UNIVERSITY OF NOTTINGHAM
10 0758808 2
ROM THE
KU-393-105

DRUG DELIVERY

Wiley Series in Drug Discovery and Development

Binghe Wang, Series Editor

Computer Applications in Pharmaceutical Research and Development
Edited by Sean Ekins

Glycogen Synthase Kinase-3 (GSK-3) and Its Inhibitors: Drug Discovery and Development
Edited by Ana Martinez, Ana Castro, and Miguel Medina

Aminoglycoside Antibiotics: From Chemical Biology to Drug Discovery
Edited by Dev P. Arya

Drug Transporters: Molecular Characterization and Role in Drug Disposition
Edited by Guofeng You and Marilyn E. Morris

Drug-Drug Interactions in Pharmaceutical Development
Edited by Albert P. Li

Dopamine Transporters: Chemistry, Biology, and Pharmacology
Edited by Mark L. Trudell and Sari Izenwasser

Carbohydrate-Based Vaccines and Immunotherapies
Edited by Zhongwu Guo and Geert-Jan Boons

ABC Transporters and Multidrug Resistance
Edited by Ahcène Boumendjel, Jean Boutonnat, and Jacques Robert

Drug Design of Zinc-Enzyme Inhibitors: Functional, Structural, and Disease Applications
Edited by Claudiu T. Supuran and Jean-Yves Winum

Kinase Inhibitor Drugs
Edited by Rongshi Li and Jeffrey A. Stafford

Evaluation of Drug Candidates for Preclinical Development: Pharmacokinetics, Metabolism, Pharmaceutics, and Toxicology
Edited by Chao Han, Charles B. Davis, and Binghe Wang

HIV-1 Integrase: Mechanism and Inhibitor Design
Edited by Nouri Neamati

Carbohydrate Recognition: Biological Problems, Methods, and Applications
Edited by Binghe Wang and Geert-Jan Boons

Chemosensors: Principles, Strategies, and Applications
Edited by Binghe Wang and Eric V. Anslyn

Medicinal Chemistry of Nucleic Acids
Edited by Li He Zhang, Zhen Xi, and Jyoti Chattopadhyaya

Plant Bioactives and Drug Discovery: Principles, Practice, and Perspectives
Edited by Valdir Cechinel Filho

Dendrimer-Based Drug Delivery Systems: From Theory to Practice
Edited by Yiyun Cheng

Cyclic-Nucleotide Phosphodiesterases in the Central Nervous System: From Biology to Drug Discovery
Edited by Nicholas J. Brandon and Anthony R. West

Drug Transporters: Molecular Characterization and Role in Drug Disposition, Second Edition
Edited by Guofeng You and Marilyn E. Morris

Drug Delivery: Principles and Applications, Second Edition
Edited by Binghe Wang, Longqin Hu, and Teruna J. Siahaan

DRUG DELIVERY

Principles and Applications

Second Edition

Edited by

BINGHE WANG
Georgia State University, Atlanta, GA, USA

LONGQIN HU
Rutgers - The State University of New Jersey, Piscataway, NJ, USA

TERUNA J. SIAHAAN
University of Kansas, Lawrence, KS, USA

Published by John Wiley & Sons, Inc., Hoboken, New Jersey
Published simultaneously in Canada

For general information on our other products and services or for technical support, please contact our Customer Care Department within the United States at (800) 762-2974, outside the United States at (317) 572-3993 or fax (317) 572-4002.

Wiley also publishes its books in a variety of electronic formats. Some content that appears in print may not be available in electronic formats. For more information about Wiley products, visit our web site at www.wiley.com.

Library of Congress Cataloging-in-Publication Data

Names: Wang, Binghe, 1962– editor. | Hu, Longqin, editor. | Siahaan, Teruna, editor.
Title: Drug delivery : principles and applications / edited by Binghe Wang, Longqin Hu, Teruna J. Siahaan.
Description: Second edition. | Hoboken, New Jersey : John Wiley & Sons Inc., 2016. | Revision of: Drug delivery / Binghe Wang, Teruna Siahaan, Richard Soltero. 2005. | Includes bibliographical references and index.
Identifiers: LCCN 2015039604 (print) | LCCN 2015041751 (ebook) | ISBN 9781118833360 (cloth) | ISBN 9781118833230 (Adobe PDF) | ISBN 9781118833308 (ePub)
Subjects: LCSH: Drug delivery systems. | Pharmaceutical chemistry.
Classification: LCC RS199.5 .W36 2016 (print) | LCC RS199.5 (ebook) | DDC 615.1/9–dc23
LC record available at http://lccn.loc.gov/2015039604

Set in 10/12pt Times by SPi Global, Pondicherry, India

Printed in the United States of America

10 9 8 7 6 5 4 3 2 1

CONTENTS

LIST OF CONTRIBUTORS

Norah Albekairi, Department of Obstetrics & Gynecology, University of Texas Medical Branch, Galveston, TX, USA

Shariq Ali, Department of Obstetrics & Gynecology, University of Texas Medical Branch, Galveston, TX, USA

Dewey H. Barich, Department of Pharmaceutical Chemistry, The University of Kansas, Lawrence, KS, USA

Barlas Büyüktimkin, Department of Pharmaceutical Chemistry, The University of Kansas, Lawrence, KS, USA

Anna M. Calcagno, Department of Pharmaceutical Chemistry, The University of Kansas, Lawrence, KS, USA

Xiaoyuan Chen, Laboratory of Molecular Imaging and Nanomedicine (LOMIN), National Institute of Biomedical Imaging and Bioengineering (NIBIB), National Institutes of Health (NIH), Bethesda, MD, USA

Yuan Cheng, Department of Pharmaceutical Chemistry, Macromolecule Vaccine Stabilization Center, The University of Kansas, Lawrence, KS, USA

Sanaalarab Al Enazy, Department of Obstetrics & Gynecology, University of Texas Medical Branch, Galveston, TX, USA

Laird Forrest, Department of Pharmaceutical Chemistry, The University of Kansas, Lawrence, KS, USA

Chris V. Galliford, Department of Chemistry, Purdue University, West Lafayette, IN, USA

Xiangming Guan, Department of Pharmaceutical Sciences, College of Pharmacy, South Dakota State University, Brookings, SD, USA

Xiaowen Guan, Department of Pharmaceutical Sciences, School of Pharmacy and Pharmaceutical Sciences, University at Buffalo, State University of New York, Buffalo, NY, USA

Chao Han, Biologics Clinical Pharmacology, Janssen R&D LLC, Spring House, PA, USA

Anthony J. Hickey, RTI International, Research Triangle Park, NC, USA

Longqin Hu, Department of Medicinal Chemistry, Ernest Mario School of Pharmacy, Rutgers, The State University of New Jersey, Piscataway, NJ, USA

Krishnaveni Janapareddi, Department of Pharmaceutics, University College of Pharmaceutical Sciences, Kakatiya University, Warangal, Telangana, India; Department of Pharmaceutics and Medicinal Chemistry, Thomas J Long School of Pharmacy and Health Sciences, University of the Pacific, Stockton, CA, USA

Bhaskara R. Jasti, Department of Pharmaceutics and Medicinal Chemistry, Thomas J Long School of Pharmacy and Health Sciences, University of the Pacific, Stockton, CA, USA

Irina Kalashnikova, Department of Obstetrics & Gynecology, University of Texas Medical Branch, Galveston, TX, USA

Krizia Karry, School of Engineering, Rutgers, The State University of New Jersey, Piscataway, NJ, USA

Paul Kiptoo, Department of Pharmaceutical Chemistry, The University of Kansas, Lawrence, KS, USA

Jeffrey P. Krise, Department of Pharmaceutical Chemistry, The University of Kansas, Lawrence, KS, USA

Meng Li, Pharmacokinetics and Pharmacometrics, Drug Disposition, Safety and Animal Research, Sanofi US, Bridgewater, NJ, USA

Xiaoling Li, Department of Pharmaceutics and Medicinal Chemistry, Thomas J Long School of Pharmacy and Health Sciences, University of the Pacific, Stockton, CA, USA

Gang Liu, State Key Laboratory of Molecular Vaccinology and Molecular Diagnostics & Center for Molecular Imaging and Translational Medicine, School of Public Health, Xiamen University, Xiamen, China

Wansheng Jerry Liu, Fox Rothschild LLP, Lawrenceville, NJ, USA

Philip S. Low, Department of Chemistry, Purdue University, West Lafayette, IN, USA

Dan Menasco, Department of Chemistry and Biochemistry, University of South Carolina, Columbia, SC, USA

Bozena Michniak-Kohn, Ernest Mario School of Pharmacy, Center for Dermal Research, and The New Jersey Center for Biomaterials, Rutgers, The State University of New Jersey, Piscataway, NJ, USA

C. Russell Middaugh, Department of Pharmaceutical Chemistry, Macromolecule Vaccine Stabilization Center, The University of Kansas, Lawrence, KS, USA

Donald W. Miller, Department of Pharmacology and Therapeutics, Kleysen Institute for Advanced Medicine, University of Manitoba, Winnipeg, Manitoba, Canada

Marilyn E. Morris, Department of Pharmaceutical Sciences, School of Pharmacy and Pharmaceutical Sciences, University at Buffalo, State University of New York, Buffalo, NY, USA

Eric J. Munson, Department of Pharmaceutical Chemistry, The University of Kansas, Lawrence, KS, USA

Ngoc H. On, Department of Pharmacology and Therapeutics, Kleysen Institute for Advanced Medicine, University of Manitoba, Winnipeg, Manitoba, Canada

Anil K. Philip, Department of Pharmaceutics, School of Pharmacy, College of Pharmacy and Nursing, University of Nizwa, Nizwa, Sultanate of Oman

Tannaz Ramezanli, Ernest Mario School of Pharmacy, Center for Dermal Research, and The New Jersey Center for Biomaterials, Rutgers, The State University of New Jersey, Piscataway, NJ, USA

Charles M. Roth, Department of Chemical & Biochemical Engineering, Department of Biomedical Engineering, Rutgers, The State University of New Jersey, Piscataway, NJ, USA

Erik Rytting, Department of Obstetrics & Gynecology, University of Texas Medical Branch, Galveston, TX, USA

Neha Sahni, Department of Pharmaceutical Chemistry, Macromolecule Vaccine Stabilization Center, The University of Kansas, Lawrence, KS, USA

Can Sarisozen, Center for Pharmaceutical Biotechnology and Nanomedicine, Northeastern University, Boston, MA, USA

Kishore Shah, Polytherapeutics Inc., Bridgewater, NJ, USA

Shahnam Sharareh, Fox Rothschild LLP, Lawrenceville, NJ, USA

Teruna J. Siahaan, Department of Pharmaceutical Chemistry, The University of Kansas, Lawrence, KS, USA

John Stewart Jr., Department of Pharmaceutical Chemistry, The University of Kansas, Lawrence, KS, USA

Hongying Su, Department of Chemical Engineering, Kunming University of Science and Technology, Kunming, China; State Key Laboratory of Molecular Vaccinology and Molecular Diagnostics & Center for Molecular Imaging and Translational Medicine, School of Public Health, Xiamen University, Xiamen, China

Zhizhi Sun, Department of Pharmacology and Therapeutics, Kleysen Institute for Advanced Medicine, University of Manitoba, Winnipeg, Manitoba, Canada

Kayann Tabanor, Department of Pharmaceutical Chemistry, The University of Kansas, Lawrence, KS, USA

Vladimir P. Torchilin, Center for Pharmaceutical Biotechnology and Nanomedicine, Northeastern University, Boston, MA, USA; Department of Biochemistry, Faculty of Science, King Abdulaziz University, Jeddah, Saudi Arabia

David B. Volkin, Department of Pharmaceutical Chemistry, Macromolecule Vaccine Stabilization Center, The University of Kansas, Lawrence, KS, USA

Binghe Wang, Department of Chemistry, Georgia State University, Atlanta, GA, USA

Guijun Wang, Department of Chemistry and Biochemistry, Old Dominion University, Norfolk, VA, USA

Qian Wang, Department of Chemistry and Biochemistry, University of South Carolina, Columbia, SC, USA

Qingping Wang, DMPK, Safety and Animal Research, Sanofi US, Waltham, MA, USA

Christine Xu, Pharmacokinetics and Pharmacometrics, Drug Disposition, Safety and Animal Research, Sanofi US, Bridgewater, NJ, USA

Qiuhong Yang, Department of Pharmaceutical Chemistry, The University of Kansas, Lawrence, KS, USA

Vinith Yathindranath, Department of Pharmacology and Therapeutics, Kleysen Institute for Advanced Medicine, University of Manitoba, Winnipeg, Manitoba, Canada

Mark T. Zell, Pfizer Global Research and Development, Ann Arbor Laboratories, Ann Arbor, MI, USA

Yun Zeng, State Key Laboratory of Molecular Vaccinology and Molecular Diagnostics & Center for Molecular Imaging and Translational Medicine, School of Public Health, Xiamen University, Xiamen, China; Sichuan Key Laboratory of Medical Imaging, North Sichuan Medical College, Nanchong, China

Zheng Zhang, The New Jersey Center for Biomaterials, Rutgers, The State University of New Jersey, Piscataway, NJ, USA

Sarah K. Zingales, Department of Chemistry and Physics, Armstrong State University, Savannah, GA, USA

PREFACE

Drug delivery is an integral part of drug discovery and development. To set the context for the issues related to drug delivery, it is important to look at the big picture as well. It has been 12 years since the last edition of this book was published. As a personal journey, it is amazing to see how different the pharmaceutical industry is now when compared to what it was 12 years ago. Much has changed. In size, the 2003 US pharmaceutical market was $220B (IMS Health, Doug Long); and the 2014 size was $337B (Pharmarceutical Commerce, November 20, 2014). This is an increase of over 50%. While the United States remains the largest market for pharmaceuticals, the rest of the world is changing too, and new and emerging markets are becoming increasingly important. For example, China was ranked no. 10 in pharmaceutical market size in 2003; and in 2013, it had surpassed all European countries in becoming no. 3 behind the United States and Japan (China's Pharmaceutical Industry—Poised for the Giant Leap, KPMG 2011). In 2014, China's pharmaceutical sales were estimated to be about $100B ("The Next Phase: Opportunities in China's Pharmaceuticals Market" by Yvonne Wu, Deloitte, 2011), and China is poised to overtake Japan in 2016 in becoming the no. 2 market with an estimated size of about $150B (Mind power Solutions, March 2012). In terms of business models, the industry is becoming more global. Outsourcing of research work to new and emerging markets such as China and India and going from a model of mostly in-house research and discovery to a heavy emphasis on in-licensing represent a trend. In terms of the science, there is more of an emphasis on biologics with antibodies and antibody conjugates leading the way.

All these changes will affect how we formulate ideas in drug discovery and development, and thus drug delivery issues as a result. For example, the growing weight of the emerging markets means that intellectual property (IP) protection in countries such as China is becoming increasingly important. In addition, the IP rights of the prodrug in relation to its parent drug may be viewed differently in various countries. With all these changes, some of the drug discovery and development issues have changed too. The "silo" structure, prevalant in the pharmaceutical industry decades ago, is no longer the norm. Team-based discovery and development efforts are more of the norm. The feverish feast with combinatorial chemistry has been replaced by a more rational approach of using diversity in chemistry to address drug discovery problems, with special emphasis on "drug-like" properties. Even with all those changes, some of the drug delivery issues remain the same. Thus the Preface for the first edition is reprinted after this as well. With the new edition, the basic flow of the book has not changed. However, new chapters have been added and old chapters have been updated. Among the new chapters added, there is a special emphasis on delivery to specific organs or sites. For example, we have added chapters on "Intracellular Delivery and Disposition of Small Molecular Weight Drugs" by Jeff Krise and "Intracellular Delivery of Proteins and Peptides" by Can Sarisozen and Vladimir P. Torchilin. These additions largely reflect the tremendous progress made in recent years on the understanding of intracellular trafficking of drugs. We have added chapters on (i) "Transdermal Delivery of Drugs Using Patches and Patchless Delivery Systems" by Bozena Michniak-Kohn and colleagues; (ii) "Nanoparticles as Drug Delivery Vehicles" by Qian Wang and colleagues; (iii) "Evolution of Controlled Drug Delivery Systems" by Xiaoling Li and colleagues, (iv) "Targeted Delivery of Drugs to the Colon" by Anil K. Philip and Sarah K. Zingales; (v) "Protein and Peptide Conjugates for Targeting Therapeutics and Diagnostics to Specific Cells" by Teruna Siahaan and colleagues; (vi) "Drug Delivery to the Lymphatic System" by Qiuhong Yang and Laird Forrest; (vii) "The Development of Cancer Theranostics: a New Emerging Tool towards Personalized Medicine" by Chen and colleagues; (viii) "Vaccine Delivery: Current Routes of Administration and Novel Approaches" by David Volkin and colleagues; and (ix) "Delivery of Genes and Oligonucleotides" by Charles M. Roth. These additions reflect our thinking that specialized delivery to sites beyond the general circulation is a major challenge. The additions of theranostics, nanoparticles in drug delivery, and genes and oligonucleotides largely were based on the tremendous development in these areas since the previous edition.

We hope that this new edition will provide valuable information for students and professionals alike, and welcome suggestions and participations in future revisions.

Binghe Wang, Longqin Hu, and Teruna J. Siahaan
September 2015

1

FACTORS THAT IMPACT THE DEVELOPABILITY OF DRUG CANDIDATES

CHAO HAN[1] AND BINGHE WANG[2]

[1] *Biologics Clinical Pharmacology, Janssen R&D LLC, Spring House, PA, USA*
[2] *Department of Chemistry, Georgia State University, Atlanta, GA, USA*

1.1 CHALLENGES FACING THE PHARMACEUTICAL INDUSTRY

Drug discovery and development is a long, arduous, and expensive journey. It was estimated that the total cost of developing a new drug in the US pharmaceutical industries was well over a billion dollars in the 2000s, and this figure has been increasing [1, 2]. This figure may be slightly better for biotechnology-based research and development (R&D) [1]. The entire process may take up to 14 years [1, 3]! Yet, only 2 out of 10 marketed drugs would return revenues that match or exceed R&D costs according to a recent analysis [4]. There has been a tremendous amount of pressure on the industry to maximize efficiency, shorten development time, and reduce the cost during discovery and development. In order to accomplish such objectives, one needs to analyze the entire drug discovery and development process so as to identify steps where changes can be made to increase efficiency.

The entire endeavor of developing a new drug from an idea to the market is generally divided into several stages: target identification, hit identification/discovery, hits' optimization, lead selection and further optimization, candidate identification, and preclinical and clinical development [5]. Among these, each stage has

Drug Delivery: Principles and Applications, Second Edition. Edited by Binghe Wang, Longqin Hu, and Teruna J. Siahaan.

many interrelated aspects and components. A target is identified in early discovery when there is sufficient evidence to suggest a relationship between the intervention of a target and treatment of the disease or conditions. Tens of thousands new molecules are then synthesized and screened against the target to identify a few molecules (hits) with desired biological activities. Analogs of these selected molecules are then made and screened further for improved activities and drug-like properties. Optimization results in identifying a small number of compounds for testing in pharmacological and other models. Those active compounds (leads) are further optimized for their biopharmaceutical properties, and the most drug-like compound(s) (drug candidates, only 1–2 in most cases) are then selected for further preclinical and clinical development. The drug discovery and development path with an emphasis on the discovery stages is schematically illustrated in Figure 1.1.

Having been through the screening and optimization processes, however, of those drug candidates with most drug-like properties, only about 40% successfully make their way into the evaluations in humans (first-in-human or FIH clinical trial) [6]. Unfortunately, data from historical average reveals an almost 90% overall attrition rate in clinical development [6]. In another word, only one molecule successfully makes into the market from 10 compounds tested in humans. Results from another statistical analysis gave a similar success rates for new chemical entities or new molecular entities (NCEs/NMEs) for which an investigational new drug (IND) application or a biologic license application (BLA) was filed in almost four decades [7], and the figure has not changed much [8]. This high attrition rate obviously does not meet the needs of long-term success desired by both the pharmaceutical industry and health care system.

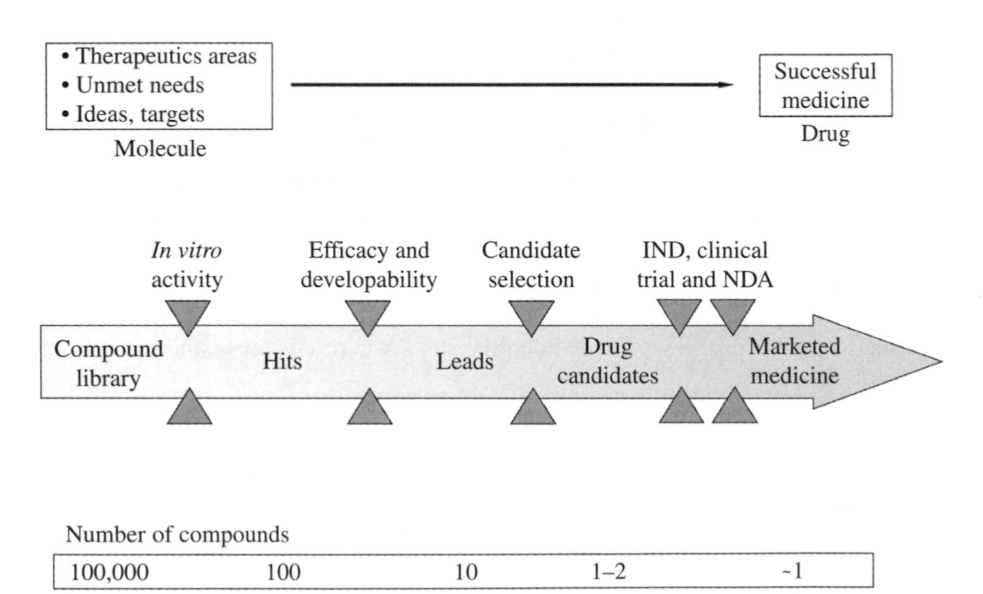

FIGURE 1.1 A schematic illustration of the drug discovery and development process with the estimated number of compounds shown for each step.

Prentis et al. [9] analyzed many factors that potentially were attributable for such a high attrition rate based on the data from seven UK-based pharmaceutical companies from 1964 through 1985. The results from this statistical analysis revealed that a 39% failure was due to poor pharmacokinetic properties in man, 29% was due to a lack of clinical efficacy, 21% was due to toxicity and adverse effects, and about 6% was due to commercial limitations. Although not enough information was available in a great detail, it is believed that some intrinsic relations of these factors existed. For instance, toxicity or lack of efficacy can be precipitated by undesired drug metabolism and pharmacokinetic (DMPK) properties of the molecule. Based on the assumptions that most failures was not due to the lack of "biologic activities" per se as defined by *in vitro* testing, there has been a drive to incorporate the evaluation of drug delivery properties, which may potentially precipitate developmental failures, into the early drug discovery and candidate selection processes with the intention of reducing the proportion of late stage failures, which is obviously most costly.

Rapid development in biology, and in rational and structure-based chemical design in addition to new technologies such as generation of diversity libraries, automation in high throughput screening, and advanced instrumentation in bioanalysis have significantly accelerated lead identification and discovery process [10, 11] for a given target. In light of these scientific and technical advances and under the pressure to reduce the cost and shorten the time of discovery and development, many major organizations in the pharmaceutical industry went through rapid and drastic changes from the late 1990s to early 2000s. A conference entitled "Opportunities for integration of pharmacokinetics, pharmacodynamics, and toxicokinetics in rational drug development" [12] was a landmark of this fundamental change in the pharmaceutical industry [13]. The developability concept was introduced to pharmaceutical R&D with an organizational and functional integration in early drug discovery and development [14]—optimization of DMPK properties of drug candidates in conjunction with toxicology and pharmaceutical development. These changes were successful in addressing some of the specific causes of the attrition. Early investment in optimizing absorption, distribution, metabolism, and elimination (ADME) in drug discovery [15] has successfully reduced attrition rate due to poor human pharmacokinetics from about 39% in the previous survey [9] to approximately 10% in the year of 1991–2000 [16]. A top cause of failures appeared to have shifted to toxicology related. Furthermore, failures due to other reasons, such as the lack of clinical efficacy, remain to be a major issue.

Being encouraged by the successes in addressing ADME issues early on in the discovery and preclinical development, R&D in pharmaceutical industry bolstered the number of drug candidates entering into clinical trials during the early 2000s. Unfortunately, this did not make the expected positive impact on the output in terms of the number of new medicines into the market. The success rate, instead, fell to approximate 5% in the year of 2006–2008 [17]. Thus there is a need of improved understanding of disease mechanism(s) and issues in drug delivery. It shouldn't be forgotten that waves of mergers and acquisitions aim at boosting R&D performance in the pharmaceutical industry apparently failed to effectively address the issues either [18].

Nevertheless, the march goes on. A fairly recent analysis indicated that the number of approved new drugs from pharmaceutical companies has essentially been relatively constant during the past 60 years [19]. Over a thousand new drugs had been approved by the US Food and Drug Administration (FDA) in this period of time. There is no doubt that these medicines helped enormously in treating diseases, managing health conditions, and improving the quality of life. Indeed, life expectancy and cancer survival rate improved due to new treatments [20, 21]. Death rates in cardiovascular diseases decreased significantly [22]. Average cholesterol level in adults in the United States fell to the ideal level—below 200 mg/dl [23, 24]. The most striking example was the dramatic drop in HIV/AIDS death rate since the approvals of antiretroviral treatments [25]. These testimonial facts are the demonstration of the value of pharmaceutical R&D of new medicine.

Since the first therapeutic monoclonal antibody—muromonab-CD3 (Orthoclone OKT3®)—was approved by the US FDA in 1986 [26], more than 30 therapeutic monoclonal antibodies have been approved, and probably hundreds based on the same platform of therapeutics are under clinical development. This class of molecules mimics the human immune system and very specifically intervene cell membrane-bound or soluble targets by antagonizing (a few agonists too) the pathway or neutralizing the ligand [27]. Monoclonal antibody therapeutics along with other biologics such as recombinant or fusion proteins are commonly referred as large molecule to differentiate from synthetic drugs or small molecules. Based on an analysis [8] of the data up to 2004, clinical approval success rate for large molecule therapeutics more than doubled that for small molecules. An in-depth survey on only monoclonal antibody-based therapeutics reveals similar encouraging trend [28]. The discovery and development of biologics are seeing rapid growth. It is expected that the top list of sales will be dominated by biologics in a few years according to Slatko's analysis based on the observations made in 2010 [29].

Taking advantage of the high specificity of a monoclonal antibody as a guided carrier to deliver chemotherapeutic agent specifically to the tumor cells was truly an innovation in drug delivery. This class of coupled molecules is commonly referred as antibody–drug conjugate (ADC) (for a thorough review, see Zolot et al. [30]). The very first ADC was approved for acute myeloid leukemia in 2000 (although it was withdrawn in 2010 based on US FDA's recommendations) [31]. Shortly in the next few years a CD30-specific ADC, brentuximab vedotin, was approved by the US FDA in 2011. Trastuzumab emtansine was approved in 2013. Through conjugation of anti-human HER2 receptor antibody with mertansine, a tubulin inhibitor, Trastuzumab was created as a unique ADC [32]. Because the monoclonal antibody targets HER2, and HER2 is over-expressed in certain cancer cells, the cytotoxic toxin is delivered specifically to tumor cells such as in breast cancer [33]. It has been proven to be a very successful drug delivery strategy for cancer therapies.

Over the past several decades, the never-ending endeavors conjointly by pharmaceutical, academic, and regulatory scientists and researchers have been devoted to finding more effective and safer medicines for treating variety of diseases. The journey has been focused more closely on understanding the biology, learning the etiology, finding the right molecule, and delivering the molecule to the right target. Many successful stories and good lessons learned undoubtedly demonstrate that drug delivery has been playing a critical role. The developability of drug candidates is an assemblage of assessments that programmatically ensure and optimize drug delivery. The concept has not changed although the domains of developability have been continuously extending along with the development of technologies and advance in sciences. The evaluation of developability mostly involves the integration of research activities in functional areas such as DMPK, pharmaceutical development, safety assessment, and process chemistry into drug discovery and development process in very early stages of discovery. The inputs from other functional areas as well as those from clinical, regulatory, commercial, and marketing groups in the early stage help to minimize costly mistakes in late stages of development and have become more and more important to the success of the drug discovery and development process. Developability is an overall evaluation of the drug-like properties of an NCE/NME. Many of the changes in the pharmaceutical industry have been driven by the concept of ensuring developability. These changes, in other words, the integration of the sciences and strategies in multifunctional areas in drug discovery and development, are to ensure that the NCEs/NMEs of interest will have the best possibility of success in every step toward the final goal.

In the next few sections, examples of some factors that are often examined for developability and their intrinsic relationship are briefly discussed. This is, of course, not a comprehensive coverage of the developability. However, we do hope that this section will put various chapters in perspective and allow readers to see individual sections in the context of an integrated drug discovery and developmental process.

1.2 FACTORS THAT IMPACT DEVELOPABILITY

In most pharmaceutical companies, many efforts have been made to create a clear framework for selecting compound(s) with minimal ambiguity for further progression. Such a framework is not a simple list of the factors that impact the quality of a drug-like molecule and can vary from company to company [34]. This framework, which is more often referred as "developability criteria," is a comprehensive summary of the characteristics, properties, and qualities of the NCE/NME(s) of interest, which normally consist of preferred profiles with a minimally acceptable range. A preferred profile describes the optimal goal for selection and further progression of a candidate, whereas the minimum range gives the acceptable properties for a compound that is not ideal but may succeed. Molecules that do not meet the criteria will not be further progressed. Such criteria cover all the functional areas in drug development. Some of the major developability considerations are briefly described as examples in the following paragraphs.

1.2.1 Commercial Goal

It is self-evident that we are in a business world! Generally speaking a product needs to bring value to the health care system and be profitable to the manufacturer to be viable. Therefore, early input from commercial, marketing, and medical outcome professionals is very important for setting up a projective product profile, which profoundly affects the development of the developability criteria for intended therapeutics. In general, this portfolio documents the best possible properties of the product and minimum acceptable ranges that may succeed based on the studies of market desires. These studies should be suggestive based on the results from professional analyses of health care needs, potential market, and existing leading products for the same, similar, or related indications. The following aspects need to be well thought out and fully justified before the commencement of a project: (i) therapeutic strategy, (ii) dose form and regimen, and (iii) the best possible safety profile such as therapeutic window, potential drug–drug interactions, and any other potential adverse effects. Using the development of an anticancer agent, as an example, for therapeutic strategy selection, one may consider the choice of developing a chemotherapeutic (directly attacking the cancer cells) versus anti-angiogentic agent (depriving cancer cells of their nutrients) and in combination versus stand-alone therapy. In deciding the optimal dose form and regimen, one may consider the following: whether an oral or iv or both formulation should be developed, should it be once daily or in a different dose interval, and would projected dose regimen be acceptable or convenient for the patients. The results from such an analysis form the frameworks for developing the developability criteria and become the guidelines of setting up the criterion for each desired property. For example, pharmacokinetic properties such as half-life and oral bioavailability of a drug candidate will have direct impact on developing a drug that is to be administered orally once a day.

1.2.2 The Chemistry Efforts

Medicinal chemistry is always the starting point and a driver of small-molecule drug discovery programs. In a large pharmaceutical R&D organization, early discovery of bioactive compounds (hits) can be carried out either by high throughput screening of compound libraries or by rational design, or a combination of both. Medicinal chemists will then use the structural information of the pharmacophore thus identified to optimize the structures. Chemical tractability needs to be examined carefully at the very beginning when a new chemical series is identified. Chemistry space around the core structure for modification is closely studied. Upon a thorough examination of a small number of compounds, an initial exploratory *structure–activity relationship* (SAR) or quantitative SAR (QSAR) should be developed. Rheault and colleagues [35] described an example of how to establish and explore SAR around a pharmacophore in the discovery of a potent and oral bioavailable BRAF inhibitor. In this example, numerous substructural changes were made leading to the most potent compounds while considering the other properties such as the pharmacokinetics and metabolism. Such efforts are normally made in parallel with several different chemical series. It is important for medicinal chemists that many different SARs are being considered, developed, and integrated into their efforts at the same time, which provide more opportunities to avoid other undesirable properties unrelated to their intended biological activities. Such factors, again, may include potential *CYP*450 inhibition, permeability, selectivity, stability, and solubility, etc.

Structural novelty of the compounds (in other words, can this piece of art be protected in a patent?), complexity of synthetic routes, scalability (can the syntheses be scaled up to industrial production scale?) and the cost of starting materials (cost of goods at the end of the game), potential environmental concerns, and toxic intermediate issues will all need to be closely examined at early stages of the drug discovery and development processes. It is never too early to have those thoughts and to put them into actions.

1.2.3 Biotechnology in the Discovery of Medicine

Comparing to medicinal chemistry efforts in the processes of searching a bioactive molecule, the initiation of a biotechnology-based project is more specific and target driven. The biologic activity of large molecule therapeutics is generally believed to be more specific; therefore, there are fewer unexpected off-target effects and potential toxicity issues, which can be a major advantage. Yet, many different hurdles have to be overcome.

The issues with large-molecule products during early development are similar by nature. Thus monoclonal antibodies are used as an example here. The discovery of hybridoma technology by Köhler and Milstein in 1975 was a milestone in the development of monoclonal antibodies in immunology and biomedicine [36]. The Nobel Prize in Physiology or Medicine in 1984 was awarded jointly to Niels K. Jerne, Georges J.F. Köhler, and César Milstein "for theories concerning the specificity in development and control of the immune system and the discovery of principle for production of monoclonal antibodies" [37]. It is fascinating to see how this discovery has changed the face of immunology and biomedicine nowadays [38].

Monoclonal antibodies can be made fully humanized with current technology. Several molecular and cellular biology techniques have been established to generate human monoclonal antibodies [39]. In addition to affinity maturation, engineering and selection processes for the desired specificity and binding affinity, and protein sequence and amino acid residue that may affect the stability and other physicochemical properties of the molecule are important factors in protein engineering of the molecule. The selection of a production platform and/or cell line for a stable and high-yield production of selected antibody is also a very important developability criterion that has to be considered much early on.

Immunogenicity of protein-based therapeutics has been one of the major safety concerns besides its potentially negative impact on the pharmacokinetics and pharmacodynamics. This aspect has been largely addressed by using fully human products [40]. The immunogenicity of a candidate in animal species used in pharmacology and toxicology models is also a very important factor although the occurrence and its impact are in general not predictable for humans [41]. Successful preclinical pharmacology and toxicology programs are the very first step of preclinical development. The importance of drug delivery has been exhibited even in early preclinical development for large molecule as well. Taking immunogenicity as an example, it may interfere with the investigation of pharmacokinetics and safety assessments in animal species, which may severely hinder the molecule being developed further to FIH.

Antibodies largely undergo protein catabolism leading to their eventual elimination, rather than being metabolized by the *CYPs* or other enzymes. FcRn (neonatal Fc receptor or Brambell [42] receptor) plays an important role in protecting antibodies from proteolysis in the lysosomes. That explains the long half-life of most therapeutic antibodies as well as endogenous ones. Transporters are rarely involved for large molecule's absorption and excretion. It may have less of concerns for drug metabolism-based drug–drug interactions [43]. However, the potential of drug–drug interactions should still be programmatically evaluated [44], especially when a cytokine modulator is being developed since certain soluble cytokines may play a role in regulating the expressions of *CYP* enzymes and transporters. The effects of cytokines, such as interleukin-6 and tumor necrosis factor alpha, on *CYP* modulation and possible mechanisms have been studied [45].

With the introductions to medicinal chemistry- and biotechnology-based drug discovery and early development described already, it should be relatively easy to appreciate the complexities of the factors that may affect drug developability directly and indirectly for ADCs. On top of those factors that have to be considered and evaluated for a small-molecule drug and those for the development of a monoclonal antibody, the linker between the two molecules in terms of chemical type and relative stability in a biological environment is also a key factor that has to be fine-tuned before making an ADC work [46].

1.2.4 Target Validation in Animal Models

Although drug discovery efforts almost always start with *in vitro* testing nowadays, it is well recognized that promising results from *in vitro* testing do not always translate into *in vivo* efficacy. There are numerous reasons that could lead to this discrepancy, some of which are well understood and others are not. Therefore, target validation in animal models before clinical trials in human is a critical step. Before a drug candidate is fully assessed for its safety and brought to a clinical test, demonstration of efficacy of a biologically active compound (e.g., active in an enzyme inhibition assay) in pharmacological models (*in vivo*, if available) is considered as a milestone in the path of discovering a drug candidate. This is sometimes also called proof of mechanism (PoM). Many cases exemplify the challenges and importance of pharmacological models. For example, inhibitors of integrin receptor $\alpha_v\beta_3$ have been shown to inhibit endothelial cell growth, which implies their potential as being clinically beneficial for an anti-angiogenic target for cancer treatment [47]. However, the proposed mechanism did not work in animal models although compounds were found very active *in vitro* [48, 49]. What has been recognized is that integrin receptor $\alpha_v\beta_3$ may not be the exclusive pathway that tumor cell growth depends on. Inhibition of this pathway may induce or shift to a compensatory pathway(s) for angiogenesis.

Advances in mathematical modeling have been providing very useful testing environments and have generated very useful data. Anticancer drugs, for example, may be tested in animal xenograft models. Biomarkers and antitumor efficacy data with the pharmacokinetic information could be modeled for prediction of clinical drug exposure and efficacy [50]. Knowing the limitation of animal models, the information derived from such *in vitro* and *in vivo* experiments and from mathematical modeling is invaluable for target validation and, furthermore, to provide guidance for dose selections in clinical studies. Also, it should be mentioned here that most biologic therapeutics, such as monoclonal antibodies, are very specific to human target and may not cross-react with that in animal species. This property sometimes paradoxically limits the use of preclinical animal models. Therefore, the availability of directly relevant information from preclinical species may be limited for these types of drug candidates. Nonhuman primates are often used. The development of human transgenic animals has been providing very relevant research tools. For example, hFcRn transgenic mice may predict the pharmacokinetic behavior of human monoclonal antibodies very well [51].

Ideally, an *in vivo* model should comprise all biochemical, cellular, and physiological complexities as in a real-life system, which may predict the behavior of a potential drug candidate in human much better than an *in vitro* system. In order to have a biological hypothesis tested in the system with validity, a molecule has to be evaluated in many other aspects. Knowing the pharmacokinetic parameters such as absorption, distribution, and metabolism in the animal species that is used in the pharmacological model becomes critical. Basic pharmacokinetic parameters will be briefly described in the following paragraph and discussed in detail in several chapters in the book. The importance of drug delivery is demonstrated as early as in an animal model that serves as an early milestone in preclinical drug development.

The pharmacokinetics/pharmacodynamics relationship, systemic and tissue levels of drug exposure, frequency of dosing, which allows the drug to demonstrate efficacy, and the strength of efficacy are all very important factors that may affect the future development of an NCE/NME. These are all factors that are directly or indirectly related to the topic and, therefore, have to be fully considered for drug delivery.

1.2.5 Drug Metabolism and Pharmacokinetics

The importance of DMPK in drug discovery and development is reflected in the statistics of attrition rates [9]. Most of the changes in the industry during early 2000s have happened in the areas of DMPK [13] and proven to be effective in reducing attrition [16]. The overall goal of DMPK in drug discovery and development is to predict the pharmacokinetic behaviors of a drug candidate in humans. Nevertheless, the focus could vary at different stages of the process. PK parameters in animal species that will be used in pharmacological (as briefed in the previous paragraphs) and safety assessment models provide very important insights (systemic and tissue exposures) for those studies. The results from pharmacokinetic studies in several animal species generate the data for physiologically-based models or interspecies allometric scaling [52, 53] to predict basic pharmacokinetic behaviors of a product in human. Assays using human tissues, cells, and genetically engineered cell lines provide a tremendous amount of information before a molecule can be tested in clinical studies. Optimizing DMPK developability factors are immensely beneficial for finding the candidate(s) with best the potential for success [54].

Desirable (or undesirable) biological effects of a drug *in vivo* normally are directly related to its exposure. One of the following factors, namely, total systemic exposure, maximum concentration, or duration of the concentration above a certain level, is usually used as a parameter that is correlated with the efficacy and/or adverse effects [55]. The exposure is governed at a given dose by (i) the ability for the body to remove the drug as a xenobiotic and (ii) the route via which the drug is delivered. Blood or plasma clearance (CL) is often used as a measurement of the capability to eliminate a drug molecule from the systemic circulation. Low-to-moderate clearance molecule is desirable in most situations unless a fast-action and short-duration drug is being designed [56]. Biologics such as monoclonal antibodies generally have much lower clearance when compared to small-molecule drugs. Since endosomal proteolysis of monoclonal antibody is protected by its binding to the FcRn receptors [42], the half-life of a therapeutic monoclonal antibody is normally 2–3 weeks. Monthly or even longer dosing interval thus are possible.

A drug can be directly delivered into the systemic circulation by several methods. However, for convenience and many other reasons, oral dosage forms are preferred in many situations. Therefore, oral bioavailability of the compound is one of the very important developability criteria for oral drug delivery. Many factors affect the oral bioavailability of a drug. Orally delivering a biologic therapeutic protein is still quite challenging due to the digestive system. Subcutaneous or intramuscular delivery is the commonly used route of administration in addition to intravenous infusion. The understanding of the mechanisms and factors affecting subcutaneous absorption is still primary. These factors will be discussed in detail in several of the following chapters. In addition to clearance and bioavailability, other major pharmacokinetic parameters that should be evaluated are also discussed in related chapters.

Volume of distribution is a conceptual pharmacokinetic parameter that measures the extent of a drug distributed into tissues. A well-known parameter, elimination half-life, can be derived from clearance and volume of distribution. It is a very important developability criterion, which warrants desired dose regimen. It should be noted here that a discussion of half-life has to be in the context of pharmacologically relevant concentration. A purely mathematically derived half-life is sometimes pharmacologically irrelevant. Some more definitive explanation and comprehensive discussion of the major pharmacokinetic parameters and their biological relevance have been extensively reviewed [57, 58].

These parameters should be examined across several different preclinical species to reliably predict the behavior in human. However, with therapeutic monoclonal antibodies, although available data usually are limited to only one relevant animal species, the predictability has been impressively good and reliable [59]. The pharmacokinetics and pharmacodynamics topics will be discussed in several related chapters in this book.

Inhibition and induction of drug metabolizing enzymes [60], P-glycoprotein (P-gp) substrate property [61, 62], plasma protein binding and binding kinetics [63, 64], metabolic stability in the microsomes or hepatocytes from different species including humans [65], metabolic pathway, and the metabolite(s) identified [66] are all very important developability measurements in the assessment of safety, potential drug–drug interaction, and predictability. These factors need to be optimized and carefully examined against developability criteria. Drug metabolism-related issues are outlined and discussed in Chapter 9. The impact of transporter including efflux transporter in drug delivery and the models used to study and address these issues are discussed in Chapters 5 and 7.

1.2.6 Preparation for Pharmaceutical Products

Before the early 1990s, the issues of solid state, salt form, aqueous solubility, and dosing formulation for agents used in pharmacological, pharmacokinetic, and toxicological studies have not been brought to full attention. However, an inappropriate salt version or solid form may precipitate potential drug delivery and stability problems (both physicochemically and chemically) during formulation and pharmaceutical engineering. Now it has been realized that the investigation of physicochemical properties of an NCE/NME against developability criteria should start early in the R&D processes. Chapter 3 and several other chapters discuss these physicochemical properties that have major impact on drug delivery.

Aqueous solubility is one of the most important physicochemical properties. It is believed that a drug has to be in solution to be absorbed [67]. From a pharmaceutical development point of view, solid-state form is another important factor that affects solubility and dissolution rate, and eventually the developability. Solid-state form is the determinant of, to some extent, physicochemical stability, intellectual property, and formulation scalability; this factor ought to be carefully examined and optimized. Changes in crystallinity from different chemical processes, in some cases, result in a big difference in bioavailability when the drug is delivered by a solid-dosage formulation.

Many of the earlier-described properties could change when salt version and form change. The salt with the best solubility, dissolution rate (therefore, could result in best bioavailability if by solid dose), stability, and other properties such as moisture absorption should be selected before a molecule enters full development [68]. *In situ* salt screening is a new technology to select the right salt form for drug candidate [69]. For instance, the HCl salt [70, 71] used to be almost the default version for a weak base; however, it has been shown in many cases not to be the best. Application of these screening processes in early drug development is one of the major steps in integrating pharmaceutical development into drug discovery and development.

Preclinical safety assessment (toxicology) is another functional area, which in itself stands to serve as a big milestone in drug discovery and development. NCE/NMEs have to be evaluated for their genetic toxicity as well as acute, short-term, and long-term toxicity when appropriate. The results are crucial for further development of the molecule in FIH clinical study and beyond. Although the principles and importance of toxicology will not be discussed in this book, many efforts in DMPK and pharmaceutics are to assure drug delivery in the animal species used in safety assessment programs. Metabolism profiles of a drug candidate in the species used in the toxicology studies are to be compared with that from human tissues for major difference. The profiles are also examined for potential active/toxic metabolite(s).

It should be noted that process chemistry and biologics production are large functional areas that can have major impacts on a drug's developability, but will not be covered in this book. Although developability criteria in this area will not be discussed here, it is important to point out that it is essential that collaboration with these areas is considered early on in order to define the best strategy for drug delivery.

1.3 REMARKS ON DEVELOPABILITY

The concept of ensuring developability in drug discovery and development represents an integration and synchronization of all functional areas that impact efficiency, and thus the quality and quantity as well as timelines for drug development. Coordination of these multifunctional, interlinked, parallel ongoing scientific and technological research activities is a new challenge to the management of a drug discovery and development enterprise nowadays.

The developability concept was adapted and executed much earlier and more rigorously by larger pharmaceutical companies than their smaller counterparts. However, an analysis by Munos [19] of pharmaceutical innovations in the past 60 years suggested a trend that smaller companies may have outperformed larger companies in their NME/NCE outputs. The underlying reasons for this difference are not clear, especially about whether it was due to a difference in the directions of innovation investments and/or the impact from heightened safety concerns of regulators. There were also not enough data to make the comparison on final approvals. Nonetheless, it was probably more certain that the way in which developability criteria are being adapted and applied was somewhat different. A recent publication clearly indicated varying organization-dependent criteria in different companies [34]. It is reasonable to expect a more focused and objective-driven process in smaller biotechs; wherease larger pharmaceutical companies may use more compartmentalized and criterion-driven development processes. In another word, the question of how we achieve our goals should be asked conjunctively with the questions of how likely it will be to achieve the goals, knowing that the risks, resources, and timelines have to be balanced in practice. Developability is about an in-depth understanding of the molecule regardless of the size of the company or number of molecules in the pipeline.

It is interesting to note some "exceptions" to the commonly accepted developability criteria in the recent history of drug development. In those exceptions the candidate had been successfully developed and even became a blockbuster although the molecule was inborn with some strongly undesired properties. One of the examples would be atorvastatin. The molecule had very limited bioavailability in preclinical species (e.g., ~5% in rats) due to the interplay between transporters and drug metabolizing enzymes in the intestines and liver [72]. Thus, if the preclinical bioavailability criteria used by most preclinical development organizations were applied to atorvastatin, it would not have been selected for development and, therefore, would not have made it to a top-selling drug at all. We learned that in human clinical studies, the bioavailability of atorvastatin was not very high either (14%) [73]. Another story is about a recent-approved cancer drug—dabrafenib (GSK2118436). Inhibition and induction of major *CYP* enzymes are serious concerns for potential drug–drug interactions. Drug candidates usually are deselected for that reason. If the concerns of *CYP*3A4 induction plus the inhibition of several other *CYPs* (public data in *gsk media*) [74] were used as a litmus test, GSK2118436 would not have been selected for development and, therefore, there would not be the successful story of dabrafenib and trametinib combination therapy for melanoma [75]. The successful stories, or hypothetical arguments if one would, tell us that the developability should never be simply an artificially defined bar for a candidate to jump over. It is a complex process that requires judgment calls based on the full understanding of the properties of a molecule.

1.4 DRUG DELIVERY FACTORS THAT IMPACT DEVELOPABILITY

Delivery of a pharmaceutical agent to the systemic circulation and consequently to the site of action to produce the desired pharmacological effect is the ultimate goal of drug delivery. The developability of a drug candidate from drug delivery perspectives has become the core of developability criterion in drug development. As discussed in the previous sections, many other factors in developability criteria are closely related to drug delivery; these thoughts and practices are applicable from research laboratory all the way to clinical trials and from early discovery to post market development. In order to accomplish the task, one has to overcome numerous barriers that hinder drug delivery.

As the nature of a biological system, multiple or redundant mechanisms may exist to protect the system from exposures to almost any foreign substance while preserving the ability of nutrients uptake. The physiological arrangements and the chemical and biochemical barriers associated with the physiological structures form the first line of defense. Any drug, delivered by any route, will almost certainly encounter some of these barriers before reaching the site of action. These barriers, as well as their physiological and biochemical functions, and their role in drug delivery, will be discussed in detail in Chapter 2. In the first several chapters, general concepts that are directly related to drug delivery, principles in evaluation of drug delivery, along with come common approaches to study drug delivery from anatomical to cellular level are introduced and discussed in sequence from Chapters 3 to 8.

Earlier in this chapter we touched on some conventional routes of drug delivery, such as intravenous injection. Specific factors associated with different routes of drug delivery, such as the first-pass effects following oral administration are discussed in Chapter 9. How a drug molecule interacts with these barriers is very much determined by the intrinsic properties of the molecule. The intrinsic properties are, in another word, the physicochemical and biochemical characteristics of a molecule. In Chapter 3, the physicochemical properties and their implication in formulation and drug delivery will be extensively discussed.

Development of pharmacokinetics and pharmacodynamics relationship by mathematical modeling of the interactions of a drug molecule with the entire biological system is important to the prediction of drug concentrations in the systemic circulation, and, therefore, the pharmacological responses. Better understanding of the system will allow a pharmaceutical scientist to utilize the system and manipulate the system for the purpose of drug delivery. Chapter 4 discusses the basic principles and topics in pharmacokinetics and pharmacodynamics. Approaches in drug delivery based on the understanding pharmacokinetic principles are essential in pharmaceutical development.

Developability in drug delivery is an overall assessment of all the important factors. For example, in oral drug delivery [76] solubility is important because a drug molecule has to be dissolved to be absorbed. Some lipophilicity is essential for the molecule to cross the cell membrane by diffusion. In order to finally reach systemic circulation, the molecule has to survive various chemical and biochemical attacks in the gastrointestinal system and the liver. A flow chart describing sequentially the factors that can impact drug delivery is illustrated in Figure 1.2. The order in which these factors are listed could also be the order of logical thinking when one plans to tackle an oral drug delivery problem, and could be a reference point for other routes of delivery too.

It is believed that permeability and metabolic stability of a drug molecule are two major factors in drug delivery or in the prediction of a drug's absorption [77] when the molecule is in solution. Permeability can be further divided into passive diffusion and transporter-mediated processes. Metabolism of a drug molecule in the liver and intestine can be evaluated by *in vitro* experimental methods. In many cases, *in vitro* metabolism (intrinsic clearance) can be used to predict *in vivo* metabolic clearance successfully [78]. It is obvious that when efflux transporters, such as P-gp, are

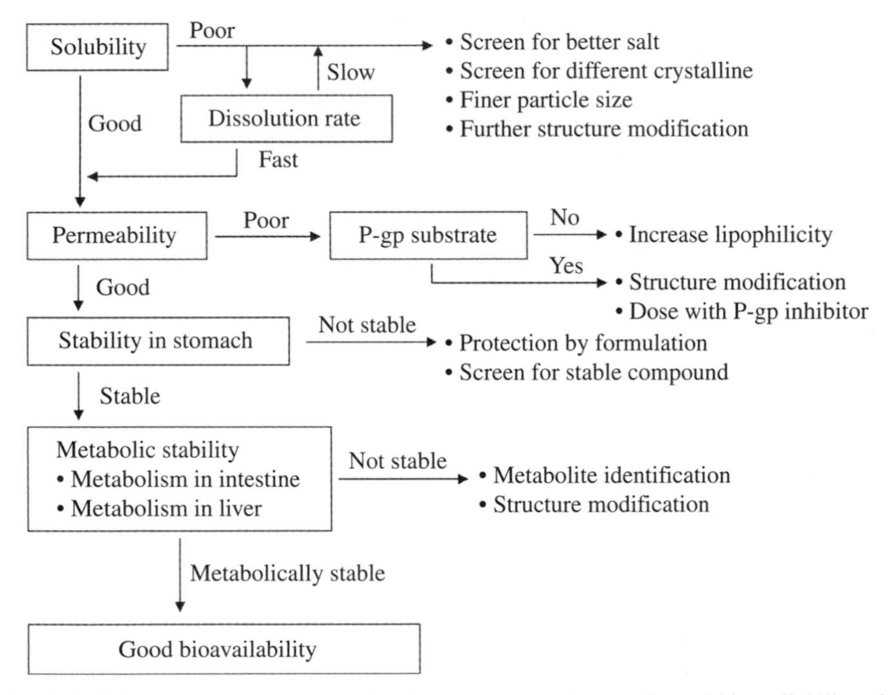

FIGURE 1.2 The evaluation steps of various factors that impact the oral bioavailability of a drug candidate.

involved, the predictability of the *in vivo* clearance using metabolic intrinsic clearance becomes uncertain [79]. A more in-depth understanding of drug transporters and their function in combination with our knowledge on drug metabolism will help predict oral absorption [80, 81]. Transporter-related drug delivery issues as well as *in vivo* and *in vitro* models used to address these issues are discussed in Chapter 5.

In addition to parental delivery of a therapeutic agent, many other routes of drug delivery are developed for convenience, safety, specific targeting, and delivery of special agents. First-pass metabolism is especially applicable to oral drug delivery, and will be discussed in Chapter 9. Several other "unconventional" routes for drug delivery such as pulmonary (Chapter 10) and transdermal absorption (Chapter 11) are discussed together with strategies in development and technical challenges to be considered. Although this book does not cover most routes of delivery individually, the philosophy and logical thinking discussed should be generally applicable to the development of other route of delivery. Figure 1.2 provides, for example, thinking paths in addressing an issue for oral bioavailability. The discussions are further projected into several of later chapters on controlled target-specific drug delivery (Chapters 15–22). Targeting specific organ or tumor tissues through different technologies and potential personalized drug delivery are discussed in Chapters 17–22. Several chapters provided a number of technical approaches to improve drug delivery. Physicochemical approaches by formulation include controlled release (Chapter 15), prodrug approaches (Chapter 12), liposome vehicles (Chapter 13), and nanoparticles (Chapter 14).

It was discussed previously in this chapter that the discovery and development of biologic therapeutics have seen increasing attention and proven to be successful in recent years. Biologic therapeutics are expected to be dominant in the market in the future [29]. Unfortunately, the delivery of biologics had been mostly limited to those by parental injection. A large body of contents related to the delivery of biologic therapeutics or macromolecular drugs are newly added into this edition. Formulations for delivery of vaccines are discussed in Chapter 24. Cutting-edge researches in delivery systems for gene therapy are specifically reviewed and discussed in Chapter 25. It is known that the distribution of large molecules into intracellular space is limited. New developments in sciences and technology focused on intracellular delivery of protein and peptides are introduced in Chapter 23.

The goal of this book is to provide readers with a basic understanding of all the major issues in drug delivery. In this edition, new developments in drug delivery sciences and technology are captured in addition to updates made to those already included in the last edition. A much more detailed examination of various topics can also be found in the references cited in this chapter and the specific discussions in the relevant chapters.

REFERENCES

1. DiMasi, J. A.; Grabowski, H. G. *Manage. Decis. Econ.* 2007, **28**(4–5), 469–479.
2. US PHARM. *Profile-Biopharmaceutical Research Industry*; Pharmaceutical Research and Manufacturers of America: Washington, DC, 2013, http://www.phrma.org (accessed September 30, 2015).

3. DiMasi, J. A. *Clin. Pharmacol. Ther.* 2001, **69**, 286–296.
4. Vernon, J. A.; Golec, J. H.; DiMasi, J. A. *Health Econ.* 2010, **19**(8), 1002–1005.
5. Kuhlmann, J. *Int. J. Clin. Pharmacol. Ther.* 1997, **35**, 541–552.
6. Venkatesh, S.; Lipper, R. A. *J. Pharm. Sci.* 1999, **89**, 145–154.
7. DiMasi, J. A. *Clin. Pharmacol. Ther.* 2001, **69**, 297–307.
8. DiMasi, J. A.; Feldman, L.; Seckler, A.; Wilson, A. *Clin. Pharmcol. Ther.* 2010, **87**(3), 272–277.
9. Prentis, R. A.; Lis, Y.; Walker, S. R. *Br. J. Clin. Pharmacol.* 1988, **25**, 387–396.
10. Drews, J. *Science* 2000, **287**, 1960–1964.
11. Hopfinger, A. J.; Duca, J. S. *Curr. Opin. Chem. Biol.* 2000, **11**, 97–103.
12. Yacobi, A.; Skelly, J. P.; Shah, V. P.; Benet, L. Z. *Integration of Pharmacokinetics, Pharmacodynamics, and Toxicokinetics in Rational Drug Development*; Plenum Press: New York and London, 1993.
13. Lesko, L. J.; Rowland, M.; Peck, C. C.; Blaschke, T. F. *Pharm. Res.* 2000, **17**, 1335–1344.
14. Railkar, A. S.; Sandhu, H. K.; Spence, E.; Margolis, R.; Tarantino, R.; Bailey, C. A. *Pharm. Res.* 1996, **13**, S-278.
15. Davis, C. B. Introduction. In: *Evaluation of Drug Candidates for Preclinical Development—Pharmacokinetics, Metabolism, Pharmaceutics, and Toxicology*; Han, C.; Davis, C. B.; Wang, B. (Eds.). John Wiley & Sons, Inc.: Hoboken, NJ, 2010, pp 1–7.
16. Kola, I.; Landis, J. *Nat. Rev. Drug Discov.* 2004, **3**(8), 711–715.
17. Arrowsmith, J. *Nat. Rev. Drug Discov.* 2012, **11**(1), 17–18.
18. Lamattina, J. *Nat. Rev. Drug Discov.* 2011, **10**(8), 559.
19. Munos, B. *Nat. Rev. Drug Discov.* 2009, **8**(12), 959–968.
20. Sun, E. *J. Clin. Oncol.* 2008, **26**(suppl 15), Abstract 6616, http://meeting.ascopubs.org/cgi/content/short/26/15_suppl/6616 (accessed November 2015).
21. Lichtenberg, F. R. National Bureau of Economic Research Working Paper. 2004, http://www.nber.org/papers/w10328.pdf (accessed September 30, 2015).
22. Go, A. S.; Mozaffarian, D.; Roger, V. L.; et al. *Circulation* 2013, **127**(1), e6–e245.
23. Schober, S. E.; Carroll, M. D.; Lacher, D. A.; Hirsch, R. NCHS Data Brief 2007, No. 2, 1–8.
24. Carroll, M. D.; Kit, B. K.; Lacher, D. A.; Shero, S. T.; Mussolino, M. E. *JAMA* 2012, **308**(15), 1545–1554.
25. Hoyert, D. L.; Xu, J. *Natl. Vital Stat. Rep.* 2012, **61**(6), 1–51.
26. Hooks, M. A.; Wade, C. S.; Millikan, W. J. *Pharmacotherapy* 1991, **11**(1), 26–37.
27. Waldmann, T. A. *Nat. Med.* 2003, **9**(3), 269–277.
28. Nelson, A. L. *Nat. Rev. Drug Discov.* 2010, **9**(10), 767–774.
29. Slatko, J. *Med. Ad. News* 2010, **29**(7), 1,6–7.
30. Zolot, R. S.; Basu, S.; Million, R. P. *Nat. Rev. Drug Discov.* 2013, **12**(4), 259–260.
31. US FDA. Mylotarg (gemtuzumab ozogamicin): Market Withdrawal. http://www.fda.gov/Safety/MedWatch/SafetyInformation/SafetyAlertsforHumanMedicalProducts/ucm216458.htm (accessed September 30, 2015).
32. LoRusso, P. M.; Weiss, D.; Guardino, E.; Girish, S.; Sliwkowski, M. X. *Clin. Cancer Res.* 2011, **17**(20), 6437–6447.

33. Verma, S.; Miles, D.; Gianni, L.; et al. *N. Engl. J. Med.* 2012, **367**(19), 1783–1791.
34. Leeson, P. D.; St-Gallay, S. A. *Nat. Rev. Drug Discov.* 2011, **10**(10), 749–765.
35. Rheault, T.; Stellwagen, J.; Adjabeng, G.; et al. *Med. Chem. Lett.* 2013, **4**(3), 358–362.
36. Köhler, G.; Milstein, C. *Nature* 1975, **256**, 495–497.
37. Nobelprize.org. The Nobel Prize in Physiology or Medicine 1984, http://www.nobelprize.org/nobel_prizes/medicine/laureates/1984/ (accessed September 30, 2015).
38. Alkan, S. S. *Nat. Rev. Immunol.* 2004, **4**, 153–156.
39. Wang, S. *Antibody Tech. J.* 2011, **1**, 1–4.
40. Kuester, K.; Kloft, C. Part II, Chapter 3. Pharmacokinetics of monoclonal antibodies. In: *Pharmacokinetics and Pharmacodynamics of Biotech Drugs*; Meibohm, B. (Ed.). Wiley-VCH: Weinheim, Germany, 2006, pp 45–91.
41. Pendley, C.; Schantz, A.; Wagner, C. *Curr. Opin. Mol. Ther.* 2003, **5**(2), 172–179.
42. Brambell, F. W. R.; Halliday, R.; Morris, I. G. *Nature* 1964, **203**(4952), 1352–1355.
43. Zhou, H.; Mascelli, M. A. *Ann. Rev. Pharmacol. Toxicol.* 2011, **51**, 359–372.
44. Zhang, L.; Zhang, Y.; Zhao, P.; Huang, S.-M. *AAPS J.* 2009, **11**(2), 300–306.
45. Morgan, E. T.; Goralski, K. B.; Piquette-Miller, M.; et al. *Drug Metab. Dispos.* 2008, **36**(2), 205–216.
46. Nolting, B. *Methods. Mol. Biol.* 2013, **1045**, 71–100.
47. Eliceiri, B. P.; Cheresh, D. A. *J. Clin. Invest.* 1999, **103**, 1227–1230.
48. Miller, W. H.; Alberts, D. P.; Bhatnger, P. K.; et al. *J. Med. Chem.* 2000, **43**, 22–26.
49. Carron, C. P.; Meyer, D. M.; Pegg, J. A.; et al. *Cancer Res.* 1998, **58**, 1930–1935.
50. Luo, F. R.; Yang, Z.; Camuso, A.; et al. *Clin. Cancer Res.* 2006, **12**(23), 7180–7186.
51. Tam, S. H.; McCarthy, S. G.; Brosnan, K.; Goldberg, K. M.; Scallon, B. J. *MAbs* 2013, **5**(3), 1–9.
52. Mahmood, I. *Am. J. Ther.* 2002, **9**, 35–42.
53. Mahmood, I.; Balian, J. D. *Clin. Pharmacokinet.* 1999, **36**, 1–11.
54. Lin, J. H. *Ernst Schering Res. Found. Workshop* 2002, **37**, 33–47.
55. Woodnutt, G. *J. Antimicrob. Chemother.* 2000, **46**, 25–31.
56. Bodor, N.; Buchwald, P. *Med. Res. Rev.* 2000, **20**, 58–101.
57. Benet, L. Z. *Eur. J. Respir. Dis.* 1984, **65**, 45–61.
58. Benet, L. Z.; Zia-Amirhosseini, P. *Toxicol. Pathol.* 1995, **23**, 115–123.
59. Han, C.; Zhou, H. *Ther. Deliv.* 2011, **2**(3), 359–368.
60. Lin, J. H. *Curr. Drug Metab.* 2000, **1**, 305–331.
61. van Asperen, J.; van Tellingen, O.; Beijnen, J. H. *Pharm. Res.* 1998, **37**, 429–435.
62. Wacher, V. J.; Salphati, L.; Benet, L. Z. *Adv. Drug Deliv. Rev.* 2001, **46**, 89–102.
63. Tawara, S.; Matsumoto, S.; Kamimura, T.; Goto, S. *Antimicrob. Agents Chemother.* 1992, **36**, 17–24.
64. Talbert, A. M.; Tranter, G. E.; Holmes, E.; Francis, P. L. *Anal. Chem.* 2002, **74**, 446–452.
65. Rodrigues, A. D.; Wong, S. L. *Adv. Pharmacol.* 2000, **43**, 65–101.
66. Baillie, T. A.; Cayen, M. N.; Fouda, H.; et al. *Toxicol. Appl. Pharmacol.* 2002, **182**, 188–196.
67. Rowland, M.; Tozer, T. N. *Clinical Pharmacokinetics. Concepts and Applications*; Lippincott Williams & Wilkins: Philadelphia, 1995; Chapter 9, pp 119–136.

68. Morris, K. R.; Fakes, M. G.; Thakur, A. B.; et al. *Int. J. Pharm.* 1994, **105**, 209–217.
69. Tong, W.-Q.; Whitesell, G. *Pharm. Dev. Technol.* 1998, **3**, 215–223.
70. Berge, S. M.; Bighley, L. D.; Monkhouse, D. C. *J. Pharm. Sci.* 1977, **66**, 1–19.
71. Gould, P. L. *Int. J. Pharm.* 1986, **33**, 201–217.
72. Lau, Y. Y.; Okochi, H.; Huang, Y.; Benet, L. Z. *Drug Metab. Dispos.* 2006, **34**(7), 1175–1181.
73. Pfizer. Lipitor Physician Prescribing Information. 2012, http://labeling.pfizer.com/ShowLabeling.aspx?id=587 (accessed September 30, 2015).
74. GlaxoSmithKline Inc. Product Monograph: PrTAFINLAR™, Dabrafenib (as Dabrafenib Mesylate). 2015, http://gsk.ca/english/docs-pdf/product-monographs/tafinlar.pdf (accessed November 2015).
75. Flaherty, K. T.; Infante, J. R.; Daud, A.; et al. *N. Engl. J. Med.* 2012, **367**, 1694–1703.
76. Martinez, M. N.; Amidon, G. L. *J. Clin. Pharmacol.* 2002, **42**, 620–643.
77. Lipinski, C. A.; Lombardo, F.; Dominy, B. W.; Feeney, P. J. *Adv. Drug Deliv. Rev.* 1997, **23**, 3–25.
78. Houston, J. B. *Biochem. Pharmacol.* 1994, **47**, 1469–1479.
79. Wu, C. Y.; Benet, L. Z.; Hebert, M. F.; et al. *Clin. Pharmacol. Ther.* 1995, **58**, 492–497.
80. Benet, L. Z.; Cummins, C. L. *Adv. Drug Deliv. Rev.* 2001, **50**, S3–S11.
81. Polli, J. W.; Wring, S. A.; Humpherys, J. E.; et al. *JEPT* 2001, **299**, 620–628.

2

PHYSIOLOGICAL, BIOCHEMICAL, AND CHEMICAL BARRIERS TO ORAL DRUG DELIVERY

Paul Kiptoo, Anna M. Calcagno, and Teruna J. Siahaan

Department of Pharmaceutical Chemistry, The University of Kansas, Lawrence, KS, USA

2.1 INTRODUCTION

The development of orally bioavailable drugs is challenging due to the presence of the intestinal mucosa barrier [1–3]. In addition, various potential therapeutic agents derived from biotechnology research (peptidomimetics, peptides, proteins, oligonucleotides) present even bigger challenges than small-molecule drugs. These challenges for oral delivery are due to at least two factors: (1) the presence of physical, physiological, and biochemical barriers of the intestinal mucosa and (2) the unfavorable physicochemical properties of drugs [1, 4]. Oral delivery of small and large molecules has been the focus of many review articles. This chapter focuses on various aspects of the intestinal mucosa barrier and potential solutions to overcome these barriers [1–3].

Many biological barriers protect the human body and segregate each organ according to its functions. These barriers protect organs or tissues from pathogenic invaders, including toxins, viruses, and bacteria. The skin is the largest barrier to protect the body from its surrounding. A more specific protector for the brain is the blood–brain barrier (BBB), which prevents unwanted molecules from entering the brain from the blood stream. Similarly, the intestinal mucosa barrier filters out the unwanted pathogens (i.e., virus and bacteria) and toxins from food and prevents them from getting into the blood stream. Unfortunately, these barriers also become

Drug Delivery: Principles and Applications, Second Edition. Edited by Binghe Wang, Longqin Hu, and Teruna J. Siahaan.

barricades for delivering therapeutic agents orally or to the brain to treat brain diseases. Thus, understanding the structure and biological properties of the intestinal mucosa barrier and the physicochemical properties of drugs that can cross the intestinal mucosa barrier are valuable for designing ways to increase the oral bioavailability of the potential drugs. Here, the properties of the intestinal mucosa barrier and types of molecules that can penetrate the barrier will be discussed in greater detail.

Several different obstacles must be overcome for the delivery of drugs through the intestinal mucosa barrier. Although there are differences between the intestinal mucosa barrier and BBB, they have some fundamental similarities. The drug has to have optimal physicochemical properties for it to cross the intestinal mucosa barrier and enter into the systemic circulation. To cross the intestinal mucosa barrier, the drug has to enter the cell cytoplasm via partition into the cell membranes. The presence of efflux pumps can prevent membrane partition of the drug molecules. From the cytoplasm, the drug has to cross basolateral cell membranes before entering the systemic circulation. During this process, the drug also must overcome the biochemical barriers such as metabolizing enzymes that can degrade the drug and prevent it from crossing the basolateral membranes to enter the systematic circulation. To enhance drug delivery to cross these various barriers, many of these factors can be taken into consideration when designing drugs with good intestinal mucosa absorption characteristics.

2.2 PHYSIOLOGICAL BARRIERS TO DRUG DELIVERY

An aqueous mucus layer covers the luminal side of the gastrointestinal tract, and the mucus that is composed of glycoproteins is secreted by the goblet cells. The mucus traps water molecules with a turnover rate of 12–24 hours. A drug molecule has to penetrate the mucus layer with a thickness of 100–150 μm before crossing the epithelial cell layer of the intestinal mucosa barrier [5]. This mucus layer acts as a filter for molecules with molecular weights of 600–800 Da. Drug penetration through this mucus and unstirred water layer is the rate-limiting step before the drug reaches the surface of the epithelial cells of the enterocytes [5, 6].

Under the mucus layer is a single layer of columnar epithelial cells that are joined together by tight intercellular junctions to form a barrier against systemic delivery for orally administered drugs [1, 4, 6]. This layer of cells is composed of enterocytes, goblet cells, endocrine cells, and paneth cells. The amount of goblet cells differs from the small intestine to the distal colon: only 10% of the cells in the small intestine are goblet cells, whereas this percentage increases to 24% in the distal colon. The gastric epithelium from the proximal to distal stomach has four regions including the nonglandular stratified squamous, the cardiac, the glandular proper gastric (fundic), and the pyloric regions. Each region has a different physiological function. The toughness of the stratified squamous region allows it to resist food abrasion, while the cardiac region is responsible for the production of mucus and bicarbonate. Pepsinogen and hydrochloric acid are secreted from the proper gastric region.

The final section is associated with the release of gastrin and pepsinogen [7]. Both villi and crypts are lined with the epithelial cell layer. The microvilli amplify the intestinal surface area for nutrient absorption into the systemic circulation while crypts are responsible for cell renewal [5, 8].

The basolateral side of epithelial cell layer sits on the lumen of the gastrointestinal tract, and the lamina propria supports the epithelial layer [9]. The lamina propria has many components, including smooth muscle cells, nerve cells, lymph vessels, and blood vessels. It serves as a bridge between the food side of the intestinal mucosa and systemic circulation to provide nutrition. The lymphatic system for circulation of immune cells is connected to the intestinal mucosa at the lamina propria. The nerve cells found at the lamina propria link it to the nervous system. The contractility of the intestinal mucosa is regulated by the muscularis mucosa at the mucosa's deepest layer [9].

A drug molecule can cross the intestinal mucosa via several different mechanisms, depending upon its physiochemical properties. Hydrophobic drugs that can partition into the cell membranes are more likely to cross the intestinal mucosa through the transcellular pathway (Fig. 2.1, Pathway A). Drugs that cross

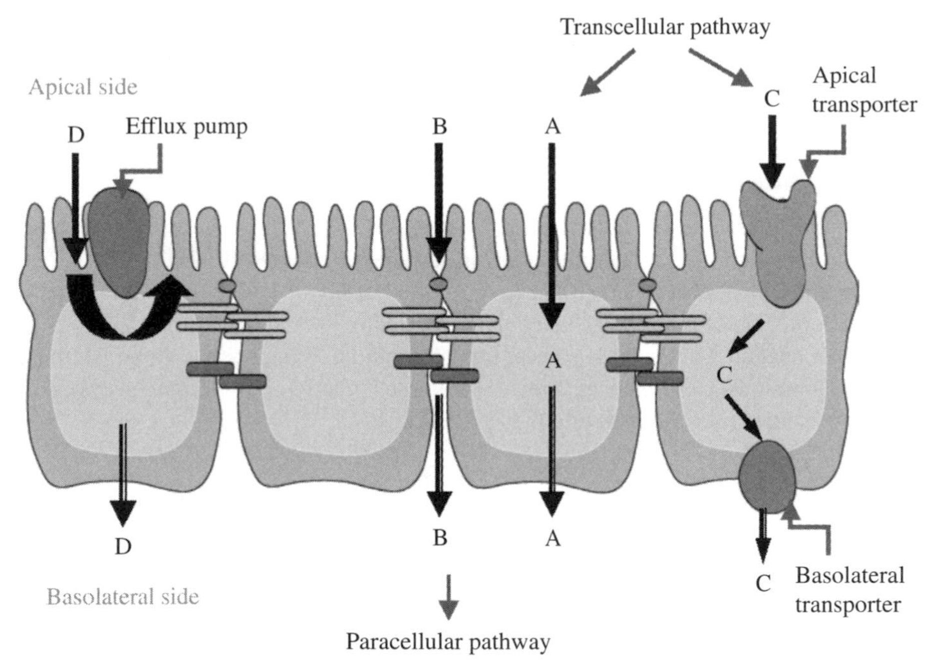

FIGURE 2.1 Several possible mechanisms of drug transport through the intestinal mucosa barrier. Pathway A is the passive transcellular route in which a drug permeates the cell passively by partitioning to cell membranes at both apical (AP) and basolateral (BL) sides. Pathway B is called the paracellular route where the drug passively diffuses between cells at the intercellular junctions. Pathway C is the active transport route where the drug is recognized by transporters, which shuttle the drug from the AP to BL sites. Pathway D is the route where the drug permeation is inhibited by efflux pumps; the efflux pumps expel the drug molecule from the cell membranes during the cell membrane partition process.

via the transcellular pathway normally have a good balance between optimal hydrophobicity and solubility. In contrast, hydrophilic drugs cannot partition into the cell membranes so they cannot cross the intestinal mucosa via Pathway A. For hydrophilic molecules, they must use the paracellular pathway (Fig. 2.1, Pathway B), but only molecules with a hydrodynamic radius less than 11 Å can pass through this pathway [10]. The presence of tight junctions restricts the permeation of molecules through the paracellular pathway [11, 12]. The tight junctions are constructed by protein–protein interactions that connect the membranes of adjacent cells [11, 13–15]. Pathway B is the most likely route of transport for peptides and proteins; however, the large size of peptides and proteins prevents their penetration through the tight junctions. Receptor-mediated endocytosis pathway (Pathway C) is another way for drug molecules to cross the intestinal mucosa barrier. With this pathway, the drug has to be recognized by the transporter on the apical side of the intestinal mucosa barrier. For example, peptide transporters have been shown to improve the oral bioavailability of a drug by conjugating the drug to an amino acid [16–18]. Finally, the intestinal mucosa has efflux pumps (Fig. 2.1, Pathway D) to protect the barrier from unwanted molecules such as toxins to cross this barrier [19]. These pumps recognize molecules with certain features and expel them from the epithelial cell membranes of the intestinal mucosa barrier [20].

2.2.1 Paracellular Pathway

The paracellular pathway is a channel in between cells that is connected with intercellular junctions between neighboring cell membranes (Fig. 2.2) [1, 15, 21, 22]. The tortuous channel of intercellular junctions spans about 80-nm long and runs the entire lateral side of the cells [10]. This pathway allows small molecules and ions to cross the intestinal mucosa barrier. The intercellular junction has three regions: (1) tight junctions (*zonula occludens*), (2) adherens junctions (*zonula adherens*), and (3) desmosomes (*macula adherens*) [1, 15, 21, 22].

2.2.1.1 Tight Junctions The tight junctions (*zonula occludens*) are found at the most apical region of the epithelial cells of the intestinal mucosa barrier. The tight junction region brings the membranes of the opposing cells in close contact, which often is referred to as the "kiss" region. The kiss region can be seen by freeze-fracture electron microscopy as branching fibrils that circumscribe the cells [23]. The tight junction region functions as the gate of the intercellular junctions to prevent molecules from freely passing through the paracellular pathway (Fig. 2.2) [23]. The kiss region is populated with proteins that form cell–cell adhesion among neighboring cells, causing cell surface polarity. This produces the fence function, which restricts free diffusion of lipids and proteins from the apical plasma membrane to the basolateral surface [24, 25]. Thus, paracellular permeation of a drug through the intercellular junctions is regulated by the pore size of the tight junctions. The villus tips have the smallest pores while the crypt regions have the largest pores for the percolation of

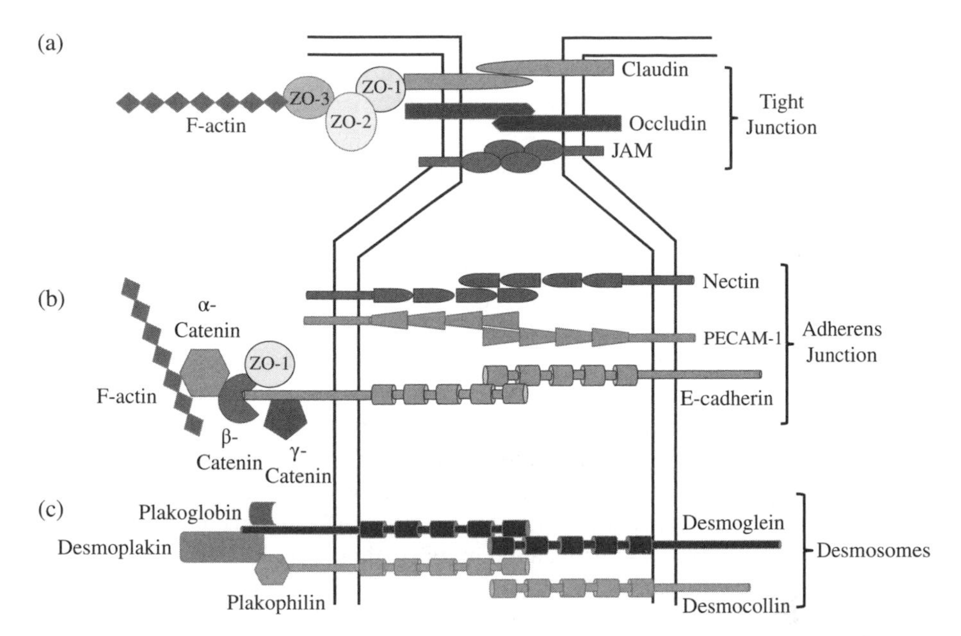

FIGURE 2.2 The intercellular junction is mediated by protein–protein interactions at different regions of intercellular junctions, including (a) tight junction (*zonula occludens*), (b) adherens junction (*zonula adherens*), and (c) desmosomes (*macula adherens*).

molecules through the tight junctions [7]. The integrity of the tight junctions is calcium dependent, and removal of calcium causes a rearrangement of the tight junction proteins [26–28]. It is possible that removal of calcium disrupts the interactions and integrity of calcium-dependent proteins such as cadherins at the adherens junctions. A number of cytokines and growth factors have also been shown to decrease the barrier function of the tight junctions [29].

Protein–protein interactions at the tight junctions are mediated by occludins, claudins (claudin-3 and -5), and junctional adhesion molecules (i.e., JAM-A, -B, and -C), and these proteins are involved in both the gate and fence functions [11, 12, 30–33]. Occludins and claudins have a similar general structure with some distinct differences. Occludins and claudins have four transmembrane domains, two extracellular domains that form loops (loop-1 and -2), and a cytoplasmic carboxyl tail [34]. The extracellular loops as well as the cytoplasmic domain of claudins and occludins play a vital role in creating cell–cell contact [25]. Both occludins and claudins interact with the cytoplasmic proteins called zonula occludin-1 (ZO-1), ZO-2, or ZO-3 that belong to the membrane-associated guanylate kinase (MAGUK) family. ZO-1 interacts with the C-terminal cytoplasmic domain of occludins via the guanylate kinase-like (GUK) domain while it interacts with claudin via its PDZ1 domain [35, 36]. ZO-1 also interacts with JAM-1 through the PDZ3 domain. ZO-1 stabilizes the tight junction by cross-linking occludins and claudins to the actin cytoskeleton. The phosphorylated occludins and claudins are the main forms located at the tight junctions to maintain tight junction integrity. Dephosphorylation of occludins and claudins can cause their relocation from the tight junctions' cell surface into the cytoplasmic intracellular compartments [37–40]. This relocation can loosen the tight junctions to increase the penetration of molecules through the tight junctions.

2.2.1.2 Adherens Junctions Adherens junctions are found below the tight junctions and mediate cell–cell adhesion through protein–protein interactions [1, 41]. The adherens junctions are formed prior to the formation of the tight junctions [42]. All three regions of the intercellular junctions work together to make up the transepithelial electrical resistance (TEER) across paracellular pathway [43]. The TEER value is a measure of the integrity of the intercellular junction, and it is a reciprocal of the ability of the ion and small molecule to permeate through the intercellular junctions [44]. The increase in resistance can be correlated to the increase in the number of strands found in the tight junctions of the intestinal mucosa barrier. The TEER values are different in different regions of the human intestine with the jejunum displaying $20\,\Omega\,cm^2$ and the large intestine showing $100\,\Omega\,cm^2$ [45]. Within the adherens junctions, the perijunctional actin–myosin II ring encircles the epithelial cells, impacting the solute permeation in this region [43].

Cell–cell adhesion within the adherens junction is controlled by calcium-dependent E-cadherin and calcium-independent nectin and platelet endothelial cell adhesion molecule-1 (PECAM-1) [1, 15, 46, 47]. E-cadherin is a glycoprotein with an extracellular domain, one single transmembrane domain, and a cytoplasmic domain. The extracellular domain is divided into five repeats (EC1–EC5); each repeat unit has about 110 amino acid residues [48]. Cadherin molecules interact with each other to form cell–cell adhesion in many different ways [15]. For example, E-cadherin protrudes from the cell surface as a dimer (*cis*-dimer), and this *cis*-dimer interacts with another *cis*-dimer from the opposing cell to form a *trans*-dimer. Calcium ions have been shown to be involved in forming a rod-like structure of E-cadherin to induce *trans*-homophilic interactions of E-cadherins. However, structural studies for E- and C-cadherins indicate that there are various ways that cadherins form homophilic interactions [49, 50]. The highly conserved cytoplasmic domain of cadherins is necessary for the adhesion property, and it interacts with α- and β-catenins, which link the cadherins to the actin cytoskeleton [51].

2.2.1.3 Desmosomes The last region of the paracellular pathway is the desmosome, which is located nearest the basolateral membrane surface of the enterocyte. The desmosome is connected by protein–protein interactions of desmocollins (i.e., Dscs: Dsc-1, -2, and -3) and desmogleins (i.e., Dsgs: Dsg-1, -2, -3, and -4), which belong to the calcium-dependent cadherin family such as E- and N-cadherin [52–55]. The difference between Dscs or Dsgs with E-cadherin is in the structure of the cytoplasmic domain. Both Dscs and Dsgs have intracellular cadherin-like sequence (ICS) and intracellular anchor (IA) [52–55]. In addition, Dsgs have proline-rich linker (IPL), repeat unit domain (RUD), and desmoglein terminal domain (DTC). The number of repeats in RUD is different in Dsg-1 (5 repeats) and in Dsg-3 (2 repeats).

2.2.2 Transcellular Pathway

A drug with the appropriate physiochemical characteristics can traverse through the cell by passive diffusion. These optimal characteristics can partly be identified as the rule of five. In the case of peptides, peptidomimetics, and proteins, their physicochemical properties may not be suitable for permeation through the cell membrane via the transcellular pathway. The drug molecules must pass through the lipid bilayers that make up the membranes as the rate-limiting barrier to the passive flow of molecules. The resistance across the transcellular path can be described as resistors in a series arrangement, where the apical and basolateral membranes act as the two resistors [43]. The outer region of the bilayers is surrounded by a large body of water molecules, and it is embedded with proteins (e.g., receptors, enzymes, and transporters) and carbohydrates. The polar head groups of the membranes are in the next region of the cellular barrier, and this region has the highest molecular density, making passive diffusion of drug molecules through the membranes difficult. The third region contains nonpolar tails of phospholipids in the inner membrane region, which have a hydrophobic characteristic [56]. Both bilayer density and hydrophobicity select molecules with optimal physicochemical properties (e.g., size, shape, hydrogen-bonding potential, and hydrophobicity) that can partition into and penetrate the cell membranes.

After the drug partitions into the membranes, it must also enter the cytosol before exiting through basolateral membranes. The cytosol contains various compartments and drug-metabolizing enzymes to trap and/or degrade (metabolize) the drug molecules. It has been shown that basic drugs can be sequestered in the endosomes/lysosomes due to their low pH, and as a result they cannot escape to endosomes to cross the basolateral membranes. Finally, metabolism can change the drug's physicochemical properties from hydrophobic to hydrophilic metabolites that can be trapped in cytosol or lysosomes. Thus, metabolism of the drug can lower the amount of drug molecules that cross basolateral membranes of intestinal mucosa into the systemic circulation.

2.3 BIOCHEMICAL BARRIERS TO DRUG DELIVERY

There is great interindividual variability in the drug metabolism processes as a result of differing enzyme activity (inhibition or induction), genetic polymorphisms, or even disease state [57]. Enzymes found within the intestine are from mammalian and bacteria-associated sources. The mammalian enzymes are located within the lumen and in the enterocytes. The enzymes from microflora are found in the ileum and colon [9]. The focus of this discussion will center only on degradation by the mammalian enzymes.

2.3.1 Metabolizing Enzymes

The first metabolic barrier that drug molecules encounter is a mixture of hydrochloric acid and proteolytic pepsins within the lumen of the stomach. The acidic conditions at pH 2–5 can cause hydrolysis of peptides and proteins, especially when they contain the aspartic acid residue [5]. The luminal enzymes at the upper small intestine function as the second barrier [5]. In addition, several proteolytic enzymes are found at the lumen of the duodenum, including trypsin, chymotrypsin, elastase, and carboxypeptidase A and B, and their highest activity is found at pH 8. These enzymes degrade 30–40% of large proteins and small peptides within 10 minutes [5].

The major enzymatic barrier is peptidases found within the brush border and in the cytosol of the enterocytes [5]. There is an increase in brush border peptidase activity from the upper duodenum to the lower ileum. These enzymes degrade smaller peptides ranging from di- to tetrapeptides. The brush border peptidases are selective for tripeptides, while the cytosolic proteases have selectivity for dipeptides [5]. The metabolic enzyme activity decreases along the intestine to a nearly negligible rate within the colon, while the drug permeation at the colon epithelium remains good [2]. This indicates that the colon is a good target region for drug delivery to avoid enzyme degradation. The intestinal surface pH on the brush border is 5.5–6.0, and it is more acidic than the lumen pH [58]. The enterocytes have an intercellular pH of 7.0–7.2, and the gastrointestinal pH is also changed during the fasted and fed states [59].

The proximal small intestine shows the greatest metabolic activity due to its large surface area and the plethora of intestinal enzymes and transporters [9]. The intestine has Phase I and II enzymes, and CYP superfamily are the most notable Phase I enzymes. The concentration of P-450 enzymes in the intestinal walls is approximately 20 times less than that found within the liver; however, their drug metabolism activity is comparable to that found in the liver [57, 60]. The highest activity of the P-450 enzymes is displayed in the proximal part of the gastrointestinal tract, and away from the proximal part P-450 enzyme activity decreases [57]. The highest concentration of P-450 enzymes is found in the villus tips of the upper and middle third of the intestine [56]. There is intra- and interindividual variability in an enzyme activity; this is due to the exposure of the enterocytes to external stimuli such as food and drugs that can either induce or inhibit these enzymes. These intestinal P-450 enzymes are more responsive to inducers or stimulators than are their hepatic counterparts [56]. Although the blood flow to the intestine is lower than to the liver, the villus tip has a large surface area where the enzyme can interact freely with its substrate to allow extensive drug metabolisms [61]. Metabolic activity has been shown to be route dependent, and the drug metabolism is normally greater with oral administration than with intravenous administration [57, 62]. In this case, the intestinal metabolism occurs during the initial absorption of the drug across the intestinal barrier, and the metabolism is lower with recirculation of the drug. The major factor that influences route-dependent metabolism is the residence time of the drug within the enterocyte. The residence time can be lengthened when there is drug trapping in cytoplasm and/or lowering of the blood flow. Conversely, the residence time can be shortened due to basolateral clearance by basolateral transporters [57].

The CYP1, CYP2, and CYP3 subfamilies are mostly involved in xenobiotic metabolism. Each subfamily of isoenzymes has its own drug substrate specificity. CYP1A1, CYP2C, CYP2D6, and CYP3A4 enzymes are found within the human small intestine [9]. The characterization of CYP2D6 is difficult because it has numerous polymorphic forms [63]. The CYP3A4 is the most abundant P-450 enzyme subfamily; it makes up 70% of the CPY in the intestine [63]. There are structural similarities between the intestinal and hepatic CYP enzymes; however, they appear to be independently regulated [9]. Food interactions have been shown to affect the regulation of the intestinal CYP enzymes. Grapefruit juice inhibits CYP3A, while grilled and smoked foods induce CYP1A1 activity [9]. Variations in the enzyme population can also affect the degradation of drug molecules including peptide and protein drugs.

Phase II enzymes are referred to as metabolizing and conjugating enzymes found in the intestine. Phase II enzymes such as glucuronyltransferase, *N*-acetyltransferase, sulfotransferase, and glutathione-*S*-transferase have high activity at the intestinal mucosa barrier [9]. The enzymes can form drug conjugates within the cell to become substrates for the multidrug resistance-associated protein (MRP) family of transporters [64]. The MRP family are ATP-dependent transporters that excrete the substrate into the lumen of the intestinal mucosa.

2.3.2 Transporters and Efflux Pumps

The presence of transporters has been found in the intestinal barrier, and some of these transporters recognize di- and tripeptides. Peptides can be transported through the brush border membrane in a carrier-mediated and pH-dependent fashion [5]. Peptide transport into the cell is energy-dependent and saturable, which are the characteristics of receptor-mediated transport.

Although most transporters are situated on the apical membrane, some of them are also located on the basolateral membrane surface. The Na^+/A amino acid transporter, Na^+/ASC amino acid transporter, GLUT2 hexose transporter, and the Na^+-independent folic acid transporter are examples of such basolateral transporters [56]. PepT1 is an apical H^+/dipeptide transporter, and it is most abundant in the villus tip [18, 65]. The transporter population increases from the duodenum to the ileum, and the expression of this transporter increases during starvation. At the basolateral membrane side, PepT2 transporters act as the H^+/dipeptide transporter to allow the substrates to exit the enterocyte [56, 66].

P-glycoprotein (Pgp) serves as an efflux pump found at the brush borders of the villus tips of the small and large intestine [67–69]. The efflux pump can prevent the membrane partition of small molecules (e.g., natural products, fluorescent dyes, and anticancer drugs) and peptides. The hydrophobic aromatic and tertiary amine serve as signature recognition for this efflux pump in drug molecules. The expression of Pgp increases from the stomach to the colon [7]. Pgp has broad substrate recognition with a wide range of molecular structures, and the substrate affinity varies as a function of intestinal region [56, 68]. A common feature of the substrates is hydrophobicity. As mentioned previously, the efflux pumps assist the intestinal metabolism by returning the drug to the lumen, allowing the metabolizing enzymes to work on the drug another time as well as preventing product inhibition by removing primary metabolites that have been formed [60, 68]. This interaction is enhanced due to the colocalization of the CYP3A enzymes and Pgp on the apical membrane as well as the overlap in substrate specificities and shared inducers and inhibitors [60, 68]. There are several inhibitors of Pgp including GF-120918, cyclosporine A (CsA), and PCS833, and these inhibitors can enhance the biological barrier transport of drugs that are substrates for Pgp. Grapefruit juice also interferes with the transport mediated by Pgp; however, not all substrates for the CYP3A enzyme behave as substrates for Pgp [67, 70, 71]. Pgp receptors are also expressed in other biological barriers (i.e., liver, kidney, pancreas, and capillary endothelium of the brain) and function as a defense mechanism against xenobiotics [72].

2.4 CHEMICAL BARRIERS TO DRUG DELIVERY

The chemical structure of a drug determines its solubility and barrier permeability profiles and, in turn, the effective concentration at the intestinal lumen influences the rate and extent of intestinal absorption [73]. Unfavorable physiochemical properties have been shown to be the limiting factors in oral absorption of peptides and peptidomimetics [74–76]. As an example, the structural factors that affect the permeation of peptides will be described here.

2.4.1 Hydrogen-Bonding Potential

Hydrogen-bonding potential to water molecules has been shown to be an important factor in the permeation of peptides. Studies *in vivo* and in various cell culture models of the intestinal mucosa and BBB indicate that desolvation or hydrogen-bonding potential regulates the permeation of peptides [73, 76–78]. The energy needed to desolvate the polar amide bonds in the peptide to allow it to enter and traverse the cell membrane is the principle behind the concept of hydrogen-bonding potential. For small organic molecules, the octanol-water partition coefficient is the best predictor of cell membrane permeation with a sigmoidal relationship [77]. However, this is not the case with peptides; the desolvation energy or hydrogen-bonding potential is a better predictor for membrane permeation of peptides. Burton et al. have reported partition coefficients of model peptides in *n*-octanol/Hanks' balanced salt solution (HBSS), isooctane/HBSS, and heptane/ethylene glycol systems [79]. It was found that measuring the partition coefficient of peptide in heptane/ethylene glycol correlates well with the hydrogen-bonding potential. This method is simpler and more direct than the method that uses the difference in partition coefficients between octanol/HBSS and isooctane/HBSS [79].

The predictive rule of five by Lipinski has been used to predict the transport of a molecule through membranes of biological barriers; this rule is based on the H-bond potential, Log *P* value, and molecular weight [80, 81]. Molecules with lower potential H-bonds (e.g., 2 H-bonds) to water have higher membrane permeation than those with higher H-bonds (e.g., 8 H-bonds). A drug molecule with a molecular weight higher than 500 Da with more than five hydrogen bond donors and ten hydrogen bond acceptors is less likely to cross the biological barriers. It is predicted, that even if it is a hydrophobic molecule with a molecular weight higher than 500 Da (e.g., >800 Da), it will have difficulty in crossing the biological membranes of the intestinal mucosa.

2.4.2 Other Properties

In the case of peptides, other properties such as size, charge, and hydrophilicity influence the peptide membrane partition and permeation [74, 76]. A change in hydrophilicity of a peptide may alter its route of permeation; as the lipophilicity increases, the peptide permeation shifts from paracellular to transcellular pathways. Studies using Caco-2 cell monolayers confirm that drug permeation via the paracellular path is size dependent, and this highlights the sieving abilities of the intercellular junctions [76]. Although the paracellular path is negatively charged, the effect of charge on paracellular permeation of molecules is not well understood. One study suggests that a positive net charge on a peptide produces the best paracellular permeation, but another study suggests that a −1 or −2 charge is most effective in paracellular transport [74]. It has also been suggested that the effect of charge is negligible as the molecular size of the peptide increases [74].

2.5 DRUG MODIFICATIONS TO ENHANCE TRANSPORT ACROSS BIOLOGICAL BARRIERS

Several methods have been explored to improve drug permeation across biological barriers [1, 18, 76, 82–86]. One method is by chemical modification of drug entities such as prodrug and peptidomimetic. Another method is designing a formulation that enhances the drug permeation through the biological barriers.

2.5.1 Prodrugs and Structural Modifications

A prodrug approach has been utilized to optimize drug solubility and transport as discussed in more detail in Chapter 12 [87, 88]. A prodrug is defined as a chemical derivative that is inactive pharmacologically until it is converted *in vivo* to the active drug moiety [87, 88]. A targeted prodrug design has emerged in which prodrugs have been used to target membrane transporters or enzymes [83, 84]. This method improves the oral drug absorption or site-specific drug delivery. Extensive knowledge about the structure, distribution within biological barriers, and substrate specificities of the transporter is needed for using it as a target for drug delivery.

Prodrug strategies have been very successful with small molecules; however, their use in peptides has not been widely implemented. The cyclic peptide prodrug approach has been shown to improve peptide membrane permeation [76]. In this method, the N- and C-termini of the peptide are connected via a linker to form a cyclic peptide. The linker can be cleaved by esterase to release the linear peptide. The formation of a cyclic peptide prodrug increases the conformational rigidity, improves the intramolecular hydrogen bonding, and lowers the hydrogen-bonding potential to water molecules as a solvent. In addition, the lipophilicity of the cyclic prodrug increases, which shifts its transport from paracellular to transcellular [75]. It has also been reported that cyclic peptides are less susceptible to amino- and carboxypeptidases than linear peptides because the amino and carboxy terminals are protected from these enzymes [76].

Peptide structural modification has been applied to improve peptide membrane permeation. Metabolism of peptides can occur in various regions along the route to oral absorption, and inhibition of this degradation is advantageous in enhancing drug delivery. To improve enzymatic stability, peptides have been converted to peptidomimetics. In this case, the peptide bond is converted to its bioisostere that is stable to proteolytic enzymes [86]. Other structural modification strategies to improve membrane permeation of peptides include lipidization, halogenation, glycosylation, cationization, and conjugation to polymers [85].

2.5.2 Formulations

Intestinal absorption of drug molecules can be improved by designing an optimal formulation [86, 89]. For peptides, several methods to enhance drug absorption have been suggested, including addition of ion-pairing and complexation molecules, nonsurfactant membrane permeation enhancers, surfactant adjuvants, or combinations of these additives [86]. Addition of perturbants of tight junctions such as cytoskeletal agents, oxidants, hormones, calcium chelators, and bacterial toxins into the formulation has been investigated to improve drug permeation [89]. Another novel delivery system involves the use of mucoadhesives to enhance drug delivery because of their long retention time at the targeted mucosal membrane; lectins have been identified as potential carriers for peptides in an oral mucoadhesive system [4]. For peptides, their coadministration with inhibitors of metabolizing enzymes has also been suggested to increase oral absorption [86, 90, 91].

Modulations of protein–protein interactions in the intercellular junctions have been shown to increase paracellular permeation of molecules through the biological barriers [1]. Peptides derived from tight junction proteins such as occludin and claudin can modulate the intercellular junctions of the biological barriers. A 29-mer peptide (C1C2) derived from claudin-1 can lower the TEER values of cell monolayers as well as increase the permeation of paracellular permeation of 10 kDa protein, which was labeled with fluorescein isothiocyanate (FITC) [92]. This peptide enhances the brain delivery of opioids such as tetrodotoxin and opioid peptide *in vivo* [92]. A claudin hexapeptide (DFNYNP) can disrupt and weaken the tight junctions of cell monolayers. This disruption is because of internalization of claudins into the vehicles in cell cytoplasmic domain [93]. An occludin peptide called OCC1 can lower the TEER values of Caco-2 cell monolayers, and it increases permeation of ^{14}C-mannitol across the cell monolayers [94]. HAV- and ADT-peptides derived from the EC1 domain of E-cadherin can also modulate the intercellular junctions of MDCK cell monolayers [95, 96]. These peptides lowered the TEER values of MDCK cell monolayers and also enhanced the ^{14}C-mannitol transport across the monolayers. The HAV hexapeptide (Ac-SHAVSS-NH_2) can improve brain delivery of ^{14}C-mannitol, ^{3}H-daunomycin (anticancer agent), and Gd-DTPA (MRI contrast agent) using the *in-situ* rat brain perfusion model [97]. These peptides can also increase the brain delivery of Gd-DTPA, a near IR dye R800, and an 800cw polyethylene glycol (25 kDa) through the BBB in *in vivo* balb/c mice [98]. These results indicate the possibility of delivering drug and diagnostic molecules through the biological barriers *in vivo* by modulating the cell–cell adhesion at the intercellular junctions.

2.6 CONCLUSIONS

The absorption of orally administered drug molecules depends on the successful passage of the molecules through several barriers to drug delivery. The gastrointestinal epithelial layer is a formidable obstacle to the passage of drugs. The drug molecules can pass either between the cells or through the cells, depending on their physiochemical properties. Recent studies have shown that metabolism within the intestine forms a major obstruction to drug absorption. The concerted activity of these drug-metabolizing enzymes and efflux systems confounds the problem. Although many challenges exist for traversing the intestinal epithelial layer, pharmaceutical scientists and medicinal chemists are overcoming them with innovative methods to optimize pharmacological activity and enhance drug delivery.

ACKNOWLEDGMENT

Research in Dr. Siahaan's laboratory on the modulation of intercellular junctions has been funded by R01-NS075374 grant from the National Institutes of Health.

REFERENCES

1. Laksitorini M., Prasasty V. D., Kiptoo P. K., Siahaan T. J., *Ther Deliv* 2014, **5**, 1143–1163.
2. Lipka E., Crison J., Amidon G. L., *J Control Release* 1996, **39**, 121–129.
3. Yu L. X., Lipka E., Crison J. R., Amidon G. L., *Adv Drug Deliv Rev* 1996, **19**, 359–376.
4. Kompella U. B., Lee V. H., *Adv Drug Deliv Rev* 2001, **46**, 211–245.
5. Fricker G., Drewe J., *J Pept Sci* 1996, **2**, 195–211.
6. Turner J. R., *Nat Rev* 2009, **9**, 799–809.
7. Martinez M., Amidon G., Clarke L., Jones W. W., Mitra A., Riviere J., *Adv Drug Deliv Rev* 2002, **54**, 825–850.
8. Fricker G., Miller D. S., *Pharmacol Toxicol* 2002, **90**, 5–13.
9. Doherty M. M., Charman W. N., *Clin Pharmacokinet* 2002, **41**, 235–253.
10. Adson A., Raub T. J., Burton P. S., Barsuhn C. L., Hilgers A. R., Audus K. L., Ho N. F., *J Pharm Sci* 1994, **83**, 1529–1536.
11. Anderson J. M., Van Itallie C. M., *Cold Spring Harb Perspect Biol* 2009, **1**, a002584.
12. Anderson J. M., Van Itallie C. M., *Curr Biol* 2008, **18**, R941–R943.

13. Van Itallie C. M., Aponte A., Tietgens A. J., Gucek M., Fredriksson K., Anderson J. M., *J Biol Chem* 2013, **288**, 13775–13788.
14. Van Itallie C. M., Fanning A. S., Holmes J., Anderson J. M., *J Cell Sci* 2010, **123**, 2844–2852.
15. Zheng K., Trivedi M., Siahaan T. J., *Curr Pharm Des* 2006, **12**, 2813–2824.
16. Jensen E., Bundgaard H., *Acta Pharm Nord* 1991, **3**, 147–150.
17. Landowski C. P., Sun D., Foster D. R., Menon S. S., Barnett J. L., Welage L. S., Ramachandran C., Amidon G. L., *J Pharmacol Exp Ther* 2003, **306**, 778–786.
18. Gupta D., Varghese Gupta S., Dahan A., Tsume Y., Hilfinger J., Lee K. D., Amidon G. L., *Mol Pharm* 2013, **10**, 512–522.
19. Varma M. V., Ambler C. M., Ullah M., Rotter C. J., Sun H., Litchfield J., Fenner K. S., El-Kattan A. F., *Curr Drug Metab* 2010, **11**, 730–742.
20. Shepard R. L., Cao J., Starling J. J., Dantzig A. H., *Int J Cancer* 2003, **103**, 121–125.
21. Van Itallie C. M., Anderson J. M., *Methods Mol Biol* 2011, **762**, 1–11.
22. Deli M. A., *Biochim Biophys Acta* 2009, **1788**, 892–910.
23. Yap A. S., Mullin J. M., Stevenson B. R., *J Membr Biol* 1998, **163**, 159–167.
24. Cereijido M., Valdes J., Shoshani L., Contreras R. G., *Annu Rev Physiol* 1998, **60**, 161–177.
25. Mitic L. L., Anderson J. M., *Annu Rev Physiol* 1998, **60**, 121–142.
26. Gonzalez-Mariscal L., Betanzos A., Nava P., Jaramillo B. E., *Prog Biophys Mol Biol* 2003, **81**, 1–44.
27. Gonzalez-Mariscal L., Chavez de Ramirez B., Cereijido M., *J Membr Biol* 1985, **86**, 113–125.
28. Rothen-Rutishauser B., Riesen F. K., Braun A., Gunthert M., Wunderli-Allenspach H., *J Membr Biol* 2002, **188**, 151–162.
29. Walsh S. V., Hopkins A. M., Nusrat A., *Adv Drug Deliv Rev* 2000, **41**, 303–313.
30. Tsukita S., Furuse M., Itoh M., *Curr Opin Cell Biol* 1999, **11**, 628–633.
31. Cummins P. M., *Mol Cell Biol* 2012, **32**, 242–250.
32. Piontek J., Winkler L., Wolburg H., Muller S. L., Zuleger N., Piehl C., Wiesner B., Krause G., Blasig I. E., *FASEB J* 2008, **22**, 146–158.
33. Schneeberger E. E., Lynch R. D., *Am J Physiol Cell Physiol* 2004, **286**, C1213–C1228.
34. Lapierre L. A., *Adv Drug Deliv Rev* 2000, **41**, 255–264.
35. Anderson J. M., Fanning A. S., Lapierre L., Van Itallie C. M., *Biochem Soc Trans* 1995, **23**, 470–475.
36. Anderson J. M., Van Itallie C. M., *Am J Physiol* 1995, **269**, G467–G475.
37. Rao R., *Ann N Y Acad Sci* 2009, **1165**, 62–68.
38. Suzuki T., Elias B. C., Seth A., Shen L., Turner J. R., Giorgianni F., Desiderio D., Guntaka R., Rao R., *Proc Natl Acad Sci U S A* 2009, **106**, 61–66.
39. Jain S., Suzuki T., Seth A., Samak G., Rao R., *Biochem J* 2011, **437**, 289–299.
40. Van Itallie C. M., Tietgens A. J., LoGrande K., Aponte A., Gucek M., Anderson J. M., *J Cell Sci* 2012, **125**, 4902–4912.
41. Hartsock A., Nelson W. J., *Biochim Biophys Acta* 2008, **1778**, 660–669.

42. Mitic L. L., Van Itallie C. M., Anderson J. M., *Am J Physiol Gastrointest Liver Physiol* 2000, **279**, G250–G254.
43. Madara J. L., *Annu Rev Physiol* 1998, **60**, 143–159.
44. Burton P. S., Goodwin J. T., Vidmar T. J., Amore B. M., *J Pharmacol Exp Ther* 2002, **303**, 889–895.
45. Hilgendorf C., Spahn-Langguth H., Regardh C. G., Lipka E., Amidon G. L., Langguth P., *J Pharm Sci* 2000, **89**, 63–75.
46. Shore E. M., Nelson W. J., *J Biol Chem* 1991, **266**, 19672–19680.
47. Takeichi M., *Annu Rev Biochem* 1990, **59**, 237–252.
48. Chothia C., Jones E. Y., *Annu Rev Biochem* 1997, **66**, 823–862.
49. Koch A. W., Bozic D., Pertz O., Engel J., *Curr Opin Struct Biol* 1999, **9**, 275–281.
50. Boggon T. J., Murray J., Chappuis-Flament S., Wong E., Gumbiner B. M., Shapiro L., *Science* 2002, **296**, 1308–1313.
51. Yap A. S., Brieher W. M., Gumbiner B. M., *Annu Rev Cell Dev Biol* 1997, **13**, 119–146.
52. Delva E., Tucker D. K., Kowalczyk A. P., *Cold Spring Harb Perspect Biol* 2009, **1**, a002543.
53. Dusek R. L., Godsel L. M., Chen F., Strohecker A. M., Getsios S., Harmon R., Muller E. J., Caldelari R., Cryns V. L., Green K. J., *J Invest Dermatol* 2007, **127**, 792–801.
54. Dusek R. L., Godsel L. M., Green K. J., *J Dermatol Sci* 2007, **45**, 7–21.
55. Getsios S., Amargo E. V., Dusek R. L., Ishii K., Sheu L., Godsel L. M., Green K. J., *Differentiation* 2004, **72**, 419–433.
56. Martinez M. N., Amidon G. L., *J Clin Pharmacol* 2002, **42**, 620–643.
57. Agoram B., Woltosz W. S., Bolger M. B., *Adv Drug Deliv Rev* 2001, **50** Suppl 1, S41–S67.
58. Tsuji A., Tamai I., *Pharm Res* 1996, **13**, 963–977.
59. Charman W. N., Porter C. J., Mithani S., Dressman J. B., *J Pharm Sci* 1997, **86**, 269–282.
60. Engman H. A., Lennernas H., Taipalensuu J., Otter C., Leidvik B., Artursson P., *J Pharm Sci* 2001, **90**, 1736–1751.
61. Wacher V. J., Salphati L., Benet L. Z., *Adv Drug Deliv Rev* 2001, **46**, 89–102.
62. Cong D., Doherty M., Pang K. S., *Drug Metab Dispos* 2000, **28**, 224–235.
63. Benet L. Z., Izumi T., Zhang Y., Silverman J. A., Wacher V. J., *J Control Release* 1999, **62**, 25–31.
64. Suzuki H., Sugiyama Y., *Eur J Pharm Sci* 2000, **12**, 3–12.
65. Behrens I., Kamm W., Dantzig A. H., Kissel T., *J Pharm Sci* 2004, **93**, 1743–1754.
66. Daniel H., *Ann N Y Acad Sci* 2000, **915**, 184–192.
67. Wagner D., Spahn-Langguth H., Hanafy A., Koggel A., Langguth P., *Adv Drug Deliv Rev* 2001, **50** Suppl 1, S13–S31.
68. Benet L. Z., Cummins C. L., *Adv Drug Deliv Rev* 2001, **50** Suppl 1, S3–S11.
69. Tournier N., Decleves X., Saubamea B., Scherrmann J. M., Cisternino S., *Curr Pharm Des* 2011, **17**, 2829–2842.
70. Mandery K., Balk B., Bujok K., Schmidt I., Fromm M. F., Glaeser H., *Eur J Pharm Sci* 2012, **46**, 79–85.

71. Kane G. C., Lipsky J. J., *Mayo Clin Proc* 2000, **75**, 933–942.
72. Zhang Y., Benet L. Z., *Clin Pharmacokinet* 2001, **40**, 159–168.
73. Goodwin J. T., Conradi R. A., Ho N. F., Burton P. S., *J Med Chem* 2001, **44**, 3721–3729.
74. Pauletti G. M., Okumu F. W., Borchardt R. T., *Pharm Res* 1997, **14**, 164–168.
75. Okumu F. W., Pauletti G. M., Vander Velde D. G., Siahaan T. J., Borchardt R. T., *Pharm Res* 1997, **14**, 169–175.
76. Pauletti G. M., Gangwar S., Siahaan T. J., Jeffrey A., Borchardt, R. T., *Adv Drug Deliv Rev* 1997, **27**, 235–256.
77. Burton P. S., Conradi R. A., Ho N. F., Hilgers A. R., Borchardt R. T., *J Pharm Sci* 1996, **85**, 1336–1340.
78. Chikhale E. G., Ng K. Y., Burton P. S., Borchardt R. T., *Pharm Res* 1994, **11**, 412–419.
79. Burton P. S., Conradi R. A., Hilgers A. R., Ho N. F., Maggiora L. L., *J Control Release* 1992, **19**, 87–97.
80. Lipinski C., Hopkins A., *Nature* 2004, **432**, 855–861.
81. Lipinski C. A., Lombardo F., Dominy B. W., Feeney P. J., *Adv Drug Deliv Rev* 2001, **46**, 3–26.
82. Gupta D., Gupta S. V., Lee K. D., Amidon G. L., *Mol Pharm* 2009, **6**, 1604–1611.
83. Gupta S. V., Gupta D., Sun J., Dahan A., Tsume Y., Hilfinger J., Lee K. D., Amidon G. L., *Mol Pharm* 2011, **8**, 2358–2367.
84. Han H. K., Amidon G. L., *AAPS PharmSci* 2000, **2**, E6.
85. Witt K. A., Gillespie T. J., Huber J. D., Egleton R. D., Davis T. P., *Peptides* 2001, **22**, 2329–2343.
86. Aungst B. J., *J Pharm Sci* 1993, **82**, 979–987.
87. Stella V. J., Nti-Addae K. W., *Adv Drug Deliv Rev* 2007, **59**, 677–694.
88. Bundgaard H., Falch E., Jensen E., *J Med Chem* 1989, **32**, 2503–2507.
89. Lutz K. L., Siahaan T. J., *J Pharm Sci* 1997, **86**, 977–984.
90. Mizuma T., Ohta K., Koyanagi A., Awazu S., *J Pharm Sci* 1996, **85**, 854–857.
91. Gotoh S., Nakamura R., Nishiyama M., Quan Y. S., Fujita T., Yamamoto A., Muranishi S., *J Pharm Sci* 1996, **85**, 858–862.
92. Zwanziger D., Hackel D., Staat C., Bocker A., Brack A., Beyermann M., Rittner H., Blasig I. E., *Mol Pharm* 2012, **9**, 1785–1794.
93. Zwanziger D., Staat C., Andjelkovic A. V., Blasig I. E., *Ann N Y Acad Sci* 2012, **1257**, 29–37.
94. Wong V., Gumbiner B. M., *J Cell Biol* 1997, **136**, 399–409.
95. Makagiansar I. T., Avery M., Hu Y., Audus K. L., Siahaan T. J., *Pharm Res* 2001, **18**, 446–453.
96. Sinaga E., Jois S. D., Avery M., Makagiansar I. T., Tambunan U. S., Audus K. L., Siahaan T. J., *Pharm Res* 2002, **19**, 1170–1179.
97. Kiptoo P., Sinaga E., Calcagno A. M., Zhao H., Kobayashi N., Tambunan U. S., Siahaan T. J., *Mol Pharm* 2011, **8**, 239–249.
98. On N. H., Kiptoo P., Siahaan T. J., Miller D. W., *Mol Pharm* 2014, **11**, 974–981.

3

PHYSICOCHEMICAL PROPERTIES, FORMULATION, AND DRUG DELIVERY

Dewey H. Barich[1], Mark T. Zell[2], and Eric J. Munson[1]

[1] *Department of Pharmaceutical Chemistry, The University of Kansas, Lawrence, KS, USA*
[2] *Pfizer Global Research and Development, Ann Arbor Laboratories, Ann Arbor, MI, USA*

3.1 INTRODUCTION

The goal of drug formulation and delivery is to administer a drug at a therapeutic concentration to a particular site of action for a specified period of time. The design of the final formulated product for drug delivery depends upon several factors. First, the drug must be administered using a narrow set of parameters that are defined by the therapeutic action of the drug. Some of the parameters include the site of action (either targeted to a specific region of the body or systemic), the concentration of the drug at the time of administration, the duration that the drug must remain at a therapeutic concentration, and initial release rate of the drug for oral/controlled release systems. Second, the drug must remain physically and chemically stable in the formulation for at least 2 years. Third, the choice of delivery method must reflect the preferred administration route for the drug, such as oral, parenteral, and transdermal.

A complete knowledge of the relevant therapeutic and physicochemical properties of the drug is required to determine the proper formulation and delivery method of a drug. For example, the physicochemical properties of the drug strongly influence the choice of delivery methods. Figure 3.1 shows the interdependence of the three main

Drug Delivery: Principles and Applications, Second Edition. Edited by Binghe Wang, Longqin Hu, and Teruna J. Siahaan.

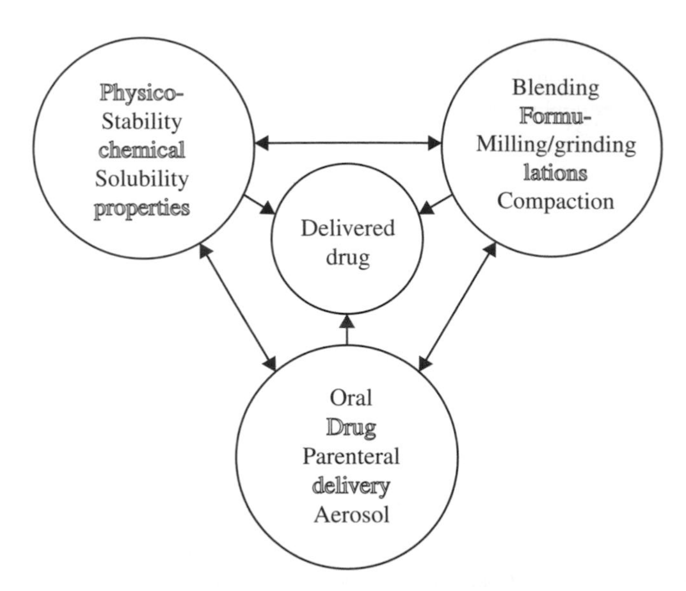

FIGURE 3.1 Schematic diagram showing the interdependence of physicochemical properties, formulation, and drug delivery.

topics covered in this chapter. This creates a problem in dividing this chapter into specific sections, as a discussion of the important physicochemical properties of a drug will be different for oral administration of a solid tablet compared with parenteral administration of a drug in solution. For this reason, we have chosen to take a broad approach in the physicochemical properties section of discussing the basic physicochemical properties that are determined for almost all drugs. A similar approach has been taken in the formulation and delivery sections.

This chapter is divided into three sections. In the physicochemical section, the two most relevant physicochemical properties to drug delivery—solubility and stability—are discussed. In addition to providing a basic understanding of the importance of solubility and stability to drug delivery, methods to enhance solubility and physical and chemical stability are described. The second section focuses on the processes required for the proper drug formulation. Since most drugs are administered in the solid state, the formulation process for tablets is described in detail. The last section is a discussion of some of the basic drug delivery methods, with an emphasis on the physicochemical properties that impact those methods.

3.2 PHYSICOCHEMICAL PROPERTIES

The most important goal in the delivery of a drug is to bring the drug concentration to a specific level and maintain it at that level for a specific duration of time. Stability and solubility are two key physicochemical properties that must be considered when designing a successful drug formulation. Many challenges must be overcome to formulate a product that has sufficient chemical and physical stability to avoid degradation during the shelf life of the product, yet has sufficient solubility (and dissolution rate) to reach the required therapeutic level.

The physicochemical properties of the drug in both solution and solid state play a critical role in drug formulation. The solid-state form of the drug is often preferred, because it is often more chemically stable, easier to process, and more convenient to administer than liquid formulations. However, if the drug is in the solid state, it must dissolve before it can be therapeutically active, and once it is in solution, it must be both sufficiently soluble and chemically stable. For these reasons it is critical to determine the physicochemical properties of the drug both in solution and in the solid state.

There are several parameters that affect the solubility and chemical stability of a drug in solution. The pH of the solution can dramatically affect both the solubility and chemical stability of the drug. Buffer concentration/composition and ionic strength can also have an effect, especially on chemical stability. The hydrophobic/hydrophilic nature of the drug influences solubility. A typical characterization of a drug will start with a study of the chemical stability of the drug as a function of pH. The structure of the degradation products will be characterized to determine the mechanism of the degradation reaction.

In the solid state, the form of the drug will affect both the solubility and the physicochemical stability of the drug. A full characterization of the drug in the solid state will often include a determination of the melting point and heat of fusion using differential scanning calorimetry, loss of solvent upon heating using thermogravimetric analysis, and a characterization of the molecular state of the solid using diffraction and spectroscopic techniques.

In the following two sections, solubility and stability will be discussed as they relate to drug formulation. In the solubility section, the emphasis is on methods to increase solubility. In the stability section, the emphasis is on describing the types of reactions that lead to decreased stability.

3.2.1 Solubility

A drug must be maintained at a specific concentration to be therapeutically active. In many cases the drug solubility is lower than the required concentration, in which case the drug is no longer effective [1]. There is a trend in new drug molecules toward larger molecular weights, which often leads to lower solubility. The ability to formulate a soluble form of a drug is becoming both more important and more challenging. This has resulted in an extensive research on methods to increase drug solubility.

Solubility is affected by many factors. One of the most important factors is pH. Other factors that affect the solubility of the drug include temperature, hydrophobicity of the drug, solid form of the drug, and the presence of complexing agents in solution.

For drugs with low solubility, special efforts must be made to bring the concentration into the therapeutically active range. In this section some of the common methods to increase solubility will be discussed: salt versus free form, inclusion compounds, prodrugs, solid form selection, and dissolution rate. It should be noted that efforts to increase solubility also have an influence (often negative) on the stability of a compound. For this reason the most soluble form is often not the first choice when formulating the drug.

3.2.1.1 Salt versus Free Forms One of the easiest ways to increase the solubility of a therapeutic agent is to make a corresponding salt form of the drug. The salt form must be made from either the free acid or free base. Carboxylic acids are the most common acidic functional groups found in drug molecules, while amines are the most common basic groups. An important consideration in the choice of salt versus free forms of a drug is that the pH changes depending upon the location in the intestinal tract. In the stomach, the pH is typically 1–3 and changes to 6–8 in the small intestine. Since the majority of adsorption occurs in the small intestine, it is often desirable to have the maximum solubility at neutral to basic pH values. In general, the acid form of a drug will be ionized at intestinal pH values and therefore more soluble, whereas the basic form will be unionized and less soluble. Salts are typically more soluble than the free forms, although this often comes with increased hydrophilicity and possible decrease in chemical stability due to increased moisture sorption.

Usually the choice of salt versus free forms is made based upon the physicochemical properties of each individual compound. However, some generalizations can be made. Free acid forms of a drug usually have adequate solubility and dissolution rates at pH values found in the intestine, and salts of weak bases may be preferred to the free forms because of greater solubility and dissolution rates. It should also be noted that the counter ion can have a dramatic effect upon the solubility and/or stability of the drug. Salt form screening is routinely performed on compounds to determine the counter ion that possesses the best combination of solubility and stability.

3.2.1.2 Inclusion Compounds Another method for improving solubility is to create an inclusion compound between the drug molecule and a host molecule. To be effective, the host/guest inclusion compound must have a higher solubility than the individual drug molecule. An inclusion complex of a drug is usually not crystalline and thus should have a higher solubility than a crystalline material. Cyclodextrins complexed to drugs are an example of inclusion compounds commonly used in pharmaceutics.

Cyclodextrins are nonreducing cyclic oligosaccharides made up of six to eight glucopyranose molecules. This class of molecules has a unique structure that is often represented as a tapered doughnut (with the opening at one side larger than the other). The guest molecule then fits inside this cavity and is much less likely to crystallize. Such complexes are also used to improve drug stability by reducing interactions between the drug and its environment. Chemically modified cyclodextrins, which exhibit different stabilizing effects than the natural forms, are also used. They also increase the solubility of insoluble drugs by complexing the drug with the cyclodextrin, generating a metastable form of the drug. Two examples of drugs whose solubility is enhanced by cyclodextrins are prednisolone [2] and prostaglandin E_1 [3]. Figure 3.2 shows an example demonstrating the improvement in solubility provided by sulfobutyl ether-β-cyclodextrin (Captisol™) for prednisolone [4].

3.2.1.3 Prodrugs Prodrugs are chemically modified forms of the drug that commonly contain an additional functional group (e.g., an ester group) designed to enhance solubility, stability, and/or transport across a biological membrane. Once the

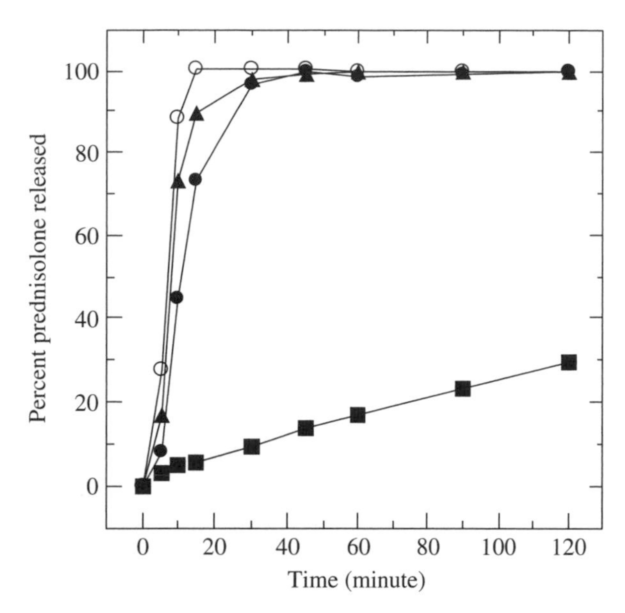

FIGURE 3.2 Plot of percent prednisolone released versus time for different complexes of cyclodextrin and prednisolone. Prednisolone : $(SBE)_{7m}$-β-CD at a 1 : 1 molar ratio (●), Prednisolone : $(SBE)_{7m}$-β-CD at a 1 : 2 molar ratio (○), Prednisolone : HP-β-CD at a 1 : 2 molar ratio (▲), Prednisolone : sugar (■). Okimoto et al. [4], pp. 1562–1568, figure 2. Reproduced with permission of Springer.

prodrug is inside the body, the additional functional group is cleaved off, either hydrolytically or enzymatically, leaving the drug so that it may fulfill its therapeutic function. Examples of prodrugs (given as prodrug[drug]) that improve solubility include fosphenytoin[phenytoin] [5–7], valacyclovir[acyclovir] [8–10], and capecitabine[5-fluorouracil] [11]. More discussion of prodrugs can be found in Chapter 12.

3.2.1.4 Solid Form Selection A drug can exist in multiple forms in the solid state. If the two forms have the same molecular structure but different crystal packing, then they are polymorphs. Pseudopolymorphs (or solvatomorphs) differ in the level of hydration/solvation between forms. Polymorphs and pseudopolymorphs in principle will have a different solubility, melting point, dissolution rate, etc. While less thermodynamically stable polymorphs have higher solubilities, they also have the potential to convert to the more thermodynamically stable form. This form conversion can lead to reduced solubilities for the formulated product. One example is the case of Ritonavir, a protease inhibitor compound used to treat AIDS. Marketed by Abbott Labs as Norvir, this compound began production in a semisolid form and an oral liquid form. In July 1998, dissolution tests of several new batches of the product failed. The problem was traced to the appearance of a previously unknown polymorph (Form II) of the compound. This form is thermodynamically more stable than Form I and, therefore, less soluble. In this case the solubility is at least a factor of two below that of Form I [12]. The discovery of this new polymorph ultimately led to a temporary withdrawal of the solid form of Norvir from the market and a search for a new formulation.

3.2.1.5 Dissolution Rate While not directly related to solubility, the ability to rapidly reach the therapeutic concentration may be useful for fast-acting therapeutic agents, and compensate for drugs that may have sufficient solubility but are metabolized/excreted too quickly to reach the desired concentration. An example of a method to enhance dissolution is the WOWTAB® technology by Astellas Pharma, Japan [13].

3.2.2 Stability

Formulation scientists must consider two types of stability: chemical and physical. Physical stability is the change in the physical form of the drug, for example, an amorphous form changing into a crystalline form. The chemical composition remains the same as it was prior to crystallization, but the drug now has different physical properties. Chemical stability is a change in the molecular structure through a chemical reaction. Hydrolysis and oxidation are two common chemical degradation pathways.

3.2.2.1 Physical Stability Physical stability can refer to molecular-level changes, such as polymorphic changes, or macroscopic changes, such as dissolution rate or tablet hardness. At the molecular level, form changes include amorphous to crystalline, changes in crystalline form (polymorphism), and changes in solvation state (solvatomorphism). The impact of polymorphic changes on the solubility of Ritonavir was discussed in the previous section. In general, a metastable solid form may convert to a more thermodynamically stable form, and it is usually desirable to market the most stable form if possible to avoid such transformations. The presence of seed crystals of the more stable form may initiate or accelerate the conversion from the metastable form to the more stable form. In addition, the presence of solvents, especially water, may cause formation of a solvate with significantly different physicochemical properties. Desolvation is also a possible reaction. For drug formulations, the choice of salt forms (hydrates, solvates, and polymorphs) plays a role in identifying the most suitable form for the pharmaceutical product. Polymorphism in drug formulations makes the characterization of polymorphic forms very important. This is most commonly done with X-ray powder diffraction or solid-state NMR spectroscopy.

When improvements in physical stability of a product are needed, choices must be made based upon the nature of the problem and the desired goal. One of the first choices made is to use the most stable polymorph of the drug. This may involve an extensive polymorph screening effort to attempt to find the most stable polymorph. If the most stable polymorph is undesirable for some reason (e.g., solubility issues), then avoiding contamination of the desired polymorph with seeds of the most stable polymorph becomes very important. In a product that uses an amorphous form of a drug, it is critical to inhibit crystallization to avoid dramatic changes in stability and solubility.

3.2.2.2 Chemical Stability Chemical degradation of the drug includes reactions such as hydrolysis, dehydration, oxidation, photochemical degradation, or reaction with excipients. The constant presence of water and oxygen in our environment means that exposure to moisture or oxygen can affect the chemical stability of a compound. Chemical stability is very important not only because a sufficient amount of the drug is needed at the time of administration for therapeutic purposes but also because chemical degradation products may adversely affect the properties of the formulated product, and may even be toxic.

Determining how a drug degrades and what factors affect degradation is very important in pharmaceutical product development. The importance of reaching (or avoiding) the activation barrier of a particular chemical process makes temperature one of the most important variables in this area [14]. A second factor in drug degradation is pH. The degradation rate depends on the pH of the formulation and/or the compartments of the body in which the drug is present. Many drug degradation pathways are catalyzed by either hydronium or hydroxide ions, reiterating the important role of water [14]. Described below (with an example or two) are several degradation reactions including hydrolysis, dehydration, oxidation, photodegradation, isomerization, racemization, decarboxylation, and elimination.

Hydrolysis is one of the most common drug degradation reactions. In hydrolysis reactions the drug reacts with water to form two degradation products. The two most common types of hydrolysis reactions encountered in pharmaceutical chemistry are the hydrolysis of ester or amide functional groups. Esters hydrolyze to form carboxylic acids and alcohols, while amides form carboxylic acids and amines. For example, the ester bond in aspirin is hydrolyzed to produce salicylic acid and acetic acid, while the amide bond is hydrolyzed in acetaminophen [15–17].

Dehydration reactions are another common degradation pathway. Ring closures are a fairly common type of dehydration, as is seen for both lactose [18, 19] and glucose [20–22]. Both of these compounds dehydrate to form 5-(hydroxymethyl)-2-furfural. Batanopride is another example of a compound that can undergo a dehydration reaction [23].

Elimination degradation pathways are also possible. Decarboxylation, in which a carboxylic acid releases a molecule of CO_2, occurs for p-aminosalicylic acid [24]. Oxidation is very common as well, largely due to the presence of oxygen during manufacture and/or storage. Several examples can be found in Yoshioka and Stella [14]. Isomerization and racemization reactions are other degradation pathways. Some examples of compounds that undergo isomerization reactions are amphotericin B [25] and tirilazad [26].

Photodegradation of pharmaceuticals has been known for decades. A complication encountered when studying photodegradation reactions is that there are many degradation pathways that each have the potential to yield different products. When oxidizers are present, photodegradation can accompany oxidation.

There are several options available to improve the stability of drugs. One is the use of cyclodextrins, in which the formation of the inclusion complex produces a more stable form of the drug. Examples of cyclodextrins inhibiting drug degradation include tauromustine [27], mitomycin C [28], and thymoxamine [29]. Another possibility is to generate a prodrug that has increased stability compared to the parent compound. Examples of prodrugs that enhance stability include [prodrug (drug)] enalapril (enalaprilat) [30–32] and dipivefrin (epinephrine) [33].

3.3 FORMULATIONS

Formulation is the stage of product manufacturing in which the drug is combined with various excipients to prepare a dosage form for delivery of the drug to the patient. Excipients are defined by IPEC-Americas [34] as "... substances other than the pharmacologically active drug or prodrug which are included in the manufacturing process or are contained in a finished pharmaceutical product dosage form." These include binders to form a tablet, aggregates to keep the tablet together, disintegrants to aid dissolution once the drug is administered, and coloring or flavoring agents. Excipients help keep the drug in the desired form until administration, aid in delivering the drug, control the release rate of the drug, or make the product more appealing in some way to the patient.

Formulation is dictated by the physicochemical properties of the drug and excipients. Each drug delivery method has specific formulation issues. As previously mentioned, the solid dosage form is the most convenient and most preferred means of administering drugs, and therefore the discussion here will focus on solid dosage forms. The vast majority of solid dosage forms are tablets, which are produced by compression or molding. Powders are the most common form of both the drug and the excipients prior to processing. The process of creating tablets from bulk materials has a number of steps. Some of these are discussed later.

3.3.1 Processing Steps

First, milling is often used to ensure that the particle size distribution is adequate for mixing. Milling both reduces the particle size and produces size and shape uniformity. There are several milling options, though perhaps the most common is the ball mill, in which balls are placed inside a hard cylindrical container along with the bulk drug. The cylinder is then turned horizontally along its long axis to cause the balls to repeatedly tumble over one another, thereby breaking the drug particles into smaller pieces.

Next, the drug and excipients must be blended or mixed together. It is very important at this stage that the bulk properties of the materials be conducive to mixing. This means that the materials must have good flowability characteristics. Lubricants such as magnesium stearate may be added to improve the flowability of the formulation.

Once the formulation has been blended, it must then be compressed into a tablet. Flowability remains important at this stage of the processing because a uniform dose of the blended ingredient mixture must be delivered to the tableting machine. Poor flowability results in poor tablet weight reproducibility. Lubricants are needed to ensure that the tablet can be removed intact from the die once it has been compressed. Finally, the tablet may require a coating. This could be as simple as a flavor coating, or it could be an enteric coating designed to avoid an upset stomach by delaying dissolution until the tablet enters the small intestine.

3.3.2 Influence of Physicochemical Properties on Drugs in Formulations

Most of the processing steps depend at least indirectly upon the physicochemical properties of the drug. Particle size, shape, and morphology often are determined by the solid form of the drug and the conditions from which the drug is crystallized. Aspirin, for example, can have multiple crystal morphologies depending upon the conditions of recrystallization [35]. Processing can also result in changes in the form of the drug. Amorphous drug formation, changes in the polymorphic form of the drug, or the production of crystal defects can all have a negative effect upon the solubility and stability of the drug [36]. Drug–excipient interactions can affect both solubility and stability. These interactions impact the physical properties of the drug by altering its chemical nature by reactions such as desolvation, or the Maillard reaction (also known as the browning reaction based on the color of the products).

Physicochemical changes in the form of the drug at the formulation and processing stages are almost always undesirable. Such changes can be very costly if found only toward the end of product development. Thus, many times it is desirable to perform preformulation studies to determine the optimum form for delivery [14, 37].

3.3.3 Other Issues

New excipients are needed in the industry, as not all formulation needs are satisfied by currently known excipients. This situation is likely to worsen over time as new products, each with potentially unique requirements, are brought to the development stage. Despite this need, the introduction of new excipients is becoming more difficult [38] because new excipients face similar regulatory requirements as new drugs themselves. Difficulty in satisfying different nations' regulatory requirements on excipients sometimes makes it more difficult for companies to make a single product that can be marketed in different countries.

3.4 DRUG DELIVERY

For many drugs, the therapeutic nature of the drug dictates the method of administration. For example, oral drug delivery may be the most logical choice for gastrointestinal diseases. If drug release is systemic, then the choice of method often relies on the physicochemical and therapeutic properties of the drug. Transdermal drug delivery, although having the advantage of being noninvasive, has several criteria that must be met by the drug in order to be delivered properly, such as high potency, ready permeability through the stratum corneum, and non-irritation.

In drug delivery, the three most important questions are as follows: When is the drug delivered? Where is the drug delivered? How is the drug delivered? For this reason the rest of the section is divided into three parts that address the when, where, and how of drug delivery.

3.4.1 Duration of Release

The goal of drug delivery is to maintain the drug at the appropriate therapeutic level for a specified period of time. There are several methods to achieve this goal, some of which are demonstrated in Figure 3.3. The first is the administration of a single dose, with immediate release of drug to the site of action. This method is useful for acute therapeutic treatment requiring a short period of action. For chronic problems, the goal is to maintain the drug at the therapeutic level for a sustained period of time. Multiple-dose administration is one method for providing sustained therapeutic levels of drug. However, there are many disadvantages to multiple-dose therapy, including variations in drug levels during the treatment period and requiring patient compliance with dosage regimen requirements. To avoid this problem, non-immediate release devices are used to deliver the drug over an extended period of time. Non-immediate release devices have three types of release mechanisms: delayed release, prolonged release, and controlled release. Delayed release allows multiple doses to be incorporated into a single dosage form, alleviating the problem of frequent dosing and patient noncompliance. The prolonged release device extends the release of the drug, for example, by slowing the dissolution rate of the drug compared with an immediate release device. The controlled release device meters out the drug to maintain a constant release rate throughout the desired dosage period. In the prolonged and controlled release dose, there is usually an initial release of drug to bring the drug into the therapeutic window, followed by additional drug that is released over a longer period of time. Nonimmediate release devices maintain a more consistent level of drug than multiple doses while retaining the advantage of requiring fewer doses, which helps with patient compliance. The disadvantage of non-immediate

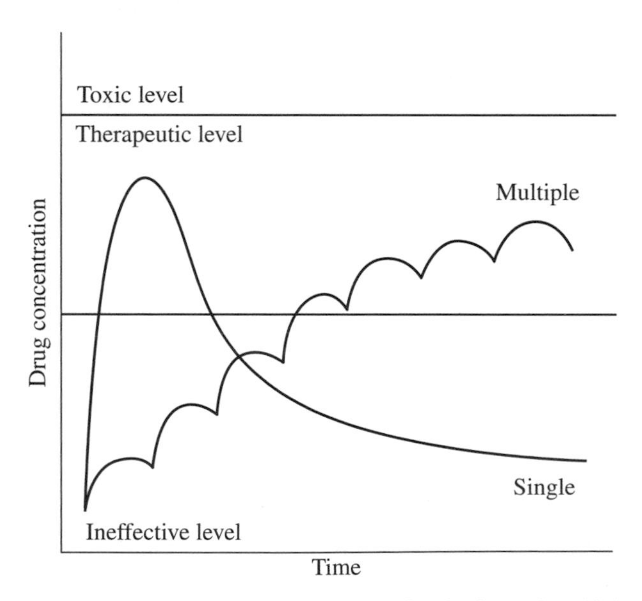

FIGURE 3.3 Plot of drug concentration versus time for single- and multiple-dose therapy.

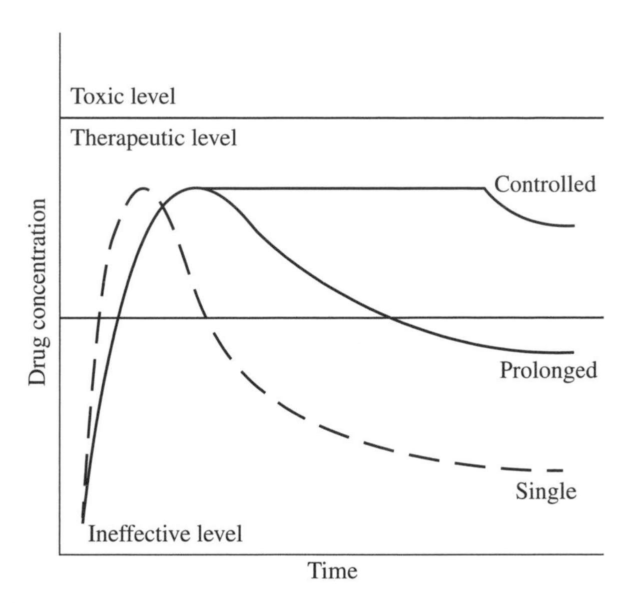

FIGURE 3.4 Plot of drug concentration versus time for nonimmediate release systems.

release delivery devices is the inability to stop delivery if adverse reactions are observed in the patient. The concentration characteristics of different non-immediate release systems are shown in Figure 3.4.

Non-immediate or sustained release devices can be divided into two categories. The first is a reservoir device, where the drug is loaded into the reservoir as either a solid or a liquid. Drug release occurs by diffusion through either a semipermeable membrane or a small orifice. Lasers are commonly used to generate uniform orifices through which the drug will diffuse. Osmotic pressure is commonly used to provide the driving force for drug dispersion. The second is a matrix diffusion device, where the drug is dispersed evenly in a solid matrix. Polymers are commonly used as the matrix. Drug delivery is accomplished by either dissolution of the matrix, with corresponding release of drug, or diffusion of the drug from the insoluble matrix.

The physicochemical properties of the drug are critical in the design of the dosage form. Solubility, stability, and pH can strongly affect whether a drug can be delivered effectively from a controlled delivery device. Because sustained release devices often contain multiple doses that if released immediately would reach toxic levels, the physicochemical properties and formulation process may have to be more tightly controlled compared with immediate release systems.

3.4.2 Site of Administration

Targeted drug delivery is often used if the desired site of action is located in a diseased organ or tissue, and release of the drug systemically would produce toxic or deleterious effects. One approach to targeted drug delivery is to place the delivery device adjacent to the site of action, which is especially applicable if the device is controlled release. The other approach is to design the drug so that it has a particular receptor that is found only within the targeted tissue.

3.4.3 Methods of Administration

There is an increasing number of delivery systems that are available for drug delivery. The drug delivery method is chosen based upon the physicochemical properties of the drug, the desired site of action, the duration of action, and the biological barriers (including rapid drug metabolism) that must be overcome to deliver the drug. Some of the most common delivery methods are tablets (oral), parenteral, transdermal, and aerosol. The advantages and disadvantages of each of these methods are described in the text that follows.

3.4.3.1 Oral Administration The oral drug delivery method is the most common and usually the most preferred drug delivery method by both the formulator and the patient for reasons discussed earlier in this chapter. If the oral delivery method is not chosen, it is primarily due to incompatibilities with the physicochemical properties, site of action, or a biological barrier. Disadvantages of oral drug administration form include the low pH of the gastric juices, the first-pass effect of the liver, oral metabolism, and that some patients may have difficulty swallowing the dosage form.

3.4.3.2 Parenteral Administration Parenteral dosage forms include a wide variety of delivery routes, including injections, implants, and liposomes. The advantages of parenteral delivery systems are that they avoid first-pass effects, oral metabolism, and the harsh chemical environment of the stomach's gastric juices. The disadvantage is that the delivery mechanism is invasive.

3.4.3.3 Transdermal Administration Transdermal drug delivery systems have several advantages over other drug delivery methods. These include avoiding gastrointestinal drug adsorption, first-pass effects, replacement of oral administration, and oral metabolism. It also provides for multiday therapy from a single dose, quick termination of drug administration, and rapid identification of the medication. The biggest disadvantage of transdermal delivery systems is that only relatively potent drugs are suitable for administration in this manner. Other disadvantages include drug irritation of the skin and adhering the system to the skin.

3.4.3.4 Aerosol Administration Aerosols can be used for nasal, oral, and topical drug delivery. For topical delivery, aerosols have the advantages of convenient use, protection from air and moisture, and maintaining sterility of the dosage form. Metered dose inhalers (MDIs) are used for oral and nasal delivery of drugs. MDIs are used most effectively for the treatment of asthma and are being developed for the delivery of insulin [39]. They have the advantages of avoiding first-pass effects and degradation within the GI tract, and rapid onset of action. Some of the disadvantages of aerosol delivery systems for oral and nasal delivery include particle size uniformity of drug for proper delivery.

3.4.3.5 Other Delivery Methods In addition to the methods described earlier, suspensions, emulsions, ointments, and suppositories are all effective drug delivery methods. New delivery methods are continually being developed as many of the new drugs have low solubilities and stabilities, requiring improved methods for delivery of these drugs. Improvements in the delivery methods for peptides and proteins are necessary as they continue to be developed as drugs.

3.5 CONCLUSION

Designing a successful drug delivery method for a new therapeutic agent requires a thorough understanding of the physicochemical properties of the drug. If all of the relevant physicochemical properties are not determined, the drug may not be correctly formulated, resulting in product failure at scale up, or even after the drug is on the market. In this chapter we have tried to explain some of the relevant physicochemical properties that must be considered in the proper formulation of a drug method.

REFERENCES

1. Lee, T. W.-L.; Robinson, J. R. Controlled drug delivery systems. In *Remington: The Science and Practice of Pharmacy*; 20th ed.; Gennaro, A. R., Ed.; University of Sciences in Philadelphia: Philadelphia, PA, 2000, p 903–929.
2. Rao, V. M.; Haslam, J. L.; Stella, V. J. *J. Pharm. Sci.* 2001, **90**, 807–816.
3. Uekama, K.; Hieda, Y.; Hirayama, F.; Arima, H.; Sudoh, M.; Yagi, A.; Terashima, H. *Pharm. Res.* 2001, **18**, 1578–1585.
4. Okimoto, K.; Miyake, M.; Ohnishi, N.; Rajewski, R. A.; Stella, V. J.; Irie, T.; Uekama, K. *Pharm. Res.* 1998, **15**, 1562–1568.
5. Stella, V. J.; Martodihardjo, S.; Katsuhide, T.; Venkatramana, M. R. *J. Pharm. Sci.* 1998, **87**, 1235–1241.
6. Stella, V. J. *Adv. Drug Del. Rev.* 1996, **19**, 311–330.
7. Boucher, B. A. *Pharmacotherapy* 1996, **16**, 777–791.
8. Acosta, E. P.; Fletcher, C. V. *Ann. Pharmacother.* 1997, **31**, 185–191.
9. Perry, C. M.; Faulds, D. *Drugs* 1996, **52**, 754–772.
10. Shinkai, I.; Ohta, Y. *Bioorg. Med. Chem.* 1996, **4**, 1–2.
11. Budman, D. R. *Invest. New Drugs* 2000, **18**, 355–363.
12. Bauer, J.; Spanton, S.; Henry, R.; Quick, J.; Dziki, W.; Porter, W.; Morris, J. *Pharm. Res.* 2001, **18**, 859–866.
13. Agarwal, V.; Kothari, B. H.; Moe, D. V.; Khankari, R. K. Drug delivery: Fast-dissolve systems. In *Encyclopedia of Pharmaceutical Technology*; Swarbrick, J., Ed.; Informa Healthcare Inc.: New York, 2006, p 1104–1114.
14. Yoshioka, S.; Stella, V. J. *Stability of Drugs and Dosage Forms*; Kluwer Academic/Plenum: New York, 2000.
15. Edwards, L. J. *Trans. Faraday Soc.* 1950, **46**, 723–735.

16. Garrett, E. R. *J. Am. Chem. Soc.* 1957, **79**, 3401–3408.
17. Koshy, K. T.; Lach, J. L. *J. Pharm. Sci.* 1961, **50**, 113–118.
18. Brownley, C. A., Jr.; Lachman, L. *J. Pharm. Sci.* 1964, **53**, 452–454.
19. Koshy, K. T.; Duvall, R. N.; Troup, A. E.; Pyles, J. W. *J. Pharm. Sci.* 1965, **54**, 549–554.
20. Wolfrom, M. L.; Schuetz, R. D.; Calvalieri, L. F. *J. Am. Chem. Soc.* 1948, **70**, 514–517.
21. Taylor, R. B.; Jappy, B. M.; Neil, J. M. *J. Pharm. Pharmacol.* 1972, **24**, 121–129.
22. Taylor, R. B.; Sood, V. C. *J. Pharm. Pharmacol.* 1978, **30**, 510–511.
23. Nassar, M. N.; House, C. A.; Agharkar, S. N. *J. Pharm. Sci.* 1992, **81**, 1088–1091.
24. Jivani, S. G.; Stella, V. J. *J. Pharm. Sci.* 1985, **74**, 1274–1282.
25. Hamilton-Miller, J. M. T. *J. Pharm. Pharmacol.* 1973, **25**, 401–407.
26. Snider, B. G.; Runge, T. A.; Fagerness, P. E.; Robins, R. H.; Kaluzny, B. D. *Int. J. Pharm.* 1990, **66**, 63–70.
27. Loftsson, T.; Baldvinsdottir, J. *Acta Pharm. Nord.* 1992, **4**, 129–132.
28. van der Houwen, O. A. G. J.; Teeuwsen, J.; Bekers, O.; Beijnen, J. H.; Bult, A.; Underberg, W. J. M. *Int. J. Pharm.* 1994, **105**, 249–254.
29. Musson, D. G.; Evitts, D. P.; Bidgood, A. M.; Olejnik, O. *Int. J. Pharm.* 1993, **99**, 85–92.
30. Allen, L. V., Jr.; Erickson, M. A., III. *Am. J. Health Sys. Pharm.* 1998, **55**, 1915–1920.
31. Nahata, M. C.; Morosco, R. S.; Hipple, T. F. *Am. J. Health Sys. Pharm.* 1998, **55**, 1155–1157.
32. Stanisz, B. *J. Pharm. Biomed. Anal.* 2003, **31**, 375–380.
33. Jarho, P.; Jarvinen, K.; Urtti, A.; Stella, V. J.; Jarvinen, T. *Int. J. Pharm.* 1997, **153**, 225–233.
34. IPEC-Americas. What are pharmaceutical excipients? http://ipecamericas.org/about/faqs (accessed November 16, 2015).
35. Byrn, S. R.; Pfeiffer, R. R.; Stowell, J. G. *Solid-State Chemistry of Drugs*; 2nd ed.; SSCI: West Lafayette, IN, 1999.
36. Zell, M. T.; Padden, B. E.; Grant, D. J. W.; Schroeder, S. A.; Wachholder, K. L.; Prakash, I.; Munson, E. J. *Tetrahedron* 2000, **56**, 6603–6616.
37. Gibson, M., Ed. *Pharmaceutical Preformulation and Formulation: A Practical Guide from Candidate Drug Selection to Commercial Dosage Form*; Interpharm Press: Englewood, CO, 2002.
38. Moreton, R. C. Excipients to the Year 2000. In *Excipients and Delivery Systems for Pharmaceutical Formulations*; Karsa, D. R.; Stephenson, R. A., Eds.; The Royal Society of Chemistry: Cambridge, UK, 1995, p 12–22.
39. Owens, D. R.; Zinman, B.; Bolli, G. *Diabet. Med.* 2003, **20**, 886–898.

4

TARGETED BIOAVAILABILITY: A FRESH LOOK AT PHARMACOKINETIC AND PHARMACODYNAMIC ISSUES IN DRUG DISCOVERY AND DEVELOPMENT

CHRISTINE XU
Pharmacokinetics and Pharmacometrics, Drug Disposition, Safety and Animal Research, Sanofi US, Bridgewater, NJ, USA

4.1 INTRODUCTION

Drug delivery strategies, including the design and choice of target and biological molecular platforms are intended to improve drug efficacy and safety to enhance the overall therapeutic index of new or existing drug. Pharmacokinetic (PK) and pharmacodynamic (PD) properties of drugs play an important role in the drug delivery; these properties are a critical part of drug discovery and development process. It has been increasingly recognized that early optimization for key parameters such as absorption, distribution, metabolism, and excretion (ADME) of a drug candidate during the drug discovery process is important to reduce the failure rate during the development stage. With the rapid increases in cost and duration for drug discovery and development, the critical decisions are being made at every stage of drug development. PK–PD evaluations and analyses can identify the key

Drug Delivery: Principles and Applications, Second Edition. Edited by Binghe Wang, Longqin Hu, and Teruna J. Siahaan.

"drug-like" ADME properties and establish the PK–PD relationship of drug efficacy and safety, which help in decision making.

4.2 TARGET BIOAVAILABILITY

In assessing pharmacokinetic (PK) and pharmacodynamic (PD) issues in drug delivery, it is necessary to carefully consider the definition of bioavailability. The concept of "target bioavailability," a term that extends the idea that the true bioavailability of a drug, is the fraction of the administered dose that reached the site of action.

One way of viewing the many processes involved in the delivery of a drug from its site of administration to its site of action is to consider each barrier along the delivery path. One or more mechanism(s) may be affecting the rate and extent of drug that reaches the site of action during these processes (Fig. 4.1). The process depicted in Figure 4.1 is a representative example for oral administration and is not meant to be exhaustive. The potential barriers that a compound must pass to reach the systemic circulation are traditionally thought to contribute to the bioavailability of a compound (Fig. 4.1). Since the target bioavailability is the fraction of the administered dose that reached the site of action. Additional barriers need to be overcome after the drug leaves the bloodstream to reach the site of action (Fig. 4.1). The importance of various barriers depends on the site of administration and the physicochemical characteristics of the drug.

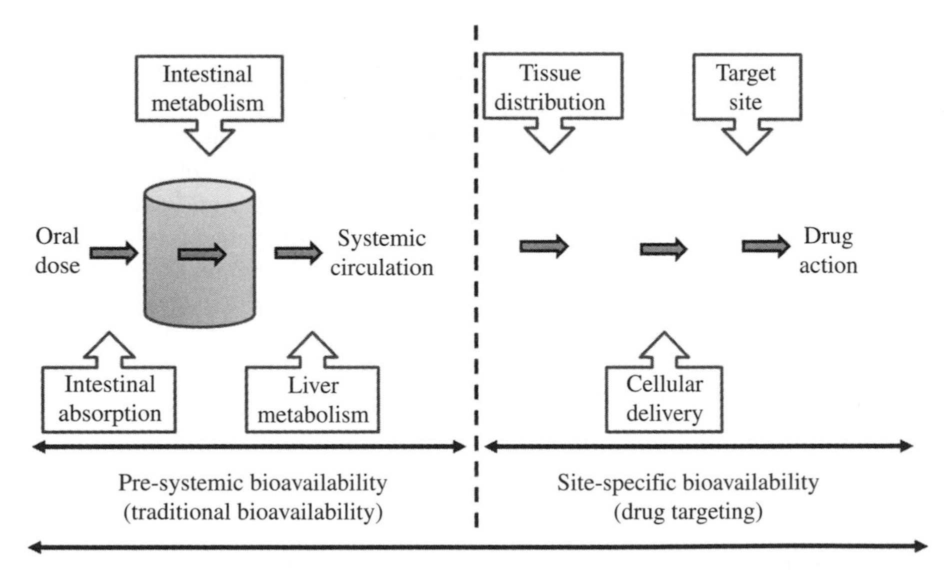

FIGURE 4.1 A schematic representation of the potential biological barriers that an orally administered compound must pass before reaching the site of action.

4.3 DRUG DELIVERY TRENDS AND TARGETS RELATED TO PK AND PD

The traditional drug delivery technology and device include different routes: oral drug, pulmonary, transdermal, and mucosal delivery. These approaches include prodrug, biopolymers, drug carriers (e.g., liposomes), nanoparticles, and controlled drug release systems. They have been specially devised to optimize drug absorption processes (para- and transcellular permeation, active transport, lymphatic uptake, and absorption rate control).

While pharmaceutical industry has allocated a large portion of resources to the evaluation and application of existing mature technologies using traditional approaches, recently, the target drug delivery highlights biological and molecular approaches. The biologics approaches include peptides, recombinant proteins, monoclonal antibody (mAb), and oligonucleotide with special emphasis on the high selective targets. Major pharmaceutical companies have acquired or increased biological molecular platforms in the pipelines. Because of predicable PK and clearance mechanism, as well as the maturation of mAb and molecular biotechnologies, mAb is one of the fast-growing drug delivery areas for new drug development (Table 4.1). For the past two decades, there was a growth of the antibody–drug conjugates (ADCs) in the platform in many biopharmaceutical companies. ADC was introduced as a "magic bullet" targeted drug delivery about a century ago [1]. In this targeted drug delivery hypotheses, there are two critically important components: (1) a selective compound (or agent such as antibodies) for targeting and (2) a toxin (or drug) combined in one unit, so that the toxin or drug finds its way only to disease-causing cells or pathogenic tissues. Therefore, such a targeted drug delivery system exhibits low toxicity to nontargeted tissues in the body and enhances the benefit and risk index. Recent clinical success and approval of ADCs (Seattle Genetics' Adcetris (brentuximab vedotin; approved August 2011), and Roche/Genentech's Kadcyla (adotrastuzumab emtansine; approved February 2013)) have renewed interest and resurgence in targeted drug delivery for oncology area [2, 3].

4.4 PK–PD IN DRUG DISCOVERY AND DEVELOPMENT

The definitions have been refined over the centuries, often, it is simply said that "pharmacokinetics is what the body does to the drug; PD is what the drug does to the body." More specifically, PK describes the time course of plasma and/or tissue drug concentrations. PK studies rely on the measurement of the active drug ingredient or therapeutic moiety and/or its metabolite(s) in an accessible biologic fluid such as blood, plasma, or urine. The definition of PD in a broad term intends to describe all aspects of the pharmacological actions, pathophysiological effects, and therapeutic responses of both beneficial and adverse for the active drug ingredient, therapeutic moiety, and/or its metabolites(s) on various systems of the body from subcellular effects to clinical outcomes [4].

TABLE 4.1 Summary of the Number of Clinical Trials Listed on ClinicalTrials.gov Organized as the Three Major Drug Delivery Categories (Updated on September 10, 2013)

	All	Phase I	Phase II	Phase III	Phase IV
Drug delivery technology and system					
Device	15,729	1280	1996	1816	2149
Dosage form	13,614	3629	4288	2590	1536
Drug delivery system	4094	769	1148	792	536
Formulation	3446	1522	828	672	305
Liposome	624	192	307	109	46
Transdermal	601	103	125	171	127
Formulation comparison	1470	681	272	321	140
Route	1293	370	355	269	166
Sustained release	375	63	90	124	56
Lipid formulation	139	52	33	18	25
Nanoparticles	142	48	81	10	2
Aerosol and inhalation	152	36	34	39	19
Prodrugs	129	53	53	15	10
Colloid	178	6	30	23	28
Subtotal	37,738				
Biological molecule platform/technologies					
Antibody	7532	2058	3128	1511	481
Biologics and vaccines	4048	1410	1411	932	410
Peptide	2205	623	671	291	226
Recombinant proteins	673	261	227	130	37
Antibody conjugates	415	82	125	164	42
Antisense	119	67	59	13	0
Oligonucleotide	112	49	46	13	3
siRNA	72	19	18	17	5
Aptamer	22	10	9	4	1
Subtotal	14,104				

PK analysis and PK–PD relationships often involve mathematical models. The objectives of PK studies are usually to establish the relationship between dose and concentrations (systemic exposure) and to obtain important information on drug absorption and disposition [4]. PK modeling has been used to quantitatively describe the concentration (plasma, serum, blood, urine)–time course of active drug ingredients in the body. PK–PD relationships and modeling build the bridge between PK and PD, which links the concentration–time profile (PK) to the physiological or pharmacological effects and clinical outcomes (PD).

The basic concepts of PK–PD modeling are illustrated in Figure 4.2 [5]. As shown in Figure 4.2, the PK model relates dose to plasma concentrations (C_p). A rate-limiting step for the drug efficacy is the drug distribution to the site of action. The term "biophase" is used to describe drug permeation to receptors in such sites. A modeling approach has been developed by Sheiner et al. [6] for drugs that have delayed response; here, a hypothetical effect-compartment is included to mathematically link

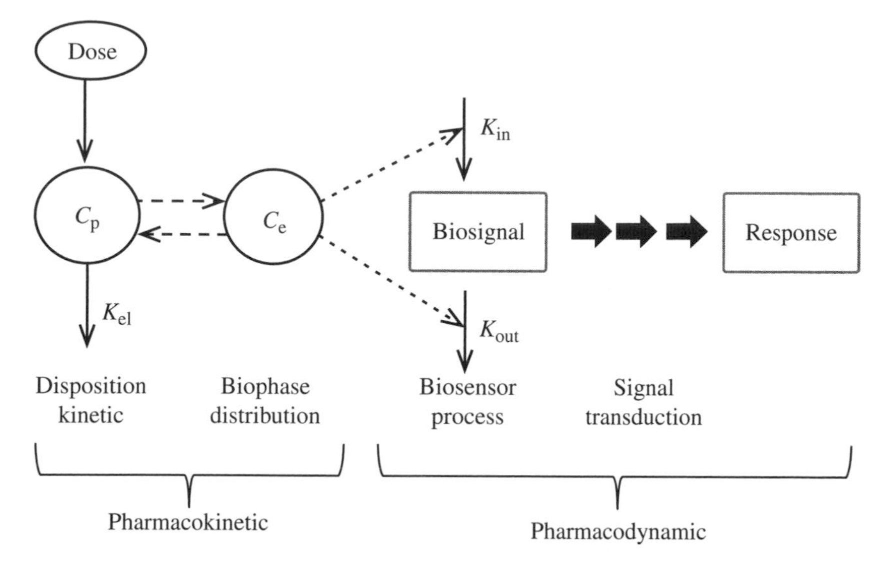

FIGURE 4.2 The drug action model in terms of basic components of pharmacokinetic and pharmacodynamic models. Adapted from Jusko et al. [5], pp. 5–8. Reproduced with permission of Springer.

plasma concentrations and drug effects at the site of action. In the effect compartment PK–PD model, the biophase or the effect site concentrations (C_e) is assumed to be proportional to the plasma concentrations (C_p). The PD model then translates the drug concentration at biophase/the effect site (C_e) into effects as the pharmacological responses (R). In some cases, C_p at steady state is in a direct correspondence with effect, which leads to fairly simple mathematical models that can relate dose to effect (e.g., linear, log-linear, E_{max}, and sigmoid E_{max}). In other cases where this simple direct relationship does not exist or where the link from PK to PD is slow relative to changes in dose, the time-varying nature of the system is usually modeled as a biophase or a biosensing or transduction kinetics problem.

Beginning in the 1990s, pharmacometrics began to emerge as a mathematical and statistical approach to characterize, understand, and predict a drug's PK, PD, diseases progress, and clinical trials at both population and individual levels. Pharmacometrics including, but not limited to, PK–PD modeling can be defined as the science of computer-based modeling and simulation techniques. It has been used to characterize and predict interactions between drug pharmacology, disease pathogenesis, and intrinsic/extrinsic patient factors. Pharmacometrics can be established with different degrees of complexity, ranging from simply data-driven and descriptive (nonmechanistic or empirical) to complex and completely mechanistic [7]. Modeling and simulation have been used for a variety of purposes such as candidate selection, first in human dose prediction, analyzing the exposure–response or dose–exposure–outcome relationships from the reported results of completed clinical trials. They also have been used to simulate subsequent clinical trials and predict responses among individual subgroups of the population not included in the clinical trial. In addition, the combination of sparse sampling strategies, population PK modeling, and population PK–PD modeling often evaluate the relevant covariates. The covariates assessment can be used to (i) predict exposure–response relationships of efficacy and safety in subgroups that are not formally studied in clinical trials and (ii) to estimate the effect size on the PK in subgroups such as body weight, age, sex, and genotype.

The process of drug development can be viewed as iterative learn–confirm cycles (Fig. 4.3) [8]. The model-based drug development facilitates the development of a comprehensive knowledge base through continuous integration of knowledge generated along the development path and provides a quantitative tool for informed decision making. The learning activity involves utilization of knowledge available in the form of previous information, experimental data, and logical assumptions to construct an appropriate mathematical model. Simulation of outcomes with associated uncertainty based on this model provides rational criteria for go/no go decisions for further subsequent studies. Once the confirmatory data from these studies become available to further corroborate the predicted outcome, the model can be further refined and updated for future use [9].

Under the traditional drug development paradigm, advances have been made to optimize the ADME properties of new medical product at the early stage. PK/PD modeling plays an important role in optimization of ADME properties, thus contributes to the reduction of the attrition rate due to ADME issues in the later-stage drug development. In the past decades, the overall goals of PK and PD have shifted to the model-based drug development. Although the majorities of the PK–PD models are data driven using empirical functions to characterize data, the field of PK/PD modeling has merged to the employment of a diverse array of the complex and completely mechanistic models. A mechanistically based approach is the preferred methodology for exposure–response model building as it reflects the underlying physiological processes for the observed pharmacologic response and allows integrating information from different stages of development [10].

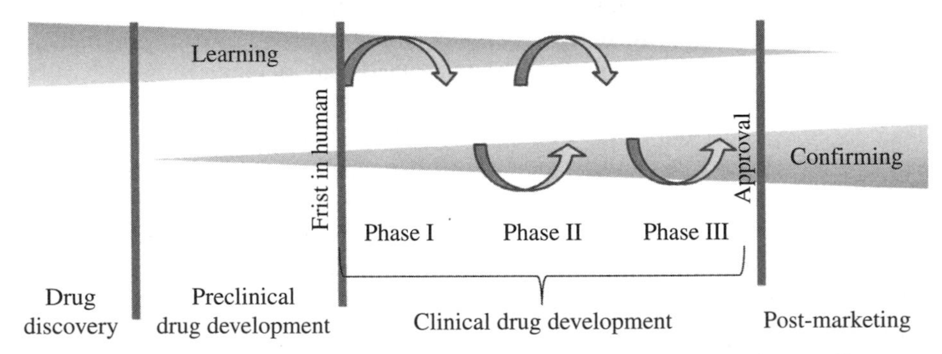

FIGURE 4.3 Model-based drug development is a new paradigm that embraces all aspects of drug development from drug discovery to post-marketing. Adapted from Sheiner [8], pp. 285–291.

4.5 SOURCE OF VARIABILITY OF DRUG RESPONSE

Effective drug delivery to the site of action is dependent on many factors that influence the PK and PD of drug. The intrinsic and extrinsic factors that contribute to the interindividual PK variability are

Intrinsic factors:

- Age, body weight, gender, and race
- Renal function
- Hepatic function
- Disease
- Pregnancy/lactation
- Pharmacogenetics

Extrinsic factors:

- Drug–drug interaction
- Smoking/diet
- Food effect
- Environment
- Formulation

The variability of the PK would lead to variability in the PD responses. However, it is possible that total PD variability is not fully accounted by in the variability seen in the PK measurement. Thus, it is important to conduct quantitative analysis for the source of variability of PD as well as PK.

Compared with the descriptive (nonmechanistic or empirical) models, physiology-based PK (PBPK) models is particularly useful to understand each "location" of drug concentrations that can be influenced by multiple mechanisms that affect drug disposition. As shown in Figure 4.4, PBPK models represent a subtype of mechanism-based models for individualization. These models distinguish between drug-specific and biological system-specific parameters. Drug-specific parameters describe the interaction between the drug and the biological system, whereas system-specific parameters describe the functioning of the underlying biological system. PBPK models are used primarily for characterizing and predicting drug exposure at different sites for different populations by dividing the biological system into a number of compartments, each of which represents a different organ or tissue. The various organ and tissue compartments are interconnected through the arterial and venous blood circulation.

PBPK models are increasingly employed to provide a mechanistic explanation of the impact of intrinsic and extrinsic variables on exposure–exposure relationships for both desired and undesired drug effects. The effect of individual or combined intrinsic/extrinsic factors on drug exposure is projected via both drug-dependent and

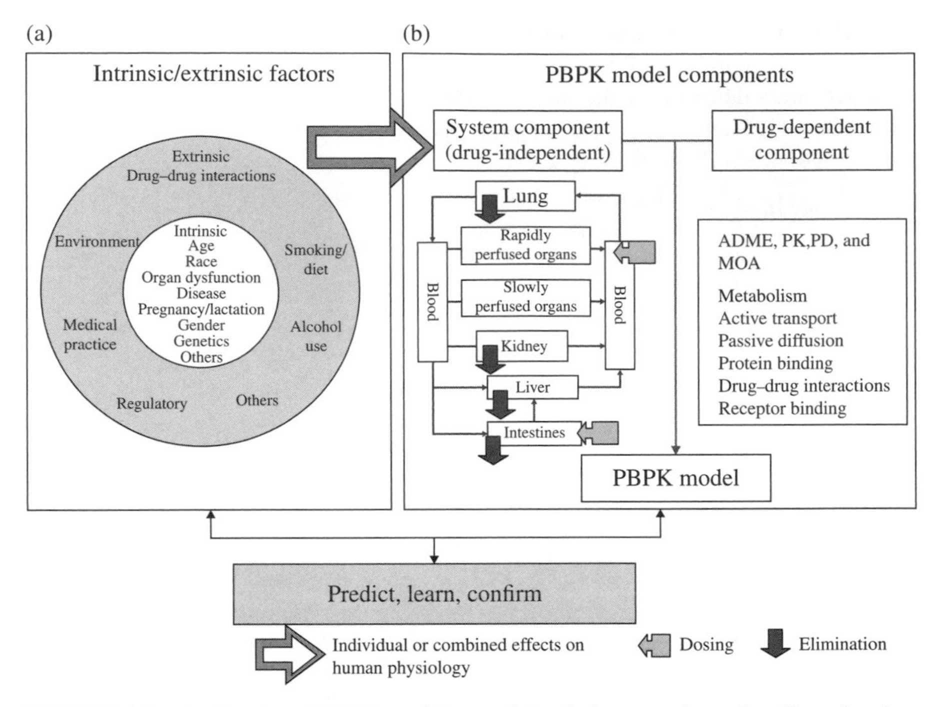

FIGURE 4.4 Application of PBPK modeling and simulation to evaluate the effect of various extrinsic and intrinsic factors on drug exposure and response: (a) Intrinsic and extrinsic patient factors that can affect drug exposure and response and (b) components of PBPK modeling (drug-dependent component and drug-independent (system) component) [11]. Zhao et al. [11], pp. 259–267, figure 1. Reproduced with permission of John Wiley & Sons.

drug-independent (system) components of the PBPK model. These are consequently valuable tools for guiding the establishment of a rational dosing regimen, particularly in clinically understudied patient populations, such as children, the elderly, patients with renal or hepatic impairment, and patients receiving concomitant drug therapy who could be at risk for adverse drug reactions.

While it is increasingly recognized that it is important to measure/predict the drug concentrations at the "delivery site" and/or biophase, collection of samples from the "delivery sites" such as intestinal mucous membrane, liver, brain, or tumor tissues are often constrained by technical and ethical complexities. PBPK models can be utilized to predict the drug level at the delivery site and/or biophase of drug concentrations. The major challenges in modeling these systems are to obtain the parameters of biological system-dependent mechanism process in the model to adequately predict the observations. There have been remarkable advances in the knowledge of system pharmacology, physiology of the regulation, and expression of many of the important factors in drug response. Many of these parameters affecting drug delivery are discussed in other chapters in this book, including drug metabolism, membrane permeability, receptor affinity for receptor-mediated delivery, and efflux transport.

In addition to PBPK, population PK and population PK–PD approaches are widely used to identify and estimate the variability attributed by factors such as body weight, age, sex, genotype, and disease baseline characteristics.

4.6 RECENT DEVELOPMENT AND ISSUES OF BIO-ANALYTICAL METHODOLOGY

It is widely accepted that bioanalysis is an integral part of the PK/PD characterization of a novel chemical entity from the time of its discovery, various stages of drug development, to its market authorization.

Often the drug concentration of real interest and drug action is not that in the blood; but rather, the drug concentration in the biophase or close to the site of action. Therefore, the distribution of a compound to the biophase is important in the assessment of targeted bioavailability. In order to quantitatively assess the various mechanisms that may influence the distribution of drug from the blood to the site of action [12], collecting samples from target tissue/organ have become increasingly common in the clinical studies. Specialized sampling techniques are developed to collect tumor tissues from cancer patients, liver biopsy samples from hepatitis B virus (HBV) and hepatitis C virus (HCV) infection patients, and central nervous system (CNS) fluids via microdialysis. New optical methodologies like fluorescence and bioluminescence are of interest due to low cost and versatility with a number of solutes [13]. Novel detection systems, such as the ultrasensitive accelerator mass spectroscopy may allow measuring the concentration at the targeted site of action. New advancements in imaging techniques such as magnetic resonance imaging, X-ray computed tomography, and positron emission tomography have also enable their wide applications in the clinical research of drug delivery/drug distribution [14].

Therapeutic antibody assessment in biofluids requires fit-for-purpose bioanalytical methods. The traditional and reference bioanalytical method is the immunoassay, most as enzyme-linked immunosorbent assay (ELISA—an analytical technique involving antibody, antigen, and an enzyme used to quantify proteins, antigens, or antibody). Liquid chromatography coupled to (tandem) mass spectrometry (LC–MS/MS) is likely considered as the most attractive and flexible analytical technique to cope with the high demands related to the rapid growth of biopharmaceuticals [15].

Bioanalysis of ADCs is challenging compared with small-molecule drugs and protein therapeutics [16] because ADCs have complex molecular structures, combining the molecular characteristics of small-molecule drugs with those of protein therapeutics. In addition, ADCs are heterogeneous because conjugating the drug to the antibody results in a range of drug-to-antibody ratios with the cytotoxic drug covalently bound at multiple sites in the antibody. Both ligand-binding methods and LC–MS/MS methods are used to quantify ADCs. There are some limitations to note with either approach. In the case of LC–MS/MS, an *a priori* postulated form of the drug released by the ADC in circulation is measured; however, this may not correspond to the major form(s) of the drug released. For ligand-binding assays, the calibration standard curve is composed of a reference standard that may not represent the changing drug-to-antibody ratios *in vivo* over time. Some methods developed specifically for ADCs combine aspects of ligand binding and mass spectrometry, such as affinity capture LC–MS. Another hybrid method involves affinity capture and LC–MS/MS to measure antibody-conjugated drug [17].

4.7 MECHANISTIC PK–PD MODELS

Omalizumab (Xolair®) is an anti-IgE mAb to treat allergic asthma, and the second generation of IgE was called "HAE1." Omalizumab preclinical and clinical data were used as an example of learn–confirm cycles of mechanism-based PK/PD model for HAE1 (Fig. 4.5) [18]. In this case, the *in vitro* binding data for HAE1 and omalizumab were compared; HAE1 has been shown to have 23-fold higher affinity than omalizumab. Therefore, HAE1 has expanded patient population compared to omalizumab, and it includes patients that have higher baseline IgE levels. The higher affinity helps reduce the dose levels or increase dosing intervals compared to current drugs. The PK/PD model incorporated many parameters including (i) PK of anti-IgE mAb, (ii) the reversible binding with free IgE, and (iii) the disposition of free IgE and anti-IgE mAb : IgE complexes. Free IgE clears rapidly from the circulation by endocytosis and binding to Fcε receptors. The clearance of anti-IgE mAb is controlled by nonspecific IgG clearance and reversible binding to free IgE. The nonspecific IgG clearance includes the protection from endocytosis by recycling through the IgG salvage receptor, FcRn. The anti-IgE mAb can bind to free IgE and form small and inert complexes. These complexes can be cleared Fcγ receptors on the

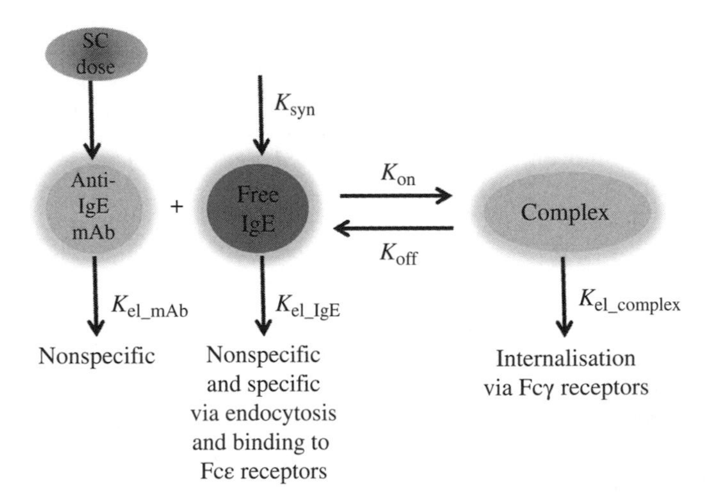

FIGURE 4.5 Mechanism-based model of PK–PD for an anti-IgE monoclonal antibody. Putnam et al. [18], figure 2. Reproduced with permission of Springer.

reticuloendothelial system. In the Phase I studies, the preliminary exposure–response model was used to simulate Phase I dose studies, which was based on IgE suppression duration at below 10 IU/ml.

The results from omalizumab clinical studies indicate that free IgE can be used as a PD biomarker to develop HAE1. In Phase I clinical trial for HAE1, the objective was to develop dosing regimen that is convenient to achieve IgE suppression at the target level of approximately 10 IU/ml for asthma patients. The goal is to significantly lower the exacerbation rates compared to the data from omalizumab clinical studies.

In the next step, Phase I data were used to refine the mechanism-based PK/PD model. The objectives were to enable multi-dose HAE1 exposure–response predictions and provide study design and decision-making support for Phase II studies in asthma patients. Subsequent dose selection for the Phase II dose-ranging trial was accomplished by using a clinical trial simulation platform based on the exposure–response model. The clinical trial simulation results provided the probability of treated subjects achieving the target-free IgE level and attaining the clinical response as a function of dose for each clinical trial scenario. The doses were selected based on the simulation studies; for this, two independent methods were used including dose–response nonlinear fitting and linear mixed modeling. The following criteria were selected: (i) 15 mg for "minimal effect," (b) 45 and 90 mg and two intermediate doses, and (iii) 180 mg as a high dose that is near the top of the dose–response curve (Fig. 4.6).

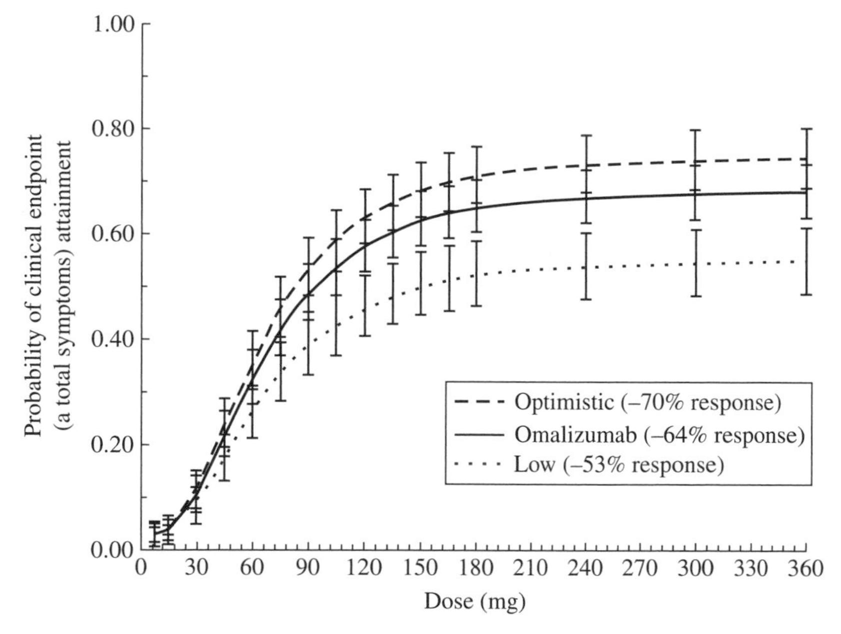

FIGURE 4.6 Simulated probability between clinical response (Δ total symptom score) and dose for three different cases: (1) an optimistic response rate (70%), (2) a response rate similar to that expected for omalizumab (64%), and (3) a low response rate (50%). Putnam et al. [18], figure 5. Reproduced with permission of Springer.

Finally, the development of HAE1 was largely supported and accelerated by the mechanism of PK–PD modeling and simulation, which enable data-driven decision making as well as confirmation of projections of new incoming data.

4.8 SUMMARY

For the drug delivery, one shall take into consideration the true bioavailability at the site of action as well as the various pharmacokinetic (PK) and pharmacodynamic (PD) issues. In recent decade, there is trend of increase in the target drug delivery that highlights biological and molecular approaches, comparing the traditional drug delivery technology and device. PK–PD modeling is a tool that is increasingly used throughout the drug discovery and element continuum to support fast and rationale decision making. Thereby, it has the potential to accelerate and increase the cost effectiveness of the drug development process. The merging pharmacometric approaches have integrated the pharmacokinetic, pharmacology, diseases progress, and intrinsic/extrinsic factors. The novel bioanalytical methods allow measure and the pharmacometric approaches including PBPK models that allow to estimate/predict the drug concentration at the site of action. There are many factors contributing to PK and PD variability; population PK and population PK–PD modeling can identify, characterize, and quantify the specific attributers. PBPK modeling is also a useful tool to quantify drug-dependent and drug-independent (system) components.

REFERENCES

1. Ehrlich, P. *Fortschritte der Medizin* 1897, 15, 41–43.
2. Ho, R. J. Y.; Chien, J. *J. Pharm. Sci.* 2014, 103, 71–77.
3. Ho, R. J. Y.; Chien, J. Y. *J. Pharm. Sci.* 2012, 101, 2668–2674.
4. Derendorf, H.; Lesko, L. J.; Chaikin, P.; Colburn, W. A.; Lee, P.; Miller, R.; Powell, R.; Rhodes, G.; Stanski, D.; Venitz, J. *J. Clin. Pharmacol.* 2000, 40, 1399–1418.
5. Jusko, W. J.; Ko, H. C.; Ebling, W. F. *J. Pharmacokinet. Biopharm.* 1995, 23, 5–8.
6. Sheiner, L. B.; Stanski, D. R.; Vozeh, S.; Miller, R. D.; Ham, J. *Clin. Pharmacol. Ther.* 1979, 25, 358–371.
7. Lesko, L. J.; Schmidt, S. *Clin. Pharmacol. Ther.* 2012, 92, 458–466.
8. Sheiner, L. B. *Clin. Pharmacol. Ther.* 1997, 61, 275–291.
9. Suryawanshi, S.; Zhang, L.; Pfister, M.; Meibohm, B. *Expert Opin. Drug Discov.* 2010, 5, 311–321.
10. Levy, G. *Clin. Pharmacol. Ther.* 1994, 56, 356–358.
11. Zhao, P.; Zhang, L.; Grillo, J. A.; Liu, Q.; Bullock, J. M.; Moon, Y. J.; Song, P.; Brar, S. S.; Madabushi, R.; Wu, T. C.; Booth, B. P.; Rahman, N. A.; Reynolds, K. S.; Berglund, E. G.; Lesko, L. J.; Huang, S. M. *Clin. Pharmacol. Ther.* 2011, 89, 259–267.
12. Fischman, A.; Alpert, N.; Rubin, R. *Clin. Pharmacokinet.* 2002, 41, 581–602.
13. Rudin, M.; Weissleder, R. *Nat. Rev. Drug Discov.* 2003, 2, 123–131.

14. Lappin, G.; Garner, R. C. *Nat. Rev. Drug Discov.* 2003, 2, 233–240.
15. van den Broek, I.; Niessen, W. M. A.; van Dongen, W. D. *J. Chromatogr. B* 2013, 929, 161–179.
16. Gorovits, B.; Alley, S. C.; Bilic, S.; Booth, B.; Kaur, S.; Oldfield, P.; Purushothama, S.; Rao, C.; Shord, S.; Siguenza, P. *Bioanalysis* 2013, 5, 997–1006.
17. Kaur, S. *Bioanalysis* 2013, 5, 981–983.
18. Putnam, W.; Li, J.; Haggstrom, J.; Ng, C.; Kadkhodayan-Fischer, S.; Cheu, M.; Deniz, Y.; Lowman, H.; Fielder, P.; Visich, J.; Joshi, A.; Jumbe, N. S. *AAPS J.* 2008, 10, 425–430.

5

THE ROLE OF TRANSPORTERS IN DRUG DELIVERY AND EXCRETION

MARILYN E. MORRIS AND XIAOWEN GUAN

Department of Pharmaceutical Sciences, School of Pharmacy and Pharmaceutical Sciences, University at Buffalo, State University of New York, Buffalo, NY, USA

5.1 INTRODUCTION

Therapeutic agents or other xenobiotic compounds can exert their pharmacological or toxicological activities only when sufficient concentrations of the compounds are present at the site of action and upon binding to the targeted receptors or enzymes. Therefore, the ability of drug molecules to cross biological membranes represents an important determinant of drug absorption, distribution, metabolism, elimination, and, ultimately, their therapeutic or toxic effects. It is known that the complex biological membrane system is not pure lipid bilayers, but lipid bilayers embedded with numerous proteins, including transporters. The fundamental role of transporters is to maintain the cellular homeostasis and to provide physiological function for movement of endogenous substances, such as amino acids, glucose, and hormones; however, many of these transporters are also involved in transporting therapeutic agents. Thus, for a large number of drug molecules, their ability to pass through biological membranes is not only solely determined by their physiochemical parameters such as lipophilicity but also governed by drug transporter activities.

Drug transporters are classified into two families: the ABC (ATP-binding cassette) and SLC (solute carrier) families. As the name implies, ABC transporters are primary active transporters that utilize ATP as the source of energy. ABC transporters are efflux transporters that transport substrates out of the cell. Members of ABC

Drug Delivery: Principles and Applications, Second Edition. Edited by Binghe Wang, Longqin Hu, and Teruna J. Siahaan.

family that are important in drug excretion include P-glycoprotein (P-gp), multidrug resistance-associated proteins (MRPs), and breast cancer resistant protein (BCRP). In addition, these ABC transporters are also known to transport a number of clinically important anticancer agents, thereby limiting intracellular accumulation of these drugs and resulting in inefficient tumor cell killing, a phenomenon known as multidrug resistance (MDR) that is attributed to overexpression of these transporters on tumor cells [1–4]. In contrast SLC transporter families are secondary active transporters that use the concentration gradient of a cosubstrate such as protons or sodium ions to drive their transport. SLC transporters are mostly influx transporters that transport substrates from the extracellular lumen into the cell; exceptions to this are monocarboxylate transporters (MCTs) 1 and 4 and multidrug and toxin extrusion (MATE) transporters that can function as efflux transporters.

Localization of drug transporters in particular organs and tissues and in polarized membrane domains (apical versus basolateral membrane (BLM)) governs their functional role in the absorption, intestinal, hepatobiliary and renal excretion, and tissue distribution of a variety of endogenous and exogenous compounds [5]. The expression of drug transporters in the luminal membrane of the blood–brain barrier (BBB), blood–testis barrier, and placenta suggests central roles of these transporters in regulating the entry of potentially harmful compounds into these tissues [5]. Considering the impact of drug transporters on drug disposition and the wide substrate specificity of many drug transporters, adverse drug interactions due to inhibition or induction of drug transporters by coadministered drugs, ingested dietary, or environmental compounds could be expected, and have been reported in a number of animal and clinical studies [6]. On the other hand, these transporter-based interactions may also result in a beneficial outcome and improvement of therapeutic efficacy of a particular drug of interest. For instance, the poor bioavailability of some anticancer agents could be improved by inhibiting intestinal P-gp or other efflux transporters, and transporters can therefore be targeted as a drug delivery mechanism. To appreciate the importance of the transporters in drug delivery and excretion, an understanding of the molecular and functional characteristics of drug transporters, such as their tissue distribution and the impact of transporters on drug disposition, is essential and therefore will be the focus of this chapter.

5.2 DRUG TRANSPORT IN ABSORPTION AND EXCRETION

5.2.1 Intestinal Transport

For orally administered drugs, the US Food and Drug Administration (FDA) has incorporated the Biopharmaceutics Classification System (BCS) [7] in which solubility and permeability are used as determining factors for oral absorption. These factors will highly impact the influence of transporter-mediated absorption since highly permeable drugs will predominantly undergo passive diffusion in the intestine, and solubility will determine drug concentrations and the potential saturation of drug transporters at high drug concentrations. However, other factors including transporter expression and capacity, as well as substrate specificity of the drug need to be considered. For those compounds with low permeability, uptake transporters will be important for their intestinal absorption (Classes 3 and 4). However efflux transporters may also contribute significantly to bioavailability for drugs in Classes 2, 3, and 4, resulting in decreased intestinal absorption. On the other hand, for those compounds with high permeability and high solubility, transporter-mediated absorption is likely not important since these compounds will be taken up into enterocytes by passive absorption and intracellular concentrations will be high, likely resulting in saturation of efflux transporters. High drug concentrations will result in saturation of influx transporters resulting in reduced bioavailability, whereas saturation of efflux transporters can enhance the extent of drug absorption [6]. As a modification of BCS, Wu and Benet have proposed the Biopharmaceutics Drug Disposition Classification System (BDDCS) [8] to include both transporter and metabolism effects on the overall bioavailability of drugs.

Various members of both families of drug transporters, ABC and SLC transporters, are expressed throughout the intestinal tract (Fig. 5.1a). Although mRNA expression of numerous transporters has been identified in the gastrointestinal (GI) tract, only a few transporters have been found to play a crucial role in intestinal drug absorption. One of the reasons for this is that many transporters show overlapping substrate specificity, which makes it difficult to attribute the effect of one particular intestinal transporter to changes in drug absorption. Additionally, transporter-mediated changes in both bioavailability and clearance can occur, both resulting in altered plasma concentrations following oral administration of the drug; the contribution of each of these processes may not be known unless additional studies after intravenous (i.v.) administration of the drug are performed. Nevertheless, the use of probe substrates, specific inhibitors and inducers of transporters, transfected cell lines, small interference RNAs (siRNAs), and knockout animals has provided useful tools to assess the effect of a specific transporter *in vitro* and *in vivo*.

The high expression levels and transport capacities of the influx transporters peptide transporter (PEPT1) and MCT1 have resulted in their use as drug delivery targets in order to increase oral drug absorption, which will be discussed later in this chapter. Efflux transporters can limit drug bioavailability, and the interplay between transporters and drug metabolizing enzymes (DMEs) can contribute to another level of complexity with first-pass metabolism and transporter-mediated efflux in the gut contributing to low oral drug bioavailability.

5.2.2 Hepatic Transport

As the major clearance organ in the body, the liver can metabolize drugs or excrete the parent drug into the bile. Following the uptake of xenobiotics into hepatocytes via the sinusoidal (basolateral) membrane, compounds may undergo metabolic modifications, or the parent compound, as well as the formed metabolites, may be excreted into the bile across the canalicular membrane or effluxed back into the blood via the sinusoidal membrane (Fig. 5.1b). The relatively small surface area of the canalicular membrane (10–15% of the hepatocyte surface area) in contrast to the sinusoidal membrane (at least 70%) and small intracanalicular fluid volume suggests that carrier-mediated transport may significantly contribute to the biliary excretion of both

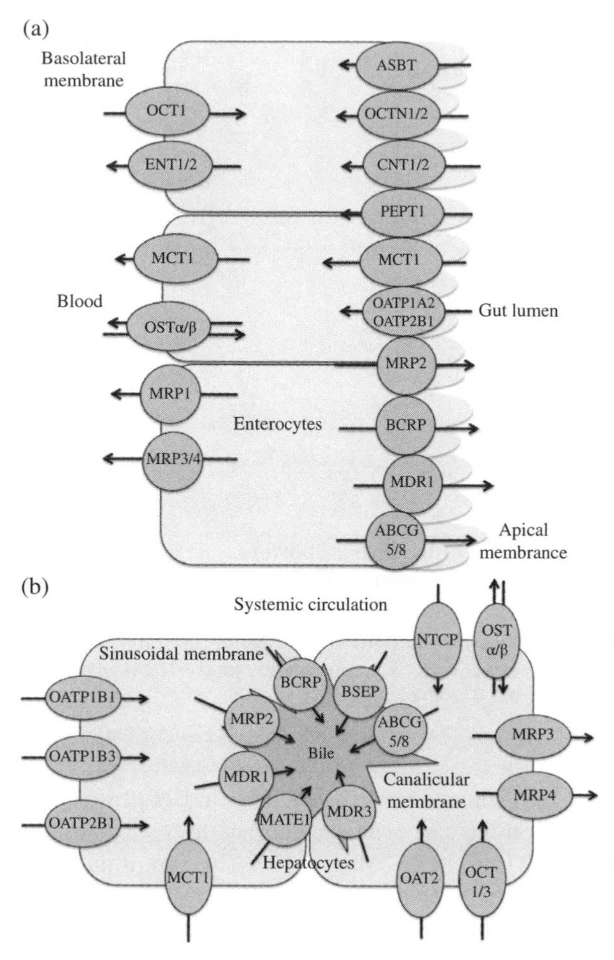

FIGURE 5.1 Membrane transporters for drugs and endogenous substances. Localization of transporters on the plasma membrane of enterocytes, hepatocytes, and kidney proximal tubules are presented. (a) **Intestinal transporters.** On the apical (luminal) membrane of enterocytes, several solute carrier (SLC) transporters are presented. These include ileal apical sodium/bile acid cotransporter (ASBT), carnitine/organic cation transporters (OCTN1/2), concentrative nucleoside transporters (CNT1/2), peptide transporter 1 (PEPT1), monocarboxylate transporter 1 (MCT1), and organic anion transporting polypeptide transporters (OATP1A2 and 2B1). The apical ATP-binding cassette, ABC transporters, include multidrug resistance-associated protein 1 (MRP1), breast cancer resistance protein (BCRP), P-glycoprotein (P-gp; MDR1), and ABC subfamily G members 5/8 (ABCG5/8). The transporters on the BLM of enterocytes consist of organic cation transporter 1 (OCT1), equilibrative nucleotide transporters (ENT1/2), MCT1, heteromeric organic solute transporters (OSTα-OSTβ), MRPs 1, 3, and 4. (b) **Hepatic transporters.** The uptake transporters on the basolateral (sinusoidal) membrane of hepatocytes include sodium/taurocholate cotransporting peptide (NTCP), OSTα-OSTβ, organic anion transporters (OAT1 and OAT2), OCT1/3, MCT1 and OATP1B1, 1B3, and 2B1. The efflux transporters present on the BLM of hepatocytes include MRP3 and 4. The efflux transport across canalicular membrane of hepatocytes includes BCRP, MRP2, MDR1 and 3, multidrug and toxin extrusion transporter (MATE1), ABCG5/8 and bile salt export pump (BSEP).

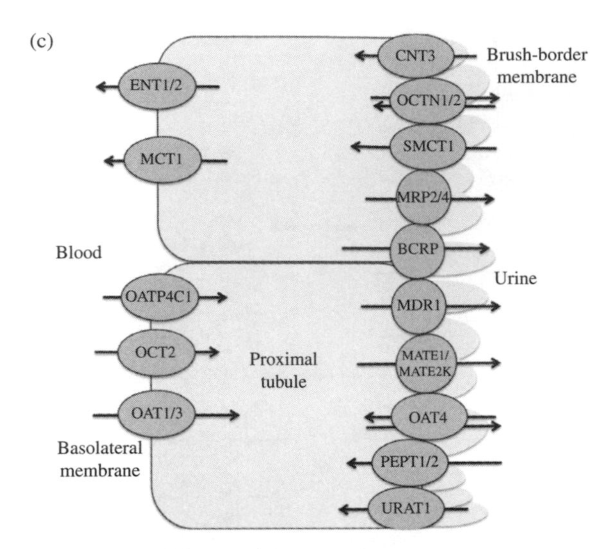

FIGURE 5.1 (*Continued*) (c) **Renal transporters.** The uptake transporters on the apical (brush-border) membrane of proximal tubules include CNT3, OCTN1/2, sodium-dependent monocarboxylate transporter 1 (SMCT1), OAT4, PEPT1/2 and urate transporter 1 (URAT1). The efflux transporters on the brush border membrane of proximal tubules contain MRP2 and 4, BCRP, P-gp, and MATE 1 and 2K. The transporters on the BLM of proximal tubules consist of ENT1/2, MCT1, OATP4C1, OCT2, and OAT1/3.

endogenous and exogenous compounds [9, 10]. Indeed, many active transporters have been identified on the canalicular membrane to mediate this process [10–12], including P-gp, BCRP, and MRP2 (Fig. 5.1b). The contribution of P-gp in biliary secretion has been demonstrated for several drugs including doxorubicin [13]. MRP2 is known for its significant role in the biliary excretion of drug conjugates, whereas BCRP is also involved in biliary excretion of drug conjugates as well as some unconjugated drugs. Furthermore, since the liver represents the primary site for drug metabolism by Phase I and Phase II DMEs, the uptake of drugs into the hepatocytes is necessary for their access to DMEs. For drugs that are hydrophilic, uptake transporters localized on the sinusoidal membrane such as organic anion-transporting polypeptides (OATPs) are important determinants in the overall distribution of drugs into the liver (Fig. 5.1b). However, only when the uptake of drugs by transporters is rate-limiting relative to the metabolism will it affect the overall drug elimination from the liver. The relevance of drug metabolism that is limited by an uptake transporter has been demonstrated for the OATP1B1-mediated uptake of various statin drugs including pravastatin [14, 15]. In addition, the effect of OATP1B1 polymorphisms and OATP1B1 inhibitors, as observed in several clinical studies, has further proven the important role of OATP1B1 on the disposition of statins in the liver [16, 17]. Other uptake transporters present on the sinusoidal membrane include MCT1, which may be involved in the uptake of L-lactic acid into the liver for gluconeogenesis, especially after exercise [18].

5.2.3 Renal Transport

The kidneys represent another important site for the elimination of a large number of xenobiotic compounds. Renal clearance is a dynamic process involving glomerular filtration, transporter-mediated renal tubular secretion and reabsorption, and passive reabsorption. Unbound small-molecule drugs are filtered at the glomerulus, and for a lipophilic drug, passive reabsorption from proximal tubule back into the systemic circulation will result in minimal renal clearance. On the other hand, if the drug is hydrophilic, the filtered drug will be excreted into the urine unless the drug is actively reabsorbed back into the systemic circulation by a transporter-mediated mechanism. Drugs can also be secreted from the blood into the lumen of the proximal tubules, resulting in a renal clearance greater than the glomerular filtration rate. There are two major carrier systems responsible for the renal secretion of drugs: organic anion transporters (OATs) and organic cation transporters (OCTs) [19]. In addition, several ATP-dependent transporters, including P-gp and MRPs, as well as the SLC transporters from the MATE family are present on the brush border membrane and are responsible for the efflux of drugs out of proximal tubule cells into the urine [20–22] (Fig. 5.1c). Other SLC transporters present on the brush border membrane are responsible for the active reabsorption of endogenous compounds including glucose, amino acids, and phosphate, as well as drugs including the drug of abuse γ-hydroxybutyrate (GHB) [23]. These transporter-mediated processes are saturable, leading to nonlinear pharmacokinetics of drugs. Also, the most common drug–drug interaction (DDI) involving renal clearance is due to inhibition of active secretion, such as that occurring with penicillin and probenecid [21].

5.2.4 BBB Transport

The BBB comprises endothelial cells that form tight junctions around the capillaries and serves as a protective barrier for the brain against xenobiotics or toxins. Transport of essential polar nutrient molecules (e.g., amino acids and glucose) into the brain requires uptake transporters. There is considerable interest in the targeted delivery of drugs into the brain to treat neuronal diseases and brain tumors. One of the prime examples for such a delivery strategy has exploited the L-type amino acid transporter (LAT1) in enhancing levodopa uptake into the brain for Parkinson's disease [24, 25]. Additionally, inhibition of efflux transporters present at the BBB (e.g., P-gp and BCRP) has been evaluated as a potential alternative strategy to enhance penetration of drugs (e.g., vinblastine, digoxin, colchicine, paclitaxel, and nelfinavir) into the brain by decreasing drug efflux [26].

5.3 ABC (ATP-BINDING CASSETTE) TRANSPORTER FAMILY

5.3.1 P-Glycoprotein (ABCB1)

MDR1, also commonly known as P-gp, has been extensively studied since its discovery by Juliano and Ling in 1976 from the plasma membrane of Chinese hamster ovary cells [27]. The molecular structure of P-gp was predicted to have two homologous halves;

each consisting of six transmembrane domains (TMDs) and a hydrophilic nucleotide domain for ATP binding that is located intracellularly [28, 29]. At least three proposed mechanisms by which P-gp performs its transport function have been hypothesized. These include classical, "hydrophobic vacuum cleaner" and "flippase" models. In the classical model, P-gp acts as a pore-forming protein and directly expels substrates from the cytoplasm into the extracellular space [22]. In contrast, in the "hydrophobic vacuum cleaner" and "flippase" models, substrates are removed directly from the cell membrane by P-gp through either direct binding to P-gp or binding of substrates in the inner leaflet of the plasma membrane, respectively [2, 30].

One of the distinctive features of P-gp is its broad substrate specificity (Table 5.1). P-gp substrates encompass many therapeutically administered drug classes including anticancer agents, cardiovascular drugs, HIV protease inhibitors, immunosuppressants, antibiotics, steroids, and cytokines. P-gp substrates are commonly hydrophobic, positively charged or neutral compounds with a planar structure [2, 33]; however, negatively charged compounds, such as methotrexate and phenytoin, can also serve as substrates under certain circumstances [34–36]. The mechanism by which P-gp recognizes such a wide range of structurally unrelated chemical entities still remains an enigma, but could be partly due to the multiple drug-binding sites present in the TMDs of the protein, and the relative hydrophobicity of substrate in the lipid membrane may facilitate the transport by P-gp even in the absence of high-affinity binding [37–40]. The relative lipophilicity of P-gp substrates also suggests that these substrates can passively diffuse across membranes and, therefore, transport by P-gp must be efficient to have an impact on drug absorption, clearance, or tissue distribution [6]. In addition, P-gp substrates are often substrates of other transporters and DMEs, so that it is difficult to identify interactions solely due to P-gp.

P-gp is found in a variety of tissues such as the liver, the kidney, the GI tract, the BBB, blood–testis barrier, and the adrenal glands, as well as in MDR tumor cells. At the subcellular level, P-gp has been shown to be predominantly located on the apical surface of the epithelial (or endothelial) cells with a specific barrier function, such as the endothelial cells of the blood capillaries in the brain, the canalicular membrane of the hepatocytes, the brush border membrane of renal proximal tubules, and the luminal membrane of the enterocytes in the colon and jejunum [41–43]. The polarized expression of P-gp in the excretory organs (liver, kidney, and intestine) and blood–tissue barriers, together with its ability to transport a wide variety of chemicals, indicates that P-gp may play an important role in protecting the body or certain tissues (such as brain and testis) from the insult of ingested toxins and toxic metabolites, by actively excreting these toxic agents into bile, urine, and intestine, or by restricting their entry into the brain and other pharmacological sanctuaries. P-gp is also present in placental trophoblasts, from the first trimester of pregnancy to full term, indicating it may be also involved in the protection of the developing fetus [2].

A wide range of P-gp inhibitors that are as chemically diverse as the substrates have also been identified (Table 5.1). Interestingly, a number of pharmaceutical excipients (e.g., cremophor EL, Tween 80, and polyethylene glycols) [44, 45] and dietary compounds in a variety of natural products (e.g., flavonoids [46–49], curcumin [50], and piperine [51]) have been also shown to inhibit P-gp. The potential

TABLE 5.1 ABC Transporters

Transporter	HUGO Symbol	Tissue Localization	Drug Substrates	Inhibitors
MDR1 (P-gp)	ABCB1	Liver, intestine, kidney, adrenal gland, blood–brain barrier, placenta	Anthracyclines, vinca alkaloids, epipodophyllotoxins, paclitaxel, topotecan, mitoxantrone, HIV protease inhibitors, digoxin, Rhodamine123, methotrexate [3]	Verapamil, diltiazem, trifluoperazine, quinidine, reserpine, cyclosporin A, valinomycin, terfenidine, PSC833, VX710, PAK-104P, GF120918, LY335979, XR9576 [3]
MRP1	ABCC1	Ubiquitous	Aflatoxin B1, doxorubicin, etoposide, vincristine, methotrexate, and various lipophilic glutathione, glucuronide, and sulfate conjugates	MK571, cyclosporin A, VX710, PA-104P [3]
MRP2 (cMOAT)	ABCC2	Liver, intestine, kidney	Glutathione conjugates, glucuronides, sulfate conjugates, methotrexate, temocaprilat, CPT11 carboxylate, SN38 carboxylate, cisplatin, pravastatin, PAH, vinblastine	MK571, benzbromarone [31]
MRP3	ABCC3	Liver, intestine, kidney, adrenal gland	Glutathione conjugates, glucuronides, sulfate conjugates, methotrexate, monoanionic bile acids (taurocholate, glycocholate), vincristine, etoposide	MK571
MRP4	ABCC4	Kidney, liver, brain	Azidothymidin, adefovir, ganciclovir, loop diuretics, thiazides, angiotensin II receptor antagonists [32]	MK571 [31]
BCRP (ABCP, MXR)	ABCG2	Placenta, liver, intestine apical membrane	Anthracyclines, epipodophyllotoxins, camptothecins or their active metabolites, mitoxantrone, bisantrene, methotrexate, flavopiridol, zidovudine, lamivudine	FTC, GF120918, Ko-134

applications of P-gp inhibitors in enhancing oral bioavailability, drug delivery to the brain, and restoring MDR-related tumor responsiveness (a phenomenon resulting from overexpression of the efflux transporters in tumor cells) to chemotherapeutic agents have been extensively studied both *in vitro* and *in vivo*. Many P-gp inhibitors have undergone clinical testing for their ability to restore tumor responsiveness to chemotherapeutic agents; however, the toxicities associated with the high concentrations of these inhibitors required for significant P-gp inhibition have prevented their clinical use. The newly developed second and third generations of P-gp inhibitors such as PSC833 [52], GF120918 [53], LY335979 [54], and XR9576 [55] have very high potency and low toxicity, and clinical trials using these agents as chemosensitizers have produced some promising results [56–59].

Development of Mdr1 knockout rodent models has provided convincing evidence concerning the functional role of P-gp in drug disposition. Significant P-gp-mediated effects on absorption and intestinal excretion have also been observed in a number of studies. Using Mdr1 knockout mouse models, the importance of P-gp in the bioavailability of a number of orally administered drugs including paclitaxel [60], digoxin [61], grepafloxacin [62], vinblastine [13], and HIV protease inhibitors [63] has been demonstrated. The clinical relevance of these observations in animal studies has been demonstrated in several human studies (e.g., talinolol, a β_1-adrenergic receptor blocker [64], and digoxin [65]). Following intravenous (i.v.) administration, the intestinal secretion of talinolol was shown to occur against a concentration gradient (5.5 (lumen):1 (blood)), indicating the involvement of an active process. In addition, the secretion rate of talinolol in the presence of a simultaneous intraluminal perfusion of R-verapamil, a known P-gp inhibitor, dropped to 29–59% of the values obtained in the absence of R-verapamil [64]. Similar results have also been obtained for digoxin with concomitant oral administration of clarithromycin [65]. With P-gp induction due to the administration of rifampin, intestinal secretion of talinolol was also increased significantly in human subjects, and the increased secretion correlated with a 4.2-fold increase in intestinal P-gp expression [66]. Together these results suggest the significance of P-gp in oral drug bioavailability.

The presence of P-gp on the canalicular membrane of hepatocytes suggests a role in the biliary excretion of drugs. The administration of P-gp inhibitors, cyclosporin A or its analogue PSC833, resulted in a decrease in the biliary excretion of both colchicine and doxorubicin *in vivo* [67, 68]. Inhibition with erythromycin decreased P-gp-mediated biliary excretion of fexofenadine in an isolated perfused rat liver study [69]. In an *Mdr1a* (–/–) knockout mouse model, the biliary excretion of unchanged doxorubicin decreased from 13.3% of the dose in wild-type mice to only 2.4% in knockout mice after a 5 mg/kg i.v. dose [13]. However, other studies have failed to find significant effects on P-gp-mediated biliary excretion in *Mdr1a* (–/–) knockout mice with paclitaxel, digoxin, and vinblastine. One possible explanation of these conflicting results is the presence of alternative transport processes responsible for the secretion of these substrates into the bile. P-gp may act in concert with other transporters in excreting certain substrates into bile, and the loss of P-gp function could be compensated by other transport processes under certain circumstances. Indeed, it has been shown that *Mdr1b* expression in the liver and kidney was consistently increased in *Mdr1a* (–/–) knockout mice compared to the wild-type animals, indicating that the loss of *Mdr1a* function could be compensated by *Mdr1b* protein for their common substrates [70]. Other canalicular membrane transporters may also exhibit overlapping substrate specificity for certain P-gp substrates. Nevertheless, in humans the P-gp inhibitors verapamil and quinidine produced significant reduction in the biliary clearance of digoxin, (a non-metabolized P-gp substrate), as well as an increase in digoxin plasma concentration, indicating the clinical significance of this transporter-mediated interaction. Additionally, in humans, the renal clearance of digoxin was decreased by 20% ($p<0.01$) by the concomitant use of the P-gp inhibitor, itraconazole. Since digoxin is mainly excreted unchanged into urine, this reduction is most likely mediated by the inhibition of P-gp [71]. Similarly, the renal clearance of quinidine was also decreased by 50% ($p<0.001$) by itraconazole in a double-blind, randomized crossover study, and inhibition of P-gp is most likely the underlying mechanism involved [72]. Taken together, these studies demonstrated that P-gp also significantly contributes to the renal excretion of its substrates.

5.3.2 Multidrug Resistance-Associated Proteins (ABCC)

The family of multidrug resistance-associated proteins (MRPs), consisting of nine members, is another group of ABC transporters involved in drug transport. Among the family members, MRP2, in particular, plays a role in drug transport, while MRPs 1 and 3 have recently been identified as emerging members of MRPs with clinical importance [31]. MRPs 1–3 have similar topology, containing a typical ABC core structure of two segments with each consisting of six TMDs and an ATP-binding domain, similar to P-gp, and an extra N-terminal segment of five TMDs linked to the core structure through an intracellular loop [73].

The first member of this family, MRP1, was cloned in 1992 from the resistant human lung cancer cell line [74], which does not overexpress P-gp [75–78]. Subsequent transfection studies demonstrated that overexpression of this 190 kDa membrane protein is able to confer multidrug resistance (MDR) against a number of natural product anticancer agents such as the anthracyclines, vinca alkaloids, and epipodophyllotoxins, by extruding these cytotoxic agents from cells and thus lowering their intracellular concentrations [79–82]. Later on, MRP2 and other members were also identified and characterized to varying extents [83–92]. MRP3 is the most closely related member to MRP1 with 58% amino acid identity, followed by MRP2 (49%) [93]. Substrates of MRP1 include amphiphilic anions, preferentially lipophilic compounds conjugated with glutathione (e.g., leukotriene C4 and dinitrophenylglutathione), glucuronide (e.g., bilirubin and 17β-estradiol), or sulfate [94]. Some unconjugated amphiphilic anions such as methotrexate and Fluo-3, a penta-anionic fluorescent dye, can also be transported by MRP1 in an unchanged form [95, 96]. In addition to the anionic compounds, MRP1 can also transport unconjugated amphiphilic cations or neutral compounds, such as anthracyclines, etoposide, and vinca alkaloids but requires reduced glutathione (GSH) as a cotransporting substrate [97–100], one of the distinctive property of MRP1-3-mediated transport. Both MRP2 and MRP3 share similar substrate specificity with MRP1. They can also transport conjugates of lipophilic substances with glutathione, glucuronide, and sulfate such as the glutathione S-conjugate leukotriene C4, bilirubin glucuronide, and anticancer agents such as methotrexate, vincristine, and etoposide [94]. However, the substrate specificity of these three MRP isoforms is not identical, and for their common substrates, the transporting efficiency by these isoforms varies substantially; [94] there are also substrates that can be recognized by one isoform but not the others.

MRP1 is ubiquitously expressed throughout the body, and its distribution is solely confined to the BLMs with low expression in the liver [73, 101, 102]. It is has been well established that the involvement of MRP1 in transporting many exogenous and endogenous toxic compounds is important for its protective role in preventing tissue or organ exposure to these toxins such as proinflammatory cysteinyl leukotriene C_4 (LTC_4) [103]. MRP1 is also known for its cellular defense against oxidative stress by extruding oxidized glutathione (GSSG) out of the cells [104]. Interestingly, although GSH is a co-substrate for MRPs 1–3, it also has been demonstrated to play an important role in modulating MRP1-mediated transport [104]. Drugs such as verapamil and many of the dietary flavonoids have been shown to act as modulators for stimulating MRP1-mediated GSH transport via increases in the apparent affinity of MRP1 [105, 106]. Furthermore, depletion of intracellular GSH by buthionine sulfoximine (BSO), an inhibitor of GSH synthesis, can increase the intracellular accumulation of many anticancer agents in MRP1-overexpressing cells [107, 108], suggesting that alteration of intracellular GSH concentration may be one of the reasons for MRP1-mediated MDR. In addition to its role in transport, recent studies have also suggested an involvement of MRP1 in modulating the biological activities of its endogenous toxins, mainly lipid metabolites such as LTC_4, via altering the expression of other proinflammatory molecules, leading to reduced inflammatory responses as observed in Mrp1 knockout mice [103, 109]. Despite the physiological role of MRP1, it has been reported that mice with disrupted Mrp1 (*Mrp1* (–/–)) are viable, fertile, and have no physiological or histological abnormalities, indicating that Mrp1 may not be essential for normal mouse physiology, and that these mice are a useful model for studying Mrp1-mediated effects.

The expression of MRP1 found in cells of blood–tissue barriers such as choroid blood–CSF barrier in the BLM of the choroid plexus suggests its role in protecting some tissues or organs from exposure to toxic substances. Indeed, numerous studies in *Mrp1* (–/–) knockout mice have demonstrated that *Mrp1* (–/–) knockout significantly increases sensitivity to a cytotoxic agent etoposide with increased bone marrow toxicity [110] and increased CSF exposure [111]. A similar observation has also been reported with a therapeutic dose of vincristine, which normally does not exhibit bone marrow toxicity and gastrointestinal damage, causing extensive damage to these tissues in both *Mrp1* (–/–) and *Mdr1a/1b* (–/–) knockout mice, indicating that Mrp1, Mdr1-type P-gp, and probably other related efflux transporters work in concert as detoxifying mechanisms to protect tissue from damage induced by toxic agents [112].

Similar to MRP1, both MRP2 and MRP3 have also been shown to confer MDR to several anticancer drugs *in vitro* [113–116]. The clinical relevance for

MRP1-mediated MDR has been a topic of extensive investigation, and there is increasing evidence suggesting that overexpression of MRP1 might represent a poor prognostic factor in various cancers particularly in neuroblastoma [103, 117–127]. The clinical relevance of MRP2- and MRP3-mediated MDR is currently unknown.

MRP2 transports organic anions and is the only MRP member that is localized exclusively on the apical membrane of polarized cells including hepatocytes, intestinal epithelial cells, and renal proximal tubule cells [128]. In addition to sharing similar substrate spectrum with MRP1, MRP2 can also transport many of the phase II drug conjugated metabolites [129]. MRP2 deficiency in humans is associated with Dubin–Johnson syndrome (DJS), a hereditary disorder characterized by mild conjugated hyperbilirubinemia and pigment disposition in the liver due to impairment in an MRP2-mediated transport [130–133]. Two naturally occurring mutant rat strains, GY/TR$^-$ and EHBR rats from the Wistar and Sprague-Dawley rat colonies, respectively, also lack Mrp2 expression and are considered animal models for DJS in humans [83, 84, 134, 135]. Furthermore utilization of these Mrp2-deficient rats has been valuable in the functional characterization of MRP2.

While MRP2 is localized on the brush border membrane of organs, other MRPs, including MRP 1, 3, 4, 5, and 6, are present on the BLM. An exception is MRP4, which is present on the BLM of hepatocytes but on the brush border of renal proximal tubules and brain capillary endothelial cells [21, 31]. Currently, there are no selective inhibitors available for MRPs 1–4. MK-571, a quinolone derivative (developed as leukotriene D4 receptor antagonists), represents a widely used nonselective inhibitor that targets MRPs 1, 2, 3, and 4 [104]; however, it also inhibits OATP1B1, 1B3, and 2B1 [136].

MRP2 is best known for its significant role in the biliary excretion of drug glucuronide and sulfate conjugates and divalent bile acids. MRP2 is also involved in the renal secretion of some drug substrates. Unlike MRP1, MRP2 shows extensive overlap in substrates with P-gp. Studies have shown that MRP2 is involved in the biliary excretion of doxorubicin (a substrate for P-gp as well), pravastatin, and valsartan, which further suggests that P-gp and MRP2 collectively are involved in biliary excretion and oral drug absorption for many therapeutic agents [6, 137–139]. Using Mrp2-deficient rats, it has been shown that the AUC (0–6 hour) of ^{14}C-temocapril was dramatically increased and the biliary clearance, as measured by total radioactivity, was markedly decreased in EHBR rats compared with the control rats after i.v. administration. Since the active metabolite temocaprilat accounted for over 95% of the total radioactivity, again these data indicate that Mrp2 plays a central role in the biliary excretion of the metabolites of this drug [140]. Interestingly, it was shown that the biliary excretion of irinotecan, its active metabolite SN-38, and its glucuronide conjugate can be substantially decreased by probenecid, an MRP2 inhibitor, with concomitant elevation of plasma concentrations of these compounds in normal rats, resulting in decreased GI toxicity [141].

MRP3 may be responsible for the development of MDR against etoposide, teniposide, and vincristine [115, 142]. The primary role of MRP3 is to transport drug metabolites that have been glucuronidated in the liver (e.g., morphine-3-glucuronide, bilirubin-glucuronide, etoposide-glucuronide, and acetaminophen-glucuronide)

[32, 143] and efflux them out of hepatocytes into the systemic circulation via the sinusoidal membrane. It has been shown that the expression of MRP3 in the liver is highly variable and inducible [144]. Interestingly, MRP3 is significantly upregulated in the liver of Mrp2-deficient rats and in patients with DJS or primary biliary cirrhosis [142, 145], indicating that MRP3 may serve as a compensatory mechanism to remove the conjugates from hepatocytes through the sinusoidal membrane in conditions where MRP2-mediated biliary excretion is impaired [94].

In the kidney, secretion of anions across the brush border membrane into the urine is mainly mediated by MRP2 and MRP4. Based on recent findings that MRP4 protein expression was reported to be fivefold higher than MRP2 and MRP4 exhibits higher affinity toward anions than MRP2 in human renal cortex [21, 146], it was suggested that MRP4 may play a much greater role in anion secretion in the kidney than MRP2. Significant accumulation of antiviral drugs (e.g., adefovir and tenofovir) [147], diuretics [148], and antibiotics [149] in renal cells was observed in Mrp4 knockout mice. Studies in human kidney revealed that MRP4 is primarily involved in excreting many antiviral drugs such as adefovir and cidofovir [150]. Interestingly, it was found that cidofovir in particular is not transported by MRP4 and, therefore, further suggested that absence of MRP4-mediated secretion of antiviral drugs such as cidofovir may in part be one of the reasons for the nephrotoxicity seen with some of the antiviral drugs [6, 147].

5.3.3 Breast Cancer Resistance Protein (ABCG2)

BCRP was initially cloned from a doxorubicin-resistant breast cancer cell line (MCF-7/AdrVp) selected with a combination of adriamycin and verapamil [151]. Two other groups also independently identified this transporter from human placenta [152] and human colon carcinoma cells (S1-M1-80) [153] and named the protein ABCP (ABC transporter in placenta) and mitoxantrone resistance (MXR)-associated protein, respectively. As its name implies, BCRP exhibits MDR in various tumors. Significant and variable expression of BCRP has been reported in human tumors such as acute leukemia and breast cancer; however, the contribution of this efflux transporter to clinical MDR needs to be further investigated [154–160]. In contrast to P-gp and MRP1 or MRP2, BCRP is a “half ABC transporter” that consists of only six TMDs and one ATP-binding site [152]. As a “half transporter,” BCRP most likely forms a homodimer or homomultimers to transport its substrates out of the cells utilizing the energy derived from ATP hydrolysis [3, 151, 161–163].

BCRP shows considerable overlap in substrate specificity with P-gp and MRP1 or MRP2, although the binding affinity of substrates to these transporters may vary substantially [3]. Interestingly, an expression of BCRP in normal tissues is similar to P-gp. Because BCRP shares similar substrates with P-gp and is often colocalized with P-gp, specific BCRP inhibitors and knockout animals are valuable for attributing drug transport to BCRP in drug disposition. Nevertheless, it is thought that it is the functional redundancy of BCRP with other transporters like P-gp that makes the occurrence of BCRP-related DDI relatively rare clinically [164]. Constitutive expression of BCRP was detected in various human tissues such as placenta (syncytiotrophoblast), liver (canalicular membrane of hepatocytes), small and large intestine (apical membrane of the epithelium), ducts and lobules of the breast tissue, BBB, and kidney (proximal tubules) [152, 164–167]. By analogy with P-gp, it is reasonable to speculate that one, if not the major physiological function of BCRP, is to protect the body from exposure to toxic endogenous or exogenous compounds. And indeed it has been shown that some physiological roles of BCRP include effluxing porphyrins from hematopoietic cells and hepatocytes and secreting urate into urine [168, 169].

BCRP can limit drug bioavailability either by decreasing intestinal uptake or by hepatobiliary drug excretion. A study conducted by Jonker et al. [170], using *Bcrp1* (–/–) knockout mice, strongly supported this speculation. In this study, the authors demonstrated that the oral bioavailability of topotecan increased about sixfold in *Bcrp1* (–/–) mice compared to the wild-type mice. Additionally, in humans, it has been shown that the apparent oral bioavailability of topotecan was significantly increased from 40.0 to 97.1% following the coadministration of GF120918 [171]. As it is known that topotecan is only a weak substrate of P-gp [171], this observed change most likely resulted from the inhibition of BCRP. In addition, oral coadministration of rosuvastatin and BCRP inhibitors (e.g., atazanavir/ritonavir) significantly increased rosuvastatin plasma concentrations in humans, which further suggested that the interaction resulted from inhibition of BCRP-mediated intestinal efflux and biliary excretion [172]. It should also be noted that the expression of BCRP and MRP2 in the human intestine is even higher than P-gp [173], and, therefore, it may be possible that the contribution of BCRP to the intestinal secretion and oral absorption of xenobiotics can be comparable with, if not greater than, that of P-gp.

The expression of BCRP on the luminal side of the intestinal epithelial cells and canalicular membrane of the hepatocytes suggests that BCRP may also play a significant role in the biliary excretion, thus limiting the entry of xenobiotic toxins into the systemic circulation or facilitating drug elimination. Unlike MRP2, it was found that BCRP plays an important role in biliary excretion of sulfate conjugates along with unconjugated drugs [174, 175]. In mice lacking Bcrp1, a decrease in biliary excretion of acetaminophen sulfate and other sulfate conjugated drug was observed while the same effect is not seen in mice lacking Mrp2, suggesting that sulfate conjugates are predominately exported by BCRP into bile [175–177]. Clinically relevant changes in plasma concentrations of BCRP substrates, such as rosuvastatin, can occur due to genetic polymorphisms in BCRP [178].

In the brain, BCRP also plays a protective role at the BBB. Through liquid chromatography-tandem mass spectrometry quantitative membrane transporter expressions were determined in human brain, and it was found that BCRP was the most abundant protein among the other transporters, being 1.33-fold higher than P-gp [164, 179]. In comparison to mice, BCRP protein level is 1.85-fold higher in humans, and it is thought that the relative contribution of human BCRP in BBB transport may be underestimated when extrapolating from mouse studies [164, 179]. Knockout of both P-gp and BCRP in a mouse model has shown to increase the accumulation of numerous anticancer drugs in the brain [164, 180–182]; however, knockout of only one transporter may have limited effects for substrates that exhibit overlapping substrate specificity for these two transporters. This functional relevance of BCRP in brain penetration of drugs has provided insight on targeting inhibitors of both P-gp and BCRP as a potential strategy to enhance anticancer, antiviral, and antiepileptic drugs delivery to the brain. As for the role of BCRP in the renal elimination of drugs, the expression level of BCRP/Bcrp1 is high in rodents but low in humans and, therefore, may not be important for the renal secretion of drugs in humans [183].

5.3.4 Other ABC Transporters

MDR3 (ABCB3) is the other human P-gp isoform with virtually identical molecular structure to that of the human *MDR1* and mouse *mdr1b* genes [184]. MDR3 is mainly present in the canalicular membrane of liver hepatocytes and functions as an ATP-dependent phosphatidylcholine translocator. It has been shown that MDR3 is also capable of transporting several cytotoxic drugs such as digoxin, paclitaxel, and vinblastine, but with a low efficiency. Bile salt export pump (BSEP, ABCB11) is almost exclusively present in the liver and localized to the canalicular microvilli and subcanalicular vesicles of the hepatocytes, functioning as a major bile salt efflux protein in mammalian livers [185]. At this time, it is generally believed that both MDR3 and BSEP may not play significant roles in terms of drug disposition. Additionally, other members of the ABCG family, related to BCRP, including ABCG1 and 3 are involved in cholesterol transport and ABCG5 and 8 in cholesterol and phytosterol transport [186].

5.4 SLC (SOLUTE CARRIER) TRANSPORTER FAMILY

5.4.1 Organic Anion Transporting Polypeptides (SLCO)

OATPs are part of SLCO solute carrier family that are involved in sodium-independent transport of large-molecular-weight amphipathic organic anions (>300 Da) such as bile acids and cationic and neutral compounds [187]. OATPs are predicted to have 12 TMDs and a large extracellular loop at the fifth position [188]. OATPs are antiporters, and their mechanism of transport most likely involves using intracellular glutathione concentrations as a driving force [187, 189]. Because members of the OATP family are poorly conserved and orthologs for human OATPs may not be present in rodents, the relevance of extrapolation from animal models to understand OATP-mediated drug disposition in humans is limited [169].

OATP1B1 and OATP1B3 are exclusively expressed at the sinusoidal membrane of the liver where they serve to transport substrates into hepatocytes for metabolism or biliary excretion. Generally, OATP1B3 shares many substrates with OATP1B1, with a few exceptions (Table 5.2). In the OATP family, OATP1B1 is probably best known for its clinically significant DDIs reported for HMG-CoA reductase inhibitors (e.g., rosuvastatin and cerivastatin) when given with OATP1B1 inhibitors [17, 190–194]. While OATP1B3 plays a minor role in the uptake of drugs into hepatocytes, as compared to OATP1B1, it often acts as compensatory transporter for OATP1B1

TABLE 5.2 Solute Carrier SLC Transporters

Transporter	HUGO Symbol	Tissue Localization	Substrates	Inhibitors
OATP1A2	SLCO1A2	Brain, intestine	Digoxin, erythromycin, fexofenadine, imatinib, levofloxacin, methotrexate, ouabain, pitavastatin, rocuronium, rosuvastatin, saquinavir, thyroxine, unoprostone [6, 190]	Grapefruit juice, rifampin, ritonavir, saquinavir, verapamil [6, 190]
OATP1B1	SLCO1B1	Liver	Atorvastatin, atrasentan, benzylpenicillin, bosentan, caspofungin, cerivastatin, enalapril, fexofenadine, fluvastatin, methotrexate, olmesartan, pravastatin, repaglinide, rifampin, rosuvastatin, simvastatin, SN-38, valsartan [6, 190]	Atorvastatin, clarithromycin, cyclosporine, erythromycin, gemfibrozil, paclitaxel, rifampin, ritonavir, saquinavir, tacrolimus, telmisartan [6, 190]
OATP1B3	SLCO1B3	Liver	Atrasentan, bosentan, digoxin, docetaxel, enalapril, erythromycin, fexofenadine, fluvastatin, imatinib, methotrexate, olmesartan, ouabain, paclitaxel, pitavastatin, pravastatin, rifampin, rosuvastatin, telmisartan, SN-38, thyroxine, valsartan [6, 190]	Clarithromycin, cyclosporine, erythromycin, rifampin, ritonavir [6, 190]
OATP2B1	SLCO2B1	Intestine, liver	Atorvastatin, benzylpenicillin, bosentan, fexofenadine, fluvastatin, glibenclamide, pravastatin, rosuvastatin, unoprostone [6, 190]	Cyclosporine, gemfibrozil, rifampin [6, 190]
OAT1	SLC22A6	Kidney	Acyclovir, adefovir, cidofovir, ciprofloxacin, lamivudine, methotrexate, penicillins, tenofovir, zidovudine [6]	Probenecid

(*Continued*)

TABLE 5.2 ***(Continued)***

Transporter	HUGO Symbol	Tissue Localization	Substrates	Inhibitors
OAT3	SLC22A8	Kidney	Bumetanide, cefaclor, ceftizoxime, furosemide, NSAIDs, penicillins [6]	Probenecid
OCT1	SLC22A1	Liver	Metformin, lamivudine, oxaliplatin [6]	Disopyramide, quinidine [6]
OCT2	SLC22A2	Kidney	Cisplatin, lamivudine, metformin, oxaliplatin, procainamide [6]	Cetirizine, cimetidine, quinidine [6]
MATE1	SLC47A1	Kidney, liver	Cephalexin, cephradine, creatinine, cimetidine, Metformin, MPP, oxaliplatin procainamide, TEA [6, 20]	Cimetidine, quinidine, procainamide [6]
MATE2-k	SLC47A2	Kidney	Cimetidine, creatinine, Metformin, MPP, oxaliplatin, TEA [20]	Cimetidine, quinidine, procainamide
MCT1	SLC16A1	Ubiquitous	Atorvastatin, salicylic acid, pravastatin, valproic acid, γ-hydroxybutyric acid, XP13512	Dietary flavonoids, L-lactate, CHC
PepT1	SLC15A1	Intestine, kidney, liver, pancreas	Captopril, cefadroxil, cephalexin, enalapril, valacyclovir, valganciclovir, midodrine	Glycyl-proline, zinc [6]

[187]. In addition to transporting a wide spectrum of substrates, OATP1B1 is also involved in transporting numerous endogenous substrates including bilirubin and estrogen conjugates. OATP1B1 inhibitors such as gemfibrozil, cyclosporine, and rifampin (act as both substrates and inhibitors for OATP1B1) are also commonly used in the clinic. It is important to note that cyclosporine and rifampin are inhibitors of multiple transporters and DMEs and, therefore, specific transporter attribution for a clinical DDI is often difficult to ascertain based on interaction studies. Interestingly, many single nucleotide polymorphisms (SNPs) have been identified in the *SLCO1B1* gene in various populations and shown to significantly contribute to the reduction in the uptake of drugs into the liver, resulting in decreased hepatic clearance of statins and many other OATP1B1 substrates (e.g., repaglinide [195], atrasentan [196], olmesartan [197], and SN38 [198], the active metabolite of irinotecan). Along with OATP1B1 polymorphisms, hepatic uptake of substrates such as bosentan has been speculated to be rate-limiting in their overall hepatic clearance [199]; thus, inhibition of OATP1B1-mediated uptake may represent the predominant mechanism in some clinically significant DDIs.

Several clinical studies have demonstrated the importance of OATP1B1 in the disposition of statins. Following coadministration of gemfibrozil with cerivastatin, a significant increase in cerivastatin exposure was shown to increase the risk for rhabdomyolysis in healthy volunteers [200], and it was later determined that the interaction was due to inhibition of OATP1B1 and CYP2C8 by gemfibrozil [17]. Further demonstrating the significant role of OATP1B1 in the hepatic uptake of statins, the concomitant administration of gemfibrozil also resulted in increasing plasma concentration of other statins such as rosuvastatin, which undergoes minimal metabolism [17, 190, 191]. Studies examining the effect of OATP1B1 polymorphisms on the pharmacokinetics of statin drugs reported that individuals with SLCO1B1*15 expression have an increased exposure to statins including pravastatin [15], pitavastatin [201], simvastatin acid [202], atorvastatin [203], and rosuvastatin [204], compared to subjects with the wild-type allele. This increase in statin exposure was shown to be associated with myopathy observed in individuals with the SLCO1B1*15 allele [16, 205]. Collectively, these results clearly provide evidence of the significant role of OATP1B1 in hepatic uptake and hepatic clearance of drugs.

OATP isoforms involved in the intestinal absorption of drugs include OATP1A2 and 2B1 [206]. In addition to its expression in the intestine, OATP1A2 is also expressed at the BBB where it is involved in BBB transport of drugs and endogenous molecules [6, 207]. Unlike OATP1B1 and 1B3, OATP2B1 is more widely expressed; it is found in intestine (apical membrane of intestinal cells) and liver (sinusoidal membrane of hepatocytes) [208]. The role of OATP1A2 in intestinal absorption has been demonstrated with the interaction between grapefruit juice and fexofenadine. Fexofenadine is also a substrate for P-gp, and grapefruit juice can also inhibit P-gp. Interestingly, it was shown *in vitro* that the inhibition of P-gp was observed at grapefruit juice concentrations greater than 20%, whereas at lower concentrations (<5%) [209], grapefruit juice is a good inhibitor of OATP-mediated fexofenadine transport [210]. In clinical studies, coadministration of grapefruit juice with fexofenadine decreased the plasma concentrations of fexofenadine by 63%, suggesting that it is likely that a decrease in bioavailability of fexofenadine is mediated by inhibition of OATP1A2 [211, 212].

OATP4C1 is the only member of the OATP family found in the human kidney, where it is localized to the BLM of proximal tubules [213]. Substrates for OATP4C1 include digoxin, thyroid hormones, methotrexate, and antidiabetic drug sitagliptin, and OATP4C1 may be involved in the renal secretion of these drugs from the blood into the kidney [213, 214].

5.4.2 Organic Anion Transporters (SLC22A)

In addition to MRPs and OATPs, OATs are also involved in transporting organic anions, and OATs are known particularly for their significant role in the renal secretion and clearance of many anionic drugs. Like the other SLC22 family members, OATs are predicted to consist of 12 α-helical transmembrane-spanning domains [215]. OAT1 and OAT3 are highly expressed in the kidney on the BLM of proximal tubule cells. OAT1 and OAT3 are also expressed in various other tissues such as the choroid plexus [19, 216]. OAT1 and OAT3 function as organic anion/dicarboxylate exchangers that are involved in facilitating anionic drug transport into the proximal tubule cells from the blood through the utilization of an intracellular α-ketoglutarate gradient [217]. Substrates for OAT1 and OAT3 include β-lactam antibiotics, nonsteroidal antiinflammatory drugs (NSAIDs), and other structurally distinct classes of type I organic anions (monovalent or selected divalent anions that are <500 Da) [215, 217, 218] (Table 5.2). Despite overlapping substrate specificity between OAT1 and OAT3, OAT3 prefers more bulky amphipathic anions such as HMG-CoA reductase inhibitors (statins) and some cationic drugs such as H_2 receptor antagonists [217].

In addition to OAT1 and OAT3, OAT4 is also found in the kidney where it is localized to the apical membrane of proximal tubule cells. OAT4 is involved in both renal secretion (e.g., glutarate) and reabsorption (e.g., loop diuretic torasemide, methotrexate, urate, and prostaglandin) of drugs and many endogenous substrates; however, its primary role is likely the renal reabsorption for many of its substrates [219]. Unlike OAT1 and OAT3, OAT4 has a much narrower substrate specificity.

High expression of OAT1 and OAT3 in the kidney and their broad substrate spectrum have resulted in these transporters playing a significant role in the renal excretion of numerous therapeutic agents. Numerous earlier studies, reported from the 1960s to the 1990s, have demonstrated that coadministration of the nonspecific OAT inhibitor, probenecid, with various penicillin derivatives (e.g., piperacillin, nafcillin, and ticarcillin), statins, antiviral drugs (e.g., acyclovir, cidofovir, ganciclovir), and diuretics (e.g., furosemide) decrease the renal clearance of these drugs in humans [21]. The decrease in furosemide renal clearance when probenecid was coadministered also resulted in a reduced diuretic effect, which was associated with the decreased furosemide concentrations in tubular fluid [220, 221]. Because probenecid is also an MRP2 inhibitor and MRP2 is known to mediate the renal secretion of anionic drugs, but is present at the apical membrane of proximal tubule cells, it is likely that probenecid effects on the renal clearance for some of these anionic drugs can be attributed to both inhibition of OATs and MRP2. Furthermore, since some of the antiviral drugs such as acyclovir, cidofovir, ganciclovir, which are renally secreted mainly by OATs and MRP2, can cause nephrotoxicity, OAT inhibition could represent a potential strategy to prevent renal accumulation of these drugs and, hence, minimize the nephrotoxicity associated with these agents.

Hepatic transport of organic anions across hepatocytes can be mediated by OAT2, the major OAT on the sinusoidal membrane of hepatocytes. OAT2 is also expressed on the BLM of proximal tubule cells, but to a much lesser extent [222]. Inhibition of OAT2-mediated hepatic uptake of theophylline by erythromycin has been reported to produce a 25% decrease in theophylline concentrations [223, 224].

5.4.3 Organic Cation Transporters (SLC22)

OCT1-3 are electrogenic uniporters that use the negative membrane potential to facilitate diffusion of organic cations across membranes in both directions [225]. OCT isoforms show both species- and tissue-specific distribution (e.g., OCT2 is the major OCT in human kidney while Oct1 is the major OCT in rodent kidney) [226]. Interestingly, all three OCT isoforms exhibit genetic polymorphisms [227–229], but OCT2 displays the greatest number of genetic variants [228], and it is clinically most relevant [230–232].

Substrates for OCTs are often small-molecular-weight and hydrophilic organic cations with distinctive and diverse structure, hence the term "polyspecific" [233]. However, they can also transport weak bases and some neutral compounds as well [19]. The "polyspecific" property of OCTs could be due to the transporters' large binding site that allows for interaction of different substrates with their binding domains and tissue-specific distribution of OCT isoforms in the body [233, 234]. OCT1 is predominantly expressed on the sinusoidal membrane of the hepatocytes in the liver and on the BLM of intestinal cells [226, 235]. OCT1 is also present on the apical membrane of proximal tubule cells but at a much lower amount [226, 235]. OCT2 is mainly expressed on the BLM of proximal tubule cells [236]. OCT3 shows a much broader tissue distribution and is present in the brain, heart, skeletal muscle, placenta, and the liver, with high expression in placenta and brain [19]. In contrast to OCT1 and OCT2, OCT3 is involved in transporting mostly endogenous substrates such as monoamines [237]. OCT3 is known for its central role in eliminating catecholamines from the fetal blood circulation [238].

With tissue-specific distribution of OCT1 in the liver and OCT2 in the kidney, OCT1 and OCT2 have been shown to play significant roles in the disposition of drugs such as metformin and cisplatin, and potentially in the tissue-specific toxicity observed in humans [239, 240]. Using both single and double Oct1/2 knockout mouse models, studies have characterized the pharmacological and physiological roles of OCTs. Studies in *Oct1* (–/–) mice showed significant decreases in tetraethyl ammonium (TEA) and metformin concentrations in the liver and significant increases in renal excretion of TEA and metformin as compared to wild-type mice [241, 242]. With *Oct1/2* (–/–) double knockout mice, renal excretion of TEA was completely abolished and the plasma concentrations of TEA were significantly increased. Because of the differences in human and murine OCT isoform tissue distribution, these data most likely reflect the effect of human OCT2 in the kidney [243].

Clinically, OCT-mediated transport of metformin and cisplatin has been reported to be associated with drug interactions and toxicity [239, 240]. Metformin, used in the treatment of diabetes, is a substrate for both OCT1 and OCT2. The role of OCT1 in the transport of metformin has been studied both *in vitro* [242] and *in vivo* [242, 244], and OCT1 is the major isoform involved in metformin uptake into the liver. In *Oct1* (–/–) knockout mice the distribution of metformin in the liver was greater than 30 times lower than in wild-type mice [242] and blood lactate concentrations were significantly increased in wild-type mice [244], suggesting an important role of OCT1 for metformin uptake into the liver and in the toxicity of metformin. It was also demonstrated that OCT2 polymorphisms and coadministration of the OCT inhibitor cimetidine, with metformin, resulted in decreased renal clearance of metformin, which can be attributed mainly to decreased OCT2- mediated renal secretion [231].

OCT2-mediated transport of the anticancer agent cisplatin into proximal tubule cells has been identified as an important factor in cisplatin-induced nephrotoxicity observed clinically [240]. Mice with deletion of Oct1 and Oct2 exhibit significantly impaired renal secretion of cisplatin and reduced cisplatin-induced tubular toxicity [245]. Collectively, these results clearly indicate the importance of OCT2 in the renal excretion and, hence, the toxicity associated with cisplatin. Therefore, aside from MRP inhibition, OCT2 inhibition may provide another therapeutic strategy to mitigate nephrotoxicity associated with cisplatin and other drugs.

Little is known about the role of OCTs in intestinal excretion of xenobiotics although Oct1 knockout (Oct1 –/–) mouse models lack excretion of TEA into the lumen of intestine as well as uptake of TEA into the hepatocytes, suggesting a role for Oct1 in mediating these transport processes [246]. This is most likely due to the differences in tissue distribution of OCT1 between humans and rodents; rat Oct1 mRNA has been detected in the liver, kidney, and intestine [247], whereas human OCT1 is primarily expressed in the liver. Although human OCT1 and OCT3 mRNA levels are detected in jejunum and colon [248], the importance of OCTs in the intestinal transport of organic cations has not been demonstrated.

The carnitine/organic cation transporter (OCTN) family consists of two members: OCTN1 and OCTN2. In addition to OATs and OCTs, OCTNs are also part of subfamily of SLC22A superfamily. OCTN1 was first discovered in 1997, followed by OCTN2 in HEK293 cells [249, 250]. They consist of 11–12 putative membrane-spanning domains. Interestingly, OCTN1 was found to have a nucleotide-binding site motif; thus, the reason why it was named OCTN [249–251]. Unlike OCTs, OCTNs are uniquely involved in transporting zwitterionic carnitine [251]. OCTNs are known to be important in the placenta for the transport of carnitine to the fetus [6, 238]. They are also involved in transporting xenobiotics, particularly cationic compounds [251].

OCTN1 is an organic cation proton exchanger, and its mRNA and/or protein expression have been found in various tissues including gut, placenta, skeletal muscle, bone marrow, and trachea; however, it is mainly expressed on the apical membrane of renal tubular cells [251, 252]. The physiological role of OCTN1 has not been well established although it was found to preferentially transport the antioxidant ergothioneine with relatively high affinity [251, 253]. In addition, OCTN1 is capable of transporting drugs including quinidine, pyrilamine, verapamil, TEA, oxaliplatin, acetylcholine, and the anticholinergic drugs ipratropium and tiotropium [19, 251, 254, 255] (Table 5.2). OCTN1 polymorphisms were found to be associated with autoimmune diseases such as rheumatoid arthritis and Crohn's disease. Interestingly, it has been reported that OCTN1-mediated renal secretion of gabapentin in individuals with OCTN1 polymorphisms was decreased, suggesting relevance of OCTN1 in renal secretion of gabapentin as well as other xenobiotics [256]. Although many studies have provided evidence of the role of OCTN1, more studies are needed to further confirm the significance of OCTN1 in drug transport [251].

OCTN2 is a sodium-dependent, high-affinity transporter, but it can also function as sodium-independent transporter for TEA, choline, verapamil, and pyrilamine [252]. OCTN2 mRNA and/or protein expression have been found ubiquitously throughout the body (e.g., liver, kidney, skeletal muscle, heart, brain, placenta, and intestine) but high expression of OCTN2 is found on the apical membrane of renal tubular cells [251, 252]. OCTN2 is primarily responsible for the transport of carnitine in the body for β-oxidation of long-chain fatty acids in mitochondria [251]. Mutations in OCTN2 cause primary systemic carnitine deficiency syndrome, which leads to various symptoms including cardiomyopathy, skeletal muscle weakness, male infertility, and fatty liver [251]. Many drug substrates for OCTN2 are also transported by OCTN1 (Table 5.2). However, the anticholinergic drugs ipratropium and tiotropium are transported mainly by OCTN2 [6]. Inhibitors of OCTN generally inhibit both OCTN1 and OCTN2, and many of those inhibitors are also substrates, including valproate and verapamil [19, 251]. Using human placenta brush border membrane vesicles, it was demonstrated that many anticonvulsants including valproate inhibit OCTNs, and it has been suggested that the inhibition of OCTNs could be responsible, at least in part, for the teratogenicity and other side effects associated with anticonvulsant drugs [6, 257].

5.4.4 Multidrug and Toxin Extrusion Transporters (SLC47A)

MATEs are cation transporters, with transport mediated by proton exchange. There are two members of the MATE family identified: MATE1 and the splice-variant MATE2-K. MATE1 is localized predominantly on the apical membrane of proximal tubule cells and is also expressed in liver (canalicular membrane), adrenal gland, testis, and skeletal muscle [258]. MATE2-K is human specific, present only in the kidney on the apical side [259]. MATE1 and MATE2-K share similar substrate specificities [260]; however, it was demonstrated that the zwitterions cephalexin and cephradine are MATE1 specific [261].

The unidirectional transport involved in the renal secretion of organic cations across proximal tubule cells, from the blood to the urine, was further elucidated following the discovery and characterization of MATEs. It is now known that MATEs on the apical membrane of renal tubular cells act in concert with OCT2, which is expressed on the basolateral side, to transport organic cationic drugs. OCT2 and MATE1 double-transfected MDCK cells have been a useful *in vitro* system for studying renal secretion of cationic drugs in that it mimics the vectorial transport in human kidney [262]. The results of many of the DDI studies using double-transfected MDCK cells agreed well with findings from knockout mouse models. Metformin has been characterized as a substrate for MATEs in addition to OCT1 and OCT2, and cimetidine inhibits MATEs as well as OCTs. To understand the pharmacokinetic role of MATE1 *in vivo*, a murine Mate1 knockout mouse model has been developed. Following i.v. administration of metformin to *Mate* (–/–) mice, the renal excretion of metformin decreases significantly, which further suggested that in addition to Oct2, Mate1 is also important in the renal elimination of metformin [263].

It was also demonstrated that MATE1 has a protective role in preventing nephrotoxicity associated with many of platinum-based drugs. Cisplatin is a substrate for OCT2 but not MATEs and, therefore, accumulates in proximal tubule cells leading to nephrotoxicity [264–266], while another platinum-based drug, oxaliplatin, which is a substrate for both OCT2 and MATEs, is not associated with nephrotoxicity. This suggests that transport of oxaliplatin by MATE1 and MATE2-K allows for its efficient secretion into the urine, thus reducing the potential for renal toxicity [266]. Furthermore carboplatin and nedaplatin, which are neither MATE nor OCT2 substrates, do not manifest any nephrotoxicity [266]. Together these results indicate that the substrate specificity of platinum-based drugs determines their nephrotoxic features, and that MATE-mediated efflux of oxaliplatin has a protective role in preventing nephrotoxicity.

In addition, the presence of MATE1 on the canalicular membrane of hepatocytes suggests its potential role in biliary excretion of cationic drugs. Both the renal clearance and biliary excretion of metformin decreased with pyrimethamine (a potent MATE inhibitor) [267], suggesting a role of MATE in biliary excretion. In the *Mate1* (–/–) knockout mouse model, elevated hepatic metformin concentrations are present and associated with the extent of lactic acidosis [268]. However, because metformin is almost completely excreted unchanged in urine, it is very likely that biliary excretion of metformin by MATE1 is minimal. Nevertheless, the presence of MATE1 on the canalicular membrane of hepatocytes suggests that it may play a role in biliary excretion of some endogenous compounds or drugs.

5.4.5 Monocarboxylate Transporters (SLC16 and SLC5)

Two families of MCTs; namely, the proton-dependent transporters (MCTs, SLC16A) and the sodium-dependent transporters (SMCTs, SLC5A) have been identified. The proton-dependent MCT family is comprised of fourteen isoforms, with seven of the members functionally characterized to date [269, 270]. Only the first four members of MCTs (MCT1-MCT4) exhibit proton-linked symport of various monocarboxylates (e.g., lactate, pyruvate, and ketone bodies) and are important for cellular metabolism, as well as transport of therapeutic agents (e.g., salicylate, valproic acid, and atorvastatin) [18, 270]. MCT proteins contain 12 TMDs with C- and N-termini within the cytoplasm and an intracellular loop between TMDs 6 and 7 [271]. The ubiquitous tissue localization of MCTs that includes the liver, kidney, intestine, and brain suggests their potential role in drug pharmacokinetics and pharmacodynamics. MCT1 is by far most studied in terms of its role in drug disposition. In particular, the impact of MCT1 on drug disposition has been characterized for GHB, a drug of abuse that is also marketed as Xyrem® in the United States for the treatment of narcolepsy [272] and in Europe for the treatment of alcohol withdrawal [273]. MCT6, MCT8, and MCT10 have been identified to transport diuretics [274], thyroid hormones [275], and aromatic amino acids [276], respectively. Interestingly, unlike the other MCTs, MCT6 does not transport short-chain monocarboxylates; substrates that have been identified are mainly pharmaceutical agents such as bumetanide [274].

MCTs are important for the renal transport of endogenous and exogenous substrates. MCT1 is predominantly expressed on the BLM of rat kidney membrane vesicles [23, 277] and human kidney HK-2 cells [278]. Although GHB was found to be a substrate for MCTs 1, 2, and 4, studies performed with silencing RNA for MCT1, 2, and 4 in HK-2 cells showed that GHB is primarily transported by MCT1 [278]. This suggests that MCT1 most likely plays a role in the renal handling of GHB. Indeed *in vitro* and *in vivo* studies using MCT inhibitors such as L-lactate [23], α-cyano-4-hydroxycinnamate (CHC) [23, 279], and flavonoids (e.g., quercetin and luteolin) [280] supported previous findings that MCTs are involved in the active renal reabsorption of GHB in the kidney. Coadministration of L-lactate with GHB resulted in an increase in GHB renal elimination from 63 to 118 ml/hour/kg in rats [281]. In addition, the renal clearance of GHB significantly increased by more than threefold with luteolin administration (10 mg/kg) in rats [280]. Collectively, these data suggest that GHB is a substrate for MCTs, and the administration of MCT inhibitors can increase its renal elimination resulting in increased total clearance and decreased plasma concentrations.

In the intestine, MCT1 has been shown to be the predominant isoform among the proton-dependent MCTs, and it is expressed on the apical membrane of intestinal cells [282]. Previously, our laboratory has carried out transport studies in Caco-2 cells, known to express MCT1, 2, and 4, and demonstrated that GHB and D-lactate uptake occurred in a pH- and concentration-dependent manner, suggesting that GHB undergoes saturable intestinal absorption mediated, at least in part, by MCTs [277]. In addition to GHB, cefdinir [283], carindacillin [284], salicylic acid [285], pravastatin [286], and atorvastatin [287] were also shown to be transported by MCTs in various intestinal cell lines and intestinal membrane vesicles. In particular, MCT1 has been reported to play a role in the intestinal uptake of carindacillin (prodrug of β-lactam antibiotic) in rat intestinal brush-border membrane vesicles, and it is speculated that the enhanced exposure of carindacillin may be attributed to MCT-mediated uptake [284]. Furthermore, due to the high transport capacity of MCT1 and its high expression throughout the intestine, MCT1 is an important target for oral drug delivery. The oral delivery of gabapentin has been enhanced by designing gabapentin as a prodrug, XP13512 (Horizant®, XenoPort) that targets MCT1 [288]. MCT1-mediated uptake of XP13512 was confirmed in Caco-2, HEK, and MDCK cell lines [288]. The oral bioavailability of gabapentin increased from 25 to 84% with XP13512 administration in monkeys [288], and in a human clinical trial, XP13512 showed dose-proportional increases in gabapentin plasma concentrations [289]. In 2012, the FDA approved XP13512 under the trade name Horizant for postherpetic neuralgia [290].

MCTs are also widely expressed in rat, mouse, and human brain (e.g., MCTs 1, 2, and 4), with MCT1 present at the BBB [291, 292]. The physiological role of MCT1 at the BBB is to mediate transport of metabolic monocarboxylates (e.g., lactate and ketone bodies) across the BBB. The importance of MCT1-mediated BBB transport has been examined in several *in vitro* and *in vivo* studies. Tsuji et al. were probably the first to examine MCT-mediated uptake of acidic drugs such as valproic acid, benzoic acid, and various β-lactam antibiotics (e.g., benzylpenicillin, propicillin, and cefazolin) across the BBB in rat brain [293]. Later Tsuji et al. reported the uptake of various statins including simvastatin and atorvastatin into bovine brain capillary endothelial cells in a pH-dependent and carrier-mediated manner [294]. Furthermore, *in vivo* brain perfusion studies demonstrated that statin (e.g., simvastatin and lovastatin) penetration into the brain correlated well with sleep disturbance—a side effect of statins [295]. An *in situ* rat brain perfusion study showed saturable transport of GHB into the brain [296], and a more recent study showed that the transport of GHB in rat (RBE4) and human (hCMEC/D3) brain endothelial cells was significantly inhibited by CHC, an MCT inhibitor, with GHB uptake in these cell lines exhibiting Michaelis–Menten kinetics [297]. High doses of L-lactate can decrease the MCT-mediated BBB uptake of GHB, indicating the potential for inhibition of brain uptake [297]. Taken together, a greater understanding of the role of MCTs in drug transport across the BBB may provide additional insight in optimizing drug delivery into the brain.

The SLC5 transporter gene family consists of 12 members, with the cotransport of sodium representing the driving force for membrane transport. Two members, SLC5A8 and SLC5A12, transport physiological monocarboxylates including lactate, pyruvate, and butyrate and are known as SMCT1 and SMCT2, respectively. SMCT1 and SMCT2 are localized predominantly in the kidney and GI tract, although SMCT1 is also present on neuronal membranes in the brain. Due to their tissue localization, these transporters play an important role in the reabsorption of L-lactate in the kidney and nicotinamide in the intestine. Their role in drug transport is less well understood, although GHB is a substrate for SMCT1. Due to their low capacity for transport, compared with MCT1, SMCTs have not been utilized as a drug delivery target [298–300].

5.4.6 Peptide Transporters (SLC15A)

The proton-coupled oligopeptide transporter (POT; SLC15) superfamily consists of four members, peptide transporters 1 and 2 (PEPT1 and PEPT2) and peptide/histidine transporters 1 and 2 (PHT1 and PHT2), which are responsible for absorption and secretion of a wide range of di- and tripeptides as well as peptidomimetic drugs via proton-mediated symport [301]. Members of the POT family share similar topology, 12 putative α-helical TMDs with N- and C-termini embedded intracellularly [302, 303]. PEPT1 is a low-affinity and high-capacity transporter, and it is primarily expressed on the apical membrane of enterocytes of small intestine, particularly in the duodenum. PEPT1 is also highly expressed in renal proximal tubules, hepatic bile duct, and pancreas, with lower expression in other tissues [304]. Compared with PEPT1, PEPT2 is a high-affinity and low-capacity transporter, and it is more ubiquitously expressed throughout the body; high expression of PEPT2 was detected on the apical membrane of proximal tubule cells [305]. In contrast to the PEPT transporters, PHTs have not been as extensively studied and little is known about their physiological role, although PHT1 has been implicated in the efflux of peptide and neuroactive peptides out of the brain [306]. PHT1 is primarily expressed throughout the brain, whereas PHT2 is widely expressed in various tissues [304].

Overall, the primary role of PEPT1, PEPT2, and PHT1 seems to be the absorption of peptide nutrients, whereas the role of PHT2 needs to be elucidated. Various studies have suggested that the role of PEPT2 could be expanded to include the elimination of neuropeptides from the brain [307, 308]. Substrates for PHT1 include histidine, with relatively high affinity, and carnosine [309]. PEPTs transport endogenous di- and -tripeptides, as well as several drugs including zanamivir prodrug, oseltamivir analogue, bisphosphates, and angiotensin-converting enzyme (ACE) inhibitors.

PEPT1 has been targeted for the oral drug delivery for some of the antiviral and vasopressor drugs that exhibit poor bioavailability. Since PEPT1 is a high-capacity transporter present on the apical membrane of enterocytes in the duodenum, adding chemical moieties (e.g., valine or glycine) to make the drug molecules mimic peptides (peptide-based prodrug) has been exploited in targeting drug uptake in the small intestine via PEPT1-mediated transport. In addition, developing a good understanding of the structural requirements for PEPT1 is also a key aspect in screening and determining potential applicability of transporters for drug delivery. Currently, three marketed drugs have utilized PEPT1-mediated transport for their absorption across the intestinal barrier. These include valacyclovir (Valtrex®, GlaxoSmithKline), valganciclovir (Valcyte®, Roche), and midodrine (ProAmatine®, Shire plc). Both valacyclovir and valganciclovir are antiviral drugs with the amino acid valine as their promoiety of acyclovir and ganciclovir, respectively [310, 311]. The low oral bioavailability of acyclovir and ganciclovir is primarily due to the high hydrophilicity of the drugs. After transport across the intestinal membrane by PEPT1, valacyclovir and valganciclovir are readily hydrolyzed back to their parent drugs. Comparing oral bioavailability from clinical trials, oral bioavailability for valganciclovir improved from 12 to 20% (acyclovir) to 54% [25, 312] and even higher for valganciclovir from 6 (ganciclovir) to 61% [25]. The increased bioavailability of these prodrugs translates to higher plasma concentrations of the parent drug and increased antiviral activity [312, 313]. Midodrine is a prodrug with a glycine promoiety attached to desglymidodrine (DMAE), a selective α1-receptor agonist used for orthostatic hypotension, and utilizes PEPT1 to improve oral absorption [25, 314, 315].

In targeting an intestinal uptake transporter such as PEPT1, it is also important to consider the effect of metabolizing enzymes, such as CYP3A4, or other transporters, such as the efflux transporters P-gp and MRP2, on the overall bioavailability [316–318]. Many HIV protease inhibitors given orally such as saquinavir and lopinavir are substrates for P-gp, MRP1, and MRP2. To improve bioavailability of saquinavir and lopinavir through exploiting PEPT1-mediated transport, the peptide prodrug derivative with valine–valine (Val–Val) or glycine–valine (Gly–Val) promoiety was developed [316, 319, 320]. It was demonstrated that introducing Val–Val or Gly–Val promoieties to these protease inhibitors not only targeted PEPT1 uptake but also reduced the affinity and capacity of the prodrugs for efflux transporters, compared with the parent compound [316, 319, 320]. Furthermore, to exploit an intestinal transporter for drug delivery requires an understanding of its substrate specificity, inhibitors, and changes in its expression with disease, age, and diet, since these factors will impact the overall prodrug oral bioavailability.

5.4.7 Other SLC Transporters

SLC transporters currently are divided into 53 families, many of which transport endogenous substrates. The nucleoside transporters consist of two families: the concentrative nucleoside transporters (CNT, SLC28) and the equilibrative nucleoside transporters (ENT, SLC29). These transporters are important for the disposition of purines and pyrimidines, as well as nucleoside analogues including zidovudine, Ara-C, and gemcitabine. As well, some of the bile acid transporters may be involved in drug transport, including the sodium taurocholate cotransporting polypeptide (NTCP, SLC10A1) present on the sinusoidal membrane of hepatocytes.

5.5 CONCLUSIONS

The ongoing molecular and functional characterization of transporters has increased our understanding concerning how these transporters control the passage of a diverse range of substrates through biological membranes. The characterization of their tissue localization and their function has suggested that transport proteins may significantly impact the absorption, elimination, and distribution of xenobiotic compounds. In addition, identifying transporter protein features, such as their relative transporter expression level, substrate specificity, and transporter capacity, has resulted in their application in drug delivery. The generation of knockout mice lacking a specific transporter or transporters and the identification of specific inhibitors and siRNA studies have greatly enhanced our ability to understand the physiological and pharmacological functions of these transporters. It has been clearly demonstrated by the studies presented here, as well as others, that these transporters play an essential role in intestinal absorption, biliary excretion, and renal secretion and contribute to the barrier functions between the blood and various tissues such as brain, testis, and placenta. The potential importance of transporters for oral drug delivery in the small intestine and for transport into specific tissue sites, such as the brain, remains an area for further exploration, as our molecular and functional understanding of transporters increases.

ACKNOWLEDGMENT

We acknowledge research support through NIH grant DA023223.

REFERENCES

1. Gottesman, M. M.; Pastan, I. *J Biol Chem* 1988, **263** (25), 12163–6.
2. Gottesman, M. M.; Pastan, I. *Annu Rev Biochem* 1993, **62**, 385–427.
3. Litman, T.; Druley, T. E.; Stein, W. D.; Bates, S. E. *Cell Mol Life Sci* 2001, **58** (7), 931–59.
4. Hipfner, D. R.; Deeley, R. G.; Cole, S. P. *Biochim Biophys Acta* 1999, **1461** (2), 359–76.
5. You, G.; Morris, M. E. Overview of Drug Transporter Families. In *Drug Transporters*. John Wiley & Sons, Inc.: Hoboken, NJ, 2006; pp 1–10.
6. Morris, M. E.; Morse, B. L. Membrane Drug Transporters. In *Foye's Principles of Medicinal Chemistry*, 7th edition, Lemke, T. L.; Williams, D. A.; Roche, V. F., and William Zito, S., Eds. Lippincott Williams & Wilkins: Baltimor, MD/Philadelphia, PA, 2011.
7. Amidon, G. L.; Lennernas, H.; Shah, V. P.; Crison, J. R. *Pharm Res* 1995, **12** (3), 413–20.
8. Wu, C. Y.; Benet, L. Z. *Pharm Res* 2005, **22** (1), 11–23.
9. Ayrton, A.; Morgan, P. *Xenobiotica* 2001, **31** (8–9), 469–97.
10. Keppler, D.; Arias, I. M. *FASEB J* 1997, **11** (1), 15–18.
11. Hooiveld, G. J.; van Montfoort, J. E.; Meijer, D. K.; Muller, M. *Eur J Pharmaceut Sci* 2000, **12** (1), 13–30.
12. Kim, R. B. *Toxicology* 2002, **181–182**, 291–7.
13. van Asperen, J.; van Tellingen, O.; Beijnen, J. H. *Drug Metab Dispos* 2000, **28** (3), 264–7.
14. Yamazaki, M.; Akiyama, S.; Nishigaki, R.; Sugiyama, Y. *Pharm Res* 1996, **13** (10), 1559–64.
15. Nishizato, Y.; Ieiri, I.; Suzuki, H.; Kimura, M.; Kawabata, K.; Hirota, T.; Takane, H.; Irie, S.; Kusuhara, H.; Urasaki, Y.; Urae, A.; Higuchi, S.; Otsubo, K.; Sugiyama, Y. *Clin Pharmacol Ther* 2003, **73** (6), 554–65.
16. Romaine, S. P.; Bailey, K. M.; Hall, A. S.; Balmforth, A. J. *Pharmacogenomics J* 2010, **10** (1), 1–11.
17. Shitara, Y.; Hirano, M.; Sato, H.; Sugiyama, Y. *J Pharmacol Exp Ther* 2004, **311** (1), 228–36.
18. Halestrap, A. P.; Meredith, D. *Pflugers Arch* 2004, **447** (5), 619–28.
19. Koepsell, H. *Mol Aspects Med* 2013, **34** (2–3), 413–35.
20. Terada, T.; Inui, K. *Biochem Pharmacol* 2008, **75** (9), 1689–96.
21. Masereeuw, R.; Russel, F. G. *Pharmacol Ther* 2010, **126** (2), 200–16.
22. Sharom, F. J. *Essays Biochem* 2011, **50** (1), 161–78.
23. Wang, Q.; Darling, I. M.; Morris, M. E. *J Pharmacol Exp Ther* 2006, **318** (2), 751–61.
24. Pinho, M. J.; Serrao, M. P.; Gomes, P.; Hopfer, U.; Jose, P. A.; Soares-da-Silva, P. *Kidney Int* 2004, **66** (1), 216–26.
25. Rautio, J.; Kumpulainen, H.; Heimbach, T.; Oliyai, R.; Oh, D.; Jarvinen, T.; Savolainen, J. *Nat Rev Drug Discov* 2008, **7** (3), 255–70.
26. Begley, D. J. *Curr Pharm Des* 2004, **10** (12), 1295–312.

27. Juliano, R. L.; Ling, V. *Biochim Biophys Acta* 1976, **455** (1), 152–62.
28. Germann, U. A. *Eur J Cancer* 1996, **32A** (6), 927–44.
29. Ambudkar, S. V.; Lelong, I. H.; Zhang, J.; Cardarelli, C. O.; Gottesman, M. M.; Pastan, I. *Proc Natl Acad Sci U S A* 1992, **89** (18), 8472–6.
30. Higgins, C. F.; Gottesman, M. M. *Trends Biochem Sci* 1992, **17** (1), 18–21.
31. Hillgren, K. M.; Keppler, D.; Zur, A. A.; Giacomini, K. M.; Stieger, B.; Cass, C. E.; Zhang, L.; International Transporter Consortium. *Clin Pharmacol Ther* 2013, **94** (1), 52–63.
32. Kock, K.; Brouwer, K. L. *Clin Pharmacol Ther* 2012, **92** (5), 599–612.
33. Kusuhara, H.; Suzuki, H.; Sugiyama, Y. *J Pharm Sci* 1998, **87** (9), 1025–40.
34. de Graaf, D.; Sharma, R. C.; Mechetner, E. B.; Schimke, R. T.; Roninson, I. B. *Proc Natl Acad Sci U S A* 1996, **93** (3), 1238–42.
35. Norris, M. D.; De Graaf, D.; Haber, M.; Kavallaris, M.; Madafiglio, J.; Gilbert, J.; Kwan, E.; Stewart, B. W.; Mechetner, E. B.; Gudkov, A. V.; Roninson, I. B. *Int J Cancer* 1996, **65** (5), 613–19.
36. Potschka, H.; Loscher, W. *Epilepsia* 2001, **42** (10), 1231–40.
37. Martin, C.; Berridge, G.; Higgins, C. F.; Mistry, P.; Charlton, P.; Callaghan, R. *Mol Pharmacol* 2000, **58** (3), 624–32.
38. Dey, S.; Ramachandra, M.; Pastan, I.; Gottesman, M. M.; Ambudkar, S. V. *Proc Natl Acad Sci U S A* 1997, **94** (20), 10594–9.
39. Shapiro, A. B.; Ling, V. *Eur J Biochem* 1997, **250** (1), 130–7.
40. Shapiro, A. B.; Fox, K.; Lam, P.; Ling, V. *Eur J Biochem* 1999, **259** (3), 841–50.
41. Thiebaut, F.; Tsuruo, T.; Hamada, H.; Gottesman, M. M.; Pastan, I.; Willingham, M. C. *Proc Natl Acad Sci U S A* 1987, **84** (21), 7735–8.
42. Cordon-Cardo, C.; O'Brien, J. P.; Casals, D.; Rittman-Grauer, L.; Biedler, J. L.; Melamed, M. R.; Bertino, J. R. *Proc Natl Acad Sci U S A* 1989, **86** (2), 695–8.
43. Thiebaut, F.; Tsuruo, T.; Hamada, H.; Gottesman, M. M.; Pastan, I.; Willingham, M. C. *J Histochem Cytochem* 1989, **37** (2), 159–64.
44. Friche, E.; Jensen, P. B.; Sehested, M.; Demant, E. J.; Nissen, N. N. *Cancer Commun* 1990, **2** (9), 297–303.
45. Hugger, E. D.; Audus, K. L.; Borchardt, R. T. *J Pharm Sci* 2002, **91** (9), 1980–90.
46. Zhang, S.; Morris, M. E. *J Pharmacol Exp Ther* 2003, **304** (3), 1258–67.
47. Ferte, J.; Kuhnel, J. M.; Chapuis, G.; Rolland, Y.; Lewin, G.; Schwaller, M. A. *J Med Chem* 1999, **42** (3), 478–89.
48. Conseil, G.; Baubichon-Cortay, H.; Dayan, G.; Jault, J. M.; Barron, D.; Di Pietro, A. *Proc Natl Acad Sci U S A* 1998, **95** (17), 9831–6.
49. de Wet, H.; McIntosh, D. B.; Conseil, G.; Baubichon-Cortay, H.; Krell, T.; Jault, J. M.; Daskiewicz, J. B.; Barron, D.; Di Pietro, A. *Biochemistry* 2001, **40** (34), 10382–91.
50. Romiti, N.; Tongiani, R.; Cervelli, F.; Chieli, E. *Life Sci* 1998, **62** (25), 2349–58.
51. Bhardwaj, R. K.; Glaeser, H.; Becquemont, L.; Klotz, U.; Gupta, S. K.; Fromm, M. F. *J Pharmacol Exp Ther* 2002, **302** (2), 645–50.
52. Twentyman, P. R. *Biochem Pharmacol* 1992, **43** (1), 109–17.
53. Hyafil, F.; Vergely, C.; Du Vignaud, P.; Grand-Perret, T. *Cancer Res* 1993, **53** (19), 4595–602.

54. Dantzig, A. H.; Shepard, R. L.; Cao, J.; Law, K. L.; Ehlhardt, W. J.; Baughman, T. M.; Bumol, T. F.; Starling, J. J. *Cancer Res* 1996, **56** (18), 4171–9.

55. Roe, M.; Folkes, A.; Ashworth, P.; Brumwell, J.; Chima, L.; Hunjan, S.; Pretswell, I.; Dangerfield, W.; Ryder, H.; Charlton, P. *Bioorg Med Chem Lett* 1999, **9** (4), 595–600.

56. Advani, R.; Saba, H. I.; Tallman, M. S.; Rowe, J. M.; Wiernik, P. H.; Ramek, J.; Dugan, K.; Lum, B.; Villena, J.; Davis, E.; Paietta, E.; Litchman, M.; Sikic, B. I.; Greenberg, P. L. *Blood* 1999, **93** (3), 787–95.

57. Advani, R.; Fisher, G. A.; Lum, B. L.; Hausdorff, J.; Halsey, J.; Litchman, M.; Sikic, B. I. *Clin Cancer Res* 2001, **7** (5), 1221–9.

58. Chico, I.; Kang, M. H.; Bergan, R.; Abraham, J.; Bakke, S.; Meadows, B.; Rutt, A.; Robey, R.; Choyke, P.; Merino, M.; Goldspiel, B.; Smith, T.; Steinberg, S.; Figg, W. D.; Fojo, T.; Bates, S. *J Clin Oncol* 2001, **19** (3), 832–42.

59. Thomas, H.; Coley, H. M. *Cancer Control* 2003, **10** (2), 159–65.

60. Sparreboom, A.; van Asperen, J.; Mayer, U.; Schinkel, A. H.; Smit, J. W.; Meijer, D. K.; Borst, P.; Nooijen, W. J.; Beijnen, J. H.; van Tellingen, O. *Proc Natl Acad Sci U S A* 1997, **94** (5), 2031–5.

61. Mayer, U.; Wagenaar, E.; Beijnen, J. H.; Smit, J. W.; Meijer, D. K.; van Asperen, J.; Borst, P.; Schinkel, A. H. *Br J Pharmacol* 1996, **119** (5), 1038–44.

62. Yamaguchi, H.; Yano, I.; Saito, H.; Inui, K. *J Pharmacol Exp Ther* 2002, **300** (3), 1063–9.

63. Kim, R. B.; Fromm, M. F.; Wandel, C.; Leake, B.; Wood, A. J.; Roden, D. M.; Wilkinson, G. R. *J Clin Invest* 1998, **101** (2), 289–94.

64. Gramatte, T.; Oertel, R. *Clin Pharmacol Ther* 1999, **66** (3), 239–45.

65. Drescher, S.; Glaeser, H.; Murdter, T.; Hitzl, M.; Eichelbaum, M.; Fromm, M. F. *Clin Pharmacol Ther* 2003, **73** (3), 223–31.

66. Westphal, K.; Weinbrenner, A.; Zschiesche, M.; Franke, G.; Knoke, M.; Oertel, R.; Fritz, P.; von Richter, O.; Warzok, R.; Hachenberg, T.; Kauffmann, H. M.; Schrenk, D.; Terhaag, B.; Kroemer, H. K.; Siegmund, W. *Clin Pharmacol Ther* 2000, **68** (4), 345–55.

67. Speeg, K. V.; Maldonado, A. L. *Cancer Chemother Pharmacol* 1994, **34** (2), 133–6.

68. Speeg, K. V.; Maldonado, A. L.; Liaci, J.; Muirhead, D. *Hepatology* 1992, **15** (5), 899–903.

69. Milne, R. W.; Larsen, L. A.; Jorgensen, K. L.; Bastlund, J.; Stretch, G. R.; Evans, A. M. *Pharm Res* 2000, **17** (12), 1511–15.

70. Schinkel, A. H.; Smit, J. J.; van Tellingen, O.; Beijnen, J. H.; Wagenaar, E.; van Deemter, L.; Mol, C. A.; van der Valk, M. A.; Robanus-Maandag, E. C.; te Riele, H. P.; Berns, A. J.; Borst, P. *Cell* 1994, **77** (4), 491–502.

71. Jalava, K. M.; Partanen, J.; Neuvonen, P. J. *Ther Drug Monit* 1997, **19** (6), 609–13.

72. Kaukonen, K. M.; Olkkola, K. T.; Neuvonen, P. J. *Clin Pharmacol Ther* 1997, **62** (5), 510–17.

73. Borst, P.; Evers, R.; Kool, M.; Wijnholds, J. *Biochim Biophys Acta* 1999, **1461** (2), 347–57.

74. Cole, S. P.; Bhardwaj, G.; Gerlach, J. H.; Mackie, J. E.; Grant, C. E.; Almquist, K. C.; Stewart, A. J.; Kurz, E. U.; Duncan, A. M.; Deeley, R. G. *Science* 1992, **258** (5088), 1650–4.

75. Marsh, W.; Sicheri, D.; Center, M. S. *Cancer Res* 1986, **46** (8), 4053–7.
76. Marsh, W.; Center, M. S. *Cancer Res* 1987, **47** (19), 5080–6.
77. McGrath, T.; Center, M. S. *Biochem Biophys Res Commun* 1987, **145** (3), 1171–6.
78. McGrath, T.; Latoud, C.; Arnold, S. T.; Safa, A. R.; Felsted, R. L.; Center, M. S. *Biochem Pharmacol* 1989, **38** (20), 3611–19.
79. Cole, S. P.; Sparks, K. E.; Fraser, K.; Loe, D. W.; Grant, C. E.; Wilson, G. M.; Deeley, R. G. *Cancer Res* 1994, **54** (22), 5902–10.
80. Grant, C. E.; Valdimarsson, G.; Hipfner, D. R.; Almquist, K. C.; Cole, S. P.; Deeley, R. G. *Cancer Res* 1994, **54** (2), 357–61.
81. Zaman, G. J.; Flens, M. J.; van Leusden, M. R.; de Haas, M.; Mulder, H. S.; Lankelma, J.; Pinedo, H. M.; Scheper, R. J.; Baas, F.; Broxterman, H. J. *Proc Natl Acad Sci U S A* 1994, **91** (19), 8822–6.
82. Kruh, G. D.; Chan, A.; Myers, K.; Gaughan, K.; Miki, T.; Aaronson, S. A. *Cancer Res* 1994, **54** (7), 1649–52.
83. Ito, K.; Suzuki, H.; Hirohashi, T.; Kume, K.; Shimizu, T.; Sugiyama, Y. *Am J Physiol* 1997, **272** (1 Pt 1), G16–22.
84. Buchler, M.; Konig, J.; Brom, M.; Kartenbeck, J.; Spring, H.; Horie, T.; Keppler, D. *J Biol Chem* 1996, **271** (25), 15091–8.
85. Paulusma, C. C.; Bosma, P. J.; Zaman, G. J.; Bakker, C. T.; Otter, M.; Scheffer, G. L.; Scheper, R. J.; Borst, P.; Oude Elferink, R. P. *Science* 1996, **271** (5252), 1126–8.
86. Taniguchi, K.; Wada, M.; Kohno, K.; Nakamura, T.; Kawabe, T.; Kawakami, M.; Kagotani, K.; Okumura, K.; Akiyama, S.; Kuwano, M. *Cancer Res* 1996, **56** (18), 4124–9.
87. Hopper, E.; Belinsky, M. G.; Zeng, H.; Tosolini, A.; Testa, J. R.; Kruh, G. D. *Cancer Lett* 2001, **162** (2), 181–91.
88. Lee, K.; Belinsky, M. G.; Bell, D. W.; Testa, J. R.; Kruh, G. D. *Cancer Res* 1998, **58** (13), 2741–7.
89. Lee, K.; Klein-Szanto, A. J.; Kruh, G. D. *J Natl Cancer Inst* 2000, **92** (23), 1934–40.
90. Kool, M.; de Haas, M.; Scheffer, G. L.; Scheper, R. J.; van Eijk, M. J.; Juijn, J. A.; Baas, F.; Borst, P. *Cancer Res* 1997, **57** (16), 3537–47.
91. Kool, M.; van der Linden, M.; de Haas, M.; Baas, F.; Borst, P. *Cancer Res* 1999, **59** (1), 175–82.
92. Chen, Z. S.; Hopper-Borge, E.; Belinsky, M. G.; Shchaveleva, I.; Kotova, E.; Kruh, G. D. *Mol Pharmacol* 2003, **63** (2), 351–8.
93. Leslie, E. M.; Deeley, R. G.; Cole, S. P. *Toxicology* 2001, **167** (1), 3–23.
94. Konig, J.; Nies, A. T.; Cui, Y.; Leier, I.; Keppler, D. *Biochim Biophys Acta* 1999, **1461** (2), 377–94.
95. Zeng, H.; Chen, Z. S.; Belinsky, M. G.; Rea, P. A.; Kruh, G. D. *Cancer Res* 2001, **61** (19), 7225–32.
96. Keppler, D.; Cui, Y.; Konig, J.; Leier, I.; Nies, A. *Adv Enzyme Regul* 1999, **39**, 237–46.
97. Loe, D. W.; Almquist, K. C.; Deeley, R. G.; Cole, S. P. *J Biol Chem* 1996, **271** (16), 9675–82.
98. Loe, D. W.; Deeley, R. G.; Cole, S. P. *Cancer Res* 1998, **58** (22), 5130–6.
99. Renes, J.; de Vries, E. G.; Nienhuis, E. F.; Jansen, P. L.; Muller, M. *Br J Pharmacol* 1999, **126** (3), 681–8.

100. Evers, R.; de Haas, M.; Sparidans, R.; Beijnen, J.; Wielinga, P. R.; Lankelma, J.; Borst, P. *Br J Cancer* 2000, **83** (3), 375–83.

101. Borst, P.; Evers, R.; Kool, M.; Wijnholds, J. *J Natl Cancer Inst* 2000, **92** (16), 1295–302.

102. Evers, R.; Zaman, G. J.; van Deemter, L.; Jansen, H.; Calafat, J.; Oomen, L. C.; Oude Elferink, R. P.; Borst, P.; Schinkel, A. H. *J Clin Invest* 1996, **97** (5), 1211–18.

103. Cole, S. P. *Annu Rev Pharmacol Toxicol* 2014, **54**, 95–117.

104. Keppler, D. *Handb Exp Pharmacol* 2011, (201), 299–323.

105. Loe, D. W.; Deeley, R. G.; Cole, S. P. *J Pharmacol Exp Ther* 2000, **293** (2), 530–8.

106. Leslie, E. M.; Deeley, R. G.; Cole, S. P. *Drug Metab Dispos* 2003, **31** (1), 11–15.

107. Schneider, E.; Yamazaki, H.; Sinha, B. K.; Cowan, K. H. *Br J Cancer* 1995, **71** (4), 738–43.

108. Versantvoort, C. H.; Broxterman, H. J.; Bagrij, T.; Scheper, R. J.; Twentyman, P. R. *Br J Cancer* 1995, **72** (1), 82–9.

109. Wijnholds, J.; Evers, R.; van Leusden, M. R.; Mol, C. A.; Zaman, G. J.; Mayer, U.; Beijnen, J. H.; van der Valk, M.; Krimpenfort, P.; Borst, P. *Nat Med* 1997, **3** (11), 1275–9.

110. Lorico, A.; Rappa, G.; Finch, R. A.; Yang, D.; Flavell, R. A.; Sartorelli, A. C. *Cancer Res* 1997, **57** (23), 5238–42.

111. Wijnholds, J.; deLange, E. C.; Scheffer, G. L.; van den Berg, D. J.; Mol, C. A.; van der Valk, M.; Schinkel, A. H.; Scheper, R. J.; Breimer, D. D.; Borst, P. *J Clin Invest* 2000, **105** (3), 279–85.

112. Johnson, D. R.; Finch, R. A.; Lin, Z. P.; Zeiss, C. J.; Sartorelli, A. C. *Cancer Res* 2001, **61** (4), 1469–76.

113. Cui, Y.; Konig, J.; Buchholz, J. K.; Spring, H.; Leier, I.; Keppler, D. *Mol Pharmacol* 1999, **55** (5), 929–37.

114. Zeng, H.; Bain, L. J.; Belinsky, M. G.; Kruh, G. D. *Cancer Res* 1999, **59** (23), 5964–7.

115. Kool, M.; van der Linden, M.; de Haas, M.; Scheffer, G. L.; de Vree, J. M.; Smith, A. J.; Jansen, G.; Peters, G. J.; Ponne, N.; Scheper, R. J.; Elferink, R. P.; Baas, F.; Borst, P. *Proc Natl Acad Sci U S A* 1999, **96** (12), 6914–19.

116. Kawahara, M.; Sakata, A.; Miyashita, T.; Tamai, I.; Tsuji, A. *J Pharm Sci* 1999, **88** (12), 1281–7.

117. Tada, Y.; Wada, M.; Migita, T.; Nagayama, J.; Hinoshita, E.; Mochida, Y.; Maehara, Y.; Tsuneyoshi, M.; Kuwano, M.; Naito, S. *Int J Cancer* 2002, **98** (4), 630–5.

118. Laupeze, B.; Amiot, L.; Drenou, B.; Bernard, M.; Branger, B.; Grosset, J. M.; Lamy, T.; Fauchet, R.; Fardel, O. *Br J Haematol* 2002, **116** (4), 834–8.

119. Ito, K.; Fujimori, M.; Nakata, S.; Hama, Y.; Shingu, K.; Kobayashi, S.; Tsuchiya, S.; Kohno, K.; Kuwano, M.; Amano, J. *Oncol Res* 1998, **10** (2), 99–109.

120. Filipits, M.; Suchomel, R. W.; Dekan, G.; Haider, K.; Valdimarsson, G.; Depisch, D.; Pirker, R. *Clin Cancer Res* 1996, **2** (7), 1231–7.

121. Nooter, K.; Brutel de la Riviere, G.; Look, M. P.; van Wingerden, K. E.; Henzen-Logmans, S. C.; Scheper, R. J.; Flens, M. J.; Klijn, J. G.; Stoter, G.; Foekens, J. A. *Br J Cancer* 1997, **76** (4), 486–93.

122. Ota, E.; Abe, Y.; Oshika, Y.; Ozeki, Y.; Iwasaki, M.; Inoue, H.; Yamazaki, H.; Ueyama, Y.; Takagi, K.; Ogata, T. *Br J Cancer* 1995, **72** (3), 550–4.

123. Oshika, Y.; Nakamura, M.; Tokunaga, T.; Fukushima, Y.; Abe, Y.; Ozeki, Y.; Yamazaki, H.; Tamaoki, N.; Ueyama, Y. *Mod Pathol* 1998, **11** (11), 1059–63.
124. Sugawara, I.; Yamada, H.; Nakamura, H.; Sumizawa, T.; Akiyama, S.; Masunaga, A.; Itoyama, S. *Int J Cancer* 1995, **64** (5), 322–5.
125. Young, L. C.; Campling, B. G.; Voskoglou-Nomikos, T.; Cole, S. P.; Deeley, R. G.; Gerlach, J. H. *Clin Cancer Res* 1999, **5** (3), 673–80.
126. Haber, M.; Smith, J.; Bordow, S. B.; Flemming, C.; Cohn, S. L.; London, W. B.; Marshall, G. M.; Norris, M. D. *Journal Clin Oncol* 2006, **24** (10), 1546–53.
127. Pajic, M.; Murray, J.; Marshall, G. M.; Cole, S. P.; Norris, M. D.; Haber, M. *Pharmacogenet Genomics* 2011, **21** (5), 270–9.
128. Nies, A. T.; Keppler, D. *Pflugers Arch* 2007, **453** (5), 643–59.
129. Zamek-Gliszczynski, M. J.; Hoffmaster, K. A.; Nezasa, K.; Tallman, M. N.; Brouwer, K. L. *Eur J Pharm Sci* 2006, **27** (5), 447–86.
130. Kartenbeck, J.; Leuschner, U.; Mayer, R.; Keppler, D. *Hepatology* 1996, **23** (5), 1061–6.
131. Paulusma, C. C.; Kool, M.; Bosma, P. J.; Scheffer, G. L.; ter Borg, F.; Scheper, R. J.; Tytgat, G. N.; Borst, P.; Baas, F.; Oude Elferink, R. P. *Hepatology* 1997, **25** (6), 1539–42.
132. Kajihara, S.; Hisatomi, A.; Mizuta, T.; Hara, T.; Ozaki, I.; Wada, I.; Yamamoto, K. *Biochem Biophys Res Commun* 1998, **253** (2), 454–7.
133. Toh, S.; Wada, M.; Uchiumi, T.; Inokuchi, A.; Makino, Y.; Horie, Y.; Adachi, Y.; Sakisaka, S.; Kuwano, M. *Am J Hum Genet* 1999, **64** (3), 739–46.
134. Jansen, P. L.; Peters, W. H.; Lamers, W. H. *Hepatology* 1985, **5** (4), 573–9.
135. Kuipers, F.; Enserink, M.; Havinga, R.; van der Steen, A. B.; Hardonk, M. J.; Fevery, J.; Vonk, R. J. *J Clin Invest* 1988, **81** (5), 1593–9.
136. Letschert, K.; Faulstich, H.; Keller, D.; Keppler, D. *Toxicol Sci* 2006, **91** (1), 140–9.
137. Vlaming, M. L.; Mohrmann, K.; Wagenaar, E.; de Waart, D. R.; Elferink, R. P.; Lagas, J. S.; van Tellingen, O.; Vainchtein, L. D.; Rosing, H.; Beijnen, J. H.; Schellens, J. H.; Schinkel, A. H. *J Pharmacol Exp Ther* 2006, **318** (1), 319–27.
138. Yamazaki, M.; Akiyama, S.; Ni'inuma, K.; Nishigaki, R.; Sugiyama, Y. *Drug Metab Dispos* 1997, **25** (10), 1123–9.
139. Yamashiro, W.; Maeda, K.; Hirouchi, M.; Adachi, Y.; Hu, Z.; Sugiyama, Y. *Drug Metab Dispos* 2006, **34** (7), 1247–54.
140. Ishizuka, H.; Konno, K.; Naganuma, H.; Sasahara, K.; Kawahara, Y.; Niinuma, K.; Suzuki, H.; Sugiyama, Y. *J Pharmacol Exp Ther* 1997, **280** (3), 1304–11.
141. Horikawa, M.; Kato, Y.; Sugiyama, Y. *Pharm Res* 2002, **19** (9), 1345–53.
142. Konig, J.; Rost, D.; Cui, Y.; Keppler, D. *Hepatology* 1999, **29** (4), 1156–63.
143. Borst, P.; de Wolf, C.; van de Wetering, K. *Pflugers Arch* 2007, **453** (5), 661–73.
144. Lang, T.; Hitzl, M.; Burk, O.; Mornhinweg, E.; Keil, A.; Kerb, R.; Klein, K.; Zanger, U. M.; Eichelbaum, M.; Fromm, M. F. *Pharmacogenetics* 2004, **14** (3), 155–64.
145. Hirohashi, T.; Suzuki, H.; Sugiyama, Y. *J Biol Chem* 1999, **274** (21), 15181–5.
146. Smeets, P. H.; van Aubel, R. A.; Wouterse, A. C.; van den Heuvel, J. J.; Russel, F. G. *J Am Soc Nephrol* 2004, **15** (11), 2828–35.
147. Imaoka, T.; Kusuhara, H.; Adachi, M.; Schuetz, J. D.; Takeuchi, K.; Sugiyama, Y. *Mol Pharmacol* 2007, **71** (2), 619–27.

148. Ci, L.; Kusuhara, H.; Adachi, M.; Schuetz, J. D.; Takeuchi, K.; Sugiyama, Y. *Mol Pharmacol* 2007, **71** (6), 1591–7.
149. Hasegawa, M.; Kusuhara, H.; Adachi, M.; Schuetz, J. D.; Takeuchi, K.; Sugiyama, Y. *J Am Soc Nephrol* 2007, **18** (1), 37–45.
150. Cundy, K. C. *Clin Pharmacokinet* 1999, **36** (2), 127–43.
151. Doyle, L. A.; Yang, W.; Abruzzo, L. V.; Krogmann, T.; Gao, Y.; Rishi, A. K.; Ross, D. D. *Proc Natl Acad Sci U S A* 1998, **95** (26), 15665–70.
152. Allikmets, R.; Schriml, L. M.; Hutchinson, A.; Romano-Spica, V.; Dean, M. *Cancer Res* 1998, **58** (23), 5337–9.
153. Miyake, K.; Mickley, L.; Litman, T.; Zhan, Z.; Robey, R.; Cristensen, B.; Brangi, M.; Greenberger, L.; Dean, M.; Fojo, T.; Bates, S. E. *Cancer Res* 1999, **59** (1), 8–13.
154. Ross, D. D.; Karp, J. E.; Chen, T. T.; Doyle, L. A. *Blood* 2000, **96** (1), 365–8.
155. Kanzaki, A.; Toi, M.; Nakayama, K.; Bando, H.; Mutoh, M.; Uchida, T.; Fukumoto, M.; Takebayashi, Y. *Jpn J Cancer Res* 2001, **92** (4), 452–8.
156. Sargent, J. M.; Williamson, C. J.; Maliepaard, M.; Elgie, A. W.; Scheper, R. J.; Taylor, C. G. *Br J Haematol* 2001, **115** (2), 257–62.
157. Faneyte, I. F.; Kristel, P. M.; Maliepaard, M.; Scheffer, G. L.; Scheper, R. J.; Schellens, J. H.; van de Vijver, M. J. *Clin Cancer Res* 2002, **8** (4), 1068–74.
158. van der Kolk, D. M.; Vellenga, E.; Scheffer, G. L.; Muller, M.; Bates, S. E.; Scheper, R. J.; de Vries, E. G. *Blood* 2002, **99** (10), 3763–70.
159. Sauerbrey, A.; Sell, W.; Steinbach, D.; Voigt, A.; Zintl, F. *Br J Haematol* 2002, **118** (1), 147–50.
160. Steinbach, D.; Sell, W.; Voigt, A.; Hermann, J.; Zintl, F.; Sauerbrey, A. *Leukemia* 2002, **16** (8), 1443–7.
161. Ozvegy, C.; Litman, T.; Szakacs, G.; Nagy, Z.; Bates, S.; Varadi, A.; Sarkadi, B. *Biochem Biophys Res Commun* 2001, **285** (1), 111–17.
162. Honjo, Y.; Hrycyna, C. A.; Yan, Q. W.; Medina-Perez, W. Y.; Robey, R. W.; van de Laar, A.; Litman, T.; Dean, M.; Bates, S. E. *Cancer Res* 2001, **61** (18), 6635–9.
163. Kage, K.; Tsukahara, S.; Sugiyama, T.; Asada, S.; Ishikawa, E.; Tsuruo, T.; Sugimoto, Y. *Int J Cancer* 2002, **97** (5), 626–30.
164. Schnepf, R.; Zolk, O. *Expert Opin Drug Metab Toxicol* 2013, **9** (3), 287–306.
165. Poguntke, M.; Hazai, E.; Fromm, M. F.; Zolk, O. *Expert Opin Drug Metab Toxicol* 2010, **6** (11), 1363–84.
166. Maliepaard, M.; Scheffer, G. L.; Faneyte, I. F.; van Gastelen, M. A.; Pijnenborg, A. C.; Schinkel, A. H.; van De Vijver, M. J.; Scheper, R. J.; Schellens, J. H. *Cancer Res* 2001, **61** (8), 3458–64.
167. Meyer zu Schwabedissen, H. E.; Kroemer, H. K. *Handb Exp Pharmacol* 2011, (201), 325–71.
168. van Herwaarden, A. E.; Schinkel, A. H. *Trends Pharmacol Sci* 2006, **27** (1), 10–16.
169. International Transporter Consortium; Giacomini, K. M.; Huang, S. M.; Tweedie, D. J.; Benet, L. Z.; Brouwer, K. L.; Chu, X.; Dahlin, A.; Evers, R.; Fischer, V.; Hillgren, K. M.; Hoffmaster, K. A.; Ishikawa, T.; Keppler, D.; Kim, R. B.; Lee, C. A.; Niemi, M.; Polli, J. W.; Sugiyama, Y.; Swaan, P. W.; Ware, J. A.; Wright, S. H.; Yee, S. W.; Zamek-Gliszczynski, M. J.; Zhang, L. *Nat Rev Drug Discov* 2010, **9** (3), 215–36.

170. Jonker, J. W.; Buitelaar, M.; Wagenaar, E.; Van Der Valk, M. A.; Scheffer, G. L.; Scheper, R. J.; Plosch, T.; Kuipers, F.; Elferink, R. P.; Rosing, H.; Beijnen, J. H.; Schinkel, A. H. *Proc Natl Acad Sci U S A* 2002, **99** (24), 15649–54.

171. Kruijtzer, C. M.; Beijnen, J. H.; Rosing, H.; ten Bokkel Huinink, W. W.; Schot, M.; Jewell, R. C.; Paul, E. M.; Schellens, J. H. *J Clin Oncol* 2002, **20** (13), 2943–50.

172. Busti, A. J.; Bain, A. M.; Hall, R. G., 2nd; Bedimo, R. G.; Leff, R. D.; Meek, C.; Mehvar, R. *J Cardiovasc Pharmacol* 2008, **51** (6), 605–10.

173. Taipalensuu, J.; Tornblom, H.; Lindberg, G.; Einarsson, C.; Sjoqvist, F.; Melhus, H.; Garberg, P.; Sjostrom, B.; Lundgren, B.; Artursson, P. *J Pharmacol Exp Ther* 2001, **299** (1), 164–70.

174. Suzuki, M.; Suzuki, H.; Sugimoto, Y.; Sugiyama, Y. *J Biol Chem* 2003, **278** (25), 22644–9.

175. Funk, C. *Expert Opin Drug Metab Toxicol* 2008, **4** (4), 363–79.

176. Zamek-Gliszczynski, M. J.; Nezasa, K.; Tian, X.; Kalvass, J. C.; Patel, N. J.; Raub, T. J.; Brouwer, K. L. *Mol Pharmacol* 2006, **70** (6), 2127–33.

177. Zamek-Gliszczynski, M. J.; Hoffmaster, K. A.; Tian, X.; Zhao, R.; Polli, J. W.; Humphreys, J. E.; Webster, L. O.; Bridges, A. S.; Kalvass, J. C.; Brouwer, K. L. *Drug Metab Dispos* 2005, **33** (8), 1158–65.

178. Giacomini, K. M.; Balimane, P. V.; Cho, S. K.; Eadon, M.; Edeki, T.; Hillgren, K. M.; Huang, S. M.; Sugiyama, Y.; Weitz, D.; Wen, Y.; Xia, C. Q.; Yee, S. W.; Zimdahl, H.; Niemi, M.; International Transporter Consortium. *Clin Pharmacol Ther* 2013, **94** (1), 23–6.

179. Uchida, Y.; Ohtsuki, S.; Katsukura, Y.; Ikeda, C.; Suzuki, T.; Kamiie, J.; Terasaki, T. *J Neurochem* 2011, **117** (2), 333–45.

180. de Vries, N. A.; Zhao, J.; Kroon, E.; Buckle, T.; Beijnen, J. H.; van Tellingen, O. *Clin Cancer Res* 2007, **13** (21), 6440–9.

181. Oostendorp, R. L.; Buckle, T.; Beijnen, J. H.; van Tellingen, O.; Schellens, J. H. *Invest New Drugs* 2009, **27** (1), 31–40.

182. Tang, S. C.; Lankheet, N. A.; Poller, B.; Wagenaar, E.; Beijnen, J. H.; Schinkel, A. H. *J Pharmacol Exp Ther* 2012, **341** (1), 164–73.

183. Huls, M.; Brown, C. D.; Windass, A. S.; Sayer, R.; van den Heuvel, J. J.; Heemskerk, S.; Russel, F. G.; Masereeuw, R. *Kidney Int* 2008, **73** (2), 220–5.

184. Lincke, C. R.; Smit, J. J.; van der Velde-Koerts, T.; Borst, P. *J Biol Chem* 1991, **266** (8), 5303–10.

185. Stieger, B. *Handb Exp Pharmacol* 2011, (201), 205–59.

186. Velamakanni, S.; Wei, S. L.; Janvilisri, T.; van Veen, H. W. *J Bioenerg Biomembr* 2007, **39** (5–6), 465–71.

187. Hagenbuch, B.; Stieger, B. *Mol Aspects Med* 2013, **34** (2–3), 396–412.

188. Hagenbuch, B.; Meier, P. J. *Biochim Biophys Acta* 2003, **1609** (1), 1–18.

189. Li, L.; Lee, T. K.; Meier, P. J.; Ballatori, N. *J Biol Chem* 1998, **273** (26), 16184–91.

190. Kalliokoski, A.; Niemi, M. *Br J Pharmacol* 2009, **158** (3), 693–705.

191. Schneck, D. W.; Birmingham, B. K.; Zalikowski, J. A.; Mitchell, P. D.; Wang, Y.; Martin, P. D.; Lasseter, K. C.; Brown, C. D.; Windass, A. S.; Raza, A. *Clin Pharmacol Ther* 2004, **75** (5), 455–63.

192. Kostapanos, M. S.; Milionis, H. J.; Elisaf, M. S. *Am J Cardiovasc Drugs* 2010, **10** (1), 11–28.

193. Neuvonen, P. J.; Niemi, M.; Backman, J. T. *Clin Pharmacol Ther* 2006, **80** (6), 565–81.

194. Lau, Y. Y.; Huang, Y.; Frassetto, L.; Benet, L. Z. *Clin Pharmacol Ther* 2007, **81** (2), 194–204.

195. Niemi, M.; Backman, J. T.; Kajosaari, L. I.; Leathart, J. B.; Neuvonen, M.; Daly, A. K.; Eichelbaum, M.; Kivisto, K. T.; Neuvonen, P. J. *Clin Pharmacol Ther* 2005, **77** (6), 468–78.

196. Katz, D. A.; Carr, R.; Grimm, D. R.; Xiong, H.; Holley-Shanks, R.; Mueller, T.; Leake, B.; Wang, Q.; Han, L.; Wang, P. G.; Edeki, T.; Sahelijo, L.; Doan, T.; Allen, A.; Spear, B. B.; Kim, R. B. *Clin Pharmacol Ther* 2006, **79** (3), 186–96.

197. Suwannakul, S.; Ieiri, I.; Kimura, M.; Kawabata, K.; Kusuhara, H.; Hirota, T.; Irie, S.; Sugiyama, Y.; Higuchi, S. *J Hum Genet* 2008, **53** (10), 899–904.

198. Xiang, X.; Jada, S. R.; Li, H. H.; Fan, L.; Tham, L. S.; Wong, C. I.; Lee, S. C.; Lim, R.; Zhou, Q. Y.; Goh, B. C.; Tan, E. H.; Chowbay, B. *Pharmacogenet Genomics* 2006, **16** (9), 683–91.

199. Treiber, A.; Schneiter, R.; Hausler, S.; Stieger, B. *Drug Metab Dispos* 2007, **35** (8), 1400–7.

200. Backman, J. T.; Kyrklund, C.; Neuvonen, M.; Neuvonen, P. J. *Clin Pharmacol Ther* 2002, **72** (6), 685–91.

201. Chung, J. Y.; Cho, J. Y.; Yu, K. S.; Kim, J. R.; Oh, D. S.; Jung, H. R.; Lim, K. S.; Moon, K. H.; Shin, S. G.; Jang, I. J. *Clin Pharmacol Ther* 2005, **78** (4), 342–50.

202. Pasanen, M. K.; Neuvonen, M.; Neuvonen, P. J.; Niemi, M. *Pharmacogenet Genomics* 2006, **16** (12), 873–9.

203. Pasanen, M. K.; Fredrikson, H.; Neuvonen, P. J.; Niemi, M. *Clin Pharmacol Ther* 2007, **82** (6), 726–33.

204. Lee, E.; Ryan, S.; Birmingham, B.; Zalikowski, J.; March, R.; Ambrose, H.; Moore, R.; Lee, C.; Chen, Y.; Schneck, D. *Clin Pharmacol Ther* 2005, **78** (4), 330–41.

205. Group, S. C.; Link, E.; Parish, S.; Armitage, J.; Bowman, L.; Heath, S.; Matsuda, F.; Gut, I.; Lathrop, M.; Collins, R. *N Engl J Med* 2008, **359** (8), 789–99.

206. Konig, J.; Seithel, A.; Gradhand, U.; Fromm, M. F. *Naunyn Schmiedebergs Arch Pharmacol* 2006, **372** (6), 432–43.

207. Kullak-Ublick, G. A.; Hagenbuch, B.; Stieger, B.; Schteingart, C. D.; Hofmann, A. F.; Wolkoff, A. W.; Meier, P. J. *Gastroenterology* 1995, **109** (4), 1274–82.

208. Tamai, I.; Nezu, J.; Uchino, H.; Sai, Y.; Oku, A.; Shimane, M.; Tsuji, A. *Biochem Biophys Res Commun* 2000, **273** (1), 251–60.

209. Dresser, G. K.; Bailey, D. G.; Leake, B. F.; Schwarz, U. I.; Dawson, P. A.; Freeman, D. J.; Kim, R. B. *Clin Pharmacol Ther* 2002, **71** (1), 11–20.

210. Perloff, M. D.; von Moltke, L. L.; Greenblatt, D. J. *J Clin Pharmacol* 2002, **42** (11), 1269–74.

211. Dresser, G. K.; Kim, R. B.; Bailey, D. G. *Clin Pharmacol Ther* 2005, **77** (3), 170–7.

212. Bailey, D. G. *Br J Clin Pharmacol* 2010, **70** (5), 645–55.

213. Mikkaichi, T.; Suzuki, T.; Onogawa, T.; Tanemoto, M.; Mizutamari, H.; Okada, M.; Chaki, T.; Masuda, S.; Tokui, T.; Eto, N.; Abe, M.; Satoh, F.; Unno, M.; Hishinuma, T.; Inui, K.; Ito, S.; Goto, J.; Abe, T. *Proc Natl Acad Sci U S A* 2004, **101** (10), 3569–74.

214. Chu, X. Y.; Bleasby, K.; Yabut, J.; Cai, X.; Chan, G. H.; Hafey, M. J.; Xu, S.; Bergman, A. J.; Braun, M. P.; Dean, D. C.; Evers, R. *J Pharmacol Exp Ther* 2007, **321** (2), 673–83.

215. You, G. *Med Res Rev* 2004, **24** (6), 762–74.

216. Alebouyeh, M.; Takeda, M.; Onozato, M. L.; Tojo, A.; Noshiro, R.; Hasannejad, H.; Inatomi, J.; Narikawa, S.; Huang, X. L.; Khamdang, S.; Anzai, N.; Endou, H. *J Pharmacol Sci* 2003, **93** (4), 430–6.

217. Burckhardt, B. C.; Burckhardt, G. *Rev Physiol Biochem Pharmacol* 2003, **146**, 95–158.

218. Wright, S. H. *Toxicol Appl Pharmacol* 2005, **204** (3), 309–19.

219. Rizwan, A. N.; Burckhardt, G. *Pharm Res* 2007, **24** (3), 450–70.

220. Walshaw, P. E.; McCauley, F. A.; Wilson, T. W. *Clin Invest Med* 1992, **15** (1), 82–7.

221. Honari, J.; Blair, A. D.; Cutler, R. E. *Clin Pharmacol Ther* 1977, **22** (4), 395–401.

222. Miyazaki, H.; Sekine, T.; Endou, H. *Trends Pharmacol Sci* 2004, **25** (12), 654–62.

223. Jonkman, J. H.; Upton, R. A. *Clin Pharmacokinet* 1984, **9** (4), 309–34.

224. Upton, R. A. *Clin Pharmacokinet* 1991, **20** (1), 66–80.

225. Koepsell, H.; Endou, H. *Pflugers Arch* 2004, **447** (5), 666–76.

226. Gorboulev, V.; Ulzheimer, J. C.; Akhoundova, A.; Ulzheimer-Teuber, I.; Karbach, U.; Quester, S.; Baumann, C.; Lang, F.; Busch, A. E.; Koepsell, H. *DNA Cell Biol* 1997, **16** (7), 871–81.

227. Leabman, M. K.; Huang, C. C.; DeYoung, J.; Carlson, E. J.; Taylor, T. R.; de la Cruz, M.; Johns, S. J.; Stryke, D.; Kawamoto, M.; Urban, T. J.; Kroetz, D. L.; Ferrin, T. E.; Clark, A. G.; Risch, N.; Herskowitz, I.; Giacomini, K. M.; Pharmacogenetics Of Membrane Transporters Investigators. *Proc Natl Acad Sci U S A* 2003, **100** (10), 5896–901.

228. Leabman, M. K.; Huang, C. C.; Kawamoto, M.; Johns, S. J.; Stryke, D.; Ferrin, T. E.; DeYoung, J.; Taylor, T.; Clark, A. G.; Herskowitz, I.; Giacomini, K. M.; Pharmacogenetics of Membrane Transporters Investigators. *Pharmacogenetics* 2002, **12** (5), 395–405.

229. Sakata, T.; Anzai, N.; Kimura, T.; Miura, D.; Fukutomi, T.; Takeda, M.; Sakurai, H.; Endou, H. *J Pharmacol Sci* 2010, **113** (3), 263–6.

230. Filipski, K. K.; Mathijssen, R. H.; Mikkelsen, T. S.; Schinkel, A. H.; Sparreboom, A. *Clin Pharmacol Ther* 2009, **86** (4), 396–402.

231. Song, I. S.; Shin, H. J.; Shim, E. J.; Jung, I. S.; Kim, W. Y.; Shon, J. H.; Shin, J. G. *Clin Pharmacol Ther* 2008, **84** (5), 559–62.

232. Fujita, T.; Urban, T. J.; Leabman, M. K.; Fujita, K.; Giacomini, K. M. *J Pharm Sci* 2006, **95** (1), 25–36.

233. Koepsell, H. *Biol Chem* 2011, **392** (1–2), 95–101.

234. Ciarimboli, G. *Expert Opin Drug Metab Toxicol* 2011, **7** (2), 159–74.

235. Zhang, L.; Dresser, M. J.; Gray, A. T.; Yost, S. C.; Terashita, S.; Giacomini, K. M. *Mol Pharmacol* 1997, **51** (6), 913–21.

236. Motohashi, H.; Sakurai, Y.; Saito, H.; Masuda, S.; Urakami, Y.; Goto, M.; Fukatsu, A.; Ogawa, O.; Inui, K. *J Am Soc Nephrol* 2002, **13** (4), 866–74.

237. Grundemann, D.; Schechinger, B.; Rappold, G. A.; Schomig, E. *Nat Neurosci* 1998, **1** (5), 349–51.

238. Ganapathy, V.; Prasad, P. D. *Toxicol Appl Pharmacol* 2005, **207** (2 Suppl), 381–7.

239. Somogyi, A.; Stockley, C.; Keal, J.; Rolan, P.; Bochner, F. *Br J Clin Pharmacol* 1987, **23** (5), 545–51.

240. Iwata, K.; Aizawa, K.; Kamitsu, S.; Jingami, S.; Fukunaga, E.; Yoshida, M.; Yoshimura, M.; Hamada, A.; Saito, H. *Clin Exp Nephrol* 2012, **16** (6), 843–51.

241. Jonker, J. W.; Wagenaar, E.; Van Eijl, S.; Schinkel, A. H. *Mol Cell Biol* 2003, **23** (21), 7902–8.

242. Wang, D. S.; Jonker, J. W.; Kato, Y.; Kusuhara, H.; Schinkel, A. H.; Sugiyama, Y. *J Pharmacol Exp Ther* 2002, **302** (2), 510–15.

243. Jonker, J. W.; Schinkel, A. H. *J Pharmacol Exp Ther* 2004, **308** (1), 2–9.

244. Wang, D. S.; Kusuhara, H.; Kato, Y.; Jonker, J. W.; Schinkel, A. H.; Sugiyama, Y. *Mol Pharmacol* 2003, **63** (4), 844–8.

245. Ciarimboli, G.; Deuster, D.; Knief, A.; Sperling, M.; Holtkamp, M.; Edemir, B.; Pavenstadt, H.; Lanvers-Kaminsky, C.; am Zehnhoff-Dinnesen, A.; Schinkel, A. H.; Koepsell, H.; Jurgens, H.; Schlatter, E. *Am J Pathol* 2010, **176** (3), 1169–80.

246. Jonker, J. W.; Wagenaar, E.; Mol, C. A.; Buitelaar, M.; Koepsell, H.; Smit, J. W.; Schinkel, A. H. *Mol Cell Biol* 2001, **21** (16), 5471–7.

247. Grundemann, D.; Gorboulev, V.; Gambaryan, S.; Veyhl, M.; Koepsell, H. *Nature* 1994, **372** (6506), 549–52.

248. Seithel, A.; Karlsson, J.; Hilgendorf, C.; Bjorquist, A.; Ungell, A. L. *Eur J Pharm Sci* 2006, **28** (4), 291–9.

249. Tamai, I.; Yabuuchi, H.; Nezu, J.; Sai, Y.; Oku, A.; Shimane, M.; Tsuji, A. *FEBS Lett* 1997, **419** (1), 107–11.

250. Tamai, I.; Ohashi, R.; Nezu, J.; Yabuuchi, H.; Oku, A.; Shimane, M.; Sai, Y.; Tsuji, A. *J Biol Chem* 1998, **273** (32), 20378–82.

251. Tamai, I. *Biopharm Drug Dispos* 2013, **34** (1), 29–44.

252. Koepsell, H.; Lips, K.; Volk, C. *Pharm Res* 2007, **24** (7), 1227–51.

253. Nakamura, T.; Yoshida, K.; Yabuuchi, H.; Maeda, T.; Tamai, I. *Biol Pharm Bull* 2008, **31** (8), 1580–4.

254. Yabuuchi, H.; Tamai, I.; Nezu, J.; Sakamoto, K.; Oku, A.; Shimane, M.; Sai, Y.; Tsuji, A. *J Pharmacol Exp Ther* 1999, **289** (2), 768–73.

255. Jong, N. N.; Nakanishi, T.; Liu, J. J.; Tamai, I.; McKeage, M. J. *J Pharmacol Exp Ther* 2011, **338** (2), 537–47.

256. Urban, T. J.; Brown, C.; Castro, R. A.; Shah, N.; Mercer, R.; Huang, Y.; Brett, C. M.; Burchard, E. G.; Giacomini, K. M. *Clin Pharmacol Ther* 2008, **83** (3), 416–21.

257. Wu, S. P.; Shyu, M. K.; Liou, H. H.; Gau, C. S.; Lin, C. J. *Epilepsia* 2004, **45** (3), 204–10.

258. Otsuka, M.; Matsumoto, T.; Morimoto, R.; Arioka, S.; Omote, H.; Moriyama, Y. *Proc Natl Acad Sci U S A* 2005, **102** (50), 17923–8.

259. Masuda, S.; Terada, T.; Yonezawa, A.; Tanihara, Y.; Kishimoto, K.; Katsura, T.; Ogawa, O.; Inui, K. *J Am Soc Nephrol* 2006, **17** (8), 2127–35.

260. Astorga, B.; Ekins, S.; Morales, M.; Wright, S. H. *J Pharmacol Exp Ther* 2012, **341** (3), 743–55.

261. Motohashi, H.; Inui, K. *AAPS J* 2013, **15** (2), 581–8.

262. Sato, T.; Masuda, S.; Yonezawa, A.; Tanihara, Y.; Katsura, T.; Inui, K. *Biochem Pharmacol* 2008, **76** (7), 894–903.

263. Tsuda, M.; Terada, T.; Mizuno, T.; Katsura, T.; Shimakura, J.; Inui, K. *Mol Pharmacol* 2009, **75** (6), 1280–6.

264. Yokoo, S.; Yonezawa, A.; Masuda, S.; Fukatsu, A.; Katsura, T.; Inui, K. *Biochem Pharmacol* 2007, **74** (3), 477–87.

265. Katsuda, H.; Yamashita, M.; Katsura, H.; Yu, J.; Waki, Y.; Nagata, N.; Sai, Y.; Miyamoto, K. *Biol Pharm Bull* 2010, **33** (11), 1867–71.

266. Yonezawa, A.; Inui, K. *Biochem Pharmacol* 2011, **81** (5), 563–8.

267. Ito, S.; Kusuhara, H.; Kuroiwa, Y.; Wu, C.; Moriyama, Y.; Inoue, K.; Kondo, T.; Yuasa, H.; Nakayama, H.; Horita, S.; Sugiyama, Y. *J Pharmacol Exp Ther* 2010, **333** (1), 341–50.

268. Toyama, K.; Yonezawa, A.; Masuda, S.; Osawa, R.; Hosokawa, M.; Fujimoto, S.; Inagaki, N.; Inui, K.; Katsura, T. *Br J Pharmacol* 2012, **166** (3), 1183–91.

269. Jackson, V. N.; Halestrap, A. P. *J Biol Chem* 1996, **271** (2), 861–8.

270. Halestrap, A. P.; Price, N. T. *Biochem J* 1999, **343** (Pt 2), 281–99.

271. Poole, R. C.; Sansom, C. E.; Halestrap, A. P. *Biochem J* 1996, **320** (Pt 3), 817–24.

272. Mamelak, M.; Scharf, M. B.; Woods, M. *Sleep* 1986, **9** (1 Pt 2), 285–9.

273. Gallimberti, L.; Spella, M. R.; Soncini, C. A.; Gessa, G. L. *Alcohol* 2000, **20** (3), 257–62.

274. Murakami, Y.; Kohyama, N.; Kobayashi, Y.; Ohbayashi, M.; Ohtani, H.; Sawada, Y.; Yamamoto, T. *Drug Metab Dispos* 2005, **33** (12), 1845–51.

275. Friesema, E. C.; Ganguly, S.; Abdalla, A.; Manning Fox, J. E.; Halestrap, A. P.; Visser, T. J. *J Biol Chem* 2003, **278** (41), 40128–35.

276. Kim, D. K.; Kanai, Y.; Chairoungdua, A.; Matsuo, H.; Cha, S. H.; Endou, H. *J Biol Chem* 2001, **276** (20), 17221–8.

277. Morris, M. E.; Felmlee, M. A. *AAPS J* 2008, **10** (2), 311–21.

278. Wang, Q.; Lu, Y.; Morris, M. E. *Pharm Res* 2007, **24** (6), 1067–78.

279. Wang, Q.; Morris, M. E. *Drug Metab Dispos* 2007, **35** (8), 1393–9.

280. Wang, Q.; Morris, M. E. *Drug Metab Dispos* 2007, **35** (2), 201–8.

281. Morris, M. E.; Hu, K.; Wang, Q. *J Pharmacol Exp Ther* 2005, **313** (3), 1194–202.

282. Gill, R. K.; Saksena, S.; Alrefai, W. A.; Sarwar, Z.; Goldstein, J. L.; Carroll, R. E.; Ramaswamy, K.; Dudeja, P. K. *Am J Physiol Cell Physiol* 2005, **289** (4), C846–52.

283. Tsuji, A.; Tamai, I.; Nakanishi, M.; Terasaki, T.; Hamano, S. *J Pharm Pharmacol* 1993, **45** (11), 996–8.

284. Li, Y. H.; Tanno, M.; Itoh, T.; Yamada, H. *Int J Pharm* 1999, **191** (2), 151–9.

285. Neuhoff, S.; Ungell, A. L.; Zamora, I.; Artursson, P. *Eur J Pharm Sci* 2005, **25** (2–3), 211–20.

286. Tamai, I.; Takanaga, H.; Maeda, H.; Ogihara, T.; Yoneda, M.; Tsuji, A. *Pharm Res* 1995, **12** (11), 1727–32.

287. Wu, X.; Whitfield, L. R.; Stewart, B. H. *Pharm Res* 2000, **17** (2), 209–15.

288. Cundy, K. C.; Annamalai, T.; Bu, L.; De Vera, J.; Estrela, J.; Luo, W.; Shirsat, P.; Torneros, A.; Yao, F.; Zou, J.; Barrett, R. W.; Gallop, M. A. *J Pharmacol Exp Ther* 2004, **311** (1), 324–33.

289. Cundy, K. C.; Sastry, S.; Luo, W.; Zou, J.; Moors, T. L.; Canafax, D. M. *J Clin Pharmacol* 2008, **48** (12), 1378–88.

290. Jeffrey, S. FDA Approves Gabapentin Enacarbil for Postherpetic Neuralgia. http://www.medscape.com/viewarticle/765233 (accessed September 23, 2015).

291. Pierre, K.; Pellerin, L. *J Neurochem* 2005, **94** (1), 1–14.

292. Vijay, N.; Morris, M. E. *Curr Pharm Des* 2014, **20** (10), 1487–98.

293. Kang, Y. S.; Terasaki, T.; Tsuji, A. *J Pharmacobiodyn* 1990, **13** (2), 158–63.

294. Tsuji, A.; Saheki, A.; Tamai, I.; Terasaki, T. *J Pharmacol Exp Ther* 1993, **267** (3), 1085–90.

295. Saheki, A.; Terasaki, T.; Tamai, I.; Tsuji, A. *Pharm Res* 1994, **11** (2), 305–11.

296. Bhattacharya, I.; Boje, K. M. *J Pharmacol Exp Ther* 2004, **311** (1), 92–8.

297. Roiko, S. A.; Felmlee, M. A.; Morris, M. E. *Drug Metab Dispos* 2012, **40** (1), 212–18.

298. Srinivas, S. R.; Gopal, E.; Zhuang, L.; Itagaki, S.; Martin, P. M.; Fei, Y. J.; Ganapathy, V.; Prasad, P. D. *Biochem J* 2005, **392** (Pt 3), 655–64.

299. Rodriguez, A. M.; Perron, B.; Lacroix, L.; Caillou, B.; Leblanc, G.; Schlumberger, M.; Bidart, J. M.; Pourcher, T. *J Clin Endocrinol Metab* 2002, **87** (7), 3500–3.

300. Cui, D.; Morris, M. E. *Drug Metab Dispos* 2009, **37** (7), 1404–10.

301. Kennedy, D. J.; Leibach, F. H.; Ganapathy, V.; Thwaites, D. T. *Pflugers Arch* 2002, **445** (1), 139–46.

302. Covitz, K. M.; Amidon, G. L.; Sadee, W. *Biochemistry* 1998, **37** (43), 15214–21.

303. Herrera-Ruiz, D.; Knipp, G. T. *J Pharm Sci* 2003, **92** (4), 691–714.

304. Herrera-Ruiz, D.; Wang, Q.; Gudmundsson, O. S.; Cook, T. J.; Smith, R. L.; Faria, T. N.; Knipp, G. T. *AAPS PharmSci* 2001, **3** (1), E9.

305. Smith, D. E.; Pavlova, A.; Berger, U. V.; Hediger, M. A.; Yang, T.; Huang, Y. G.; Schnermann, J. B. *Pharm Res* 1998, **15** (8), 1244–9.

306. Lindley, D. J.; Carl, S. M.; Mowery, S. A.; Knipp, G. T. *Rev Mex Cienc Farm* 2011, **42** (4), 57–65.

307. Shen, H.; Smith, D. E.; Keep, R. F.; Brosius, F. C., 3rd. *Mol Pharm* 2004, **1** (4), 248–56.

308. Shen, H.; Smith, D. E.; Keep, R. F.; Xiang, J.; Brosius, F. C., 3rd. *J Biol Chem* 2003, **278** (7), 4786–91.

309. Yamashita, T.; Shimada, S.; Guo, W.; Sato, K.; Kohmura, E.; Hayakawa, T.; Takagi, T.; Tohyama, M. *J Biol Chem* 1997, **272** (15), 10205–11.

310. de Vrueh, R. L.; Smith, P. L.; Lee, C. P. *J Pharmacol Exp Ther* 1998, **286** (3), 1166–70.

311. Sugawara, M.; Huang, W.; Fei, Y. J.; Leibach, F. H.; Ganapathy, V.; Ganapathy, M. E. *J Pharm Sci* 2000, **89** (6), 781–9.

312. Perry, C. M.; Faulds, D. *Drugs* 1996, **52** (5), 754–72.

313. Cocohoba, J. M.; McNicholl, I. R. *Ann Pharmacother* 2002, **36** (6), 1075–9.

314. Cruz, D. N. *Expert Opin Pharmacother* 2000, **1** (4), 835–40.

315. Tsuda, M.; Terada, T.; Irie, M.; Katsura, T.; Niida, A.; Tomita, K.; Fujii, N.; Inui, K. *J Pharmacol Exp Ther* 2006, **318** (1), 455–60.

316. Agarwal, S.; Boddu, S. H.; Jain, R.; Samanta, S.; Pal, D.; Mitra, A. K. *Int J Pharm* 2008, **359** (1–2), 7–14.

317. Carl, S. M.; Herrera-Ruiz, D.; Bhardwaj, R. K.; Gudmundsson, O.; Knipp, G. T. Mammalian Oligopeptide Transporters. In *Drug Transporters: Molecular Characterization and Role in Drug Disposition*, You, G.; Morris, M. E., Ed. John Wiley & Sons, Inc.: Hoboken, NJ, 2006.

318. Lindley, D. J.; Carl, S. M.; Herrera-Ruiz, D.; Pan, L. F.; Karpes, L. B.; Goole, J. M. E.; Gudmundsson, O. S.; Knipp, G. T. Drug Transporters and Their Role in Absorption and Disposition of Peptides and Peptide-Based Pharmaceuticals. In *Oral Bioavailability: Basic Principles, Advanced Concepts, and Applications*, Hu, M.; Li, X., Ed. John Wiley & Sons, Inc.: Hoboken, NJ, 2011.

319. Jain, R.; Agarwal, S.; Mandava, N. K.; Sheng, Y.; Mitra, A. K. *Int J Pharm* 2008, **362** (1–2), 44–51.

320. Jain, R.; Agarwal, S.; Majumdar, S.; Zhu, X.; Pal, D.; Mitra, A. K. *Int J Pharm* 2005, **303** (1–2), 8–19.

6

INTRACELLULAR DELIVERY AND DISPOSITION OF SMALL-MOLECULAR-WEIGHT DRUGS

Jeffrey P. Krise

Department of Pharmaceutical Chemistry, The University of Kansas, Lawrence, KS, USA

6.1 INTRODUCTION

The term "drug delivery" encompasses a broad array of pharmaceutical research, as the reader will certainly appreciate upon reading the various chapters contained within this book. In a simplistic and broad sense, drug delivery represents the science related to getting a drug from the site of administration to the intended site of action in the body (i.e., the receptor). Traditionally this term broadly includes everything from drug formulation to tissue-targeting strategies. This chapter will focus on the intracellular aspects of drug delivery pertaining to small-molecular-weight drugs that enter and exit cells by passive diffusion-type mechanisms.

For drugs that have intracellular targets, the diffusion and transport within the cell is fundamentally important because it truly represents the final stage of drug delivery. In this chapter I will provide an overview as to why the intracellular distribution of a drug is an important therapeutic consideration. These discussions will be specifically directed toward the influences of this on drug activity and pharmacokinetics. I will provide some background with regard to methods typically employed to study the intracellular distribution of a drug. Finally, I will review the major drug sequestering compartments that have been identified to date, which include the mitochondria, lysosomes, and the nucleus. I will provide some examples of drugs that accumulate

Drug Delivery: Principles and Applications, Second Edition. Edited by Binghe Wang, Longqin Hu, and Teruna J. Siahaan.

within these compartments, with a focus on those where the intracellular distribution was purposely modified in an effort to optimize activity and/or pharmacokinetics.

6.2 THE RELATIONSHIP BETWEEN THE INTRACELLULAR DISTRIBUTION OF A DRUG AND ITS ACTIVITY

Up until fairly recently, it was often incorrectly assumed that if a drug was shown to significantly accumulate within a cell that harbors the cognate receptor, then a subsequent therapeutic response could be anticipated, assuming that the drug was not significantly deactivated through metabolism or degradation. While drug accumulation within drug target-containing cells is certainly a necessary step in drug delivery, it often does not constitute the final step. This is because our cells are extensively compartmentalized with over 50% of the total cell volume comprising membrane-bound compartments (i.e., organelles) that can potentially provide additional barriers between a drug and its intended target.

So even when various stages of drug delivery are distilled down to just those events occurring within the boundaries of the plasma membrane, the scenario can be quite complex. Successful drug delivery within a cell ultimately depends on the degree of co-localization between the drug and the drug target. Much is known about drug target localization in cells; they are typically not evenly distributed but instead are often confined to a single organelle population (i.e., mitochondria, cytosol, and nucleus). On the contrary, much less is known about how drugs themselves localize and distribute within cells.

It is useful to consider hypothetical scenarios pertaining to the intracellular drug distribution. The best possible scenario would be for a drug to *concentrate*, in its free form, only in the organelle that houses the intended target. Under these circumstances the potency of the drug would be maximized and, importantly, the potential for off-target interactions or side effects would be minimized. On the other end of the spectrum, it is possible that the drug could specifically localize and concentrate in an organelle that is separate from the one containing the drug target. Under these circumstances the only anticipated drug effect would result from off-target interactions. As will be discussed in detail later, this conceptual understanding has prompted efforts to design drugs with appropriate and optimized intracellular distribution profiles to improve potency and reduce side effects.

6.3 THE RELATIONSHIP BETWEEN THE INTRACELLULAR DISTRIBUTION OF A DRUG AND ITS PHARMACOKINETIC PROPERTIES

A second important potential therapeutic consequence linked with the intracellular localization of drugs has to do with its effects on pharmacokinetic properties. The most obvious influence would pertain to the apparent volume of distribution (V_D) of a drug. The volume of distribution of a drug relates the total amount of drug in the body with the concentration of the drug found in the blood. The lowest possible value for the volume of distribution is the volume of the blood. There is no upper limit for this volume term, and for many drugs the volume of distribution far exceeds the total volume of the body.

Drugs that possess relatively large volumes of distribution are often presumed to have a high affinity for tissues and/or adipose tissue, but the precise basis for high volume of distribution is typically not understood. The volume of distribution term is practically important because it, along with the clearance term (Cl), dictates the elimination half-life, which dictates the dosing frequency and time required to reach steady state in multiple-dosing situations. The relationship between volume of distribution, clearance and half-life is shown as follows:

$$t_{1/2} = \frac{0.693 \times V_{\mathrm{D}}}{\mathrm{Cl}}$$

The nonconcentrative distribution of a drug in a cell should not theoretically impact the volume of distribution of a drug. In other words, two drugs that have the same steady-state accumulation in cells and tissues but have stark differences in intracellular distribution should have the same pharmacokinetic distribution properties, assuming the kinetics of uptake and release from the cells and tissues is the same. In contrast to this, in some cases the accumulation of a drug within a specific organelle can be classified as concentrative. Under these circumstances the drug reaches levels in an organelle that are significantly greater than those found in surrounding compartments. In such cases the intracellular compartments can be responsible for driving the distribution of drug into the tissues.

As will be discussed with specific examples further, the most notable example of this phenomenon occurs with weakly basic drugs that are substrates for extensive entrapment within lysosomes. Accordingly, it has been shown that weakly basic drugs have, on average, much higher volumes of distribution than do weakly acidic or neutral drugs [1].

6.4 OVERVIEW OF APPROACHES TO STUDY INTRACELLULAR DRUG DISPOSITION

Despite the potential therapeutic importance of understanding the intracellular disposition of a drug, there is little to no information regarding this available for most drugs on the market. The main reason for this, in my opinion, is that it is perceived to be experimentally challenging to characterize this. It is true that such evaluations can be challenging, mostly because they are not amenable to high-throughput platforms. However, as will be discussed later, more recent approaches have been described that help decrease this barrier. I have superficially categorized some of the main approaches that have been most routinely used along with their pros and cons. A more thorough discussion of the specific approaches for evaluating the intracellular distribution of a drug can be found in several research reports and review articles on this topic [2–9].

6.4.1 Fluorescence Microscopy

Fluorescence microscopy is, by far, the easiest method to assess the intracellular distribution of a drug. Obviously, a major drawback is that it requires that the drug be intrinsically fluorescent, which very few drugs are. Examples of fluorescent drugs include the anticancer agents such as doxorubicin and daunorubicin as well as antimalarial drugs such as quinacrine. The intracellular distribution of these drugs has received considerable attention for this reason.

In order to determine which organelle(s) the fluorescent drug is associated with, the fluorescent drug is often incubated with cells in conjunction with additional fluorescent organelle vital stains to allow for co-localization studies. There are organelle-specific fluorescent vital stains available for a variety of organelles and cellular components including the nucleus, lysosomes, endoplasmic reticulum, and mitochondria that are commercially available through vendors such as Molecular Probes (Invitrogen, Life Technologies). These probes are usually available with different fluorescence properties to allow differentiation from the fluorescence properties of the drug. These studies are best performed with live cells but can be performed with cells fixed with cross-linking reagents such as formaldehyde, but it should be confirmed that fixation does not alter the intracellular distribution of the drugs being investigated.

The fluorescence of a drug can also be exploited in the characterization of drug associated with organelles in a heterogeneous cell lysate. For example, a unique analytical approach for evaluating the accumulation of the fluorescent drug doxorubicin in lysosomes was described by Chen and coworkers [10]. While intact, the lysosomes of the cells were labeled with fluorescent nanospheres using a pulse-chase technique. The authors utilized capillary electrophoresis with laser-induced fluorescence and dual-channel detection to identify lysosomes containing both fluorescent nanospheres and doxorubicin. The authors were able to identify heterogeneity among lysosomes in their doxorubicin content, an observation not readily achievable using alternative techniques.

For nonfluorescent drugs it is potentially possible to covalently attach a "fluorescent tag" to aid in the analysis of intracellular distribution using fluorescence microscopy. I, however, do not favor such an approach. It is often the case that the fluorescent tag is as large as or larger than the drug requiring evaluation. It would be very difficult to imagine a scenario whereby the addition of a fluorophore to a nonfluorescent drug wouldn't dramatically alter its intracellular distribution.

6.4.2 Organelle Isolation

It is possible to assess the intracellular distribution of a drug by isolating the cellular organelle(s) of interest, followed by subsequent extraction and quantitation of the drug from the isolated fraction using conventional approaches (i.e., LC–MS/MS). There have been two general types of assessments that utilize purified organelles. The first is that the organelles are isolated from cells that have been pre-exposed to drug and the second approach involves isolation of the organelles prior to drug exposure.

The positive aspect of this approach is that it is the most direct approach, at least in principle. The disadvantages of this approach are significant, however. Even the most rapid organelle isolation protocol takes significant amounts of time that potentially allows for the drug to diffuse out of the organelle. It is often quite difficult to obtain pure organelle fractions, which is particularly true when isolating lysosomes. The isolation of relatively pure nuclei and mitochondria is more realistic because of their differences in size and density relative to other organelles. In addition, it is possible that the integrity of the organelle lipid bilayer can be compromised during the isolation, resulting in leakage of luminal contents. Great care must be taken that proteins/enzymes that create the driving force for organelle-specific accumulation remain active to ensure drug retention. Finally, if one is attempting to estimate the concentration of a drug in the organelle, one must accurately determine not only the amount of drug contained within the isolated fraction but also the volume of the isolated organelle fraction, which is typically estimated from electron micrographs of the isolated fraction. This step is very difficult to do with any accuracy and it is time-consuming.

6.4.3 Indirect Methods

A number of indirect approaches have been described for investigating the intracellular accumulation of drugs in cells. The basis for such approaches can vary. Some assays are fluorescence based and rely on the assumption that if a drug accumulates within an organelle, then it will cause some change in the fluorescence of an organelle-specific vital stain. In the general design, a fluorescent organelle-specific vital stain is incubated with cells and its cellular accumulation and/or distribution is monitored using fluorescence plate readers and/or microscopy. Subsequently, the test drug is added and perturbations to the fluorescence of the vital stain provide indirect evidence that the drug has accumulated in this compartment. The advantage of this approach is that it does not require that the test drug be fluorescent and it can be high throughput. The disadvantage is that the drugs must cause some phenotypic change in the organelle in order to detect accumulation. Accordingly, concentrations of drugs employed often exceed what might be therapeutically relevant. In principle, this type of assay could be performed with almost any organelle for which there exists a fluorescence vital stain. In practice, however, this assay has mainly been employed to assess drug accumulation in lysosomes as will be discussed later.

The previously discussed approach will give some indication of whether or not a drug has accumulated within an organelle, but it is not particularly useful in estimating the concentration of the drug in an organelle. For this purpose we and others have proposed an approach by which the total cellular accumulation of the drug could be measured before and after conditions are applied to the cell that are known to specifically dissipate organelle accumulation. For the evaluation of mitochondrial accumulation, reagents such as FCCP have been used [11]. For lysosomal sequestration, inhibitors of the vacuolar proton ATPase (i.e., concanamycin A, bafilomycin) or ionophores (nigericin, monensin) and buffering agents such as chloroquine have been used. For DNA binding the use of competitive DNA binders has been explored [12]. The differences in the total amount of drug associated with cells with or without treatment can provide a good and relatively fast estimate regarding the total amount of drug accumulating in a specific organelle in a given population of cells. Then, using published estimates for the percentage of the total cell volume that is comprised by the organelle of interest, one can arrive at the concentration of the drug in the organelle.

In contrast to the aforementioned approaches, the same concept has been explored without the use of the chemicals. Hu and colleagues have selected for a human osteosarcoma cell line that was devoid of mitochondrial DNA and that that could not perform electron transport or oxidative phosphorylation [13]. Using this cell line and matched wild-type cells, the authors conducted parallel IC_{50} evaluations of a series of drugs and were able to conclude based on these readouts which of the drugs were potential substrates for mitochondrial sequestration and which ones were not.

6.5 THE ACCUMULATION OF DRUGS IN MITOCHONDRIA, LYSOSOMES, AND NUCLEI

Many drugs have been shown to specifically accumulate in organelles of mammalian cells. Discussion in this chapter will focus only on small-molecular-weight drugs that have some capacity to enter cells and organelles by passive diffusion. For simplicity, this chapter will not specifically cover therapeutically important macromolecules and polar small-molecular-weight drugs that have been shown to accumulate specifically in endocytic compartments of cells following receptor-mediated and/or nonspecific fluid-phase endocytosis. It is assumed that such examples will be addressed in other chapters of this book.

Previous studies on the intracellular distribution of drugs in cells have led to the identification of three compartments that, in my opinion, comprise the most common sites for intracellular distribution. These compartments comprise the nucleus, the mitochondria, and the lysosomes. For each of these compartments I will generally describe the mechanism for organelle accumulation and provide examples of drugs that have had documented accumulation within them. The examples that I provide are not meant to be all-inclusive; I will instead focus on those examples that reveal interesting information regarding the structure–localization relationship for accumulation or those examples where the organelle accumulation was intentional to optimize an activity and/or pharmacokinetics.

6.5.1 Mitochondrial Accumulation of Drugs

Mitochondria play a central role in human health, and many drugs have therapeutic or toxicological effects involving this organelle or its constituents (see Szewczyk and Wojtczak for a review of the mitochondria as a pharmacological target [14]).

As discussed in this review, there are numerous drugs that have been shown to have effects on mitochondrial structure and function, and from this one could infer that those drugs have gained some degree of access to these mitochondrial targets. A summary of the structure–localization relationship for these drugs has been previously reviewed [15]. This chapter, however, will not attempt to examine all of these drugs. Instead, this chapter will focus on documented examples whereby a drug was either deliberately designed or modified in an attempt to enhance mitochondrial accumulation for the purpose of increasing interactions with mitochondrial targets.

The majority of attempts made to increase the mitochondrial accumulation of drugs have attempted to exploit the unique membrane potential associated with this compartment. The membrane potential across the inner membrane of the mitochondria is negative and ranges from 120 to 170 mV. This negative potential can drive the accumulation of cationic drugs into this space (see Fig. 6.1). The Nernst equation can be used to predict the equilibrium distribution of cations inside and outside the inner mitochondrial membrane (see equation in Fig. 6.1). According to the Nernst equation and the existing membrane potential associated with the mitochondria, it is anticipated that under equilibrium conditions the concentration of cationic drugs inside the mitochondria could be 100–1000 times greater than they are in the cell cytosol [16]. Every 61.5 mV increase in the membrane potential results in a 10-fold increase in the cation concentration in the mitochondria [17].

We will first focus on the attributes of molecules historically shown to accumulate in the mitochondria according to the negative membrane potential. The most important requirement is that the molecule must have a significant net cationic charge at physiological pH, or cytosolic pH to be more specific. However, it is important to note that not every cationic molecule would be expected to reach equilibrium levels dictated by the Nernst equation in a reasonable time frame. Specifically, those cationic molecules that have low permeability across the inner mitochondrial membrane would be expected to require exceedingly long times to reach equilibrium.

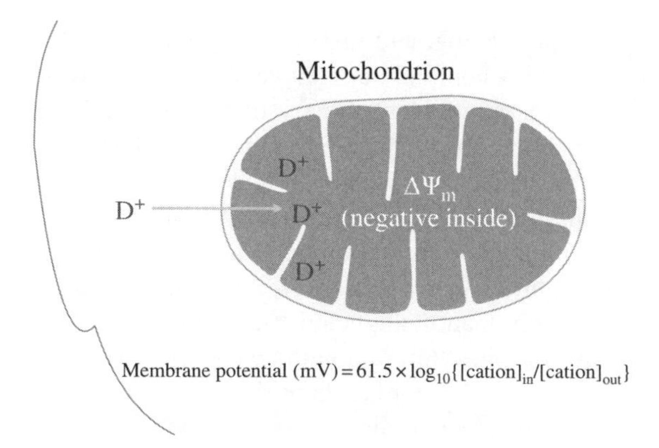

FIGURE 6.1 Diagram illustrating the accumulation of cationic drugs (D^+) in the inner mitochondrial space. The Nernst equation that relates equilibrium concentrations (inside vs. outside) of the cation with membrane potential is shown.

Alternatively, those that possess greater permeability in their cationic form will reach equilibrium much faster, potentially in therapeutically relevant timescales. Molecules that have a delocalized positive charge and/or those cations that are effectively shielded and lipophilic have been shown to possess the ability to cross membranes in the charged state. Rhodamine dyes (i.e., Rhodamine 123) and phosphonium salts (i.e., methyltriphenylphosphonium) are examples of delocalized and shielded cations, respectively, whose accumulation has been used to measure mitochondrial membrane potential.

From a drug delivery perspective, there has been an increasing desire to exploit this mitochondrial accumulation pathway to enhance the local concentrations of drugs that have targets in this space or to deliver agents with more general properties such as antioxidants. The "mitochondriotropic" properties can conceivably be designed into new drugs under development or they can be affixed to existing drugs, but the latter scenario is most common. Most examples of this mitochondrial targeting strategy start with a drug that is fairly "tolerant" to significant chemical modifications without dramatic changes to the biological activity. The idea then is to attach a molecule with known mitochondrial concentrative capacity, and the expectation is that once derivatized the overall conjugate will adopt the mitochondriotropic character similar to that of the molecule that has been attached to the drug. There are some important considerations and limitations to this approach. Importantly, the overall conjugate must have net cationic character. For example, if the parent drug had overall anionic charge at physiological pH and a single cationic mitochondrial targeting appendage was added, the approach would fail because the conjugate would be considered to be zwitterionic overall and the net charge would be zero.

Commonly employed mitochondrial targeting groups are single or repeating cyclic or acyclic guanidinium [18] or triphenylphosphonium moieties [19] (see Fig. 6.2). Such moieties have been used to target both antioxidants and anticancer agents, as will be discussed later.

There has been much therapeutic interest in targeting antioxidants to mitochondria to mitigate oxidative damage that is associated with neurodegenerative disorders, ischemia-reperfusion injury, aging, and inflammatory damage [20, 21]. Smith and colleagues have made a mitochondrial targeted version of the antioxidant Vitamin E by covalently coupling the antioxidant moiety of this molecule with a lipophilic triphenylphosphonium cation (the conjugate was named "TPPB" [11]). The authors studied the mitochondrial uptake of this molecule and its antioxidant effects in isolated mitochondria and in cultured cells. TPPB was shown to reach concentrations within the inner mitochondria that were several hundred fold greater than cytosolic concentrations. The authors showed that TPPB protected mitochondria from oxidative damage far more effectively than nontargeted Vitamin E itself. Similar approaches using covalent addition of triphenylphosphonium were used to target the antioxidants ubiquinol [22] and lipoic acid [23] to mitochondria.

In addition to antioxidants, there is also a keen interest in mitochondrial targeting of anticancer drugs [8, 24]. Some of this is based on the observation that cancer cells have a greater negative membrane potential across the inner mitochondrial membrane than do normal cells [25, 26]. Sibrian-Vazquez have synthesized porphyrin

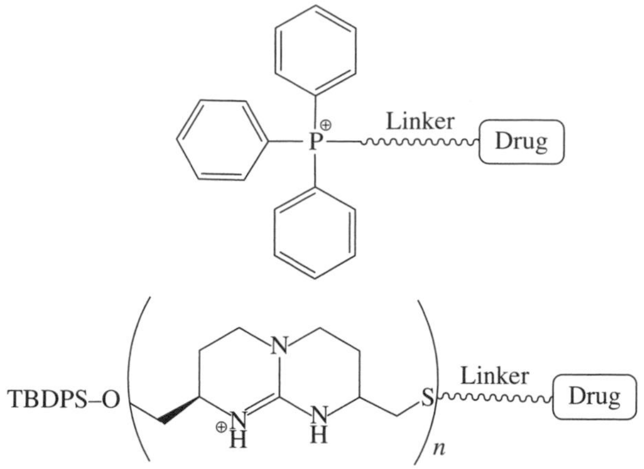

FIGURE 6.2 Structures of common moieties that are covalently attached to drugs to promote mitochondrial accumulation.

derivatives containing guanidine or a biguanidine in an effort to target these agents to the mitochondria as part of an effort toward the development of photodynamic therapy photosensitizers with increased biological efficacy in treating cancer [18]. The authors used fluorescence microscopy to show a significant degree of co-localization of the conjugates with fluorescent mitochondrial vital stains. Lim and colleagues have also exploited the more negative membrane potential associated with tumor cells with their work on the synthesis of substituted rosamines with substituted delocalized aromatic amines [27]. The authors were able to show cytotoxicity and exclusive mitochondrial accumulation with selected derivatives. There has been additional work on targeting ceramides to mitochondria through the addition of pyridinium rings [28]. These agents have been shown to accumulate in mitochondria and have cytotoxic effects that rely on the negative mitochondrial membrane potential [29].

The molecular chaperone Hsp90 has also emerged as an attractive target for anticancer agents. The Hsp90 homolog TRAP1 resides in the mitochondria and plays an important role in regulating mitochondrial integrity, protecting against oxidative stress and inhibiting cell death [30]. Several drugs targeting mitochondrial Hsp90 and TRAP1 have shown good selectivity toward cancer cells, and approaches for targeting these drugs to the mitochondria have received considerable attention [19]. Gamitrinibs are the first mitochondria-targeted small molecules that inhibit TRAP1 and Hsp90 inside the mitochondria [31, 32]. The prototype Hsp90 inhibitor to these drugs is geldanamycin, and the covalently linked mitochondrial targeting motif consisted of cyclic gaunidinium or triphenylphosphonium groups coupled to the 17-position of the geldanamycin backbone through stable linkers. Mechanistically, Gamitrinib triggered acute mitochondrial dysfunction in cultured cells with loss of organelle inner membrane potential and release of cytochrome-c in the cytosol.

It is important to note that despite the initial success of this approach with Hsp90 inhibitors described earlier, it may be unrealistic to assume that a similar approach could be easily applied to other therapeutics. Geldanamycin is fairly unique in that substitutions off of its 17-position, regardless of their size, do not dramatically effect binding affinity with Hsp90 [33]. Few other drugs display such tolerance to synthetic modification.

6.5.2 Lysosomal Accumulation of Drugs

Lysosomotropism or being lysosomotropic can have varied meanings. Almost all membrane permeable drugs would be expected to have some interactions with lysosomes, but only a subset of these drugs will be the focus of this discussion. I will not cover drugs that merely have the capacity to reach lysosomes, even if they are capable of causing profound effects on the organelle structure and function. Examples of drugs in this category include molecules like inhibitors of the lysosomal vacuolar proton ATPase (i.e., concanamycin A and bafilomycin) or molecules that are thought to intercalate with lysosomal membranes and modify their structure (i.e., steroids such as progesterone). This discussion will also not concentrate on macromolecules or small-molecular-weight polar compounds that are considered to have limited membrane permeability that have been shown to concentrate in lysosomes through fluid phase or receptor-mediated endocytosis. This section will only focus on those relatively small-molecular-weight weakly basic drugs that are significantly concentrated in lysosomes by a pH partition–type mechanism. I, therefore, limit my use of the term "lysosomotropic" to this class of molecules.

The mechanism for the lysosomal sequestration of drugs is referred to as ion trapping or pH partitioning. I use the terms interchangeably but prefer the term "pH partitioning" because ion trapping connotes some degree of irreversibility, which is not the case. The driving force for lysosomal sequestration is provided by the pH gradient that exists across the limiting lysosomal lipid bilayer (see Fig. 6.3). The pH inside lysosomes is relatively low (~pH 4.5) and is maintained by the vacuolar proton ATPase [34]. The low pH is required for the optimal activity of resident lysosomal hydrolases.

To understand the process of ion trapping let us consider an example in which we have a cell that has an extracellular and cytosolic pH of 7.4 and the pH within the lysosomes is 4.4. Let's assume that this cell is exposed to a weakly basic drug having a pK_a of 7.4 in its extracellular fluid, and that this drug is considered to be completely membrane impermeable when it is ionized and freely membrane permeable when unionized. When initially outside the cell, this drug will exist 50% in its membrane impermeable ionized form and 50% in its membrane-permeable unionized form. Initially the concentration of the drug in the cell cytosol will be zero, and the extracellular unionized drug will diffuse into the cytosol down its concentration gradient. The drug that has diffused into the cell cytosol will also exist 50/50 (unionized/ionized) because of the relationship between the pK_a of the drug and the pH of the cytosol. The unionized drug in the cell cytosol is free to diffuse across the limiting lipid bilayers of all cellular organelles. However, when this weakly basic drug diffuses into the lumen of acidic organelles such as lysosomes, the drug will now exist predominantly (99.9%) in the membrane impermeable ionized form. Only a

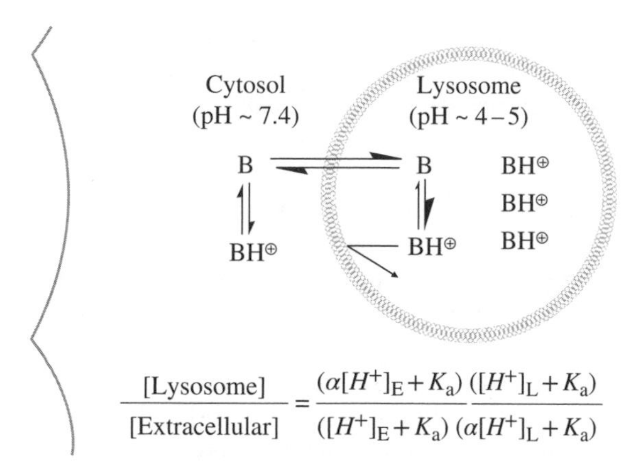

FIGURE 6.3 Diagram illustrating the pH partitioning-based mechanism for accumulation of weakly basic drugs (B) in the acidic lysosomes. The equation at the bottom of the figure represents the lysosome to extracellular space steady-state concentration ratio for a base. The dissociation constant for the conjugate acid of the weak base is denoted as K_a and $[H^+]$ is the proton concentration (subscript E represents extracellular and L represents lysosomal). The ratio of permeabilities of the ionized base to that of the unionized base in the lysosomal lipid bilayer is denoted by the α term.

very small percentage (0.01%) of the drug will exist in the membrane permeable unionized form that is in equilibrium with the unionized drug that is in the cell cytosol. The unionized drug in the cell cytosol, therefore, will initially be in great excess compared to the scant levels present in the lysosomal lumen. This chemical potential will drive diffusion of the unionized species from the cytosol to the lysosomal lumen, and in the lysosomal lumen the drug will exist predominantly in the ionized membrane impermeable form, which is unable to establish equilibrium with the ionized drug in the cytosol. These iterative steps of diffusion and ionization will continue until the concentration of the unionized species in the lysosomes is exactly the same as the concentration of the unionized species in the cytosol and extracellular space. Under this condition of steady state the total concentration of base (ionized and unionized) will be substantially greater than the total concentration of base in the cytosol. A simplified diagrammatic illustration of this scenario is shown in Figure 6.3.

It is important to note that this mechanism for drug accumulation will apply to any biological compartment for which a pH gradient exists. For example, in human cells the *trans*-Golgi apparatus and endosomes have luminal pH values that are below the pH of the cytosol. Accordingly, these organelles, in addition to lysosomes, will accumulate weakly basic drugs. However, the degree of accumulation is directly related to the magnitude of the pH differential. Based on this notion and the fact that lysosomes are, by far, the most acid intracellular compartment it has been common, for simplicity sake, to only refer to lysosomes when discussing pH partitioning. Nevertheless, it is important to realize that this is not factually correct.

An elegant theoretical commentary on lysosomotropism was published by Christian de Duve and colleagues in the 1970s [35], which was foundational to this field of research. De Duve was awarded the Nobel Prize for his work pertaining to the identification and characterization of lysosomes. In their commentary, an equation relating the steady-state concentration ratio of a weakly basic drug inside lysosomes compared with the extracellular space was derived. This equation is shown in Figure 6.3.

According to this equation, the maximum possible concentration ratio for a weakly basic substance in lysosomes (relative to the extracellular space) is equal to the ratio of the hydrogen ion concentration in the lysosomes relative to the fluid surrounding the cells. For example, if the extracellular pH was 7.4 and the pH of lysosomes was 4.4, the maximal lysosomal concentration ratio would be 1000. There are two drug-related terms that can theoretically influence the steady-state accumulation of weakly basic drugs in lysosomes. The first is the weak base pK_a and the second variable is termed alpha (α), which represents the ratio of permeabilities of the weakly basic drug across the lysosomal lipid bilayer in its ionized and unionized forms.

The pK_a of the weakly basic drug is predicted to impact the steady-state lysosomal accumulation ratio. Molecules with pK_a values of 6 or below will not be considered to be lysosomotropic because they will not be sufficiently ionized in lysosomes to drive the accumulation. Low pK_a molecules can never reach the maximal steady-state accumulation ratio of 1000. Molecules with pK_a values of 8 or greater can theoretically achieve thousand-fold higher levels in lysosomes relative to the extracellular space. However, molecules with a pK_a high above 8 will start to become less lysosomotropic in a therapeutically relevant timescale because they are predicted to take exceedingly long times to reach steady state. Specifically, the permeation rate was predicted to decrease by a factor of 10 for each 1 unit increase in drug pK_a. Interestingly, de Duve's theoretical calculations predict that the time to reach one half of steady state for molecules with a pK_a of 12 will be over one and a half years. In contrast, bases with pK_a values near 7 are predicted to take 10 minutes to reach one half steady-state levels. Using a series of weakly basic structural isomers that varied in pK_a from 4 to 9, we experimentally investigated the influence of pK_a on lysosomal accumulation [36]. Consistent with theoretical predictions, we found that molecules with pK_a values below 6 were not appreciably lysosomotropic. The molecule with a pK_a value of 9 had the greatest degree of lysosomal accumulation. Using this relationship between pK_a and lysosomal accumulation we have performed a proof of concept that manipulation of a drug's pK_a would allow for the modulation of lysosomal entrapment.

The term alpha (α) refers to the ratio of the permeability coefficients for the ionized base divided by that of the unionized base [35]. This term can theoretically vary from zero to one. Molecules with an alpha of zero will be the most lysosomotropic and will reach a theoretical maximal steady-state ratio of accumulation dictated by the pH differential. However, as the alpha parameter increases in magnitude the maximal steady-state accumulation ratio of a drug in the lysosomes significantly decreases. For example, a drug with an alpha value equal to one will have equal membrane permeability regardless of ionization. Under these circumstances, extensive protonation of the drug in lysosomes would not result in entrapment. We have experimentally estimated alpha values for a series of drugs and model compounds and correlated this value with the lysosome-to-cytosol concentration ratio [37]. As a proxy for lysosomal membrane permeability we measured the octanol/water partition coefficients for drugs as a function of pH to obtain estimates for the Log*D* of the ionized and unionized species and took the ratio to obtain alpha. As anticipated, we found that compounds with lower alpha values accumulated to a greater extent in lysosomes relative to compounds with higher alpha values. We also examined how alpha correlated with mitochondrial versus lysosomal accumulation of weakly basic drugs. We found that molecules with low alpha values accumulated almost exclusively in lysosomes whereas molecules with high alpha values accumulated within mitochondria. Based on our evaluations, we observed that alpha appeared to correlate with the degree of charge delocalization present in a molecule. Weak bases with fixed localized positive charge tended to have very low alpha values, whereas those with delocalized charges tended to have higher alpha values near one. As previously discussed, we think that this is due to differences in how tightly the molecules bind water, which subsequently relates to the energy required to pass into a nonpolar lipid-like environment. A theoretical relationship illustrating the combined influence of pK_a and alpha on lysosomal trapping is shown in Figure 6.4.

The relative lipophillicity (i.e., Log *P*) does not influence the predicted steady-state accumulation ratios for weakly basic drugs. However, the relative lipophillicity will most definitely influence the time to reach steady state and the propensity to exhibit lysosomal trapping in a therapeutically relevant timescale, as will be discussed later.

Having reviewed some of the background and principles behind lysosomal trapping I will now summarize some of the work that has been done where specific attempts have been made to either increase or decrease lysosomal trapping of drugs for the optimization of activity and/or pharmacokinetics.

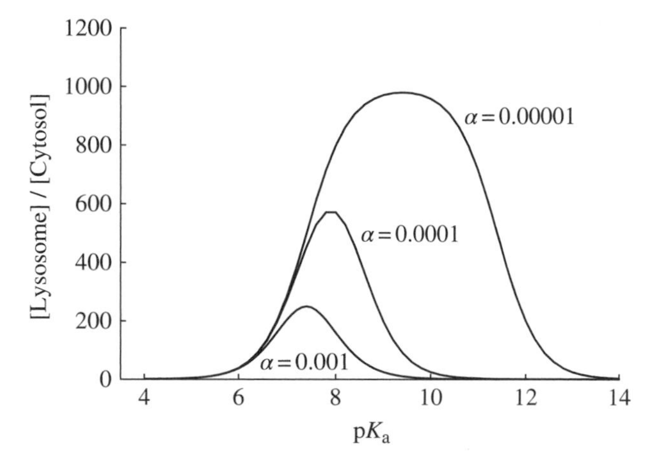

FIGURE 6.4 Theoretical relationship between weak base pK_a and the alpha permeability parameter. The relationships were derived using the equation shown in Figure 6.3, with indicated values for alpha.

6.5.2.1 Influence of Lysosomotropism on Drug Activity Some of the earliest work to exploit lysosomal trapping came from scientists at Merck who synthesized a series of varying chain length alkyl amines that were designed to act as detergents in the lysosomes of cells for potential use as anticancer agents [38, 39]. These detergents were designed to specifically dissolve the lysosomal lipid bilayer, which would in turn release lysosomal enzymes into the cell cytosol, thereby killing the cell. These molecules were specifically designed to have pK_a values that varied between 5 and 7, so that they would only become amphiphilic detergents when they were protonated, following accumulation in the acidic lysosomes. The compounds were found to be toxic to a variety of cells containing lysosomes but not to red blood cells that are devoid of lysosomes. It was later shown that toxicity of the lysosomotropic detergents correlated with the growth rate of cells, and such agents were preferentially toxic to rapidly growing cells [40].

There has also been a great interest in the exploitation of lysosomotropism to decrease the escape of various enveloped virus, toxins, and other membrane impermeable cargo from lysosomes [41–45]. Work suggests that lysosomotropic amines can suppress major histocompatibility complex (MHC) class II antigen presentation [46] and decrease iron release from lysosomes [47]. Groups have also used the lysosomotropic agents to partially reverse anticancer drug resistance through a mechanism speculated to encompass impairment of lysosomal vesicle-mediated exocytosis of the anticancer agents [48–51].

It is not exactly clear how lysosomotropic amines are effective in decreasing lysosomal egress. One explanation is that the amines, through a buffering effect, transiently increase the pH of lysosomes. This pH increase could decrease the activity of lysosomal hydrolases involved in lysosomal escape pathways. Alternatively, it is thought that the lysosomotropic amines can impact the lysosomal membrane such that membrane fusion and fission events required for the vesicle-mediated trafficking becomes impaired. Consistent with the later notion, the weakly basic drug primaquine has been shown to specifically impair budding events at the Golgi [52].

Ceramidase inhibitors have been synthesized and evaluated as anticancer agents in a variety of cell lines [53, 54]. In this work it was clear that lysosomal targeting was considered in the design of the inhibitors. Agents with lysosomotropic properties had much greater cellular accumulation in cultured cells and showed increased potency relative to neutral, nontargeted agents. Inhibitors with basic groups were shown to increase lysosomal pH and promote the release of lysosomal enzymes into the cytosol in prostate cancer cells [55].

Lysosomotropic properties have also been exploited and optimized in the development of drugs useful in the treatment of osteoporosis [56]. The lysosomal protease cathepsin K is involved in the process of bone reabsorption, and efforts are underway to develop potent and specific inhibitors against this enzyme. In an elegant study, Falgueyret and colleagues have synthesized and evaluated both lysosomotropic and less lysosomotropic inhibitors of cathepsin K [56, 57]. Interestingly, molecules with lysosomotropic properties had far greater activity in whole cell enzyme occupancy assays compared to activity evaluations conducted with isolated enzymes. This was pronounced for lysosomotropic inhibitors. Fortuitously, these compounds were inherently fluorescent, and their cellular distribution could be monitored using two-photon confocal fluorescence microscopy. Microscopic evaluations demonstrated that compounds localized in lysosomes as was confirmed with colocalization studies with the lysosomal vital stain LysoTracker Red (LTR). The lysosomotropic inhibitors of cathepsin K had much improved pharmacokinetic properties in rats. The lysosomotropic inhibitors had high tissue to plasma concentration ratios in organs rich in lysosomes including the liver, lung, and kidneys. In addition, the half-life of lysosomotropic inhibitors was sevenfold longer than nonlysosomotropic inhibitors. One noteworthy obstacle with this approach is that basic cathepsin K inhibitors were also concentrated in the lysosomes of cells other than osteoclasts, and lysosomotropism tended to pronounce their potential for off-target interactions with other members of the cathepsin family (i.e., cathepsin B, S, and L) [58].

The osteoclast resorbs bone by forming a ring of close contact with the bone surface within the confines of which it secretes protons and lysosomal enzymes, thus forming an extracellular digestive hemivacuole. In this hemivacuole, the mineral component of bone is dissolved by protons, and the organic component is digested by cathepsin K. Interestingly, lysosomotropism was shown to occur in the hemivacuole surrounding the osteoclast [59]. This space was shown to be capable of trapping lysosomotropic drugs, thereby increasing their potency and prolonging their duration of action relative to non-lysosomotropic inhibitors that had equal binding affinity to cathepsin K.

While the most obvious exploitation of lysosomotropism is centered on drug targets contained in lysosomes, this lysosomotropic behavior can be exploited for targets that are outside lysosomes. We have described an approach for increasing the selectivity of anticancer agents to tumor cells that involved lysosomal trapping of drugs [60–63]. We referred to this approach as "intracellular distribution-based targeting." This approach represents a new concept in delivery-based approach for increasing the selective toxicity toward cancer cells. Traditional drug delivery approaches rely on the premise that more drug must be delivered to the cancer cells relative to the normal cells. This approach exploits differences in intracellular distribution behavior that exists for some drugs in normal cells versus cancer cells. In order for a drug to exert its cytotoxic effects it must not only enter the cell, but it must also sufficiently concentrate in the specific subcellular compartment that houses the target. For most anticancer drugs their targets reside in the nucleus or cytosol, which are contiguous compartments for small-molecular-weight drugs due to the large pore and abundance of the nuclear pore complexes.

We and others have shown that many, but not all, anticancer cell lines have defective acidification of lysosomes for reasons that have not been fully elucidated [64–67]. Based on this notion, we envisioned that weakly basic anticancer agents that are exquisite substrates for lysosomal trapping would achieve very low concentrations in the cell cytosol and nucleus of normal cells, thus having reduced capacity to interact with targets and exert cytotoxic effects. Alternatively, when present in cancer cells with elevated lysosomal pH, the lysosomotropic drug would be relatively less concentrated in lysosomes and would exist to a greater degree in the cell cytosol and nucleus. Accordingly we envisioned that a lysosomotropic anticancer agent would possess increased selectivity against tumor cells compared to nonlysosomotropic anticancer agents (see Fig. 6.5 for a diagrammatic illustration of this concept).

Our initial evaluations of this targeting concept were conducted in cultured cells that had variable lysosomal pH. The model cancer drugs were inhibitors of the heat shock protein 90 (Hsp90) with or without lysosomotropic properties. As mentioned earlier, the prototypical Hsp90 inhibitor geldanamycin is amenable to diverse chemical derivatization at the 17-position with little effect on binding to Hsp90 [33]. We, therefore, selected inhibitors that had variations at the 17-position that made them either weakly basic (i.e., lysosomotropic) or neutral (non-lysosomotropic). These studies revealed that weakly basic inhibitors had additional selective toxicity toward cells with elevated lysosomal pH that was not the case for non-lysosomotropic inhibitors [60].

We further investigated this strategy in non-tumor-bearing mice with and without elevated lysosomal pH, again using lysosomotropic and nonlysosomotropic Hsp90 inhibitors [63]. We found that mice with elevated lysosomal pH were far more susceptible to the toxic effects of lysosomotropic Hsp90 inhibitors. As a control, we have shown that non-lysosomotropic inhibitors had toxicities that did not vary with alterations in lysosomal pH. More recently, we have synthesized a series of Hsp90 derivatives with variable pK_a in the 17-position, and we have shown that the

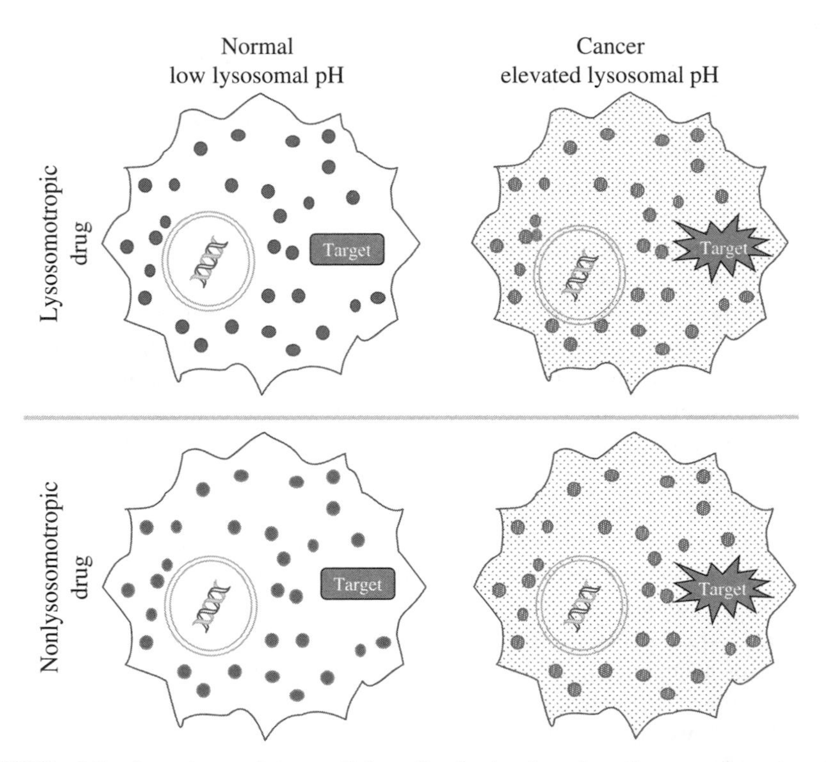

FIGURE 6.5 Overview of intracellular distribution-based anticancer drug targeting platform.

degree of selectivity toward tumor cells increased as the pK_a was changed to values that were most conducive to lysosomal trapping [61].

Kang and colleagues have similarly exploited this difference in lysosomal pH between normal cells and cancer cells to aid in the development of fluorescent probes that could selectively identify cancer cells [68]. The authors have developed a fluorophore (3,6-bis-(1-methyl-4-vinylpyridinium)carbazole diiodide (BMVC)) whose quantum yield increases by a hundred fold when bound to DNA. The authors show that in normal cells these lysosomotropic fluorophores are not particularly fluorescent because they are trapped in lysosomes and unavailable to bind to DNA. However, in cancer cells with elevated lysosomal pH the fluorophores escape the lysosomes and bind with the mitochondrial and/or nuclear DNA and become highly fluorescent.

6.5.2.2 Influence of Lysosomotropism on Pharmacokinetics There is no question that the pharmacokinetics of drugs can be highly influenced by lysosomal trapping. Specifically the volume of distribution and, therefore, half-life of a drug will be expected to be influenced by lysosomal trapping (see previous discussion).

It is important to realize that the human body comprises many single cells, and understanding drug accumulation at this level provides foundational insight into whole animal drug distribution. To illustrate the potential importance of lysosomal trapping on pharmacokinetic distribution I will utilize a hypothetical example that compares the apparent volume of distribution of two drugs that are identical in all aspects, except that one is lysosomotropic and the other is not (see Fig. 6.6). Each drug is to be administered by continuous IV infusion such that the steady-state plasma levels are equivalent, which is consistent with both drugs having identical clearance values. Assume that both drugs are freely permeable across all biological membranes, and they do not bind with endogenous molecules or membranes (i.e., free fraction in the blood and tissues are both equal to one). In this example we will assume that the steady-state plasma concentration is 1 mg/l. Under these assumptions the volume of distribution of the nonlysosomotropic drug would be roughly equal to the volume of the total body water, which is approximately 42 l for a 70 kg male. For simplicity, I will refer to this as 50 l.

In the case of the lysosomotropic drug, the apparent volume of distribution at steady state should be larger because the drug is concentrated within lysosomes. To estimate how much larger the volume of distribution will be, we need to estimate the volume of the lysosomal compartment and the relative concentration of drug inside the lysosomes. Our body is estimated to contain 15 trillion cells. The average volume of a human liver cell is 3.4×10^{-9} cm^3. This equates to approximately 51 l of cells in the human body. Assuming that lysosomes comprise 1% of the total cell volume, the volume of lysosomes in the human body can be estimated to be 0.51 l in total, which I will round to 0.5 l. If the lysosomal to extracellular pH differential is equal to 3 units, then the total drug concentration in the lysosomes will be 1000 times greater than extracellular concentrations. If the free drug is assumed to be evenly distributed throughout the entire apparent volume of distribution, the physiological volume estimate of the lysosomes must be increased by thousand-fold. This gives

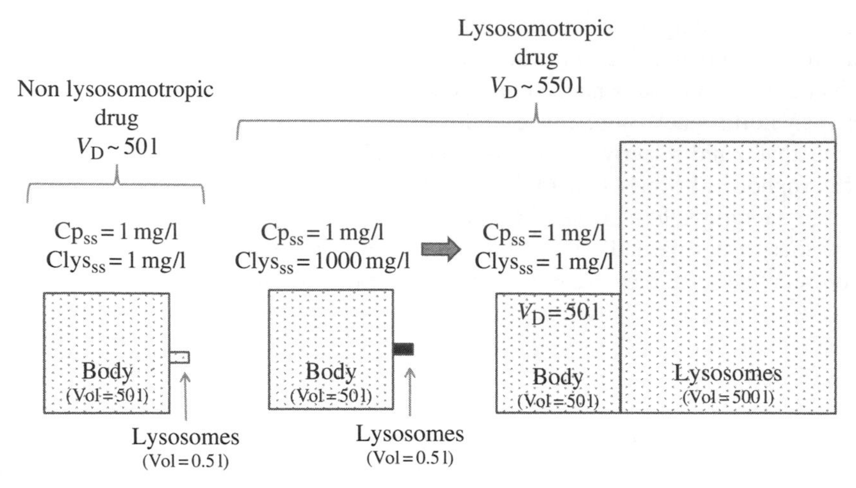

FIGURE 6.6 Hypothetical example illustrating the potential influence of lysosomotropism on the apparent volume of distribution of a drug. See text for details.

a new lysosomal volume estimate of 500 l. One can now see that under these assumptions, lysosomal partitioning of drugs can account for apparent volumes of distribution that are approximately 10-fold greater in magnitude in comparison with non-lysosomotropic counterparts. This, as previously discussed, can translate into elimination half-lives that are also 10-fold greater for the lysosomotropic drug, assuming clearance is the same. These predictions are in fairly close agreement with similar predictions made by Hung and colleagues [69]. In their work they predicted that the unbound concentrations of drugs in tissues would be over seven times greater than the perfusate concentrations for lysosomotropic drugs (i.e., propranolol), and the ratio would be one to one for non-lysosomotropic drugs such as antipyrine.

It is important to emphasize that when measured experimentally, lysosomal trapping occurs at levels much greater than basic pH partitioning theory would predict [4]. Accordingly, it is possible that lysosomal trapping could have an even more significant impact on the volume of distribution and half-life of drugs than theoretically predicted. MacIntyre and Cutler [70], in *in vitro* and *in vivo* studies in rats, have shown that hepatocytes exposed to chloroquine at therapeutic levels had tissue-to-plasma accumulation ratios that were nearly 800. The authors have shown that this accumulation was consistent with lysosomal trapping. They showed that the accumulation ratio remained constant at various therapeutic concentrations of CQ but that it significantly decreased when concentrations of CQ increased, likely owing to the fact that high concentrations raise lysosomal pH. Similarly, Cramb has shown that the apparent concentration of the lysosomotropic drug propranolol in a variety of cultured cells is over 1000 times that in the extracellular media when incubated at therapeutically relevant concentrations [71]. It was speculated that the lysosomal trapping may prolong the activity of this drug, following abrupt cessation of therapy. Ishizaki and colleagues have measured the accumulation of imipramine in isolated lysosomes and have shown that accumulation was approximately 140 times greater than pH partitioning theory predicted [72]. The authors speculate that binding of the drug to lysosomal molecules and/or aggregation could contribute to this. We have also quantified quinacrine accumulation in lysosomes of cultured cells and showed that it accumulated much higher than theory would predict [4].

As previously discussed, having a large volume of distribution can prolong half-life, and this could be advantageous from the point of view of having a long dosing interval. However, it is possible that extensive lysosomal trapping could present a barrier to the development and evaluation of a drug. For example, in collaboration with scientists at GSK, we had shown that lysosomal trapping of melanocortin receptor agonists in the liver following oral dosing was responsible for virtually undetectable levels in the blood [73]. It was shown that less lysosomotropic versions had improved detectability in plasma.

Considering how strongly lysosomal trapping can impact the pharmacokinetics of drugs, it is becoming increasingly apparent that drug interactions involving lysosomes need to be carefully considered as an important source of pharmacokinetic variability. Daniel and coworkers have investigated a new class of pharmacokinetic drug–drug interactions involving lysosomes [74, 75]. Most of their studies were done using lysosomotropic antidepressant/antipsychotic drugs that were co-administered. Their investigations were focused on potential short-term interactions that might occur immediately after the administration of the drugs. The authors speculated that one lysosomotropic drug, if presented at a high enough concentration, could temporarily raise lysosomal pH and, therefore, decrease the lysosomal accumulation of a second drug. The authors provided evidence to support this interaction both *in vitro* and *in vivo*. The authors speculate that this interaction could cause a redistribution of lysosomotropic drugs from lysosome-rich tissues (i.e., lung, liver, kidney), to organs whose lysosomal abundance is less (i.e., heart, skeletal muscle, and brain).

Some high throughput-type assays that are designed to determine whether or not a molecule is lysosomotropic rely on this type of drug–drug interaction [76]. In this assay, the cells are cultured in multiwell format and exposed to the lysosomotropic fluorophore LTR. In untreated cells, the lysosomal pH will be normal and low, and the amount of LTR accumulated in the cells will be at its greatest. However, cells exposed to high concentrations of lysosomotropic drugs might have elevated lysosomal pH and, therefore, will accumulate less LTR. Molecules that can decrease LTR accumulation are classified as being lysosomotropic. This assay has also been developed and evaluated using hepatocytes grown in culture [77].

We have explored potential drug–drug interactions involving lysosomes from a different perspective, which we believe has more widespread therapeutical relevance. We are focused on the more long-term effects of drugs on lysosome structure and function that would occur at therapeutically relevant concentrations. Under these conditions most drugs would not be expected to significantly raise lysosomal pH. Instead, we question whether or not one drug could impact the volume of the lysosomal compartment. Toward this end, we have shown that numerous weakly basic drugs from a variety of therapeutic classes can cause an appreciable increase (two- to threefold) in the apparent volume of the lysosomes in a concentration- and time-dependent manner [78]. In these conditions the presence of the first drug can dramatically increase the cellular accumulation of a secondarily administered lysosomotropic drug. We have investigated the mechanism for this effect, and we have shown that this is due to changes in the vesicle-mediated trafficking associated with lysosomes [79]. Specifically, the amine-containing drugs that perpetrate the interaction do so by increasing the flux of membrane and volume into the lysosomes through the induction of autophagy. In addition, the perpetrating amines have also been shown to decrease the efficiency of vesicle-mediated flux out of the lysosomes. We have also examined the SAR of drugs that perpetrate this effect and found that both the propensity to be lysosomotropic and having a capacity to intercalate within lipid bilayer membranes are most important [80]. So far this drug interaction pathway has not been extensively studied in animals or humans. However, we believe that future studies pertaining to this drug–drug interaction pathway involving lysosomes could shed light on a novel basis for variability in drug pharmacokinetics.

6.5.3 Nuclear Accumulation of Drugs

For small-molecular-weight drugs, the focus of this chapter, the nuclear membrane does not provide a true physical barrier that can limit entry. The nuclear membrane contains numerous, relatively large diameter, nuclear pore complexes that allow for free diffusion of small-molecular-weight drugs and ions, even those with high polarity. Accordingly, for small-molecular-weight drugs the nuclear space can be considered to be continuous with the cell cytosol.

The nucleus represents an important site of accumulation for many drugs. The mechanism for nuclear accumulation is fundamentally different than the type of accumulation that occurs in the lysosomes and mitochondria. Generally speaking, drugs that accumulate in the nucleus do so because of their high affinity for DNA (see Fig. 6.7). Once complexed to a macromolecule resident of the nucleus the complex is too large to freely diffuse out through the nuclear pore complexes, resulting in the accumulation of the drug.

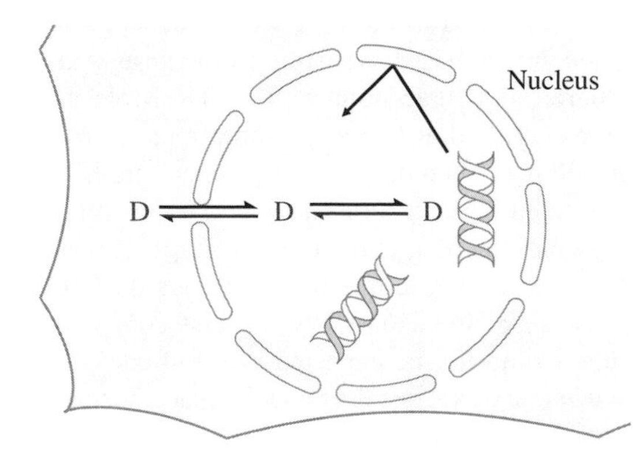

FIGURE 6.7 Diagram illustrating the nuclear accumulation of drugs in the nucleus. Small-molecular-weight drugs can freely diffuse into the nucleus through nuclear pore complexes. Once bound to resident macromolecules (i.e., DNA), the complexed drug is too large to diffuse out.

It is likely that similar mechanisms for organelle accumulation occur for drugs capable of binding to other macromolecules in organelles other than the nucleus. However, this phenomenon is most commonly associated with the nucleus because of the high concentration of DNA.

Drug binding to DNA has shown to result in very significant nuclear accumulation of drugs. However, it is important to realize that this type of organelle accumulation would not likely be "exploitable" from a drug delivery perspective. This is because the free, unbound, drug in the nucleus would be expected to be the same in the cytosol and in the nucleus.

The nuclear accumulation of fluorescent DNA intercalating drugs, in particular antitumor agents, has been previously shown. For example adriamycin and daunorubicin are highly fluorescent and have been shown to localize in the nucleus of cultured cells [81]. The nuclear accumulation of other fluorescent anticancer agents has been investigated including mitoxantrone [82]. Interestingly, the nuclear accumulation of adriamycin has been shown to decrease after the acquisition of the multidrug resistance phenotype [83].

Lansiaux and colleagues have conducted some elegant intracellular distribution studies on a number of furamidine derivatives that have been shown to have antimicrobial and antitumor properties [12, 84]. The main cellular target of furamidine and related diphenylfuran analogues is considered to be DNA. Fortuitously, these drugs have strong fluorescence, which enabled the investigators the opportunity to observe how structure influenced intracellular distribution and uptake. The authors independently assessed DNA-binding affinity of the derivatives through assessment of their effects on the DNA melting temperature. Interestingly, dicationic molecules such as DB75 exhibited high DNA-binding affinity (ΔT_m 24.7°C) and had predominantly nuclear accumulation, whereas monocationic derivatives such as DB607 had less affinity with DNA (ΔT_m 5.2°C) and were shown to preferentially accumulate in the cell cytoplasm.

In general, molecules with nuclear accumulation according to this mechanism are relatively planar (allowing intercolation in the DNA grooves) and have one or multiple cationic charges, which facilitate electrostatic interactions with the DNA.

6.6 SUMMARY AND FUTURE DIRECTIONS

For over 100 years it has been known that small-molecular-weight fluorescent compounds have the propensity to accumulate within distinct regions of cells. These so-called organelle vital stains have been used extensively by cell biologists in the characterization of organelle structure and function in both healthy and diseased cells. It is only fairly recently that the intracellular distribution of drugs has received much attention, and in most of these instances this has been limited to the few drugs that are intrinsically fluorescent.

As discussed in this chapter, there have been a number of elegant experimental evaluations of the structure–localization relationship for organelle-specific accumulation. In addition, researchers have begun to group molecules that have been shown to have organelle-specific accumulation and identified common features that allow for prediction of cellular distribution. Mathematical models have been developed that incorporate intracellular distribution phenomenon and have been used to model the dynamics of drug accumulation and trafficking within and through a cell [85, 86]. With these advances, the pharmaceutical industry is gaining the necessary insight and tools to facilitate investigations pertaining to the intracellular distribution of drugs under development.

It is increasingly realized that the intracellular distribution profile of a drug can have a profound impact on the activity, pharmacokinetics, and toxicity profile of a drug. From my experience in consulting with scientists involved in drug discovery, both in industry and academia, there is often a significant discrepancy between the *in vitro* activity of a drug using isolated targets and the activity observed in cell-based assays. I believe that in many instances such deviations can be rationally attributed to the intracellular distribution of the drug. In addition the intracellular distribution behavior can have a profound impact on pharmacokinetic distribution properties. A more comprehensive understanding of this may allow for the design of drugs with optimized half-life and potentially novel mechanisms for pharmacokinetic variability and drug–drug interaction pathways. Finally, although not the focus of this chapter, there is a growing interest and potential concern about the effects that drugs can have on organelle structure and function when they are concentrated in these environments. *In vitro* toxicity screens for off-target interactions are often conducted at some range in concentration that is thought to be reflective of therapeutic levels. *In vivo* microdialysis studies have suggested that many drugs reach levels in the body in the nanomolar to low micromolar range. However, we now know that drugs that are substrates for accumulation in organelles such as lysosomes can potentially reach concentrations inside the lumen of the organelle that are a thousand-fold or higher in concentration than in other compartments. Under such high concentrations it is possible that unanticipated effects on organelle structure and function could occur. These could contribute to unwanted toxicity, but interesting to consider, could also be responsible for some of the desired therapeutic effects on some drugs. Future research will be needed to answer these questions.

REFERENCES

1. Meanwell NA (2011) Improving drug candidates by design: a focus on physicochemical properties as a means of improving compound disposition and safety. *Chemical Research in Toxicology* **24**: 1420–1456.
2. Kaufmann AM, Krise JP (2007) Lysosomal sequestration of amine-containing drugs: analysis and therapeutic implications. *Journal of Pharmaceutical Sciences* **96**: 729–746.
3. Reijngoud DJ, Tager JM (1977) The permeability properties of the lysosomal membrane. *Biochimica et Biophysica Acta* **472**: 419–449.
4. Duvvuri M, Krise JP (2005) A novel assay reveals that weakly basic model compounds concentrate in lysosomes to an extent greater than pH-partitioning theory would predict. *Molecular Pharmaceutics* **2**: 440–448.

5. Solheim AE, Seglen PO (1983) Structural and physical changes in lysosomes from isolated rat hepatocytes treated with methylamine. *Biochimica et Biophysica Acta* 763: 284–291.
6. Lloyd JB (2000) Lysosome membrane permeability: implications for drug delivery. *Advanced Drug Delivery Reviews* 41: 189–200.
7. Morjani H, Millot JM, Belhoussine R, Sebille S, Manfait M (1997) Anthracycline subcellular distribution in human leukemic cells by microspectrofluorometry: factors contributing to drug-induced cell death and reversal of multidrug resistance. *Leukemia: Official Journal of the Leukemia Society of America, Leukemia Research Fund, UK* 11: 1170–1179.
8. Rosania GR (2003) Supertargeted chemistry: identifying relationships between molecular structures and their sub-cellular distribution. *Current Topics in Medicinal Chemistry* **3**: 659–685.
9. Chu X, Korzekwa K, Elsby R, Fenner K, Galetin A, et al. (2013) Intracellular drug concentrations and transporters: measurement, modeling, and implications for the liver. *Clinical Pharmacology and Therapeutics* **94**: 126–141.
10. Chen Y, Walsh RJ, Arriaga EA (2005) Selective determination of the doxorubicin content of individual acidic organelles in impure subcellular fractions. *Analytical Chemistry* **77**: 2281–2287.
11. Smith RA, Porteous CM, Coulter CV, Murphy MP (1999) Selective targeting of an antioxidant to mitochondria. *European Journal of Biochemistry/FEBS* **263**: 709–716.
12. Lansiaux A, Tanious F, Mishal Z, Dassonneville L, Kumar A, et al. (2002) Distribution of furamidine analogues in tumor cells: targeting of the nucleus or mitochondria depending on the amidine substitution. *Cancer Research* **62**: 7219–7229.
13. Hu Y, Moraes CT, Savaraj N, Priebe W, Lampidis TJ (2000) Rho(0) tumor cells: a model for studying whether mitochondria are targets for rhodamine 123, doxorubicin, and other drugs. *Biochemical Pharmacology* **60**: 1897–1905.
14. Szewczyk A, Wojtczak L (2002) Mitochondria as a pharmacological target. *Pharmacological Reviews* **54**: 101–127.
15. Horobin RW, Trapp S, Weissig V (2007) Mitochondriotropics: a review of their mode of action, and their applications for drug and DNA delivery to mammalian mitochondria. *Journal of Controlled Release* **121**: 125–136.
16. Murphy MP (1997) Selective targeting of bioactive compounds to mitochondria. *Trends in Biotechnology* **15**: 326–330.
17. Rottenberg H (1979) The measurement of membrane potential and deltapH in cells, organelles, and vesicles. *Methods in Enzymology* **55**: 547–569.
18. Sibrian-Vazquez M, Nesterova IV, Jensen TJ, Vicente MG (2008) Mitochondria targeting by guanidine- and biguanidine-porphyrin photosensitizers. *Bioconjugate Chemistry* **19**: 705–713.
19. Kang BH (2012) TRAP1 regulation of mitochondrial life or death decision in cancer cells and mitochondria-targeted TRAP1 inhibitors. *BMB Reports* **45**: 1–6.
20. Milagros Rocha M, Victor VM (2007) Targeting antioxidants to mitochondria and cardiovascular diseases: the effects of mitoquinone. *Medical Science Monitor: International Medical Journal of Experimental and Clinical Research* **13**: RA132–145.
21. Murphy MP, Smith RA (2007) Targeting antioxidants to mitochondria by conjugation to lipophilic cations. *Annual Review of Pharmacology and Toxicology* **47**: 629–656.

22. Adlam VJ, Harrison JC, Porteous CM, James AM, Smith RA, et al. (2005) Targeting an antioxidant to mitochondria decreases cardiac ischemia-reperfusion injury. *FASEB Journal: Official Publication of the Federation of American Societies for Experimental Biology* **19**: 1088–1095.
23. Brown SE, Ross MF, Sanjuan-Pla A, Manas AR, Smith RA, et al. (2007) Targeting lipoic acid to mitochondria: synthesis and characterization of a triphenylphosphonium-conjugated alpha-lipoyl derivative. *Free Radical Biology & Medicine* **42**: 1766–1780.
24. Dias N, Bailly C (2005) Drugs targeting mitochondrial functions to control tumor cell growth. *Biochemical Pharmacology* **70**: 1–12.
25. Hockenbery DM (2010) Targeting mitochondria for cancer therapy. *Environmental and Molecular Mutagenesis* **51**: 476–489.
26. Ralph SJ, Low P, Dong L, Lawen A, Neuzil J (2006) Mitocans: mitochondrial targeted anti-cancer drugs as improved therapies and related patent documents. *Recent Patents on Anti-Cancer Drug Discovery* **1**: 327–346.
27. Lim SH, Wu L, Burgess K, Lee HB (2009) New cytotoxic rosamine derivatives selectively accumulate in the mitochondria of cancer cells. *Anti-Cancer Drugs* **20**: 461–468.
28. Szulc ZM, Bielawski J, Gracz H, Gustilo M, Mayroo N, et al. (2006) Tailoring structure-function and targeting properties of ceramides by site-specific cationization. *Bioorganic & Medicinal Chemistry* **14**: 7083–7104.
29. Beckham TH, Lu P, Jones EE, Marrison T, Lewis CS, et al. (2013) LCL124, a cationic analog of ceramide, selectively induces pancreatic cancer cell death by accumulating in mitochondria. *The Journal of Pharmacology and Experimental Therapeutics* **344**: 167–178.
30. Kang BH, Altieri DC (2009) Compartmentalized cancer drug discovery targeting mitochondrial Hsp90 chaperones. *Oncogene* **28**: 3681–3688.
31. Kang BH, Siegelin MD, Plescia J, Raskett CM, Garlick DS, et al. (2010) Preclinical characterization of mitochondria-targeted small molecule hsp90 inhibitors, gamitrinibs, in advanced prostate cancer. *Clinical Cancer Research: An Official Journal of the American Association for Cancer Research* **16**: 4779–4788.
32. Kang BH, Tavecchio M, Goel HL, Hsieh CC, Garlick DS, et al. (2011) Targeted inhibition of mitochondrial Hsp90 suppresses localised and metastatic prostate cancer growth in a genetic mouse model of disease. *British Journal of Cancer* **104**: 629–634.
33. Tian ZQ, Liu Y, Zhang D, Wang Z, Dong SD, et al. (2004) Synthesis and biological activities of novel 17-aminogeldanamycin derivatives. *Bioorganic & Medicinal Chemistry* **12**: 5317–5329.
34. Forgac M (1989) Structure and function of vacuolar class of ATP-driven proton pumps. *Physiological Reviews* **69**: 765–796.
35. de Duve C, de Barsy T, Poole B, Trouet A, Tulkens P, et al. (1974) Commentary. Lysosomotropic agents. *Biochemical Pharmacology* **23**: 2495–2531.
36. Duvvuri M, Konkar S, Funk RS, Krise JM, Krise JP (2005) A chemical strategy to manipulate the intracellular localization of drugs in resistant cancer cells. *Biochemistry* **44**: 15743–15749.
37. Duvvuri M, Gong Y, Chatterji D, Krise JP (2004) Weak base permeability characteristics influence the intracellular sequestration site in the multidrug-resistant human leukemic cell line HL-60. *The Journal of Biological Chemistry* **279**: 32367–32372.

38. Firestone RA, Pisano JM, Bonney RJ (1979) Lysosomotropic agents. 1. Synthesis and cytotoxic action of lysosomotropic detergents. *Journal of Medicinal Chemistry* **22**: 1130–1133.

39. Miller DK, Griffiths E, Lenard J, Firestone RA (1983) Cell killing by lysosomotropic detergents. *The Journal of Cell Biology* **97**: 1841–1851.

40. Wilson PD, Hreniuk D, Lenard J (1991) A relationship between multidrug resistance and growth-state dependent cytotoxicity of the lysosomotropic detergent N-dodecylimidazole. *Biochemical and Biophysical Research Communications* **176**: 1377–1382.

41. Helenius A, Marsh M, White J (1982) Inhibition of Semliki forest virus penetration by lysosomotropic weak bases. *The Journal of General Virology* **58** Pt 1: 47–61.

42. Marnell MH, Stookey M, Draper RK (1982) Monensin blocks the transport of diphtheria toxin to the cell cytoplasm. *The Journal of Cell Biology* **93**: 57–62.

43. Gupta DK, Gieselmann V, Hasilik A, von Figura K (1984) Tilorone acts as a lysosomotropic agent in fibroblasts. *Hoppe-Seyler's Zeitschrift fur physiologische Chemie* **365**: 859–866.

44. Schwartz AL, Hollingdale MR (1985) Primaquine and lysosomotropic amines inhibit malaria sporozoite entry into human liver cells. *Molecular and Biochemical Parasitology* **14**: 305–311.

45. Alcami A, Carrascosa AL, Vinuela E (1989) The entry of African swine fever virus into Vero cells. *Virology* **171**: 68–75.

46. Schultz KR, Gilman AL (1997) The lysosomotropic amines, chloroquine and hydroxychloroquine: a potentially novel therapy for graft-versus-host disease. *Leukemia & Lymphoma* **24**: 201–210.

47. Komarov AM, Hall JM, Chmielinska JJ, Weglicki WB (2006) Iron uptake and release by macrophages is sensitive to propranolol. *Molecular and Cellular Biochemistry* **288**: 213–217.

48. Shiraishi N, Akiyama S, Kobayashi M, Kuwano M (1986) Lysosomotropic agents reverse multiple drug resistance in human cancer cells. *Cancer Letters* **30**: 251–259.

49. Klohs WD, Steinkampf RW (1988) The effect of lysosomotropic agents and secretory inhibitors on anthracycline retention and activity in multiple drug-resistant cells. *Molecular Pharmacology* **34**: 180–185.

50. Zamora JM, Pearce HL, Beck WT (1988) Physical-chemical properties shared by compounds that modulate multidrug resistance in human leukemic cells. *Molecular Pharmacology* **33**: 454–462.

51. Klugmann FB, Decorti G, Crivellato E, Candussio L, Mallardi F, et al. (1990) Effect of lysosomotropic and membrane active substances on adriamycin uptake and histamine release. *Anticancer Research* **10**: 1571–1577.

52. Hiebsch RR, Raub TJ, Wattenberg BW (1991) Primaquine blocks transport by inhibiting the formation of functional transport vesicles. Studies in a cell-free assay of protein transport through the Golgi apparatus. *The Journal of Biological Chemistry* **266**: 20323–20328.

53. Szulc ZM, Mayroo N, Bai A, Bielawski J, Liu X, et al. (2008) Novel analogs of D-e-MAPP and B13. Part 1: synthesis and evaluation as potential anticancer agents. *Bioorganic & Medicinal Chemistry* **16**: 1015–1031.

54. Bielawska A, Bielawski J, Szulc ZM, Mayroo N, Liu X, et al. (2008) Novel analogs of D-e-MAPP and B13. Part 2: signature effects on bioactive sphingolipids. *Bioorganic & Medicinal Chemistry* **16**: 1032–1045.
55. Holman DH, Turner LS, El-Zawahry A, Elojeimy S, Liu X, et al. (2008) Lysosomotropic acid ceramidase inhibitor induces apoptosis in prostate cancer cells. *Cancer Chemotherapy and Pharmacology* **61**: 231–242.
56. Black WC, Percival MD (2006) The consequences of lysosomotropism on the design of selective cathepsin K inhibitors. *Chembiochem: A European Journal of Chemical Biology* **7**: 1525–1535.
57. Falgueyret JP, Desmarais S, Oballa R, Black WC, Cromlish W, et al. (2005) Lysosomotropism of basic cathepsin K inhibitors contributes to increased cellular potencies against off-target cathepsins and reduced functional selectivity. *Journal of Medicinal Chemistry* **48**: 7535–7543.
58. Desmarais S, Black WC, Oballa R, Lamontagne S, Riendeau D, et al. (2008) Effect of cathepsin k inhibitor basicity on in vivo off-target activities. *Molecular Pharmacology* **73**: 147–156.
59. Fuller K, Lindstrom E, Edlund M, Henderson I, Grabowska U, et al. (2010) The resorptive apparatus of osteoclasts supports lysosomotropism and increases potency of basic versus non-basic inhibitors of cathepsin K. *Bone* **46**: 1400–1407.
60. Duvvuri M, Konkar S, Hong KH, Blagg BS, Krise JP (2006) A new approach for enhancing differential selectivity of drugs to cancer cells. *ACS Chemical Biology* **1**: 309–315.
61. Ndolo RA, Luan Y, Duan S, Forrest ML, Krise JP (2012) Lysosomotropic properties of weakly basic anticancer agents promote cancer cell selectivity in vitro. *PloS One* **7**: e49366.
62. Ndolo RA, Jacobs DT, Forrest ML, Krise JP (2010) Intracellular distribution-based anticancer drug targeting: exploiting a lysosomal acidification defect associated with cancer cells. *Molecular and Cellular Pharmacology* **2**: 131–136.
63. Ndolo RA, Forrest ML, Krise JP (2010) The role of lysosomes in limiting drug toxicity in mice. *The Journal of Pharmacology and Experimental Therapeutics* **333**: 120–128.
64. Altan N, Chen Y, Schindler M, Simon SM (1998) Defective acidification in human breast tumor cells and implications for chemotherapy. *The Journal of Experimental Medicine* **187**: 1583–1598.
65. Kokkonen N, Rivinoja A, Kauppila A, Suokas M, Kellokumpu I, et al. (2004) Defective acidification of intracellular organelles results in aberrant secretion of cathepsin D in cancer cells. *The Journal of Biological Chemistry* **279**: 39982–39988.
66. Schindler M, Grabski S, Hoff E, Simon SM (1996) Defective pH regulation of acidic compartments in human breast cancer cells (MCF-7) is normalized in adriamycin-resistant cells (MCF-7adr). *Biochemistry* **35**: 2811–2817.
67. Gong Y, Duvvuri M, Krise JP (2003) Separate roles for the Golgi apparatus and lysosomes in the sequestration of drugs in the multidrug-resistant human leukemic cell line HL-60. *The Journal of Biological Chemistry* **278**: 50234–50239.
68. Kang CC, Huang WC, Kouh CW, Wang ZF, Cho CC, et al. (2013) Chemical principles for the design of a novel fluorescent probe with high cancer-targeting selectivity and sensitivity. *Integrative Biology : Quantitative Biosciences from Nano to Macro* **5**: 1217–1228.

69. Hung DY, Chang P, Weiss M, Roberts MS (2001) Structure-hepatic disposition relationships for cationic drugs in isolated perfused rat livers: transmembrane exchange and cytoplasmic binding process. *The Journal of Pharmacology and Experimental Therapeutics* **297**: 780–789.
70. MacIntyre AC, Cutler DJ (1988) Role of lysosomes in hepatic accumulation of chloroquine. *Journal of Pharmaceutical Sciences* **77**: 196–199.
71. Cramb G (1986) Selective lysosomal uptake and accumulation of the beta-adrenergic antagonist propranolol in cultured and isolated cell systems. *Biochemical Pharmacology* **35**: 1365–1372.
72. Ishizaki J, Yokogawa K, Ichimura F, Ohkuma S (2000) Uptake of imipramine in rat liver lysosomes in vitro and its inhibition by basic drugs. *The Journal of Pharmacology and Experimental Therapeutics* **294**: 1088–1098.
73. Gong Y, Zhao Z, McConn DJ, Beaudet B, Tallman M, et al. (2007) Lysosomes contribute to anomalous pharmacokinetic behavior of melanocortin-4 receptor agonists. *Pharmaceutical Research* **24**: 1138–1144.
74. Daniel WA, Wojcikowski J (1997) Interactions between promazine and antidepressants at the level of cellular distribution. *Pharmacology & Toxicology* **81**: 259–264.
75. Daniel WA, Wojcikowski J, Palucha A (2001) Intracellular distribution of psychotropic drugs in the grey and white matter of the brain: the role of lysosomal trapping. *British Journal of Pharmacology* **134**: 807–814.
76. Lemieux B, Percival MD, Falgueyret JP (2004) Quantitation of the lysosomotropic character of cationic amphiphilic drugs using the fluorescent basic amine Red DND-99. *Analytical Biochemistry* **327**: 247–251.
77. Kazmi F, Hensley T, Pope C, Funk RS, Loewen GJ, et al. (2013) Lysosomal sequestration (trapping) of lipophilic amine (cationic amphiphilic) drugs in immortalized human hepatocytes (Fa2N-4 cells). *Drug Metabolism and Disposition: The Biological Fate of Chemicals* **41**: 897–905.
78. Funk RS, Krise JP (2012) Cationic amphiphilic drugs cause a marked expansion of apparent lysosomal volume: implications for an intracellular distribution-based drug interaction. *Molecular Pharmaceutics* **9**: 1384–1395.
79. Logan R, Kong A, Krise JP (2013) Evaluating the roles of autophagy and lysosomal trafficking defects in intracellular distribution-based drug-drug interactions involving lysosomes. *Journal of Pharmaceutical Sciences* **102**: 4173–4180.
80. Logan R, Kong AC, Axcell E, Krise JP (2014) Amine-containing molecules and the induction of an expanded lysosomal volume phenotype: a structure-activity relationship study. *Journal of Pharmaceutical Sciences* **103**: 1572–1580.
81. Egorin MJ, Hildebrand RC, Cimino EF, Bachur NR (1974) Cytofluorescence localization of adriamycin and daunorubicin. *Cancer Research* **34**: 2243–2245.
82. Smith PJ, Sykes HR, Fox ME, Furlong IJ (1992) Subcellular distribution of the anticancer drug mitoxantrone in human and drug-resistant murine cells analyzed by flow cytometry and confocal microscopy and its relationship to the induction of DNA damage. *Cancer Research* **52**: 4000–4008.
83. Marquardt D, Center MS (1992) Drug transport mechanisms in HL60 cells isolated for resistance to adriamycin: evidence for nuclear drug accumulation and redistribution in resistant cells. *Cancer Research* **52**: 3157–3163.

84. Lansiaux A, Dassonneville L, Facompre M, Kumar A, Stephens CE, et al. (2002) Distribution of furamidine analogues in tumor cells: influence of the number of positive charges. *Journal of Medicinal Chemistry* **45**: 1994–2002.
85. Baik J, Rosania GR (2013) Modeling and simulation of intracellular drug transport and disposition pathways with virtual cell. *Journal of Pharmaceutics & Pharmacology* **1**: 1–17.
86. Min KA, Zhang X, Yu JY, Rosania GR (2014) Computational approaches to analyse and predict small molecule transport and distribution at cellular and subcellular levels. *Biopharmaceutics & Drug Disposition* **35**: 15–32.

7

CELL CULTURE MODELS FOR DRUG TRANSPORT STUDIES

Irina Kalashnikova, Norah Albekairi, Shariq Ali, Sanaalarab Al Enazy, and Erik Rytting

Department of Obstetrics & Gynecology, University of Texas Medical Branch, Galveston, TX, USA

7.1 INTRODUCTION

High-throughput screening of pharmaceutical lead compounds can reduce the time, resources, and cost of identifying the most promising drug candidates at an early stage of the drug development process. Provided that a cell culture model can adequately predict permeability, biotransformation, or toxicity in a cost-effective manner, the use of such models can reduce the number of animal experiments that might have been required to narrow down a list of new compounds having the most preclinical promise. The elucidation of specific molecular mechanisms taking place at the cellular level represents another advantage of cell culture models [1].

This chapter will provide an overview of cell culture models that may be used to predict drug transport across relevant biological barriers. General considerations regarding the utility of such a model are discussed, followed by a brief description of physiological cell barrier properties and cell culture models that have been utilized or proposed to understand mechanisms of drug transport in the intestinal epithelium, the blood–brain barrier (BBB), nasal and pulmonary epithelium, ocular epithelial and endothelial barriers, the placenta, and renal epithelium. Examples of new developments regarding three-dimensional (3D) models are

Drug Delivery: Principles and Applications, Second Edition. Edited by Binghe Wang, Longqin Hu, and Teruna J. Siahaan.

also presented. Nevertheless, this is not a comprehensive overview of all possible cell culture models applicable to drug transport. Readers are referred to other sources for information on cell culture models of the skin and liver, for example see Refs. 2–6.

7.2 GENERAL CONSIDERATIONS

Cell culture models generally fall into one of two categories: primary cells and immortalized cell lines. Primary cultures are isolated from human or animal tissues, and cells of interest can be disaggregated by enzymatic or mechanical techniques (see Fig. 7.1) [7–9]. Cells capable of proliferation can be removed and redistributed onto new cell growth surfaces as a subculture. Additional subculturing (passaging) of cells can give rise to a cell line. Most primary cell cultures are not stable after a certain number of passages, and some will not proliferate at all. Senescence of the cell line can be caused by shortening of telomeres on the chromosomes. Upon reaching a critical telomere length, the cells will not continue to divide. Accordingly, transfection of the telomerase gene hTRT may be one means to produce an immortalized cell line [9]. Although immortalized cell lines can be generated by chemical or viral methods, some (often malignant) cells are spontaneously continuous [10].

Although immortalized cell lines are generally easier to work with, primary cells have certain advantages. In comparison to immortalized cell lines derived from malignant cells, primary cultured cells may be more representative of normal cell behavior *in vivo*. A major disadvantage of primary cell culture is the need for repeated isolation procedures due to the fact that most primary cells cannot be continuously subcultured to a substantial cell passage number. Differences in isolation conditions or heterogeneous tissue sources can lead to large variability in results obtained using

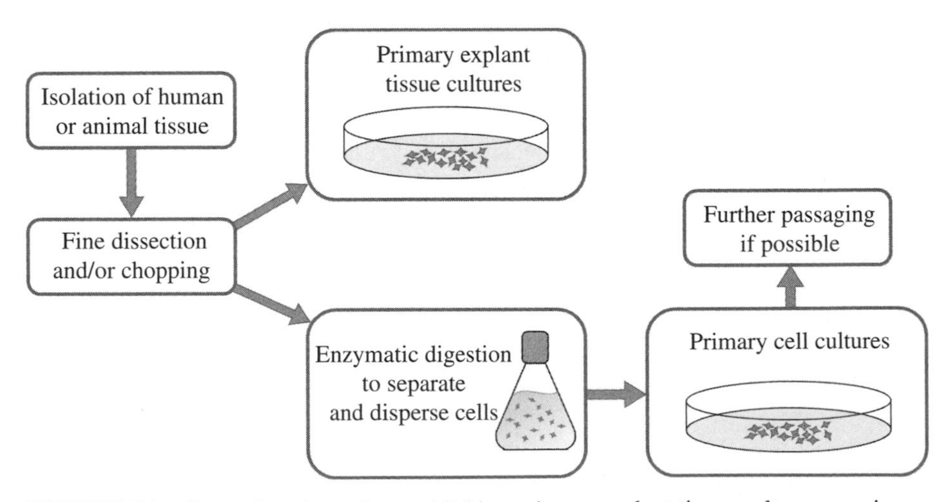

FIGURE 7.1 General methods for establishing primary explant tissue cultures or primary cell culture models isolated from human or animal tissues.

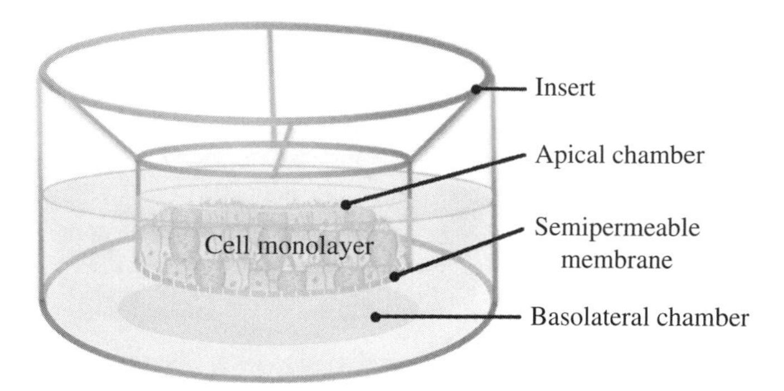

FIGURE 7.2 Inserts with semipermeable membranes can be used to study directional drug transport across cell monolayers. Samples can be taken from both the apical and basolateral sides of the cells.

primary cells. Although immortalized cell lines are more convenient to maintain, the abnormal nature of cell lines derived from malignant sources may limit the extrapolation of some data to normal cell populations. For example, certain immortalized cell lines may not express all relevant transporters [11]. Although the homogeneous nature of an immortalized cell line may lead to a smaller degree of variability in one's experimental results, it is important to note that from a pharmacogenetic standpoint, an immortalized cell line represents $n = 1$ individual and, therefore, cannot reflect the genetic variability of a population as well as might be achieved by isolating multiple sets of primary cells [12].

A number of criteria should be taken into account when selecting an appropriate model. For example, in addition to the aforementioned general advantages and disadvantages of primary versus immortalized cell lines, one should consider the passage number of the cell line and any associated limits on cell line stability, the suitability of the cell model to reflect *in vivo* barrier properties, reproducibility, the ease with which transport studies can be conducted, and the expression of relevant transporter proteins [1, 13]. A number of experimental set-ups permitting the feasibility of drug permeability studies across a cell monolayer have been described previously [1, 14, 15]. One example is the Transwell® insert, depicted in Figure 7.2. A comprehensive discussion of considerations for the selection of an appropriate experimental transport system—namely, cell type, matrix, filter support, medium or buffer composition, stirring, solute properties, and assay conditions—is provided by Ho et al. [16].

7.3 INTESTINAL EPITHELIUM

7.3.1 The Intestinal Epithelial Barrier

Assessing the absorption and bioavailability of orally administered drugs is an important step in drug development. By using cell culture models to understand the oral bioavailability of drugs, costs are significantly reduced and, as a result, drugs

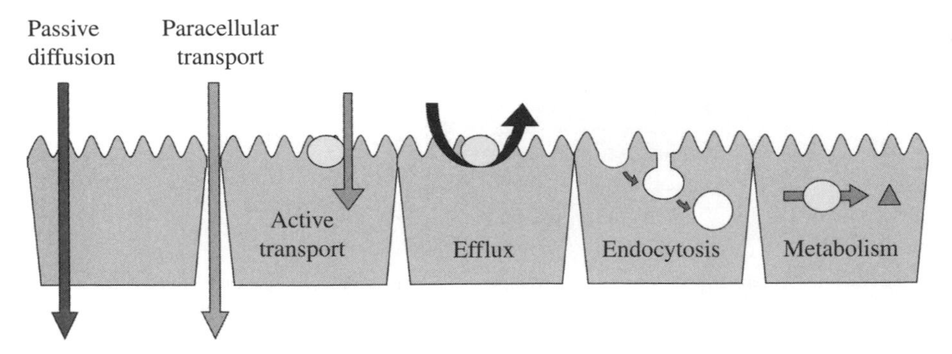

FIGURE 7.3 General mechanisms of drug transport across a cellular barrier. It should be noted that active, facilitative, and efflux transporter proteins may be present in both apical and basolateral membranes. Metabolizing enzymes within cells may in effect limit the transport of drugs by means of biotransformation to metabolites with altered or no pharmacological activity.

can be screened at an earlier stage of development. The reason that a cell culture model is a valid way to assess drug permeability is because the intestinal epithelium is the limiting barrier for the permeation and absorption of drugs [17].

The intestinal mucosa comprises a monolayer of many different cell types. These cells include enterocytes, goblet cells, lymphocytes, M cells, and crypt cells; each cell type has a unique function. Of these cells, enterocytes are the most numerous and form the basis of most cell culture models [18]. Enterocytes are coated with microvilli on the apical surface, which allows for greater surface area and, thus, a higher ability to interact with the luminal contents of the bowel.

The determinants of absorption of drugs through the intestinal epithelium are the permeability of the drug, solubility, and stability. The apparent permeability (P_e) quantifies the rate at which molecules can cross the epithelium [19]. (Readers are referred to previous examples in the literature that demonstrate the calculation of permeability from drug transport experiments [12, 15].) There are many mechanisms by which drugs can permeate through the intestinal epithelium (see Fig. 7.3). The paracellular pathway involves passage through tight junctions that exist between cells, namely, occludins and claudins [20]. Molecules can also cross transcellularly by passive diffusion through the lipid membrane. This pathway of absorption is characteristic of lipophilic molecules [21]. The transcellular pathway includes active and facilitated transport, both of which use specific proteins in the cell membranes to allow passage of molecules into and through the cell. A number of influx and efflux transporters (e.g., P-glycoprotein) exist in the membrane that also regulate the passage of certain substrates.

7.3.2 Intestinal Epithelial Cell Culture Models

To date, a number of *in vitro* models of the intestinal epithelium have been investigated, all of which may possess different characteristics and, thus, have different potential applications in the study of drug transport. A number of reviews have been published comparing the usefulness of these models in ranking the order of drug permeability as well as correlations to the *in vivo* situation.

Primary cell culture techniques for the intestinal epithelium have been developed. However, these models typically cannot form monolayers and are not polarized. Immortalized cell lines have the advantage of forming monolayers and expressing a phenotype similar to the mature intestinal epithelium. The most widely used model is the Caco-2 cell line. This cell line originated from adenocarcinoma of the colon, but it resembles the small intestine in terms of transporter expression. Once this cell line reaches confluence in culture, it spontaneously changes the phenotype to mature epithelium [17]. The expression of transporters in this model is actually less than found *in vivo* [19]. Because of these factors, its usefulness is limited to studying the transport of drugs that display high passive transport across the epithelium [20]. The intercellular pore size of Caco-2 cells is 4.5 Å, which is smaller than the normal intestinal pore size of 8–13 Å [22]. Depending on the mechanism of transport, it can be useful for specific transport studies [19]. Many transporters present in the intestinal epithelium are expressed by Caco-2 cells, including MDR1 (P-glycoprotein) and ABCG2 (breast cancer-resistant protein, BCRP). Several metabolic enzymes are also expressed, including CYP1A. CYP3A4 can also be induced in this cell line by administration of vitamin D3 [22]. A disadvantage of this cell line, however, is that it is heterogeneous. It also lacks many characteristics of the *in vivo* intestinal epithelium. To that end, other models involving the Caco-2 system have been developed. For instance, there exists a subclone called TC7, which is useful in that it expresses many brush border enzymes [17].

Other cell lines that can mimic the intestinal epithelium are the Madin-Darby canine kidney epithelial cell line (MDCK) and the 2/4/A1 cell line. MDCK cells are derived from the renal epithelium of dogs and are useful for studying membrane permeability. It is similar to Caco-2 in its predictive power of the behavior of drugs *in vivo*. The 2/4/A1 cell line, however, is derived from the rat small intestine. It has a better predictive capability than the Caco-2 cell line, even though it lacks several transporters. It is useful for soluble compounds and demonstrates passive diffusion characteristics similar to intestinal epithelium *in vivo* [19].

Coculture models, which demonstrate more features of the heterogeneous cell population of the intestinal epithelium, can provide more insight into drug behavior in the intestine. Examples of this include the Caco-2/HT29 model, which produces mucus, and the Caco-2/Raji-B coculture model, which is meant to mimic follicle-associated epithelium. A triple culture of Caco-2/HT29/Raji-B cells has also been developed [17].

7.4 THE BLOOD–BRAIN BARRIER

7.4.1 The Blood–Brain Endothelial Barrier

The blood–brain barrier (BBB) is composed of the brain microvascular endothelial cells (BMECs) and is a part of the neurovascular unit. Perivascular cells (pericytes and astrocytes) contribute to the basement membrane, surrounding the capillaries and containing structural proteins, collagens, and specialized proteins [23–25]. This barrier functions as an interface between the brain parenchyma and the systemic circulation. The BBB is largely impermeable to the transfer of most compounds; for example, clinical trials have shown that 98% of drugs intended to treat cancers in the central nervous system had inadequate BBB permeability [26–28].

Several features of the brain endothelium contribute to its functional barrier and limit the permeability of many drugs. These include tight interendothelial junctions, which limits paracellular permeability; few pinocytic vesicles and a low level of endocytosis and transcytosis; an enzymatic barrier, which includes acetylcholinesterase, alkaline phosphatase, γ-glutamyl transpeptidase, monoamine oxidases, and drug metabolizing enzymes; and efflux transporters [29]. The tight junctions are characterized with high transendothelial electrical resistance values in the range of 1500–2000 $\Omega\,cm^2$, and selective transporter proteins within the BBB include efflux proteins such as P-glycoprotein (ABCB1, MDR1), BCRP (ABCG2), multidrug resistance associated proteins (MRPs, ABCC1-6), and organic anion transporters (OATs), but carrier-mediated solute carriers and endocytosis mechanisms are also present [26, 27, 29–31].

7.4.2 BBB Cell Culture Models

Since there is a large market for central nervous system drugs to treat or prevent a number of disorders (e.g., epilepsy, multiple sclerosis, Alzheimer's disease, inflammation, edema, hypoxia, ischemia, glaucoma, Parkinson's disease, depression, and HIV), several BBB models—ranging from in silico to *in vivo*—have been developed to predict the pharmacodynamic and pharmacokinetic behavior of many different classes of drugs [25, 26, 32–37]. Primary cell culture models of the BBB are based on the isolation of BMECs from various species. Bovine BMECs (BBMECs) and porcine BMECs (PBMECs) are most commonly used [38]. Primary cells can also be cocultured with astroglial cells, neurons, pericytes, and/or astrocytes in order to approximate *in vivo* conditions more closely [39–42]. Additional modifications to mimic *in vivo* conditions include the use of a plate viscometer or a cone-plate apparatus to build a dynamic model, and materials such as collagen, hydrogels, extracellular matrix proteins, or polypropylene can be used to create a three-dimensional (3D) scaffold or a hollow organ-like structure to improve *in vitro* modeling of the BBB. Contact coculture and 3D dynamic models closely mimic the *in vivo* BBB, and high transendothelial electrical resistance values have been obtained with such models (up to 1650 $\Omega\,cm^2$) [25, 29, 31, 40, 41, 43].

Several immortalized cell line models of the BBB have been developed, including cell lines of human origin (HMEC-1, hCMEC/D3, and TY08), rat origin (RBE4, GP8, GPNT, and TR-BBB), and murine origin (bEnd.3-5 and TM-BBB) [25, 29, 31, 40, 42, 44, 45]. Readers are referred to recently published examples of culturing protocols for the hCMEC/D3 and bEnd.5 cell lines [23]. As was the case for primary cell culture models, the addition of astrocytes, pericytes, neurons, and glial cells can improve barrier function (increases in transendothelial electrical resistance values, expression of transporters, induction of relevant enzymes, and decreases in passive and paracellular permeability) [29–31, 39–41, 45, 46]. For example, coculture of hCMEC/D3 cells with astrocytes resulted in significant increases in transendothelial electrical resistance values. However, exposing the cells to flow-based shear stress resulted in the most substantial increases in transendothelial electrical resistance, with values exceeding 1000 Ω cm^2 [31]. The development of human pluripotent stem cell–derived BMECs has been described recently, which may be useful in future drug screening applications [40].

7.5 NASAL AND PULMONARY EPITHELIUM

7.5.1 The Respiratory Airway Epithelial Barrier

The pulmonary system is the site of gas exchange between the blood and air. It can be categorized into the upper respiratory tract (including the nasal cavity, pharynx, and larynx) and the lower respiratory tract (including the trachea, primary bronchi, bronchioles, and the alveolar region) [47]. The airway diameter decreases progressively along its length until it ends with the alveolar sacs, the distal, functional respiratory units where gas exchange occurs. Healthy human lungs contain between 300 and 500 million alveoli [48].

The entire respiratory system is lined by a continuous layer of epithelial cells (see Fig. 7.4). This epithelial layer plays a number of vital roles: (i) it forms a physical protective barrier, (ii) it facilitates mucociliary clearance, (iii) it secretes protective substances including antimicrobial agents, (iv) it helps regenerate and repair other pulmonary components, and (v) it plays a role in inflammatory and immune responses [47, 50]. The epithelial cell type changes along the respiratory system to accommodate specific functions, which are morphologically and functionally unique to each region.

The epithelium of the upper respiratory tract is composed of stratified squamous cells, and the nasal cavity is composed of transitional nonciliated epithelial cells. This region has traditionally been considered of low interest for drug delivery because of the small surface area and low vascularity. Within the nasal cavity, however, is the olfactory epithelium, located on the upper part of the nasal cavity, which represents approximately 8% of the total nasal surface area [51]. It is composed of pseudostratified ciliated cells, supporting cells, basal cells, Bowman's glands, and olfactory sensory neurons. These olfactory neurons are unique in that they are directly in contact with the nasal cavity environment, and their axons are in contact with the olfactory bulb in the brain. Therefore, the delivery of drugs to the brain is possible by means of intranasal administration [52].

The respiratory epithelium of the pharynx and the larynx consists of ciliated and nonciliated columnar cells, goblet cells, and basal cells. Most of the luminal surfaces of the nasal mucosa are covered with mucus, produced by goblet cells within the surface of the epithelium. This mucus plays an important role in filtering particles from the inhaled air. It also contains several isoforms of cytochrome P-450 enzymes, which are responsible for metabolizing many drugs. The mucus is constantly cleared toward the nasopharynx by the beating cilia (mucociliary clearance), and then cleared through the digestive tract via the esophagus [53].

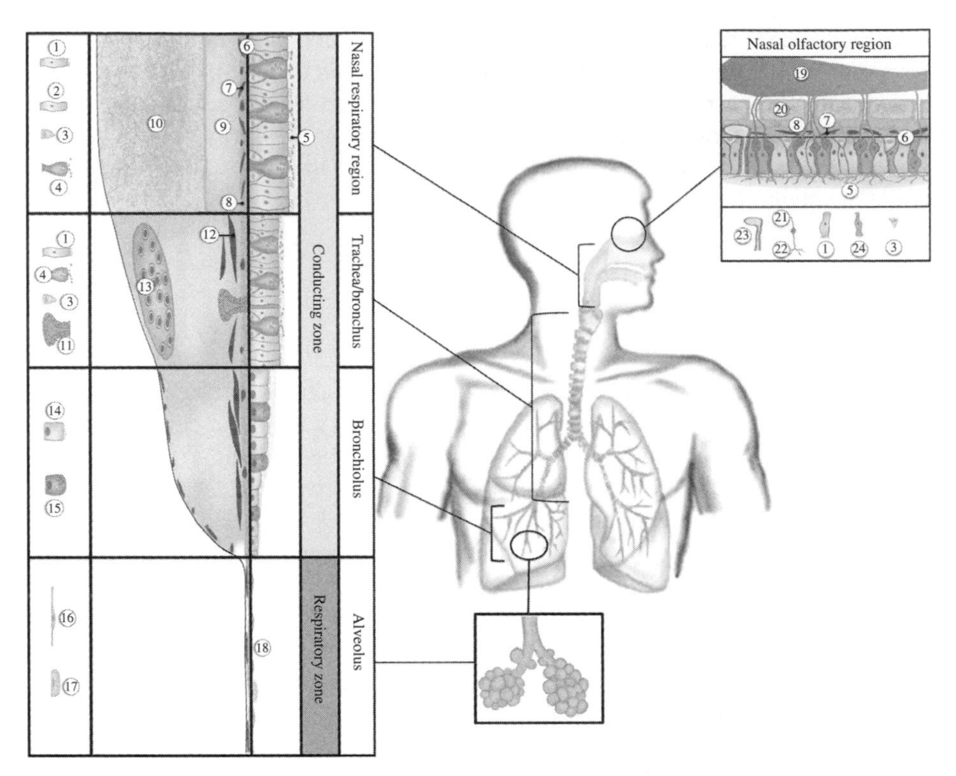

FIGURE 7.4 Cell types lining the respiratory system. Key: (1) ciliated columnar cells, (2) non-ciliated columnar cells, (3) basal cells, (4) goblet cells, (5) mucus, (6) basement membrane, (7) blood vessels, (8) lamina propria, (9) fibro-cartilaginous layer, (10) bone, (11) gland, (12) smooth muscle cells, (13) cartilage, (14) ciliated cuboidal cells, (15) Clara cells, (16) type I alveolar cells, (17) type II alveolar cells, (18) surfactant, (19) olfactory bulb, (20) cribriform plate, (21) receptor cell, (22) receptor cell cilia, (23) Bowman's gland, (24) supporting cell. Adapted from Klein et al. [49], pp. 1516–1534. (*See insert for color representation of the figure.*)

The epithelium in the lower respiratory tract consists of columnar and pseudostratified cells, secretory goblet cells, basal cells, and ciliated cells for mucociliary clearance. In the bronchioles, the epithelium consists of cuboidal cells with short cilia and secretory Clara cells. The distal respiratory tract consists mainly of the alveolar epithelium that is composed of type I and type II alveolar cells. Type II alveolar cells produce surfactant and serve as progenitors for type I alveolar cells [50, 51, 53].

7.5.2 The Nasal Epithelial Barrier and Cell Culture Models

Intranasal drug delivery is noninvasive, results in rapid onset of action, and avoids first-pass metabolism [54]. Several *in vivo* animal models have been evaluated for intranasal delivery; however, these models have failed to correlate well with results in humans. Therefore, human nasal *in vitro* models have been established as a substitute for such animal models [55].

Primary culture of human nasal epithelial cells provides a promising tool for studying drug transport, metabolism, toxicology, and electrophysiology. These cultures form confluent monolayers in 6–8 days, with differentiated goblet and ciliated cells. The primary human nasal cell can be obtained by several methods. Nevertheless, these methods are limited by intensive labor, limited passage number, heterogeneity, risk of pathogenic contamination, and donor variability. Several factors affect primary nasal epithelial cell culture. For example, prolonged culturing can lead to loss of cilia and supplementary extracellular matrix can affect cell differentiation [56–62].

Some of the limitations of primary culture may be overcome by the use of cell lines. The most common cell line of human origin is RPMI 2650, derived from human nasal anaplastic squamous cell carcinoma of the nasal septum [56, 57]. Nonhuman nasal cell lines include bovine turbinate cells and NAS 2BL (derived from rat nasal squamous carcinoma). Although RPMI 2650 cells may be useful for metabolism studies, they do not form monolayers spontaneously, and they lack goblet and ciliated cells, which makes them unsuitable for drug permeation studies. Nevertheless, it has been reported that the use of RPMI 2650 cells under air–liquid interface culture conditions can promote the formation of a monolayer with sufficient transepithelial electrical resistance (TEER) to perform drug permeation studies [63, 64]. A 3D reconstructed nasal mucosa was recently developed, which also shows promise for drug transport studies [65, 66]. Although Calu-3 cells are not of nasal origin, but are derived from human lung adenocarcinoma, these cells have also been investigated as a screening tool for nasal drug delivery. At an air–liquid interface, Calu-3 cells form a polarized monolayer with suitable TEER values and a uniform mucus layer [61, 64, 67–70]. Additional studies are needed to correlate *in vitro* and *in vivo* results to promote the use of such *in vitro* models in high throughput drug candidate screening.

7.5.3 The Airway Epithelial Barrier and Cell Culture Models

A variety of epithelial cell line models have been developed to specifically study basic cellular pathways, toxicology, and drug transport within the airways. A number of protocols have been developed for culturing primary human airway cells *in vitro* [49, 71, 72]. Primary human cells can be cultured from normal or disease-derived cells. In fact, comparing cultures derived from normal versus diseased cells provides a methodology for probing the underlying pathways that contribute to disease. Primary bronchial epithelial cell monolayers derived from asthmatic donors have been shown to retain their phenotype even after several passages [73]. To better replicate the *in vivo* airway epithelium, complex *in vitro* models have been established, which include coculture techniques and 3D mucus-secreting spheroids of primary bronchial epithelial cells [74–76]. Precultured cell systems are commercially available, such as EpiAirway™ and MucilAir™. The EpiAirway model is prefilled with primary cells of human tracheal or bronchial epithelium, and the MucilAir model contains primary cells of the human respiratory tract. In both models, the cells are well differentiated, form beating cilia, secrete mucus, and form tight junctions. In addition, they have a long life span and are suitable for uptake and transport studies [17, 48, 49].

Three immortalized cell lines most widely used for investigations of the airway epithelium include Calu-3, 16HBE14o-, and BEAS-2B cells [77]. The Calu-3 adenocarcinoma cell line generates a confluent monolayer with tight junctions and mucus production. Transport studies can be performed after 6–8 days in submerged culture and after 10–14 days in air–interface culture. 16HBE14o- is an immortalized epithelial cell line derived from healthy human bronchial epithelium. The cells form tight junctions, retain differentiated epithelia with apical microvilli, and have no mucus secretion. The 16HBE14o- cell line has been used to study oxidative stress and inflammation, and transport studies can be performed after 6–9 days in submerged culture or after 6–7 days using air–interface culture. Both Calu-3 and 16HBE14o- cells express P-glycoprotein. Like 16HBE14o- cells, BEAS-2B cells are viral transformed human bronchial epithelial cells. However, BEAS-2B cells do not form tight junctions [49, 78–80].

7.5.4 The Alveolar Epithelial Barrier and Cell Culture Models

There are two types of alveolar epithelial cells: type I and type II. Type I alveolar epithelial cells are squamous cells which form the air–blood barrier. These cells cover 95% of the alveolar surface. Type II alveolar epithelial cells are cuboidal, produce surfactant and proinflammatory mediators, and are the progenitors of type I alveolar cells. A population of alveolar macrophages is also found in the interstitial spaces; these macrophages clear pathogens and debris deposited in the air spaces [47, 48, 81, 82].

Primary human alveolar epithelial cells (hAEpC) are derived from type II alveolar cells isolated from human lung biopsies. After 8–9 days in culture, tight junctions are formed and most of the cells are flattened, displaying a type I alveolar cell-like phenotype. To more closely resemble the air–blood barrier, cocultures of hAEpC cells with macrophages, dendritic cells, and/or primary human pulmonary microvascular endothelial cells are possible [83–88]. An innovative alveolar "lung-on-a-chip" model has been described, which provides structural, functional, and mechanical properties of the human alveolar–capillary interface (see Fig. 7.5). The alveolar cells are separated from the endothelial cells by a flexible membrane coated with fibronectin. Nutrients are supplied through microfluidic flow of medium and immune cells, and the flexible membrane allows for the cyclical stretching of alveolar tissue during respiration [87, 88].

The A549 alveolar epithelial cell line derived from a human adenocarcinoma exhibits properties consistent with type II alveolar cells, such as the presence of lamellar bodies and surfactant proteins [89]. However, A549 cells do not differentiate to a type I alveolar cell-like phenotype, and they do not form tight junctions, which limits the use of this cell line for transport studies [90]. NCI-H441 cells are of human distal lung epithelial origin and are more suitable for transport studies. NCI-H441

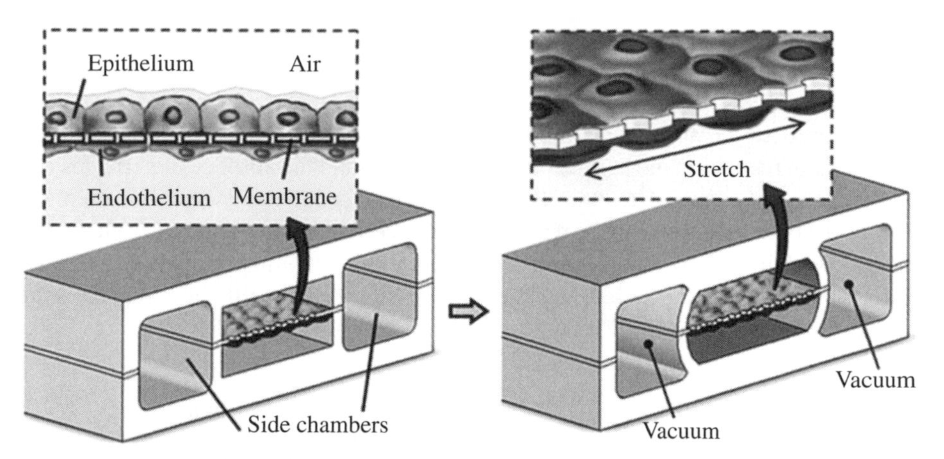

FIGURE 7.5 *In vitro* model of the alveolar–capillary barrier. Cells representing the airway epithelium and pulmonary endothelium are grown on opposite sides of a flexible membrane. Vacuum is applied to the side chambers to mimic physiological breathing, causing mechanical stretching of the flexible membrane. Huh et al. [87], pp. 1662–1668. Reprinted with permission of the AAAS.

cells are derived from human lung papillary adenocarcinoma and in some aspects are representative of the bronchiolar pulmonary epithelium and type II alveolar epithelial cells. In culture, this cell line forms confluent monolayers with tight junctions, with peak TEER values around $1000\,\Omega\,cm^2$. Drug transporter activity and expression is similar to that seen in primary alveolar epithelial cells, including the presence of P-glycoprotein and a number of organic cation transporters [89].

7.6 THE OCULAR EPITHELIAL AND ENDOTHELIAL BARRIERS

7.6.1 The Corneal and Retinal Barriers

The cornea and the conjunctiva are the two major components of the anterior barrier of the eye. The cornea is a clear, avascular layer of epithelial cells that is only 520 μm thick. This layer includes superficial cells, wing cells, basal cells, Bowman's layer, stroma, Descemet's membrane, and endothelium. The tight junctions of the corneal epithelium prevent the paracellular passage of hydrophilic molecules, but passive diffusion is possible for lipophilic compounds. The pores in the intercellular spaces are negatively charged at physiological pH due to the carboxylic groups on the tight junction proteins. The conjunctival layer comprises mucus-producing epithelial cells. It acts as a protective surface that maintains the tear film by production of mucus glycoproteins. The conjunctiva is more permeable to hydrophilic compounds and macromolecules than the cornea. Nevertheless, drug delivery across the conjunctiva is considered nonproductive absorption because most of the drug is lost to the blood rather than reaching a site of action within the aqueous humor [91–94].

The blood–retinal barrier (BRB) consists of outer and inner layers. The retinal pigment epithelial cells form the outer layer, and the retinal vascular endothelial cells form the inner layer. The BRB resembles the BBB in that it limits the transport of molecules between the neural retina and the systemic circulation. Pericytes are important in regulating the permeability of the BRB, and the absence of pericytes can cause vascular leakage. The inner BRB is essential for intact vision and is involved in the transport of nutrients to the retina and the excretion of waste products. Occludin, claudin, and junctional adhesion molecules maintain tight junctions between the retinal vascular endothelial cells. The polarized retinal pigment epithelium separates the neural retina from the choroid and also has tight junctions [95–98].

7.6.2 Cell Culture Models of Ocular Epithelium and Endothelium

Primary cell culture models of the corneal epithelium derived from rabbits have been described for use in permeability studies [95]. Primary human corneal epithelial cells (HCEpiC) have been utilized to characterize the expression of monocarboxylate and efflux transporters [99, 100]. Immortalized cell models of the human corneal epithelium that have been utilized for studying transport include HCE cells, tet HPV16-E6/E7 transduced HCE cells, and HCLE cells [95, 101]. Primary models of the conjunctival epithelium include cells isolated from rabbits, which have shown TEER values similar to that of excised rabbit conjunctiva [95]. An immortalized human conjunctival epithelial cell line (HCjE) has been used to investigate drug transporter expression [101].

Primary cell culture models of the retinal pigment epithelium include cells isolated from both human and animal sources, but to avoid the challenges associated with interspecies data extrapolation, the use of human primary cells is preferred. The human immortalized ARPE-19 cell line has been used extensively to characterize drug transport across the retinal pigment epithelium [95]. A primary bovine cell model of the retinal capillary endothelium can be difficult to grow as a tight monolayer. Human primary retinal endothelial cells (hREC) can also be isolated and have been utilized in permeability studies to assess monolayer integrity [102]. An immortalized cell line of the retinal capillary endothelium (TR-iBRB) was derived from a transgenic rat, but the cells do not form a tight barrier [95]. A telomerase-immortalized cell line (HREC-hTERT) has been derived from human primary cells and may serve as an appropriate model of the human BRB [103].

7.7 THE PLACENTAL BARRIER

7.7.1 The Syncytiotrophoblast Barrier

The placenta is the interface between the mother and the fetus during pregnancy. It serves many functions, such as gas and nutrient exchange, waste elimination, and the metabolism of some drugs. For instance, the oxygen diffusing capacity of the placenta influences gas exchange between the fetal and maternal circulations. The human syncytiotrophoblast cell layer is replenished by precursor cytotrophoblast cells and has the dual functionality of a transport barrier and a hormone producer [104, 105]. It is the rate-limiting barrier for exchange between the fetal and the maternal circulations, and it is responsible for the regulation of the transport of toxins, xenobiotics, waste products, and nutrients [106]. The mechanism of transport across the placenta varies depending on the size of the molecule. In general, many molecules having molecular weights less than 1000 Da can freely cross the placental barrier by passive diffusion. Other transplacental transport processes exist, namely, facilitated diffusion (not requiring metabolic energy) and active transport (requiring metabolic energy) [107]. Syncytiotrophoblast cells are polarized, with basolateral (facing the fetus) and apical (facing the mother) plasma membranes. The polarity of the syncytiotrophoblast layer has an important role in regulating the movement of nutrients and waste products between the maternal and fetal circulations [105, 106]. Transporters in the maternal side include P-glycoprotein (MDR1), BCRP, MRP2, OCTN2, SERT, and NET, while transporters expressed on the fetal side include MRP1, MRP3, MRP5, OATP-B, OAT4, OCT3, and MDR3 [108].

7.7.2 Trophoblast Cell Culture Models

In vitro models to elucidate drug transport and uptake in placental trophoblast cells include isolated tissue explants, primary cells, and immortalized cell lines. The use of placental tissue explants, membrane vesicles, and primary cytotrophoblast cells can provide information regarding placental drug uptake, but these systems do not form confluent monolayers sufficient to carry out transport studies [14, 109]. Immortalized trophoblast cell culture models include JEG-3 cells, JAr cells, and BeWo cells. JEG-3 and JAr cells secrete placental hormones can syncytialize in culture and can be used for drug uptake studies. However, they are generally considered unsuitable for polarized transport experiments [110]. It should be noted that two clones of the BeWo cell line (b24 and b30) form confluent monolayers amenable to drug transport studies, but the original BeWo cell line available through the American Type Culture Collection does not have this same monolayer-forming ability [12, 14]. BeWo b30 cells have been utilized in a number of drug transport and metabolism studies [111, 112]. In fact, recent comparisons have demonstrated very good correlation between *in vitro* permeability data from BeWo b30 cell transport experiments and measurements of drug transfer across *ex vivo* dually perfused human placental lobules [12, 113].

7.8 THE RENAL EPITHELIUM

7.8.1 The Renal Epithelial Barrier

The kidneys are highly perfused; blood flow to the kidneys is approximately 25% of total cardiac output. Each kidney contains more than 1 million nephrons. Within a single nephron, blood undergoes glomerular filtration. Most substances having a molecular weight below 5000 Da pass through the glomerulus, whereas most proteins remain in the capillaries. The plasma ultrafiltrate moves from the glomerulus into the proximal tubule, Henle's loop, the distal tubule, and finally as urine into the collecting duct. Tubular secretion and reabsorption processes take place along this pathway to maintain fluid and electrolyte balance and concentrate the urine. A number of nutrients and ions are reabsorbed from the filtrate into the blood. Osmotic pressure promotes the reabsorption of water from the filtrate back into the renal capillaries. The renal tubules are lined with epithelial cells, which contain a number of transporter proteins that participate in tubular secretion mechanisms. The apical (brush border) membrane faces the urine, and basolateral infoldings of the epithelium interact with the renal capillaries. Certain drugs (ionized compounds in particular) may be actively secreted against a concentration gradient out of the blood and into the filtrate, whereas other drugs (e.g., unionized lipophilic compounds) may diffuse across the tubular epithelium and be reabsorbed into the blood. [114, 115].

7.8.2 Renal Epithelial Cell Culture Models

A number of cell culture models have been described for studying tubular secretion and reabsorption. This paragraph will introduce primary cell models, and the following paragraph will discuss some of the advantages and limitations of certain cell line models. Terryn et al. have isolated primary proximal tubule cells from mice [116]. Markadieu et al. have described a unique method to isolate primary cultured distal tubule cells from mice expressing enhanced green fluorescent protein under the parvalbumin promoter. Parvalbumin is expressed in the distal convoluted tubule. Fluorescent parvalbumin–containing tubules were separated and a cellular monolayer with tight junctions was established [117]. Primary proximal tubular cells from porcine kidney were confirmed to express several important transporters, although some transporters that were expressed in freshly isolated cells were down-regulated in culture [118]. Taub has reported the use of hormonally defined serum-free media to promote the differentiation of primary rabbit kidney tubule epithelial cells without fibroblast overgrowth. The resulting cells had a polarized morphology and tight junctions [119]. A similar approach was applied to the isolation of primary human proximal tubule cells by a Percoll gradient, followed by culture in hormonally defined serum-free media, which allowed the formation of a tight monolayer for up to three passages [120]. Another technique for separating primary human proximal tubule cells involves the use of antibodies to CD10 and CD13 as markers of proximal tubule epithelial cells in conjunction with fluorescence activating cell sorting (FACS) [121]. Primary cultured human proximal tubule cells express a number of transporters relevant to drug disposition, and they can be cultured on Millicell® filter inserts or on Transwell inserts [122]. Jang et al. cultured primary human proximal tubule cells on a microfluidic device whereon the monolayer is exposed to fluid shear stress. This resulted in enhanced cell polarization and formation of cilia compared to the growth of cells on Transwell inserts [123].

Cell lines as models for renal tubular epithelial cells include: MDCK cells (canine); OK cells (opossum); LLC-PK_1 (porcine); PKSV-PCT, PKSV-PR, and mpkCCD$_{cl4}$ (murine); and HKC, HK-2, Caki-1, and RPTEC/TERT1 cells (human). MDCK cells represent the distal renal epithelium and express P-glycoprotein, monocarboxylic acid, organic cation, and peptide transporters, as well as alkaline phosphatase, glutathione S-transferase, and sulfotransferase. When grown on semipermeable inserts, MDCK cells are polarized and form tight junctions. MDCK cells have been used as a model of intestinal drug absorption because, similar to Caco-2 cells, MDCK permeability values correlate with drug absorption, and MDCK cells grow faster than Caco-2 cells [124]. OK opossum kidney cells have been shown to reflect the *in vivo* paracellular permeability properties of rat proximal tubules [125]. In LLC-PK$_1$ proximal tubule cells, apical SGLT and basolateral GLUT are highly expressed, and the cells have a moderate expression of OCTs and OATs [126]. Chassin et al. have described the derivation of the proximal tubule cell lines PKSV-PCT and PKSV-PR from transgenic mice, as well as the mpkCCD$_{cl4}$ collecting duct cell line [127]. HKC-8 cells express the Na^+-HCO_3^- cotransporter (NBC-1) with transport activity similar to that of intact proximal tubules [128]. HK-2 cells express SLC16A1 (MCT) and SLCO4C1 (OATP4C1), MDR1, and MRPs, but they do not express the SLC22 transporters OAT1, OAT3, and OCT2, nor do they express ABCG2 (BCRP) [129]. The TEER of Caki-1 cells was stable between passage numbers 8 and 71 and was representative of a leaky epithelium. Although they retain activity of certain transport proteins and metabolizing enzymes, in comparison to native kidney cells, Caki-1 cells have high expression of MRP3, MRP4, and MRP6, and low levels of MDR1 [129, 130]. RPTEC/TERT1 cells display intact vectorial transport and hormonal response similar to primary cells, but this cell line is genomically stable for up to 90 population doublings [131].

7.9 3D *IN VITRO* MODELS

Because two-dimensional cell monolayers cannot fully represent the 3D nature and extracellular matrices of tissues *in vivo*, the development of 3D *in vitro* models can provide a cellular microenvironment that preserves cell–cell interactions and tissue architecture. 3D cellular models can provide environmental cues affecting cell signaling, differentiation, and morphology [132–134]. A number of cellular 3D models have been developed, including models of liver, breast, cardiac, muscle, bone, corneal tissue, and tumors. 3D tumor models include multicellular spheroids, hollow fibers, and multicellular layers. Spheroids, for example, mimic the heterogeneity of tumor cells, having a hypoxic, necrotic region in the core [133].

3D tissue constructs can be developed by top-down or bottom-up approaches. In a top-down approach, cells are seeded in a biodegradable scaffold, but it can be difficult to control cell alignment or cell–cell interactions by this method. Four types of bottom-up approaches include cellular layer-by-layer, extracellular matrix-assisted cellular layer-by-layer, cell accumulation, and inkjet printing of cells and polymers. One cellular layer-by-layer approach utilizes temperature-sensitive polymeric culture dishes to harvest a layer of cells without using proteolytic enzymes. Cell sheets can then be stacked to create the 3D model. In the extracellular matrix-assisted layer-by-layer technique, a film of fibronectin and gelatin is placed on cell monolayers, onto which a second cell seeding can be performed. This process can be repeated to generate the desired 3D cellular structure. The cell accumulation method involves coating single cells with fibronectin and gelatin, which promotes cell–cell adhesion [134]. In one example of a 3D model, Astashkina et al. used a hyaluronic acid hydrogel to encapsulate an intact proximal tubule, that is, not just the epithelial cells. These 3D organoid proximal tubule cultures were more sensitive to toxic insult than 2D LLC-PK_1 or HEK293 cells [132, 135].

7.10 CONCLUSIONS

The application of cell culture models for predicting drug transport has facilitated an accelerated pace of understanding the biochemical processes affecting the permeability of molecules across cellular barriers. Besides serving as a tool for screening passive permeability and studying uptake and efflux mechanisms, many of these same cell lines also play a number of roles in various pharmacological and toxicological assays. It is interesting to review the previous edition of this chapter, which contained expressions of hope for future developments, including a more convenient model of the BBB and 3D cellular models. As we appreciate and welcome these advances, which have been realized during the past decade, we look forward to future refinements and opportunities to improve our understanding of drug transport and improve our clinical approaches to drug delivery.

REFERENCES

1. Audus, K. L.; Bartel, R. L.; Hidalgo, I. J.; Borchardt, R. T. *Pharm Res* 1990, **7**, 435–451.
2. Gotz, C.; Pfeiffer, R.; Tigges, J.; Blatz, V.; Jackh, C.; Freytag, E. M.; Fabian, E.; Landsiedel, R.; Merk, H. F.; Krutmann, J.; Edwards, R. J.; Pease, C.; Goebel, C.; Hewitt, N.; Fritsche, E. *Exp Dermatol* 2012, **21**, 358–363.
3. Suhonen, M. T.; Pasonen-Seppanen, S.; Kirjavainen, M.; Tammi, M.; Tammi, R.; Urtti, A. *Eur J Pharm Sci* 2003, **20**, 107–113.
4. Sahi, J.; Grepper, S.; Smith, C. *Curr Drug Discov Technol* 2010, **7**, 188–198.
5. Swift, B.; Pfeifer, N. D.; Brouwer, K. L. *Drug Metab Rev* 2010, **42**, 446–471.
6. Malinen, M. M.; Palokangas, H.; Yliperttula, M.; Urtti, A. *Tissue Eng Part A* 2012, **18**, 2418–2425.
7. Forrest, I. A.; Murphy, D. M.; Ward, C.; Jones, D.; Johnson, G. E.; Archer, L.; Gould, F. K.; Cawston, T. E.; Lordan, J. L.; Corris, P. A. *Eur Respir J* 2005, **26**, 1080–1085.
8. Petroff, M. G.; Phillips, T. A.; Ka, H.; Pace, J. L.; Hunt, J. S. *Methods Mol Med* 2006, **121**, 203–217.
9. Freshney, R. I. Basic Principles of Cell Culture. In *Culture of Cells for Tissue Engineering*; Vunjak-Novakovic, G., Freshney, R. I., Eds.; John Wiley & Sons, Inc.: Hoboken, NJ, 2006, 3–22.

10. Soule, H. D.; Maloney, T. M.; Wolman, S. R.; Peterson, W. D., Jr.; Brenz, R.; McGrath, C. M.; Russo, J.; Pauley, R. J.; Jones, R. F.; Brooks, S. C. *Cancer Res* 1990, **50**, 6075–6086.
11. Serrano, M. A.; Macias, R. I.; Briz, O.; Monte, M. J.; Blazquez, A. G.; Williamson, C.; Kubitz, R.; Marin, J. J. *Placenta* 2007, **28**, 107–117.
12. Poulsen, M. S.; Rytting, E.; Mose, T.; Knudsen, L. E. *Toxicol In Vitro* 2009, **23**, 1380–1386.
13. Gumbleton, M.; Audus, K. L. *J Pharm Sci* 2001, **90**, 1681–1698.
14. Bode, C. J.; Jin, H.; Rytting, E.; Silverstein, P. S.; Young, A. M.; Audus, K. L. *Methods Mol Med* 2006, **122**, 225–239.
15. Tavelin, S.; Grasjo, J.; Taipalensuu, J.; Ocklind, G.; Artursson, P. *Methods Mol Biol* 2002, **188**, 233–272.
16. Ho, N. F. H.; Raub, T. J.; Burton, P. S.; Barsuhn, C. L.; Adson, A.; Audus, K. L.; Borchardt, R. T. Quantitative Approaches to Delineate Passive Transport Mechanisms in Cell Culture Monolayers. In *Transport Processes in Pharmaceutical Systems*; Amidon, G. L., Lee, P. I., Topp, E. M., Eds.; Marcel Dekker: New York, 2000, 219–316.
17. Sarmento, B.; Andrade, F.; da Silva, S. B.; Rodrigues, F.; das, N. J.; Ferreira, D. *Expert Opin Drug Metab Toxicol* 2012, **8**, 607–621.
18. Hidalgo, I. J. *Curr Top Med Chem* 2001, **1**, 385–401.
19. Fagerholm, U. *J Pharm Pharmacol* 2007, **59**, 905–916.
20. Baumgart, D. C.; Dignass, A. U. *Curr Opin Clin Nutr Metab Care* 2002, **5**, 685–694.
21. Volpe, D. A. *Future Med Chem* 2011, **3**, 2063–2077.
22. Sun, H.; Chow, E. C.; Liu, S.; Du, Y.; Pang, K. S. *Expert Opin Drug Metab Toxicol* 2008, **4**, 395–411.
23. Czupalla, C. J.; Liebner, S.; Devraj, K. *Methods Mol Biol* 2014, **1135**, 415–437.
24. Vandenhaute, E.; Dehouck, L.; Boucau, M. C.; Sevin, E.; Uzbekov, R.; Tardivel, M.; Gosselet, F.; Fenart, L.; Cecchelli, R.; Dehouck, M. P. *Curr Neurovasc Res* 2011, **8**, 258–269.
25. Naik, P.; Cucullo, L. *J Pharm Sci* 2012, **101**, 1337–1354.
26. Geldenhuys, W. J.; Allen, D. D.; Bloomquist, J. R. *Expert Opin Drug Metab Toxicol* 2012, **8**, 647–653.
27. Adkins, C. E.; Mittapalli, R. K.; Manda, V. K.; Nounou, M. I.; Mohammad, A. S.; Terrell, T. B.; Bohn, K. A.; Yasemin, C.; Grothe, T. R.; Lockman, J. A.; Lockman, P. R. *Front Pharmacol* 2013, **4**, 136.
28. Pardridge, W. M. *Drug Discov Today* 2007, **12**, 54–61.
29. Wilhelm, I.; Fazakas, C.; Krizbai, I. A. *Acta Neurobiol Exp* 2011, **71**, 113–128.
30. Kusuhara, H.; Sugiyama, Y. *NeuroRx* 2005, **2**, 73–85.
31. Weksler, B.; Romero, I. A.; Couraud, P. O. *Fluids Barriers CNS* 2013, **10**, 16–10.
32. Zlokovic, B. V. *Neuron* 2008, **57**, 178–201.
33. Kaur, C.; Ling, E. A. *Curr Neurovasc Res* 2008, **5**, 71–81.
34. Desai, B. S.; Monahan, A. J.; Carvey, P. M.; Hendey, B. *Cell Transplant* 2007, **16**, 285–299.
35. Eugenin, E. A.; Clements, J. E.; Zink, M. C.; Berman, J. W. *J Neurosci* 2011, **31**, 9456–9465.

36. Marchi, N.; Granata, T.; Ghosh, C.; Janigro, D. *Epilepsia* 2012, **53**, 1877–1886.
37. Grieshaber, M. C.; Flammer, J. *Surv Ophthalmol* 2007, **52** Suppl 2, S115–S121.
38. Kuhnline Sloan, C. D.; Nandi, P.; Linz, T. H.; Aldrich, J. V.; Audus, K. L.; Lunte, S. M. *Annu Rev Anal Chem (Palo Alto Calif)* 2012, **5**, 505–531.
39. Fletcher, N. F.; Callanan, J. J. Cell Culture Models of the Blood–Brain Barrier: New Research. In *The Blood–Brain Barrier: New Research*; Montenegro, P. A., Juarez, S. M., Eds.; Nova Science Publishers: Hauppauge, NY, 2012.
40. Lippmann, E. S.; Al Ahmad, A.; Palecek, S. P.; Shusta, E. V. *Fluids Barriers CNS* 2013, **10**, 2–10.
41. Cohen-Kashi, M. K.; Cooper, I.; Teichberg, V. I. *Brain Res* 2009, **1284**, 12–21.
42. Li, G.; Simon, M. J.; Cancel, L. M.; Shi, Z. D.; Ji, X.; Tarbell, J. M.; Morrison, B., III; Fu, B. M. *Ann Biomed Eng* 2010, **38**, 2499–2511.
43. Lu, J. *J Exp Integr Med* 2012, **2**, 39–43.
44. Sano, Y.; Kashiwamura, Y.; Abe, M.; Dieu, L. H.; Huwyler, J. Ã.; Shimizu, F.; Haruki, H.; Maeda, T.; Saito, K.; Tasaki, A. *Clin Exp Neuroimmunol* 2013, **4**, 92–103.
45. Vernon, H.; Clark, K.; Bressler, J. P. *Methods Mol Biol* 2011, **758**, 153–168.
46. Terasaki, T.; Ohtsuki, S.; Hori, S.; Takanaga, H.; Nakashima, E.; Hosoya, K. *Drug Discov Today* 2003, **8**, 944–954.
47. Chang, M. M.-J.; Shih, L.; Wu, R. Pulmonary Epithelium: Cell Types and Functions. In *The Pulmonary Epithelium in Health and Disease*; Proud, D., Ed.; John Wiley & Sons, Ltd: Chichester, 2008.
48. Berube, K.; Prytherch, Z.; Job, C.; Hughes, T. *Toxicology* 2010, **278**, 311–318.
49. Klein, S. G.; Hennen, J.; Serchi, T.; Blomeke, B.; Gutleb, A. C. *Toxicol In Vitro* 2011, **25**, 1516–1534.
50. Toppila-Salmi, S.; Renkonen, J.; Joenvaara, S.; Mattila, P.; Renkonen, R. *Curr Opin Allergy Clin Immunol* 2011, **11**, 29–32.
51. Ali, J.; Ali, M.; Baboota, S.; Sahani, J. K.; Ramassamy, C.; Dao, L.; Bhavna. *Curr Pharm Des* 2010, **16**, 1644–1653.
52. Ugwoke, M. I.; Agu, R. U.; Verbeke, N.; Kinget, R. *Adv Drug Deliv Rev* 2005, **57**, 1640–1665.
53. Harkema, J. R.; Carey, S. A.; Wagner, J. G. *Toxicol Pathol* 2006, **34**, 252–269.
54. Pires, A.; Fortuna, A.; Alves, G.; Falcao, A. *J Pharm Pharm Sci* 2009, **12**, 288–311.
55. Werner, U.; Kissel, T. *Pharm Res* 1996, **13**, 978–988.
56. Merkle, H. P.; Ditzinger, G.; Lang, S. R.; Peter, H.; Schmidt, M. C. *Adv Drug Deliv Rev* 1998, **29**, 51–79.
57. Yoon, J. H.; Kim, K. S.; Kim, S. S.; Lee, J. G.; Park, I. Y. *Ann Otol Rhinol Laryngol* 2000, **109**, 594–601.
58. Yoo, J. W.; Kim, Y. S.; Lee, S. H.; Lee, M. K.; Roh, H. J.; Jhun, B. H.; Lee, C. H.; Kim, D. D. *Pharm Res* 2003, **20**, 1690–1696.
59. Cho, H. J.; Termsarasab, U.; Kim, J. S.; Kim, D. D. *J Pharm Invest* 2010, **40**, 321–332.
60. Gray, T.; Koo, J. S.; Nettesheim, P. *Toxicology* 2001, **160**, 35–46.
61. Cho, H. J.; Balakrishnan, P.; Shim, W. S.; Chung, S. J.; Shim, C. K.; Kim, D. D. *Int J Pharm* 2010, **400**, 59–65.

62. Lee, M. K.; Yoo, J. W.; Lin, H.; Kim, Y. S.; Kim, D. D.; Choi, Y. M.; Park, S. K.; Lee, C. H.; Roh, H. J. *Drug Deliv* 2005, **12**, 305–311.
63. Bai, S.; Yang, T.; Abbruscato, T. J.; Ahsan, F. *J Pharm Sci* 2008, **97**, 1165–1178.
64. Harikarnpakdee, S.; Lipipun, V.; Sutanthavibul, N.; Ritthidej, G. C. *AAPS PharmSciTech* 2006, **7**, E12.
65. Wengst, A.; Reichl, S. *Eur J Pharm Biopharm* 2010, **74**, 290–297.
66. Reichl, S.; Becker, K. *J Pharm Pharmacol* 2012, **64**, 1621–1630.
67. Foster, K. A.; Avery, M. L.; Yazdanian, M.; Audus, K. L. *Int J Pharm* 2000, **208**, 1–11.
68. Berger, J. T.; Voynow, J. A.; Peters, K. W.; Rose, M. C. *Am J Respir Cell Mol Biol* 1999, **20**, 500–510.
69. Teijeiro-Osorio, D.; Remunan-Lopez, C.; Alonso, M. J. *Biomacromolecules* 2009, **10**, 243–249.
70. Amoako-Tuffour, M.; Yeung, P. K.; Agu, R. U. *Acta Pharm* 2009, **59**, 395–405.
71. Wicks, J.; Haitchi, H. M.; Holgate, S. T.; Davies, D. E.; Powell, R. M. *Thorax* 2006, **61**, 313–319.
72. Larsen, K.; Malmstrom, J.; Wildt, M.; Dahlqvist, C.; Hansson, L.; Marko-Varga, G.; Bjermer, L.; Scheja, A.; Westergren-Thorsson, G. *Respir Res* 2006, **7**, 11.
73. Bucchieri, F.; Puddicombe, S. M.; Lordan, J. L.; Richter, A.; Buchanan, D.; Wilson, S. J.; Ward, J.; Zummo, G.; Howarth, P. H.; Djukanovic, R.; Holgate, S. T.; Davies, D. E. *Am J Respir Cell Mol Biol* 2002, **27**, 179–185.
74. Deslee, G.; Dury, S.; Perotin, J. M.; Al Alam, D.; Vitry, F.; Boxio, R.; Gangloff, S. C.; Guenounou, M.; Lebargy, F.; Belaaouaj, A. *Respir Res* 2007, **8**, 86.
75. Wu, X.; Peters-Hall, J. R.; Bose, S.; Pena, M. T.; Rose, M. C. *Am J Respir Cell Mol Biol* 2011, **44**, 914–921.
76. Kunz-Schughart, L. A.; Freyer, J. P.; Hofstaedter, F.; Ebner, R. *J Biomol Screen* 2004, **9**, 273–285.
77. Forbes, I. I. *Pharm Sci Technol Today* 2000, **3**, 18–27.
78. Atsuta, J.; Sterbinsky, S. A.; Plitt, J.; Schwiebert, L. M.; Bochner, B. S.; Schleimer, R. P. *Am J Respir Cell Mol Biol* 1997, **17**, 571–582.
79. Manford, F.; Tronde, A.; Jeppsson, A. B.; Patel, N.; Johansson, F.; Forbes, B. *Eur J Pharm Sci* 2005, **26**, 414–420.
80. Forbes, B.; Ehrhardt, C. *Eur J Pharm Biopharm* 2005, **60**, 193–205.
81. Mühlfeld, C.; Ochs, M. Functional Aspects of Lung Structure as Related to Interaction with Particles. In *Particle-Lung Interactions*, 2nd ed.; Gehr, P., Mühlfeld, C., Rothen-Rutishauser, B., Blank, F., Eds.; Informa Healthcare: New York, 2010, 1–16.
82. Möller, W.; Kreyling, W. G.; Schmid, O.; Semmler-Behnke, M.; Schulz, H. Deposition, Retention and Clearance, and Translocation of Inhaled Fine and Nano-Sized Particles. In *Particle-Lung Interactions*, 2nd ed.; Gehr, P., Mühlfeld, C., Rothen-Rutishauser, B., Blank, F., Eds.; Informa Healthcare: New York, 2010, 79–107.
83. Sakagami, M. *Adv Drug Deliv Rev* 2006, **58**, 1030–1060.
84. Lehmann, A. D.; Daum, N.; Bur, M.; Lehr, C. M.; Gehr, P.; Rothen-Rutishauser, B. M. *Eur J Pharm Biopharm* 2011, **77**, 398–406.
85. Alfaro-Moreno, E.; Nawrot, T. S.; Vanaudenaerde, B. M.; Hoylaerts, M. F.; Vanoirbeek, J. A.; Nemery, B.; Hoet, P. H. *Eur Respir J* 2008, **32**, 1184–1194.

86. Hermanns, M. I.; Unger, R. E.; Kehe, K.; Peters, K.; Kirkpatrick, C. J. *Lab Invest* 2004, **84**, 736–752.
87. Huh, D.; Matthews, B. D.; Mammoto, A.; Montoya-Zavala, M.; Hsin, H. Y.; Ingber, D. E. *Science* 2010, **328**, 1662–1668.
88. Choe, M. M.; Tomei, A. A.; Swartz, M. A. *Nat Protoc* 2006, **1**, 357–362.
89. Salomon, J. J.; Muchitsch, V. E.; Gausterer, J. C.; Schwagerus, E.; Huwer, H.; Daum, N.; Lehr, C. M.; Ehrhardt, C. *Mol Pharm* 2014, **11**, 995–1006.
90. Foster, K. A.; Oster, C. G.; Mayer, M. M.; Avery, M. L.; Audus, K. L. *Exp Cell Res* 1998, **243**, 359–366.
91. Kompella, U. B.; Kadam, R. S.; Lee, V. H. *Ther Deliv* 2010, **1**, 435–456.
92. Chang, J. N. *Handbook of Non-Invasive Drug Delivery Systems*, 1st ed.; Elsevier: Burlington, 2010.
93. Holland, E. J.; Mannis, M. J.; Lee, W. B. *Ocular Surface Disease: Cornea, Conjunctiva and Tear Film*; Elsevier Health Sciences: Philadelphia, 2013.
94. Ye, T.; Yuan, K.; Zhang, W.; Song, S.; Chen, F.; Yang, X.; Wang, S.; Bi, J.; Pan, W. *Asian J Pharm Sci* 2013, **8**, 207–217.
95. Hornof, M.; Toropainen, E.; Urtti, A. *Eur J Pharm Biopharm* 2005, **60**, 207–225.
96. Mannermaa, E.; Vellonen, K. S.; Urtti, A. *Adv Drug Deliv Rev* 2006, **58**, 1136–1163.
97. Hosoya, K.; Tachikawa, M. *Adv Exp Med Biol* 2012, **763**, 85–104.
98. Wisniewska-Kruk, J.; Hoeben, K. A.; Vogels, I. M.; Gaillard, P. J.; Van Noorden, C. J.; Schlingemann, R. O.; Klaassen, I. *Exp Eye Res* 2012, **96**, 181–190.
99. Vellonen, K. S.; Mannermaa, E.; Turner, H.; Hakli, M.; Wolosin, J. M.; Tervo, T.; Honkakoski, P.; Urtti, A. *J Pharm Sci* 2010, **99**, 1087–1098.
100. Vellonen, K. S.; Hakli, M.; Merezhinskaya, N.; Tervo, T.; Honkakoski, P.; Urtti, A. *Eur J Pharm Sci* 2010, **19**, 241–247.
101. Xu, S.; Flanagan, J. L.; Simmons, P. A.; Vehige, J.; Willcox, M. D.; Garrett, Q. *Mol Vis* 2010, **16**, 1823–1831.
102. Rangasamy, S.; Srinivasan, R.; Maestas, J.; McGuire, P. G.; Das, A. *Invest Ophthalmol Vis Sci* 2011, **52**, 3784–3791.
103. Kashyap, M. V.; Ranjan, A. P.; Shankardas, J.; Vishwanatha, J. K. *In Vivo* 2013, **27**, 685–694.
104. Carter, A. M. *Physiol Rev* 2012, **92**, 1543–1576.
105. Kitano, T.; Iizasa, H.; Hwang, I. W.; Hirose, Y.; Morita, T.; Maeda, T.; Nakashima, E. *Biol Pharm Bull* 2004, **27**, 753–759.
106. Lager, S.; Powell, T. L. *J Pregnancy* 2012, **2012**, 179827.
107. Rytting, E.; Ahmed, M. S. Fetal Drug Therapy. In *Clinical Pharmacology During Pregnancy*; Mattison, D. R., Ed.; Elsevier: Amsterdam, 2013, 55–72.
108. Vahakangas, K.; Myllynen, P. *Br J Pharmacol* 2009, **158**, 665–678.
109. Sastry, B. V. *Adv Drug Deliv Rev* 1999, **38**, 17–39.
110. Prouillac, C.; Lecoeur, S. *Drug Metab Dispos* 2010, **38**, 1623–1635.
111. Avery, M. L.; Meek, C. E.; Audus, K. L. *Placenta* 2003, **24**, 45–52.
112. Rytting, E.; Bryan, J.; Southard, M.; Audus, K. L. *Biochem Pharmacol* 2007, **73**, 891–900.
113. Li, H.; van Ravenzwaay, B.; Rietjens, I. M.; Louisse, J. *Arch Toxicol* 2013, **87**, 1661–1669.

114. Koeppen, B. M.; Stanton, B. A. *Berne & Levy Physiology*, 6th ed.; Mosby Elsevier: Philadelphia, 2010.

115. Pandit, N. K. *Introduction to the Pharmaceutical Sciences*; Lippincott Williams & Wilkins: Baltimore, 2007.

116. Terryn, S.; Jouret, F.; Vandenabeele, F.; Smolders, I.; Moreels, M.; Devuyst, O.; Steels, P.; Van Kerkhove, E. *Am J Physiol Renal Physiol* 2007, **293**, F476–F485.

117. Markadieu, N.; San Cristobal, P.; Nair, A. V.; Verkaart, S.; Lenssen, E.; Tudpor, K.; van Zeeland, F.; Loffing, J.; Bindels, R. J.; Hoenderop, J. G. *Am J Physiol Renal Physiol* 2012, **303**, F886–F892.

118. Schlatter, P.; Gutmann, H.; Drewe, J. *Eur J Pharm Sci* 2006, **28**, 141–154.

119. Taub, M. *Methods Mol Biol* 2005, **290**, 231–247.

120. Vesey, D. A.; Qi, W.; Chen, X.; Pollock, C. A.; Johnson, D. W. *Methods Mol Biol* 2009, **466**, 19–24.

121. Van der Hauwaert, C.; Savary, G.; Gnemmi, V.; Glowacki, F.; Pottier, N.; Bouillez, A.; Maboudou, P.; Zini, L.; Leroy, X.; Cauffiez, C.; Perrais, M.; Aubert, S. *PLoS One* 2013, **8**, e66750.

122. Lash, L. H.; Putt, D. A.; Cai, H. *Toxicology* 2006, **228**, 200–218.

123. Jang, K. J.; Mehr, A. P.; Hamilton, G. A.; McPartlin, L. A.; Chung, S.; Suh, K. Y.; Ingber, D. E. *Integr Biol (Camb)* 2013, **5**, 1119–1129.

124. Volpe, D. A. *J Pharm Sci* 2008, **97**, 712–725.

125. Liang, M.; Ramsey, C. R.; Knox, F. G. *Kidney Int* 1999, **56**, 2304–2308.

126. Kobayashi, M.; Shikano, N.; Nishii, R.; Kiyono, Y.; Araki, H.; Nishi, K.; Oh, M.; Okudaira, H.; Ogura, M.; Yoshimoto, M.; Okazawa, H.; Fujibayashi, Y.; Kawai, K. *Nucl Med Commun* 2010, **31**, 141–146.

127. Chassin, C.; Bens, M.; Vandewalle, A. *Cell Biol Toxicol* 2007, **23**, 257–266.

128. Hara, C.; Satoh, H.; Usui, T.; Kunimi, M.; Noiri, E.; Tsukamoto, K.; Taniguchi, S.; Uwatoko, S.; Goto, A.; Racusen, L. C.; Inatomi, J.; Endou, H.; Fujita, T.; Seki, G. *Pflugers Arch* 2000, **440**, 713–720.

129. Jenkinson, S. E.; Chung, G. W.; van Loon, E.; Bakar, N. S.; Dalzell, A. M.; Brown, C. D. *Pflugers Arch* 2012, **464**, 601–611.

130. Glube, N.; Giessl, A.; Wolfrum, U.; Langguth, P. *Nephron Exp Nephrol* 2007, **107**, e47–e56.

131. Wieser, M.; Stadler, G.; Jennings, P.; Streubel, B.; Pfaller, W.; Ambros, P.; Riedl, C.; Katinger, H.; Grillari, J.; Grillari-Voglauer, R. *Am J Physiol Renal Physiol* 2008, **295**, F1365–F1375.

132. Astashkina, A. I.; Mann, B. K.; Prestwich, G. D.; Grainger, D. W. *Biomaterials* 2012, **33**, 4700–4711.

133. Elliott, N. T.; Yuan, F. *J Pharm Sci* 2011, **100**, 59–74.

134. Matsusaki, M.; Case, C. P.; Akashi, M. *Adv Drug Deliv Rev* 2014, **74**, 95–103.

135. Astashkina, A. I.; Mann, B. K.; Prestwich, G. D.; Grainger, D. W. *Biomaterials* 2012, **33**, 4712–4721.

8

INTELLECTUAL PROPERTY AND REGULATORY ISSUES IN DRUG DELIVERY RESEARCH

SHAHNAM SHARAREH AND WANSHENG JERRY LIU

Fox Rothschild LLP, Lawrenceville, NJ, USA

8.1 INTRODUCTION

There is no secret that the market success of the innovative pharmaceutical companies in the past 50 years, at least partially, depended on their successful exclusivity strategies. This has been even more evident in the recent years as many blockbuster drugs have lost or are losing patent protection, and the revenues generated from such drugs begin to dwindle. Having faced fierce generic competitions and lack of new blockbuster drug portfolios, many of the big pharmaceutical companies have started a paradigm shift of extending their products' life cycle by repurposing or developing a generic version of their own products.

Drug delivery research perhaps offers one alternative approach toward improving clinical outcomes while maintaining some degree of market control. Drug delivery research has evolved significantly in the past 30 years. Providing shelf-stable immediate release formulation some 40 years ago may now seem trivial compared to today's techniques for smart drug delivery systems that seem to match the characteristics of an active ingredient with a patient's specific needs to maximize clinical outcomes. Even though discovery of active ingredients, chemical or biological, still plays a critical role in the review and approval process, developing an appropriate delivery system to not only enhance clinical efficacy but also maintain

Drug Delivery: Principles and Applications, Second Edition. Edited by Binghe Wang, Longqin Hu, and Teruna J. Siahaan.

a competitive edge in the marketplace has become an integral part of the pharmaceutical industry's product strategy.

Any drug delivery system must first undergo safety and efficacy testing before entering the market. Still, reaching the market is not necessarily guaranteed. In fact, a successful project is merely an illusion if a company does not develop proper intellectual property and regulatory strategies to maximize the life cycle of a given product line. The Food and Drug Administration (FDA) regulations and patent laws place substantial barriers to market entry for innovative drug companies to achieve market competitiveness.

In view of such barriers and the rapid rise in costs of drug development, innovative drug companies are more in need of developing strong intellectual property strategies to offset the costs associated with the development stages. The competitive nature of the industry demands the players to be savvy and efficient in handling their regulatory and intellectual property matters. The scientific and legal issues that arose during recent litigation cases involving drugs such as Aciphex®, Prilosec®, Zyprexa®, Prozac®, and Buspar® further call for strategic planning of drug product life-cycle management. Therefore, scientific, legal, and practical market considerations must carefully be weighed to best protect the competitive edge. For such reasons, it is important that pharmaceutical scientists obtain a stronger understanding of the patent principles and regulatory strategies to maximize the life cycle of a given product line.

This chapter discusses the basis of patent laws and FDA regulations that drug manufacturers must be mindful of before developing a patent strategy.

8.2 PHARMACEUTICAL PATENTS

A patent is a legal instrument that grants a federal recognized exclusivity over a new, unobvious invention having industrial utility. Just as the deed for a real property protects the holder from trespassers, a patent protects a patent owner from other people's impermissible making or using the patented invention for a 20-year term. A pharmaceutical product can be protected by different types of patents.

These include patents that protect a new active ingredient and its formulations, collectively called "product patents." The same product can also be protected by one or more process patents to cover processes of, for example, using or making the product. The rights obtained from a patent are defined in the respective claims of the patent.

In product patents, the claims may be directed to an active pharmaceutical substance as a new chemical entity, a formulation containing such entity, and even an article that contains either, such as a kit or a customized delivery container. Although the patent laws or regulations may differ from one country to another, as different countries may impose certain limitations on the scope of protection, by and large, most countries having a patent system substantially follow a similar set of rules.

Process patents are those that claim processes of making or using a product. For example, the indication approved on the label of a pharmaceutical product may provide the rationale for a method of use (or method of treatment) patent. The same is applicable to patents directed to processes of manufacturing drugs.

Formulation and drug delivery system patents are product patents that claim pharmaceutical dosage forms. They are also referred to as "composition of matter" patents. Composition of matter patents can take a variety of claim structures and may include new formulations for an old drug, such as a delayed or controlled release version, transdermal patches, liposomal, or polymeric delivery systems.

8.3 STATUTORY REQUIREMENTS FOR OBTAINING A PATENT

Article I, section 8 of the US Constitution provides that "[C]ongress shall have power … To promote the progress of science and useful arts, by securing for limited times to authors and inventors the exclusive right to their respective writings and discoveries." Section 101 of the US Patent Act provides that "whoever invents or discovers any new and useful process, machine, manufacture, or composition of matter, or any new and useful improvement thereof, may obtain a patent therefor, subject to the conditions and requirements of this title." Therefore, to be patentable a pharmaceutical invention must be of statutory subject matter, new or novel, nonobvious, and useful with industrial applicability. These requirements are substantially similar in all global existing patent regimes.

8.3.1 Patentable Subject Matter

Among patentability requirements, the patentable subject matter requirement may vary from country to country. Different countries include or exclude certain subject matter from patentability even if the invention as a whole is still novel, nonobvious, and has industrial applicability. As the main restriction, the laws of nature, physical phenomena, abstract ideas, and mathematical formula are generally not acceptable subject matters. However, a process may be patentable if "(1) it is tied to a particular machine or apparatus, or (2) it transforms a particular article into a different state or thing" [1].

In recent years, what is viewed to be a patentable subject matter has been the issue of much controversy. In *Association for Molecular Pathology v. Myriad Genetics, Inc.*, 133 S.Ct. 2017 (2013) [2], the US Supreme Court ruled that even though a complementary DNA sequence may qualify as a patentable subject matter, the natural or isolated DNA cannot, because "the portion of the DNA isolated from its natural state sought to be patented is identical to that portion of the DNA in its natural state." In *Mayo Collaborative Services v. Prometheus Laboratories, Inc.*, 132 S.Ct. 1289 (2012) [3], a process that established a correlation between blood test results and patient health in determining an appropriate dosage of a specific medication for the patient was held to be nonpatentable subject matter by the Supreme Court, because the correlation was deemed to be a law of nature.

In the field of drug delivery, systems used for delivering a DNA may be patentable so long as the inventive concept is not the DNA, rather the delivery platform. Similarly, a drug delivery system delivering the DNA or methods of treating a condition using such DNA can still qualify as a patentable subject matter.

8.3.2 Novelty

The novelty or "newness" requirement is based on the underlying principle that an invention cannot be patented if certain public disclosures are available about that invention. Accordingly, if a claimed invention is not exactly the same as previously disclosed, then it is considered novel. In the US Patent system, 35 U.S.C. § 102 sets forth the novelty requirement for patentability. A claim in a patent application is viewed to be not novel, or be anticipated, "if each and every limitation is found either expressly or inherently in a single prior art reference" [4]. The teachings of the prior art do not need to be explicit, but "anticipation by inherent disclosure is appropriate only when the reference discloses prior art that must *necessarily* include the unstated limitation..." [5]. Therefore, new undiscovered properties of an old formulation may not necessarily warrant a new product patent, if the product itself was already known in the art. "The discovery of a new property or use of a previously known composition, even when that property and use are unobvious from the prior art, cannot impart patentability to the known composition" (noting that "a new use of a known composition ... may be patentable as a process") [6].

The America Invents Act ("AIA") of September 16, 2011 brought various changes to the novelty rules of the US Patent Act, particularly relevant to determination of novelty and nonobviousness requirements. AIA changed the US Patent system from a "first-to-invent" system to "first-inventor-to-file" system. Thus, patent publications, printed publications, public uses, and sale of the invention that predates a patent application's effective filing date are considered prior art against the patent application. Any technical disclosure prior to filing a patent application must therefore be carefully considered so as not to create self-imposed prior art.

Historically, the general approach in the pharmaceutical patents was to draft patent applications very broadly to capture all possible applications and variations of a technology in the first application. However, such strategy requires rethinking in view of the evolution of sciences and the growth of the body of prior art. Due to the pressure of cost in today's industry, pharmaceutical patent drafting cannot afford a defeat under the novelty requirement. For such reasons a robust strategy is paramount to avoid inadvertent novelty problems.

8.3.3 Nonobviousness

For an invention to be patentable, it must not only be novel but also be nonobvious over the teachings of prior art. Under 35 U.S.C. § 103, nonobviousness is determined by assessing whether the invention to be patented would have been obvious "to one of ordinary skill in the art at the time the invention was filed." This determination may appear formulaic but is complex and fact-sensitive in nature. A nonobviousness analysis requires that the claimed invention be compared to the teachings of prior art; and if the differences in the new invention would have been obvious to a person having ordinary skill in the same technology, then the claimed invention is said to be obvious.

This analysis was articulated in the Supreme Court case of *Graham v. John Deer Co.*, 383 U.S. 1 (1966) [7]. In *Graham,* the Supreme Court stated that the question of obviousness must be resolved on the basis of the following factual inquiries: (i) determining the scope of the claimed invention and content of the prior art, (ii) ascertaining the differences between the claimed invention and the prior art, and (iii) resolving the level of ordinary skill in the pertinent art. Evidence, referred to as "secondary considerations," such as commercial success, long-felt but unsolved needs, failure of others and unexpected results, can also be used to further justify nonobviousness.

In the landmark case of *KSR International Co. v. Teleflex Inc.*, 550 U.S. 398 (2007) [8], the US Supreme Court further clarified the obviousness analysis that should be used in patent cases. The Supreme Court rejected the use of a "rigid and mandatory formula" developed in Federal Circuit in determining obviousness of a patent claim. The Supreme Court mandated that instead of using "expansive and flexible approach," a more commonsense approach should be considered on a case-by-case basis. Accordingly, patentable improvements should be "more than the predictable use of prior art elements according to their established functions" [9]. Such a standard solidified a lowered threshold for establishing a *prima facie* case of obviousness. *KSR*'s impact on the pharmaceutical industry was projected to be substantial. Reasonable minds differ on the degree of this impact, depending on which side of the industry they are standing.

Drug delivery research mainly falls into the pharmaceutical formulation inventions and are generally directed to particular forms of delivery, such as a capsule, a tablet with specific delivery characteristics, a polymeric matrix containing drug, smart ionic delivery systems, solutions, suspensions, liposomal preparation, and the like. All of such formulations are designed for delivery of an active chemical or biological ingredient that is placed therein. Perhaps the biggest challenge in obtaining a patent for a pharmaceutical formulation in drug delivery research is the nonobviousness determination, because post–*KSR* law still does not follow a uniform approach in determining obviousness of composition patents. In some cases, Federal Circuit leans toward KSR's approach; in others, the Court maintains a showing of more affirmative reasoning for any modifications. Moreover, *post facto* determination of obviousness for a drug formulation is marred by improper hindsight of the scientists in the field. Therefore, biased conclusions after the fact can taint the nonobviousness assessment.

For example, in *Bayer Schering Pharma Ag v. Barr Laboratories*, 575 F.3d 1341 (Fed. Cir. 2009) [10], the Federal Circuit did not determine whether there was a reason for the patent holder to combine micronized drospirenone with the normal pill formulation, yet the Court found that the patent was obvious because micronization process in normal pill formulation was indeed "obvious to try" and well within the skills of an ordinary formulation scientist.

In another scenario, and perhaps the most telling of Federal Circuit's disagreements with the *KSR* decision, the Court relied on a "lead compound test" to maintain validity of a patent. In *Unigene Labs. and Upsher-Smith Labs v. Apotex*, 655 F.3d 1352 (Fed. Cir. 2011) [11], the Federal Circuit affirmed the lower court and provided an analysis that was not necessarily consistent with the KSR rationale. These approaches pose the question as to whether *KSR* can uniformly be applied in patent-sensitive industries such as pharmaceutical industry.

Historically, in many chemical cases prior to *KSR*, the obviousness determination followed a legal test called "lead compound test." According to this legal test, the obviousness analysis begins with the closest prior art compound. The differences between the closest prior art compound and the claimed invention are then alleviated by the teachings of prior art. Such analysis had not *per se* existed in patent disputes involving a drug delivery composition. However in *Unigene*, the Federal Circuit expanded the same principles in the lead compound test to pharmaceutical compositions. Referring to it as a "reference composition test," in *Unigene*, the reference composition was alleged to have been modified with a pH adjuster instead of a preservative. The challenger took the position that such modification would have been obvious to those of ordinary skill in the art at the time of the filing of the patent application.

However, the Federal Circuit rejected such a position and rationalized that under the reference composition test, one of ordinary skill in the art must first identify a particular starting point composition and then provide motivation to develop a bioequivalent one. In short, it would be essential to explain why one of ordinary skill in the art would modify the reference composition as alleged. Therefore, the Court ruled that the state of art would not cause a person of ordinary skill to replace a preservative in referenced composition with a pH adjuster in the normal course of research and development; and absent a reason substituting a pH adjuster in place of a preservative, the claimed invention would not have been obvious.

Despite gradual evolution of the patent pharmaceutical formulation laws, the obviousness analysis remains fact intensive and challenging. As a formulation scientist develops new ways to administer products, such new ways must show certain benefits, superiority, or unexpected results to improve the chances of patentability over the existing products or prior art. Simply changing the active ingredient in a similar formulation designs may not warrant patentability if the modifications are deemed obvious in view of those of ordinary skill in the art. Therefore, in any endeavor related to drug delivery, researchers should at least consider a showing of clinical advantage and superior or unexpected results to justify patentability over older versions of formulations.

8.4 PATENT PROCUREMENT STRATEGIES

More than ever, intellectual property strategies must include not only patents but also trade secrets as a mode to protect novel ideas and know-how's. This inherently creates a tension in certain research activities, such as in university or public settings, when researchers are balancing patenting and the need for publications. In the university research settings, for example, a central organization commonly referred to as Office of Technology Transfer is in charge of identifying research with potential commercial interest and developing strategies to exploit it commercially. For pharmaceutical and drug delivery research, patent procurement strategy is an integral part of any commercial drug formulation research projects.

There are many methods of developing a patent portfolio and managing the patent term for a given platform. It typically begins with the filing of a provisional patent application to reserve additional time to further develop the research. Once an affirmative decision to pursue the technology is made, a utility patent application for examination of the invention should be filed. It would then continue by developing follow-on continuation applications, reissue, or design patent applications to cover an approved drug product. AIA substantially changed the US patent laws and established a patent system where patent publications, printed publications, public uses, and sale of the invention that predates a patent's effective filing date are considered prior art, even if the publication is based on the inventor's own work.

A provisional patent application is a type of application that gives inventors additional time to develop the invention, and the application can be updated on or before the 1-year anniversary date with newly developed data when the provisional application is converted to a utility application or an international (or PCT) application. This step, or commonly referred to in the industry as the "conversion time," is the only step when a patent application can be changed. Future follow-on applications can be modified to contain new data, but the new data cannot enjoy the priority date protecting them from inventors' own work or patents. Second generation, or "continuation-in-part," applications are also part of the mechanisms available to cover modifications and improvements in the technology. For example, a class of lead drug candidate molecules might be claimed in an initial provisional or utility patent application, but more specific inventions can be claimed in later unrelated applications. As the pharmaceutical data for such members becomes available, different branches of the research could support entire new patent family.

At the outset every filed application would be examined or prosecuted at the respective patent offices around the world. Upon meeting statutory requirements and approvals, the granted patents undergo a patent term adjustment that accounts for any lost term due to patent office delays that extend the period of patent. Granted patents may further be narrowed in scope or reexamined for post-grant protection.

8.5 REGULATORY REGIME

In the United States, before any drug formulation reaches the market, it must meet the safety and efficacy criteria by the Food, Drug, and Cosmetics Act (FDCA) as enforced by the FDA. FDA sets the regulatory standards and barriers to market entry. Accordingly, no drug formulation may enter the market without prior FDA review and approval.

Since the passage of the Drug Price Competition and Patent Term Restoration Act, also known as the Hatch-Waxman Act of 1984, pharmaceutical companies have had three drug approval pathways to clear a drug product through the FDA. Congress in its passage of the Hatch-Waxman Act intended to strike a balance between innovation and easier access to medication, a balance to incentivize continued innovation by brand pharmaceutical companies, while encouraging the generic manufacturers to challenge the status quo.

The new drug application (NDA) approval process under the FDCA essentially requires applicants to substantiate safety and efficacy of a drug product through well-controlled human clinical trials. This path is typically referred to in the art as section 505(b)(1)[1] application, which applies to all new pharmaceutical products. An NDA is a comprehensive document submitted to the FDA through which a sponsor requests the FDA to approve a new pharmaceutical product for sale and marketing in the United States.

Before a sponsor can even begin human clinical trials it must first submit an investigational new drug application (IND) to the FDA. IND is merely a request to obtain FDA's permission for interstate shipment of an unapproved drug product. Once the IND is submitted to the FDA, the sponsor must wait 30 days before initiating any human studies.

The data gathered during the animal studies and human clinical trials of an IND become a part of the NDA. The equivalent application for biological drugs is referred to as the Biologics License Application (BLA). The NDA and BLA are comprehensive packages setting forth the manufacturing, chemistry, stability testing data, packaging, and the proposed labeling of a new drug product based on the results of animal and human clinical data obtained during the developmental stage of the product. Establishing that the drug or biologics product is both safe and effective for its proposed use is the fundamental premise of any NDA or BLA. Noncompliance of an approved drug or biologics product with established proper procedures can lead to the issuance of warning letters or initiation of other FDA enforcement actions against the manufacturer of the product.

The second path under the Hatch-Waxman Act is under section 505(j) of the Act. The abbreviated NDA (ANDA) process is an administrative mechanism introduced by the Hatch-Waxman action for a rapid approval of a generic pharmaceutical product. Under this path, an applicant needs to show that its proposed product is bioequivalent to a drug product on the market, or called "a reference drug." Prior to the Hatch-Waxman Act, generic drugs had to undergo the same safety and efficacy testing as required of innovative drug products. ANDA is a faster and substantially cheaper process, allowing generic manufacturers to rely on the safety data of an already approved drug.

The third path is a hybrid form of sections 505(b)(1) and 505(j), known as section 505(b)(2). Under this path, historically referred to as a "paper NDA," an NDA is filed; however, the sponsor of the application can include information obtained from studies not conducted by or for the sponsor of the 505(b)(2) application. This approach is particularly useful in the field when approvals of new formulations of an older active ingredient are sought. It is generally a faster-approval process, especially for those that represent some limited changes from an existing or approved drug product. This path is ideal for such developmental projects as (i) seeking a new indication for an older formulation, (ii) changing the dosage form of a previously approved drug, and (iii) introducing new strengths, dosing regimen, route of administration, new combination products, new active ingredients, and prodrugs of an existing drug.

Prodrugs are viewed as new molecular entities. However, in certain cases such a new molecular entity may have been studied by a party other than the sponsor or the

[1] Referring to the section of the FDCA authorizing the process.

applicant who is submitting the premarket application. In other scenarios, there may be published information available that is pertinent to the new molecular entity or an active metabolite of an approved drug. Therefore, a 505(b)(2) application could be the preferred path toward approval.

Many other factors can play a role in determination of a proper regulatory path. To name a few, careful consideration of factors such as the disease to be treated, the exclusivity strategies, and commercial-scale manufacturing methodologies may prove significant in establishing realistic expectation of a drug development project. Accordingly, it is advisable that any long-term drug development strategy should prudently include consideration of the intellectual property and regulatory issues to maximize the life cycle of a product.

8.6 FDA MARKET EXCLUSIVITIES

Under the Hatch-Waxman Act, the FDA cannot approve an ANDA before relevant patent issues are resolved. The Act mandates that an NDA applicant submit to the FDA its patent information, covering its product within a time limit but during the pendency of its NDA. This information is made available in a publication called "Approved Drug Products with Therapeutic Equivalence Evaluations" (commonly known as the "Orange Book"). The Orange Book compiles information of all drug products marketed in the United States and approved by the FDA on the basis of safety and efficacy. It also contains all patent numbers that cover the active ingredient, formulation, and other articles for which an NDA was submitted. An NDA holder may also submit patent information that covers the method of using the drug that was the subject of the NDA. The list in the Orange Book is collectively referred to in the industry as the Orange Book patents. These include all product and method of use patents that the NDA filer believes could reasonably be asserted against potential generic drug competitors.

When a generic drug company files an ANDA application to seek approval of a generic version of a drug product already approved under an NDA, it must submit a patent certification on each patent listed in the Orange Book by the NDA filer. Although the Act provides a safe harbor against patent infringement, allowing bioequivalence testing to be performed prior to patent expiration to allow generics to reach the market quickly, it does not allow automatic market entry of the generic product until all patent issues have undergone the statutory dispute resolution process.

These procedures include an ANDA filer's independent review of all Orange Book and non-Orange Book patents relevant to the product. The Act provides four types of patent certifications: Paragraph I through Paragraph IV. Upon comprehensive review of all the relevant patents, the ANDA filer determines which type of certification is required for a given listed patent. Under Paragraph I, Paragraph II, and Paragraph III certifications, ANDA filers do not challenge any of the listed patents covering the subject or referenced product. As such, multiple generics can enter the market at the same time, causing substantial erosion on market control of the branded product.

However, by submitting a Paragraph IV certification, the ANDA filer effectively challenges at least one of the listed Orange Book patents. A Paragraph IV certification is considered an act of infringement as a matter of law. The patentee can in return initiate a legal action against the generic manufacturer for patent infringement. The filing of the action triggers an automatic 30-month stay on an FDA approval of the ANDA. As such, only the first ANDA filer gets to enjoy a limited 6-month (180 days) market exclusivity upon the first commercial marketing of its product, if successful in the patent challenge. During this 180-day exclusivity period, the FDA will not approve any other generic drug application. In the scenario when the ANDA filer loses the patent challenge or settles the litigation, it may be forced to amend its certification to a paragraph III certification. If the ANDA filer does not enter the market on a timely fashion, it may forfeit its exclusivity right.

Over the years, there were incidences that NDA filers entered into certain agreements with the ANDA filers in which they employed a strategy to temporarily keep low-cost generic drugs off the market. The strategy is commonly referred to as "pay-for-delay," which has been a top priority for the Federal Trade Commission (FTC) attempting to make it illegal for many years, as it was perceived by the FTC to be anticompetitive amounting to a violation of antitrust laws. The issue came culminated in a recent legal battle, wherein the Supreme Court held that FTC can sue parties in a pay-for-delay deal for potential antitrust violations if the facts support such intent [12].

In an attempt to level the field, the Hatch-Waxman Act provides a patent restoration process under which the patent holder of an approved product can request for restoring any patent term lost while the drug was under regulatory approval process. Despite certain limitations, a patent term can be extended up to 5 years or for a maximum period of 14 years after the date of approval of the NDA for the drug product [13].

In addition to the patent term restoration, FDA provides additional 5 years of data exclusivity when the NDA is filed for a product that contains a new chemical entity that has not been previously approved for any therapeutic use. Commonly referred to as the 5-year NCE exclusivity, the NDA filer enjoys an absolute exclusivity during which the FDA will not approve a second application covering the same chemical entity, unless a new NDA filer submits its own safety and efficacy data in a new NDA application. The NCE exclusivity applies to such NDAs regardless of whether there is a patent listed in the Orange Book covering the product.

Typically employed as a life-cycle strategy, should an NDA be approved for a new use or a new formulation under the 505(b)(2) route, the FDA provides a 3-year limited exclusivity only toward the approved new uses or formulations. Changes in an approved drug product that affect its active ingredient(s), strength, dosage form, stereoisomeric forms, route of administration, or conditions of use that require further clinical investigations for approval typically qualify for the application for an up to 3-year term. These FDA exclusivities provide market protection independent of any patents protecting the product or listed in the Orange Book to claim the approved product.

There are also other FDA laws and regulations applicable to new uses or new formulations of a drug that provide certain market exclusivities for an NDA filer. These include orphan drug exclusivities for an orphan disease (7 years), pediatric (6 months) exclusivities for additional pediatric studies, and most recently biologics and biosimilar type exclusivities. The 6-month exclusivity based on pediatric studies of a drug adds to the end of an NDA holder's patent term, which encourages the NDA holders to conduct such studies.

8.7 REGULATORY AND PATENT LAW LINKAGE

Manufacturers of drug products with a potential long-market life should consider managing patent term and regulatory exclusivities as part of their strategies to maximize their product's life cycle and market success in the face of scientific uncertainty. As such, any comprehensive drug development strategy calls for a robust plan for available exclusivities. In the United States, the Hatch-Waxman Act has created a flexible Pharmaceutical Regulatory–Patent Law linkage regime that requires all participants of the industry to take prudent measures before approval of their respective premarket drug application in order to sustain their market competitiveness.

By the way of example, any novel NCE can not only be covered by a patent with possible patent term extension but, if approved, it can also enjoy 5-year data exclusivity under the FDA regulations. Patents directed to formulations and methods of use of a particular therapy place additional barriers for generic competition in the market. Pediatric exclusivities or supplemental indications can further extend the market control for additional periods. Finally, a drug delivery strategy must include protecting follow-on and improvement products. Therefore, pursuing new indications with new dosage forms, or even new molecular entities that are derived from an already approved product should be explored as a means to further expand the economic benefits.

The linking of the patent and regulatory regime is an excellent deriver for the industry's expansion as a whole, thereby neutralizing the financial risks facing the pharmaceutical companies. It offers an extended legal protection on drugs at all stages of development and commercialization. Yet, it is complex and demands extensive preplanning. Thus, aside from good science, a better understanding of related legal issues is paramount for effectively navigating in the sea of uncertainty.

REFERENCES

1. In *re Bilski*, 130 S.Ct., 3218, 3226 (2010).
2. In *Association for Molecular Pathology v. Myriad Genetics, Inc.*, 133 S.Ct., 2017 (2013).
3. In *Mayo Collaborative Services v. Prometheus Laboratories, Inc.*, 132 S.Ct., 1289 (2012).
4. In *Celeritas Techs. Ltd. v. Rockwell Int'l Corp.*, 150 F.153d 1354 (Fed. Cir. 1998).
5. In *Transclean Corp. v. Bridgewood Servs., Inc.*, 290 F.3d 1364 (Fed. Cir. 2002).
6. In *re Spada*, 911 F.2d 705 (Fed. Cir. 1990).
7. In *Graham v. John Deer Co.*, 383 U.S., 1 (1966).

8. In *KSR International Co. v. Teleflex Inc.*, 550 U.S., 398 (2007).
9. In *KSR International Co. v. Teleflex Inc.*, 550 U.S., 417 (2007).
10. In *Bayer Schering Pharma Ag v. Barr Laboratories*, 575 F.3d 1341 (Fed. Cir. 2009).
11. In *Unigene Labs. and Upsher-Smith Labs v. Apotex*, 655 F.3d 1352 (Fed. Cir. 2011).
12. In *FTC v. Actavis, Inc.*, 133 S.Ct., 2223 (2013).
13. *35 U.S.C § 156(c)(3).*

9

PRESYSTEMIC AND FIRST-PASS METABOLISM

QINGPING WANG[1] AND MENG LI[2]

[1] *DMPK, Safety and Animal Research, Sanofi US, Waltham, MA, USA*

[2] *Pharmacokinetics and Pharmacometrics, Drug Disposition, Safety and Animal Research, Sanofi US, Bridgewater, NJ, USA*

9.1 INTRODUCTION

The choice of drug administration route is based on the consideration of the site of action, the clinical condition of the patients, and the rate and extent of drug absorption. Oral route is the most common and preferred method of administration for small-molecule drugs. Once disintegrated and dissolved in the gastrointestinal (GI) fluid, a drug molecule needs to traverse the biological barriers of the intestine and liver where it may be subject to "first-pass" loss by the sequential actions of intestinal and hepatic metabolism prior to entering systemic circulation to exert pharmacological effect. The process is referred to as first-pass effect. The fraction of the oral dose, which reaches systemic circulation/target site as intact drug, is referred to as oral bioavailability (F). The oral drug bioavailability is determined by the product of the fraction of dose absorbed from the GI tract (F_a) and the fraction of the absorbed dose that escapes the presystemic metabolism at the intestine (F_G) and liver (F_H), respectively, when drug passes through the elimination organs for the first time.

To understand the first-pass events, several aspects of GI physiology need to be taken into account. The intestine can be broadly divided into small and large intestine. Small intestine (duodenum and jejunum) is the major site for oral drug absorption

Drug Delivery: Principles and Applications, Second Edition. Edited by Binghe Wang, Longqin Hu, and Teruna J. Siahaan.

due to its large absorptive surface area composed of villi and microvilli. Absorption across the gut wall is primarily mediated by paracellular and passive transcellular transport. Enterocytes, the intestinal absorptive cells lining the gut lumen, are not inert cells, but possess significant levels of drug-metabolizing enzymes as well as influx and efflux transporters, which may play important roles in limiting systemic availability of a drug by metabolism and active transport. As the intestine and liver are two first-pass organs arranged in series anatomically, drug absorbed unchanged from gut lumen may also undergo significant first-pass metabolism in the liver as it transits through the liver via the portal vein.

Various factors controlling the extent of intestinal absorption and first-pass metabolism of oral drugs can be ascribed to a combination of drug properties: GI anatomy and physiology characteristics. While factors that limit drug release and dissolution in the gut lumen can be addressed by formulation improvement, drugs that are metabolically labile are likely to exhibit low and variable oral bioavailability due to first-pass metabolism. In addition, first-pass organs are also recognized as common sites for drug–drug interactions (DDIs). Drugs altering the ability of the intestine and liver in the first-pass handling may cause significant changes in the pharmacokinetics (PK) of other affected drug. For instance, concomitant administration of cytochrome P450 (CYP) 3A inducers or inhibitors can alter systemic exposure of orally dosed CYP3A substrates by modulating intestinal first-pass metabolism [1, 2]. Such interactions may result in either subtherapeutic efficacy or increased incidences of adverse effects. Therefore, a thorough understanding of the underlying biochemical basis of presystemic metabolic processes in both intestine and liver is essential in the discovery and development of oral drug therapy.

This chapter will review the current literature of enzyme systems responsible for first-pass metabolism in the intestine and liver, and provide an updated overview of methodology in assessing first-pass effect utilizing *in vitro*, *in vivo*, and *in silico* approaches. Our opinion on the potential strategies to improve oral bioavailability will also be offered.

9.2 HEPATIC FIRST-PASS METABOLISM

The liver is the primary site of metabolism for both endogenous substances and xenobiotics. Hepatocytes, which constitute more than 90% of cells in the liver, abundantly express a large repertoire of enzymes for phase I and II metabolism. Phase I metabolism is the process of functionalization in which functional groups such as hydroxyl, carboxylic, amino, and thiol groups are introduced to the parent drug structure via oxidation, reduction, and hydrolysis. Phase II metabolism is the process of conjugation in which highly polar conjugate is formed by the addition of polar moiety to either phase I metabolite or the parent drug to facilitate further biliary or urinary excretion. The common phase II reactions include glucuronidation, sulfation, glutathione conjugation, and acetylation. The liver not only plays a central role in the systemic elimination of many therapeutic drugs but also presents itself as an important first-pass metabolic barrier in limiting the systemic availability of orally

TABLE 9.1 Hepatic and Intestinal Expression of Major Human Xenobiotic-Metabolizing Enzymes

Enzymes	Expressed in Liver	Expressed in Intestine
Cytochrome P450 (CYP)	1A2, 2A6, 2B6, 2C9, 2C19, 2D6, 2E1, 3A4, 3A5	2C9, 2C19, 2J2, 2D6, 3A4, 3A5
Glucuronosyl transferase (UGT)	1A1, 1A3, 1A4, 1A6, 1A9, 2A3, 2B4, 2B7, 2B10, 2B11, 2B15	1A1, 1A6, 1A8, 1A10, 2B7, 2B15, 2B17
Sulfotransferase (SULT)	1A1, 1B1, 1E1, 2A1	1A1, 1A3, 1B1, 1E1, 2A1
Flavin-containing monooxygenase (FMO)	FMO1, FMO3	FMO1
Carboxylesterase (CES)	CES1, CES2	CES2

administered drugs during absorption. The major enzymes involved in the hepatic first-pass metabolism include CYPs, UDP-glucuronosyltransferases (UGTs), and sulfotransferases (SULTs) (Table 9.1).

9.2.1 Hepatic Enzymes

9.2.1.1 Cytochrome P450 Enzymes The CYP enzymes comprise a superfamily of heme proteins, members of which are present in all branches of the phylogenetic tree from archaebacteria to higher mammals. The human genome encodes 57 individual CYP isoforms that are categorized to 18 different families and 43 subfamilies on the basis of amino acid sequence homology (http://drnelson.uthsc.edu/human.P450.table.html). CYPs are membrane proteins found in either the endoplasmic reticulum (ER) or mitochondrial membrane. The catalytic activity of CYPs requires electron transport via either reduced nicotinamide adenine dinucleotide phosphate (NADPH)-CYP reductase, adrenodoxin, or the small heme protein cytochrome b5. The CYP catalytic cycle has been extensively studied [3], and most of the intermediate states are now fairly well characterized, although the precise nature of the ultimate oxygenating species is still an area of some controversy and debate. CYPs are predominantly expressed in the liver; appreciable expression is also found in the intestine, skin, kidney, and brain. In human liver, the abundance of major CYP enzymes has been reported [4]. As shown in Figure 9.1a, CYP3A is by far the most important and abundant isoform (29% of total hepatic CYPs) of these enzymes and has the broadest range of known substrates [3, 5]. CYP2C and 1A2 represent 18 and 13% of total hepatic P450 content, respectively, followed by minor contribution of CYP2E1, 2A6, and 2D6. Of these, CYP3A, CYP2D, and CYP2C subfamilies are the principle drug-metabolizing enzymes, responsible for the metabolism of approximately 51, 25, and 15% of therapeutic drugs (Fig. 9.1b), respectively [3].

CYP enzymes carry out diverse metabolic reactions including hydroxylation, N-, S-, and O-dealkylation, expoxidation, N- and S-oxidation, deamination, and dehalogenation. For example, metabolism of debrisoquine, an antihypertensive drug, is mediated by CYP2D6 through hydroxylation at the 4-position to form (+)-(*S*)-4-hydroxydebrisoquine [6]. A large variety of intrinsic and extrinsic factors have been shown to influence the

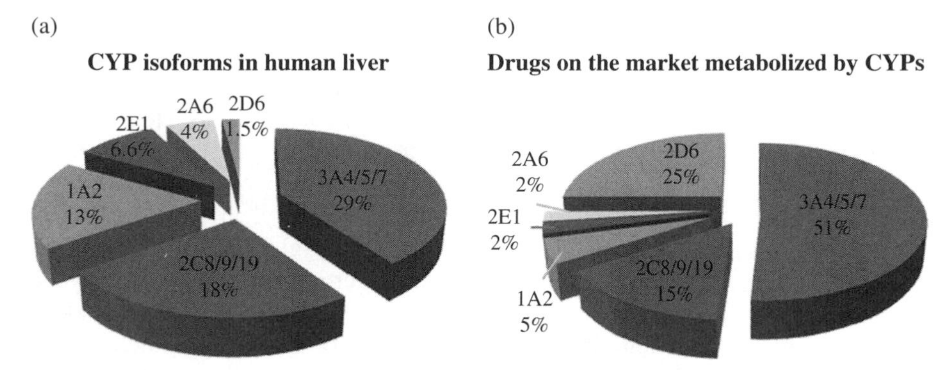

FIGURE 9.1 Relative contribution of (a) CYP isoforms in human liver and (b) of CYP isoforms to the total CYP-mediated metabolism of marketed drugs.

expression and functional activity of CYPs such as age, gender, diet, and disease status. Many CYP inducers and inhibitors have also been identified to modulate enzyme activities. The presence of genetic polymorphism is known in a number of human CYP genes. To date, the best-characterized polymorphic CYPs include CYP2C9, 2C19, 2D6, and 3A5. These predominant polymorphic CYPs are responsible for approximately 40% of human P450-dependent drug metabolism and are linked clinically to the abolished, quantitatively or qualitatively altered drug metabolism [7]. For instance, normal metabolizers of typical CYP2D6 substrates have at least one functional copy of CYP2D6*1 or CYP2D6*2, whereas poor metabolizers with deletions or mutations in both alleles exhibit defective expression of functional CYP2D6 [8, 9]. In contrast, ultrarapid metabolizers have multiple copies of functional CYP2D6*2, resulting from a heritable gene amplification [10]. CYP2D6 polymorphism has led to the withdrawal of a number of clinically used drugs (e.g., debrisoquine and perhexiline) and the vigilant use of other known CYP2D6 substrates due to safety concerns associated with highly variable drug exposure.

As the liver contains the greatest abundance of drug-metabolizing CYPs, CYP substrates are susceptible to efficient hepatic first-pass metabolism, which leads to low bioavailability. For example, buspirone, cyclosporine, lovastatin, saquinavir, and verapamil exhibit poor oral bioavailability (<30%), mainly due to the extensive CYP3A-mediated first-pass metabolism [11].

9.2.1.2 Glucuronosyl Transferases UDP-glucuronosyl transferases (UGTs) are the second most relevant enzymes involved in first-pass drug metabolism based on the number of drug substrates [12]. UGTs catalyze the transfer of glucuronic acid moiety to nucleophilic substrate from the cofactor uridine-5′-diphospho-α-D-glucuronic acid (UDPGA) [13]. UGTs are ER membrane–bound proteins; their active site is exposed on the luminal side of the ER. UGT superfamily constitutes a large multiplicity of isoforms. Of them, the subfamily of UGT1A and UGT2B is mainly involved in drug metabolism. UGTs exhibit broad tissue distribution with liver as the major site of expression; extrahepatic expression is found in the intestine, kidney, lung, skin, brain, prostate, and thymus [14, 15]. In human liver, the major drug-metabolizing UGTs are

UGT1A1, 1A3, 1A4, 1A6, 1A9, 2B4, 2B7, and 2B15, responsible for approximately 35% of the drugs metabolized by phase II metabolism including analgesics, antiviral drugs, nonsteroidal anti-inflammatory drugs, and anticonvulsants [16, 17].

Although CYPs are quantitatively the most important enzyme system in drug metabolism, a large variety of drugs exist that are at least partially metabolized by glucuronidation [18]. This is explained by the diversity of functional groups to which glucuronic acid can be coupled. For example, the hydroxyl groups in phenols, alcohols, hydroxylamines, and hydroxylamides form *O*-glucuronides; carboxylic acids form acyl-glucuronides; primary, secondary, tertiary aromatic, heterocyclic amines or amides, and sulfonamides form *N*-glucuronides; thiols and thioacids lead to *S*-glucuronides; and a few strongly acidic carbons form *C*-glucuronides. A case in point is flavopiridol, an anticancer agent, which is primarily metabolized by UGT1A1-mediated glucuronidation pathway [19, 20]. Drugs that undergo considerable biliary excretion as glucuronide conjugates may be subject to enterohepatic recirculation (EHR), in which the glucuronide metabolites are hydrolyzed by glucuronidases and reabsorbed in the intestine. For example, ezetimibe, a cholesterol-lowering drug, forms two *O*-glucuronides, which are excreted mainly in the bile and undergo EHR [21].

The occurrence of genetic variation has been identified in multiple UGT isoforms including UGT1A1, 1A6, 1A7, 2B4, 2B7, and 2B15 [22, 23]. Of them, UGT1A1 polymorphism appears to be the most clinically relevant, with over 50 genetic variants identified to date [24]. Deficiency in UGT1A1 expression and activity in bilirubin glucuronidation lead to hyperbilirubinemia, namely, Crigler–Najjar (rare) and Gilbert's (more common) syndromes. As UGT-mediated glucuronide conjugation is one of the most important detoxification pathways of phase II drug metabolism, defective glucuronidation may result in either direct toxicity of the substrates or enhanced substrate bioactivation to toxic-reactive intermediates. SN-38, the active metabolite of irinotecan (Fig. 9.2), is an inhibitor of topoisomerase I and is primarily eliminated via glucuronidation mediated by UGT1A1. Generic polymorphisms in UGT1A1 (e.g., UGT1A1*28 variant) is linked to the increased toxicities of irinotecan, which largely arise from the inefficient or reduced glucuronidation of SN-38 [25]. The polymorphic diversity in UGTs may also influence flavopiridol disposition and toxicity in a similar manner to irinotecan. The systemic glucuronidation of flavopiridol was reported to be inversely associated with the risk of diarrhea occurrence [26].

Extensive first-pass glucuronidation acts as a metabolic barrier impacting oral drug bioavailability of UGT substrates and may result in their reduced efficacies. Raloxifene, an estrogen receptor modulator used for the prevention of osteoporosis, is known to undergo extensive presystemic glucuronidation to form 6-β- and 4′-β-*O*-glucuronides by UGT1A1, 1A8, and 1A10, which at least partially account for its low oral bioavailability of 2% [27].

9.2.1.3 Sulfotransferase Sulfotransferases (SULTS) are cytosolic enzymes and catalyze sulfate conjugation by transferring of the sulfuryl group of 3′-phospho-adenosine-5′-phosphosulfate (PAPS) to the substrates [28]. Similar to those of UGTs,

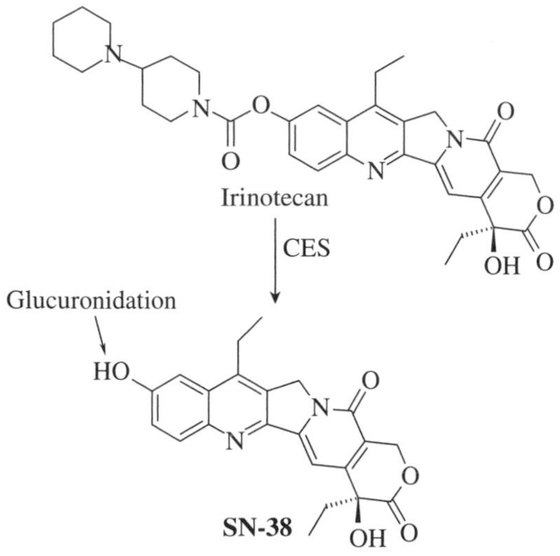

Irinotecan

CES

Glucuronidation

SN-38

FIGURE 9.2 Structure of irinotecan and its active metabolite SN-38.

SULT substrates include alcohols, phenols, and amine-containing compounds. In contrast to the low-affinity, high-capacity reactions of glucuronidation, sulfation reactions are characterized by high affinity (low K_m) and low capacity due to the limited PAPS availability. In human, 11 SULT isoforms have been identified that mediate the sulfation of both endogenous substrates and xenobiotics [18, 29, 30]. SULT1 and SULT2 families are the main enzymes for drug metabolism, with SULT1A1 being the most important enzyme to catalyze the sulfation of a broad variety of structurally diverse xenobiotics [31]. The abundance of SULT expression in human liver has been determined by quantitative immunoblotting [32]. SUL1A1 is the main hepatic SULT isoforms, accounting for 53% of total SULT content. Expression of SULT2A1 (27%), SULT1B1 (14%), and SULT1E1 (6%) is also presented in the liver, albeit at a lower abundance [32].

Many drugs undergo sulfation in parallel to glucuronidation. For instance, the major metabolic pathways of traxoprodil, an *N*-methyl-D-aspartate receptor antagonist, are direct O-glucuronidation and O-sulfation in poor CYP2D6 metabolizers [33]. Extensive sulfation in the liver and/or intestine can lead to low systemic availability for many phenolic drugs including acetaminophen, albuterol, isoproterenol, phenylephrine, and terbutaline. The absolute oral bioavailability of phenylephrine was approximately 38%, and the substantial loss of the drug during absorption resulted from predominant first-pass conjugation [34].

9.2.1.4 Other Enzymes In addition to the enzyme systems mentioned earlier, the following hepatic enzyme families can also play significant roles in the first-pass metabolism.

The flavin-containing monooxygenases (FMOs) are expressed in the ER and are capable of carrying out a variety of oxidative biotransformation. Various aspects of this enzyme family have recently been reviewed [35, 36]. Typical FMO-mediated reactions are the oxidation of nitrogen or sulfur heteroatoms. Of the six families of FMOs, FMO3 is the prominent form present in an adult human liver and is associated with the bulk of FMO-mediated drug metabolism including drugs such as cimetidine, tamoxifen, itopride, and sulindac [37]. FMO3 is highly polymorphic; genetic deficiency in FMO3 is associated with trimethylaminuria disorder [38].

Esterases are hydrolytic enzymes located in the cytoplasm and ER and catalyze the hydrolysis of endogenous substances and xenobiotics that contain ester, amide, or thioester groups. Of them, carboxylesterases (CESs), particularly CES1 and CES2, play important roles in the hydrolytic biotransformation and prodrug activation for a broad variety of therapeutic drugs [39]. Ester prodrugs such as anti-influenza drug oseltamivir and anticancer drug irinotecan (Fig. 9.2) are specifically designed to enhance oral bioavailability and are extensively converted to the active metabolites by esterase-mediated hydrolysis, following absorption from the GI tract [40, 41]. The human liver predominantly expresses CES1 with smaller quantities of CES2, whereas the small intestine contains CES2 with virtually no CES1. Although there is a high degree of sequence homology between CES1 and CES2, they display distinct substrate specificity. Drugs such as clopidogrel and oseltamivir are specifically hydrolyzed by human CES1, whereas irinotecan and prasugrel are preferentially hydrolyzed by CES2 [42, 43].

9.3 INTESTINAL FIRST-PASS METABOLISM

While the intestine was historically regarded as an absorptive site, accumulating *in vitro* evidence has demonstrated that the intestinal mucosa is rich with phase I and II enzymes implicated in drug biotransformation, most notably CYPs, UGTs, and SULTs (Table 9.1) [11, 44, 45]. A series of preclinical and human studies have further proven the significant contributions of intestinal mucosa to the *in vivo* first-pass metabolism of drugs, including CYP3A substrates cyclosporine [1, 46, 47], midazolam [48, 49], verapamil [2, 50, 51], and felodipine [52] as well as UGTs and/or SULTs substrates raloxifene and terbutaline [53, 54]. These findings and other biochemical evidences have collectively established intestine as the most important extrahepatic site of drug metabolism where metabolic transformation may occur concurrently with drug absorption from the intestinal lumen.

9.3.1 Intestinal Enzymes

9.3.1.1 Cytochrome P450 Enzymes Since the discovery of CYP3A in human intestinal mucosa by Watkins and coworkers [55, 56], considerable efforts have been focused on the characterization of CYP contents in human intestine. While CYP1A1, 1B1, 2C, 2D6, 2E1, 3A4, and 3A5 mRNA expression was identified in human enterocytes, only CYP3A and CYP2C were detected at the protein level in human intestinal microsomes [57]. In a more recent study, mucosal CYP expression was profiled in microsomal preparations of small intestine from 31 human donors by immunoblotting [58]. Analogous to the liver, the major P450s in the small intestine are CYP3As, accounting for 82% of the total enteric CYP content. The expression of other CYP isoforms is also found in the small intestine to a minor extent, with CYP2C9 contributing about 15% followed by CYP2C19 (2%), 2J2 (1.4%), and 2D6 (0.7%), respectively [55, 58, 59]. Studies examining CYP catalytic activity of isoform-specific substrate probes have demonstrated similar K_m and comparable intrinsic clearance (per picomole CYP) between intestinal CYP3A, 2C9, 2C19, 2D6 and their hepatic counterparts [59–61].

In contrast to the homogenous CYP expression in the liver, intestinal expression of CYPs exhibits regional heterogeneity along the length of the intestinal tract. CYP3A expression is generally the highest in the proximal region of the intestine and declines distally toward ileum [57, 62]; mean (range) expression of 30.6 (<3.0–91), 22.6 (<2.1–98), and 16.6 (<1.9–60) pmol/mg microsomal protein was found in the duodenum, distal jejunum, and distal ileum, respectively [59]. CYP expression also varies on the crypt–villus axis, with the highest level found in the mature enterocytes lining of the villus tips [63]. This local variation is associated with the differentiation of cell function during cell migration from crypt toward apical tip.

There are large differences in the CYP expression in human intestine among individuals as assessed by both direct *in vitro* biochemical analysis and clinical studies. In an analysis of 20 duodenal pinch biopsies from healthy adult volunteers, an 8- or 11-fold variation was noted in CYP3A4 mRNA or immunoreactive CYP3A4 protein in small bowel biopsy samples [64]. Paine et al. also reported a 17-, 5-, and 9-fold variation in the enteric protein expression of CYP3A4, 3A5, and 2C9, respectively [58]. This marked interindividual variability may originate from the presence of genetic polymorphisms. The small intestine, CYP3A5 expression has been shown to exhibit polymorphic patterns similar to that found in the liver [58]. Differences in the response to CYP regulation by enzyme inhibitors or inducing agents can also contribute to the variation in constitutive CYP expression. Gut wall represents a highly sensitive site for CYP modulation because of its unique anatomical location and exposure to high luminal concentration of interacting drugs. Repeated daily administration of 600 mg rifampin increased intestinal CYP3A expression by 4.4 ± 2.7-fold as compared with the control [65]. Coadministration of rifampin and phenytoin was shown to enhance intestinal first-pass extraction of cyclosporine and verapamil via the preferential induction of intestinal CYP3A [1, 2, 47, 66]. Concurrent intake of grapefruit juice (GFJ) selectively inhibits intestinal CYP3A and was shown to increase the systemic exposure of oral alfentanil and cyclosporine, but had no effect on hepatic CYP enzymes and thus on the pharmacokinetics (PK) of alfentanil and cyclosporine administered by the intravenous route [67, 68]. The large interindividual variability of enteric CYP expression may contribute to the variable oral bioavailability of their substrates, particularly for those drugs whose intestinal metabolism is predominated by the CYP isoforms exhibiting highly variable expression levels. It is also noteworthy that the gut wall first-pass extraction may be more susceptible to enzyme saturation in comparison to liver due to the high drug concentration in the gut lumen.

Although the relative microsomal CYP3A-specific contribution to the intestine doubles that of the liver, the total CYP3A content of the entire human small intestine is estimated to be approximately 1% of that of the liver, taken into account of the microsomal protein content and organ weight [59]. Nevertheless, clinical evidence has shown that both liver and intestine contributed comparably to the first-pass extraction for a number of CYP3A substrates including midazolam and cyclosporine [47, 49, 69]. It was postulated that this may be ascribed to a complex interplay of intestinal absorptive transport and metabolism, differential protein binding and blood flow in the two first-pass organs, as well as the anterior placement of the intestine in the sequential first-pass process.

9.3.1.2 UDP-Glucuronosyltransferase The major UGT forms identified in the human intestine are UGT1A1, UGT1A6, UGT1A8, UGT1A10, UGT2B7, UGT2B15, and UGT2B17 [15, 45]. Of them, UGT1A8 and UGT1A10 exhibit tissue-specific expression found in the intestinal tract, but not in the liver [70–72]. UGT expression appears to be uniform along the intestinal tract. Using various intestinal tissue preparations, *in vitro* studies have demonstrated the gut UGT activities for the metabolism of acetaminophen [73], estradiol and 17β-estradiol [74], ethinyl estradiol [75], morphine [76, 77], propofol [78], and raloxifene [27]. Few *in vivo* studies have been conducted to directly evaluate the impact of intestinal drug glucuronidation. Extrahepatic formation of propofol glucuronide was observed in anhepatic patients, which was likely attributed to intestinal and renal glucuronidation [79]. Further studies are warranted to determine the quantitative contribution of intestinal glucuronidation to *in vivo* drug metabolism.

The absolute abundance of selected UGTs in the intestine was characterized in a recent study employing stable isotope-labeled peptide standards and liquid chromatography–tandem mass spectrometry (LC–MS/MS). Of the three individual donors examined, enteric expression of UGT1A1 and 1A6 was approximately 35% of that of the liver on the basis of per milligram microsomal protein [80]. Studies examining the relative importance of intestinal and hepatic glucuronidation *in vitro* have shown substrate-dependent results [81–85]. Compared to the liver, lower intestinal intrinsic clearance of glucuronidation ($CL_{int.UGT}$) was observed for mycophenolic acid, gemfibrozil, and diclofenac, whereas higher $CL_{int.UGT}$ by intestine was noted for raloxifene and troglitazone. In the latter case, intestine-specific UGT1A8 and 1A10 are determined as the main UGT isoforms in raloxifene and troglitazone glucuronidation and may be significant contributors to their intestinal first-pass metabolism *in vivo* [27, 83].

Large interindividual variability has been observed for intestinal UGTs. Strassburg et al. reported a polymorphic expression pattern for most of the intestinal UGT1A and 2B isoforms, signifying their essential roles to the interindividual variability of first-pass glucuronidation [86]. Similar to CYPs, intestinal expression and activities of UGTs could be altered by extrinsic factors such as induction and inhibition. Inducible human UGT enzymes include UGT1A1, 1A3, 1A6, 1A9, and 2B7. Coregulation of UGTs with CYP enzymes is observed for many UGT inducers, including carbamazepine, rifampin, and phenobarbital via nuclear receptor pathways [45].

9.3.1.3 Sulfotransferase Sulfation is the second most clinically relevant phase II metabolic pathway aside from glucuronidation. *In vitro* studies showed the extensive metabolism of ethinylestradiol to ethinylestradiol sulfate by human jejunal mucosa [75]. Higher activity of terbutaline sulfation was noted in the human intestinal mucosa isolated from duodenum, ileum, and colon, as compared to the liver [53]. In a human study, terbutaline metabolism was administration route dependent. Terbutaline administered intravenously appeared largely unchanged in urine, whereas terbutaline was mostly excreted as sulfate conjugate following oral administration [87]. These early research indicated a key role of the gut wall in the first-pass sulfation of orally administered drugs.

Using immunoblotting and immunohistochemistry techniques, Teubner and coworkers reported the cytosolic expression of SULT1A1, 1A3, 1B1, 1E1, and 2A1 in the intestinal tract [88]. SULT1A1, 1A3, and 1B1 were found to be predominantly expressed in the differentiated enterocytes with the highest level in the ileum [88]. SULT expression profile in small intestine was also determined utilizing quantitative immunoblotting approach [32]. In contrast to the liver where SULT1A1 is predominantly expressed, SULT1B1 and 1A3 are the major intestinal SULT isoforms, accounting for 36 and 31% of the total immunoquantified SULTs, followed by SULT1A1 (19%), SULT1E1 (8%), and SULT2A1 (6%), respectively [32]. Collectively, the small intestine represents the major expression site for SULT1A3, SULT1B1, and SULT1E1, whereas the liver is the major tissue for SULT1A1 and SULT2A1 expression. The small intestine exhibits a higher total expression of SULTs than the liver (7800 vs. 5960 ng/mg cytosol protein) [32]. The substantial quantities of SULTs in the small intestine are indicative of their significant roles in determining the extent of first-pass metabolism for SULT substrates. Substantial variation of enteric SULT expression was also observed in the same study [32], which may be ascribed to SULT genetic variations and differences in gene regulation.

9.3.1.4 Other Enzymes Similar to the liver, the intestine is known to possess hydrolytic activities. Ester-containing drugs and prodrugs can be hydrolyzed by esterases present in the intestinal mucosa. For example, irinotecan was shown to undergo conversion to its active metabolite SN-38 in intestinal S9 fraction [89]. The CES isoform expressed in human small intestine is mainly CES2, which plays an important role in the inactivation of a variety of structurally diverse drugs and activation of prodrugs [39].

The intestinal bacterial microflora may also contribute to the metabolism of drugs whose conjugated metabolites go through significant biliary excretion. Bacterial β-glucuronidase and sulfatase facilitate enterhepatic recirculation by catalyzing the hydrolysis of drug conjugates once secreted into the small intestine (e.g., morphine) [90].

9.3.2 Interplay of Intestinal Enzymes and Transporters

In addition to drug-metabolizing enzymes, the significance of intestinal transporters (solute carrier and ATP-binding cassette transporters) has been increasingly recognized as important determinants of oral drug PK [91, 92]. Efflux transporter P-glycoprotein (P-gp), expressed on the apical membrane of enterocytes, is the most investigated transporter and has been shown to share a significant overlap in substrate specificity with CYP3A. It is hypothesized that these two proteins may act in concert to prolong the residence time in enterocytes and enhance CYP3A-mediated drug metabolism. In support of the hypothesis, *in vitro* drug metabolism studies in CYP3A4-transfected Caco-2 cells demonstrated a decrease of metabolism for dual CYP3A and P-gp substrate K77 in the presence of selective P-gp inhibitor GG918 [93]. Consistent *in vivo* evidence for P-gp modulation of K77 intestinal metabolism was also obtained from rat single-pass intestinal perfusion model [94]. Although there is still considerable controversy around this hypothesis and its clinical relevance, those results support the investigation and incorporation of synergistic interactions of drug-metabolizing enzymes and transporters to improve the precision of *in vitro–in vivo* prediction of oral first-pass metabolism.

9.4 PREDICTION OF FIRST-PASS METABOLISM

Knowledge of the absorption, distribution, metabolism, and excretion (ADME) processes of drug disposition is essential to the compound selection and optimization process in drug discovery. It is also important in the effective management of clinical drug interactions and the understanding of interindividual pharmacokinetic variability during drug development. Various *in vitro*, *in vivo*, and *in silico* methods have been developed and employed to predict the *in vivo* intestinal and hepatic first-pass metabolism in human.

9.4.1 *In vivo* Assessment of First-Pass Metabolism

The absolute net oral bioavailability of a drug can be determined by comparing dose-normalized exposure, following intravenous or oral drug administration of the drug either in a crossover or parallel group design; the latter assumes negligible interindividual variability (Eq. 9.1). The respective values of F_G and F_H (Eq. 9.2) are estimated with a few underlying assumptions that the drug exhibits linear kinetics, has negligible extrahepatic metabolism besides gut, and undergoes complete gut absorption ($F_a = 1$).

$$F = \left(\frac{\text{AUC}_{\text{oral}}}{\text{AUC}_{\text{iv}}}\right) \times \left(\frac{\text{Dose}_{\text{iv}}}{\text{Dose}_{\text{oral}}}\right) \quad (9.1)$$

$$F = F_a \times F_G \times F_H \quad (9.2)$$

Animal studies are routinely used in the preclinical stage to characterize the PK and systemic availability of candidate compounds. Results of these studies have been traditionally used in human PK prediction using allometric scaling approach.

The allometry method is based on the empirical principle that drug clearance and volume of distribution across species correlate with body weight or size. It has become increasingly recognized that many other factors such as the marked species differences in metabolizing enzymes should be investigated and accounted for better human PK prediction. Although animal models such as rat intestinal perfusion model have been utilized to investigate the mechanistic aspects of intestinal metabolism/ transport, there is a tendency to assess the extent of intestinal metabolism in human rather than using animal PK data extrapolation. The species difference of intestinal enzyme expression and metabolic activities remain largely unknown between human and common preclinical animal species. It was reported that *in vitro* intestinal metabolic activities for midazolam and nisoldipine exhibit a large species difference in a rank order of monkey ~ human > rat [95]. Recently, species difference of intestinal metabolism was investigated in intestinal microsomes obtained from rats, dogs and humans using 43 substrate probes of human CYP3A, 2C, 2D6, 1A2 and 2J2. Intestinal CL_{int} values of CYP3A substrates in rat and dog appeared to be lower for most of the compounds and showed moderate correlation with those of human [96]. In another study examining *in vitro* and *in vivo* intestinal glucuronidation activities for 17 UGT substrates in human, rat, dog, and monkey [97], $CL_{int,UGT}$ of intestinal microsomes showed a good correlation between human and animals, while the values tended to be lower in human than in laboratory animals. *In vivo*, the fraction of dose absorbed into portal vein (F_aF_G) in human was shown to correlate with that in dogs and monkeys, but not with rat [97]. Future studies to clarify the quantitative species differences in gut metabolism will be critical in evaluating appropriate animal model for the prediction of human intestinal metabolism and intestinal drug interactions.

Information on the direct assessment of first-pass metabolism in human, in particular the intestinal first-pass, is scarce due to the technical and ethical complexities in such studies. Patient studies utilizing the anhepatic phase during liver transplant have demonstrated the importance of small intestine first-pass metabolism for CYP3A substrates [46, 48]. Concurrent oral administration of GFJ was shown to selectively inhibit intestinal CYP3A *in vivo*, while had no impact on the hepatic enzyme [52, 68, 98]. Furanocoumarins were identified as the active GFJ ingredients responsible for CYP3A inhibition [99]. Given that GFJ was also reported to inhibit intestinal transporters *in vitro* [100], the observed effect may reflect its modulation of first-pass effect via inhibiting both intestinal transport and metabolism. Because of the complete intestinal CYP3A inactivation with appropriate GFJ regimen, the use of GFJ may provide a feasible clinical approach to the determination of intestinal CYP3A-mediated first-pass metabolism in human.

9.4.2 *In vitro* Assessment of First-Pass Metabolism

Given the challenges associated with *in vivo* assessment, *in vitro* investigation is vital to characterizing first-pass metabolism by gut wall and liver. Key *in vitro* characterizations for bioavailability prediction include the determination of compound solubility, membrane permeability and metabolism. Compound solubility is usually evaluated at pH 7.4 in buffer (i.e., 100 mM phosphate buffer). Membrane permeability is often determined by measuring apical to basolateral transport across polarized monolayers of human colon carcinoma–derived (Caco-2) cells or Madin–Darby canine kidney (MDCK) cells. Alternatively, the passive membrane permeability can be determined using artificial membrane (i.e., parallel artificial membrane permeability assay, PAMPA).

There are several *in vitro* systems available to evaluate hepatic metabolism with varying degrees of success, each of which has its own merits and limitations [101]. Microsomal preparation is the most widely used *in vitro* model for its simplicity, high levels of metabolizing enzyme activity and commercial availability. Characterization of CYP and UGT-mediated metabolism in liver microsomes or alamethicin-activated liver microsomes requires the addition of corresponding cofactors, namely NADPH or UDPGA. The major limitations of microsomes are the lack of intact cell membrane and full complement of enzymes/cofactors, which may lead to biased estimate of CL_{int} for certain compounds. The involvements of specific isoforms in drug metabolism can be determined using commercially available recombinant human enzymes. Additionally, S9 supernatant, which contains both cellular cytosolic and ER membrane-bound enzymes, are also used for investigating metabolic reaction by extramicrosomal enzymes. Besides subcellular fractions, primary cells (i.e., cryopreserved and to a less extent freshly isolated hepatocytes) are also often used to characterize drug metabolism and offer unique advantage to mimic the metabolic profile similar to that found *in vivo* [102]. HepaRG cells, a recently established hepatocyte model, are terminally differentiated hepatic cells derived from a human hepatic progenitor cell line that retains many characteristics of primary human hepatocytes and express phase I and phase II enzymes at levels comparable to those in human [103]. HepaRG cells represent a promising alternative to primary human hepatocytes for xenobiotic metabolism and toxicity studies. Other available *in vitro* systems of intact tissues such as precision-cut liver slices (PCLS) preserve the innate tissue structural integrity and are highly suitable for characterizing drug metabolism–transport interplay and enzyme regulation involving multicellular processes [104].

For intestinal metabolism, the predictive *in vitro* methods are not as well characterized and widely available as for the liver. Intestinal microsomes have been successfully used to predict intestinal first-pass metabolism for CYP3A and UGT substrates [105, 106]. Similar to hepatocytes, intestinal epithelial cells (enterocytes) can be isolated by specialized techniques to obtain functional enterocytes. Higher metabolic activity was observed in enterocytes isolated by enterocyte elution compared to mucosal scraping method [61]. The use of primary enterocytes has only been examined in limited studies and requires further validation as a reproducible and quantitative tool for intestinal metabolism evaluation [107]. Equivalent to liver slices, precision-cut intestinal slices (PCIS) represent a more complete model of intestinal physiology compared to microsomes [108]. PCIS can be prepared from all regions of intestine to investigate the regional heterogeneity of intestinal metabolic activity. Although its application in metabolism research is limited by its availability and short-term viability (8–24 hours incubation), PCIS have shown high activities of drug-metabolizing enzymes and are useful *in vitro* tool for drug metabolism, transport, and drug–drug interaction studies [109].

Common procedure of metabolic stability determination involves steps of xenobiotics incubation with *in vitro* human matrix systems such as microsomes or hepatocytes, followed by reaction quenching and sample preparation by centrifugation. The resultant supernatant is analyzed for substrate depletion or metabolite formation using liquid chromatography mass spectrometry (LC–MS). These processes can be adapted to multiwell plate format for rapid high-throughput automation [110]. Major advancements in the selectivity and sensitivity of LC–MS have propelled its applications in the development of high-throughput metabolism assays. Many labs have recently started using various mass spectrometry methodologies for simultaneous screening of the metabolic stability of the parent compound and metabolite profiling within a single LC/MS analysis [111–114]. The approach allows efficient and reliable identification of compounds with adequate metabolic stability for further development.

9.4.3 *In vitro–in vivo* Prediction

It is a routine practice to scale *in vitro* metabolism data to predict human clearance of new chemical entities in drug discovery. Due to the complexity of the physiological processes involved in the intestinal and hepatic extraction, an appropriate mechanistic model is often required to integrate *in vitro* CL_{int} from intestine or liver preparations with relevant physiological factors for the quantitative prediction of the extent of *in vivo* first-pass metabolism. Commonly used *in vitro* systems for hepatic clearance estimation include liver microsomes, hepatocytes and recombinant enzymes. Among various scale-up methods for hepatic CL_{int} and F_H prediction, the conventional well-stirred model, which incorporates hepatic blood flow (Q_H) and plasma protein binding (f_u) in the model prediction, has shown reasonable predictability of the *in vivo* outcome (Eq. 9.3) [104].

$$F_H = \frac{Q_H}{Q_H + f_u \times CL_{int.u}} \tag{9.3}$$

The well-stirred model assumes an instantaneous drug distribution and a homogeneous distribution of metabolic enzymes in all regions of the organ, which are not valid in the case of intestine. Moreover, drug is delivered from gut lumen during intestinal absorption, rather than via the intestinal blood flow. Given the physiological complexities unique to the intestine, more sophisticated physiologically based models have emerged as useful tools to predict intestinal first-pass metabolism. These methods take into account drug properties such as permeability and metabolism, as well as GI physiology including enterocytic blood flow, the distributional heterogeneity of intestinal metabolic enzyme and drug transporters in the models [115, 116]. The Q_{Gut} model is one of the simplest form of physiologically based models, which allows prediction of F_G using *in vitro* determined intestinal CL_{int} and membrane permeability [116, 117]. The suitability of the Q_{Gut} model to predict F_G for drugs with diverse clearance and membrane permeability characteristics was investigated in a recent study. F_G of drugs with low intestinal extraction was generally well predicted, but the prediction for drugs with high intestinal extraction (F_G<0.5) was considerably less accurate [106]. These results indicate the necessity of incorporating additional complexities of the underlying physiological and biochemical processes in the model before *in vivo* intestinal first-pass metabolism can be quantitatively predicted from *in vitro* data for a broad range of drugs [118]. Further research efforts in the further development and refinement of physiologically based models are warranted.

9.4.4 *In Silico* Approach

A high CYP-mediated clearance is of particular concern as it would not only necessitate high and frequent dose regimen but also carry the risk of large variation in exposure. Therefore, improving metabolic stability through structural modification in the early discovery stage is the main focus of medicinal chemists. High-resolution X-ray crystal structures of human CYP1A2, 2A6, 2B6, 2C8, 2C9, 2C19, 2D6, 2E1, and 3A4/5 are now available. One general observation with these and other animal and bacterial P450 structures is that the flexibility of P450 active sites in accommodating very diverse interactions with different substrates. Therefore, the structure of a particular P450/ligand complex is of limited utility in understanding the binding mode(s) of structurally distinct ligands to the same P450 enzyme [119]. Thus, *in silico* methods (computer-based prediction systems or expert systems) are useful tools to predict sites of metabolism [120, 121]. For example, MetaSite, a commercial computational software from Molecular Discovery is used to predict and rank metabolic soft spot(s) in compound structure for CYP-mediated phase I metabolism [122]. Recently, Trunzer and coworkers described a robust metabolic soft spot identification procedure combining *in silico* predictions of the most likely metabolites by MetaSite and mass spectrometric confirmation [123]. Among the set of 18 marketed drugs and 95 Novartis compounds undergoing primarily CYP-mediated metabolic biotransformation, MetaSite's first-rank order predictions of the soft spot were experimentally confirmed for about 55% of the compounds. The secondary- and tertiary-rank order predictions of MetaSite were detected for another 29% of the compounds. Therefore, this automatic and high-throughput reprioritization of likely soft spot may increase the accuracy in the identification of metabolic liable sites to guide structural modification in drug discovery.

9.5 STRATEGIES FOR OPTIMIZATION OF ORAL BIOAVAILABILITY

Oral bioavailability depends on a number of key factors, primarily aqueous solubility/dissolution, membrane permeability, first-pass metabolism, and active transport. As such, optimization of bioavailability is an integral part of drug discovery and development. In the hit to lead process, high-throughput *in vitro* screening assays are set up to efficiently assess the basic compound properties of metabolic stability and permeability. Structure-ADME property relationship is investigated to aid the early effort of medicinal chemistry for the development of chemical series with adequate balance of *in vitro* pharmacology and ADME properties. A general screening tree is

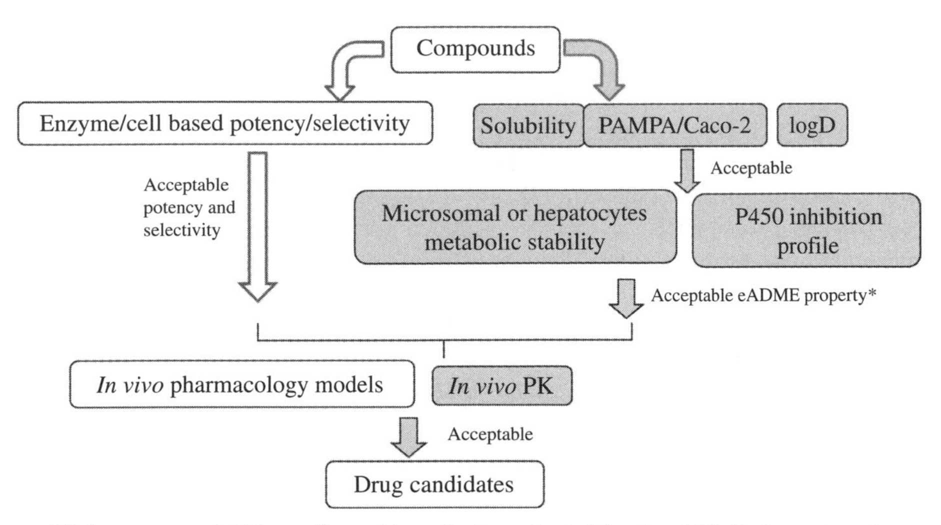

FIGURE 9.3 Integrated ADME-based screen strategies during the candidate identification phase of drug discovery.

shown in Figure 9.3, which illustrates parallel ADME and *in vitro* pharmacology screens that can be implemented to rapidly generate key data to support decision-making. During the lead to drug candidate optimization stage, various high-throughput and customized *in vitro* assays, *in vivo* studies, and *in silico* methods as described in Section 9.4 can be employed to identify and characterize the root causes of low and/or erratic bioavailability, and the strategies to improve the oral bioavailability can be specifically developed to correct the liability. While preserving the core series structure, compound structure can be further modified to improve ADME properties. Issues with compound dissolution and chemical stability can be addressed with appropriate formulation design. Mechanism of carrier protein–mediated active transport in the intestine and liver can be investigated by transporter expressing cell lines or membrane vesicles. Prodrug approaches have been also utilized to overcome compound deficiency in both solubility and permeability. For example, prodrugs with ionizable promoieties such as phosphate can increase the solubility up to orders of magnitudes. Ester prodrugs for carboxylic acids, phenols, and alcohols can increase permeability of the compounds by masking the polar functional groups. Overall, an integrated approach should be considered in the optimization process based on a thorough understanding of the ADME principles governing the absorption process.

9.6 SUMMARY

First-pass metabolism is a key factor in determining systemic availability of orally administered drugs. Over recent years, major advance has been made in our understanding of the biochemical and physiological processes determining first-pass metabolism. *In vitro* characterizations of intestinal expression and catalytic activities of CYPs, UGTs, and SULTs have highlighted the intestine as the most important extrahepatic site of metabolism and brought increased awareness of the significance of intestinal first-pass. A variety of *in vitro* methods utilizing subcellular fractions, intact tissue preparations, primary cells, and cell lines have been developed, optimized, and validated as essential tools for studying first-pass metabolism. While various mechanistic models are being developed for the quantitative prediction of intestinal first-pass metabolism, there are still remaining gaps to be elucidated with respect to the appropriate integration of drug properties and GI physiology in the model development. Further efforts in elucidating the interplay of multiple processes in drug absorption will be crucial to the accurate prediction of first-pass metabolism and to improve our understanding of interindividual variability of absorption for oral drug delivery optimization.

REFERENCES

1. Hebert, M. F., Roberts, J. P., Prueksaritanont, T., and Benet, L. Z. *Clin Pharmacol Ther.* 1992, **52**, 453–457.
2. Fromm, M. F., Busse, D., Kroemer, H. K., and Eichelbaum, M. *Hepatology.* 1996, **24**, 796–801.
3. Guengerich, F. P. *Chem Res Toxicol.* 2008, **21**, 70–83.
4. Shimada, T., Yamazaki, H., Mimura, M., Inui, Y., and Guengerich, F. P. *J Pharmacol Exp Ther.* 1994, **270**, 414–423.
5. Danielson, P. B. *Curr Drug Metab.* 2002, **3**, 561–597.
6. Fuhr, U., Jetter, A., and Kirchheiner, J. *Clin Pharmacol Ther.* 2007, **81**, 270–283.
7. McGraw, J., and Waller, D. *Expert Opin Drug Metab Toxicol.* **8**, 371–382.
8. Gonzalez, F. J., Skoda, R. C., Kimura, S., Umeno, M., Zanger, U. M., Nebert, D. W., Gelboin, H. V., Hardwick, J. P., and Meyer, U. A. *Nature.* 1988, **331**, 442–446.
9. Gough, A. C., Miles, J. S., Spurr, N. K., Moss, J. E., Gaedigk, A., Eichelbaum, M., and Wolf, C. R. *Nature.* 1990, **347**, 773–776.
10. Johansson, I., Lundqvist, E., Bertilsson, L., Dahl, M. L., Sjoqvist, F., and Ingelman-Sundberg, M. *Proc Natl Acad Sci U S A.* 1993, **90**, 11825–11829.
11. Shen, D. D., Kunze, K. L., and Thummel, K. E. *Adv Drug Deliv Rev.* 1997, **27**, 99–127.
12. Wienkers, L. C., and Heath, T. G. *Nat Rev Drug Discov.* 2005, **4**, 825–833.
13. Tukey, R. H., and Strassburg, C. P. *Annu Rev Pharmacol Toxicol.* 2000, **40**, 581–616.
14. Collier, A. C., Ganley, N. A., Tingle, M. D., Blumenstein, M., Marvin, K. W., Paxton, J. W., Mitchell, M. D., and Keelan, J. A. *Biochem Pharmacol.* 2002, **63**, 409–419.
15. Fisher, M. B., Paine, M. F., Strelevitz, T. J., and Wrighton, S. A. *Drug Metab Rev.* 2001, **33**, 273–297.
16. Evans, W. E., and Relling, M. V. *Science.* 1999, **286**, 487–491.
17. Kiang, T. K., Ensom, M. H., and Chang, T. K. *Pharmacol Ther.* 2005, **106**, 97–132.
18. McCarver, D. G., and Hines, R. N. *J Pharmacol Exp Ther.* 2002, **300**, 361–366.

19. Hagenauer, B., Salamon, A., Thalhammer, T., Kunert, O., Haslinger, E., Klingler, P., Senderowicz, A. M., Sausville, E. A., and Jager, W. *Drug Metab Dispos.* 2001, **29**, 407–414.
20. Ramirez, J., Iyer, L., Journault, K., Belanger, P., Innocenti, F., Ratain, M. J., and Guillemette, C. *Pharm Res.* 2002, **19**, 588–594.
21. Ghosal, A., Hapangama, N., Yuan, Y., Achanfuo-Yeboah, J., Iannucci, R., Chowdhury, S., Alton, K., Patrick, J. E., and Zbaida, S. *Drug Metab Dispos.* 2004, **32**, 314–320.
22. Guillemette, C. *Pharmacogenomics J.* 2003, **3**, 136–158.
23. Burchell, B. *Am J Pharmacogenomics.* 2003, **3**, 37–52.
24. Tukey, R. H., Strassburg, C. P., and Mackenzie, P. I. *Mol Pharmacol.* 2002, **62**, 446–450.
25. Fujita, K., and Sparreboom, A. *Curr Clin Pharmacol.* 2010, **5**, 209–217.
26. Innocenti, F., Stadler, W. M., Iyer, L., Ramirez, J., Vokes, E. E., and Ratain, M. J. *Clin Cancer Res.* 2000, **6**, 3400–3405.
27. Kemp, D. C., Fan, P. W., and Stevens, J. C. *Drug Metab Dispos.* 2002, **30**, 694–700.
28. Kauffman, F. C. *Drug Metab Rev.* 2004, **36**, 823–843.
29. Gamage, N., Barnett, A., Hempel, N., Duggleby, R. G., Windmill, K. F., Martin, J. L., and McManus, M. E. *Toxicol Sci.* 2006, **90**, 5–22.
30. James, M. O., and Ambadapadi, S. *Drug Metab Rev.* 2013, **45**, 401–414.
31. Dong, D., Ako, R., and Wu, B. *Expert Opin Drug Metab Toxicol.* 2012, **8**, 635–646.
32. Riches, Z., Stanley, E. L., Bloomer, J. C., and Coughtrie, M. W. *Drug Metab Dispos.* 2009, **37**, 2255–2261.
33. Johnson, K., Shah, A., Jaw-Tsai, S., Baxter, J., and Prakash, C. *Drug Metab Dispos.* 2003, **31**, 76–87.
34. Hengstmann, J. H., and Goronzy, J. *Eur J Clin Pharmacol.* 1982, **21**, 335–341.
35. Krueger, S. K., and Williams, D. E. *Pharmacol Ther.* 2005, **106**, 357–387.
36. Phillips, I. R., and Shephard, E. A. *Trends Pharmacol Sci.* 2008, **29**, 294–301.
37. Cashman, J. R. *Pharmacogenomics.* 2002, **3**, 325–339.
38. Krueger, S. K., Williams, D. E., Yueh, M. F., Martin, S. R., Hines, R. N., Raucy, J. L., Dolphin, C. T., Shephard, E. A., and Phillips, I. R. *Drug Metab Rev.* 2002, **34**, 523–532.
39. Satoh, T., and Hosokawa, M. *Annu Rev Pharmacol Toxicol.* 1998, **38**, 257–288.
40. Shi, D., Yang, J., Yang, D., LeCluyse, E. L., Black, C., You, L., Akhlaghi, F., and Yan, B. *J Pharmacol Exp Ther.* 2006, **319**, 1477–1484.
41. Humerickhouse, R., Lohrbach, K., Li, L., Bosron, W. F., and Dolan, M. E. *Cancer Res.* 2000, **60**, 1189–1192.
42. Tang, M., Mukundan, M., Yang, J., Charpentier, N., LeCluyse, E. L., Black, C., Yang, D., Shi, D., and Yan, B. *J Pharmacol Exp Ther.* 2006, **319**, 1467–1476.
43. Williams, E. T., Jones, K. O., Ponsler, G. D., Lowery, S. M., Perkins, E. J., Wrighton, S. A., Ruterbories, K. J., Kazui, M., and Farid, N. A. *Drug Metab Dispos.* 2008, **36**, 1227–1232.
44. Lin, J. H., Chiba, M., and Baillie, T. A. *Pharmacol Rev.* 1999, **51**, 135–158.
45. Tripathi, S. P., Bhadauriya, A., Patil, A., and Sangamwar, A. T. *Drug Metab Rev.* 2013, **45**, 231–252.

46. Kolars, J. C., Awni, W. M., Merion, R. M., and Watkins, P. B. *Lancet.* 1991, **338**, 1488–1490.
47. Wu, C. Y., Benet, L. Z., Hebert, M. F., Gupta, S. K., Rowland, M., Gomez, D. Y., and Wacher, V. J. *Clin Pharmacol Ther.* 1995, **58**, 492–497.
48. Paine, M. F., Shen, D. D., Kunze, K. L., Perkins, J. D., Marsh, C. L., McVicar, J. P., Barr, D. M., Gillies, B. S., and Thummel, K. E. *Clin Pharmacol Ther.* 1996, **60**, 14–24.
49. Thummel, K. E., O'Shea, D., Paine, M. F., Shen, D. D., Kunze, K. L., Perkins, J. D., and Wilkinson, G. R. *Clin Pharmacol Ther.* 1996, **59**, 491–502.
50. Sandstrom, R., Karlsson, A., Knutson, L., and Lennernas, H. *Pharm Res.* 1998, **15**, 856–862.
51. von Richter, O., Greiner, B., Fromm, M. F., Fraser, R., Omari, T., Barclay, M. L., Dent, J., Somogyi, A. A., and Eichelbaum, M. *Clin Pharmacol Ther.* 2001, **70**, 217–227.
52. Edgar, B., Bailey, D., Bergstrand, R., Johnsson, G., and Regardh, C. G. *Eur J Clin Pharmacol.* 1992, **42**, 313–317.
53. Pacifici, G. M., Eligi, M., and Giuliani, L. *Eur J Clin Pharmacol.* 1993, **45**, 483–487.
54. Kosaka, K., Sakai, N., Endo, Y., Fukuhara, Y., Tsuda-Tsukimoto, M., Ohtsuka, T., Kino, I., Tanimoto, T., Takeba, N., Takahashi, M., and Kume, T. *Drug Metab Dispos.* 2011, **39**, 1495–1502.
55. Watkins, P. B., Wrighton, S. A., Schuetz, E. G., Molowa, D. T., and Guzelian, P. S. *J Clin Invest.* 1987, **80**, 1029–1036.
56. Kolars, J. C., Schmiedlin-Ren, P., Schuetz, J. D., Fang, C., and Watkins, P. B. *J Clin Invest.* 1992, **90**, 1871–1878.
57. Zhang, Q. Y., Dunbar, D., Ostrowska, A., Zeisloft, S., Yang, J., and Kaminsky, L. S. *Drug Metab Dispos.* 1999, **27**, 804–809.
58. Paine, M. F., Hart, H. L., Ludington, S. S., Haining, R. L., Rettie, A. E., and Zeldin, D. C. *Drug Metab Dispos.* 2006, **34**, 880–886.
59. Paine, M. F., Khalighi, M., Fisher, J. M., Shen, D. D., Kunze, K. L., Marsh, C. L., Perkins, J. D., and Thummel, K. E. *J Pharmacol Exp Ther.* 1997, **283**, 1552–1562.
60. von Richter, O., Burk, O., Fromm, M. F., Thon, K. P., Eichelbaum, M., and Kivisto, K. T. *Clin Pharmacol Ther.* 2004, **75**, 172–183.
61. Galetin, A., and Houston, J. B. *J Pharmacol Exp Ther.* 2006, **318**, 1220–1229.
62. Thorn, M., Finnstrom, N., Lundgren, S., Rane, A., and Loof, L. *Br J Clin Pharmacol.* 2005, **60**, 54–60.
63. Murray, G. I., Barnes, T. S., Sewell, H. F., Ewen, S. W., Melvin, W. T., and Burke, M. D. *Br J Clin Pharmacol.* 1988, **25**, 465–475.
64. Lown, K. S., Kolars, J. C., Thummel, K. E., Barnett, J. L., Kunze, K. L., Wrighton, S. A., and Watkins, P. B. *Drug Metab Dispos.* 1994, **22**, 947–955.
65. Greiner, B., Eichelbaum, M., Fritz, P., Kreichgauer, H. P., von Richter, O., Zundler, J., and Kroemer, H. K. *J Clin Invest.* 1999, **104**, 147–153.
66. Rowland, M., and Gupta, S. K. *Br J Clin Pharmacol.* 1987, **24**, 329–334.
67. Kharasch, E. D., Walker, A., Hoffer, C., and Sheffels, P. *Clin Pharmacol Ther.* 2004, **76**, 452–466.
68. Ducharme, M. P., Warbasse, L. H., and Edwards, D. J. *Clin Pharmacol Ther.* 1995, **57**, 485–491.

69. Masica, A. L., Mayo, G., and Wilkinson, G. R. *Clin Pharmacol Ther.* 2004, **76**, 341–349.
70. Strassburg, C. P., Oldhafer, K., Manns, M. P., and Tukey, R. H. *Mol Pharmacol.* 1997, **52**, 212–220.
71. Strassburg, C. P., Manns, M. P., and Tukey, R. H. *J Biol Chem.* 1998, **273**, 8719–8726.
72. Ohno, S., and Nakajin, S. *Drug Metab Dispos.* 2009, **37**, 32–40.
73. Rogers, S. M., Back, D. J., and Orme, M. L. *Br J Clin Pharmacol.* 1987, **23**, 727–734.
74. Radominska-Pandya, A., Little, J. M., Pandya, J. T., Tephly, T. R., King, C. D., Barone, G. W., and Raufman, J. P. *Biochim Biophys Acta.* 1998, **1394**, 199–208.
75. Back, D. J., Bates, M., Breckenridge, A. M., Ellis, A., Hall, J. M., Maciver, M., Orme, M. L., and Rowe, P. H. *Br J Clin Pharmacol.* 1981, **11**, 275–278.
76. Pacifici, G. M., Bencini, C., and Rane, A. *Xenobiotica.* 1986, **16**, 123–128.
77. Cappiello, M., Giuliani, L., and Pacifici, G. M. *Eur J Clin Pharmacol.* 1991, **41**, 345–350.
78. Raoof, A. A., van Obbergh, L. J., de Ville de Goyet, J., and Verbeeck, R. K. *Eur J Clin Pharmacol.* 1996, **50**, 91–96.
79. Veroli, P., O'Kelly, B., Bertrand, F., Trouvin, J. H., Farinotti, R., and Ecoffey, C. *Br J Anaesth.* 1992, **68**, 183–186.
80. Harbourt, D. E., Fallon, J. K., Ito, S., Baba, T., Ritter, J. K., Glish, G. L., and Smith, P. C. *Anal Chem.* 2012, **84**, 98–105.
81. Cubitt, H. E., Houston, J. B., and Galetin, A. *Pharm Res.* 2009, **26**, 1073–1083.
82. Bernard, O., and Guillemette, C. *Drug Metab Dispos.* 2004, **32**, 775–778.
83. Watanabe, Y., Nakajima, M., and Yokoi, T. *Drug Metab Dispos.* 2002, **30**, 1462–1469.
84. Bowalgaha, K., and Miners, J. O. *Br J Clin Pharmacol.* 2001, **52**, 605–609.
85. Trdan Lusin, T., Trontelj, J., and Mrhar, A. *Drug Metab Dispos.* **39**, 2347–2354.
86. Strassburg, C. P., Kneip, S., Topp, J., Obermayer-Straub, P., Barut, A., Tukey, R. H., and Manns, M. P. *J Biol Chem.* 2000, **275**, 36164–36171.
87. Davies, D. S., George, C. F., Blackwell, E., Conolly, M. E., and Dollery, C. T. *Br J Clin Pharmacol.* 1974, **1**, 129–136.
88. Teubner, W., Meinl, W., Florian, S., Kretzschmar, M., and Glatt, H. *Biochem J.* 2007, **404**, 207–215.
89. Ahmed, F., Vyas, V., Cornfield, A., Goodin, S., Ravikumar, T. S., Rubin, E. H., and Gupta, E. *Anticancer Res.* 1999, **19**, 2067–2071.
90. Ouellet, D. M., and Pollack, G. M. *Drug Metab Dispos.* 1995, **23**, 478–484.
91. Giacomini, K. M., Huang, S. M., Tweedie, D. J., Benet, L. Z., Brouwer, K. L., Chu, X., Dahlin, A., Evers, R., Fischer, V., Hillgren, K. M., Hoffmaster, K. A., Ishikawa, T., Keppler, D., Kim, R. B., Lee, C. A., Niemi, M., Polli, J. W., Sugiyama, Y., Swaan, P. W., Ware, J. A., Wright, S. H., Yee, S. W., Zamek-Gliszczynski, M. J., and Zhang, L. *Nat Rev Drug Discov.* 2010, **9**, 215–236.
92. Estudante, M., Morais, J. G., Soveral, G., and Benet, L. Z. *Adv Drug Deliv Rev.* 2012, **65**, 1340–1356.
93. Cummins, C. L., Jacobsen, W., and Benet, L. Z. *J Pharmacol Exp Ther.* 2002, **300**, 1036–1045.
94. Cummins, C. L., Salphati, L., Reid, M. J., and Benet, L. Z. *J Pharmacol Exp Ther.* 2003, **305**, 306–314.

95. Komura, H., and Iwaki, M. *J Pharm Sci.* 2008, **97**, 1775–1800.
96. Nishimuta, H., Nakagawa, T., Nomura, N., and Yabuki, M. *Xenobiotica.* 2013, **43**, 948–955.
97. Furukawa, T., Naritomi, Y., Tetsuka, K., Nakamori, F., Moriguchi, H., Yamano, K., Terashita, S., Tabata, K., and Teramura, T. *Xenobiotica.* 2014, **44**, 205–216.
98. Paine, M. F., and Oberlies, N. H. *Expert Opin Drug Metab Toxicol.* 2007, **3**, 67–80.
99. Paine, M. F., Widmer, W. W., Hart, H. L., Pusek, S. N., Beavers, K. L., Criss, A. B., Brown, S. S., Thomas, B. F., and Watkins, P. B. *Am J Clin Nutr.* 2006, **83**, 1097–1105.
100. Dresser, G. K., Bailey, D. G., Leake, B. F., Schwarz, U. I., Dawson, P. A., Freeman, D. J., and Kim, R. B. *Clin Pharmacol Ther.* 2002, **71**, 11–20.
101. Costa, A., Sarmento, B., and Seabra, V. *Expert Opin Drug Metab Toxicol.* 2014, **10**, 103–119.
102. Hewitt, N. J., Lechon, M. J., Houston, J. B., Hallifax, D., Brown, H. S., Maurel, P., Kenna, J. G., Gustavsson, L., Lohmann, C., Skonberg, C., Guillouzo, A., Tuschl, G., Li, A. P., LeCluyse, E., Groothuis, G. M., and Hengstler, J. G. *Drug Metab Rev.* 2007, **39**, 159–234.
103. Antherieu, S., Chesne, C., Li, R., Camus, S., Lahoz, A., Picazo, L., Turpeinen, M., Tolonen, A., Uusitalo, J., Guguen-Guillouzo, C., and Guillouzo, A. *Drug Metab Dispos.* 2012, **38**, 516–525.
104. Obach, R. S. *Curr Top Med Chem.* 2011, **11**, 334–339.
105. Nishimuta, H., Sato, K., Yabuki, M., and Komuro, S. *Drug Metab Pharmacokinet.* 2011, **26**, 592–601.
106. Gertz, M., Harrison, A., Houston, J. B., and Galetin, A. *Drug Metab Dispos.* 2010, **38**, 1147–1158.
107. Bonnefille, P., Sezgin-Bayindir, Z., Belkhelfa, H., Arellano, C., Gandia, P., Woodley, J., and Houin, G. *Fundam Clin Pharmacol.* 2011, **25**, 104–114.
108. de Graaf, I. A., Olinga, P., de Jager, M. H., Merema, M. T., de Kanter, R., van de Kerkhof, E. G., and Groothuis, G. M. *Nat Protoc.* 2010, **5**, 1540–1551.
109. Groothuis, G. M., and de Graaf, I. A. *Curr Drug Metab.* 2013, **14**, 112–119.
110. Tong, X. S., Xu, S., Zheng, S., Pivnichny, J. V., Martin, J., and Dufresne, C. *J Chromatogr B Analyt Technol Biomed Life Sci.* 2006, **833**, 165–173.
111. Kantharaj, E., Tuytelaars, A., Proost, P. E., Ongel, Z., Van Assouw, H. P., and Gilissen, R. A. *Rapid Commun Mass Spectrom.* 2003, **17**, 2661–2668.
112. Gu, M., and Lim, H. K. *J Mass Spectrom.* 2001, **36**, 1053–1061.
113. O'Connor, D., Mortishire-Smith, R., Morrison, D., Davies, A., and Dominguez, M. *Rapid Commun Mass Spectrom.* 2006, **20**, 851–857.
114. Carlson, T. J., and Fisher, M. B. *Comb Chem High Throughput Screen.* 2008, **11**, 258–264.
115. Fan, J., Chen, S., Chow, E. C., and Pang, K. S. *Curr Drug Metab.* 2010, **11**, 743–761.
116. Yang, J., Jamei, M., Yeo, K. R., Tucker, G. T., and Rostami-Hodjegan, A. *Curr Drug Metab.* 2007, **8**, 676–684.
117. Rostami-Hodjegan, A., and Tucker, G. T. *Hepatology.* 2002, **35**, 1549–1550; author reply 1550–1541.
118. Gertz, M., Houston, J. B., and Galetin, A. *Drug Metab Dispos.* 2011, **39**, 1633–1642.
119. Johnson, E. F., Connick, J. P., Reed, J. R., Backes, W. L., Desai, M. C., Xu, L., Estrada, D. F., Laurence, J. S., and Scott, E. E. *Drug Metab Dispos.* 2014, **42**, 9–22.

120. Kirchmair, J., Williamson, M. J., Tyzack, J. D., Tan, L., Bond, P. J., Bender, A., and Glen, R. C. *J Chem Inf Model.* **52**, 617–648.
121. de Groot, M. J., Kirton, S. B., and Sutcliffe, M. J. *Curr Top Med Chem.* 2004, **4**, 1803–1824.
122. Cruciani, G., Carosati, E., De Boeck, B., Ethirajulu, K., Mackie, C., Howe, T., and Vianello, R. *J Med Chem.* 2005, **48**, 6970–6979.
123. Trunzer, M., Faller, B., and Zimmerlin, A. *J Med Chem.* 2009, **52**, 329–335.

10

PULMONARY DRUG DELIVERY: PHARMACEUTICAL CHEMISTRY AND AEROSOL TECHNOLOGY

Anthony J. Hickey
RTI International, Research Triangle Park, NC, USA

10.1 INTRODUCTION

Delivery of drugs as aerosols to the lungs requires consideration of a number of disciplines including physical chemistry, device design and engineering, aerosol physics, and involves physiological/anatomical and pharmacological strategies [1]. The multidisciplinary nature of the approach required to develop effective therapies is daunting but reflects a high standard of achievement in both science and technology. More than four generations of scientists, engineers, and clinicians have expended their energies in pursuit of the most effective, safe, and elegant solution to the problems of pulmonary drug delivery. In terms of numbers of researchers and quantities of resources there has been nothing to match the activity since 1990.

At the end of the 1980s the aerosol products that were available were a reflection of approximately 30 years of research and development [2]. During this period it was clear that the propellant-driven metered-dose inhaler (pMDI) had revolutionized asthma therapy, largely due to the invention of the metering valve and actuator combination [3]. In contrast, dry powder inhalers (DPIs) were very primitive and somewhat ineffective systems that deservedly did not compete with pMDIs in terms of physician and patient acceptance [4, 5]. Consequently, they were not a great commercial success. Nebulizers had been available for a number of years and had a

Drug Delivery: Principles and Applications, Second Edition. Edited by Binghe Wang, Longqin Hu, and Teruna J. Siahaan.

significant role in acute care [6, 7]. Some questions were being asked about the susceptibility of nebulizer performance to operating conditions against the backdrop of few, if any, manufacturer specifications and little government regulation.

A growing movement for change occurred throughout the late 1980s. Occurrences in the 1990s resulted in dramatic developments in the nature of aerosol products driven by restrictions on the use of chlorofluorocarbon (CFC) propellants and also by upsurge in novel biotherapeutic agents, which were intended for lung delivery [8]. Alternative, nonozone depleting, propellant systems appeared to replace CFC products extending the application of pMDI technology [9]. A host of new DPIs were under development based on a greater understanding of the requirements for adequate aerosol dispersion [4, 5]. Very few of these products have appeared commercially. Finally a range of more efficient nebulizers and hand-held aqueous aerosol delivery systems were developed bringing a standard of performance to the systems that had previously been absent [8, 10].

In the following sections the performance of recently developed devices will be contrasted with that of earlier systems in order to emphasize the progress that has been made. The conclusion will focus on reasonable expectations of the future and indicate areas in which further fundamental observations may help in the design and production of new aerosol delivery systems.

10.2 AEROSOL TECHNOLOGY

The devices that are employed to deliver drugs to the lungs may be divided into three categories: pMDIs, DPIs, and nebulizers [11]. Each of these systems delivers aerosols by a different principle, and the chemistry associated with the product varies significantly between them.

10.2.1 Particle Production

Before drug particles can be incorporated into aerosol products they must be prepared in size ranges and with structures suitable for delivery to the lungs. A variety of methods have been developed for the preparation of particles [12, 13]. The most common method of particle size reduction is the one in which bulk drug product, most of which is prepared by conventional crystallization/precipitation followed by drying techniques, is air jet milled at high pressures [12]. Attrition of particles leads to the micronization of the product. In recent years methods have been used that combine the solution conversion to solid with size reduction or crystal engineering. The first of these methods—spray drying—involves forcing the drug solution through a nozzle at high pressure into a drying airstream from which small particles can be recovered suitable for inclusion in aerosol products [13, 14]. These particles occasionally have unique properties that will be discussed later in the text. Supercritical fluid manufacture is a particle construction or crystal engineering method [13]. In its simplest form this involves dispersing drug in a supercritical fluid (usually carbon dioxide), and by controlling the conditions of temperature, pressure, volume, or the presence of an antisolvent the drug maybe crystallized to form morphologically well-defined particles.

10.2.2 Propellant-Driven Metered-Dose Inhalers

Drugs delivered from pMDIs are initially prepared as suspensions or solutions in a selected propellant [15, 16]. Often other components are added to aid in suspension of particles or drug dispersion into solution. These additives may be cosolvents, such as ethanol, or surfactants, such as oleic acid.

The original pMDIs employed chlorofluorocarbon (CFC) propellants 11, 114, and 12. (e.g., beclomethasone dipropionate (BDP) products Vanceril and Beclovent) [17]. In recent years these propellants have been replaced with hydrofluoroalkane (HFA) propellants 134a or 227 (e.g., BDP product Q_{var}) due to atmospheric ozone depletion associated with CFCs [18, 19].

Containers are filled on a large scale with drug formulation by a variety of techniques, based on high pressure or low temperature to control the state of the propellant during filling [20]. Valves are crimped on the opening to the container either before (pressure filling) or after (cold filling) propellant filling occurs.

The principle of aerosol delivery from pMDIs is based on the following sequence of events [16, 21]. A small volume of a homogeneous dispersion of the drug, in solution or suspension, in high vapor pressure propellant or propellant blend from a reservoir is isolated. The small volume container (the metering chamber) is opened through an actuator nozzle. The metering chamber filling and opening to atmosphere are achieved by means of a metering valve. Once opened to atmosphere the high vapor pressure contents of the metering valve immediately begin to equilibrate with atmospheric pressure. This has the effect of propelling the contents rapidly through the nozzle that causes shear and droplet formation. Throughout this process the propellant is evaporating propelling, shearing, and ultimately reducing the size of the droplets produced. The components of an MDI are shown in Figure 10.1a.

10.2.3 Dry Powder Inhalers

DPIs have been through a number of evolutionary changes over the last 40 years [4, 5]. All approved inhalers have been passive in the sense that they employ the patient's inspiratory flow as the means of dispersion and entrainment of the aerosol into the lungs. The majority of powder products are blends of respirable drug particles and large lactose carrier particles [22–24]. Early designs employed a unit dose gelatin capsule metering system (Rotahaler, Spinhaler). This has to some extent been superceded by multiple unit dose blister discs (Diskhaler) and strips/rolls (Diskus) or reservoir powder devices (Turbuhaler, Clickhaler). The general principle of powder delivery is shown in Figure 10.1b. A powder bed is exposed to a shearing air supply (usually the inspiratory airflow) that entrains particles. A blend will employ the fluidizing effects of large lactose particles to help disperse the respirable particles associated with their surfaces [24, 25]. The small drug particles will be carried to the lungs of the patient, while the large carrier particles will be deposited in the mouthpiece of the inhaler or the oropharynx of the patient.

A variety of mechanisms for assisting with the dispersion of the powder have been adopted including impellors (Spiros), compressed air assist (Nektar), vibration (Oriel, Microdose), impact hammers (3M, DelSys) [26, 27].

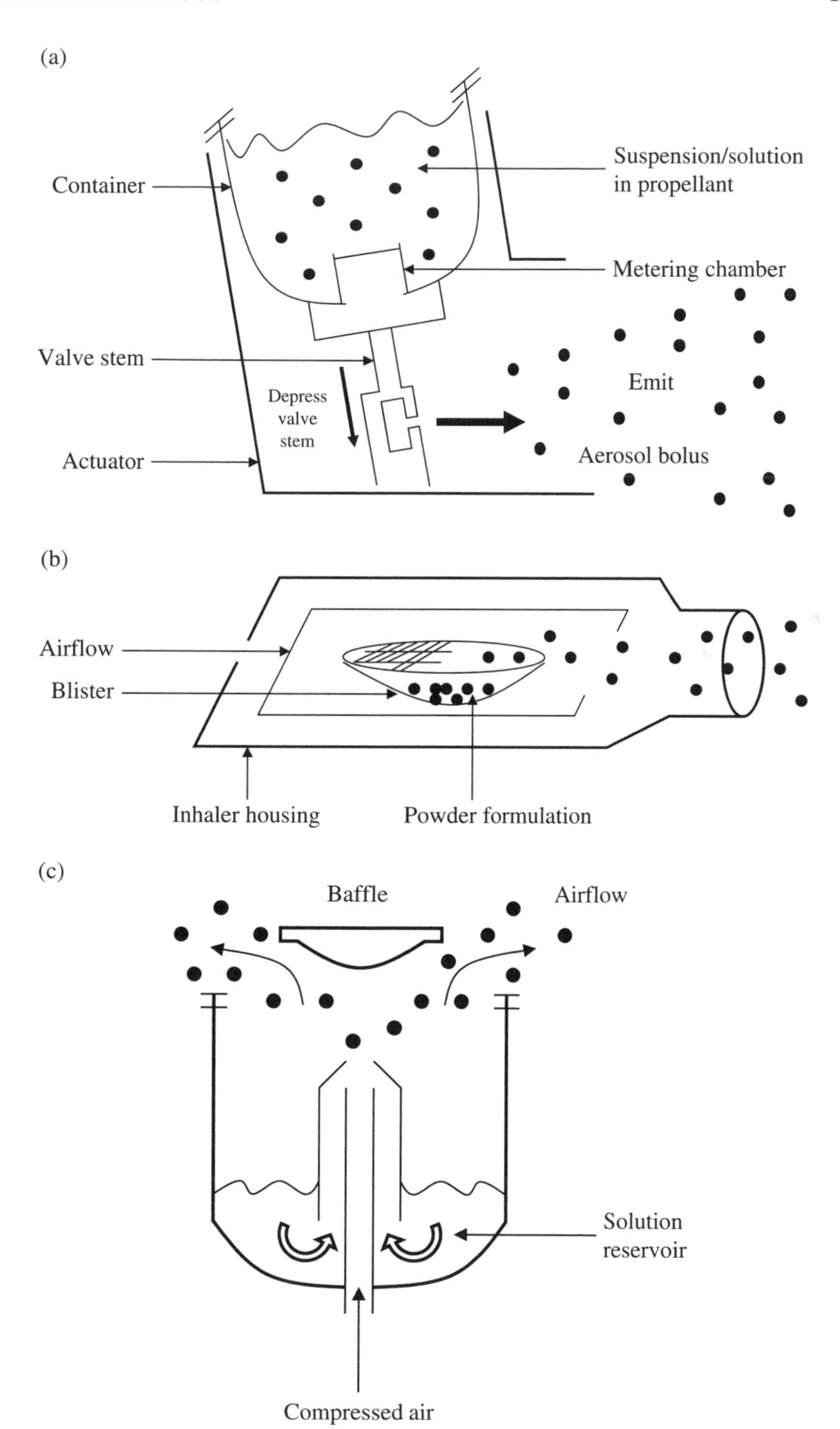

FIGURE 10.1 (a) Schematic diagram of a propellant-driven metered-dose inhaler. (b) Schematic diagram of a dry powder inhaler. (c) Schematic diagram of an air-jet nebulizer.

10.2.4 Nebulizer

Nebulizers are among the oldest devices used for the delivery of therapeutic agents [28]. They employ energy from compressed gas or piezoelectric ceramics to generate droplets of water-containing drug. The principle of air jet dispersion is shown in Figure 10.1c. Drug solution (or occasionally suspension) is drawn from a reservoir through a capillary tube by Venturi (Bernoulli) effect. In principle, a low-pressure region is created at the exit from the capillary tube by passing compressed gas at high velocity over the tube thereby drawing liquid into the air where droplets are formed. Large droplets are projected onto a baffle where they are collected, and small ones pass around the baffle and are delivered to the patient's lungs on their inspiratory flow.

In the last decade vibrating mesh systems (eFlow, Pari; Aeroneb, Aerogen; and iNeb, Phillips) have become popular because of their ability to generate high-concentration aerosols in small volumes of air that requires a shorter duration of treatment and improves patient compliance [29]. These systems have facilitated breath control delivery of aerosols by a variety of methods notably the Akita (InaMed) system and novel systems such as the tPAD device (Parion).

10.3 DISEASE THERAPY

The lungs have been a route of drug delivery for millennia. Modern medical applications of aerosol delivery can be traced to the development of the pMDI in the middle of the last century [30].

Initially drugs for asthma therapy were the prominent therapeutic category of interest. With increased understanding of pulmonary biology and the pathogenesis of disease, agents such as proteins and peptides, to achieve local and systemic therapeutic effect [31], and antimicrobials for infectious disease therapy [32] have been studied. There are a number of classical texts that outline the pharmacology of the lungs including Goodman and Gilman [33, 34] and the medicinal chemistry of drugs [35]. These general references have been drawn on for the following discussion. Thorough descriptions of agents delivered to the lungs may also be found in the literature [36, 37].

10.3.1 Asthma

The first drugs developed for asthma therapy were used to bronchodilate patients by acting on the sympathetic or parasympathetic receptors. These agents fall broadly into two categories: β-adrenergic agonists (BAAs) and anticholinergics. Subsequently, other agents were added that acted on other manifestations of the disease. Notably, steroids acted on the underlying inflammation. Figure 10.2 depicts the action of some of the commonly used drugs to treat asthma by therapeutic category [38].

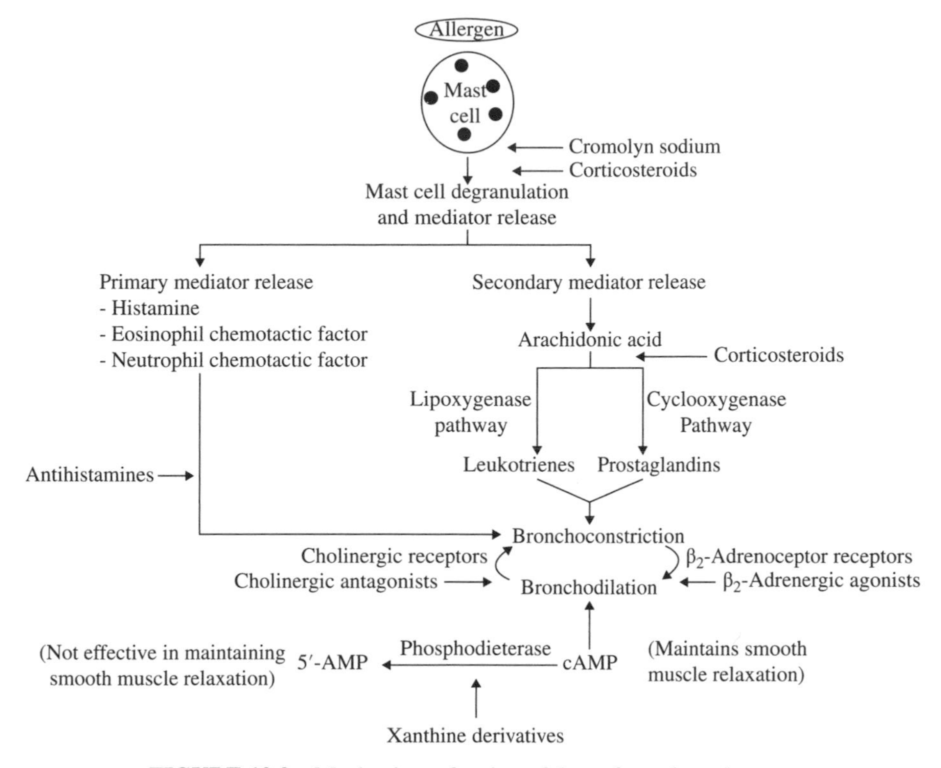

FIGURE 10.2 Mechanism of action of drugs for asthma therapy.

10.3.1.1 β-Adrenergic Agonists The receptors for BAA are distributed throughout the lung but occur in increasing numbers as you approach the periphery. Consequently, aerosol BAAs must be delivered to the periphery of the lungs to exhibit their pharmacodynamic effect.

A number of BAAs were evaluated approximately 50 years ago for their ability to induce bronchodilation. The molecules are analogs of epinephrine, a very short-acting bronchodilator. Epinephrine is currently marketed in an over-the-counter product Primatene Mist (Whitehall-Robins, Richmond, VA). The safety of this product relates to its very short duration of action in the lung following delivery. However, early examples of nonspecific agents such as isoproterenol were longer acting in the lungs and at other sites capable of inducing serious, in some cases life-threatening, side effects. In the 1960s more specific β_2 adrenergic agonists were developed, the most notable of which was albuterol (salbutamol, 3M, GSK, Schering Plough (SP is now Merck)), which targeted airway receptors and exhibited reduced systemic side effects. A series of short-acting agents following the same general structure (shown in Fig. 10.3a) were developed including terbutaline (AstraZeneca) and fenoterol (Boehringer Ingelheim).

As the chemistry of these drugs was understood they could be modified to achieve several things including local metabolism to an active agent, increased residence time at the receptor (extend duration of action), and selection for receptor binding to reduce toxicity. The first prodrugs were bitolterol and bambuterol, both of which are

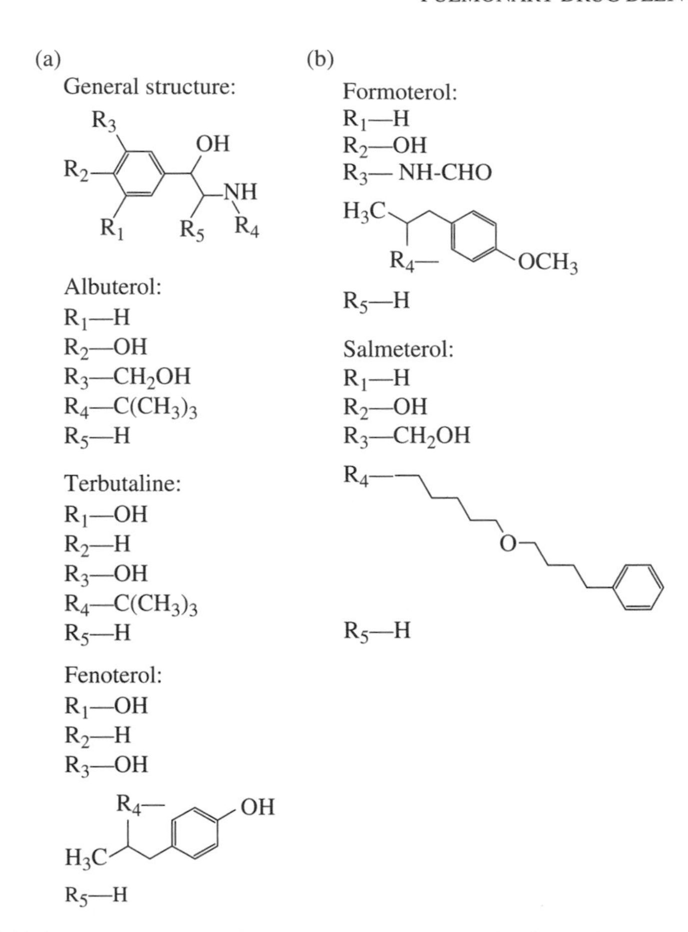

FIGURE 10.3 (a) Short-acting β-agonists and (b) long-acting β-agonists.

metabolized to active adrenergic agonists. Two long-acting agents have been developed based on increased residence time at the receptor: salmeterol (GSK) and formoterol (BI) (shown in Fig. 10.3b). It was noted that selected isomeric forms of BAAs may exhibit lower toxicity and enhanced efficacy. The most prominent example of this is Levalbuterol (Xopenex, Sepracor).

10.3.1.2 Anticholinergics Anticholinergic agents act on the central airways, the location of the majority of receptors. A small number of anticholinergic agents have been developed in the last 30 years. Ipratropium (BI) was the first of these agents to be approved, and it was followed by oxitropium, see Figure 10.4. Asthma was the target disease state for these agents. The success of tiotropium in the treatment of chronic obstructive pulmonary disease, for which there has previously been no aerosol therapy, has opened up the opportunity for a new generation of anticholinergics.

Ipratropium:

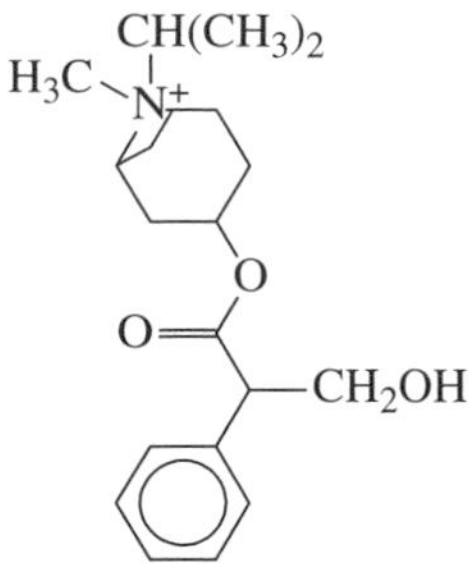

FIGURE 10.4 Anticholinergic.

10.3.1.3 Corticosteroids Corticosteroids have been known to be therapeutically beneficial in the treatment of asthma for some time. They have the significant advantage that they treat the underlying inflammation of the lungs that causes the disease. However, the systemic use of corticosteroids is associated with substantial systemic side effects. Consequently, courses in orally ingested steroids are given infrequently.

Steroids, which are active following oral ingestion such as prednisone, prednisolone, and dexamethasone, cause significant systemic immunosuppression with consequent risks to the health of the patient. Since the inflammation associated with asthma is localized in the lungs, local topical delivery with aerosols has the advantage of requiring low doses sufficient to achieve local therapeutic effect but resulting in small circulating concentrations of drug capable of causing systemic side effects. The first drug in this class was beclomethasone dipropionate (GSK, 3M, Schering Plough (SP now Merck)). This was followed by triamcinolone acetonide (Aventis, drug now owned by Abbott). While these agents exhibited some preferential local effects the introduction of budesonide (AstraZeneca), flunisolide and fluticasone (GSK) gave maximum local effect and minimum systemic effects based on the local binding of the steroids in the lungs. Some examples of steroid structures are shown in Figure 10.5.

10.3.1.4 Cromones Cromones (cromolyn sodium and nedocromil sodium, Aventis; Fig. 10.6) are a unique class of compounds that are known to cause mast cell stabilization, thereby preventing histamine release involved in local hypersensitivity of the lungs. In addition, these agents are implicated in preventing the release of other inflammatory mediators and the sensitivity of myelinated nerves in the airways. The major action of these compounds is still obscure but they have found a particular application in exercise-induced asthma.

10.3.2 Emphysema

A number of agents have been employed for the treatment of emphysema, associated with the action of elastase in the lungs. The most prominent example of a drug for the treatment of emphysema is α1-antitrypsin [39]. Others include peptidyl carbamates

General structure:

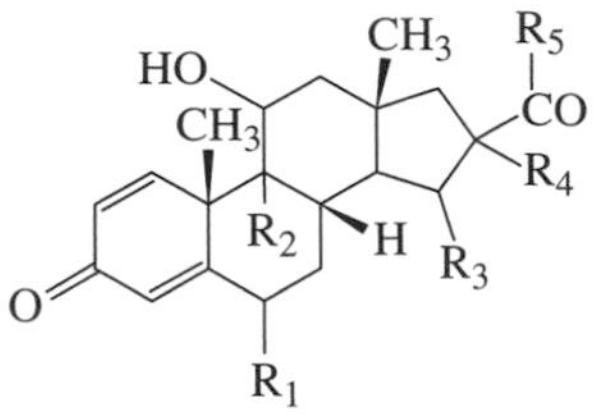
HO
CH3
CH3
R5
CO
R4
R2
H
R3
O
R1

Beclomethasone dipropionate:

R_1—H
R_2—Cl
R_3—CH_3
R_4—$COOC_2H_5$
R_5—$CH_2COOC_2H_5$

Triamcinolone acetonide:

R_1—H
R_2—F

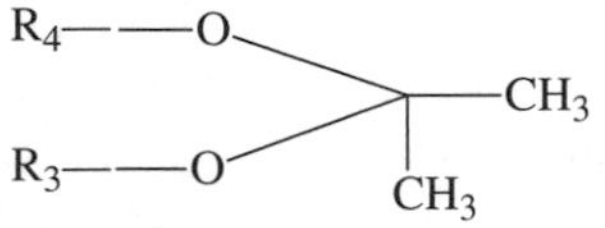
R4——O
CH3
R3——O
CH3

R_5—CH_2OH

Budesonide:

R_1—H
R_2 - H

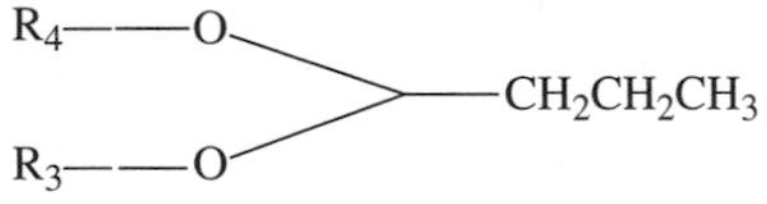
R4——O
CH2CH2CH3
R3——O

R_5—CH_2OH

Fluticasone propionate:

R_1—F
R_2—F
R_3—CH_3
R_4—$COOC_2H_5$
R_5—SCH_2F

FIGURE 10.5 Steroids.

and Eglin C. The use of these drugs reduces the free elastase in the lungs thereby preventing the structural remodeling of the lungs, which gives rise to poor gaseous exchange and severe disability associated with the disease. Some individuals exhibit a genetic predisposition for this disease, others exhibit onset as the result of a history of smoking.

Cromolyn sodium:

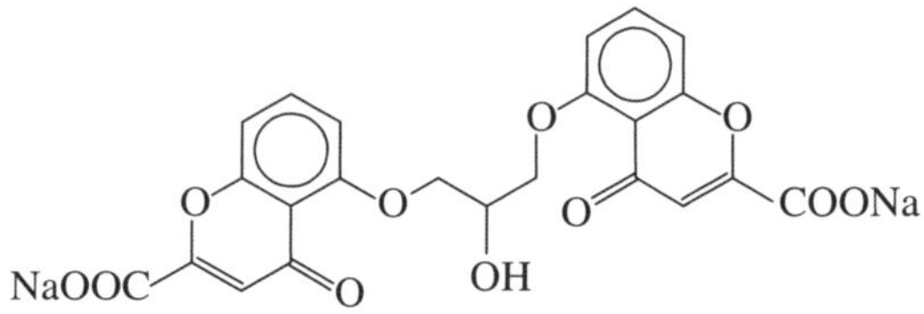

Nedocromyl:

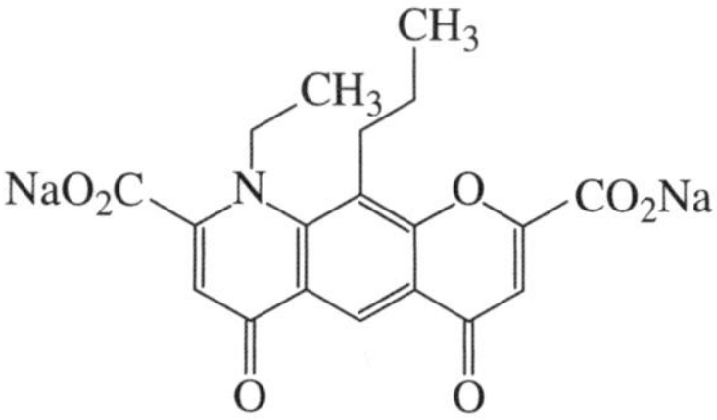

FIGURE 10.6 Cromones.

10.3.3 Cystic Fibrosis

Cystic fibrosis (CF) is characterized by an imbalance in airway chloride ion concentrations due to the absence of CF transmembrane regulator (CFTR) receptor. Three approaches have been taken to using aerosols to treat this disease [40]. The first involves the delivery of recombinant human DNase to cleave the tangle of leukocyte DNA, which results from cell infiltrate into the lungs thereby reducing the viscosity of the mucus layer in the lungs and facilitating expectoration. A second aerosol is employed to deliver an antibacterial agent tobramycin (shown in Fig. 10.7), which acts against *Pseudomonas aeruginosa* that grows on mucus plaques in the lungs of patients. Finally nucleic acid is employed to correct the genetic imbalance in CF expression and thereby correct chloride ion transport [41].

10.3.4 Other Locally Acting Agents

Amikacin and amphotericin B have both been prepared in liposomal formulations for different reasons. Amikacin can be targeted to macrophages for the treatment of intracellular microorganisms such as *Mycobacterium avium* complex (MAC) [42, 43]. Amphotericin B exhibits increased solubility in liposomes capable of delivering a dose to the lungs for the treatment of Aspergillosis [44].

Pentamidine (shown in Fig. 10.7) and its analogs have been delivered to the lungs for the treatment of *Pneumocystis carinii* pneumonia, a secondary infection associated with Acquired Immunodeficiency Disease Syndrome (AIDS) [45]. The advent of drugs to treat human immunodeficiency viral (HIV) infections has reduced the need for pentamidine therapy.

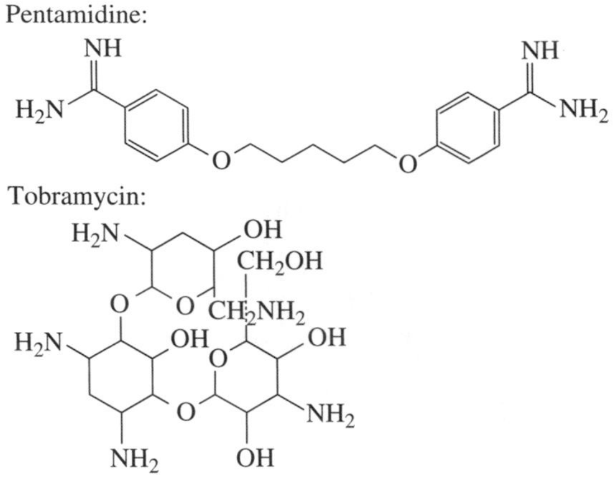

Pentamidine:

Tobramycin:

FIGURE 10.7 Antimicrobials.

10.3.5 Systemically Acting Agents

A variety of systemically acting agents have been evaluated for delivery via the lungs. Among these are insulin, leuprolide acetate, calcitonin, parathyroid hormone, and growth hormone [46].

Each of these agents targets a different disease state. Insulin is employed in the control of diabetes. Leuprolide acetate is employed for the treatment of prostate cancer and endometriosis. Calcitonin and parathyroid hormones are used to prevent osteoporosis. Growth hormone, as its name implies, controls deficiencies in the normal growth of children.

The developments of the last 10 years in inhaled products have been reviewed elsewhere [47]. In the period since the original publication of this chapter, inhaled insulin had been considered both a success and failure by industry analysts. In 2006 Exubera (Pfizer/Nektar Therapeutics) was launched with great fanfare and to the satisfaction of all who had been involved in the development of inhaled therapies in the period from the late 1980s onward. In 2007 a business decision was made to withdraw the product, due largely to poor initial sales. The evidence that inhaled insulin achieved the desired therapeutic outcome is clear. Whether it might have achieved greater market share over time is now simply the subject of speculation. Nevertheless, there continues to be interest in inhaled insulin, and it is probable that a new product will be introduced in the coming years.

Also in the period since the last publication, a new treatment for migraine headaches, hydroxyergotamine, has been approved as a pMDI product.

10.4 FORMULATION VARIABLES

The aforementioned agents are delivered from a variety of dosage forms. Their ompatibility with various solvents (propellants, alcohol, water); liquid- (glycerol, vethylene glycol, oleic acid, sorbitan trioleate, lecithin); or solid-phase excipients (lactose) is the key to the chemical and physical stability of the products. *The Handbook of Pharmaceutical Excipients* lists most of these materials [48].

10.4.1 Excipients

10.4.1.1 Propellant-Based Systems Propellants are relatively inert materials in which most drugs exhibit limited solubility. Consequently, there are two favored approaches to incorporating drug into propellants. The first is to use surfactants to disperse respirable particles in a suspension. Three excipients were approved for this purpose in CFC propellants: oleic acid (Fig. 10.8), sorbitan trioleate, and lecithin (Fig. 10.9). HFA systems are much more limited in their use of excipients, and oleic acid is the only excipient currently employed in these systems. The second approach is to use the cosolvent, ethanol, to bring the drug into solution and thereby achieve a molecularly homogeneous distribution in the propellant.

Suspension formulations are prepared to achieve a controlled flocculation, which will allow ease of redispersion upon shaking [49]. A number of physicochemical phenomena may occur to disrupt the stability of the product. Moisture ingress to the container (and association with the drug particles) will lead to hydrolysis for molecules that are susceptible to this mechanism of degradation. Because of the inert hydrophobic nature of propellant this can be controlled to some extent. The presence of moisture will also give rise to interactions between particles, which may result in an irreversible aggregation. Related examples of aggregation may come about due to

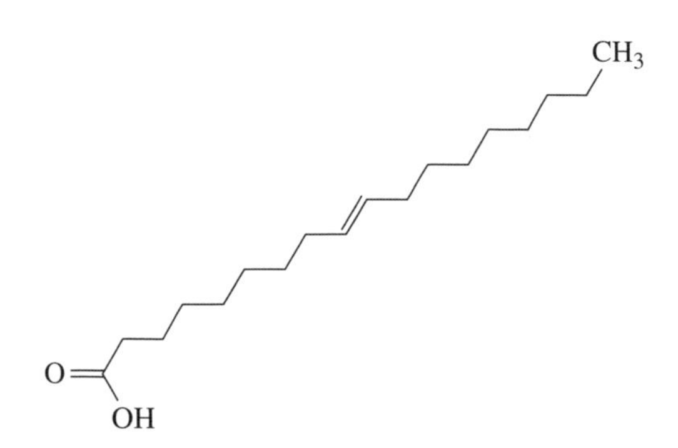

FIGURE 10.8 Fatty acid (oleic acid).

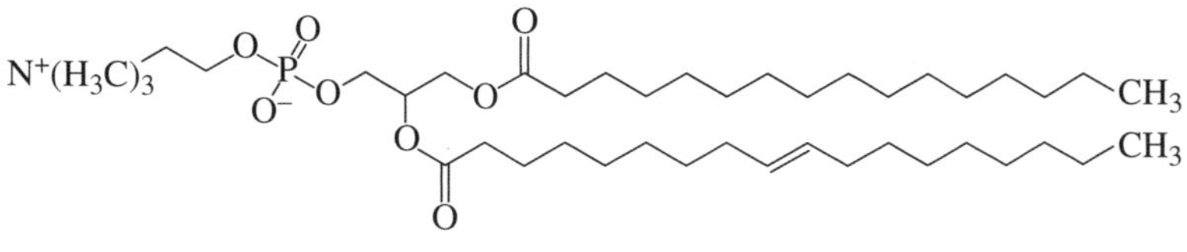

FIGURE 10.9 Lecithin.

electrostatic effects. The presence of small quantities of moisture and a low but defined solubility of drug in propellant may also result in particle growth by Ostwald ripening. Aggregation and particle growth will change the dose delivery and proportion of particles in the respirable range, which in turn will impact on lung deposition and therapeutic effect.

Solution formulations generally require a semipolar solvent to achieve dissolution, and these cosolvents are more susceptible to moisture uptake. Since the drug is molecularly dispersed, side groups are exposed that may be susceptible to rapid degradation.

In both solution and suspension formulations the propellant can potentially leach extractables (nitrosamines, elastomers) from the gaskets in the containers, which may result in instability.

10.4.1.2 Dry Powder Systems Dry powder systems are occasionally prepared from pure drug substance. More frequently blends with lactose (Fig. 10.10) are prepared. The lactose blends consist of respirable drug particles and large (~50–150 μm) excipient particles [50–52]. The excipient is included as a diluent to aid in dispensing the drug and as a fluidizing agent to assist dispersion. Tertiary blends have been prepared in which small lactose particles (<50 μm) have been used to aid flow and dispersion of drug from the inhaler. Other sugars have been evaluated as excipients (e.g., mannitol and trehalose), but none has yet been approved by the FDA for marketed products.

Some spray-dried particles have been prepared using other generally regarded as safe (GRAS) substances such as lecithin, human serum albumin, polylysine, and polyarginine, but these have also not yet been approved for delivery to the lungs.

10.4.1.3 Nebulizer Solutions Solutions intended for delivery from nebulizers are subject to the rules guiding general solution chemistry. Susceptibility to hydrolysis would mitigate against solution formulation. Light and heat may accelerate degradation. The presence of complexing agents will influence solubility. Using concentrations near the solubility limit of the drug may result in precipitation (pentamidine).

Nebulizer solutions are now prepared as sterile product due to the degradation of drug and potential for nosocomial infections from contaminating microorganisms [53].

Historically preservatives such as benzalkonium chloride were employed, but this approach is no longer favored due to adverse events associated with these agents.

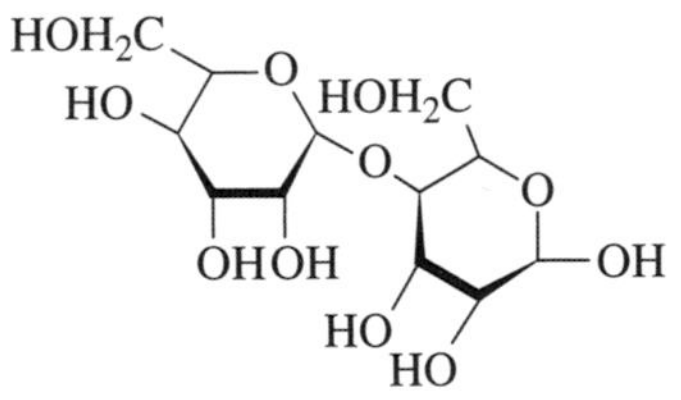

FIGURE 10.10 Lactose.

10.4.2 Interactions

The combination of drugs and excipients often results in unforeseen interactions, which are detrimental to the product and potentially to the health of the patient. These interactions may result in physical or chemical changes, which impact upon the stability, efficacy, and toxicity of the product. It would be impossible to include a comprehensive review of all of the circumstances where such interactions could occur. However, examples are given below for each of the major delivery systems.

10.4.2.1 Propellant-Driven Metered Dose Inhalers The incorporation of surfactants in pMDIs contributes to the suspension stability and lubricates the valve to facilitate the delivery of consistent doses throughout the lifespan of the product. It has been noted in a model system that side-chain interactions may occur, which will ultimately influence the physicochemical properties of the product [54]. In this context disodium fluorescein, a suitable model hygroscopic material for certain drugs, was shown to interact with selected fatty acids in nonaqueous solution. This interaction was concentration dependent and potentially initiated several interfacial phenomena ranging from adsorption to precipitation [55]. It was noted that a specific interaction occurred between the phenolic sodium of the disodium fluorescein and the carboxylate group of the fatty acid, which, in the extreme, could result in the production of the salt form of the fatty acid and a monoanion of fluorescein. This indirect observation was superceded by studies of a variety of drugs and their interaction with approved excipients such as oleic acid and sorbitan trioleate [56]. In these studies while examining albuterol (salbutamol) and two salt forms of isoproterenol (isoprenaline), it was postulated that a proton exchange occurred between the adsorbed surfactant and drug, which resulted in a charge effect capable of contributing to the stability of the suspension. A reduction in susceptibility to hygroscopic growth, both on storage and in transit through the airways, was an incidental effect of the adsorption/association of the surfactant with the surface of model drug particles [57, 58].

Another clear effect of the presence of surfactant in a nonaqueous suspension of particles is the potential for changes in crystal habit [59] or Ostwald ripening [60].

10.4.2.2 Dry Powder Systems Dry powder formulations are susceptible to a number of potential interactions. Since there is currently only one approved excipient, the drugs have to be compatible with lactose [4, 5]. In addition, dry powders are prone to moisture sorption, which can give rise to chemical degradation by hydrolysis or physical instability due to capillary forces [61]. As other excipients, such as lecithin, are explored as excipients in dry powder product hydrophobic effect may become the source of aggregation, and the stability of the excipient itself may become a source of concern.

10.4.2.3 Nebulizer Solutions The potential interactions that can occur in nebulizer solutions are subject to simple solution chemistry. A few excipients are employed in nebulizer products, notably sodium chloride to achieve isotonicity with body fluids. There are some rare examples in which interactions with sodium chloride and other solutes influence nebulizer performance. In studies of the delivery of amiloride hydrochloride and trisodium uridine triphosphate (UTP) both used in the treatment of CF an interaction occurred in which a precipitate of the two drugs was formed in the ratio 3 amiloride: 1 UTP [62]. The presence of sodium chloride suppressed the solubility of amiloride further by the common ion effect.

Suspension nebulizer formulations exhibit a complex behavior in which the particle or aggregate size in suspension may influence the droplet size delivered from an air jet nebulizer, and as the size of the particles approaches and exceeds the droplet size produced, size selectivity in dispersion of the particles may occur [63]. Ultrasonic nebulizers are also susceptible to particle size in dispersion [64].

10.4.3 Stability

The potential for drugs to interact with solvent, excipient, packaging materials, atmospheric moisture, oxygen, light, and heat contributes to the overall stability of the final product. These factors have been recognized by international regulatory bodies. Notable among these is the US Food and Drug Administration (FDA). Clear guidance has been given for the chemistry, manufacturing, and section of any submission to new drug approval [65, 66].

The performance of pMDIs and DPIs is scrutinized in the context of efficiency and reproducibility of dose delivery, particle size, and distribution under a range of storage conditions, with respect to temperature and humidity, for extended periods of time, up to 2 years [67]. Since the nebulizer products do not bring the device in contact with the drug until the point of use, a slightly different approach is taken to their approval. Recommended devices and conditions of operation for the delivery of a particular drug must now be stated. The solution formulation is then viewed as a sterile parenteral product and requires concomitant testing.

10.5 REGULATORY CONSIDERATIONS

The last decade has seen an increasing interest in the development of generic aerosol products. However, the regulatory guidance has been insufficient to render this a straightforward task [68]. The adoption of quality by design and statistical process and quality control approaches by the regulators at the beginning of this period gave a more scientifically rigorous foundation to product development activities [69]. However, many years passed without product-specific guidance documents. In 2013, two important guidances—one on albuterol-HFA 134a [70] and the other on fluticasone propionate-salmeterol xinafoate dry powders [71]—were published by the US FDA. These are important new developments that should facilitate the rapid development of new generic products.

10.6 FUTURE DEVELOPMENTS

In the relatively near future the focus for aerosol delivery of drugs may shift from the physics and chemistry associated with the product development to the biology of identifying new drug candidates and targets. The driving force for such a shift is multifaceted. As device technology is refined and the limits to efficiency and reproducibility of drug delivery are approached the ability to improve disease therapy will be refocused on the desired pharmacological effect. In this context the desire to localize receptor, enzyme, and transport targets will become stronger, and the notion of combining therapies to achieve a particular goal will be more widespread. This opens up a significant opportunity for those involved in quantitative structure–activity relationship research and bioinformatics to search existing chemical libraries and reexamine the use of previously overlooked potential drugs. It may be predicted that this approach will lead to improved disease management, new generations of drugs with desired efficacy, and reduced toxicity. To facilitate these developments government agencies and industries are opening their drug libraries to searches using up-to-date tools in pharmacoinformatics to allow rapid and inexpensive new target identification. This is an important new approach in the "big data" environment that has been created by continuing developments in information technology and computer capacity.

10.7 CONCLUSION

The importance of generating drug particles in the 1–5 μm size range is central to the efficient delivery of drugs. A variety of devices and mechanisms have been developed over the last half century to achieve this goal.

A steady increase in new aerosol products has occurred in the last 10 years. Notably, the majority of these products have come from large pharmaceutical companies with a history of achievement in this field. It can only be hoped that the other products currently in development will soon be commercially available.

As it becomes increasingly probable that delivery of drugs as aerosols can be achieved readily the focus can shift to the nature of the therapeutic agent and its physical and chemical stability in the dosage forms. New chemical entities can be considered for delivery to the lungs to facilitate the control of pulmonary diseases or disease that may be treated by pulmonary drug delivery.

The chemical composition and structure of drug and excipient particles play a significant role in the success of therapeutic agents in a number of diseases for which the lungs are a target or portal. Consideration of these issues is essential to promoting effective disease management. There are some specific sources of drug instability, degradation, and physical properties that detract from product performance. If these issues are given consideration, the likelihood of developing a commercially viable, new therapy that will meet stringent regulatory requirements is very good.

In the last decade a sense of urgency has driven a desire for rapid product development. The pharmaceutical industry appears to be in a period of metamorphosis that has political, economic, sociological, and scientific origins. Adopting the tools of the information age in the search for new therapeutic targets and drugs is one aspect of this change. Bringing efficient but scientifically rigorous practical approaches including a range of high throughput technologies, rapid screening techniques, and statistical methodologies throughout the development process should serve the need to develop new and inexpensive drugs in the context of a changing global and local health care environment.

REFERENCES

1. A.J. Hickey, *Inhalation Aerosols: Physical and Biological Basis for Therapy*, Second Edition, Vol. **221**, Marcel Dekker, Inc., New York, 2007.
2. D. Ganderton, T.C. Jones, *Drug Delivery to the Respiratory Tract*, John Wiley & Sons, Inc., New York, 1988.
3. C. Thiel, From Susie's question to CFC free: an inventor's perspective on forty years of MDI development and regulation, in: R.N. Dalby, P.R. Byron, S.J. Farr (Eds.), *Respiratory Drug Delivery V*, Interpharm Press, Inc., Phoenix, AZ, 1996, pp. 115–123.
4. C.A. Dunbar, A.J. Hickey, P. Holzner, Dispersion and characterization of pharmaceutical dry powder aerosols, *KONA* **16** (1998) 7–44.
5. M.J. Telko, A.J. Hickey, Dry powder inhaler formulation. *Respir Care* **50** (2005) 1209–1227.
6. H. Matthys, D. Köhler, Pulmonary deposition of aerosols by different mechanical devices, *Respiration* **48** (1985) 269–276.
7. M.M. Clay, D. Pavia, S.P. Newman, T.L. Lennard-Jones, S.W. Clarke, Assessment of jet nebulisers for lung aerosol therapy, *Lancet* **2** (1983) 592–594.
8. C. Dunbar, A.J. Hickey, A new millennium for inhaler technology, *Pharm Tech* **21** (1997) 116–125.
9. R. Davies, C. Leach, B. Lipworth, R. Shaw, Asthma management with HFA-BDP (Qvar), *Hosp Med* **60** (1999) 263–270.
10. T.M. Crowder, M.D. Louey, V.V. Sethuraman, H.D.C. Smyth, A.J. Hickey, 2001: an odyssey in inhaler formulations and design, *Pharm Tech* **25** (2001) 99–113.
11. A.J. Hickey, Summary of common approaches to pharmaceutical aerosol administration, in: A.J. Hickey (Ed.), *Pharmaceutical Inhalation Aerosol Technology*, Second Edition, Marcel Dekker, Inc., New York, 2003, pp. 385–421.
12. A.J. Hickey, D. Ganderton, *Pharmaceutical Process Engineering*, Second Edition, Vol. **195**, Informa Healthcare, New York, 2010, pp. 136–154.
13. M. Sacchetti, M.M.V. Oort, Spray-drying and supercritical fluid particle generation techniques, in: A.J. Hickey (Ed.), *Inhalation Aerosols: Physical and Biological Basis for Therapy*, Second Edition, Vol. **221**, Informa Healthcare, New York, 2007, pp. 307–346.
14. R. Vehring, Pharmaceutical particle engineering via spray drying, *Pharm Res* **25** (2007) 999–1022.

15. K. Johnson, Interfacial phenomena and phase behavior in metered dose inhaler formulations, in: A.J. Hickey (Ed.), *Inhalation Aerosols: Physical and Biological Basis for Therapy*, Second Edition, Vol. **221**, Informa Healthcare, New York, 2007, pp. 347–371.
16. H.D.C. Smyth, A.J. Hickey, R.M. Evans, Aerosol generation from propellent-driven metered dose inhalers, in: A.J. Hickey (Ed.), *Inhalation Aerosols: Physical and Biological Basis for Therapy*, Second Edition, Vol. **221**, Informa Healthcare, New York, 2007, pp. 399–416.
17. T.S. Purewal, D.J. Grant, *Metered Dose Inhaler Technology*, CRC Press, Boca Raton, FL, 1997.
18. H.D.C. Smyth, V.P. Beck, D. Williams, A.J. Hickey, The influence of formulation and spacer device on the *in vitro* performance of solution chlorofluorocarbon-free propellant-driven metered dose inhalers, *AAPS PharmSciTech* **5** (2004) E7.
19. H.D.C. Smyth, A.J. Hickey, Multimodal particle size distributions emitted from HFA-134a solution pressurized metered-dose inhalers, *AAPS PharmSciTech* **4** (2003) E38.
20. C. Sirand, J.-P. Varlet, A.J. Hickey, Aerosol-filling equipment for the preparation of pressurized pack pharmaceutical formulations, in: A.J. Hickey (Ed.), *Pharmaceutical Inhalation Aerosol Technology*, Second Edition, Marcel Dekker, Inc., New York, 2003, pp. 311–343.
21. C.A. Dunbar, A.P. Watkins, J.F. Miller, An experimental investigation of the spray issued from a pMDI using laser diagnostic techniques, *J. Aerosol Med* **10** (1997) 351–368.
22. A.J. Hickey, H.M. Mansour, M.J. Telko, Z. Xu, H.D. Smyth, T. Mulder, R. McLean, J. Langridge, D. Papadopoulos, Physical characterization of component particles included in dry powder inhalers. I. Strategy review and static characteristics, *J Pharm Sci* **96** (2007) 1282–1301.
23. A.J. Hickey, H.M. Mansour, M.J. Telko, Z. Xu, H.D. Smyth, T. Mulder, R. McLean, J. Langridge, D. Papadopoulos, Physical characterization of component particles included in dry powder inhalers. II. Dynamic characteristics, *J Pharm Sci* **96** (2007) 1302–1319.
24. Z. Xu, H.M. Mansour, T. Mulder, R. McLean, J. Langridge, A.J. Hickey, Heterogeneous particle deaggregation and its implication for therapeutic aerosol performance, *J Pharm Sci* **99** (2010) 3442–3461.
25. Z. Xu, H.M. Mansour, A.J. Hickey, Particle interactions in dry powder inhaler unit processes: a review, *J Adhes Sci Technol* **25** (2011) 451–482.
26. A.J. Hickey, T.M. Crowder, *Next Generation Dry Powder Inhalation Delivery Systems, in Inhalation Aerosols: Physical and Biological Basis for Therapy*, Second Edition, Vol. **221**, Informa Healthcare, New York, 2007, pp. 445–460.
27. T.M. Crowder, M.J. Donovan, Science and technology of dry powder inhalers, in: H.D.C. Smyth, A.J. Hickey (Eds.), *Controlled Pulmonary Delivery*, Springer, New York, 2012 203–222.
28. A.J. Hickey, R.W. Niven, Atomization and nebulizers, in: A.J. Hickey (Ed.), *Inhalation Aerosols: Physical and Biological Basis for Therapy*, Second Edition, Vol. **221**, Informa Healthcare, New York, 2007, pp. 253–283.
29. B.J. Greenspan, Ultrasonic and electrohydrodynamic methods for aerosol generation, in: A.J. Hickey (Ed.), *Inhalation Aerosols Physical and Biological Basis for Therapy*, Second Edition, Vol. **221**, Informa Healthcare, New York, 2007, pp. 285–306.
30. R. Dalby, J. Suman, Inhalation therapy: technological milestones in asthma treatment, *Adv Drug Deliver Rev* **55** (2003) 779–791.

31. A. Adjei, J. Garren, Pulmonary delivery of peptide drugs: effect of particle size on bioavailability of leuprolide acetate in healthy male volunteers, *Pharm Res* **7** (1990) 565–569.
32. D.W. Williams, The application of aerosolized antimicrobial therapies in lung infections, in: A.J. Hickey (Ed.), *Pharmaceutical Inhalation Aerosol Technology*, Second Edition, Marcel Dekker, Inc., New York, 2003, pp. 473–488.
33. L.S. Goodman, A.G. Gilman, L.E. Limbird, J.G. Hardman, A.G. Gilman, *Goodman & Gilman's The Pharmacological Basis of Therapeutics*, McGraw-Hill Co., New York, 2001.
34. B.G. Katzung, *Basic and Clinical Pharmacology (Lange Series)*, Lange Medical Books/ McGraw-Hill, New York, 2000.
35. A. Gringauz, *Introduction to Medicinal Chemistry: How Drugs Act and Why*, Wiley-VCH, New York, 1997.
36. P.A. Crooks, A.M. Al-Ghananeem, Drug targeting to the lung: chemical and biochemical considerations, in: A.J. Hickey (Ed.), *Pharmaceutical Inhalation Aerosol Technology*, Second Edition, Marcel Dekker, Inc., New York, 2003, pp. 89–154.
37. J.A. Bernstein, H. Amin, S.J. Smith, Therapeutic uses of lung aerosols, in: A.J. Hickey (Ed.), *Inhalation Aerosols: Physical and Biological Basis for Therapy*, Second Edition, Vol. **221**, Informa Healthcare, New York, 2007, pp. 219–252.
38. D. Olivieri, P.J. Barnes, S.S. Hurd, G.C. Folco, *Asthma Treatment A Multidisciplinary Approach*, Vol. **229**, Plenum Press, New York, 1992.
39. R.M. Smith, Aerosol delivery of alpha1 proteinase inhibitor: a paradigm for delivery of therapeutic proteins to the lung, in: A.L. Adjei, P.K. Gupta (Eds.), *Inhalation Delivery of Therapeutic Peptides and Proetins*, Marcel Dekker, Inc., New York, 1997, pp. 331–354.
40. L. Garcia-Contreras, A.J. Hickey, Pharmaceutical and biotechnological aerosols for cystic fibrosis therapy, *Adv Drug Deliver Rev* **54** (2002) 1491–1504.
41. J. Hanes, M. Dawson, Y.-E. Har-el, J. Suh, J. Fiegel, Gene delivery to the lung, in: A.J. Hickey (Ed.), *Pharmaceutical Inhalation Aerosol Technology*, Second Edition, Marcel Dekker, Inc., New York, 2003, pp. 489–539.
42. M. Ausborn, B.V. Wichert, M.T. Carvajal, R.W. Niven, D.M. Soucy, R.J. Gonzalez-Rothi, X.Y. Gao, H. Schreier, Amikacin liposomes for inhalation therapy: *in vitro* efficacy against Mycobacterium avium intracellulare in alveolar macrophages, *Proc Program Int Symp Controlled Release Bioact Mater* **18** (1991) 371–372.
43. B.V. Wichert, R.J. Gonzalez-Rothi, L.E. Straub, B.M. Wichert, H. Schreier, Amikacin liposomes: characterization, aerosolization, and *in vitro* activity against Mycobacterium avium intracellulare in alveolar macrophages, *Int J Pharm* **78** (1992) 227–235.
44. A. BitMansour, J.M.Y. Brown, Prophylactic administration of liposomal amphotericin B is superior to treatment in a murine model of invasive aspergillosis after hematopoietic cell transplantation, *J Infect Dis* **186** (2002) 134–137.
45. A.J. Hickey, A.B. Montgomery, Aerosolized pentamidine for treatment and prophylaxis of pneumocystis carinii pneumonia in patients with acquired immunodeficiency syndrome, in: A.J. Hickey (Ed.), *Pharmaceutical Inhalation Aerosol Technology*, Second Edition, Marcel Dekker, Inc., New York, 2003, pp. 459–472.
46. P.R. Byron, J.S. Patton, Drug delivery via the respiratory tract, *J Aerosol Med* **7** (1994) 49–75.
47. A.J. Hickey, Back to the future: inhaled drug products, *J Pharm Sci* **102** (2013) 1165–1172.
48. A. Wade, P.J. Weller, *Handbook of Pharmaceutical Excipients*, American Pharmaceutical Association, Washington, DC, 1994.

49. S.R.P. da Rocha, B. Bharatwaj, S. Saiprasad, Science and technology of pressurized metered dose inhalers, in: H.D.C. Smyth, A.J. Hickey (Eds.), *Controlled Pulmonary Drug Delivery*, Springer, New York, 2012, pp. 165–201.
50. Z. Xu, H.M. Mansour, T. Mulder, R. McLean, J. Langridge, A.J. Hickey, Dry powder aerosols generated by standardized entrainment tubes from drug blends with lactose monohydrate: 1. Albuterol sulfate and disodium cromoglycate, *J Pharm Sci* **99** (2010) 3398–3414.
51. Z. Xu, H.M. Mansour, T. Mulder, R. McLean, J. Langridge, A.J. Hickey, Dry powder aerosols generated by standardized entrainment tubes from drug blends with lactose monohydrate: 2. Ipratropium bromide monohydrate and fluticasone propionate, *J Pharm Sci* **99** (2010) 3415–3429.
52. H.M. Mansour, Z. Xu, A.J. Hickey, Dry powder aerosols generated by standardized entrainment tubes from alternative sugar blends: 3. Trehalose dihydrate and D-mannitol carriers, *J Pharm Sci* **99** (2010) 3430–3441.
53. M.S. Simberkoff, M.R. Santos, Prevention of community-acquired and nosocomial pneumonia, *Curr Opin Pulm Med* **2** (1996) 228–235.
54. A.J. Hickey, G.V. Jackson, F.J.T. Fildes, Preparation and characterization of disodium fluorescein powders in association with lauric and capric acids, *J Pharm Sci* **77** (1988) 804–809.
55. D.J. Cooney, A.J. Hickey, Preparation of disodium fluorescein powders in association with lauric and capric acids, *J Pharm Sci* **92** (2003) 2341–2344.
56. S.J. Farr, L. McKenzie, J.G. Clarke, Drug-surfactant interactions in apolar systems: relevance to the optimized formulations of suspension MDIs, in: P.R. Byron, R.N. Dalby, S.J. Farr (Eds.), *Respiratory Drug Delivery IV*, Interpharm Press, Inc., Buffalo Grove, IL, 1994, pp. 221–230.
57. A.J. Hickey, P.R. Byron, Effect of a non-hygroscopic surfactant coating upon fluorescein absorption from the respiratory tract, *Drug Des Deliv* **2** (1987) 35–39.
58. A.J. Hickey, I. Gonda, W.J. Irwin, F.J.T. Fildes, Effect of hydrophobic coating on the behavior of a hygroscopic aerosol powder in an environment of controlled temperature and relative humidity, *J Pharm Sci* **79** (1990) 1009–1014.
59. K.A. Fults, I.F. Miller, A.J. Hickey, Effect of particle morphology on emitted dose of fatty acid-treated disodium cromoglycate powder aerosols, *Pharm Dev Technol* **2** (1997) 67–79.
60. E.M. Phillips, Crystal growth-formulation dependence and early detection, *J Biopharm Sci* **3** (1992) 11–18.
61. N.M. Concessio, A.J. Hickey, Assessment of micronized powders dispersed as aerosols, *Pharm Tech* **18** (1994) 88–98.
62. R.J. Pettis, M.R. Knowles, K.N. Olivier, A.J. Hickey, Ionic interaction of amiloride and uridine 5′-triphosphate in nebulizer solutions, *J Pharm Sci* **93** (2003) 2399–2406.
63. A.J. Hickey, K. Kuchel, L.E. Masinde, Comparative efficiency of solution and suspension output from jet nebulizers, in: P.R. Byron, R.N. Dalby, S.J. Farr (Eds.), *Repiratory Drug Delivery IV*, Interpharm Press, Buffalo Grove, IL, 1994, pp. 259–263.
64. R.N. Dalby, A.J. Hickey, S.L. Tiano, Medical devices for the delivery of therapeutic aerosols to the lungs, in: A.J. Hickey (Ed.), *Inhalation Aerosols, Physical and Biological Basis for Therapy*, Second Edition, Vol. **221**, Informa Healthcare, New York, 2007, pp. 417–444.

65. USDHHS, Draft Guidance for Industry—Nasal Spray and Inhalation Solution, Suspension and Spray Drug Products Chemistry, Manufacturing, and Controls, Food and Drug Administration, Center for Drug Evaluation and Research, May 26, 1999.
66. USP24, <601> Aerosols, metered dose inhalers, and dry powder inhalers, in: *The United States Pharmacopoeia and National Formulary*, United States Pharmacopoeial Convention, Rockville, MD, 2000, pp. 1895–1912.
67. L.D. Jones, P. McGlynn, L. Bovet, A.J. Hickey, Analysis and physical stability of pharmaceutical aerosols, *Pharm Tech* **24** (2000) 40–54.
68. US Food and Drug Administration, Draft Guidance for the Industry, Metered Dose Inhaler (MDI) and Dry Powder Inhaler (DPI) Chemistry, Manufacturing and Controls Documentation, November 1998.
69. A.J. Hickey, D. Ganderton, *Pharmaceutical Process Engineering*, Second Edition, Informa Healthcare, New York, 2010, pp. 193–206.
70. US Food and Drug Administration, Draft Guidance on Albuterol Sulfate, recommended April, revised June 2013.
71. US Food and Drug Administration, Draft Guidance on Fluticasone Propionate; Salmeterol Xinafoate, September 2013.

11

TRANSDERMAL DELIVERY OF DRUGS USING PATCHES AND PATCHLESS DELIVERY SYSTEMS

Tannaz Ramezanli[1,2,3], Krizia Karry[4], Zheng Zhang[3], Kishore Shah[5], and Bozena Michniak-Kohn[1,2,3]

[1] *Ernest Mario School of Pharmacy, Rutgers, The State University of New Jersey, Piscataway, NJ, USA*
[2] *Center for Dermal Research, Rutgers, The State University of New Jersey, Piscataway, NJ, USA*
[3] *The New Jersey Center for Biomaterials, Rutgers, The State University of New Jersey, Piscataway, NJ, USA*
[4] *School of Engineering, Rutgers, The State University of New Jersey, Piscataway, NJ, USA*
[5] *Polytherapeutics Inc., Lakewood, NJ, USA*

11.1 INTRODUCTION

Transdermal drug delivery (TDD) refers to a dosage form where a compound diffuses across the skin and enters the bloodstream with the intention of therapeutic action not in skin but in another part of the body. In the past three decades TDD systems have provided an attractive alternative to oral drug delivery and hypodermic injections [1, 2]. Systemic delivery of compounds via their application on the skin is desirable for many reasons [2, 3]:

- Increased patient compliance: less invasive compared to hypodermal injections and also suitable for people who have difficulty swallowing tablets.
- Relatively large and accessible surface area for absorption (1–2 m^2).

Drug Delivery: Principles and Applications, Second Edition. Edited by Binghe Wang, Longqin Hu, and Teruna J. Siahaan.

- Bypassing first-pass metabolism and therefore increase in bioavailability.
- The potential for sustained or controlled release products, which are favorable for the drugs with short biological half-lives.
- Ease of termination of the treatment by removing the TDD system from the skin.

The stratum corneum (SC) is the most outer layer of the skin and the main barrier to drugs crossing the skin. This layer is composed of lipids arranged in bilayer structures in the interstitial space between groups of anucleated cells known as corneocytes [4]. In conventional TDD design, there are ideal properties required for candidate drugs to be considered for this route of delivery [1, 3]. The molecular weight should be less than a few hundred Dalton, octanol–water partition coefficients should be between 10 and 1000, the required dose should be less than 10 mg/day, and absolute solubility should be greater than 1 mg/ml. Hence, the first generation of TDD systems were designed for transport of small, lipophilic, low-dose drugs [1]. One also realizes that there are limitations for traditional passive TDD dosage forms, and these are listed below:

- Inability to deliver ionic or large-molecular-weight drugs.
- Inability to deliver drugs in a pulsatile fashion.
- Possibility of skin irritation and allergic reaction.
- Therapeutic drug concentrations in the plasma cannot be obtained with many drugs.

Some of the TDD limitations can be overcome using passive and active penetration enhancement techniques. In this chapter we will discuss patch and patchless delivery systems and several novel approaches in the TDD delivery.

11.2 TRANSDERMAL PATCH DELIVERY SYSTEMS

11.2.1 Definition and History of Patches

A transdermal patch is an adhesive film/membrane/matrix-containing medication that is placed on the skin. These patches are designed to transfer specific doses of drug at specific rates through intact skin into the systemic circulation to produce therapeutic effects.

The first skin patch was developed by Michaels et al. [5] in 1970s. This new drug delivery system was approved by the US Food and Drug Administration (FDA) in 1979. It was a 3-day patch designed to deliver scopolamine to the bloodstream to prevent motion sickness. In 1981 the second transdermal patch system was approved by the FDA and this utilized nitroglycerin for angina.

Transdermal patches have introduced new options for existing therapeutics. Depending on the solute, the duration of patch application on the skin can last from 1 to 7 days [6]. This delivery system is particularly useful for those drugs that have poor oral uptake, interactions with stomach acid, high metabolism by liver, and need frequent administration.

Table 11.1 depicts examples of some FDA-approved drugs formulated in transdermal patches and their applications. The nicotine patch is currently the best-selling transdermal patch in the United States and assists patients with cessation of smoking. Opioid patches are another example of widely used transdermal patches for chronic pain relief. For example, fentanyl and buprenorphine patches are administered to provide round-the-clock relief in severe chronic pain.

Despite the successes with transdermal patches, there are still only a few drugs that can be formulated into passive TDD systems. As an example, the drug with the smallest molecular weight formulated in a patch system currently on the market is nicotine (162 Da) and the largest is oxybutynin (359 Da).

11.2.2 Anatomy and Designs of Patches

The patch structure is composed of a liner, backing, drug compartment, and an adhesive layer. The liner is the protective layer at the bottom of the patch that is removed by the patient before applying the patch on the skin to enable drug release. The backing is the top layer of the patch usually chosen for appearance of the patch but more importantly for the occlusion [7]. Polyester, polyethylene, and polyolefin are materials often used for the patch backing. An adhesive layer is found in all transdermal patches. This overlay may or may not contain drug located on top of the liner and usually functions to secure the patch on the skin [8]. There are two types of adhesive layers: the face and the peripheral. In the face adhesive type, which is more common, a layer of adhesive covers the whole face of the patch, but in the peripheral type the adhesive is placed around the outer edge of the patch, usually surrounding the drug compartment [9]. Most of the transdermal patches in the market consist of multilayers. They can be categorized into two groups of reservoir-type and matrix-type designs (see Fig. 11.1).

11.2.2.1 Reservoir In this type of patch the drug is dissolved/suspended in liquid or gel compartment that acts as drug reservoir. The reservoir also contains many excipients. Two other components in this system are the rate-controlling membrane and the adhesive. Drug release rate is controlled by a membrane placed between the reservoir and adhesive layer. The drug can reach the skin once it permeates through the membrane and the adhesive layer [7]. This system possesses more complexity and results in a good control over delivery rate. However, there is a possibility of dose-dumping with reservoir patches. If reservoir patch fails in any way, it has the potential to release high concentrations of drug on to the skin. Thus, this type of patch should not be cut or altered to avoid destruction of the rate-controlling membrane. The fentanyl patch, for example, was initially designed as a reservoir patch, and due to some manufacturing defects fentanyl gel in the reservoir could leak out and come into direct contact with the skin. This resulted in millions of recalls of opioid pain patches. Fentanyl patch has now been redesigned as a matrix-type patch described later. Estraderm®, Nicoderm®, and Transderm®Scop are some examples of the reservoir patches in the market.

TABLE 11.1 Examples of Transdermal Patches Approved in the United States

Drug Generic Name	Trade Name	Manufacturer	Approved Year	System Type	Indication
Scopolamine	Transderm®Scop	Novartis	1979	Reservoir	Motion sickness
Nitroglycerin	Transderm®Nitro	Novartis	1981	Reservoir	Angina pectoris
Clonidine	Catapres-TTS®	Boehringer Ingelheim	1984	Reservoir	Hypertension
Estradiol	Estraderm®	Novartis	1986	Reservoir	Menopausal symptoms
Fentanyl	Duragesic®	Ortho McNeil Janssen	1990	Reservoir	Chronic pain
Nicotine	Nicoderm®	ALZA	1991	Reservoir	Smoking cessation
Testosterone	Testoderm-TTS®	ALZA	1993	Matrix	Hormone replacement therapy in men
Estradiol-norethindrone	Combipatch®	Novartis	1998	Drug-in adhesive	Menopausal symptoms
Lidocaine	Lidoderm®	Endo Pharmaceuticals	1999	Drug-in adhesive	Local anesthesia
Ethinyl estradiol-norelgestromin	Ortho Evra®	Ortho-McNeil Pharmaceutical	2001	Matrix	Contraception
Oxybutynin	Oxytrol®	Watson pharma	2003	Drug-in adhesive	Overactive bladder
Methylphenidate	Daytrana®	Noven Pharmaceuticals	2006	Matrix	Attention-deficit hyperactivity disorder
Selegiline	Emsam®	Somerset Pharmaceuticals	2006	Drug-in adhesive	Depression disorder
Rivastigmine	Exelon®	Novartis	2007	Matrix	Dementia
Granisetron	Sancuso®	Prostrakan	2008	Matrix	Chemotherapy-induced nausea and vomiting

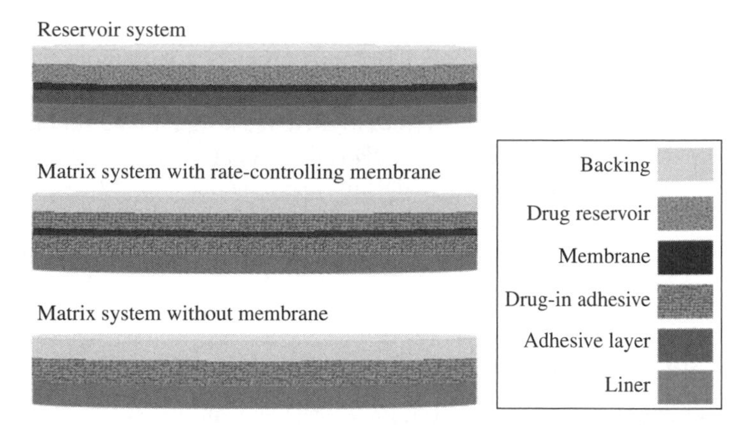

FIGURE 11.1 Popular designs of transdermal patches. (*See insert for color representation of the figure.*)

11.2.2.2 Matrix In this system the active ingredient is dissolved/suspended in a semisolid polymer matrix surrounded by an adhesive layer, or the matrix consists of the drug incorporated in an adhesive layer, and this layer is responsible for releasing the drug as well as sticking the patch on to the skin. Drug-in adhesive patches are designed as single- or multilayer patches. In the multilayer system a membrane is placed between two layers of drug-in adhesive patches [10]. The matrix may also contain excipients, such as penetration enhancers, solubilizers, etc. [9]. Matrix-type patches were introduced after reservoir-type patches, and in the absence of rate-controlling membrane, drug delivery rate is governed by skin permeability. In the matrix system, the drug is evenly distributed throughout the patch. Hence, there is less risk of accidental overdose and less potential for abuse with matrix-type patches than the reservoir system [11]. Matrix-type patches can be easier to manufacture but have limited flexibility in their design [6]. Oxytrol® and Emsam® are examples of matrix-type patches where drug is incorporated in the adhesive layer. In some other designs like Exelon®, there is a separate adhesive layer below drug matrix compartment. Fentanyl extended-release transdermal system is an example of matrix-type patch with rate-controlling membrane currently in the market.

11.2.2.3 Vapor Patch This system is the most recent design of patch where the adhesive layer is responsible for keeping the various layers of patch together and in addition releasing vapor. One example of this system on the market is patches which release essential oils used to relieve congestion and improve the quality of sleep [10].

11.3 PATCHLESS TRANSDERMAL DRUG DELIVERY SYSTEMS

As discussed earlier in the chapter, the TDD poses several advantages over traditional oral routes for the delivery of drugs with high first-pass metabolism, for example, drugs with poor bioavailability [12]. However, assuring that these drugs travel through the skin layers is no easy task. In the previous section the utilization of patches for the TDD was discussed. Here we will focus on patchless transdermal delivery where the aims are to enhance skin permeability by disrupting the microstructure of the SC and/or by employing external forces to drive the drug into the skin layers.

Regardless of the system in use, the TDD is dependent on drug diffusion through the skin layers (i.e., the SC, epidermis, dermis, and hypodermis) [12]. So in order to develop such capable systems different mechanisms of action need to be employed and assessed. These mechanisms have been conveniently summarized into three categories [1].

11.3.1 First-Generation Systems

These TDD systems employ little to no use of chemical enhancers. Conventional patches are the most common examples; however, their use is limited to lipophilic delivery of candidates of low molecular weight that are efficacious at low doses.

In recent years there has been an increased need for drug selectivity so as to target specific diseases and conditions via "engineered" mechanisms of action [13]. Unfortunately, this selectivity requires coupling of drug molecules with long carbon chains that increase their molecular size and thus impair their delivery through the skin [14]. Furthermore, since first-generation systems rely solely on passive drug diffusion, these systems have fallen behind if not coupled with other physical aids such as those described later.

11.3.2 Second-Generation Systems

These systems will be the focus of this section as they can be aptly called "modified patchless TDD systems." Specifically, second-generation systems focus on reversible changes to the SC by introducing external forces to drive transport of the drug into the skin layers and/or by enhancing skin permeability via physicochemical interactions of the drug with inactive molecules. Below is a list of the most common second-generation systems. The list is not comprehensive, but it will give an idea of the systems available and their designs.

Chemical penetration enhancers (CPEs) are contained in the majority of novel topical and transdermal formulations. For the most part, these are amphiphilic molecules that disrupt the lipid bilayer structure by introducing nanometer-sized defects that disorganize the molecular packaging of the tissue [1]. Upon disorganizing the tissues, they make way for transient pathways within the skin layers that enable formulations to have much higher drug permeation rates compared to the pure drug without CPEs. For example, researchers evaluated the kinetics and rates of drug diffusion of two common nonsteroidal anti-inflammatory drugs (NSAIDs) through human cadaver skin [15]. They found that after time, the tissue became saturated and did not allow the passage of drug. However, with formulations containing the chemical enhancer *N*-methyl-2-pyrrolidone (NMP) there was a 16-fold increase in permeation of ibuprofen compared to the formulation without the CPE [15].

There are many CPEs available, but the most widely used are poly(ethylene glycol) [16], SEPA (2-*n*-nonyl-1,3-dioxolane) [17], sodium lauryl sulfate (SLS) [18], and Azone® (1-docecylazacycloheptan-2-one) [19, 20]. There have also been an increased number of studies where biomolecules and/or combinations of CPEs have proved very successful in increasing the skin permeability so as to achieve the TDD [12, 16].

A disadvantage of these enhancers is that because of the augmented skin permeability there have been reports of increased skin irritation [21]. However, very interestingly, these skin irritation levels have been correlated to the absorption of the amide I band and its permeation effectiveness to the symmetric CH_2 stretching via Fourier Transform Infrared Spectroscopy (FTIR) [22]. Thus, this nondestructive assessment has allowed CPEs to be screened in a more efficient manner.

Supramolecular structures are large structures comprised of many small molecules that are held together by weak hydrogen bonds and/or strong covalent bonds. They may assemble spontaneously due to low-activation energies within the system, that is, emulsions, or they may be fabricated in the laboratory to have specific sizes and structures (e.g., vesicular systems [23], nanospheres [24], and cyclodextrin complexes [25]). As can be predicted, the drug structure plays a vital role in drug encapsulation and subsequent penetration efficiency. Henceforth, scientists have developed supramolecular structures that incorporate (i) cyclodextrins to improve the bioavailability of poorly soluble drugs, with (ii) liposomes (vesicles) to accommodate hydrophobic drugs within its aqueous compartment. In one case, this combination ultimately resulted in a formulation with negligible skin irritation and improved drug deposition [26]. In addition, seeking out these desirable traits, researchers have also employed chemical principles by designing spatial drug structures utilizing metal ions platinum and palladium II—elements known to form right angles upon complexion [27].

Another approach that has been investigated is that *using prodrugs*. This approach is typically followed when the parent drug molecule exhibits very poor diffusivity across the SC, as is the case for most hydrophilic molecules. *Prodrugs* are inactive molecules that contain the inactivated drug within a functional chemical structure that when enzymatically cleaved, that is, hydrolyzed, activates the drug moiety [28]. One major advantage of using a prodrug is that its lipophilicity can be "adjusted" by manipulating the nature and length of its side chains—where longer chains lead to more hydrophobic (and larger) molecules. A hydrophilic molecule is usually a poor transdermal drug candidate; however, when reacted with an ester or an amide in an esterification or amide formation reaction, the lipophilicity of the drug will increase such that percutaneous diffusion is now positive [28, 29].

Two other advantages of the prodrug approach are that (i) it is applicable to all delivery systems (oral, systemic, parenteral, ocular, etc.) and (ii) since the approach is to increase drug penetration and not skin permeability, prodrugs rarely cause any skin irritation [1, 28]. Nevertheless, there are also some disadvantages to prodrugs, the first one being its poor acceptability by the FDA—this is because approval for both the inactive and active drug molecules is required [1]. Hence, these systems typically have longer clinical trials. With extended trials and a need for thorough assessment of side products after percutaneous activation, this approach has not seen much success in the market.

Even with these implications in July 2007 the FDA approved the inactive prodrug Vyvanse™ (lisdexamfetamine dimesylate) for the treatment of attention-deficit hyperactivity disorder (ADHD) in children. It proved to be a high-revenue drug with almost two million prescriptions within its first 8 months of availability [30]. Explicitly, the inactivated drug has been covalently bonded to aminoacid l-lysine, which when metabolized releases and activates the drug moiety [30].

As can be predicted, some of these second-generation systems may exist in combination with each other, for example, vesicles containing CPEs [31], prodrug enhancer [32], and vesicles with prodrugs [28]. Furthermore, even with all these approaches, there still are molecules that resist drug diffusion unless driven by physical methods into the skin layers.

Iontophoresis is a physical process that involves the use of an electric or magnetic field to drive diffusive transport of a drug into the SC and percutaneous layers. It is a low-voltage (≤10 V), noninvasive method, primarily focused on the transdermal delivery of charged lipophilic drug molecules [33]. Unparalleled by most, these systems control drug delivery rate by the applied electrical current. Even more, there is proportionality between the release rate of the drug and the current. So, extended or complex systemic profiles of the drug can be achieved very easily by changing the amperage or volts of the device [1].

As with prodrugs, since there is no direct change to the skin barrier, skin irritation is usually low for iontophoretic devices. Its major disadvantages have to do with the fact that its ideal candidates are charged drugs for which stability becomes an issue [34]. These charged drugs convectively move through the skin layers by the induction of a mechanism similar to electrophoresis with continuous current or even voltage pulses [34].

Another disadvantage correlates with the fact that the electrical properties of the skin are inevitably variable within and between subjects. The skin's pH, hydration, elasticity, perspiration, time of day/year, and even emotional state have been studied as important factors [35]. In particular, changes in the electrical conductance of the skin due to experienced levels of anxiety have been leveraged as adequate tools to assess different interviewing protocols [36]. Specifically, in a study conducted in 1988 researchers found that there existed a directly proportional relationship between the anxiety experienced by an interviewed student and the skin's conductance [36]. Because conductance is the reciprocal of resistivity, it can be assumed that at higher levels of anxiety there is a decreased resistance from the skin and thus higher permeability to drug molecules [35, 37].

11.3.3 Third-Generation Systems

Third-generation systems make use of specific (mostly) physical technology to directly permeate or disrupt the SC and epidermis [1]. The following is a list of the most commonly used third-generation systems.

Electroporation makes use of high-voltage (≥50 V) short pulses to disrupt the lipid bilayer structure within the SC and drive drug diffusion into transient nanosized pores deep in the skin layers [33, 38]. The mechanism of action is not exactly understood, but it has been hypothesized that the potential difference in charges between the device and the skin leads to reduced skin resistivity by inducing conformational changes and/or reorientations of the lipids in the SC [37]. Another theory suggests that the reduction in skin resistance, and thus an increase in drug permeation, is caused by the alteration of the skin's ion content upon applying an electric charge [35]. All in all, there are numerous instances in which electroporation has proved advantageous over first- and second-generation systems, even more when it was combined with the former [39].

Compared with iontophoresis, electroporation is associated with much higher voltages—sometimes 10-fold of iontophoresis [33]. At 80 V, a linear increase in percutaneous delivery was observed for an electroporation-calcein dye system [40]. Nevertheless, at higher voltages (>100 V) and high pulsation frequencies (<10 s) the system reached a plateau for which no linear dependency of voltage-to-active diffusion was apparent [40]. Hence, a thorough evaluation of both applied voltage and pulsation frequency/spacing need to be assessed in the early development stages of an electrophoretic system.

Sonophoresis (phonophoresis, ultrasound) makes use of low-frequency ultrasonic waves (20 kHz) to disrupt the lipid bilayers and induce microvibrations in the tissues [41]. These vibrations generate heat that in turn increases the kinetic energy of the drug molecules and drive convective transport of the entities into pores within the skin layers. Compared to other third-generation systems, sonophoresis has been around for a long time. For example, in ultrasound imaging of fetuses in the mother's womb. Nonetheless, its applications for transdermal delivery of macromolecules have started to be explored only a decade ago.

Purposely, because of the technique's disruptive nature, it allows for efficient delivery of peptides and proteins—a challenging task for all other systems (except microneedles, which shall be discussed later). In a recent study, subcutaneous insulin injections were compared to an ultrasonic transdermal delivery system. It was found that the ultrasonic systemic led to increased bioavailability and faster systemic delivery [42].

Thermal ablation works very similar to sonophoresis. In this technique, however, the skin is heated so as to induce formation of micropores in the SC. These pores then allow passage of the drugs into deeper skin layers. One of its disadvantages is that depending on the patient's sensitivity and fat deposition the heating may affect cells, cutaneous nerves, and even some pain sensors such as free nerve endings within the hypodermis [43].

Microdermabrasion is a technique in which the skin's surface layers are removed or wounded before drug application [44]. Since the skin's resistance is reduced, drugs are then able to penetrate more efficiently into subcutaneous layers and reach the systemic circulation. In cosmetic applications, the SC is thought of a reservoir of sunspots, blemishes, and fine wrinkles, while the viable epidermis is the "elixir of healthy cells and beautiful skin." Although those assumptions may not be necessarily true, it is safe to state that for TDD applications, it is has been necessary to remove or scar the deeper epidermis layer so as to ensure adequate insulin delivery [44]. Unfortunately, even though the skin scarring is temporal, the immediate pain and discomfort experienced by the patients has been an issue that has made microdermabrasion a slow-advancing transdermal system for drug delivery.

Microneedles are the best example of third-generation systems. They disrupt the SC lipid bilayer and create channels that allow for an efficient drug delivery, including macromolecules and even vaccines. These will be further explored in latter sections.

11.4 RECENT ADVANCES IN TRANSDERMAL DRUG DELIVERY

So far we have discussed different designs and techniques used for patch and patchless TDD. Some of the penetration enhancement techniques have also been used in the design of transdermal patches. For example, Zecuity® is an iontophoretic transdermal patch approved by the FDA in 2013 for treatment of migraine headaches. A microchip inside the patch is preprogrammed to apply low voltage and deliver the drug sumatriptan across the skin.

The third generation of the TDD has opened new avenues for delivery of hydrophilic and/or large-molecular-weight compounds into the skin. In the past few years, there has been huge number of studies on skin delivery of peptides, vaccines, and siRNAs.

On the other hand, ease of manufacturing and low cost of first- and second-generation chemical enhancers of the TDD make them remain popular for delivery of small lipophilic drugs. In this section the use of microneedles for vaccine delivery and a new polymer-based patchless system called PharmaDur® are explored.

11.4.1 Frontier in Transdermal Drug Delivery: Transcutaneous Immunization via Microneedle Techniques

The human skin is not only a protective barrier but also an immunological organ that contains immunocompetent cells, that is, Langerhans cells in epidermis and dermal dendritic cells in dermis. These cells have on their surface high amounts of the major histocompatibility complex class II molecules (MHC2), which confer the immunocompetent cells with the capability of detecting and capturing antigens (e.g., vaccine macromolecules) [45]. Subsequently, the immune response of the Langerhans cells and dermal dendritic cells is elicited. This will lead to the migration of these antigen-presenting cells (APCs) into the draining lymph nodes [46]. The helper T-lymphocytes recognize the MHC2 and get activated, releasing cytokines that further stimulate the activity of macrophages, killer T cells, and antibody-producing B cells.

Based on this immune function of the skin, various transcutaneous immunization techniques have been developed. Tremendous efforts and progress have been devoted and achieved in electroporation, iontophoresis, sonophoresis, jet injections, patch formulations, microneedle techniques, nanoparticles, and liquid-based vesicles (for a recent review, see Ref. [47]). Here, we will focus on the transcutaneous immunization via microneedle techniques. The most recent research papers, ongoing clinical trials, and FDA-approved microneedle-based product for transcutaneous immunization are reviewed herein.

Microneedle techniques are based on an array of micrometer-sized needles that can create a transport pathway large enough for proteins and nanoparticles, but small enough to avoid pain [47]. Recent review papers have classified four types of microneedle techniques with respect to delivery mechanism: (i) solid microneedles for pretreatment of the skin to increase permeability, (ii) microneedles coated with drug that dissolve in the skin, (iii) polymer microneedles encapsulating the drug that fully dissolve in the skin, and (iv) hollow microneedles for infusing the drug into the skin [48]. Figure 11.2 illustrates the scheme of these four types of microneedle techniques.

Among the aforementioned types of microneedles, the first group represents *solid microneedles* that are used to pretreat the skin and, therefore, increase the permeability of the vaccines that are applied topically later. In this case, the vaccines are not directly delivered by the microneedles; instead, the microneedles penetrate through the SC and then are removed from the skin, resulting in holes or pores for vaccines to be transported through. It successfully enhanced the skin permeation of low molecular weight drugs [48]. However, for vaccine delivery, solid microneedles that are used to pretreat the skin have been proven not as effective as intradermal injection in a number of studies [48–53], probably due to inefficient delivery of vaccine from the topical formulations even though micrometer-sized holes are introduced by the microneedles before applying the topical formulation [48].

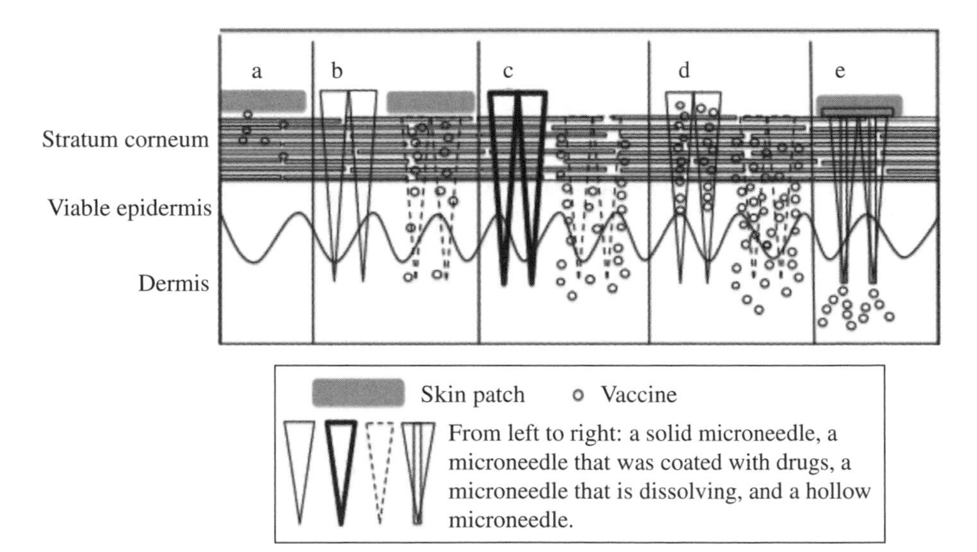

FIGURE 11.2 Different designs of microneedle technology for vaccine delivery: (a) via a skin patch without using microneedles; (b) via a skin patch after treatment with solid microneedles; (c) via coated microneedles that were coated with the drug; (d) via dissolving microneedles in which the drug was loaded; (e) via hollow microneedles in combination with a skin patch.

Hollow microneedles technique is the most successfully translated toward routine clinical practice; different from the traditional microneedle arrays, most hollow microneedles vaccination studies have been carried out with 1.5 mm-long, small-gauge hypodermic needles mounted on a syringe. More detailed information on current clinical trials using hollow microneedles for transcutaneous immunization can be found at http://clinicaltrials.gov. It is worthwhile noting that Intanza® (developed by Sanofi Pasteur, the vaccine division of Sanofi) is the first clinically applied microneedle vaccine against influenza: the European commission and FDA have approved the release of Intanza on the marget in 2009 and 2011, respectively [54].

For the remaining two types of microneedle techniques, the vaccines are delivered via the microneedle arrays. Briefly, the *coated microneedles* technique is the most extensively used for skin delivery of vaccines. *Dissolving microneedles* technique represents a fast-growing aspect in the recent years, as the worry on the "broken tips of the microneedles" and the "bolus-type of release" of the vaccines is circumvented. The selected recent research papers reporting the use of coated and dissolving microneedles in transdermal vaccine delivery are listed in Table 11.2.

TABLE 11.2 The Application of Coated and Dissolving Microneedle Techniques in Transdermal Vaccine Delivery

Features	References
Coated microneedles: Microneedles coated with vaccines that dissolve in the skin	
Category of vaccines: Influenza (whole inactivated influenza virus, and virus-like particle [55]); Calmette–Guerin vaccine; DNA vaccine encoding hepatitis C virus protein; Hepatitis B surface antigen; measles vaccine.	For early publications, see reviews [48, 56, 57]. A number of most recent publications (2012–2013) are listed [55, 58–75]
Coating methods: Direct coating; coated together with external excipients to ensure homogeneity coating and stability; layer-by-layer coating of "smart" polymers and DNA vaccine [58, 59].	
Dissolving microneedles: Polymer microneedles encapsulating the vaccines that fully dissolve in the skin	
The design of the dissolving microneedles utilizes the capability of polymeric drug delivery system, resulting in more homogeneous drug loading and higher stability of the vaccine moiety. In addition, the controlled release with tunable release rates of the loaded vaccines, rather than the bolus-type of release for coated microneedles, has been generally considered as an advantage of dissolving microneedles. For a recent review, see Ref. [76].	Ganterz® AN-139; PLGA microparticles [77]; PLGA nanoparticles loaded in PMVE/MA [78]; sodium chondroitin sulfate [79]; sodium alginate [80]; liposome loaded in PVP-K30 [81]; CMC [82]; PVP [83]; PVA-PVP [84]; PEGDA [85]; sodium hyaluronate [86].

11.4.2 Patchless Transdermal Delivery: The PharmaDur "Virtual Patch"

Inherent skin barrier function and limited surface area of plastic patches present a significant obstacle to the delivery of high dosages of a drug transdermally. Patchless TDD in the form of topically applied gel, cream, lotion, or solution can to some extent overcome the limitation of skin barrier by application of the drug formulation to a much larger area of skin than would be possible with plastic patches. However, retention of such applied semisolid dosage form on the skin is a challenge because of its facile removal from the skin surface by flaking, perspiration, or wiping by contact with hands or articles of clothing. A graft copolymer delivery system that is designed to address these and other issues has been commercially available since 2004 under the trade name PharmaDur (www.pharmadur.com; Polytherapeutics, Inc.). PharmaDur is a graft copolymer (CAS nomenclature: poly(*N*,*N*-dimethylacrylamide-co-acrylic acid-co-polystyrene ethyl methacrylate)) having a combination of hydrophilic and hydrophobic structural moieties.

When a dermatological vehicle, for example, cream, lotion, or a gel, formulated with PharmaDur polymer is applied to the skin, it forms an imperceptible and invisible hydrogel film (*A Virtual Patch*; Fig. 11.3). The PharmaDur polymer film is nontacky, nongreasy, has a very smooth feel, does not exhibit flaking or skin stretchiness, and is breathable. It is characterized by excellent bioadhesion to skin and exhibits controlled release of both hydrophobic and hydrophilic molecules [87]. The polymer film and the drug contained therein are retained on the skin for 24+ hours unless the film is physically washed off. Thus, PharmaDur Virtual Patch combines benefits of both the plastic and semisolid dosage forms. PharmaDur film on skin serves as matrix for the drug and other nonvolatile excipients of the product formulation.

Furthermore, studies have shown the PharmaDur polymer to act synergistically with skin penetration enhancers in maximizing the rate of the TDD.

For example, Michniak et al. [88] have reported the effect of PharmaDur polymer on transdermal permeation of caffeine in the presence of enhancer oleic acid. Skin permeation rate of caffeine with and without oleic acid and the PharmaDur polymer was studied using human cadaver skin and Franz diffusion cells. It was found that the

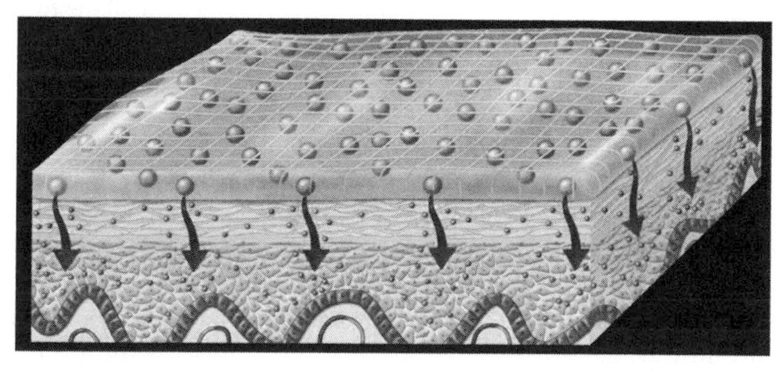

FIGURE 11.3 Schematic diagram of PharmaDur Virtual Patch. (Shah et al. [87].)

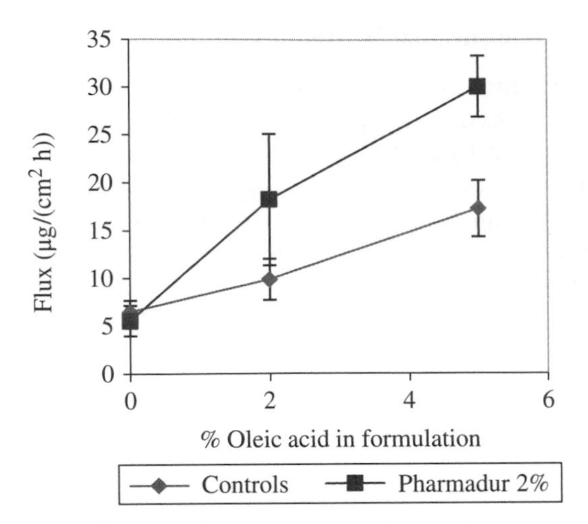

FIGURE 11.4 Comparison of flux of caffeine with oleic acid formulations (controls) and with added 2% PharmaDur polymer. (Michniak et al. [88].)

presence of the PharmaDur polymer almost doubled the skin permeation rate of caffeine compared to that of the corresponding control formulations consisting of caffeine and oleic acid (Fig. 11.4).

Transdermal delivery of NSAID diclofenac for systemic therapeutic activity using PharmaDur Virtual Patch has been previously reported. Seven hydroalcoholic gel formulations of diclofenac sodium were prepared containing the PharmaDur polymer and an enhancer consisting of *N*-methyl 2-pyrrolidone (NMP, Pharmasolve™, ISP) or dimethyl sulfoxide USP (DMSO, Gaylord Chemicals) or a combination of the two enhancers in different amounts.

The skin permeation studies were done utilizing Vertical Franz cells and human cadaver skin, using phosphate buffer of pH 7.4 as receptor phase. The donor compartment was left uncovered for uniform drying of the formulation on skin surface. Commercially available topical diclofenac product Voltaren® Gel (Novartis) was used as a control product in all the experiments. Average Voltaren diclofenac flux (J), studied over 10 different donor skins, was 1.16 μg/(cm² h). In order to compare the flux of diclofenac from formulations tested over different donor skins, flux values for all the formulations were normalized with respect to Voltaren diclofenac $J = 1.16$ μg/(cm² h). Diclofenac J was found to increase with an increase in the concentration of the enhancer. NMP was found to be a more effective enhancer than DMSO.

Formulation # 7 represents a mixed enhancer composition containing 15% NMP and 25% DMSO. An average value of diclofenac J for this composition, which was studied over six different donor skins, was found to be 12 ± 2.7 μg/(cm² h). Comparative diclofenac skin permeation profiles for the formulation # 7 and Voltaren Gel are shown in Figure 11.5.

Calculated steady-state blood plasma concentration of the diclofenac resulting from application of the formulation # 7 to a patient's skin area of 400 cm² would be 272 ng/ml (using reported [87] clearance value of 252 ml/(h kg)). Since the minimum effective plasma concentration of diclofenac is known to be 50 ng/ml [89], a

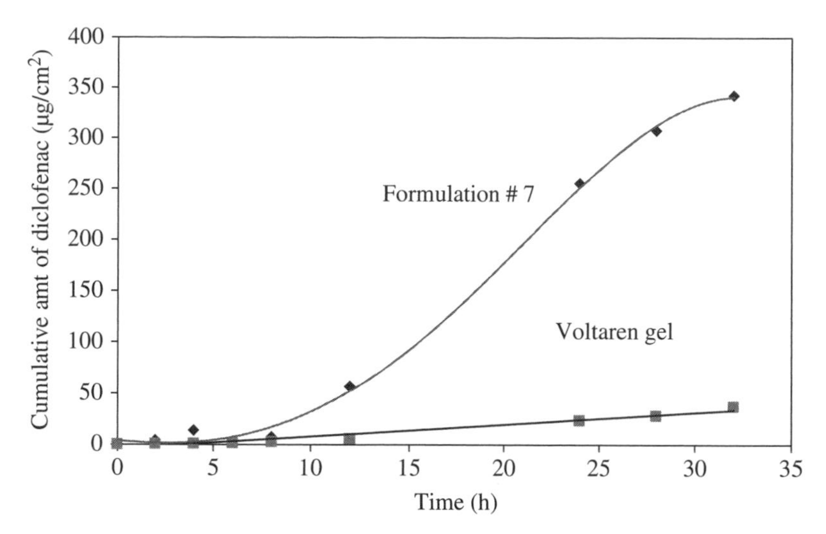

FIGURE 11.5 Comparative diclofenac skin permeation profiles.

PharmaDur Virtual Patch can transdermally provide systemic anti-inflammatory and analgesic activity all day long. Key advantages of such transdermal administration of diclofenac include sustained therapeutic activity and minimization of gastrointestinal tract irritation. In addition, systemic toxicity risks (e.g., cardiovascular) may be minimized by avoiding high drug concentrations in the blood in comparison with those attained by oral drug administration.

These findings suggest that PharmaDur Virtual Patch can be useful in transdermal delivery of drugs requiring higher dosages or those that have low skin permeability. Limitations of this approach are that it is not suitable for drugs having a narrow therapeutic window, or for those drugs that that have high skin permeability and need a membrane-controlled release.

11.5 SUMMARY

Even though TDD systems have been very popular since 1979, when the FDA approved the first patch for the treatment of motion sickness, the technology has been limited by the molecular size and lipophilicity of the drugs. Nonetheless, recent advances have propelled a new trend based on patchless transdermal systems that focus on increasing skin permeability via physicochemical interactions and external driving forces to allow delivery of small drugs to peptides and macromolecules. Future of the transdermal industry is very bright; new device-based transdermal products are expected in the near future with programmable systems that allows drug-on-demand and variable-rate delivery. Third generation of penetration enhancement systems has overcome skin barriers and decreased limitations existed for traditional TDD. However, proper adhesion of the patch onto the skin and lack of skin irritation still remain the main concerns for transdermal products.

REFERENCES

1. Prausnitz MR, Langer R. Transdermal drug delivery. *Nat Biotechnol.* 2008; **26**(11):1261–8.
2. Tanner T, Marks R. Delivering drugs by the transdermal route: review and comment. *Skin Res Technol.* 2008;**14**(3):249–60.
3. Naik A, Kalia YN, Guy RH. Transdermal drug delivery: overcoming the skin's barrier function. *Pharmaceutical Science and Technology Today.* 2000;**3**(9):318–26.
4. Walters KA, Roberts MS. The structure and function of skin. *Drugs Pharm Sci.* 2002;**119**:1–40.
5. Michaels A, Chandrasekaran S, Shaw J. Drug permeation through human skin: theory and invitro experimental measurement. *AIChE J.* 1975;**21**(5):985–96.
6. Prausnitz MR, Mitragotri S, Langer R. Current status and future potential of transdermal drug delivery. *Nat Rev Drug Discov.* 2004;**3**(2):115–24.
7. Wokovich AM, Prodduturi S, Doub WH, Hussain AS, Buhse LF. Transdermal drug delivery system (TDDS) adhesion as a critical safety, efficacy, and quality attribute. *Eur J Pharm Biopharm.* 2006;**64**(1):1–8.
8. Lee M, Phillips J. Transdermal patches: high risk for error. *Drug Top.* 2002;**146**(7):54–5.
9. Allen LV, Popovich NG, Ansel HC. *Ansel's Pharmaceutical Dosage Forms and Drug Delivery Systems.* Philadelphia, PA: Lippincott Williams & Wilkins; 2005.
10. Patel D, Patel N, Parmar M, Kaur N. Transdermal drug delivery system: review. *Int J Pharm Toxicol Res.* 2011;**1**:61–80.
11. Margetts L, Sawyer R. Transdermal drug delivery: principles and opioid therapy. *Contin Educ Anaesth Crit Care Pain.* 2007;**7**(5):171–6.
12. Desai P, Patlolla RR, Singh M. Interaction of nanoparticles and cell-penetrating peptides with skin for transdermal drug delivery. *Mol Membr Biol.* 2010;**27**(7):247–59.
13. Mencher SK, Wang LG. Promiscuous drugs compared to selective drugs (promiscuity can be a virtue). *BMC Clin Pharmacol.* 2005;**5**:3.
14. Verma DD, Verma S, Blume G, Fahr A. Particle size of liposomes influences dermal delivery of substances into skin. *Int J Pharm.* 2003;**258**(1–2):141–51.
15. Akhter SA, Barry BW. Absorption through human skin of ibuprofen and flurbiprofen; effect of dose variation, deposited drug films, occlusion and the penetration enhancer N-methyl-2-pyrrolidone. *J Pharm Pharmacol.* 1985;**37**(1):27–37.
16. Gannu R, Vishnu YV, Kishan V, Rao YM. In vitro permeation of carvedilol through porcine skin: effect of vehicles and penetration enhancers. *PDA J Pharm Sci Technol.* 2008;**62**(4):256–63.
17. Parhi R, Suresh P, Mondal S, Kumar PM. Novel penetration enhancers for skin applications: a review. *Curr Drug Deliv.* 2012;**9**(2):219–30.
18. Som I, Bhatia K, Yasir M. Status of surfactants as penetration enhancers in transdermal drug delivery. *J Pharm Bioallied Sci.* 2012;**4**(1):2–9.
19. Degim IT, Uslu A, Hadgraft J, Atay T, Akay C, Cevheroglu S. The effects of Azone and capsaicin on the permeation of naproxen through human skin. *Int J Pharm.* 1999;**179**(1):21–5.
20. Norlen L, Engblom J. Structure-related aspects on water diffusivity in fatty acid-soap and skin lipid model systems. *J Control Release.* 2000;**63**(1–2):213–26.

21. Kanikkannan N, Singh M. Skin permeation enhancement effect and skin irritation of saturated fatty alcohols. *Int J Pharm.* 2002;**248**(1–2):219–28.
22. Karande P, Jain A, Ergun K, Kispersky V, Mitragotri S. Design principles of chemical penetration enhancers for transdermal drug delivery. *Proc Natl Acad Sci U S A.* 2005; **102**(13):4688–93.
23. Sinico C, Fadda AM. Vesicular carriers for dermal drug delivery. *Expert Opin Drug Deliv.* 2009;**6**(8):813–25.
24. Zhang Z, Tsai PC, Ramezanli T, Michniak-Kohn BB. Polymeric nanoparticles-based topical delivery systems for the treatment of dermatological diseases. *Wiley Interdiscip Rev Nanomed Nanobiotechnol.* 2013;**5**(3):205–18.
25. Maestrelli F, Gonzalez-Rodriguez ML, Rabasco AM, Mura P. Preparation and characterisation of liposomes encapsulating ketoprofen-cyclodextrin complexes for transdermal drug delivery. *Int J Pharm.* 2005;**298**(1):55–67.
26. Kaur N, Puri R, Jain SK. Drug-cyclodextrin-vesicles dual carrier approach for skin targeting of anti-acne agent. *AAPS PharmSciTech.* 2010;**11**(2):528–37.
27. Leininger S, Olenyuk B, Stang PJ. Self-assembly of discrete cyclic nanostructures mediated by transition metals. *Chem Rev.* 2000;**100**(3):853–908.
28. Huttunen KM, Rautio J. Prodrugs—an efficient way to breach delivery and targeting barriers. *Curr Top Med Chem.* 2011;**11**(18):2265–87.
29. Paudel KS, Milewski M, Swadley CL, Brogden NK, Ghosh P, Stinchcomb AL. Challenges and opportunities in dermal/transdermal delivery. *Ther Deliv.* 2010;**1**(1):109–31.
30. PNA Holdings LLC. Vyvanse Approved to Treat ADHD in Adults. Basingstoke/Philadelphia, PA; 2008. Available from Drugs.com (accessed on October 6, 2015; cited January 30, 2014).
31. Manconi M, Sinico C, Caddeo C, Vila AO, Valenti D, Fadda AM. Penetration enhancer containing vesicles as carriers for dermal delivery of tretinoin. *Int J Pharm.* 2011; **412**(1–2):37–46.
32. Bando H, Takagi T, Yamashita F, Takakura Y, Hashida M. Theoretical design of prodrug-enhancer combination based on a skin diffusion model: prediction of permeation of acyclovir prodrugs treated with 1-geranylazacycloheptan-2-one. *Pharm Res.* 1996;**13**(3):427–32.
33. Banga AK, Prausnitz MR. Assessing the potential of skin electroporation for the delivery of protein- and gene-based drugs. *Trends Biotechnol.* 1998;**16**(10):408–12.
34. Pikal MJ. The role of electroosmotic flow in transdermal iontophoresis. *Adv Drug Deliv Rev.* 2001;**46**(1–3):281–305.
35. Prausnitz MR. The effects of electric current applied to skin: a review for transdermal drug delivery. *Adv Drug Deliv Rev.* 1996;**18**:395–425.
36. Stem DE, Bozman CS. Respondent anxiety reduction with the randomized response technique. *Adv Consum Res.* 1988;**15**:595–9.
37. Allenby AC, Fletcher J, Schock C, Tees TFS. The effect of heat, pH and organic solvents on the electrical impedance and permeability of excised human skin. *Br J Dermatol.* 1969;**81**:31–9.
38. Vanbever R, Lecouturier N, Preat V. Transdermal delivery of metoprolol by electroporation. *Pharm Res.* 1994;**11**(11):1657–62.
39. Charoo NA, Rahman Z, Repka MA, Murthy SN. Electroporation: an avenue for transdermal drug delivery. *Curr Drug Deliv.* 2010;**7**(2):125–36.

40. Pliquett U, Weaver JC. Transport of a charged molecule across the human epidermis due to electroporation. *J Control Release*. 1996;**38**:1–10.

41. Escobar-Chavez JJ, Bonilla-Martinez D, Villegas-Gonzalez MA, Rodriguez-Cruz IM, Dominguez-Delgado CL. The use of sonophoresis in the administration of drugs throughout the skin. *J Pharm Pharm Sci*. 2009;**12**(1):88–115.

42. Park EJ, Dodds J, Smith NB. Dose comparison of ultrasonic transdermal insulin delivery to subcutaneous insulin injection. *Int J Nanomedicine*. 2008;**3**(3):335–41.

43. Lee JW, Gadiraju P, Park JH, Allen MG, Prausnitz MR. Microsecond thermal ablation of skin for transdermal drug delivery. *J Control Release*. 2011;**154**(1):58–68.

44. Andrews S, Lee JW, Choi SO, Prausnitz MR. Transdermal insulin delivery using microdermabrasion. *Pharm Res*. 2011;**28**(9):2110–18.

45. Nestle FO, Nickoloff BJ. Dermal dendritic cells are important members of the skin immune system. *Adv Exp Med Biol*. 1995;**378**:111–16.

46. Saeki H, Moore AM, Brown MJ, Hwang ST. Cutting edge: secondary lymphoid-tissue chemokine (SLC) and CC chemokine receptor 7 (CCR7) participate in the emigration pathway of mature dendritic cells from the skin to regional lymph nodes. *J Immunol*. 1999;**162**(5):2472–5.

47. Matsuo K, Hirobe S, Okada N, Nakagawa S. Frontiers of transcutaneous vaccination systems: novel technologies and devices for vaccine delivery. *Vaccine*. 2013;**31**(19):2403–15.

48. Kim YC, Park JH, Prausnitz MR. Microneedles for drug and vaccine delivery. *Adv Drug Deliv Rev*. 2012;**64**(14):1547–68.

49. Dean CH, Alarcon JB, Waterston AM, Draper K, Early R, Guirakhoo F, et al. Cutaneous delivery of a live, attenuated chimeric flavivirus vaccine against Japanese encephalitis (ChimeriVax)-JE) in non-human primates. *Hum Vaccin*. 2005;**1**(3):106–11.

50. Mikszta JA, Alarcon JB, Brittingham JM, Sutter DE, Pettis RJ, Harvey NG. Improved genetic immunization via micromechanical disruption of skin-barrier function and targeted epidermal delivery. *Nat Med*. 2002;**8**(4):415–19.

51. Ding Z, Verbaan FJ, Bivas-Benita M, Bungener L, Huckriede A, van den Berg DJ, et al. Microneedle arrays for the transcutaneous immunization of diphtheria and influenza in BALB/c mice. *J Control Release*. 2009;**136**(1):71–8.

52. Laurent PE, Bourhy H, Fantino M, Alchas P, Mikszta JA. Safety and efficacy of novel dermal and epidermal microneedle delivery systems for rabies vaccination in healthy adults. *Vaccine*. 2010;**28**(36):5850–6.

53. Slutter B, Bal SM, Ding Z, Jiskoot W, Bouwstra JA. Adjuvant effect of cationic liposomes and CpG depends on administration route. *J Control Release*. 2011;**154**(2):123–30.

54. Kis EE, Winter G, Myschik J. Devices for intradermal vaccination. *Vaccine*. 2012;**30**(3):523–38.

55. Quan FS, Kim YC, Song JM, Hwang HS, Compans RW, Prausnitz MR, et al. Long-term protective immunity from an influenza virus-like particle vaccine administered with a microneedle patch. *Clin Vaccine Immunol*. 2013;**20**(9):1433–9.

56. Hirobe S, Okada N, Nakagawa S. Transcutaneous vaccines—current and emerging strategies. *Expert Opin Drug Deliv*. 2013;**10**(4):485–98.

57. Gratieri T, Alberti I, Lapteva M, Kalia YN. Next generation intra- and transdermal therapeutic systems: using non- and minimally-invasive technologies to increase drug delivery into and across the skin. *Eur J Pharm Sci*. 2013;**50**(5):609–22.

58. DeMuth PC, Moon JJ, Suh H, Hammond PT, Irvine DJ. Releasable layer-by-layer assembly of stabilized lipid nanocapsules on microneedles for enhanced transcutaneous vaccine delivery. *ACS Nano*. 2012;**6**(9):8041–51.
59. DeMuth PC, Min Y, Huang B, Kramer JA, Miller AD, Barouch DH, et al. Polymer multilayer tattooing for enhanced DNA vaccination. *Nat Mater*. 2013;**12**(4):367–76.
60. Choi HJ, Bondy BJ, Yoo DG, Compans RW, Kang SM, Prausnitz MR. Stability of whole inactivated influenza virus vaccine during coating onto metal microneedles. *J Control Release*. 2013;**166**(2):159–71.
61. Kim YC, Song JM, Lipatov AS, Choi SO, Lee JW, Donis RO, et al. Increased immunogenicity of avian influenza DNA vaccine delivered to the skin using a microneedle patch. *Eur J Pharm Biopharm*. 2012;**81**(2):239–47.
62. Pearton M, Saller V, Coulman SA, Gateley C, Anstey AV, Zarnitsyn V, et al. Microneedle delivery of plasmid DNA to living human skin: formulation coating, skin insertion, and gene expression. *J Control Release*. 2012;**160**(3):561–9.
63. Song JM, Kim YC, Eunju O, Compans RW, Prausnitz MR, Kang SM. DNA vaccination in the skin using microneedles improves protection against influenza. *Mol Ther*. 2012;**20**(7):1472–80.
64. Kang SM, Kim MC, Compans RW. Virus-like particles as universal influenza vaccines. *Expert Rev Vaccines*. 2012;**11**(8):995–1007.
65. Kim MC, Song JM, Eunju O, Kwon YM, Lee YJ, Compans RW, et al. Virus-like particles containing multiple M2 extracellular domains confer improved cross-protection against various subtypes of influenza virus. *Mol Ther*. 2013;**21**(2):485–92.
66. Wang BZ, Gill HS, Kang SM, Wang L, Wang YC, Vassilieva EV, et al. Enhanced influenza virus-like particle vaccines containing the extracellular domain of matrix protein 2 and a Toll-like receptor ligand. *Clin Vaccine Immunol*. 2012;**19**(8):1119–25.
67. Weldon WC, Zarnitsyn VG, Esser ES, Taherbhai MT, Koutsonanos DG, Vassilieva EV, et al. Effect of adjuvants on responses to skin immunization by microneedles coated with influenza subunit vaccine. *PLoS One*. 2012;**7**(7):e41501.
68. Choi HJ, Yoo DG, Bondy BJ, Quan FS, Compans RW, Kang SM, et al. Stability of influenza vaccine coated onto microneedles. *Biomaterials*. 2012;**33**(14):3756–69.
69. Kim YC, Yoo DG, Compans RW, Kang SM, Prausnitz MR. Cross-protection by co-immunization with influenza hemagglutinin DNA and inactivated virus vaccine using coated microneedles. *J Control Release*. 2013;**172**(2):579–88.
70. Andrianov AK, Mutwiri G. Intradermal immunization using coated microneedles containing an immunoadjuvant. *Vaccine*. 2012;**30**(29):4355–60.
71. Vrdoljak A, McGrath MG, Carey JB, Draper SJ, Hill AV, O'Mahony C, et al. Coated microneedle arrays for transcutaneous delivery of live virus vaccines. *J Control Release*. 2012;**159**(1):34–42.
72. Koutsonanos DG, Vassilieva EV, Stavropoulou A, Zarnitsyn VG, Esser ES, Taherbhai MT, et al. Delivery of subunit influenza vaccine to skin with microneedles improves immunogenicity and long-lived protection. *Sci Rep*. 2012;**2**:357.
73. Chen X, Fernando GJ, Raphael AP, Yukiko SR, Fairmaid EJ, Primiero CA, et al. Rapid kinetics to peak serum antibodies is achieved following influenza vaccination by dry-coated densely packed microprojections to skin. *J Control Release*. 2012;**158**(1):78–84.

74. Kommareddy S, Baudner BC, Bonificio A, Gallorini S, Palladino G, Determan AS, et al. Influenza subunit vaccine coated microneedle patches elicit comparable immune responses to intramuscular injection in guinea pigs. *Vaccine*. 2013;**31**(34):3435–41.
75. Edens C, Collins ML, Ayers J, Rota PA, Prausnitz MR. Measles vaccination using a microneedle patch. *Vaccine*. 2013;**31**(34):3403–9.
76. Hong X, Wei L, Wu F, Chen L, Liu Z, Yuan W. Dissolving and biodegradable microneedle technologies for transdermal sustained delivery of drug and vaccine. *Drug Des Devel Ther*. 2013;**7**:945–52.
77. Demuth PC, Garcia-Beltran WF, Ai-Ling ML, Hammond PT, Irvine DJ. Composite dissolving microneedles for coordinated control of antigen and adjuvant delivery kinetics in transcutaneous vaccination. *Adv Funct Mater*. 2013;**23**(2):161–72.
78. Zaric M, Lyubomska O, Touzelet O, Poux C, Al-Zahrani S, Fay F, et al. Skin dendritic cell targeting via microneedle arrays laden with antigen-encapsulated poly-D,L-lactide-co-glycolide nanoparticles induces efficient antitumor and antiviral immune responses. *ACS Nano*. 2013;**7**(3):2042–55.
79. Naito S, Ito Y, Kiyohara T, Kataoka M, Ochiai M, Takada K. Antigen-loaded dissolving microneedle array as a novel tool for percutaneous vaccination. *Vaccine*. 2012;**30**(6): 1191–7.
80. Demir YK, Akan Z, Kerimoglu O. Sodium alginate microneedle arrays mediate the transdermal delivery of bovine serum albumin. *PLoS One*. 2013;**8**(5):e63819.
81. Guo L, Chen J, Qiu Y, Zhang S, Xu B, Gao Y. Enhanced transcutaneous immunization via dissolving microneedle array loaded with liposome encapsulated antigen and adjuvant. *Int J Pharm*. 2013;**447**(1–2):22–30.
82. Lee JW, Park JH, Prausnitz MR. Dissolving microneedles for transdermal drug delivery. *Biomaterials*. 2008;**29**(13):2113–24.
83. Sullivan SP, Koutsonanos DG, Del Pilar Martin M, Lee JW, Zarnitsyn V, Choi SO, et al. Dissolving polymer microneedle patches for influenza vaccination. *Nat Med*. 2010;**16**(8):915–20.
84. Chu LY, Prausnitz MR. Separable arrowhead microneedles. *J Control Release*. 2011;**149**(3):242–9.
85. Kochhar JS, Zou S, Chan SY, Kang L. Protein encapsulation in polymeric microneedles by photolithography. *Int J Nanomedicine*. 2012;**7**:3143–54.
86. Matsuo K, Hirobe S, Yokota Y, Ayabe Y, Seto M, Quan YS, et al. Transcutaneous immunization using a dissolving microneedle array protects against tetanus, diphtheria, malaria, and influenza. *J Control Release*. 2012;**160**(3):495–501.
87. Shah KR. PharmaDur® bioadhesive delivery system. In: Wille JJ, editor. *Skin Delivery Systems; Transdermals, Dermatologicals, and Cosmetic Actives*. Ames, IA: Blackwell; 2006: 211–222.
88. Michniak B, Thakur R, Shah K. Evaluation of Novel Polymer for Transdermal Permeation of Caffeine in Presence of Different Concentrations of Chemical Enhancer. Poster # 323, Proceedings of Controlled Release Society, 32nd Annual Meeting, Miami Beach, FL, June 18–22; 2005.
89. Toshiaki N, Akira K, Kiyoshi S, Koichi T, Koichi M, Kazuhiko S, et al. Percutaneous absorption of diclofenac in rats and humans: aqueous gel formulation. *Int J Pharm*. 1988;**46**(1):1–7.

12

PRODRUG APPROACHES TO DRUG DELIVERY

LONGQIN HU

Department of Medicinal Chemistry, Ernest Mario School of Pharmacy, Rutgers, The State University of New Jersey, Piscataway, NJ, USA

12.1 INTRODUCTION

For drugs to produce their desired pharmacological action, they often have to overcome many hurdles before they can reach the desired site of action. These hurdles include the intestinal barrier, the blood–brain barrier (BBB), and metabolic reactions that could render them inactive. These three subjects are covered in Chapters 2, 3, and 9, respectively, and therefore will not be discussed here in detail. Most drugs distribute randomly throughout the body, and the amount of drugs reaching the site of action is relatively small. For an effective amount to reach the site of action and not cause severe systemic side effects, a drug must possess certain physicochemical properties that make it conducive to penetration through various biological membranes (i.e., sufficiently bioavailable), to avoid metabolic inactivation by various enzymes, and to avoid retention in body depot tissues that could lead to undesirable long-lasting effects. These desired physicochemical properties are not always present in pharmacologically active compounds.

With the advance of new technologies, more and more compounds are being identified with extremely potent *in vitro* activity but are found to be inactive *in vivo*. They may have the optimal configuration and conformation needed to interact with their target receptor or enzyme, but they do not necessarily possess the best molecular form and physicochemical properties needed for their delivery to the site of action.

Drug Delivery: Principles and Applications, Second Edition. Edited by Binghe Wang, Longqin Hu, and Teruna J. Siahaan.

Some of the problems often encountered include (i) limited solubility and poor chemical stability preventing the drug from being adequately formulated, (ii) low or variable bioavailability due to incomplete absorption across biological membranes or extensive first-pass metabolism, and (iii) lack of site specificity. Further structural modifications are often performed but do not always solve all the problems. Another approach that is often effective in solving some of these delivery problems is the design of prodrugs by attaching a promoiety to the active drug [1–3]. This chapter will focus on the various prodrug approaches that have been used to overcome many of the pharmaceutical and pharmacokinetic barriers that hinder the optimal delivery of the active drug.

12.2 BASIC CONCEPTS: DEFINITION AND APPLICATIONS

A prodrug by definition is inactive or much less active and has to be converted to the active drug within the biological system. There are a variety of mechanisms by which a prodrug can be activated. These include metabolic activation mediated by enzymes present in the biological system as well as the less common, simple chemical means of activation such as hydrolysis.

Prodrugs occur in nature. One example is proinsulin, which is synthesized in the beta cells of the pancreas and undergoes a series of proteolysis to form the mature insulin (A-B chains) and the connecting C-peptide. In addition to the known biological functions of insulin, the C-peptide that was originally thought of only as a linker facilitating the efficient assembly, folding, and processing of insulin has recently been shown to have its own biological activity on microvascular blood flow and tissue health [4]. Most synthetic prodrugs are prepared by attachment of the active drug through a metabolically labile linkage to another molecule, the "promoiety." The promoiety is not necessary for activity but may impart some desirable properties to the drug, such as increased lipid or water solubility or site specificity. Advantages that can be gained with such a prodrug include increased bioavailability, alleviation of pain at the site of injection, elimination of an unpleasant taste, decreased toxicity, decreased metabolic inactivation, increased chemical stability, and prolonged or shortened duration of action.

12.2.1 Increasing Lipophilicity to Increase Systemic Bioavailability

This is the most successful application of prodrugs. Because of the lipid bilayer nature of biological membranes, the rate of passive drug transport is affected by both lipophilicity and aqueous solubility (also called "hydrophilicity"). The rate of passive diffusion across the biological membrane will increase exponentially with increasing lipophilicity and then level off at higher lipophilicity. This is due to the fact that an increase in lipophilicity is usually accompanied by a decrease in water solubility and will eventually decrease the flux over the membrane due to poor water solubility. The design of prodrugs aims to achieve a balance between lipophilicity and aqueous solubility in order to improve passive drug transport across various biological membranes. According to the empirical Lipinski's rule of five [5], in order to have good membrane permeability, drugs should have a relatively low molecular weight (≤500) and be relatively nonpolar with a cLog$P$ between −1 and 5. This means drugs should partition between an aqueous and a lipid phase in favor of the lipid phase but, at the same time, possess certain water solubility. The majority of effective oral drugs obey this empirical rule.

Since most drugs are either weak acids or weak bases, they are often given and present in the salt form under relevant physiological conditions. Therefore, dissociation constants also affect the membrane permeability and thus bioavailability. It is generally accepted that the neutral, unionized, and thus most lipophilic form of an acidic or basic drug is absorbed far more efficiently than the ionized species. The distribution between the ionized and neutral form is pH dependent. The effective partition coefficient for a dissociative system (Log$D$) gives the correct description of such complex partitioning equilibria.

$$\text{Log}D = \text{Log}P_{\text{HA}} - \log\left(1 + 10^{(\text{pH} - \text{p}K_{\text{a}})}\right) \quad \text{for an acid}$$

$$\text{Log}D = \text{Log}P_{\text{B}} - \log\left(1 + 10^{(\text{p}K_{\text{a}} - \text{pH})}\right) \quad \text{for a base}$$

where P_{HA} and P_{B} are the intrinsic partition coefficients of the weak acid and weak base, respectively. Programs such as ACD/Log$P$ and cLog$P$ are available to calculate with reasonable accuracy the Log$P$ and Log$D$ values using a structure-fragment approach as well as internal structure databases. To illustrate the principles discussed in this chapter, examples will be given with their Log$P$ and/or Log$D$ values calculated using the Advanced Chemistry Development (ACD) software.

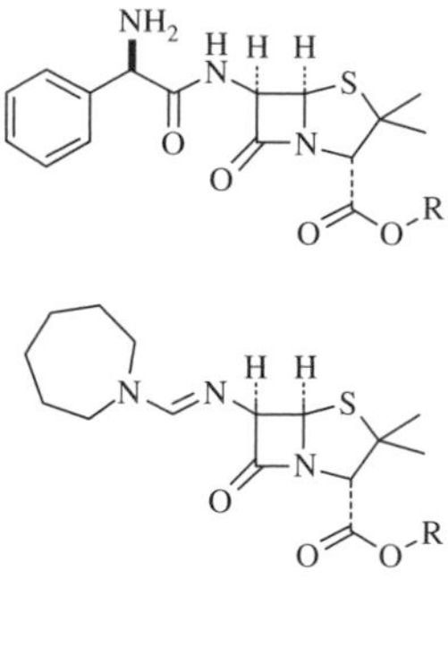

Compound	R	ACD/Log$P$	ACD/Log$D$ pH 7	ACD/Log$D$ pH 1
(**1**) Ampicillin	R = H	1.35 ± 0.32	−1.54	−1.72
(**2**) Bacampicillin	R = —$CH(CH_3)OCOOC_2H_5$	2.17 ± 0.89	1.99	−0.93
(**3**) Pivampicillin	R = —$CH(CH_3)OCOC(CH_3)_3$	2.55 ± 0.88	2.37	−0.55
(**4**) Talampicillin	R 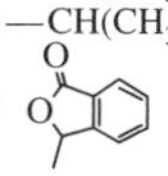 =	2.79 ± 0.88	2.61	−0.31
(**5**) Mecillinam	R = H	1.49 ± 0.87	−1.01	−1.58
(**6**) Pivmecillinam	R = —$CH(CH_3)OCOC(CH_3)_3$	3.45 ± 0.92	1.77	0.35

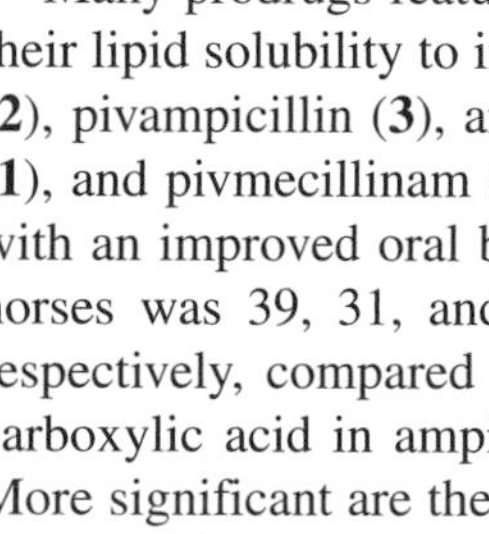

Many prodrugs feature the addition of a hydrophobic group in order to increase their lipid solubility to improve their gastrointestinal (GI) absorption. Bacampicillin (**2**), pivampicillin (**3**), and talampicillin (**4**) are more lipophilic esters of ampicillin (**1**), and pivmecillinam (**6**) is a more lipophilic ester prodrug of mecillinam (**5**), all with an improved oral bioavailability. For example, absolute oral bioavailability in horses was 39, 31, and 23% for bacampicillin, pivampicillin, and talampicillin, respectively, compared with only 2% for ampicillin sodium [6]. Esterification of carboxylic acid in ampicillin (**1**) resulted in an increase of 0.8–1.4 unit in Log$P$. More significant are the increases in Log$D$ values for the prodrugs when ionization of the amino group is taken into consideration; as much as a 4-unit difference in

Log*D* is estimated at pH values in the intestine where the prodrugs are believed to be absorbed. Other prodrugs of antibiotics include esters of carbenicillin (for urinary tract infection), cefotiam, and erythromycin.

Enalapril (**8**) is an ester prodrug of enalaprilat (**7**). The latter binds tightly to the angiotensin-converting enzyme (ACE) but is transported with low efficacy by the peptide carrier in the GI tract. The prodrug enalapril has a higher affinity for the peptide carrier [7] and is much better absorbed, with about 60% oral bioavailability [8, 9]. As a matter of fact, all ACE inhibitors except captopril and lisinopril are administered as prodrugs; other commercialized ACE inhibitor prodrugs include perindopril, quinapril, ramipril, cilazapril, benazepril, spirapril, imidapril, and trandolapril, all based on esterification of the same carboxylic acid group [10]. The esters are hydrolyzed *in vivo*, after absorption, to the corresponding active but poorly absorbed dicarboxylate forms.

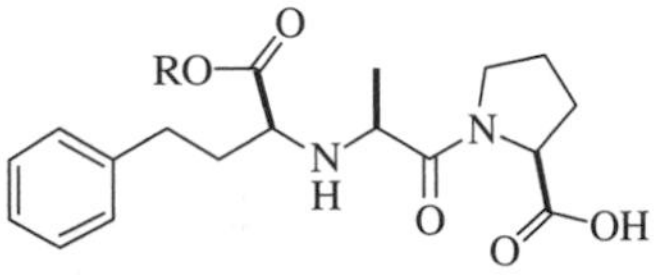

		ACD/Log*P*	ACD/Log*D*	
			pH 7	pH 1
(**7**) Enalaprilat	R = H	2.10 ± 0.57	−1.45	−0.92
(**8**) Enalapril	R = Et	2.98 ± 0.58	−0.12	−0.10

Valacyclovir (**10**) [11, 12] and famciclovir (**12**) are ester prodrugs of acyclovir (**9**) and penciclovir (**11**), respectively, for the treatment of viral infections. Both acyclovir and penciclovir exhibit site-specific conversion to the active triphosphate species by viral thymidine kinase. They show remarkable antiviral selectivity and specificity. However, their oral bioavailability is quite low, 15–20% of an oral dose being absorbed in humans for acyclovir and 5% for penciclovir [13]. Both valacyclovir and famciclovir have no intrinsic antiviral activity, and both are rapidly hydrolyzed to acyclovir and penciclovir by esterases present in the liver and gut wall. Valacyclovir displays a mean absolute bioavailability of 54%, a threefold increase in oral bioavailability over acyclovir, while famciclovir has an absolute bioavailability of 77% in humans [14–16]. Famciclovir's better bioavailability could be explained by the increase in lipophilicity; the high oral bioavailability of valacyclovir was also partly attributed to the involvement of an active transport mechanism through PEPT1 [17]. Therefore, in addition to increasing lipophilicity, prodrug design can utilize active transport mechanisms as a means of enhancing bioavailability.

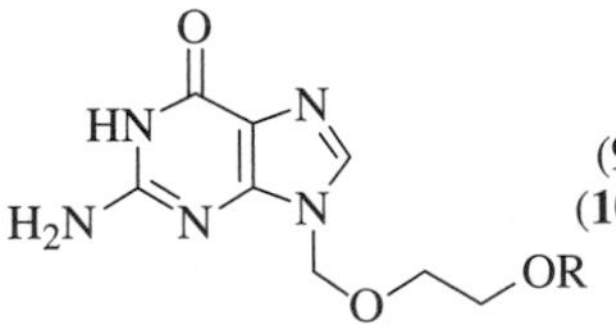

		ACD/Log*P*	ACD/Log*D*	
			pH 7	pH 1
(**9**) Acyclovir	R = H	−1.76 ± 0.49	−1.76	−3.76
(**10**) Valacyclovir	R = L-Valyl-	0.04 ± 0.58	−0.78	−4.76

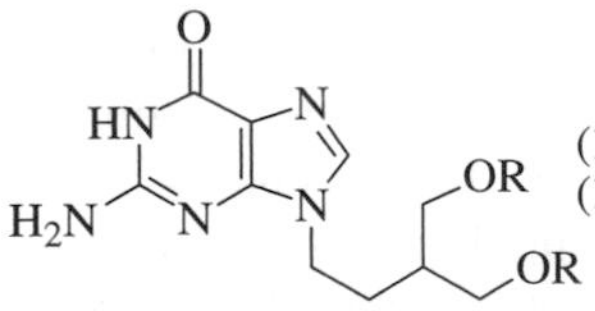

(**11**) Penciclovir	R = H	−2.03 ± 0.58	−2.03	−4.44
(**12**) Famciclovir	R = CH_3CO—	−0.09 ± 0.27	−0.09	−3.27

12.2.2 Sustained-Release Prodrug Systems

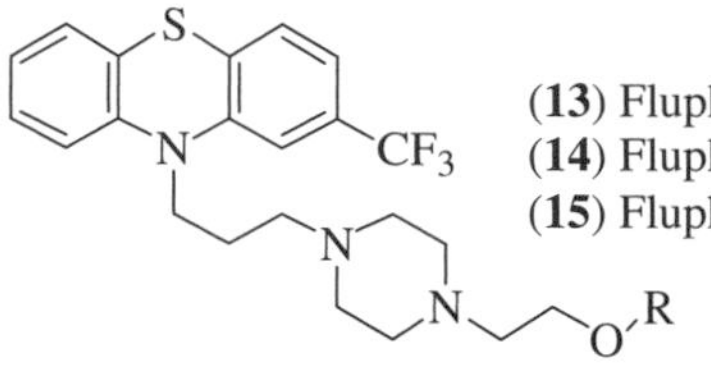

		ACD/Log*P*
(**13**) Fluphenazine	R = H	4.84 ± 0.46
(**14**) Fluphenazine enanthate	R = —$CO(CH_2)_5CH_3$	8.03 ± 0.45
(**15**) Fluphenazine decanoate	R = —$CO(CH_2)_8CH_3$	9.63 ± 0.45

Antipsychotic drugs are the mainstay treatment for schizophrenia and similar psychotic disorders. Long-acting depot injections of antipsychotic drugs are extensively used as a means of long-term maintenance treatment. The duration of action for many antipsychotic drugs with a free hydroxyl group can be considerably prolonged by the preparation of long-chain fatty acid esters with very high Log*P* values (usually 7 or above). Fluphenazine enanthate (**14**) and fluphenazine decanoate (**15**) were the first of these esters to appear in clinical use and are longer acting, with fewer side effects than the parent drug. The ability to treat patients with a single intramuscular injection every 1–2 weeks with the enanthate or every 2–3 weeks with the decanoate esters means that problems associated with patient compliance with the drug regimens and with drug malabsorption can be reduced [18]. Esterification of antipsychotic drugs with decanoic acid yields very lipophilic prodrugs that are dissolved in a light vegetable oil such as Viscoleo or sesame oil. Intramuscular injection creates an oily depot from which the prodrug molecules slowly diffuse into the systemic circulation, where they are hydrolyzed quickly by esterases to the active moieties. These depot forms allow these drugs to be given only once or twice a month, permitting the long-term treatment of schizophrenia. Antipsychotic drugs available in depot formulation include fluphenazine (**13**), flupenthixol, haloperidol, and zuclopenthixol in their enanthate or decanoate esters.

Anabolic steroids such as nandrolone and testosterone, anti-inflammatory glucocorticoids such as methylprednisolone, and contraceptives such as estradiol and levonorgestrel all have slow-release formulations of their ester prodrugs in the market.

A more recent example of a prodrug designed for prolonged release of the active parent drug molecule is lisdexamfetamine (**17**) for attention-deficit hyperactivity disorder (ADHD) [19]. The L-lysyl amide of d-amphetamine (**16**) is highly water soluble, readily absorbed through PEPT1-mediated active transport in the small intestines, and undergoes enzymatic hydrolysis by an unidentified enzyme in the red blood cells. The prodrug **17** demonstrated a long duration of effect allowing for once-a-day dosing [19].

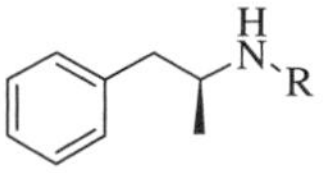

			ACD/Log*D*	
		ACD/Log*P*	pH 7	pH 1
(**16**) d-Amphetamine	R = H	1.79 ± 0.20	−0.92	−1.31
(**17**) Lisdexamfetamine	R 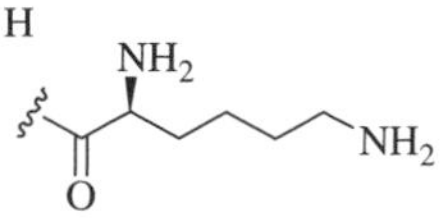 = (L-lysyl)	1.06 ± 0.52	−2.38	−3.04

12.2.3 Improving Gastrointestinal Tolerance

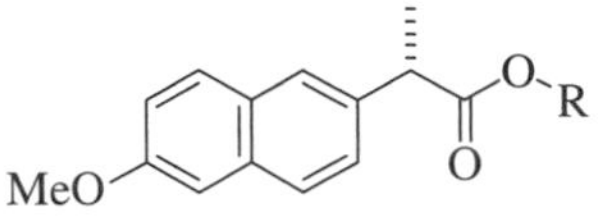

(**18**) Naproxen (R = H)
(**19**) R = —$(CH_2)_n$—N(morpholino)O

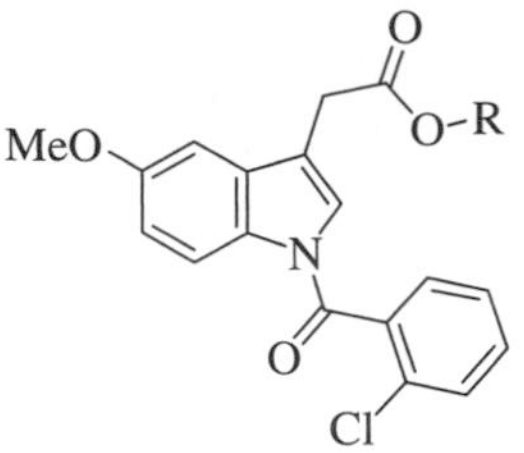

(**20**) Indomethacin (R = H)
(**21**) R = —$(CH_2)_n$—N(morpholino)O

Temporary masking of carboxylic acid groups in nonsteroidal anti-inflammatory drugs (NSAIDs) was proposed as a promising means of reducing GI toxicity resulting from direct mucosal contact mechanisms. Morpholinoalkyl esters (**19** and **21**, HCl salts) of naproxen (**18**) and indomethacin (**20**) were evaluated *in vitro* and *in vivo* for their potential use as prodrugs for oral delivery [20]. The prodrugs were freely soluble in simulated gastric fluid and pH 7.4 phosphate buffer and showed a minimum of a 2000-fold increase in solubility over the parent drugs. The prodrugs were more lipophilic than the parent drugs and were quantitatively hydrolyzed to their respective parent drugs *in vivo*. The prodrugs were 30–36% more bioavailable orally than the parent drugs following a single dose in rats. They were significantly less irritating to gastric mucosa than the parent drugs, following a single dose as well as chronic oral administration in rats.

12.2.4 Improving Taste

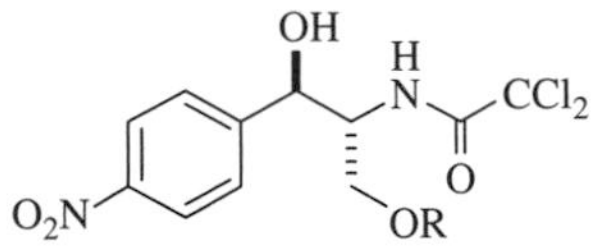

		ACD/Log*P*	ACD/Log*D*	
			pH 7	pH 1
(**22**) Chloramphenicol	R = H	1.02 ± 0.32	1.02	1.02
(**23**) Chloramphenicol palmitate	R = —$CO(CH_2)_{14}CH_3$	9.92 ± 0.76	9.92	9.92
(**24**) Chloramphenicol sodium succinate	R = —$COCH_2CH_2COO^-Na^+$	2.29 ± 0.85	–0.34	2.29

Oral drugs with markedly bitter taste may lead to poor patient compliance if administered as a solution or syrup. The prodrug approach has been used to improve the taste for chloramphenicol (**22**), clindamycin, erythromycin, and metronidazole [21]. A prodrug such as chloramphenicol palmitate (**23**), with Log*P* of around 10, does not dissolve in an appreciable amount in the mouth and, therefore, does not interact with the taste receptors.

12.2.5 Diminishing Gastrointestinal Absorption

Many prodrugs have been evaluated in this context for colon-specific drug delivery. Colon targeting is of value for the topical treatment of diseases of the colon such as Crohn's disease, ulcerative colitis, and colorectal cancer. Sustained colonic release of drugs can be useful in the treatment of nocturnal asthma, angina, and arthritis. Prodrugs have been designed to pass intact and unabsorbed from the upper GI tract and undergo biotransformation in the colon, releasing the active drug molecule. Prodrug activation can be carried out by microflora and distinct enzymes present in the colon (i.e., azoreductase, glucuronidase, glycosidase, dextranase, esterase, nitroreductase, and cyclodextranase) [22, 23]. Balsalazide (**25**), ipsalazide (**26**), olsalazine (**27**), and sulfasalazine (**28**) are azo-containing prodrugs developed for colon-specific delivery of an anti-inflammatory agent in the treatment of inflammatory bowel disease. As shown in Scheme 12.1, they can undergo azoreduction in the colon to release the active 5-amino salicylic acid (5-ASA or mesalazine, **29**). Other prodrugs evaluated for colon-specific delivery include conjugates of amino acids, glucuronide, glycoside, dextran, and cyclodextrin [24].

12.2.6 Increasing Water Solubility

Poorly water-soluble lipophilic drugs also have difficulty getting absorbed as discussed earlier. The prodrug approach has been applied to circumvent problems of low aqueous solubility by the introduction of an ionizable functional group such as phosphate esters, amino acid esters, and hemiesters of dicarboxylic acids, allowing various salts of such prodrugs to be formed. Prodrugs can also be used to increase water solubility in order to increase the amount of drug that will reach the systemic circulation through parenteral administration. Examples include chloramphenicol sodium succinate (**24**), hydrocortisone sodium succinate, methylprednisolone sodium succinate, betamethasone sodium phosphate, clindamycin phosphate, fosphenytoin sodium, prednisolone phosphate, fosaprepitant (**97**), and fosamprenavir.

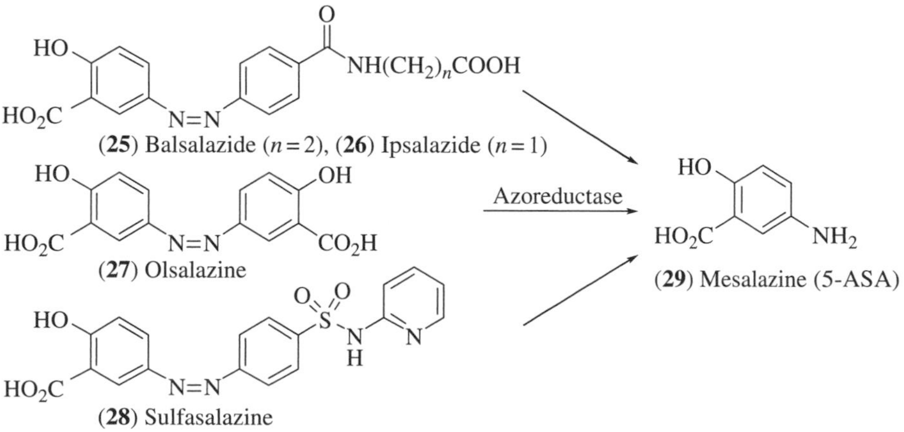

(**25**) Balsalazide ($n = 2$), (**26**) Ipsalazide ($n = 1$)

(**27**) Olsalazine

(**28**) Sulfasalazine

Azoreductase

(**29**) Mesalazine (5-ASA)

SCHEME 12.1 Activation of balsalazide (**25**), ipsalazide (**26**), olsalazine (**27**), and sulfasalazine (**28**) to form mesalazine (**29**) by microbial azoreductases in the colon.

In addition to the use of ionizable groups, disruption of the crystal lattice can also result in a significant increase in aqueous solubility, as illustrated by the antiviral agent vidarabine (**30**). The 5′-formate ester derivative (**31**) of vidarabine is 67-fold more soluble in water than vidarabine itself and has been attributed to disruption of the strong intermolecular interactions in the crystal, as indicated by the 85°C drop in the melting point [25].

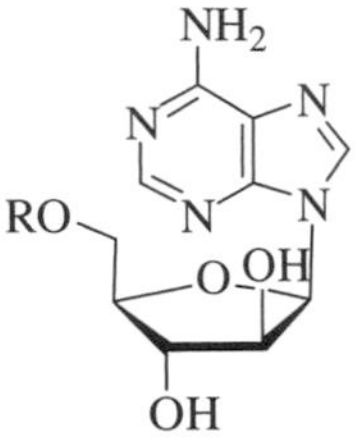

	R	ACD/Log*P*	m.p.	Solubility $_{H_2O}$ (25°C)
(**30**) Vidarabine	R = H	−1.46 ± 0.47	260°C	0.0018 M
(**31**) 5′-formate	R = C(=O)–H	−0.36 ± 0.60	175°C	0.12 M

12.2.7 Tissue Targeting and Activation at the Site of Action

Prodrugs can be designed to target-specific tissues. This is especially useful in improving the therapeutic effectiveness and decreasing the systemic toxicity of anticancer agents in the treatment of cancer. Anticancer agents are usually highly cytotoxic with a very small therapeutic index; their therapeutic effectiveness is often limited by their dose-limiting side effects. Here, several strategies for targeting chemotherapeutic agents to cancers will be briefly discussed to illustrate the applications of prodrugs. For details, please refer to Chapter 17 on metabolic activation and drug targeting.

12.2.7.1 Tumor Hypoxia and Bioreductive Activation of Anticancer Prodrugs Solid tumors often contain regions, which are subject to chronic or transient deficiencies of blood flow and, therefore, to the development of chronic or acute hypoxia owing to the primitive state of tumor vasculature [26]. Hypoxic cells in a solid tumor frequently constitute 10–20% and occasionally over half of the total viable tumor cell population. Agents that are active against proliferating cells are relatively ineffective against these hypoxic tumor cells, which are not actively replicating at the time of treatment but are capable of commencing proliferation at a later time and causing the tumor to regrow. Hypoxic cells also may be resistant to conventional chemotherapy due to pharmacodynamic considerations [26]. To produce a therapeutic response, appropriate drug concentrations must be reached. Drugs that have physicochemical properties not conducive to diffusion into tumor tissue, or that are unstable or metabolized rapidly, may not reach chronically hypoxic tumor cells located in regions of severe vascular insufficiency. Therefore, the presence of hypoxic cells in solid tumors is an obstacle to effecting a cure.

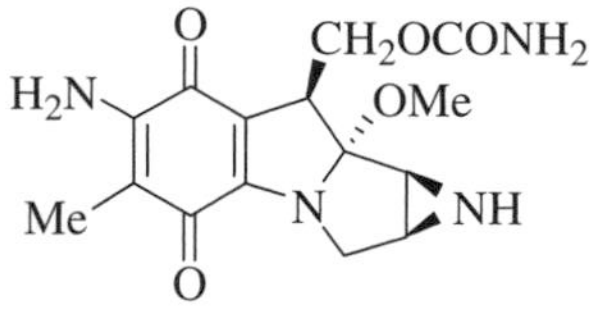

(**32**) Mitomycin C

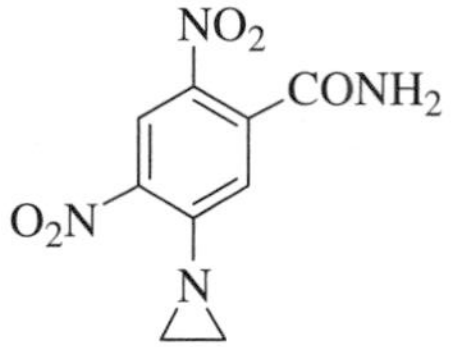

(**33**) CB1954

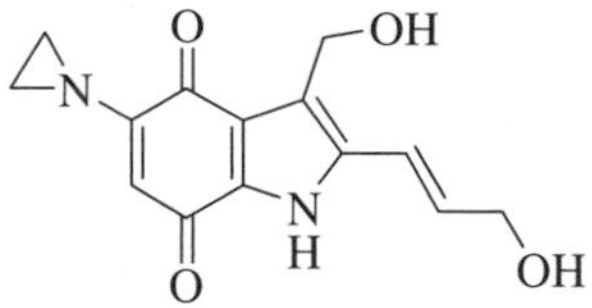

(**34**) EO9

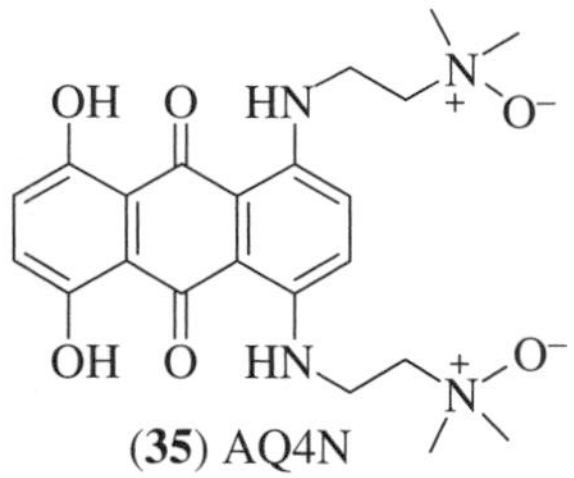

(**35**) AQ4N

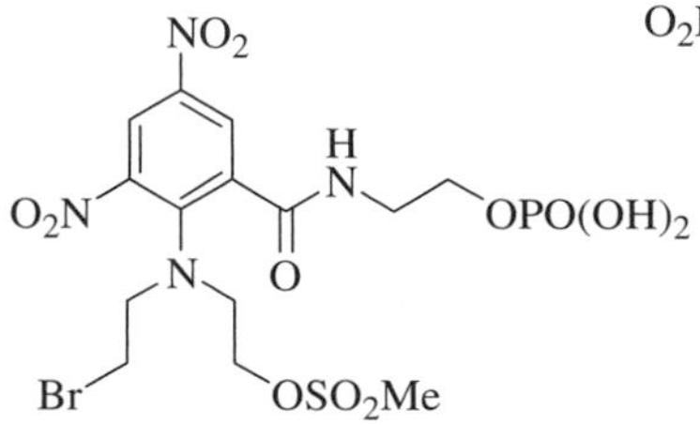

(**36**) PR-104

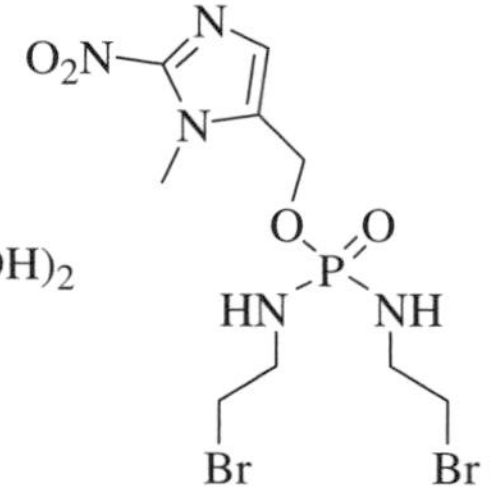

(**37**) TH-302

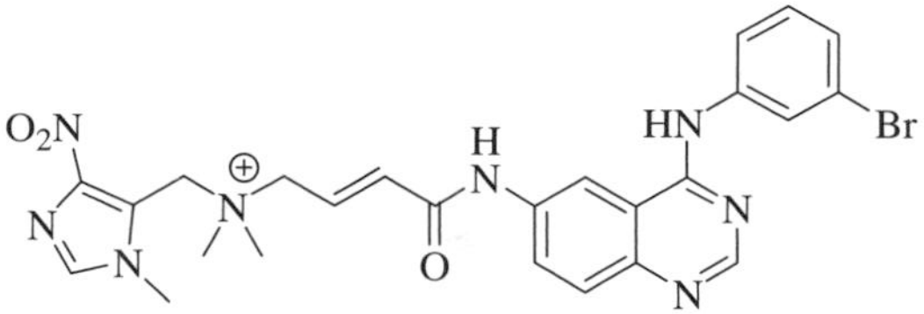

(**38**) SN29966

Since hypoxic cells located remotely from the vascular supply of a tumor mass may have a greater capacity for reductive reactions than their normal, well-oxygenated counterparts, hypoxia could provide an opportunity for the design of selective cancer chemotherapeutic agents that could be reductively activated in these hypoxic cells [26]. Several classes of agents are presently known, which exhibit preferential cytotoxicity toward hypoxic cells through reductive activation. They include nitro compounds, quinones, and aromatic *N*-oxides such as mitomycin C (**32**), CB1954 (**33**), EO9 (**34**), AQ4N (**35**), PR-104 (**36**), and TH-302 (**37**) [27, 28]. Several of these including PR-104, TH-302, and EO9 are currently undergoing clinical evaluation. Besides the DNA-damaging agents employed in these prodrugs, an irreversible pan-ErbB multikinase inhibitor has also recently been incorporated into a hypoxia-activated prodrug SN29966 (**38**) [29].

12.2.7.2 Activation of Prodrugs by Tissue- or Tumor-Specific Enzymes Investigation of the biochemistry and molecular biology of cancer has also identified several reductive or proteolytic enzymes that are unique to tumors or tissues and could be used as potential therapeutic targets or prodrug-converting enzymes for novel cancer therapy. These include DT-diaphorase [30], prostate-specific antigen (PSA) [31], plasminogen activator [32], and members of matrix metalloproteinases [33].

One such example is the peptide doxorubicin conjugate, glutaryl-Hyp-Ala-Ser-Chg-Gln-Ser-Leu-Dox, L-377202 (**39**), which was reported to have the profile of physical and biological properties needed for further clinical development [34]. Conjugate **39** was found to have a greater than 20-fold selectivity against PSA-secreting LNCaP cells relative to non-PSA-secreting DuPRO cells. In nude mouse xenograft studies, it reduced PSA levels by 95% and tumor weight by 87% at 21 μmol/kg, a dose below its maximal tolerated dose (MTD). On the basis of these results, this conjugate was selected for further studies in clinical trials to assess its ability to inhibit human prostate cancer cell growth and tumorigenesis. It was believed that PSA cleavage in and around prostate cancer cells would release, as shown in Scheme 12.2, dipeptide–doxorubicin conjugate (**40**), which would be further cleaved by aminopeptidases to the cytotoxic Leu-doxorubicin (**41**) and doxorubicin (**42**). The same approach has been used to deliver the phosphoramide mustard alkylating agent, thapsigargin, and vinblastine. Scheme 12.3 shows the peptide-linked amino-cyclophosphamides (**43a–c**) designed for selective PSA activation to release 4-aminocylophosphamide (**44**), which spontaneously degrades to the active alkylating agent, phosphoramide mustard (**45**) [35].

12.2.7.3 Antibody- or Gene-Directed Enzyme Prodrug Therapy Besides targeting hypoxic tumor cells and using tumor- or tissue-specific enzymes like PSA to activate prodrugs, other specific enzymes can be delivered to tumor tissues using antibodies or expressed by tumor cells through gene therapy and can be used as prodrug-converting enzymes. These strategies are called "antibody-directed enzyme prodrug therapy" (ADEPT) or "gene-directed enzyme prodrug therapy" (GDEPT). In these approaches, an enzyme is delivered site specifically by chemical conjugation

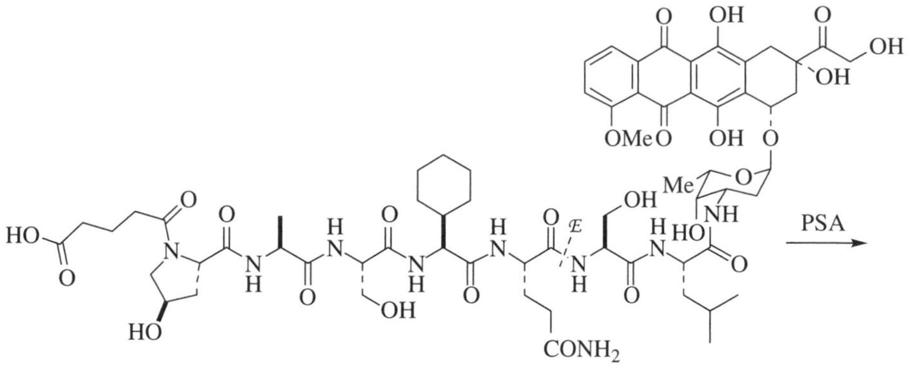

(**39**) L-377202

H-Ser-Leu-Doxorubicin → H-Leu-Doxorubicin → Doxorubicin
(**40**) (**41**) (**42**)

SCHEME 12.2 Activation of the peptide doxorubicin conjugate **39** by PSA.

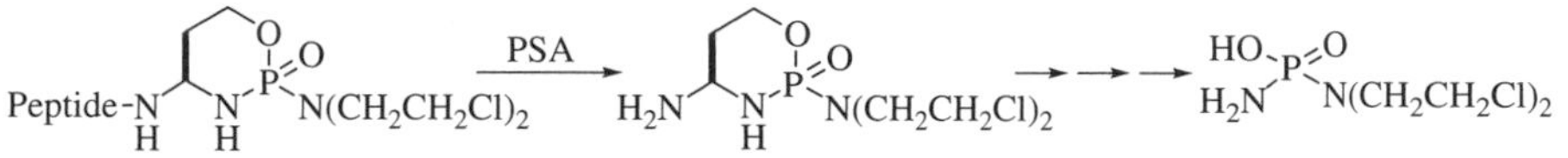

43a–c —PSA→ **44** → → → **45**

Peptdie = *Succinyl*-Ser-Lys-Leu-Gln-
Succinyl-His-Ser-Ser-Lys-Leu-Gln-
Glutaryl-Hyp-Ala-Ser-Chg-Gln-

SCHEME 12.3 Activation of the peptide-linked amino-cyclophosphamides (**43a–c**) by PSA to release 4-aminocylophosphamide (**44**).

or genetic fusion to a tumor-specific antibody or by enzyme gene delivery systems into tumor cells. This is followed by the administration of a prodrug, which is selectively activated by the delivered enzyme at the tumor cells. A number of these systems are in development and have been reviewed [36]. Many viral and nonviral delivery systems have been used for gene delivery. When the gene for GDEPT is delivered by a viral vector, it is known as virus-directed enzyme prodrug therapy (VDEPT). Many enzymes of nonmammalian origin (bacterial, viral, and yeast) have been used to achieve selective activation of prodrugs and the exogenous enzyme is delivered. The main disadvantage of using a nonhuman enzyme is its potential immunogenicity. Among the enzymes under evaluation is a bacterial nitroreductase from *Escherichia coli*. This is an FMN-containing flavoprotein capable of reducing certain aromatic nitro groups to the corresponding amines or hydroxylamines in the presence of a cofactor NADH or NADPH. The nitroaromatics that were found to be good substrates of *E. coli* nitroreductase include dinitroaziridinylbenzamide CB1954 (**33**), dinitrobenzamide mustards SN 23862 (**46**), 4-nitrobenzylcarbamates (**47**), and nitrophenyl phosphoramides (**48** and **49**) [37, 38].

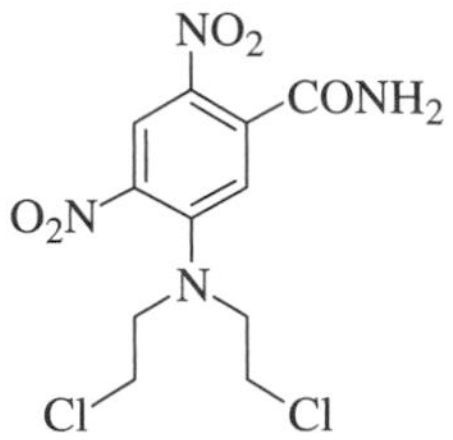

(**46**) SN 23862

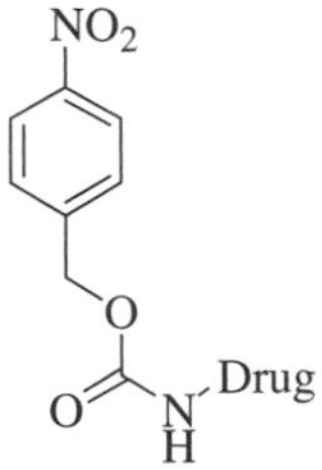

(**47**) 4-Nitrobenzylcarbamates

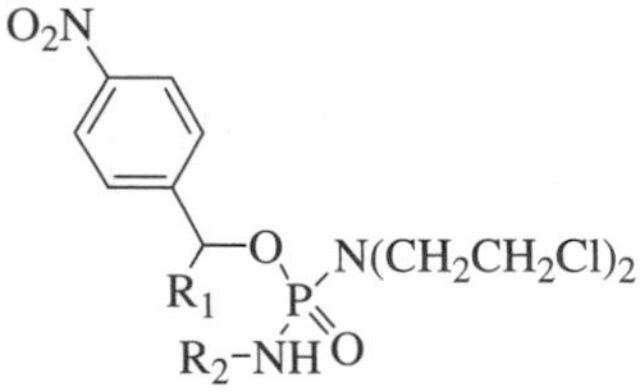

Nitrophenyl phosphoramides
(**48**) $R_1 = R_2 = H$ (LH7)
(**49**) $R_1, R_2 = CH_2CH_2$

12.2.7.4 Tumor-Specific Transporters Antibody–drug conjugates would have to overcome problems inherent in proteins such as susceptibility toward proteolytic cleavage and high immunogenicity; the latter could lead to an antibody response against the conjugate, thereby precluding further use. To increase the selectivity of chemotherapeutic agents, considerable efforts have also been made to identify biochemical characteristics unique to malignant tumor cells that could be exploited in a therapeutic intervention. The small and nonimmunogenic, tumor-specific molecules like folic acid are among the promising alternatives to antibody molecules as targeting agents for drug delivery. Folate conjugates of radiopharmaceuticals, MRI contrast agents, antisense oligonucleotides and ribozymes, proteins and protein toxins, immunotherapeutic agents, liposomes with entrapped drugs, and plasmids have all been successfully delivered to folate receptor-expressing cells [39].

12.3 PRODRUG DESIGN CONSIDERATIONS

Medicinal chemists often encounter a situation where a compound has adequate pharmacological activity but an inadequate pharmacokinetic profile (i.e., absorption, distribution, metabolism, and excretion). Prodrugs can be designed to improve physicochemical properties, resulting in an improvement in pharmacokinetic properties as well as pharmaceutical properties. The pharmaceutical properties that could be improved, as discussed earlier, include drug product stability, taste and odor, pain on injection, and GI irritation. Despite the many benefits, there are also concerns unique to prodrug design. Of particular concern is the fact that toxicological studies might not be relevant for human use of the drug because of differences in the rate and/or extent of formation of the active moiety–metabolic aspects. Experiments should thus be designed early to address these concerns. As examples of interspecies differences, the pivaloyloxyethyl ester of methyldopa was essentially hydrolyzed presystematically to pivalic acid and methyldopa at the same rate in human, dog, and rat, while the succinimidoethyl derivative was hydrolyzed faster in rat than in man and dog [40]. This suggests that the succinimidoethyl ester of methyldopa was more resistant to extrahepatic esterase action in man and dog but not in rat. For different ester prodrugs of dyphylline, the relative rates of release were 1.3–13 times faster in rabbit plasma than in human plasma [41].

The bond between the active moiety (parent drug) and the promoiety plays a major role in determining the pharmacokinetic properties of a prodrug. Knowledge about the nature of the bond and the promoiety may help explain the nature of the biotransformation process and its location in specific tissues or cells. The study of the fate in the body of the promoiety is particularly important from the safety point of view and should be investigated just as thoroughly as that of the active moiety. In some cases, the fate of the released carrier moiety is well known, such as the esters of methanol or ethanol; no extra study is needed during drug development. In other cases, additional pharmacokinetic investigations may be necessary.

Rational design of a prodrug should begin with identification of the problem(s) encountered with the delivery of the parent compound/drug and the physicochemical properties needed to overcome the delivery problem(s). Only then can the appropriate promoiety be selected to construct a prodrug with the proper physicochemical properties that can be effectively transformed to the active drug in the desired biological compartment.

The most important requirement in prodrug design is naturally the adequate reconversion of the prodrug to the active drug *in vivo* at the intended compartment. This prodrug–drug conversion may take place before absorption (e.g., in the GI system), during absorption (e.g., in the GI wall or in the skin), after absorption, or at the specific site of drug action. It is important that the conversion be essentially complete because the intact prodrug, being usually inactive, represents an unavailable drug. However, the rate of conversion would depend on the specific goal of the prodrug design. A prodrug designed to overcome poor solubility for an intravenous drug formulation should be converted very quickly to the active moiety after injection. If the objective of the prodrug is to produce sustained drug action through rate-limiting conversion, the rate of conversion should not be too fast.

Prodrugs can be designed to use a variety of chemical and enzymatic reactions to achieve cleavage to generate their active drug at the desired rate and place. The design is often limited by the availability of a suitable functional group in the active drug for the attachment of a promoiety. Table 12.1 lists some of the common reversible prodrug forms for various functional groups that are often present in biologically active substances.

The most common prodrugs are those that require hydrolytic cleavage, but reductive and oxidative reactions have also been used for the *in vivo* regeneration of the active drug. Besides using the various enzyme systems for the necessary activation of prodrugs, the buffered and relatively constant physiological pH may be used to trigger their release.

Enzymes considered important to orally administered prodrugs are found in the GI walls, liver, and blood. In addition, enzyme systems present in the gut microflora may be important in metabolizing prodrugs before they reach the intestinal cells. In

TABLE 12.1 Reversible Prodrug Forms for Various Functional Groups Present in Biologically Active Substances

Functional Group	Reversible Prodrug Forms
O OH	O O R; O R O O R′; O R O O O O R′; O N R H; O O O O O O; R O; $-CH_2OH$
—OH	O O R; O O O R; O O N R H; O O P OH OH; R O O O R′; O R
—SH	O S R; R O S O R′; S S R
$-NH_2$	O N R H; O N O R H; O R O N O O R′ H; O N P OH OH H
	R N R′; R N R′ H R″; Ar N=N Ar; $Ar-NO_2$
—N	R O $\overset{+}{N}$ O R′
=O	N R; N OH; N O—R; O O R; OR OR′; O N H; S N H
=N OH	N O—R; N O O R
P O OH	P O O R; P O R O O R′; P O R O O O R′
O_n X NH Acidic	O_n X N N R R′; O_n X N OH; O_n X N O P OH OH; O_n X N O O P OH OH X = C, n = 1 X = S, n = 2
XH NH_2	X N H X = O or S
OH OH	O; N O R

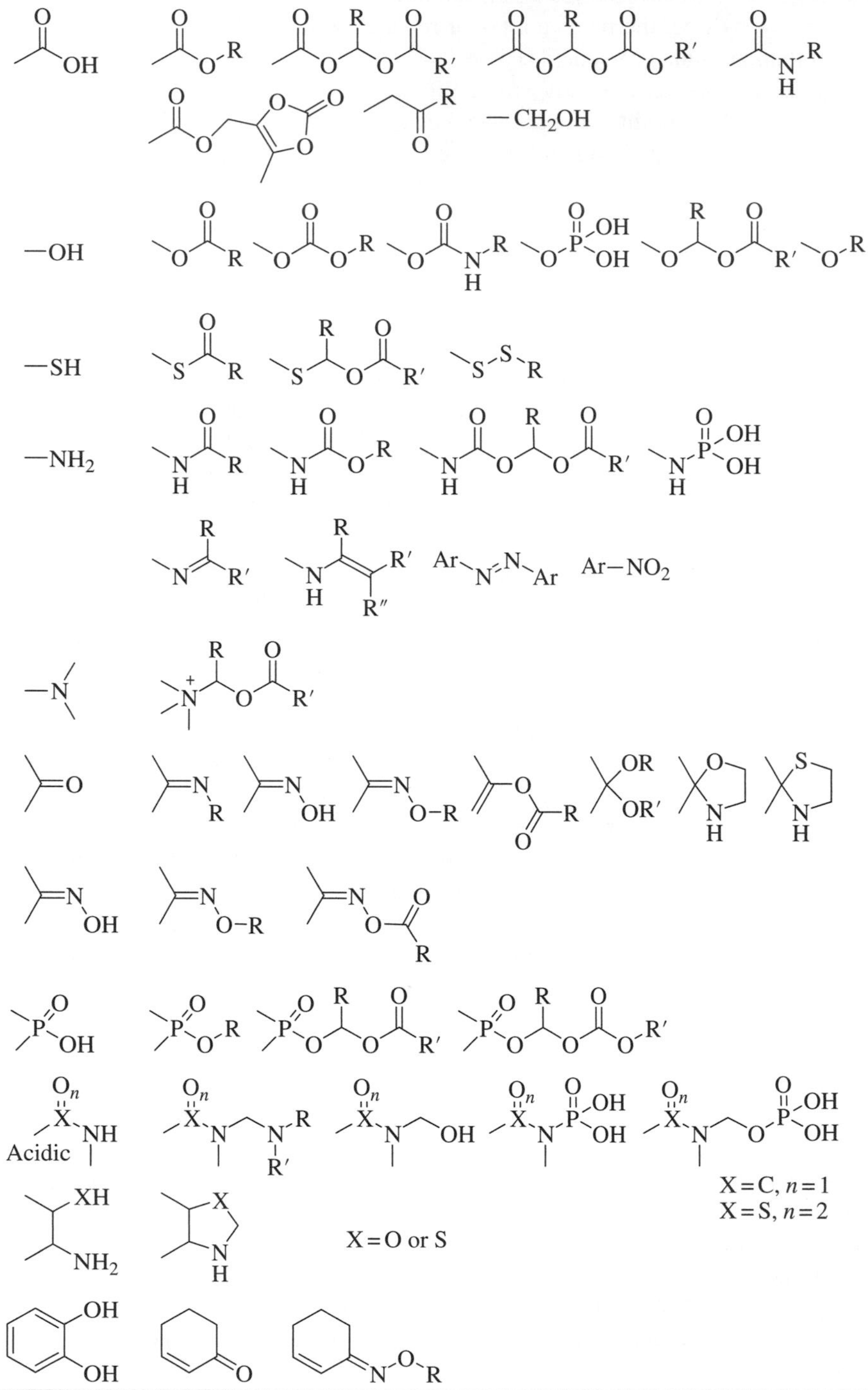

addition, site-specific delivery can be accomplished by exploiting enzymes that are present specifically or at high concentrations in the targeted tissues relative to non-target tissues. A number of enzymes can also be delivered to targeted tissues through antibodies or gene delivery approaches for the activation of subsequently administered prodrugs, as discussed earlier in this chapter.

12.4 PRODRUGS OF VARIOUS FUNCTIONAL GROUPS

12.4.1 Prodrugs of Compounds Containing —COOH or —OH

12.4.1.1 Carboxylic Acid Esters Due to the presence of a wide variety of esterases in various body tissues, it is not surprising that esters are the most common prodrugs used to improve GI absorption. By appropriate esterification of molecules containing a carboxylic acid or hydroxyl group, it is possible to obtain derivatives with almost any desirable hydrophilicity, lipophilicity, and *in vivo* lability. It should be noted that enzyme-catalyzed ester hydrolysis is quite different from nonenzymatic ester hydrolysis in terms of electronic and steric requirements in the substrates. Enzymatic reactions are more likely influenced by steric rather than electronic effects. Experimental determination should be performed to evaluate the rate of cleavage under incubation, with plasma or a homogenate from the intended tissue or organ where the prodrug would be activated. It should also be kept in mind that there are significant interspecies variations in the enzyme's expression level and catalytic capacity.

As shown in Scheme 12.4, esters in the form of **50** can be used as prodrugs for acid drugs (**51**), and the alcohol would serve as a promoiety. Esters in the form of **52** can be used as prodrugs for alcohol drugs (**53**), and here the acid would serve as a promoiety. The acid promoiety can be amino acid like in the experimental anticancer prodrug brivanib alaninate (**55**) to increase aqueous solubility [42]. Upon esterase hydrolysis, it releases a dual tyrosine kinase inhibitor BMS-540215 (**54**) active against both vascular endothelial growth factor receptor (VEGFR) and fibroblast growth factor receptor (FGFR) tyrosine kinases [43].

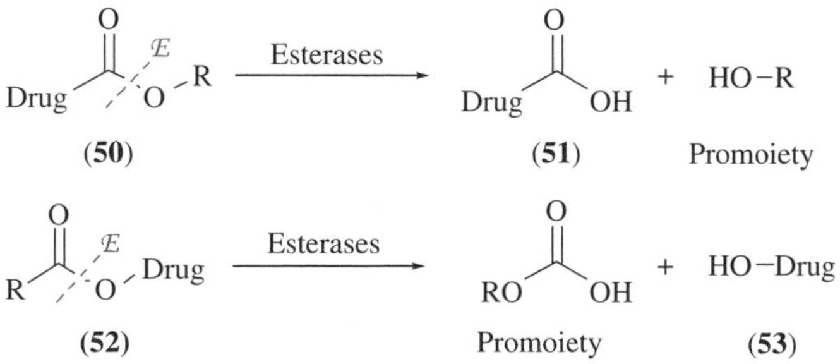

Drug–C(=O)–O–R (**50**) —Esterases→ Drug–C(=O)–OH (**51**) + HO–R (Promoiety)

R–C(=O)–O–Drug (**52**) —Esterases→ RO–C(=O)–OH (Promoiety) + HO–Drug (**53**)

SCHEME 12.4 Activation of ester prodrugs by esterases.

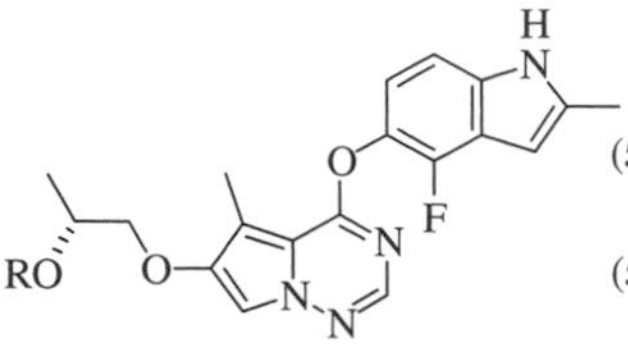

		ACD/Log*P*	ACD/Log*D* pH 7	ACD/Log*D* pH 1
(**54**) BMS-540215	R = H	2.93 ± 1.55	−2.93	2.92
(**55**) Brivanib alaninate	R 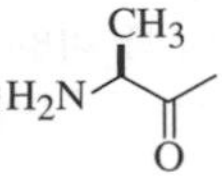 = H_2N (CH_3, O)	2.86 ± 1.56	2.15	−0.25

(5-Methyl-2-oxo-1,3-dioxol-4-yl)methyl ester (medoxomil ester) has been successfully used to prepare prodrugs of such carboxylic acid–containing drugs as olmesartan (**56**) and azilsartan (**58**), both of which are angiotensin II receptor antagonists used to treat hypertension. Unlike ACE inhibitors, olmesartan and azilsartan's activity in lowering blood pressure is independent of angiotension II synthesis. They act by blocking the binding of angiotensin II to the AT_1 receptors in vascular muscle, which directly causes vasodilation and decreases the release of vasopressin and aldosterone [44]. The medoxomil ester prodrugs **57** and **59** have shown potent and long-lasting antihypertensive activity after oral administration. Their activation is through rapid enzymatic hydrolysis in the intestine, liver, and plasma. Multiple enzymes including plasma albumin and carboxymethylenebutenolidase homolog (CMBL, an intestinal and liver hydrolase) are known to activate the medoxomil prodrugs in human [45]. The oral bioavailability of olmesartan medoxomil (**57**) in humans is about 26% [46] while that for azilsartan medoxomil (**59**) is around 60% [47].

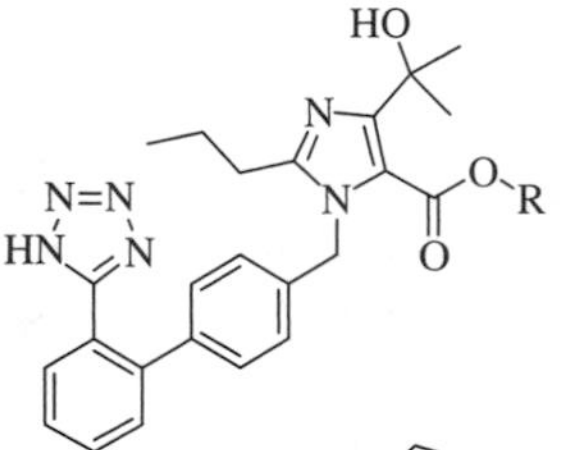

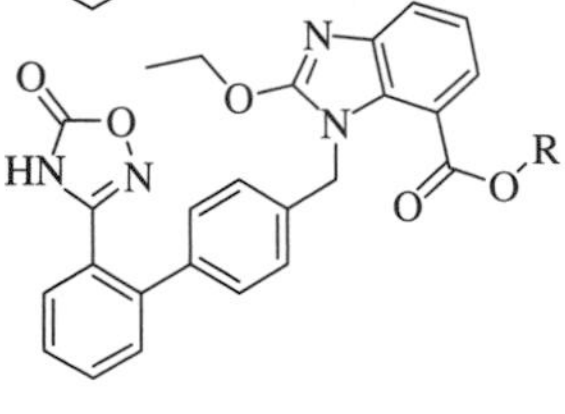

		ACD/Log*P*	ACD/Log*D* pH 7	ACD/Log*D* pH 1
(**56**) Olmesartan	R = H	2.88 ± 0.51	−0.78	0.16
(**57**) Olmesartan medoxomil	R = 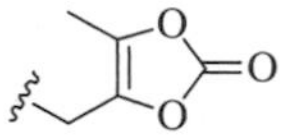	2.80 ± 0.63	0.86	0.12
(**58**) Alisartan	R = H	5.63 ± 1.02	2.42	3.21
(**59**) Alisartan medoxomil	R = 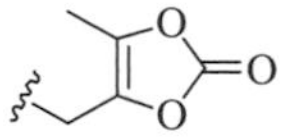	5.64 ± 1.04	5.34	3.25

12.4.1.2 Phosphates Phosphates or phosphonooxymethyl derivatives have also been used successfully to increase aqueous solubility of alcohol- and phenol-containing drugs. Examples include prednisolone phosphate, betamethasone phosphate, clindamycin phosphate, fosamprenavir, and fospropofol (**61**). The active parent drug molecules are rapidly released from these phosphate prodrugs by endogenous phosphates such as alkaline phosphatase. This enzyme is present in the GI tract, plasma, and various tissues and organs. The prodrugs can be administered orally or by parental injections. When administered orally, the prodrugs are used to increase oral dose or mask unpleasant taste and are converted to the parent drugs before or during the absorption process. Phosphate can be directly attached to a free hydroxyl group such as those in prednisolone, betamethasone, clindamycin, and amprenavir. In the case of sterically hindered hydroxyl group such as those in propofol and camptothecin, the phosphate can be attached to the hydroxylmethyl ether so that the phosphate is further away from the bulky functional groups in the drug molecule to facilitate bioconversion.

The widely used anesthetic propofol (**60**) is formulated as an oil-in-water emulsion because of its high lipophilicity and has such drawbacks as pain on injection, emulsion instability, and hyperlipidemia. The prodrug fospropofol (**61**) is significantly more water soluble and is formulated as a purely aqueous solution, avoiding some of the problems associated with propofol. After intravenous administration, fospropofol releases propofol rapidly upon hydrolysis by alkaline phosphatase present in the plasma and tissues [48].

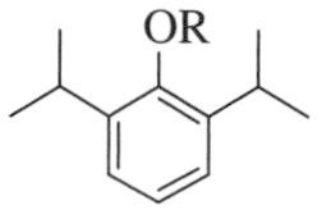

			ACD/Log$D$	
		ACD/Log$P$	pH 7	pH 1
(**60**) Propofol	R = H	3.66 ± 0.25	3.66	3.66
(**61**) Fospropofol	R 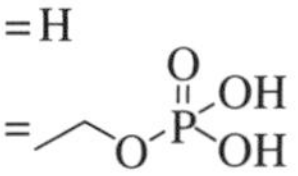 = $CH_2OP(O)(OH)_2$	1.73 ± 0.30	−2.44	1.65

12.4.1.3 α-Acyloxyalkyl, Carbonate, or Alkoxycarbonyloxyalkyl Esters Both the acyl and the alcohol portion surrounding the cleavable ester bond affect the enzyme-catalyzed ester hydrolysis. Sometimes because of steric hindrance in the active drug, direct ester formation with the existing functional group might not produce a prodrug that is sufficiently labile *in vivo*. This problem can be solved by designing the so-called cascade prodrugs containing double esters using α-acyloxyalkyl, carbonate, or alkoxycarbonyloxyalkyl esters (**62**, **63**, or **64**), where the terminal ester group is accessible for enzymatic cleavage (Scheme 12.5). Cascade prodrugs are those

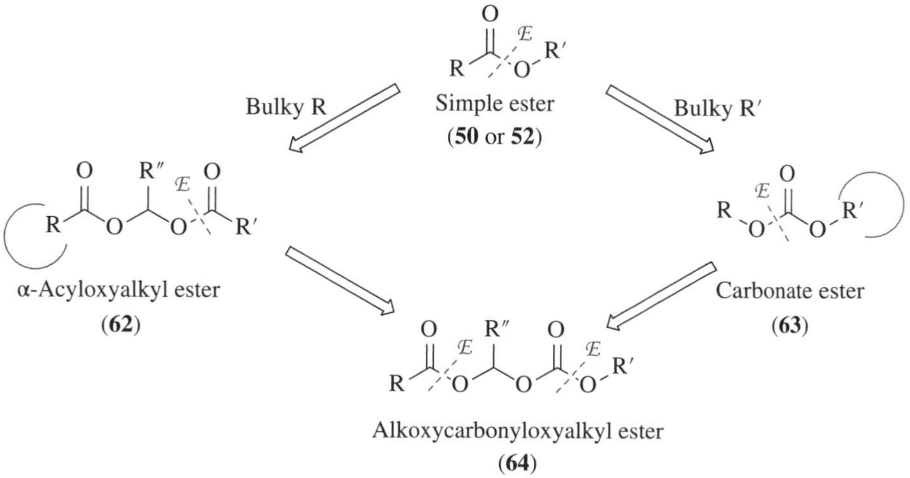

Simple ester (**50** or **52**)
Bulky R
Bulky R′
α-Acyloxyalkyl ester (**62**)
Carbonate ester (**63**)
Alkoxycarbonyloxyalkyl ester (**64**)

SCHEME 12.5 Design of cascade prodrugs containing double esters using α-acyloxyalkyl (**62**), carbonate (**63**), or alkoxycarbonyloxyalkyl (**64**) esters.

prodrugs that require a sequence of two or more reactions for drug release and activation, usually triggered by a first enzymatic-catalyzed reaction followed by a spontaneous chemical release/activation step(s). A number of such examples are known, including several prodrugs of β-lactam antibiotics, corticosteroids, and angiotensin II receptor antagonists. The 2-carboxylic acid on the thiazolidine ring of β-lactam antibiotics is required for antibacterial activity, providing an ideal site for attaching a promoiety in the design of ester prodrugs. But, because of steric hindrance, simple esters of this carboxylic acid group would resist enzymatic hydrolysis. Thus, a number of cascade prodrugs were made to extend the chain and render the terminal ester group easily accessible to hydrolytic enzymes. Examples of α-acyloxyalkyl ester prodrugs include bacampicillin (**2**), pivampicillin (**3**), pivmecillinam (**6**), and cefuroxime axetil (**65**).

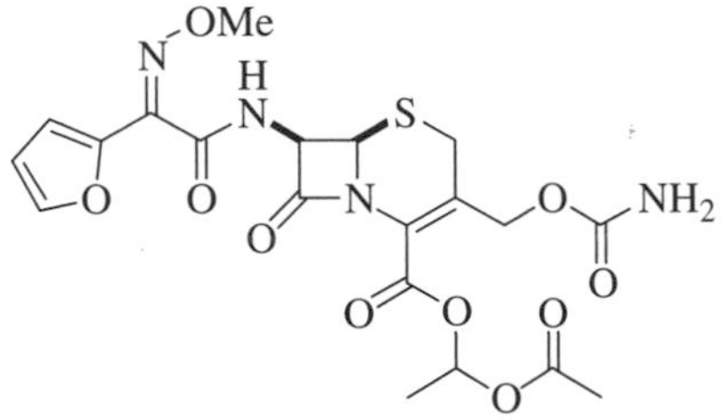

(**65**) Cefuroxime axetil

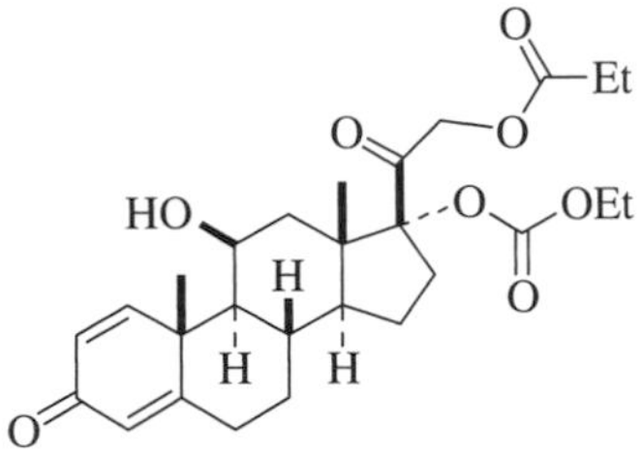

(**66**) Prednicarbate

Prednicarbate (**66**) is an example of a carbonate prodrug of corticosteroids, while candesartan cilexetil (**67**) is a racemic mixture of an alkoxycarbonyloxyalkyl ester of candesartan (**69**) with a chiral center at the carbonate ester group. Following oral administration, candesartan cilexetil (**67**) undergoes rapid and complete hydrolysis during absorption from the GI tract to form, as shown in Scheme 12.6, the active drug candesartan (**69**), which is an achiral selective AT1 subtype angiotensin II receptor antagonist [49].

The α-acyloxylalkyl esters have also been extended to include the phosphate group, phosphonic acids, and phosphinic acids. One such example is fosinopril (**71**), an inhibitor of the angiotensin-converting enzyme (ACE), where the phosphinic acid is

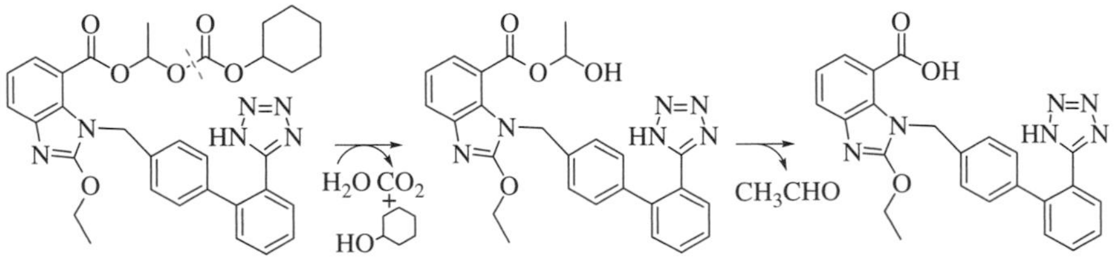

(**67**) Candesartan cilexetil
ACD/Log*P* = 7.430 ± 1.003

(**68**)

(**69**) Candesartan
ACD/Log*P* = 4.651 ± 0.930
ACD/Log*D* = 0.54 at pH 7

SCHEME 12.6 Activation of candesartan cilexetil (**67**).

O-α-acyloxyalkylated to increase lipophilicity to provide better absorption. Another such example is adefovir dipivoxil (**73**), a bis-(pivaloyloxymethyl) ester of adefovir, used for the treatment of hepatitis B with greatly improved oral bioavailability of 30% as compared to 2% for the parent [50]. Activation of adefovir dipivoxil by esterases releases adefovir and two equivalents each of pivalic acid and formaldehyde (Scheme 12.7). The release of pivalic acid has an impact on carnitine homeostasis and is a major toxicological concern of adefovir dipivoxil [51]. The alternative is the corresponding dicarbonate esters like those in tenofovir disoproxil (**75**), a prodrug of tenofovir with an (*R*)-α-methyl substitution. Tenofovir disoproxil releases isopropanol, carbon dioxide, formaldehyde, and tenofovir upon esterase-catalyzed hydrolysis and is indicated for the treatment of both HIV-1 and hepatitis B virus infections [50].

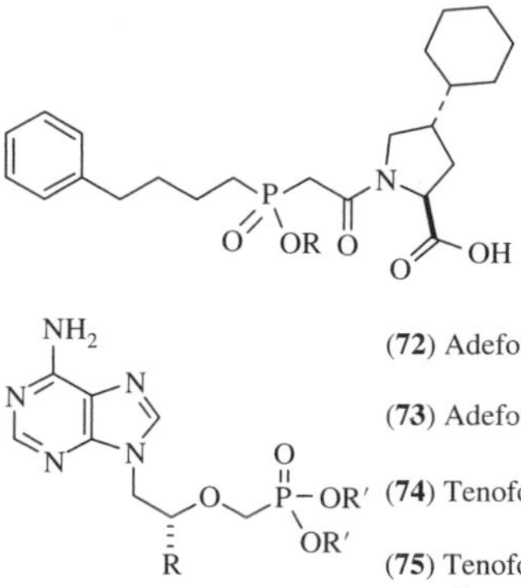

Compound	Substituents	ACD/Log*P*	ACD/Log*D* pH 7	ACD/Log*D* pH 1
(**70**) Fosinoprilat	R = H	3.07 ± 0.66	−1.93	3.06
(**71**) Fosinopril	R = 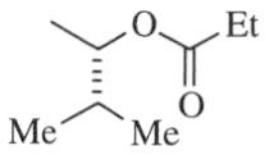	5.81 ± 0.68	2.65	5.81
(**72**) Adefovir	R = H R′ = H	−0.49 ± 1.62	−4.19	−2.94
(**73**) Adefovir dipivoxil	R = H R′ 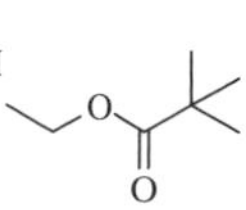 =	2.27 ± 0.64	2.27	−0.17
(**74**) Tenofovir	R = CH_3 R′ = H	−0.14 ± 0.62	−3.84	−2.59
(**75**) Tenofovir disoproxil	R = CH_3 R′ 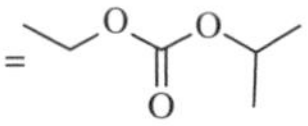 =	1.61 ± 0.75	1.60	−0.85

O-α-acyloxyalkyl ethers are also a useful prodrug type for compounds containing phenol group. Such derivatives (**76**) are hydrolyzed by a sequential reaction involving formation of an unstable hemiacetal intermediate (**77**) as shown in Scheme 12.8.

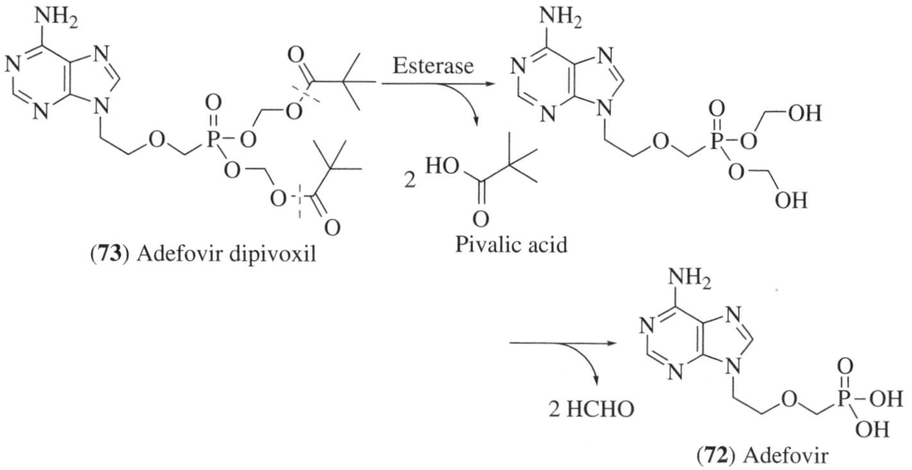

(**73**) Adefovir dipivoxil → Esterase → 2 Pivalic acid → 2 HCHO → (**72**) Adefovir

SCHEME 12.7 Activation of adefovir dipivoxil (**73**).

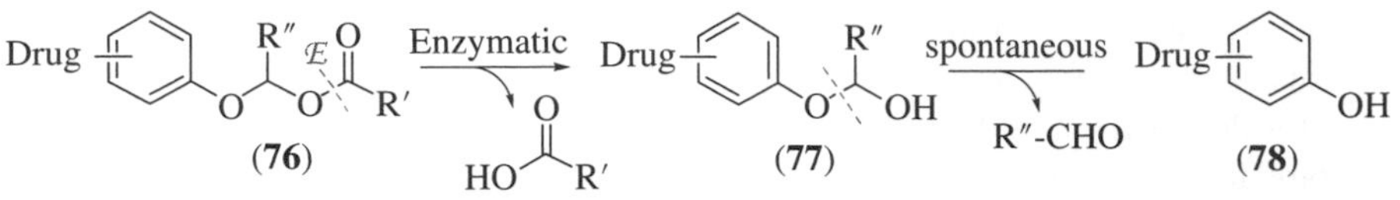

Drug (76) R″ E O O O R′ Enzymatic → HO R′ Drug (77) R″ O OH spontaneous → R″-CHO Drug (78) OH

SCHEME 12.8 Sequential activation of *O*-α-acyloxyalkyl ether prodrugs of phenol-containing drugs.

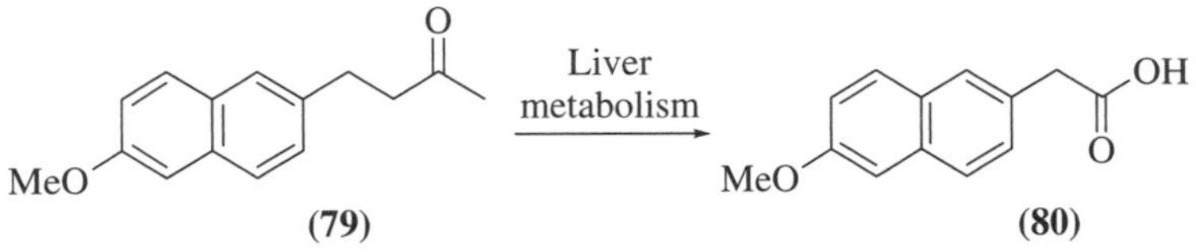

MeO (79) O Liver metabolism → MeO (80) OH O

SCHEME 12.9 Metabolic activation of nabumetone (**79**).

These kinds of ethers might be better prodrugs than normal phenol esters because they are more stable against chemical hydrolysis, but they are still susceptible to enzymatic hydrolysis by human plasma esterases.

12.4.1.4 Ketones and Alcohols Carboxylic acids have also been masked as ketones and alcohols, which would require oxidation to convert to the active acid drugs. Nabumetone (**79**) is a nonacidic prodrug, a nonsteroidal anti-inflammatory drug (NSAID) [52]. After absorption, nabumetone undergoes extensive metabolism, the main circulating active form is 6-methoxy-2-naphthylacetic acid (**80**), a potent COX-2 inhibitor (Scheme 12.9). Since nabumetone is not acidic and the active acid metabolite does not undergo enterohepatic circulation, nabumetone does not cause gastric irritation and is one of the most widely used NSAIDs in the United States.

12.4.2 Prodrugs of Compounds Containing Amides, Imides, and Other Acidic NH

12.4.2.1 Mannich Bases Mannich base prodrugs could enhance the delivery of their parent drugs through the skin because of their enhanced water solubility as well as enhanced lipid solubility. *N*-Mannich bases, or *N*-acyl *gem*-diamines (**85**), are generally formed, as shown in Scheme 12.10, by reaction of an acidic NH compound (**84**) with an aldehyde, usually formaldehyde, and a primary or secondary aliphatic amine (**81**). Aromatic amines do not usually undergo this reaction. Mannich base prodrugs are regenerated by chemical hydrolysis without enzymatic catalysis in the reverse direction of their formation [53].

Transformation of an amide to an *N*-Mannich base introduces a readily ionizable amino functional group (**85** ⇌ **86**) that would allow the preparation of sufficiently stable derivatives with greatly enhanced water solubility at slightly acidic pH. Clinically useful *N*-Mannich base prodrugs include rolitetracycline and hetacillin. The highly water-soluble rolitetracycline (**88**) is an *N*-Mannich base of tetracycline

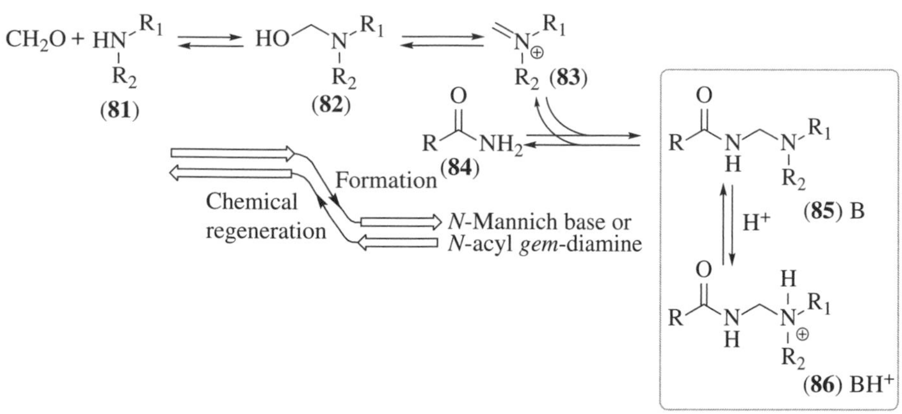

SCHEME 12.10 The formation and activation of Mannich base prodrugs.

(**87**) with pyrrolidine and is decomposed to tetracycline quantitatively with a half-life of 40 minutes at pH 7.4 and 35°C [54]. Since the decomposition of *N*-Mannich bases does not rely on enzymatic catalysis, the rate of hydrolysis is the same in plasma and in buffer. Hetacillin (**89**) is an example of a cyclic *N*-Mannich base-type prodrug, which is formed by condensation of ampicillin with acetone. The prodrug is readily converted back to the active ampicillin and acetone, with a half-life of 15–20 minutes at pH 4–8 and 35°C [55, 56]. The advantage of hetacillin is its higher stability in concentrated aqueous solutions as compared to ampicillin, which has a more nucleophilic amine that would react with the strained β-lactam ring.

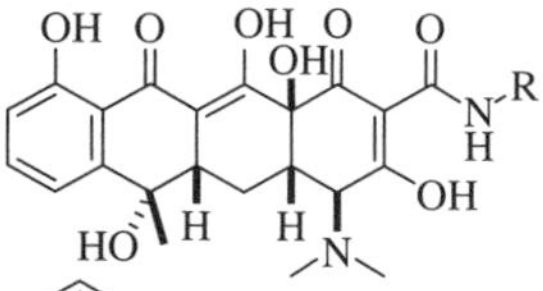

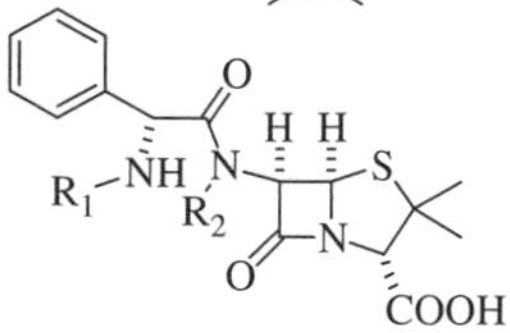

		ACD/Log*P*	ACD/Log*D* pH 7	ACD/Log*D* pH 1
(**87**) Tetracycline	R = H	−1.19 ± 0.75	−4.78	−4.29
(**88**) Rolitetracycline	R = —CH_2—N(pyrrolidine)	0.15 ± 0.75	−3.02	−4.84
(**1**) Ampicillin	$R_1 = R_2 = H$	1.35 ± 0.32	−1.54	−1.72
(**89**) Hetacillin	$R_1, R_2 = >C(CH_3)_2$	2.30 ± 0.90	−1.61	−0.29

12.4.2.2 N-α-Acyloxyalkyl Derivatives *N*-α-Acyloxyalkylation has become a commonly used approach to obtain prodrug forms of various NH-acidic drug substances such as carboxamides, carbamates, ureas, and imides. This is because *N*-α-acyloxyalkyl derivatives (**90**) combine high *in vitro* stability with enzymatic lability. The derivatives are cascade or double prodrugs. The regeneration of the parent drug occurs via a two-step mechanism, the enzymatic cleavage of the ester group followed by spontaneous decomposition of the *N*-α-hydroxyalkyl intermediate (**91**) (Scheme 12.11). The usefulness of this approach depends on the stability of the *N*-α-acyloxyalkyl derivative **90**, its susceptibility to esterase-catalyzed hydrolysis, and the rate of decomposition of the intermediate **91**. *N*-α-acyloxyalkyl derivatives of imides

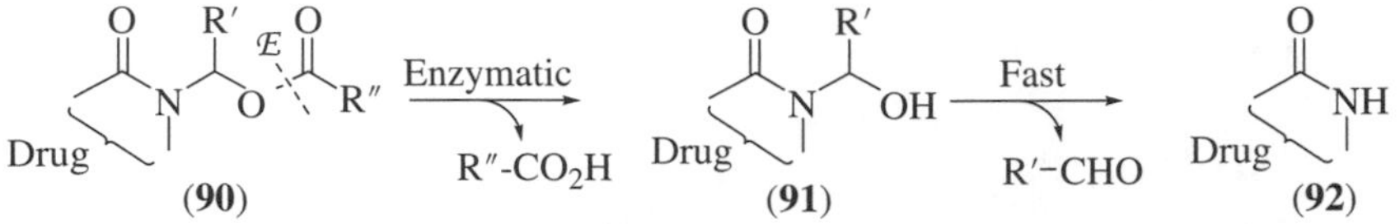

SCHEME 12.11 Sequential enzymatic and chemical activation of prodrugs in the form of *N*-α-acyloxyalkyl derivatives.

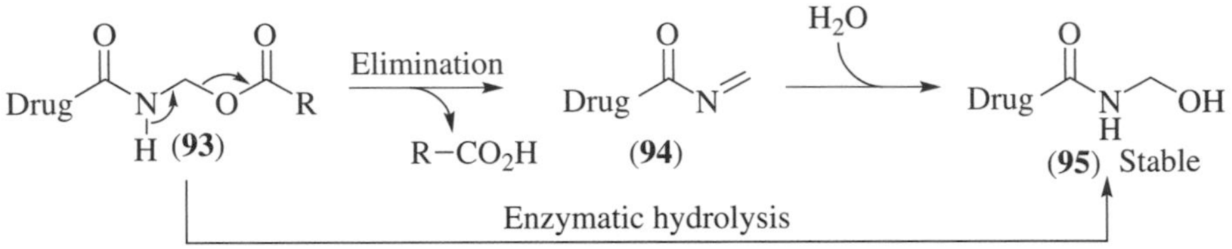

SCHEME 12.12 Decomposition of *N*-α-acyloxyalkyl derivatives (**93**) of primary amides through an elimination–addition mechanism.

and secondary amides, as well as ring structures containing such moieties, showed normal ester stability [57]. To make such *N*-α-acyloxyalkyl derivatives useful as prodrugs, the α-hydroxyalkyl intermediate (**91**) formed after the enzyme-catalyzed hydrolysis must decompose quickly to release the original drug molecule (**92**). The rate of the chemical decomposition step was found to correlate with the pK_a of the acidic NH group; a pK_a of less than 10.5 is required for instantaneous decomposition of *N*-hydroxymethyl derivatives.

However, *N*-α-acyloxyalkyl derivatives (**93**) of primary amides, and other primary amide–type structures such as carbamates and sulfonamides, are extremely unstable in aqueous solution and quickly undergo decomposition to the corresponding *N*-hydroxymethyl derivatives, which are stable. Such derivatives of simple primary amides decompose by an elimination–addition mechanism involving a reactive *N*-acylimine intermediate (**94**) (Scheme 12.12). For imides and secondary amides, their inability to form an *N*-acylimine is believed to contribute to the stability of their *N*-α-acyloxyalkyl derivatives (**90**). At pH 4 and 37°C, half-lives of hydrolysis of *N*-α-acyloxyalkyl derivatives (**93**) range from 1 to 90 minutes, whereas at pH 7.4 the half-lives of hydrolysis are less than 1 minutes [58]. The resulting *N*-hydroxymethyl derivatives (**95**) are rather stable; the half-life for the decomposition of *N*-(hydroxymethyl) benzamide is 183 hours at pH 7.4 and 37°C [58]. However, aldehydes other than formaldehyde can be used to from *N*-α-hydroxyalkyl derivatives that are more unstable than *N*-hydroxymethyl analogs. For example, the half-life for *N*-(α-hydroxybenzyl)benzamide is only 6.5 minutes at pH 7.4 and 37°C. The use of aldehydes other than formaldehyde may further expand the applicability of this approach to simple amides with pK_a above 11 [58].

12.4.2.3 ***N-α-Phosphoryl and N-α-Phosphoryloxyalkyl Derivatives*** Transformation of an acidic amide or imide to phosphoramide or phosphoryloxyalkyl derivatives leads to prodrugs with greatly enhanced aqueous solubility for parental administration. Fosaprepitant (**97**) is a prodrug of aprepitant, an antiemetic for intravenous administration, used in combination with other antiemetics for the prevention of nausea and vomiting caused by chemotherapy [59]. After intravenous infusion, fosaprepitant is quickly converted by alkaline phosphatase or a phosphoramidase to aprepitant (**96**), a substance P antagonist acting on the neurokinin 1 (NK1) receptor. Plasma concentration of the prodrug drops below the limit of detection (10 ng/ml) within 30 minutes after infusion ($t_{1/2}$ ~ 2.3 minutes) [60].

Similar to acyloxyalkyl derivatives, *N*-α-phosphoryloxyalkyl derivatives have been used successfully to convert an acidic NH to a prodrug. One such example is fosphenytoin (**99**), the phosphoryloxymethyl prodrug of phenytoin (**98**) with enhanced water solubility. The activation of fosphenytoin is triggered by the alkaline phosphatase–catalyzed hydrolysis on the terminal phosphate monoester, followed by rapid decomposition of *N*-hydroxymethyl phenytoin [61].

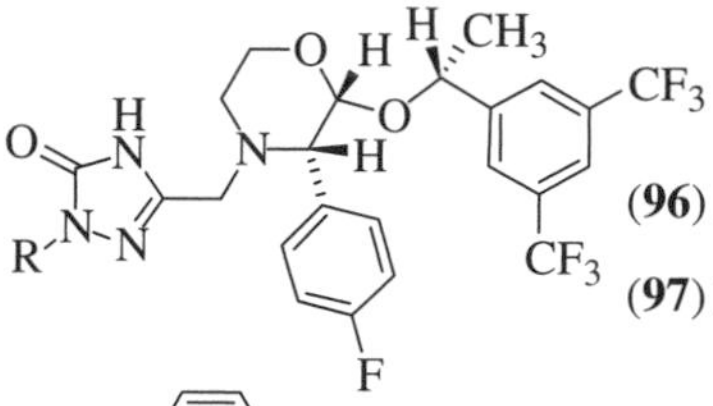

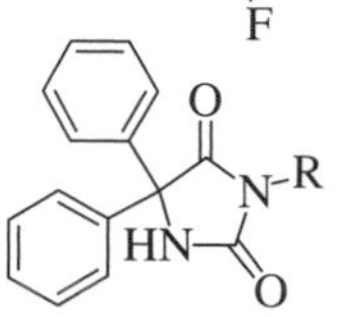

		ACD/Log*P*	ACD/Log*D* pH 7	ACD/Log*D* pH 1
(**96**) Aprepitant	R = H	4.99 ± 0.65	4.93	2.46
(**97**) Fosaprepitant	R = —PO_3H_2	1.97 ± 0.68	–2.09	–0.70
(**98**) Phenytoin	R = H	1.42 ± 0.37	1.41	–1.42
(**99**) Fosphenytoin	R = —$CH_2OPO_3H_2$	–1.06 ± 0.74	–5.73	–1.14

12.4.3 Prodrugs of Amines

The presence of an amino group can affect a drug's physicochemical and biological properties in several ways. These include (i) intermolecular or intramolecular aminolysis leading to reactive and/or potentially toxic substances, (ii) solubility problems when the drug is present with another ionizable functionality such as COOH (zwitterionic nature under physiological pH), potentially limiting its dissolution rate and/or its passive permeability, and (iii) terminal free amino acid groups providing recognition sites for proteolytic enzymes, such as aminopeptidase and trypsin, present in the GI tract lumen, the brush border region, and the cytosol of the intestinal mucosa cells. For all these reasons, prodrug approaches have been advocated for improving *in vivo* behavior of active compounds containing amino groups.

12.4.3.1 Amides Because of the relatively high stability of amides *in vivo*, N-acylation of amines was formerly of limited use in prodrug design. Only a few examples of simple amide prodrugs are known that are sufficiently labile *in vivo*; these include the *N*-L-isoleucyl derivative of dopamine [62], *N*-acetylmethionyl derivative of dopamine [63], the lysinyl conjugate of d-amphetamine (lisdexamfetamine, **17**) [19], and the *N*-glycyl derivative midodrine [64, 65]. With the use of proteases as prodrug-converting enzymes, amines can be coupled to peptide carboxylates, resulting in amide bonds cleavable by proteases (e.g., **39**).

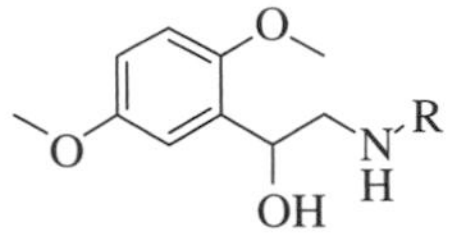

		ACD/Log*P*	ACD/Log*D* pH 7	ACD/Log*D* pH 1
(**100**) Desglymidodrine	R = H	0.38 ± 0.28	−1.14	−3.42
(**101**) Midodrine	R = —$COCH_2NH_2$	−0.32 ± 0.60	−0.83	−2.72

Midodrine (**101**) is a glycinamide prodrug, and the therapeutic effect of orally administered midodrine is due to the major metabolite desglymidodrine (**100**)—an α-agonist formed by deglycination of midodrine. Midodrine is rapidly absorbed after oral administration. The plasma level of the prodrug peaks after about half an hour and declines with a half-life of approximately 25 minutes, while the metabolite reaches peak blood concentrations about 1–2 hours after a dose of midodrine and has a half-life of about 3–4 hours. The absolute bioavailability of midodrine (measured as desglymidodrine) is 93% and is not affected by food. Approximately the same amount of desglymidodrine is formed after intravenous and oral administration of midodrine. Midodrine has been used successfully in the treatment of neurogenic orthostatic hypotension and dialysis hypotension. It acts through vasoconstriction of the arterioles and the venous capacitance vessels, thereby increasing peripheral vascular resistance and augmenting venous return, respectively. The prodrug is a unique agent in the armamentarium against orthostatic hypotension since it has minimal cardiac and CNS effects [64, 65].

12.4.3.2 Carbamates There are only a few prodrugs using carbamate as the site of cleavage. One example of a carbamate prodrug is irinotecan, where the phenolic hydroxyl group of the highly cytotoxic topoisomerase I inhibitor SN-38 forms a carbamate with 1,4′-bipiperidine promoiety. Aromatic amines or amidines have been successfully converted to carbamate prodrugs. Capecitabine (**102**) is a clinically used prodrug of 5-FU in the form of carbamate. As shown in Scheme 12.13, the activation of capecitabine follows a pathway with three enzymatic steps and two intermediary metabolites. It is first converted to 5′-deoxy-5-flurocytidine (5′-dFCR, **103**) by carboxyesterases most likely in the liver, then to 5′-deoxy-5-flurouridine (5′-dFUR, **104**) by cytidine deaminase, and finally to 5-fluorouracil (5-FU, **105**) by thymidine or uridine phosphorylase. The last two steps can occur in normal or tumor tissues, where 5-FU inhibits DNA synthesis and slows cellular growth [66].

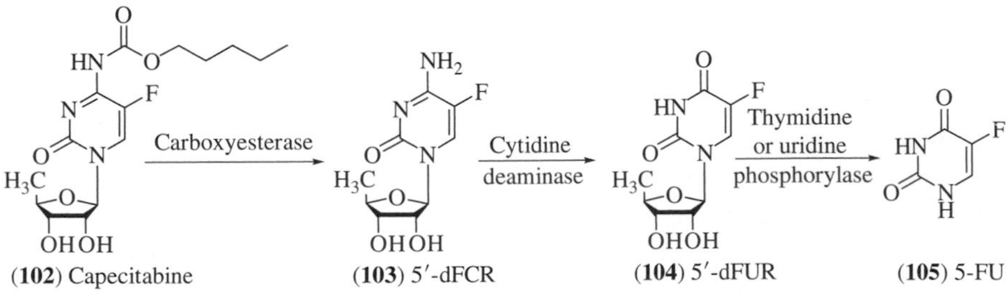

(**102**) Capecitabine —Carboxyesterase→ (**103**) 5′-dFCR —Cytidine deaminase→ (**104**) 5′-dFUR —Thymidine or uridine phosphorylase→ (**105**) 5-FU

SCHEME 12.13 Activation of capecitabine (**102**).

Dabigatran etexilate (**107**) can be viewed as a carbamate prodrug of benzamidine-containing anticoagulant thrombin inihitor dabigatran (**106**) although strictly speaking dabigatran is not an amine drug. Nevertheless, *N*-alkoxycarbonyl derivative of amidine together with the ethyl ester is responsible for the improved lipophilicity although the oral bioavailability of the prodrug is only approximately 7% but can be increased up to 75% through formulation [67].

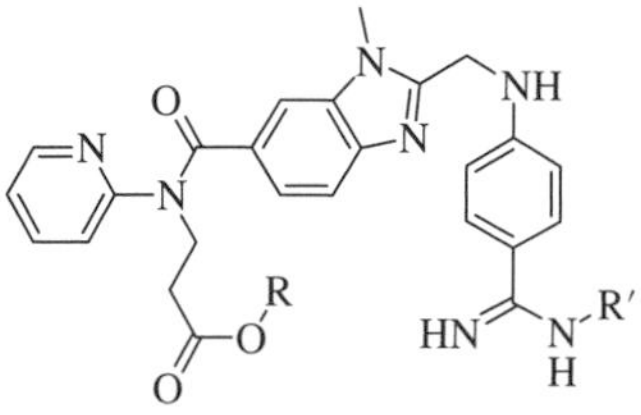

	ACD/Log*P*	ACD/Log*D* pH 7	ACD/Log*D* pH 1
(**106**) Dabigatran R = R′ = H	1.02 ± 0.63	−1.48	−3.36
(**107**) Dabigatran etexilate R = —CH_2CH_3, R′ =	4.80 ± 0.69	4.76	0.43

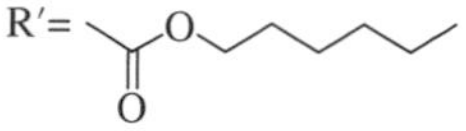

The nonsedating antihistamine loratidine (**108**) is a carbamate drug that is metabolized to the active metabolite, desloratidine (**109**) by an oxidative process involving cytochrome P450 isozyme CYP2D6 and CYP3A4 (Scheme 12.14). Desloratidine has a longer half-life than loratidine; it is also a more potent H_1 antagonist and more potent inhibitor of histamine release. Thus, desloratidine may account for most of the effects of loratidine, which led some to call loratidine a "prodrug." Strictly speaking, loratidine is not a prodrug of desloratidine because loratidine is also a potent H_1 antagonist and was certainly not designed to be a prodrug. However, this example does suggest that carbamate could be used to obtain prodrugs of amine-containing drugs.

12.4.3.3 N-α-Acyloxyalkoxycarbonyl Derivatives Carbamates are of limited use in prodrug design due to their general resistance to enzymatic hydrolytic cleavage *in vivo* as can be seen in the previous loratidine example. The introduction of an enzymatically labile ester group in the carbamate structure could render them sensitive to esterase-catalyzed hydrolysis, leading to the activation and release of an amine drug. Thus, *N*-α-acyloxyalkoxycarbonyl derivatives (**110**) of primary and secondary amines may be readily transformed, as shown in Scheme 12.15, to the parent amine (**113**) *in vivo* [68, 69]. Esterase-catalyzed hydrolysis of the ester moiety in these

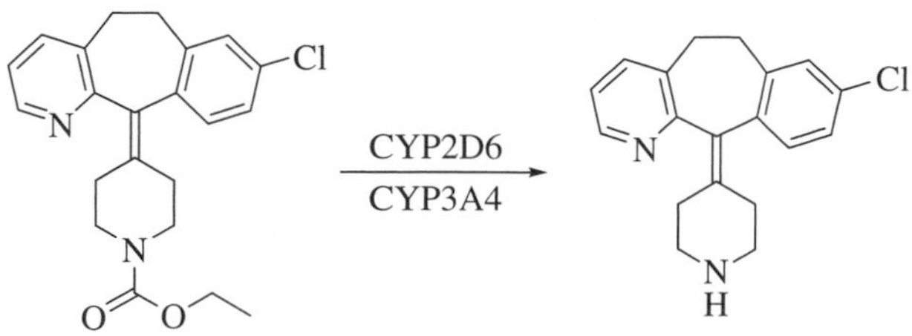

(**108**) Loratidine — CYP2D6 / CYP3A4 → (**109**) Desloratidine

SCHEME 12.14 Activation of loratidine (**108**).

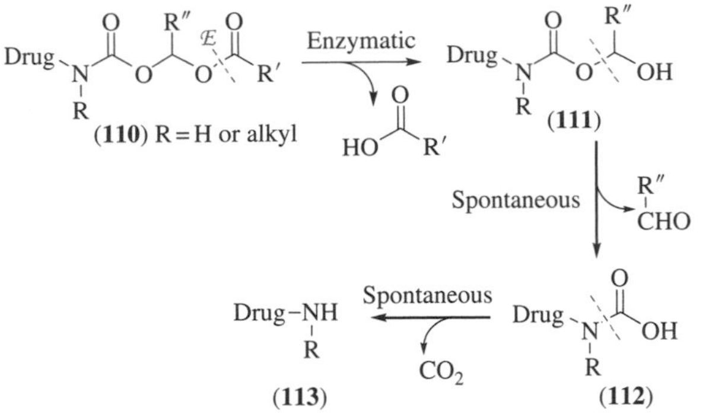

SCHEME 12.15 Activation of prodrugs of primary and secondary amines in the form of *N*-α-acyloxyalkoxycarbonyl derivatives (**110**).

derivatives leads to an unstable α-hydroxyalkoxycarbonyl intermediate (**111**), which spontaneously decomposes into the parent amine via a labile carbamic acid (**112**). These α-acyloxyalkyl carbamate derivatives are neutral and combine high stability in aqueous solution with high susceptibility to enzymatic reconversion to the active agent triggered by hydrolysis of the terminal ester functions; they may be promising reversible prodrugs for amino-containing compounds. Gabapentin enacarbil (**115**) is such a prodrug of gabapentin (**114**)—an anticonvulsant and analgesic agent. It was designed to increase oral bioavailability over gabapentin (from 25 to 84% in monkeys), and it releases rapidly gabapentin in intestinal and liver tissues [70]. It was recently approved for the treatment of restless legs syndrome and postherpetic neuralgia.

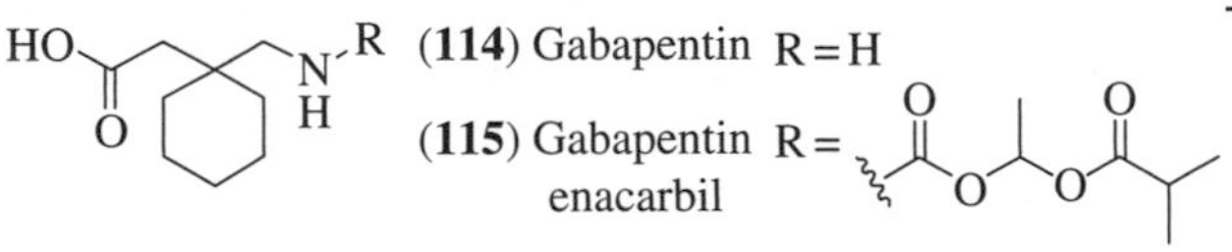

	ACD/Log*P*	ACD/Log*D* pH 7	ACD/Log*D* pH 1
(**114**) Gabapentin R = H	1.08 ± 0.24	−1.42	−2.02
(**115**) Gabapentin enacarbil	2.66 ± 0.36	0.40	2.66

This approach has also been applied to peptides and peptidomimetics in order to improve their unfavorable physicochemical characteristics (e.g., size, charge, and hydrogen-bonding potential), which prevent them from permeating biological barriers such as the intestinal mucosa, and by their lack of stability against enzymatic degradation [71, 72]. Many of the structural features of a peptide, such as the N-terminal amino group, the C-terminal carboxyl group, and the side chain carboxyl, amino, and hydroxyl groups, which bestow upon the molecule affinity and specificity for its pharmacological receptor, severely restrict its ability to permeate biological barriers and render it as a substrate of proteases. Bioreversible cyclization of the peptide backbone is one of the most promising new approaches in the development of peptide prodrugs. Cyclization of the peptide backbone enhances the

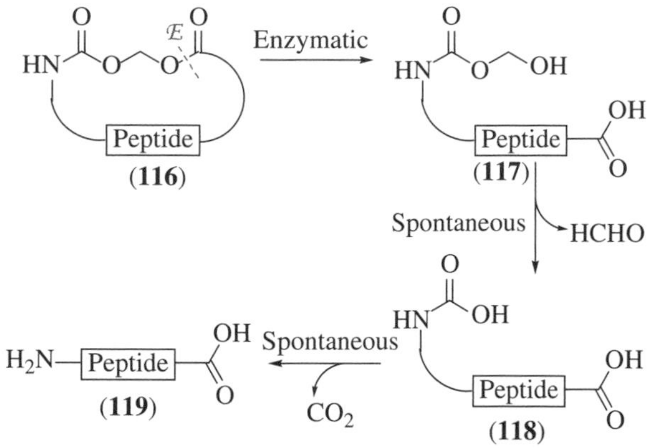

SCHEME 12.16 Activation of prodrugs in the form of a cyclized peptide via an α-acyloxyalkoxy promoiety.

extent of intramolecular hydrogen bonding and reduces the potential for intermolecular hydrogen bonding to aqueous solvent. Linking the N-terminal amino group to the C-terminal carboxyl group via an α-acyloxyalkoxy promoiety, as in (**116**), is an interesting approach that has been shown to work on a number of model peptides (Scheme 12.16). These cyclic prodrugs were designed to be susceptible to esterase-catalyzed hydrolysis (slow step), leading to a cascade of chemical reactions resulting in the generation of the linear peptide. In pH 7.4 buffer at 37°C, the cyclic prodrugs (**116**) were shown to degrade quantitatively to their corresponding linear peptides (**119**). In human plasma, the rates of hydrolysis of cyclic prodrugs were significantly faster than in buffer and were inhibited by paraoxon, a potent esterase inhibitor. In comparison to the linear peptides, the cyclic prodrugs were at least 70 times more permeable in cell culture models of the intestinal mucosa.

12.4.3.4* N-*Mannich Bases *N*-Mannich bases have been used successfully to obtain prodrugs of amide- (see Section 13.4.2.1) as well as amine-containing drugs. Due to their rapid cleavage, with half-lives between 10 and 40 minutes at physiological pH and a pronounced decrease in their basicity of 3–4 pK_a units [73], salicylamide *N*-Mannich bases (**122**) were evaluated as prodrug forms for primary and secondary amines (Scheme 12.17). In this case, the amide part of a Mannich base is the promoiety. To improve their stability *in vitro* and avoid stability-associated formulation problems, the hydroxyl group of the salicylamide *N*-Mannich bases (**122**) can be blocked by O-acyloxymethylation. *O*-acyloxymethylated derivatives (**120**) were much more stable in acidic and neutral aqueous solutions than the parent salicylamide *N*-Mannich base (**122**) and could be readily converted to the latter in the presence of human plasma by enzymatic hydrolysis. In addition to providing an *in vitro* stabilizing effect, the concept of O-acyloxymethylation makes it possible to obtain prodrug derivatives of a given amine drug with varying physicochemical properties of importance for drug delivery, such as lipophilicity and water solubility. This can simply be effected by the selection of an appropriate α-acyloxymethyl group [74].

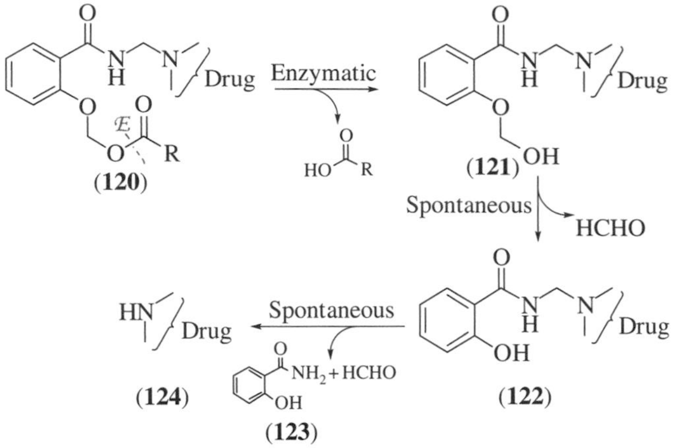

SCHEME 12.17 Enzyme-triggered activation of *O*-acyloxymethyl ether derivatives of salicylamide *N*-Mannich bases as prodrugs of amines.

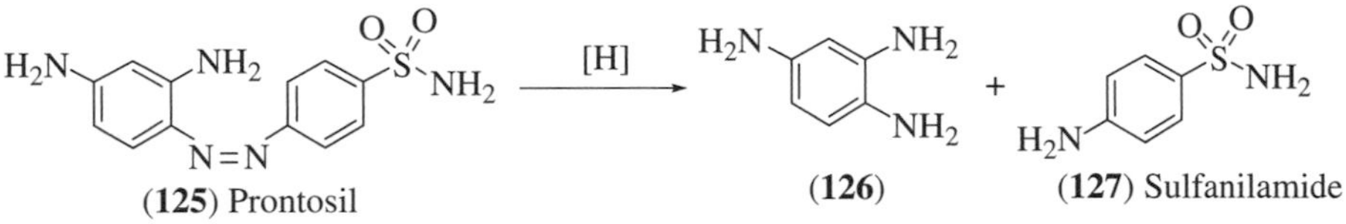

SCHEME 12.18 Activation of prontosil (**125**) to sulfanilamide (**127**).

12.4.3.5 Azo Prodrugs Amines have been incorporated into an azo linkage to form prodrugs that can be activated through azo reduction. In fact, sulfa drugs were discovered because of prontosil (**125**), an inactive azo dye that was converted *in vivo* to the active sulfanilamide (**127**) (Scheme 12.18). Clinically useful balsalazide (**25**), olsalazine (**27**), and sulfasalazine (**28**) are azo prodrugs of mesalazine (**29**). They are converted *in vivo* by bacterial azo reductases in the gut to the active 5-aminosalicylic acid (5-ASA or mesalazine, **29**), which is responsible for their anti-inflammatory activity in the treatment of ulcerative colitis, as discussed earlier.

12.4.3.6 Schiff Base Prodrugs Amines can form reversible Schiff bases with aldehydes and ketones. Although they are of limited use in small-molecule prodrugs, Schiff bases have been used to conjugate amine-containing drugs to polymers with carbonyl groups as macromolecular prodrugs for slow release and targeting. Doxorubicin was conjugated to polyethylene glycol (PEG) through a Schiff base linkage, and the resulting conjugate **128** was found to release doxorubicin under the lysosomal acidic conditions *in vitro* and very slowly under physiological conditions. Moreover, the conjugate showed strong cytotoxic activity similar to that of free doxorubicin against lymphocytic leukemia cells *in vitro* [75].

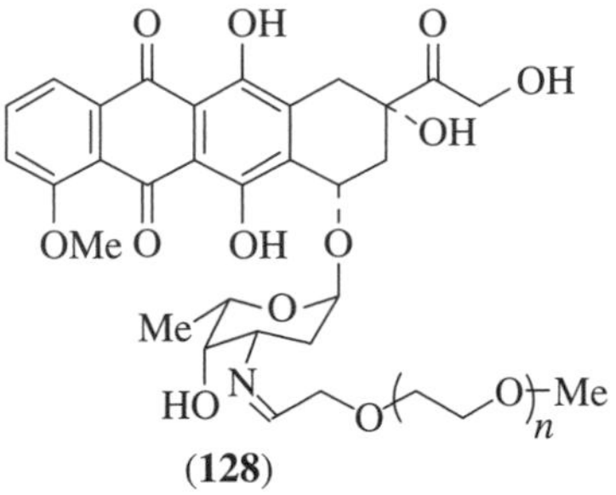

(**128**)

12.4.4 Prodrugs for Compounds Containing Carbonyl Groups

12.4.4.1 Schiff Bases and Oximes Schiff bases and oximes formed from ketones or aldehydes with amines or hydroxyl amines are chemically reversible under acidic or basic conditions. They could be used as prodrugs of compounds containing either an amine or carbonyl functionality.

Oximes of enones (**130**, **132**, and **133**) have been used as prodrugs of contraceptive norethindrone (**129**) and levonorgestrel (**131**). The oximes are highly bioavailable and are converted *in vivo* through chemical hydrolysis to their corresponding active drugs [76, 77].

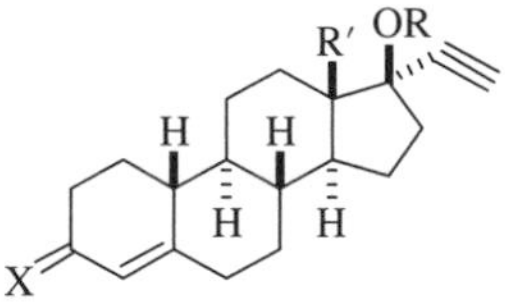

Compound	Substituents	ACD/Log*P*
(**129**) Norethindrone	R = H, R′ = Me, X = O	3.38 ± 0.35
(**130**) Norethindrone-3-oxime	R = H, R′ = Me, X = N—OH	3.87 ± 0.61
(**131**) Levonorgestrel	R = H, R′ = Et, X = O	3.92 ± 0.35
(**132**) Levonorgestrel-3-oxime	R = H, R′ = Et, X = N—OH	4.40 ± 0.61
(**133**) Norgestimate	R = Ac, R′ = Et, X = N—OH	5.00 ± 0.61

A more recent application of oxime derivatives as prodrugs is the design of cascade prodrugs of dopamine agonists for the treatment of Parkinson's disease. As shown in Scheme 12.19, enones such as *S*-(−)-6-(*N*,*N*-di-*n*-propylamino)-3,4,5,6,7,8-hexahydro-2*H*-naphthalen-1-one (**135**) can be oxidized *in vivo* to catecholamines such as (−)-5,6-dihydroxy-2-(*N*,*N*-di-*n*-propylamino)tetralin ((−)-5,6-diOH-DPAT, **136**) which are known as "mixed dopamine D_1/D_2 agonists" with potential utility in

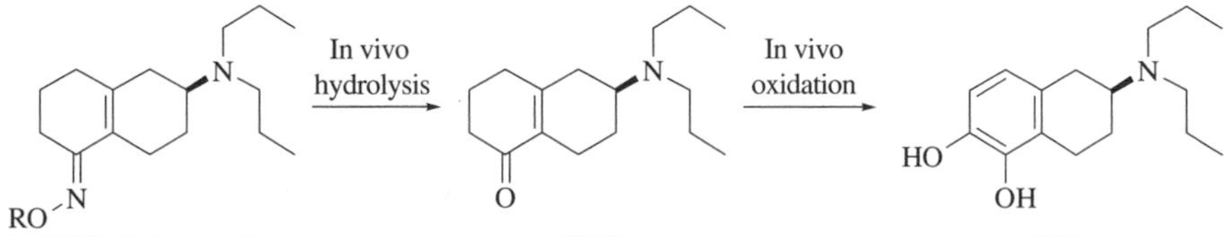

(**134**) Oxime prodrug → In vivo hydrolysis → (**135**) → In vivo oxidation → (**136**) (−)-5,6-diOH-DPAT

SCHEME 12.19 Proposed activation of an enone oxime prodrug to a catecholamine.

the treatment of Parkinson's disease [78]. Upon oral administration of catecholamines, the phenol and catechol moieties are rapidly metabolized to an extent that limits the therapeutic usefulness of these compounds. Enones such as **135** are prodrugs of such catecholamines and have been shown to improve their bioavailability and extend the duration of action. Compound **135** was found to be efficacious *in vivo* in models for Parkinson's disease in the rat. To potentially further increase the usefulness of enone **135**, a number of oxime ethers and oxime esters (**134**) were prepared as potential cascade prodrugs [79]. It was found that the unsubstituted oxime and the acetyl-oxime induced a pronounced and long-lasting effect *in vivo*. The oxime derivatives were readily hydrolyzed under acidic and alkaline conditions. The fact that these oximes as well as **135** were inactive at the dopamine receptor, yet induced dopamine D_1 and D_2 receptor-related effects *in vivo*, suggested that they were acting as prodrugs and were being converted *in vivo* to the active species (**136**).

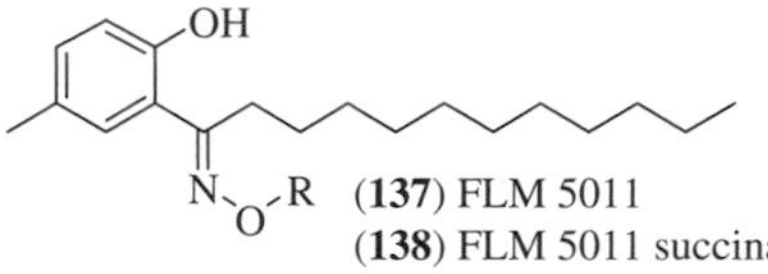

	R	ACD/Log*P*	ACD/Log*D* pH 7	ACD/Log*D* pH 1
(**137**) FLM 5011	R = H	7.71 ± 0.55	7.70	7.70
(**138**) FLM 5011 succinate	R = $COCH_2CH_2COOH$	7.71 ± 0.60	5.23	7.71

Oximes can be acylated to make prodrugs, as in the earlier example and in the case of FLM 5011 (**137**), which is a strongly lipophilic, poorly water-soluble, lipoxygenase inhibitor. The water solubility was improved by using the succinate monoester prodrug **138**. The bioavailability of FLM 5011 in rabbits after oral administration was markedly increased by its prodrug [80].

12.4.4.2 Enol Esters Enol esters are rather stable, bioreversible derivatives of ketones and may be useful as prodrugs of agents containing enolizable carbonyl groups. As shown in Scheme 12.20, 6′-acetylpapaverine enol esters (**139**), prepared by acylation of the appropriate Li enolate with the respective anhydride, were hydrolyzed to 6′-acetylpapaverine (**140**) by esterases present in rat and human plasma, rat liver, and brain tissue supernatants. The intermediate 6′-acetylpapaverine cyclizes rapidly to coralyne (**141**), which has antitumor activity but has difficulty passing through the blood–brain barrier due to the presence of the positive charge. 6′-Acetylpapaverine enol esters are neutral and stable in aqueous solution and are potential prodrugs for enhancing delivery of coralyne to brain tissues [81].

12.4.4.3 Oxazolidines The kinetics of hydrolysis of several oxazolidines derived from tris and various aldehydes and ketones was investigated to explore their suitability as prodrug forms for β-aminoalcohols such as (−) ephedrine (**142**) and for carbonyl-containing substances [82–84].

Oxazolidines were easily and completely hydrolyzed at pH 1–11 at 37°C. The hydrolysis rates were subject to general acid–base catalysis by buffer substances

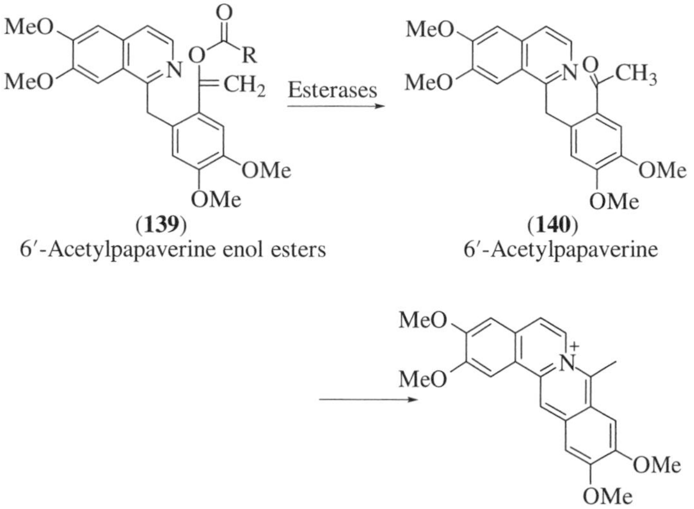

(**139**)
6′-Acetylpapaverine enol esters

(**140**)
6′-Acetylpapaverine

(**141**) Coralyne

SCHEME 12.20 Activation of an enol ester prodrug to form the positively charged coralyne in two steps.

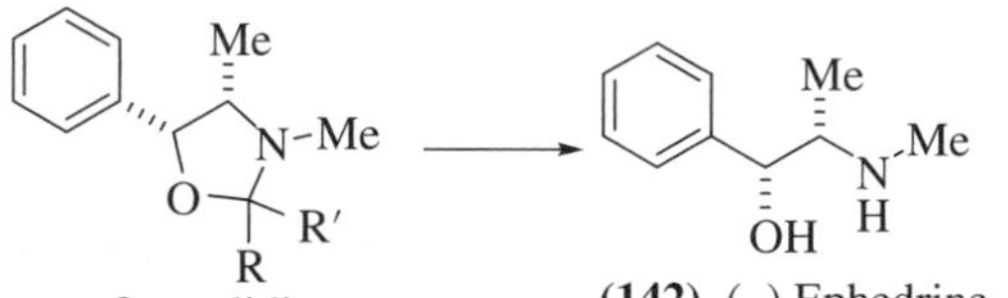

Oxazolidines
(**143**)

(**142**) (–) Ephedrine

R, R′	$t_{1/2}$ (second)
H, H	5
Et, H	18
Ph, H	300
2-HOPh, H	5
tBut, H	1800
Me, Me	280
Cyclohexylidene	360

SCHEME 12.21 Stability and chemical activation of oxazolidines at pH 7.4 and 37°C.

and depended strongly on pH. Most oxazolidines showed sigmoidal pH-rate profiles with maximum rates at pH 7–7.5. At pH 7.4 and 37°C, the half-lives of hydrolysis for the various ephedrine oxazolidines (**143**) ranged from 5 seconds to 30 minutes (Scheme 12.21). The reaction rates in neutral and basic solutions decreased with increasing steric effects of the substituents derived from the carbonyl component and decreased with increasing basicity of oxazolidines. Oxazolidines are weaker bases (pK_a 5.2–6.9) than the parent β-amino alcohol and are more lipophilic at physiological pH. Thus, oxalolidines can be considered as potentially useful prodrugs for drugs containing a β-amino alcohol moiety or carbonyl groups [82, 83]. Molecular complexation with cyclodextrins might be able to enhance the stability of oxazolidine prodrugs to make them potentially more useful [85, 86].

The stability characteristics of various *N*-acylated oxazolidines were also studied in an attempt to develop approaches, which may solve the stability problems associated with the use of oxazolidines as prodrug forms. The *N*-acylated oxazolidines, including a carbamate derivative, were in fact found to be highly stable in an aqueous solution, but they also proved to be resistant to hydrolysis by plasma enzymes. The latter limits the use of *N*-acylated oxazolidines in prodrug design [84].

12.4.4.4 Thiazolidines

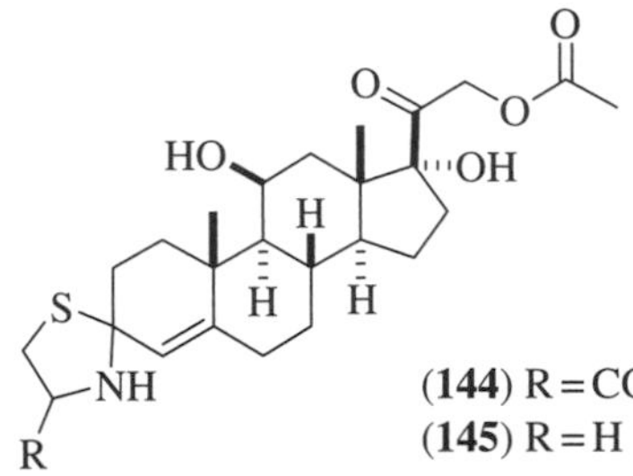

(**144**) R = CO_2Et
(**145**) R = H

Thiazolidines of some α,β-unsaturated 3-ketone steroids including progesterone, testosterone, and hydrocortisone (**144, 145**) were prepared from the reaction with cysteine alkyl esters and cysteamines as potential prodrugs [87, 88]. The thiazolidines readily reverted to their parent steroidal ketones, thus meeting the requirements for a prodrug. Most of the thiazolidines were more lipophilic than their parent steroids, thereby imparting the desired change in the physicochemical properties to the derivatives of the steroids. Thus, they can function as bioreversible derivatives of the parent steroids, cysteines, and cysteamines. Cysteine derivatives are particularly attractive as promoieties due to the release of cysteine as the by-product upon activation. Both cysteines and cysteamines were also used as chemoprotective agents against side effects of chemotherapy and radiation therapy. Thus, thiazolidines of cysteines and cysteamines could be used as prodrugs for chemoprotection during chemotherapy and radiation therapy.

12.5 DRUG RELEASE AND ACTIVATION MECHANISMS

Most prodrugs rely on enzymatic hydrolysis by esterases or proteases and, to a less extent, on chemical hydrolysis to achieve a one-step cleavage of the promoiety and the release of the original active drug (Fig. 12.1a(i)). These systems, as well as other one-step activation mechanisms (Fig. 12.1a(ii–iv)), are simple and, in many cases, sufficient in achieving useful regeneration rates of the active agent. Otherwise, a cascade release/activation mechanism can be incorporated by taking advantage of autodegradation or intramolecular cyclization reactions to effect the release and activation of a prodrug. Some of the cascade strategies that have been employed in the design of prodrugs are shown in Figure 12.1b and will be briefly discussed here.

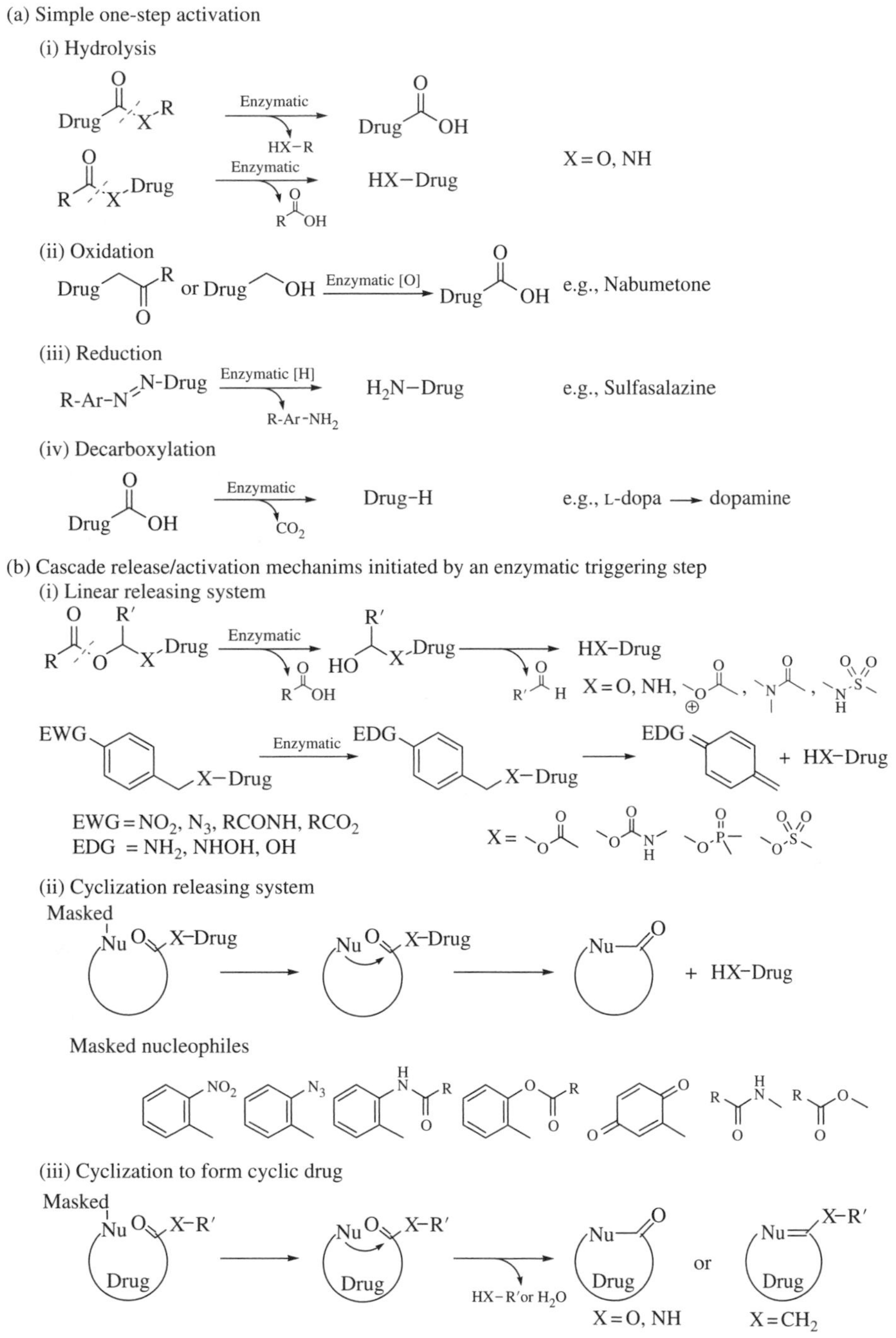

(a) Simple one-step activation

(i) Hydrolysis

Drug–C(=O)–X–R —Enzymatic, −HX−R→ Drug–C(=O)–OH

R–C(=O)–X–Drug —Enzymatic, −R–C(=O)–OH→ HX−Drug

X = O, NH

(ii) Oxidation

Drug–CH₂–C(=O)–R or Drug–CH₂–OH —Enzymatic [O]→ Drug–C(=O)–OH e.g., Nabumetone

(iii) Reduction

R-Ar–N=N-Drug —Enzymatic [H], −R-Ar-NH_2→ H_2N−Drug e.g., Sulfasalazine

(iv) Decarboxylation

Drug–C(=O)–OH —Enzymatic, −CO_2→ Drug-H e.g., L-dopa → dopamine

(b) Cascade release/activation mechanims initiated by an enzymatic triggering step

(i) Linear releasing system

R–C(=O)–O–CH(R′)–X-Drug —Enzymatic, −R–C(=O)–OH→ HO–CH(R′)–X-Drug —(−R′–CHO)→ HX-Drug

X = O, NH, (acyloxy, ⊕), (N-methyl amide), (N-H sulfonamide)

EWG–C_6H_4–CH_2–X−Drug —Enzymatic→ EDG–C_6H_4–CH_2–X−Drug → EDG=(quinone methide) + HX−Drug

EWG = NO_2, N_3, RCONH, RCO_2

EDG = NH_2, NHOH, OH

X = (ester), (carbamate, N-H), (phosphate), (sulfonate)

(ii) Cyclization releasing system

Masked Nu (ring)–C(=O)–X−Drug → Nu (ring)–C(=O)–X−Drug → Nu–C(=O) (ring) + HX−Drug

Masked nucleophiles

NO_2, N_3, NH–C(=O)R, O–C(=O)R, quinone (O, O), R–C(=O)–NH–, R–C(=O)–O–

(iii) Cyclization to form cyclic drug

Masked Nu (Drug ring)–C(=O)–X−R′ → Nu (Drug ring)–C(=O)–X−R′ —(−HX−R′ or H_2O)→ Nu–C(=O) (Drug ring) or Nu=C–X−R′ (Drug ring)

X = O, NH X = CH_2

FIGURE 12.1 Examples of drug-release/activation mechanisms.

12.5.1 Cascade Release Facilitated by Linear Autodegradation Reactions

A number of examples are known for the release of an active drug facilitated by a linear autodegradation process. This can be achieved through chemically unstable intermediates such as α-hydroxy amines, amides, and esters. Many of the double prodrugs discussed earlier belong to this category.

Another interesting linear autodegradation process often used in prodrug design involves an electron "push and pull" mechanism through a conjugative aromatic ring, which is linked to a good leaving group such as an ester in the *para* benzylic position. In such an approach, an electron-donating amino or hydroxy group is masked as an electron-withdrawing nitro group or an amide or ester group in the prodrug. Upon unmasking via reduction or hydrolysis, the resulting electron-donating group will be able to push electrons through the conjugative system to the *para* position, leading to the cleavage of the benzylic carbon–oxygen bond. The rate of this cleavage is not enzyme dependent, but rather relies on the electron-pushing capability of the unmasked electron-donating group and the electron-pulling ability of the leaving group. The formation of the negatively charged species can serve as the activation mechanism of the drug. This approach has recently been used in our effects to develop a novel and superior class of nitroaryl phosphoramides as potential prodrugs for nitroreductase-mediated enzyme prodrug therapy [37].

Several nitroaryl phosphoramides were designed and synthesized, each with a strategically placed nitro group on the benzene ring in the *para* position to the benzylic carbon (Scheme 12.22). Compound **146** is a cyclophosphamide analog with the cyclophosphamide ring fused with a benzene ring and a nitro group placed in the *para* position to the benzylic carbon. Compound **49** is a 4-nitrophenyl substituted cyclophosphamide analog, and **48** is an acyclic nitrobenzyl phosphoramide mustard (LH7). The nitro group is a strong electron-withdrawing group (Hammett σ_p electronic parameter = 0.78) and is converted to an electron-donating hydroxylamino group ($\sigma_p = -0.34$) upon nitroreductase reduction. This large difference in electronic effect ($\Delta\sigma_p = 1.12$) is exploited to effect the formation of the highly cytotoxic phosphoramide mustard or like reactive species. After reduction by nitroreductase (NTR), the resulting hydroxylamines **147**, **149**, and **151** relay their electrons to the *para* position and facilitate the cleavage of the benzylic C—O bond, producing the anionic cytotoxic species phosphoramide mustard **153** or like reactive species **148** and **150**. Structurally, the phosphoramide portion in **148** and **150** closely resembles phosphoramide mustard **153**, the reactive alkylating agent produced following the metabolic activation of cyclophosphamide in the liver, and could also be the ultimate cytotoxic alkylating agent.

Phosphoramide mustard is the proven cytotoxic metabolite of cyclophosphamide, a successful clinical anticancer prodrug that requires cytochrome P450 activation in the liver. These nitroaryl phosphoramides in combination with nitroreductase could effectively move the site of activation from the liver in the case of cyclophosphamide into nitroreductase-expressing tumor tissues. All compounds were shown to be excellent substrates of *E. coli* nitroreductase, but with varying degrees of cytotoxicity against nitroreductase-expressing V79 and SKOV3 cells. Compounds *cis*-**49** and

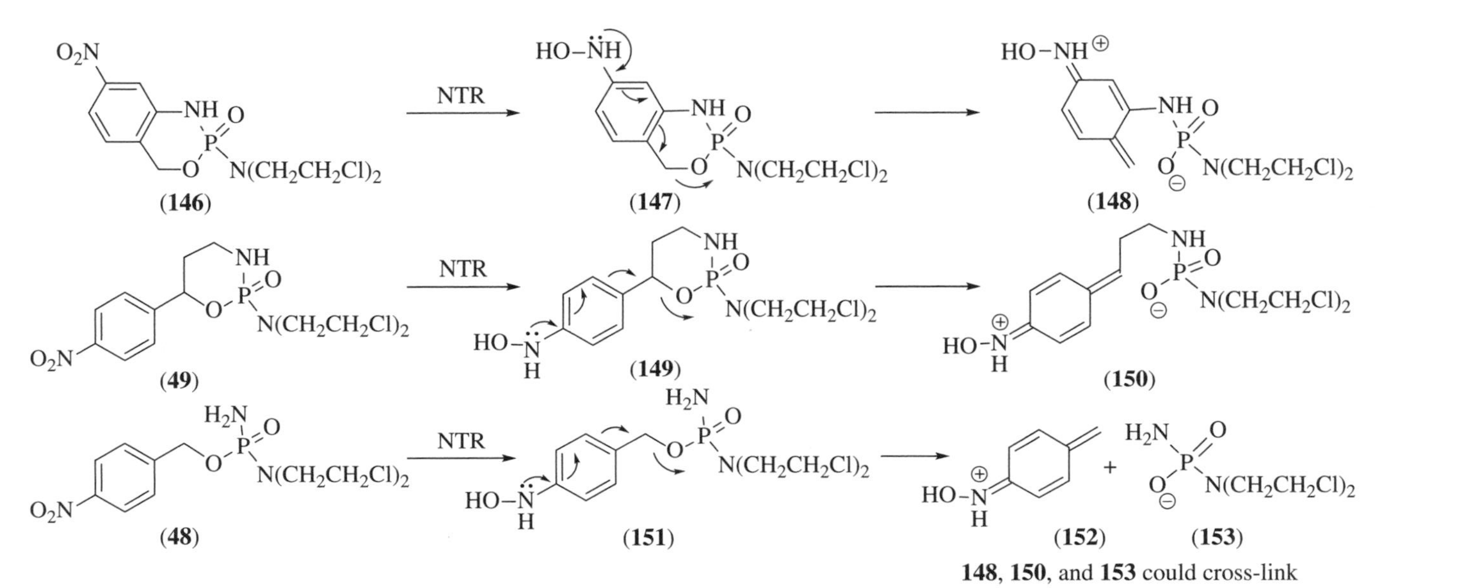

SCHEME 12.22 Reductive activation of cyclic and acyclic nitroaryl phosphoramides by nitroreductases (NTR) followed by 1,6-elimination.

trans-**49**, the best of the cyclic series, were over 22,000× more cytotoxic in nitroreductase-expressing Chinese hamster V79 cells and **48**, the acyclic compound LH7, was 167,500× more cytotoxic in the same cell line, with an IC_{50} as low as 0.4 nM upon 72-hour drug exposure. This level of activity is about 100× more active and 27× more selective than CB1954. Even when the V79 cells were exposed to each test compound for 1 hour before the media were replaced with nondrug–containing fresh media, the IC_{50} was 10 nM, which was about 30× lower than that of CB1954 (**33**). The high selectivity of *cis*-**49**, *trans*-**49**, and **48** was reproduced in SKOV3 human ovarian carcinoma cells infected with an adenovirus expressing *E. coli* nitroreductase. Enzyme kinetic analysis indicates that compound **48** was a much better substrate of *E. coli* nitroreductase with a specificity constant 20× that of CB1954.

12.5.2 Cascade Release Facilitated by Intramolecular Cyclization Reactions

Intramolecular reactions are usually thermodynamically favored over intermolecular reactions because they have lower activation energy and therefore more stable transition states. The lower activation energy is attributed to a better entropic situation. When a reaction is performed between two different reactants, the two molecules need to collide in a specific orientation and the reaction leads to a decrease in the number of molecules, resulting in an increase in order and therefore a loss of entropic energy. In an intramolecular reaction, the two reaction centers, for example, a nucleophile and an electrophile, are both present within the same molecule and are in a good position to interact and form the cyclic product. The positioning of the two reaction centers within the same molecule is very important for the reaction to take place. Steric factors resulting in less flexible molecules and better positioning of the two reaction centers would lead to increased reaction rates. Generally speaking, intramolecular reactions leading to the formation of 5- or 6-membered rings are much more favorable and occur at faster rates.

A series of alkylaminoalkyl carbamates of 4-hydroxyanisole (**154**) were evaluated as prodrugs of the melanocytotoxic phenol (**155**) that could be activated through intramolecular cyclization (Scheme 12.23) [89]. The carbamates were relatively stable at low pH but released 4-hydroxyanisole cleanly in a nonenzymatic fashion at a pH 7.4 at rates that were structure dependent. A detailed study of the *N*-methyl-*N*-[2-(methylamino)ethyl]carbamate showed that generation of the parent phenol followed first-order kinetics with $t_{1/2}$ = 36.3 minutes at pH 7.4, 37°C and was

MeO–C₆H₄–O–C(=O)–N(R)–$(CH_2)_n$NHR′ (**154**) —pH 7.5, 37°C→ MeO–C₆H₄–OH (**155**) + R–N–C(=O)–N–R′ cyclic urea with $(CH_2)_n$ (**156**)

n	R, R′	$t_{1/2}$(minute)
2	Me, Me	36.3
2	Et, Et	118
2	Me, H	304
2	H, Me	335
2	H, H	724
3	H, H	910
3	Me, Me	942

SCHEME 12.23 Activation of alkylaminoalkyl carbamates of phenolic drugs via cyclization.

accompanied by the formation of *N*,*N*-dimethylimidazolidinone (**156**). In comparison, the related derivative with three methylene units between the two N atoms releases the phenol at a much slower rate with $t_{1/2}$ = 942 minutes. These basic carbamates are examples of cyclization-activated prodrugs in which generation of the active drug is not linked to enzymatic cleavage but rather depends solely upon a predictable, intramolecular cyclization–elimination reaction.

The terminal amino group in the earlier system could be masked as an amide, thus avoiding the stability problem encountered when using the basic carbamates as prodrugs. Unmasking of the amide could be catalyzed by a specific protease *in vivo* at certain target sites, thus achieving target specificity. Another alternative is to use *o*-nitroaromatic, as in compounds **157** (Scheme 12.24), which could be converted to a nucleophilic aromatic amine upon bioreduction in hypoxic tumor tissues or by other reductases delivered to targeted cells [90–92]. Kinetic analysis of the cyclization activation process indicates that the addition of two methyl groups α to the ester carbonyl would restrict the rotational freedom of the ground-state molecule and promote the cyclization reaction. The nitro group can be reduced *in vivo* to a nucleophilic aromatic amine with a low pK_a (<4), which would be present in neutral nucleophilic form under most physiological conditions. At pH 7.4, 37°C, the amines **158** cyclized quickly to the lactam **159**, releasing the anticancer drug floxuridine (FUDR) (**160**) [90]. For tumor targeting purposes, subsequent drug release after initial specific enzymatic activation should be very fast (preferably <1 minutes) in order to prevent the active drug from escaping the targeted site.

Reduction of quinone propionic esters or amides **161** bearing three Me groups in the so-called trialkyl lock positions (*o*-, β-, β-positions) or hydrolysis of the corresponding phenolic esters **162** has been shown to undergo spontaneous lactonization with the release of alcohol or amine, respectively (Scheme 12.25) [93–95]. Several amides **161** were synthesized and tested as model redox-sensitive cascade prodrugs of amines. The reduction of model amide prodrugs (**161**) generated hydroxy amide intermediates **163**, the lactonization of which resulted in amine release. The half-lives for appearance of the product lactone **164** from these intermediates ranged from 1.4 to 3.4 minutes at pH 7.4 and 37°C. With such rapid lactonization rates, it is believed that reduction would be the rate-limiting step in the two-step conversion of the prodrugs to amines [95]. Comparison of the observed rates of lactonization at pH 7.5 and 30°**C** for three hydroxy amides obtained from the hydrolysis of phenolic ester prodrugs (**162**) allowed an estimate of the extent of rate enhancement provided

(157) → [H] → (158) → pH 7.4, 37°C → (159) + FUDR (160)

n	R	$t_{1/2}$ (minute)
0	H	14
0	Me	~0
1	H	4
1	Me	2

SCHEME 12.24 Two-step activation of *o*-nitrophenylalkanoate prodrugs via reduction and cyclization.

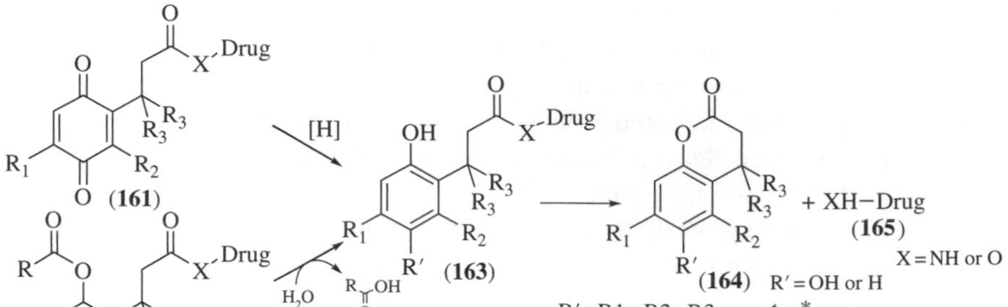

(161) (162) [H] H_2O (163) (164) R′ = OH or H + XH–Drug (165) X = NH or O

R′	R1	R2	R3	$t_{1/2}$*
H	Me	Me	Me	65.4 seconds
H	H	H	Me	10.5 hours
H	H	H	H	19.4 days

* with p-MePhNH$_2$ as a model amine, at pH 7.5, 30°C

SCHEME 12.25 Two-step activation of quinone propionic ester and amide prodrugs via reduction and cyclization.

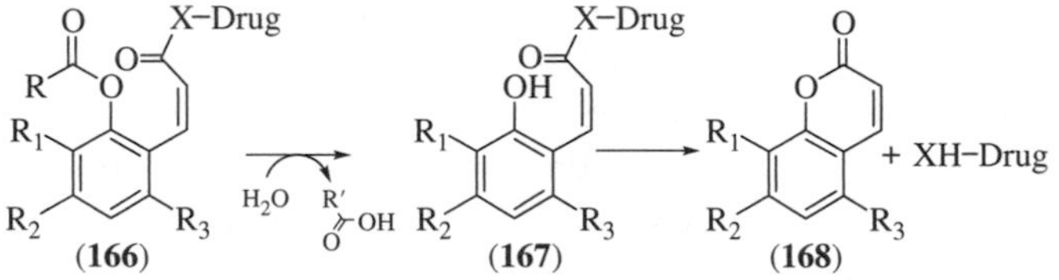

(166) H_2O (167) (168) + XH–Drug

R_1	R_2	R_3	$t_{1/2}$(minute)*
Me	H	Me	4.8
Me	Me	H	6.3
H	H	H	32.5

* with PhCH$_2$NHMe as a model amine, at pH 7.4, 37°C

SCHEME 12.26 Two-step activation of a coumarin-based prodrug system.

by the addition of a partial or total trimethyl lock for the hydroxy amide lactonization reaction under near-physiological conditions [94]. The half-life for the hydroxy amide with a full trimethyl lock was 65 seconds, a rate enhancement of 2.54×10^4 as compared to the corresponding hydroxy amide without the three methyl groups.

Still another intramolecular cyclization system is the coumarin-based prodrug system **166** that can be used for bioreversible derivatization of amine and alcohol drugs and the preparation of cyclic peptide prodrugs (Scheme 12.26) [96, 97]. This system takes advantage of the known facile lactonization of coumarinic acid and its derivatives. Such a system can be used for the development of esterase- and phosphatase-sensitive prodrugs of amines and alcohols. Esterase-sensitive prodrugs of a number of model amines prepared by using this system readily released the amines upon incubation in the presence of porcine liver esterase, with $t_{1/2}$ values ranging from 100 seconds to 35 minutes [97].

12.5.3 Cascade Activation through Intramolecular Cyclization to Form Cyclic Drugs

Pilocarpine (**171**) is used as a topical miotic for controlling elevated intraocular pressure associated with glaucoma. The drug presents significant delivery problems due to its low ocular bioavailability (1–3% or less) and its short duration of action. The poor bioavailability was partly attributed to its poor permeability across the

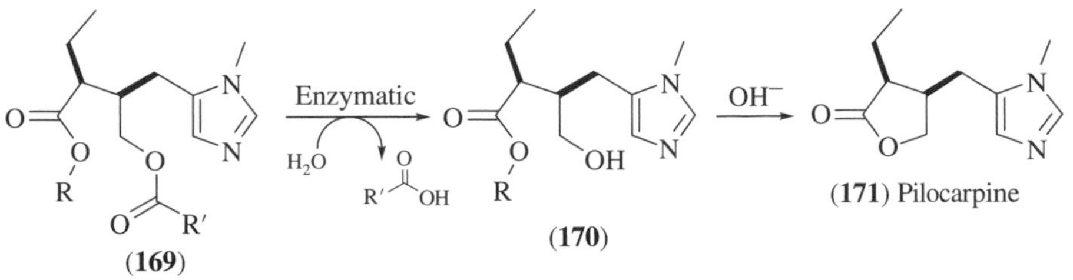

SCHEME 12.27 Activation of a pilocarpine prodrug via esterase-mediated hydrolysis and cyclization.

corneal membrane due to its low lipophilicity. Because of the low bioavailability, a large ophthalmic dose is required to enable an effective amount of pilocarpine to reach the inner eye receptors and reduce the intraocular pressure over a suitable duration. This in turn gives rise to concerns about systemic toxicity, since most of the applied drug is then available for systemic absorption from the nasolacrimal duct. The systemic absorption of pilocarpine may lead to undesired side effects, for example, in those patients who display sensitivity to cholinergic agents. Upon instillation of pilocarpine into the eye, the intraocular pressure is reduced for only about 3 hours. As a consequence, the frequency of administration is at an inconvenient 3–6 times per day.

To improve the ocular bioavailability and prolong the duration of action, various pilocarpic acid mono- and diesters were evaluated as prodrugs for pilocarpine. As shown in Scheme 12.27, the pilocarpic acid monoesters (**170**) undergo a quantitative cyclization to pilocarpine (**171**) in aqueous solution, the rate of cyclization being a function of the polar and steric effects within the alcohol portion of the esters. At pH 7.4 and 37°C, half-lives ranging from 30 to 1105 minutes were observed for the various esters. A main drawback of these monoesters is their poor solution stability, but this problem was overcome by esterification of the free hydroxy group. A number of pilocarpic diesters (**169**) were highly stable in aqueous solution (shelf lives were estimated to be more than 5 years at 20°C) and, most significantly, were readily converted to pilocarpine under conditions simulating those occurring *in vivo* through a cascade process involving rapid enzymatic hydrolysis of the *O*-acyl bond followed by spontaneous lactonization of the intermediate pilocarpic acid monoester (**170**). Both the pilocarpic acid monoesters and, in particular, diesters enhanced the ocular bioavailability of pilocarpine and significantly prolonged the duration of its activity following topical instillation, as determined by a miosis study in rabbits [98].

Derivatives of *N*-alkylbenzophenones and peptidoaminobenzophenones undergo hydrolysis, with the subsequent intramolecular condensation that results in the formation of the 1,4-benzodiazepine hypnotics. A number of such compounds, for example, rilmazafone (**172**), were suggested to have more beneficial pharmacological properties in comparison to standard benzodiazepines. Rilmazafone is a ring-opened derivative of 1,4-benzodiazepine (Scheme 12.28) and was developed in Japan as an orally active sleep inducer. Rilmazafone is exclusively metabolized by aminopeptidases in the small intestine to the labile desglycylated metabolite **173** and then to its cyclic form **174**. The concentration of **174** in the systemic plasma

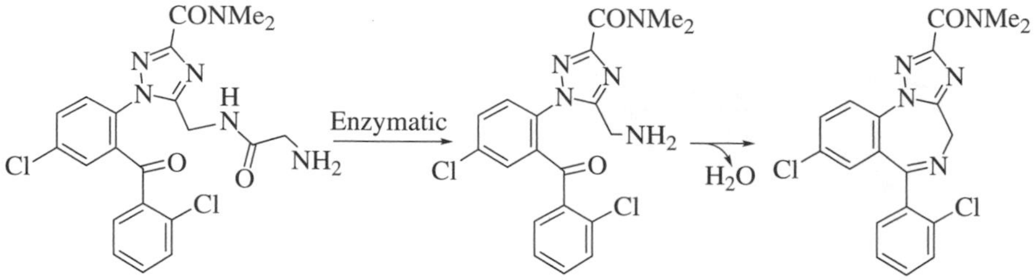

(**172**) Rilmazafone (**173**) (**174**)

SCHEME 12.28 Two-step activation of rilmazafone (**172**) via enzyme-mediated hydrolysis and cyclization.

(i.e., bioavailability) after oral administration of rilmazafone has been reported to be higher than that observed after administration of **174** due to the lower hepatic extraction of **173** than **174** [99].

12.6 PRODRUGS AND INTELLECTUAL PROPERTY RIGHTS—TWO COURT CASES

The primary purpose for developing prodrugs is, of course, not to circumvent intellectual property rights but to obviate certain disadvantages that may have precluded an active agent from being used in clinical applications. Therefore, if undesirable properties of a drug molecule cannot be overcome by conventional changes in the pharmaceutical formulation or route of administration, the method of choice is to use one of the prodrug approaches discussed earlier. The parent drug usually came first and was followed by the prodrug. The prodrug is inactive or much less active, and it was the parent drug that would ultimately act in the human body.

If a prodrug is sufficiently distinct from the parent drug and if it possesses unexpected improved properties over the parent drug, the prodrug can be patented. Chapter 8 focuses on the intellectual property issues related to drug delivery. Here only a brief discussion involving prodrugs is presented. In many cases, the patent will be granted to the same inventor or company that developed the parent drug. Even if a third party is granted a patent on the prodrug, the prodrug patent would, of course, not prevent the user of the parent drug from continuing to use it. However, an interesting question arises when a third party decides to manufacture, use, import, or sell the prodrug form of a patented drug: whether the owner of a patent covering the parent drug can object to the use of the prodrug by third parties (irrespective of whether the prodrug has been patented). The answer to this question may have to be determined on a case-by-case basis in the courts. The following two cases, though they do not fully address this question, do show the potential legal ramifications.

The first case relates to a decision by the British House of Lords on the previously discussed hetacillin (**89**) [100]. The question was whether the British patent covering the antibiotic ampicillin was infringed by the importation and use of the antibiotic hetacillin in England. The House of Lords ruled that the prodrug hetacillin infringed

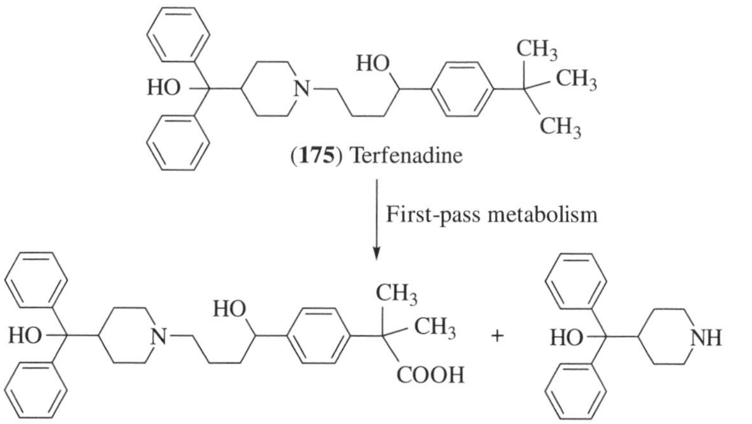

(**175**) Terfenadine

First-pass metabolism

(**176**) Fexofenadine

(**177**) Inactive metabolite

SCHEME 12.29 First-pass metabolism of terfenadine (**175**) to fexofenadine (**176**).

the patent pertaining to ampicillin. It reasoned that when the prodrug came into contact with water in the gullet, it underwent a chemical reaction and became the active substance ampicillin. The court believed that there was no therapeutic or other added value associated with the use of hetacillin. It was decided that this was an infringement, despite the existence of more than insignificant structural diversity between the claims and the infringing product. It was important to note that the court believed the prodrug did not add any value to the known invention [100].

The second case relates to the relationship between terfenadine (**175**) and its active metabolite, now known as fexofenadine (or Allegra, **176**) (Scheme 12.29). As a now discontinued "second-generation" antihistamine, terfenadine itself was active and developed as a histamine H_1 receptor antagonist; thus, it was not a prodrug in the strict sense. But, this court case does have ramifications for prodrug design. Terfenadine was successfully marketed for a long period of time, without knowledge that terfenadine was, in fact, acting *in vivo* through its active metabolite, fexofenadine. On the basis of a mass balance study using ^{14}C-labeled terfenadine, oral absorption of terfenadine was estimated to be at least 70%. However, terfenadine is not normally detectable in plasma at levels greater than 10 ng/ml; it undergoes extensive (99%) first-pass metabolism to two primary metabolites, an active acid metabolite, fexofenadine, and an inactive dealkylated metabolite. This led some to refer to terfenadine as a "prodrug" of fexofenadine. While the drug-active metabolite relationship was unknown, both to the patent owners and to the public, a new patent application was filed 7 years later covering the compound fexofenadine (formerly known as "MDL 16,455") [101, 102]. Therefore, the owner of both patents and the public realized only later that the compound fexofenadine was in fact the active drug all along.

After expiration of the terfenadine patent, the owner of the metabolite patent believed that the marketing of terfenadine-containing products infringed upon the substance and use claims of the later-filed metabolite patent. This led the owner of the metabolite patent to file infringement lawsuit in Germany, the United Kingdom, and the United States against generic companies that had launched terfenadine-containing drug products. In these cases, it was the alleged infringers, not the plaintiffs that received sympathy from the courts. The infringement lawsuits were regarded as attempts to extend the monopoly of a lapsed patent.

It is clear from the earlier discussion that prodrugs can be patented if they are designed to add value to, not to circumvent, a known invention, are sufficiently distinct from the parent drug, and possess unexpected improved properties compared to the parent drug. But one should bear in mind the potential legal ramifications arising from working on prodrugs of the patented inventions of others.

REFERENCES

1. Bundgaard, H. *Drug Future* 1991, **16**, 443–458.
2. Balant, L. P.; Doelker, E. Metabolic considerations in prodrug design. In *Burger's Medicinal Chemistry and Drug Discovery*, *Vol.* **1** (Ed. Wolff, M. E.), John Wiley & Sons, New York, 1995, pp. 949–982.
3. Larsen, C. S.; Ostergaard, J. Design and application of prodrugs. In *Textbook of Drug Design and Discovery*, 3rd ed. (Eds. Krogsgaard-Larsen, P.; Liljefors, T.; Madsen, U.), Taylor & Francis, London and New York, 2002, pp. 410–458.
4. Hills, C. E.; Brunskill, N. J. *Rev. Diabet. Stud.* 2009, **6**, 138–147.
5. Lipinski, C. A.; Lombardo, F.; Dominy, B. W.; Feeney, P. J. *Adv. Drug Deliv. Rev.* 1997, **23**, 3–25.
6. Ensink, J. M.; Vulto, A. G.; van Miert, A. S.; Tukker, J. J.; Winkel, M. B.; Fluitman, M. A. *Am. J. Vet. Res.* 1996, **57**, 1021–1024.
7. Friedman, D. I.; Amidon, G. L. *Pharm. Res.* 1989, **6**, 1043–1047.
8. Ribeiro, W.; Muscara, M. N.; Martins, A. R.; Moreno, H. J.; Mendes, G. B.; de Nucci, G. *Eur. J. Clin. Pharmacol.* 1996, **50**, 399–405.
9. Todd, P. A.; Goal, K. L. *Drugs* 1992, **43**, 346–383.
10. Menard, J.; Patchett, A. A. Angiotensin-converting enzyme inhibitors. In *Drug Discovery and Design*, *Vol.* **56** (Ed. Scolnick, E. M.), Academic Press, London, Boston, and New York, 2001, pp. 13–75.
11. Beauchamp, L. M.; Orr, G. F.; De Miranda, P.; Burnette, T.; Krenitsky, T. A. *Antivir. Chem. Chemother.* 1992, **3**, 157–164.
12. Crooks, R. J.; Murray, A. *Antivir. Chem. Chemother.* 1994, **5** (suppl.), 31–37.
13. Boyd, M. R.; Safrin, S.; Kern, E. R. *Antivir. Chem. Chemother.* 1993, **4** (suppl.), 3–11.
14. Shinkai, I.; Ohta, Y. *Bioorg. Med. Chem.* 1996, **4**, 1–2.
15. Pue, M. A.; Pratt, S. K.; Fairless, A. J.; Fowles, S.; Laroche, J.; Georgiou, P.; Prince, W. *J. Antimicrob. Chemother.* 1994, **33**, 119–127.
16. Cirelli, R.; Herne, K.; McCrary, M.; Lee, P.; Tyring, S. K. *Antiviral Res.* 1996, **29**, 141–151.
17. Han, H. K.; Oh, D. M.; Amidon, G. L. *Pharm. Res.* 1998, **15**, 1382–1386.
18. Kane, J. M. *J. Clin. Psychiatry* 1984, **45**, 5–12.
19. Pennick, M. *Neuropsychiatr. Dis. Treat.* 2010, **6**, 317–327.

20. Tammara, V. K.; Narurkar, M. M.; Crider, A. M.; Khan, M. A. *Pharm. Res.* 1993, **10**, 1191–1199.
21. Waller, D. G.; George, C. F. *Br. J. Clin. Pharmacol.* 1989, **28**, 497–507.
22. Sinha, V. R.; Kumria, R. *Pharm. Res.* 2001, **18**, 557–564.
23. Yang, L. B.; Chu, J. S.; Fix, J. A. *Int. J. Pharm.* 2002, **235**, 1–15.
24. Chourasia, M. K.; Jain, S. K. *J. Pharm. Pharm. Sci.* 2003, **6**, 33–66.
25. Repta, A. J.; Rawson, B. J.; Shaffer, R. D.; Sloan, K. B.; Bodor, N.; Higuchi, T. *J. Pharm. Sci.* 1975, **64**, 392–396.
26. Kennedy, K. A.; Teicher, B. A.; Rockwell, S.; Sartorelli, A. C. *Biochem. Pharmacol.* 1980, **29**, 1–8.
27. Siim, B. G.; Atwell, G. J.; Anderson, R. F.; Wardman, P.; Pullen, S. M.; Wilson, W. R.; Denny, W. A. *J. Med. Chem.* 1997, **40**, 1381–1390.
28. Wilson, W. R. Tumor hypoxia: challenges for cancer chemotherapy. In *Cancer Biology and Medicine, Vol.* **3** (Eds. Waring, M. J., Ponder, B. A. J.), Kluwer Academic Publishers, Lancaster, 1992, pp. 87–131.
29. Lu, G.-L.; Ashoorzadeh, A.; Anderson, R. F.; Patterson, A. V.; Smaill, J. B. *Tetrahedron* 2013, **69**, 9130–9138.
30. Rauth, A. M.; Goldberg, Z.; Misra, V. *Oncol. Res.* 1997, **9**, 339–349.
31. Ast, G. *Curr. Pharm. Des.* 2003, **9**, 455–466.
32. Eisenbrand, G.; Lauck-Birkel, S.; Tang, W. C. *Synthesis* 1996, 1246–1258.
33. Ray, J. M.; Stetler-Stevenson, W. G. *Eur. Respir. J.* 1994, **7**, 2062–2072.
34. DeFeo-Jones, D.; Garsky, V. M.; Wong, B. K.; Feng, D. M.; Bolyar, T.; Haskell, K.; Kiefer, D. M.; Leander, K.; McAvoy, E.; Lumma, P.; Wai, J.; Senderak, E. T.; Motzel, S. L.; Keenan, K.; Van Zwieten, M.; Lin, J. H.; Freidinger, R.; Huff, J.; Oliff, A.; Jones, R. E. *Nat. Med.* 2000, **6**, 1248–1252.
35. Jiang, Y.; Hu, L. *Bioorg. Med. Chem.* 2013, **21**, 7507–7514.
36. Melton, R. G.; Knox, R. J. in *Enzyme-prodrug strategies for cancer therapy*, Kluwer Academic/Plenum, New York, 1999, p. 272.
37. Hu, L. Q.; Yu, C. Z.; Jiang, Y. Y.; Han, J. Y.; Li, Z. R.; Browne, P.; Race, P. R.; Knox, R. J.; Searle, P. F.; Hyde, E. I. *J. Med. Chem.* 2003, **46**, 4818–4821.
38. Jaberipour, M.; Vass, S. O.; Guise, C. P.; Grove, J. I.; Knox, R. J.; Hu, L.; Hyde, E. I.; Searle, P. F. *Biochem. Pharmacol.* 2010, **79**, 102–111.
39. Leamon, C. P.; Low, P. S. *Drug Discov. Today* 2001, **6**, 44–51.
40. Vickers, S.; Duncan, C. A.; White S, D.; Breault, G. O.; Royds, R. B.; de Schepper, P. J.; Tempero, K. F. *Drug Metab. Dispos.* 1978, **6**, 640–646.
41. Huang, H. P.; Ayres, J. W. *J. Pharm. Sci.* 1988, **77**, 104–109.
42. Cai, Z.-W.; Zhang, Y.; Borzilleri, R. M.; Qian, L.; Barbosa, S.; Wei, D.; Zheng, X.; Wu, L.; Fan, J.; Shi, Z.; Wautlet, B. S.; Mortillo, S.; Jeyaseelan, R.; Kukral, D. W.; Kamath, A.; Marathe, P.; D'Arienzo, C.; Derbin, G.; Barrish, J. C.; Robl, J. A.; Hunt, J. T.; Lombardo, L. J.; Fargnoli, J.; Bhide, R. S. *J. Med. Chem.* 2008, **51**, 1976–1980.
43. Huynh, H.; Ngo, V. C.; Fargnoli, J.; Ayers, M.; Soo, K. C.; Koong, H. N.; Thng, C. H.; Ong, H. S.; Chung, A.; Chow, P.; Pollock, P.; Byron, S.; Tran, E. *Clin. Cancer Res.* 2008, **14**, 6146–6153.

44. Navar, L. G.; Kobori, H.; Prieto, M. C.; Gonzalez-Villalobos, R. A. *Hypertension* 2011, **57**, 355–362.
45. Ishizuka, T.; Fujimori, I.; Kato, M.; Noji-Sakikawa, C.; Saito, M.; Yoshigae, Y.; Kubota, K.; Kurihara, A.; Izumi, T.; Ikeda, T.; Okazaki, O. *J. Biol. Chem.* 2010, **285**, 11892–11902.
46. Lee, B.; Kang, M.; Choi, W.; Choi, Y.; Kim, H.; Lee, S.; Lee, J.; Choi, Y. *Arch. Pharm. Res.* 2009, **32**, 1629–1635.
47. Angeli, F.; Verdecchia, P.; Pascucci, C.; Poltronieri, C.; Reboldi, G. *Expert Opin. Drug Metab. Toxicol.* 2013, **9**, 379–385.
48. Pergolizzi, J. V.; Gan, T. J.; Plavin, S.; Labhsetwar, S.; Taylor, R. *Anesthesiol. Res. Pract.* 2011, **2011**, 458920.
49. Easthope, S. E.; Jarvis, B. *Drugs* 2002, **62**, 1253–1287.
50. De Clercq, E.; Holý, A. *Nat. Rev. Drug Discov.* 2005, **4**, 928–940.
51. Brass, E. P. *Pharmacol. Rev.* 2002, **54**, 589–598.
52. Mangan, F. R.; Flack, J. D.; Jackson, D. *Am. J. Med.* 1987, **83**, 6–10.
53. Bundgaard, H.; Johansen, M. *Int. J. Pharm.* 1981, **9**, 7–16.
54. Vej-Hansen, B.; Bundgaard, H. *Arch. Pharm. Chem. Sci. Ed.* 1979, **7**, 341–353.
55. Durbin, A. K.; Rydon, H. N. *Chem. Commun.* 1970, 1249–1250.
56. Schwartz, M. A.; Hayton, W. L. *J. Pharm. Sci.* 1972, **61**, 906–909.
57. Bundgaard, H.; Nielsen, N. M. *Acta Pharm. Suec.* 1987, **24**, 233–246.
58. Bundgaard, H.; Johansen, M. *Int. J. Pharm.* 1984, **22**, 45–56.
59. Ruhlmann, C. H.; Herrstedt, J. *Expert Rev. Anticancer Ther.* 2012, **12**, 139–150.
60. Lasseter, K. C.; Gambale, J.; Fin, B.; Bergman, A.; Constanzer, M.; Dru, J.; Han, T. H.; Majumdar, A.; Evans, J. K.; Murphy, M. G. *J. Clin. Pharmacol.* 2007, **47**, 834–840.
61. Dhareshwar, S. S.; Stella, V. J. *J. Pharm. Sci.* 2008, **97**, 4184–4193.
62. Biel, J. H.; Somani, P.; Jones, P. H.; Minard, F. N.; Goldberg, L. I. *Biochem. Pharmacol.* 1974, **23**, 748–750.
63. Yoshikawa, M.; Nishiyama, S.; Takaiti, O. *Hypertens. Res.* 1995, **18**, S211–S213.
64. Cruz, D. N. *Expert Opin. Pharmacother.* 2000, **1**, 835–840.
65. McClellan, K. J.; Wiseman, L. R.; Wilde, M. I. *Drugs Aging* 1998, **12**, 76–86.
66. Walko, C. M.; Lindley, C. *Clin. Ther.* 2005, **27**, 23–44.
67. Eisert, W. G.; Hauel, N.; Stangier, J.; Wienen, W.; Clemens, A.; van Ryn, J. *Arterioscler. Thromb. Vasc. Biol.* 2010, **30**, 1885–1889.
68. Gogate, U. S.; Repta, A. J.; Alexander, J. *Int. J. Pharm.* 1987, **40**, 235–248.
69. Gogate, U. S.; Repta, A. J. *Int. J. Pharm.* 1987, **40**, 249–255.
70. Cundy, K. C.; Branch, R.; Chernov-Rogan, T.; Dias, T.; Estrada, T.; Hold, K.; Koller, K.; Liu, X.; Mann, A.; Panuwat, M.; Raillard, S. P.; Upadhyay, S.; Wu, Q. Q.; Xiang, J.-N.; Yan, H.; Zerangue, N.; Zhou, C. X.; Barrett, R. W.; Gallop, M. A. *J. Pharmacol. Exp. Ther.* 2004, **311**, 315–323.
71. Pauletti, G. M.; Gangwar, S.; Siahaan, T. J.; Aube, J.; Borchardt, R. T. *Adv. Drug Deliv. Rev.* 1997, **27**, 235–256.
72. Song, X. P.; Xu, C. R.; He, H. T.; Siahaan, T. J. *Bioorg. Chem.* 2002, **30**, 285–301.
73. Johansen, M.; Bundgaard, H. *Int. J. Pharm.* 1980, **7**, 119–127.

74. Bundgaard, H.; Klixbull, U.; Falch, E. *Int. J. Pharm.* 1986, **29**, 19–28.
75. Ohya, Y.; Kuroda, H.; Hirai, K.; Ouchi, T. *J. Bioact. Compat. Polym.* 1995, **10**, 51–66.
76. Juchem, M.; Pollow, K.; Elger, W.; Hoffmann, G.; Moebus, V. *Contraception* 1993, **47**, 283–294.
77. Li, Q. G.; Huempel, M. *Eur. J. Drug Metab. Pharmacokinet.* 1992, **17**, 281–291.
78. Venhuis, B. J.; Wikström, H. V.; Rodenhuis, N.; Sundell, S.; Wustrow, D.; Meltzer, L. T.; Wise, L. D.; Johnson, S. J.; Dijkstra, D. *J. Med. Chem.* 2002, **45**, 2349–2351.
79. Venhuis, B. J.; Dijkstra, D.; Wustrow, D.; Meltzer, L. T.; Wise, L. D.; Johnson, S. J.; Wikstroem, H. V. *J. Med. Chem.* 2003, **46**, 4136–4140.
80. Tscheuschner, C.; Neubert, R.; Fuerst, W.; Luecke, L.; Fries, G. *Pharmazie* 1993, **48**, 681–684.
81. Repta, A. J.; Patel, J. P. *Int. J. Pharm.* 1982, **10**, 29–42.
82. Bundgaard, H.; Johansen, M. *Int. J. Pharm.* 1982, **10**, 165–175.
83. Johansen, M.; Bundgaard, H. *J. Pharm. Sci.* 1983, **72**, 1294–1298.
84. Buur, A.; Bundgaard, H. *Arch. Pharm. Chem. Sci. Ed.* 1987, **15**, 76–86.
85. Bakhtiar, R.; Hop, C. E. C. A.; Walker, R. B. *Rapid Commun. Mass Spectrom.* 1997, **11**, 598–602.
86. Walker, R. B.; Dholakia, V. N.; Brasfield, K. L.; Bakhtiar, R. *Gen. Pharmacol.* 1998, **30**, 725–731.
87. Bodor, N.; Sloan, K. B.; Little, R. J.; Selk, S. H.; Caldwell, L. *Int. J. Pharm.* 1982, **10**, 307–321.
88. Bodor, N.; Sloan, K. B. *J. Pharm. Sci.* 1982, **71**, 514–520.
89. Saari, W. S.; Schwering, J. E.; Lyle, P. A.; Smith, S. J.; Engelhardt, E. L. *J. Med. Chem.* 1990, **33**, 97–101.
90. Liu, B.; Hu, L. Q. *Bioorg. Med. Chem.* 2003, **11**, 3889–3899.
91. Hu, L.; Liu, B.; Hacking, D. R. *Bioorg. Med. Chem. Lett.* 2000, **10**, 797–800.
92. Jiang, Y.; Zhao, J.; Hu, L. *Tetrahedron Lett.* 2002, **43**, 4589–4592.
93. Carpino, L. A.; Triolo, S. A.; Berglund, R. A. *J. Org. Chem.* 1989, **54**, 3303–3310.
94. Amsberry, K. L.; Borchardt, R. T. *J. Org. Chem.* 1990, **55**, 5867–5877.
95. Amsberry, K. L.; Borchardt, R. T. *Pharm. Res.* 1991, **8**, 323–330.
96. Shan, D. X.; Nicolaou, M. G.; Borchardt, R. T.; Wang, B. H. *J. Pharm. Sci.* 1997, **86**, 765–767.
97. Liao, Y.; Wang, B. *Bioorg. Med. Chem. Lett.* 1999, **9**, 1795–1800.
98. Bundgaard, H.; Falch, E.; Larsen, C.; Mosher, G. L.; Mikkelson, T. J. *J. Med. Chem.* 1985, **28**, 979–981.
99. Koike, M.; Futaguchi, S.; Takahashi, S.; Sugeno, K. *Drug Metab. Dispos.* 1988, **16**, 609–615.
100. House of Lords. *Beecham Group v. Bristol Laboratories of 30 03* 1978, RPC 153.
101. House of Lords. *Merrell Dow Pharmaceuticals Inc v. H.N. Norton & Co Ltd* 1996, RPC 76.
102. U.S. District Court for the District of Colorado. *Marion Merrell Dow v. Geneva Pharmaceuticals, 877 F. Supp. 531* 1994.

13

LIPOSOMES AS DRUG DELIVERY VEHICLES

GUIJUN WANG
Department of Chemistry and Biochemistry, Old Dominion University, Norfolk, VA, USA

13.1 INTRODUCTION

Liposomes are spherically shaped vesicles composed of hydrophobic lipid bilayers encapsulating hydrophilic compartments. They can be unilamellar and multilamellar vesicles (Fig. 13.1). Liposomes are typically formed by biocompatible phospholipids and their analogs. Depending on the structures of the lipids and the methods of formation, the size of liposomes can range from 100 nm to 50 μm. Since the discovery of liposomes by Bangham in 1965 [1], extensive studies in the area of liposomes as biomembrane models and later on as drug delivery vehicles have been carried out. Liposomes are considered as ideal vehicles for drug delivery since both hydrophilic and hydrophobic therapeutic agents can be encapsulated in either the cavity or within the bilayer (Fig. 13.1), and their structures mimic biological membranes.

Liposome-based nanocarriers have been the most successful drug delivery systems so far with more than a dozen liposome-formulated drug delivery systems approved by the US Food and Drug Administration (FDA) for clinical use, yet many others are in different stages of clinical trials [2–10]. The drug delivery field using liposome or liposome complexes is extremely active with a large number of publications and reports appearing each year. Therefore, it is not surprising that there are also many excellent reviews [3–9] on the various aspects of liposome-based nanocarriers. As a result, this chapter will not be a comprehensive review of liposomal drug

Drug Delivery: Principles and Applications, Second Edition. Edited by Binghe Wang, Longqin Hu, and Teruna J. Siahaan.

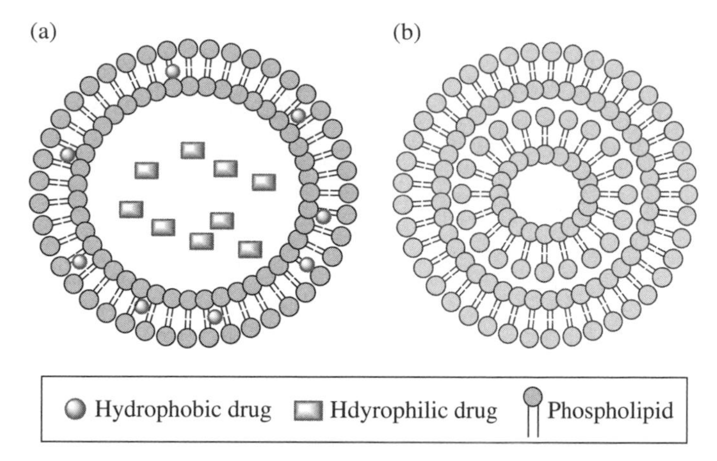

FIGURE 13.1 Schematic representations of unilamellar (a) and multilamellar (b) liposomes.

carriers. Instead, the emphasis is on the structure and function of the lipid components and their influence on drug delivery outcomes. To design and achieve multifunctional drug delivery systems, it is necessary to integrate knowledge from multiple research fields such as chemistry, biology, toxicology, materials science, and medicinal chemistry. The author wishes to address the fundamental aspects of liposomes from a chemistry perspective and hope this can provide useful insights to researchers in various disciplines.

In this chapter, the current status of liposome-based drug delivery vehicles will be briefly reviewed. Then recent development of liposomes in triggered delivery and targeted delivery will be updated.

13.2 CURRENTLY APPROVED LIPOSOMAL DRUGS IN CLINICAL APPLICATIONS

Since the discovery of liposome in 1965, many generations of liposomes have been developed, from the conventional ones to stealth liposomes, to targeted and triggered release systems. After five decades of research, a reasonable understanding of the behaviors of conventional and stealth liposome–trapped drugs *in vivo* has been achieved. Currently many liposomal drugs have been approved by the FDA for treating various conditions from cancer to infectious diseases [4–12]. Liposomal delivery systems are able to protect the cargos from undesired degradation before reaching target sites and reduce systemic toxicity of drugs such as anticancer and antifungal agents, and therefore improve therapeutic index.

Anticancer agent doxorubicin and antifungal agent amphotericin B (Fig. 13.2) are among the most extensively investigated drugs, and several liposomal formulations of these agents are currently in clinical use. Amphotericin B is a polyene antifungal agent, which binds to ergosterol of fungal cells and forms a pore in the cell membrane; it has severe and potentially lethal side effects due its interactions with mammalian

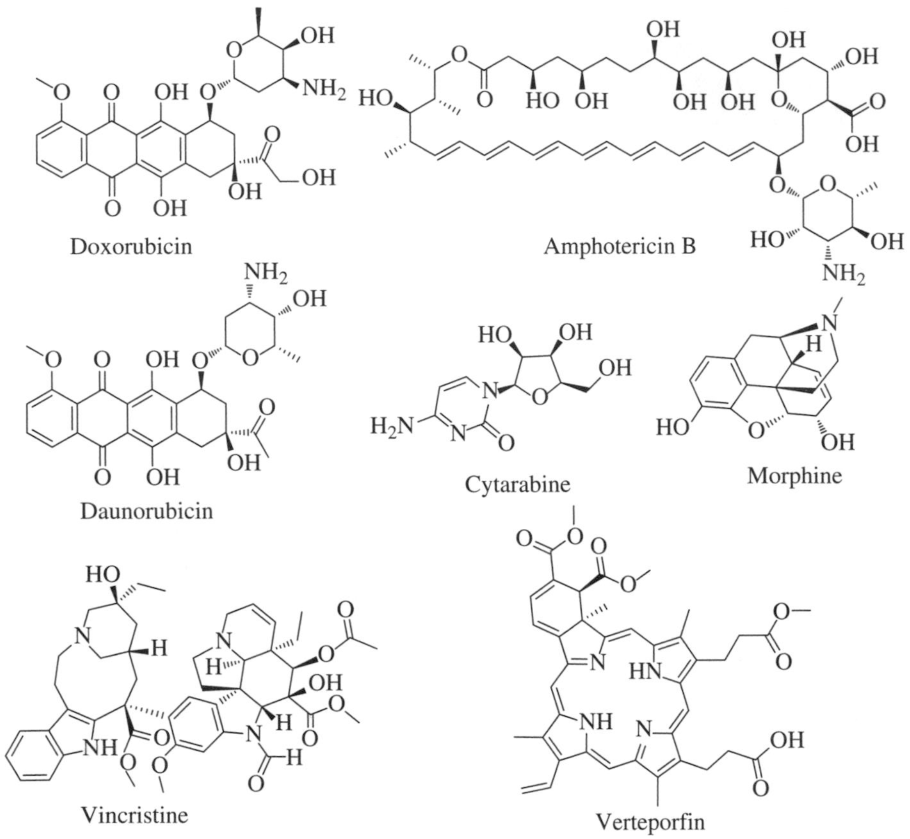

Doxorubicin

Amphotericin B

Daunorubicin

Cytarabine

Morphine

Vincristine

Verteporfin

FIGURE 13.2 Chemical structures of some small molecule drugs that have been successfully formulated as liposomal drugs and approved by the FDA.

cell membrane in a similar fashion. Several liposomal amphotericin B formulations have been approved for severe fungal infections. These include Ambisome, Abelcet, and Amphotec. For doxorubicin, the first liposomal drug, Doxil, uses PEGylated liposomes and was approved in 1995 by the US FDA for treating Kaposi's sarcoma, ovarian, and breast cancer. Afterward, more than a dozen liposomal drugs including Lipodox and Mycet have been approved and brought to clinical use. Other anticancer drugs formulated with liposomes approved for clinical use include vincristine, daunorubicin, and cytarabine. There are also other anticancer liposomal drugs including platinum-based anticancer agents in various stages of clinical trials. In different therapeutic categories, approved liposomal drugs include morphine for severe pain and verteporfin for macular degeneration. All currently approved liposomal drugs are composed of the so-called conventional and stealth liposomes. These liposomal drugs rely on passive targeting mechanism to accumulate at the disease site. The structures of the common lipids used for these drug delivery systems are shown in Figure 13.3.

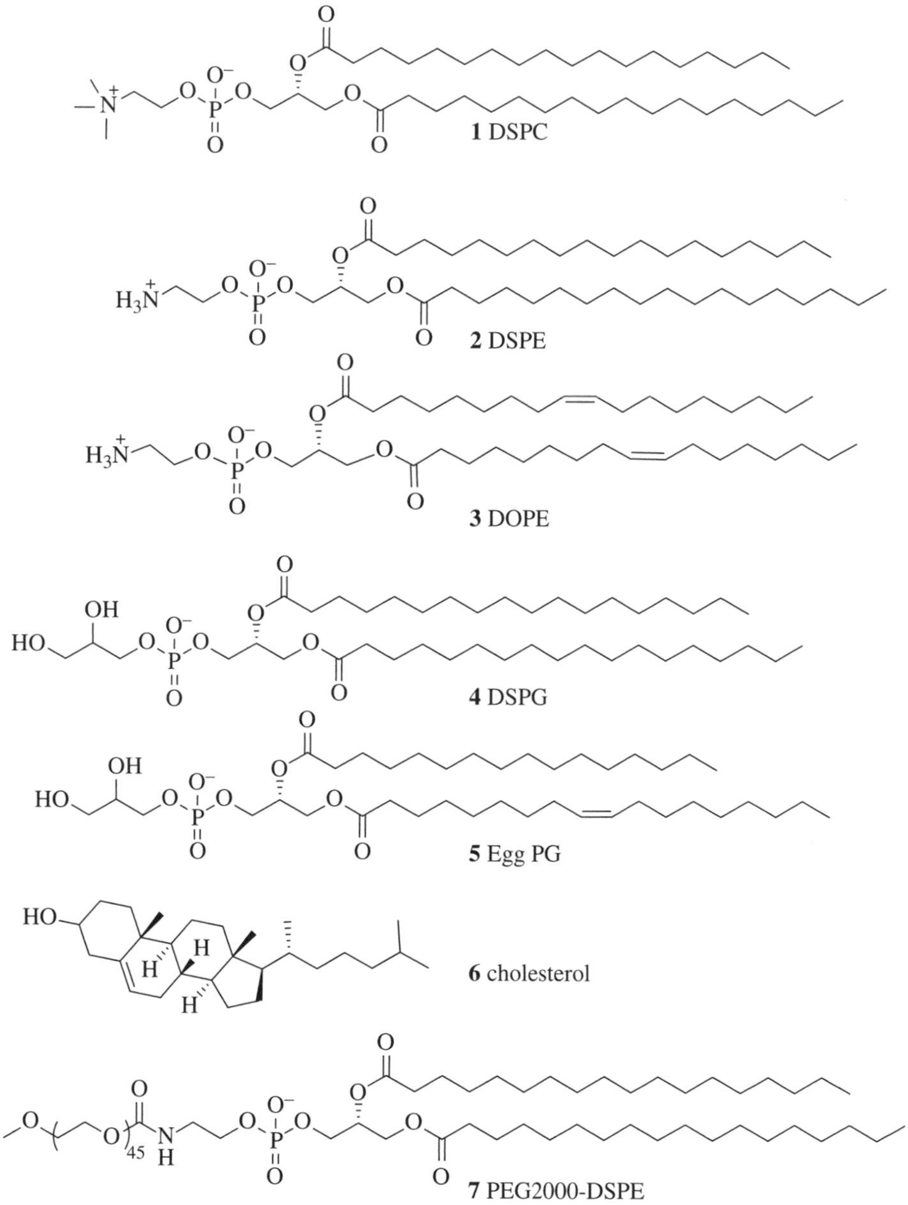

1 DSPC

2 DSPE

3 DOPE

4 DSPG

5 Egg PG

6 cholesterol

7 PEG2000-DSPE

FIGURE 13.3 General structures of several phospholipids commonly used in liposomal drug delivery systems: distearoylphosphatidylcholine (DSPC) **1**, distearoylphosphatidylethanolamine (DSPE) **2**, dioleoylphosphatidylethanolamine (DOPE) **3**, distearoylphosphatidylglycerol (DSPG) **4**, egg PG (main component L-α-phosphatidylglycerol **5**), cholesterol **6**, PEGylated DSPE **7**.

13.3 CONVENTIONAL AND STEALTH LIPOSOMES

Three representative structures of typical liposomal drug delivery vehicles are shown in Figure 13.4. The first-generation liposomes are the conventional ones or "classical liposomes," which are composed of a mixture of typical phospholipids and cholesterol found in biomembrane. They typically suffer from rapid clearance and instability in the blood stream. Various strategies of improving the liposome stability have been developed. Many efforts have been devoted to the modifications of structures and compositions of the lipids. Changing components of the lipid mixtures can also lead to the change of stability. It has been found that increasing the cholesterol content improves stability significantly. A different but more effective approach is by coating the liposome surface with an inert hydrophilic polymer, which creates a steric shield to make it "invisible" and thus to reduce blood protein adsorption and be "invisible" to the blood stream. Such efforts led to the introduction of "stealth liposome" [6, 13]. The most successful strategy in improving stability and increasing circulation time is coating liposome with polyethylene glycol (PEG) of various chain lengths; these PEG-coated liposomes are referred to as stealth liposome. Besides PEG, other polymers have also been studied for this purpose.

The application of polymer-coated liposomes, especially PEGylated liposomes, has become very popular in the research fields of liposomal drug delivery or gene delivery systems. Despite the success of several conventional and stealth liposomal drugs, there are still problems with achieving desired therapeutic index, and desirable outcome is not a routine occurrence. One main problem is the nonspecific interactions with normal cells. Another problem is the bioavailability of the entrapped drugs if they are not sufficiently unloaded upon reaching the target sites. There has been a high level of activities in searching for "ideal" drug delivery vehicles. Triggered release and targeted delivery systems, which employ specific physiochemical properties of disease sites, can potentially overcome the various issues of the first-generation liposomal drugs. Next section discusses issues related to triggered release.

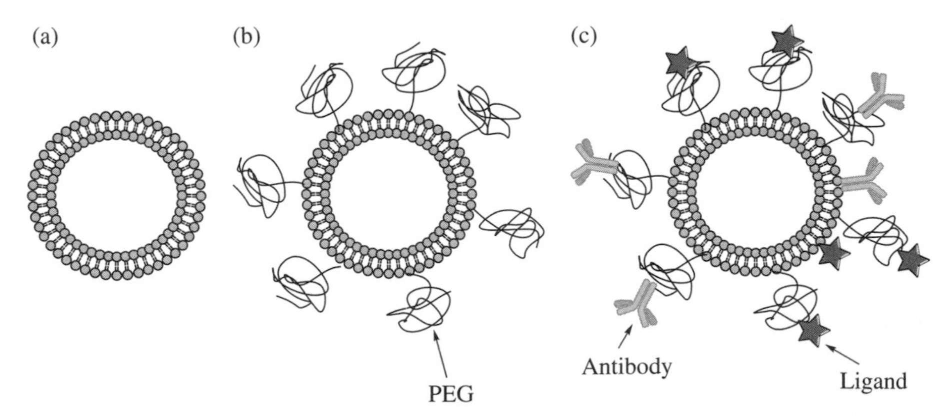

FIGURE 13.4 Representations of conventional liposome (a), stealth liposome (b), and targeted liposome (c) using antibody or ligand. The targeting components can be attached to the end of polyethylene glycol (PEG) or directly to the lipid head group.

13.4 STIMULI-RESPONSIVE LIPOSOMES OR TRIGGERED-RELEASE LIPOSOMES

Conventional liposomes typically suffer from rapid clearance from the blood stream. Upon intravenous administration, liposomes are rapidly taken up by the mononuclear phagocyte system (MPS) also known as reticuloendothelial system (RES) and cleared from the blood stream. The rapid uptake by MPS, predominantly in the liver and spleen, reduces the distribution in other tissues of the body and may cause toxicities to MPS organs. Stealth liposomes, such as PEGylated liposomes, reduce uptake by the RES systems, improve circulation time, and increase accumulation at the tumor site. Despite all these advantages, there are still other issues such as timely unloading of the cargo, which hinder the application of liposomes in drug delivery.

Since the encapsulated therapeutic agents are not bioavailable until they are released, it is important to be able to control the rate of release and the half-life. Many different types of triggered systems have been designed and studied. Based on the location of the trigger release event, these are broadly categorized as remote exogenous triggering and *in vivo* local endogenous triggering. Remote triggering methods utilize external stimuli such as light, heat, ultrasound, electromagnetic field [6, 9, 14], etc. These environmental stimuli are applied to the localized areas where liposomal drugs are targeted. Internal biological triggering systems utilize the specific properties of the local disease sites, either by responding to different acidity or by interacting with certain enzymes that are specific to or overexpressed by the targeted cells.

This section will examine several triggered release mechanisms from the perspective of the lipid bilayers' physical and chemical properties. These include temperature-, pH-, and light-sensitive liposomes as well as enzyme-triggered systems. The general mechanisms of triggered release liposomal systems are based on the change of structure and/or conformation of the lipid components of the vesicles, thus perturbing the stability of bilayer packing and allowing the contents to be released.

13.4.1 General Mechanism of Triggered Release

Before discussing various types of triggered release systems, it is necessary to review the basic biophysical properties of lipids and their assemblies. Depending on the structures and shapes of lipids, they can self-assemble and form different types of organized structures. A few representative lipid structures are shown in Figure 13.5. For phospholipids such as distearoylphosphatidylcholine (DSPC) or other phosphatidylcholines, their headgroups and two lipid tails adopt a conformation that occupy a cylindrical shape and the assembled lipid bilayers lead to spherical vesicles or liposomes (Fig. 13.5a). If the lipid has a large headgroup and a small tail such as a lysolipid, which only has one hydrophobic lipid tail, then the lipid adopts a cone-shaped structure and in turn forms spherical micelles (Fig. 13.5b). If the phospholipids have a relatively small headgroup and two hydrophobic tails such as DSPE or other phosphatidylethanolamine, they would have a truncated cone shape, and tend to form inverted hexagonal micelles with the hydrophilic headgroups inside the cavity and the tails outside (Fig. 13.5c). Typically the liposomes used for drug encapsulation are

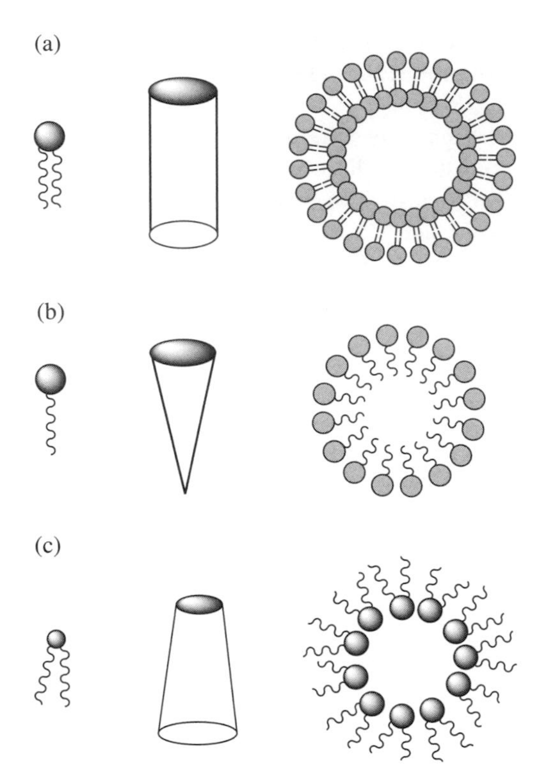

FIGURE 13.5 The basic self-assembling properties of lipids. Depending on the structures of the lipids, they can form lamellar structures as in (a), micelles as in (b), or inverted hexagons (micelles) as in (c).

composed of a mixture of different phospholipids and cholesterol [15]. The stability, structure, and shape of liposomes depend on the nature of the lipids and the composition of the mixture. Inclusion of a certain amount of cholesterol in the vesicles results in improved stability of the liposome.

In the various triggered systems, controlled release can be achieved by manipulating the lipid structure and fluidity. Specifically, structural changes can lead to changes in the packing mode of the lipids and, thus, content release. For example, protonation often results in repulsion of the headgroup and thus destabilization of the bilayer membrane. The use of lysolipids (mono-tailed lipids), however, will lead to enhanced sensitivity to stimuli and, thus, rapid release of the content. Below we discuss various types of triggered release systems.

13.4.2 Thermo-Sensitive Liposomes

Thermo- or temperature-sensitive liposomes (TSLs) have been extensively studied during the past decade, especially for anticancer drug delivery [16–37]. Several of them have progressed to clinical trials, but currently there are no approved clinical uses yet. Noninvasive-controlled heating of local area can be advantageous in cancer treatment. TSLs with doxorubicin (Doxil) have been shown to be effective in treating multiple drug-resistant (MDR) cancer cells. A thermo-sensitive liposomal formulation of doxorubicin (Thermodox) had progressed to phase III clinical trial in 2013 [19, 20]. Temperature-sensitive liposomes release their encapsulated drugs at their melting phase transition temperature, while the encapsulated drugs remain mostly dormant at normal body temperature.

The principle of the TSLs is to utilize the phase transition properties of the lipid bilayer upon heating to slightly above the body temperature, such as 39–42°C. The lipid bilayer changes from a well-packed gel-like solid phase to a more fluid-like disordered liquid phase. The melting temperature (T_m) of the lipid bilayer depends on the structures and physical properties of the phospholipids and interactions among the lipid chains. The gel-like solid phase is not permeable, but the fluid phase is permeable for substances trapped in the liposome and thus allows these trapped substances to leak out above T_m. Therefore, liposomes with the encapsulated drugs are not bioavailable until the temperature is elevated to the phase transition temperature of the bilayer. This method can achieve controlled and targeted delivery, especially for accessible solid tumors.

By selecting the appropriate lipids or lipid mixtures, the melting temperatures of the lipid bilayers can be altered. Unsaturated bonds and shorter tails all are likely to decrease the phase transition temperatures. Figure 13.6 shows the structures of several lipids used often in constructing thermo-sensitive liposomes. Incorporation of various lysolipids has been shown to modulate the transition temperature. The thermo-sensitive liposomes formed by lysolipids or other designed phospholipid derivatives are likely to produce effective drug delivery systems. Besides using special types of lipids to achieve hyperthermia-triggered delivery, thermo-sensitive polymers have been used to coat liposomes, and temperature-sensitive peptides have also been incorporated into lipid bilayers [25, 30].

The method of heating to achieve hyperthermia has also evolved over the years. Typically a heat source such as high intensity focused ultra sound (HIFU) allows noninvasive heating to establish hyperthermia (39–45°C) in targeted tissue. Recently, magnetic resonance-guided high intensity focused ultrasound (MR-HIFU) has been shown to trigger drug release from thermo-sensitive liposomes in a noninvasive, localized, and controlled manner. Such practice has greatly advanced the field of thermo-sensitive liposomal drug delivery [27, 31].

13.4.3 pH-Sensitive Liposomes

Among the many triggered release systems that are being evaluated, pH-sensitive liposomes are particularly interesting, because the acidity of many pathological sites have been well characterized, and often a pH gradient exists between normal tissues and diseased sites [38–41]. Many tumor cells have acidic microenvironment that can be exploited for the design of pH-responsive drug delivery systems. The extracellular pH values in cancerous tissues are lower (5.7–7.0) than normal blood pH of 7.4. pH gradients can also be found between the extracellular environment and intracellular

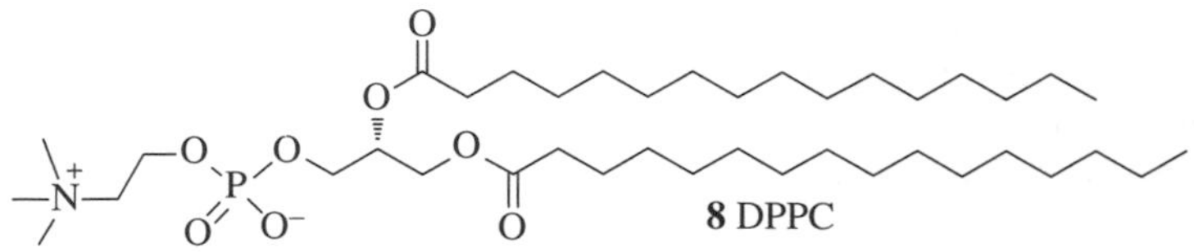

8 DPPC

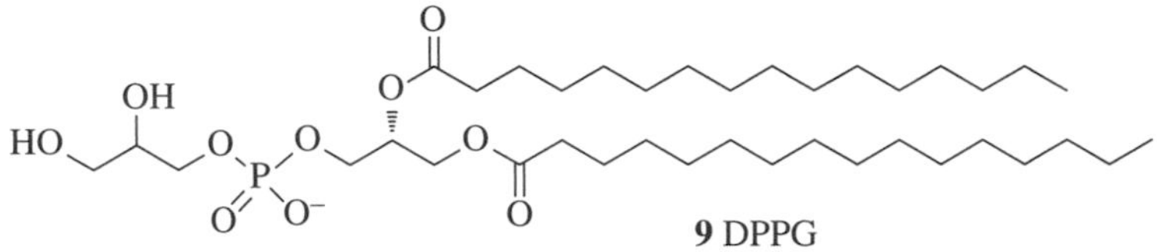

9 DPPG

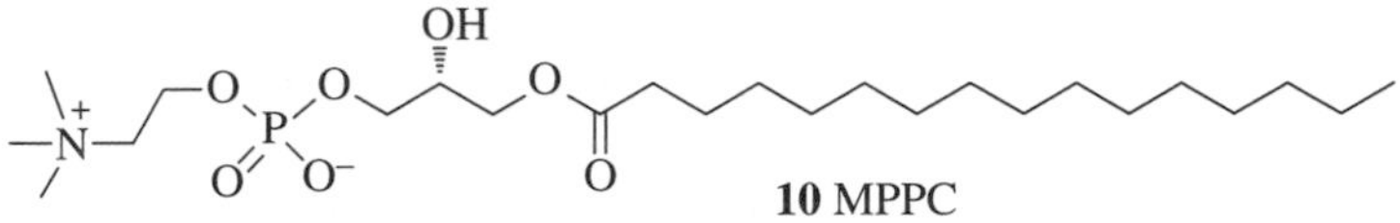

Lyso-PC

10 MPPC

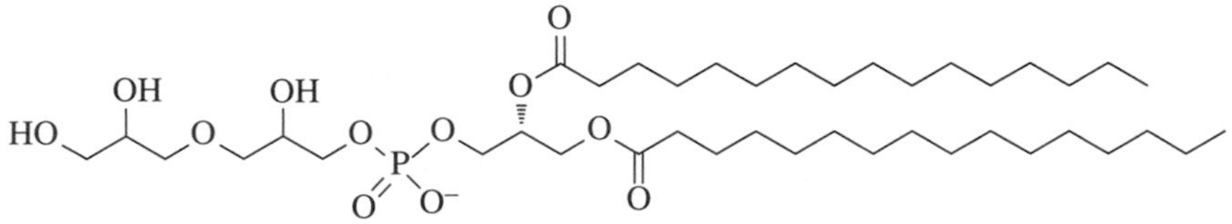

11 DPPGOG

FIGURE 13.6 Structures of lipids utilized in thermo-sensitive liposomes.

compartments such as endosomes and lysosomes (pH 4.5–6.5). This pH gradient is important for drug carriers that can be taken up by endocytosis and trapped within endosome and lysosomes.

There are several different types of pH-sensitive liposomes. The first class utilizes ionizable functional groups that will protonate or deprotonate at different pH values. These include amino groups, carboxylic acids, imidazoles, pyridines [39–44], etc. The structures of some of these lipids are shown in Figure 13.7. The most extensively studied pH-sensitive systems in this class utilize the amino group of PE (**12**), especially dioleoylphosphatidylethanolamine (DOPE). Under acidic environment, the amino group (RNH_2) is protonated (RNH_3^+), which decreases the stability of the lipid bilayer and allows the trapped drugs to be released. Systems using other ionizable groups follow similar principles. RCOOH is converted to the corresponding $RCOO^-$ anion at high pH, and the reprotonation of the carboxylate changes the stability of the assemblies depending on the lipids attached. Besides incorporating pH-sensitive structures into lipids, strategies of coating liposomes with polymers containing pH-responsive functional groups have also been explored for acid-triggered liposomes [44].

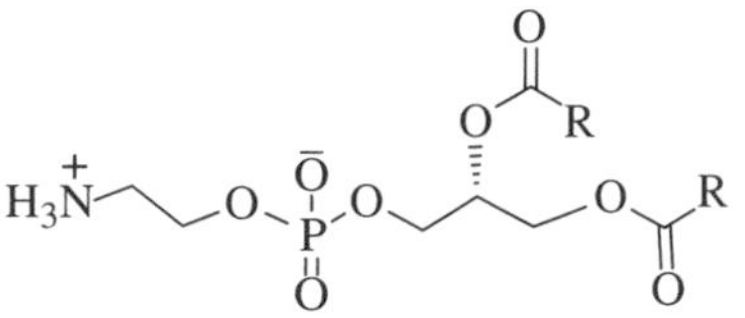

12 PE

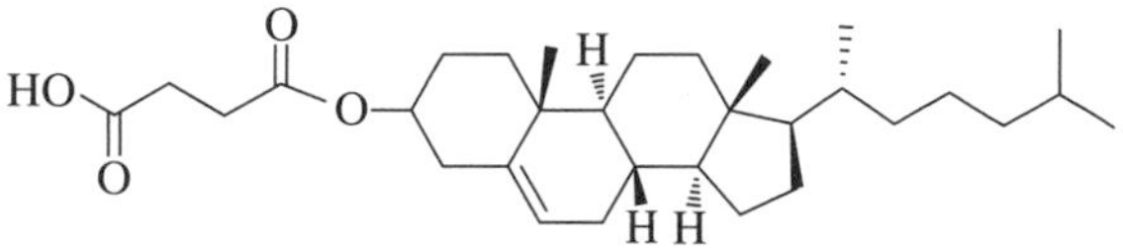

13 Cholesteryl hemisuccinate

FIGURE 13.7 Structures of lipids utilized in pH-sensitive liposomes.

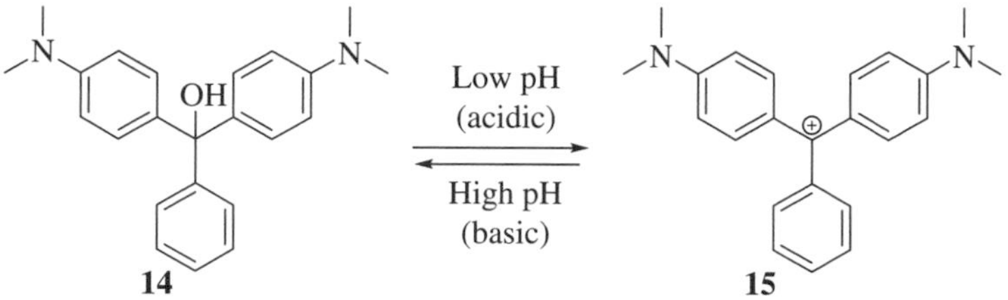

Low pH (acidic)

High pH (basic)

14 **15**

FIGURE 13.8 Conversion of neutral malachite green carbinol base to the cationic form at low pHs.

A recent example incorporated a new type of pH-triggered mechanism into liposomes. As shown in Figure 13.8, instead of directly changing the ionization states of the lipids, a pH-responsive compound (**14**) is incorporated into the lipid bilayer. Under acidic conditions, the compound dehydrates and forms the corresponding carbocation (**15**), which is more polar and hydrophilic than the neutral alcohol and consequently results in the disordering of the lipid bilayer [45]. This moiety can be introduced noncovalently to the lipid bilayer and used for the modulation of different liposomal systems.

Recently yet another class of pH-sensitive drug delivery systems incorporates carbon dioxide precursor in their structures [46]. Under acidic conditions, the bicarbonate ions react with acid and produce CO_2 bubbles. Ammonium bicarbonate and sodium bicarbonate have both been used for this strategy, and they are shown to be biocompatible. Chen et al. reported the use of ammonium bicarbonate–encapsulated liposomes for doxorubicin delivery [46]. At mild acidic condition or an elevated temperature (such as 42°C), decomposition of ammonium bicarbonate generates CO_2 bubbles and thus creates permeable defects in the lipid bilayer, which results in the rapid release of DOX. Another interesting pH-responsive lipid (Fig. 13.9) [47] changes its conformation through flipping from the diequatorial structure **16** to the diaxial chain arrangement in **17** upon protonation. The conformation change is presumably caused by the formation of intramolecular hydrogen bonds with the protonated amino group. This flip of conformation causes the disruption of the lipid bilayer and allows cargo release.

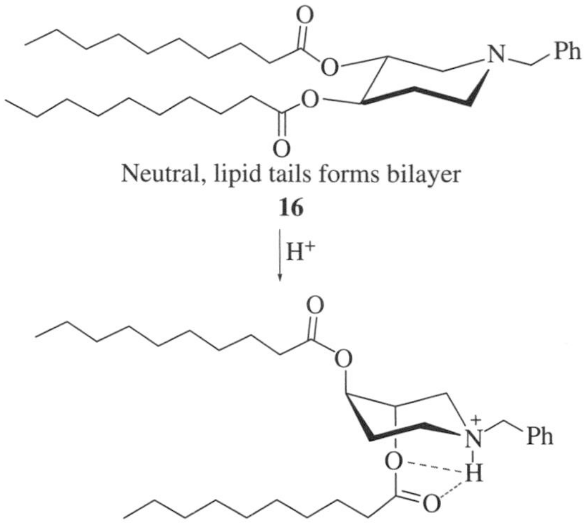

Neutral, lipid tails forms bilayer

16

H^+

Change of shape, size, and charge
lipid phase separation,

17

FIGURE 13.9 Conformational conversion upon pH changes.

The second main class of pH-responsive liposomes utilizes acid-labile functional groups, which cleave upon exposure to an acidic environment [48–66]. Functional groups that have been used in such applications include hydrazones [48, 49], *cis*-aconityl linkages [50, 51], acetals [52–54], orthoesters [55–59], vinyl ethers [60–63], phenyl-substituted vinyl ethers [64–66], etc. Several examples of lipids containing acid-labile functional groups are shown in Figure 13.10, including the aconityl lipid **18**, ortho ester **19**, and vinyl ether **20**. Figure 13.11 shows the cleavage mechanism of an orthoester lipid under acidic conditions. Figure 13.12 shows the mechanism of pH-triggered cleavage of a PEGylated phenyl-substituted vinyl ether [64]. These compounds contain chemical bonds covalently cleaved under mild acidic environment. Often they decompose to form a lipid, such as PE or lysolipids, which will not form stable lamellar structures. These changes typically cause dramatic interruption of the lipid assembly and in turn the bilayer stability. For the pH-sensitive liposomes containing chemically cleavable bonds, they lead to more pronounced changes to the bilayer structures comparing to simple change of ionization states.

13.4.4 Photo-Triggered Liposomes

In comparison to the extensive studies on pH-responsive liposomes, there are relatively fewer studies on photo-triggered liposomal drug delivery systems [67, 68]. One reason is that the application of light externally is not as convenient as the *in vivo* internal triggers. Also a suitably designed system containing photo-responsive functional groups must be custom synthesized. A related field of well-studied “light-responsive liposomes” is for the delivery of photosensitizers in photodynamic therapy (PDT) [11].

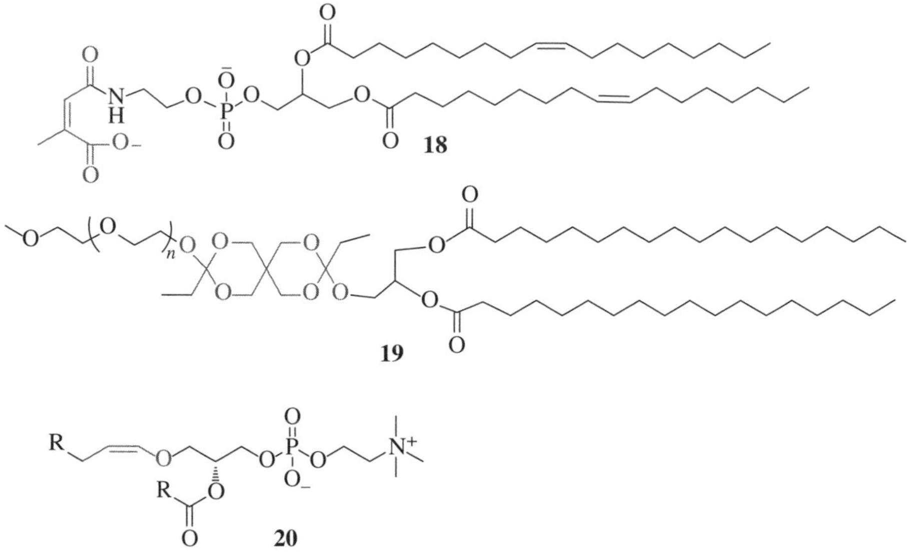

FIGURE 13.10 Structures of some representative acid-labile lipids.

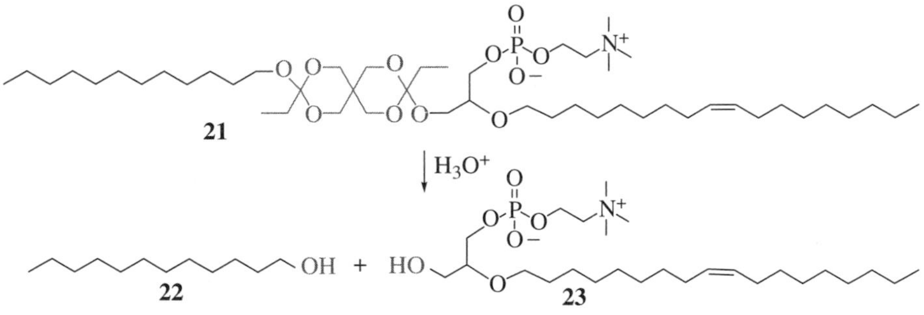

FIGURE 13.11 The acid hydrolysis of the diorthoester **21**.

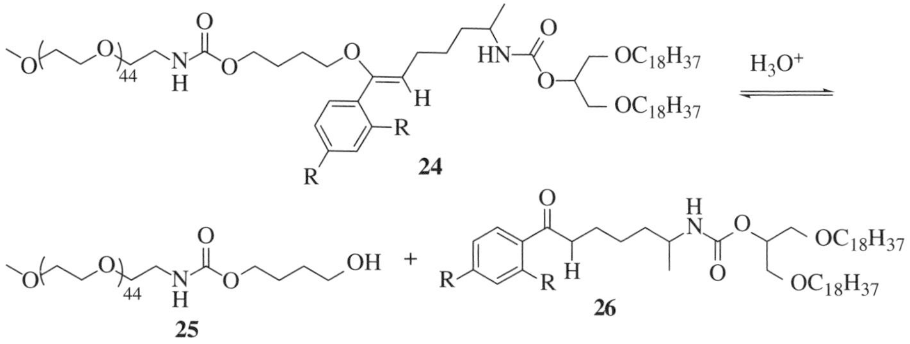

FIGURE 13.12 The hydrolysis of a pH-sensitive PEG–phenyl-substituted vinyl ether (PIVE)-linked lipid conjugate.

In PDT, ideally irradiation of tumors after systematic administration of a photosensitizer will destroy tumor cells without affecting normal tissues; however, photosensitizers often do not selectively accumulate at tumor sites; therefore, PDT may cause unwanted side effects. Liposomes have been used to effectively deliver the photosensitizer and achieve targeting to tumor cells thus reducing the toxicity of nonspecific interactions [11]. These are different comparing to photo-triggered liposomes in the sense that the lipid components usually don't contain photo-sensitive or photo-cleavable functional groups.

Photo-responsive functional groups have been used in the construction of photo-triggered liposomes. Several types of photo-triggered chemical reactions can be used in liposome design. These include photoisomerization, photopolymerization, and "uncaging" of photo-labile "caged" compounds [69–83]. Upon light irradiation, the liposomes containing these different types of lipids will have changes to the bilayer stability and permeability, which result in drug release. For the first class, the photo-polymerizable functional groups such as acrylate and diacetylenes have been incorporated in the lipid tails or headgroups [73, 74]. Upon polymerization, the properties of the lipid bilayer change and thus interrupt the normal bilayer permeability. This seems a little contradictory to the situation of polymerized vesicles, which typically have increased stability. The rationale is that certain domains of the lipid bilayers are crosslinked, which in turn form hydrophobic domains that destabilize the lamellar assemblies leading to pore formation and thus drug release. The second class contains photo-isomerizable functional groups. Such systems include stilbene derivatives, azobenzenes [69–71], and spiropyrans [72]. Figure 13.13 shows the isomerization of some of these functional groups upon light irradiation. It is obvious to see that upon isomerization the lipid packing would change accordingly, leading to the destabilization of the vesicles.

The third class utilizes photo-labile linkers in the lipid headgroups or other regions. Upon UV irradiation, these photo-labile functional groups are cleaved, leading to structural and property changes of the corresponding lipids and liposomes. Photo-cleavable groups, such as those photo-labile protecting groups used in solid-phase synthesis, and other "caged" compounds used in biological studies can be used in the design of photo-triggerable liposomes [75, 76]. Figure 13.14 shows some examples of the photo-cleavable groups, their activation wavelengths, and photolytic

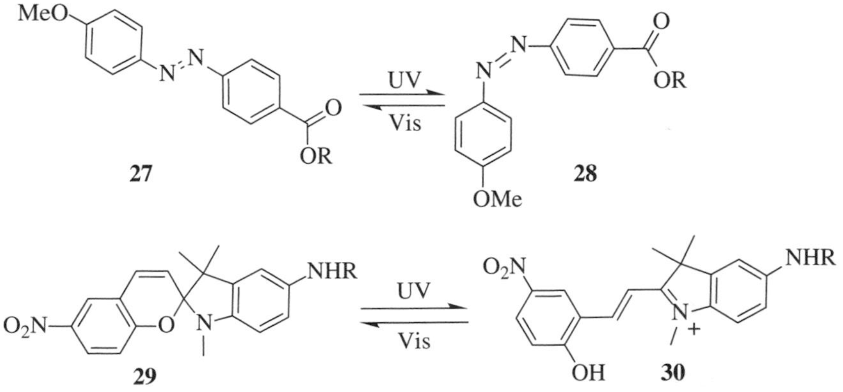

FIGURE 13.13 Photo-isomerizations of azobenzene and spriropyran derivatives.

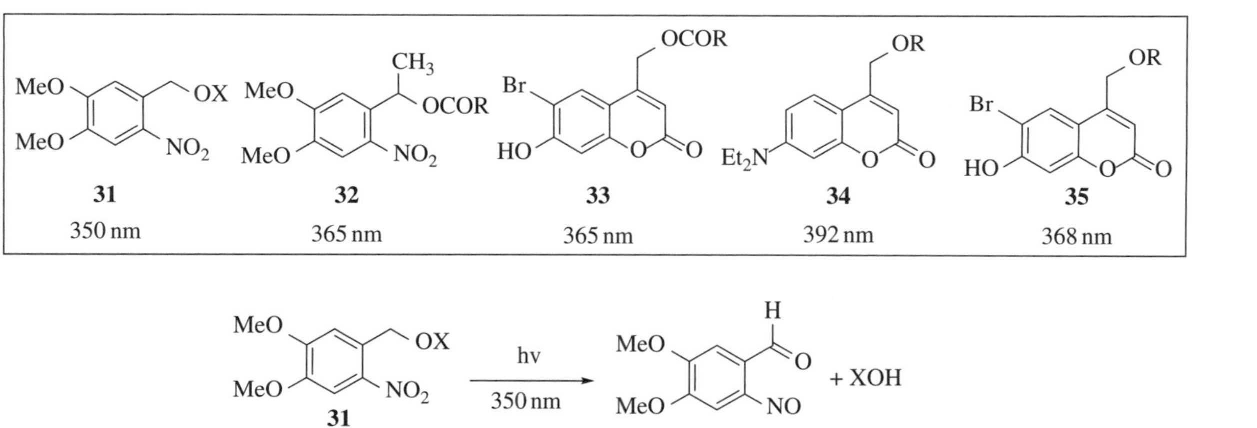

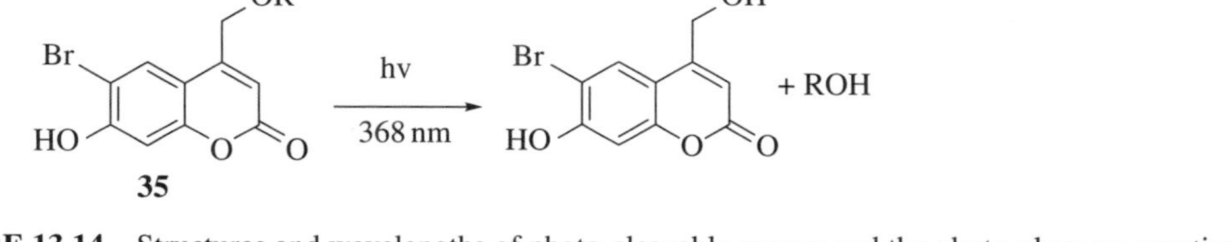

FIGURE 13.14 Structures and wavelengths of photo-cleavable groups and the photo-cleavage reactions.

reactions. The most extensively studied systems utilized nitrophenyl groups (**31**, **32**) and coumarin derivatives (**34–36**). The majority of these chromophores are cleavable by UV light at longer wavelength than 350 nm.

Figure 13.15 shows the structures of several photo-triggerable lipids containing various photo-responsive functional groups. Compound **36** contains polymerizable diacetylene groups and has been studied as photo-triggered system for doxorubicin delivery [73, 74]. Nitrophenyls are particularly useful as photo-labile protecting groups, and their various analogs (**37**, **39**, **40**) have been used in the design of photo-responsive liposomes [77–81]. Coumarin derivatives (**38**) [82, 83] are generally not toxic and can be uncaged by exposure to light in the near IR region by a two-photon absorption mechanism, which may be more applicable to *in vivo* drug delivery

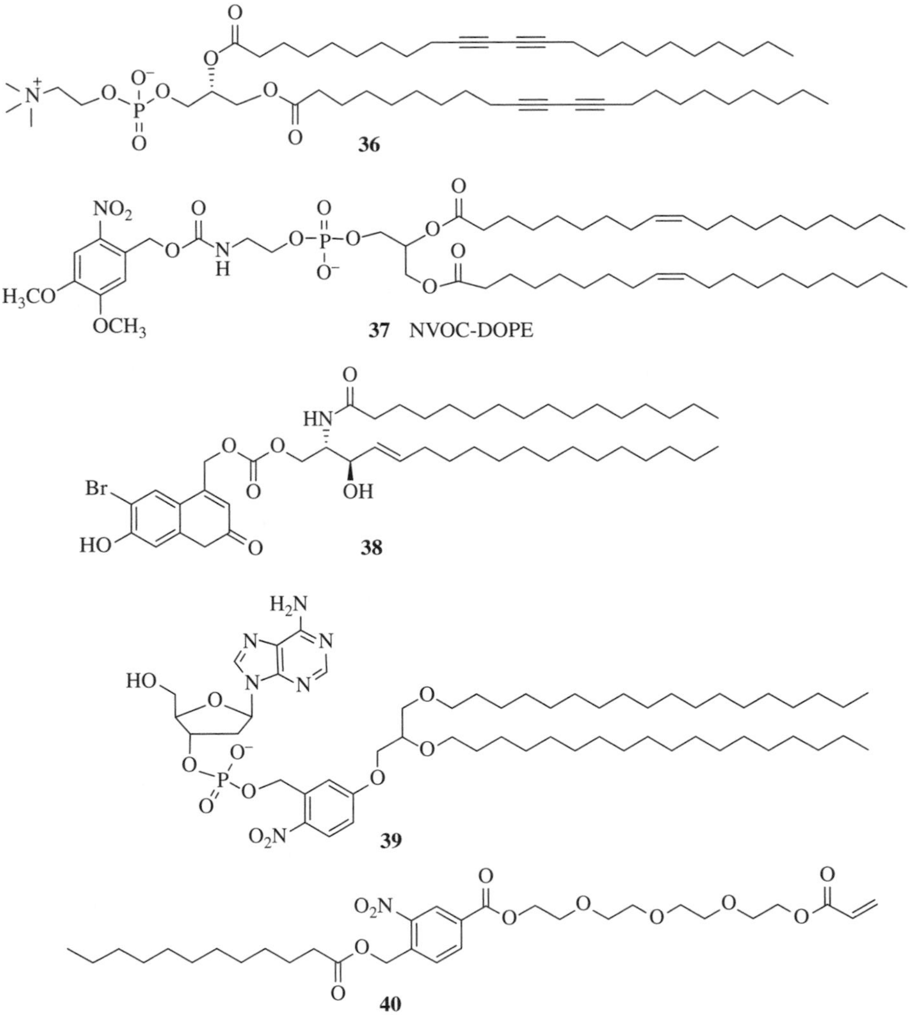

36

37 NVOC-DOPE

38

39

40

FIGURE 13.15 Structures of lipids containing photo-triggerable functional groups.

systems. Compound **40** includes both a nitrophenyl group and a polymerizable methacrylate group in the molecule and has been used in controlled photo-triggered release from the polymerized vesicles, which maintains the shape integrity upon UV irradiation. Such results allow stepwise controlled delivery [79].

13.4.5 Triggered Release Controlled by Enzymes

Enzyme-triggered liposomal drug delivery systems are highly effective [84, 85]. Disease-associated enzymes can be directly utilized for the design of the lipids and liposomes, and no other stimuli are required. Therefore, enzyme-triggered release could be biocompatible, noninvasive, and with good *in vivo* performance. The requirement, however, is to understand the biological pathways of the tumor or disease sites and use suitable functional groups that can react with the enzyme and cause phase changes or perturbations of the carrier, thereby releasing the trapped drugs.

Sometimes the enzyme-triggered delivery systems are called targeted delivery systems in the literature. In this chapter, we differentiate these two types of mechanisms though sometimes they are used simultaneously in one system. Enzyme-triggered systems typically result in the destabilization of the vesicle, which then lead to controlled release of the liposome cargo; while targeted delivery systems recognize the disease-specific receptors, antigens, etc. and aim to "guide" the carriers to preferably localize at the disease sites.

Several different types of enzymes, including lipases, proteases, glycosidases, and oxidoreductases, have been exploited for enzyme-triggered drug delivery. One main strategy is to incorporate the required functional groups in the liposomal components, which destabilizes the lipid bilayers upon reaction with the respective enzymes. A few examples are discussed here.

Using enzymes for tumor site–specific triggered drug delivery is a valid *in vivo* approach, which does not need any stimuli except the interaction of the particular enzymes with the delivery system. Typically two strategies have been used in the design of enzymatically triggered drug release from liposomes. The first type utilizes the cleavage of the lipid conjugate, resulting in the generation of a lysolipid or fusogenic lipid that destabilizes liposome. The second one utilizes the prodrug concept, which incorporates the drug into the design of the delivery system. Upon activation by the enzyme, the liposome is cleaved at where the drug is attached, which in turn destabilizes the liposome and releases the active drug.

An example of the first type of triggered release systems responds to redox state changes. Recently, a quinone-based reduction-sensitive liposome system was reported as a triggered release drug delivery system [86–88]. As shown in Figure 13.16, upon exposure to disulfide, the quinone moiety in **41** is converted to the dihydroquinone form in **42**, leading to cyclization and release of DOPE, which doesn't form stable lamellar structures.

Hydrolytic enzymes can cleave the corresponding ester or amide bonds in lipids and disturb bilayer integrity and thus release the contents [89–92]. For example, hydrolysis of the ester functional group in a lipid to form a lysolipid will interrupt the bilayer integrity. Secretory phospholipases A2 (sPLA2) are a family of lipases that catalyze the

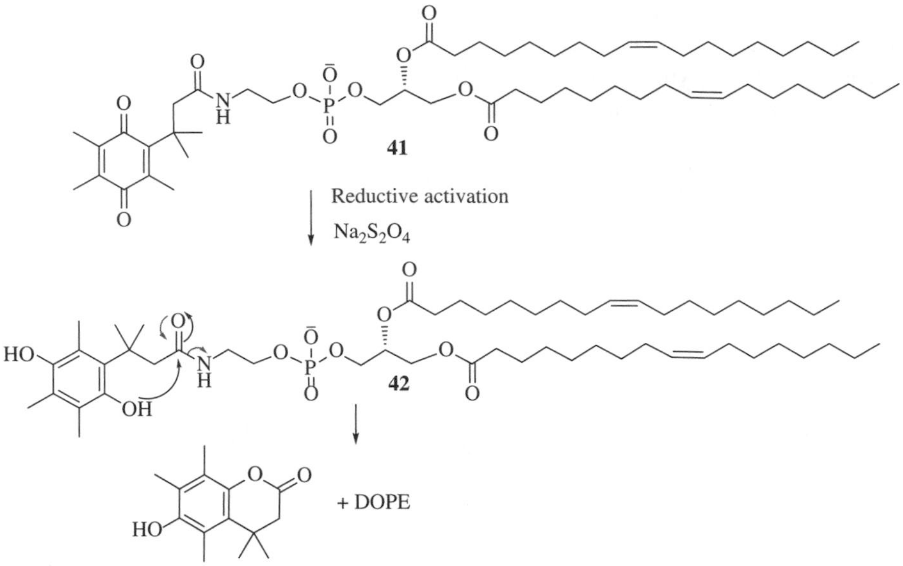

FIGURE 13.16 A reduction-responsive liposomal system.

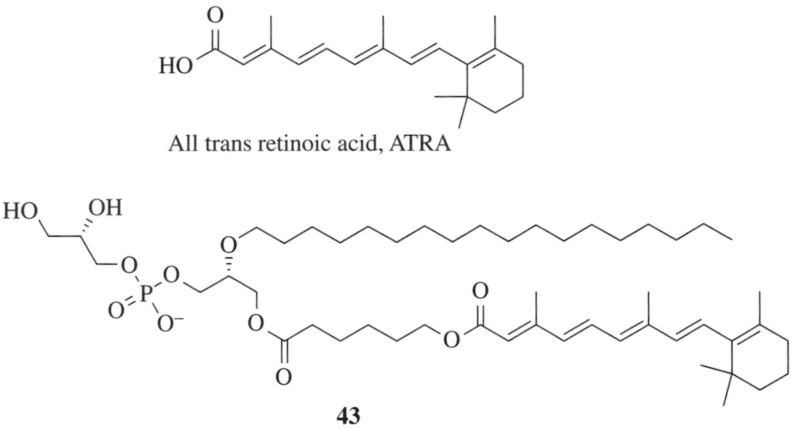

FIGURE 13.17 A prodrug that forms vesicle and can be activated by sPLA2.

hydrolysis of phospholipids in the *sn*-2 position, generating lysophospholipids and fatty acids. It has been shown that certain subtypes of sPLA2 are overexpressed in cancer cells. This has been employed in triggered release design. For example, retinoids have been incorporated into a lipase-activated drug delivery system in a prodrug form [91]. Retinoids such as all-trans retinoic acid (ATRA) have a broad spectrum of biological functions, including anticancer effect. Various methods of administration have been explored. Oral administration is limited by the low bioavailability of ATRA, and intravenous administration, however, is hindered by the low water solubility of ATRA. The ATRA-containing phospholipid prodrug **43** is shown in Figure 13.17. Upon the action of sPLA2, the prodrug liposomes were degraded and cytotoxic-free ATRA was released.

Among proteases, matrix metalloproteinases (MMPs) are over expressed in cancerous tissues and can be used for site-specific triggered release of liposomal drugs. Various proteases including MMPs have been studied and exploited in triggered release applications [93–97]. However the substrates for proteases are quite different to those of lipases. Typically tailor-made lipopeptides need to be designed and synthesized, and the structures must contain a functional group or moiety recognizable by the particular enzyme. These substrate lipopeptides then are incorporated into the liposome membrane.

13.5 TARGETED LIPOSOMAL DELIVERY

Since the first liposomal anticancer drug Doxil was approved in 1995, targeted drug delivery has been a subject of intense study [98–114]. Although many antibody-based targeted systems have reached clinical trials, no targeted or triggered release systems have been approved yet. In targeted drug delivery systems, disease-specific ligands or antibody, which can recognize tumorous cell specially or preferentially over normal tissue, are introduced to the liposomes in order to enhance the accumulation at tumor or disease sites, and reduce nonspecific circulation in other cells. The aforementioned strategy is sometimes referred to as "active targeting." This is in contrast to the fact that most of the current liposomal systems use "passive" targeting mechanism, in which the circulating drugs migrate and accumulate at the disease sites [99, 103–106]. For anticancer drugs, many of the current delivery systems assume that liposomal drugs accumulate at tumor sites due to the enhanced permeation and retention (EPR) effect.

Targeted drug deliveries to specific disease sites and then releasing the drugs at desirable rates at the intended sites have been the goals for drug delivery studies. Active ligand-targeted delivery systems to tumor site is highly complex and requires more extensive studies before they can reach the expected therapeutic potentials. It has been found that the majority of targeted delivery systems deliver their cargos by a nontargeted pathway [106]. The ligands are supposed to "guide" or direct the liposomal drugs to cancerous cells and increase the concentration site-specifically and, therefore, reducing toxicity. However, targeted liposomal and other nanoparticle delivery systems often enter the bloodstream in a similar fashion as nontargeted delivery systems [106]. Nonetheless, targeted drug delivery systems provide benefits of targeting cell internalization and tissue retention once they reach the targeted sites. Many targeted systems have shown to be safe and efficacious in preclinical models [103, 105]. Currently about 13 ligand-targeted nanoparticle systems have progressed into clinical trials [105], although a great deal of effort needs to be directed toward the understanding of the pharmacokinetics of the delivery systems.

Targeted moieties can be conjugated to phospholipids through their amino or carboxyl functional groups using standard peptide coupling chemistry. Some examples of the lipids employed for targeted drug delivery are shown in Figure 13.18. Folate receptors have been found to be over expressed by many forms of cancer [98, 104a]. Therefore, folate was conjugated to distearoylphosphatidylethanolamine (DSPE) through PEG linkage and studied in various drug delivery models. The folate-targeted

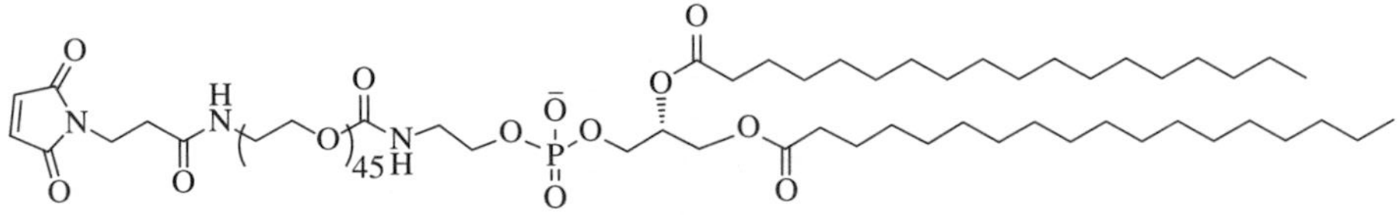

44 DSPE-PEG(2000) Maleimide

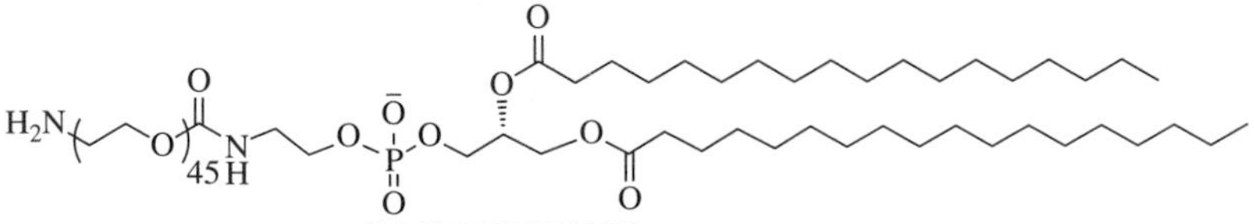

45 DSPE-PEG(2000)

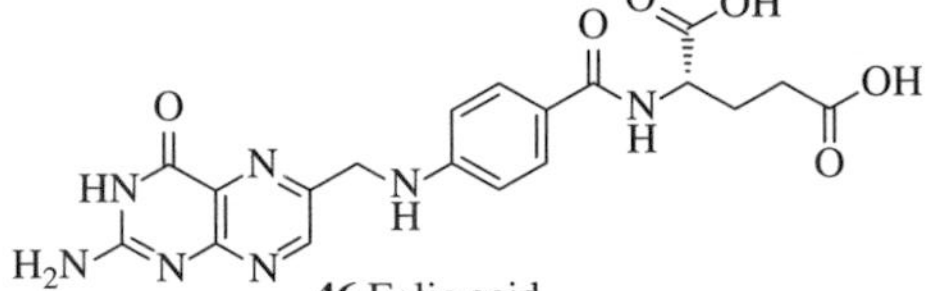

46 Folic acid

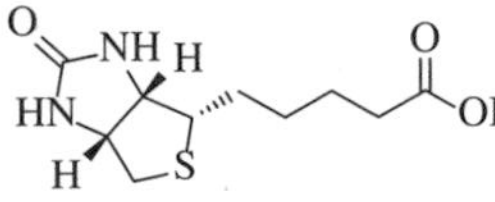

47 Biotin

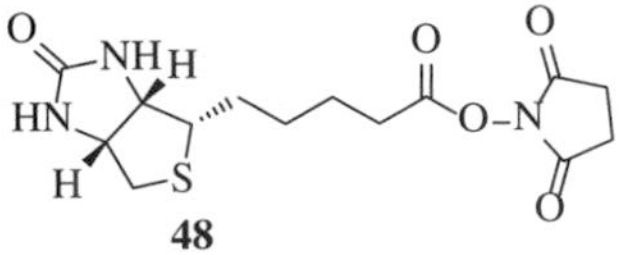

48

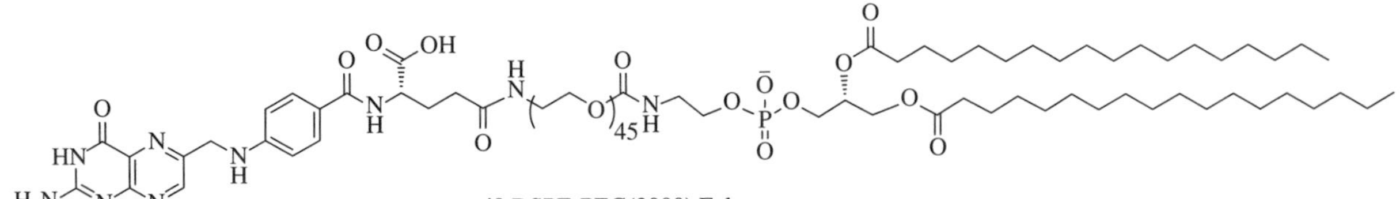

49 DSPE-PEG(2000) Folate

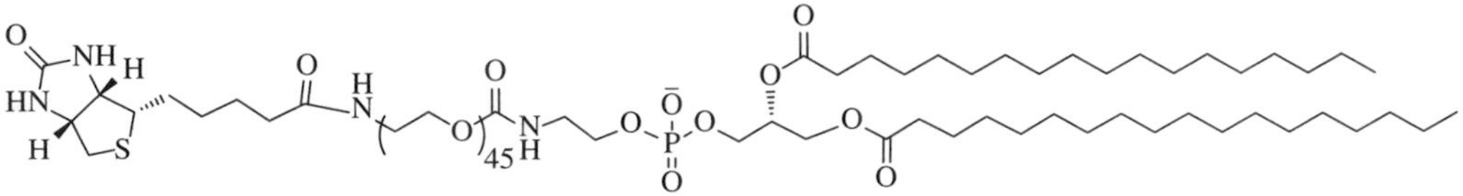

50 DSPE-PEG(2000) Biotin

FIGURE 13.18 Structures of PEGylated lipids that are useful for conjugation with ligands or antibodies, and the structures of two targeted lipids.

lipid **49** can be obtained by conjugating the PEG-DSPE **45** with folic acid **46**; and the biotinylated lipid **50** can be synthesized by coupling the activated biotin **48** with **45** [108]. The ligand-targeted approach may be advantageous in reversing multidrug resistance in tumor cells, because targeted delivery systems may be able to circumvent the drug-efflux pumps after receptor-mediated endocytosis. In contrast, nontargeted systems typically reach tumor cells by passive diffusion and presumably through the EPR mechanism [99–101, 104b, 107].

Although the majority of the targeted drug delivery systems are developed for anticancer therapy [98–112], examples of noncancer-targeted systems have also been reported. These include using multivalent glycosides targeting lectins [113] and for anticoagulation therapy by targeting thrombin [114]. Recently a thrombin-targeted drug delivery system that exhibits sustained local inhibition of thrombin has been reported [114]. In this system, a direct thrombin inhibitor (PPACK) was attached to the surface of the liposome, and the PPACK-liposomes were found to be able to bind and inhibit thrombin and prevent further clot formation. Figure 13.19 shows the coupling of PPACK **51** to a PEGylated DSPE carboxylic acid **52** and the formation of the functionaliszed lipid **53**. The liposomal formulation had a long-lasting multivalent anticlotting surface over a newly forming clot. The system does not require systematic administration of thrombin inhibitor. This has potential for site-specific treatment in an acute thrombosis.

13.6 HYBRID LIPOSOME DRUG DELIVERY SYSTEM

To achieve sustained release and to deliver multiple drugs in a controlled manner, liposomes have been combined with many other drug delivery vehicles to prepare multifunctional smart hybrid drug delivery systems [115–125]. Liposomes combined with polymers have been utilized and studied extensively. One interesting drug delivery system that may hold promise in future drug delivery is the combination of hydrogel with liposomes [115, 116]. Both encapsulation of liposomes in the gel matrix and encapsulating hydrogels inside liposomes have been explored. The advantage of using a combination of delivery methods is that upon the triggered release of liposomes, the content in the gels can be released slowly into the tumor site. Therefore, the system is more advantageous in maintaining an effective therapeutic concentration at the target site without the need for repeated administration of the drugs. This could be potentially useful in modulating proper pharmacokinetics of the drugs. Another advantage is that possible drugs or therapeutic agents with very different physical properties can be encapsulated in one delivery vehicle and applied simultaneously. This approach could increase the effectiveness of combination therapy.

There is also a growing trend of combining diagnosis with drug delivery. The so called "theranostic" systems can incorporate imaging probes into the liposomal formulations together with other triggered or targeted components [121–125]. Such approaches can be very important in nanomedicine.

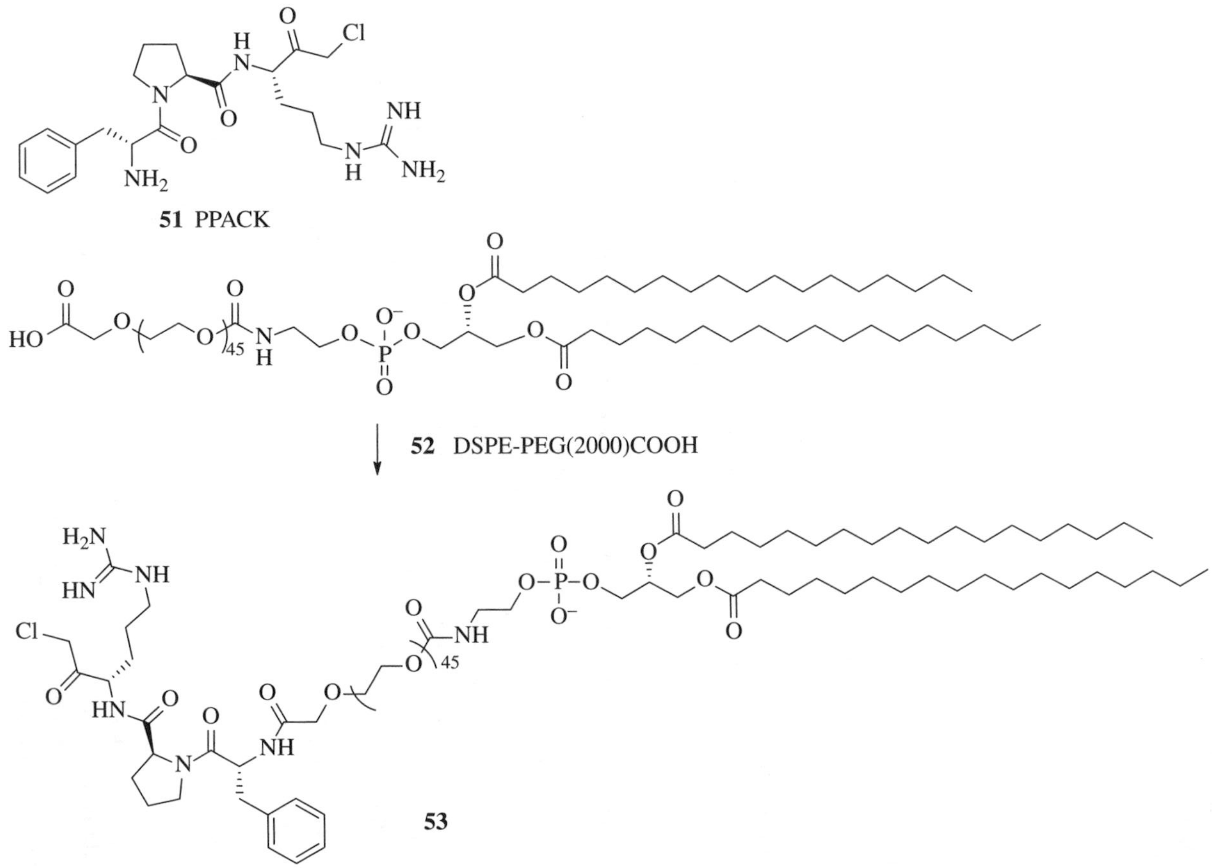

51 PPACK

52 DSPE-PEG(2000)COOH

53

FIGURE 13.19 The conjugation of PPACK to PEG-DSPE-COOH.

13.7 CONCLUSIONS AND FUTURE PERSPECTIVES

Liposomes as drug delivery vehicles will find broader applications with the study of tailor-designed lipids in conjunction with the therapeutic agents to be delivered. It is important that for certain types of drugs, the liposomes used should also be tailored to match the physical and chemical properties of the agents.

Triggered release strategies using thermosensitive, pH-sensitive, photo-sensitive, and enzyme-responsive liposomes are expected to gain more applications in future. Furthermore, triggered release mechanisms can also be used in conjunction with other mechanisms including targeted delivery. However, when adding more functionality and building complex delivery systems, one must consider how the system will perform under physiological conditions, their biocompatibility, degradability, and other pharmacokinetic issues.

The hybrid of liposomes with other nanocarriers such as polymers, inorganic nanoparticle complexes, and compartmentalized deliveries will find broader application in the future, though many therapeutic hurdles still need to be resolved. The payload may be adjusted based on the need for the therapy, and sustained release and triggered release should also be balanced based on the types of biological targets. The combinations of liposomes with hydrogels may also find greater therapeutic applications in future.

Finally theranostic agents, which combine therapeutics with diagnostics, are also gaining great momentum. Such approaches can take advantage of the availability of different nanocarriers with smart design to achieve both diagnosis and treatment (see Chapter 22 for a more in-depth discussion). Liposomes as basic systems, together with more complex vehicles, will continue to have an important impact on the future of nanomedicine.

REFERENCES

1. (a) Bangham, A. D.; Standish, M. M.; Watkins, J. C. *J. Mol. Biol.* 1965, **13**, 238–252. (b) Bangham, A. D.; Standish, M. M.; Weissman, G. *J. Mol. Biol.* 1965, **13**, 253–259.
2. Lian, T.; Ho, R. J. Y. *J. Pharm. Sci.* 2001, **90**, 667–680.
3. (a) Allen T. M.; Cullis, P. R. *Science* 2004, **303**, 1818–1822. (b) Allen, T. M.; Cullis, P. R. *Adv. Drug Deliv. Rev.* 2013, **65**, 36–48.
4. Torchilin, V. P. *Nat. Rev. Drug Discov.* 2005, **4**, 145–160.
5. Zhang, Y.; Chan, H. F.; Leong, K. W. *Adv. Drug Deliv. Rev.* 2013, **65**, 104–120.
6. Sen, K.; Mandal, M. *Int. J. Pharm.* 2013, **448**, 28–43.
7. Tacar, O.; Sriamornsak, P.; Dass, C. R. *J. Pharm. Pharmacol.* 2013, **65**, 157–170.
8. Lopez-Davila, V.; Seifalian, A. M.; Loizidou, M. *Curr. Opin. Pharmacol.* 2012, **12**, 414–419.
9. Yin, Q.; Shen, J.; Zhang, Z.; Yu, H.; Li, Y. *Adv. Drug Deliv. Rev.* 2013, **65**, 1699–1715.
10. Pisano, C.; Cecere, S. C.; Di Napoli, M.; Cavaliere, C.; Tambaro, R.; Facchini, G.; Scaffa, C.; Losito, S.; Pizzolorusso, A.; Pignata, S. *J. Drug Deliv.* 2013, 898146–898158.
11. (a) Derycke, A. S. L.; de Witte, P. A. M. *Adv. Drug Deliv. Rev.* 2004, **56**, 17–30. (b) Sharmana, W. M.; van Liera, J. E.; Allen, C. M. *Adv. Drug Deliv. Rev.* 2004, **56**, 53–76.

12. Hyodo, K.; Yamamoto, E.; Suzuki, T.; Kikuchi, H.; Asano, M.; Ishihara, H. *Biol. Pharm. Bull.* 2013, **36**, 703–707.
13. (a) Immordino, M. L.; Dosio, F.; Cattel, L. *Int. J. Nanomedicine* 2006, **1**, 297–315. (b) Lasic, D. D.; Needham, D. *Chem. Rev.* 1995, **95**, 2601–2628.
14. (a) Bibi, S.; Lattmann, E.; Mohammed, A. R.; Perrie, Y. *J. Microencapsul.* 2012, **29**, 262–276. (b) Mellal, D.; Zumbuehl, A. *J. Mater. Chem. B* 2014, **2**, 247–252. (c) Sawant, R. R.; Torchilin, V. P. *Soft Matter* 2010, **6**, 4026–4044.
15. Cui, Z.-K.; Lafleur, M. *Colloids Surf. B* 2014, **114**, 177– 185.
16. Needham, D.; Dewhirst, M. W. *Adv. Drug Deliv. Rev.* 2001, **53**, 285–305.
17. Gruell, H.; Langereis, S. *J. Control. Release* 2012, **161**, 317–327.
18. May, J. P.; Li, S.-D. *Expert Opin. Drug Deliv.* 2013, **10**, 511–527.
19. Chang, H.-I.; Yeh, M.-K. *Int. J. Nanomedicine* 2012, **7**, 49–60.
20. Ta, T.; Porter, T. M. *J. Control. Release* 2013, **169**, 112–125.
21. Park, S. M.; Kim, M. S.; Park, S.-J.; Park, E. S.; Choi, K.-S.; Kim, Y.; Kim, H. R. *J. Control. Release* 2013, **170**, 373–379.
22. Tagami, T.; May, J. P.; Ernsting, M. J.; Li, S.-D. *J. Control. Release* 2012, **161**, 142–149.
23. Gasselhuber, A.; Dreher, M. R.; Rattay, F.; Wood, B. J.; Haemmerich, D. *PLoS One* 2012, **7**, e47453.
24. Staruch, R. M.; Ganguly, M.; Tannock, I. F.; Hynynen, K.; Chopra, R. *Int. J. Hyperth.* 2012, **28**, 776–787.
25. Al-Ahmady, Z. S.; Al-Jamal, W. T.; Bossche, J. V.; Bui, T. T.; Drake, A. F.; Mason, A. J.; Kostarelos, K. *ACS Nano* 2012, **6**, 9335–9346.
26. Djanashvili, K.; ten Hagen, T. L. M.; Blange, R.; Schipper, D.; Peters, J. A.; Koning, G. A. *Bioorg. Med. Chem.* 2011, **19**, 1123–1130.
27. de Smet, M.; Heijman, E.; Langereis, S.; Hijnen, N. M.; Gruell, H. *J. Control. Release* 2011, **150**, 102–110.
28. Tagami, T.; Ernsting, M. J.; Li, S. D. *J. Control. Release* 2011, **152**, 303–309.
29. Tagami, T.; Ernsting, M. J.; Li, S. D. *J. Control. Release* 2011, **154**, 290–297.
30. Kono, K.; Ozawa, T.; Yoshida, T.; Ozaki, F.; Ishizaka, Y.; Maruyama, K.; Kojima, C.; Harada, A.; Aoshima, S. *Biomaterials* 2010, **31**, 7096–7105.
31. Negussie, A. H.; Yarmolenko, P. S.; Partanen, A.; Ranjan, A.; Jacobs, G.; Woods, D.; Bryant, H.; Thomasson, D.; Dewhirst, M. W.; Wood, B. J., Dreher, M. R. *Int. J. Hyperth.* 2011, **27**, 140–155.
32. Poon, R. T. P.; Borys, N. *Future Oncol.* 2011, **7**, 937–945.
33. Pradhan, P.; Giri, J.; Rieken, F.; Koch, C.; Mykhaylyk, O.; Doblinger, M.; Banerjee, R.; Bahadur, D.; Plank, C. *J. Control. Release* 2010, **142**, 108–121.
34. Lindner, L. H.; Eichhorn, M. E.; Eibl, H.; Teichert, N.; Schmitt-Sody, M.; Issels, R. D., Dellian, M. *Clin. Cancer Res.* 2004, **10**, 2168–2178.
35. Langereis, S.; Keupp, J.; van Velthoven, J. L. J.; de Roos, I. H. C.; Burdinski, D.; Pikkemaat, J. A.; Gruell, H. *J. Am. Chem. Soc.* 2009, **131**, 1380–1381.
36. Dicheva, B. M.; Koning, G. A. *Expert Opin. Drug Deliv.* 2014, **11**, 83–100.
37. Dicheva, B. M.; ten Hagen, T. L. M.; Li, L.; Schipper, D.; Seynhaeve, A. L. B.; van Rhoon, G. C.; Eggermont, A. M. M.; Lindner, L. H.; Koning, G. A. *Nano Lett.* 2013, **13**, 2324–2331.

38. Yatvin, M. B.; Kreutz, W.; Horwitz, B. A.; Shinitzky, M. *Science* 1980, **210**, 1253–1255.
39. Gao, W.; Chan, J. M.; Farokhzad, O. C. *Mol. Pharm.* 2010, **7**, 2469–2478.
40. Liu, X.; Huang, G. *Asian J. Pharm. Sci.* 2013, **8**, 319–328.
41. Simões, S.; Moreira, J. N.; Fonseca, C.; Düzgüneş, N.; de Lima, M. C. *Adv. Drug Delivery Rev.* 2004, **56**, 947–965.
42. Roux, E.; Passirani, C.; Scheffold, S.; Benoit, J. P.; Leroux, J. C. *J. Control. Release* 2004, **94**, 447–451.
43. Ishida, T.; Okada, Y.; Kobayashi, T.; Kiwada, H. *Int. J. Pharm.* 2006, **309**, 94–100.
44. Yeo, Y.; Xu, P. S.; Bajaj, G.; Shugg, T.; Van Alstine, W. G. *Biomacromolecules* 2010, **11**, 2352–2358.
45. Liu, Y.; Gao, F. P.; Zhang, D.; Yun-Shan Fan, Y. S.; Chen, X. G.; Wang, H *J. Control. Release* 2014, **173**, 140–147.
46. Chen, K.-J.; Liang, H.-F.; Chen, H.-L.; Wang, Y.; Cheng, P.-Y.; Liu, H.-L.; Xia, Y.; Sung, H.-W. *ACS Nano* 2013, **7**, 438–446.
47. Samoshin, A. V.; Veselov, I. S.; Chertkov, V. A; Yaroslavov, A. A.; Grishina, G. V.; Samoshina, N. M.; Samoshin, V. V. *Tetrahedron Lett.* 2013, **54**, 5600–5604.
48. Sawant, R. M.; Hurley, J. P.; Salmaso, S.; Kale, A.; Tolcheva, E.; Levchenko, T. S. Torchilin, V. P. *Bioconjug. Chem.* 2006, **17**, 943–949.
49. Kale, A. A.; Torchilin, V. P. *Bioconjug. Chem.* 2007, **18**, 363– 370.
50. (a) Remenyi, J.; Balazs, B.; Toth, S.; Falus, A.; Toth, G.; Hudecz, F. *Biochem. Biophys. Res. Commun.* 2003, **303**, 556–561. (b) Reddy, J. A.; Low, P. S. *J. Control. Release* 2000, **64**, 27–37.
51. Hu, F. Q.; Zhang, Y. Y.; You, J.; Yuan, H.; Du, Y. Z. *Mol. Pharm.* 2012, **9**, 2469–2478.
52. Gillies, E. R.; Goodwin, A. P.; Frechet, J. M. J. *Bioconjug. Chem.* 2004, **15**, 1254–1283.
53. Knorr, V.; Allmendinger, L.; Walker, G. F.; Paintner, F. F.; Wagner, E. *Bioconjug. Chem.* 2007, **18**, 1218–1225.
54. Knorr, V.; Russ, V.; Allmendinger, L.; Ogris, M.; Wagner, E. *Bioconjug. Chem.* 2008, **19**, 1625–1634.
55. Masson, C.; Garinot, M.; Mignet, N.; Wetzer, B.; Mailhe, P.; Scherman, D.; Bessodes, M. *J. Control. Release* 2004, **99**, 423–434.
56. Huang, Z. H.; Guo, X.; Li, W. J.; MacKay, J. A.; Szoka, F. C. *J. Am. Chem. Soc.* 2006, **128**, 60–61.
57. Chen, H. G.; Zhang, H. Z.; McCallum, C. M.; Szoka, F. C.; Guo, X. *J. Med. Chem.* 2007, **50**, 4269–4278.
58. Guo, X.; MacKay, J. A.; Szoka, F. C., Jr. *Biophys. J.* 2003, **84**, 1784–1795.
59. Bruyere, H.; Westwell, A. D.; Jones, A. T. *Bioorg. Med. Chem. Lett.* 2010, **20**, 2200–2203.
60. Boomer, J. A.; Inerowicz, H. D.; Zhang, Z.-Y.; Bergstrand, N.; Edwards, K.; Kim, J.-M.; Thompson, D. H. *Langmuir* 2003, **19**, 6408–6415.
61. Shin, J.; Shum, P.; Thompson, D. H. *J. Control. Release* 2003, **91**, 187–200.
62. Xu, Z.; Gu, W.; Chen, L.; Gao, Y.; Zhang, Z.; Li, Y. *Biomacromolecules* 2008, **9**, 3119–3126.
63. Boomer, J. A.; Qualls, M. M.; Inerowicz, H. D.; Haynes, R. H.; Patri, V. S.; Kim, J. M.; Thompson, D. H. *Bioconjug. Chem.* 2009, **20**, 47–59.

64. Kim, H. K.; Thompson, D. H.; Jang, H. S.; Chung, Y. J.; Van den Bossche, J. *ACS Appl. Mater. Interfaces* 2013, **5**, 5648–5658.
65. Shin, J.; Shum, P.; Grey, J.; Fujiwara, S.; Malhotra, G. S.; Gonzalez-Bonet, A.; Hyun, S.-H.; Moase, E.; Allen, T. M.; Thompson, D. H. *Mol. Pharm.* 2012, **9**, 3266–3276.
66. Kim, J. M.; Van den Bossche, J.; Hyun, S.-H.; Thompson, D. H. *Bioconjug. Chem.* 2012, **23**, 2071–2077.
67. Carmen Alvarez-Lorenzo, C.; Bromberg, L.; Concheiro, A. *Photochem. Photobiol.* 2009, **85**, 848–860.
68. Yavlovich, A.; Smith, B.; Gupta, K.; Blumenthal, R.; Puri, A. *Mol. Membr. Biol.* 2010, **27**, 364–381.
69. Liu, X. M.; Yang, B.; Wang, Y. L.; Wang, J. Y. *Biochim. Biophys. Acta Biomembr.* 2005, **1720**, 28–34.
70 (a) Sasaki, Y.; Iwamoto, S.; Mukai, M.; Kikuchi, J. *J. Photochem. Photobiol. A: Chemistry* 2006, **183**, 309–314. (b) Liang, X.; Yue, X.; Dai, Z.; Kikuchi, J. *Chem. Commun.* 2011, **47**, 4751–4753.
71. Backus, E. H. G.; Kuiper, J. M.; Engberts, J. B. F. N.; Poolman, B.; Bonn, M. *J. Phys. Chem. B* 2011, **115**, 2294–2302.
72. Tangso, K. J.; Fong, W.-K.; Darwish, T.; Kirby, N.; Boyd, B. J.; Hanley, T. L. *J. Phys. Chem. B* 2013, **117**, 10203–10210.
73. (a) Wang, G.; Hollingsworth, R. I. *Adv. Mater.* 2000, **12**, 871–874. (b) Nie, X.; Wang, G. *J. Org. Chem.* 2006, **71**, 4734–4741.
74. Yavlovich, A.; Singh, A.; Blumenthal, R.; Puri, A. *Biochim. Biophys. Acta* 2011, **1808**, 117–126.
75. Lee, H. M.; Larson, D. R.; Lawrence, D. S. *ACS Chem. Biol.* 2009, **4**, 409–427.
76. Mayer, G.; Heckel, A. *Angew. Chem. Int. Ed.* 2006, **45**, 4900–4921.
77. Chandra, B.; Subramaniam, R.; Mallik, S.; Srivastava, D. K. *Org. Biomol. Chem.* 2006, **4**, 1730–1740.
78. Thompson, S.; Dessi, J.; Self, C. H. *Biochem. Biophys. Res. Commun.* 2008, **366**, 526–531.
79. Dong, J.; Zeng, Y.; Xun, Z.; Han, Y.; Chen, J.; Li, Y.-Y.; Li, Y. *Langmuir* 2012, **28**, 1733–1737.
80. Sun, Y.; Yan, Y.; Wang, Y.; Wang, M.; Chen, C.; Xu, H.; Lu, J. R. *ACS Appl. Mater. Interfaces* 2013, **5**, 6232–6236.
81. Fang, N.-C.; Cheng, F.-Y.; Ho, J. A.; Yeh, C.-S. *Angew. Chem. Int. Ed.* 2012, **51**, 8806–8810.
82. Seo, H. J.; Kim, J.-C. *J. Nanosci. Nanotechnol.* 2011, **11**, 10262–10270.
83. Carter Ramirez, D. M.; Kim, Y. A.; Bittman, R.; Johnston, L. J. *Soft Matter* 2013, **9**, 4890–4899.
84. Andresen, T. L.; Thompson, D. H.; Kaasgaard, T. *Mol. Membr. Biol.* 2010, **27**, 353–363.
85. Andresen, T. L.; Jensen, S. S.; Jorgensen, K. *Prog. Lipid Res.* 2005, **44**, 68–97.
86. Shirazi, R. S.; Ewert, K. K.; Silva, B. F. B.; Leal, C.; Li, Y.; Safinya, C. R. *Langmuir* 2012, **28**, 10495–10503.
87. Ong, W.; Yang, Y.; Cruciano, A. C.; McCarley, R. L. *J. Am. Chem. Soc.* 2008, **130**, 14739–14744.

88. Goldenbogen, B.; Brodersen, N.; Gramatica, A.; Loew, M.; Liebscher, J.; Herrmann, A.; Egger, H.; Budde, B.; Arbuzova, A. *Langmuir* 2011, **27**, 10820–10829.
89. Andresen, T. L.; Davidsen, J.; Begtrup, M.; Mouritsen, O. G.; Jørgensen, K. *J. Med. Chem.* 2004, **47**, 1694–1703.
90. Linderoth, L.; Fristrup, P.; Hansen, M.; Melander, F.; Madsen, R.; Andresen, T. L.; Peters, G. H. *J. Am. Chem. Soc.* 2009, **131**, 12193–12200.
91. Pedersen, P. J.; Adolph, S. K.; Subramanian, A. K.; Arouri, A.; Andresen, T. L.; Mouritsen, O. G.; Madsen, R.; Madsen, M. W.; Peters, G. H.; Clausen, M. H. *J. Med. Chem.* 2010, **53**, 3782–3792.
92. (a) Foged, C.; Nielsen, H. M.; Frokjaer, S. *Int. J. Pharm.* 2007, **331**, 160–166. (b) Zhu, G.; Mock, J. N.; Aljuffali, I.; Cummings, B. S.; Arnold, R. D. *J. Pharm. Sci.* 2011, **100**, 3146–3159.
93. de la Rica R., Aili, D.; Stevens, M. M. *Adv. Drug Deliv. Rev.* 2012, **64**, 967–978.
94. Sarkar, N.; Banerjee, J.; Hanson, A. J.; Elegbede, A. I.; Rosendahl, T.; Krueger, A. B.; Banerjee, A. L.; Tobwala, S.; Wang, R.; Lu, X.; Mallik, S.; Srivastava, D. K. *Bioconjug. Chem.* 2008, **19**, 57–64.
95. Romberg, B.; Flesch, F. M.; Hennink, W. E.; Storm, G. *Int. J. Pharm.* 2008, **355**, 108–113.
96. Elegbede, A. I.; Banerjee, J.; Hanson, A. J.; Tobwala, S.; Ganguli, B.; Wang, R.; Lu, X.; Srivastava, D. K.; Mallik, S. *J. Am. Chem. Soc.* 2008, **130**, 10633–10642.
97. Yingyuad, P.; Mevel, M.; Prata, C.; Furegati, S.; Kontogiorgis, C.; Thanou, M.; Miller, A. D. *Bioconjug. Chem.* 2013, **24**, 343–362.
98. Lu, Y. J.; Low, P. S. *Adv. Drug Deliv. Rev.* 2002, **54**, 675–693.
99. Zhao, G.; Rodriguez, B. L. *Int. J. Nanomedicine* 2013, **8**, 61–71.
100. Medina, O. P., Zhu, Y., Kairemo, K. *Curr. Pharm. Des.* 2004, **10**, 2981–2989.
101. Steichen, S. D.; Caldorera-Moore, M.; Peppas, N. A. *Eur. J. Pharm. Sci.* 2013, **48**, 416–427.
102. Gao, J.; Chen, H.; Song, H.; Su, X.; Niu, F.; Li, W.; Li, B.; Dai, J.; Wang, H.; Guo, Y. *Mini-Rev. Med. Chem.* 2013, **13**, 2026–2035.
103. Noble, G. T.; Stefanick, J. F.; Ashley, J. D.; Kiziltepe, T.; Bilgicer, B. *Trends Biotechnol.* 2014, **32**, 32–45.
104. (a) Zhao, X. B.; Lee, R. J. *Adv. Drug Deliv. Rev.* 2004, **56**, 1193–1204. (b) Maruyama, K. *Adv. Drug Deliv. Rev.* 2011, **63**, 161–169.
105. van der Meel, R.; Vehmeijer, L. J. C.; Kok, R. J.; Storm, G.; van Gaal, E. V. B. *Adv. Drug Deliv. Rev.* 2013, **65**, 1284–1298.
106. Kwon, I. K.; Lee, S. C.; Han, B.; Park, K. *J. Control. Release* 2012, **164**, 108–114.
107. Danhier, F.; Feron, O.; Preat, V. *J. Control. Release* 2010, **148**, 135–146.
108. Liu, D.; Liu, F.; Liu, Z.; Wang, L.; Zhang, N. *Mol. Pharm.* 2011, **8**, 2291–2301.
109. Takara, K.; Hatakeyama, H.; Ohga, N.; Hida, K.; Harashima, H. *Int. J. Pharm.* 2010, **396**, 143–148.
110. Li, X. M.; Ding, L. Y.; Xu, Y. L.; Wang, Y. L.; Ping, Q. N. *Int. J. Pharm.* 2009, **373**, 116–123.
111. Manjappa, A. S.; Chaudhari, K. R.; Venkataraju, M. P.; Dantuluri, P.; Nanda, B.; Sidda, C.; Sawant, K. K.; Murthy, R. S. *J. Control. Release* 2011, **150**, 2–22.

112. Danhier, F.; Breton, A. L.; Preat, V. *Mol. Pharm.* 2012, **9**, 2961–2973.

113. (a) Qu, B.; Li, X.; Li, X.; Hai, L.; Guan, M.; Wu, Y. *Eur. J. Med. Chem.* 2013, **72C**, 110–118. (b) Aleandri, S.; Casnati, A.; Fantuzzi, L.; Mancini, G.; Rispoli, G.; Sansone, F. *Org. Biomol. Chem.* 2013, **11**, 4811–4817.

114. Palekar, R. U.; Myerson, J. W.; Schlesinger, P. H.; Sadler, J. E.; Pan, H.; Wickline, S. A. *Mol. Pharm.* 2013, **10**, 4168–4175.

115. Hong, J. S.; Stavis, S. M.; De Paoli Lacerda, S. H.; Locascio, L. E.; Raghavan, S. R.; Gaitan, M. *Langmuir* 2010, **26**, 11581–11588.

116. Tsumoto, K.; Oohashi, M.; Tomita, M. *Colloid Polym. Sci.* 2011, **289**, 1337–1346.

117. Li, S. H.; Goins, B.; Zhang, L. J.; Bao, A. D. *Bioconjug. Chem.* 2012, **23**, 1322–1332.

118. Hosta-Rigau, L.; Chung, S. F.; Postma, A.; Chandrawati, R.; Staedler, B.; Caruso, F. *Adv. Mater.* 2011, **23**, 4082–4087.

119. Chandrawati, R.; Hosta-Rigau, L.; Vanderstraaten, D.; Lokuliyana, S. A.; Stadler, B.; Albericio, F.; Caruso, F. *ACS Nano* 2010, **4**, 1351–1361.

120. Shahin, M.; Soudy, R.; Aliabadi, H. M.; Kneteman, N.; Kaur, K.; Lavasanifar, A. *Biomaterials* 2013, **34**, 4089–4097.

121. Mura, S.; Nicolas, J.; Couvreur, P. *Nat. Mater.* 2013, **12**, 991–1003.

122. Lehner, R.; Wang, X.; Marsch, S.; Hunziker, P. *Nanomedicine: NBM* 2013, **9**, 742–757.

123. (a) Svenson, S. *Mol. Pharm.* 2013, **10**, 848–856. (b) Lo, S.-T.; Kumar, A.; Hsieh, J.-T.; Sun, X. *Mol. Pharm.* 2013, **10**, 793–812.

124. Al-Jamal, W. T.; Kostarelos, K. *Acc. Chem. Res.* 2011, **44**, 1094–1104.

125. Shi, J.; Xiao, Z.; Kamaly, N.; Farokhzad, O. C. *Acc. Chem. Res.* 2011, **44**, 1123–1134.

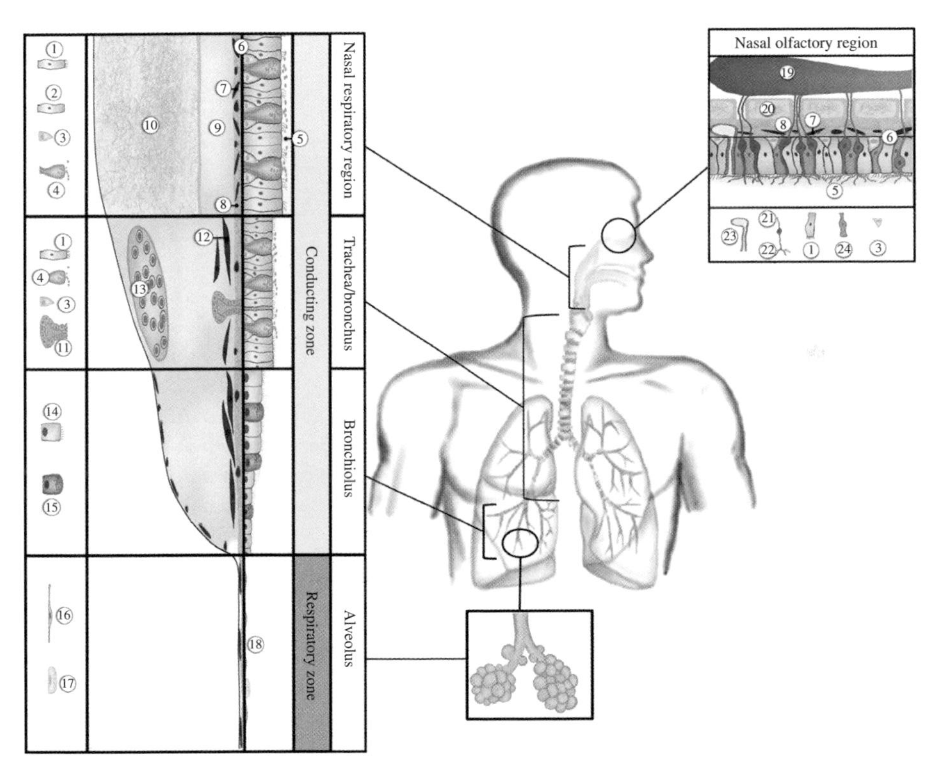

FIGURE 7.4 Cell types lining the respiratory system. Key: (1) ciliated columnar cells, (2) non-ciliated columnar cells, (3) basal cells, (4) goblet cells, (5) mucus, (6) basement membrane, (7) blood vessels, (8) lamina propria, (9) fibro-cartilaginous layer, (10) bone, (11) gland, (12) smooth muscle cells, (13) cartilage, (14) ciliated cuboidal cells, (15) Clara cells, (16) type I alveolar cells, (17) type II alveolar cells, (18) surfactant, (19) olfactory bulb, (20) cribriform plate, (21) receptor cell, (22) receptor cell cilia, (23) Bowman's gland, (24) supporting cell. Adapted from Klein et al. [49], pp. 1516–1534.

Drug Delivery: Principles and Applications, Second Edition. Edited by Binghe Wang, Longqin Hu, and Teruna J. Siahaan.

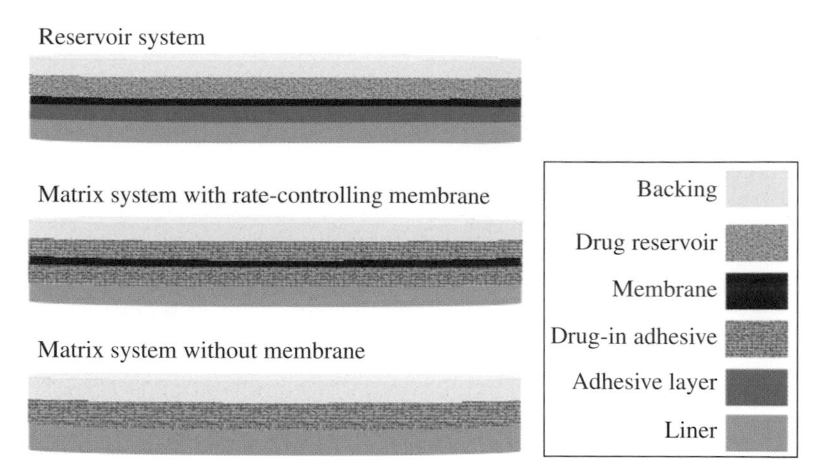

FIGURE 11.1 Popular designs of transdermal patches.

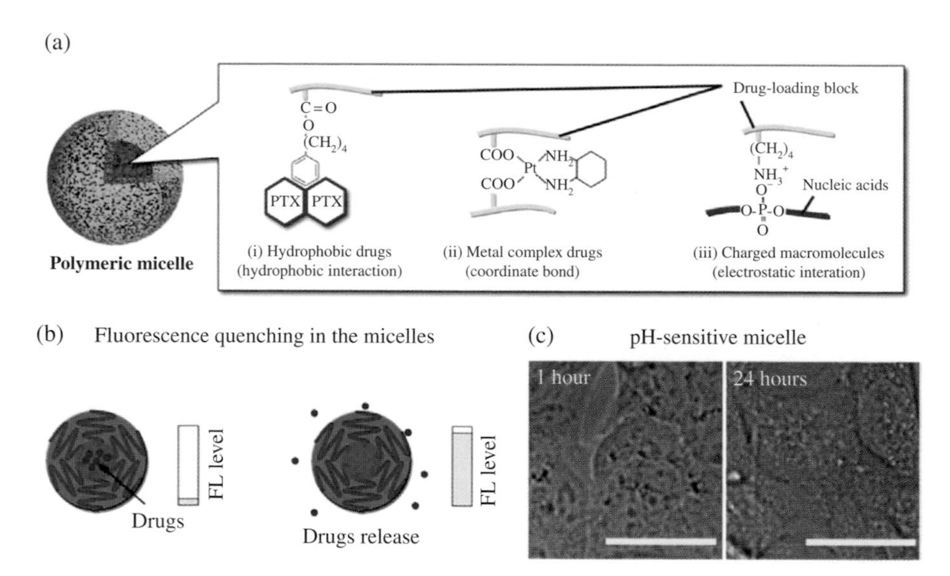

FIGURE 14.3 Potential mechanisms for drug encapsulation and release. (a) Coordination with the backbone of the polymer through either hydrophobic (i), negatively charged moieties (ii), or positively charged systems (iii) affords the direct connectivity required to both stabilize the payload, but also the DDV core. Miyata et al. [14], pp. 227–234. Reproduced with permission of Elsevier. (b) Fluorescent drugs or imaging agents undergo quenching upon encapsulation within the core, but their subsequent release restores their fluorescent properties. Bae and Kataoka [15], pp. 768–784. Reproduced with permission of Elsevier. (c) A pH-sensitive micelle demonstrates its environmental sensitivity through the release of its fluorescent payload upon a decrease in intracellular pH. Adapted from Bae and Kataoka [15]; Nishiyama et al. [16,17].

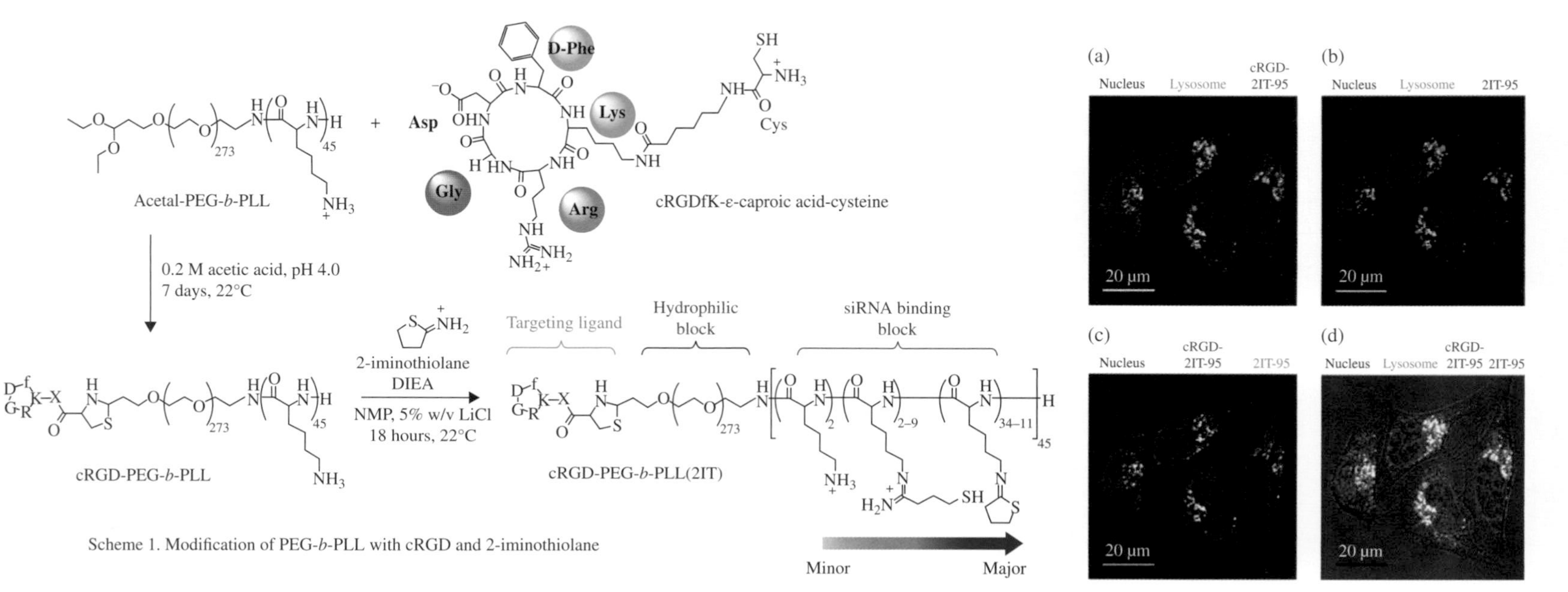

FIGURE 14.5 Effect of multivalency and DDV subcellular distribution. Introducing the cRGD peptide (Scheme 1), enhanced uptake and a broader subcellular distribution was observed. Adapted from Christie et al. [9], pp. 5174–5189. Reproduced with permission of American Chemical Society.

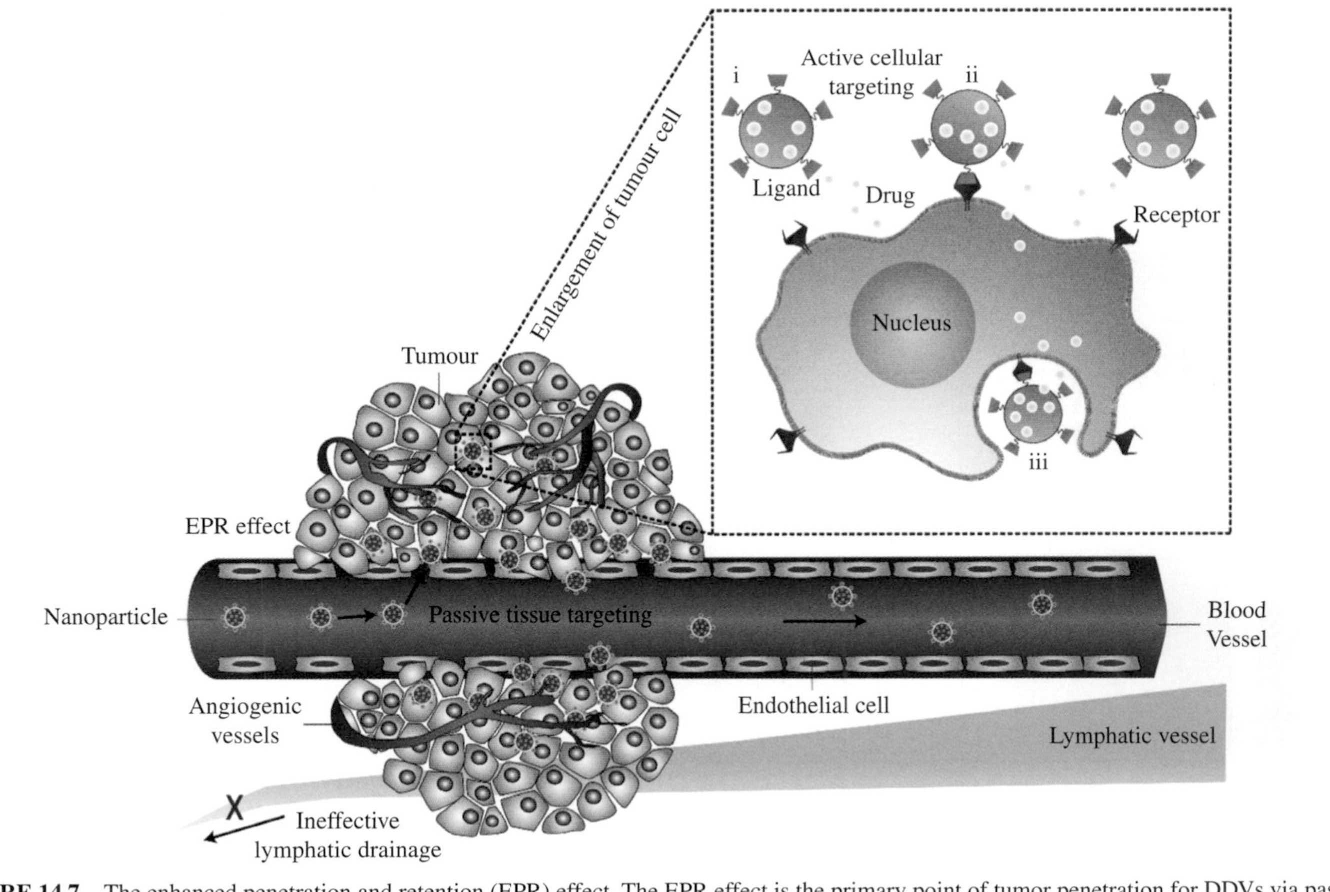

FIGURE 14.7 The enhanced penetration and retention (EPR) effect. The EPR effect is the primary point of tumor penetration for DDVs via passive targeting. Due to the irregular tumor vasculature and ineffective lymphatic drainage (RES), nanoparticles extravasate through the leaky system and enter the tumor environment. This resolves to expose the tumors cell surface and promote active DDV targeting (see *inset*) and ultimately, payload distribution [44]. Reprinted by permission from Macmillan Publishers Ltd: *Nature Nanotechnology* [44], copyright 2007.

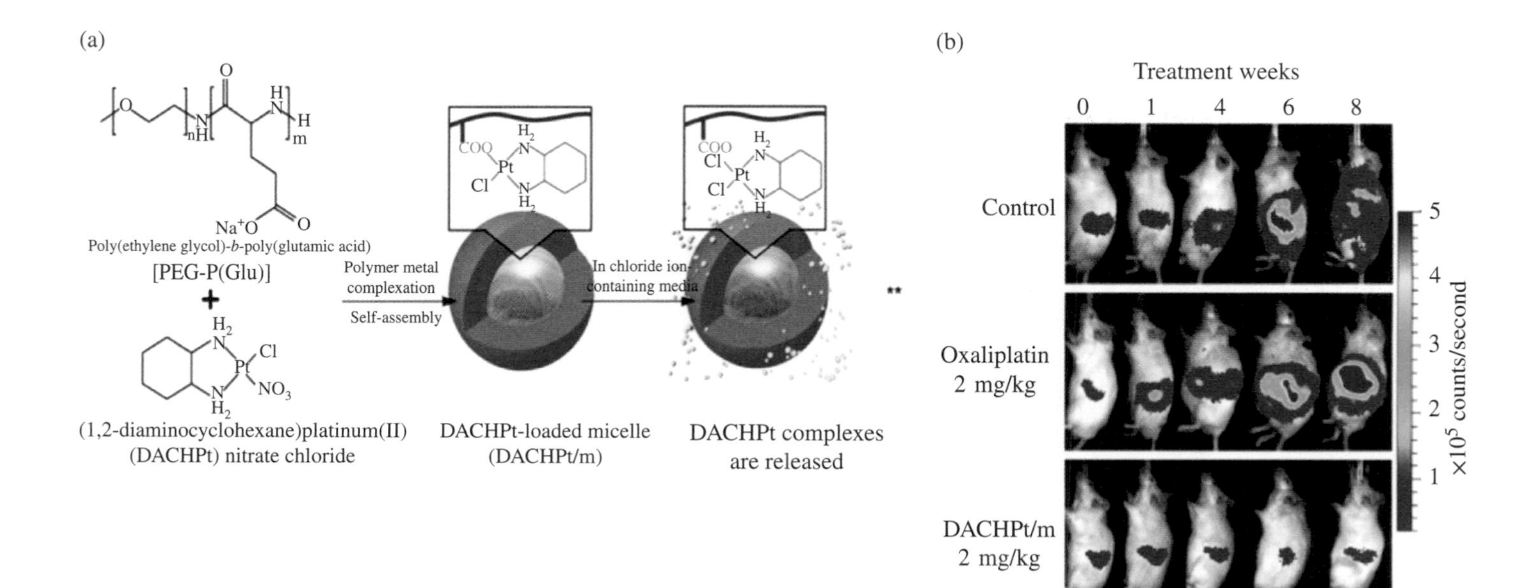

FIGURE 14.11 Preparation of DACHPt/m. (a) Self-assembly of the metal-polymer complex DACHPt and PEG-P(Glu). (b) Pancreatic tumor activity as demonstrated by bioluminescence signal (PNAS is not responsible for the accuracy of this translation). Cabral et al. [68], pp. 11397–11402. Reproduced with permission of PNAS.

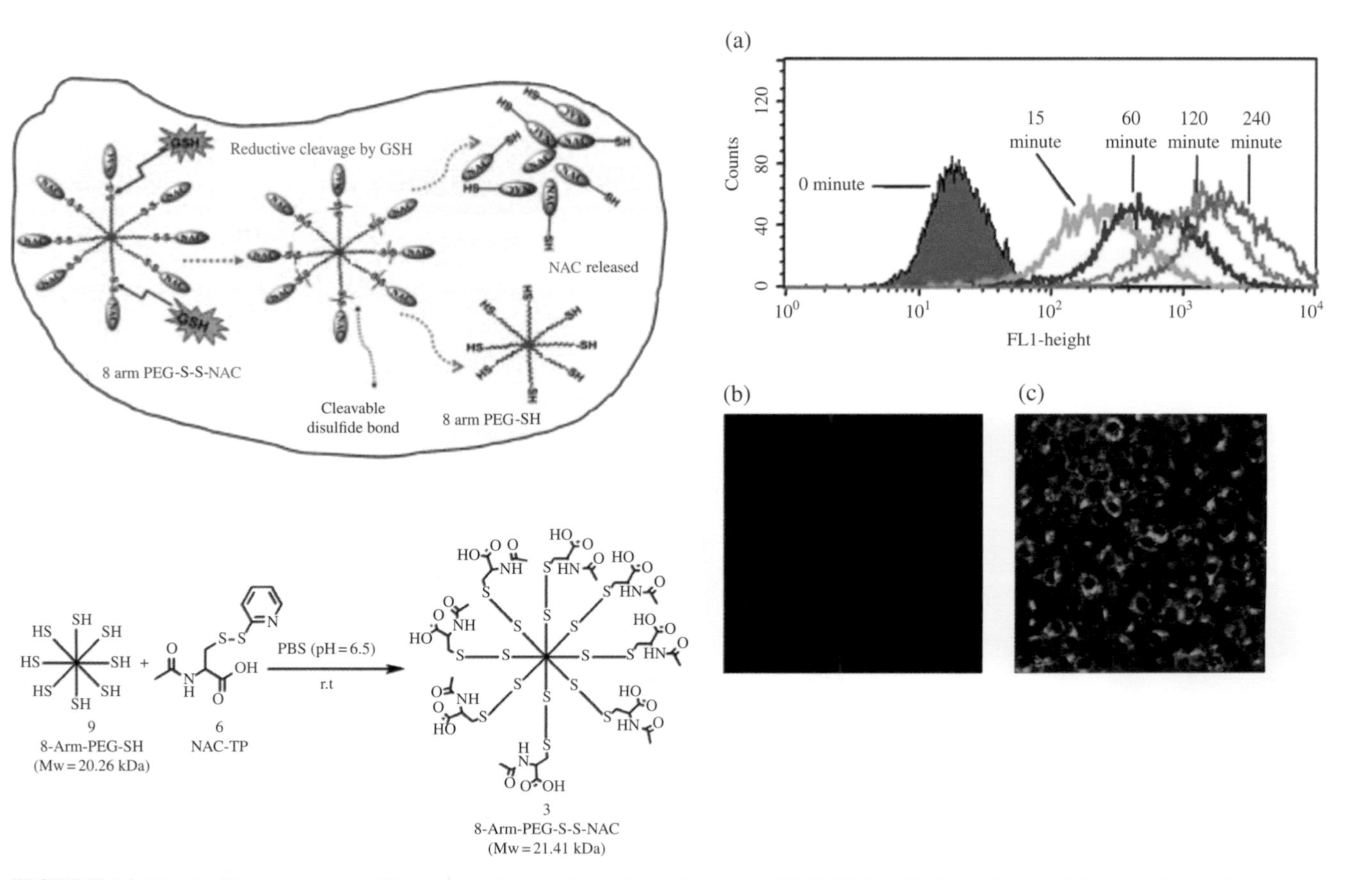

FIGURE 14.16 (a) Flow cytometry illustrating the rapid uptake of the 6-arm-PEG-FITC DDV. (b) Confocal image of negative control BV2 cells. (c) Confocal image of BV2 cells with 6-arm-PEG-FITC DDV. Navath et al. [75], pp. 447–456. Reproduced with permission of Elsevier.

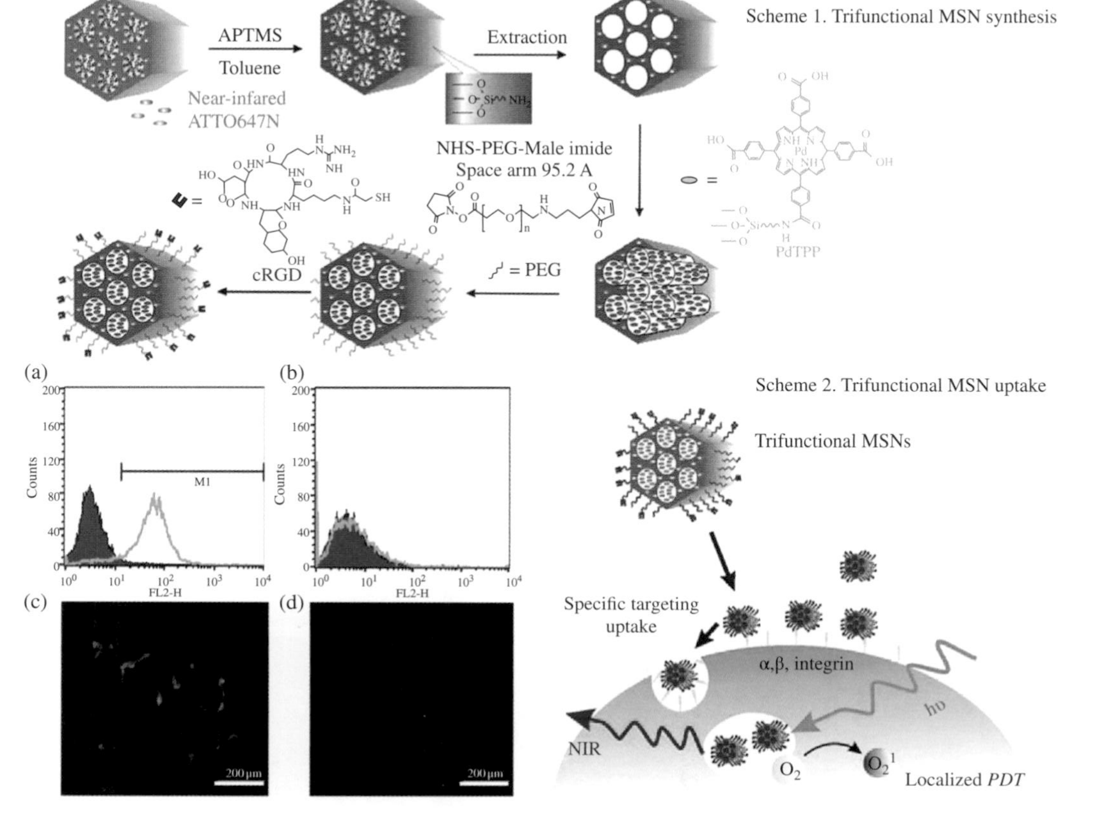

FIGURE 14.20 (Scheme 1) Synthesis of the trifunctional theranostic MSN incorporated with an NIR ATTO 647N contrast agent, cRGD targeting peptide, and palladium-based photosensitizer (Pd-TPP) for photodynamic therapy (PDT). (Scheme 2) Graphical representation of the trifunctional MSN binding to integrin receptors via the targeting cRGD peptide followed by cellular internalization, and Pd-TPP activation by PDT. (a) Flow cytometric analysis of positive control cells (U87-MG), which overexpress $\alpha_v\beta_3$ integrin receptors. (b) MCF-7 cells that are $\alpha_v\beta_3$ null for integrin receptors act as a negative control. (c and d) Immunofluorescent staining of $\alpha_v\beta_3$ integrin expression for U87-MG and MCF-7 cells, respectively. Reproduced ("Adapted" or "in part") from [89] with permission from the Royal Society of Chemistry.

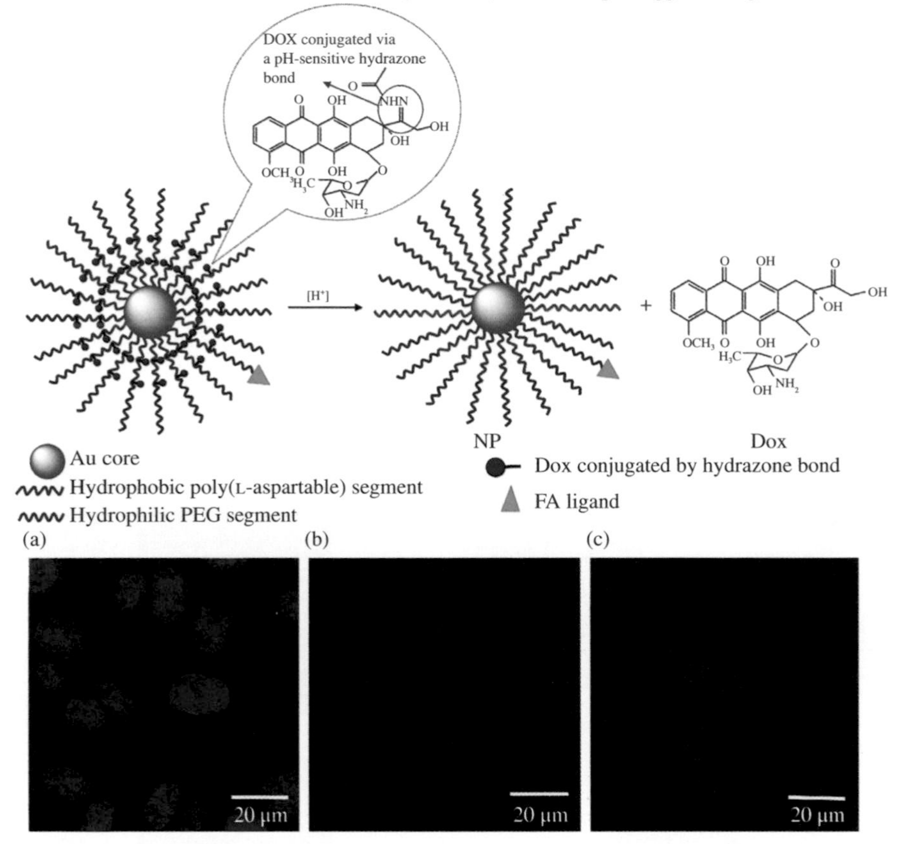

FIGURE 14.22 (Scheme 1) General schematic of the AuNP modified with Asp-PEG-FA with an acid-cleavable linker capable of releasing its fluorescent DOX payload at low pH. (a) Positive control: Free DOX in 4TI cells. (b) Au-P(LA-DOX)-*b*-PEG-OH in 4TI cells. (c) Au-P(LA-DOX)-*b*-PEG-FA in 4TI cells. Prabaharan et al. [92], pp. 6065–6075. Reproduced with permission of Elsevier.

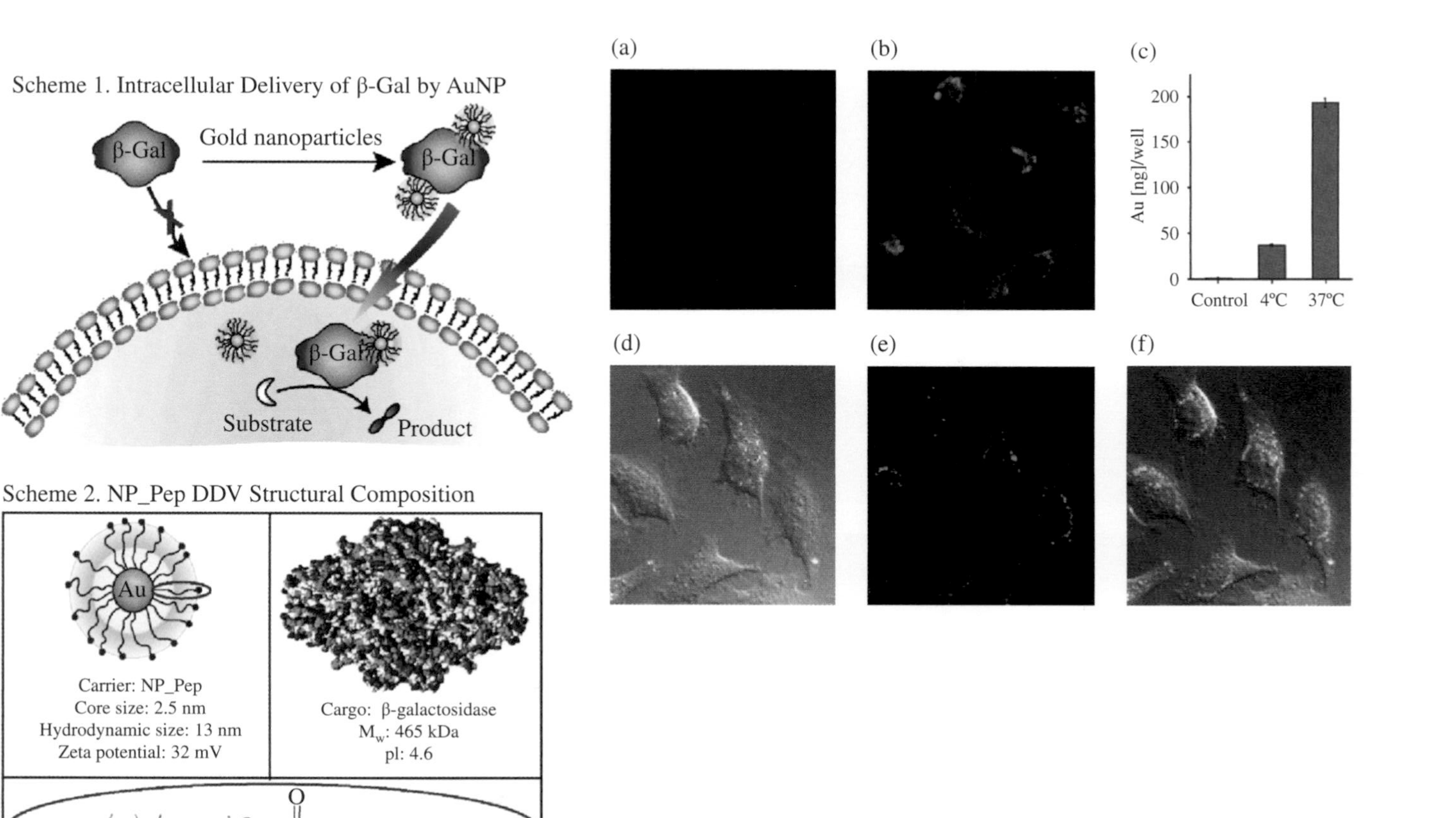

FIGURE 14.23 Fluorescence micrograph of HeLa cells transfected with FITC-β-Gal in (a) absence or (b) presence of NP_Pep. (c) ICP-MS measurements after NP_Pep/β-Gal treatment. (d–f) Confocal images of HeLa cells after protein transfection: (d) bright field, (e) fluorescence, and (f) merged. Reprinted (adapted) with permission from [93]. Copyright 2010 American Chemical Society.

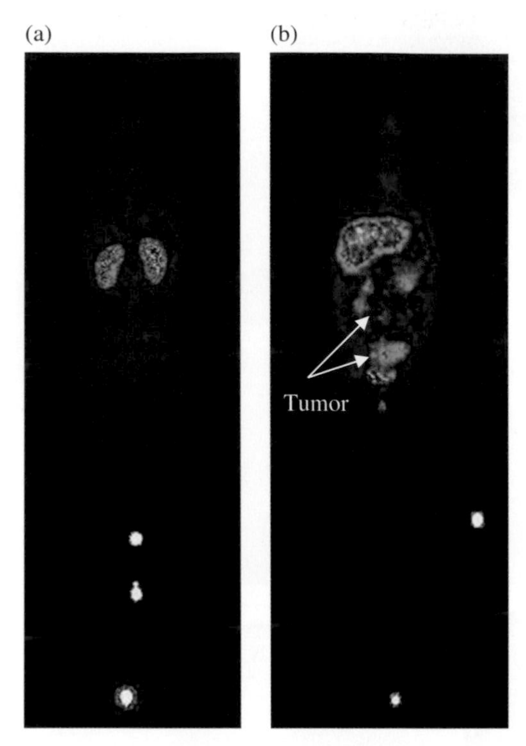

FIGURE 19.5 Anterior SPECT images of two patients receiving ^{111}In-DTPA-folate: (a) Image of a female patient without cancer and (b) image of a female patient with Stage IIIc ovarian carcinoma.

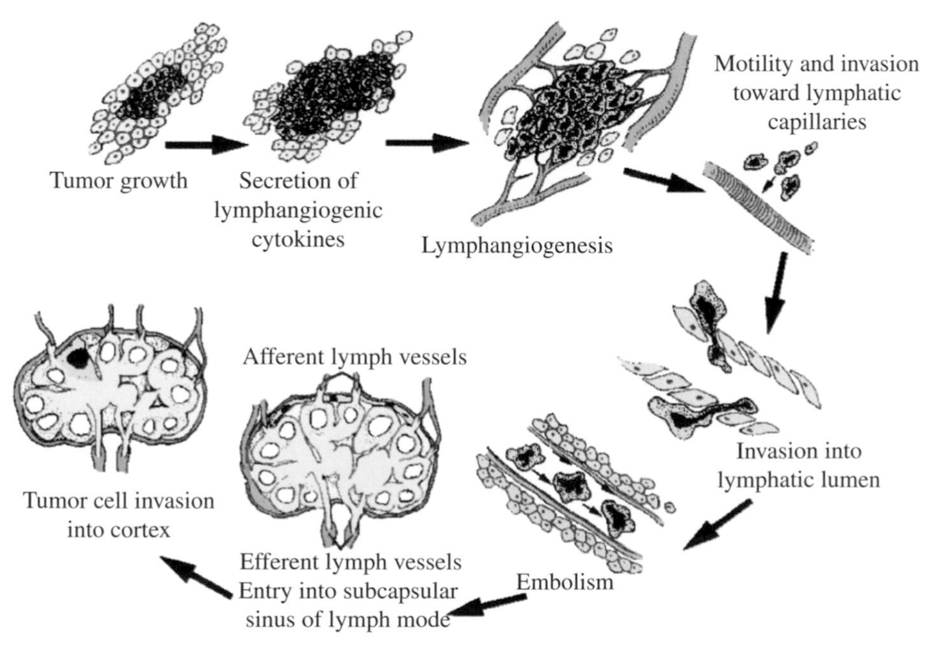

FIGURE 21.3 Tumor-associated lymphatic vessels serve as a route for lymph node metastasis. Nathanson [113], pp. 413–423. Reproduced with permission of John Wiley & Sons.

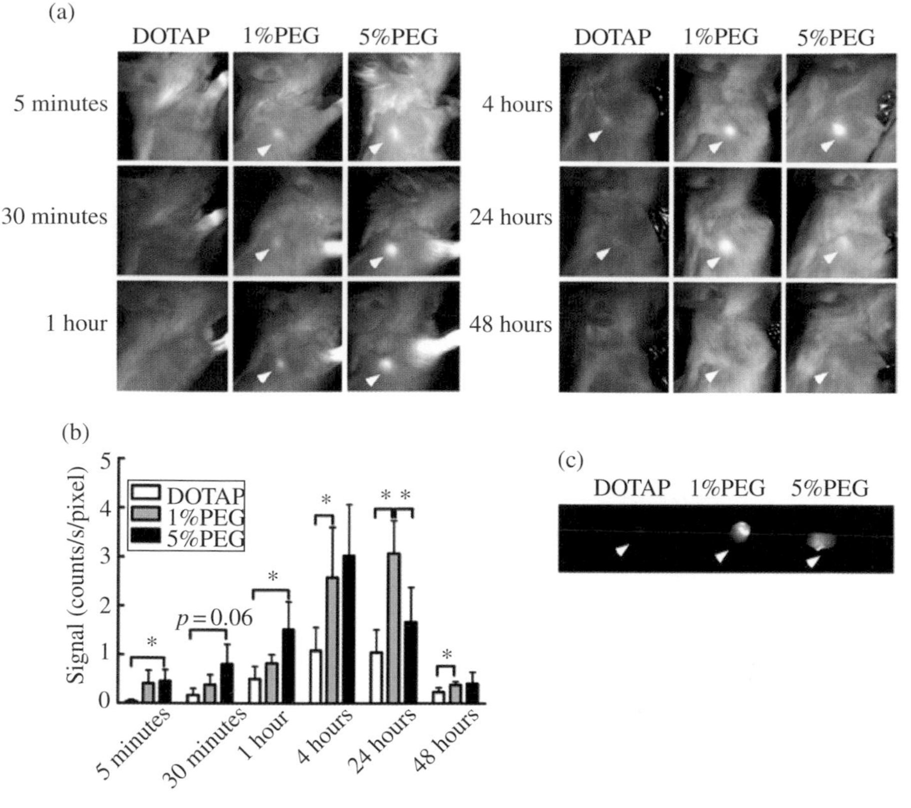

FIGURE 21.9 The lymphatic trafficking of fluorescent liposomes after SC injection into forepaws of Balb/c mice. (a) The accumulation of different liposomes in draining lymph nodes (LNs) from 5 minutes to 48 hours post injection. (b) The fluorescent intensity of LN images measured with the maestro software. Data are shown as mean ± SE ($n = 5$), $*p < 0.05$; $**p < 0.01$. (c) The fluorescent image of draining LNs at 24 hours. White arrows indicate draining LNs. Zhuang et al. [27], pp. 135–142. Reproduced with permission of Elsevier.

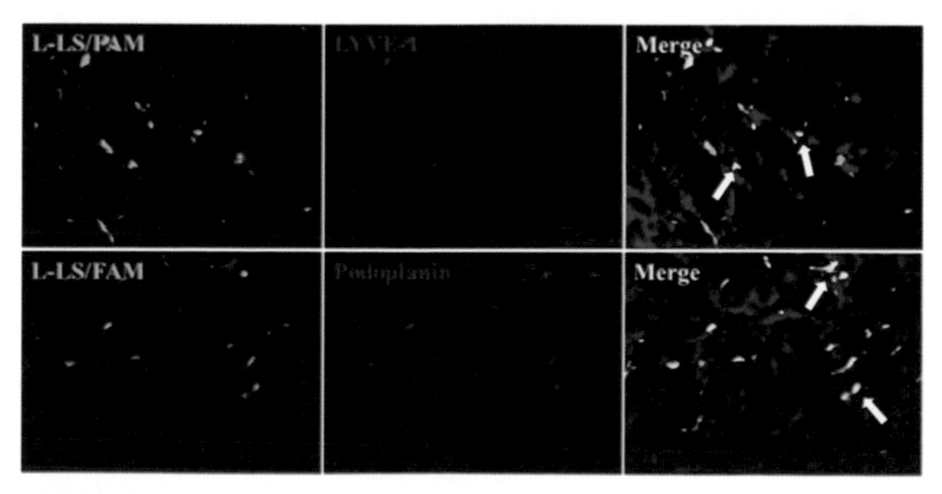

FIGURE 21.13 Specific targeting of LyP-1-PEG liposomes to breast tumor lymphatics. Fluorescent LyP-1-PEG liposomes were colocalized with lymphatic vessel markers (shown for LYVE-1 and podoplanin, arrows indicate the colocalization) in metastatic LNs, indicating the specific binding ability of LyP-1-PEG liposomes to tumor lymphatics. Yan et al. [22], p. 415103. Reproduced with permission of IOP Publishing.

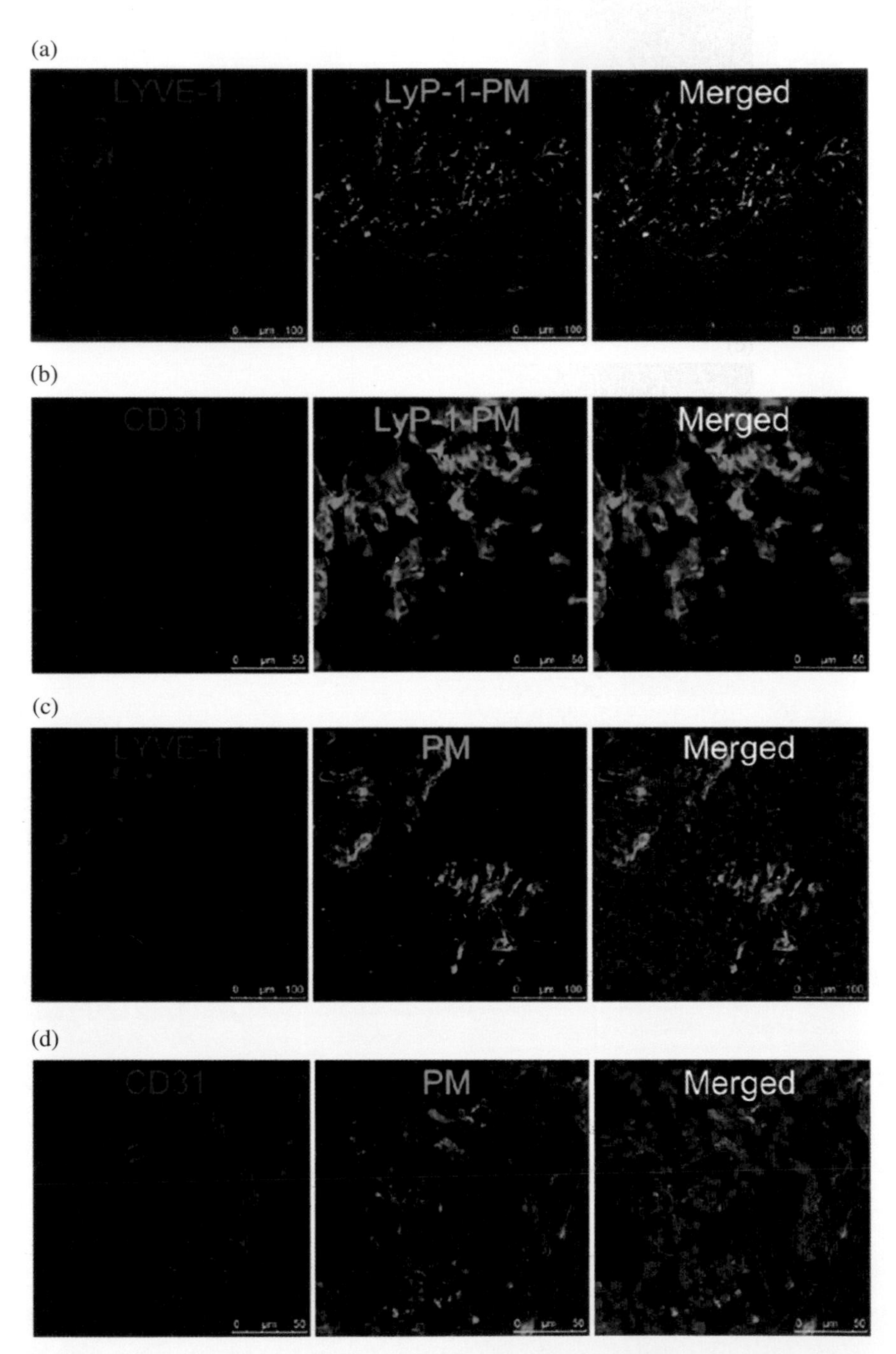

FIGURE 21.19 The targeted delivery of LyP-1-PM to tumor lymphatics in MDA-MB-435S tumor-bearing nude mice by immunofluorescence technique. LyP-1-PM colocalized with the (a) lymphatic endothelial marker (LYVE-1), but not with (b) blood vessel markers (CD31). On the contrary, PM had good colocalization with (d) CD31, but not with (c) LYVE-1. Nuclei were counterstained with Hoechst 33258 (blue). Wang et al. [62], pp. 2646–2657. Reproduced with permission of American Chemical Society.

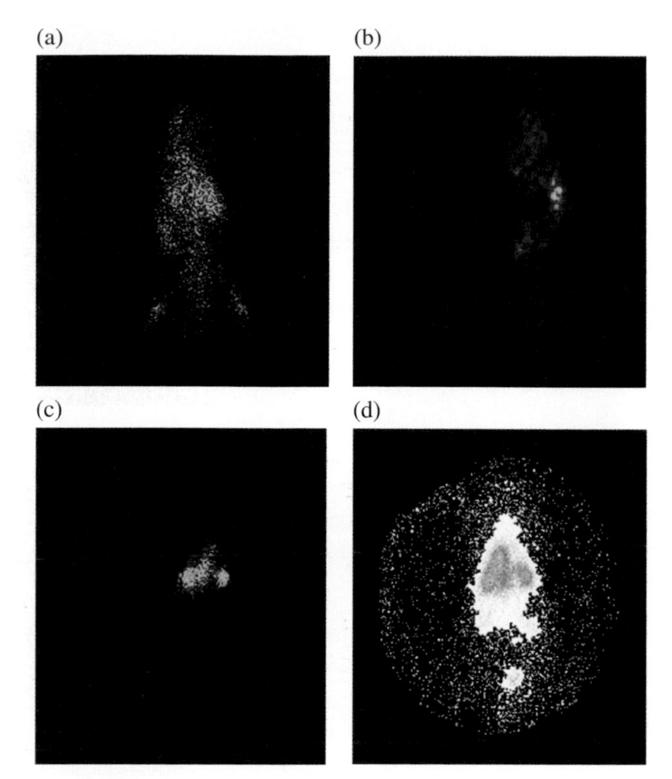

FIGURE 21.23 Gamma scintigraphy photographs of rats receiving ^{99m}Tc-amikacin-loaded SLNs (a) IV after 0.5 hour, (b) IV after 6 hours, (c) pulmonary after 0.5 hour, and (d) pulmonary after 6 hours. Varshosaz et al. [165], p. 8.

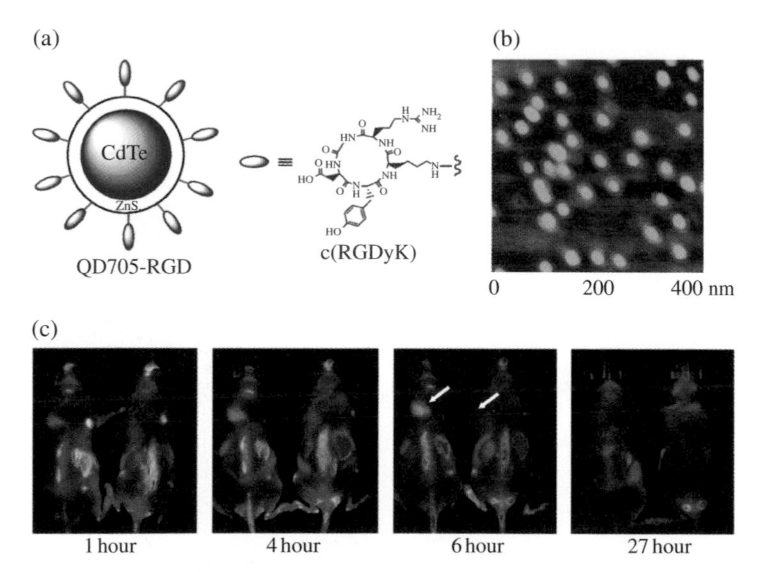

FIGURE 22.3 RGD peptide-labeled QD705 for NIR fluorescence imaging of tumor vasculature: (a) Schematic illustration of the peptide-labeled QDs (QD705-RGD), (b) AFM of QD705-RGD deposited on a silicon wafer and (c) *in vivo* NIR fluorescence imaging of U87MG tumor-bearing mice injected with QD705-RGD (left) and QD705 (right), the tumor signal intensity reached its maximum at 6-hour postinjection with QD705-RGD (pointed by white arrows). Cai et al. [59], pp. 669–676. Reproduced with permission of American Chemical Society.

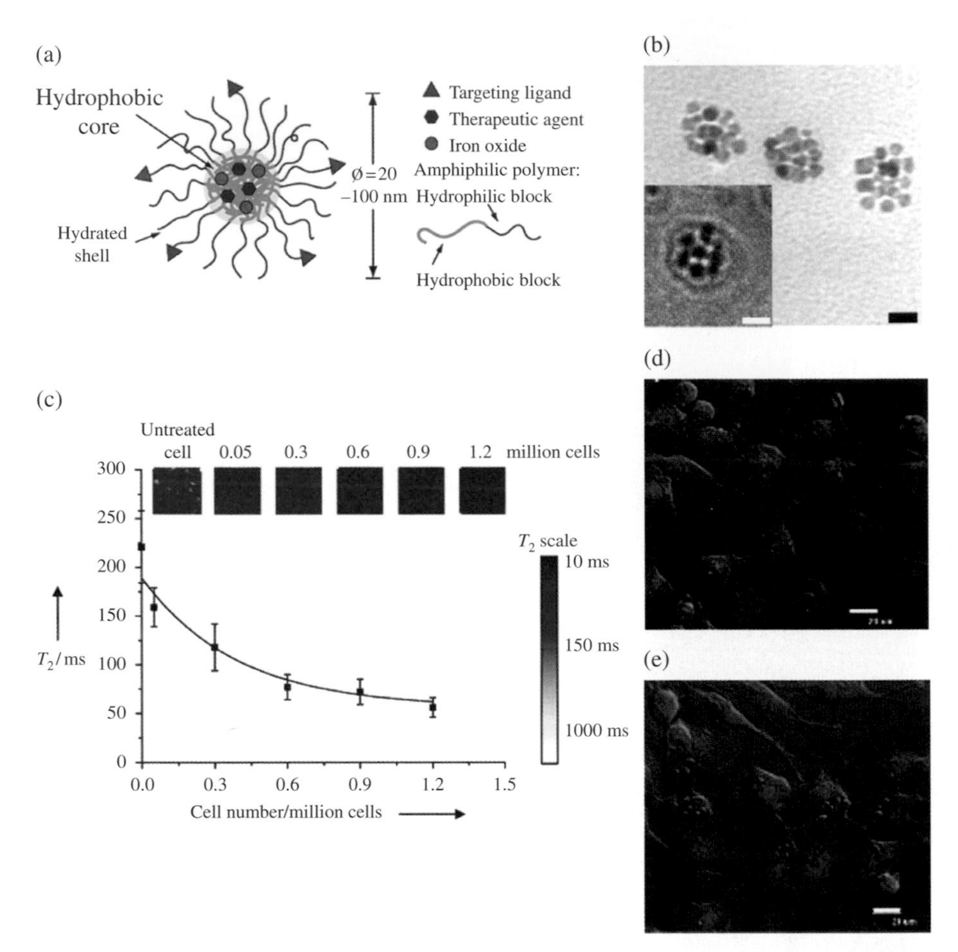

FIGURE 22.4 (a) Schematic representation of doxorubicin (Dox) and Fe_3O_4 nanoparticle-loaded PEG-PLA micelles, with cRGD peptide conjugated on the micelle surface. (b) TEM image of cRGD-DOXO-SPIO-loaded polymeric micelles (scale bars: 20 nm). (c) T_2 values of SLK cells treated with 16% cRGD-DOXO-SPIO micelles as a function of cell number. (d, e) Confocal laser scanning microscopy of SLK cells treated with 10 and 16% cRGD-DOXO-SPIO micelles. Nasongkla et al. [104], pp. 2427–2430. Reproduced with permission of American Chemical Society.

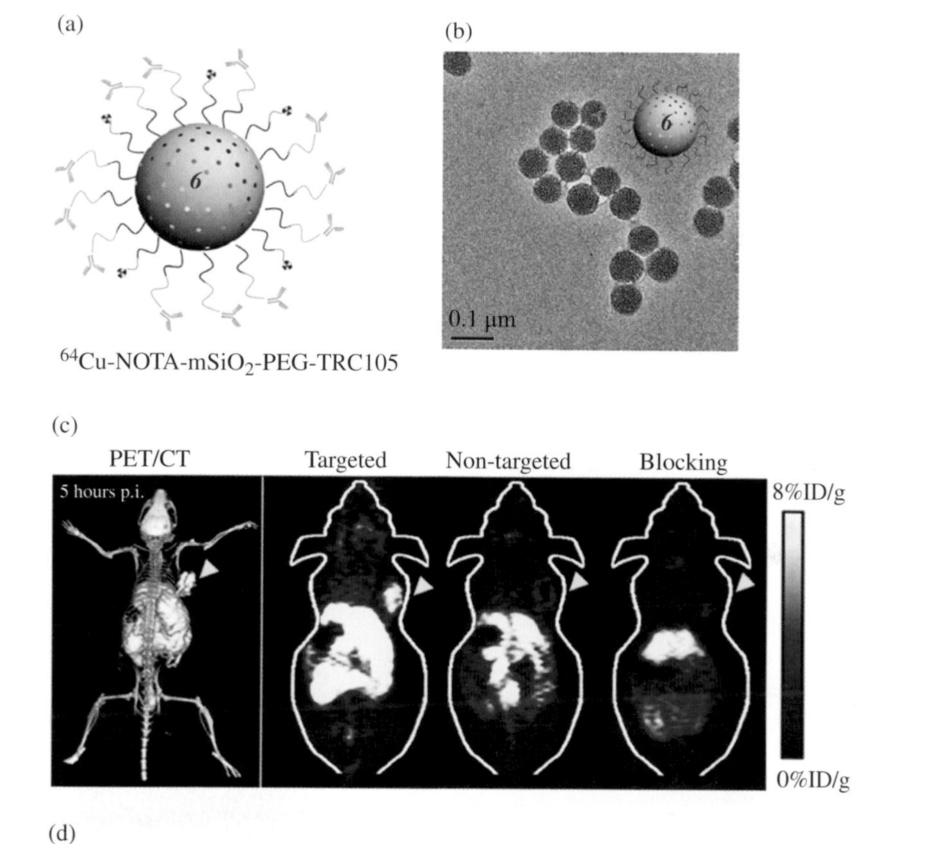

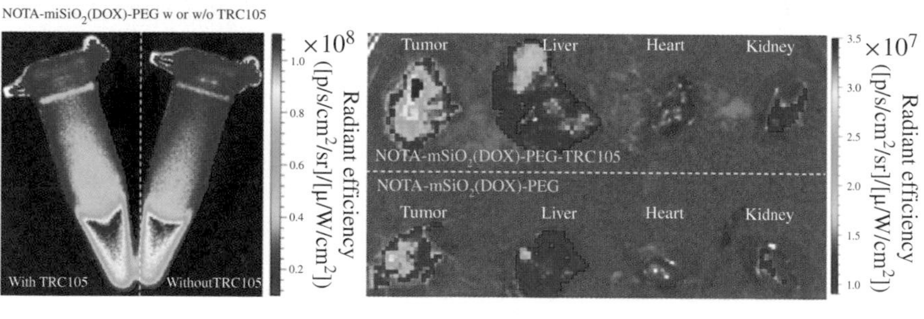

FIGURE 22.5 (a) Schematic illustration of ^{64}Cu-NOTA-mSiO$_2$-PEG-TRC105 nanoconjugate. (b) TEM image of NOTA-mSiO$_2$-PEG-TRC105 in PBS solution. (c) Representative PET/CT and PET images of mice at 5 hours postinjection. Tumors are indicated by yellow arrowheads. (d) Fluorescence images of DOX-loaded nanocomposite in PBS solution and *ex vivo* optical image of major organs at 0.5 hour after intravenous injection of DOX-loaded nanocomposite. Chen et al. [122], pp. 9027–9039. Reproduced with permission of American Chemical Society.

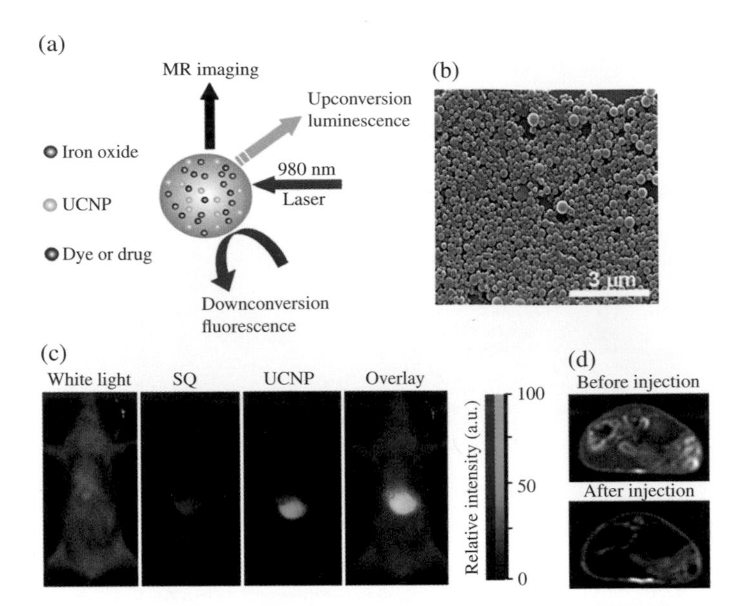

FIGURE 22.6 (a) Schematic illustration for UC-IO@Polymer nanocomposite. (b) SEM image of UC-IO@Polymer. (c) UCL (green) and FL (red) images of Squaraine (SQ) dye–loaded nanocomposites (UC-IO@Polymer-SQ) obtained by the Maestro *in vivo* imaging system. (d) T_2-weighted MR images of mice before and post injection of UC-IO@Polyme-SQ. Xu et al. [113], pp. 9364–9373. Reproduced with permission of Elsevier.

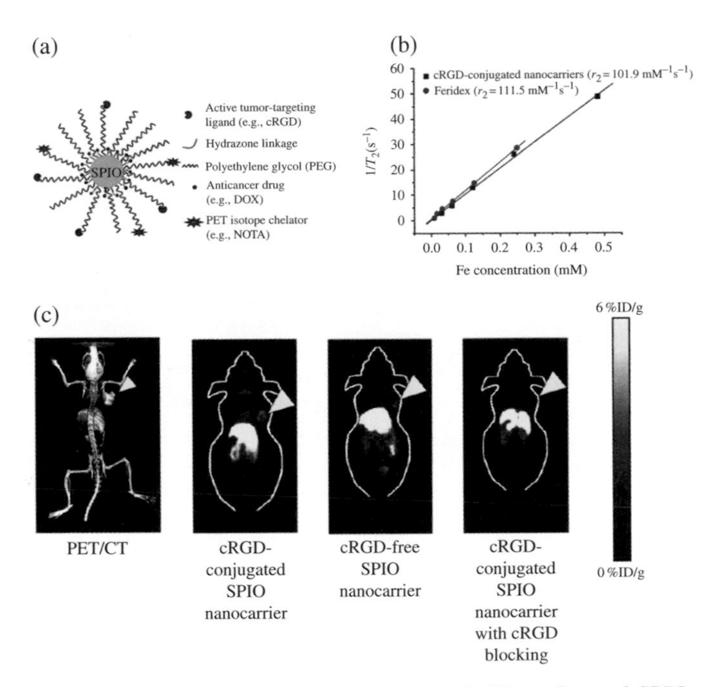

FIGURE 22.7 (a) Illustration of the multifunctional cRGD-conjugated SPIO nanocarriers for combined tumor-targeting drug delivery and PET/MR imaging. (b) T_2 relaxation rates ($1/T_2$, s^{-1}) as a function of iron concentration (mM) for both cRGD-conjugated SPIO nanocarriers and commercial Feridex. (c) PET images of U87MG tumor-bearing mice at 0.5 hour post injection of ^{64}Cu-labeled SPIO nanocarriers (cRGD-conjugated, cRGD-free, and cRGD-conjugated with a blocking dose of cRGD). Yang et al. [160], pp. 4151–4160. Reproduced with permission of Elsevier.

14

NANOPARTICLES AS DRUG DELIVERY VEHICLES

Dan Menasco and Qian Wang
Department of Chemistry and Biochemistry, University of South Carolina, Columbia, SC, USA

14.1 INTRODUCTION

Modern chemotherapeutics are challenged with the difficult task of maintaining superior pharmacokinetic efficacy by narrowing their targeting windows while simultaneously preventing their toxic side effects. Hence, the amelioration of errant targeting and undesirable side effects necessitates a nonconventional strategy for drug delivery. Ideally, to maximize the therapeutic efficacy of succedent targeting modalities, the highest degree of ligand targeting must be accompanied by both the delivery of optimal therapeutic dosage and clandestinely overcoming biological barrier(s) that prevent delivery of the therapeutic payload. Delivery devices that incorporate these strategic principles require a multidisciplinary approach; thus, motivation for designing nonconventional drug delivery vehicles (DDVs) has engendered the convergence of materials engineering and biological chemistry. Through the union of these disciplines, a new scientific frontier emerged: cancer nanotechnology [1].

Although significant advances in the etiology of cancer have progressed substantially in the past 30 years, morbidity rates reveal the existence of a large disparity between these advances and current chemotherapeutics [2, 3]. The variance between the clinic and lab lies within the inability of traditional chemotherapeutics, while still the clinical standard, to effectively mitigate the deleterious effects of off-target specificity and poor penetration of biophysical barriers [4, 5]. Even largely successful

Drug Delivery: Principles and Applications, Second Edition. Edited by Binghe Wang, Longqin Hu, and Teruna J. Siahaan.

therapeutic regimens such as the coadministration of trastuzumab (Herceptin), a monoclonal antibody that targets erbB2 breast cancer, with taxanes augments the incidence of breast cancer–derived brain metastasis due to Herceptin's poor blood–brain barrier penetration [6]. To overcome such obstacles, DDV technology has focused on the development of injectable nanoparticle, which refines the ability of traditional anticancer agents to precisely target malignant phenotypes and bypass previous insoluble biological barriers. These injectable DDVs are highly attractive therapies and have gone on to elicit exciting new cancer therapeutics and diagnostic capabilities.

14.1.1 General DDV Properties

As we will see in this chapter, the ability of nanoparticle therapeutics to selectively transport, target, and release its payload will depend on their size, surface characteristics, and ligand display; the aggregate of which will directly impact their targeting abilities and *in vivo* biodistribution. Generally, nanoparticle DDVs are manufactured with great precision and to exact nanoscale-specification composed of polymer, lipid, silicates, or other amphipathic materials. The core of the DDV is designed to act as a repository where chemotherapeutics, DNA/RNA, or an imaging agent can be stored. Whereas the core is the cargo hold of the DDV system, the exterior is responsible for *in vivo* navigation and avoiding both innate and adaptive immune detection. The ultimate destination, through systemic delivery, of the DDV is realized by presenting multiple targeting ligands such as antibodies, peptides, or other substrates specific for cell surface proteins or epitopes, to the surface of the nanoparticle.

Nanoparticle DDVs are tenable to the violent conditions within the circulatory environment, which extend their half-lives; they possess high surface-to-volume ratios capable of providing a diverse array of multivalent ligand displays, which can allow for enhanced immune-stealth, targeting, cellular internalization, and delivery of their therapeutic payload; and they can be prepared from a wide range of inexpensive organic, inorganic, and biological materials used to encapsulate or display therapeutics or contrast-imaging agents (Fig. 14.1) [7–9]. All of these robust attributes usually can be incorporated in a self-assembled fashion within the range of a 10–1000 nm (in diameter) nanoparticle [10].

Put into perspective, Paclitaxel (PXL), a first-line anticancer agent, targets and immobilizes microtubular growth within cells. This action serves to arrest cell trafficking and mitotic separation of chromosomes and ultimately induces apoptosis. Unfortunately, the therapeutic value of PXL is diminished due to its inability to target tissue-specific malignancies, but more importantly because of its poor *in vivo* solubility. Some detergent excipients, such as cremophor, were used to improve PXL solubility (c-PXL); however, due to its (cremophor) acute toxicity and that it required specialized equipment for systemic delivery, an additional formulation was sought [11]. Moreover, since PXL's therapeutic value is highest during certain phases of the cell cycle, longer infusion times (prolonged circulation) are required to capture more cells during the susceptible cell phase.

To improve solubility and prolong the circulation of PXL, human serum albumen-bound PXL, a polymeric nanoparticle, or n-PXL (Abraxane), was developed. Approved by the FDA in 2005, it was generated by mixing 3–4% human serum albumen with

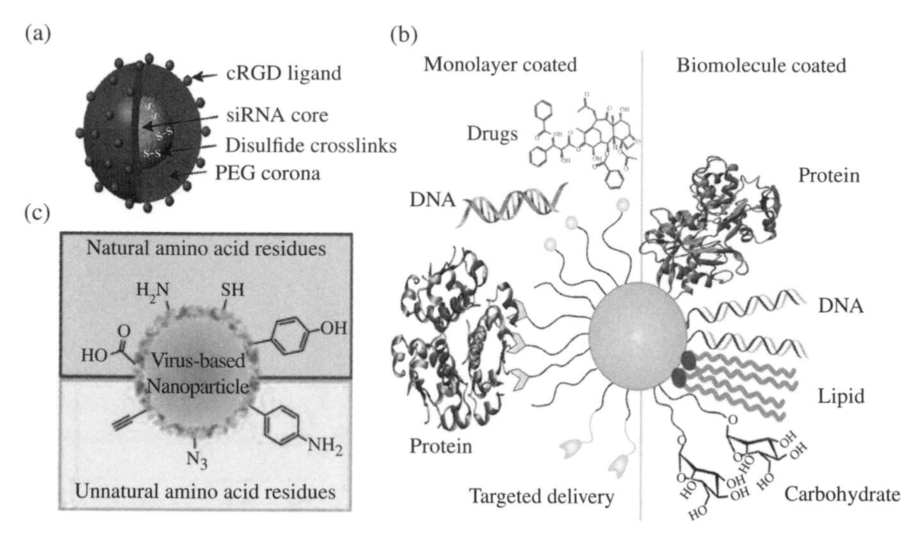

FIGURE 14.1 Schematic representation of general DDV strategies. (a) Organic polymeric micelle. (Reprinted (adapted) with permission from [7]. Copyright (2012) American Chemical Society). (b) Inorganic nanoparticle. Smith et al. [8], pp. 620–626. Reproduced with permission of Elsevier. (c) Biological virus-based nanocarriers and their potential functionalization and cargo. Christie et al. [9], pp. 5174–5189. Reproduced with permission of American Chemical Society.

PXL, resulting in a 130–150 nm nanoparticle colloidal suspension [11]. A study for patients with metastatic breast cancer with previous c-PXL treatment revealed a substantial increase in response rates from 19% with c-PXL to 33% with n-PXL [12]. The development of nanoparticles as potential medical device has not only improved the pharmacokinetic parameters of PXL but also demonstrably improved clinical outcomes. This transformation, among others, has allowed fresh perspective to burgeon in to multiple nanotherapeutical and imaging devices. DDV systems have proven versatile in their abilities to promote targeted therapeutics, diagnostics, and provide insight into the mechanisms of cellular uptake (Fig. 14.2) [13]. These strategies depend on the direct control of the DDV influence on cell and tissue uptake. Mechanisms that support uptake revolve around the direct interaction between the cell and DDV surface. Electrostatic, small molecule or peptide directed, and immune-conjugate cell/DDV relationships enable unique intracellular penetrating strategies, whereas other extracellular methods offer direct identification of cell surface receptors or biomarkers.

14.1.2 The DDV Core: Therapeutic Loading, Release, and Sensing

From a clinical standpoint, a DDV must remain intact during systemic delivery, maintain cargo integrity during transport, and foment its release under circumstances that permit direct targeting to tissues or organs. Hence, the payload must be efficiently loaded to ensure vehicle stability and prevent off-target delivery; however, loading must be coupled with an intrinsic mechanism that destabilizes the DDV for cargo release. Therefore, time-dependent drug biodistribution, along with encapsulation and timed release, must be governed by the characteristics designed into the DDV system (Fig. 14.3) [14, 15].

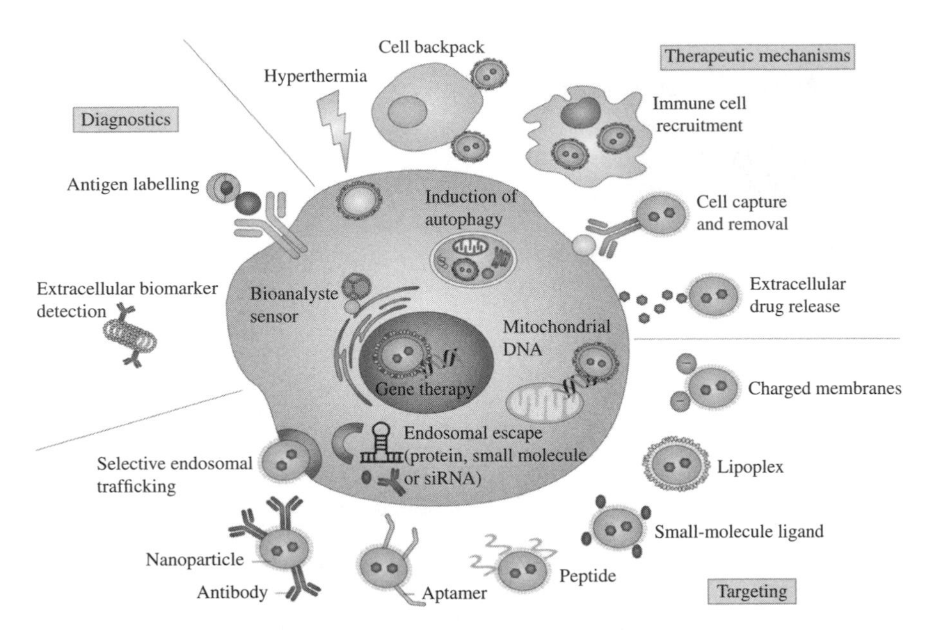

FIGURE 14.2 DDV strategies for cellular targeting and penetration. The potential for DDV systems to correctly target, penetrate, and deliver its payload relies upon the selection of secondary targeting and its concomitant avidity. Without cell-surface targeting, therapeutic payloads suffer indiscriminant distribution. This figure offers a diverse summary of various potential approaches for cellular targeting, diagnosis, and therapeutic mechanisms [13]. Reprinted by permission from Macmillan Publishers Ltd: *Nature Reviews Cancer*, 13, copyright 2012.

Loading and encapsulation can be manipulated by functionalizing a DDV to the payload. In consideration to polymer-based nanocarriers, termed "polymeric micelles," these systems have developed an attractive and diverse set of drug-core loading capabilities. These DDVs are composed of amphiphilic self-assembling block copolymers. Each block contains hydrophobic drug-loading domain and an exterior shell-forming hydrophilic section. The partition coefficient of the hydrophobic block can be altered through size and functional group distribution to promote drug encapsulation through spontaneous self-assembly [18]. Moreover, integrating functional groups such as carboxylic acids into the hydrophobic block can introduce ionizable groups capable of coordinating platinum-based therapies such as cisplatin (Fig. 14.3a) [16, 17].

Organelle-directed drug release could be achieved by customizing polymer DDVs by direct covalent attachment of drugs via cleavable linkages using hydrazone, ketal, and acetal functionality [15, 19]. These linkages are susceptible to the acidic environments within the lysosome/late-endosome (pH $<$ 5.5) or tumor microenvironments (pH 6.7) and can promote their release upon contact with these specific environments, whereas they are more stable at physiological pH (Fig. 14.4b and c) [21–23]. Intracellular release can also be achieved through linkers responsive to the redox environment of the cytosol. Disulfide connections can be used to attach drugs to the micelles hydrophobic block or to promote micelle stabilization via crosslinking. Upon exposure to the reductive environment of the endosome, reduction of the disulfide tether would release the drugs and/or destabilize the nanocarrier (Fig. 14.4a) [20].

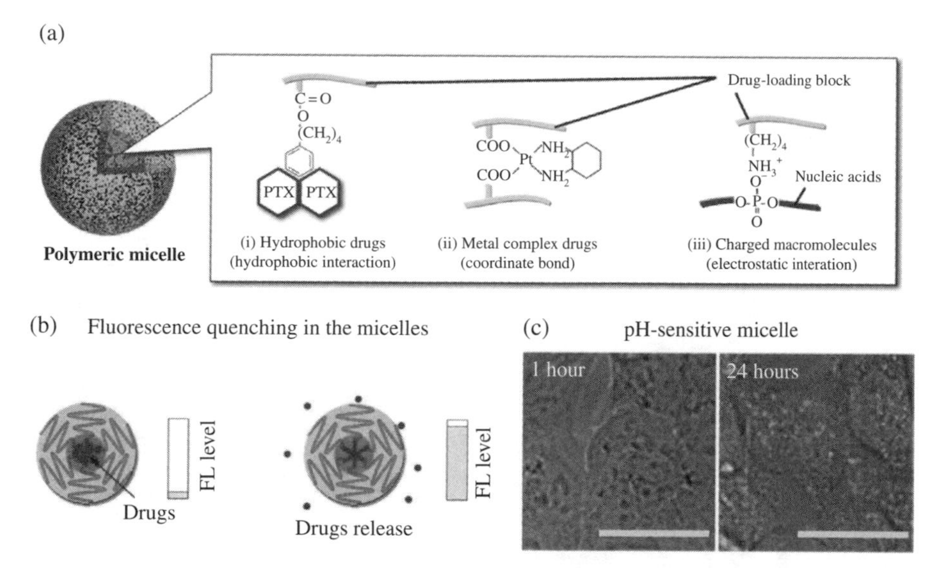

FIGURE 14.3 Potential mechanisms for drug encapsulation and release. (a) Coordination with the backbone of the polymer through either hydrophobic (i), negatively charged moieties (ii), or positively charged systems (iii) affords the direct connectivity required to both stabilize the payload, but also the DDV core. Miyata et al. [14], pp. 227–234. Reproduced with permission of Elsevier. (b) Fluorescent drugs or imaging agents undergo quenching upon encapsulation within the core, but their subsequent release restores their fluorescent properties. Bae and Kataoka [15], pp. 768–784. Reproduced with permission of Elsevier. (c) A pH-sensitive micelle demonstrates its environmental sensitivity through the release of its fluorescent payload upon a decrease in intracellular pH. Adapted from Bae and Kataoka [15]; Nishiyama et al. [16,17]. (*See insert for color representation of the figure.*)

Lastly, charge–charge or electrostatic interactions can stabilize pDNA and siRNA through the interaction of their negatively charged backbone and polycation block copolymers [14]. Small interfering RNA, or siRNA, is used to suppress transcriptional events in the milieu of eukaryotes; however, delivery of naked siRNA is highly susceptible to degradation *in vivo*. Complexation of siRNA to polycation block copolymers, polyion complex (PIC), circumvents their eminent degradation by DNAses and reduces exposure to physiological ionic strength (0.9% saline or 150 mM NaCl). However, release is accomplished during endosomal sorting where a concomitant decrease in pH protonates the “low” pKa-bearing amines of the PIC, disrupting the electrostatic relationship between nucleic acid and polymer [14].

The usage of disulfide crosslinking agents can complement PIC micelles, which can introduce an environmental sensing element, stabilize micelle formation, and allow for higher accumulation of complexed siRNA [24]. Antibodies tethered to siRNA can carry less than 10 siRNA moieties, whereas a PIC-micelle DDV can house up to 2000 moieties [25, 26]. Paramount to this capability is that neither the type nor the physical amount of the therapeutic payload affects the pharmacokinetic parameters or the *in vivo* biodistribution of the DDV; these parameters are mutually exclusive. Comparatively, direct conjugation to antibodies is physically limited to a number of available sites and can occlude or interfere with antigenic determination.

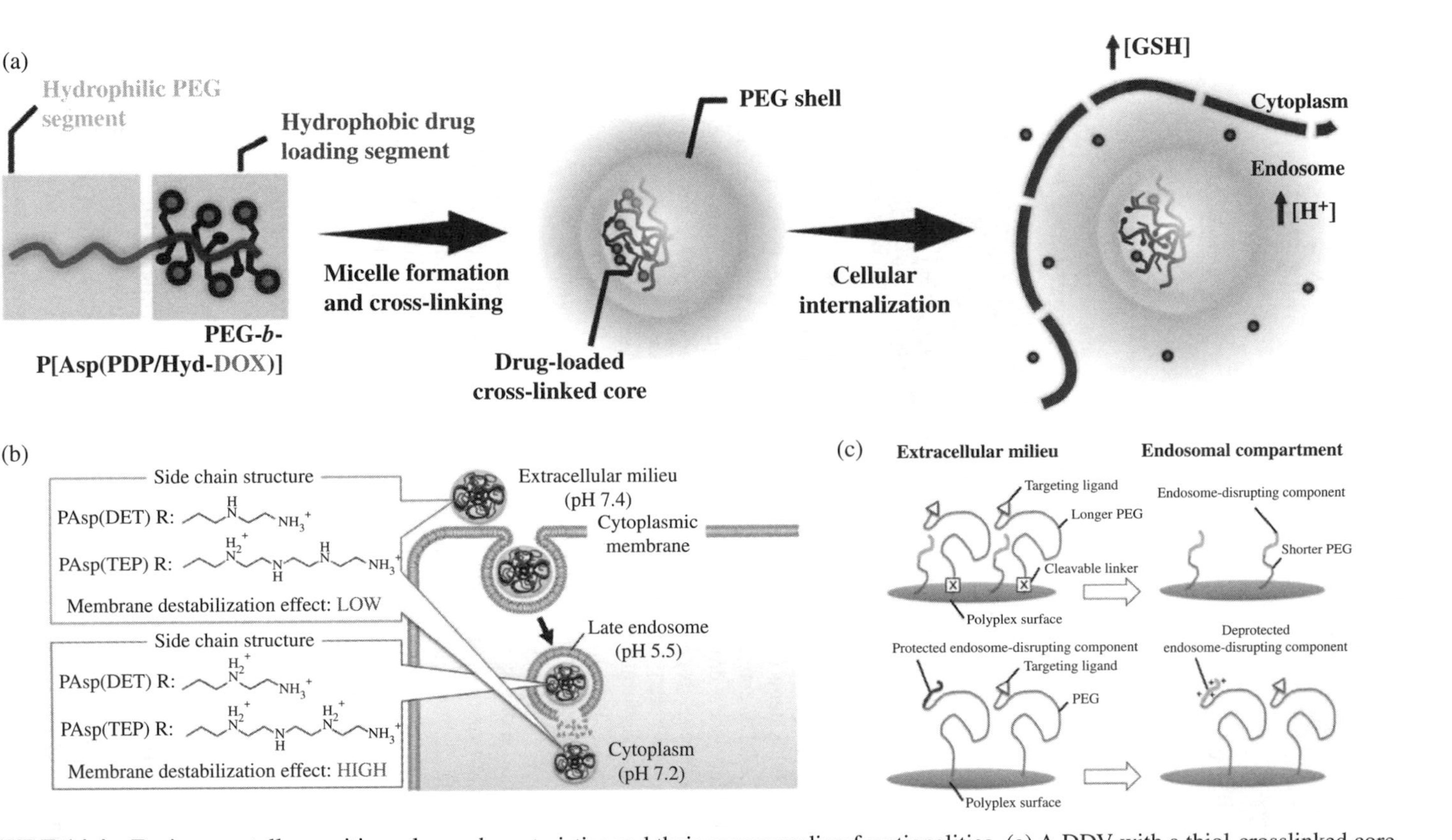

FIGURE 14.4 Environmentally sensitive release characteristics and their corresponding functionalities. (a) A DDV with a thiol-crosslinked core and doxorubicin–hydrazine linkage demonstrate dual release functionalities: upon cellular internalization, the disulfides reduce due to the increase in glutathione, destabilizing the core. As the core destabilizes, the drop in pH severs the DOX–hydrazine linkage, releasing the drug payload [20]. (b) The ability to maintain appropriate endosomal release, and therefore membrane destabilization, can be controlled by the number of mono and diprotonated moieties producing cationic charges [21]. (c) A catch-and-release function whereby the target ligand catches its cell-surface receptor is then dethatched upon entering the endosomal compartment revealing the endosomal-escape component. Conversely, both functionalities are exposed to the milieu; however, the endosomal-disrupting function is protected until the proper pH is obtained within the endosome. Adapted from Miyata et al. [21], pp. 2562–2574; Lai et al. [20], pp. 1650–1661.

14.1.3 DDV Targeting: Ligand Display

The ability to physically localize antineoplastics to a specific tissue, cell type, or disease state while maintaining low to zero off-target interactions is the primary objective in anticancer therapy. Novartis' Gleevic, a tyrosine kinase inhibitor used to mediate chronic myelogenous leukemia (CML), illustrates a remarkable case in which a one-ligand, one-target scenario results in the direct assault of aberrant tumor growth. By directly targeting the mutated kinase, which is consistently generated by a translocated chromosome, Gleevic has increased the 5-year survival rate from 33 to 89% in CML patients [27]. However, direct molecular recognition at both the cellular and subcellular level, especially of "Gleevic" magnitude, remains difficult.

A multitude of factors including both genetic and environmental, govern malignant proliferation and the ability to establish therapeutic resistance. The complex nature of the tumor microenvironment as well as its distinguished architecture and transport systems actually provide nanomedicine with a unique opportunity. Tailoring DDV morphology, targeting capabilities, and payload release characteristics to this unique environment can advance conventional therapies past the sentinel biological systems that are managed within malignancies. For example, the tumor microenvironment uses soluble factors (VGEF), specific cell markers (various clusters of differentiation), and multiple classes of overexpressed receptors to generate a complex extracellular/intracellular communication network. DDV infiltration and exploitation of this network is achieved by both primary and secondary targeting. Primary targeting, or "passive targeting," is organ specific and solely relies upon extended circulation and avoiding both renal filtration and the RES. Indirect tumor localization via passive targeting is primarily accomplished via the enhanced penetration and retention effect (EPR) effect.

Conversely, a secondary targeting, also known as "active" targeting, allows for direct ligand–receptor interactions thus providing a chemical handle for developing tumor, and therefore cellular specificity. These handles, such as short peptides, cRGD (Fig. 14.5) and YIGSR, can be appended to DDVs and directed toward integrin- or laminin-binding domains overexpressed in some tumors in addition to pendent folic acids for direct folate-receptor targeting [9, 28]. Likewise, liposomal DDVs have incorporated surface-bearing antibodies to generate immunomicelles, which are specific for their complementary epitopes [29, 30]. The most attractive and versatile asset of nanoparticle DDVs is their diverse ligand-displaying capabilities via secondary targeting (Fig. 14.6) [31, 32]. Specifically, both the avidity and the density of DDV ligand-targeting moieties can be appropriately tailored to its cellular target [33, 34]. Thus, the distribution of valency, or multivalency, will have a strong impact on both affinity and cell specificity of ligand-receptor interactions [35, 36]. As opposed to the monovalency of many chemotherapeutics, biological systems govern recognition, adhesion, and signal transduction through multi- or polyvalent interactions [37]. Although high-fidelity ligands are generally included in the repertoire of secondary targeting, the multimerization of DDVs with multiple low-affinity ligands can also lead to high avidities [38, 39].

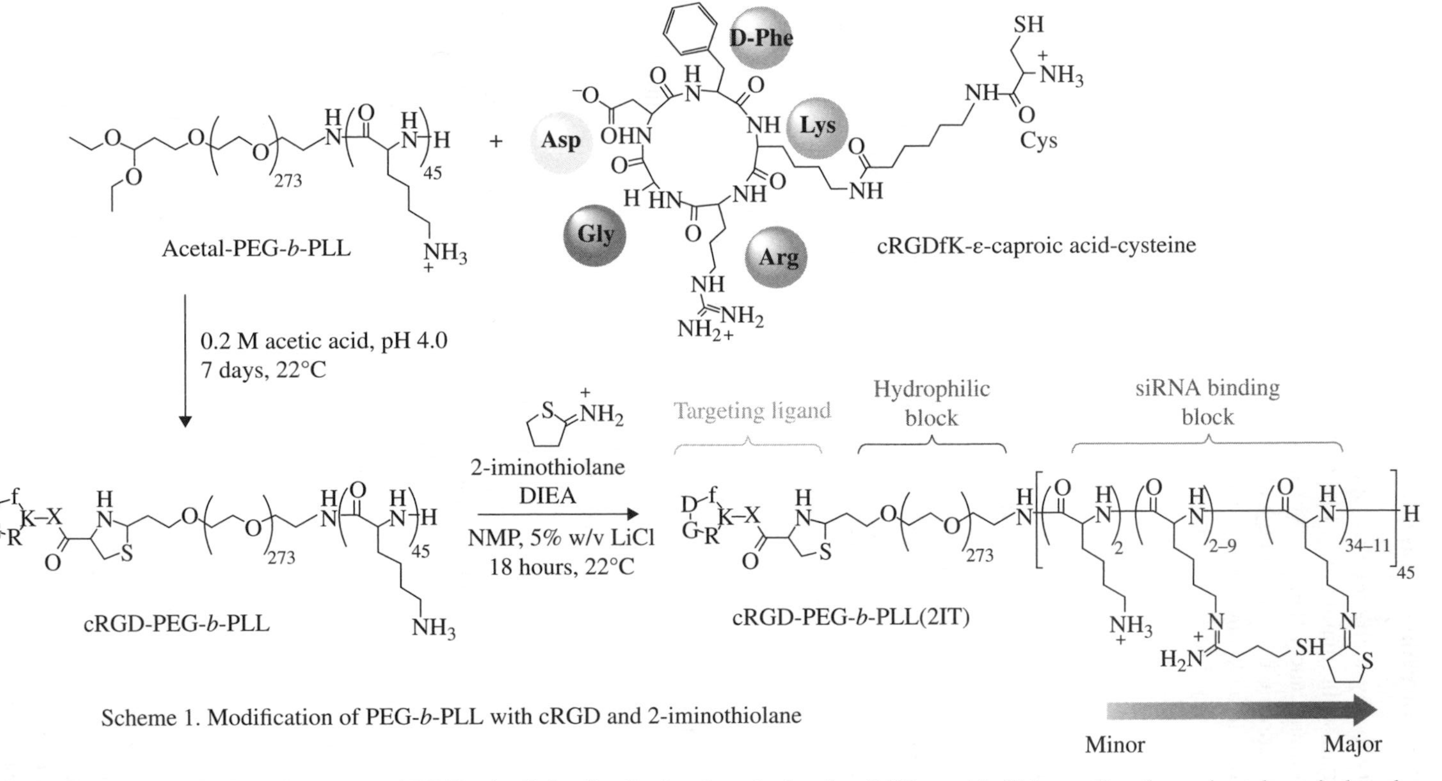

Scheme 1. Modification of PEG-*b*-PLL with cRGD and 2-iminothiolane

FIGURE 14.5 Effect of multivalency and DDV subcellular distribution. Introducing the cRGD peptide (Scheme 1), enhanced uptake and a broader subcellular distribution was observed. Adapted from Christie et al. [9], pp. 5174–5189. Reproduced with permission of American Chemical Society. (*See insert for color representation of the figure.*)

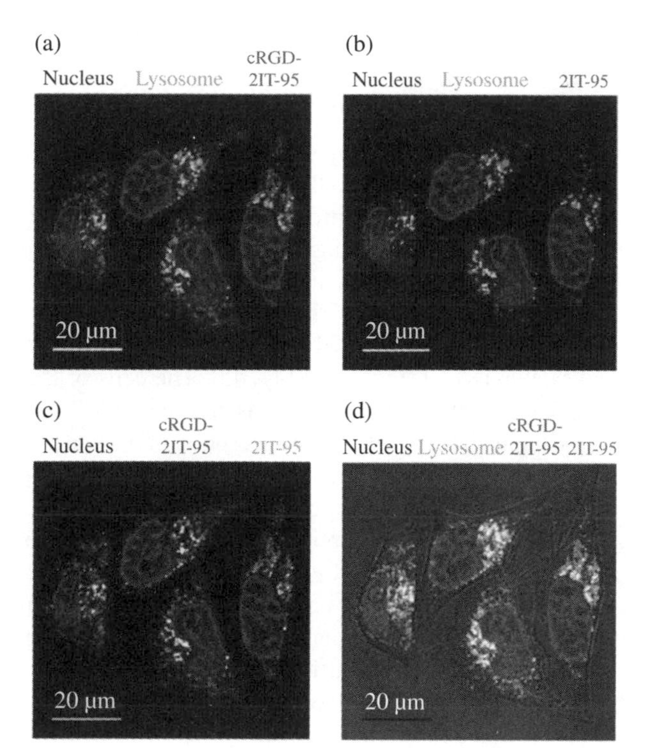

FIGURE 14.5 (*Continued*)

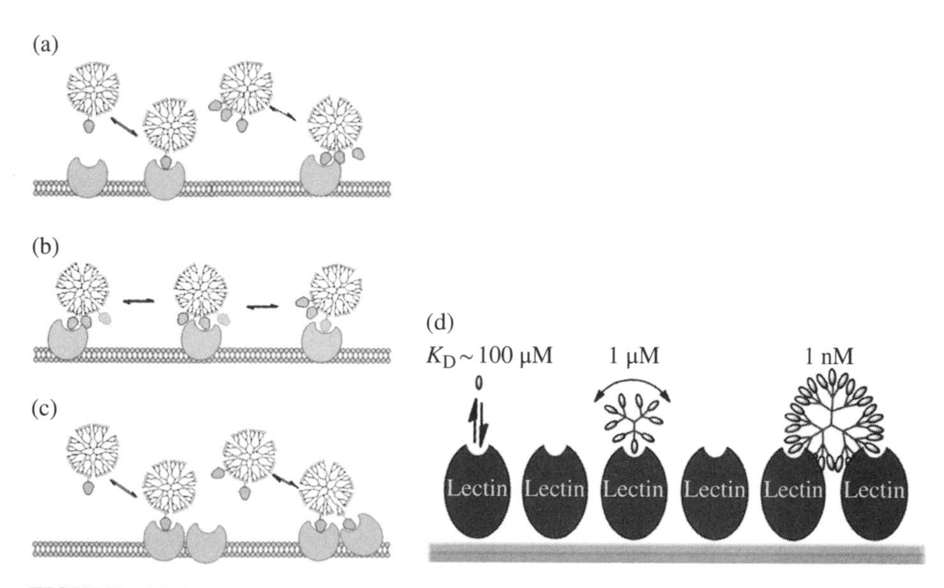

FIGURE 14.6 Ligands and DDV multivalency. Multivalent binding mechanisms. (a) Effective concentration increases the chances of binding [31]. (b) Statistical rebinding is higher for multivalent conjugates if the original interaction dissociates [31]. (c) The chelate effect allows for multiple interactions through one conjugate. (a–c) van Dongen et al. [31], pp. 3215–3234, figure 7. Reproduced with permission of American Chemical Society. (d) Binding avidity is also increased through multivalent displays. Lallana et al. [32], pp. 902–921, figure 4. Reproduced with permission of Springer.

14.1.4 DDV Size and Surface: Clearance and the EPR Effect

Effective nanocarriers, or nanoparticles, typically range from 40 to 200 nm in diameter. The lower end of this diameter is based on the kidney's glomerular capillary wall and its ability to sieve macromolecular particulates less than 5 nm during first-pass circulation [40–42]. Particles of increasing diameter, within the micrometer range, will accumulate primarily in Kupffer cells in the liver or in microcapillary beds within the lung and spleen sinusoids [42, 43]. Within these connective tissues, nanoparticle clearance from circulation is regulated by the phagocytotic action of the reticuloendothelial system (RES) (Fig. 14.7) [44, 45]. The cells of the RES network mediate particle uptake through a process known as opsonization. This process begins immediately upon systemic delivery where the adsorption of plasma proteins (or opsonins) to the nanoparticle surface elicits their accumulation within cells of the RES [43, 46]. While many of the opsonins are known, such as IgG, IgM, and proteins of the complement system, the recognition of foreign nanoparticles depends on their surface characteristics and their interaction with the local environment [47].

Processes to prevent opsonization, or enhance immune stealth, have focused on reducing hydrophobicity through steric stabilization and neutralizing ionic surface properties of nanoparticles [48–50]. A particularly successful process involves coating or grafting nanoparticles with polyethylene glycol (PEG), which prevents complete protein adsorption and extends circulation. However, PEGylation tends to interfere with cargo release kinetics and necessitates a specific formulation to reduce any systemic overexposure [51]. Hence, the combination of surface properties and size strongly influences nanoparticle DDV fate.

Large nanoparticle diameters have resulted in their sweeping removal via the RES, and sometimes occlusion of vasculature. In contrast their narrow size distribution, 40–200 nm, enables DDVs the distinct advantage of permeating the extensive "leaky" vasculature network that solid tumors produce. Construction of new capillary networks from preexisting blood vasculature occurs through the activation of a process called angiogenesis. However, in solid tumors, angiogenesis is a process of gross architectural mismanagement where large gaps or fenestrations are striated across the vessels, which range from 10 to 1000 nm. This not only serves as an access point for DDVs, but it also generates irregular flow rates around and within the tumor and restricts free-drug penetration [52].

These abnormal networks are genuine to the tumors' pathophysiology, whose metabolic requirements must be met to ensure their rapid expansion. In order to fulfill this rate of expansion, cells located in the tumor parenchyma establish these aberrant networks [53]. By emitting soluble growth factors such as vascular endothelial growth factor (VEGF) and basic fibroblast growth factor (bFGF) tumor progression is facilitated and the rate of apoptosis is shunted [54].

14.2 ORGANIC DDVs

14.2.1 Polymer-Based Nanocarriers

Amphiphilic-based polymers are currently the most commanding materials used in DDV formulation. They have been used to generate polymeric micelles [55], dendrimers [56], polyplexes [57], polymer–drug conjugates [58], and polymer–protein

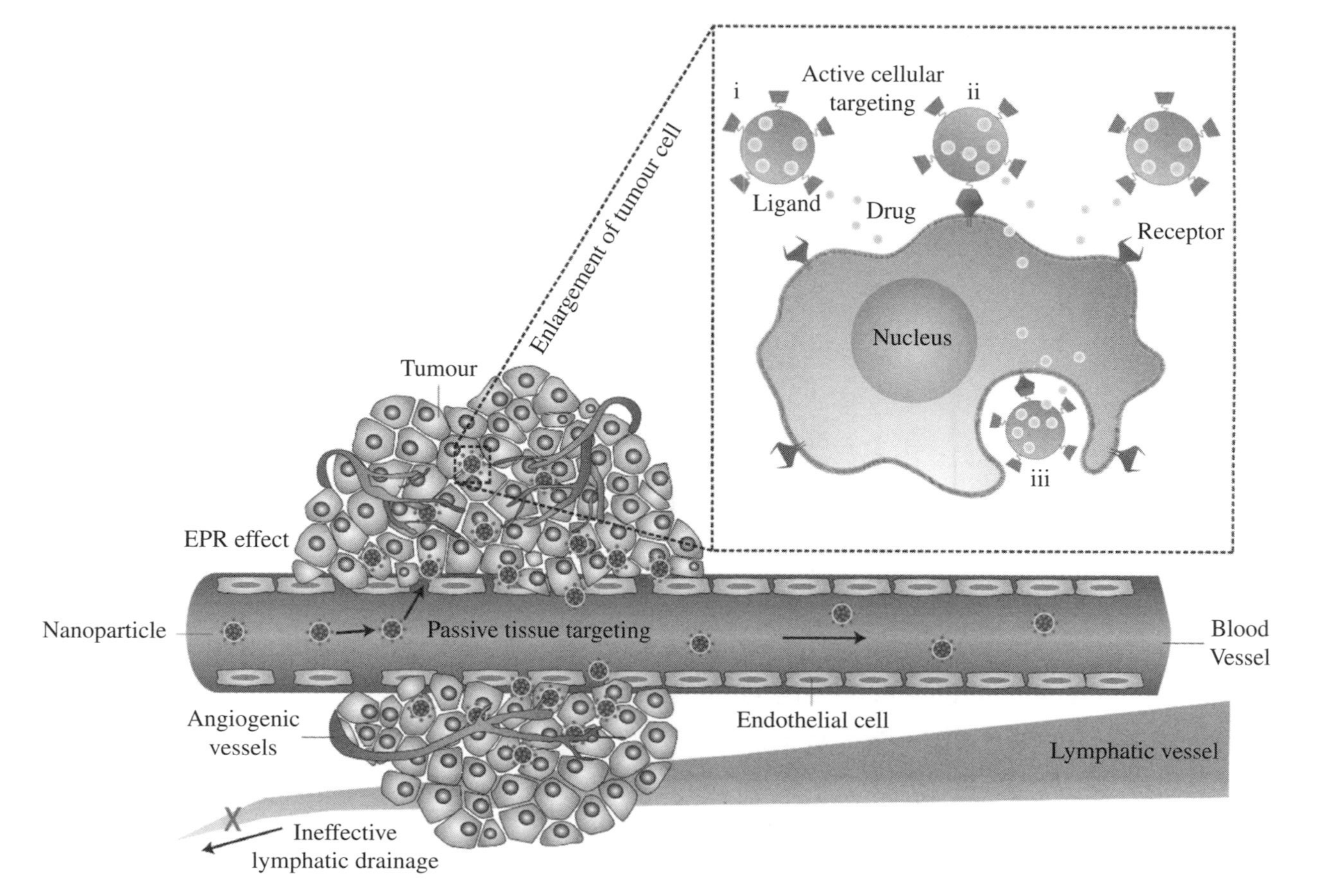

FIGURE 14.7 The enhanced penetration and retention (EPR) effect. The EPR effect is the primary point of tumor penetration for DDVs via passive targeting. Due to the irregular tumor vasculature and ineffective lymphatic drainage (RES), nanoparticles extravasate through the leaky system and enter the tumor environment. This resolves to expose the tumors cell surface and promote active DDV targeting (see *inset*) and ultimately, payload distribution [44]. Reprinted by permission from Macmillan Publishers Ltd: *Nature Nanotechnology* [44], copyright 2007. (*See insert for color representation of the figure.*)

conjugates [59]. The demand for these materials relies upon both their diverse physical and chemical properties unique to biological systems. While the polymer–drug and polymer–protein conjugates have advanced clinical exposure (polyglutamate-PXL and PEG-IFNα 2a), their systems have remained the more simplistic and consist of a tripartite system: the therapeutic, the linker, and the polymer (Fig. 14.8) [60]. Advanced formulations of polymer systems have introduced more sophisticated functionalities by chemically diversifying their macromolecular composition; branched [61] and graft polymers [62] and dendrimers [63], along with block and triblock copolymers, provide the foreshadowing toward an increased biocompatibility, environmental response/stimuli for payload release, immune stealth, internal or external payload, and controlled assembly.

14.2.2 Polymeric Micelles

Polymeric micelles are composed of amphipathic co-block unimers that self-assemble due to their opposing cumulative dipoles. The difference in the cumulative dipole of each unimer, and its concentration in the bulk or the critical micelle concentration (CMC), alters the general solubility of the co-block polymer. Subsequently, when the unimers begin to reach the CMC, the surface free energy begins to drop because the interface of the unimers hydrophobic block with the

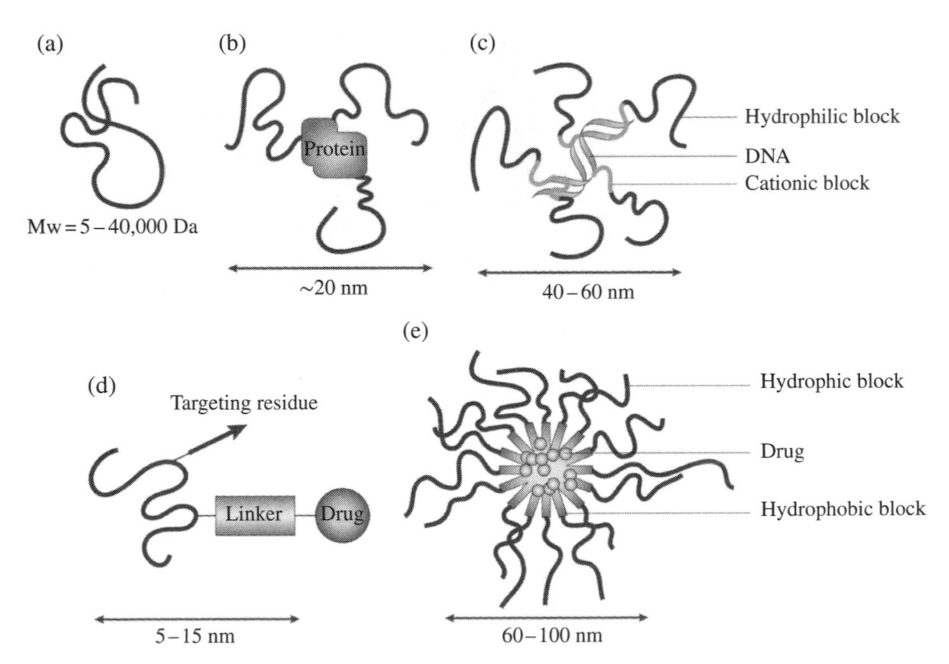

FIGURE 14.8 Schematic representation of organic-based DDVs and their attributes: (a) Polymeric drug or sequestrant, (b) polymer–protein conjugate, (c) polyplex: polymer–DNA complex, (d) polymer–drug conjugate, and (e) polymeric micelle. Reprinted by permission from Macmillan Publisher Ltd: *Nature Reviews Drug Discovery* [60], copyright 2003.

aqueous bulk decreases thus driving in micelle formation. The corona, or the hydrophilic shell of the co-block polymer, is thus exposed to the bulk medium, and further stabilizes the hydrophobic segments within the core [64]. Furthermore, the CMC is also metric of thermodynamic stability: low threshold CMC values resist dilution upon systemic delivery and ensure their structural integrity for extended biodistribution. The use of amphiphilic block copolymers can promote encapsulation of hydrophobic drugs, contrast or imaging agents, and both pDNA and siRNA due to their distinct hydrophilic and hydrophobic segments to produce micelles (Fig. 14.9) [15]. However, direct conjugation of a drug to the polymer scaffold is the most widely accepted approach as it asserts direct control over the spatial distribution of the resulting self-assembled micelle. Chief among the long list of polymers used for polymeric micelle drug delivery are the biocompatible, biodegradable PEG, and the core-forming poly(L-amino acids) (PLAA). The diversity of PLAAs such as L-Lysine, L-histidine, L-aspartic, and L-glutamic acid renders the segment water soluble and nontoxic, and provides versatile functionality capable of complexing charged anticancer payloads like cisplatin and its homologs or siRNA/pDNA systems [60].

Synthesis of PLAA unimers can be completed by ring-opening polymerization (ROP) with α-amino acid-*N*-carboxyanhydrides (NCAs) and phosgene. NCA polymerization can be initiated by a variety of nucleophiles such as primary amines or alkoxides and is generally referred to as the Fuchs–Farthing method [13]. This process uses β-benzyl protected amino acids, such as β-benzyl L-aspartate (β-BLA), to form activated monomers of β-benzyl-L-aspartate–*N*-carboxyanhydride (BLA-NCA). This ring-opening chain growth process takes place upon the addition of the macrointiator α-methoxy-ω-amino PEG. Post polymerization, the deprotection of the benzyl-protecting group then affords for either direct functionalization or complexation of the PLAA side chains with drugs [13]. The DOX-conjugated PEG-block-poly(α,β-aspartic acid), (PEG-*b*-P(Asp-Hyd-DOX)), block copolymer was one of the earliest chemotherapeutic delivery vehicles to hold therapeutic potential. For the PEG-*b*-P(Asp-Hyd-DOX) micelle, DOX was covalently attached to the γ-carboxylic acid of the PAsp through a pH-dependent, cleavable hydrozone linkage of DOX (Fig. 14.10) [65]. The conjugation of DOX to the interior was found to promote micelle formation by increasing the hydrophobic nature of the

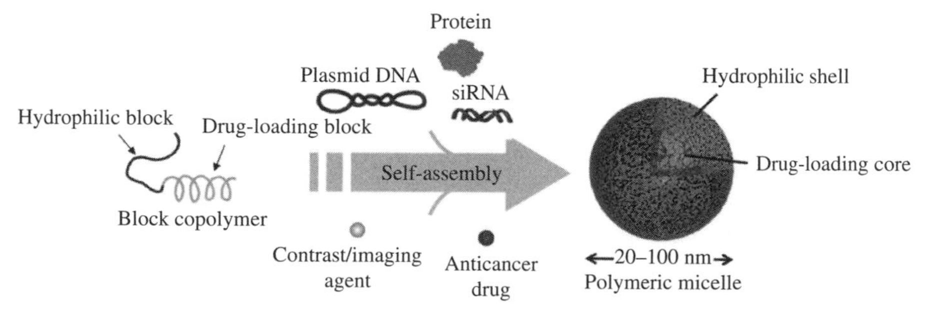

FIGURE 14.9 Self-assembly block copolymer micelles and their potential payloads. Bae and Kataoka [15], pp. 768–784. Reproduced with permission of Elsevier.

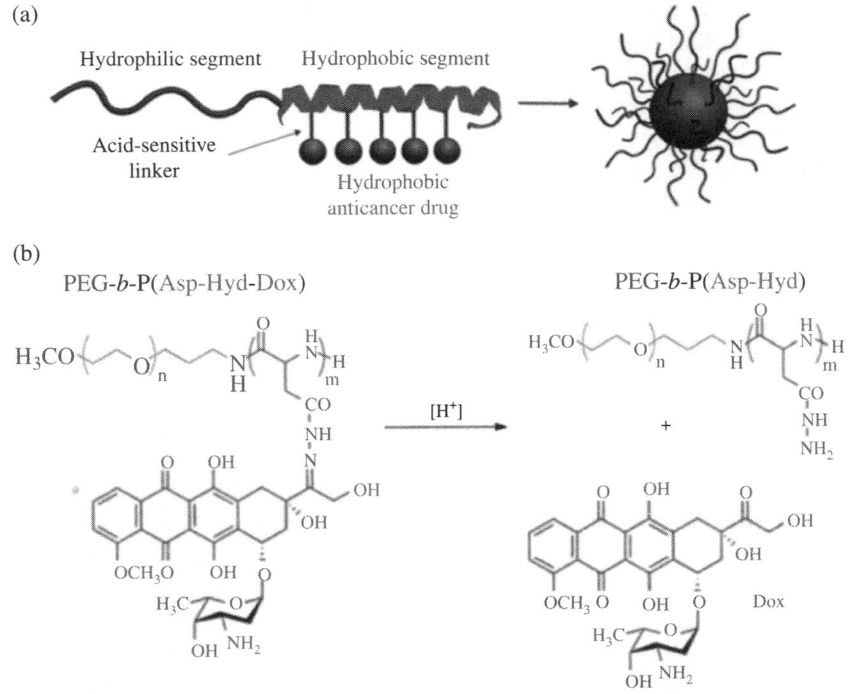

FIGURE 14.10 (a) The self-assembly of an environmentally sensitive polymeric micelle using hydrazine-linked doxorubicin tethered to an amphiphilic block copolymer. (b) A simple schematic illustrating the release of doxorubicin upon contact with an acidic environment. Reproduced ("Adapted" or "in part") from [65] with permission of the Royal Society of Chemistry.

PAsp core segment. Additional drug loading, via entrapment, and micelle stability were further enhanced due to the increased π–π interactions of conjugated and entrapped DOX [66].

Ultimately, the PEG-*b*-P(Asp-Hyd-DOX) vehicle performance was superior to that of free DOX, and its variant was the first nanocarrier introduced to the clinic in 2001 [67]. Currently under phase I clinical evaluation against intractable pancreatic cancer are (1,2-diaminocyclohexane)platinum(II) loaded micelles (DACHPt/m). The polymer–metal complexation is achieved through the interaction of the gamma-carboxylate of PEG-*b*-poly glutamic acid block copolymers (PEG-*b*-P(Glu)) and DACHPt/m (Fig. 14.11a). Its concomitant release is actuated through ligand substitution of chloride for DACHPt/m. This formulation was evaluated against a transgenic mouse model that possessed a viable immune system, which spontaneously developed bioluminescent pancreatic adenocarcinoma using two elastase 1-promoted (EL1) gene constructs: SV40 T and firefly luciferase (EL1-luc/TAg). Repeated systemic administration of DACHPt/m dramatically suppressed pancreatic malignancies and reduced metastatic incidence prolonging rodent survival.

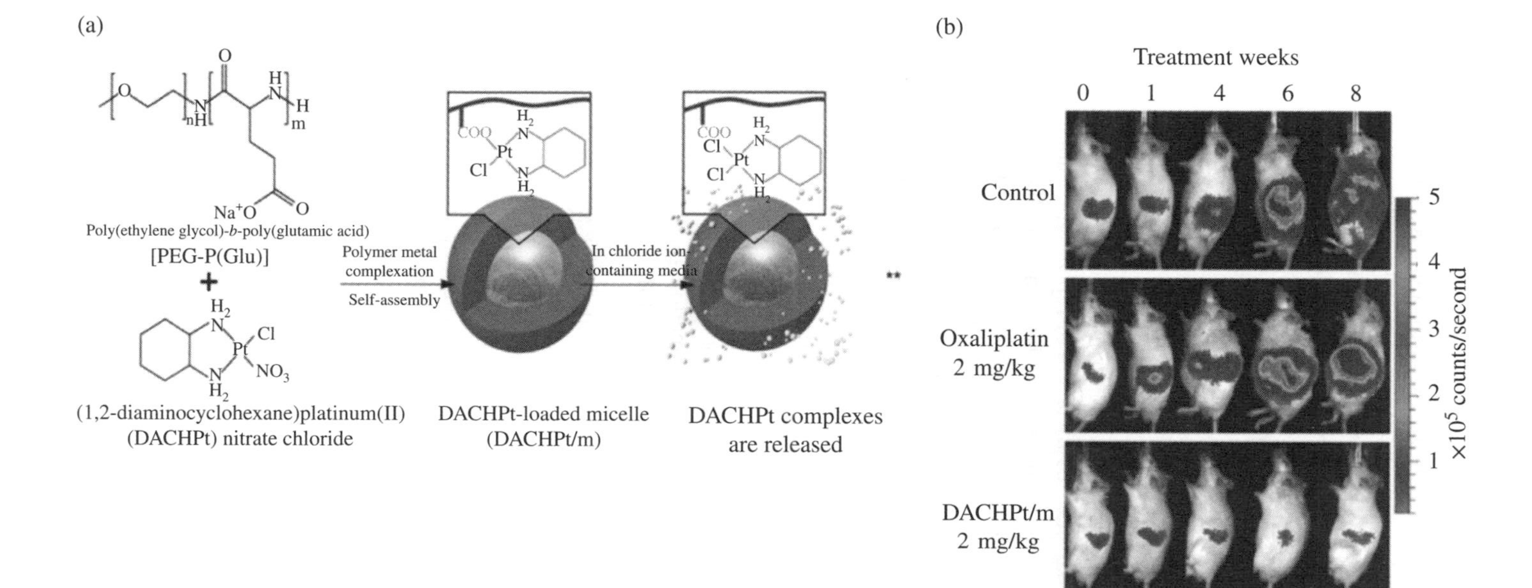

FIGURE 14.11 Preparation of DACHPt/m. (a) Self-assembly of the metal-polymer complex DACHPt and PEG-P(Glu). (b) Pancreatic tumor activity as demonstrated by bioluminescence signal (PNAS is not responsible for the accuracy of this translation). Cabral et al. [68], pp. 11397–11402. Reproduced with permission of PNAS. (*See insert for color representation of the figure.*)

This is especially novel regarding the constitutive oncogenic activity of SV40 T in the EL1-luc/TAg mice. Hence, the extended survival past the treatment window, 8 weeks with one administration per week, can only be supported by a long circulating nanocarrier. The DACHPt/m, therefore, not only maximizes each dosage through enhanced tumor–drug accumulation but also allows for fewer systemic administrations and potential side effects [68]. Notably, PXL, doxorubicin (DOX), and CDDC polymeric drug vehicles have all generated a measured amount of success in PLAA systems and have passed from their prospective development phases to different stages of clinical trials.

14.2.3 Dendrimers

Contrary to polymeric micelles, dendrimers are highly branched monodisperse segments that are characterized by layers of repeating monomers that emanate from a singular chemical moiety. These three-dimensional (3D) tree-like structures afford precision tuning regarding their size, morphology, and extended valency. Capable of both passive and active targeting, their ability to provide a structured system with extended multivalent ligand displays using small-molecule ligands and/or imaging moieties is perhaps their most formidable feature (Fig. 14.12) [69].

The hierarchal arrangement of "trees" or dendrons that are radially fixed from the dendrimers core is controlled by the multiplicity of its core. Each dendron is then categorized into three different segments: the core, its branches, and its end groups or the periphery [56]. Each fork or branching point of each dendron is referred to as a "generation," for example, G0, G1, and G3 (Fig. 14.13). Divergent dendrimer synthesis constructs these generations, or rungs, in a stepwise fashion where one monomer unit is conjugated to a multifunctional core moiety to produce the first-generation dendrimer. Successive additions of the monomer units to previous generations will result in the parent dendrimer [56].

One of the first dendrimer DDVs synthesized by the divergent method was poly(amidoamine) (PAMAM) dendrimers (Fig. 14.14a) [70]. Malik and coworkers generated a PAMAM with 3.5 generations with surface pendent cisplatin (20–25 wt%) capable of *in vitro* timed release. More importantly they reported enhanced solubilization of cisplatin and demonstrated *in vivo* antitumor activity against melanoma-burdened mice, whereas free cisplatin, even at the maximum tolerated dose, showed no activity [71]. Although divergent synthesis was productive and readily used commercially, purification of monodisperse products proved difficult and extenuating.

As an alternative, convergent synthesis was developed (Fig. 14.14b) [70]. This method begins with the construction of the individual dendrons, which is followed by the coupling to a focal or centralized unit to produce the parent dendrimer. This method leads to highly monodisperse systems, which are easily purified. Hence, dendrimer micelles have the distinct advantage of maintaining their structure at any concentration as opposed to polymeric micelles and their CMCs. Drug or guest molecules can be both covalently displayed, as were the PAMAM–cisplatin vehicles, or encapsulated within the dendrimer core. Encapsulation, however, is strongly dependent upon the drug/guest and the architectural features formed within the dendrimer. Additionally,

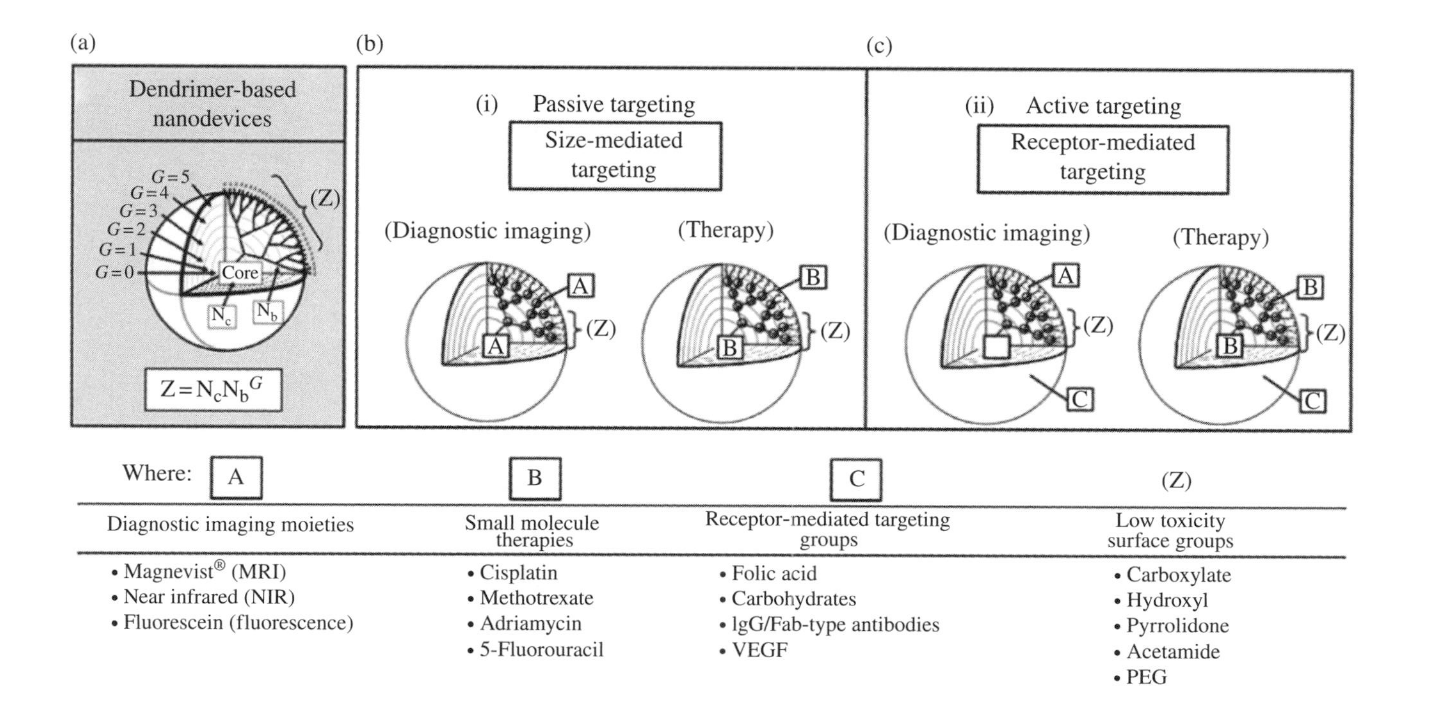

FIGURE 14.12 (a) An illustration demonstrating dendrimer topological architecture and hierarchical assembly. (b) Dendrimer passive targeting strategies and applications involving categories A, B, and Z. (c) Active targeting applications for dendrimer DDVs with all four divisions: A, B, C, and Z. (Z) A list of potential surface handles used to provide immune stealth or direct dendrimer-ligand bioconjugation. Menjoge et al. [69], pp. 171–185. Reproduced with permission of Elsevier.

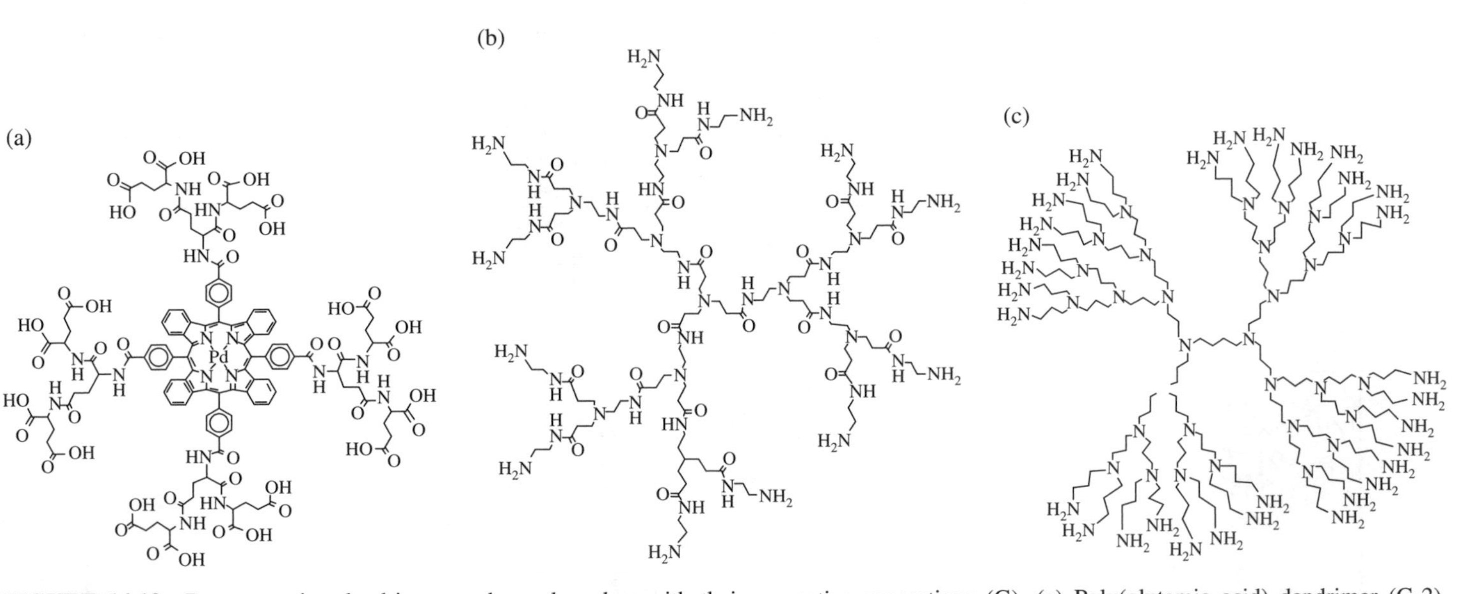

FIGURE 14.13 Representative dendrimers a, b, c, d, and e, with their respective generations (G). (a) Poly(glutamic acid) dendrimer (G-2), (b) polyamidoamine (PAMAM) dendrimer (G-2), (c) polypropyleneimine (PPI) dendrimer (G-3),

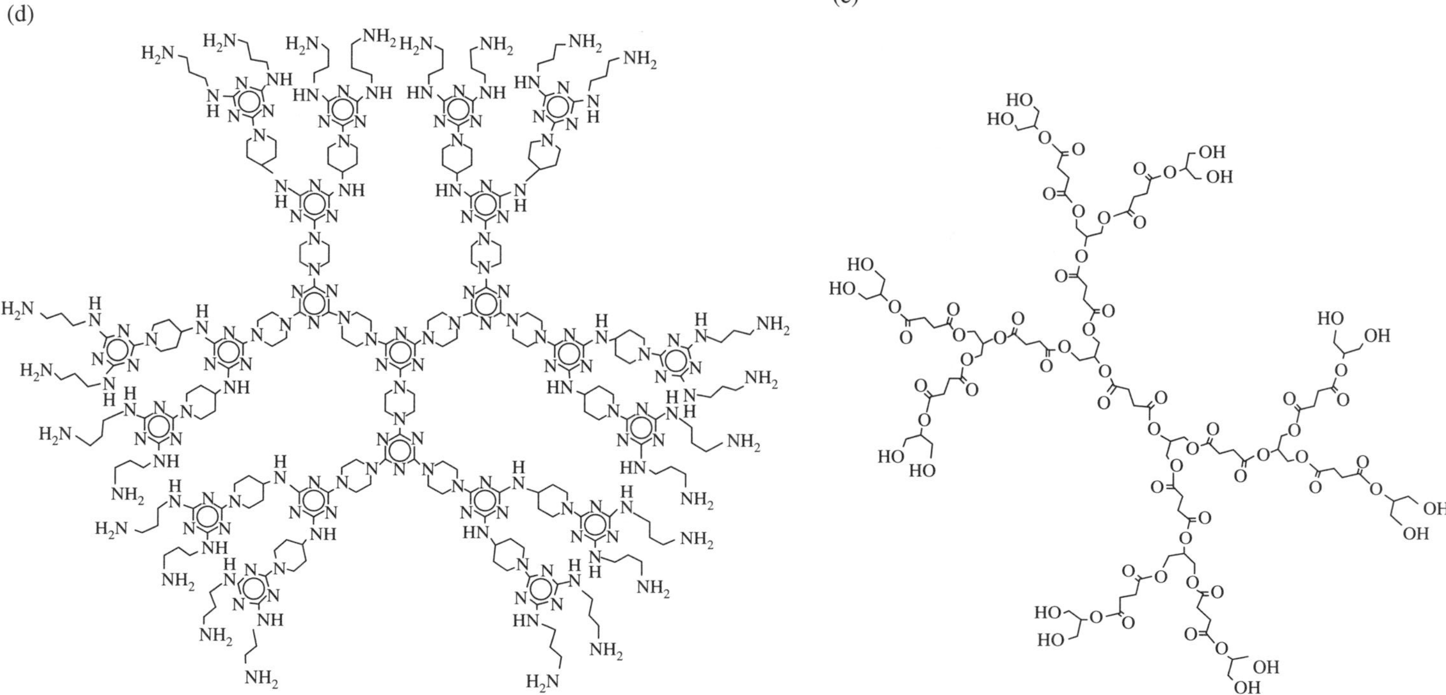

FIGURE 14.13 (*Continued*) (d) polymelamine dendrimer (G-3), and (e) polyester dendrimer (G-2). Reprinted by permission from Macmillan Publishers Ltd: *Nature Biotechnology* [56], copyright 2005.

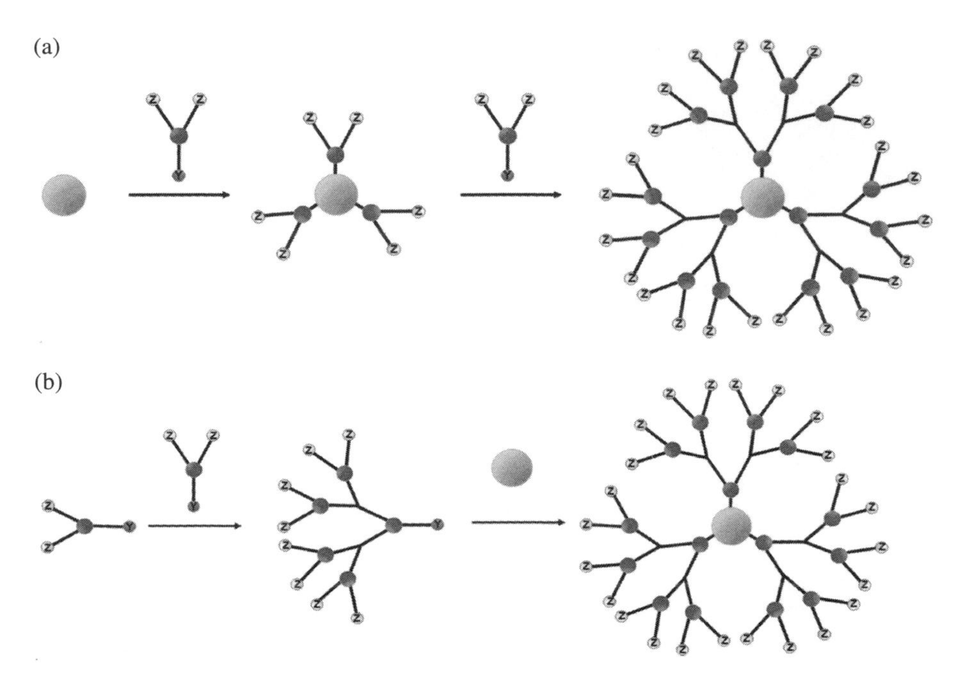

FIGURE 14.14 (a) Divergent synthesis: Starting with the reactive initiator core, activated moiety Y produced the first-generation branched dendrimer. Iterative addition of the branched monomer produces additional generations, which are terminated with functional group Z. Reprinted (adapted) with permission from [70]. Copyright 2009 American Chemical Society. (b) Convergent synthesis: The focal point Y of the branched monomer is chemically reactive to functional group Z. Iterative coupling of the branched monomer to another produced a parent dendron. The dendrimer is then assembled through addition of the reactive initiator core, which reacts with focal point Y of the parent dendron. Reprinted (adapted) with permission from [70]. Copyright 2009 American Chemical Society.

release of the drug payload is passively controlled by forces such as hydrogen bonding, hydrophobicity, electrostatic interactions, and steric interactions, which are governed at the interface of the aqueous milieu and the dendrimer surface [70].

Jain and coworkers showed that by endcapping PAMAM G4 dendrimers with 5 kDa PEG markedly increased 5-fluorouracil (5-FU) loading and reduced both its *in vivo* release rate when compared with uncoated PAMAM-NH_2 G4 dendrimers. These results strongly indicate that by endcapping with PEG, both the release rate of 5-FU and its loading efficacy can be managed [72]. Jin et al. reasoned that a cap or pH-sensitive release motif may be able to prevent premature release of the drug cargo. They constructed a PAMAM G4 dendrimer, but used both poly(2-(*N*,*N*-diethylamino)ethyl methacrylate) (PDEA) and methoxy-PEG (mPEG) to generate a core-shell (PPD) dendrimer. With the core PAMAM parent dendrimer, the pH-responsive PDEA withheld loaded 5-FU at pH 7.4 while displaying full release at pH 6.5 (Fig. 14.15a). Furthermore, the mPEG chains further enhanced *in vivo* circulation taking full advantage of the tumors RES [73].

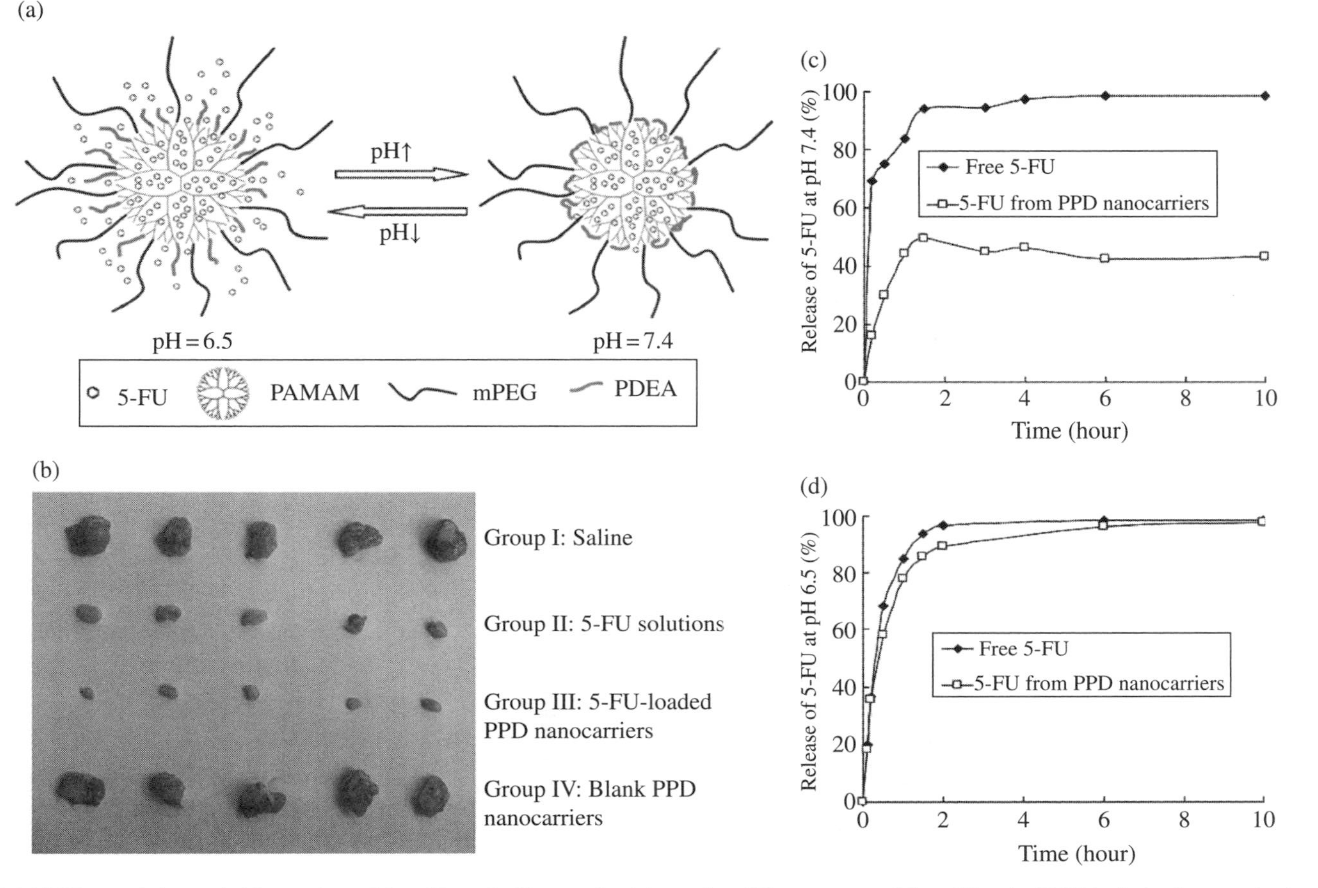

FIGURE 14.15 (a) Schematic illustration of the effect of pH upon dendrimer. As pH becomes weakly acidic, the PDEA chains become hydrophilic and extend, releasing the pyrimidine analog, 5-fluorouracil. (b) Surgically resected mouse tumors of the saline (negative control), 5-FU (positive control), PPD-5-FU, and PPD-empty groups. (c and d) Plots demonstrating the release characteristics of the 5-FU-PPD dendrimer at both physiological and endosomal pH of 7.4 and 6.5, respectively. Jin [73], pp. 378–384. Reproduced with permission of Elsevier.

Their PDD nanodevice displayed high tumor targeting and precise pH-directed release with advanced circulation in tumor-bearing mice (Fig. 14.15c and d).

Although Bradshaw and coworkers successfully functionalized their PDD dendrimer with sensing modalities, more sophisticated and facile means are available to intermediate sustained circulation and drug release. The simple conjugation or covalent modification to the peripheral units of dendrimer vehicles allows for the multivalent display of drugs, ligands, or fluorescent probes and contrast agents. Each potentially capable of environmental sensing through tethered degradable linkages, such as pH or redox-sensitive drug–dendrimer connectivities. Still, with optimal release in mind, mobilizing these attributes into successful DDVs has been challenging.

Recently, a new approach was developed for clinically viable therapeutics. Usually, amide, ester, or hydrazone bonds are used to append drug or imaging systems to the periphery of dendrimers for their endosomal or lysosomal hydrolytic release [69]. The anti-inflammatory N-acetylated cysteine (NAC) linked to PAMAM DDVs provides the means for rapid intracellular release by glutathione promoted cleavage, followed by excretion of the PAMAM–NAC dendrimer conjugates. This proof-of-principle release mechanism revealed that approximately 60% of its NAC payload was released within the first hour under cellular GSH levels, whereas no release was detected under physiological conditions similar to plasma GSH levels [74]. The authors went on to examine the relationship between the polymer scaffolding and the release profile of NAC by substituting PAMAM chains for PEG (Fig. 14.16) [75]. Using an octavalent-PEG core linked to the GSH-sensitive NAC linker, it was found that the DDV retained the same release characteristics at a pH of 6.5 and 7.4 as the PAMAM–NAC vehicle. Together, these studies highlight the salient relationship between drug/guest placement for dendrimer-like ligand-based targeting and the concomitant ligand-release characteristics within biological samples.

14.3 INORGANIC DDVs: METAL- AND SILICA-BASED SYSTEMS

Inorganic nanoparticles have been the focus of material science for decades. Their use in thin films and composites is well documented; however, within the last 20 years their intrinsic properties have transformed drug delivery [76]. Their magnetic, optical, electronic, and catalytic properties as well as their ability to act as a template material have all seen wide-ranging use under the auspice of biological applications [77–80]. Inorganic nanoparticles are typically composed of a core-shell motif, where the core can be composed of metals such as silver, gold, iron oxide, silica, or quantum dots (QDs) and tailored from 1.5 to 10 mm in diameter [81]. Their external shells can be composed of polymers, drugs, or biomacromolecules such as proteins, antibodies, or other biological polymers suited for *in vivo* applications (Fig. 14.17) [82]. The preparation of inorganic DDVs can extend to a variety of synthetic strategies including surface coating, direct synthesis, or through bioconjugation techniques that are centered on metals or silica complexes [83]. However, the most common techniques for surface modification of inorganic DDVs employ either physical or chemisorption, with the latter being preferred.

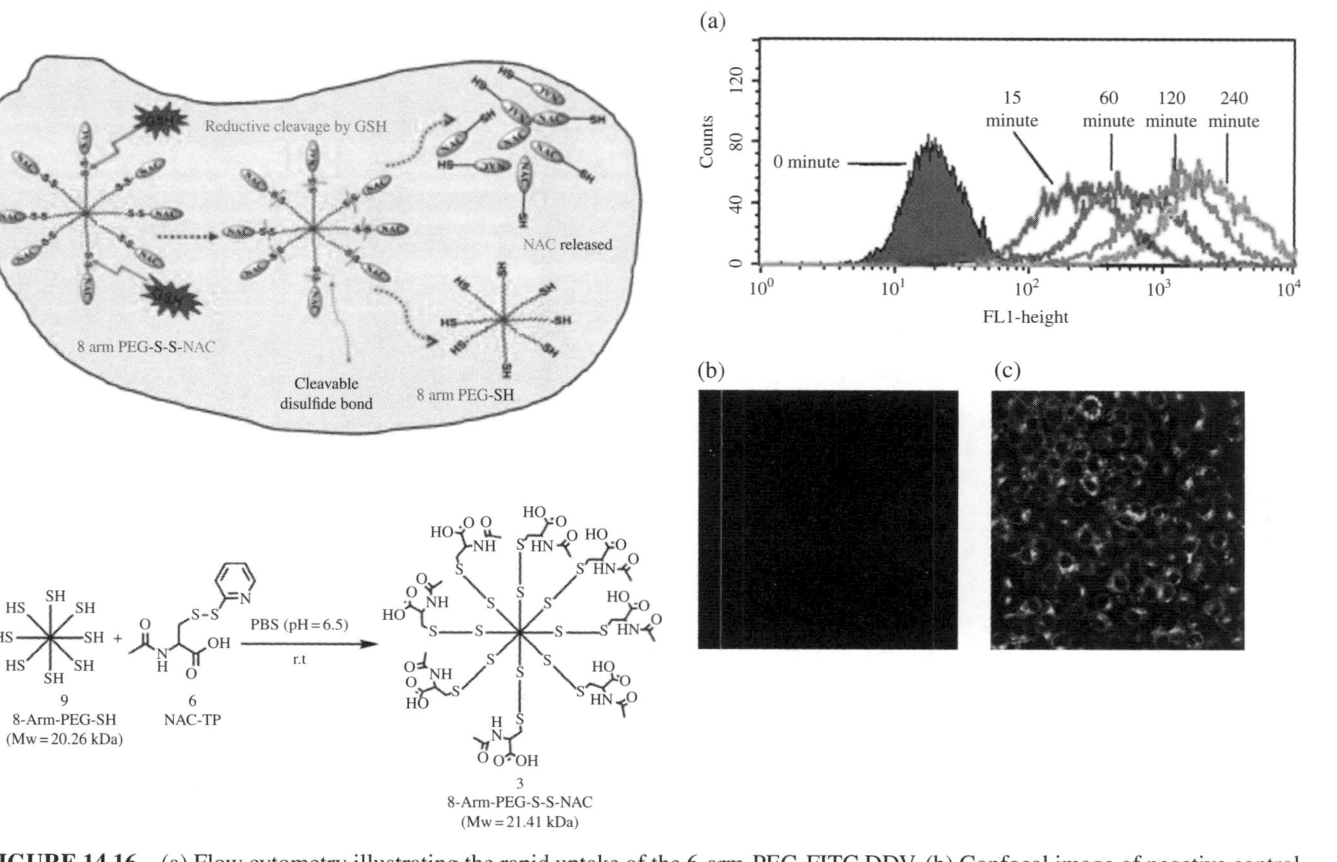

FIGURE 14.16 (a) Flow cytometry illustrating the rapid uptake of the 6-arm-PEG-FITC DDV. (b) Confocal image of negative control BV2 cells. (c) Confocal image of BV2 cells with 6-arm-PEG-FITC DDV. Navath et al. [75], pp. 447–456. Reproduced with permission of Elsevier. (*See insert for color representation of the figure.*)

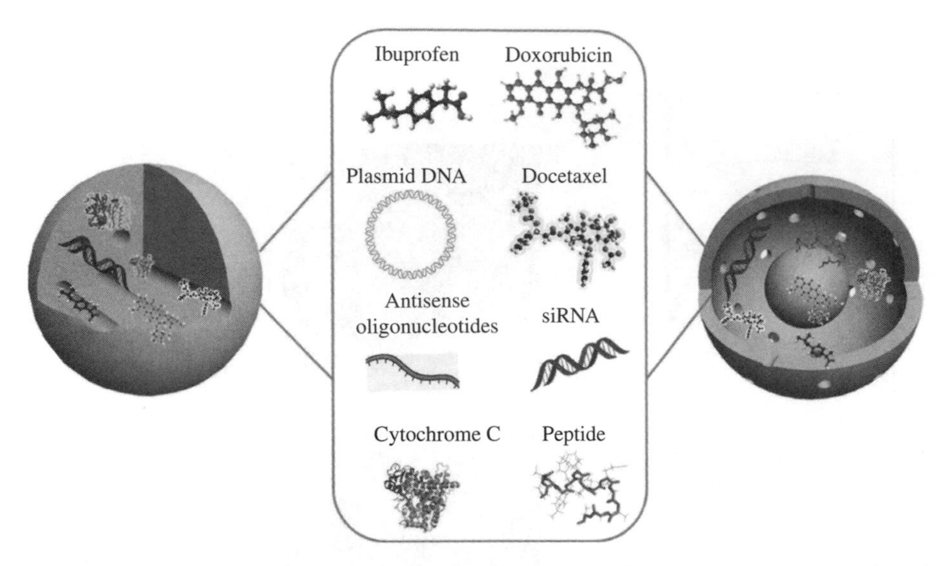

FIGURE 14.17 Mesoporous silica nanoparticles are a multifaceted DDV system capable of transporting NSAIDs (ibuprofen), anticancer therapeutics (doxorubicin), and therapeutic biopolymers such as pDNA and cytochrome C. Tang et al. [76], pp. 1504–1534. Reproduced with permission of John Wiley & Sons.

14.3.1 Inorganic DDVs: Mesoporous Silica Nanoparticles

Among potential inorganic DDVs mesoporous silica nanoparticles (MSNs) have raised great interest. With their inexpensive building blocks and tailorable architectural properties, MSPs are capable of supporting large pore volumes and high specific surface areas necessary for a capable and diverse drug delivery or bioimaging platform [84]. Moreover, MSNs possess dual functional surfaces, both exterior and interior, which can further diversify its targeting and delivery modalities. For drug delivery applications, precise control over diameter, loading capacity, and targeting geometry must be demonstrated. Hence, the synthesis of MSNs must account for each of these metrics. The Mobile Oil Company developed the first class of MSNs known as the M41S phase. These materials were created to advance the control of zeolite molecular sieves, which were limited to pore sizes around 1.5 nm [85]. The M41S phase materials proved superior, and their pores could be controlled ranging from 2 to 10 nm with amorphous characteristics. The assembly of the more notable MCM-41, lesser-known MCM-48, and MCM-50 silicates, was controlled using an ionic-surfactant structure-directing agent (SDA) that generated supramolecular aggregates with uniform diameters and pore sizes [85]. Their assembly is promoted through the liquid crystal-templating effect of the SDA, cetyltrimethylammonium bromide (CTAB), to generate micellular structures at concentrations above their CMC. Upon the addition of tetraethyl orthosilicate (TEOS) or sodium metasilicate (Na_2SiO_3), the mesostructured composite assembles during the condensation of the silicates under alkaline conditions. The final MSN is generated upon removal of the surfactant SDA by extraction or calcination (Fig. 14.18).

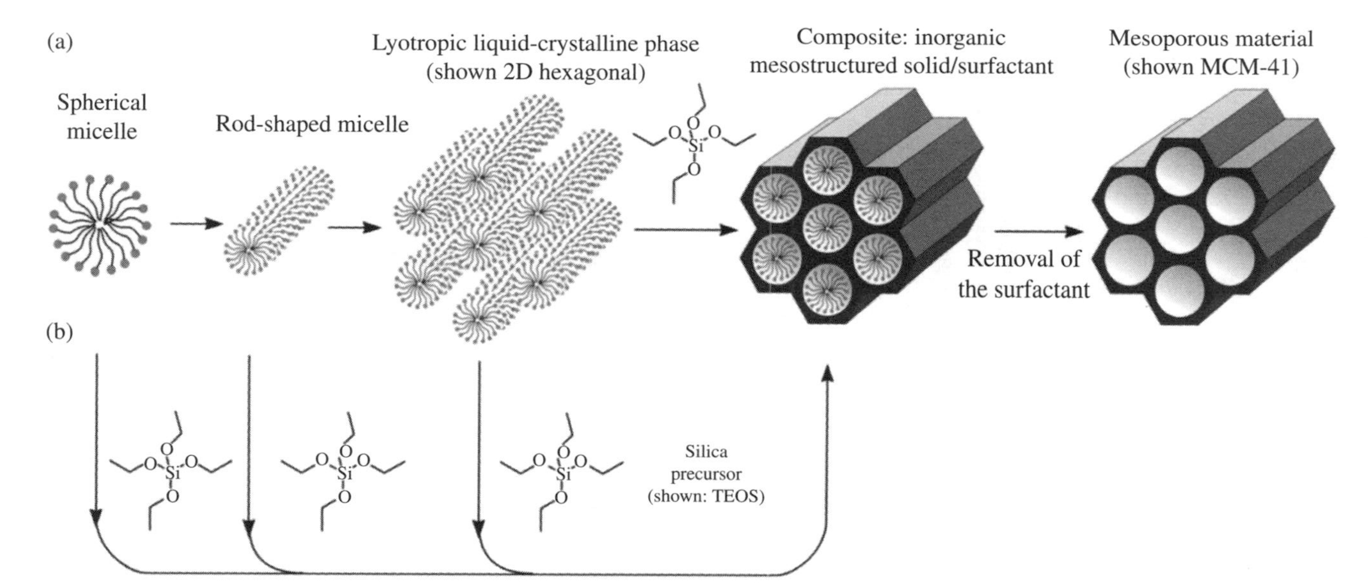

FIGURE 14.18 Scheme showing the formation of mesoporous silica using structure-directing agents (SDAs): (a) the liquid-crystal template mechanism and (b) the cooperative liquid-crystal template mechanism. Tang et al. [76], pp. 1504–1534. Reproduced with permission of John Wiley & Sons.

However, the most powerful facet of MSN DDVs is their ability to combine the highly structured inorganic building blocks of a single material with multiple organic functionalizations. Common techniques for surface modification of MSN DDVs employ either physical or chemisorption—with the latter being preferred. Among the covalent modifications are grafting, co-condensation, and periodic mesoporous organosilicas (PMO). Both grafting and co-condensation methods are reference methods of subsequent modification to the interior surface of the mesoporous organosilanes. Grafting refers to the direct modification of the type $(R'O)_3SiR$ and to some extent $ClSiR_3$. This method, however, is applied as postsynthesis and therefore can result in nonhomogeneous degree of distribution of the organosilane functionalization within the pores (where R is the organic functional group). Particularly, the grafting method tends to preferentially label the entrance of the pore, occluding the opening and eliminating its ability to occupy chemical payloads or perform any additional bioconjugation (Fig. 14.19a) [85].

Conversely the co-condensation of tetraalkoxysilanes, $(RO)_4Si$, with terminal trialkoxyorganosilanes of the type $(R'O)_3SiR$ in the presence of an SDA results in the covalently anchored residues to the inner surface of the pore walls. Because the organic functionalities are loaded with the SDA presynthesis, they're more evenly distributed within the pore structures. However, a concentration gradient of the functionalities occurs with increasing concentrations of $(R'O)_3SiR$ in the reaction mixture that generates disordered products. This, in turn, limits total functionalization to approximately 40 mol%, which can decrease pore volume and generate overcrowding of functional groups (Fig. 14.19b) [85].

The use of a bridged organosilica precursor of the type $(R'O)_3Si\text{-}R\text{-}Si(OR')_3$ is ostensibly an alternate route of functionalization that draws upon the success of co-condensation, but avoids its pitfalls through incorporation of the 3D network of the silica matrix. The preparation of these PMOs can afford large internal surface areas with a very narrow pore radii and hold the highest potential for delivery systems (Fig. 14.19c) [85]. Consequently, the biocompatibility and biodistribution parameters of MSNs have also been demonstrated through applications in both bioimaging and as a DDV [82, 86–88]. An excellent example is the trifunctionalized MSN DDV reported by Cheng et al. In their work, they incorporated, in a sequential fashion, a contrast agent for *in vitro* tracking, a chemotherapeutic, and a bidirectional targeting ligand.

This theranostic system (a combination of diagnostic and therapeutic device) was decorated with a highly specific cRGD (cyclic-Arg-Gly-Asp) peptide-targeting motif combined with a matrix-incorporated near-infrared (NIR) contrast agent within the MSN framework (Fig. 14.20) [89]. As a payload, a photosensitizer was used for photodynamic therapy. These features enabled the direct detection of high cellular penetration and cytotoxicity against cells that overexpressed the cRGD-targeted integrins. This example clearly demonstrates the ability of MSN to perform as an inorganic DDV and bioimaging/contrast agent, and its full potential as a nanotherapeutic device.

14.3.2 Inorganic DDVs: Gold Nanoparticles

Other inorganic nanocarriers have also found success as DDV systems. Using a metal core system significant efforts have been made to transport pharmaceuticals, macromolecules, DNA or RNA as payloads. Specifically, gold nanoparticles (AuNPs)

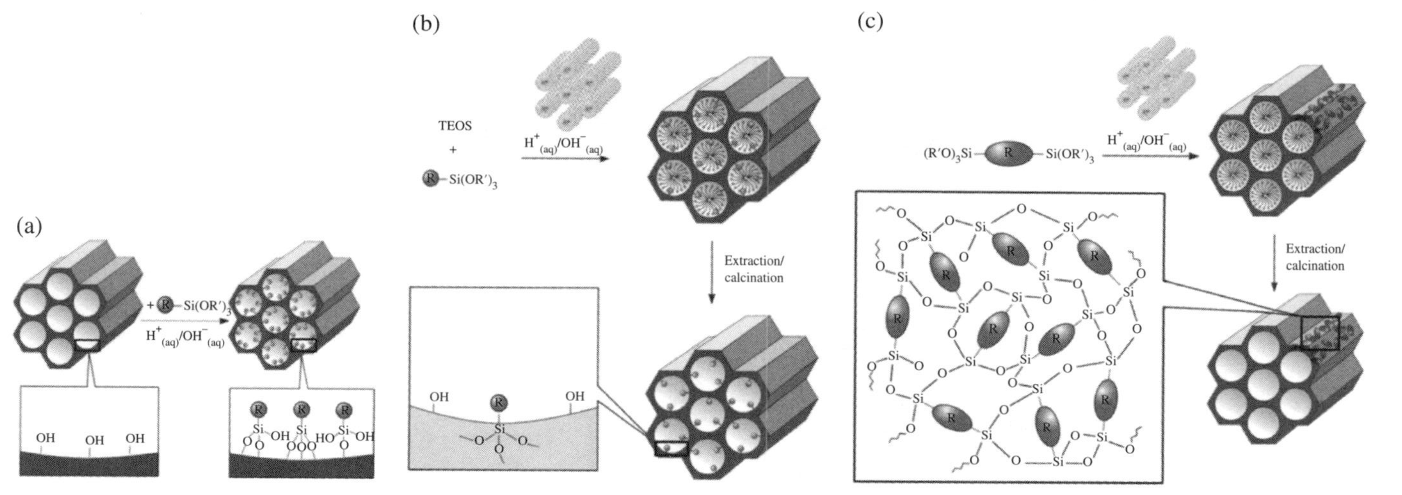

FIGURE 14.19 (a) Grafting: a post-synthesis chemical modification with organosilanes of type $(R'O)_3SiR$ (R = organic functional group). (b) The co-condensation of tetraalkoxysilanes [$(RO)_4Si$ (TEOS)] with terminal $(R'O)_3SiR$ groups and appropriate SDAs that project organic functionalities into the pore. (c) Periodic mesoporous organosilicas (PMOs): type $(R'O)_3Si$-R-$Si(OR')_3$ precursors are used to generate a 3D lattice, which are homogeneously distributed throughout the MSN's walls. Hoffmann et al. [85], pp. 3216–3251. Reproduced with permission of Wiley-VCH.

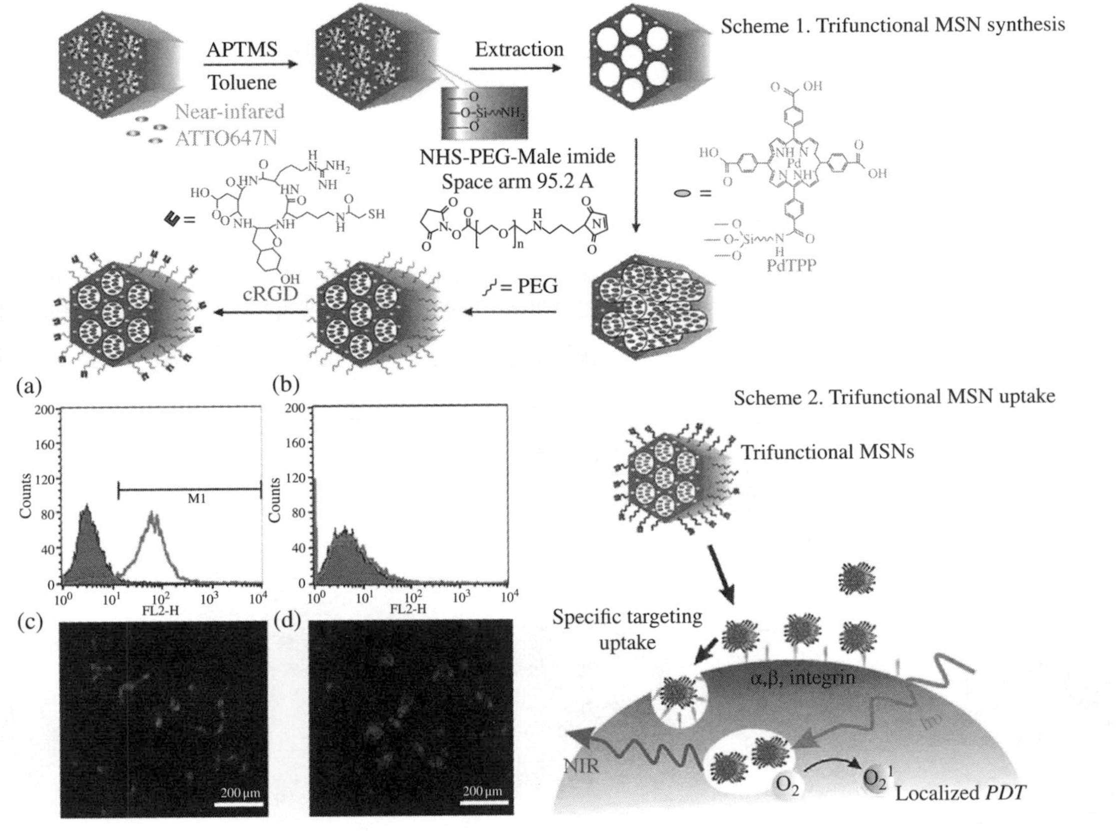

FIGURE 14.20 (Scheme 1) Synthesis of the trifunctional theranostic MSN incorporated with an NIR ATTO 647N contrast agent, cRGD targeting peptide, and palladium-based photosensitizer (Pd-TPP) for photodynamic therapy (PDT). (Scheme 2) Graphical representation of the trifunctional MSN binding to integrin receptors via the targeting cRGD peptide followed by cellular internalization, and Pd-TPP activation by PDT. (a) Flow cytometric analysis of positive control cells (U87-MG), which overexpress $\alpha_v\beta_3$ integrin receptors. (b) MCF-7 cells that are $\alpha_v\beta_3$ null for integrin receptors act as a negative control. (c and d) Immunofluorescent staining of $\alpha_v\beta_3$ integrin expression for U87-MG and MCF-7 cells, respectively. Reproduced ("Adapted" or "in part") from [89] with permission from the Royal Society of Chemistry. (*See insert for color representation of the figure.*)

make excellent scaffolds for DDVs because of their complete inertness, biocompatibility, and their range of size, 1–150 nm, which possesses excellent dispersity [7]. These characteristics allow easy AuNP fabrication, which can be customized to meet the strict biological parameters that DDV systems must adhere to for efficient targeting and payload delivery. Additionally, the morphological aspects of Au allow for high surface-to-volume ratios, which can support substantial and highly tunable functionalization of targeting ligands and chemotherapeutics [7]. Their synthesis typically follows the preparation of monolayer-protected clusters (MPC) of AuNPs via the Brust–Schiffrin reaction (Fig. 14.21).

This method uses a surfactant, tetraoctylammonium bromide (TOAB), to transfer $AuCl_4^-$ to the organic phase where it is reduced in the presence of $NaBH_4$ and alkanethiols. The nucleophilic nature of the thiol allows for the direct alkylation where further modification can be applied through the formation of mixed monolayered clusters via postfunctionalization or using place-exchange methods with specific ligands [90]. These methods differ from contemporary colloid formulations in that they can be repeatedly isolated and redissolved in traditional organic solvents without the deleterious effects of irreversible aggregation [91].

As a DDV, AuNP can act as a highly diverse scaffold for functionalization. Particularly, monolayered-protected systems, which are functionalized through thiol linkages, have seen great use as a therapeutic delivery system. Moreover, functionalizing with PEG linked to a targeting ligand enables the AuNP to minimize any interactions with the RES and, therefore, increases its potential circulation time allowing for accumulation in tumors. Prabaharan and coworkers used this approach with cysteinamine as the thiol-conjugating agent to generate amine functionalized AuNP. The amine-bearing AuNP was used as the macro initiator for the ROP of β-benzyl L-aspartate *N*-carboxyanhydride (BLA-NCA)—the Fuchs–Farthing method—followed by PEGylation and subsequent conjugation of folic acid as the tumor-targeting moiety.

Finally, DOX was coupled to the inner poly (L-aspartate) segment using the pH-sensitive hydrozone linkage to generate the final AuNP: AuP(LA-DOX)-b-PEG-FA. This synthetic approach combines the complexity of PLAA construction (BLA-NCA polymerization) used for organic DDVs and the diversity inherent to AuNP to generate an estimable delivery device with an average diameter of 34 nm and a narrow polydispersity index of 0.029 (Fig. 14.22) [92]. Its release mechanism was verified as the release of DOX occurred at a pH of the endosomal compartments (5–6) and stabilized under a physiological pH of 7.4. Moreover, when measured

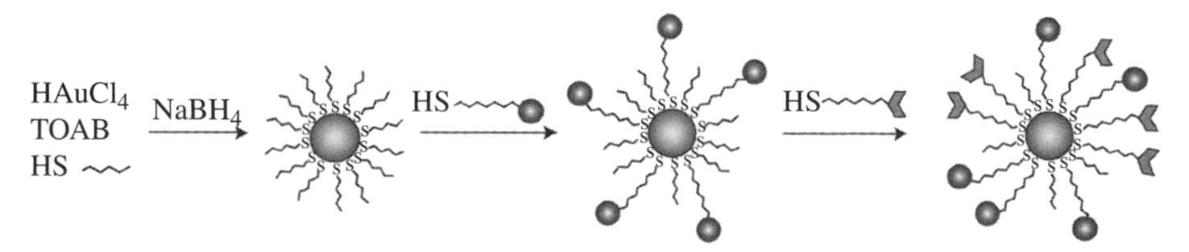

FIGURE 14.21 Preparation of a monolayer-protected cluster (MPC) using the Brust–Schiffrin reaction. Rana et al. [7], pp. 200–216. Reproduced with permission of Elsevier.

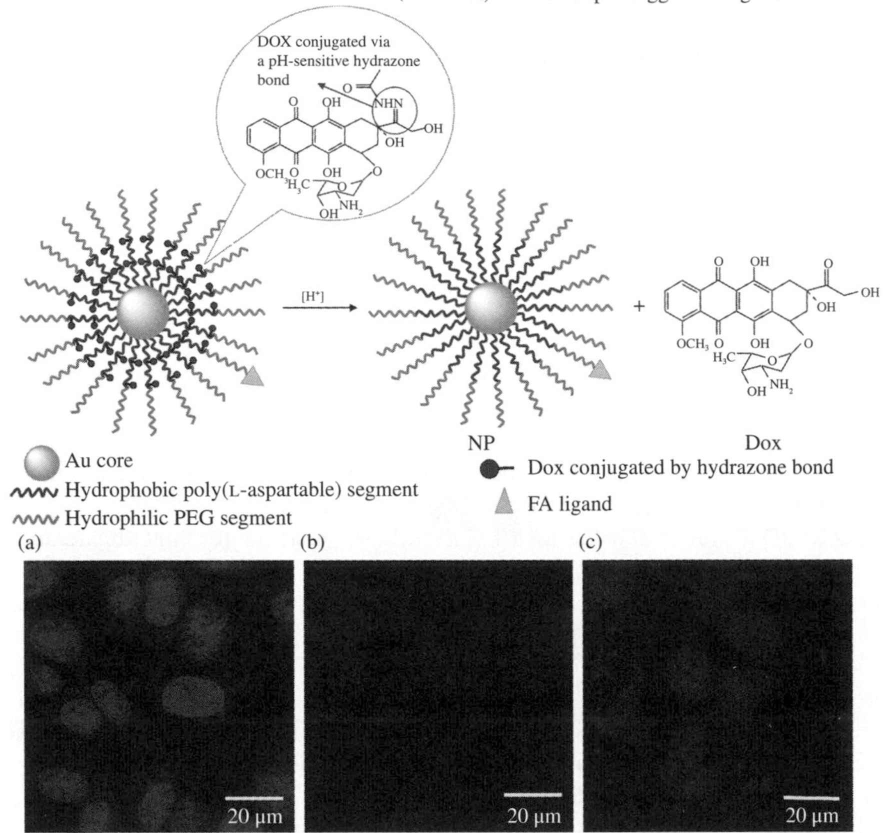

FIGURE 14.22 (Scheme 1) General schematic of the AuNP modified with Asp-PEG-FA with an acid-cleavable linker capable of releasing its fluorescent DOX payload at low pH. (a) Positive control: Free DOX in 4TI cells. (b) Au-P(LA-DOX)-*b*-PEG-OH in 4TI cells. (c) Au-P(LA-DOX)-*b*-PEG-FA in 4TI cells. Prabaharan et al. [92], pp. 6065–6075. Reproduced with permission of Elsevier. (*See insert for color representation of the figure.*)

against AuP(LA-DOX)-b-PEG, AuP(LA-DOX)-b-PEG-FA showed an increase in cellular targeting and uptake in folate receptor positive 4TI cells (Fig. 14.22) [92]. This particular DDV illustrates the diversity and ease at which AuNP can be projected into other DDV frameworks to yield facile and effective delivery systems. Delivery of proteins into cellular or subcellular stations is an onerous task due to their large size and variable charge density. The ability to transport biomacromolecules across cell membranes, with retention of their activity and structure preservation, presents an extremely attractive means for biological probing and medicinal therapies. Demonstrably, Ghosh et al. fabricated an AuNP with a core diameter of 2.5 nm incorporated with a three-tier structural domain: core-stabilizing alkyl domains, a tetraethylene glycol (TET) corona, and pendent peptide recognition ligands, HLRL (Scheme 1 in Fig. 14.23), to promote protein surface recognition, and endosomal escape via the proton sponge effect (Fig. 14.23) [93]. With a strongly cationic

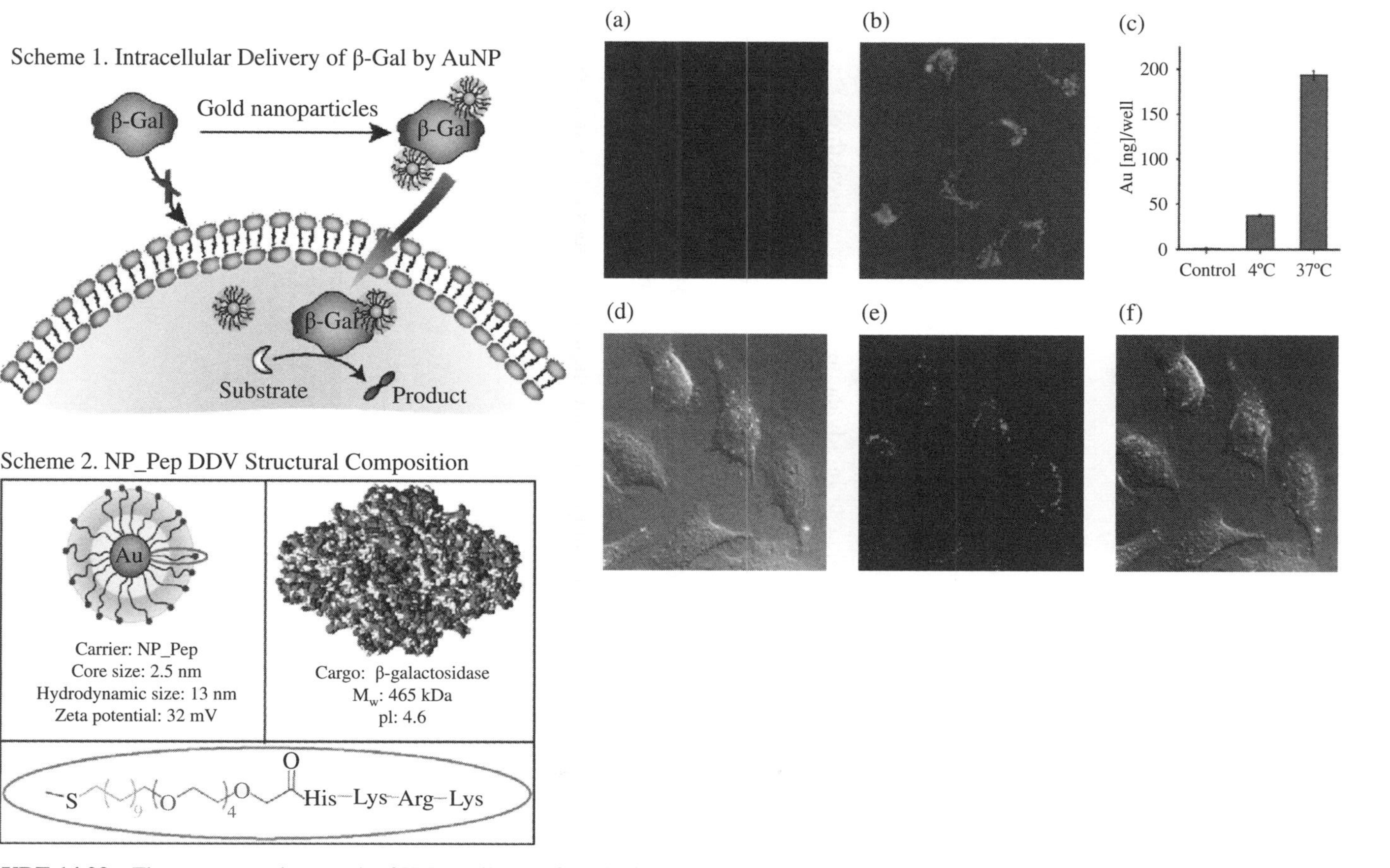

FIGURE 14.23 Fluorescence micrograph of HeLa cells transfected with FITC-β-Gal in (a) absence or (b) presence of NP_Pep. (c) ICP-MS measurements after NP_Pep/β-Gal treatment. (d–f) Confocal images of HeLa cells after protein transfection: (d) bright field, (e) fluorescence, and (f) merged. Reprinted (adapted) with permission from [93]. Copyright 2010 American Chemical Society. (*See insert for color representation of the figure.*)

surface, the AuNP easily promoted β-galatosidase complexation and was subsequently challenged against HeLa cells. The bioactivity of β-galatosidase was assessed *in vitro* post transduction using appropriate substrate concentrations of X-gal. Significantly, blue precipitate was observed indicating the colorless X-gal substrate had been successfully hydrolyzed to its blue product, thus the catalytic activity of β-galatosidase had been preserved [93].

14.4 CONCLUSION

Drug delivery vehicles (DDVs) have been used to combat some of the most challenging hurdles of modern cancer therapies. Their modular design–bearing functionalities have improved upon the pharmacological action of clinical therapeutics through extending their systemic circulation, refining their targeting, and implemented programmed payload delivery through environmental sensing and remote actuating components.

These technologies, driven by the amalgamation of cancer biology and nanomaterials engineering, have proven valuable in advancing significant breakthroughs in both the biology of cancer and its therapies. To improve the DDV platform additional exploration of DDV vectors is underway. Using plant viruses as scaffold for *in vivo* imaging or endogenous protein-shell DDVs to enhance immune-stealth represent other pathways to combat cancer [94–96]. These tour-de-force devices, along with those described in this chapter, possess the greatest potential to harness the next generation of personalized medicine. Each patient can potentially have his or her treatment designed around the unique circumstances of their diagnosis. However, more work is needed to viably and safely translate these techniques into a successful clinical regimen where the inherent activities of DDVs can be realized.

REFERENCES

1. Wang, A.Z., Langer, R., & Farokhzad, O.C. Nanoparticle delivery of cancer drugs. *Annu. Rev. Med.* **63**, 185–198 (2012).
2. Petros, R.A. & DeSimone, J.M. Strategies in the design of nanoparticles for therapeutic applications. *Nat. Rev. Drug Discov.* **9**, 615–627 (2010).
3. Sanhai, W.R., Sakamoto, J.H., Canady, R., & Ferrari, M. Seven challenges for nanomedicine. *Nat. Nanotechnol.* **3**, 242–244 (2008).
4. Ferrari, M. Cancer nanotechnology: opportunities and challenges. *Nat. Rev. Cancer* **5**, 161–171 (2005).
5. Allen, T.M. Ligand-targeted therapeutics in anticancer therapy. *Nat. Rev. Cancer* **2**, 750–763 (2002).
6. Church, D.N. et al. Extended survival in women with brain metastases from HER2 overexpressing breast cancer. *Am. J. Clin. Oncol.* **31**, 250–254 (2008).
7. Rana, S., Bajaj, A., Mout, R., & Rotello, V.M. Monolayer coated gold nanoparticles for delivery applications. *Adv. Drug Deliv. Rev.* **64**, 200–216 (2012).

8. Smith, M.T., Hawes, A.K., & Bundy, B.C. Reengineering viruses and virus-like particles through chemical functionalization strategies. *Curr. Opin. Biotechnol.* **24**, 620–626 (2013).
9. Christie, R.J. et al. Targeted polymeric micelles for siRNA treatment of experimental cancer by intravenous injection. *ACS Nano* **6**, 5174–5189 (2012).
10. Stolnik, S., Illum, L., & Davis, S.S. Long circulating microparticulate drug carriers. *Adv. Drug Deliv. Rev.* **64**, 290–301 (2012).
11. Yared, J.A. & Tkaczuk, K.H.R. Update on taxane development: new analogs and new formulations. *Drug Des. Devel. Ther.* **6**, 371–384 (2012).
12. Gradishar, W.J. et al. Phase III trial of nanoparticle albumin-bound paclitaxel compared with polyethylated castor oil-based paclitaxel in women with breast cancer. *J. Clin. Oncol.* **23**, 7794–7803 (2005).
13. Schroeder, A. et al. Treating metastatic cancer with nanotechnology. *Nat. Rev. Cancer* **12**, 39–50 (2012).
14. Miyata, K., Christie, R.J., & Kataoka, K. Polymeric micelles for nano-scale drug delivery. *React. Funct. Polym.* **71**, 227–234 (2011).
15. Bae, Y. & Kataoka, K. Intelligent polymeric micelles from functional poly(ethylene glycol)-poly(amino acid) block copolymers. *Adv. Drug Deliv. Rev.* **61**, 768–784 (2009).
16. Nishiyama, N. et al. Preparation and characterization of self-assembled polymer-metal complex micelle from cis-dichlorodiammineplatinum(II) and poly(ethylene glycol)-poly(alpha,beta-aspartic acid) block copolymer in an aqueous medium. *Langmuir* **15**, 377–383 (1999).
17. Nishiyama, N. et al. Novel cisplatin-incorporated polymeric micelles can eradicate solid tumors in mice. *Cancer Res.* **63**, 8977–8983 (2003).
18. Kowalczuk, A. et al. Loading of polymer nanocarriers: factors, mechanisms and applications. *Prog. Polym. Sci.* **39**, 43–86 (2014).
19. Duong, H.T.T., Marquis, C.P., Whittaker, M., Davis, T.P., & Boyer, C. Acid degradable and biocornpatible polymeric nanoparticles for the potential codelivery of therapeutic agents. *Macromolecules* **44**, 8008–8019 (2011).
20. Lai, T.C., Cho, H., & Kwon, G.S. Reversibly core cross-linked polymeric micelles with pH- and reduction-sensitivities: effects of cross-linking degree on particle stability, drug release kinetics, and anti-tumor efficacy. *Polym. Chem.* **5**, 1650–1661 (2014).
21. Miyata, K., Nishiyama, N., & Kataoka, K. Rational design of smart supramolecular assemblies for gene delivery: chemical challenges in the creation of artificial viruses. *Chem. Soc. Rev.* **41**, 2562–2574 (2012).
22. Carmeliet, P. & Jain, R.K. Angiogenesis in cancer and other diseases. *Nature* **407**, 249–257 (2000).
23. Pillay, C.S., Elliott, E., & Dennison, C. Endolysosomal proteolysis and its regulation. *Biochem. J.* **363**, 417–429 (2002).
24. Kakizawa, Y., Harada, A., & Kataoka, K. Environment-sensitive stabilization of core-shell structured polyion complex micelle by reversible cross-linking of the core through disulfide bond. *J. Am. Chem. Soc.* **121**, 11247–11248 (1999).
25. Bartlett, D.W. & Davis, M.E. Physicochemical and biological characterization of targeted, nucleic acid-containing nanoparticles. *Bioconjug. Chem.* **18**, 456–468 (2007).
26. Song, E.W. et al. Antibody mediated in vivo delivery of small interfering RNAs via cell-surface receptors. *Nat. Biotechnol.* **23**, 709–717 (2005).

27. Druker, B.J. et al. Five-year follow-up of patients receiving imatinib for chronic myeloid leukemia. *N. Engl. J. Med.* **355**, 2408–2417 (2006).
28. Sarfati, G., Dvir, T., Elkabets, M., Apte, R.N., & Cohen, S. Targeting of polymeric nanoparticles to lung metastases by surface-attachment of YIGSR peptide from laminin. *Biomaterials* **32**, 152–161 (2011).
29. Torchilin, V.P., Lukyanov, A.N., Gao, Z.G., & Papahadjopoulos-Sternberg, B. Immunomicelles: targeted pharmaceutical carriers for poorly soluble drugs. *Proc. Natl. Acad. Sci. U. S. A.* **100**, 6039–6044 (2003).
30. Kirpotin, D.B. et al. Antibody targeting of long-circulating lipidic nanoparticles does not increase tumor localization but does increase internalization in animal models. *Cancer Res.* **66**, 6732–6740 (2006).
31. van Dongen, M.A., Dougherty, C.A., & Holl, M.M.B. Multivalent polymers for drug delivery and imaging: the challenges of conjugation. *Biomacromolecules* **15**, 3215–3234 (2014).
32. Lallana, E., Fernandez-Trillo, F., Sousa-Herves, A., Riguera, R., & Fernandez-Megia, E. Click chemistry with polymers, dendrimers, and hydrogels for drug delivery. *Pharm. Res.* **29**, 902–921 (2012).
33. Hong, S. et al. The binding avidity of a nanoparticle-based multivalent targeted drug delivery platform. *Chem. Biol.* **14**, 107–115 (2007).
34. Montet, X., Funovics, M., Montet-Abou, K., Weissleder, R., & Josephson, L. Multivalent effects of RGD peptides obtained by nanoparticle display. *J. Med. Chem.* **49**, 6087–6093 (2006).
35. Kiessling, L.L., Gestwicki, J.E., & Strong, L.E. Synthetic multivalent ligands as probes of signal transduction. *Angew. Chem. Int. Ed.* **45**, 2348–2368 (2006).
36. Kiessling, L.L., Gestwicki, J.E., & Strong, L.E. Synthetic multivalent ligands in the exploration of cell-surface interactions. *Curr. Opin. Chem. Biol.* **4**, 696–703 (2000).
37. Fasting, C. et al. Multivalency as a chemical organization and action principle. *Angew. Chem. Int. Ed.* **51**, 10472–10498 (2012).
38. Davis, M.E., Chen, Z., & Shin, D.M. Nanoparticle therapeutics: an emerging treatment modality for cancer. *Nat. Rev. Drug Discov.* **7**, 771–782 (2008).
39. Carlson, C.B., Mowery, P., Owen, R.M., Dykhuizen, E.C., & Kiessling, L.L. Selective tumor cell targeting using low-affinity, multivalent interactions. *ACS Chem. Biol.* **2**, 119–127 (2007).
40. Venturoli, D. & Rippe, B. Ficoll and dextran vs. globular proteins as probes for testing glomerular permselectivity: effects of molecular size, shape, charge, and deformability. *Am. J. Physiol. Ren. Physiol.* **288**, F605–F613 (2005).
41. Choi, H.S. et al. Renal clearance of quantum dots. *Nat. Biotechnol.* **25**, 1165–1170 (2007).
42. Mitragotri, S. & Lahann, J. Physical approaches to biomaterial design. *Nat. Mater.* **8**, 15–23 (2009).
43. Moghimi, S.M. & Patel, H.M. Serum-mediated recognition of liposomes by phagocytic cells of the reticuloendothelial system—the concept of tissue specificity. *Adv. Drug Deliv. Rev.* **32**, 45–60 (1998).
44. Peer, D. et al. Nanocarriers as an emerging platform for cancer therapy. *Nat. Nanotechnol.* **2**, 751–760 (2007).

45. Illium, L. et al. Blood clearance and organ deposition of intravenously administered colloidal particles—the effects of particle-size, nature and shape. *Int. J. Pharm.* **12**, 135–146 (1982).
46. Goppert, T.M. & Muller, R.H. Polysorbate-stabilized solid lipid nanoparticles as colloidal carriers for intravenous targeting of drugs to the brain: comparison of plasma protein adsorption patterns. *J. Drug Target.* **13**, 179–187 (2005).
47. Vonarbourg, A., Passirani, C., Saulnier, P., & Benoit, J.P. Parameters influencing the stealthiness of colloidal drug delivery systems. *Biomaterials* **27**, 4356–4373 (2006).
48. Owens, D.E. & Peppas, N.A. Opsonization, biodistribution, and pharmacokinetics of polymeric nanoparticles. *Int. J. Pharm.* **307**, 93–102 (2006).
49. Li, S.D. & Huang, L. Nanoparticles evading the reticuloendothelial system: role of the supported bilayer. *Biochim. Biophys. Acta Biomembr.* **1788**, 2259–2266 (2009).
50. Claesson, P.M., Blomberg, E., Froberg, J.C., Nylander, T., & Arnebrant, T. Protein interactions at solid-surfaces. *Adv. Colloid Interface Sci.* **57**, 161–227 (1995).
51. Li, S.D. & Huang, L. Stealth nanoparticles: high density but sheddable PEG is a key for tumor targeting. *J. Control. Release* **145**, 178–181 (2010).
52. Munn, L.L. Aberrant vascular architecture in tumors and its importance in drug-based therapies. *Drug Discov. Today* **8**, 396–403 (2003).
53. Fang, J., Nakamura, H., & Maeda, H. The EPR effect: unique features of tumor blood vessels for drug delivery, factors involved, and limitations and augmentation of the effect. *Adv. Drug Deliv. Rev.* **63**, 136–151 (2011).
54. Narang, A.S. & Varia, S. Role of tumor vascular architecture in drug delivery. *Adv. Drug Deliv. Rev.* **63**, 640–658 (2011).
55. Torchilin, V.P. Micellar nanocarriers: pharmaceutical perspectives. *Pharm. Res.* **24**, 1–16 (2007).
56. Lee, C.C., MacKay, J.A., Frechet, J.M., & Szoka, F.C. Designing dendrimers for biological applications. *Nat. Biotechnol.* **23**, 1517–1526 (2005).
57. Lai, T.C., Kataoka, K., & Kwon, G.S. Pluronic-based cationic block copolymer for forming pDNA polyplexes with enhanced cellular uptake and improved transfection efficiency. *Biomaterials* **32**, 4594–4603 (2011).
58. Duncan, R., Vicent, M.J., Greco, F., & Nicholson, R.I. Polymer-drug conjugates: towards a novel approach for the treatment of endrocine-related cancer. *Endocr. Relat. Cancer* **12**, S189–S199 (2005).
59. Duncan, R. Polymer conjugates as anticancer nanomedicines. *Nat. Rev. Cancer* **6**, 688–701 (2006).
60. Duncan, R. The dawning era of polymer therapeutics. *Nat. Rev. Drug Discov.* **2**, 347–360 (2003).
61. Stiriba, S.E., Kautz, H., & Frey, H. Hyperbranched molecular nanocapsules: comparison of the hyperbranched architecture with the perfect linear analogue. *J. Am. Chem. Soc.* **124**, 9698–9699 (2002).
62. Bury, K. & Neugebauer, D. Novel self-assembly graft copolymers as carriers for anti-inflammatory drug delivery. *Int. J. Pharm.* **460**, 150–157 (2014).
63. Astruc, D., Boisselier, E., & Ornelas, C. Dendrimers designed for functions: from physical, photophysical, and supramolecular properties to applications in sensing, catalysis, molecular electronics, photonics, and nanomedicine. *Chem. Rev.* **110**, 1857–1959 (2010).

64. Adams, M.L., Lavasanifar, A., & Kwon, G.S. Amphiphilic block copolymers for drug delivery. *J. Pharm. Sci.* **92**, 1343–1355 (2003).
65. Delplace, V., Couvreur, P., & Nicolas, J. Recent trends in the design of anticancer polymer prodrug nanocarriers. *Polym. Chem.* **5**, 1529–1544 (2014).
66. Kataoka, K., Harada, A., & Nagasaki, Y. Block copolymer micelles for drug delivery: design, characterization and biological significance. *Adv. Drug Deliv. Rev.* **47**, 113–131 (2001).
67. Lavasanifar, A., Samuel, J., & Kwon, G.S. Poly(ethylene oxide)-block-poly(L-amino acid) micelles for drug delivery. *Adv. Drug Deliv. Rev.* **54**, 169–190 (2002).
68. Cabral, H. et al. Targeted therapy of spontaneous murine pancreatic tumors by polymeric micelles prolongs survival and prevents peritoneal metastasis. *Proc. Natl. Acad. Sci. U. S. A.* **110**, 11397–11402 (2013).
69. Menjoge, A.R., Kannan, R.M., & Tomalia, D.A. Dendrimer-based drug and imaging conjugates: design considerations for nanomedical applications. *Drug Discov. Today* **15**, 171–185 (2010).
70. Medina, S.H. & El-Sayed, M.E.H. Dendrimers as carriers for delivery of chemotherapeutic agents. *Chem. Rev.* **109**, 3141–3157 (2009).
71. Malik, N., Evagorou, E.G., & Duncan, R. Dendrimer-platinate: a novel approach to cancer chemotherapy. *Anticancer Drugs* **10**, 767–776 (1999).
72. Bhadra, D., Bhadra, S., Jain, S., & Jain, N.K. A PEGylated dendritic nanoparticulate carrier of fluorouracil. *Int. J. Pharm.* **257**, 111–124 (2003).
73. Jin, Y.G. et al. A 5-fluorouracil-loaded pH-responsive dendrimer nanocarrier for tumor targeting. *Int. J. Pharm.* **420**, 378–384 (2011).
74. Kurtoglu, Y.E. et al. Poly(amidoamine) dendrimer-drug conjugates with disulfide linkages for intracellular drug delivery. *Biomaterials* **30**, 2112–2121 (2009).
75. Navath, R.S., Wang, B., Kannan, S., Romero, R., & Kannan, R.M. Stimuli-responsive star poly(ethylene glycol) drug conjugates for improved intracellular delivery of the drug in neuroinflammation. *J. Control. Release* **142**, 447–456 (2010).
76. Tang, F.Q., Li, L.L., & Chen, D. Mesoporous silica nanoparticles: synthesis, biocompatibility and drug delivery. *Adv. Mater.* **24**, 1504–1534 (2012).
77. Caruso, F., Caruso, R.A., & Mohwald, H. Nanoengineering of inorganic and hybrid hollow spheres by colloidal templating. *Science* **282**, 1111–1114 (1998).
78. Kamat, P.V. Photophysical, photochemical and photocatalytic aspects of metal nanoparticles. *J. Phys. Chem. B* **106**, 7729–7744 (2002).
79. Tartaj, P., Morales, M.D., Veintemillas-Verdaguer, S., Gonzalez-Carreno, T., & Serna, C.J. The preparation of magnetic nanoparticles for applications in biomedicine. *J. Phys. D. Appl. Phys.* **36**, R182–R197 (2003).
80. Solanki, A., Kim, J.D., & Lee, K.B. Nanotechnology for regenerative medicine: nanomaterials for stem cell imaging. *Nanomedicine* **3**, 567–578 (2008).
81. De, M., Ghosh, P.S., & Rotello, V.M. Applications of nanoparticles in biology. *Adv. Mater.* **20**, 4225–4241 (2008).
82. Chen, F. et al. In vivo tumor targeting and image-guided drug delivery with antibody-conjugated, radio labeled mesoporous silica nanoparticles. *ACS Nano* **7**, 9027–9039 (2013).

83. Erathodiyil, N. & Ying, J.Y. Functionalization of inorganic nanoparticles for bioimaging applications. *Acc. Chem. Res.* **44**, 925–935 (2011).
84. Lu, J., Liong, M., Li, Z., Zink, J.I., & Tamanoi, F. Biocompatibility, biodistribution, and drug-delivery efficiency of mesoporous silica nanoparticles for cancer therapy in animals. *Small* **6**, 1794–1805 (2010).
85. Hoffmann, F., Cornelius, M., Morell, J., & Froba, M. Silica-based mesoporous organic–inorganic hybrid materials. *Angew. Chem. Int. Ed.* **45**, 3216–3251 (2006).
86. Li, X., Xie, Q.R., Zhang, J.X., Xia, W.L., & Gu, H.C. The packaging of siRNA within the mesoporous structure of silica nanoparticles. *Biomaterials* **32**, 9546–9556 (2011).
87. Kim, J. et al. Multifunctional uniform nanoparticles composed of a magnetite nanocrystal core and a mesoporous silica shell for magnetic resonance and fluorescence imaging and for drug delivery. *Angew. Chem. Int. Ed.* **47**, 8438–8441 (2008).
88. Xia, T.A. et al. Polyethyleneimine coating enhances the cellular uptake of mesoporous silica nanoparticles and allows safe delivery of siRNA and DNA constructs. *ACS Nano* **3**, 3273–3286 (2009).
89. Cheng, S.-H. et al. Tri-functionalization of mesoporous silica nanoparticles for comprehensive cancer theranostics—the trio of imaging, targeting and therapy. *J. Mater. Chem.* **20**, 6149 (2010).
90. Fan, J., Chen, S.W., & Gao, Y. Coating gold nanoparticles with peptide molecules via a peptide elongation approach. *Colloids Surf. B Biointerfaces* **28**, 199–207 (2003).
91. Templeton, A.C., Wuelfing, M.P., & Murray, R.W. Monolayer protected cluster molecules. *Acc. Chem. Res.* **33**, 27–36 (2000).
92. Prabaharan, M., Grailer, J.J., Pilla, S., Steeber, D.A., & Gong, S.Q. Gold nanoparticles with a monolayer of doxorubicin-conjugated amphiphilic block copolymer for tumor-targeted drug delivery. *Biomaterials* **30**, 6065–6075 (2009).
93. Ghosh, P. et al. Intracellular delivery of a membrane-impermeable enzyme in active form using functionalized gold nanoparticles. *J. Am. Chem. Soc.* **132**, 2642–2645 (2010).
94. Chen, L.M., Zhao, X., Lin, Y., Huang, Y.B., & Wang, Q. A supramolecular strategy to assemble multifunctional viral nanoparticles. *Chem. Commun.* **49**, 9678–9680 (2013).
95. Suthiwangcharoen, N. et al. Facile co-assembly process to generate core-shell nanoparticles with functional protein corona. *Biomacromolecules* **15**, 948–956 (2014).
96. Suthiwangcharoen, N. et al. M13 bacteriophage-polymer nanoassemblies as drug delivery vehicles. *Nano Res.* **4**, 483–493 (2011).

15

EVOLUTION OF CONTROLLED DRUG DELIVERY SYSTEMS

KRISHNAVENI JANAPAREDDI*, BHASKARA R. JASTI, AND XIAOLING LI

Department of Pharmaceutics and Medicinal Chemistry, Thomas J Long School of Pharmacy and Health Sciences, University of the Pacific, Stockton, CA, USA

15.1 INTRODUCTION

The use of drugs in crude form derived from natural sources such as plants, animals, and minerals existed in the early human history. The earliest reports of medicinal preparations date back to 1550 B.C. as seen in Ebers Papyrus with more than 800 formulas and 700 drugs [1, 2]. "De materia medica" written by Dioscorides, a Greek physician and botanist of A.D. 50–70, was the basis for the development of pharmaceutical botany or pharmacognosy. Claudius Galen (A.D. 129–200), a Greek pharmacist–physician was the author of 500 treatise in medicine, and he formulated many preparations of natural origin by mixing or melting of ingredients. The field of pharmaceutical preparations was once known as *Galenic Pharmacy*. In Asia, the use of animal fats in medicinal preparations to prolong the action was reported in 168 B.C. in Chinese "Recipes for Fifty-Two Ailment." Indian Charaka Samhita (Ayurveda), a compendium containing detailed description of various drugs, formulations, and their pharmacological actions was dated between 200 B.C. and A.D. 200. Chinese Compendium of Materia Medica (Bencao Gangmu) was a collection of 1892 herbal medicines and preparations by Li Shizhen in 1593. A review of history of medicine

*Current affiliation: Department of Pharmaceutics, University College of Pharmaceutical Sciences, Kakatiya University, Warangal, Telangana, India

Drug Delivery: Principles and Applications, Second Edition. Edited by Binghe Wang, Longqin Hu, and Teruna J. Siahaan.

revealed the need to have active ingredients in proper form for administration. In the modern age, this proper form is called a dosage form or a drug delivery system.

Dosage form is a pharmaceutical preparation that contains active ingredient and excipients. The excipients in the preparations are included to impart unique properties and characteristics for the manufacturing process and finished products. These properties and characteristics include bulkiness or size of the product, flowability of intermediate materials, palatability, stability, solubility, disintegration time, dissolution of a product, etc. Most disease states demand immediate action of the active ingredients, and the dosage forms are designed to release most or the entire active ingredient upon entering the biological system. This type of dosage form is referred as conventional dosage form, and the dosage form does not provide any control over the release rate of active ingredients at the absorption site. When a pharmaceutical preparation is designed to modulate the release rate, release patterns, and/or duration of action, the dosage forms are termed controlled or modified drug release system/dosage forms. Although modulating the drug action by using excipient could be dated back to 200 B.C., the concept of modern controlled release drug delivery system was not fully established and implemented until 1960s. A historical timeline for the major milestones of controlled or modified drug delivery system is shown in Figure 15.1.

Unmet clinical needs, establishment of drug biopharmaceutics and pharmacokinetics, availability of novel materials, use of biological molecules as therapeutic agents and introduction of new technologies enabled the development of controlled drug delivery systems. In this chapter, the evolution of controlled release drug delivery systems and technologies will be discussed based on key advancements that impact the development of controlled drug delivery: (i) biopharmaceutics and pharmacokinetics, (ii) material science, (iii) protein and nucleic acids, (iv) pathophysiology/drug delivery by spatial control, and (v) microelectronics and microfabrication technologies.

15.2 BIOPHARMACEUTICS AND PHARMACOKINETICS

The word "pharmacokinetics" was first coined by FH Dost in 1953. Pharmacokinetics, the study of time course of drug in body began in 1930s with sporadic publications on some aspects and advanced at a rapid pace from 1971 onward. Wagner and Nelson were the two pioneering scientists whose publications on pharmacokinetics gained the attention of researchers worldwide and stimulated the works on pharmacokinetics [1–7]. Riegelman was a pioneering scientist, made significant contributions to the field of Pharmacokinetics [8]. An important development in pharmaceutical sciences that has a significant impact on controlled and modified release products is the correlation of bioavailability of dosage forms and dissolution in late 1960 [9, 10]. The introduction of biopharmaceutics and pharmacokinetics in 1960s provided a foundation for the quantitative description of plasma drug concentration-time profile. The relationship between the level of drug in the plasma and its pharmacological effect as well as half-life of drug allowed pharmaceutical scientists to define the duration of action. This resulted in the design of modified or controlled release drug delivery systems that improve patient compliance.

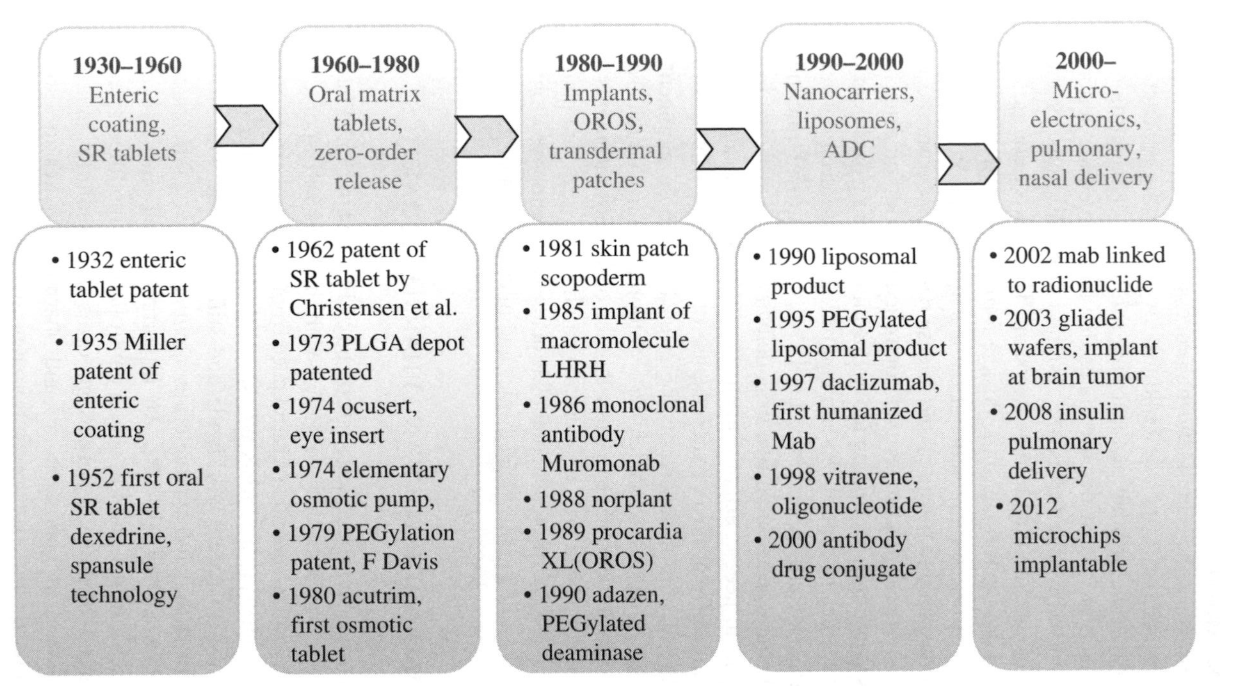

FIGURE 15.1 Milestones in controlled and modified drug delivery systems.

Understanding of mechanisms of drug release and application of mathematical models for characterizing drug release from dosage forms are essential in the design and evaluation of controlled release systems and contributed to the growth of controlled release technology. Tacheru Higuchi was the first scientist to derive an equation for quantification of drug release from the ointment films, containing dispersed drug [11]. The development of an equation for drug release modeling from matrix systems involving moving boundaries by Ping Lee in 1980 represents another key step in understanding and designing the controlled release of drugs [12, 13]. The mathematical model for describing the release of drug from a swellable, bioerodible controlled release dosage forms of various geometrical shapes provided a tool for predicting the release kinetics and designing controlled delivery systems [14–17].

The controlled release concept emerged when it became evident that there was a need to improve patient compliance by reducing the frequency of administration, especially in chronic disease states. In addition to reducing the dosing frequency, controlled drug release systems minimize the adverse reactions by maintaining the drug concentration below the minimum toxic level and above the minimum effective level between the dosing intervals. By keeping the drug concentration between the minimum toxic level and the minimum effective level for an extended period, the fluctuations in drug concentration can be minimized that prolong the duration of action.

Judah Folkman, an American scientist and pioneer in the field of controlled drug delivery is well known for his work on tumor angiogenesis. Folkman is the first scientist to suggest the zero-order delivery of drugs. He developed implants of drug enclosed in a silastic (silicon) capsule that can deliver the drug for a prolonged period of time [18]. The founding of ALZA Corporation in Palo Alto, California, in 1968, by Alejandro Zaffaroni, a synthetic chemist, was a breakthrough in commercialization of controlled drug delivery. Extensive progress was made by this company through the introduction of many controlled release products such as transdermal patches, ocular inserts, intrauterine devices, and oral osmotic drug delivery systems [19].

In 1952, Smith Kline & French introduced the first oral SR product Dexedrine employing Spansule® technology with an enormous success. This product was the result of 7 years of research by Donald R MacDonnel and his group. The product contained tiny beads of drug coated with wax of varied thickness. These beads release the drug at different time points due to the varying coating thickness [20]. Following the success of Spansule, the first oral sustained release matrix tablet made of hydrophilic polymers was patented by Christenson in 1962 [21].

The first controlled drug delivery device developed and introduced into the market by Alza in 1974 was an ophthalmic insert with trade name of Ocusert, containing pilocarpine for the treatment of glaucoma [22]. In this system, pilocarpine was allowed to diffuse through a flexible polymeric rate-controlling membrane. Intrauterine device called Progestasert, containing progesterone, a contraceptive steroid was introduced soon after. Both devices were of reservoir type with polymer as the rate-controlling membrane to achieve zero-order release.

In 1966, Sheldon J Segal developed Norplant, a subcutaneous silicone implant containing six small silicone capsules of levonorgestrel, a contraceptive steroid that releases the steroid over a period of 5 years. Norplant was introduced by Population Council in 1983 [23].

In 1989, matrix tablets for oral delivery employing swelling and gelling-controlled hydrogels of polymers were developed. In the later years, a number of oral sustained release tablets were introduced into the market owing to the simplicity in manufacturing of matrix tablets as compared to osmotic pumps [15, 24, 25].

Transdermal patches were introduced in early 1980s. Zaffaroni was issued a patent for the skin patch he developed in 1971. He called it as "bandage for administering drugs." The device was reservoir type of transdermal drug delivery system with PVA as the rate-controlling membrane. The first transdermal patch Scopoderm containing scopolamine was introduced into the market in 1979. Scopoderm was used as an antiemetic for sea sickness [26]. Later nitroglycerine patch for treating angina, nicotine patch for smoking cessation, fentanyl patch for the relief of pain, and patches containing hormones, clonidine, vitamin B_{12} were also introduced.

Development of osmotic pump capsules by Alza for the controlled delivery of drugs through oral route (OROS) started in the early 1980s and continued into 1990s. Most of the osmotic-controlled devices developed by Alza used cellulose acetate as rate-controlling membrane with zero-order drug release. First patent of osmotic delivery of drugs by oral route was issued to Theeuwes and Higuchi in 1974 [27]. Multiparticulate osmotic pump in which each particle served as an osmotic pump, known as controlled porosity osmotic pump was patented by Haslam in 1989 [28]. The drug and osmogen were coated with insoluble polymer containing pore-forming agent. The drug delivery was controlled by using particles of varied thickness similar to that of Spansule.

In addition to control the rate of release to achieve sustained pharmacological action based on pharmacokinetics, modulating the physicochemical properties of drugs and physiology or pathophysiology based on biopharmaceutics and pharmacokinetics also played an important role in the design and development of controlled drug delivery systems. Solubility is an important parameter that influences the drug bioavailability. Many of the drug candidates (40–60%) discovered through high-throughput screening and rational drug design have low bioavailability due to poor solubility, resulting in failure for advancement to the clinical trials [29–32]. Various techniques have been employed to enhance the solubility of drug with the goal of enhancing the absorption since late 1960s. These techniques include cosolvency, use of surfactants, micronization, nanonization by media milling technique [33], solid dispersions, polymeric micelles [34], and liposomal formulation [35, 36]. First-pass metabolism is also an important factor that affects bioavailability. To bypass the first-pass metabolism, drug delivery via non-oral but noninvasive routes was considered. Pulmonary administration of insulin—dry powder inhaler (Exubera)—was approved in 2014. Peptide drugs (desmopressin) for pulmonary delivery, antimigraine drugs, vaccines by nasal route and opioid analgesics by nasal route were also introduced into the market. Recently many peptide drugs were studied by nasal route for the treatment of neurodegenerative disorders. Nasal delivery has the advantage of brain targeting by direct transport of drugs through olfactory pathways to the brain. Nasal route has proved successful in the delivery of many drugs and the products such as vaccines, peptides, or drugs for systemic delivery are available in the market.

The success of controlled or modified release drug delivery technology not only created the products that provide constant or zero-order release, but the technology has also been used to reformulate the existing conventional dosage form products to modulate the time course of drug plasma level and improve or optimize therapeutic efficacy and establish proprietary position for the products after patent protection expired [37]. During the past few decades, many drugs have been reformulated to improve biopharmaceutical properties of a drug such as solubility, permeability, or life cycle management.

With the understanding of biopharmaceutical and pharmacokinetic properties of drug and drug candidates, the design of a dosage form and selection of the route of administration have evolved from an empirical or trial and error-based experiments to a rational design approach.

15.3 MATERIAL SCIENCE

Utilizing the unique characteristics of material has been an inseparable part of controlled drug delivery system design. In the early part of controlled drug delivery history, the off-the-shelf materials were used in the formulation of sustained or controlled drug delivery systems. As the needs to have materials with unique characteristics for controlled drug delivery arise, materials tailored to controlled drug delivery were developed. Therefore, discovery and synthesis of novel materials played an important role in the evolution of controlled release drug delivery systems.

Enteric coated or delayed release oral drug delivery systems are designed to deliver the drug in the small intestine either to protect the drug from stomach pH or gastric mucosa from the gastroirritant drug. One may consider that the history of using materials responding to external conditions in controlled release began with the development of enteric coating of the oral solid dosage forms. In 1884, Paul Unna developed keratin-coated pellets that remained intact in the stomach and released the drug in the intestine [38]. Grover C Miller was issued a patent in 1935 for the enteric coating of tablet. He prepared enteric coatings using a mixture of high melting point fatty acid—stearic acid, carnauba wax, and petroleum jelly with hygroscopic, effervescent, or naturally soluble substance [39]. Herman prepared enteric hard-shell capsule by adding aqueous solution of alkali metal salt (sodium carbonate) and cellulose derivative (cellulose acetate phthalate) to gelatin solution [40]. Shellac was used as an enteric coating material in the earlier years. Due to disadvantages of slow and unreliable dissolution of shellac-coated tablets, shellac has been replaced by other pH-responsive polymers such as copolymer of methacrylate and (meth)acrylic acids and cellulose derivatives in the enteric-coated controlled release products [41].

Polymers that can respond to the change of pH, temperature, and solvents such as water and are also biocompatible add a new dimension to the design of controlled drug delivery. Triblock polymers of PEG-PLGA-PEG were designed as thermally responsive material. These types of polymers form gel when injected to tissues and reverse to sol upon application of heat [42]. Polyacetal hydrogels formed from divinyl ethers and polyols were patented by Heller et al. for the applications that need to have materials with properties such as water soluble, bioerodible, and biodegradable [43]. Ping Lee contributed to the theoretical aspects of swellable and matrix systems by providing an equation for drug release from matrices. He patented diffusion-controlled matrix devices made from, swellable hydrogel matrices from cross-linked polymers of methacrylates, polyvinylpyrrolidone, and also linear polymers of high molecular weight, natural polymers [44].

One of the limitations for oral controlled release drug delivery system is short resident time of the system in the gastrointestinal (GI) tract that restricts the delivery of drug beyond 18 hours. Implantable drug delivery system can circumvent this limitation. However, the removal of implants after the depletion of drug is a serious drawback of such system. Biodegradable polymers for controlled drug delivery were developed to meet the needs for creating implantable controlled drug delivery systems. Copolymers of lactic acid and glycolic acids (PLA and PGA) initially developed for absorbable suturing [45] were further developed as rate-controlling polymers in implants/tablets as these polymers are biodegradable and biocompatible. This series of materials led to the successful development of depot systems in late 1980s. Boswell and Scribner developed polylactide biodegradable depot system containing contraceptive hormone estradiol and patented in 1973 [46]. Dunn et al. developed biodegradable *in situ* implant made of thermoplastic and thermosetting polymers such as polylactide or PLGA dissolved in solvent, which on injection set to form a gel or implant [47]. Polylactic glycolic acid polymer (PLGA) implants were studied for local slow release of chemotherapeutic agent placed at tumor site. Polyorthoethers and polyanhydrides were two of biodegradable polymers that have unique degradation properties. The degradation of these polymers occurs via surface erosion. The rate of erosion can be controlled by varying the levels of chain flexibility, thereby zero-order release from these polymers can be achieved. Gliadel wafers made of biodegradable polyanhydride copolymer, containing carmustine for the treatment of glioma were approved by FDA in 1997. PLGA was the most commonly used biodegradable polymer for implant or depot type of controlled drug delivery systems. PLGA was successfully employed for the delivery of a wide range of drugs, peptides, nucleic acids, and vaccines. Nutropin Depot—a recombinant growth hormone—was the first approved PLGA microsphere product developed by Genentech in 1999. Polycaprolactone, polyanhydride, polyortho esters, and polyphosphazenes are the other biodegradable polymers that are designed and used for drug delivery.

Other materials that have been specifically developed for controlled drug delivery include dendrimers, amphiphilic molecules, lipids, and poly cationic polymers. Some of these materials are covered in other sections of this chapter. Dendrimers are macromolecular carriers with branched tree-like structure, which gained much attention during the past decade due to their unique features [48, 49]. Both small and large molecules with amphiphilic characteristics have been designed and investigated for drug delivery [50–53]. Amphiphilic block copolymers were widely investigated, because they are capable of forming different nanostructures such as micelle, polymeric vesicles or polymersomes, nanocapsules and nanospheres [54]. Polymeric micelles were studied by a number of researchers and proved to be successful in improving the solubility of poorly soluble drugs, which is due to the larger space available to accommodate more drug and were more stable than micelles of surfactants [55, 56].

The creation of novel materials specifically for the purpose of drug delivery removes the barriers to design controlled release drug delivery system. Therefore, new materials play a crucial role in the evolution of controlled drug delivery.

15.4 PROTEINS, PEPTIDES AND NUCLEIC ACIDS

Therapeutic application of insulin to treat diabetic patients was first reported in 1922 [57, 58]. Human insulin is a peptide consisting 51 amino acids with a molecular weight of 5808 Da. In addition to the short half-life of insulin (4–5 minutes), the fluctuation of insulin level in response to blood sugar levels is desirable in the treatment. Therefore, an ideal insulin delivery system should be able to maintain a baseline level and also respond to the level of glucose. In addition, the system should be easy to administer.

For the past few decades insulin delivery systems by noninvasive routes are being investigated extensively, which includes oral, buccal, transdermal, rectal and pulmonary products. Among various alternative routes of administration, commercialization of noninvasive delivery of insulin was realized via pulmonary delivery product—Exubera by Pfizer and Nektar Therapeutics. Exubera is an inhalation product approved in 2006, but it was removed from the market in 2007 as a result of combination of multiple factors including sales, potential risk of lung cancer, and insurance reimbursement. [59, 60]. Oral administration of insulin in combination with sodium *N*-[8-(2-hydroxybenzoyl)amino] caprylate was studied by a US-based company (Emisphere) and demonstrated glucose-lowering effect [61].

Prior to 1950s, most of the drug discovery was by serendipity or from exhaustive screening for the activity. Many of these drugs entered the market without much knowledge of the mechanism of action, metabolism, and metabolites, but solely based on the efficacy in treating a disease. The drug discovery process is revolutionized in twenty-first century by the integration of technologies such as informatics, genomics, proteomics, high-throughput screening, and rational drug design. These new approaches and rational drug design enabled scientists to identify new targets at faster rate, leading to the discovery of drugs that are more specific and safe. The development of biologically active oligo nucleotides or macromolecules such as proteins, peptides, and nucleic acids as therapeutic agents has become a new frontier for drug delivery in recent years.

Delivery of macromolecular drugs such as proteins, peptides, oligonucleotides and DNA is more challenging than the small-molecule drugs, because they are less permeable across the biological membranes, more susceptible to enzymatic degradation and more vulnerable to conformational changes by unfavorable pH and temperature. These macromolecules have potential problems of degradation in the GI tract, unstable three-dimensional (3D) structures, shorter biological half-life, low solubility, and difficulty in reaching the target, posing serious challenges to the delivery via the most convenient oral route. With the exception of very few small-molecular peptides and proteins administered by oral route, macromolecular drugs are administered by parenteral route. Protection of drug from degradation, increasing the permeation, and reducing the clearance are the primary objectives in formulating and delivering these biologicals or biopharmaceuticals. The development in the drug delivery technologies attempts to address the aforementioned problems and ultimately improve therapeutic efficacy.

PEGylation of proteins discovered by Frank F Davis in 1970 was a milestone in the history of drug delivery. PEGylation of proteins increased circulation half-life of protein by grafting hydrophilic polyethylene glycol chains on protein to prevent the degradation and elimination of protein by increasing the stability of proteins. This technology enabled a number of protein drugs commercialization by parenteral route [62, 63]. First product of PEGylated protein introduced in 1990 by Enzon Corporation was Adazen (PEG deaminase) for the treatment of combined immunodeficiency [64]. Other approaches to modify the protein were also utilized in delivering protein drugs. Acylation of proteins with long alkyl chain makes the protein lipophilic. This approach has been used for the slow release of protein from polymer implant. Long-acting insulin detemir (Levemir) and long-acting GLP-1 analogue liraglutide (Victoza) are the examples of marketed products given subcutaneously, where proteins are modified with alkyl chain to become more lipophilic. Encapsulation or conjugation of biological molecules with nanocarriers such as liposomes is also a promising approach in protecting these drugs from proteases and nucleases.

Controlled release of proteins and macromolecules from biodegradable polymers was first demonstrated by Langer in 1983 [65]. First slow-releasing peptide, LHRH analogue formulated as polymeric microspheres, was approved by FDA in 1989 [66].

Delivery of gene presents another challenge in controlled drug delivery history. This category of molecules needs to be delivered into the cells to exert their pharmacological effects. Viral gene delivery systems contain modified virus, which are replication deficient but can deliver the genes to the cells efficiently. Viral vectors have the advantage of constant expression of therapeutic genes. Adenoviruses, retroviruses, and lentiviruses were used as vectors for gene delivery. However, use of viral vectors has the disadvantages of immunogenicity and toxicity. Nonviral carriers such as cationic lipids and positively charged polymers have been used in gene delivery. Positively charged lipids and polymers interact with negatively charged DNA, RNA, oligonucleotides through ionic interaction to form lipoplexes and polyplexes respectively. Cationic polymers like chitosan, polyethyleneimine, and dendrimers were investigated for the DNA delivery. However, the transfection efficiency of nonviral vectors was significantly low when compared to viral vectors.

The first clinical trial of gene therapy for the treatment of severe combined immunodeficiency (SCID) syndrome was conducted in 1990. The T lymphocytes from the patient were collected and the gene for the enzyme adenosine deaminase was introduced into T lymphocytes using retrovirus as vector. The modified T lymphocytes were reintroduced into the patient [67]. A number of clinical trials of gene therapy using viral delivery systems are in phase I and II trials [68]. Due to the potential vector-associated toxicities and other side effects like carcinoma, there were only limited number of approved gene therapy products. In 2003, China approved gendicine, a gene therapy product, a human p53–expressing adenovirus injection, for the treatment of head and neck squamous cell carcinoma [69]. Vitravene, an antisense oligonucleotide containing fomivirsen, administered by injecting intravitreally for the local treatment of cytomegalovirus retinitis in HIV patients was approved in 1998 by FDA. Glybera is another gene therapy product approved in 2012 by European Medicines Agency (EMA) for the treatment of ultrarare inherited disorder, lipoprotein lipase deficiency administered by intramuscular injection. Glybera consists of an engineered copy of the human LPL gene in a nonreplicating adeno-associated virus serotype 1 (AAV1) as vector [70].

To circumvent the degradation of macromolecules in the GI tract, delivery of the macromolecular drugs by noninvasive routes other than oral such as nasal, pulmonary, buccal, sublingual, transdermal, rectal, and vaginal is still actively being explored in both academia and industry. The advances in basic science and material science will facilitate the evolution of delivering this category of drugs.

15.5 DISCOVERY OF NEW MOLECULAR TARGETS—TARGETED DRUG DELIVERY

Delivery of drug based on the pathophysiologic mechanisms of diseases can achieve the therapeutic outcomes specific to the disease state while minimizing the unwanted effects to the normal physiological processes. In other words, if the drug could be targeted solely to the disease-related organs, tissues, cells, or molecular targets, it will not interfere with normal biological functions. The concept of targeted delivery can be traced back to 1906, a German medical scientist, Paul Ehrlich, used the term "magic bullet" to describe a drug specifically targeted to the cause of disease [71]. Targeted delivery or spatial placement of drug at the site of action became important in therapy optimization, as it enhances the drug effects at the site of action while reduces the toxic effects of drug reaching the locations other than the site of action.

Targeting is accomplished by directly delivering the drug to the site of action if possible, or using a carrier-mediated delivery system, which is directed to the site of action by attaching a ligand or antibodies specific to the receptors at the target site. During the past few decades significant improvements in the field of site-specific drug delivery took place due to the advances in biomedical research. Research in understanding disease pathophysiology, cellular, and molecular biology, identification of genes responsible for the diseases, and discovery of structures of receptors brought a paradigm shift by providing a well-defined target for targeted drug delivery. When a target is identified, various strategies are developed to target the drug to the desired organ or cells or molecule targets. The strategies developed include, use of prodrug, drug or a carrier coupled to ligands specific to receptors on cell, or conjugation of drug to a polymer for passive targeting.

Although enteric coating product could be considered as an early example of organ-level targeting, modern targeted drug delivery refers to the drug delivery to specific tissue, organ, cells or cellular organelles, and targeted molecules. The field of cancer chemotherapy could be considered as the battle front for targeted drug delivery. Over the past two decades, cancer chemotherapy has changed remarkably due to the discovery of tumor microenvironment conditions and factors related to tumor growth, angiogenesis, and the overexpression of certain receptors on cancer cells. Hiroshi Maeda discovered the enhanced permeation and retention (EPR) effect in 1986. This finding led to the passive targeting of anticancer drugs using macromolecular drug conjugates or nanocarriers. Utilizing the EPR effect to target cancer has been a successful strategy that resulted in a number of products being introduced into the market with liposomal doxorubicin (Doxil) marketed in 1995 [72, 73].

Liposomes, lipid-bilayer vesicles of phospholipid, were first described by Bingham in 1961 and were utilized to study the biological process of membranes as they are similar to the cell membrane. The potential application of liposomes as carrier for drugs was noticed by Gregoriadis G in 1971 [74–76]. Since then, researchers have been studying various liposomal drug delivery systems. The first liposomal product was introduced in 1990, and currently there are about 15 liposomal drug products including anticancer drugs doxorubicin, daunorubicin, paclitaxel, antifungal drug amphotericin, proteins, peptides, nucleic acids, and vaccines in the market.

When the overexpression of proteins on the cancer cells is identified, these proteins can serve as an ideal target for targeted drug delivery to differentiate the cancer cells from normal cells. Targeting the tumor tissue by using cancer-specific antibody began with Rituximab in 1997 and continued with the increase in the number of mAbs introduced every year. Conjugation of a chemotherapeutic agent to antibody or antibody fragment via linker can deliver the anticancer agents specifically to cancer cells and reduce the side effects. Mylotarg, the first antibody–drug conjugate (ADC) formed by coupling ozogamicin to gemtuzumab was introduced in 2000 for the treatment of acute myeloid leukemia but was withdrawn in 2010 due to the concerns on product safety and clinical benefits to the patients. Currently there are two ADCs in the market and close to 30 ADCs are in different phases of clinical trials [77, 78]. Antibody fragments are also used in place of whole antibodies to serve as specific binding moiety in targeted drug delivery systems.

During the past few decades, there is a remarkable growth in the development of carrier-mediated nanoscale systems, which include liposomes, solid lipid nanoparticles, nanostructured lipids, nanoparticles for passive targeting delivery as well as ligand modified nanocarriers, aptamers, dendrimers, polymeric micelles, and ADCs for active targeting. Utilizing the potential of albumin uptake by cancer cells via albumin receptor (gp 60)-mediated transcytosis, albumin was used as a carrier to increase drug delivery to tumors. Albumin-bound Paclitaxel nano suspension (Abraxane) was approved by FDA in 2005. Abraxane delivered higher dose of anticancer agent without the use of any organic solvent such as cremophor EL and therefore reducing the solvent-related side effects [79–81].

Metabolism and enzymes can be utilized to deliver drug at the desired time and locations [82]. Determination of the metabolic pathways of a drug is part of current drug discovery and development process. A proper understanding of drug metabolism also plays an important role in the design of drug delivery systems, especially for the selection of route of delivery and mode of delivery. Based on the drug metabolism, prodrugs were developed to improve drug bioavailability or release active parent drug at a desired location. Prodrugs have been developed to address problems such as solubility, stability, taste masking, and targeted delivery (see Chapter 12). Drug targeting to specific tissue or organ using a prodrug, which is metabolized enzymatically to an active drug only at the target organ by the presence of enzymes specific to that organ or tissue or type of cells, is an important strategy for targeting drugs to the liver, GI tract, and eye.

In addition to increasing the specificity of drug, the targeting delivery strategy can also revive those drugs or drug candidates that are rejected in the development due to their nonspecific toxicities. The compounds used in ADC products are a good example for reviving the compounds that are highly potent but lack of specificity when they were first discovered.

Targeted drug delivery relies on the discovery of disease targets or understanding of the mechanisms and pathophysiology of diseases. Therefore, the advancement in understanding diseases and their mechanisms has been a key factor in the evolution of drug delivery. As biomedical researchers unveil more mechanisms of diseases or the disease-related biomarkers, the development of targeted drug delivery systems will also advance accordingly.

15.6 MICROELECTRONICS AND MICROFABRICATION TECHNOLOGIES

Advances in chemical, mechanical, and electronic engineering allow pharmaceutical scientists to deliver drugs using microfabricated- or microelectronic-driven devices. The use of microelectronics and miniature sensors make it possible to design electronically controlled devices capable of delivering the drug in response to the disease conditions in real time as a close-looped feedback-controlled system and capable of delivering multiple drugs at programmable rate and time [83, 84]. Drug delivery can be more precisely controlled in a programmable manner using microelectronic devices. These devices can be designed to contain one or more reservoirs, which can be loaded with single or multiple drugs. The delivery of drug can be in a pulsatile, feedback-controlled, or on demand mode.

First portable infusion pump for continuous subcutaneous delivery of insulin was designed by Dean Kamen in 1976 and was introduced into the market in 1985 [85]. As the technology is evolving, a delivery system in the form of a smart pump that can calculate the dose based on the calorie intake, deliver drug according to the glucose meter coupled to the system, determine injection schedule, and set alarm at low and high sugar values came in to existance [86]. Insulin pump is effective in managing diabetes for those who require intensive insulin therapy with multiple injections in an outpatient setting. Because of the precision of rate control in delivering medications, infusion pumps are also useful in treating a variety of conditions including cancer chemotherapy.

Iontophoretic transdermal drug delivery is one of the early successful approaches in combining drug delivery and microelectronics. Iontophoresis is a noninvasive method by which transportation of ionic or nonionic drugs across the skin is enhanced under the influence of electric field. Iontophoresis technique demonstrated substantial enhancement of transdermal absorption of ionic drugs and showed potential to deliver peptides and oligonucleotides [87]. Currently, iontophoresis technique has been used to deliver lidocaine, NSAIDs, epinephrine, fentanyl and corticosteroids [88–90]. Examples of iontophoresis principle–based products in the market include lidocaine (Ionsys), for topical dermal anesthesia in superficial procedures like venipuncture, approved in 2006 [91] and sumatryptan iontophoretic system (Zecuity) for treating migraine was approved by US FDA in 2013 [92]. Iontophoretic delivery of insulin was also explored by various research groups and industries [93–97]. However, it was not yet successful in developing into a product.

Application of microelectronics in drug delivery is not limited to external devices. A capsule has been designed to release the drug from capsule precisely in the predetermined region of intestinal tract upon activation by a radio frequency signal. This delivery system has been used in the clinical trial to determine the drug absorption in different segments of the GI tract [98].

Microfabrication provides another avenue for pharmaceutical scientists to create the drug delivery systems with precision and unique architectures [99]. One of the microfabrication applications in drug delivery is creating microneedles to overcome the biological barriers such as the skin. This technology has the potential for the administration of biologicals like peptides, vaccines through transdermal route while avoiding pain and the requirement of skilled medical personnel. Both undissolvable and dissolvable microneedles have been investigated for the drug delivery of small and macromolecules for delivery of insulin [100]. Dissolvable microneedles were safe when compared to metal needles. Later many biodegradable polymers such as chondroitin sulfate, caboxymethylcellulose, and amylopectin were also employed in fabricating microneedles for immediate (bolus) and sustained release [101–103]. Preparation of microneedles by *in situ* polymerization of PVP and methacrylic acid in PDMS molds was reported by Yan [104]. MicroCor®, a biodegradable microneedle patch technology for the delivery of proteins and vaccines, is available in the market. Transdermal patch made of titanium microprojections technology, which can be applied with a reusable applicator, was developed for the delivery of parathyroid hormone and has completed phase I and II clinical trials [105].

As technologies advance and mature, many high-tech products are gradually merging into drug delivery. 3D printing technology could fabricate the drug delivery systems with unique architecture to achieve a predetermined release profile. The polymeric devices prepared by 3D printing enabled complex drug release profiles of drug by precise control over 3D position, microstructure, and composition. Wu et al. applied 3D printing technology to create resorbable devices using polyethylene oxide and polycaprolactone as matrix polymers and obtained precise, reproducible release. 3D printing can produce multiple diffusion gradients within a single device, resulting in complicated drug release profiles [106].

Technology is a driving force in advancing the development of drug delivery. The products built by incorporation of microelectronics and microfabrication represent a new generation of drug delivery system. As the application implementation of technology becomes simple and affordable, the next wave of novel drug delivery system will be integrated with microelectronics and microfabrication to leverage the unique capability of microelectronics and programming.

15.7 CONCLUSION

Optimizing drug delivery for improving the therapeutic outcome has been recognized in the early years of use of medication in disease treatment. Evolution of drug delivery has been a result of advancement in medical research, pharmaceutical sciences, material sciences and microelectronic engineering. Pharmacokinetics and biopharmaceutics provided a foundation for the design and development of drug delivery systems. Discovery of disease pathophysiology paved the road for rational design of drug delivery systems. New materials and new technologies elevated the development of drug delivery system to a new level. Increased number of macromolecules and poorly soluble drugs posed a new direction for drug delivery. As human genome is demystified and pathophysiology of many diseases is revealed, development of controlled release systems took place at a rapid pace. There is still ample scope to advance in near future to address the clinical needs and be part of revolution in medicine such as personalized medicine.

REFERENCES

1. Bender, G. A. *Great moments in pharmacy*. Detroit: Northwood Institute Press, 1966; pp 8–10.
2. Von Klein, C. H. *The medical features of the Papyrus Ebers*. Chicago, Press of the American Medical Association, 1905. (Republished by Forgotten Books, 2013).
3. Nelson, E.; Schaldemose, I. *J Am Pharm Assoc* 1959, **48** (9), 489–495.
4. Nelson, E. *J Am Pharm Assoc* 1960, **49** (1), 54–56.
5. Wagner, J. G. *J Pharm Sci* 1961, **50** (5), 359–387.
6. Wagner, J. *J Pharmacokinet Biopharm* 1976, **4** (5), 395–425.
7. Wagner, J. G. *Pharmacol Ther* 1981, **12** (3), 537–562.
8. Riegelman, S.; Crowell, W. J. *J Am Pharm Assoc* 1958, **47** (2), 127–133.
9. Hanano, M. *Chem Pharm Bull* 1967, **15** (7), 994–1001.
10. Shaw, T. R.; Raymond, K.; Howard, M. R.; Hamer, J. *BMJ* 1973, **4** (5895), 763–766.
11. Higuchi, T. *J Pharm Sci* 1961, **50**, 874–875.
12. Lee, P. I. *J Pharm Sci* 1984, **73** (10), 1344–1347.
13. Lee, P. I. *J Control Release* 1985, **2**, 277–288.
14. Hopfenberg, H. B.; In Paul, D. R.; Harris, F. W. Eds. *Controlled Release of Polymeric Formulations*. ACS Symposium Series 33, Washington, DC: American Chemical Society, 1976; pp 26–31.
15. Peppas, N. A.; Colombo, P. *J Control Release* 1997, **45** (1), 35–40.
16. Kaunisto, E.; Marucci, M.; Borgquist, P.; Axelsson, A. *Int J Pharm* 2011, **418** (1), 54–77.
17. Lao, L. L.; Peppas, N. A.; Boey, F. Y. C.; Venkatraman, S. S. *Int J Pharm* 2011, **418** (1), 28–41.

18. Folkman, J.; Long, D. M. *J Surg Res* 1964, **4** (3), 139–142.
19. Peppas, N. A. *Adv Drug Deliv Rev* 2013, **65** (1), 5–9.
20. Ullyot, G. B.; Ullyot, B. H.; Slater, L. B. *Bull Hist Chem* 2000, **25** (1), 16–21.
21. Christenson, G. L.; Dale, L. B. U.S. Patent 3,065,143 A 1962.
22. Ness, R. A. U.S. Patent 3,618,604 1971.
23. Watkins, E. S. *J Womens Hist* 2010, **22** (3), 88–111.
24. Colombo, P.; Manna, A. L.; Conte, U. U.S. Patent 4,839,177 A 1989.
25. Losi, E.; Bettini, R.; Santi, P.; Sonvico, F.; Colombo, G.; Lofthus, K.; Colombo, P.; Peppas, N. A. *J Control Release* 2006, **111** (1–2), 212–218.
26. Zaffaroni, A. U.S. Patent 3,797,494 A 1973.
27. Higuchi, T.; Theeuwes, F. U.S. Patent 3,845,770 A 1974.
28. Haslam, J. L.; Rork, G. S. U.S. Patent 4,880,631 1989.
29. Lipper, R. A.; Higuchi, W. I. *J Pharm Sci* 1977, **66** (2), 163–164.
30. Lipinski, C. A. *J Pharmacol Toxicol Methods* 2000, **44** (1), 235–249.
31. Lipinski, C. A.; Lombardo, F.; Dominy, B. W.; Feeney, P. J. *Adv Drug Deliv Rev* 2001, **46** (1–3), 3–26.
32. Kennedy, T. *Drug Discov Today* 1997, **2** (10), 436–444.
33. Junghanns, J. U.; Müller, R. H. *Int J Nanomedicine* 2008, **3** (3), 295–309.
34. Werner, M. E.; Cummings, N. D.; Sethi, M.; Wang, E. C.; Sukumar, R.; Moore, D. T.; Wang, A. Z. *Int J Radiat Oncol Biol Phys* 2013, **86** (3), 463–468.
35. Fan, Y.; Zhang, Q. *Asian J Pharm Sci* 2013, **8** (2), 81–87.
36. Yang, T.; Cui, F. D.; Choi, M. K.; Cho, J. W.; Chung, S. J.; Shim, C. K.; Kim, D. D. *Int J Pharm* 2007, **338** (1–2), 317–326.
37. Rowland, M.; Noe, C. R.; Smith, D. A.; Tucker, G. T.; Crommelin, D. J.; Peck, C. C.; Rocci, M. L., Jr.; Besancon, L.; Shah, V. P. *J Pharm Sci* 2012, **101** (11), 4075–4099.
38. Thompson, H. O.; Lee, C. O. *J Am Pharm Assoc* 1945, **34** (5), 135–138.
39. Miller, C. G. U.S. Patent 2,011,587 A 1935.
40. Herman, B. H. U.S. Patent 2,491,475 A 1949.
41. Savage, G. V.; Rhodes, C. T. *Drug Dev Ind Pharm* 1995, **21** (1), 93–118.
42. Jeong, B.; Bae, Y. H.; Kim, S. W. *J Biomed Mater Res* 2000, **50** (2), 171–177.
43. Heller, J.; Penhale, D. W. H. U.S. Patent 4,713,441 A 1987.
44. Lee, P. I. Google Patent CA 1,246,447 A1 1988.
45. Emil, S. E.; Albert, P. R. U.S. Patent 3,297,033 A 1967.
46. Boswell, G.; Scribner, R. U.S. Patent 3,773,919 A 1973.
47. Dunn, R. L.; English, J. P.; Cowsar, D. R.; Vanderbilt, D. P. U.S. Patent 5,990,194 A 1990.
48. Tomalia, D. A.; Baker, H.; Dewald, J.; Hall, M.; Kallos, G.; Martin, S.; Roeck, J.; Ryder, J.; Smith, P. *Polym J* 1985, **17** (1), 117–132.
49. Cheng, Y.; Xu, Z.; Ma, M.; Xu, T. *J Pharm Sci* 2008, **97** (1), 123–143.
50. Torchilin, V. P. *J Control Release* 2001, **73** (2–3), 137–172.
51. Adams, M. L.; Lavasanifar, A.; Kwon, G. S. *J Pharm Sci* 2003, **92** (7), 1343–1355.
52. Haag, R. *Angew Chem Int Ed* 2004, **43** (3), 278–282.
53. Rösler, A.; Vandermeulen, G. W. M.; Klok, H. A. *Adv Drug Deliv Rev* 2012, **64**, 270–279.

54. Letchford, K.; Burt, H. *Eur J Pharm Biopharm* 2007, **65** (3), 259–269.
55. Kwon, G. S.; Kataoka, K. *Adv Drug Deliv Rev* 1995, **16** (2–3), 295–309.
56. Kwon, G. S.; Okano, T. *Adv Drug Deliv Rev* 1996, **21** (2), 107–116.
57. Bliss, M. *The Discovery of Insulin*. Chicago, IL: University of Chicago Press, 2007.
58. Feudtner, C. *N Engl J Med* 2008, **358** (9), 975–976.
59. Mack, G. S. *Nat Biotechnol* 2007, **25** (12), 1331–1332.
60. Heinemann, L. *J Diabetes Sci Technol* 2008, **2** (3), 518–529.
61. Kidron, M.; Dinh, S.; Menachem, Y.; Abbas, R.; Variano, B.; Goldberg, M.; Arbit, E.; Bar-On, H. *Diabet Med* 2004, **21** (4), 354–357.
62. Davis, F. F.; Van Es, T.; Palczuk, N. C. U.S. Patent 4,179,337 A 1979.
63. Davis, F. F. *Adv Drug Deliv Rev* 2002, **54** (4), 457–458.
64. Veronese, F. M.; Pasut, G. *Drug Discov Today* 2005, **10** (21), 1451–1458.
65. Langer, R. *Pharmacol Therap* 1983, **21** (1), 35–51.
66. Langer, R. *Acc Chem Res* 1999, **33** (2), 94–101.
67. Kohn, D. B. *Pediatr Res* 2000, **48** (5), 578–578.
68. Sheridan, C. *Nat Biotechnol* 2011, **29** (2), 121–128.
69. Xin, H. *Science* 2006, **314** (5803), 1233.
70. Yla-Herttuala, S. *Mol Ther* 2012, **20** (10), 1831–1832.
71. Strebhardt, K.; Ullrich, A. *Nat Rev Cancer* 2008, **8** (6), 473–480.
72. Matsumura, Y.; Maeda, H. *Cancer Res* 1986, **46** (12 Part 1), 6387–6392.
73. Maeda, H. *J Control Release* 2012, **164** (2), 138–144.
74. Gregoriadis, G.; Leathwood, P. D.; Ryman, B. E. *FEBS Lett* 1971, **14** (2), 95–99.
75. Gregoriadis, G.; Ryman, B. E. *Biochem J* 1972, **128** (4), 142–143.
76. Gregoriadis, G.; Perrie, Y. *Liposomes eLS*. Chichester: John Wiley & Sons, Ltd, 2010.
77. Flygare, J. A.; Pillow, T. H.; Aristoff, P. *Chem Biol Drug Des* 2013, **81** (1), 113–121.
78. Panowksi, S.; Bhakta, S.; Raab, H.; Polakis, P.; Junutula, J. R. *mAbs* 2014, **6** (1), 34–45.
79. Desai, N. P.; Tao, C.; Yang, A.; Louie, L.; Zheng, T.; Yao, Z.; Soon-Shiong, P.; Magdassi, S. U.S. Patent 5,916,596 A 1999.
80. Green MR, Manikhas GM, Orlov S, Afanasyev B, Makhson AM, Bhar P, Hawkins MJ. Abraxane®, a novel Cremophor®-free, albumin-bound particle form of paclitaxel for the treatment of advanced non-small-cell lung cancer. *Annals of Oncology*. 2006; **17** (8): 1263–1268.
81. Kratz, F. *J Control Release* 2008, **132** (3), 171–183.
82. Krishna, D. R.; Klotz, U. *Clin Pharmacokinet* 1994, **26** (2), 144–160.
83. Guo, X. D.; Prausnitz, M. R. *Expert Rev Med Devices* 2012, **9** (4), 323–326.
84. Mura, S.; Nicolas, J.; Couvreur, P. *Nat Mater* 2013, **12** (11), 991–1003.
85. Alsaleh, F. M.; Smith, F. J.; Keady, S.; Taylor, K. M. *J Clin Pharm Ther* 2010, **35** (2), 127–138.
86. Kemper, S. *Code name Ginger: the story behind Segway and Dean Kamen's quest to invent a new world*. Boston, MA: Harvard Business School Press, 2003.
87. Costello, C. T.; Jeske, A. H. *Phys Ther* 1995, **75** (6), 554–563.
88. Anderson, C. R.; Morris, R. L.; Anderson, C. J.; Grace, L. A. U.S. Patent 6,745,071 B1 2004.

89. Zempsky, W. T.; Sullivan, J.; Paulson, D. M.; Hoath, S. B. *Clin Ther* 2004, **26** (7), 1110–1119.
90. Batheja, P.; Thakur, R.; Michniak, B. *Expert Opin Drug Deliv* 2006, **3** (1), 127–138.
91. Herwadkar, A.; Banga, A. K. *Ther Deliv* 2012, **3** (3), 339–355.
92. Vikelis, M.; Mitsikostas, D. D.; Rapoport, A. M. *Pain Manag* 2014, **4** (2), 123–128.
93. Stephen, R. L.; Petelenz, T. J.; Jacobsen, S. C. *Biomed Biochim Acta* 1984, **43** (5), 553–558.
94. Brandon, E. F.; Raap, C. D.; Meijerman, I.; Beijnen, J. H.; Schellens, J. H. *Toxicol Appl Pharmacol* 2003, **189** (3), 233–246.
95. Pillai, O.; Borkute, S. D.; Sivaprasad, N.; Panchagnula, R. *Int J Pharm* 2003, **254** (2), 271–280.
96. Pillai, O.; Nair, V.; Panchagnula, R. *Int J Pharm* 2004, **269** (1), 109–120.
97. Banga, A. K. *Expert Opin Drug Deliv* 2009, **6** (4), 343–354.
98. Wilding, I.; Hirst, P.; Connor, A. *Pharm Sci Technol Today* 2000, **3** (11), 385–392.
99. Santini, J. T.; Cima, M. J.; Langer, R. *Nature* 1999, **397** (6717), 335–338.
100. Ito, Y.; Hagiwara, E.; Saeki, A.; Sugioka, N.; Takada, K. *Eur J Pharm Sci* 2006, **29** (1), 82–88.
101. Lee, J. W.; Park, J. H.; Prausnitz, M. *Biomaterials* 2008, **29** (13), 2113–2124.
102. Ito, Y.; Yoshimitsu, J.; Shiroyama, K.; Sugioka, N.; Takada, K. *J Drug Target* 2006, **14** (5), 255–261.
103. Sullivan, S. P.; Murthy, N.; Prausnitz, M. *Adv Mater* 2008, **20** (5), 933–938.
104. Yan, L.; Raphael, A. P.; Zhu, X.; Wang, B.; Chen, W.; Tang, T.; Deng, Y.; Sant, H. J.; Zhu, G.; Choy, K. W.; Gale, B. K.; Prow, T. W.; Chen, X. *Adv Healthc Mater* 2014, **3** (4), 462–462.
105. Ameri, M.; Fan, S.; Maa, Y.-F. *Pharm Res* 2010, **27** (2), 303–313.
106. Wu, B. M.; Borland, S. W.; Giordano, R. A.; Cima, L. G.; Sachs, E. M.; Cima, M. *J Control Release* 1996, **40** (1–2), 77–87.

16

PATHWAYS FOR DRUG DELIVERY TO THE CENTRAL NERVOUS SYSTEM

NGOC H. ON, VINITH YATHINDRANATH, ZHIZHI SUN, AND DONALD W. MILLER

Department of Pharmacology and Therapeutics, Kleysen Institute for Advanced Medicine, University of Manitoba, Winnipeg, Manitoba, Canada

16.1 INTRODUCTION

The approval rate for drugs to treat central nervous system (CNS) disorders is significantly lower than other therapeutic classes [1]. Despite aggressive efforts, many new compounds targeting the CNS are not effective in delivering a safe and efficacious dose to the brain [1, 2]. While there are many potential reasons for the discrepancy in CNS drug approval compared to other therapeutic classes, one challenging aspect of CNS drug development is the requirement of the compound to cross the blood–brain barrier (BBB) in sufficient amounts for therapeutic response. The normally restrictive nature of the BBB poses a well-known and formidable obstacle for the delivery of water-soluble drugs and large therapeutic macromolecules to the brain [3, 4]. In general, molecules greater than 400 Da molecular weight are considered too large to cross the BBB. However, misconceptions concerning the ability of small molecules to penetrate the BBB and the effectiveness of small molecules in treating CNS-related pathologies have contributed to an underappreciation of the importance of drug delivery to the brain. It has been estimated that less than 2% of the compounds in the current drug discovery pipeline have sufficient BBB permeability to actually enter the brain in therapeutically relevant amounts [5]. Furthermore, CNS-related disorders, many of which are increasing in prevalence, including neurodegenerative diseases

Drug Delivery: Principles and Applications, Second Edition. Edited by Binghe Wang, Longqin Hu, and Teruna J. Siahaan.

(e.g., Alzheimer's), psychiatric disorders (schizophrenia and depression), stroke, traumatic brain injury, and brain tumors, remain largely unresponsive to current small-molecule therapy [2, 5]. Thus the search for effective treatments for CNS diseases must concentrate not only on the drug target but also on efficient delivery of the drug to the site of action within the brain.

This chapter will review the basic cellular and molecular aspects of the BBB and blood–cerebrospinal fluid barrier (BCSFB) and discuss various approaches for enhancing drug delivery to the brain. The advantages and limitations of each approach are discussed along with the emerging areas of interest within each respective CNS drug delivery pathway.

16.1.1 Cellular Barriers to Drug Delivery in the CNS

There are two cellular barriers that separate the brain extracellular (EC) fluid from the blood (see Fig. 16.1). The first and largest interface is the BBB. The BBB consists of a continuous layer of endothelial cells, surrounded by astrocyte foot processes, and scattered pericytes. The astrocytes and pericytes provide structural support for the brain capillary endothelial cells. In addition, the classic experiments of Stuart and Wiley demonstrate the importance of the astrocytes in releasing chemical factors that regulate brain capillary endothelial cell permeability [7]. However, it is the brain capillary endothelial cells that provide the cellular anatomical barrier separating the blood from the EC fluid of the brain. Unlike capillary endothelial cells found in the peripheral tissue, the brain capillary endothelial cells have complex tight junctions, low endocytic activity, and an absence of fenestrations [8]. These anatomical features prevent the passage of most polar and hydrophilic solutes from the blood into the brain. While lipophilic solutes can passively diffuse across the BBB, the expression of numerous efflux transporters including breast cancer resistance protein (BCRP), P-glycoprotein (P-gp), and multidrug resistance-associated protein (MRP) within the brain microvessel endothelial cells [9, 10] acts to restrict the passage of a variety of endogenous solutes as well as many small molecule drugs from the blood to the brain [11–15].

The second cellular barrier involved in regulation of solute entry into the brain is the BCSFB. The BCSFB is a composite barrier made up of the choroid plexuses and the arachnoid membranes of the circumventricular organs. Unlike the brain capillaries that form the BBB, the capillaries found in the circumventricular organs are fenestrated and lack tight junctions between the adjacent endothelial cells (see Fig. 16.1) [16, 17]. In the case of the BCSFB the choroid plexus epithelial cells that line the ventricles provide the cellular barrier separating the blood from the CSF in the brain [16, 17]. The epithelial cells in the choroid plexus have complex tight junctions on their apical (CSF) side that prevent paracellular diffusion of polar, hydrophilic compounds into the CSF. These tight junctions formed by the epithelial cells in the choroid plexus are slightly more permeable than those found in the capillary endothelial cells of the BBB [17, 18], but are still sufficient to prevent paracellular diffusion of polar, hydrophilic compounds into the CSF. Like their brain capillary endothelial cell counterparts, the epithelial cells of the choroid plexus also express a wide variety of solute transporters

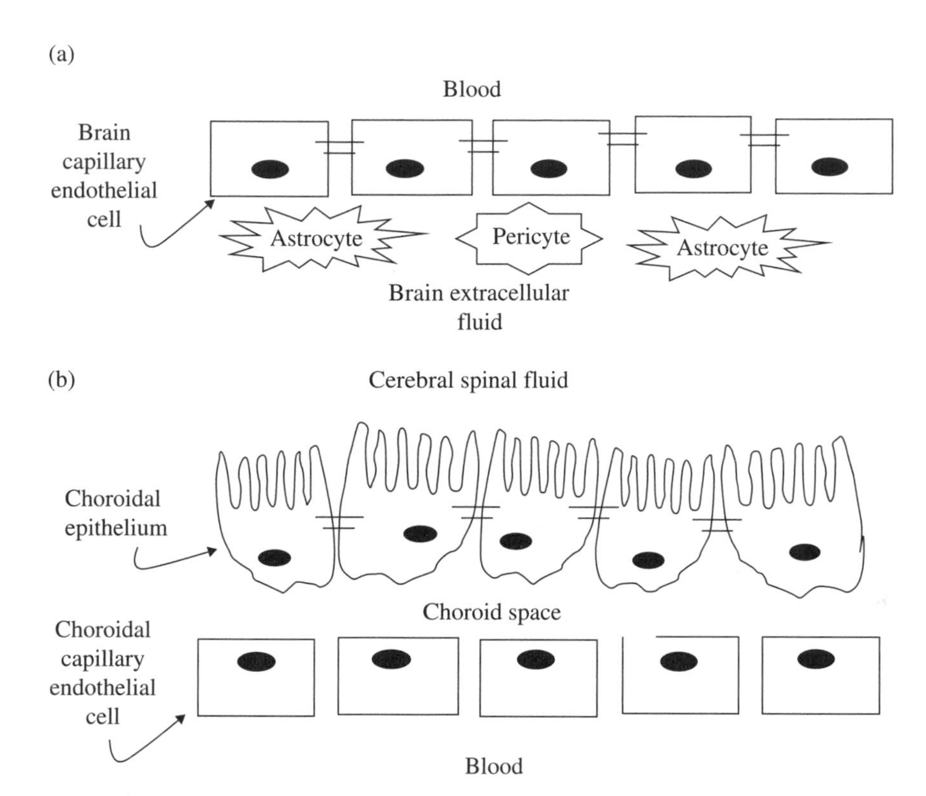

FIGURE 16.1 (Panel a): Schematic representation of the cellular components of the blood–brain barrier (BBB). (Panel b): Blood–cerebrospinal fluid barrier (BCSFB). The BBB consists of continuous-type endothelial cells with complex tight junctions to limit paracellular diffusion. The astrocytes and pericytes located in close proximity to the brain endothelial cells release various endogenous factors that modulate endothelial cell permeability. In contrast, the choroid endothelial cells are fenestrated and the BCSFB properties are provided by the tight junctions formed between the choroid epithelial cells. Zhang and Miller [6]. Reproduced with permission of John Wiley & Sons, Inc.

and drug efflux transporters that help regulate brain EC fluid content [19]. However, while the efflux transporters in the BBB contribute to reductions in the brain penetration of solutes, the efflux transporters expressed in the BCSFB result in the secretion of solutes into the CSF. This is an important distinction as CSF concentrations of drug have been used as a surrogate marker for brain concentration of drug in clinical trials, and depending on the compound of interest and its transporter liabilities can result in overestimations of brain penetration [20].

Although the apical membrane of the epithelial cells forming the BCSFB has numerous microvilli, the total surface area is still substantially smaller than that of the BBB [21]. It has been estimated that in the human brain there are approximately 100 billion capillaries with a total surface area of 20 m^2 [22]. Given the density of the capillary network in the brain, and the close proximity of neuronal cells to these capillaries, most drug delivery approaches have focused on either circumventing the BBB or

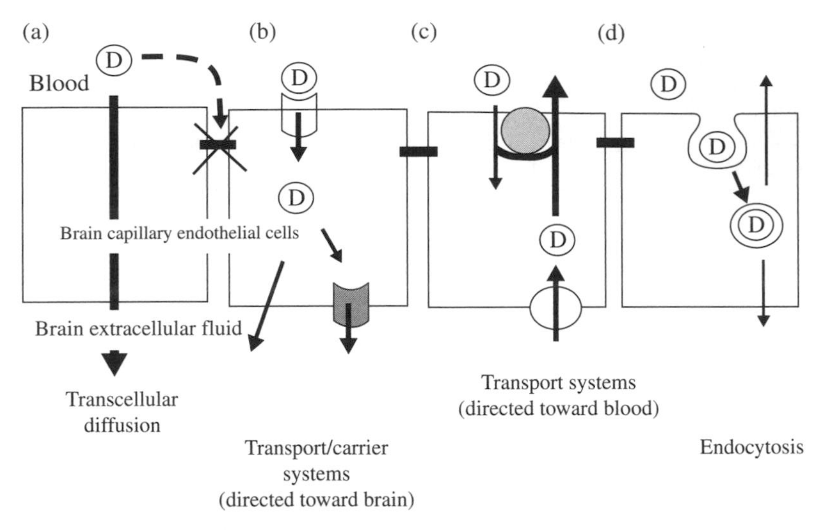

FIGURE 16.2 Potential mechanisms for drug (Ⓓ) movement across the BBB. Routes of passage include passive diffusion through the brain capillary endothelial cells (a), utilization of inwardly directed transport or carrier systems expressed on the brain capillary endothelial cells (b), utilization of outwardly directed efflux transport systems (c), or utlization in various endocytic vesicular transport processes occurring within the brain capillary endothelial cells (d). Zhang and Miller [6]. Reproduced with permission of John Wiley & Sons.

increasing drug permeability through the BBB. The various routes of solute/drug passage across the BBB are discussed later and shown schematically in Figure 16.2.

16.1.2 General Approaches for Increasing Brain Penetration of Drugs

There are three general pathways that have been used for delivering drugs to the CNS. One approach is to circumvent the difficulties associated with drug permeability in the BBB and/or BCSFB completely by the central administration of the drug directly into the brain. The remaining pathways focus on exploiting one or more of the various routes for solute/drug passage across the BBB (see Fig. 16.2). This may involve temporary disruption of tight junction complexes between the brain capillary endothelial cells to allow therapeutic agents to enter the brain through an enhanced paracellular diffusion pathway. Alternative approaches involving chemical modifications of the drug molecule or drug carrier to improve BBB permeability have also been used. Each of these strategies are discussed later.

16.2 CIRCUMVENTING THE CNS BARRIERS

Despite recent advances in CNS drug formulations and delivery platforms, many potential efficacious drug molecules still cannot penetrate the brain at relevant therapeutic concentrations due to the restrictive nature of the BBB. Thus, one strategy to increase drug delivery into the brain is to circumvent the problem of limited BBB penetration entirely through direct administration of the drugs into the brain via transcranial delivery method. The transcranial delivery approach is invasive, requiring a craniotomy in which a small hole is drilled into the skull, and it encompasses three basic delivery methods including intracerebroventricular (ICV) injection, intracerebral (IC) administration, with or without convection-enhanced diffusion (CED), and/or implantation of sustained release dosage forms into the brain [23]. A major advantage of using these techniques is the ability to achieve a much higher concentration of drug in the brain parenchyma and the CSF, without the observed systemic side effects [24]. Furthermore, these direct delivery methods also enable both small- as well as large-therapeutic drug molecules in various formulations to penetrate the brain and their target sites. However, one major drawback of this invasive technique is the limited diffusion ability of drug from the injection site. The low diffusion rate of drug away from the injection site requires a high concentration of drug in order to achieve and maintain therapeutic drug levels in large-tissue region [24]. This can also lead to dose-limiting toxicity.

16.2.1 Intracerebroventricular Injection

One approach to bypass the BBB is to introduce the therapeutic agents directly into the cerebral ventricles. The rationale is that the ventricles of the brain with its CSF contents can interact with the interstitial fluid of the brain, thereby allowing the diffusion of drugs into the brain parenchyma. The extent to which drugs can diffuse into the brain parenchyma following intracerebroventricular injection (ICV) is a concern. While, ICV injection of glial cell line-derived neurotrophic factor (GDNF), a neurotrophic factor for dopaminergic neurons, showed improvements in a toxin-induced rodent model of Parkinson's disease, no therapeutic response was observed following ICV injection of GDNF in Parkinsonian patients [25]. The main reason for the failure of clinical response to ICV administration of therapeutic agents is the low diffusion capabilities of the drug molecules into the brain parenchyma [26]. This is especially the case for large hydrophilic drug molecules. Indeed, the rate of CSF turnover in the human brain is much faster than the rate of parenchymal drug diffusion from the ependymal surface [23]. As a result, the drug is more likely to distribute to the general circulation than to the target sites within the brain [23].

Because of these limitations, the pharmacological effect of ICV injection is significant only if the target receptor for the drug is localized near the ependymal surface of the brain. Applications using ICV delivery of drugs is best suited for the treatment of primary or metastatic brain tumors localized in the meninges and the subarachnoid area of the brain where the drugs in the CSF can more easily distribute [27]. An additional consideration with the ICV delivery method is the possible development subependymal astrogliodic reactions, especially for prolonged drug treatment through the ICV route [23].

16.2.2 Intracerebral Administration

In an attempt to overcome the limited diffusional capabilities of ICV drug delivery, direct administration of drug to the brain can be done through IC administration with or without CED, or by the implantation of degradable or nondegradable polymers into the brain. Intracerebral injection requires the insertion of a catheter into the brain, with a potential risk of hemorrhage and infection at the catheter insertion site [28]. CED involves the use of positive hydrostatic pressure to infuse drug directly into the brain cerebral region thereby bypassing the BBB. The use of a positive pressure gradient enables the administered drug to penetrate further into the target tissue [28]. Drug is administered through the implanted catheter using an infusion pump to generate a continuous positive pressure gradient. Brain-related uses of CED have been mostly limited to experimental treatment of brain tumors [28]. While the CED approach has shown positive responses in preclinical brain tumor experiments, its clinical effectiveness is modest at best [28]. In addition to the development of astrogliosis due to the hydrostatic pressure of the infusion into the brain, the use of CED delivery, like all the direct delivery methods previously discussed, is costly and requires surgical intervention for catheter and/or implant placement.

16.2.3 Intranasal Delivery Route

An alternative approach to bypassing the BBB is through intranasal (IN) delivery of drug. The unique anatomical and physiological structures of the olfactory region of the nasal passages provide both intraneuronal and extraneuronal pathways into the CNS that avoid the BBB. The olfactory bulb is the portion of the brain tasked with receiving and sorting sensory information pertaining to odor and smell. It receives sensory input from the olfactory receptor neurons within the olfactory epithelium in the nasal cavity that have axonal projections through the ethmoid bone to the olfactory bulb. The sensory information delivered to the olfactory bulb is transmitted to other brain centers through interneurons within the olfactory bulb. The IN route utilizes the neuronal projections from the epithelial cells in the nasal cavity for drug delivery to the brain and CSF [29]. Thus it has the advantage over other direct brain administration routes in that it is less invasive, requiring only instillation of drugs to the olfactory epithelium where the ciliated epithelium provides a large surface area for drug entry into the CNS [30]. Once the drug reaches the olfactory epithelium, IN delivery utilizes two distinct pathways for drug penetration into the brain. The intraneuronal pathway involves internalization of the drug into the olfactory receptor neurons and retrograde axonal transport to the olfactory bulb [31]. This route is the slower pathway taking hours to days to reach the olfactory bulb depending on the physical-chemical properties of the drug [31]. For the extraneuronal pathway, the molecules diffuse through the perineural space surrounding the neuron to the olfactory bulb [31]. Once the drug diffuses into the olfactory bulb, it enters the nearby brain regions by EC diffusion process.

The IN delivery had been successfully applied in rats to deliver insulin-like growth factor-I (IGF-I)—a 7.65 kDa protein for the treatment of Alzheimer's disease and stroke—because of its ability to promote neuronal survival [32]. However, there are certainly limitations to this technique. The nasal mucosa is highly vascularized, receiving its blood supply from the nearby carotid arteries [33]. Thus, there is the potential for systemic absorption of the drug in the nasal mucosa, as well as the respiratory mucosa, prior to reaching the olfactory epithelium in the back of the nasal cavity. Due to the anatomical features of the nasal cavity, delivery of drug to the olfactory epithelium can be difficult and variable [33]. For these reasons, the most suitable therapeutic agents for IN delivery to the brain are large macromolecules that would have minimal systemic absorption within the nasal and respiratory epithelium in the nasal cavity.

Studies in animal models using IN brain delivery of peptide and protein therapeutics, either with or without permeability enhancers, have been positive. However, clinical demonstration of the effectiveness of this route for drug delivery to the brain has been difficult, with the main obstacle being limited penetration of drug to small, localized areas of the brain. The feasibility of this approach will depend on the results of ongoing clinical trials.

16.3 TRANSIENT BBB DISRUPTION

The BBB is a complex network composed of brain capillary endothelial cells, pericytes, astrocytic endfeet, and neuronal cells, which collectively function as a protective barrier limiting the paracellular diffusion of molecules and solutes from the blood to the brain. Modifying the integrity of the tight junctions causing a controlled and transient disruption of the BBB can also increase drug delivery to the brain. The various approaches for transient BBB disruption and the known mechanisms involved are discussed later.

16.3.1 Osmotic BBB Disruption

The concept of osmotic BBB disruption was introduced in 1972 when Rapoport et al. [34] observed the presence of Evans Blue, which extensively binds to albumin, in the CNS following an intraarterial infusion of hypertonic arabinose. This observation suggested that a hypertonic solution increased the BBB permeability by inducing the shrinkage of endothelial cells, thereby disrupting the tight junctions between the brain capillary endothelial cells [35]. Other hypertonic solutions including lactamide, saline, urea, radiographic contrast agent, and mannitol can also be used to transiently disrupt the BBB. The osmotic disruption technique is the most established method for transient BBB disruption. This technique has been extensively used in preclinical rodent models to deliver drugs to the brain [35] and clinically in the treatment of brain tumors, which have poor prognosis and limited chemotherapeutic response due to restricted BBB penetration [36].

The most common osmotic agent used for BBB disruption is mannitol (1.4–1.8 M), infused in a retrograde fashion in the carotid artery, followed by the infusion of the therapeutic agent. The magnitude of BBB disruption produced by osmotic agents has been established using various permeability markers including contrast-enhanced CT or magnetic resonance imaging (MRI) agents as well as fluorescently labeled dyes including Evans blue and horseradish peroxidase (HRP) [37]. The disruption of the BBB following administration of osmotic agent results in increased accumulation of both small and large molecules to the brain. Studies in rats using contrast-enhanced

MRI suggest that the time course of BBB disruption is rapid, occurring within minutes following the injection of mannitol with the maximal exposure occurring at 60 minutes [34]. The duration of BBB opening is variable depending on the animal model used [38]. However, the maximal time for restoration of BBB properties following osmotic disruption in animal studies is approximately 6 hours [34]. While the BBB opening should be sufficient in duration to deliver drug to the brain, the potential for toxicity from BBB disruption increases the longer the BBB is compromised.

Osmotic disruption of the BBB has also been examined in humans for increasing the delivery of chemotherapeutic agents to the brain for the treatment of brain tumors. This technique had been shown to increase the survival rate of the brain tumor patients, particularly with primary CNS lymphoma and malignant gliomas [39]. However, the main drawback of this technique is the long recovery period associated with the disruption, which allows macrophages as well as other small and large molecules enter the CNS inducing a transient increase in intracranial pressure [35]. The rise in intracranial pressure associated with osmotic disruption is a major contraindication for brain tumor treatments. Other inherent factors that can undermine the success of this procedure include the hemodynamic variability of each patient, the anesthetic agent being used, as well as the infusion rate of the hypertonic solutions [35, 39]. An additional confounding factor, especially in the treatment of brain tumors, is the variable BBB disruption seen upon repeated exposure to osmotic disruption and/or radiation treatment [35].

16.3.2 Pharmacological Disruption of the BBB

Both the luminal and abluminal plasma membranes of brain endothelial cells possess numerous receptors that when activated by various endogenous ligands including leukotrienes, histamines, arachidonic acid, and bradykinin are known to modulate the permeability of the BBB [40]. Most of these endogenous vasoactive compounds have a very short response half-life due to their degradation by proteases as well as tachyphylaxis at the receptor signal transduction level. Furthermore, many of these endogenous ligands also produce undesired off-target effects due to activation of their receptors at different organ and tissue sites. Despite these limitations, modulation of BBB permeability through pharmacological targeting of receptors within the brain microvasculature represents a valid approach for enhancing drug delivery to the brain.

16.3.2.1 Bradykinin Analogs Pharmacological activation of bradykinin receptors on endothelial cells has been shown to be a valid pharmacological target for producing transient disruption of the BBB. This was done through the design and synthesis of a more selective bradykinin B_2 receptor agonist, Cereport (Labradimil or RMP-7) [40]. The half-life of Cereport is two to three times greater than bradykinin due to the substitution of (2-thienyl)-Ala for a Phe at position 5 as well as substitution of methyl-Tyr for Phe at position 7 [41]. These substitutions make Cereport less susceptible to proteolytic cleavage. *In vitro* studies with rat brain microvessel endothelial cells, guinea pig ileum, and rat uterus demonstrate that Cereport is selective for B_2

receptors and acts as a B_2 receptor agonist. While Cereport has lower affinity than bradykinin for the B_2 receptor, the increased half-life results in a greater potency than bradykinin in cell-based and *in vivo* systems [42].

Cereport increases BBB permeability by interfering with the tight junctions formed between the endothelial cells without affecting the vasculature of the brain itself [41]. The increased BBB permeability associated with Cereport occurs rapidly, within minutes following the infusion of the compound. *In vivo* work in rat and dog confirmed this rapid enhancement of BBB permeability as it occurred only 5 minutes following the administration of Cereport [41, 43]. Furthermore, this disruption only lasted for a brief period of time since the BBB properties restored back to its normal condition within 2–5 minutes following the termination of the infusion. Another important pharmacodynamic property of Cereport is the occurrence of tachyphylaxis usually within minutes, following the continuous or multiple administrations of the compound [44]. A continuous infusion of bradykinin or Cereport induces a spontaneous restoration of the barrier properties occurring within 10–20 minutes, with the barrier returning to baseline permeability within 30–60 minutes [41]. Studies have shown two consecutive administrations of Cereport separated by a 15-minute interval induced a complete tachyphylaxis, while a 60-minute interval induced a smaller magnitude of disruption as compared with single infusion of the compound [41].

Transient disruption of the BBB mediated by Cereport to improve drug delivery to the brain has been investigated in several CNS disorders including brain tumors, neuropathic pain, and Parkinson's disease [44–46]. The majority of studies both preclinical and clinical have focused on brain tumor applications for Cereport [41, 47]. In rodent brain tumor models, Cereport has been used successfully to increase the delivery of chemotherapeutic agents to the brain [47]. Given the positive safety profile of Cereport and its ability to transiently enhance permeability and drug delivery in rodent brain tumor models, clinical trials were initiated in glioma patients. An intracarotid infusion of Cereport showed a significant enhancement of Ga-EDTA at the tumor region of the glioma patients [48]. The increased permeability enhancement observed in clinical trials was mostly confined to the tumor vasculature, with minimal impact on the integrity of the barrier in normal brain tissue [41]. This nonuniform disruption of the BBB mediated by Cereport may help explain how despite favorable preclinical results, the clinical trials with Cereport for enhanced brain delivery of chemotherapeutics have failed to provide significant increase in treatment response [49]. Another possible explanation for the failure of Cereport in clinical trials may be the varying level of bradykinin B_2 receptor expression in glioma patients eliciting variable amounts of BBB and blood–tumor barrier (BTB) disruption [50]. Together these studies point to the need for a more controlled method of transient BBB disruption that can influence permeability within both tumor and nontumor regions of the brain.

16.3.2.2 Alkylglycerols Transient disruption of the BBB has also been demonstrated through pharmacological administration of various alkylglycerols [51]. Indeed, disruption of BBB mediated by short-chain alkylglycerols had been shown to increase the extraluminal accumulation of a macromolecule, fluorescein isothiocyanate (FITC)-dextran 40,000 in the brain in a concentration-dependent manner [52, 53]. The opening of the BBB induced by akylglycerols was rapid and reversible with the baseline permeability being restored within 120 minutes following treatment [51]. Studies have also been performed evaluating the effectiveness of alkylglycerols in enhancing chemotherapeutic delivery in a rodent gliobastoma brain tumor model. While there was little if any brain penetration of the chemotherapeutics under control conditions, treatment with alkylglycerols resulted in dose-dependent increases in chemotherapeutic drug distribution into the brain [53]. The magnitude of increase in methotrexate delivery to the tumor region using the highest doses of alkylglycerols was comparable to that observed with osmotic disruption, and was significantly greater than that observed with bradykinin analog, Cereport [52, 53]. Although human clinical trials have not been initiated, studies in rat brain tumor model show no long-term signs of toxicity associated with these alkylglycerols, suggesting that these agents can be effective in the treatment of brain tumors.

Treatment with alkylglycerols results in increased brain accumulation of hydrophilic solutes and macromolecules, suggesting that the modulation of tight junctions and paracellular diffusion routes are involved. While the exact molecular mechanism remains unknown, there are several findings that support pharmacological targeting of a receptor within the brain vasculature. First, the effects on BBB permeability observed with alkylglycerols are concentration dependent. Second, both the magnitude and duration of BBB disruption show structural dependency. Using a series of alkylmono, di-, and triglycerols, Erdlenburch and colleagues found that BBB disruption increased as a function of chain length of the alkyl group with the pentyl and hexylglycerols providing the best response [54]. Furthermore the effects on BBB permeability were best with the monoglycerols as the addition of di- and triglycerols resulted in diminished disruption [54]. The concentration and structure dependency of the alkylglycerol on BBB disruption suggests that the activation of receptor sites within the brain microvessel cells mediates the effects of alkylglycerol molecules on BBB permeability.

16.3.2.3 Lysophosphatidic Acid Another compound that had been shown to have an impact on the vascular permeability of the BBB is lysophosphatidic acid (LPA). LPA is a phospholipid found in most cell types including neurons, Schwann cells, adipocytes, and fibroblasts [55]. It is secreted by activated platelets and found in numerous bodily fluids including serum, saliva, follicular fluid, and malignant effusions. LPA is capable of producing a variety of responses in mammalian cells by binding to their receptors on the surface of plasma membrane and initiating their response through G-protein-coupled receptor (GPCR) signaling pathways. There are currently at least six different types of LPA receptors (LPAR1–LPAR6) [56, 57]. Within the brain, LPAR1 is the most highly expressed, although LPAR2 and LPAR3 are also present [56].

An important cellular target of LPA is the vascular endothelium where LPA-induced changes in mitogenesis, cell migration [58], and permeability [59, 60] occur. Studies of LPA on brain microvessel endothelial cell permeability support an enhancement of permeability with increases in both transcellular electrical resistance

(TEER) and flux of vascular markers observed in primary cultured brain endothelial cells and *in situ* cranial window preparations [60, 61]. Recent studies by On et al. [36] demonstrated that intravenous (i.v.) injections of LPA produced dose-dependent disruption of BBB integrity. Modulation of BBB permeability with LPA was rapid in onset (within 3 minutes) and short in duration, with complete restoration of BBB integrity observed within 20 minutes [36]. Furthermore, the ability to enhance brain accumulation of both small and large hydrophilic permeability markers as opposed to more lypophilic imaging agents suggested an increased paracellular diffusion route through altered tight junction between the brain endothelial cells [36].

There are several advantages to targeting LPA receptors for pharmacological modulation of BBB permeability. One is the fast onset of BBB disruption observed following LPA exposure and the relatively short time course for restoration of BBB integrity. A second advantage is the wide range of molecules that could be delivered with this approach. Additionally, unlike Cereport that appears to have nonuniform disruption of BBB, LPA produced similar magnitudes of BBB opening throughout all areas of the brain examined [36]. While there are multiple LPA receptors expressed within the brain microvasculature, there is the possibility that targeting selected LPA receptors with more selective pharmacological agents would further refine the extent of BBB modulation achievable. Thus there is reason to be guardedly optimistic about the potential for modulation of BBB permeability through phospholipid receptors on the brain microvasculature.

16.3.2.4 Cadherin-Binding Peptides The most recent pharmacological agents to be used experimentally to modulate BBB permeability are the cadherin-binding peptides. The adherens junction in brain microvessel endothelial cells is primarily composed of cadherin proteins. The binding of these proteins on adjacent brain microvessel endothelial cells forms a homolytic dimer within the cell junction that limits the paracellular passage of solutes with diameter greater than 11 Å or approximately 500 Da [62, 63]. The cadherin protein has an EC domain with five tandem repeated units (EC-1–EC-5), and highly conserved regions of His-Ala-Val (HAV) are involved in the formation of the dimer. Synthetic peptides based on the HAV region sequence have also been shown to inhibit the interactions between the E-cadherin molecules and prevent the aggregation of bovine brain microvessel endothelial cells in a concentration-dependent manner [64]. Cell culture studies using a HAV-based peptide having the amino acid sequence of Ac-SHAVSS-NH_2 reported an increased paracellular diffusion of radiolabeled mannitol and decreased transepithelial electrical resistance (TEER) [65]. In the *in situ* brain perfusion model, HAV peptide administered into the perfusate increased the brain delivery of ^{14}C-mannitol and ^{3}H-daunomycin [66].

Recently On et al. [62] have shown that i.v. injections of HAV peptide in mice produced increased BBB permeability to a wide variety of permeability markers. Administration of HAV peptide resulted in significant dose-dependent increases in BBB permeability that were apparent within 3 minutes. The BBB permeability enhancement was completely abolished when the mice were given HAV peptide at 1 hour prior to the injection of the contrast agent, indicating that BBB integrity was completely restored back to normal within 1 hour [62]. Like other pharmacological BBB-disrupting agents, the effects of the HAV peptide were most apparent for small hydrophilic permeability markers. However, increased brain distribution of large macromolecules as well as P-gp imaging agents were observed [62]. The advantages of this approach are the ability to further refine the BBB disruption in terms of magnitude and duration of opening through modulation of the cadherin-binding peptide. An example of this are the recent studies with cyclized cadherin-binding peptides that show large and small molecule discrimination and extended periods of BBB opening (up to 4 hours) [67]. Furthermore, unlike other pharmacological agents, the BBB disruptions produced by the cadherin-binding peptides were uniform throughout the entire brain.

16.4 TRANSCELLULAR DELIVERY ROUTES

16.4.1 Solute Carrier Transport Systems in the BBB

A broad range of endogenous solutes enter the brain through carrier-mediated transport. The brain capillary endothelial cells have several solute carriers (SLCs) or transport proteins expressed in the luminal and/or abluminal sides [68]. The function of these transporters is to move essential polar metabolites including glucose, amino acids, nucleosides, monocarboxylates, small peptides, and organic anions/cations into and out of the brain. The gene symbols of these transporters under SLC superfamily are prefixed with *SLC*. For example, the human and rodent (rat, mice) amino acid transporter, LAT1, is denoted as *SLC7A5* (SLC family 7 (amino acid transporter light chain, L System), member 5) and *Slc7a5*, respectively. Depending on the transporter location in the BBB and BCSFB, solutes can be translocated from the blood to endothelium/epithelium and/or endothelium/epithelium to the brain/CSF. Mechanistically, the SLCs are facilitated diffusion-based carriers; thus, solutes move bidirectionally through the plasma membrane driven by the concentration gradient of the solute or cotransport of an electrolyte.

There are two major approaches for the delivery of drugs to the brain via transport-mediated routes: (i) chemical modification of the drug to optimize transport as a pseudosubstrate or (ii) conjugation of the drug to an endogenous transport substrate (prodrug or "piggyback" approach) [69]. There are several factors that can influence the carrier-mediated transport of drug molecules (pseudosubstrate or prodrug). Indeed, any drug designed to exploit SLC pathways is in direct competition with endogenous substrates and, consequently, can be inhibited by the endogenous substrates [70]. The brain capillary density (C_{cap}) of SLCs differs between types and between animal species, and the same is true for transport kinetics (Table 16.1). Hence, drug design and experimental strategies should address these potential roadblocks or discrepancies. In this section, key SLCs expressed in the brain capillary endothelial cells, their endogenous substrates, and recent advances in their usage to transport CNS therapeutic/diagnostic agents will be discussed.

16.4.1.1 Amino Acid Transporters Several amino acid transporters are expressed in the brain capillary endothelial cells. The most widely studied amino acid transporters are the Na^+-independent systems L (LAT1—*SLC7A5* and LAT2—*SLC7A8*), y^+ (*SLC7A7*), and x_c^- (*x*CT—*SLC7A11*) [79]. The system L actively transports large neutral amino acids such as L-leucine, L-phenylalanine, L-tryptophan, L-tyrosine, L-isoleucine, L-methionine, and L-valine. The system y^+ transports cationic amino acids including L-arginine, L-lysine, and L-ornithine. The substrates of system x_c^- include anionic amino acids such as L-glutamate and L-aspartate. *In vitro* studies have suggested the presence of other Na^+-dependent amino acid transporters called systems A/N (SNAT1—*SLC38A1*, SNAT2—*SLC38A2*, SNAT3—*SLC38A3*, SNAT4—*SLC38A4*, and SNAT5—*SLC38A5*), $B^{o,+}$ (*SLC7A9*), ASC (*SLC7A10*), β (transports β-amino acids such as β-alanine and taurine), and X^- (*SLC1A1*) [79]. The substrates of systems A and ASC include small neutral amino acids (e.g., L-alanine, L-serine, and L-cysteine), $B^{o,+}$ include neutral and basic amino acids (e.g., L-arginine and L-lysine), and that of X^- include anionic amino acids (e.g., L-glutamate and L-aspartate). Systems A, $B^{o,+}$, ASC, and X^- are thought to be predominantly localized at the abluminal side of the brain capillary endothelial cells, thereby facilitating amino acid efflux from brain EC fluid [79].

The system L-amino acid transporter (especially LAT1), with its wide range of substrates, is perhaps the most exploited SLC pathway for delivering therapeutic/diagnostic agents across the BBB. The system L recognizes molecules with an amine and a carboxylic acid group attached to a single carbon (characteristic of α-amino acids), and a bulky hydrophobic side group [80]. Drug molecules fulfilling these stereochemical requirements and transported by system L include agents such as L-DOPA [72], Melphalan [81], Gabapentin [82], D,L-NAM [83], and Acivicin [84]. The system L-amino acid transporter has also been used for delivering various positron emission tomography (PET) imaging agents to the brain, for example, 6-[^{18}F]fluoro-L-3,4-dihydroxy-phenylalanine (^{18}F-FDOPA) [85], L-[^{11}C-methyl]methionine (^{11}C-MET) [86], *O*-(2-[^{18}F]fluoroethyl)-L-tyrosine (^{18}F-FET) [87], and 3-[^{123}I]iodo-α-methyl-L-tyrosine (^{123}I-IMT) [88]. Diagnostic imaging of low-grade brain tumors using the SLC-based PET imaging probe, ^{11}C-MET, displayed better tumor (recurrence) differentiation than gadolinium-enhanced MRI [89]. This is because gadolinium-based MRI contrast agents do not cross (paracellularly) intact BBB in low-grade gliomas; however, ^{11}C-MET undergoes enhanced active transport due to upregulated transporter protein expression/activity within the brain tumor microvasculature [89, 90].

To date, the therapeutic/diagnostic agents shown to undergo carrier-mediated transport via system L were structural homologues (pseudosubstrates) of endogenous substrates. There are a very few reports where a drug conjugated to amino acid substrate ("piggyback"/prodrug) was shown to undergo carrier-mediated transport. Several reports have shown the ability of drug–amino acid conjugates to bind and inhibit LAT1 [91, 92]. For example, 6-mercaptopurin conjugated to L-cystine inhibited [^{14}C] L-leucine uptake in rat brain [91]. But, these studies did not present any evidence of carrier-mediated transport. Evidence for the carrier-mediated transport of a prodrug was given in a recent study with dopamine-L-phenylalanine conjugate

[72]. Dopamine-L-phenylalanine conjugate was found to have higher binding affinity toward LAT1 than L-DOPA. The rat brain uptake of dopamine-L-phenylalanine was 0.307 pmol/(mg minute), which dropped to 0.095 pmol/(mg minute) during cold (5°C) perfusion proving carrier-mediated transport [72]. Further studies on drugs with poor BBB permeation are needed to establish the effectiveness of "piggyback"/prodrug approach with LAT1 transporter.

16.4.1.2 Glucose Transporters Glucose transporters carry glucose and other hexoses (mannose and galactose) across the plasma membrane and are classified into Na^+-dependent SGLT and Na^+-independent GLUT (facilitative) families [93]. The primary glucose transporter in the BBB is GLUT1 (*SLC2A1*), which is expressed in the luminal and abluminal sides of the BBB. GLUT1 efficiently manages the high glucose demand in the brain [71]. The capillary density (C_{cap}) and the transport capacity of GLUT1 is the highest among SLCs in the BBB (Table 16.1) [68, 71].

Owing to its high transport capacity, GLUT1 transporter has been targeted for carrier-mediated transport across the BBB. However, like many of the SLCs, GLUT1 transporter has very stringent stereochemical and molecular size requirements for binding and transport [80, 94]. The majority of studies have taken the "piggyback" approach of conjugating therapeutic agents to actual GLUT1 substrates (i.e., various hexoses) [69, 95, 96]. Opioid peptide analogues (neuromodulators) linked to glucose showed an improved BBB penetration compared to unmodified peptides in mice model [95, 96]. Initially, the enhanced permeation was attributed to GLUT1-mediated transport [95]. However, later studies disproved this hypotheses and demonstrated that improved metabolic stability and bioavailability of the glycopeptides, as opposed to GLUT1-mediated transport, was the reason for enhanced BBB permeation [69, 97]. Glucosylated prodrugs of various anti-inflammatory agents have been examined for improved delivery to the brain [98]. While many studies can demonstrate increased brain delivery of glucosylated prodrugs, there is some question as to whether the compounds are actually transported by the GLUT1 transporter. In fact there are many reports where the drug–glucose conjugates inhibited GLUT1 without any evidence of being substrates for the GLUT1 transporter. While the GLUT1 transporter holds great potential in terms of its transport capacity, more studies are needed to design and identify drug candidates that have higher affinity toward the transporter than glucose.

16.4.1.3 Monocarboxylate Transporters The monocarboxylate transporter (MCT) family includes fourteen known types. Among them, MCT1–MCT4 are involved in the transmembrane transport of proton-linked monocarboxylates (e.g., L–lactate/H^+, pyruvate/H^+, and other ketone bodies) or monocarboxylic acids [99]. Among the MCTs, MCT1 (*SLC16A1*) is expressed in brain capillary endothelial cells and functions to carry out bidirectional transport (influx or efflux) of lactic acid, based on the intracellular and EC substrate concentrations and the pH gradient across the cell membrane [100]. The MCT1 transporter is known to transport exogenous substrates (with monocarboxylic acid attached to small alkyl or aryl groups) such as salicylic acid, nicotinic acid, and benzoic acid across the BBB. The 3-Hydroxy-3-methylglutaryl coenzyme A (HMG-CoA)

TABLE 16.1 The Kinetic Parameters of Selected Transporter Proteins Expressed in the Brain Capillary Endothelial Cells (BBB) for Selected Endogenous/Exogenous Substrates

Transporter (Gene)	Substrate	K_m (μM)	V_{max} [a,b]	C_{cap} (fmol/μg protein)
LAT1 (*SLC7A5*)	L-phenyl-alanine	26±6[c] [71]	22±4[a,c] [71]	3.00±0.62[c] [68], 2.19±0.21[d] [68], 0.43±0.09[e] [68]
	Dopamine-L-phenylalanine	227.9±52.0	0.00099±0.00009[a,c] [72]	
GLUT1 (*SLC2A1*)	D-glucose	11000±1400[c] [71]	1420±140[a,c] [71]	91.1±4.7[c] [68], 90±4.5[d] [68], 145±20[f] [68], 139±46[e] [68]
	Ketoprofen-D-glucose	—	0.0013±0.00018[a,c] [69]	
	Indomethacine-D-glucose	—	0.00993±0.00043[a,c] [69]	
MCT1 (*SLC16A1*)	L-lactic acid	1800±600[c] [71]	91±35[a,c] [71]	12.6±0.5[c] [68], 23.7±1.6[d] [68], 3.04±0.35[f] [68], 2.27±0.85[e] [68]
ENT1 (*SLC29A1*)	FLT	3400±200[g] [73]	169±4[b,g] [73]	0.985±0.363[d] [68], 0.568±0.134[e] [68]
ENT2 (*SLC29A2*)	FLT	2600±400[g] [73]	180±13[b,g] [73]	<0.180[e] [68]
CNT1 (*SLC28A1*)	AZT	549±98[g] [74]	26.4±6.6[b,g] [74]	<0.308[e] [68]
	Ddc	503±35[g] [74]	19.8±4.6[b,g] [74]	
	Uridine	37±7[g] [74]	20.8±1.1[b,g] [74]	
	FLT	130±10[g] [73]	52±1[b,g] [73]	
CNT2 (*SLC28A2*)	Adenosine	25±3[c] [71]	0.75±0.08[a,c] [71]	<0.14[e] [68]
	2,3-Dideoxyinosine	29.2±8.3[b,g] [75]	0.40±0.11[b,g] [75]	
CNT3 (*SLC28A3*)	FLT	110±10[g] [73]	37±1[b,g] [73]	<0.552[e] [68]
OATP1A2 (*SLCO1A2*)	Levofloxacin	136±48.0[g] [76]	0.55±0.065[b,g] [76]	—
	Methotrexate	457±118[g] [77]	0.29±0.08[b,g] [77]	
OAT3 (*SLC22A8*)	Indoxyl sulfate	298±43[a,c] [78]	3.75±0.16[a,c] [78]	1.97±0.11[d] [68], <0.348[e] [68]

AZT, 3′-azido-3′-deoxythymidine; C_{cap}, capillary concentration; ddc, 2′,3′-dideoxycytidine; FLT, 3′-deoxy-3′-[^{18}F]fluorothymidine; K_m, substrate affinity; V_{max}, maximal transport rate.

[a] nmol/minute/g.

[b] pmol/cell/minute.

[c] Rat.

[d] Mouse.

[e] Human.

[f] Marmoset.

[g] *In vitro*.

reductase inhibitors lovastatin and simvastatin are also believed to undergo MCT1-facilitated transport into the brain. In terms of substrates, the MCT1 transporter has a broad affinity for short-chain monocarboxylates (e.g., formate, oxamate, L-lactate, D-lactate, and pyruvate) and the ones that are substituted at the second or third positions with halide, hydroxyl, and carbonyl groups (e.g., *R*-, *S*-chloropropionate; D-,L-α-hydroxybutyrate; D-,L-β-hydroxybutyrate; and acetoacetate) [100].

16.4.1.4 Nucleoside Transporters Purine (adenosine, guanosine, and inosine) and pyrimidine (uridine, cytidine, and thymidine) nucleosides are the precursors of nucleotides, which in turn are the building blocks of DNA, RNA, and ATP. Nucleosides (e.g., adenosine) also serve as signaling molecules and as neuromodulators [101]. Nucleosides serve important functions in DNA and RNA synthesis as well as in signaling and neuromodulation within the brain [101]. The hydrophilic nature of these molecules is such that specialized transporters are required for nucleosides to cross the BBB. In humans, two families—namely, facilitative equilibrative (ENT, subtype: ENT1—*SLC29A1*, ENT2—*SLC29A2*, ENT3—*SLC29A3*, and ENT4—*SLC29A5*) and Na^+-dependent concentrative (CNT, subtype: CNT1—*SLC28A1*, CNT2—*SLC28A2*, and CNT3—*SLC28A3*) nucleoside transporters (NTs) are expressed in the luminal (blood to endothelium) and abluminal (endothelium to brain) side of the BBB, respectively [102, 103]. ENT1, ENT2, and ENT3 transports both purine and pyrimidine nucleosides whereas ENT4 specifically transports adenosine [103]. Among the CNT family, CNT3 transports both purine and pyrimidine nucleosides; CNT2 transports guanosine, adenosine, and uridine; and CNT1 transports adenosine and pyrimidine nucleosides [103].

The NTs have been attractive targets for therapeutic agents that are structural homologues (pseudosubstrate) of nucleoside substrates. For example, anticancer drugs gemcitabine (2′,2′-difluorodeoxycytidine), zalcitabine, and cytarabine as well as the antiviral agent, zidovudine, enter into the brain through ENT1 and CNT1 [104]. Recent studies indicate that A_1 receptor agonist, tecadenoson (structural analogue of adenosine), undergoes carrier-mediated transport via ENT1 in mice brain [105].

16.4.1.5 Organic Anion Transporting Polypeptides Organic anion-transporting polypeptides (OATPs) mediate the Na^+-independent bidirectional transport of endogenous substrates. Several studies have suggested possible exchange of intracellular bicarbonate, glutathione or glutathione conjugates for substrates, by OATPs. OATPs are grouped into six families (OATP1–OATP6), out of which OATP1 (subfamily: OATP1A, OATP1B, OATP1C, and OATP1D) is the most studied [106]. The members of OATP1A family include OATP1A2 (*SLCO1A2*), OATP1B1 (*SLCO1B1*), OATP1B3 (*SLCO1B3*), and OATP1C1 (*SLCO1C1*). In humans, OATP1A2 is expressed in the luminal side of the brain capillary endothelial cells (BBB) [107]. OATP1C1 is expressed in the basolateral side of the choroid plexus epithelial cells (BCSFB) [108].

The OATP1A2 transporter recognizes a broad range of substrates including bile salts, thyroid harmones, steroid conjugates, linear/cyclic peptides, and several xenobiotics. In contrast, the OATP1C1 transporter is more restrictive displaying high affinity for thyroxine [109]. The role of OATP1A2 in transporting useful therapeutic agents has been investigated. *In vitro* transport studies of antibacterial agent levofloxacin in *Xenopus laevis* oocyte cells expressing OATP1A2 showed high capacity [76]. Another study showed the high transport capacity of OATP1A2-expressed oocyte cells for chemotherapeutic agent methotrexate [77]. Similarly several studies using cell culture models have shown that deltrophin II [110] and D-penicillamine(2,5)encephalin [110] are transported by OATP1A2. The available evidence suggests that the OATP1A2 transporter is relatively tolerant in terms of substrate stereochemical requirements. Hence, a pseudosubstrate approach may be a promising strategy for transmembrane delivery of potential CNS drugs, via OATP1A2.

16.4.2 Adenosine Triphosphate-Binding Cassette Transport Systems in the BBB

There are 48 known adenosine triphosphate (ATP)-binding cassette (ABC) transporters (membrane proteins) in humans. They are expressed by seven gene subfamilies (*ABCA–ABCG*) [111]. Among these transporters, P-gp (*ABCB1*), MRP1-6 (*ABCC1-6*), and the BCRP (*ABCG2*) are the main efflux transporters. The efflux transporter proteins bind to ATP and use the energy derived from the latter's hydrolysis to translocate solutes across extra/intracellular membranes.

Within the BBB, P-gp is the best characterized drug efflux transporter interacting with a broad range of structurally unrelated substrates including anticancer agents (e.g., doxorubicin, vinblastine, paclitaxel, and methotrexate), natural products (e.g., curcuminoids), linear/cyclic peptides (e.g., pepstatin A, valinomycin), and most recently large macromolecules such as beta amyloid [112, 113]. Much of the understanding of P-gp's role in drug distribution to the brain has been elucidated through mouse models in which *ABCB1* has been knocked out [10]. The presence of multiple MRP drug efflux transporters within the BBB and BCSFB [114–116] and the broad overlap in substrates of the various MRP transporters have made it difficult to determine the role of MRPs in the BBB permeability of drugs. The MRP drug efflux transporters interact with a variety of organic anions including anticancer agents (e.g., doxorubicine, vinblastine, vincristine, and methotrexate) and conjugated metabolites of drugs [117]. The BCRP is the most recent ABC transporter to be identified in the BBB [118]. With a single nucleotide-binding domain and only six transmembrane spanning regions, BCRP is termed as a half transporter, requiring formation of the homodimer to function as an efflux pump [119]. The substrates of BCRP have substantial overlap with P-gp and MRP1. As a result many drugs are excluded from entry into the brain by a combination of both P-gp and BCRP activity [10].

16.4.2.1 Approaches for Altering Drug Efflux Activity in the BBB For drugs with significant drug efflux transporter liabilities, improvements in drug delivery to the brain can be achieved through minimizing the interactions of the drug with the efflux transporter. This can be done by chemically modifying the drug molecule to reduce the affinity toward the efflux transporter or through pharmacological inhibition of the drug efflux transporter activity. An example of the former is the studies by George et al. [120] in which the BBB permeation of taxane was enhanced by chemical modifications at C10 position with a carboxylic acid group (TX-67). The chemical modifications of TX-67 resulted in reduced P-gp efflux in brain endothelial cells and an increase in permeability compared to the parent drug (paclitaxel) [120]. Similar results were observed in a series of paclitaxel analogues with added carboxylic acid or amide groups at the C7 position [121]. It should be noted that structural changes to the drug that limit efflux transporter interactions may alter biological responses to the drug. In addition, such changes in the structure may unmask other transporter interactions. In the paclitaxel example discussed earlier, the increases in BBB permeability may be a net result of increased influx through carboxylic acid transporters and reduced efflux by P-gp [121].

Agents not amenable to chemical modification can evade drug efflux transporters through formulation into liposomal or polymeric drug carriers or through coadministration of agents that inhibit drug efflux transporter activity. Incorporation of drugs into liposomes or polymeric micelles protects the drug from interacting with drug efflux transporters [122]. This approach has been used to increase the delivery of a wide variety of therapeutic agents to the brain [123–125]. Coadministration of inhibitors, especially of P-gp drug efflux transporters, has also been extensively studied [126, 127]. While pharmacological inhibition of drug efflux has been used to enhance intestinal absorption and tissue distribution of drugs, complete inhibition of efflux transporter activity in the BBB can be difficult [128–131]. Case in point is the effects of pharmacological inhibition of drug efflux transporters and brain penetration of protein kinase inhibitors (Fig. 16.3). Many of the tyrosine kinase inhibitors are both P-gp and BCRP substrates thus limiting their BBB permeability and brain tumor therapy applications. While studies in both P-gp and BCRP knockout mice show substantial increases in the brain distribution of these agents compared with wild-type controls, the effects of pharmacological inhibition of efflux transporters are dependent on the kinase inhibitor examined (Fig. 16.3). Such findings suggest that the effectiveness of pharmacological modulation of drug efflux transporters at the BBB will depend extensively on the physicochemical properties of the drug.

16.4.3 Vesicular Transport in the BBB

There are two general types of vesicular transport processes: fluid-phase endocytosis and adsorptive endocytosis. While both processes require energy and can be inhibited by metabolic inhibitors, only adsorptive endocytosis involves an initial binding or interaction of the molecule with the plasma membrane of the cell. As such, vesicular transport due to adsorptive endocytosis is a saturable, ligand-selective phenomenon. In general, vesicular transport in the brain capillary endothelial cells is reduced compared to other capillary beds. However, several large macromolecules of importance for normal brain function are transported from the blood into the brain through receptor-mediated endocytosis. Thus, the design of therapeutic agents and biomacromolecules to utilize these specific receptor-mediated transport processes in the brain capillary endothelial cells represents another approach for enhancing

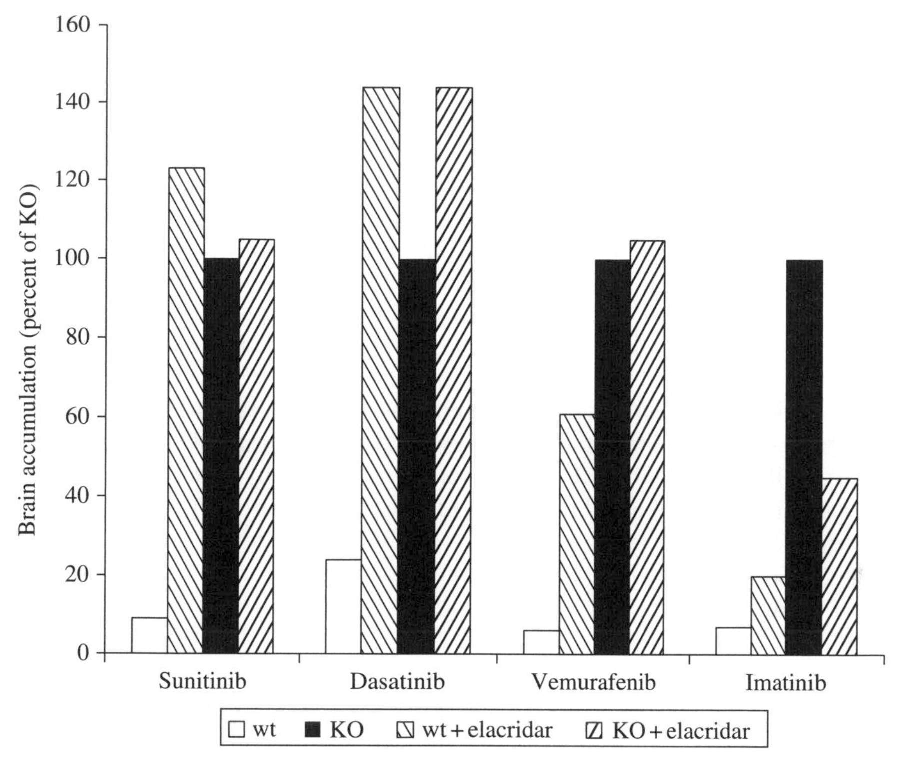

FIGURE 16.3 Effects of drug efflux transporters on the accumulation of various protein kinase inhibitors in the brain. Drug accumulation was examined in wild-type (wt) mice and triple knockout *Abcb1a/1b –/–; Abcg2 –/–* (KO) mice in the presence and absence of the drug efflux transport inhibitor, elacridar. All drugs were administered orally, and the dose of elacridar was (100 mg/kg). Data for this figure was based on previously published results [128–131].

transcellular permeability across the BBB. The method for delivery of drugs conjugated to specific ligands that undergo receptor-mediated transport across brain endothelial cells has been termed the "Trojan Horse" approach. The most well-characterized receptor-mediated vesicular transport processes used as "Trojan Horses" in the BBB are discussed in the following text.

16.4.3.1 Transferrin Receptor–Mediated Vesicular Transport Serum transferrin is a monomeric glycoprotein with a molecular weight of 80 kDa that is crucial for the transport of iron throughout the body [132]. Iron enters the cell complexed with transferrin through an endocytic process that is initiated by the binding of transferrin to its receptor on the plasma membrane [133]. The brain capillary endothelial cells have a high density of transferrin receptors on their surface compared to other types of cells [134]. The binding of transferrin to its receptor on the brain capillary endothelial cells triggers the internalization of the transferrin–iron complex. Inside the brain endothelial cell, the iron is removed from the transferrin in the endosome, and through the vesicular cell–sorting process, iron is released into the brain EC fluid, and transferrin and its receptor are recycled back to the luminal (blood) plasma membrane. The prevalence of transferrin receptors in the BBB and the resulting vesicular transport that occurs following binding to the receptor have stimulated an interest in the potential use of this transport system for targeted drug delivery to the brain. However, the biggest limitation in utilizing this transport pathway is that the transferrin itself undergoes cellular processing and is ultimately recycled back to the luminal surface of the brain capillary endothelial cell. Because of this, the use of the transferrin molecule as a drug carrier is not likely to enhance BBB permeability. To get around this issue, researchers have identified a murine monoclonal antibody to the transferring receptor, OX26, which appears to be suitable for use as a drug carrier for this transport system [135, 136]. There are three important characteristics of OX26 that make it ideal as a drug carrier. First, unlike other antibodies to the transferrin receptor, this antibody binds to the receptor and triggers endocytosis. Second, the cellular processing of the internalized antibody is such that a significant portion of the internalized OX26 actually undergoes exocytosis (release) at the abluminal (brain side) plasma membrane [135, 137]. The third important characteristic is that the OX26 antibody binds to an EC epitope on the transferrin receptor that is distinct from the transferrin ligand-binding site; thus the OX26 monoclonal antibody does not interfere with transferrin binding to its receptor on the brain endothelial cells [129]. The OX26 antibody has proven to be an effective brain delivery vector, as it has been conjugated to a variety of drugs including methotrexate [137], nerve growth factor [138], and brain-derived neurotrophic factor [139, 140].

16.4.3.2 Insulin Receptor–Mediated Vesicular Transport Insulin is a pancreatic peptide hormone with important functions in glucose regulation. The presence of insulin receptors in the CNS [141], coupled with the neurotropic and neuromodulatory actions of insulin in neuronal cells, suggests that insulin may have important functions within the brain as well [142, 143]. The finding that few neurons even express insulin mRNA136 and that brain levels of insulin are directly correlated with the concentration of the peptide in the blood suggests that insulin has a non-CNS origin. Studies demonstrating the presence of high-affinity insulin receptors on the luminal plasma membrane of brain microvessel endothelial cells and their involvement in the vesicular transport of insulin indicate that the peptide penetrates the BBB through a receptor-mediated transport process [144, 145]. Several studies support the potential use of insulin as a transport vector for the delivery of therapeutic agents and macromolecules to the brain. Studies by Kabanov et al. [146] examined the use of polymer micelles for the delivery of the antipsychotic agent, haloperidol, to the brain in mice. Conjugation of insulin to the polymer micelles improved the CNS responses to haloperidol while decreasing the deposition of the micelles in peripheral organs such as the lung and liver. These studies, together with more recent investigations in cultured brain microvessel endothelial cells demonstrating that the transport of the insulin-conjugated micelles is a saturable process inhibited by excess free insulin [147], suggest that the insulin-conjugated micelles undergo a receptor-mediated vesicular transport process in the BBB.

Insulin has also been used as a BBB transport vector for proteins. Studies by Fukuta and coworkers [148] examined HRP activity in the brain, following i.v. injections of either HRP or HRP conjugated to insulin. Those mice receiving i.v. injections of the insulin-conjugated HRP had significantly higher peroxidase activity in the brain compared with either vehicle- or HRP-treated mice [148]. Together these studies demonstrate the feasibility of the insulin receptor as a transport system into the brain. It should be noted that despite the favorable preclinical data, there is a lack of evidence for demonstrating clinical efficacy using this "Trojan Horse" technology for brain delivery of drugs.

16.4.3.3 LRP1-Mediated Vesicular Transport The low-density lipoprotein receptor–related protein 1 (LRP1/CD91/α2-macroglobulin receptor) is a member of the LDL receptor family, and it is highly expressed at both luminal and abluminal plasma membrane of brain microvessel endothelial cells [149]. Upon binding to their ligands, the receptors carry the ligand across the BBB through a receptor-mediated transcytosis process. There are two forms of LRP1 receptor: one is membrane-bound form and the other is soluble form. This section focuses mainly on the membrane-bound form of LRP1. The natural ligand of LRP1 is rather diverse, including beta amyloid, ApoE, lactoferrin, RAP, α2-macrogloblin, and many others [150]. It has been reported that LRP1 mediates cellular internalization of ligands and their transport across the BBB.

Angiopeps, a family of Kunitz domain–derived peptides, have demonstrated high transcytosis capacity in brain capillaries and substantial parenchymal brain accumulation due to binding and transport via LRP1 [151]. The angiopep peptides have been advanced as potential BBB-targeting vectors for drug delivery to the brain, especially for the delivery of chemotherapeutics for treating brain tumor. A potential advantage of angiopep-targeted brain delivery of therapeutic molecules is the ability of the targeting vector to localize in tumors with an increased LRP1 expression [152]. Particularly attractive, in addition to the high transcytosis of the peptide, is the capability to attach multiple drug molecules to the peptide. With the ability to attach up to three small molecules onto the angiopep peptide backbone, the relatively low efficiency of the transcytosis pathway can be countered.

Recently, the use of angiopep-targeting vectors to deliver nanoparticles to the brain has been reported. Using angiopep-coated nanoparticles, a significant increase in survival rate was observed in a mouse glioma model compared with both control and temozolomide treatment groups [153]. In these studies the angiopep peptide was used to deliver DNA-loaded nanoparticles to the tumor cells, taking advantage of the relatively potent activity of the gene-based therapy. Additional studies have used angiopep peptide to deliver paclitaxel-loaded nanoparticles (ANG-PEG-NP) [19]. These studies showed an enhanced tumor response to paclitaxel-loaded ANG-PEG-NP compared to paclitaxel-loaded nanoparticles without the ANG vector in a mouse glioblastoma model [154].

The most clinically advanced angiopep delivery molecule is GRN1005—a paclitaxel–angiopep-2 peptide–drug conjugate—that has successfully completed phase I clinical trial for the treatment of recurrent glioma [155]. Additional angiopep conjugations with doxorubicin have shown good response in preclinical brain tumor models [156]. However, phase II clinical trial of GRN1005 in non-small-cell lung cancer patients with brain metastases was discontinued, and it is likely not to advance past phase III due to absence of significant effect. This discrepancy may be due in part to the substantial differences in size of rodent brains used in the pre-clinical trials. Leakiness within a rodent brain tumor may result in greater brain exposure than that observed in human brain tumor patients. Currently, GRN1005 is undergoing a phase II study conducted in patients with high-grade glioma.

Besides angiopep, apoE is another ligand used for LRP1-mediated transcytosis across the BBB. Instead of using apoE as a targeting vector directly, polysorbate 80 (Tween-80) is often used. It has been reported that polysorbate 80–coated NPs are capable of delivering BBB-impermeable molecular imaging contrast agents into the brain [157]. However, the same NPs do not cross the BBB in apoE knockout mice, demonstrating polysorbate 80 can absorb apoE on NP surface and cross the BBB via receptor-mediated transcytosis [157].

16.4.3.4 Diphtheria Toxin Receptor–Mediated Vesicular Transport Another novel ligand for transport across the BBB is the cross-reacting material 197 (CRM197), a nontoxic mutant of diphtheria toxin. Like diphtheria toxin, it binds to diphtheria receptors that are expressed in the BBB and are unregulated in ischemic stroke and glioma [158, 159]. CRM197-grafted polybutylcyanoacrylate (PBCA) nanoparticles were able to traverse the monolayer of human brain microvascular endothelial cells (HBMECs) *in vitro* [160]. In addition, an increase in the grafting quantity of CRM197 enhanced the permeability and the uptake quantity of NPs by HBMECs [160]. Another study suggested that CRM197 can increase BBB permeability via upregulation of caveolin-1 protein, increased pinocytotic vesicles, and redistribution of tight junction–associated proteins in brain microvessel [161].

16.4.3.5 Considerations for Vesicular Transport in BBB Drug delivery utilizing receptor-mediated transcytosis in the BBB certainly shows great potential to enhance drug accumulation in the brain, especially for macromolecule drug delivery to the brain. However, the most limiting drawback associated with this method is its low efficiency. For example, transferrin-conjugated micelles have been reported with localization of 0.03% ID/g inside the brain [162]. By fusing with heavy chain of insulin receptor antibody, 0.02% of ID/g of human erythropoietin is taken up by the brain [163]. This represents a very small fraction of the drug reaching its intended target within the brain. While the LRP1 receptor has shown substantially better delivery efficiencies (approximately 10-fold better using angiopep-modified poly-amidoamine dendrimers), there was still only 0.25% ID/g delivered to the brain [164]. The best efficiency to date is with doxorubicin and etoposide-conjugated angiopep, which reported a delivery of 0.5% and 1% ID/g in mouse brain tissue and brain tumor tissue, respectively [156]. Macromolecules can be passively retained in tumor tissue by a process called enhanced permeability and retention (EPR) effect. These data collectively suggest that 1% of the ID/g is the upper limit for vesicular transport in intact and/or compromised BBB. With this relatively low efficiency of delivery, vesicular transport would be limited to high potency therapeutic agents.

One of the limitations in evaluating the various receptor-mediated vesicular transport routes for the delivery of drugs to the brain is a lack of standardization in terms of both the drug being delivered and the preclinical model used to examine BBB permeability. However, studies by van Rooy and colleagues provide an initial assessment of the various receptor-mediated delivery approaches by comparing the brain delivery efficiency of five targeting ligands including transferrin, RI7217 (an antibody against transferrin receptor), COG133 (an apoE-mimetic peptide-targeting LRP1), angiopep-2, and CRM197. Each targeting ligand was attached to a liposome drug carrier using similar chemistry and stoichiometry and evaluated for BBB permeability using the same *in vitro* and *in vivo models* [165]. *In vitro* experiments showed that only CRM197-modified liposomes were able to bind to murine endothelial cells (bEnd.3), while both CRM197- and RI7217-modified liposomes associated with human endothelial cells (hCMEC/D3). Interestingly, only the RI7217 was able to significantly enhance brain parenchyma uptake of liposome by 4.3-fold 6-hour post injection despite the fact that only 0.18% of injected dose accumulated in the brain after 12 hours [165]. The possible explanation for failure of CRM197-modified liposome to cross the BBB *in vivo* is low expression level of target receptor in healthy mice.

A recent review has summarized the ligand-targeted nanomedicines currently undergoing clinical trials [166]. None of the nanoparticle formulations have progressed into phase III clinical trials. This is due to the fact that comparison of PK and tissue distribution of ligand-targeted versus nontargeted nanomedicine is often inconclusive. More importantly, contribution of targeting ligand to efficacy is yet to be proven. Of the 13 ligand-targeted nanomedicine discussed, only one formulation targets brain tumor using glutathione (GSH) as the targeting ligand, with the aim to cross the BBB via glutathione transporters. This nanoparticle formulation, 2B3-101, is currently undergoing phase I/IIa trials to determine the safety and pharmacokinetics in patients with solid tumors and brain metastases or recurrent malignant glioma [166].

16.5 CONCLUSIONS

The clinical need for drugs that can effectively treat brain-related disorders is an issue that will continue to impact on health care systems throughout the world. Cellular interfaces such as the BBB and BCSFB can present substantial obstacles for the delivery of drugs to the brain. Several approaches for enhancing drug delivery to the brain have been or are currently being developed. While there is likely to be no single approach that will work, there are several pathways that may be utilized to increase drug penetration to sites within the brain.

REFERENCES

1. Alavijeh, M. S.; Chishty, M.; Qaiser, M. Z.; Palmer, A. M. *NeuroRx* 2005, **2**, 554–571.
2. Reichel, A. *Chemistry and Biodiversity* 2009, **6**, 2030–2049.
3. Loscher, W.; Potschka, H. *NeuroRx* 2005, **2**, 86–98.

4. Loscher, W.; Potschka, H. *Progress in Neurobiology* 2005, **76**, 22–76.
5. Pardridge, W. M. *Journal of Drug Targeting* 2010, **18**, 157–167.
6. Zhang, Y.; Miller, D. W. Pathways for drug delivery to the central nervous system. In *Drug Delivery Principles and Applications*; Wang, B.; Siahaan, T.; Soltero, R. A., Eds. John Wiley & Sons, Inc.: Hoboken, NJ, 2005; Vol. **1**, p 29–56.
7. Stewart, P. A.; Wiley, M. J. *Developmental Biology* 1981, **84**, 183–192.
8. Girardin, F. *Dialogues in Clinical Neuroscience* 2006, **8**, 311–321.
9. Hermann, D. M.; Bassetti, C. L. *Trends in Pharmacological Sciences* 2007, **28**, 128–134.
10. On, N. H.; Miller, D. W. *Current Pharmaceutical Design* 2014, **20**, 1499–1509.
11. Balayssac, D.; Authier, N.; Cayre, A.; Coudore, F. *Toxicology Letters* 2005, **156**, 319–329.
12. Beaulieu, E.; Demeule, M.; Ghitescu, L.; Beliveau, R. *The Biochemical Journal* 1997, **326** (Pt 2), 539–544.
13. Feng, B.; Mills, J. B.; Davidson, R. E.; Mireles, R. J.; Janiszewski, J. S.; Troutman, M. D.; de Morais, S. M. *Drug Metabolism and Disposition* 2008, **36**, 268–275.
14. Lee, C. A.; Cook, J. A.; Reyner, E. L.; Smith, D. A. *Expert Opinion on Drug Metabolism and Toxicology* 2010, **6**, 603–619.
15. Schinkel, A. H.; Wagenaar, E.; Mol, C. A.; van Deemter, L. *The Journal of Clinical Investigation* 1996, **97**, 2517–2524.
16. Davson, H.; Segal, M. B., Eds. The blood cerebral spinal fluid barrier. In *Physiology of the CSF and Blood-Brain Barrier*. CRC Press: Boca Raton, FL, 1996; 1st ed., p 30–48.
17. Segal, M. The blood-CSF barrier and chorioid plexus. In *Introduction to the Blood-Brain Barrier Methodology, Biology and Pathology*; Pardridge, W. M., Ed. Cambridge University Press: Cambridge, 1998, p 251–258.
18. Meller, K. *Cell and Tissue Research* 1985, **242**, 289–300.
19. Zhang, H.; Song, Y. N.; Liu, W. G.; Guo, X. L.; Yu, L. G. *Journal of Clinical Neuroscience* 2010, **17**, 679–684.
20. On, N. H.; Chen, F.; Hinton, M.; Miller, D. W. *Pharmaceutical Research* 2011, **28**, 2505–2515.
21. Johanson, C. E. Potential for pharmacological manipulation of the blood-cerebral spinal fluid barrier. In *Implications of the Blood-Brain Barrier and Its Manipulation*; Neuwelt, E. A., Ed. Plenum Publishing: New York, 1989; Vol. **1**, p 223–260.
22. Pardridge, W. M. *Molecular Interventions* 2003, **3**, 90–105, 151.
23. Pardridge, W. M. *Drug Discovery Today* 2007, **12**, 54–61.
24. Huynh, G. H.; Deen, D. F.; Szoka, F. C., Jr. *Journal of Controlled Release* 2006, **110**, 236–259.
25. Nutt, J. G.; Burchiel, K. J.; Comella, C. L.; Jankovic, J.; Lang, A. E.; Laws, E. R., Jr.; Lozano, A. M.; Penn, R. D.; Simpson, R. K., Jr.; Stacy, M.; Wooten, G. F. *Neurology* 2003, **60**, 69–73.
26. Gabathuler, R. *Neurobiology of Disease* 2010, **37**, 48–57.
27. Alam, M. I.; Beg, S.; Samad, A.; Baboota, S.; Kohli, K.; Ali, J.; Ahuja, A.; Akbar, M. *European Journal of Pharmaceutical Sciences* 2010, **40**, 385–403.
28. Stockwell, J.; Abdi, N.; Lu, X.; Maheshwari, O.; Taghibiglou, C. *Chemical Biology and Drug Design* 2014, **83**, 507–520.

29. Hanson, L. R.; Frey, W. H., 2nd. *BMC Neuroscience* 2008, **9** (Suppl 3), S5.
30. Mittal, D.; Ali, A.; Md, S.; Baboota, S.; Sahni, J. K.; Ali, J. *Drug Delivery* 2014, **21**, 75–86.
31. Scott, J. W.; Wellis, D. P.; Riggott, M. J.; Buonviso, N. *Microscopy Research and Technique* 1993, **24**, 142–156.
32. Thorne, R. G.; Pronk, G. J.; Padmanabhan, V.; Frey, W. H., 2nd. *Neuroscience* 2004, **127**, 481–496.
33. Dhuria, S. V.; Hanson, L. R.; Frey, W. H., 2nd. *Journal of Pharmaceutical Sciences* 2010, **99**, 1654–1673.
34. Rapoport, S. I.; Fredericks, W. R.; Ohno, K.; Pettigrew, K. D. *American Journal of Physiology* 1980, **238**, R421–431.
35. Bellavance, M. A.; Blanchette, M.; Fortin, D. *The AAPS Journal* 2008, **10**, 166–177.
36. On, N. H.; Savant, S.; Toews, M.; Miller, D. W. *Journal of Cerebral Blood Flow and Metabolism* 2013, **33**, 1944–1954.
37. Farrell, C. L.; Shivers, R. R. *Acta Neuropathologica* 1984, **63**, 179–189.
38. Rapoport, S. I. *Expert Opinion on Investigational Drugs* 2001, **10**, 1809–1818.
39. Kroll, R. A.; Pagel, M. A.; Muldoon, L. L.; Roman-Goldstein, S.; Fiamengo, S. A.; Neuwelt, E. A. *Neurosurgery* 1998, **43**, 879–886; discussion 886–879.
40. Kemper, E. M.; Boogerd, W.; Thuis, I.; Beijnen, J. H.; van Tellingen, O. *Cancer Treatment Reviews* 2004, **30**, 415–423.
41. Borlongan, C. V.; Emerich, D. F. *Brain Research Bulletin* 2003, **60**, 297–306.
42. Straub, J. A.; Akiyama, A.; Parmar, P. *Pharmaceutical Research* 1994, **11**, 1673–1676.
43. Fike, J. R.; Gobbel, G. T.; Mesiwala, A. H.; Shin, H. J.; Nakagawa, M.; Lamborn, K. R.; Seilhan, T. M.; Elliott, P. J. *Journal of Neuro-Oncology* 1998, **37**, 199–215.
44. Bartus, R. T.; Snodgrass, P.; Marsh, J.; Agostino, M.; Perkins, A.; Emerich, D. F. *The Journal of Pharmacology and Experimental Therapeutics* 2000, **293**, 903–911.
45. Borlongan, C. V.; Emerich, D. F.; Hoffer, B. J.; Bartus, R. T. *Brain Research* 2002, **956**, 211–220.
46. Emerich, D. F.; Snodgrass, P.; Pink, M.; Bloom, F.; Bartus, R. T. *Brain Research* 1998, **801**, 259–266.
47. Emerich, D. F.; Dean, R. L.; Snodgrass, P.; Lafreniere, D.; Agostino, M.; Wiens, T.; Xiong, H.; Hasler, B.; Marsh, J.; Pink, M.; Kim, B. S.; Perdomo, B.; Bartus, R. T. *The Journal of Pharmacology and Experimental Therapeutics* 2001, **296**, 632–641.
48. Black, K. L.; Cloughesy, T.; Huang, S. C.; Gobin, Y. P.; Zhou, Y.; Grous, J.; Nelson, G.; Farahani, K.; Hoh, C. K.; Phelps, M. *Journal of Neurosurgery* 1997, **86**, 603–609.
49. Prados, M. D.; Schold, S. C., Jr.; Fine, H. A.; Jaeckle, K.; Hochberg, F.; Mechtler, L.; Fetell, M. R.; Phuphanich, S.; Feun, L.; Janus, T. J.; Ford, K.; Graney, W. *Neuro-Oncology* 2003, **5**, 96–103.
50. Black, K. L.; Ningaraj, N. S. *Cancer Control* 2004, **11**, 165–173.
51. Erdlenbruch, B.; Kugler, W.; Schinkhof, C.; Neurath, H.; Eibl, H.; Lakomek, M. *Journal of Drug Targeting* 2005, **13**, 143–150.
52. Erdlenbruch, B.; Schinkhof, C.; Kugler, W.; Heinemann, D. E.; Herms, J.; Eibl, H.; Lakomek, M. *British Journal of Pharmacology* 2003, **139**, 685–694.
53. Erdlenbruch, B.; Alipour, M.; Fricker, G.; Miller, D. S.; Kugler, W.; Eibl, H.; Lakomek, M. *British Journal of Pharmacology* 2003, **140**, 1201–1210.

54. Erdlenbruch, B.; Jendrossek, V.; Eibl, H.; Lakomek, M. *Experimental Brain Research* 2000, **135**, 417–422.
55. Rivera, R.; Chun, J. *Reviews of Physiology Biochemistry and Pharmacology* 2008, **160**, 25–46.
56. Choi, J. W.; Herr, D. R.; Noguchi, K.; Yung, Y. C.; Lee, C. W.; Mutoh, T.; Lin, M. E.; Teo, S. T.; Park, K. E.; Mosley, A. N.; Chun, J. *Annual Review of Pharmacology and Toxicology* 2010, **50**, 157–186.
57. Zhao, Y.; Natarajan, V. *Cellular Signalling* 2009, **21**, 367–377.
58. Nitz, T.; Eisenblatter, T.; Psathaki, K.; Galla, H. J. *Brain Research* 2003, **981**, 30–40.
59. Sarker, M. H.; Hu, D. E.; Fraser, P. A. *Microcirculation (New York, N.Y.: 1994)* 2010, **17**, 39–46.
60. Schulze, C.; Smales, C.; Rubin, L. L.; Staddon, J. M. *Journal of Neurochemistry* 1997, **68**, 991–1000.
61. Tigyi, G.; Hong, L.; Yakubu, M.; Parfenova, H.; Shibata, M.; Leffler, C. W. *American Journal of Physiology* 1995, **268**, H2048–2055.
62. On, N. H.; Kiptoo, P.; Siahaan, T. J.; Miller, D. W. *Molecular Pharmaceutics* 2014, **11**, 974–981.
63. Zheng, K.; Trivedi, M.; Siahaan, T. J. *Current Pharmaceutical Design* 2006, **12**, 2813–2824.
64. Lutz, K. L.; Sianhaan, T. J. *Drug Delivery* 1997, **10**, 187–193.
65. Makagiansar, I. T.; Avery, M.; Hu, Y.; Audus, K. L.; Siahaan, T. J. *Pharmaceutical Research* 2001, **18**, 446–453.
66. Kiptoo, P.; Sinaga, E.; Calcagno, A. M.; Zhao, H.; Kobayashi, N.; Tambunan, U. S.; Siahaan, T. J. *Molecular Pharmaceutics* 2011, **8**, 239–249.
67. Laksitorini, M. D.; Kiptoo, P. K.; On, N. H.; Thliveris, J. A.; Miller, D. W.; Siahaan, T. J. *Journal of Pharmaceutical Sciences* 2015, **104**, 1065–1075.
68. Uchida, Y.; Ohtsuki, S.; Katsukura, Y.; Ikeda, C.; Suzuki, T.; Kamiie, J.; Terasaki, T. *Journal of Neurochemistry* 2011, **117**, 333–345.
69. Gynther, M.; Ropponen, J.; Laine, K.; Leppanen, J.; Haapakoski, P.; Peura, L.; Jarvinen, T.; Rautio, J. *Journal of Medicinal Chemistry* 2009, **52**, 3348–3353.
70. del Amo, E. M.; Urtti, A.; Yliperttula, M. *European Journal of Pharmaceutical Sciences* 2008, **35**, 161–174.
71. Pardridge, W. M. *Journal of Cerebral Blood Flow and Metabolism* 2012, **32**, 1959–1972.
72. Peura, L.; Malmioja, K.; Huttunen, K.; Leppänen, J.; Hämäläinen, M.; Forsberg, M.; Rautio, J.; Laine, K. *Pharmaceutical Research* 2013, **30**, 2523–2537.
73. Paproski, R. J.; Ng, A. M. L.; Yao, S. Y. M.; Graham, K.; Young, J. D.; Cass, C. E. *Molecular Pharmacology* 2008, **74**, 1372–1380.
74. Yao, S. Y. M.; Cass, C. E.; Young, J. D. *Molecular Pharmacology* 1996, **50**, 388–393.
75. Li, J. Y.; Boado, R. J.; Pardridge, W. M. *Journal of Pharmacology and Experimental Therapeutics* 2001, **299**, 735–740.
76. Maeda, T.; Takahashi, K.; Ohtsu, N.; Oguma, T.; Ohnishi, T.; Atsumi, R.; Tamai, I. *Molecular Pharmaceutics* 2006, **4**, 85–94.
77. Badagnani, I.; Castro, R. A.; Taylor, T. R.; Brett, C. M.; Huang, C. C.; Stryke, D.; Kawamoto, M.; Johns, S. J.; Ferrin, T. E.; Carlson, E. J.; Burchard, E. G.; Giacomini, K. M. *Journal of Pharmacology and Experimental Therapeutics* 2006, **318**, 521–529.

78. Ohtsuki, S.; Asaba, H.; Takanaga, H.; Deguchi, T.; Hosoya, K.-I.; Otagiri, M.; Terasaki, T. *Journal of Neurochemistry* 2002, **83**, 57–66.
79. Smith, Q. R. *The Journal of Nutrition* 2000, **130**, 1016.
80. Begley, D. J. *Pharmacology and Therapeutics* 2004, **104**, 29–45.
81. Cornford, E. M.; Young, D.; Paxton, J. W.; Finlay, G. J.; Wilson, W. R.; Pardridge, W. M. *Cancer Research* 1992, **52**, 138–143.
82. Su, T.-Z.; Lunney, E.; Campbell, G.; Oxender, D. L. *Journal of Neurochemistry* 1995, **64**, 2125–2131.
83. Takada, Y.; Greig, N. H.; Vistica, D. T.; Rapoport, S. I.; Smith, Q. R. *Cancer Chemotheraphy and Pharmacology* 1991, **29**, 89–94.
84. Geier, E. G.; Schlessinger, A.; Fan, H.; Gable, J. E.; Irwin, J. J.; Sali, A.; Giacomini, K. M. *Proceedings of the National Academy of Sciences of the United States of America* 2013, **110**, 5480–5485.
85. Lizarraga, K. J.; Allen-Auerbach, M.; Czernin, J.; DeSalles, A. A. F.; Yong, W. H.; Phelps, M. E.; Chen, W. *Journal of Nuclear Medicine* 2014, **55**, 30–36.
86. Galldiks, N.; Ullrich, R.; Schroeter, M.; Fink, G.; Kracht, L. *European Journal of Nuclear Medicine and Molecular Imaging* 2010, **37**, 84–92.
87. Rapp, M.; Heinzel, A.; Galldiks, N.; Stoffels, G.; Felsberg, J.; Ewelt, C.; Sabel, M.; Steiger, H. J.; Reifenberger, G.; Beez, T.; Coenen, H. H.; Floeth, F. W.; Langen, K.-J. *Journal of Nuclear Medicine* 2013, **54**, 229–235.
88. Samnick, S.; Bader, J. B.; Hellwig, D.; Moringlane, J. R.; Alexander, C.; Romeike, B. F. M.; Feiden, W.; Kirsch, C.-M. *Journal of Clinical Oncology* 2002, **20**, 396–404.
89. Ohno, M.; Narita, Y. *Japanese Journal of Clinical Oncology* 2013, **43**, 448.
90. Watanabe, M.; Tanaka, R.; Takeda, N. *Neuroradiology* 1992, **34**, 463–469.
91. Killian, D. M.; Hermeling, S.; Chikhale, P. J. *Drug Delivery* 2007, **14**, 25–31.
92. Walker, I.; Nicholls, D.; Irwin, W. J.; Freeman, S. *International Journal of Pharmaceutics* 1994, **104**, 157–167.
93. Wood, I. S.; Trayhurn, P. *The British Journal of Nutrition* 2003, **89**, 3–9.
94. Pardridge, W. M. *Journal of Neurochemistry* 1998, **70**, 1781–1792.
95. Polt, R.; Porreca, F.; Szabo, L. Z.; Bilsky, E. J.; Davis, P.; Abbruscato, T. J.; Davis, T. P.; Horvath, R.; Yamamura, H. I.; Hruby, V. J. *Proceedings of the National Academy of Sciences of the United States of America* 1994, **91**, 7114–7118.
96. Fichna, J.; Mazur, M.; Grzywacz, D.; Kamysz, W.; Perlikowska, R.; Piekielna, J.; Sobczak, M.; Salaga, M.; Toth, G.; Janecka, A.; Chen, C.; Olczak, J. *Bioorganic and Medicinal Chemistry Letters* 2013, **23**, 6673–6676.
97. Negri, L.; Lattanzi, R.; Tabacco, F.; Scolaro, B.; Rocchi, R. *British Journal of Pharmacology* 1998, **124**, 1516–1522.
98. Fan, W.; Wu, Y.; Li, X. K.; Yao, N.; Li, X.; Yu, Y. G.; Hai, L. *European Journal of Medicinal Chemistry* 2011, **46**, 3651–3661.
99. Halestrap, A. P. *Molecular Aspects of Medicine* 2013, **34**, 337–349.
100. Halestrap, A. P. *IUBMB Life* 2012, **64**, 1–9.
101. Boison, D. *Current Opinion in Pharmacology* 2008, **8**, 2–7.
102. Abbott, N. J.; Patabendige, A. A. K.; Dolman, D. E. M.; Yusof, S. R.; Begley, D. J. *Neurobiology of Disease* 2010, **37**, 13–25.

103. Parkinson, F. E.; Damaraju, V. L.; Graham, K.; Yao, S. Y. M.; Baldwin, S. A.; Cass, C. E.; Young, J. D. *Current Topics in Medicinal Chemistry* 2011, **11**, 948–972.

104. Choi, M.-K. *Archives of Pharmacal Research* 2012, **35**, 921–927.

105. Lepist, E. I.; Damaraju, V. L.; Zhang, J.; Gati, W. P.; Yao, S. Y. M.; Smith, K. M.; Karpinski, E.; Young, J. D.; Leung, K. H.; Cass, C. E. *Drug Metabolism and Disposition* 2013, **41**, 916–922.

106. Urquhart, B.; Kim, R. *European Journal of Clinical Pharmacology* 2009, **65**, 1063–1070.

107. Bronger, H.; König, J.; Kopplow, K.; Steiner, H.-H.; Ahmadi, R.; Herold-Mende, C.; Keppler, D.; Nies, A. T. *Cancer Research* 2005, **65**, 11419–11428.

108. Roth, M.; Obaidat, A.; Hagenbuch, B. *British Journal of Pharmacology* 2012, **165**, 1260–1287.

109. Westholm, D. E.; Stenehjem, D. D.; Rumbley, J. N.; Drewes, L. R.; Anderson, G. W. *Endocrinology* 2009, **150**, 1025–1032.

110. Gao, B.; Hagenbuch, B.; Kullak-Ublick, G. A.; Benke, D.; Aguzzi, A.; Meier, P. J. *Journal of Pharmacology and Experimental Therapeutics* 2000, **294**, 73–79.

111. Dean, M.; Rzhetsky, A.; Allikmets, R. *Genome Research* 2001, **11**, 1156–1166.

112. Eckford, P. D. W.; Sharom, F. J. *Chemical Reviews* 2009, **109**, 2989–3011.

113. Sharom, F. J. *Pharmacogenomics* 2008, **9**, 105–127.

114. Deeken, J. F.; Löscher, W. *Clinical Cancer Research* 2007, **13**, 1663–1674.

115. de Lange, E. C. M. *Advanced Drug Delivery Reviews* 2004, **56**, 1793–1809.

116. Zhang, Y.; Schuetz, J. D.; Elmquist, W. F.; Miller, D. W. *Journal of Pharmacology and Experimental Therapeutics* 2004, **311**, 449–455.

117. Deeley, R. G.; Westlake, C.; Cole, S. P. C. *Physiological Reviews* 2006, **86**, 849–899.

118. Cooray, H. C.; Blackmore, C. G.; Maskell, L.; Barrand, M. A. *Neuroreport* 2002, **13**, 2059–2063.

119. Natarajan, K.; Xie, Y.; Baer, M. R.; Ross, D. D. *Biochemical Pharmacology* 2012, **83**, 1084–1103.

120. Rice, A.; Liu, Y.; Michaelis, M. L.; Himes, R. H.; Georg, G. I.; Audus, K. L. *Journal of Medicinal Chemistry* 2005, **48**, 832–838.

121. Turunen, B. J.; Ge, H.; Oyetunji, J.; Desino, K. E.; Vasandani, V.; Guethe, S.; Himes, R. H.; Audus, K. L.; Seelig, A.; Georg, G. I. *Bioorganic and Medicinal Chemistry Letters* 2008, **18**, 5971–5974.

122. Miller, D. W.; Kabanov, A. V. *Colloids and Surfaces B-Biointerfaces* 1999, **16**, 321–330.

123. Huwyler, J.; Cerletti, A.; Fricker, G.; Eberle, A. N.; Drewe, J. *Journal of Drug Targeting* 2002, **10**, 73–79.

124. Pinzon-Daza, M. L.; Garzon, R.; Couraud, P. O.; Romero, I. A.; Weksler, B.; Ghigo, D.; Bosia, A.; Riganti, C. *British Journal of Pharmacology* 2012, **167**, 1431–1447.

125. Kreuter, J. *Advanced Drug Delivery Reviews* 2001, **47**, 65–81.

126. Tamaki, A.; Ierano, C.; Szakacs, G.; Robey, R. W.; Bates, S. E. *Essays in Biochemistry* 2011, **50**, 209–232.

127. Leonard, G. D.; Fojo, T.; Bates, S. E. *The Oncologist* 2003, **8**, 411–424.

128. Durmus, S.; Sparidans, R. W.; Wagenaar, E.; Beijnen, J. H.; Schinkel, A. H. *Molecular Pharmaceutics* 2012, **9**, 3236–3245.

129. Tang, S. C.; Lagas, J. S.; Lankheet, N. A. G.; Poller, B.; Hillebrand, M. J.; Rosing, H.; Beijnen, J. H.; Schinkel, A. H. *International Journal of Cancer* 2012, **130**, 223–233.
130. Lagas, J. S.; van Waterschoot, R. A. B.; van Tilburg, V. A. C. J.; Hillebrand, M. J.; Lankheet, N.; Rosing, H.; Beijnen, J. H.; Schinkel, A. H. *Clinical Cancer Research* 2009, **15**, 2344–2351.
131. Oostendorp, R. L.; Buckle, T.; Beijnen, J. H.; van Tellingen, O.; Schellens, J. H. M. *Investigational New Drugs* 2009, **27**, 31–40.
132. Aisen, P.; Listowsky, I. *Annual Review of Biochemistry* 1980, **49**, 357–393.
133. McClelland, A.; Kuhn, L. C.; Ruddle, F. H. *Cell* 1984, **39**, 267–274.
134. Jefferies, W. A.; Brandon, M. R.; Hunt, S. V.; Williams, A. F.; Gatter, K. C.; Mason, D. Y. *Nature* 1984, **312**, 162–163.
135. Pardridge, W. M.; Buciak, J. L.; Friden, P. M. *The Journal of Pharmacology and Experimental Therapeutics* 1991, **259**, 66–70.
136. Pardridge, W. M. *Peptide Drug Delivery to the Brain*; Raven Press: New York, 1991.
137. Friden, P. M.; Walus, L. R.; Musso, G. F.; Taylor, M. A.; Malfroy, B.; Starzyk, R. M. *Proceedings of the National Academy of Sciences of the United States of America* 1991, **88**, 4771–4775.
138. Friden, P. M.; Walus, L. R. *Advances in Experimental Medicine and Biology* 1993, **331**, 129–136.
139. Zhang, Y.; Pardridge, W. M. *Brain Research* 2001, **889**, 49–56.
140. Wu, D.; Pardridge, W. M. *Proceedings of the National Academy of Sciences of the United States of America* 1999, **96**, 254–259.
141. Baskin, D. G.; Wilcox, B. J.; Figlewicz, D. P.; Dorsa, D. M. *Trends in Neurosciences* 1988, **11**, 107–111.
142. Knusel, B.; Michel, P. P.; Schwaber, J. S.; Hefti, F. *The Journal of Neuroscience* 1990, **10**, 558–570.
143. Palovcik, R. A.; Phillips, M. I.; Kappy, M. S.; Raizada, M. K. *Brain Research* 1984, **309**, 187–191.
144. Miller, D. W.; Keller, B. T.; Borchardt, R. T. *Journal of Cellular Physiology* 1994, **161**, 333–341.
145. Pardridge, W. M.; Eisenberg, J.; Yang, J. *Journal of Neurochemistry* 1985, **44**, 1771–1778.
146. Kabanov, A. V.; Chekhonin, V. P.; Alakhov, V.; Batrakova, E. V.; Lebedev, A. S.; Melik-Nubarov, N. S.; Arzhakov, S. A.; Levashov, A. V.; Morozov, G. V.; Severin, E. S.; Kabanov, V. A. *FEBS Letters* 1989, **258**, 343–345.
147. Batrakova, E. V.; Han, H. Y.; Miller, D. W.; Kabanov, A. V. *Pharmaceutical Research* 1998, **15**, 1525–1532.
148. Fukuta, M.; Okada, H.; Iinuma, S.; Yanai, S.; Toguchi, H. *Pharmaceutical Research* 1994, **11**, 1681–1688.
149. Ueno, M.; Nakagawa, T.; Wu, B.; Onodera, M.; Huang, C. L.; Kusaka, T.; Araki, N.; Sakamoto, H. *Current Medicinal Chemistry* 2010, **17**, 1125–1138.
150. Sagare, A. P.; Deane, R.; Zlokovic, B. V. *Pharmacology and Therapeutics* 2012, **136**, 94–105.
151. Demeule, M.; Currie, J. C.; Bertrand, Y.; Che, C.; Nguyen, T.; Regina, A.; Gabathuler, R.; Castaigne, J. P.; Beliveau, R. *Journal of Neurochemistry* 2008, **106**, 1534–1544.

152. Bertrand, Y.; Currie, J. C.; Poirier, J.; Demeule, M.; Abulrob, A.; Fatehi, D.; Stanimirovic, D.; Sartelet, H.; Castaigne, J. P.; Beliveau, R. *British Journal of Cancer* 2011, **105**, 1697–1707.

153. Huang, S.; Li, J.; Han, L.; Liu, S.; Ma, H.; Huang, R.; Jiang, C. *Biomaterials* 2011, **32**, 6832–6838.

154. Xin, H.; Sha, X.; Jiang, X.; Zhang, W.; Chen, L.; Fang, X. *Biomaterials* 2012, **33**, 8167–8176.

155. Drappatz, J.; Brenner, A.; Wong, E. T.; Eichler, A.; Schiff, D.; Groves, M. D.; Mikkelsen, T.; Rosenfeld, S.; Sarantopoulos, J.; Meyers, C. A.; Fielding, R. M.; Elian, K.; Wang, X.; Lawrence, B.; Shing, M.; Kelsey, S.; Castaigne, J. P.; Wen, P. Y. *Clinical Cancer Research* 2013, **19**, 1567–1576.

156. Che, C.; Yang, G.; Thiot, C.; Lacoste, M. C.; Currie, J. C.; Demeule, M.; Regina, A.; Beliveau, R.; Castaigne, J. P. *Journal of Medicinal Chemistry* 2010, **53**, 2814–2824.

157. Koffie, R. M.; Farrar, C. T.; Saidi, L. J.; William, C. M.; Hyman, B. T.; Spires-Jones, T. L. *Proceedings of the National Academy of Sciences of the United States of America* 2011, **108**, 18837–18842.

158. Tanaka, N.; Sasahara, M.; Ohno, M.; Higashiyama, S.; Hayase, Y.; Shimada, M. *Brain Research* 1999, **827**, 130–138.

159. Mishima, K.; Higashiyama, S.; Asai, A.; Yamaoka, K.; Nagashima, Y.; Taniguchi, N.; Kitanaka, C.; Kirino, T.; Kuchino, Y. *Acta Neuropathologica* 1998, **96**, 322–328.

160. Kuo, Y. C.; Chung, C. Y. *Colloids and Surfaces. B, Biointerfaces* 2012, **91**, 242–249.

161. Wang, P.; Liu, Y.; Shang, X.; Xue, Y. *Journal of Molecular Neuroscience* 2011, **43**, 485–492.

162. Zhang, P.; Hu, L.; Yin, Q.; Zhang, Z.; Feng, L.; Li, Y. *Journal of Controlled Release* 2012, **159**, 429–434.

163. Boado, R. J.; Hui, E. K.; Lu, J. Z.; Pardridge, W. M. *The Journal of Pharmacology and Experimental Therapeutics* 2010, **333**, 961–969.

164. Ke, W.; Shao, K.; Huang, R.; Han, L.; Liu, Y.; Li, J.; Kuang, Y.; Ye, L.; Lou, J.; Jiang, C. *Biomaterials* 2009, **30**, 6976–6985.

165. van Rooy, I.; Mastrobattista, E.; Storm, G.; Hennink, W. E.; Schiffelers, R. M. *Journal of Controlled Release* 2011, **150**, 30–36.

166. van der Meel, R.; Vehmeijer, L. J.; Kok, R. J.; Storm, G.; van Gaal, E. V. *Advanced Drug Delivery Reviews* 2013, **65**, 1284–1298.

17

METABOLIC ACTIVATION AND DRUG TARGETING

XIANGMING GUAN

Department of Pharmaceutical Sciences, College of Pharmacy, South Dakota State University, Brookings, SD, USA

17.1 INTRODUCTION

Drug targeting through metabolic activation provides a way to deliver a drug to the desired site of drug action. It has the advantage of increasing drug's site selectivity and reducing systemic adverse effects. Since metabolic activation is an enzyme-mediated process, drug targeting through metabolic activation requires that the activating enzyme is unique to or, at least, highly enriched at the target site [1]. Successful examples have been reported in targeting a drug to the colon, kidney [1, 2], and liver [2]. Drug targeting through metabolic activation has its unique value in cancer chemotherapy, which will be the focus of this chapter.

Chemotherapy is one of the three major treatments (chemotherapy, radiotherapy, and surgical resection) in cancer therapy. However, it is often associated with severe systemic side effects due to the fact that anticancer drugs are primarily cytotoxic agents that not only kill cancer cells but also cause damage to normal cells, especially to those proliferating cells such as bone marrow and gut epithelia. As a result, the success of chemotherapy is often hampered by severe systemic side effects. Consequently, increasing tumor selectivity of the chemotherapeutic agents has been an intense research effort in improving chemotherapeutic efficacy. There are three general approaches in increasing the selectivity: (i) identify agents that will be selective in killing cancer cells than normal cells, (ii) deliver the chemotherapeutic

Drug Delivery: Principles and Applications, Second Edition. Edited by Binghe Wang, Longqin Hu, and Teruna J. Siahaan.

agent selectively (ideally specifically) to cancer cells, and (iii) mask the chemotherapeutic agent in such a way that it will be released selectively (ideally specifically) in cancer cells. Discussion of the first two approaches is beyond the scope of this chapter. In the third approach, the masked agent is called a prodrug.

By definition, a prodrug is a compound that exhibits no biological activity and will be turned into a drug with the desired pharmacological activity through, most often, an enzyme-mediated process. The prodrug approach has been applied in improving pharmaceutical properties of a drug, including solubility, bioavailability, and half-life, which is extensively covered in Chapter 12 and will not be discussed here. This chapter will focus on prodrugs that are aimed at releasing cytotoxic anticancer drugs selectively at tumor sites. Different anticancer prodrug approaches and the biochemical processes based on which the prodrugs are designed will be presented.

17.2 ANTICANCER PRODRUGS AND THEIR BIOCHEMICAL BASIS

One of the challenges in anticancer prodrug design is the identification of cancer-associated biochemical processes that can be utilized to activate or release anticancer drugs from prodrugs. The obvious advantage of designing such an anticancer prodrug is its high selectivity. Ideally, the cancer-associated biochemical processes do not or barely occur in normal/healthy cells. Two cancer-related features that have been extensively used for anticancer prodrug design are hypoxia and elevated peptidase levels associated with tumors. Anticancer prodrugs based on the biochemical processes associated with hypoxia and elevated peptidases have demonstrated high selectivity in killing cancer cells [2–8]. In addition, enzymes with elevated activities in cancer have also been exploited for anticancer prodrug design. These enzymes include β-D-glucuronidase [9], prostate-specific antigen (PSA) [2], matrix metalloproteinases (MMPs) [2], phospholipase [10], β-D-galactosidase [9], glutathione-*S*-transferase [11], prolidase [9], cytochrome P450 [12], etc. Obviously, the selective antitumor effects of the prodrugs designed based on these enzymes will depend on the difference in the activities of the enzyme between tumor sites and healthy tissues. The higher the activity at the tumor site, the better the selectivity of the prodrug. To increase the activity of the prodrug-activating enzyme at tumor sites, two major approaches have been developed: (i) deliver the enzyme to the tumor site through the aid of a tumor-specific antibody (antibody-directed enzyme prodrug therapy (ADEPT)) [9, 13–15] and (ii) increase the expression of the enzyme through delivering the gene to the tumor site (gene-directed enzyme prodrug therapy (GDEPT), also named as suicide gene therapy (SGT)) [9, 16–19]. In both ADEPT and GDEPT, the therapy involves two steps. First, an enzyme or gene encoding the enzyme is delivered to the tumor site followed by the second step—prodrug administration. Therefore, ADEPT and GDEPT are also referred as two-step therapy. On the other hand, therapy involving a prodrug alone is called prodrug monotherapy (PMT).

The development of ADEPT and GDEPT provides the possibility of delivering prodrug-activating enzyme that may not be present in humans (nonmammalian origin). Prodrugs designed based on the enzyme that is not present in humans offer much higher efficiency and selectivity, or even specificity if the enzyme can be delivered specifically to tumor sites.

17.2.1 Tumor-Activated Anticancer Prodrugs Based on Hypoxia

Hypoxia is a characteristic microenvironment in the majority of solid tumors [20]. Tumor hypoxia is a result of tumor cell rapid proliferation that exceeds the development of tumor blood supply. The diffusion limit of oxygen is about 150 μm; tumor cells beyond the distance from the blood supply will experience insufficient oxygen. This is termed as chronic or diffusion-limited hypoxia, while acute or perfusion-limited hypoxia is a case when the blood supply is temporally shut down due to the transient changes in blood flow caused by the disordered tumor vasculature [4, 20]. The extent of hypoxia in the tumor varies with a median partial pressure of oxygen (pO_2) of approximately 10 mmHg (1% O_2) compared with a pO_2 of 40–60 mmHg (5–6% O_2) in healthy subcutaneous tissues [20]. The hypoxia inducible factor (HIF)-1 is a primary transcriptional factor that regulates oxygen and the cellular response to hypoxia [6, 20].

Hypoxia is a major problem in radiotherapy and chemotherapy [6, 21, 22]. The cytotoxic effects of radiation require the presence of oxygen. In the case of chemotherapy, hypoxia reduces the distribution of chemotherapeutic agents. Further, hypoxic cells receive not only a reduced amount of oxygen but also a low supply of nutrients, which causes cells to stop or slow down their rate of progress through the cell cycle. Since most anticancer agents are more effective against rapidly proliferating cells, an anticancer agent will be far less effective in eradicating hypoxic tissues.

Although hypoxia posts a challenge in cancer treatment, the unique biochemical events associated with hypoxia provide an opportunity for hypoxia-selective antitumor prodrugs. In particular, the elevated activities of various reductive enzymes have been extensively employed for developing hypoxia-selective prodrugs. These reductive enzymes are elevated as a cellular response to hypoxic conditions [23] and are capable of reducing electron-deficient organic functional groups.

Another feature associated with hypoxia is low extracellular pH [24]. The low extracellular pH is likely to be a result of insufficient blood supply that leads to the accumulation of acidic metabolites. Anticancer prodrugs activated in a low pH environment have been explored [2, 9, 25, 26].

It is worth noting that only a small proportion of the cells in a solid tumor are likely to be chronically hypoxic. Hypoxia-selective prodrugs are not expected to be curative on their own. Other agents need to be used to eradicate aerobic cancer cells. Nevertheless, because of acute hypoxia, a significant fraction of solid tumor cells may be killed, theoretically, by hypoxia-selective prodrugs when they are experiencing transient hypoxia. In addition, a bystander effect also contributes significantly to the antitumor activity. The bystander effect is thought to occur through two major mechanisms: local bystander effect and immune-mediated bystander effect [18]. The local bystander effect involves killing of nearby cells due to diffusion of an active drug from the site of the activation or release. However, this effect will be relatively distant limited since most active drugs cannot travel far due to their high chemical reactivity. Immune-mediated bystander effect is proposed to be caused by the presence of inflammatory infiltrates, chemokines, and cytokines in regressing tumors of immunocompetent animals receiving the SGT treatment [18]. These inflammatory factors stimulate immune responses against tumor cells.

Although anticancer prodrugs designed by targeting hypoxia-related enzymes have demonstrated high tumor selectivity, toxicity associated with this approach has been observed, specifically retinal toxicity. The inner retina is vascularized, but the outer retina is avascular and relies on diffusion of O_2 from the inner retinal vessels and from the choriocapillaris. The retinal toxicity of 2-nitroimidazole alkylating agent CI-1010 and tirapazamine (TPZ) has been linked to the physiological hypoxia in the retina [27]. Nevertheless, this toxicity does not appear to be related to some other hypoxia-selective anticancer prodrugs such as the quinone porfiromycin, anthraquinone N-oxide AQ4N, and nitrogen mustard prodrugs SN 23816 and SN 25341 [27].

17.2.1.1 Tumor-Activated Anticancer Prodrug Based on Reductive Enzymes and Low Oxygen Tension Most hypoxia-elevated reductive enzymes are also present in aerobic cells. The enzymes that are involved in the reductive activation of anticancer prodrugs include NADPH-cytochrome P450 reductase, NADH-cytochrome b_5 reductase, xanthine oxidase, NADH:ubiquinone oxidoreductase, ferredoxin-$NADP^+$ reductase, and NAD(P)H:Quinone acceptor oxidoreductase (DT-diaphorase) [20, 28]. These are flavoenzymes and all catalyze a single-electron reduction except DT-diaphorase that carries out two-electron reductions. The hypoxia selectivity of targeting the single-electron reductases arises from the fact that upon a single-electron reduction, a prodrug is converted to a single-electron reduction adduct, which, in the absence of molecular oxygen, is further reduced to the active drug. However, in the presence of molecular oxygen, the one-electron adduct is oxidized back to the prodrug (Scheme 17.1). Therefore, formation of the active drug is restricted to hypoxic tissue. In the case of DT-diaphorase, a two-electron reductase, the selectivity is derived from the fact that the enzyme is found to be at high levels throughout many human solid tumors such as thyroid, adrenal, breast, ovarian, colon, and non-small-cell lung cancer [29]. It needs to be noted that bioreductive activation of a prodrug is not necessarily restricted to one enzyme. It appears that several different enzymes can participate to different extents with the various bioreductive agents (Fig. 17.1) [30]. The bioreductive organic functional groups employed for these reductases primarily include nitro groups, quinones, aromatic N-oxides, aliphatic N-oxides, sulfoxides, and transition metals [23]. Figure 17.1 listed the chemical structures of representative bioreductive prodrugs [2, 23]. Some of them have gone through clinical trials [23].

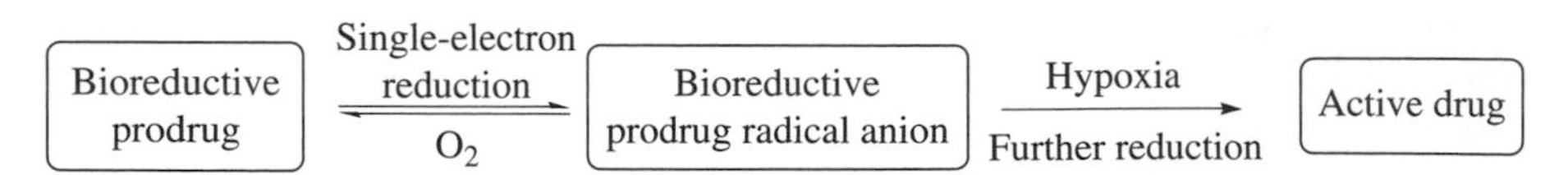

SCHEME 17.1 Bioactivation of prodrugs by a single-electron reductase.

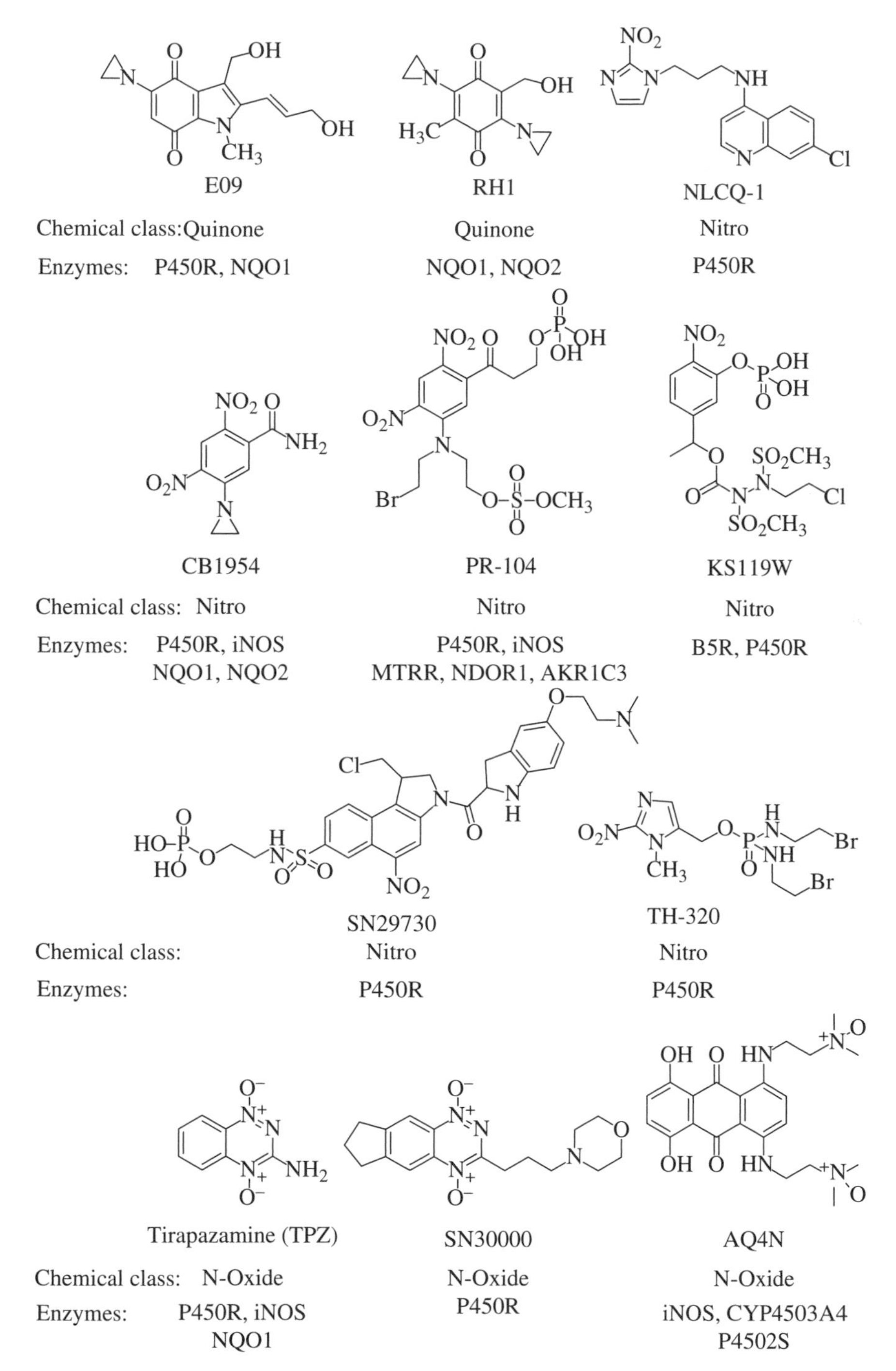

FIGURE 17.1 Chemical structures of representative bioreductive prodrugs [23]. AKR1C3, aldo-keto reductase 1C3; B5R, NADH-cytochrome *b*5 reductase; iNOS, inducible nitric oxide synthase; MTRR, methionine synthase reductase; NDOR1, NADPH-dependent diflavin oxidoreductase 1; NQO, NAD(P)H dehydrogenase; P450, cytochrome P450; P450R, NADPH-cytochrome P450 reductase.

Hypoxia-Selective Quinone-Containing Prodrugs Quinones can be reduced by a variety of reductases including single electron–donating enzymes such as NADPH:cytochrome P-450 reductase and xanthine oxidase [31–34] and two electron–donating enzyme DT-diaphorase. Hypoxia-selective quinone-containing prodrugs include some naturally occurring anticancer prodrugs, such as mitomycin (MC) and compounds designed to be hypoxia selective.

Mitomycin C (MC, Scheme 17.2) is an anticancer agent that has been shown to be more cytotoxic to hypoxic tumor cells than their aerobic counterparts. MC is considered as the prototype bioreductive alkylating agent [35]. Although the agent was not designed as a prodrug, the mode of action of this agent is thought to involve a bioreductive activation to a species (**1c**, Scheme 17.2) that alkylates DNA [36]. Similar activation occurs with structurally related compounds such as indolequinone EO9 (**2**, Scheme 17.2). Note that the aziridine ring in the structure provides an additional site that can be attacked by DNA. This group may account for the aerobic toxicity of the compound.

An elegant design of a dual action prodrug by coupling EO9 with another anticancer drug (**3**, Scheme 17.2) has been proposed [37] in which the dual action prodrug, upon reduction, releases two active anticancer drugs.

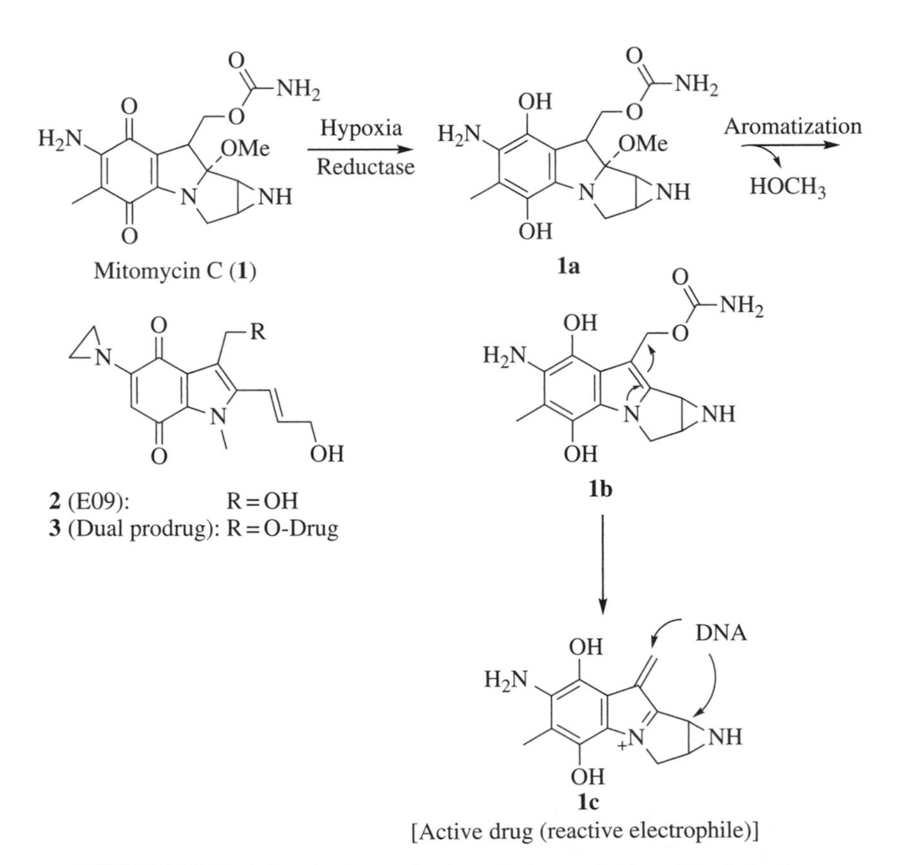

SCHEME 17.2 Bioreductive activation of mitomycin C, and its derivatives.

In a similar fashion, prodrug **4** (Scheme 17.3) is bioreductively activated in hypoxic cells to produce species **4b** and a nitrogen mustard **4c**, both of which are capable of reacting with nucleophiles, for example DNA and proteins [4]. Note that the nitrogen mustard in prodrug **4** is much less reactive than the one in **4c** due to the reduced electron density of the aromatic ring by the electron-withdrawing effect of the ester functionality. Under physiologic pH, the anionic nature of the carboxylic group increases the electron density that favors the formation of an aziridinium ring (**4d**, Scheme 17.3), a reactive electrophilic group that alkylates DNA.

Hypoxia-Selective Aromatic Nitro–Containing Prodrugs An aromatic nitro (Ar–NO_2) group can be reduced to an aromatic amine through the intermediates of a nitroso group (Ar–NO) and a hydroxylamine (Ar–NH–OH) group. The reduction turns a strong electron-withdrawing group (EWG) (—NO_2) into an electron-donating group (EDG) (—NH_2). The change in electron density, as a result of the reduction, has been utilized to turn a chemically stable compound into a reactive electrophile. The systemic toxicity of nitrogen mustard anticancer agents is derived from its indiscriminate alkylation to cancer cells and normal cells. The drug action mode of nitrogen mustard anticancer agents is illustrated in Scheme 17.4. The availability of the nitrogen lone pair electrons is crucial in determining the reactivity of the nitrogen mustard. During the formation of a reactive aziridinium ring (Scheme 17.4), the lone pair electrons aid the elimination of the chloride through a neighboring group participation effect (**5e–5f**, Scheme 17.4). The resulting aziridinium ring alkylates DNA, leading to cell killing. One of the approaches in reducing systemic toxicity of nitrogen mustards is the use of a prodrug in which the electron density of the nitrogen (or the basicity of the nitrogen) is reduced but will be regained in tumor cells. This concept has already been illustrated in Scheme 17.3 in which an ester functional group is used to reduce the basicity of the nitrogen. In the following example, a strong electron-withdrawing nitro group is employed for the same purpose. Scheme 17.4 provides a general scheme to demonstrate the activation of aromatic nitro–containing nitrogen mustard prodrugs. Note that

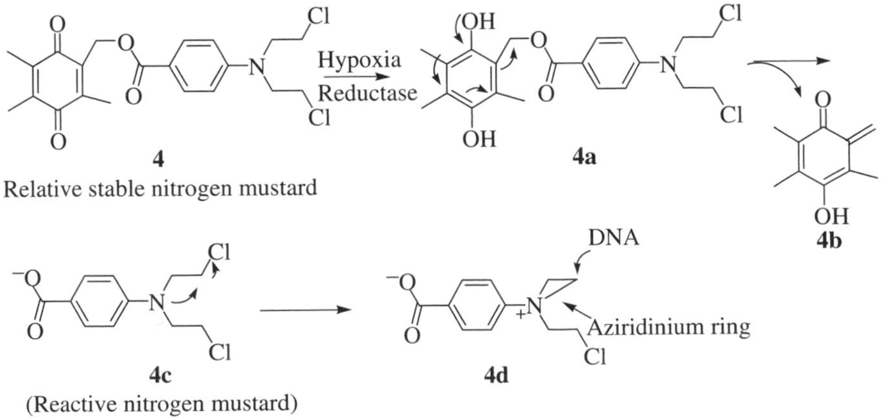

SCHEME 17.3 Bioreductive activation of prodrug **4**.

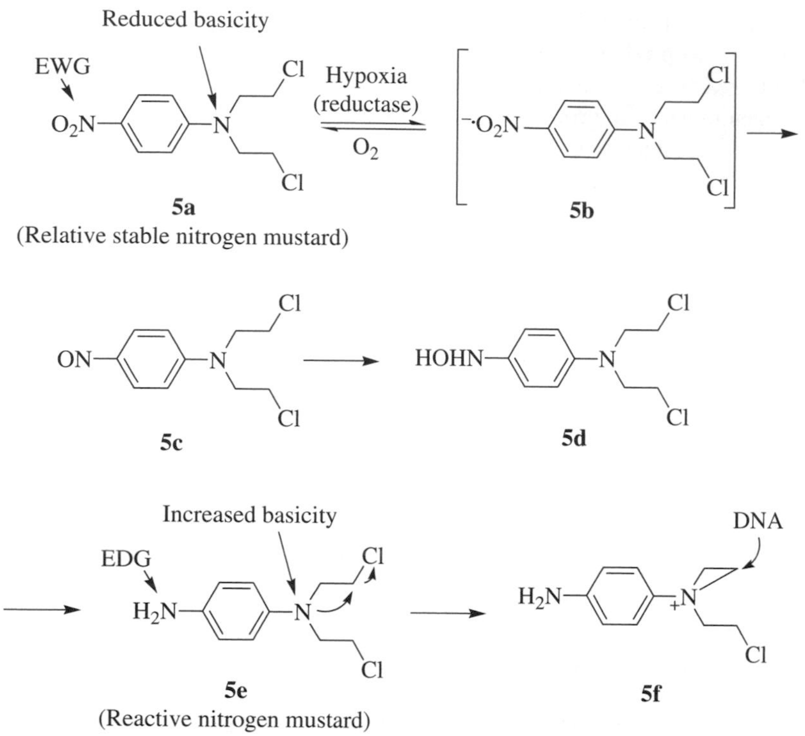

SCHEME 17.4 Bioactivation of aromatic nitro–containing hypoxia-selective nitrogen mustard prodrugs.

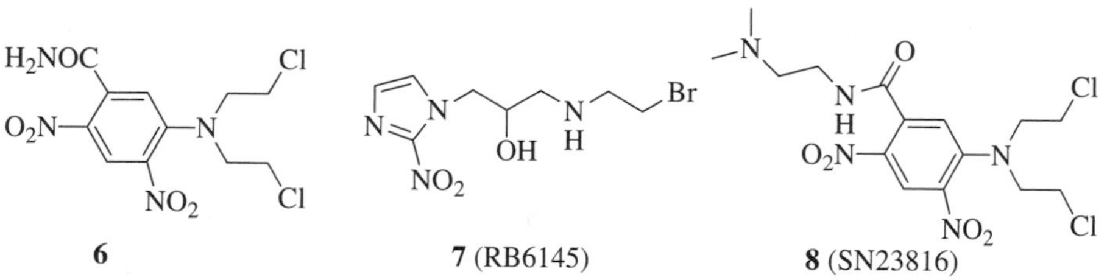

FIGURE 17.2 Representative aromatic nitro–containing hypoxia-selective nitrogen mustard prodrugs.

one electron reduction of the nitro group leads to the nitro radical anion, which, in the presence of oxygen, is reversed back to the nitro group. Under a hypoxic condition, the reduction proceeds further to a nitroso compound, hydroxylamine, and amine. Figure 17.2 shows representative hypoxia-selective aromatic nitro–containing nitrogen mustard prodrugs [27]. One of the most effective bioreductive drugs against hypoxic cells in murine tumors is the 2-nitroimidazole alkylating agent CI-1010 (PD 144872), which is the *R*-enantiomer of the racemate RB 6145 [1-[(2-bromoethyl)amino]-3-(2-nitro-1*H*-imidazole-1-yl)-2-propanol hydrobromide] (**7**). However, further clinical evaluation of this compound was terminated due to its severe retinal toxicity [27].

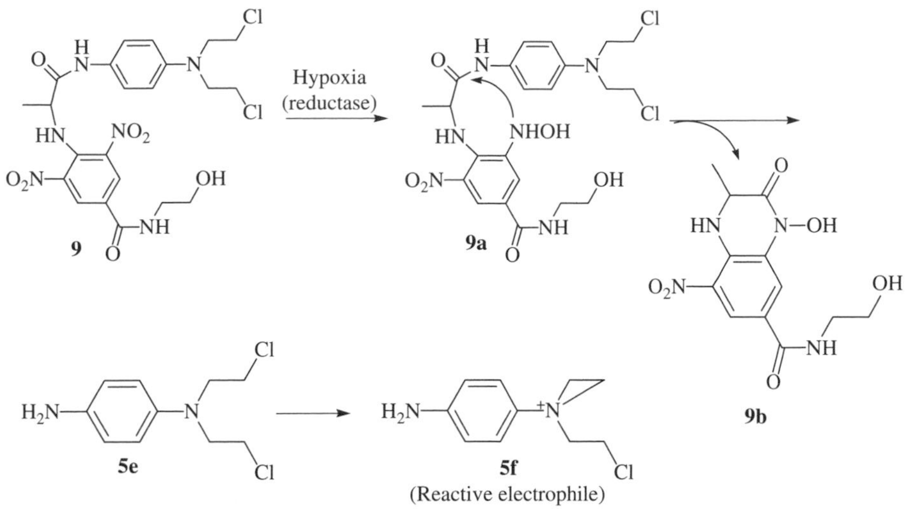

SCHEME 17.5 Bioreductive activation of prodrug **9**.

Alternatively, an increase in electron density can be achieved through the conversion of an amide group (a much less EDG) to an amino group (a strong EDG) as illustrated by prodrug **9** (Scheme 17.5) [38]. Upon reduction of the nitro group, the formed hydroxylamine (**9a**) rapidly cyclizes to release the alkylating agent **5e**.

A very elegant design using an aromatic nitro compound to release an active drug is through a reduction of the nitro group followed by a spontaneous fragmentation (1,6-elimination). Scheme 17.6 provides a general description for this approach and some of the representative prodrugs based on this design [39–42]. Most of the prodrugs via this design contain nitrobenzyl and carbamate functionalities. It has been reported that release of the active drug occurs after the formation of the hydroxylamine (**10a**, R = OH) rather than the amine (**10a**, R = H) [40].

In addition to nitrobenzyl carbamate prodrugs, some nitro heteroaromatic prodrugs have also been reported. Scheme 17.7 shows the general mode of active drug release [43–45].

Similar approach has been adopted for paclitaxel prodrugs. Paclitaxel was released upon reduction of the nitro group or aromatic azido group to an amino or hydroxylamino group and upon subsequent 1,6-elimination of a 4-amino or 4-hydroxylamino benzyloxycarbonyl moiety, respectively (Scheme 17.8) [2].

Hypoxia-Selective N-Oxide-Containing Prodrugs N-oxide compounds can be reduced by a variety of reductases. Compound **22** is a prodrug of compound **22a** (Scheme 17.9) [46], which is a structural analog of the metal-binding unit of bleomycin, an anticancer drug. The metal-binding unit is a key to the antitumor activity of bleomycin. It is believed that bleomycin forms a chelate with iron (Fe^{2+}). Five of the six coordination positions of Fe^{2+} are strongly coordinated to bleomycin. The sixth is available for coordination to oxygen. The chelate alters the redox

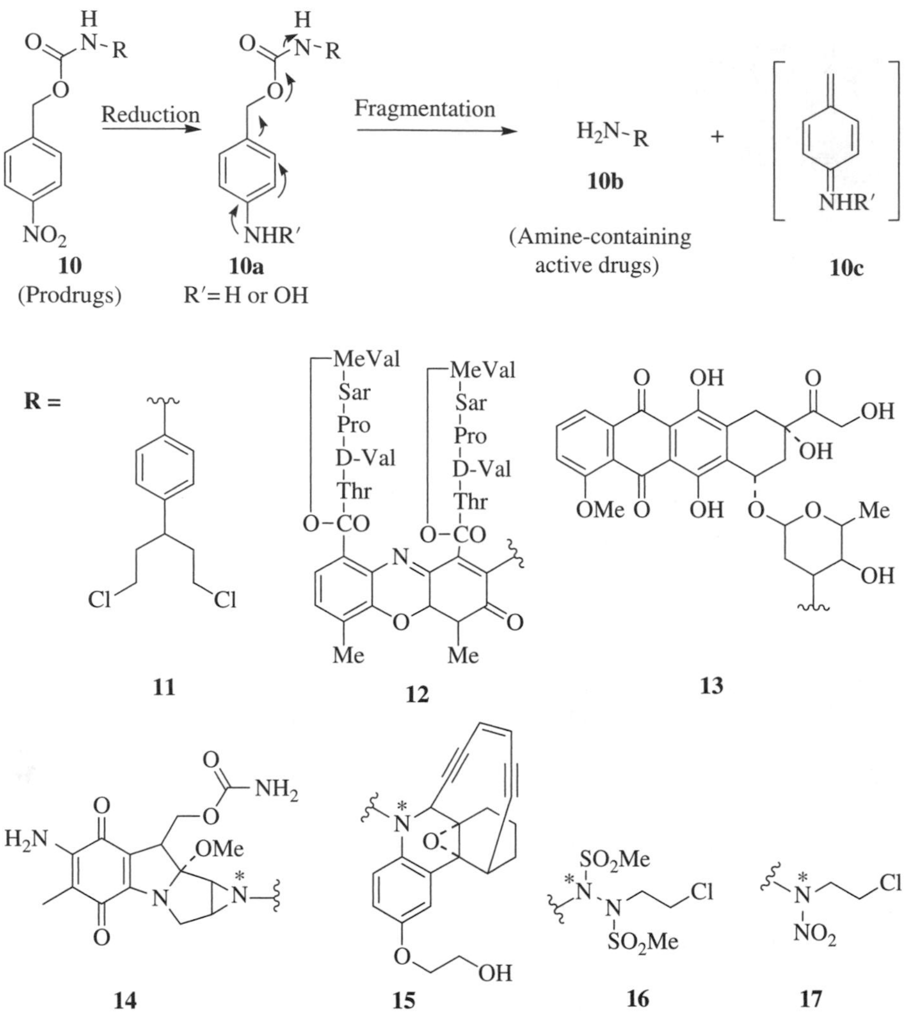

SCHEME 17.6 Bioreductive activation through a reduction of a nitro group followed by 1,6-elimination. *The nitrogen is part of the carbamate structure.

potential of iron such that bound oxygen is reduced, converting the oxygen into a reactive radical species, the hydroxyl radical, which causes cell killing through DNA degradation [36]. By converting one of the nitrogen atoms involved in metal chelating to N-oxide (prodrug **22**), the structure is not capable of metal chelating and, therefore, becomes nontoxic.

TPZ (Scheme 17.10) is a benzotriazine di-N-oxide bioreductive anticancer prodrug. TPZ is activated to a DNA-damaging oxidizing radical by cytochrome P450 reductase and one-electron reductases in the absence of oxygen [47]. TPZ has been demonstrated to be effective in killing hypoxic cells in murine tumors [47] and in sensitizing hypoxic cells to cisplatin [48, 49]. Clinical studies have showed that TPZ enhances cisplatin

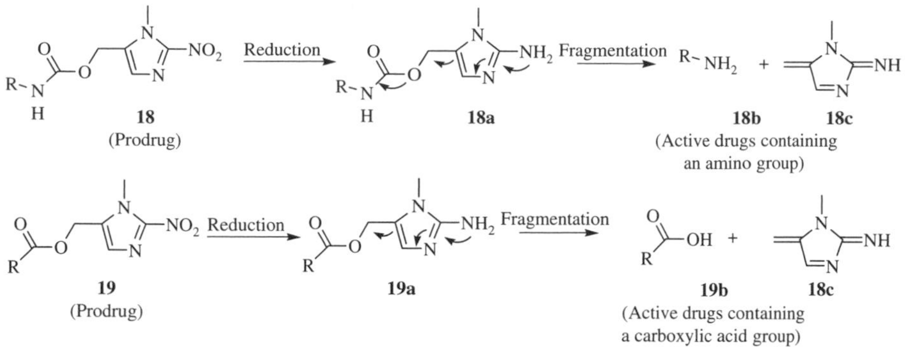

18 (Prodrug) → Reduction → 18a → Fragmentation → $R{-}NH_2$ + 18c

18b 18c (Active drugs containing an amino group)

19 (Prodrug) → Reduction → 19a → Fragmentation → RCOOH + 18c

19b 18c (Active drugs containing a carboxylic acid group)

SCHEME 17.7 Bioreductive activation of heteroaromatic prodrugs via reduction of a nitro group, followed by 1,5-elimination.

activity against non-small-cell lung cancer [50, 51]. However, like Cl-1010 (R enantiomer of **7**, Fig. 17.2) TPZ has also been shown to produce retinal toxicity [27].

The N-oxide approach can also be used in reducing the basicity of the nitrogen in nitrogen mustards. Nitromin has been used with some success for the treatment of Yoshida ascites sarcomas in rats [52]. Studies showed that nitromin (**23**, Scheme 17.11) serves as a bioreductive prodrug of a nitrogen mustard [53]. NADPH-dependent cytochrome P-450 reductase has been shown to catalyze the reduction of nitromin to its active nitrogen mustard (**23a**) [53].

AQ4N (**24**) is a prodrug of the DNA-binding agent and topoisomerase inhibitor AQ4. The less toxic prodrug is activated in hypoxic tissue through reduction of the N-oxide (Scheme 17.12) [54, 55].

Hypoxia-Selective S-Oxide-Containing Prodrugs The metabolic fate of sulfoxide (S-oxide) has been shown to be different under an aerobic or anaerobic condition. Under anaerobic conditions, sulfoxides (**25a**) can be reduced to sulfides in a reversible reaction (**25b**) (Scheme 17.13) [3]. Under aerobic conditions, sulfoxides (**25a**) are oxidized to sulfones (**25**), a process that is not reversible (Scheme 17.13). This difference has been exploited for the design of hypoxia-selective nitrogen mustard prodrugs. Since the sulfoxide group is an electron-withdrawing group, a sulfoxide nitrogen mustard prodrug is a less cytotoxic agent. This changes when, under a hypoxic condition, the sulfoxide group is converted into a sulfide, an electron-donating group that increases the electron density of the nitrogen; this in turn favors the formation of a reactive aziridinium ion (**25c**). The aziridinium ion then reacts with DNA resulting in cell death. On the other hand, under aerobic condition, a sulfoxide group is more likely to be oxidized to a corresponding sulfone, which is a stronger electron-withdrawing group and makes the formation of the reactive aziridinium ring more difficult [3]. A successful example demonstrating hypoxia selectivity of sulfoxide-containing anticancer prodrug was described by Kwon [3]. 1-[Bis(2-chloroethyl)amino-4-{4-[bis(2-chloroethyl)amino]phenyl}sulfinyl]benzene, a diphenylsulfoxide-containing nitrogen mustard prodrug, was synthesized and found to exhibit high hypoxia selectivity.

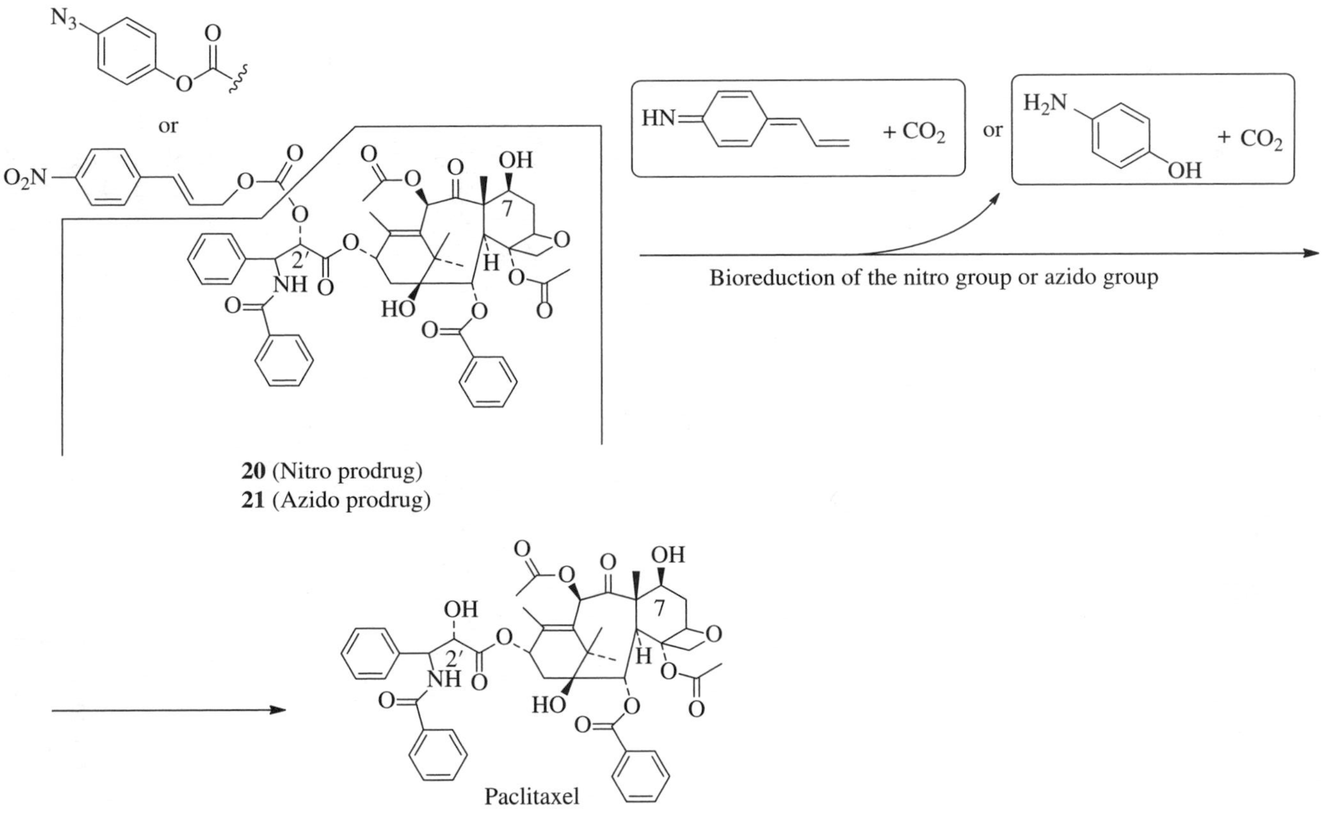

SCHEME 17.8 Bioreductive activation of nitro or azido paclitaxel prodrug [2].

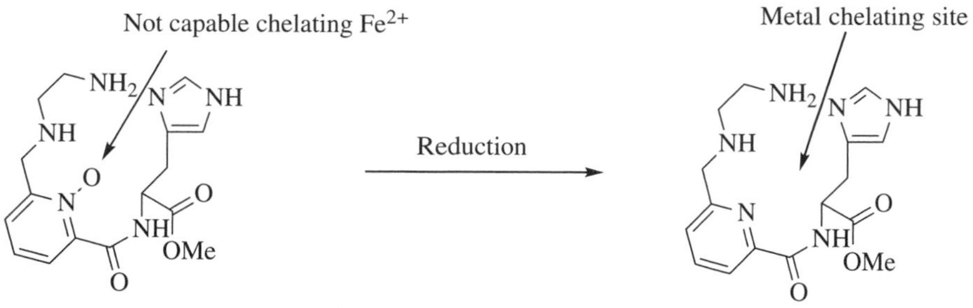

Not capable chelating Fe^{2+} | Metal chelating site

Reduction

22 → **22a**

SCHEME 17.9 Bioreductive activation of prodrug **22** to active drug **22a**.

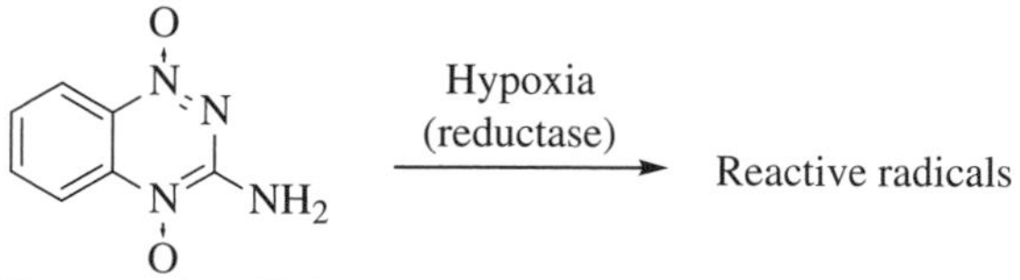

Hypoxia (reductase)

Tirapazamine (TPZ) → Reactive radicals

SCHEME 17.10 Bioreductive activation of Tirapazamine (TPZ).

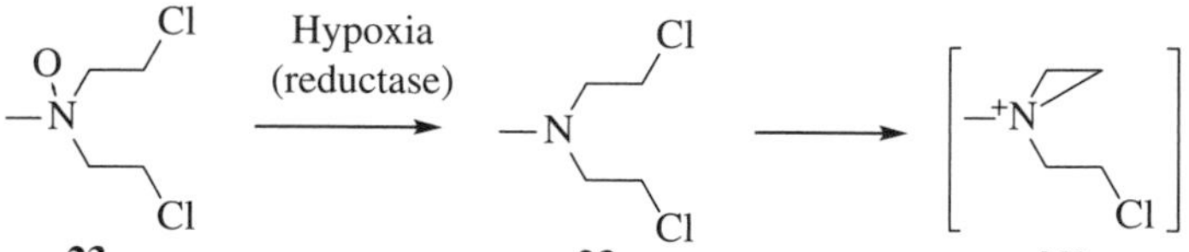

Hypoxia (reductase)

23 (Nitromin) → **23a** → **23b** (Reactive electrophile)

SCHEME 17.11 Bioreductive activation of nitromin.

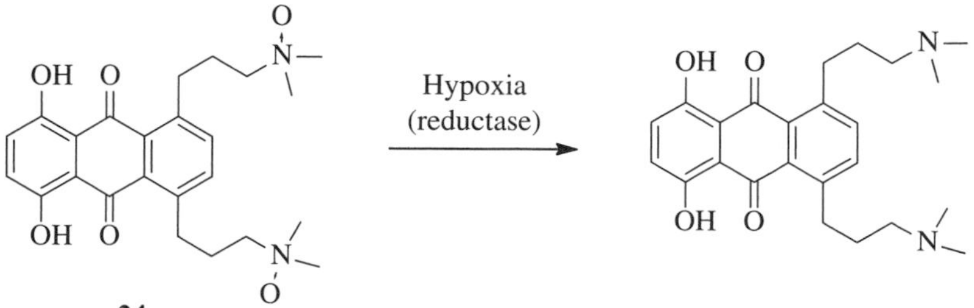

Hypoxia (reductase)

24 (AQ4N, prodrug) → AQ4 (DNA-binding agent and topoisomerase inhibitor)

SCHEME 17.12 Bioreductive activation of AQ4N.

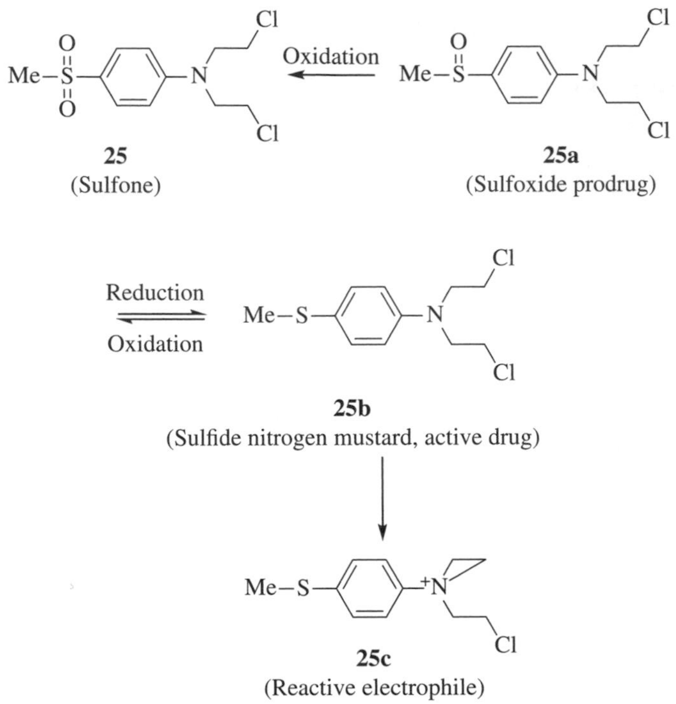

25 (Sulfone) ← Oxidation — **25a** (Sulfoxide prodrug)

Reduction / Oxidation ⇌ **25b** (Sulfide nitrogen mustard, active drug)

↓

25c (Reactive electrophile)

SCHEME 17.13 Bioreductive activation of sulfoxide-containing nitrogen mustard prodrugs.

Bioreductive Prodrugs Containing Transition Metals Metal-containing compounds comprise only a small percentage of anticancer agents. Metals exhibit the property of various redox states and can be in the form of an inert oxidized state and reduced to the reactive reductive state in a reductive environment [56]. The research focus of the redox-activatable metals has been primarily on platinum (Pt) and Ruthenium (Ru) while iron, cobalt, and copper have also been employed.

Pt is the metal atom in cisplatin and its derivatives carboplatin, oxaliplatin, nedaplatin, lobaplatin, and heptaplatin (Fig. 17.3). Cisplatin is one of the most successful anticancer drugs. Clinical application of cisplatin has been limited by the dose-limiting nephrotoxicity and other side effects. The side effects may be improved by the use of prodrugs with Pt in the more inert +IV oxidation state [Pt(IV)] that can be reduced to the reactive Pt(II). The low reactivity is due to the low-spin d6 electron configuration with octahedral geometry of Pt(IV) complexes. The configuration is relatively inert to substitution; reactions with biological nucleophiles are thus not favored compared to Pt(II) complexes, and the lifetime in biological fluids is expected to increase [56]. Potential reducing agents for Pt(IV) in the cell are glutathione, vitamin C, NAD(P)H, and cysteine-containing proteins (Scheme 17.14) [56].

The two additional axial ligands of Pt(IV) (Scheme 17.14) provide the possibility of modifying the structure to improve the pharmacokinetic property, to target the

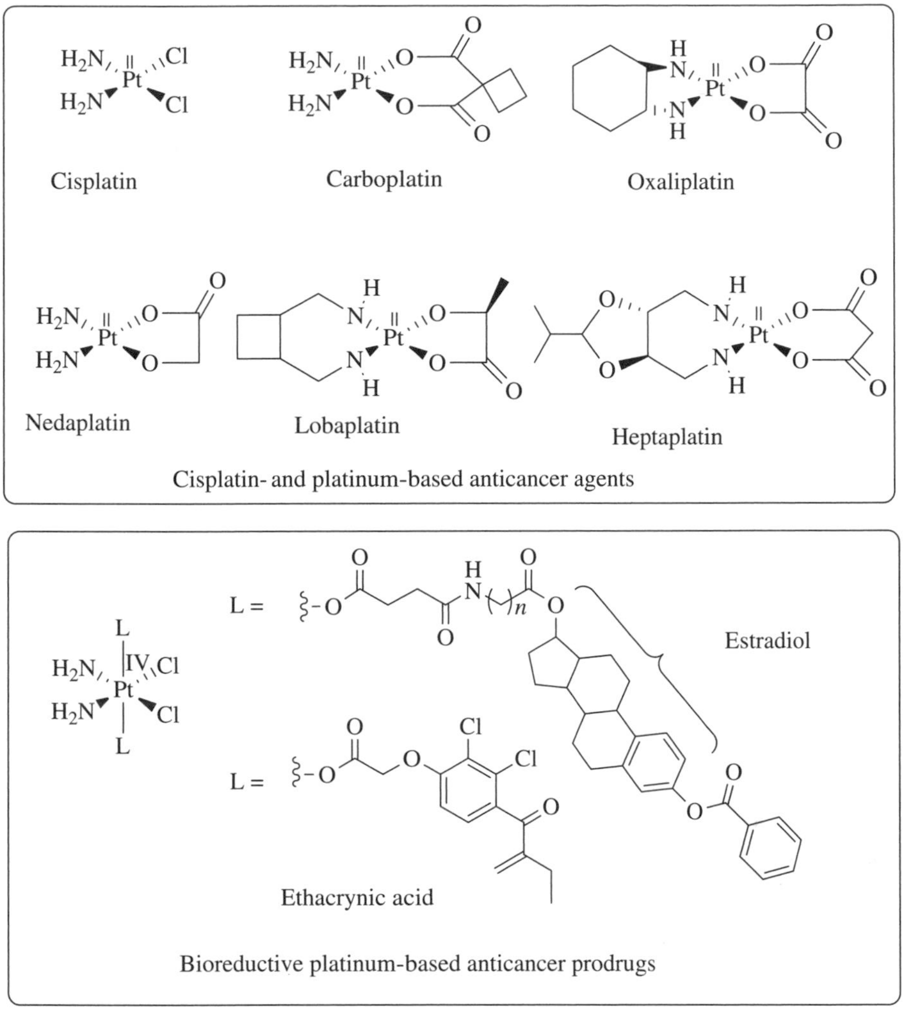

FIGURE 17.3 Chemical structures of representative platinum-based anticancer drugs and bioreductive prodrugs [56].

prodrug to tumor sites, or to provide another drug moiety for enhancing anticancer activity. Estradiol moiety in Pt–prodrug in Figure 17.3 helps the prodrug to target the estrogen-receptor-positive [ER(+)] cancers like breast and ovarian cancers and also to improve the lipophilicity and chemical stability of the prodrug. The other Pt prodrug in Figure 17.3 contains an ethacrynic acid moiety, which, upon release, inhibits glutathione *S*-transferase (GST) and helps reduce the drug resistance caused by GST.

Similar to Pt, Ru(III) complexes are bioreducible to Ru(II) complexes by vitamin C or glutathione (Scheme 17.15). Unlike Pt, both Ru(III) and Ru(II) complexes maintain the octahedral ligand set [12]. It has also been reported that

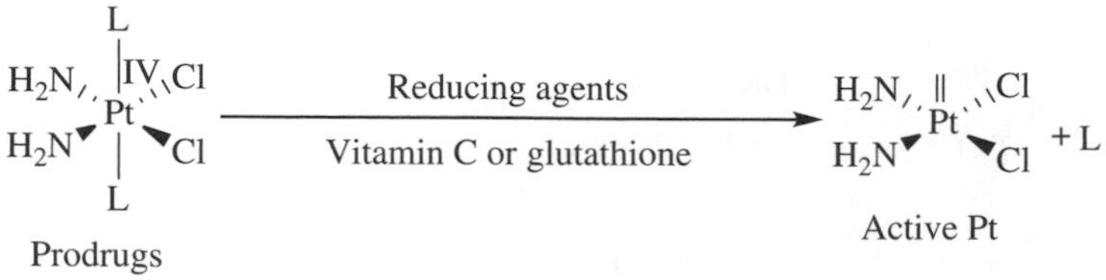

SCHEME 17.14 Bioreduction of platinum-based prodrugs. L, Ligand and can be another drug [56].

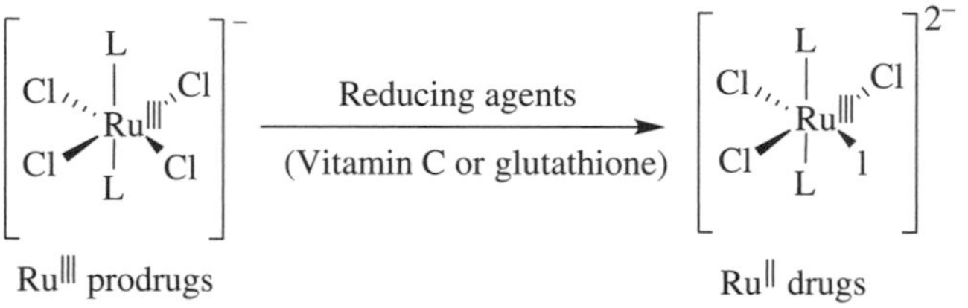

SCHEME 17.15 Bioreduction activation of Ru(III) prodrugs to Ru(II) drugs. L, Ligands [56].

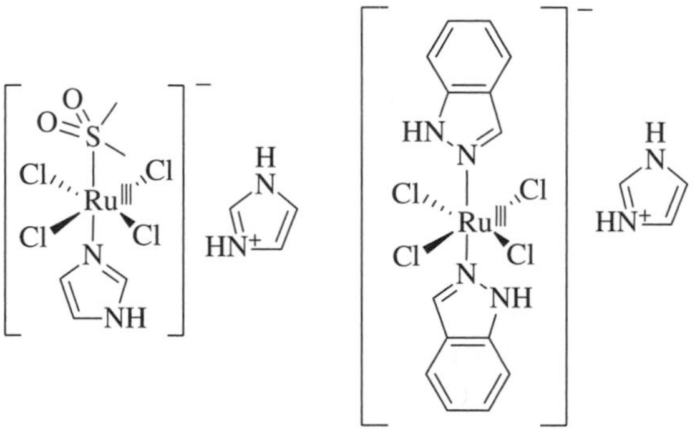

FIGURE 17.4 Chemical structures of Ru(III) anticancer prodrugs NAMI-A and KP1019.

there was more reduction of Ru(III) complexes to Ru(II) complexes in a hypoxic environment [56]. The antitumor activity of Ru(III) compounds is suggested to depend on the conversion of Ru(III) complexes to Ru(II) complexes [56]. The first Ru-based anticancer agent in clinical trials was NAMI-A, followed by KP1019 in 2003 (Fig. 17.4).

Other metals employed for bioreductive prodrugs include cobalt, copper, and iron. Co(III) complexes can be reduced to Co(II) complexes under hypoxic conditions to release their neutral ligands (Scheme 17.16). The reduction is inhibited in

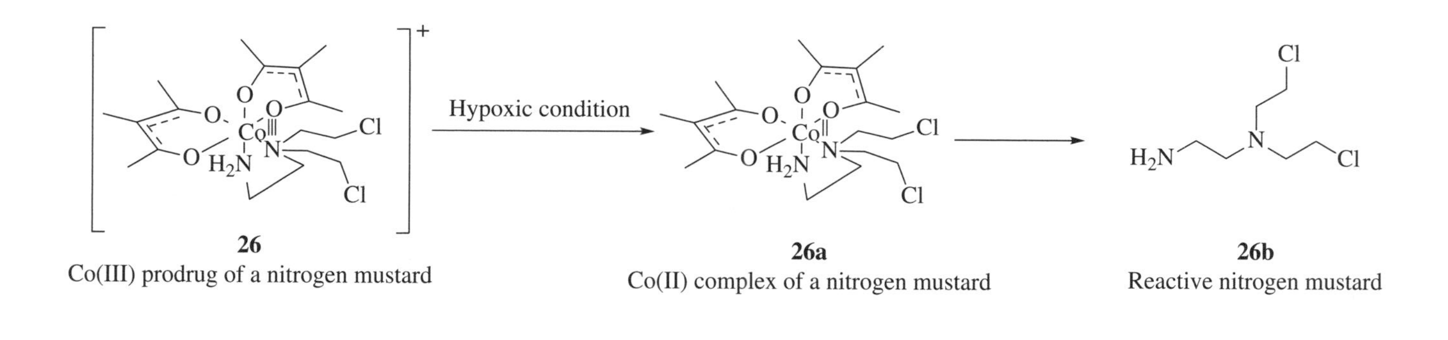

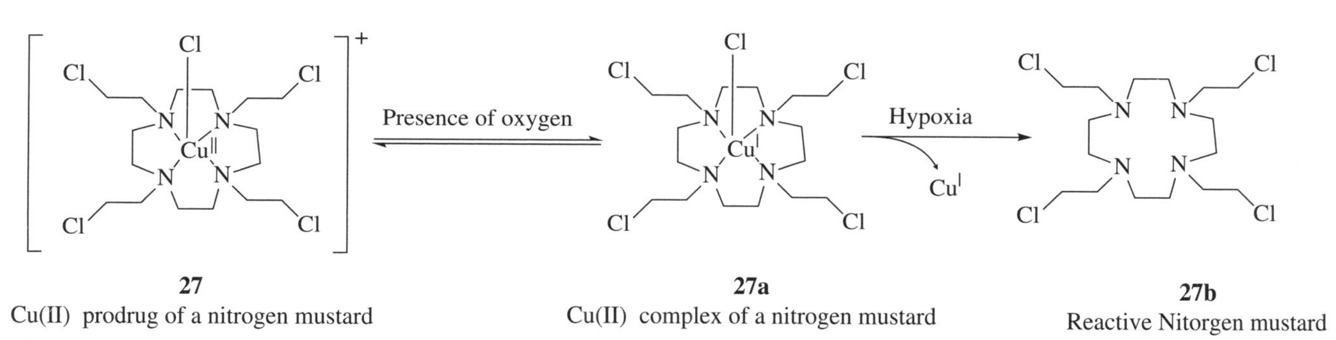

SCHEME 17.16 Bioreductive activation of Co(III) and Cu(II) nitrogen mustard prodrugs under hypoxic conditions [57, 58].

the presence of oxygen. The neutral ligands employed include nitrogen mustards [57]. Coordination of the lone pair electron of the nitrogen mustard to Co(III) reduces the reactivity of the mustard due to the fact that the lone pair electron is no longer available to undergo the neighboring group participation effect that leads to the formation of the reactive aziridinium ion. Once Co(III) is reduced to Co(II), the mustard is released (Scheme 17.16) [57]. As for Co(III), under hypoxic conditions, Cu(II) complexes can be reduced to Cu(I) complexes that release a neutral ligand as demonstrated with the Cu(II) complex with a mustard derivative of cyclen (1,4,7,10-tetraazacyclododecane) in Scheme 17.16 [58]. The Cu(II) complex showed a 24-fold increase in cytotoxicity *in vitro* under hypoxic conditions versus oxic conditions [58]. An additional advantage of copper is its radioactive isotopes (^{64}Cu and ^{67}Cu) that offer the possibility of a combination of radiotherapy and bioreductive antitumor prodrugs [12]. The selective antitumor effect associated with the reduction of Fe(III) complexes to Fe(II) complexes is believed to be associated with the production of reactive oxygen species generated from reduction that damage DNA [56].

17.2.1.2 Activation of Prodrugs Based on Low pH of Solid Tumors Lower extracellular pH has been observed in many solid tumors as a result of the limited blood flow [24]. The lower extracellular pH is believed to be caused by an insufficient clearance of acidic metabolites from chronically hypoxic cells, a phenomenon that can lower the mean extracellular pH in tumors to below circa 6.3 up to 1 pH unit lower than the intracellular pH, which is actively regulated. Prodrugs that are stable at physiological pH and can be activated by this lower extracellular pH in solid tumors have been reported. The acid labile ketal, acetal, and hydrazone functional groups are often employed for this purpose as demonstrated in Figure 17.5 [2, 59]. These two prodrugs are stable at physiological pH (7.4) but readily hydrolyzed to produce the active drug under an acidic condition. Scheme 17.17 shows the acid activation of an acetal glycoside prodrug of aldophosphamide.

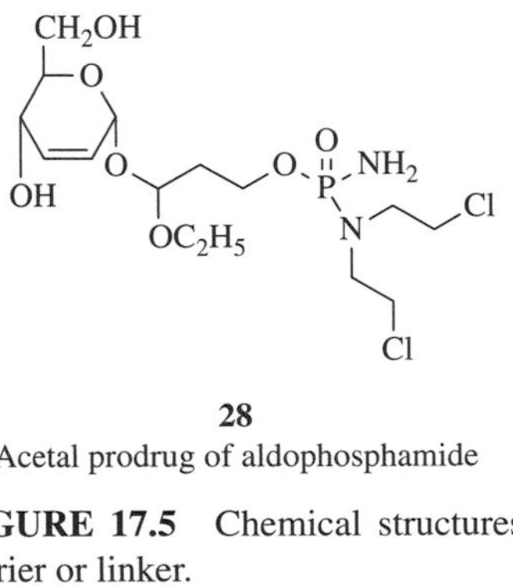

28
Acetal prodrug of aldophosphamide

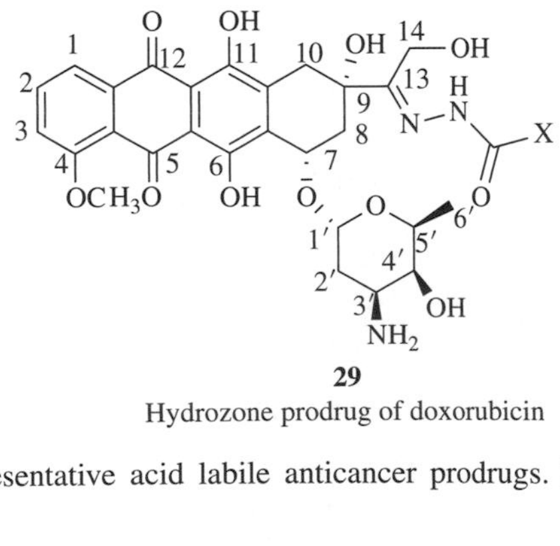

29
Hydrozone prodrug of doxorubicin

FIGURE 17.5 Chemical structures of representative acid labile anticancer prodrugs. X, carrier or linker.

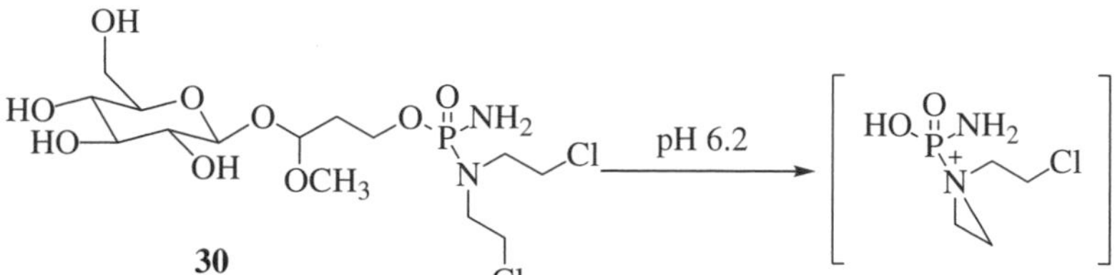

30
Acetal glycoside prodrug
of aldophosphamide

pH 6.2

30a
Reactive active drug

SCHEME 17.17 Activation of an acetal glycoside prodrug of aldophosphamide at acidic pH.

17.2.2 Tumor-Activated Prodrugs Based on Elevated Peptidases or Proteases

An increase in proteolytic enzymes such as cathepsin, MMPs, prolidase, and serine proteases represented by plasminogen activator and plasmin is often associated with tumors [2, 60–64]. These enzymes are thought to be critically involved in the events that lead to metastasis because they are capable of degrading the basement membrane and extracellular matrix around tumor tissue, allowing tumor cells to migrate and invade the surrounding stroma and endothelium. Although these enzymes are also produced by normal cells, their activity is normally tightly regulated by hormonal controls and by specific inhibitors [65, 66]. Therefore, these enzymes have been investigated as activating enzymes for anticancer prodrugs. In general, prodrugs targeting proteases contain two components—a peptide and a parent drug. The peptide chosen is usually a di-, tri-, or tetrapeptide, which should give rise to a prodrug resistant to the serum peptidase but susceptible to peptidases present around the tumor mass.

17.2.2.1 Prodrug Bond Linkage Between a Peptide and a Drug A peptide can be connected to a drug directly (Scheme 17.18) or indirectly through a link termed as a spacer (Schemes 17.19 and 17.20). The direct linkage of a peptide to a drug leads to a prodrug that can either release the parent drug or a drug that contains vestiges of the bound peptide [67–69]. In the latter case, the released drug may have impaired cytotoxic activity. An additional consideration for direct drug attachment to peptides is that the drug may reduce the hydrolysis rate of the peptide by the activating enzyme, resulting in a slow release of the active drug. This often occurs when a bulky drug is involved. The problem can be circumvented through an indirect linkage in which a self-immolative spacer is employed between the drug and the peptide. The spacer spatially separates the drug from the site of enzymatic cleavage so that the drug will not affect the hydrolysis of the peptide bond. Subsequent fragmentation of the spacer releases the active drug. Carl et al. developed one of the most commonly used

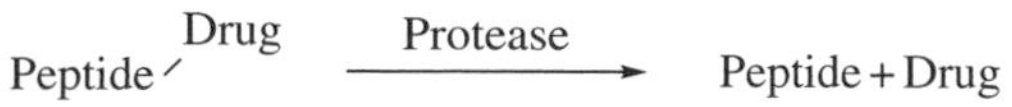

Peptide–Drug —Protease→ Peptide + Drug

SCHEME 17.18 Prodrugs with a drug directly connected to a peptide.

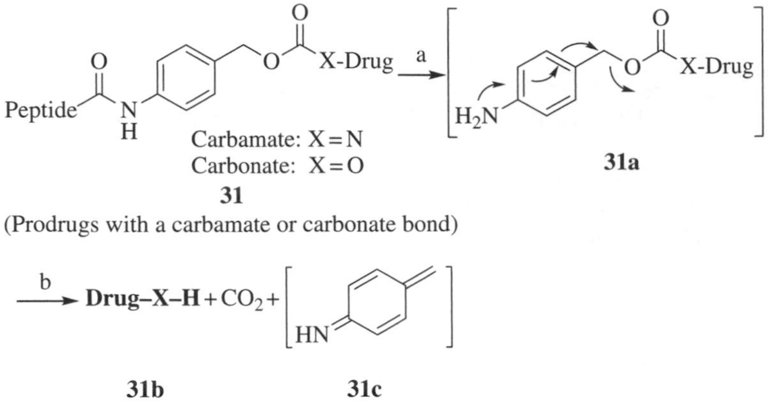

Peptide–NH–C(=O)–C₆H₄–CH₂–O–C(=O)–X-Drug —a→ [H$_2$N–C₆H₄–CH₂–O–C(=O)–X-Drug]

Carbamate: X = N
Carbonate: X = O

31
(Prodrugs with a carbamate or carbonate bond)

31a

—b→ **Drug–X–H** + CO_2 + [HN=C₆H₄=CH₂]

31b **31c**

SCHEME 17.19 Hydrolytic activation of prodrugs with a *p*-aminobenzyl alcohol as a spacer that contains a carbamate or carbonate bond. a, protease; b, 1,6 elimination.

Peptide–NH–C(=O)–C₆H₄–CH₂–O-Drug —a→ [H$_2$N–C₆H₄–CH₂–O-Drug]

32
(Prodrugs with a ether bond)

32a

—b→ **Drug-OH** + CO_2 + [HN=C₆H₄=CH₂]

32b **31c**

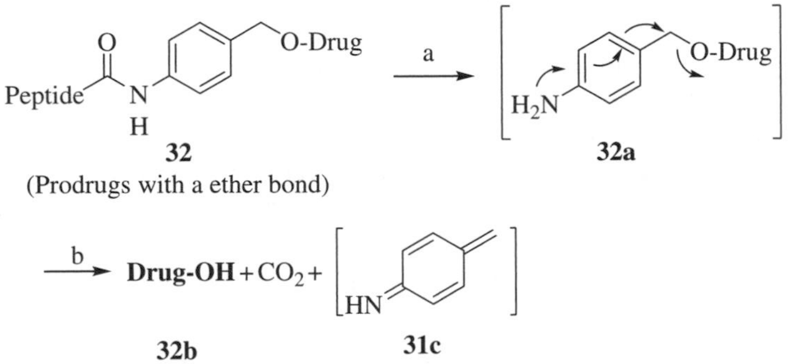

SCHEME 17.20 Hydrolytic activation of prodrugs with a *p*-aminobenzyl alcohol as a spacer that contains an ether bond. a, protease; b, 1,6 elimination.

spacers—the bifunctional *p*-aminobenzyl alcohol group—which is linked to the peptide through the amine moiety and to the drug through the alcohol moiety [70]. The alcohol moiety of the spacer can form a carbamate bond with an amino-containing drug or a carbonate bond with an alcohol-containing drug (Scheme 17.19). The formed prodrugs are activated upon protease-mediated cleavage of the amide bond, followed by a 1,6-elimination that releases the drug, carbon dioxide, and remnants of the spacer (Scheme 17.19). Since the carbonate is easily susceptible to hydrolysis, only a few alcohol-containing drugs, such as paclitaxel, can be made stable enough prodrugs through a carbonate bond. Therefore, the chemistry of this drug attachment has generally been restricted to amine-containing drugs. For alcohol-containing drugs, a recently reported approach through an ether bond has proved to be a more feasible one than the carbonate (Scheme 17.20). However, the ether bond does not undergo fragmentation as readily as the carbamate bond, and some of the prepared ether prodrugs did not undergo fragmentation to release the parent drug [71].

It is noteworthy that the length of the spacer plays an important role in the rate of enzymatic activation. This becomes especially true when using *p*-aminobenzyl alcohol carbamate as a spacer and a bulky drug is involved. As shown in Scheme 17.19, after cleavage of the amide bond, the resulting amino group of the aromatic ring is an electron-donating group and initiates an electronic cascade that leads to the expulsion of the leaving group (1,6-elimination), which releases the free drug after elimination of carbon dioxide. The 1,6-elimination of the carbamate prodrug was found to be virtually instantaneous upon unmasking of the amine group [4, 72]. Since the spacer rapidly eliminates after prodrug activation, the enzymatic activation itself determines the efficiency of drug release. This assumption is supported by the fact that prodrugs derived from bulky drugs often offer a slower drug release rate [72]. To increase enzyme activation rates, de Groot et al. propose to increase the length of the spacer between the drug and the peptide to further keep the bulky drug away from the site of the enzymatic reaction [72]. This approach results in a significant increase in enzyme activation rates and will be discussed under Section 17.2.2.2.

Other spacers such as ethylene diamine, *o*-aminobenzyl alcohol, *p*-hydroxylbenzyl alcohol, and *o*-hydroxylbenzyl alcohol have also been utilized. The application of some of these spacers will be illustrated in the examples presented later.

It is worth mentioning that the concept of using a spacer is not limited to prodrugs involving a peptide. Spacers discussed above have also been extensively employed in other prodrugs as well.

17.2.2.2 Tumor-Selective Prodrugs Activated by Plasmin Plasmin is derived from inactive plasminogen by plasminogen activators (Pas) and is involved in extracellular matrix degradation. The levels of Pas are high in many types of malignant cells and human tumors such as malignant lung [73] and colon [74]. Consequently, tumor-associated plasmin activity is highly localized [62, 73, 75].

Plasmin is a protease with specificity for arginine or lysine as amino acids participating in bond cleavage [76]. Examination of the preferred sites for plasmin cleavage of the fibrinogen molecule (physiological function of plasmin) shows that the preferred sites involve lysine linked to a hydrophobic amino acid. Thus, the choice of which peptidic sequence to use has been focused on a peptide having a hydrophobic amino acid linked to lysine [77].

The idea of selecting plasmin to activate peptide-containing prodrugs was first proposed in 1980 by Carl et al. [76] A D-Val-Leu-Lys tripeptide connected to the amino functional group AT-125 (**33**, Fig. 17.6) or phenylene diamine mustard (**34**, Fig. 17.6) was prepared and found to generate the free drug upon treatment with plasmin. Selection of the D-configuration of the N-terminal amino acid (D-Val) prevents undesired proteolysis by serum peptidases or other ubiquitous enzymes. A five- to sevenfold increase in selectivity between Pas-producing cells in comparison with low-level Pas-containing cells was demonstrated [76].

The first prodrugs of doxorubicin (Dox) for plasmin activation were designed by Chakravarty et al. [78, 79] Dox was directly linked to the tripeptide D-Val-Leu-Lys (**35**, Fig. 17.7). The formed prodrug showed a sevenfold increase in selectivity for Pas-producing cells in comparison with low Pas-containing cells, but the drug was

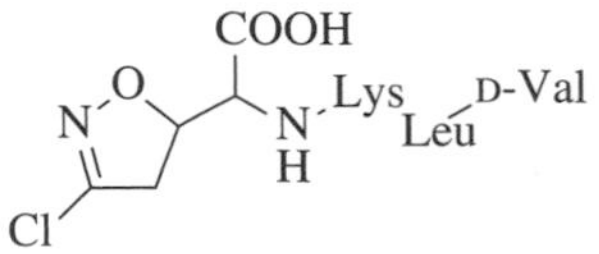

33 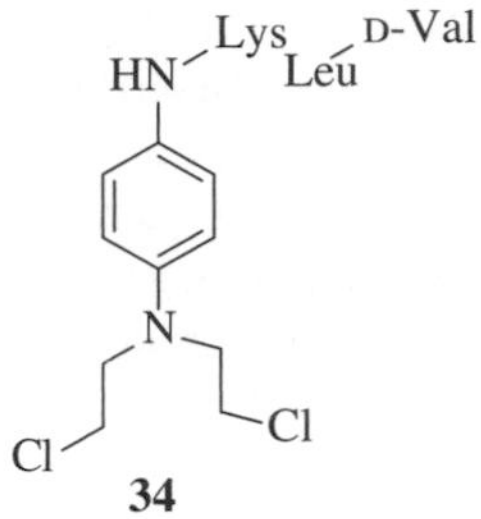 34

FIGURE 17.6 Anticancer prodrugs activated by plasmin.

very inefficiently released by plasmin, a phenomenon likely caused by steric hindrance of Dox. To overcome this problem, de Groot et al. placed a spacer between Dox and the peptide [80]. Compounds **36a–c** were prepared with a *p*-aminobenzyl alcohol as a self-immolative 1,6-elimination spacer to separate Dox or its anthracycline derivatives from the tripeptide D-Ala-Phe-Lys. All prodrugs were stable in buffer and serum for 3 days and generated the parent drugs upon incubation with plasmin. Compound **36c** demonstrated the fastest plasmin cleavage rate. Upon incubation with seven human tumor cell lines, the prodrugs showed a marked decrease of cytotoxicity in comparison with the corresponding parent drugs. *In vitro* selectivity was demonstrated by incubation of the prodrugs with Pas-transfected MCF-7 cells, in comparison with nontransfected cells. The prodrugs showed cytotoxicity similar to free Dox only in the Pas-transfected cells, while the prodrugs were much less toxic in the nontransfected cells [80]. The enzyme activation rate was even further increased when a longer spacer is placed between the drug and the peptide. de Groot et al. placed two *p*-aminobenzyl alcohol moieties between Dox and the peptide (**37**, Fig. 17.7) and found that the enzymatic release rate was increased by twofold [72]. The rate was further increased (approximately threefold) when three *p*-aminobenzyl alcohol moieties were incorporated (**38**, Fig. 17.7) [72]. Plasmin-activated prodrug of bleomycin has also been reported [2].

An elegant cyclization spacer was used in the prodrugs of *N*-nitrosourea [81]. Prodrugs **39a** and **b** contain an ethylene diamine (**39a**) or monomethylated ethylene diamine (**39b**) spacer. Upon cleavage by plasmin, a cyclization reaction led to the formation of a pentacyclic urea derivative (imidazolidin-2-one), with concomitant expulsion of the reactive electrophile (**39g**, Scheme 17.21). Compared with *p*-aminobenzyl alcohol that undergoes 1,6-elimination, cyclization with an ethylene diamine spacer occurs at a slower rate and often is a rate-determining factor in releasing the active drug.

de Groot et al. used the same approach to prepare prodrugs of paclitaxel for activation by plasmin [80]. However, most of the prepared prodrugs were either not stable or resistant to the hydrolysis by plasmin. An alternative approach using a *p*-aminobenzyl alcohol group as a spacer appeared to be successful. Prodrug **40** yielded free paclitaxel upon incubation with plasmin (Scheme 17.22). Further, prodrug **40** showed a dramatic decrease of cytotoxicity in seven human tumor cell lines in comparison with paclitaxel. A similar approach to increase the enzymatic activation

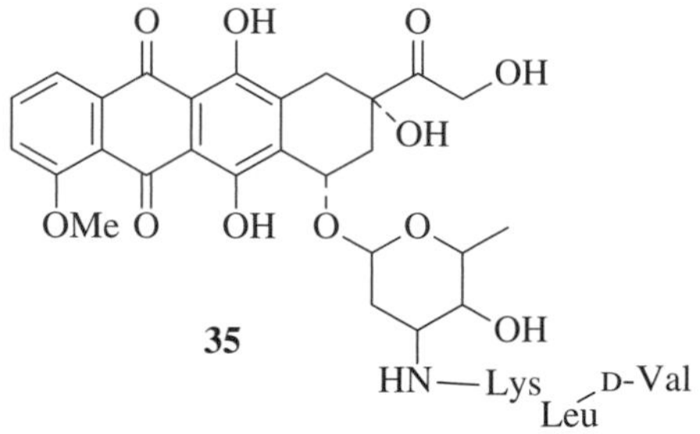

(A prodrug with a direct linkage of a peptide to doxorubicin)

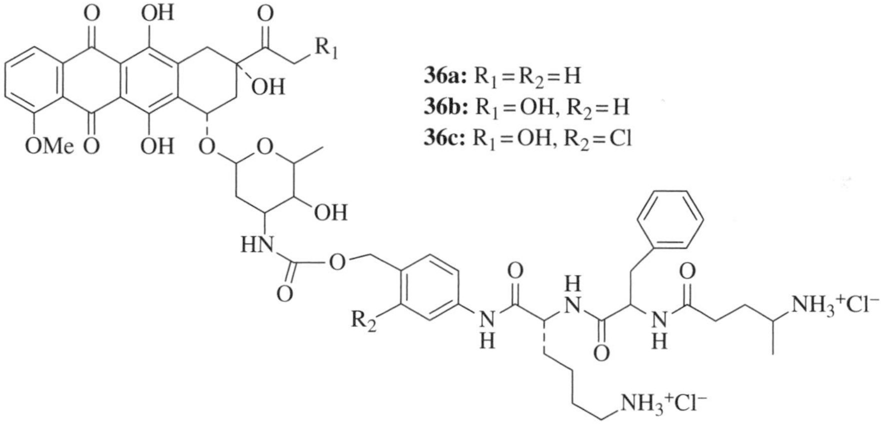

36a: $R_1 = R_2 = H$
36b: $R_1 = OH$, $R_2 = H$
36c: $R_1 = OH$, $R_2 = Cl$

(Prodrugs with one spacer between peptide and doxorubicin or its derivatives)

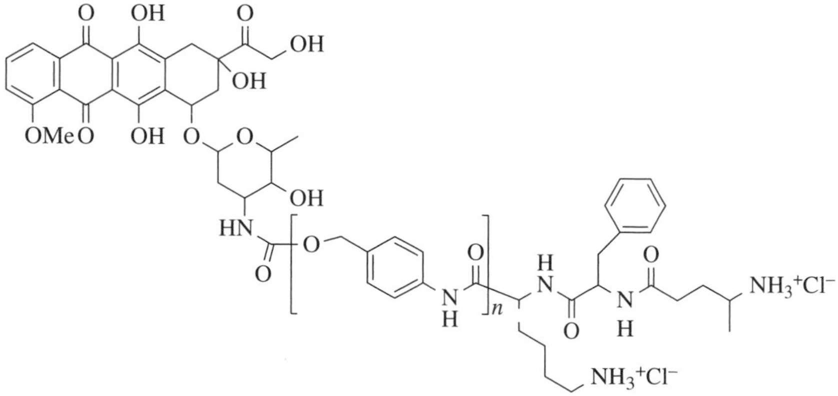

37: $n = 2$, a prodrug with two spacers between a peptide and doxorubicin
38: $n = 3$, a prodrug with three spacers between a peptide and doxorubicin

FIGURE 17.7 Prodrugs of doxorubicin and its derivatives with different number of spacers between the drug and a peptide.

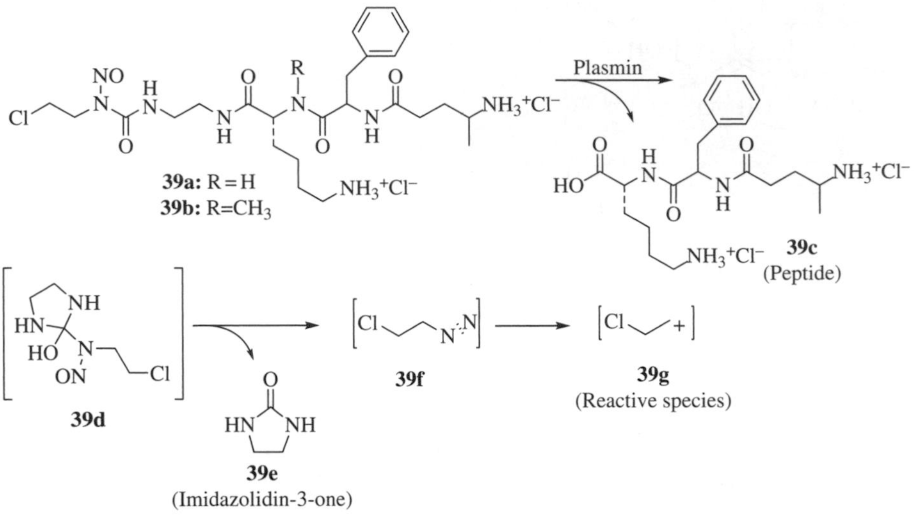

Plasmin

39a: R = H
39b: R=CH_3

39c (Peptide)

39d

39e (Imidazolidin-3-one)

39f

39g (Reactive species)

SCHEME 17.21 Hydrolytic bioactivation of prodrugs **39a** and **39b**.

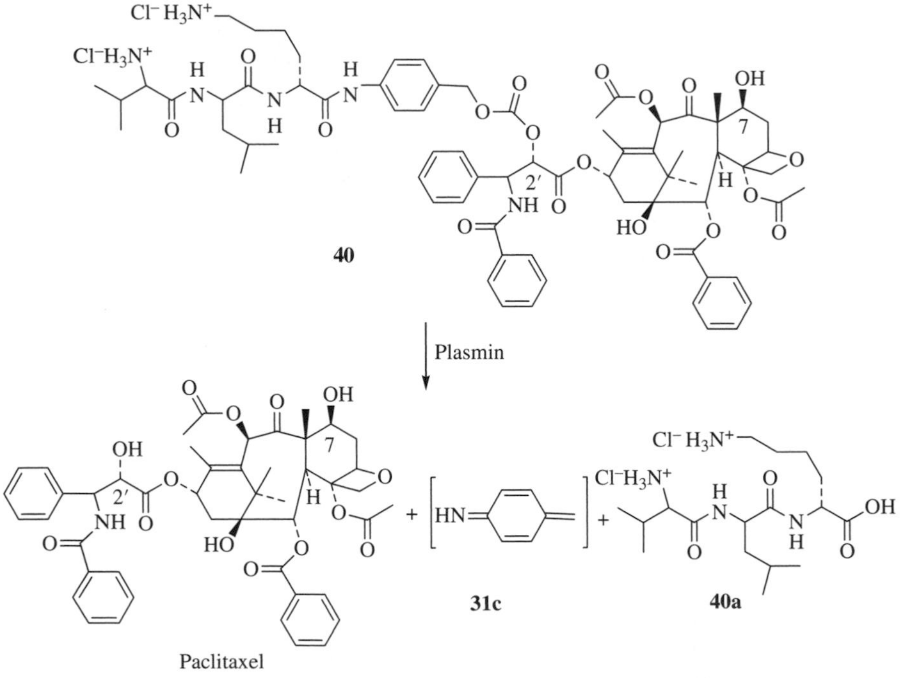

40

Plasmin

Paclitaxel

31c

40a

SCHEME 17.22 Hydrolytic bioactivation of prodrug **40** by plasmin.

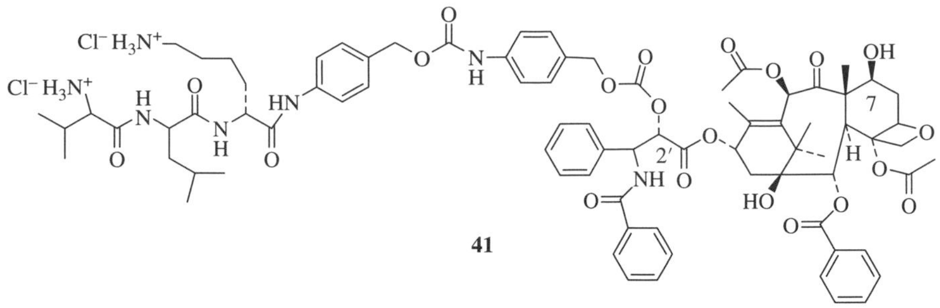

FIGURE 17.8 Prodrug of paclitaxel with two *p*-aminobenzyl alcohol moieties.

rate through an increase in the space between paclitaxel and the peptide was also conducted [72]. Upon incubation with plasmin, the prodrug (**41**, Fig. 17.8) with two *p*-aminobenzyl alcohol moieties released paclitaxel with a sixfold increase in the release rate when compared to the single-spacer-containing prodrug (**40**) [72].

17.2.2.3 Tumor-Selective Prodrugs Activated by Cathepsin Cathepsins are proteases including cysteine proteases, serine proteases, and aspartic proteases [82]. Cathepsins are overexpressed in tumors and play an important role in cancer proliferation, angiogenesis, metastasis, and invasion. The cathepsin family includes cathepsin A, B, C, D, E, F, G, H, L, K, O, S, V, and W [82]. Cathepsin B has been shown to be clinically relevant in cancer progression with studies demonstrating that cytosolic enzyme levels were 11 times higher than those in benign breast tissue specimens [83]. Patients with high intratumor cathepsin B levels suffer a significantly worse prognosis than do patients with low levels [83].

A number of prodrugs for cathepsin B activation were prepared (**42–47**, Fig. 17.9). However, not much biological information is available on these compounds. The half-lives of cathepsin B cleavage of Dox prodrug **45** and mitomycin C prodrug **47** were much shorter than the half-lives of paclitaxel prodrug **44** and **46**, indicating steric hindrance imposed by the paclitaxel part of the prodrug [4].

Toki et al. succeeded in developing cathepsin B–activated prodrugs of combretastatin A-4 and etoposide (Scheme 17.23) [71]. Combretastatin A-4 is a promising anti-angiogenic agent that inhibits the polymerization of tubulin [84]. Etoposide is a clinically approved topoisomerase inhibitor that has demonstrated utility in chemotherapeutic combinations for the treatment of leukemia, lymphoma, germ cell tumors, small-cell lung tumors, and several other carcinomas [85]. In a prodrug approach, combretastatin A-4 or etoposide was coupled to *Z*-valine-citrulline peptide, an N-protected valine-citrulline peptide, through an ether and amide bond using *p*-aminobenzyl alcohol as a spacer (Scheme 17.23). The formed prodrugs (**48** for combretastatin A-4 and **49** for etoposide) were both substrates of cathepsin B and released the active drugs upon incubation with the enzyme, suggesting that these prodrugs can be activated by cathepsin B. It is noteworthy that this is the first example demonstrating 1,6-fragmentation with an ether bond instead of a carbamate bond when *p*-aminobenzyl alcohol is used as a spacer. The prodrug **48** was less potent than

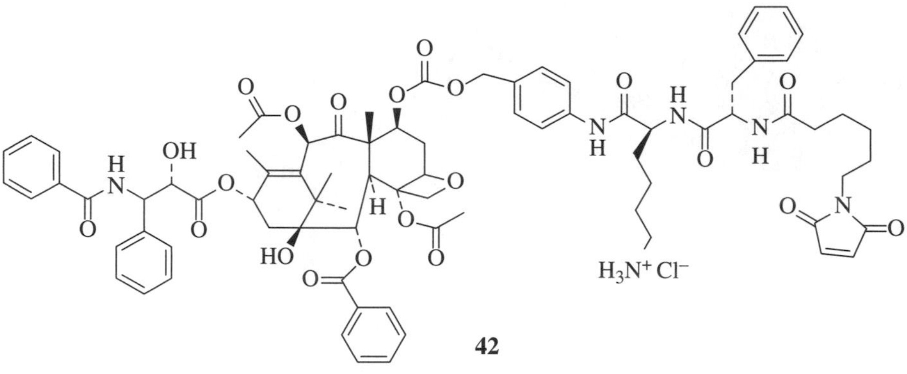

42

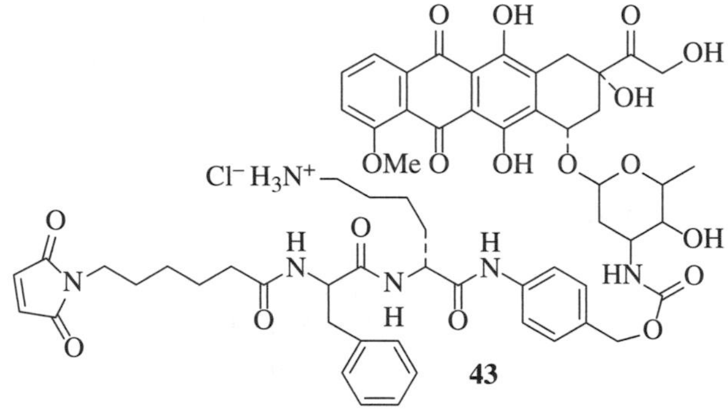

43

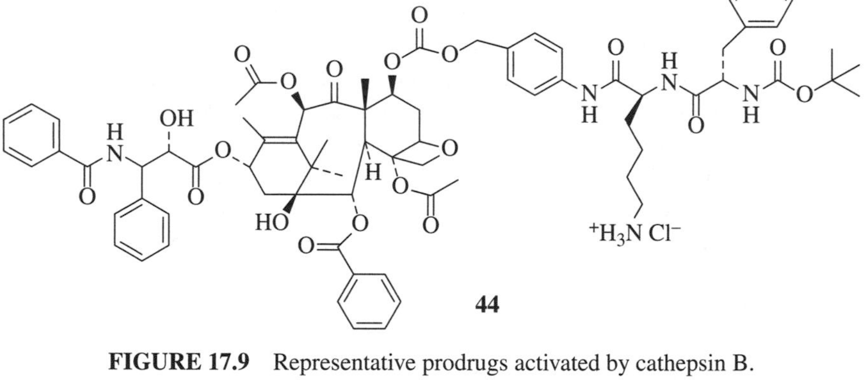

44

FIGURE 17.9 Representative prodrugs activated by cathepsin B.

the parent drug combretastatin A-4 by a factor of 13 in cell killing on L2987 human lung adenocarcinoma, while the prodrug **49** was 20–50 times less active than the parent drug etoposide in the cell lines L2987, WM266/4 (human melanoma), and IGR-39 (human melanoma), confirming that the prodrugs were much less cytotoxic [71]. However, as mentioned earlier, the fragmentation of the ether bond did not proceed as readily as the carbamate bond, which is illustrated by a prodrug derived

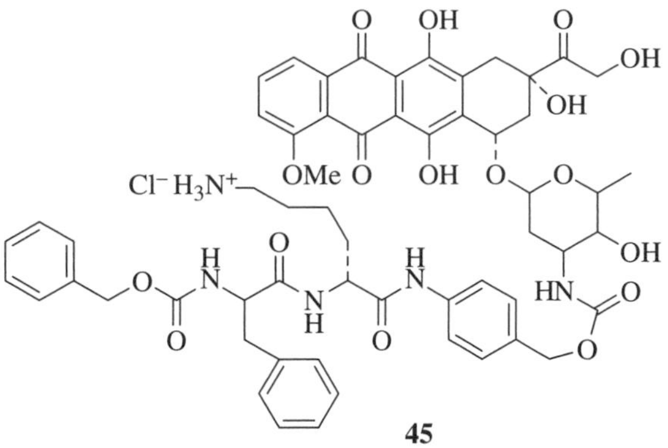

45

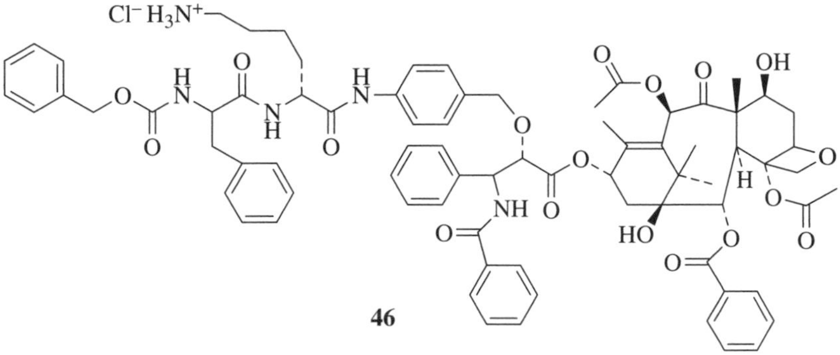

46

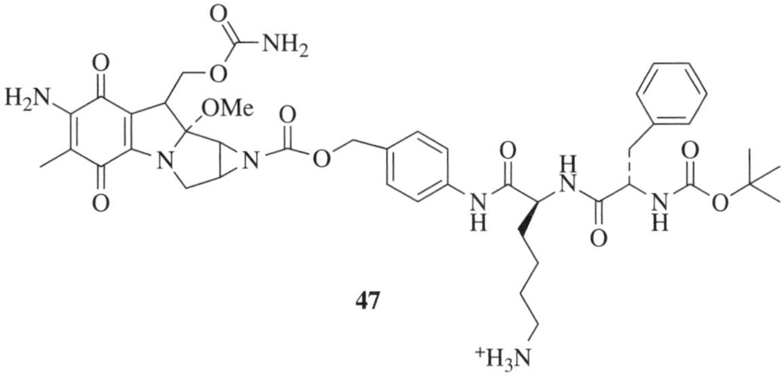

47

FIGURE 17.9 (*Continued*)

from *N*-acetylnorephedrine (Scheme 17.24). The prodrug **50** underwent hydrolysis by cathepsin B but failed to release the drug *N*-acetylnorephedrine, suggesting that the alkoxyl ether bond was not as readily cleaved as those derived from phenoxyl ether-containing drugs, such as the one in etoposide or combretastatin. This phenomenon is consistent with the fact that a phenoxyl group is a better leaving group than an alkoxyl group.

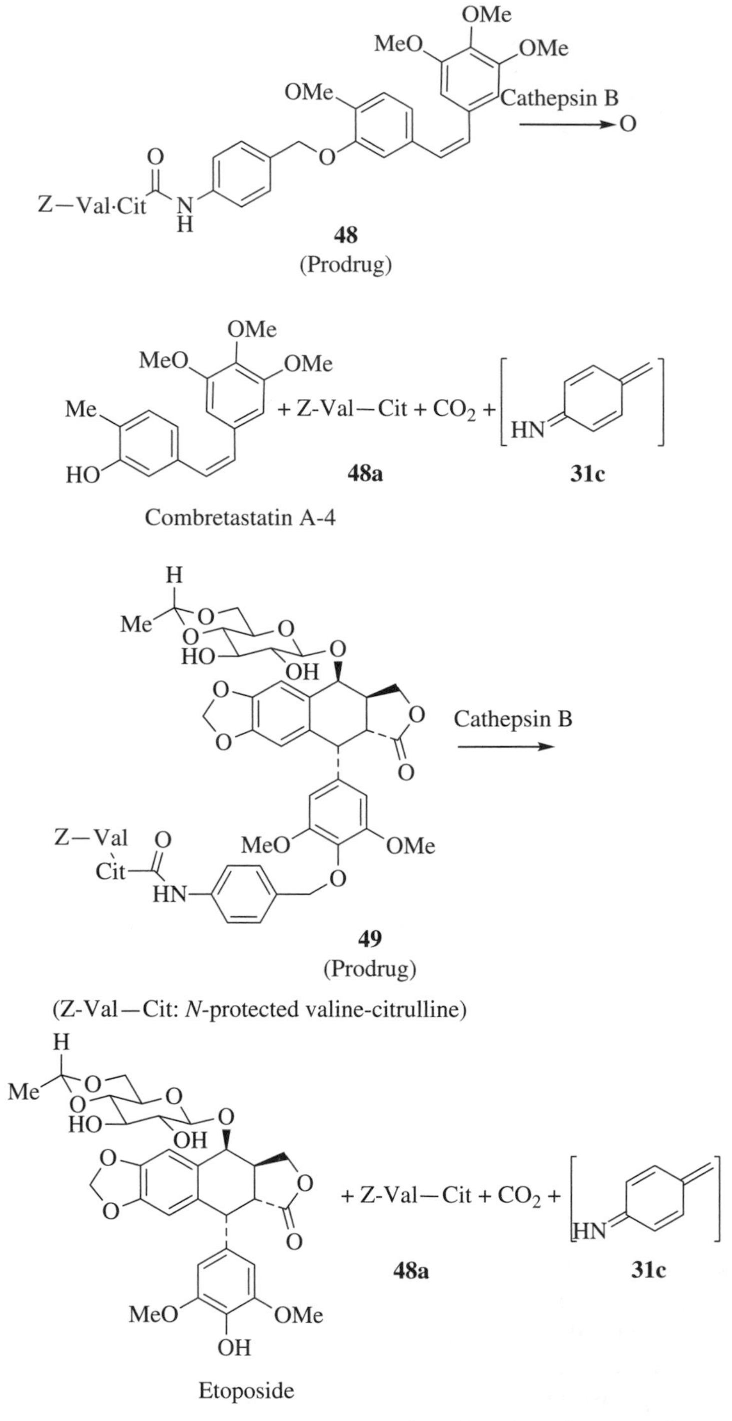

(Z-Val—Cit: *N*-protected valine-citrulline)

SCHEME 17.23 Hydrolytic bioactivation of prodrugs by cathepsin B.

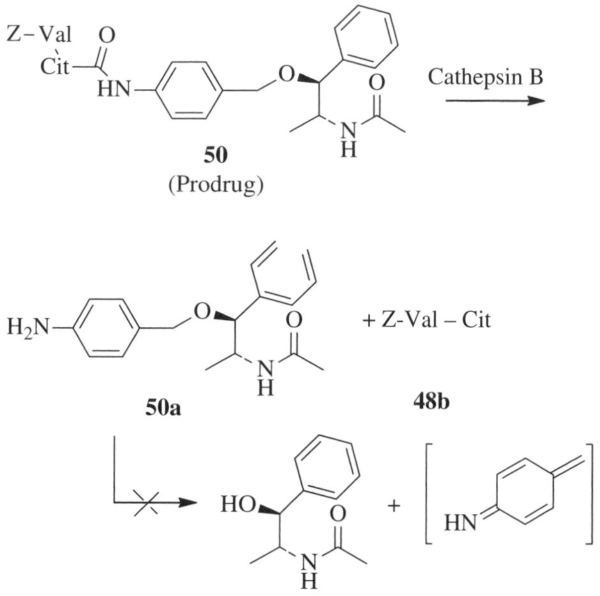

Z-Val Cit HN O O N H Cathepsin B

50 (Prodrug)

H_2N O O N H + Z-Val – Cit

50a 48b

HO O N H + [HN]

(*N*-Acetylnorephedrine) 31c

SCHEME 17.24 The hydrolyzed prodrug (**50a**) fails to undergo 1,6-fragmentation to release *N*-acetylnorephedrine.

17.2.2.4 Tumor-Selective Prodrugs Activated by MMPs MMPs are a family of structurally related zinc-containing proteases with more than 20 members [86]. An increased expression of the enzymes have been associated with cancer progression. The enzymes degrade extracellular matrix proteins and facilitates cancer cell spreading. Hu et al. reported a series of MMP-activated prodrugs by coupling Dox to peptides that are known to be substrate of MMP [87]. A general structure of the prodrugs are presented in Figure 17.10. These prodrugs are formed through linking the carboxylic acid of the peptide C-terminal directly to the amino group of Dox while the N-terminal was capped with an acetyl or other acyl groups to prevent aminopeptidase degradation before MMP cleavage. One of the prodrugs was found to be more effective than Dox with less toxicity in a mouse model [87].

MMP-activated prodrugs of paclitaxel were also reported. Yamada and colleagues reported two prodrugs of paclitaxel for MMP through coupling paclitaxel at different sites with an acetylated peptide sequence AcGly-Pro-Leu-Gly-Ile-Ala-Gly-Gln (AcGPLGIAGQ) that is known to be selectively and efficiently hydrolyzed by MMP (Fig. 17.10) [64]. However, the *in vitro* cytotoxicity of the prodrugs did not decrease as one would expect, rather the prodrugs were more cytotoxic than paclitaxel. The authors attribute this to an increase in hydrophilicity of the prodrugs.

17.2.2.5 Tumor-Selective Prodrugs Activated by Prolidase Prolidase is a cytosolic exopeptidase that cleaves imidodipeptides with C-terminal L-proline. The substrate requirements for prolidase also include the presence of a free α-amino

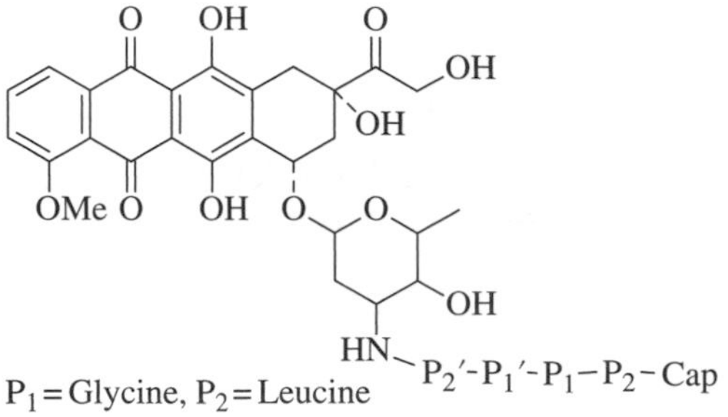

P_1 = Glycine, P_2 = Leucine

HN–P_2'-P_1'-P_1-P_2-Cap

51

MMP-activated prodrugs of Dox

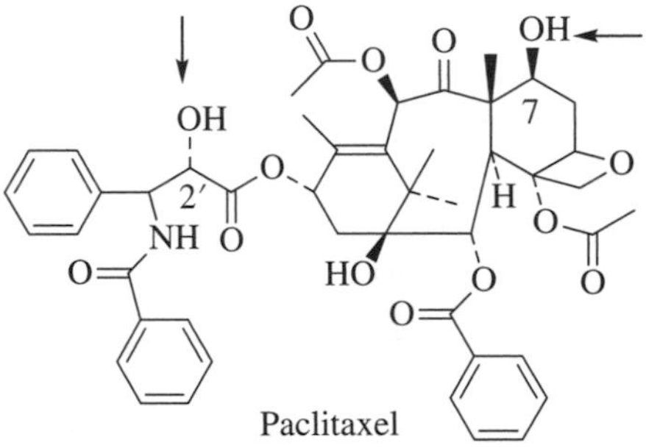

52: 2′-O-R
53: 7-O-R

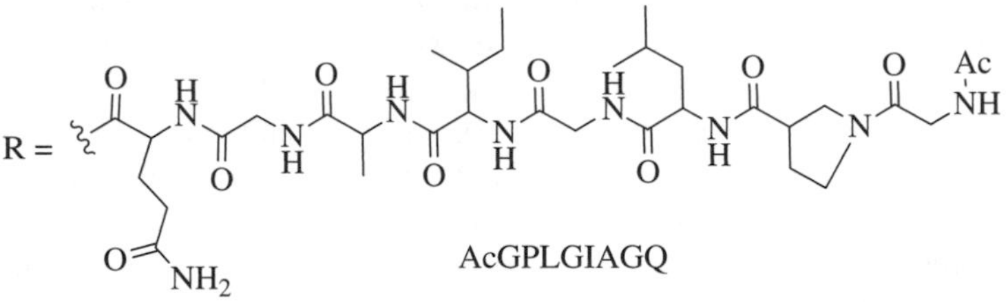

Paclitaxel

R =

AcGPLGIAGQ

MMP-activated prodrugs of Paclitaxel

FIGURE 17.10 Chemical structures of representative MMP-activated prodrugs of Dox and paclitaxel.

group. The function of the enzyme is primarily for the metabolism of proline-containing protein degradation products and the recycling of proline from imidodipeptides for resynthesis of proline-containing proteins, mainly collagen [88]. Increased prolidase activity has been found in some tumor tissues [63, 88]. Prolidase-activated prodrugs of melphalan and carmustine were reported through coupling a proline-containing dipeptide or proline with the anticancer drugs (Fig. 17.11) [63, 88]. However, in both reports, the IC_{50} values of the prodrugs are not significantly different than the corresponding anticancer drug [63, 88].

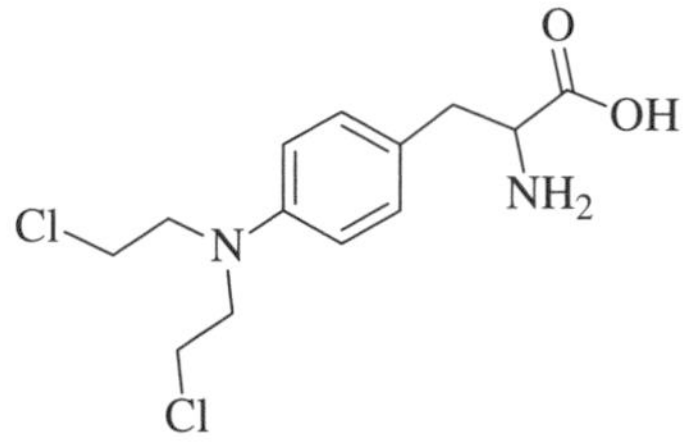

Melphalan

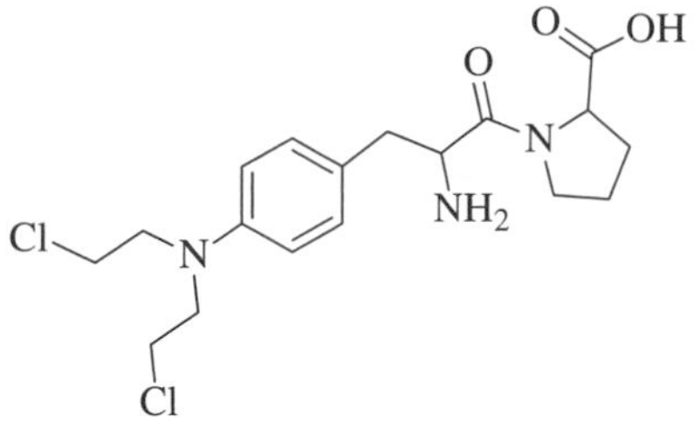

Prodrug of melphalan

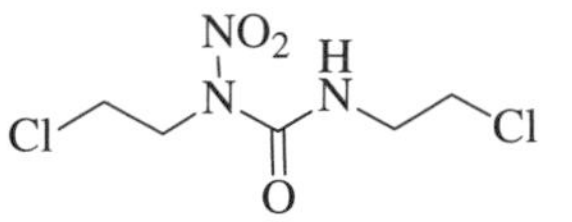

Carmustine

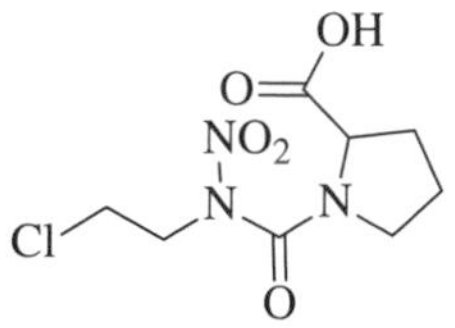

Prodrug of carmustine

FIGURE 17.11 Chemical structures of prolidase-activated prodrugs, and their corresponding anticancer drugs.

17.2.3 Tumor-Activated Prodrugs Based on Enzymes with Elevated Activity at Tumor Sites

17.2.3.1 β-Glucuronidase β-Glucuronidase is an exoglycosidase that cleaves glucuronosyl-*O* bonds [89]. Glucuronidase is intracellularly located in lysosomes in many organs and body fluids such as macrophages, most blood cells, liver, spleen, kidney, intestine, lung, muscle, bile, intestinal juice, urine, and serum [89]. There is large interindividual variability in its activity and expression. The activity of the enzyme is high in some tissues such as the liver. The rationale of selecting β-glucuronidase to activate anticancer prodrugs is based on the observation that the activity of the enzyme has been shown to be elevated in many tumors [12, 90]. Further, the activity of β-glucuronidase is high at low pH values as found in tumor tissues and low at neutral pH, thus favoring a prodrug activation in tumor tissues [9]. An additional advantage of targeting β-glucuronidase is that a glucuronide prodrug exhibits high hydrophilicity, which greatly reduces the distribution of the prodrug into cells resulting in a reduced systemic toxicity. However, the same feature also hampers the activation of a glucuronide prodrug at tumor sites. Since the enzyme is intracellularly located, a glucuronide prodrug needs to enter tumor cells in order to be activated. Bosslet et al. demonstrated that β-glucuronidase is liberated extracellularly in high concentration in necrotic areas of human cancer [91]. Therefore, necrotic areas are the areas where glucuronide prodrugs can be activated. It is worth mentioning that an increase in tissue/serum β-glucuronidase activity has also been observed in other disease states—for example, inflammatory joint disease, ichthyosiform dermatosis, some hepatic diseases, and AIDS. For example, serum β-glucuronidase activity has been reported to be 16-fold higher in HIV-infected patients than in healthy individuals [92]. The extracellular presence of β-glucuronidase in areas other than tumor sites will lead to prodrug activation and side effects.

The β-glucuronidase-mediated release approach has been used in prodrugs of a number of anticancer drugs. The prodrugs are formed by linking a drug to glucuronic acid through an anomeric ether bond directly or more often indirectly (Scheme 17.25). In the latter case, a spacer is used. In general, prodrugs with a spacer are cleaved more readily by the enzyme, especially when the parent drug molecules are bulky. Figure 17.12 shows the structures of representative prodrugs activated by β-glucuronidase. Prodrugs **55** and **56** are prodrugs of Dox. Both were much less toxic than the parent drug in the human ovarian cancer cell line (OVCAR-3), while prodrug **56**, with a spacer between the drug and glucuronyl group, was activated much faster by β-glucuronidase [93]. Similarly, daunorubicin prodrug **57** showed 200 times less toxic than the parent drug and readily cleaved by β-glucuronidase in 2 hours [94]. The spacers used most extensively are *p*- or *o*-hydroxybenzyl alcohol moieties.

Florent et al. evaluated a series of anthracyclines glucuronide prodrugs with a *p*- or *o*-hydroxylbenzyl alcohol as a spacer (Fig. 17.12) [95, 96]. From these prodrugs, compound **58d** appeared most promising for further development due to its reduced cytotoxicity and fast hydrolysis by β-glucuronidase.

The camptothecins are a class of promising anticancer agents, of which several derivatives are clinically used. Prodrug **60** (Fig. 17.13) of 9-aminocamptothecin showed a 20- to 80-fold reduced toxicity in comparison with 9-aminocamptothecin [97]. The prodrug was readily cleaved by β-glucuronidase *in vitro*.

Schmidt et al. prepared and evaluated compound **61** as a prodrug of a phenol mustard. After removal of glucuronic acid by β-glucuronidase, the spacer was eliminated through cyclization liberating the phenol mustard **61c** [98]. The prodrug was 80-fold less cytotoxic in comparison with the phenol mustard in LoVo cells. Chemically, the prodrug (**61**) was much more stable than the corresponding parent drug (**61c**) in phosphate buffer (Scheme 17.26).

Another prodrug of **61c** is compound **62**. Prodrug **62**, which contains an aromatic and aliphatic bis-carbamate spacer, is activated by β-glucuronidase [99]. A rapid cleavage of the glycosidic bond occurred ($t_{1/2}$ = 6.6 minutes) with concomitant appearance of intermediate **62c**, of which the ethylene diamine spacer cyclized with a

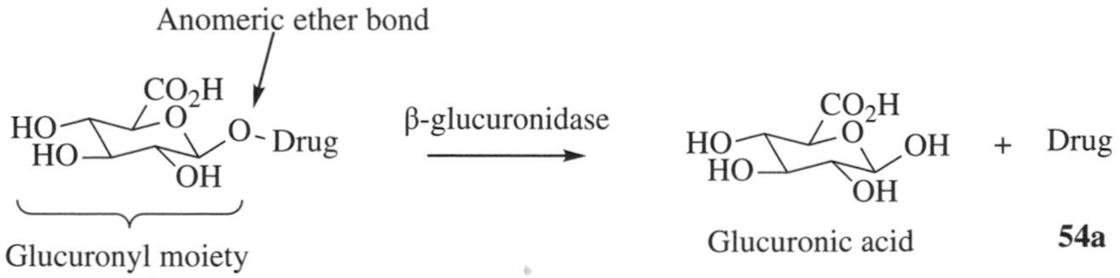

SCHEME 17.25 Bioactivation of glucuronide prodrugs by β-glucuronidase.

half-life of 2 hours (Scheme 17.27). The cytotoxicity of **62** against LoVo cells was about 50 times less than that of the corresponding phenol mustard **61c**.

A similar approach was also applied in paclitaxel prodrugs **63** and **64** [100]. After hydrolysis by β-glucuronidase, ring closure occurred resulting in the release of paclitaxel (Schemes 17.28 and 17.29). The half-lives of β-glucuronidase-mediated activation were 2 hours and 45 minutes for prodrugs **63** and **64b,** respectively. Both prodrugs were two orders of magnitude less cytotoxic than paclitaxel. Prodrug **64a** appeared as cytotoxic as paclitaxel, which was explained by the fact that the prodrug underwent spontaneous hydrolysis in buffer solution under physiological conditions.

Glucuronide prodrugs **65a, 65b,** and **66** of 5-fluorouracil exhibited half-lives of approximately 20 minutes when incubated with 25 μg/ml of β-glucuronidase and reduced (6–9 times less) cytotoxicity against LoVo cells in comparison with the parent drug 5-fluorouracil (Fig. 17.14) [101].

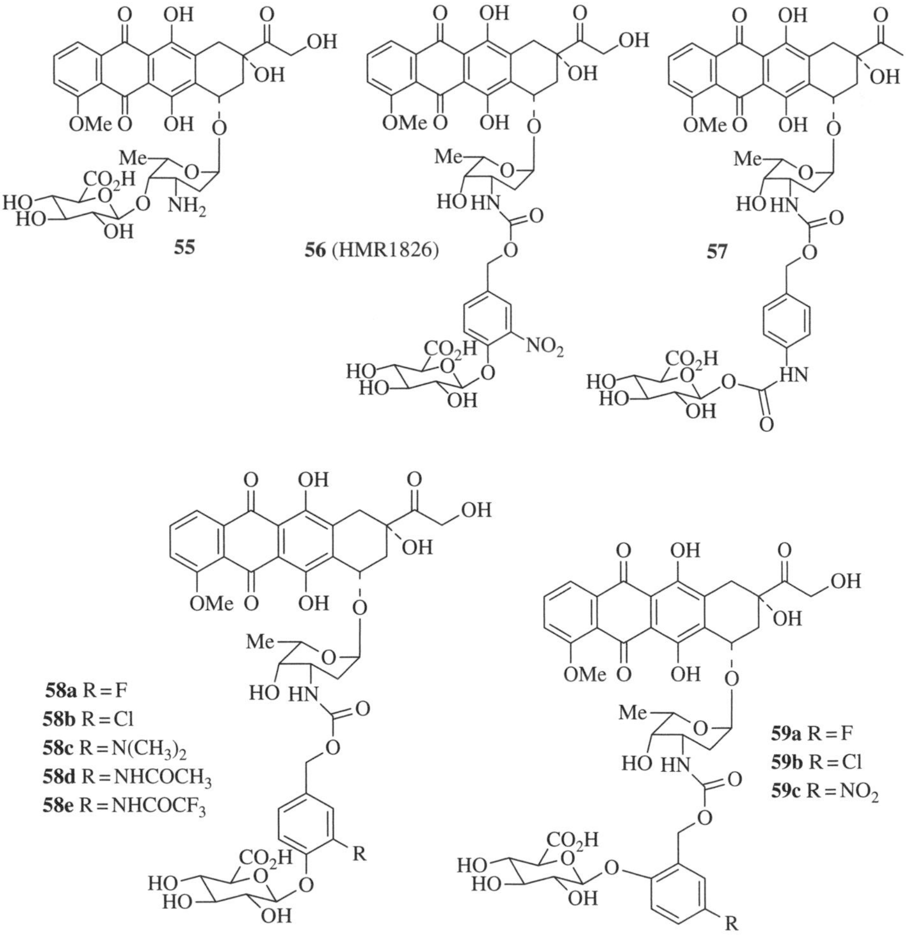

55 **56** (HMR1826) **57**

58a R = F
58b R = Cl
58c R = $N(CH_3)_2$
58d R = $NHCOCH_3$
58e R = $NHCOCF_3$

59a R = F
59b R = Cl
59c R = NO_2

FIGURE 17.12 Representative prodrugs bioactivated by β-glucuronidase.

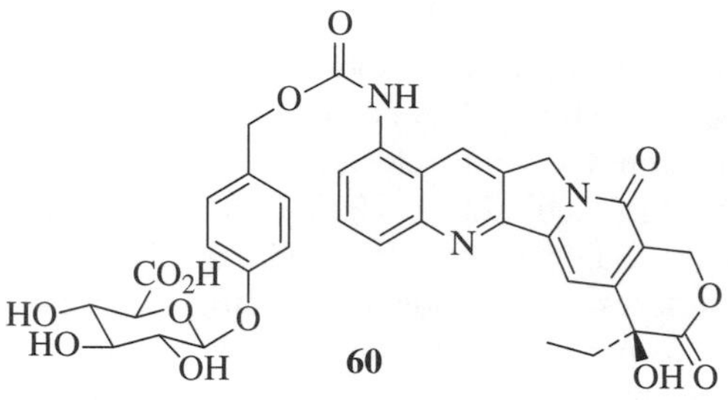

60

FIGURE 17.13 Structure of a glucuronide prodrug of 9-aminocamptothecin.

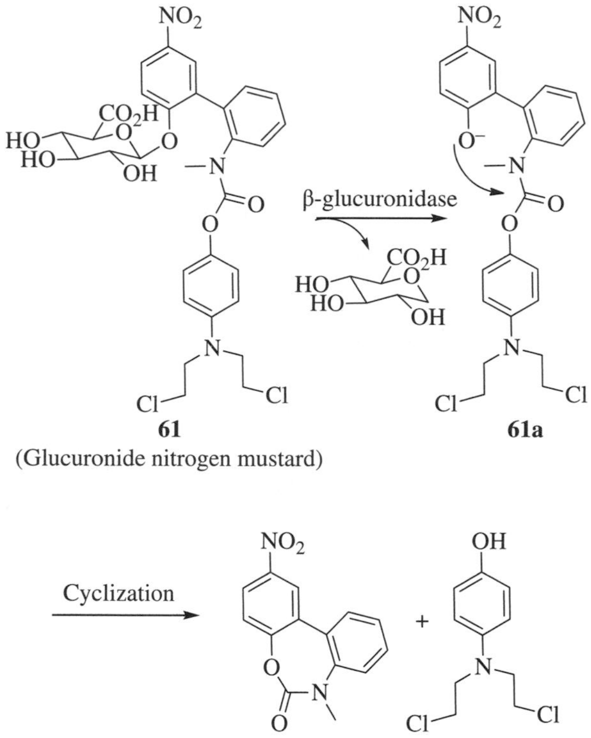

61
(Glucuronide nitrogen mustard)

61a

61b 61c

SCHEME 17.26 Activation of glucuronide nitrogen mustard prodrug **61**.

17.2.3.2 Prostate-Specific Antigen and Prostate Membrane-Specific Antigen Prostate-specific antigen (PSA) is a serine protease, which is present extracellularly in prostate cancers. The PSA is inhibited in the blood stream. As a result, active PSA is only present in prostate cancer [4, 7]. Several peptide sequences have been identified as PSA substrates, and the most widely used sequences are His-Ser-Ser-Lys-Leu-Gln and Ser-Ser-Lys-Tyr-Gln [7, 102]. A prodrug (**67**, Fig. 17.15)

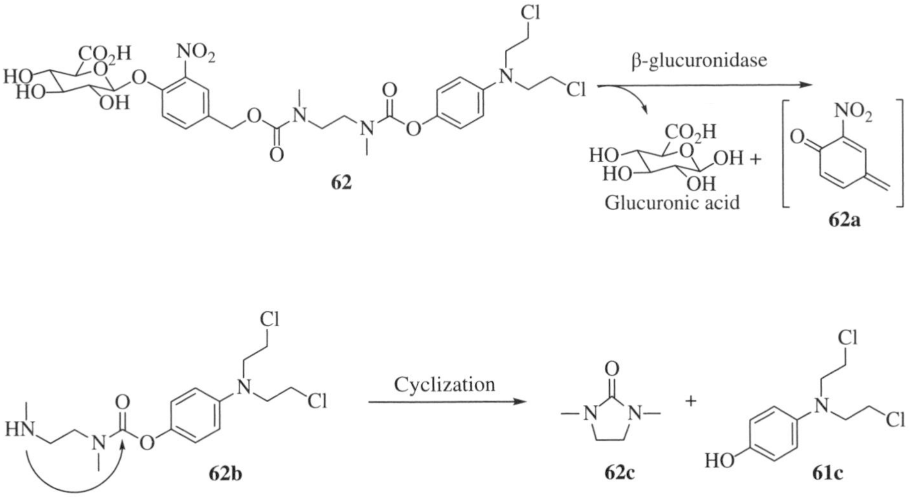

SCHEME 17.27 Glucuronidase-mediated bioactivation of prodrug **62**.

of Dox was formed by coupling the C-terminal carboxylic acid group of a heptapeptide to Dox. The sequence His-Ser-Ser-Lys-Leu-Gln was selected because of specificity and serum stability [102]. In compound **67**, an extra leucine residue was added after glutamine, which increases the distance between the peptide bond to be cleaved (leucine–glutamine) and Dox. This increased distance reduced the steric hindrance that Dox posed to the leucine–glutamine bond and facilitated the bond cleavage by PSA [68]. The prodrug **67** underwent cleavage by PSA at the leucine–glutamine bond to release not Dox but Dox–leucine, which was an anticancer agent itself. *In vitro* selectivity was demonstrated by the fact that 70 nM of the prodrug killed 50% of the PSA-producing human prostate cancer cells (LNCaP cells), whereas doses as high as 1 μM had no cytotoxic effect on PSA-nonproducing TSU human prostate cancer cells [68]. Other anticancer drugs like 5-fluorodeoxyuridine (5FudR), paclitaxel, cyclopamine, and thapsigargin have also been conjugated to these peptides [7].

PSMA is a type II membrane glycoprotein (100 kDa) and a well-known tumor antigen. The expression of PSMA in prostate cancer cells, especially the most advanced androgen-resistant prostate cancer cells, is about a 1000-fold greater than the expression in normal tissues. However, a limited PSMA substrate peptide has limited the application of the enzyme for prostate cancer–selective prodrug design [7].

17.2.3.3 P450 as Activating Enzymes for Anticancer Prodrugs Cytochrome P450 (P450) comprises a superfamily of enzymes involved in the oxidation of a great number of exogenous and endogenous compounds. Oxidation of compounds by P450 increases polarity and aids further metabolism or removal of the compound from the body. Therefore, P450 enzymes are viewed as the most important enzymes in removing exogenous compounds or toxic molecules from the body. P450 enzymes

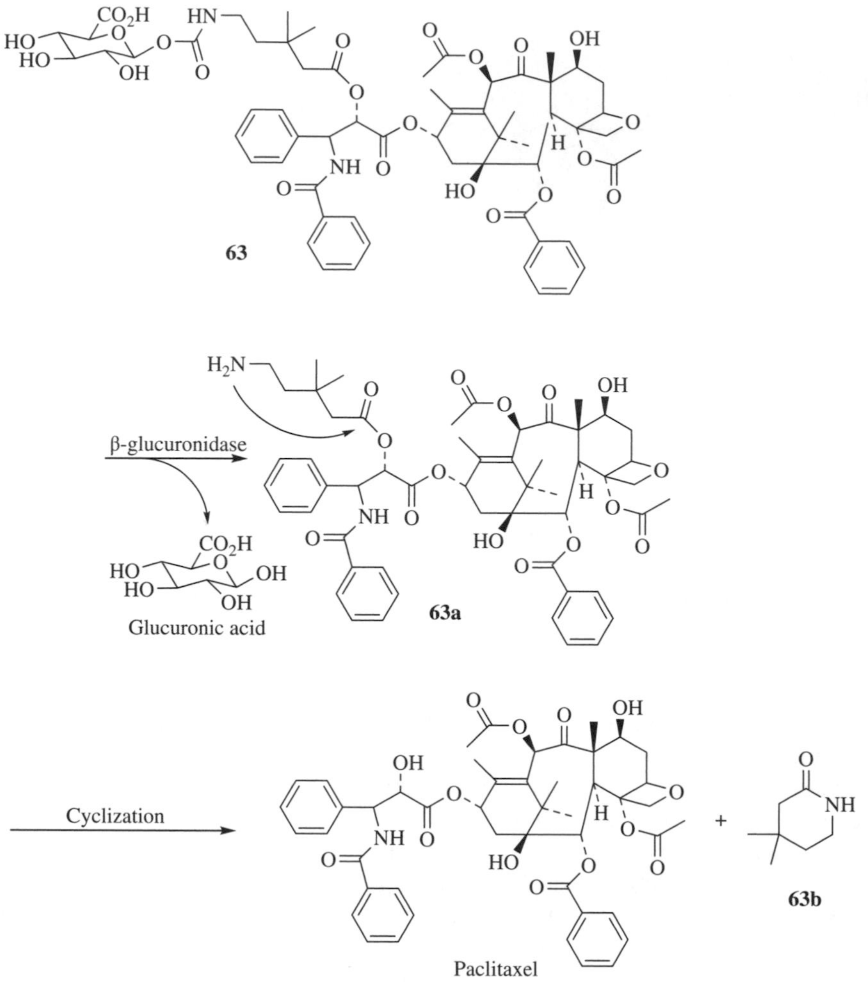

SCHEME 17.28 Bioactivation of prodrug **63** by β-glucuronidase.

are present in normal tissues with the highest levels in the liver consistent with the role of the liver as the detoxification organ. P450 enzymes are not only involved in inactivation of anticancer drugs but also in activation of several anticancer drugs such as cyclophosphamide (CYP3A4) and its isomer ifosfamide (CYP2B6), tegafur (CYP2A6), dacarbazine (DTIC) (CYP1A1 and 1A2), procarbazine (CYP1A1), flutamide (CYP1A2 and 3A4), and tamoxifen (CYP3A4, 2D6, and 2C9) [12]. As expected, P450-activated anticancer prodrugs are not highly tumor selective since most of these prodrugs are activated by the liver P450. These prodrugs will not be discussed here. It is worth noting that the levels of P450 in tumors vary: some tumors contain higher and some lower than normal tissues [12, 90]. Efforts to increase tumor selectivity of P450-activated anticancer prodrugs have been made.

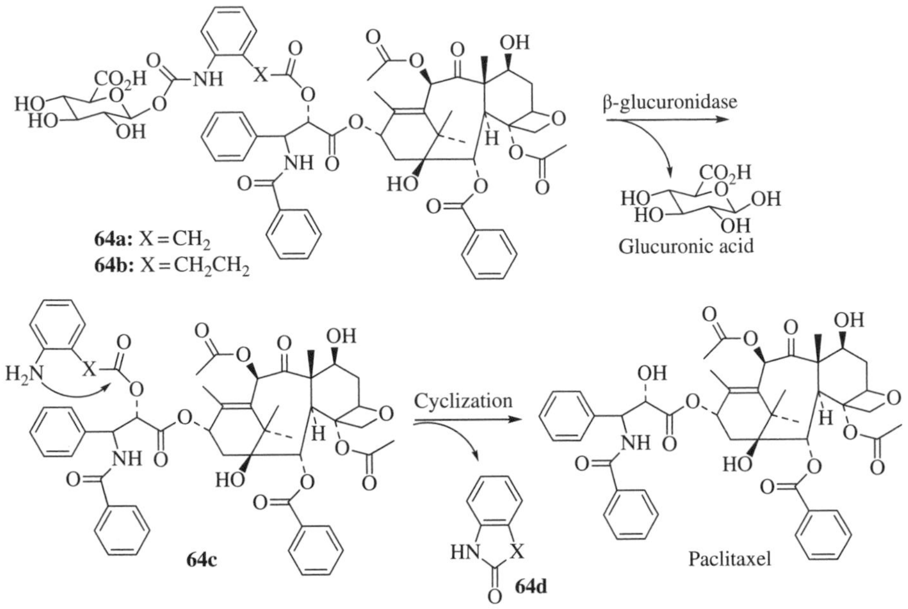

SCHEME 17.29 Bioactivation of prodrug **64a** and **64b** by β-glucuronidase.

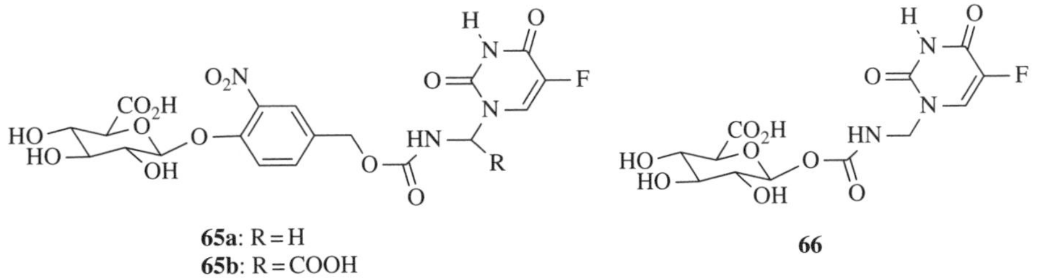

FIGURE 17.14 Structures of glucuronide prodrugs **65a, 65b,** and **66**.

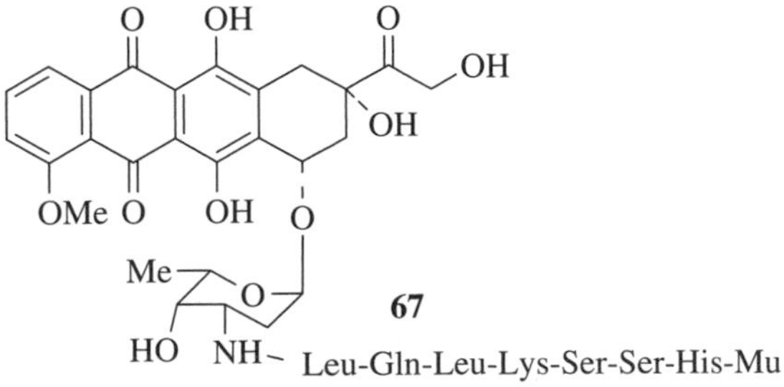

FIGURE 17.15 Structure of a glucuronide prodrug activated by prostate-specific antigen (PSA). Mu, morpholinocarbonyl.

Of particular interest is the approach to deliver a P450 enzyme gene to tumor sites through GDEPT [103–106]. In addition, efforts to increase tumor selectivity and decrease systemic toxicity through GDEPT in combination with inhibition of the liver P450 have also been reported [107, 108]. A more detailed discussion of GDEPT is presented in 17.3.

Other enzymes of mammalian origin that have been employed for anticancer prodrug activation include β-D-galactosidase [9], GST [90], phospholipase A2 [10], and cytidine deaminase [2]. Some of these enzymes will be discussed later.

17.3 ANTIBODY- AND GENE-DIRECTED ENZYME PRODRUG THERAPY

ADEPT involves two steps [14, 109]. The first step is to deliver a chosen enzyme to the surface of cancer cells by a monoclonal antibody (mAb), followed by a second step: administration of a prodrug that is activated by the enzyme to release the active parent drug. Similarly GDEPT, also named as SGT, involves two steps as well [9, 109]. In the first step, a gene encoding a chosen enzyme is transported to cancer cells. After expression of the chosen enzyme, the prodrug is administered. In ADEPT, the enzyme is attached to the surface of cells while in GDEPT, the enzyme can be expressed intracellularly or extracellularly [110, 111]. The potential advantages of extracellular expression of the enzyme are twofolds. First, it gives an improved bystander effect because the active drug will be generated in the interstitial spaces within the tumor rather than inside as with an intracellularly expressed activating enzyme. Second, the prodrug does not need to enter cells to be activated and, therefore, noncell permeable prodrugs can be used.

Enzymes being used for ADEPT or GDEPT can be divided into three major classes: (i) enzymes of nonmammalian origin that have no mammalian homologues, (ii) enzymes of nonmammalian origin with a mammalian homologue, and (iii) enzymes of mammalian origin [112]. Examples of enzymes in class I include carboxypeptidase G2 (CPG2), β-lactamase, penicillin G amidase, cytosine deaminase (CD), herpes simplex virus thymidine kinase, etc. The rationale of using such enzymes is that prodrugs designed based on these enzymes will not be activated by endogenous human enzymes. Since many of the nonmammalian enzymes are from bacteria or easily expressed in bacteria, they are available in large quantities. The main disadvantage of employing these enzymes is that they elicit immune responses in humans. The criterion for selecting enzymes in class II is that the level or activity of the endogenous counterparts should be low. Examples of class II enzymes include β-glucuronidase, β-D-galactosidase, nitroreductase (NTR), etc. One of the advantages of employing class II enzymes is that high catalytic efficiency enzymes of nonmammalian can be selected. For example, bacterial β-glucuronidase exhibits much higher turnover rate than its human counterpart. Additionally, the exogenous enzyme could differ significantly enough from the endogenous counterpart that a prodrug can be designed to be cleaved only by the exogenous enzyme. As with the class I enzyme, class II enzymes also suffer from limitations due to immunogenicity.

The major advantage of using class III enzymes is that they are much less immunogenic than bacterial or fungal enzymes. The obvious disadvantage is that the designed prodrugs can also be activated by endogenous enzymes present in healthy cells resulting in nonspecific activation and side effects. Enzymes belonging to this class include alkaline phosphatase, carboxypeptidase A, β-glucuronidase, cytochrome P450, etc.

Clearly, in both ADEPT and GDEPT, the activity of the enzyme activating a prodrug is enhanced at tumor sites leading to an increased tumor selectivity of the prodrug.

17.3.1 ADEPT

Monoclonal antibodies (mAbs) have been used to deliver chemotherapeutic drugs [113, 114], potent plant and bacterial toxins [115], and radionuclides [116] to tumor sites. A number of mAb-based drugs have been approved clinically (Rituxan, Herceptin, and Panorex), and several others are in advanced clinical trials. However, application of ADEPT for solid tumor has been limited. The major challenge for developing mAb-based drug therapy for solid tumors lies, in part, in the barriers of macromolecule penetration within the tumor masses and the heterogeneity in target antigen expression [14, 112]. The penetration issue has prompted research to find an alternative strategy that can dissociate the drug from the mAb delivery system after reaching the target. ADEPT was a result of this effort. In the ADEPT approach, a chosen enzyme, not a prodrug, is conjugated to a tumor-specific mAb or a mAb fragment to form a mAb-enzyme conjugate. After administration, the conjugate will bind to the corresponding antigen that is located on the surface of tumor cells. A clearance period is given to allow removal of nonantigen-bound mAb–enzyme conjugate from the body before a prodrug is administered. The administered prodrug will now selectively be activated by the enzyme attached to the surface of the cancer cells. The advantages of ADEPT include the following: (i) the drug is not linked to the antibody and can spread through diffusion, (ii) the drug will be continuously generated at the tumor site overcoming the delivery capacity hurdle related to the situation when a drug is delivered by linking to an antibody, and (iii) cell membrane–impermeable prodrugs can also be activated since the enzyme is present on the surface of cancer cells. This provides a way to minimize systemic cytotoxicity by developing cell membrane–impermeable prodrugs for ADEPT application. There are disadvantages of ADEPT as discussed by Bagshawe [14]. The main problem is the immunogenicity of the antibody–enzyme conjugate. Clearance of the conjugate in the blood circulation before prodrug administration is also an issue. Any presence of the conjugate in the blood circulation after prodrug administration can lead to activation of the prodrug in the circulation, resulting in systemic cytotoxicity [14]. Various ways of accelerating the clearance of the conjugate or even inactivating the enzyme of the conjugate in the circulation have been reported [14]. It was concluded that a successful ADEPT system needs to include an antibody–enzyme conjugate, an enzyme-inactivating/an enzyme-clearing agent, and a prodrug [14].

Various prodrug-activating enzymes have been employed in ADEPT. These enzymes include those that are of nonmammalian origin (carboxypeptidase G2, β-lactamase, and CD) or mammalian origin but with very low activity in serum (β-D-galactosidase and NTR) [9, 14]. A large number of prodrugs derived from anticancer drugs such as anthracyclines, mustards, methotrexate, 5-fluorouracil, etoposide, vinblastine, paclitaxel, or camptothecin have been proposed for application in ADEPT [9]. It has been suggested that an anticancer drug with a short half-life, such as nitrogen mustard, is more appropriate than an anticancer drug with a long half-life as a candidate for prodrug design for ADEPT application [14]. The reason for that is an anticancer drug with a short-life does not diffuse far to cause systemic toxicity after activation. In this chapter, ADEPT will be introduced through three representative examples with CPG2, β-lactamase, and β-D-galactosidase as activating enzymes.

17.3.1.1 ADEPT with CPG2 CPG2 is a bacterial zinc-dependent metalloproteinase [117]. This enzyme is an exoprotease that specifically cleaves terminal glutamic acid amides [112]. Niculescu-Duvaz et al. described an ADEPT system that uses a combination of mAb, CPG2, and nitrogen mustard prodrugs [117]. CPG2 is coupled to $F(ab')_2$ fragments of the mAb A5B7. A5B7 is an antihuman carcinoembryonic antigen (CEA) mouse mAb. The antibody–enzyme conjugate ($F(ab')_2$-CPG2 conjugate) was investigated for its ability to activate the prodrug (2-chloroethyl)-(2-mesyloxygen)amino-benzoyl-L-glutamic acid (CMDA) and other *N*-L-glutamyl amide nitrogen mustard prodrugs. The general structural features of the prepared *N*-L-glutamyl amide nitrogen mustard prodrugs are shown in Scheme 17.30.

The bond to be cleaved by CPG2 is the amide bond derived from L-glutamic acid. An additional advantage of including the L-glutamic acid moiety in the prodrug is the increased hydrophilicity. As mentioned earlier, in ADEPT an activating enzyme is anchored on the outer membrane of tumor cells. It will be advantageous if the prodrug

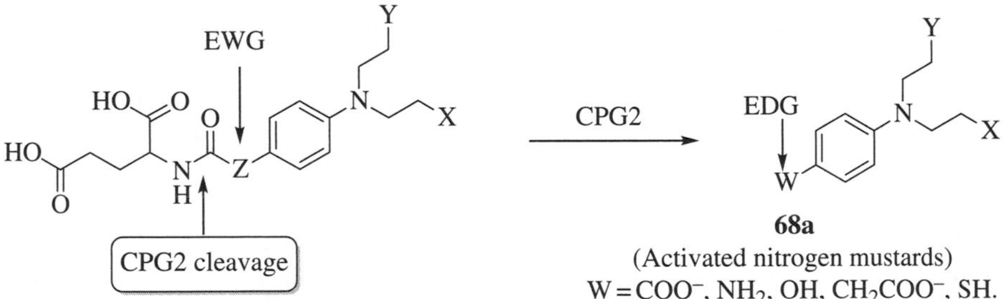

68
(*N*-L-glutamyl amide nitrogen mustard prodrugs)
X, Y = Cl, Br, I, OSO_2CH3
Z = –, O, NH_2, CH_2, S
CMDA: X = Cl, Y = OSO_2CH_3, Z = –.

68a
(Activated nitrogen mustards)
W = COO^-, NH_2, OH, CH_2COO^-, SH.

SCHEME 17.30 General structures of *N*-L-glutamyl amide nitrogen mustard prodrugs and activation of the prodrugs by CPG2.

is more hydrophilic than the parent drug since high hydrophilicity will reduce the distribution of the prodrug and systemic side effects. The original work was conducted by Springer et al. who used nude mice implanted with chemoresistant choriocarcinoma xenografts as an animal model for the study of CMDA (Scheme 17.30) [118]. It was demonstrated that 9 out of 12 mice were long-time survivors (>300 days), whereas all control mice were dead by day 111. A clinical trial was also conducted in patients with advanced colorectal carcinoma of the lower intestinal tract [13, 119]. A dose of 20,000 enzyme units/minute [2] mAb-CPG2 conjugate was administered in the first step. This conjugate dose gave tumor CPG2 levels comparable to the values found optimum in nude mice bearing xenografts. This treatment was followed 24–48 hours later by administration of a clearing agent, anti-CPG2 galactosylated Ab (220 mg/minute [2]), to help removal of the mAb-CPG2 conjugate that was not bound to the antigen. In the last step, prodrug CMDA was injected over 1–5 days up to a total dose of 1.2–10 g/minute [2]. Oral cyclosporine was coadministered to suppress the host immune response. From eight evaluable patients, there were four partial responses and one mixed response. Another clinical trial with patients with colorectal carcinoma expressing CEA using a similar approach demonstrates that the median tumor:plasma ratio of enzyme (CPG2) exceeded 10,000 : 1 at the time of prodrug administration. Enzyme concentrations in the tumor were sufficient to generate cytotoxic levels of active drug [120]. Tumor response to the ADEPT was observed [120]. Thus the results obtained are encouraging and prove the feasibility of the ADEPT at the clinical level.

17.3.1.2 ADEPT with β-Lactamase Activation of β-lactam-based prodrugs is based on the well-established β-lactam chemistry. It was demonstrated that a molecule attached to the 3′-position of cephalosphorins was eliminated through a 1,4-fragmentation reaction [112]. Scheme 17.31 shows a general scheme of this reaction, and Figure 17.16 shows the structures of the prodrugs of a vinca derivative (**70**), phenylenediamine mustard (**71**), Dox (**72**), melphalan (**73**), paclitaxel (**74**), and mitomycin (**75**) prepared based on this chemistry [112]. The first report of *in vivo* activity in a mAb–lactamase system used the β-lactamase from *Enterobacter cloacae* and a cephalosphorin-vinca alkaloid prodrug LY266070 (**70**, Fig. 17.16) [121]. Nude mice implanted with human colorectal carcinoma were used for the investigation. The tumor inhibitory effects of the ADEPT with mAb–lactamase conjugate and LY266070 were superior to prodrug alone. The effects were also superior to those obtained when the vinca alkaloid (Fig. 17.16) was attached directly to the mAb. Long-term regressions of established tumors were observed in several dosing regimens, even in animals having tumors as large as 700 mm [3] at the initiation of therapy [121].

In a related study, Kerr et al. reported that the ADEPT treatment with a combination of a mAb–lactamase conjugate and a prodrug of cephalosphorin-phenylenediamine mustard (**71**) in nude mice bearing subcutaneous 3677 human tumor xenografts produced regression in all the treated mice at doses that caused no apparent toxicity [122]. At day 120 post tumor implant, four of five mice in this treatment remained tumor free. Significant antitumor effects were even seen in mice that had large (800 mm [3]) tumors before the first prodrug treatment.

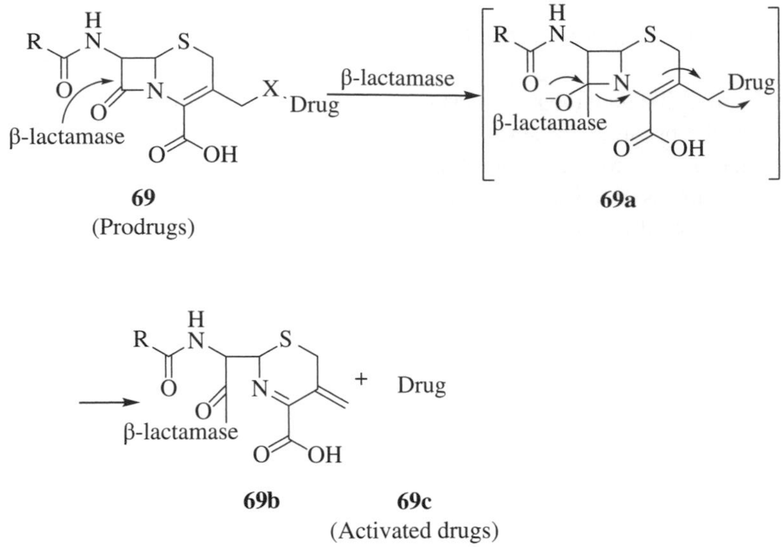

69 (Prodrugs) → β-lactamase → 69a → 69b + 69c (Activated drugs)

SCHEME 17.31 Bioactivation of prodrugs by β-lactamase.

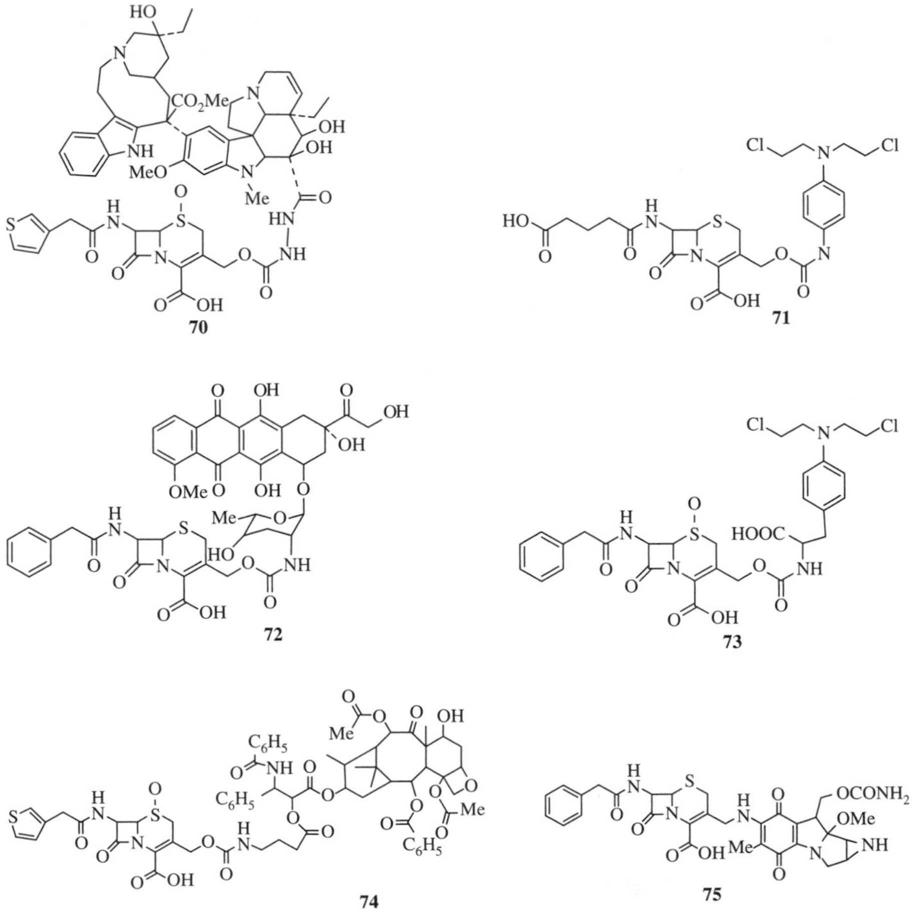

70 71 72 73 74 75

FIGURE 17.16 Structures of some representative prodrugs that are activated by β-lactamase.

17.3.1.3 ADEPT with β-D-Galactosidase β-D-galactosidase is present in humans. However, the enzyme is not detectable in serum. Further, the activity of the enzyme is rather low compared to the β-D-galactosidase from *Escherichia coli*. Moreover, addition of a galactose to a drug molecule significantly increases hydrophilicity and reduces its ability to penetrate cell membrane. Tietze's group developed a series β-D-galactose prodrugs of duocarmycin analog for application in ADEPT [9]. Scheme 17.32 demonstrates that the β-D-galactose moiety of the monoglycosidic prodrug **76** was cleaved by β-D-galactosidase or antibody–glycohydrolase conjugate. The resulting **76a** underwent rapid intramolecular cyclization (Winstein cyclization) to yield the active drug **76b**. An excellent therapeutic index was observed for **76b** with a QIC_{50} value of 4800 [9]. $QIC_{50} = IC_{50}$ (Prodrug)/IC_{50} (Drug) [9]. A minimum QIC_{50} value of 100 is suggested for an anticancer prodrug application [14].

Tietze's group also synthesized diglycosidic prodrugs as represented by **77** (Scheme 17.32) [9]. The diglycosidic prodrug exhibited less cytotoxicity than the corresponding monoglycosidic prodrug due to a reduced cellular uptake resulted from the increased hydrophilicity. Also, the cytotoxicity of **77** in the presence of β-D-galactosidase is identical to **77a** (Scheme 17.32), indicating a fast enzymatic cleavage. As a result, prodrug **77** exhibited an outstanding QIC_{50} value (6500), providing an excellent candidate for ADEPT application [9].

17.3.2 GDEPT

In GDEPT, vectors encoding the prodrug-activating enzymes are used to transduce genes into the genome of tumor cells where they lead to the expression of the exogenous enzyme [9, 17]. Successful GDEPT depends on a successful delivery of prodrug-activating enzyme gene vectors to tumor cells with high transfection efficiency, low toxicity, and low immunogenicity. There are currently two main vehicles for the delivery of gene vectors: viral and nonviral. The viral vectors can be prepared from lentivirus, poxviruses, herpes simplex virus, vaccinia virus, retroviruses, adenovirus, and adeno-associated viruses. The major advantage of viral vectors is its high transfection efficiency, but the application is limited due to safety concern and immunogenicity [17]. The safety concern has led to an increased interest in nonviral vector delivery systems. The nonviral vector delivery system includes naked DNA, cationic liposomes, polyethyleneimine, and recent biocompatible nanoparticles. Another way of delivery is to inject the vector directly to tumor (intratumor injection) [9, 15, 17, 109]. Despite various viral and nonviral vectors, only a small subpopulation of cancer cells were transfected, and a short-term expression of the gene was observed [15]. This makes the bystander effect crucial to extend cytotoxic effects from transfected cells to nontransfected neighboring cells [18].

An alternative way of using GDEPT to treat tumors is to express the suicide gene from genetically modified cells implanted near the tumor site through encapsulating the cells in microbeads. The implantation of encapsulated cells is expected to target

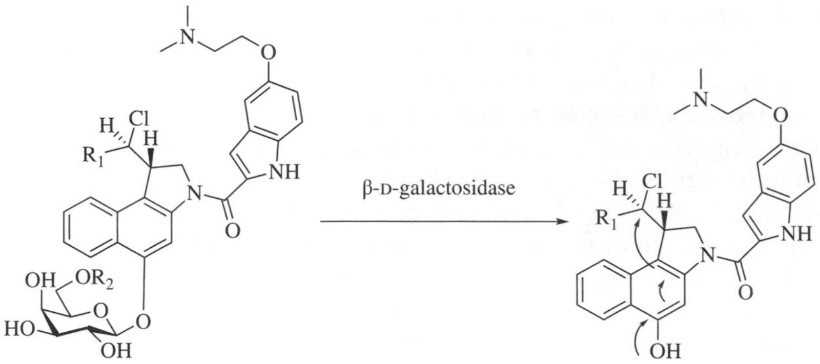

β-D-galactosidase

76: Monoglycosidic prodrug: $R_1 = Me$; $R_2 = H$.

77: Diglycosidic prodrug: $R_1 = H$, $R_2 =$

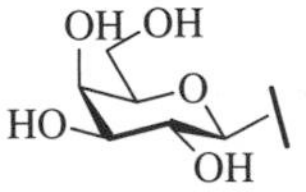

76a: $R_1 = Me$; **77a**: $R_2 = H$.

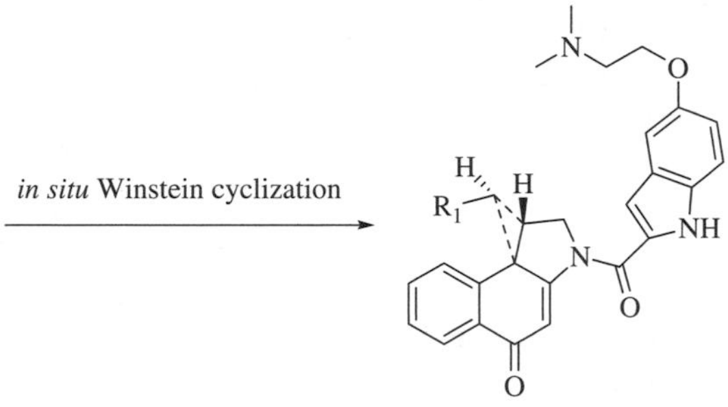

in situ Winstein cyclization

Active drugs

76b: $R_1 = Me$; **77b**: $R_2 = H$.

SCHEME 17.32 Activation of monoglycosidic or diglycosidic prodrugs by β-D-galactosidase.

primarily localized tumors, particularly inoperable tumors such as pancreatic cancer [15]. The encapsulated cells are confined physically in microbeads and protected from the patient's immune response. This also ensures a prodrug to be activated in a sustained manner that overcomes the short-term expression problem associated with other vector delivery systems. Salmons et al. encapsulated cells expressing more than one drug-activating enzyme genes (i.e., *CYP2B1* and CD) to activate multiple prodrugs (i.e., 5-FC and ifosfamide) for the treatment of solid tumors. The additive antitumor activity of two prodrugs was demonstrated in two mouse models [15].

A large number of activating enzyme genes have been employed in GDEPT [17]. The herpes simplex virus thymidine kinase gene (HSV-tk) with ganciclovir (GCV) as the prodrug and the CD of *E. coli* with 5-fluorocytosine (5-FC) as the prodrug are the most extensively studied [17]. Here, the application of GDEPT will be presented through two representative examples: the CD of *E. coli* to convert the nontoxic antifungal 5-FC into anticancer agent 5-fluorouracil (5-FU) [123] and CB1954 activated by DT-diaphorase.

CD is present in several bacteria and fungi, but not in mammalian cells [123]. CD converts 5-FC (**78**) into the anticancer drug 5-FU (Scheme 17.33). A promising approach to deliver the gene to solid tumors was described by Liu and colleagues [124]. This approach takes advantage of the hypoxic/necrotic condition in solid tumors. The bacterial genus *Clostridium* comprises a large and heterogeneous group of gram-positive, spore-forming bacteria that become vegetative and grow only in the absence (or low levels) of oxygen. Therefore, strains of these bacteria have been suggested as tools to selectively deliver the gene vector to the hypoxic and necrotic region of solid tumors [125–127]. Liu and coworkers describe, for the first time, the successful transformation of *Clostridium sporogenes*, a clostridial strain, with the *E. coli* CD gene [124]. They showed that intravenous injection of spores of *C. sporogenes* containing an expression plasmid of *E. coli* CD into tumor-bearing mice produced tumor-specific expression of CD.

More importantly, significant antitumor efficacy of systemically injected 5-FC was observed following i.v. injection of these recombinant spores. The antitumor efficacy of the prodrug 5-FC following a single i.v. injection of the recombinant spores is equivalent to or greater than that produced by maximum tolerated dose of the active drug 5-FU given by the same schedule. A major advantage of the clostridial delivery system is that not only is it tumor specific, but it has also proven itself to be safe in humans. As cited by Liu et al., Heppner and coworkers injected themselves with spores of *C. sporogenes* (a strain later named *Clostridium oncolyticum*) and experienced a mild fever as the only side effect [124]. The safety of *C. oncolyticum* was substantiated in clinical trials of cancer patients with a variety of solid tumors [128] and also in more trials in noncancer patients with a *Clostridium beijerinckii* strain [129]. An additional advantage is that injection of spores did not appear to elicit an immune response [124].

Another example of hypoxia-selective GDEPT is prodrug **79** (CB1954, Scheme 17.34). CB1954 was originally synthesized over 30 years ago. It exhibits

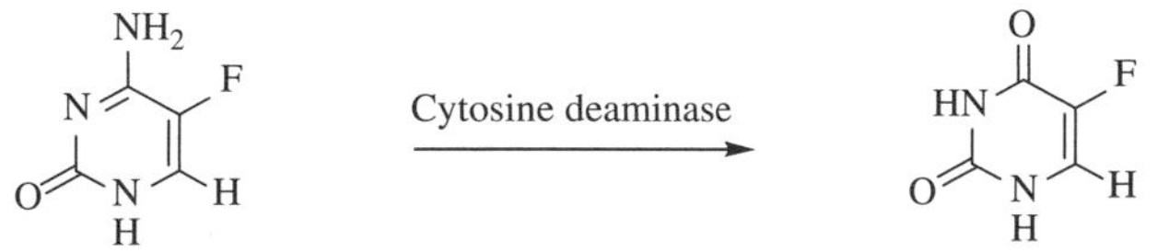

Cytosine deaminase

78

[5-Fluorocytosine, (5-FC)]

5-Fluorouracil, (5-FU)

SCHEME 17.33 Bioactivation of 5-FC by cytosine deaminase.

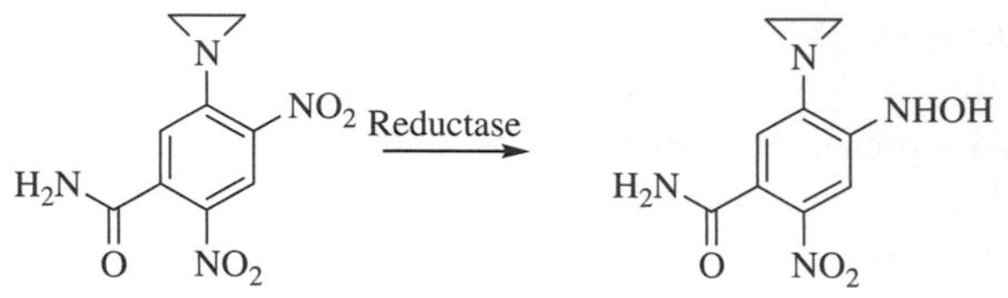

Reductase

79

5-(Aziridin-1-yl)-2,4-dinitrobenzamide, (CB 1954)

79a

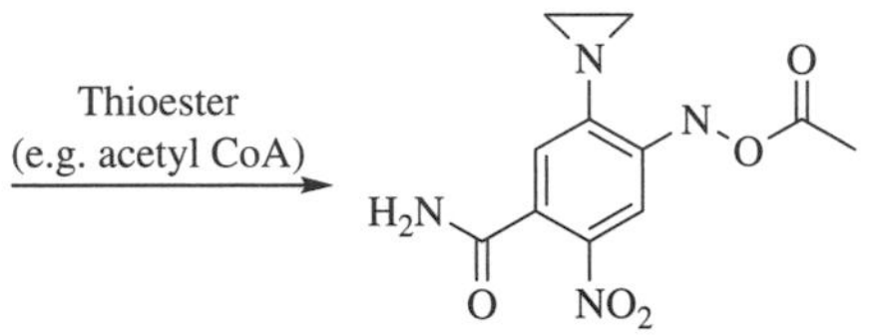

Thioester
(e.g. acetyl CoA)

79b
(DNA crosslinking agent)

SCHEME 17.34 Bioactivation of CB 1954 by reductases.

dramatic and highly specific antitumor activity against rat Walker 256 carcinoma cells [130]. The antitumor effect is due to efficient drug activation by rat DT-diaphorase [131]. CB 1954 entered clinical trials in 1970s, but little antitumor activity was observed, as human DT-diaphorase is much less active in the reduction of CB 1954 than the rat enzyme. Recent studies revealed that an amino acid difference at residue 104 between the human and rat enzyme is responsible for the catalytic difference to CB 1954 [132]. An NTR gene isolated from *E. coli* has been demonstrated to activate the prodrug CB 1954 to its toxic form approximately 90-fold more rapidly than rat DT-diaphorase, suggesting the possibility of using CB 1954 with an NTR in ADEPT [133, 134] and GDEPT [135–139].

Shibata et al. successfully transfected human HT 1080 tumor cells with the *E. coli* NTR gene [140]. The transfected human tumor cells expressed *E. coli* NTR in a time- and concentration-dependent manner under a hypoxic condition while the expression was only trace under an aerobic condition, indicating that the enzyme's expression is induced by hypoxia. No NTR was observed with wild-type HT1080 cells. The expression of NTR conferred increased sensitivity of human tumor cells to CB 1954 both *in vitro* and *in vivo*. The IC_{50} value obtained with the transfected cells was reduced by 40- to 50-fold when compared to the IC_{50} value with the wild cells in an *in vitro* experiment under a hypoxic condition. Significantly, no sensitivity difference was observed between the transfected and the wild HT 1080 cells under an aerobic condition, consistent with the notion that NTR is induced under hypoxic conditions. A significant tumor growth delay was also observed in mice implanted with transfected clones of HT 1080 cells under a hypoxic condition. Similar to the *in vitro* result, no sensitivity difference to CB 1954 was observed when the *in vivo* experiment was conducted under an aerobic condition.

17.4 SUMMARY

The search for tumor-selective prodrugs has been a long and ongoing effort. Numerous approaches have been developed based on the exploitation of the biochemical differences between cancer and normal cells. The differences can be further amplified *via* the use of antibody-directed enzyme prodrug therapy (ADEPT) and gene-directed enzyme prodrug therapy (GDEPT). Among various approaches, anticancer prodrugs based on tumor hypoxia and metastasis appear to be the most extensively explored. ADEPT and GDEPT provide the advantage of improving tumor selectivity. Nevertheless, immune response, safety, and cost are the major disadvantages of these two approaches. Although GDEPT has been successfully used in a large number of *in vitro* and *in vivo* studies, its application to cancer patients has not reached the desirable clinical significance [17]. Overall, tumor-selective anticancer prodrugs have been shown to be an effective way to improve therapeutic efficacy and reduce systemic side effects. The value of prodrug application in cancer therapy has been clearly reflected in a trend of increased approved prodrugs [2].

REFERENCES

1. Han, H. K.; Amidon, G. L. *AAPS PharmSci* 2000, **2**, E6.
2. Arpicco, S.; Dosio, F.; Stella, B.; Cattel, L. *Curr. Top. Med. Chem.* 2011, **11**, 2346.
3. Kwon, C. H. *Arch. Pharm. Res.* 1999, **22**, 533.
4. de Groot, F. M.; Damen, E. W.; Scheeren, H. W. *Curr. Med. Chem.* 2001, **8**, 1093.
5. Denny, W. A. *Eur. J. Med. Chem.* 2001, **36**, 577.
6. Nagasawa, H. *J. Pharmacol. Sci.* 2011, **115**, 446.
7. Mahato, R.; Tai, W.; Cheng, K. *Adv. Drug Deliv. Rev.* 2011, **63**, 659.
8. Seddon, B.; Kelland, L. R.; Workman, P. *Methods Mol. Med.* 2004, **90**, 515.
9. Tietze, L. F.; Schmuck, K. *Curr. Pharm. Des.* 2011, **17**, 3527.
10. Andresen, T. L.; Jensen, S. S.; Kaasgaard, T.; Jorgensen, K. *Curr. Drug Deliv.* 2005, **2**, 353.
11. Zhao, G.; Wang, X. *Curr. Med. Chem.* 2006, **13**, 1461.
12. Michael, M.; Doherty, M. M. *Expert Opin. Drug Metab. Toxicol.* 2007, **3**, 783.
13. Bagshawe, K. D. *Mol. Med. Today* 1995, **1**, 424.
14. Bagshawe, K. D. *Curr. Drug Targets* 2009, **10**, 152.
15. Salmons, B.; Brandtner, E. M.; Hettrich, K.; Wagenknecht, W.; Volkert, B.; Fischer, S.; Dangerfield, J. A.; Gunzburg, W. H. *Curr. Opin. Mol. Ther.* 2010, **12**, 450.
16. Zarogoulidis, P.; Chatzaki, E.; Hohenforst-Schmidt, W.; Goldberg, E. P.; Galaktidou, G.; Kontakiotis, T.; Karamanos, N.; Zarogoulidis, K. *Cancer Gene Ther.* 2012, **19**, 593.
17. Duarte, S.; Carle, G.; Faneca, H.; de Lima, M. C.; Pierrefite-Carle, V. *Cancer Lett.* 2012, **324**, 160.
18. Ardiani, A.; Johnson, A. J.; Ruan, H.; Sanchez-Bonilla, M.; Serve, K.; Black, M. E. *Curr. Gene Ther.* 2012, **12**, 77.

19. Vajda, A.; Marignol, L.; Foley, R.; Lynch, T. H.; Lawler, M.; Hollywood, D. *Cancer Treat. Rev.* 2011, **37**, 643.
20. Cowen, R. L.; Garside, E. J.; Fitzpatrick, B.; Papadopoulou, M. V.; Williams, K. J. *Br. J. Radiol.* 2008, **81** Spec No 1, S45.
21. Adams, G. E.; Hasan, N. M.; Joiner, M. C. *Radiother. Oncol.* 1997, **44**, 101.
22. Brown, J. M. *Cancer Res.* 1999, **59**, 5863.
23. Wilson, W. R.; Hay, M. P. *Nat. Rev. Cancer* 2011, **11**, 393.
24. Tannock, I. F.; Rotin, D. *Cancer Res.* 1989, **49**, 4373.
25. Prezioso, J. A.; Hughey, R. P.; Wang, N.; Damodaran, K. M.; Bloomer, W. D. *Int. J. Cancer* 1994, **56**, 874.
26. Jin, E.; Zhang, B.; Sun, X.; Zhou, Z.; Ma, X.; Sun, Q.; Tang, J.; Shen, Y.; Van Kirk, E.; Murdoch, W. J.; Radosz, M. *J. Am. Chem. Soc.* 2013, **135**, 933.
27. Lee, A. E.; Wilson, W. R. *Toxicol. Appl. Pharmacol.* 2000, **163**, 50.
28. Gutierrez, P. L. *Free Radic. Biol. Med.* 2000, **29**, 263.
29. Faig, M.; Bianchet, M. A.; Winski, S.; Hargreaves, R.; Moody, C. J.; Hudnott, A. R.; Ross, D.; Amzel, L. M. *Structure* 2001, **9**, 659.
30. Workman, P. *Int. J. Radiat. Oncol. Biol. Phys.* 1992, **22**, 631.
31. Powis, G. *Pharmacol. Ther.* 1987, **35**, 57.
32. Bachur, N. R.; Gordon, S. L.; Gee, M. V.; Kon, H. *Proc. Natl. Acad. Sci. U. S. A.* 1979, **76**, 954.
33. Pan, S. S.; Andrews, P. A.; Glover, C. J.; Bachur, N. R. *J. Biol. Chem.* 1984, **259**, 959.
34. Workman, P. W.; Walton, M. I. Enzyme-directed bioreductive drug development. In *Selective Activation of Drugs by Redox Processes*; Adams, G. E.; Breccia, A.; Fielden, E. M.; Wardman, P., Eds.; Plenum: New York, 1990, p 173.
35. Sartorelli, A. C. *Cancer Res.* 1988, **48**, 775.
36. Callery, P. S.; Gannett, P. M. *Cancer and Cancer Chemotherapy*; Lippincott Williams & Wilkins: Philadelphia, 2002.
37. Jaffar, M.; Naylor, M. A.; Robertson, N.; Lockyer, S. D.; Phillips, R. M.; Everett, S. A.; Adams, G. E.; Stratford, I. J. *Anticancer Drug Des.* 1998, **13**, 105.
38. Sykes, B. M.; Atwell, G. J.; Hogg, A.; Wilson, W. R.; O'Connor, C. J.; Denny, W. A. *J. Med. Chem.* 1999, **42**, 346.
39. Mauger, A. B.; Burke, P. J.; Somani, H. H.; Friedlos, F.; Knox, R. J. *J. Med. Chem.* 1994, **37**, 3452.
40. Hay, M. P.; Wilson, W. R.; Denny, W. A. *Bioorg. Med. Chem. Lett.* 1999, **9**, 3417.
41. Shyam, K.; Penketh, P. G.; Shapiro, M.; Belcourt, M. F.; Loomis, R. H.; Rockwell, S.; Sartorelli, A. C. *J. Med. Chem.* 1999, **42**, 941.
42. Reynolds, R. C.; Tiwari, A.; Harwell, J. E.; Gordon, D. G.; Garrett, B. D.; Gilbert, K. S.; Schmid, S. M.; Waud, W. R.; Struck, R. F. *J. Med. Chem.* 2000, **43**, 1484.
43. Everett, S. A.; Naylor, M. A.; Patel, K. B.; Stratford, M. R.; Wardman, P. *Bioorg. Med. Chem. Lett.* 1999, **9**, 1267.
44. Parveen, I.; Naughton, D. P.; Whish, W. J.; Threadgill, M. D. *Bioorg. Med. Chem. Lett.* 1999, **9**, 2031.
45. Hay, M. P.; Sykes, B. M.; Denny, W. A.; Wilson, W. R. *Bioorg. Med. Chem. Lett.* 1999, **9**, 2237.

46. Highfield, J. A.; Mehta, L. K.; Parrick, J.; Candeias, L. P.; Wardman, P. *J. Chem. Soc. Perkin Trans. I* 1999, 2343.
47. Brown, J. M.; Lemmon, M. J. *Cancer Res.* 1990, **50**, 7745.
48. Dorie, M. J.; Brown, J. M. *Cancer Res.* 1993, **53**, 4633.
49. Siemann, D. W.; Hinchman, C. A. *Radiother. Oncol.* 1998, **47**, 215.
50. Gatzemeier, U.; Rodriguez, G.; Treat, J.; Miller, V.; von Roemeling, R.; Viallet, J.; Rey, A. *Br. J. Cancer* 1998, **77**, Suppl 4, 15.
51. Treat, J.; Johnson, E.; Langer, C.; Belani, C.; Haynes, B.; Greenberg, R.; Rodriquez, R.; Drobins, P.; Miller, W., Jr.; Meehan, L.; McKeon, A.; Devin, J.; von Roemeling, R.; Viallet, J. *J. Clin. Oncol.* 1998, **16**, 3524.
52. Kono, O.; Terashima, H.; Azuma, S.; Murata, Y. *Gan.* 1954, **45**, 536.
53. White, I. N.; Suzanger, M.; Mattocks, A. R.; Bailey, E.; Farmer, P. B.; Connors, T. A. *Carcinogenesis* 1989, **10**, 2113.
54. Wilson, W. R.; Denny, W. A.; Pullen, S. M.; Thompson, K. M.; Li, A. E.; Patterson, L. H.; Lee, H. H. *Br. J. Cancer Suppl.* 1996, **27**, S43.
55. Raleigh, S. M.; Wanogho, E.; Burke, M. D.; McKeown, S. R.; Patterson, L. H. *Int. J. Radiat. Oncol. Biol. Phys.* 1998, **42**, 763.
56. Graf, N.; Lippard, S. J. *Adv. Drug Deliv. Rev.* 2012, **64**, 993.
57. Craig, P. R.; Brothers, P. J.; Clark, G. R.; Wilson, W. R.; Denny, W. A.; Ware, D. C. *Dalton Trans.* 2004, 611.
58. Parker, L. L.; Lacy, S. M.; Farrugia, L. J.; Evans, C.; Robins, D. J.; O'Hare, C. C.; Hartley, J. A.; Jaffar, M.; Stratford, I. J. *J. Med. Chem.* 2004, **47**, 5683.
59. Tietze, L. F.; Neumann, M.; Mollers, T.; Fischer, R.; Glusenkamp, K. H.; Rajewsky, M. F.; Jahde, E. *Cancer Res.* 1989, **49**, 4179.
60. Liotta, L. A.; Rao, C. N.; Wewer, U. M. *Annu. Rev. Biochem.* 1986, **55**, 1037.
61. Mignatti, P.; Rifkin, D. B. *Physiol. Rev.* 1993, **73**, 161.
62. Vassalli, J. D.; Pepper, M. S. *Nature* 1994, **370**, 14.
63. Mittal, S.; Song, X.; Vig, B. S.; Landowski, C. P.; Kim, I.; Hilfinger, J. M.; Amidon, G. L. *Mol. Pharm.* 2005, **2**, 37.
64. Yamada, R.; Kostova, M. B.; Anchoori, R. K.; Xu, S.; Neamati, N.; Khan, S. R. *Cancer Biol. Ther.* 2010, **9**, 192.
65. Naylor, M. S.; Stamp, G. W.; Davies, B. D.; Balkwill, F. R. *Int. J. Cancer* 1994, **58**, 50.
66. Uria, J. A.; Ferrando, A. A.; Velasco, G.; Freije, J. M.; Lopez-Otin, C. *Cancer Res.* 1994, **54**, 2091.
67. Putnam, D. A.; Shiah, J. G.; Kopecek, J. *Biochem. Pharmacol.* 1996, **52**, 957.
68. Denmeade, S. R.; Nagy, A.; Gao, J.; Lilja, H.; Schally, A. V.; Isaacs, J. T. *Cancer Res.* 1998, **58**, 2537.
69. Harada, M.; Sakakibara, H.; Yano, T.; Suzuki, T.; Okuno, S. *J. Control. Release* 2000, **69**, 399.
70. Carl, P. L.; Chakravarty, P. K.; Katzenellenbogen, J. A. *J. Med. Chem.* 1981, **24**, 479.
71. Toki, B. E.; Cerveny, C. G.; Wahl, A. F.; Senter, P. D. *J. Org. Chem.* 2002, **67**, 1866.
72. de Groot, F. M.; Loos, W. J.; Koekkoek, R.; van Berkom, L. W.; Busscher, G. F.; Seelen, A. E.; Albrecht, C.; de Bruijn, P.; Scheeren, H. W. *J. Org. Chem.* 2001, **66**, 8815.
73. Markus, G.; Takita, H.; Camiolo, S. M.; Corasanti, J. G.; Evers, J. L.; Hobika, G. H. *Cancer Res.* 1980, **40**, 841.

74. Corasanti, J. G.; Celik, C.; Camiolo, S. M.; Mittelman, A.; Evers, J. L.; Barbasch, A.; Hobika, G. H.; Markus, G. *J. Natl. Cancer Inst.* 1980, **65**, 345.
75. Campo, E.; Munoz, J.; Miquel, R.; Palacin, A.; Cardesa, A.; Sloane, B. F.; Emmert-Buck, M. R. *Am. J. Pathol.* 1994, **145**, 301.
76. Carl, P. L.; Chakravarty, P. K.; Katzenellenbogen, J. A.; Weber, M. J. *Proc. Natl. Acad. Sci. U. S. A.* 1980, **77**, 2224.
77. Cavallaro, G.; Pitarresi, G.; Licciardi, M.; Giammona, G. *Bioconjug. Chem.* 2001, **12**, 143.
78. Chakravarty, P. K.; Carl, P. L.; Weber, M. J.; Katzenellenbogen, J. A. *J. Med. Chem.* 1983, **26**, 638.
79. Chakravarty, P. K.; Carl, P. L.; Weber, M. J.; Katzenellenbogen, J. A. *J. Med. Chem.* 1983, **26**, 633.
80. de Groot, F. M.; de Bart, A. C.; Verheijen, J. H.; Scheeren, H. W. *J. Med. Chem.* 1999, **42**, 5277.
81. Eisenbrand, G.; Lauck-Birkel, S.; Tang, W. C. *Synthesis* 1996, **10**, 1247.
82. Tan, G. J.; Peng, Z. K.; Lu, J. P.; Tang, F. Q. *World J. Biol. Chem.* 2013, **4**, 91.
83. Thomssen, C.; Schmitt, M.; Goretzki, L.; Oppelt, P.; Pache, L.; Dettmar, P.; Janicke, F.; Graeff, H. *Clin. Cancer Res.* 1995, **1**, 741.
84. Horsman, M. R.; Murata, R.; Breidahl, T.; Nielsen, F. U.; Maxwell, R. J.; Stodkiled-Jorgensen, H.; Overgaard, J. *Adv. Exp. Med. Biol.* 2000, **476**, 311.
85. Hande, K. R. *Eur. J. Cancer* 1998, **34**, 1514.
86. Woessner, J. F.; Nagase, H. *Matrix Metalloproteinases and TIMPs*; 1st ed.; Oxford University Press: New York, 2000.
87. Hu, Z.; Jiang, X.; Albright, C. F.; Graciani, N.; Yue, E.; Zhang, M.; Zhang, S. Y.; Bruckner, R.; Diamond, M.; Dowling, R.; Rafalski, M.; Yeleswaram, S.; Trainor, G. L.; Seitz, S. P.; Han, W. *Bioorg. Med. Chem. Lett.* 2010, **20**, 853.
88. Bielawski, K.; Bielawska, A.; Slodownik, T.; Bolkun-Skornicka, U.; Muszynska, A. *Pharmacol. Rep.* 2008, **60**, 171.
89. Sperker, B.; Backman, J. T.; Kroemer, H. K. *Clin. Pharmacokinet.* 1997, **33**, 18.
90. Michael, M.; Doherty, M. M. *J. Clin. Oncol.* 2005, **23**, 205.
91. Bosslet, K.; Straub, R.; Blumrich, M.; Czech, J.; Gerken, M.; Sperker, B.; Kroemer, H. K.; Gesson, J. P.; Koch, M.; Monneret, C. *Cancer Res.* 1998, **58**, 1195.
92. Saha, A. K.; Glew, R. H.; Kotler, D. P.; Omene, J. A. *Clin. Chim. Acta* 1991, **199**, 311.
93. Haisma, H. J.; van Muijen, M.; Pinedo, H. M.; Boven, E. *Cell Biophys.* 1994, **24–25**, 185.
94. Leenders, R. G.; Damen, E. W.; Bijsterveld, E. J.; Scheeren, H. W.; Houba, P. H.; van der Meulen-Muileman, I. H.; Boven, E.; Haisma, H. J. *Bioorg. Med. Chem.* 1999, **7**, 1597.
95. Florent, J. C.; Dong, X.; Gaudel, G.; Mitaku, S.; Monneret, C.; Gesson, J. P.; Jacquesy, J. C.; Mondon, M.; Renoux, B.; Andrianomenjanahary, S.; Michel, S.; Koch, M.; Tillequin, F.; Gerken, M.; Czech, J.; Straub, R.; Bosslet, K. *J. Med. Chem.* 1998, **41**, 3572.
96. Desbene, S.; Van, H. D.; Michel, S.; Koch, M.; Tillequin, F.; Fournier, G.; Farjaudon, N.; Monneret, C. *Anticancer Drug Des.* 1998, **13**, 955.
97. Leu, Y. L.; Roffler, S. R.; Chern, J. W. *J. Med. Chem.* 1999, **42**, 3623.
98. Schmidt, F.; Florent, J. C.; Monneret, C.; Straub, R.; Czech, J.; Gerken, M.; Bosslet, K. *Bioorg. Med. Chem. Lett.* 1997, **7**, 1071.
99. Lougerstay-Madec, R.; Florent, J. C.; Monneret, C.; Nemati, F.; Poupon, M. F. *Anticancer Drug Des.* 1998, **13**, 995.

100. de Bont, D. B.; Leenders, R. G.; Haisma, H. J.; van der Meulen-Muileman, I.; Scheeren, H. W. *Bioorg. Med. Chem.* 1997, **5**, 405.
101. Lougerstay-Madec, R.; Florent, J. C.; Monneret, C. *J. Chem. Soc. Perkin Trans.* **1** 1999, 1369.
102. Denmeade, S. R.; Lou, W.; Lovgren, J.; Malm, J.; Lilja, H.; Isaacs, J. T. *Cancer Res.* 1997, **57**, 4924.
103. Wei, M. X.; Tamiya, T.; Chase, M.; Boviatsis, E. J.; Chang, T. K.; Kowall, N. W.; Hochberg, F. H.; Waxman, D. J.; Breakefield, X. O.; Chiocca, E. A. *Hum. Gene Ther.* 1994, **5**, 969.
104. Chen, L.; Waxman, D. J. *Cancer Res.* 1995, **55**, 581.
105. Chen, L.; Waxman, D. J.; Chen, D.; Kufe, D. W. *Cancer Res.* 1996, **56**, 1331.
106. Jounaidi, Y.; Hecht, J. E.; Waxman, D. J. *Cancer Res.* 1998, **58**, 4391.
107. Huang, Z.; Raychowdhury, M. K.; Waxman, D. J. *Cancer Gene Ther.* 2000, **7**, 1034.
108. Huang, Z.; Waxman, D. J. *Cancer Gene Ther.* 2001, **8**, 450.
109. Zarogoulidis, P.; Darwiche, K.; Sakkas, A.; Yarmus, L.; Huang, H.; Li, Q.; Freitag, L.; Zarogoulidis, K.; Malecki, M. *J. Genet. Syndr. Gene Ther.* 2013, **4**, 1.
110. Marais, R.; Spooner, R. A.; Light, Y.; Martin, J.; Springer, C. J. *Cancer Res.* 1996, **56**, 4735.
111. Marais, R.; Spooner, R. A.; Stribbling, S. M.; Light, Y.; Martin, J.; Springer, C. J. *Nat. Biotechnol.* 1997, **15**, 1373.
112. Senter, P. D.; Springer, C. J. *Adv. Drug Deliv. Rev.* 2001, **53**, 247.
113. Dubowchik, G. M.; Walker, M. A. *Pharmacol. Ther.* 1999, **83**, 67.
114. Chari, R. V. *Adv. Drug Deliv. Rev.* 1998, **31**, 89.
115. Brinkmann, U. *In Vivo* 2000, **14**, 21.
116. Illidge, T. M.; Johnson, P. W. *Br. J. Haematol.* 2000, **108**, 679.
117. Niculescu-Duvaz, I.; Cooper, R. G.; Stribbling, S. M.; Heyes, J. A.; Metcalfe, J. A.; Springer, C. J. *Curr. Opin. Mol. Ther.* 1999, **1**, 480.
118. Springer, C. J.; Bagshawe, K. D.; Sharma, S. K.; Searle, F.; Boden, J. A.; Antoniw, P.; Burke, P. J.; Rogers, G. T.; Sherwood, R. F.; Melton, R. G. *Eur. J. Cancer* 1991, **27**, 1361.
119. Bagshawe, K. D.; Sharma, S. K.; Springer, C. J.; Antoniw, P.; Boden, J. A.; Rogers, G. T.; Burke, P. J.; Melton, R. G.; Sherwood, R. F. *Dis. Markers* 1991, **9**, 233.
120. Napier, M. P.; Sharma, S. K.; Springer, C. J.; Bagshawe, K. D.; Green, A. J.; Martin, J.; Stribbling, S. M.; Cushen, N.; O'Malley, D.; Begent, R. H. *Clin. Cancer Res.* 2000, **6**, 765.
121. Meyer, D. L.; Jungheim, L. N.; Law, K. L.; Mikolajczyk, S. D.; Shepherd, T. A.; Mackensen, D. G.; Briggs, S. L.; Starling, J. J. *Cancer Res.* 1993, **53**, 3956.
122. Kerr, D. E.; Schreiber, G. J.; Vrudhula, V. M.; Svensson, H. P.; Hellstrom, I.; Hellstrom, K. E.; Senter, P. D. *Cancer Res.* 1995, **55**, 3558.
123. Yata, V. K.; Gopinath, P.; Ghosh, S. S. *Appl. Biochem. Biotechnol.* 2012, **167**, 2103.
124. Liu, S. C.; Minton, N. P.; Giaccia, A. J.; Brown, J. M. *Gene Ther.* 2002, **9**, 291.
125. Minton, N. P.; Mauchline, M. L.; Lemmon, M. J.; Brehm, J. K.; Fox, M.; Michael, N. P.; Giaccia, A.; Brown, J. M. *FEMS Microbiol. Rev.* 1995, **17**, 357.
126. Fox, M. E.; Lemmon, M. J.; Mauchline, M. L.; Davis, T. O.; Giaccia, A. J.; Minton, N. P.; Brown, J. M. *Gene Ther.* 1996, **3**, 173.

127. Lemmon, M. J.; van Zijl, P.; Fox, M. E.; Mauchline, M. L.; Giaccia, A. J.; Minton, N. P.; Brown, J. M. *Gene Ther.* 1997, **4**, 791.

128. Heppner, F.; Mose, J.; Ascher, P. W.; Walter, G. *13th Int. Cong. Chemother.* 1983, **226**, 38.

129. Fabricius, E. M.; Schneeweiss, U.; Schau, H. P.; Schmidt, W.; Benedix, A. *Res. Microbiol.* 1993, **144**, 741.

130. Cobb, L. M.; Connors, T. A.; Elson, L. A.; Khan, A. H.; Mitchley, B. C.; Ross, W. C.; Whisson, M. E. *Biochem. Pharmacol.* 1969, **18**, 1519.

131. Knox, R. J.; Boland, M. P.; Friedlos, F.; Coles, B.; Southan, C.; Roberts, J. J. *Biochem. Pharmacol.* 1988, **37**, 4671.

132. Chen, S.; Knox, R.; Wu, K.; Deng, P. S.; Zhou, D.; Bianchet, M. A.; Amzel, L. M. *J. Biol. Chem.* 1997, **272**, 1437.

133. Anlezark, G. M.; Melton, R. G.; Sherwood, R. F.; Coles, B.; Friedlos, F.; Knox, R. J. *Biochem. Pharmacol.* 1992, **44**, 2289.

134. Knox, R. J.; Friedlos, F.; Sherwood, R. F.; Melton, R. G.; Anlezark, G. M. *Biochem. Pharmacol.* 1992, **44**, 2297.

135. Bridgewater, J. A.; Springer, C. J.; Knox, R. J.; Minton, N. P.; Michael, N. P.; Collins, M. K. *Eur. J. Cancer* 1995, **31A**, 2362.

136. Clark, A. J.; Iwobi, M.; Cui, W.; Crompton, M.; Harold, G.; Hobbs, S.; Kamalati, T.; Knox, R.; Neil, C.; Yull, F.; Gusterson, B. *Gene Ther.* 1997, **4**, 101.

137. Drabek, D.; Guy, J.; Craig, R.; Grosveld, F. *Gene Ther.* 1997, **4**, 93.

138. McNeish, I. A.; Green, N. K.; Gilligan, M. G.; Ford, M. J.; Mautner, V.; Young, L. S.; Kerr, D. J.; Searle, P. F. *Gene Ther.* 1998, **5**, 1061.

139. Connors, T. A. *Gene Ther.* 1995, **2**, 702.

140. Shibata, T.; Giaccia, A. J.; Brown, J. M. *Neoplasia* 2002, **4**, 40.

18

TARGETED DELIVERY OF DRUGS TO THE COLON

Anil K. Philip[1] and Sarah K. Zingales[2]

[1] *Department of Pharmaceutics, School of Pharmacy, College of Pharmacy and Nursing, University of Nizwa, Nizwa, Sultanate of Oman*

[2] *Department of Chemistry and Physics, Armstrong State University, Savannah, GA, USA*

18.1 INTRODUCTION

There is a real need for colon-targeting therapies for both local treatment of bowel diseases and the systemic release of certain drug classes (Table 18.1) [1]. Bowel diseases such as ulcerative colitis (UC), Crohn's disease, inflammatory bowel disease (IBD), and colon cancer plague a large portion of the population, with colon cancer being the third most prevalent cancer in both men and women [2].

One factor to consider in specifically targeting the colon is the question of how well the colon can absorb different drugs. One study showed that glibenclamide was absorbed equally well in the duodenum, stomach, and colon, but the rate of absorption varied [3]. In addition, another study showed that theophylline was absorbed in the ileum, stomach, and colon equally, but the half-life of theophylline in the colon was twice as long [4]. These early findings indicated that in order to specifically target the colon, prodrug formulations would be needed, especially with drugs that show little specificity for absorption in the gastrointestinal (GI) tract [5–9]. Targeting of drugs to the colon should lower drug-induced side effects due to nonspecific uptake. In addition, if a real need arises for increased bioavailability, bypassing the first-pass metabolism can be envisioned [10].

Drug Delivery: Principles and Applications, Second Edition. Edited by Binghe Wang, Longqin Hu, and Teruna J. Siahaan.

TABLE 18.1 List of Target Sites, Disease Condition, and Drug Used for Colon Targeting

Target Sites	Disease Conditions	Drug and Active Agents
Topical action	Inflammatory bowel diseases, irritable bowel disease, and Crohn's disease.	Hydrocortisone, budenoside, prednisolone, sulfaselazine, olsalazine, mesalazine, and balsalazide
Local action	Chronic pancreatitis, pancreatectomy, cystic fibrosis, and colorectal cancer	Digestive enzyme supplements 5-Flourouracil
Systemic action	For the prevention of gastric irritation and first-pass metabolism of orally ingested drugs	NSAIDs Steroids
	Oral delivery of peptides	Insulin
	Oral delivery of vaccines	Typhoid

Reproduced with permission from *Oman Medical Journal* [1].

Another factor to consider in colon-specific delivery is transit time. As the colon is the most distal part of the GI tract, in order to have colon-specific drug release, the drug delivery system must survive through the stomach and small intestine and have quick, immediate release upon entering the colon [1]. The actual time that it takes for drugs and prodrugs to arrive at the colon affects the efficacy of the colonic drug delivery [11]. Many factors influence the transit time including food taken, GI motility, physiological factors, and any changes due to disease. Normal motility time in humans is 1–3 days for a complete transit from the mouth to the anus. In a healthy adult, the transit time from the mouth to the small intestine is 0.5–2 hours; however, the transit time could possibly increase to 6–8 hours depending on food consumption. The transit time for solid food is generally longer than that of fluid. Transit time from the stomach to the large intestine is 2–4 hours, and from the small intestine to the anus is 6–48 hours. The lengthy colonic transit time might be deemed by some as unfavorable, but it can promote the complete drug release from the delivery system [12].

The physiology of the colon lends itself to prodrug targeting. First, the colon has a much larger concentration and diversity of microbes than the rest of the GI. This microflora contains a variety of enzymes used in their metabolism that are unique to the colon while still maintaining a lower amount of hydrolytic and proteolytic enzymes [13–18]. Second, the pH of the colon ranges from 6.4 to 7.0 compared to the gastric pH of 1.0–2.5 and the small intestinal pH of 6.6–7.5 [19]. This pH profile was demonstrated by tracking radiotelemetry capsules in healthy individuals (Fig. 18.1) [19, 20]. The fact that the colon and terminal small intestine have the highest pH of the GI tract allows for targeting to those regions [21]. Third, the colon has a much higher absorption of water and thus a higher viscosity than the upper GI. This allows for targeting the colon with vehicles that can be ruptured by the intestinal pressure. In general, there are more complications with site-directed drug delivery than with site-bioactivated drug delivery, in that many unpredictable barriers are present [22, 23]. In summary, prodrug techniques typically used to target the colon include microbially activated prodrugs, which take advantage of the unique

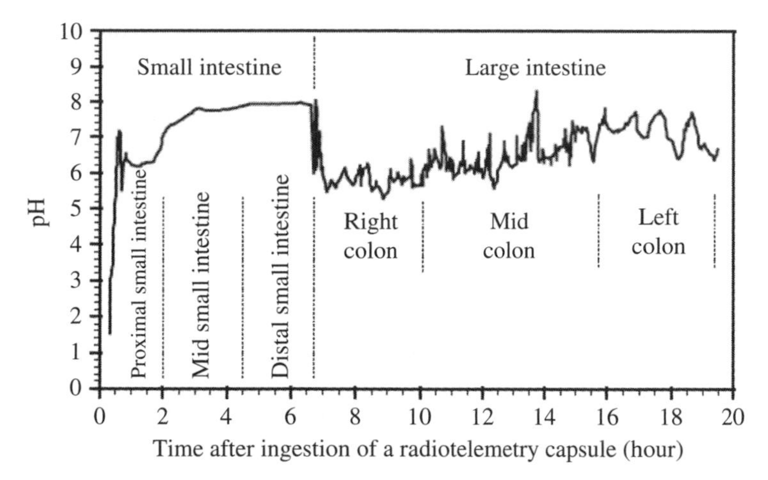

FIGURE 18.1 pH profile of a healthy GI tract. Leopold [20], pp. 157–170. Reproduced with permission of Wiley-VCH.

microflora in the colon; pH-sensitive released prodrugs, which are designed to degrade in the colonic pH range; osmotic-controlled release prodrugs, which combine the former two techniques with an internal pressure agent; pressure-controlled release prodrugs, which combine the former techniques with release from the colonic peristalsis and high viscosity, as well as novel formulations such as nanoparticles (Table 18.2) [24]. The various colon-specific delivery methods are described later.

18.2 MICROBIALLY TRIGGERED RELEASE

The most common colonic prodrug approach is to achieve site-specific drug bioactivation using the endogenous enzymes of the colonic microflora [25]. The colon is host to a large variety of microorganisms, which have various metabolic enzymes that can be exploited for drug release. The most well-explored examples are the azo-based compounds [26–28], amino acid linkers [29–33], and glucoside and glucuronide conjugates [34–39]. In addition, a drug delivery mechanism for controlling the release of an active agent with the enzyme proteinase appeared noteworthy [40].

18.2.1 Azo-Linked Compounds

Due to the unique colonic microflora and their azoreductase enzymes, the earliest prodrug approaches to colon-specific drug delivery involved using an azo linker to covalently modify an amino group on the drug [41]. Upon delivery to the colon, the azo linkage is cleaved by the azoreductases to yield two amines [42]. Since only the colon contains the bacteria with the proper digestive enzymes for drug release, the prodrug is stable for transit through the GI tract. Prontosil, a prodrug for the antibiotic sulfanilamide (Fig. 18.2a), was the first commercially available antimicrobial prodrug, which was developed in the 1930s and remained on the market until the

TABLE 18.2 Different Delivery Methods for Achieving Colon Targeting

Delivery Method	Principle	Achieved with	Examples
pH-controlled drug release	Difference in pH between the small and large intestine	pH-dependent dissolution of polymeric coatings	Enteric coatings and basic polymers
		pH-dependent polymer swelling of hydrogels	Acrylic polymers
		pH-dependent drug release from drug/ ion exchange resin complexes	Insulin + gelatin B, olsalazine + anion-exchange resin
Enzyme-controlled drug release	Degradation of dosage form components by the enzymes of the colonic microflora	Degradable prodrugs	Mono, oligo, or polymers with degradable drug-carrier bonds
		Coating materials with degradable bonds including capsule shells	Azo polymers and polymers with glycosidic bonds
		Hydrogels and matrices consisting of cross-linked, degradable polymers	Cross-linked guar, pectin, dextran, inulin, and azo polymers
		Sustained release coating materials with degradable domains (pore formers)	Ethylcellulose or Eudragit RS with galactomannans, β-cyclodextrin, glassy amylose, and inulin
Time-controlled drug release	Relatively constant transit time in the small intestine of about 3 hours	Time-dependent swellable polymers	Cellulose ethers, Eudragit coatings with sustained release
		Slow build-up of an osmotic pressure in the dosage form	COER-24™
		Polymer layers with time- dependent erosion or dissolution	Cellulose ethers; Eudragit E, chitosan (in combination with an acid in the dosage form)
Pressure-controlled drug release	Disintegration of the dosage form in the colon by intra-luminal pressure resulting from strong peristaltic waves	Thick coating consisting of water-insoluble, non-swellable polymers	Hard gelatin capsule with inner ethylcellulose coating

Reproduced with permission from Wiley-VCH Verlag GmbH & Co. KGaA [20].

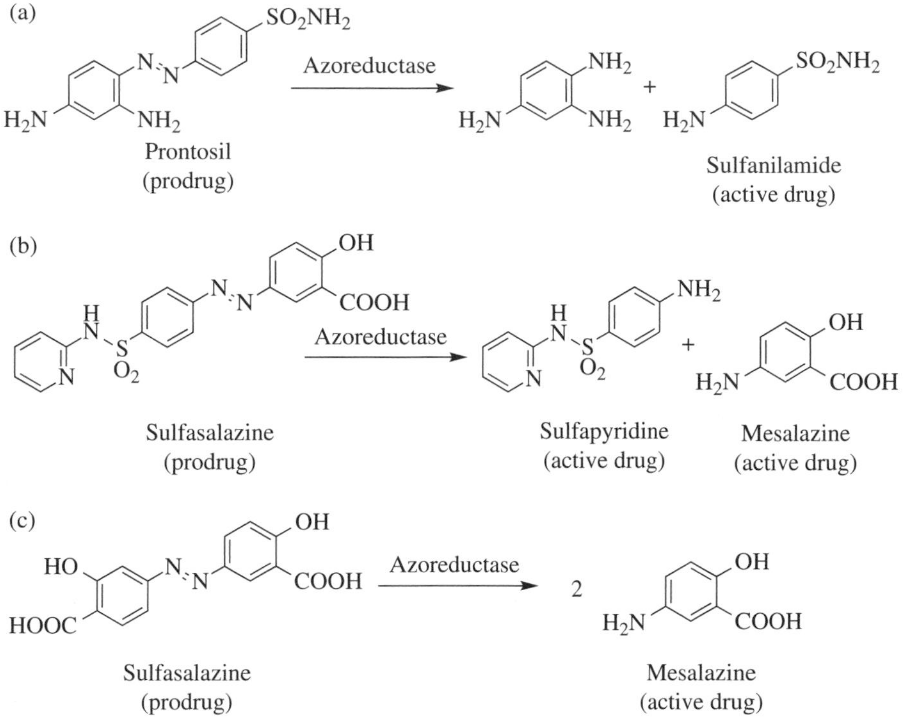

FIGURE 18.2 Azo-linked prodrugs. (a) Prontosil, the prodrug of sulfanilamide; (b) sulfasalazine, the prodrug of mesalazine and sulfapyridine; and (c) olsalazine, the prodrug dimer of mesalazine.

1960s. Almost all azo-based prodrugs are modeled on prontosil. The most common azo prodrugs are those for mesalazine (5-aminosalicylic acid, ASA, Fig. 18.2), an anti-inflammatory drug with an affinity for inflamed bowels. Sulfasalazine, developed in the 1950s, was the first such prodrug, with 5-ASA covalently linked to sulfapyridine, a sulfa drug antibiotic (Fig. 18.2b). Due to the side effects of sulfapyridine, however, other azo prodrugs for 5-ASA have been developed, including olsalazine, which is a dimerized form of mesalazine (Fig. 18.2c) and was approved by the FDA in the 1990s. In a clinical trial in 1992, olsalazine was compared to 5-ASA alone for prevention of relapse in ulcerative colitis patients. The results showed that the olsalazine treatment group had a lower rate of treatment failure (24%) and a lower relapse rate (12%) than the 5-ASA group (46% and 33%, respectively) [43]. Other methods include linking drugs to azo-linked polymers [44]. For example, azo-linked polymeric prodrugs of 5-ASA have been shown *in vitro* to release 5-ASA to the proximal colon [45]. Advantages to this prodrug strategy include site-specific delivery, stability to GI transit, the ability to link multiple drugs together for delivery, and the ability to dimerize drugs for double dosage. Disadvantages include the fact that the active drug must have an amino group for linkage and that both metabolites must be biologically active or inert upon reduction.

18.2.2 Amino Acid Conjugates

Amino acids are often used in prodrug formulations due to their structural diversity, water solubility, and chemical stability. Many anti-inflammatory drugs have been conjugated with amino acids for colonic delivery. The chemically stable amide linkages between the amino acid and the drug should be stable in the upper GI and then metabolized in the colon by the microflora. The first such prodrugs were reported in the early 1990s, with alanine, tyrosine, methionine, and glutamic acid conjugates of salicylic acid (SA), a common non-steroidal anti-inflammatory drug (NSAID), investigated as prodrugs designed to prolong the lifetime of SA in the bloodstream. In the initial animal studies of SA-ALA and SA-GLU, free SA was detected in the bloodstream 2 hours after oral administration for both, with release occurring in the hindgut and cecum, respectively [46, 47]. The SA-TYR and SA-MET conjugates, however, were metabolized in the upper GI, and thus determined unsuitable [48]. In addition, successful glycine conjugates have also been reported (Fig. 18.3): glycine-5-ASA in 2000, glycine-flurbiprofen in 2008, and glycine-ketoprofen in 2009 [29, 31, 49]. The *in vitro* revision studies for glycine–NSAID combinations demonstrated negligible release during its upper GI passage with maximum drug release in simulated colonic environment. The experimental colitis model revealed statistically negligible ulcerogenic activity ($p > 0.05$), and significant inflammatory potential for the prodrug. The histopathological features indicated that the morphological disturbances associated with acetic acid administration were corrected by treatment with glycine–NSAIDs. Although there are a variety of amino acids, only small polar or ionic amino acids have demonstrated site-specific delivery to the colon.

18.2.3 Sugar-Derived Prodrugs

Due to the glycosidase and glucuronidase activity in the colon as well as the poor absorption of glycosides and glucuronides in the upper GI, many different plant glycosides, such as glucosides, flavonoids, amygdalins, sennosides, and anthraquinones, and animal-derived glucuronides are used to make colon targeting prodrugs

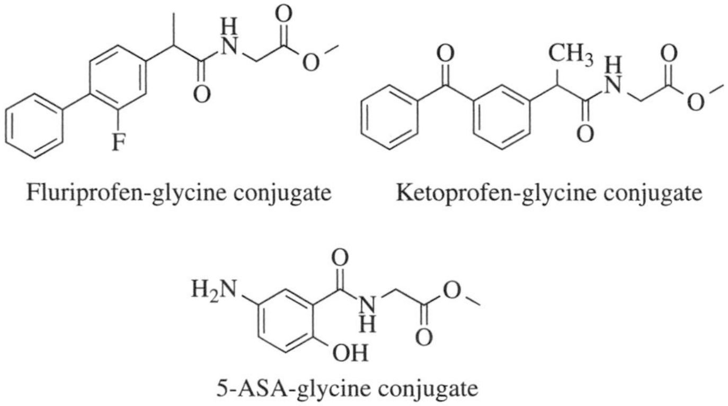

Fluriprofen-glycine conjugate

Ketoprofen-glycine conjugate

5-ASA-glycine conjugate

FIGURE 18.3 Amino acid conjugate prodrugs.

[50]. These types of prodrugs are enzymatically cleaved in the colon to yield the active drug, called an "aglycone," and the sugar moiety. Particularly, sugar conjugates of glucocorticoids have been effective treatments for inflammation of the bowel and ulcerative colitis (UC) [51, 52]. When dexamethasone- and prednisolone-glucosides were examined in an animal model for their ability to release the free glucocorticoid, roughly 60% and 15%, respectively, of an oral dose of the sugar conjugates reached the cecum, whereas less than 1% of either free glucocorticoid reached the cecum; they were absorbed in the small intestine [37]. In another animal model, budesonide-glucuronide was investigated as a treatment for ulcerative pancolitis. Animals treated with the prodrug formulation showed no signs of adrenal suppression and demonstrated accelerated healing at a fourfold lower dose compared to budesonide alone [53]. One drawback to these types of prodrugs, however, is the possibility of the release of toxic metabolites from the bacterial degradation of the sugars [54].

In addition to small sugars, many polysaccharides can also be used for drug delivery, either by covalent modification or by use as a carrier. Guar gum, a polysaccharide consisting of a 1-4-linked-β-D-mannopyranose backbone with 1-6-linked-α-D-galactopyranose side chains, has been utilized for colon targeting because of its slow drug release and the ease with which it undergoes microbial degradation [55]. Guar-gum based formulations of 5-fluoruracil [56], curcumin [57], and albendazole [58] have all been successfully released selectively in the colon *in vitro*. Pectin, a hetermeric polysaccharide, is also used for colon targeting due to its stability in the upper GI and facile degradation in the colon by pectinase enzymes released by the colonic bacteria [59]. Pectin-based formulations have been used to deliver sensitive drugs, such as proteins and polypeptides to the colon [60]. Formulations of very insoluble drugs, such as indomethacin, have also been reported using calcium pectinate, the insoluble salt of pectin, which is less water soluble, more stable at low pH, and more resistant to hydration in the GI tract than pectin [59, 61, 62]. Perhaps the most commonly utilized family of polysaccharides in drug delivery are dextrans, including cyclodextrans. Historically, drug–dextran conjugates were mainly for parenteral administration, but these conjugates remain unchanged and unabsorbed until metabolism by the bacteria of the colon, making them ideal for colon-target delivery. Many dextran prodrugs have been reported, including those of naproxen, 5-ASA, ketoprofen, methyl prednisolone, dexamethasone, and 4-biphenylylacetic acid (BPAA) [59]. Specifically, cyclodextran–BPAA conjugates released 95% of the BPAA to the colon both *in vitro* and *in vivo* [63]. In addition, dextran–chemotherapy conjugates have shown to be effective and to improve the cytotoxic effects of the chemotherapeutic agents [64]. Chitosan, a polysaccharide composed of a deacetylated ß-(1-4)-linked D-glucosamine and *N*-acetyl-D-glucosamine, is derived from the deacetylation of chitin, which is found in the exoskeleton of crabs, shrimps, and other crustaceans and cell wall of fungi [65]. The chitosan backbone contains several functional groups that can be covalently modified to synthesize chitosan prodrug derivatives and the colonic microflora can metabolism even the strongest cross-linked chitosan polymers [58]. Colon-specific chitosan hydrogel beads were able to deliver FITC-labeled bovine serum albumin as a model protein [66].

18.3 pH-SENSITIVE POLYMERS FOR TIME-DEPENDENT RELEASE

Another formulation strategy for colonic drug delivery is the use of a pH-sensitive polymer coating. These coatings can be used as a single unit or as multiparticulate systems [67]. These coatings consist of multiple, various types of polymers, mainly derivatives of acrylic acid and cellulose (Table 18.3), which each has a certain pH at which it dissolves [68–70].The pH-dependent polymers typically included in colon-specific drug delivery are insoluble at low pH and become more soluble as the pH increases. This strategy aims to protect the drug from release and/or degradation in the upper GI and allows the drug to be released into the colon specifically. There is some likelihood, however, that during this passage, body fluids might enter the polymer and trigger premature release of the drug. One strategy designed to overcome this problem is the use of a thick coat of polymer, but it too has its drawbacks; thicker coats have been shown to rupture due to the contractile activity of the stomach [68]. Other reasons for the variable results for this type of drug delivery include the pH change from the distal small intestine to the colon (it drops from 7.0 at the terminal part of the ileum to around 6.0 in the ascending colon), prolonged lag times at the ileocecal junction, and the rapid movement through the ascending colon [71]. In fact, this method can also be used to deliver drugs to the terminal part of the ileum as well as the colon.

Eudragit® L-100 coated mesalazine (5-ASA) tablets are commercially available as Claversal®, Salofalk®, and Rowasa®. In IBD patients, these 5-ASA-Eudragit tablets have been proven to deliver mesalazine efficaciously at the terminal ileum as well as proximal colon. A scintigraphic study of Claversal tablets with a group of 13 patients with Crohn's disease and ulcerative colitis showed more than 70% of presented tablets disintegrated within 3.2 hours after gastric emptying, leading to drug dispersion in the distal small intestine and proximal colon [72]. This delayed release is termed "time clock" or pulsatile drug delivery. Fairly extensive *in vitro* studies have shown that despite the diverse simulated physiological conditions, pulsatile release of salbutamol was reproducible [73]. In another study, a colon-specific pulsatile drug delivery system for theophylline was engineered. The delivery design included an insoluble hard gelatin capsule body, containing Eudragit-coated theophylline microcapsules and sealed with a hydrogel plug [74]. Another technique

TABLE 18.3 pH-Sensitive Polymers Used in Colonic Drug Delivery

Polymers	Optimum Dissolution pH
Cellulose acetate butyrate (CAB), cellulose acetate phthalate (CAP), cellulose acetate trimellitate (CAT)	≥5.5
Polyvinyl acetate phthalate (PVAP)	5.0
Hydroxypropyl methylcellulose phthalate (HPMCP)	≥5.0
Hydroxypropyl methylcellulose acetate succinate (HPMCAS)	≥5.5
Eudragit L100-55, Eudragit L30D-55	≥5.5 (Duodenum)[a]
Eudragit L-100, Eudragit L12,5	≥6.0 (Jejunum)[a]
Eudragit S-100, Eudragit S12,5, Eudragit FS30D	≥7.0 (Colon)[a]

[a] Targeted Drug Release Area.

is to add polymers that will swell at alkaline pH, which effectively builds in a lag time for the drug release [75]. This technique is designed for slower dissolution of the polymer in the small intestine, where the polymers are still compressed, and improved dissolution in the colon, where the polymers will swell. Typical systems include blends of EUDRAGIT with low pH-resistant polymers that swell in alkaline pH and are biodegradable in the colonic microflora [76, 77].

18.4 OSMOTIC RELEASE

Colon-targeted osmotic drug delivery systems resemble conventional membrane-controlled systems in their basic design: the drug to be delivered is in the core, which is surrounded by a membrane. However, for osmotic release, the membrane enables bodily fluids to enter the system and dissolve the contents. Immediately upon the dissolution of the core contents, hydrostatic pressure builds up inside the system and pushes the drug outside. The hydrostatic pressure can be controlled through osmotic agents, normally salts of organic or inorganic acids, which regulate the release of the active content for the preferred period [78–81]. In general, Biopharmaceutical Classification System (BCS) Class I drugs, which are highly permeable and highly soluble, are more suited to this type of colonic delivery than those in class III/IV, which have low permeability and/or solubility [82]. One of the most widely used osmotic delivery systems is OROS-CT™ (oral osmotic system for colon targeting), which is employed in the treatment of conditions that affect the colon such as colitis, ulcerative colitis, and Crohn's disease [83]. A microbially triggered colon-targeted osmotic pump (MTCT-OP) using the gelable character of chitosan has been reported as well, with release occurring specifically in the colon and not in the upper GI tract [84]. In another study, a microporous bilayer osmotic tablet composed of dicyclomine hydrochloride and diclofenac potassium was formulated for colon targeting. *In vitro* results indicated the pectin-based system was acid-resistant and released the drug in a steady rate up to 24 hours after delivery [85]. Another colonic delivery system has been developed for metronidazole, the preferred therapy for intestinal amoebiasis, with guar gum as the pore former in the membrane [86]. A recent report on *Sterculia urens* gum as a carrier for a colon-targeted drug delivery using an osmotic agent has shown that formulations with *Sterculia* gum alone gave premature release of drug to the upper GI and further suggested that the combination of *Sterculia* gum with chitosan/Eudragit RLPO (ammonio-methacrylate copolymer) might offer good resistance against the upper GI tract fluids and still undergo rapid enzymatic degradation once reaching the colon [87].

18.5 PRESSURE-CONTROLLED DELIVERY

The constant reabsorption of water by the colon increases its viscosity compared to that of the small intestine. The higher viscosity of colonic fluids offers a real challenge to drug dissolution in the colon. Pressure-controlled delivery systems, however, take advantage of this higher viscosity in the colon. In such systems, drug release typically

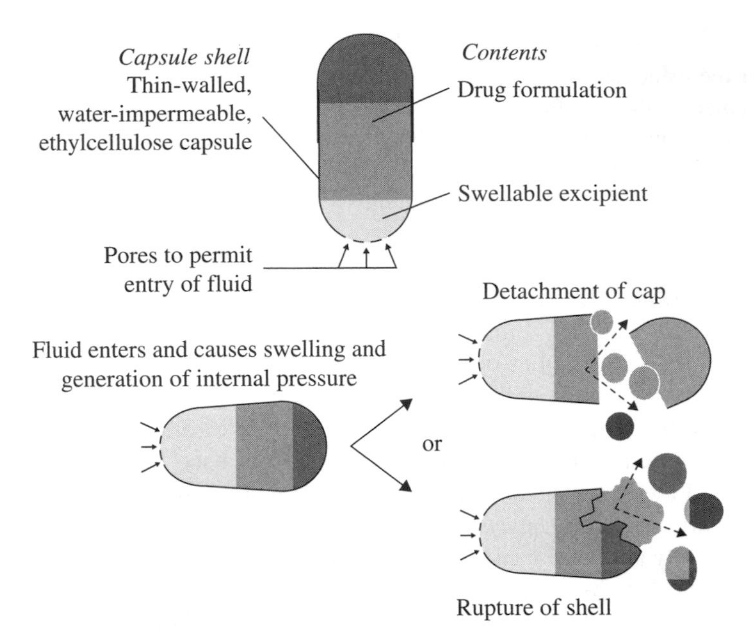

FIGURE 18.4 Representation of a pressure-controlled release system. Niwa et al. [91], pp. 83–89. Reproduced with permission of Informa Healthcare.

occurs following the disintegration of a water-insoluble polymer capsule due to pressure in the lumen of the colon. The most common system uses ethylcellulose as the encapsulating agent. In such systems, a thin-walled ethylcellulose capsule does not disintegrate in the upper GI tract transit, but ruptures in the colon due to the pressure of the colonic peristalsis [88]. In a study [89], *in vitro* drug release from three different colonic release systems were investigated for their relationship with *in vivo* drug absorption. The drug release and drug dissolution in the colonic lumen were important factors for the systemic availability of drugs from these colon drug delivery systems. A new pressure-controlled colon delivery system for managing nocturnal asthma was reported that used a film coating with Eudragit S100 over sealed conventional hard gelatin capsules containing a drug-lipid matrix of theophylline. The film-coated capsules were resistant for 2 hours in the stomach (pH 1.2) and 3 hours in phosphate buffer (pH 6.8 at 37°C) [90]. A novel capsule system has been reported, which is perforated at one end (base) with micropores to allow the aqueous fluid to enter the capsule and to help in swelling the excipient. The internal pressure created triggered the rupture of the membrane, which promoted the drug release (Fig. 18.4) [91].

18.6 NANOPARTICLE APPROACHES

There is an increasing interest in developing nanoparticles for targeting the colon [92, 93]. This interest is due in part to perplexing results in critical reviews for macro- and micromolecules [94]. Studies that have shown that some patients with IBD have

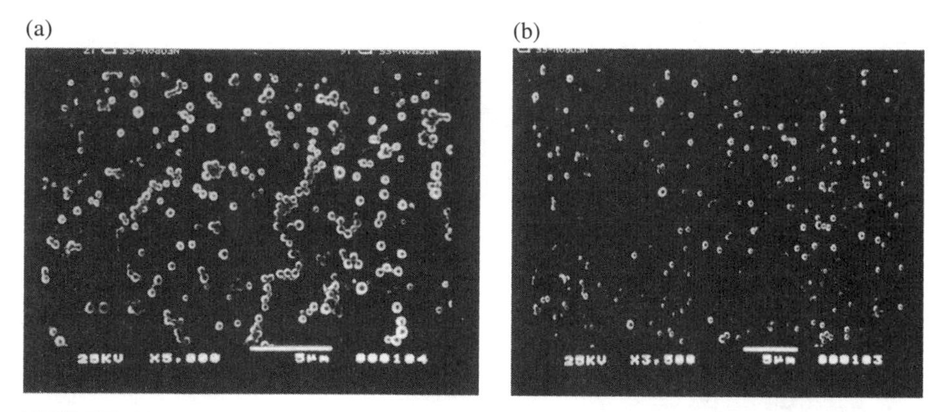

FIGURE 18.5 Scanning electron microscopic images of nanoparticles prepared by (a) PLGA (mol. wt. 20,000) or (b) PLGA (mol. wt. 5,000). Lamprecht et al. [100], pp. 775–781. Reproduced with permission of ASPET.

increased instances of diarrhea when the drug delivery agent is bigger than 200 μm. This is doubly problematic in that it increases the severity of the patients' symptoms as well as decreases the amount of the drug in circulation by increasing excretion of the drug [94, 95]. In addition, during colitis the cellular immune response is quite high, with increased number of mast cells, regulatory T-cells, neutrophils, etc., present at the site of inflammation [96]. Since these macrophages take up smaller particles, such as microspheres and nanoparticles, far better than larger ones, it can be imagined that using nanoparticles as drug delivery agents will allow accumulation of the drug at the site of inflammation [97]. This target site accumulation might also enable dose reduction of nanoparticle drug formulations [98].

Nanoparticle approaches can incorporate any of the previously mentioned colon-targeting techniques, including release by colonic microflora metabolism, pH, and pressure. For example, nanoparticles made of polysaccharide copolymers were engineered to deliver the anti-inflammatory tripeptide Lys-Pro-Val (KPV) to the colon. In this instance, the same therapeutic effect was realized *in vivo* with a 12,000-fold less dose of KPV in the nanoparticle formulation than the free KPV dose [99]. In addition, another study demonstrated that poly(lactic-*co*-glycolic acid) nanoparticles containing rolipram, an anti-inflammatory drug (Fig. 18.5), were effective for mitigating colitis *in vivo*. The nanoparticles accumulated in the inflamed tissue, which led to less adverse effects than the free rolipram, and also demonstrated long-term delayed release with a protective response compared to the relapse in symptoms seen with the treatment of free rolipram (Fig. 18.6) [100]. The use of colon cancer-targeting nanoparticles has also been of interest. For example, the use of monoclonal antibody-modified nanoparticles to target the colon was demonstrated in a study where cetuximab-modified human serum albumin (HSA)-nanoparticles were shown to differentially bind to and to accumulate in colon carcinoma cells [101]. The study by Kim et al., exploits the reduced uptake of nanoparticles by liver and spleen. This helps in solid tumor treatment as the

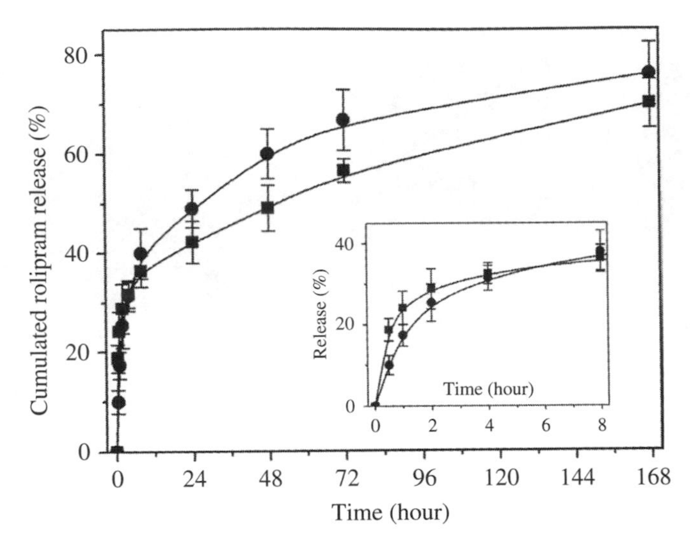

FIGURE 18.6 Release profiles of rolipram nanoparticles in phosphate buffer, pH 7.4, at 37°C during 168 hour. Lamprecht et al. [100], pp. 775–781. Reproduced with permission of ASPET.

increased nanoparticle circulation favors accumulation and may force out in tumor tissue. The study included self-assembled nanoparticles comprising hydrophobically modified glycol chitosan (HGC) as a carrier of paclitaxel (PTX), which were effective in suppressing tumor growth and reducing adverse effects than paclitaxel alone [102].

18.7 CONCLUSION

Colon-targeted drug deliveries have become valuable and efficient drug delivery options. These kinds of delivery systems offer significant therapeutic improvements for the treatment of local diseases as well as systemic delivery of certain therapeutics. Presently, several modified-release formulations are offered as colonic drug therapies. These particular formulations rely upon the colonic microflora, pH, and pressure for colon-specific release. Each of these approaches comes with advantages and drawbacks. In particular, site-specificity is generally better realized with systems that take advantage of the unique colonic bacterial enzymes. Colon-specific drug delivery is still a developing field and needs to be further explored.

ACKNOWLEDGMENT

The Research Council, Sultanate of Oman is acknowledged for the grant ORG/HSS/12/004.

REFERENCES

1. Philip, A.K.; Philip, B. *OMJ* 2010, **25**(2), 70–78.
2. American Cancer Society. *Cancer Facts & Figures 2013*, American Cancer Society, Atlanta, GA, 2013.
3. Brockmeier, D.; Grigoleit, H.G.; Leonhardt, H. *Eur. J. Clin. Pharmacol.* 1985, **29**(2), 193–197.
4. Staib, A.H.; Loew, D.; Harder, S.; Graul, E.H.; Pfab, R. *Eur. J. Clin. Pharmacol.* 1986, **30**(6), 691–697.
5. Fara, J.W. Colonic drug absorption and metabolism, In: *Novel Drug Delivery and its Therapeutic Application*, Prescott, L.F.; Nimmo, W.S. (Eds.), John Wiley and Sons, Ltd, Chichester, 1989, 103–120.
6. Laufen, H.; Wildfeuer, A. Absorption of isosorbide-5-nitrate for the gastrointestinal tract, In: *Drug Absorption at Different Regions of the Human Gastrointestinal Tract*, Rietbrock, N.; Woodcock, B.G.; Staib, A.H.; Loew, D. (Eds.), Methods in Clinical Pharmacology, Vieweg Verlag, Braunschweig, Wiesbaden, 1987, 76–81.
7. Friend, D.R. Glycosides in colonic drug delivery, In: *Oral Colon Specific Drug Delivery*, Friend, D.R. (Ed.), CRC Press, Boca Raton, FL/London, 1992, 153.
8. Watts, P.J.; Lllum, L. *Drug Dev. Ind. Pharm.* 1997, **23**(9), 893–913.
9. Tirosh, B.; Rubinstein, A. The varied mucus secretory response of the rat GI tract to cholinergic stimulus, In: *Proceedings International Symposium on Control Release of Bioactive Materials*, Controlled Release Society Inc., Minneapolis, MN, 1993, **23**, 186–187.
10. Chourasia, M.K.; Jain, S.K. *J. Pharm. Pharm. Sci.* 2003, **6**(1), 33–66.
11. Jain, A.; Gupta, Y.; Jain, S.K. *J. Pharm. Pharm. Sci.* 2007, **10**(1), 86–128.
12. McLeod, A.D.; Tozer, T.N. Kinetic perspective on colonic drug delivery, In: *Oral Colon-Specific Drug Delivery*, Friend, D.R. (Ed.), CRC Press, Boca Raton, FL, 1992, 83.
13. Sinha, V.R.; Kumria, R. *Eur. J. Pharm. Sci.* 2003, **18**, 3–18.
14. Asghar, L.F.A.; Chandran, S. *J. Pharm. Pharm. Sci.* 2006, **9**(3), 327–338.
15. Mackay, M.; Tomlinson, E. Colonic delivery of therapeutic peptides & proteins, In: *Colonic Drug Absorption and Metabolism*, Biek, P.R. (Ed.), Marcel Dekker Inc., New York, 1993, 159–176.
16. Lee, V.H.L.; Dodd-Kashi, S.; Grass, G.M.; Rubas, W. Oral route of peptide and protein drug delivery, In: *Protein and Peptide Drug Delivery*, Lee, V.H.L. (Ed.), Marcel Dekker Inc., New York, 1991, 691–740.
17. Wang, X.; Brown, I.; Khaled, D.; Mahoney, M.C.; Evans, A.J.; Conway, P.L. *J. Appl. Microbiol.* 2002, **93**(3), 390–397.
18. Freibauer J.; Gossrau, R. *Acta Histochem.* 1988, **83**(2), 207–232.
19. Evans, D.F.; Pye, G.; Bramley, R.; Clark, A.G.; Dyson, T.J.; Hardcastle, J.D. *Gut* 1988, **29**, 1035–1041.
20. Leopold, C.L. A practical approach in the design of colon-specific drug delivery systems, In: *Drug Targeting Organ-Specific Strategies*, Molema, G.; Meijer, D.K.F. (Eds.), Wiley-VCH Verlag GmbH, Weinheim/New York, 2001, 157–170.
21. Ashford, M.; Fell, J.T.; Attwood, D.; Sharma, H.; Woodhead, P.J. *Int. J. Pharm.* 1993, **95**, 193–199.

22. Huttunen, K.M.; Raunio H.; Rautio, J. *Pharmacol. Rev.* 2011, **63**(3), 750–771.
23. Johnson, R.M.; Verity, A.N. *Proc. West. Pharmacol. Soc.* 2002, **45**, 219–222.
24. Friend, D.R. *Adv. Drug Deliv. Rev.* 2005, **57**(2), 247–265.
25. Houba, P.H.J.; Boven, E.; van der Meulen M.H.; Leenders, R.G.G.; Scheeren, J.W.; Pinedo, H.M.; Haisma, H.J. *Br. J. Cancer* 2001, **84**, 550–557.
26. Roldo, M.; Barbu, E.; Brown, J.F.; Laight, D.W.; Smart J.D.; Tsibouklis, J. *Expert Opin. Drug Deliv.* 2007, **4**(5), 547–560.
27. Jain, A.; Gupta, Y.; Jain S.K. *Crit. Rev. Ther. Drug Carrier. Syst.* 2006, **23**(5), 349–400.
28. Klotz, U. *Dig. Liver. Dis.* 2005, **37**(6), 381–388.
29. Philip, A.K.; Dubey, R.K.; Pathak, K. *J. Pharm. Pharmacol.* 2008, **60**(5), 607–613.
30. Oz, H.S.; Chen, T.S.; Nagasawa, H. *Transl. Res.* 2007, **150**(2), 122–129.
31. Philip, A.K.; Dabas, S.; Pathak, K. *J. Drug Target.* 2009, **17**(3), 235–241.
32. Cassano, R., Trombino, S.; Cilea, A.; Ferrarelli, T.; Muzzalupo, R.; Picci, N. *Chem. Pharm. Bull.* 2010, **58**(1), 103–105.
33. Landowski, C.P.; Song, X.; Lorenzi, P.L.; Hilfinger, J.M.; Amidon, G.L. *Pharm. Res.* 2005, **22**(9), 1510–1518.
34. Philip, L.L.; Christopher, P.L.; Xueqin, S.; Katherine, Z.B.; Julie, M.B.; Jae, S.K.; John, M.H.; Leroy, B.T.; John, C.D.; Gordon, L.A. *J. Pharmacol. Exp. Ther.* 2005, **314**(2), 883–890.
35. Tozer, T.N.; Rigod, J.; McLeod, A.D.; Gungon, R.; Hoag, M.K.; Friend, D.R. *Pharm. Res.* 1991, **8**(4), 445–454.
36. de Graaf, M.; Pinedo, H.M.; Quadir, R.; Haisma, H.J.; Boven, E. *Biochem. Pharmacol.* 2003, **65**(11), 1875–1881.
37. Friend, D.R.; Chang, G.W. *J. Med. Chem.* 1984, **27**, 261–266.
38. Haisma, H.J.; Boven, E.; van Muijen, M.; de Jong J.; van der Vijgh W.J.; Pinedo, H.M. *Br. J. Cancer* 1992, **66**(3), 474–478.
39. Wang, S.M.; Chern, J.W.; Yeh, M.Y.; Ng, J.C.; Tung, E.; Roffler, S.R. *Cancer Res.* 1992, **52**(16), 4484–4491.
40. Windsor, B.J.; Nitin, N. *Enzyme mediated delivery system*, US patent 20090269405, October 29, 2009.
41. Liu, G.; Zhou, J.; Fu, Q.S.; Wang, J. *J. Bacteriol.* 2009, **191**(20), 6394–6400.
42. Scheline, R.R. *Pharmacol. Rev.* 1973, **25**, 451–523.
43. Courtney, M.G.; Nunes, D.P.; Bergin, C.F.; O'Driscoll, M.; Trimble, V.; Keeling, P.W.; Weir, D.G. *Lancet* 1992, **339**, 1279–1281.
44. Kopecek, J.; Kopeckova, P. *N*-(2-Hydroxypropyl) methacrylamide copolymers for colon-specific drug delivery, In: *Oral Colon-Specific Drug Delivery*, Friend, D.R. (Ed.), CRC Press, Boca Raton, FL, 1992, 189–211.
45. Schacht, E.; Gevaert, A.; Kenawy, E.R.; Molly, K.; Verstraete, W.; Adriaensens, P.; Carleer, R.; Gelan, J. *J. Control. Release* 1996, **39**, 327–338.
46. Nakamura, J.; Asai, K.; Nishida, K.; Sasaki, H. *Chem. Pharm. Bull.* 1992, **40**, 2164–2168.
47. Nakamura, J.; Tagami, C.; Nishida, K.; Sasaki, H. *J. Pharm. Pharmacol.* 1992, **44**, 295–299.
48. Nakamura, J.; Kido, M.; Nishida, K.; Sasaki, H. *Int. J. Pharm.* 1992, **87**, 59–66.
49. Jung, Y.L.; Lee, J.S.; Kim, Y.M. *J. Pharm. Sci.* 2000, **89**, 594–602.

50. Peters, T.J. *Gut* 1990, **11**, 720–725.
51. Hanauer, S.B. *Gut* 2002, **51**(4), 616.
52. Friend, D.R. *Aliment. Pharmacol. Ther.* 1998, **12**, 591–603.
53. Cui, N.; Friend, D.R.; Fedorak, R.N. *Gut* 1994, **35**, 1439–1446.
54. Brown, J.P. *Crit. Rev. Food Sci. Nutr.* 1997, **8**, 229–336.
55. Soumya, R.S.; Gosh, S.; Abraham, E.T. *Int. J. Biol. Macromol.* 2010, **46**(2), 267–269.
56. Krishnaiah, Y.S.R.; Satyanarayana, V.; Dinesh Kumar, B.; Karthikeyan, R.S. *Eur. J. Pharm. Sci.* 2002, **16**, 185–192.
57. Elias, E.J.; Anil, S.; Ahmad, S.; Daud, A. *Nat. Prod. Commun.* 2010, **5**, 915–918.
58. Krishnaiah, Y.S.; Seetha Devi, A.; Nageswara Rao, L.; Bhaskar Reddy, P.R.; Karthikeyan, R.S.; Satyanarayana, V. *J. Pharm. Pharm. Sci.* 2001, **4**, 235–243.
59. Sinha, V.R.; Kumaria, R. *Int. J. Pharm.* 2001, **224**(1–2), 19–38.
60. Liu, L.; Won, Y.J.; Cooke, P.H.; Coffin, D.R.; Fishman, M.L.; Hicks, K.B.; Ma, P.X. *Biomaterials* 2004, **25**(16), 3201–3210.
61. Liu, L.; Fishman, M.L.; Kost, J.; Hicks, K.B. *Biomaterials* 2003, **24**(19), 3333–3343.
62. Rubinstein, A.; Radai, R. Ezra, M.; Pathak, S.; Rockem, J.S. *Pharm. Res.* 1993, **10**(2), 258–263.
63. Uekama, K.; Minami, K.; Hirayama, F. *J. Med. Chem.* 1997, **40**, 2755–2761.
64. Varshosaz. J. *Expert Opin. Drug Deliv.* 2012, **9**(5), 509–523.
65. Thanou, M.; Kean, T.; Roth, S. *J. Control. Release* 2005, **103**(3), 643–653.
66. Zhang, H.; Alsarra, I.A.; Neau, S.H. *Int. J. Pharm.* 2002, **239**(1–2), 197–205.
67. Rudolph, M.W., Klein, S., Beckert, T.E., Petereit, H., Dressman, J.B. *Eur. J. Pharm. Biopharm.* 2001, **51**(3), 183–190.
68. Singh, B.N. *Recent Pat. Drug Deliv. Formul.* 2007, **1**, 53–63.
69. Yehia, S.A.; Elshafeey, A.H.; Elsayed, I. *Drug Deliv.* 2011, **18**, 620–630.
70. Milabuer, M.N.; Kam, Y.; Rubinstein, A. Orally administered drug delivery systems to the colon, In: *Oral Controlled Release Formulation Design and Drug Delivery: Theory to Practice*, Wen, H.; Park, K. (Ed.), John Wiley & Sons, Inc., Hoboken, NJ, 2010, 230.
71. Thakral, S., Thakral, N., Majumdar, D.K. *Expert Opin. Drug Deliv.* 2013, **10**(1), 131–149.
72. Healey, J.N. *Scand. J. Gastroenterol. Suppl.* 1990, **172**, 47–51.
73. Pozzi, F.; Furloni, P.; Gazzaniga, A.; Davis, S.S.; Wilding, I. *J.Control. Release* 1994, **31**, 99–108.
74. Dandagi, P.M.; Jain, S.S.; Gadad, A.P.; Kulkarni, A.R. *Int. J. Pharm.* 2007, **328**(1), 49–56.
75. Qi, M.; Wang, P.; Wu, D. *Drug Dev. Ind. Pharm.* 2003, **29**(6), 661–667.
76. Bauer, K.H. New experimental coating material for colon-specific drug delivery, In: *Drug Targeting Technology: Physical Chemical Biological Methods*, Schreier, H. (Ed.), Marcel Dekker Inc., New York, 2001, 33.
77. Nunthanid, J.; Huanbutta, K.; Luangtana-Anan, M.; Sriamornsak, P.; Limmatvapirat, S.; Puttipipatkhachorn, S. *Eur. J. Pharm. Biopharm.* 2008, **68**, 253–259.
78. Philip, A.K.; Pathak, K. *AAPS PharmSciTech* 2006, **7**(3), 1–11.
79. Philip, A.K.; Pathak, K. *PDA J. Pharm. Sci. Technol.* 2007, **61**(1), 24–36.
80. Philip, A.K.; Pathak, K.; Shakya, P. *Eur. J. Pharm. Biopharm.* 2008, **69**(2), 658–666.

81. Philip, A.K.; Pathak, K. *Drug Dev. Ind. Pharm.* 2008, **34**(7), 735–743.
82. Tannergren, C; Bergendal, A.; Lennernas, H.; Abrahamsson, B. *Mol. Pharm.* 2009, **6**(1), 60–73.
83. Guittard, G.V.; Theeuwes, F.; Wong, P.S.L. *Delivery of drug to colon by oral dosage form*, US Patent 4904474, February 27, 1990.
84. Liu, H.; Yang, X.G.; Wei, L.L.; Zhou, L.L.; Tang, R.; Pan, W.S. *Int. J. Pharm.* 2007, **332**(1–2), 115–124.
85. Chaudhary, A.; Tiwari, N.; Jain, V.; Singh, R. *Eur. J. Pharm. Biopharm.* 2011, **78**(1), 134–140.
86. Kumar, P.; Singh, S.; Mishra, B. *Chem. Pharm. Bull. (Tokyo)* 2008, **56**(9), 1234–1242.
87. Nath, B.; Nath, L.K. *PDA J. Pharm. Sci. Technol.* 2013, **67**(2), 172–184.
88. Takaya, T.; Niwa, K.; Matsuda, K.; Danno, N.; Takada, K. Evaluation of pressure controlled colon delivery capsule made of ethylcellulose, In: *Proceedings International Symposiumon Control Release of Bioactive Materials*, Controlled Release Society Inc., Minneapolis, MN, 1996, **23**, 603–604.
89. Takaya, T.; Niwa, K.; Muraoka, M.; Ogita, I.; Nagai, N.; Yano, R.; Kimura, G.; Yoshikawa, Y.; Yoshikawa, H.; Takada, K. *J. Control. Release* 1998, **50**(1–3), 111–122.
90. Barakat, N.S.; Al-Suwayeh, S.A.; Taha, EI.; Bakry Yassin, A.E. *J. Drug Target.* 2011, **19**(5), 365–372.
91. Niwa, K.; Takaya, T.; Morimoto, T.; Takada, K. *J. Drug Target.* 1995, **3**(2), 83–89.
92. Stevanovi, M.; Uskokovi, D. *Curr. Nanosci.* 2009, **5**, 1–15.
93. Millotti, G.; Schnurch, A.B. Nano and microparticles in oral delivery of macromolecular drugs, In: *Oral Delivery of Macromolecular Drugs: Barriers, Strategies and Future Trends*, Bernkop, A.S. (Ed.), Springer, New York, 2009.
94. Hardy, F.H.; Davis, S.S.; Khosla, R.; Robertson, C.S. *Int. J. Pharm.* 1988, **48**, 79–82.
95. Watts, P.J.; Barrow, L.; Steed, K.P.; Wilson, C.G.; Spiller, R.C.; Melia, C.D.; Davies, M.C. *Int. J. Pharm.* 1992, **87**, 215–221.
96. Allison, M.C.; Cornwall, S.; Poulter, L.W.; Dhillon, A.P.; Pounder, R.E. *Gut* 1988, **29**, 1531–1538.
97. Tabata, Y.; Inoue, Y.; Ikada, Y. *Vaccine* 1996, **14**, 1677–1685.
98. Stein, J.; Ries, J.; Barrett, K.E. *Am. J. Physiol.* 1998, **274**, G203–G209.
99. Laroui, H.; Dalmasso, G.; Nguyen, H.T.; Yan, Y.; Sitaraman, S.V.; Merlin, D. *Gastroenterology* 2010, **138**(3), 843–853.
100. Lamprecht, A.; Ubrich, N.; Yamamoto, H.; Schafer, U.; Takeuchi, H.; Maincent, P.; Kawashima, Y.; Lehr, C.M. *J. Pharmacol. Exp.Ther.* 2001, **299**(2), 775–781.
101. Low, K.; Wacker, M.; Wagner, S.; Langer, K.; von Briesen, H. *Nanomedicine* 2011, **7**(4), 454–463.
102. Kim, J.H.; Kim, Y.S.; Kim, S.; Park, J.H.; Choi, K.; Chung, H.; Jeong, S.Y.; Park, R.W.; Kim, I.S.; Kwon, I.C. *J. Control. Release* 2006, **111**, 228–234.

19

RECEPTOR-MEDIATED DRUG DELIVERY

Chris V. Galliford and Philip S. Low

Department of Chemistry, Purdue University, West Lafayette, IN, USA

19.1 INTRODUCTION

Most current drugs distribute nonspecifically and randomly throughout the body, entering both healthy and pathologic cells with roughly equal efficiency. Not surprisingly, when normal cells are sensitive to such drugs, their health can be compromised, leading to side effects that can limit the use of the therapeutic agents. In the case of drugs designed to promote only minor changes in cell behavior (e.g., aspirin), such side effects are usually acceptable. However, when the drug is designed to cause cell death or induce a significant change in cell behavior, toxicity to normal cells can undermine its use. The development of receptor-targeted therapeutic agents has been initiated primarily to limit the distribution of toxic drugs to only the pathologic cells, thus minimizing collateral damage to normal cells [1–3]. However, as will be noted later, receptor-mediated drug delivery can also enable otherwise membrane-impermeable drugs to enter target cells by receptor-mediated endocytosis, or it can induce desirable changes in cell behavior by activating a receptor's normal signaling pathway. Because endocytosis is intimately involved in each of the aforementioned merits of receptor-mediated drug delivery, we will begin this chapter by summarizing the basic characteristics of this process.

Endocytosis constitutes the pathway by which extracellular material is carried into a cell by membrane invagination and internalization [4–6]. Endocytosis occurs in virtually all eukaryotic cells [7, 8], and can assist in such diverse processes as

Drug Delivery: Principles and Applications, Second Edition. Edited by Binghe Wang, Longqin Hu, and Teruna J. Siahaan.

hormone signaling and removal, vitamin and mineral uptake, extracellular solute uptake, pathogen removal, and even simple membrane turnover. In fact, endocytosis is so active in some cells that the entire plasma membrane is internalized and replaced in less than 30 minutes [9].

Endocytosis can be divided into three subcategories. The first is commonly referred to as phagocytosis, or the process of cellular "eating." Phagocytosis plays a major role in host defense mechanisms by mediating the ingestion and degradation of microorganisms. Phagocytosis is also essential for tissue remodeling/differentiation and elimination of cellular debris. In contrast to other forms of endocytosis, phagocytosis is generally carried out in higher eukaryotes by professional phagocytes, such as polymorphonuclear granulocytes, monocytes, and macrophages.

The second category of endocytosis is called fluid-phase pinocytosis, which arises from entrapment of solutes by vesicles that invaginate from the cell surface. Importantly, the amount of material taken in by this route is proportional to a component's concentration in the extracellular environment. As such, pinocytosis is generally regarded as the means by which solutes enter cells non-specifically.

The third subcategory of endocytosis is referred to as "receptor-mediated." Receptors belong to a special class of cell surface proteins that utilize the endocytosis machinery to physically carry exogenous ligands into cells. When a specific cell surface receptor is overexpressed on a pathologic cell, receptor-mediated endocytosis can be exploited for targeted drug delivery.

The specific events that occur during receptor-mediated endocytosis are illustrated in Figure 19.1. Initially, exogenous ligands bind to externally oriented receptors on the cell membrane. This is a highly specific event, that is analogous to a key (ligand) inserting into a lock (receptor). Ligand binding usually occurs within minutes, but the kinetics of this event are dictated by the rate of ligand diffusion and the intrinsic affinity of the ligand for its receptor. Immediately after binding, the plasma membrane surrounding the ligand–receptor complex begins to invaginate until a distinct internal vesicle, called an "early endosome," forms within the cell [10].

Endosomal vesicles often move to their intracellular destinations along tracks of microtubules in a random, salutatory motion [11]. They may eventually interact with the trans Golgi reticulum where they are believed to fuse with membranous compartments prior to converting into late endosomes or multi-vesicular bodies. These latter compartments are capable of sorting the dissociated ligands from their empty receptors. At this juncture, there are four possible fates of the ligand and receptor:

1. Both the ligand and its receptor can be directed to the lysosomes for destruction (e.g., various hormones).
2. The ligand can be directed to a lysosome for destruction, while its receptor is recycled back to the plasma membrane to participate in another round of endocytosis (e.g., asialoglycoprotein).

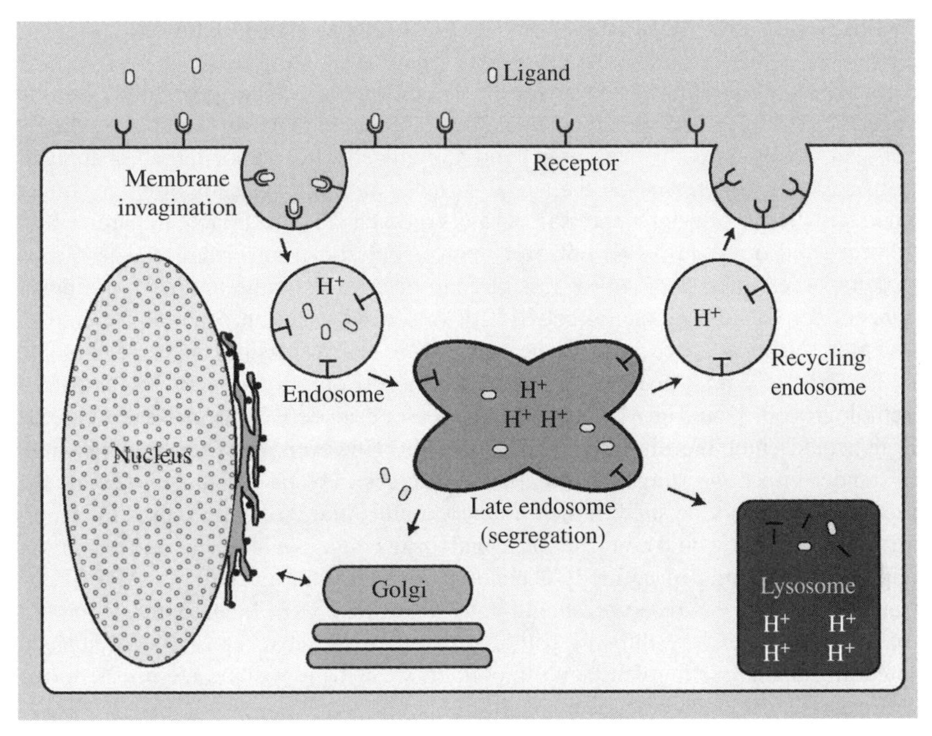

FIGURE 19.1 Receptor-mediated endocytosis. Exogenous ligands bind specifically to their cell surface receptors. The plasma membrane invaginates around the ligand–receptor complexes to form an intracellular vesicle (endosome). Intracellular trafficking of the ligand–drug conjugate may then involve passage of the conjugate through early endosome (EE) and late endosome (e.g., CURL), with ultimate translocation of the conjugate through a recycling endosome (RE) to the cell surface or directly to lysosomes (L).

3. The ligand can be transferred into the cytosol, while its receptor is recycled back to the plasma membrane to participate in another round of endocytosis (e.g., folate receptor or FR).
4. Both the ligand and its receptor can be recycled back to the plasma membrane (e.g., transferrin receptor).

Peculiarly, the fates of many receptor–ligand complexes can change from one of the aforementioned categories to another, depending on the percent occupancy of the cell surface receptor [12]. Thus, high receptor occupancy often causes a traditional recycling receptor to divert into a degradative pathway. When the same cell surface receptor is exploited for receptor-mediated drug delivery, the ligand–drug conjugate generally follows the intracellular itinerary of the free ligand. The only known exception to this rule arises when multiple ligands are attached to a single therapeutic particle (e.g., a liposome). Under these conditions, the natural endocytic pathway can be aborted, and the multivalent complex may be trafficked to lysosomes or some other unnatural destination [13–15].

19.2 SELECTION OF A RECEPTOR FOR DRUG DELIVERY

The choice of a receptor for receptor-mediated drug delivery is generally based on several criteria. First, the receptor should be present at high density on the pathologic cell, but largely absent or inaccessible on normal cells. For tumor-targeting applications, receptors expressed on the apical surfaces of epithelial cells often constitute good targets, since such receptors in normal epithelia are inaccessible to parenterally administered drugs [16–19]; however upon neoplastic transformation, these sites become accessible as a result of loss of cell polarity (also note that 80% of human cancers derive from epithelial cells [20]). A second criterion often considered in receptor selection concerns the heterogeneity in its expression on the pathologic cells. Thus, receptors that are present at high levels on only a small percentage of pathologic cells would be a less desirable target for drug delivery because the targeted drug would enter the diseased tissue unevenly. However, when drugs with large bystander effects are targeted, such disadvantages may be minimized. Third, the receptor should not be shed in measurable amounts into circulation, thereby generating a decoy that would compete for ligand–drug conjugate binding. And except for applications relating to antibody-dependent enzyme-prodrug therapy (ADEPT) or immunotherapy, the receptor should internalize and recycle in order to permit maximal drug delivery into the pathologic cell. Because receptor specificity and internalization/recycling can be so important, we will now elaborate on these two characteristics in greater detail.

19.2.1 Specificity

Perhaps the most significant advantage of receptor-mediated drug delivery lies in the researcher's ability to restrict drug deposition to tissues that express the ligand's receptor. Thus, the biodistribution of a ligand–drug conjugate should, in principle, follow the expression pattern of the ligand's receptor in the body. In our experience, this approximation is, in fact, realized if (i) the affinity of the receptor for the ligand is high, (ii) the attached drug introduces no competing affinity of its own, and (iii) the conjugate does not become entrapped in nontargeted compartments.

The ligand–drug conjugate's specificity for its receptor can and should be evaluated *in vitro* before it is tested *in vivo*. Typically, such specificity can be established by showing that (i) a ligand–drug conjugate binds to and becomes internalized by receptor-positive cell lines, (ii) association of the ligand–drug conjugate with these cells is blocked when an excess of free ligand is either pre- or co-incubated with the cells, and (iii) no measurable cell association occurs with either receptor-negative cell lines or cells from which the receptor has been removed by cleavage or transcriptional/translational suppression.

In vivo specificity can similarly be evaluated by (i) comparing uptake of the ligand–drug conjugate in a known receptor-positive tissue (e.g., tumor) with its uptake in several receptor-negative tissues (e.g., lung, liver, and heart), and (ii) examining the competitive blockade of the ligand–drug conjugate's enrichment in target tissue upon pre- or co-injection of the animal with excess free ligand. While some nonspecific retention of a conjugate in normal tissues cannot usually be avoided, in our experience, nontargeted uptake can be minimized by constructing the conjugate such that its linker and therapeutic cargo exhibit little affinity for cell surfaces on their own. In general, the more hydrophilic a conjugate is, the less it will be plagued by nonspecific tissue adsorption.

19.2.2 Receptor Internalization/Recycling

As noted earlier, following ligand binding and endocytosis, some (but not all) receptors unload their ligands and recycle back to the cell surface where they participate in another round of ligand binding and endocytosis [21–23]. In order to maximize drug delivery, it would seem intuitive to try to identify such a receptor, since recycling receptors can continually deliver ligand–drug conjugates into their target cells [21, 23]. A simple calculation will serve to emphasize the importance of this consideration. Assume that a target cell expresses 500,000 molecules of receptor and that the net efficiency of receptor unloading, endosome escape, and release of free drug in the cytosol is only approximately 1%. If the receptor cannot recycle back to the cell surface, then approximately 5000 active drug molecules would enter an average aqueous cytosolic space of approximately 0.4 picoliters per cell [24], leading to a cytosolic concentration of only 20 nM. In contrast, if the receptor recycles every 2 hours, then a controlled release formulation of the same drug delivered over a week's time span could establish a cytosolic concentration of up to 1.68 μM. Obviously, the delivered drug will need to exceed a threshold concentration (e.g., IC90) in the cell to achieve a therapeutic endpoint.

19.3 DESIGN OF A LIGAND–DRUG CONJUGATE

19.3.1 Linker Chemistry

A typical structure for a ligand–drug conjugate is cartooned in Figure 19.2. While the ligand, linker, and drug can all take on a diversity of sizes, shapes, and chemistries, a few fundamental principles can be followed to enhance therapeutic efficacy. We will

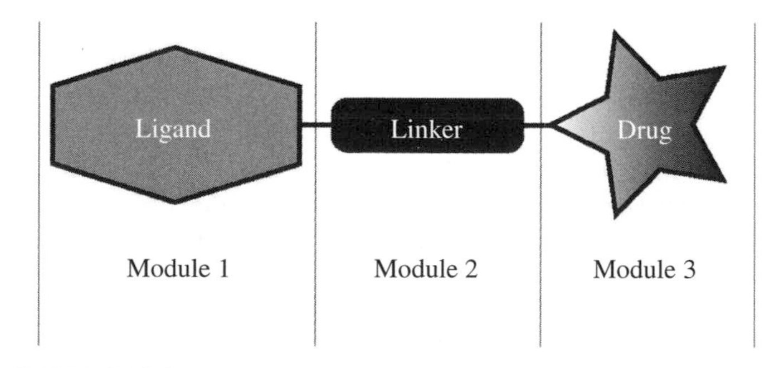

FIGURE 19.2 Structural design of a ligand-targeted drug conjugate.

begin by briefly outlining the desirable features to include in the design of a linker, and then proceed to describe the preferred characteristics of both the ligand and the drug.

Because one's freedom to change the chemistry of either ligand or drug is frequently limited by the functional roles these components must perform, the investigator's greatest creativity is often required in designing a linker that endows the conjugate with the optimal properties. Thus, not only must the linker be equipped with appropriate groups to react with available functional moieties on both the ligand and drug, but when improvements in water solubility, kidney excretion, serum protein binding, or other pharmacokinetic properties are required, the linker is often the only site where such modifications can be made.

The length of the linker can also be critical to drug delivery, since drug moieties positioned too close to a low-molecular-weight ligand can sterically reduce or even eliminate the affinity of the ligand for its receptor. Conversely, drug moieties separated too far from their targeting ligands by flexible spacers can often loop back and interact with the ligand, thereby also compromising the ligand's affinity for its receptor. Further, depending on the nature of the drug's activity, release of an intact unmodified drug may be critical to full expression of activity. Such processes have, in fact, evolved in nature to yield plant, fungal, and bacterial protein toxins of extraordinary potency [25–28], and similar release strategies have been recently shown to maximize the biological activities of ligand-targeted therapeutics (see later text).

Interestingly, knowledge gained from the study of protein toxins has proven highly useful in the design of receptor-targeted drugs. Thus, it was learned early in the characterization of protein toxins that replacement of the natural binding (B) chain with an alternative ligand would produce a powerful therapeutic agent with the new ligand's cell targeting specificity. These and related studies demonstrated that such toxins were constructed of independent binding and active domains, much like the drug conjugate cartooned in Figure 19.2. Second, it was shown that natural release of the toxic domain from the binding domain frequently involves intra-endosomal disulfide bond reduction [29–31], suggesting that disulfide bond reduction might also be exploited to release synthetic drugs from their ligands following entry into their target cells [28, 32–35].

Knowledge that certain endocytic vesicles become acidified [36, 37] has also prompted some to explore the use of acid-labile linkers in their designs of ligand–drug conjugates. As anticipated from the pH profiles of endocytic compartments, a drug attached to its ligand via an acid-sensitive linker can often be released shortly after formation of the endosome. Importantly, most of the progress in this area has come from studies of hydrazone-, acetal- and ketal-based linkers. For example, Neville et al. reported that the potency of an IgG–ketal–diphtheria toxin conjugate is increased 50-fold over its non-cleavable counterpart, and researchers at the former Wyeth–Ayerst pharmaceutical company demonstrated good activity of an anti-CD33 antibody–calicheamicin construct, featuring a cleavable hydrazine–disulfide linker in acute myelogenous leukemia patients [38, 39]. Notably, the latter construct was the first antibody–drug conjugate (ADC) to receive approval by the FDA and was marketed under the name Mylotarg® [40] until its voluntary withdrawal from the US market in 2010 amid fears over off-target cytotoxicity. Since then, two more ADCs have been approved for therapeutic use, ado-Trastuzumab emtansine (Kadcyla®) [41], and Brentuximab vedotin (Adcetris®) [42], which target HER2 and CD30, respectively. Both of these newer constructs feature a more stable thioether linker than that found in Mylotarg [43, 44] to assure that the therapeutic warhead does not release prematurely from the antibody during its prolonged circulation in the vasculature.

A third strategy for enabling the release of a drug from its targeting ligand following endocytosis consists of insertion of a peptide linker whose sequence is recognized and cleaved by endosomal/lysosomal enzymes. Indeed, the peptides Gly-Phe-Leu-Gly and citrulline–valine have been used successfully to promote lysosome-specific release of a variety of drugs [45–47], but their utility is probably limited to use with ligands that naturally target the destructive lysosomal compartment.

For any of the aforementioned release strategies, one must be concerned about the chemical nature of the liberated drug fragments. For instance, as illustrated earlier for the enzymatic techniques, the released drug fragment will contain a portion of the cleaved peptide (e.g., Phe-Gly-drug or drug-Gly-Leu if the construct design were reversed). In some cases, this added chemical baggage may affect the drug's intrinsic activity or its ability to traverse the endosomal membrane. In other cases, one of the released fragments could be toxic (e.g., hydrazine [48]). The same principle applies to all release strategies. In fact, the authors have experienced the inactivation of a potent microtubule stabilizing drug following its hydroxy esterification with a thiopropionyl linker moiety [49]. Overall, we believe that the best release strategy consists of one that discharges the drug in its original, unmodified form [49].

19.3.2 Selection of Ligands

A variety of biological ligands have been used to deliver drugs to target cells. Table 19.1 lists some of the most common ligand–receptor systems exploited to date for the delivery of therapeutic molecules. Most of these have been coupled to functionally active peptides and proteins and, in some cases, to small-molecular-weight chemotherapeutic drugs. Each receptor system has advantages and disadvantages. Unfortunately, limitations in space preclude an in-depth discussion of each system. Therefore, the reader is referred to the listed references when additional information is needed. However, for the purpose of completing this discussion on receptor-targeted drug delivery, we shall illustrate in detail how the folate-targeted drug delivery pathway has been exploited, both at the academic and clinical levels.

19.3.3 Selection of Therapeutic Drug

While the nature of the pathology often dictates the choice of the therapeutic agent, wherever multiple selections exist, a few guidelines can be beneficial. First, because receptor-mediated delivery pathways are frequently of low capacity, higher activity will likely occur with those conjugates constructed with the more potent drugs. Second, since target cell penetration is mediated by an endocytic pathway, rapid membrane permeability is often not a necessary property of the drug. In fact, the

TABLE 19.1 Ligands Frequently Used for Receptor-Mediated Drug Delivery

Ligand	Drug Payload	References
Insulin	Drugs, enzymes	[50, 51]
Epidermal growth factor	Protein toxins	[52, 53]
Transferrin	Drugs, protein toxins, and gene therapy vectors	[54]
Thyrotropin-releasing hormone	Protein toxins	[55]
Human chorionic gonadotropin	Protein toxins	[56, 57]
Leutinizing hormone	Protein toxins	[58, 59]
Interleukin-2	Protein toxins	[60, 61]
Mannose-6-phosphate	Enzymes	[62, 63]
Asialoglycoprotein	Drugs, protein toxins, and gene therapy vectors	[64]
IgG	All pharmaceutical classes	[65, 66]
Vitamin B12	Drugs, peptides	[67–70]
Glucose	Drug	[71, 72]
NGR	Protein toxin	[73, 74]
Somatostatin analogs	All pharmaceutical classes	[75–77]
DUPA and related ligands	All pharmaceutical classes	[78–83]
CCK2R ligands	All pharmaceutical classes	[84–86]
NK-1 ligands	All pharmaceutical classes	[87]
Folate	All pharmaceutical classes	See Section 19.4

more hydrophobic drugs are frequently less desirable for ligand-targeted conjugates, since they can promote nonspecific adsorption to cell surfaces and the consequent unwanted toxicity to nontargeted cells. Finally, if specificity for the pathologic cells is high, the toxicity characteristics of the free drug can be ignored, since targeted delivery can prevent uptake by the sensitive normal cells. Thus, drugs that have been discarded because of poor toxicity profiles can often be reconsidered for use in targeted drug therapies.

19.4 FOLATE-MEDIATED DRUG DELIVERY

As a more detailed example of receptor-targeted drug delivery, we have elected to elaborate on the delivery pathway that exploits the cell surface receptor for folic acid as a means of targeting drugs to receptor-expressing cells. Folates are low-molecular-weight vitamins required by all eukaryotic cells for one-carbon metabolism, DNA and protein methylation reactions, and de novo nucleotide synthesis [88, 89]. Since animal cells lack key enzymes of the folate biosynthetic pathway, their survival and proliferation are dependent on their abilities to acquire the vitamin from their diet. Thus, effective mechanisms for capturing exogenous folates are needed to sustain all higher forms of life. While most cells rely on one of two low-affinity ($K_D \sim 1$–$5\,\mu M$) membrane-spanning proteins that transport reduced folates directly into the cell (termed the proton-coupled folate transporter and the reduced folate carrier [90, 91]), a few cells

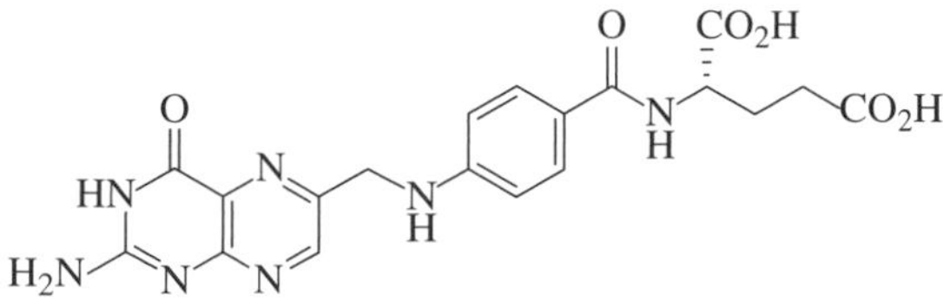

FIGURE 19.3 Structure of folic acid.

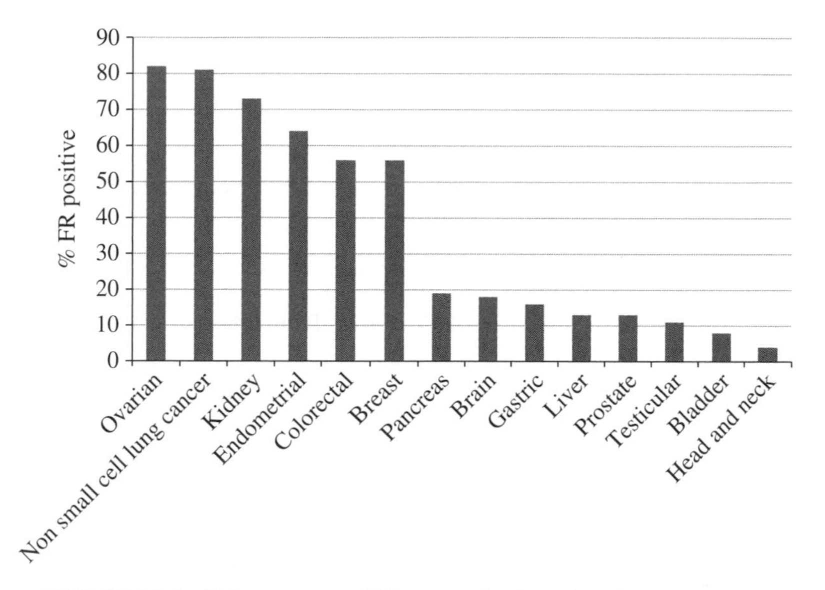

FIGURE 19.4 Folate receptor (FR) expression in various human cancers.

also express a high-affinity (K_D ~ 1 nM) receptor, generally referred to as the FR, which preferentially mediates the uptake of oxidized forms of the vitamin (e.g., folic acid) by receptor-mediated endocytosis [92, 93]. As will be explained later, attachment of folic acid (Fig. 19.3) via one of its carboxyl groups to a therapeutic or imaging agent allows targeting of the conjugate to cells that express FR [94], with no measurable uptake by cells that express only the proton coupled folate transporter or reduced folate carrier.

19.4.1 Expression of FRs in Malignant Tissues

In 1991, a clinically valuable tumor marker was purified from ovarian cancers, and sequence analysis showed that it was the receptor for folic acid [95]. Subsequent to that finding, FRs have been shown to be overexpressed on the cell surfaces of a number of different types of human cancers [95–98]. In general, the FR is upregulated in malignant tissues of epithelial origin. As detailed in Figure 19.4, FR expression has been detected at high levels in approximately 80% of ovarian and other gynecological cancers, and at high to moderate levels in the lung, kidney, endometrial, and triple-negative breast carcinomas [95–104]. Notably, the FR gene was also mapped to region 11q13: a chromosomal locus that is amplified in greater than 20% of tissue samples from breast and head/neck tumors [105, 106].

FR expression may also be dependent on the histologic classification of a tissue or cancer. For example, using a recently developed quantitative assay for measuring functional FR in cells and tissues, it was found that normal ovarian tissue and the mucinous form of ovarian cancer express very low levels of FR (~1 pmol FR/mg protein) [107]. However, serous ovarian carcinomas express large numbers of FRs, with the average expression level being 30 pmol FR/mg protein, or about 30-fold higher than normal ovary. Endometrioid carcinomas also express FR, although not at the levels seen in the serous form of ovarian cancer. Interestingly, others have observed a strong correlation between FR expression and both the histological grade and stage of the tumor [103]. In general, highly de-differentiated metastatic cancers express more FRs than their more localized, low-grade counterparts [107]. Non-small-cell lung carcinomas, however, display exactly the opposite trend, with the more differentiated grades showing the highest levels of FR expression.

19.4.2 Expression of FRs in Normal Tissues

FRs have also been detected in normal tissues, particularly those involved in the uptake or sequestration of the vitamin. For example, the choroid plexus in the brain expresses moderate levels of FR to facilitate concentration of the vitamin in the cerebrospinal fluid [108]. However, the receptor is primarily localized to the brain side of the blood–brain barrier, where it is inaccessible to blood-borne folates or folate–drug conjugates [104, 109, 110]. FR is similarly found in healthy lung tissue, but again its expression is limited to the apical surface of alveolar epithelial cells where it can only be accessed by inhaled folate conjugates [111, 112]. Expression of FR in this site could have evolved to reduce the availability of folic acid for use by inhaled bacteria that would otherwise proliferate in the lungs. A low level of FR has also been reported on the apical membrane of the intestinal brush border [113]; but as with choroid plexus and healthy lung, access to these sites requires entry into the lumen of the intestine; that is a site not generally encountered by parenterally administered drugs. Most importantly, FR is expressed at high levels in the proximal tubules of the kidney, where it is believed to capture folates (and small-molecular-weight folate–drug conjugates) as a means of preventing their excretion in the urine [114–116]. While these folate conjugates are internalized by the proximal tubule cells, most are not retained in the kidneys, but rather migrate across the kidney cell and are released from the basolateral membrane back into the blood. Finally, a different isoform of FR (i.e. FR beta) has been found on activated but not resting macrophages [2, 48, 117–132] that accumulate at sites of inflammation and autoimmune disease. Not surprisingly, expression of FR on these cells has triggered strong interest in the development of FR-targeted drugs for the treatment and imaging of inflammatory diseases. Pursuant to these interests, a large number of recent studies have focused on the identification of the autoimmune and inflammatory diseases that express sufficient FR-β for use in folate-mediated drug targeting. A list of these diseases is provided in Table 19.2. While FR has also been detected at very low levels in some normal tissues [97], its relative expression in these tissues is so low compared to

TABLE 19.2 Autoimmune and Inflammatory Diseases That Can Be Targeted with Folate Conjugates

Disease	References
Rheumatoid arthritis	[48, 125, 130]
Lupus	[132]
Atherosclerosis	[122, 128]
Asthma	[123]
Inflammatory bowel disease	[133]
encephalomyelitis	[134]
Autoimmune uveitis	[134]
Osteoarthritis	[135]
Multiple sclerosis	Unpublished data
Sarcoidosis	Unpublished data
Psoriasis	Unpublished data
Scleroderma	Unpublished data
Idiopathic pulmonary fibrosis	Unpublished data
Sjogren's syndrome	Unpublished data
Osteomyelitis	Unpublished data
Ischemia reperfusion injury	Unpublished data

many cancers and sites of inflammation that they need not be considered in the design or application of folate-targeted drugs. Consistent with this conclusion is the fact that intravenously administered folate-linked radioimaging agents in humans generally highlight only solid tumors, sites of inflammation, and healthy kidney tissue [136–138].

19.4.3 Applications of Folate-Mediated Drug Delivery

Initial studies on folate conjugate targeting were conducted with radiolabeled and fluorescent proteins (drug surrogates) covalently attached to folic acid [94]. These conjugates were shown to bind and become internalized by FR-positive cells via a nondestructive, functionally active endocytic process [94]. When FR was subsequently identified as a major tumor-associated antigen [95, 139], much effort was quickly devoted to determining the types of attached cargo that might be easily targeted with folic acid. These studies have revealed that conjugates of radiopharmaceutical agents [136, 137, 140–146], MRI contrast agents [147], low-molecular-weight chemotherapeutic agents [34, 148], antisense oligonucleotides and ribozymes [149–153], proteins and protein toxins [94, 154–158], immunotherapeutic agents [159–163], liposomes with entrapped drugs [131, 133, 164–169], drug-loaded nanoparticles [170–172], and plasmids [173–181] can all be selectively delivered to FR-expressing cancer cells. Indeed, the major limitation associated with the aforementioned targeting efforts appears to be the intrinsic permeability of the tumor. Thus, where perfusion barriers do not limit access to FR-expressing tumor cells, folate conjugate binding, FR-mediated endocytosis, and intracellular drug release are readily achievable if the fundamental principles outlined earlier are followed.

For the remainder of this chapter, we shall illustrate the techniques of folate-targeted radiodiagnostic imaging, chemotherapy, and immunotherapy. However, the reader is encouraged to review the listed references if information on other folate conjugates is desired.

19.4.3.1 Tumor-Targeting Through FR: Radiodiagnostic Imaging The field of nuclear medicine has been revitalized with the advent of tissue-specific radiopharmaceutical targeting technologies. Ligands capable of concentrating at pathologic sites have been derivatized with chelator–radionuclide complexes and then used as noninvasive probes for diagnostic imaging. Folate-targeted radiopharmaceuticals have also been explored both for the purpose of (i) developing an imaging agent for the localization, sizing, and characterization of cancers and (ii) obtaining "proof-of-principle" data supporting the ability of folic acid to deliver attached therapeutic agents to human tumors *in vivo*. Several animal models that contain tumors with FR levels similar to those found in common human cancers have been used to test uptake of folate-based radiopharmaceuticals in living organisms, including (i) nude mice with implanted human KB, MDA-231 or IGROV tumors, (ii) C57BL/6 mice implanted with 24JK tumors, (iii) Dupont's c-neu Oncomouse [182], (iv) DBA mice implanted with L1210A tumors, and (v) Balb/c mice implanted with syngeneic M109 tumors. This continually growing list of acceptable tumor models indicates that the location and nature of the tumor are relatively unimportant so long as the tumors express appreciable levels of FR.

In 1999, Phase I/II clinical studies were initiated by Endocyte, Inc., to evaluate ^{111}In-DTPA-folate [138, 145]. Patients suspected of having ovarian cancer received a 5 mCi (2 mg) intravenous dose of the radiopharmaceutical, and whole-body single-photon emission computerized tomographic (SPECT) images were taken 4 hours later to identify the location of the probe. Representative images from two enrolled patients are shown in Figure 19.5. The image displayed in Panel A shows that in a cancer-free patient, ^{111}In-DTPA-folate (a folate–drug conjugate) primarily concentrates in the FR-positive kidneys, while the remaining tissues of the body effectively clear the radiopharmaceutical by 4 hours post injection. However, as shown in panel B, the folate-targeted radiopharmaceutical accumulates in the widely disseminated malignant tissue in the peritoneal cavity of the ovarian cancer patient, in addition to the kidneys (and to some extent the liver). Taken together, and following the treatment of more than 45 subjects imaged with ^{111}In-DTPA-folate, we have noted a pattern of (i) consistent uptake of folate conjugates into malignant masses, (ii) absence of uptake into benign tumor masses, (iii) consistent uptake by the kidneys, and (iv) little or no uptake by other normal tissues. In summary, these clinical results provided an initial "proof of principle" confirming that folate-mediated tumor targeting also occurs in humans. It further suggested that folate-targeted radioimaging agents would be useful for noninvasively identifying the loci of pathologic FR-positive tissues within cancer patients.

Notably, a similar distribution pattern has more recently been observed in cancer patients imaged with a ^{99m}Tc-based radiopharmaceutical called "etarfolatide" [144, 183]. Thus, in the lung, kidney, brain, and ovarian cancer patients, etarfolatide has proven effective not only in imaging primary FR-positive tumors but also in localizing metastatic disease in proximal lymph nodes and distant tissues [184]. Moreover,

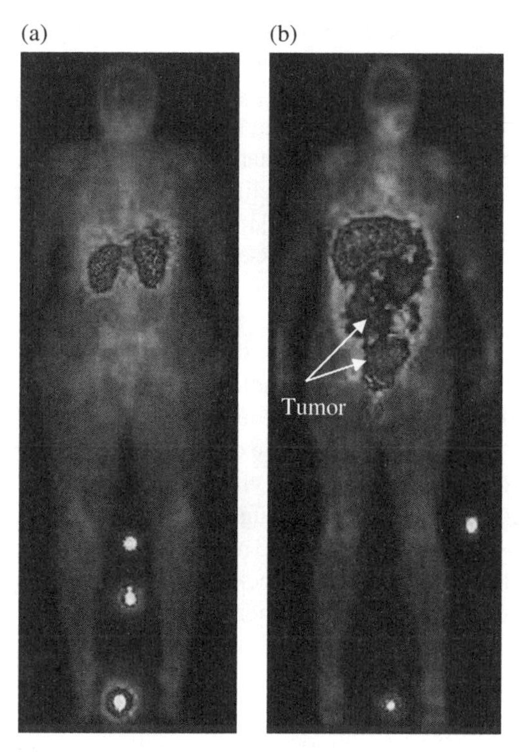

FIGURE 19.5 Anterior SPECT images of two patients receiving ^{111}In-DTPA-folate: (a) Image of a female patient without cancer and (b) image of a female patient with Stage IIIc ovarian carcinoma. (*See insert for color representation of the figure.*)

etarfolatide has also demonstrated usefulness in selecting patients with sufficient levels of FR in their tumors to respond to folate-targeted therapies [163, 185].

In inflammatory and autoimmune diseases, where FR-β-positive macrophages are known to accumulate, etarfolatide has shown efficacy in imaging the inflamed lesions in atherosclerosis, rheumatoid arthritis, ulcerative colitis, Sjogren's syndrome, osteoarthritis, sites of infection, and other autoimmune/inflammatory diseases [126, 128, 130, 133]. More recently, it has also been used to predict the level of response to therapy of a variety of inflammatory diseases to a multitude of therapeutic agents (unpublished data). In summary, folate-targeted radioimaging agents have established their utility in the detection of both FR-positive tumors and macrophages at inflamed loci.

19.4.3.2 Folate-Targeted Chemotherapy The fundamentals of folate–cytotoxin therapy were first illustrated by the targeted killing of FR-positive cells *in vitro* using folate conjugates of numerous protein synthesis-inhibiting enzymes [156]. For example, folate conjugates of cell-impermeant ribosome-inactivating proteins (e.g., momordin, saporin, gelonin, and ricin A) were all found to kill cultured FR-positive malignant cells without harming receptor-negative normal cells. The selectivity of this approach was confirmed by many important controls, which demonstrated that (i) an excess of free folic acid quantitatively blocked folate-cytotoxin cell killing, (ii) the underivatized protein was not toxic to the target cells, and (iii) pretreatment of the cells with phosphatidylinositol-specific phospholipase C, an enzyme that removes glycosylphosphatidylinositol-linked proteins (like the FR) from a cell's surface, effectively blocked folate–cytotoxin cell killing. Importantly, the same folate–cytotoxin conjugates were also shown to selectively kill only malignant cells when co-cultured in the same dish with "normal" non-transformed cells [158]. Overall, high activity (IC_{50} values of ~1 nM) was observed using a number of FR-expressing cell lines.

While it is tempting to speculate on the antitumor activity that may result from testing such folate–protein toxin conjugates *in vivo*, practical pharmaceutical considerations diminished the priority for their development. Instead, the focus shifted toward conjugates of conventional drug molecules.

Endocyte Inc. has collected data on the activities of a number of folate–drug conjugates both *in vitro* and *in vivo*. For example, folate has been conjugated to a potent small-molecular-weight DNA alkylator through a disulfide linker (folate-SS-mitomycin C), and the resultant conjugate (referred to as EC72) was found to promote target cell killing *in vitro* (IC_{50} ~ 3 nM) [186, 187]. Although the toxicity of this conjugate could be quantitatively blocked by the presence of excess free folic acid (to demonstrate FR specificity), neither EC72 nor the underivatized drug could kill all of the cancer cells *in vitro*, possibly because of an intrinsic resistance of some cells to the drug's mechanism of action. Nevertheless, EC72 was evaluated in a pilot study using Balb/c mice bearing FR-positive M109 tumors. There were two goals of the study as follows: (i) to determine if daily treatment with EC72 could prolong the lives of FR-positive tumor-bearing mice beyond that which the parent drug could do alone when tested under an identical dosing regimen and (ii) to examine the toxicological effects of EC72 on normal tissues, including the FR-positive kidneys.

As shown in Figure 19.6, all control mice died within 22 days of intraperitoneal inoculation with M109 cells, while a 39% increase in lifespan was observed for animals treated with the unmodified drug. More importantly, animals treated with the folate-targeted EC72 conjugate lived an average of 178% longer.

Following euthanasia, major organs were collected from both EC72 and parent drug-treated animals and sent to a certified pathologist for examination. The nontargeted drug-treated animals were found to suffer from massive myelosuppression (which is a major dose-limiting side effect of the parent drug), and all of the animals in this cohort died from obvious drug-related side effects. In dramatic contrast, animals treated with EC72 displayed no evidence of myelosuppression or kidney damage. Further, examination of blood collected from EC72-treated animals indicated normal blood urea nitrogen and creatinine levels following 30 consecutive daily injections with the conjugate. We concluded from these results that the use of folate–drug conjugates may be an effective form of chemotherapy that does not cause significant injury to normal tissues, including the FR-positive kidneys. While the latter conclusion was surprising, these observations do support the hypothesis that the FRs in the kidney proximal tubules function primarily to shuttle scavenged folates (or small folate–drug conjugates) back into systemic circulation rather than to deliver the conjugates into the kidney cells [115, 116].

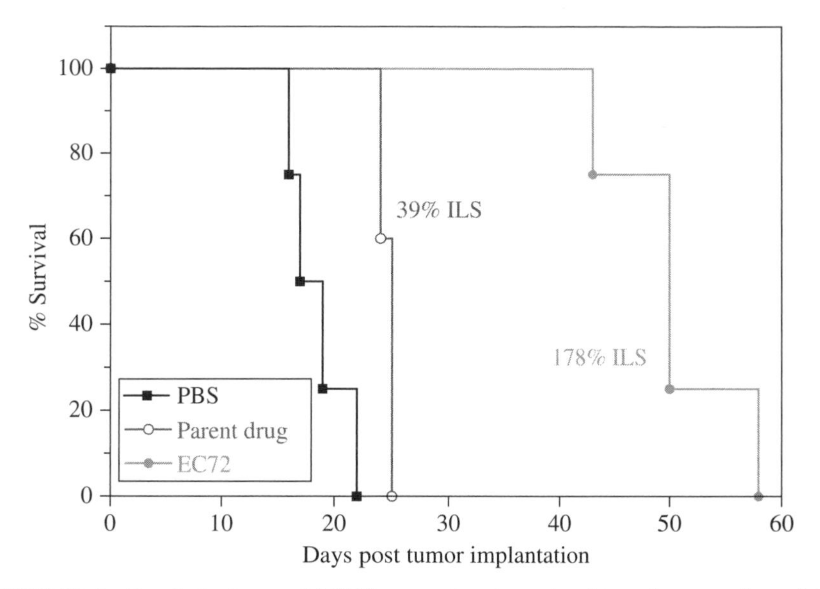

FIGURE 19.6 Survival of treated M109 tumor-bearing mice. Four days post inoculation, Balb/c mice bearing intraperitoneal M109 tumors were treated once daily with either the unmodified parent drug or the folate-derivatized drug. ILS, increased lifespan.

These initial results with a folate–cytotoxin conjugate led rapidly to the design, development, and clinical evaluation of several additional folate–chemotherapeutic agent conjugates [1]. Notably, the folate–vinblastine conjugate termed "vintafolide" recently underwent both a Phase III clinical trial for the treatment of ovarian cancer in combination with Doxil and a Phase II clinical trial for non-small-cell lung cancer in combination with docetaxel [17, 188]. Importantly, the results of a Phase II clinical trial of vintafolide plus docetaxel in non-small-cell lung cancer patients revealed that the combination therapy doubles overall survival relative to the patients treated with the docetaxel therapy alone (i.e., the current standard of care).

19.4.3.3 Folate-Targeted Immunotherapy

Intracellular Delivery Versus Cell-Surface Loading In contrast to most hormone receptors, not all FRs endocytose following ligand binding [189]. Rather, a substantial fraction of the occupied FRs remain on the cell surface as a means of storing folates for use at a later date. Since folate–drug conjugates interact with FR in the same manner as free folate [94], it was of no surprise to learn that a fraction of these conjugates also remain extracellularly bound [94, 157]. This dual destiny of occupied FR obviously allows for two distinct uses of the folate-targeting technology: (i) delivery of folate–drug conjugates into FR-expressing cells and (ii) decoration of FR-expressing cell surfaces with folate–drug conjugates. This latter application has led to the advent of folate-targeted immunotherapy [190], as outlined in the following text.

Based on the ability of folate to position an attached drug on the surface of an FR-expressing cancer cell, it was recognized that a malignant cell could be converted from its

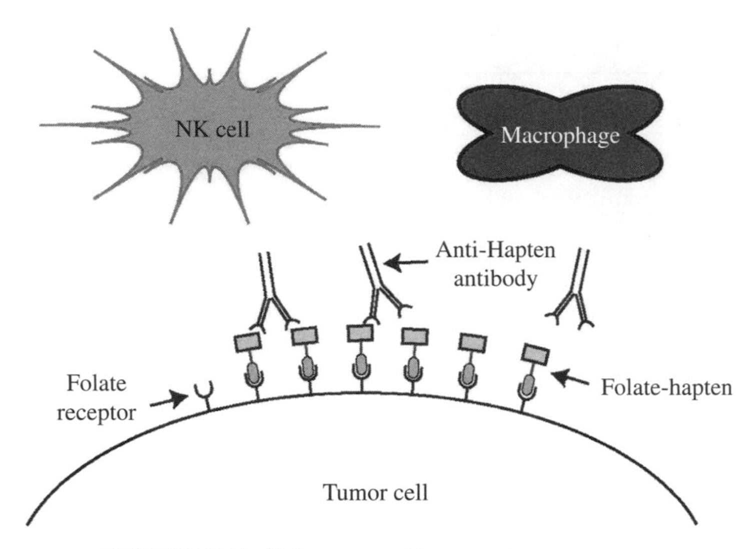

FIGURE 19.7 Folate-targeted immunotherapy strategy.

normal immunologically invisible state (the condition that permits its proliferation *in vivo*), to a state where it is vividly recognized as foreign by the immune system (and consequently subject to immune-mediated elimination) simply by targeting a highly immunogenic antigen to its cell surface. This concept is cartooned in Figure 19.7. Different from targeted chemotherapy, this treatment begins with a series of subcutaneous inoculations of a hapten-based vaccine to stimulate production of anti-hapten antibodies in the patient. After induction of an adequate antibody titer, a folate–hapten conjugate is administered to enable "marking" of all FR-expressing tumor cells with the hapten. This process rapidly promotes the tumor cell's opsonization with the previously induced endogenous anti-hapten antibodies. Mechanistically, the folate–hapten conjugate forms a molecular "bridge" between the tumor cell and the endogenous circulating anti-hapten antibody. Ultimately, this "marking" process enables Fc receptor-bearing effector cells, such as natural killer (NK) cells and macrophages, to recognize and destroy the antibody-coated tumor cells.

Although we found that this folate-targeted immunotherapy was effective in treating tumor-bearing rodents, we also noted that complete and reproducible cures were only obtained when the animals were concomitantly dosed with low levels of the cytokines: interleukin-2 and interferon-α [190]. The purpose of adding these cytokines was to stimulate those immune effector cells (like NK and macrophage cells) that are responsible for killing opsonized tumor cells. Importantly, the low cytokine dose levels applied in this technique were found not to produce effective antitumor responses in the absence of the folate–hapten conjugate. Although the exact effector mechanisms involved were never established, there was strong evidence to suggest that the cured animals developed an independent cellular immunity against the tumor cells and that this cellular immunity prevented the animals from growing tumors when rechallenged with the same malignant cells, even without additional treatment [190]. These findings also supported the notion that a long-term, perhaps T-cell-specific immunity develops during the hapten-mediated immunotherapy, and that a shift of the host's antitumor response from humoral to cellular immunity occurs. Because this therapy generated cures in multiple tumor-bearing animal models, and since the complete therapeutic regimen (vaccination and drug therapy at dose levels approaching 100-fold the human dose equivalent) did not produce test article-related toxicities in GLP toxicology studies, the strategy was approved to enter clinical trial by the FDA for the treatment of advanced ovarian and renal cell cancers. The results from a Phase I clinical trial of this targeted immunotherapy show considerable future promise [163, 191]. Further clinical evaluation should provide much information on the strengths and weaknesses of folate-targeted immunotherapies in humans.

19.5 CONCLUSIONS

Nature has designed biological membranes to serve as formidable barriers against the unwanted entry of harmful agents into cells. Life, however, cannot exist without the cell's ability to "selectively" capture and internalize certain bio-supportive molecules. Receptor-mediated endocytosis is but one mechanism by which cells retrieve required molecules from their environment, and it is also a powerful means for delivering normally impermeable molecules into target cells for medicinal purposes. The list of available ligands for use in this exploitive technique keeps growing, as does the potential for inventing alternative drug delivery strategies. It is important, however, to remember that not all ligand–receptor systems will function in the same manner. Many variations exist among cell types, including their levels of receptor expression, internalization rates, ligand affinities, intracellular compartmentalization, recycling capabilities, etc. Thus, despite recent technological advances, many challenges remain for the future development of receptor-targeted therapies. The most significant issues will undoubtedly be related to the confirmation of target tissue specificity (i.e., receptor distribution among target and nontarget tissues) as well as the design of releasable linkers for certain ligand–linker drug conjugates. Receptor-mediated delivery technology is undoubtedly influencing rational drug design, and it is our hope that the scientific and pharmaceutical community will continue to invest in this exciting new field of medicine.

ACKNOWLEDGMENTS

This work was supported in part by grants from Endocyte, Inc., and the NIH (CA89581).

REFERENCES

1. Srinivasarao, M.; Galliford, C. V.; Low, P. S. *Nat Rev Drug Discov* 2015, **14**, 203–219.
2. Feng, Y.; Shen, J.; Streaker, E. D.; Lockwood, M.; Zhu, Z.; Low, P. S.; Dimitrov, D. S. *Arthritis Res Ther* 2011, **13**, R59.
3. Low, P. S.; Kularatne, S. A. *Curr Opin Chem Biol* 2009, **13**, 256.

4. McPherson, P. S.; Kay, B. K.; Hussain, N. K. *Traffic* 2001, **2**, 375.
5. Mukherjee, S.; Ghosh, R. N.; Maxfield, F. R. *Physiol Rev* 1997, **77**, 759.
6. Schwartz, A. L. *Pediatr Res* 1995, **38**, 835.
7. Stahl, P.; Schwartz, A. L. *J Clin Invest* 1986, **77**, 657.
8. Smythe, E.; Warren, G. *Eur J Biochem* 1991, **202**, 689.
9. Marsh, M.; Helenius, A. *J Mol Biol* 1980, **142**, 439.
10. Pastan, I.; Willingham, M. C. *The Pathway of Endocytosis*; Plenum Press: New York, 1985.
11. Willingham, M. C.; Pastan, I. *Cell* 1980, **21**, 67.
12. Lai, W. H.; Cameron, P. H.; Wada, I.; Doherty, J. J., 2nd; Kay, D. G.; Posner, B. I.; Bergeron, J. J. *J Cell Biol* 1989, **109**, 2741.
13. Weflen, A. W.; Baier, N.; Tang, Q. J.; Van den Hof, M.; Blumberg, R. S.; Lencer, W. I.; Massol, R. H. *Mol Biol Cell* 2013, **24**, 2398.
14. Papademetriou, J.; Garnacho, C.; Serrano, D.; Bhowmick, T.; Schuchman, E. H.; Muro, S. *J Inherit Metab Dis* 2013, **36**, 467.
15. Ukkonen, P.; Lewis, V.; Marsh, M.; Helenius, A.; Mellman, I. *J Exp Med* 1986, **163**, 952.
16. O'Shannessy, D. J.; Somers, E. B.; Maltzman, J.; Smale, R.; Fu, Y.-S. *SpringerPlus* 2012, **1**, 1.
17. Chavakis, T.; Willuweit, A. K.; Lupu, F.; Preissner, K. T.; Kanse, S. M. *Thromb Haemost* 2001, **86**, 686.
18. Wang, L.; Madigan, M. C.; Chen, H.; Liu, F.; Patterson, K. I.; Beretov, J.; O'Brien, P. M.; Li, Y. *Gynecol Oncol* 2009, **114**, 265.
19. Kufe, D. W. *Nat Rev Cancer* 2009, **9**, 874.
20. Simons, K.; Fuller, S. D. *Annu Rev Cell Biol* 1985, **1**, 243.
21. Paulos, C. M.; Reddy, J. A.; Leamon, C. P.; Turk, M. J.; Low, P. S. *Mol Pharmacol* 2004, **66**, 1406.
22. Bandara, N. A.; Hansen, M. J.; Low, P. S. *Mol Pharm* 2014, **11**, 1007.
23. Varghese, B.; Vlashi, E.; Xia, W.; Ayala Lopez, W.; Paulos, C. M.; Reddy, J.; Xu, L. C.; Low, P. S. *Mol Pharm* 2014, **11**, 3609.
24. Hanvey, J. C.; Peffer, N. J.; Bisi, J. E.; Thomson, S. A.; Cadilla, R.; Josey, J. A.; Ricca, D. J.; Hassman, C. F.; Bonham, M. A.; Au, K. G.; Carter, S. G.; Bruckenstein, D. A.; Boyd, A. L.; Noble, S. A.; Babiss, L. E. *Science* 1992, **258**, 1481.
25. Schmidt, G.; Papatheodorou, P.; Aktories, K. *Curr Opin Microbiol* 2015, **23**, 55.
26. Weidle, U. H.; Tiefenthaler, G.; Schiller, C.; Weiss, E. H.; Georges, G.; Brinkmann, U. *Cancer Genomics Proteomics* 2014, **11**, 25.
27. Tambourgi, D. V.; van den Berg, C. W. *Mol Immunol* 2014, **61**, 153.
28. Leamon, C. P.; Pastan, I.; Low, P. S. *J Biol Chem* 1993, **268**, 24847.
29. Terada, K.; Manchikalapudi, P.; Noiva, R.; Jauregui, H. O.; Stockert, R. J.; Schilsky, M. L. *J Biol Chem* 1995, **270**, 20410.
30. Scheiber, B.; Goldenberg, H. *Arch Biochem Biophys* 1993, **305**, 225.
31. Liu, Y.; Peterson, D. A.; Kimura, H.; Schubert, D. *J Neurochem* 1997, **69**, 581.
32. McIntyre, G. D.; Scott, C. F.; Ritz, J.; Blattler, W. A.; Lambert, J. M. *Bioconjug Chem* 1994, **5**, 88.
33. Leamon, C. P.; DePrince, R. B.; Hendren, R. W. *J Drug Target* 1999, **7**, 157.

34. Ladino, C. A.; Chari, R. V. J.; Bourret, L. A.; Kedersha, N. L.; Goldmacher, V. S. *Int J Cancer* 1997, **73**, 859.
35. Liu, C.; Tadayoni, B. M.; Bourret, L. A.; Mattocks, K. M.; Derr, S. M.; Widdison, W. C.; Kedersha, N. L.; Ariniello, P. D.; Goldmacher, V. S.; Lambert, J. M.; Blattler, W. A.; Chari, R. V. *Proc Natl Acad Sci U S A* 1996, **93**, 8618.
36. Lee, R. J.; Wang, S.; Low, P. S. *Biochim Biophys Acta* 1996, **1312**, 237.
37. Yang, J.; Chen, H.; Vlahov, I. R.; Cheng, J.-X.; Low, P. S. *J Pharmacol Exp Ther* 2007, **321**, 462.
38. Neville, D. M.; Srinivasachar, K.; Stone, R.; Scharff, J. *J Biol Chem* 1989, **264**, 14653.
39. Hamann, P. R.; Hinman, L. M.; Beyer, C. F.; Lindh, D.; Upeslacis, J.; Flowers, D. A.; Bernstein, I. *Bioconjug Chem* 2002, **13**, 40.
40. Hamann, P. R.; Hinman, L. M.; Hollander, I.; Beyer, C. F.; Lindh, D.; Holcomb, R.; Hallett, W.; Tsou, H. R.; Upeslacis, J.; Shochat, D.; Mountain, A.; Flowers, D. A.; Bernstein, I. *Bioconjug Chem* 2002, **13**, 47.
41. Haddley, K. *Drugs Today* 2013, **49**, 701.
42. Perini, G. F.; Pro, B. *Biol Ther* 2013, **3**, 15–23.
43. Chari, R. V. J.; Miller, M. L.; Widdison, W. C. *Angew Chem Int Ed* 2014, **53**, 15, 3796–3827.
44. Sapra, P.; Betts, A.; Boni, J. *Expert Rev Clin Pharmacol* 2013, **6**, 541.
45. Rihova, B.; Srogl, J.; Jelinkova, M.; Hovorka, O.; Buresova, M.; Subr, V.; Ulbrich, K. *Ann N Y Acad Sci* 1997, **831**, 57.
46. Seymour, L. W.; Ulbrich, K.; Wedge, S. R.; Hume, I. C.; Strohalm, J.; Duncan, R. *Br J Cancer* 1991, **63**, 859.
47. Weinstain, R.; Segal, E.; Satchi-Fainaro, R.; Shabat, D. *Chem Commun* 2010, **46**, 553.
48. Lu, Y.; Stinnette, T. W.; Westrick, E.; Klein, P. J.; Gehrke, M. A.; Cross, V. A.; Vlahov, I. R.; Low, P. S.; Leamon, C. P. *Arthritis Res Ther* 2011, **13**, R56.
49. Vlahov, I. R.; Leamon, C. P. *Bioconjug Chem* 2012, **23**, 1357.
50. Poznansky, M. J.; Singh, R.; Singh, B. *Science* 1984, **223**, 1304.
51. Poznansky, M. J.; Hutchison, S. K.; Davis, P. J. *FASEB J* 1989, **3**, 152.
52. Shaw, J. P.; Akiyoshi, D. E.; Arrigo, D. A.; Rhoad, A. E.; Sullivan, B.; Thomas, J.; Genbauffe, F. S.; Bacha, P.; Nichols, J. C. *J Biol Chem* 1991, **266**, 21118.
53. Cawley, D. B.; Herschman, H. R.; Gilliland, D. G.; Collier, R. J. *Cell* 1980, **22**, 563.
54. Qian, Z. M.; Li, H.; Sun, H.; Ho, K. *Pharmacol Rev* 2002, **54**, 561.
55. Bacha, P.; Murphy, J. R.; Reichlin, S. *J Biol Chem* 1983, **258**, 1565.
56. Sakai, A.; Sakakibara, R.; Ishiguro, M. *J Biochem* 1989, **105**, 275.
57. Sakai, A.; Sakakibara, R.; Ohwaki, K.; Ishiguro, M. *Chem Pharm Bull (Tokyo)* 1991, **39**, 2984.
58. Singh, V.; Sairam, M. R.; Bhargavi, G. N., Akhras, R. G. *J Biol Chem* 1989, **264**, 3089.
59. Singh, V.; Curtiss, R., 3rd. *Mol Cell Biochem* 1994, **130**, 91.
60. Bacha, P.; Williams, D. P.; Waters, C.; Williams, J. M.; Murphy, J. R.; Strom, T. B. *J Exp Med* 1988, **167**, 612.
61. Shapiro, M. E.; Kirkman, R. L.; Kelley, V. R.; Bacha, P.; Nichols, J. C.; Strom, T. B. *Targeted Diagn Ther* 1992, **7**, 383.
62. Karson, E. M.; Neufeld, E. F.; Sando, G. N. *Biochemistry* 1980, **19**, 3856.

63. Sato, Y.; Beutler, E. *J Clin Invest* 1993, **91**, 1909.
64. Wu, J.; Nantz, M. H.; Zern, M. A. *Front Biosci* 2002, **7**, d717.
65. Pennell, C. A.; Erickson, H. A. *Immunol Res* 2002, **25**, 177.
66. Garnett, M. C. *Adv Drug Deliv Rev* 2001, **53**, 171.
67. Collins, D. A.; Hogenkamp, H. P.; Gebhard, M. W. *Mayo Clin Proc* 1999, **74**, 687.
68. Swaan, P. W. *Pharm Res* 1998, **15**, 826.
69. Bauer, J. A.; Morrison, B. H.; Grane, R. W.; Jacobs, B. S.; Dabney, S.; Gamero, A. M.; Carnevale, K. A.; Smith, D. J.; Drazba, J.; Seetharam, B.; Lindner, D. J. *J Natl Cancer Inst* 2002, **94**, 1010.
70. Russell-Jones, G. J.; Alpers, D. H. *Pharm Biotechnol* 1999, **12**, 493.
71. Liu, Y.; Cao, Y.; Zhang, W.; Bergmeier, S.; Qian, Y.; Akbar, H.; Colvin, R.; Ding, J.; Tong, L.; Wu, S.; Hines, J.; Chen, X. *Mol Cancer Ther* 2012, **11**, 1672.
72. Kumar, P.; Shustov, G.; Liang, H.; Khlebnikov, V.; Zheng, W.; Yang, X.-H.; Cheeseman, C.; Wiebe, L. I. *J Med Chem* 2012, **55**, 6033.
73. Di Matteo, P.; Hackl, C.; Jedeszko, C.; Valentinis, B.; Bordignon, C.; Traversari, C.; Kerbel, R. S.; Rizzardi, G. P. *Br J Cancer* 2013, **109**, 360.
74. Wang, R. E.; Niu, Y.; Wu, H.; Hu, Y.; Cai, J. *Anticancer Agents Med Chem* 2012, **12**, 76.
75. Kulkarni, H. R.; Schuchardt, C.; Baum, R. P. *Recent Results Cancer Res* 2013, **194**, 551.
76. Huo, M.; Zou, A.; Yao, C.; Zhang, Y.; Zhou, J.; Wang, J.; Zhu, Q.; Li, J.; Zhang, Q. *Biomaterials* 2012, **33**, 6393.
77. Xiao, Y.; Jaskula-Sztul, R.; Javadi, A.; Xu, W.; Eide, J.; Dammalapati, A.; Kunnimalaiyaan, M.; Chen, H.; Gong, S. *Nanoscale* 2012, **4**, 7185.
78. Barrett, J. A.; Coleman, R. E.; Goldsmith, S. J.; Vallabhajosula, S.; Petry, N. A.; Cho, S.; Armor, T.; Stubbs, J. B.; Maresca, K. P.; Stabin, M. G.; Joyal, J. L.; Eckelman, W. C.; Babich, J. W. *J Nucl Med* 2013, **54**, 380.
79. Kularatne, S. A.; Venkatesh, C.; Santhapuram, H. K.; Wang, K.; Vaitilingam, B.; Henne, W. A.; Low, P. S. *J Med Chem* 2010, **53**, 7767.
80. Banerjee, S. R.; Pullambhatla, M.; Byun, Y.; Nimmagadda, S.; Green, G.; Fox, J. J.; Horti, A.; Mease, R. C.; Pomper, M. G. *J Med Chem* 2010, **53**, 5333.
81. Kularatne, S. A.; Zhou, Z.; Yang, J.; Post, C. B.; Low, P. S. *Mol Pharm* 2009, **6**, 790.
82. Kularatne, S. A.; Wang, K.; Santhapuram, H.-K. R.; Low, P. S. *Mol Pharm* 2009, **6**, 780.
83. Chen, Y.; Dhara, S.; Banerjee, S. R.; Byun, Y.; Pullambhatla, M.; Mease, R. C.; Pomper, M. G. *Biochem Biophys Res Commun* 2009, **390**, 624.
84. Wayua, C.; Roy, J.; Putt, K. S.; Low, P. S. *Mol Pharm* 2015, **16**, 16.
85. Wayua, C.; Low, P. S. *J Nucl Med* 2015, **56**, 113.
86. Wayua, C.; Low, P. S. *Mol Pharm* 2014, **11**, 468.
87. Jahn, F.; Riesner, A.; Jahn, P.; Sieker, F.; Vordermark, D.; Jordan, K. *Int J Radiat Oncol Biol Phys* 2015, **28**, 00441.
88. Goh, Y. I.; Koren, G. *J Obstet Gynaecol* 2008, **28**, 3.
89. Crider, K. S.; Yang, T. P.; Berry, R. J.; Bailey, L. B. *Adv Nutr* 2012, **3**, 21.
90. Antony, A. C. *Blood* 1992, **79**, 2807.
91. Visentin, M.; Unal, E. S.; Najmi, M.; Fiser, A.; Zhao, R.; Goldman, I. D. *Am J Physiol Cell Physiol* 2015, **308**, 21.
92. Kamen, B. A.; Capdevila, A. *Proc Natl Acad Sci U S A* 1986, **83**, 5983.

93. Antony, A. C. *Annu Rev Nutr* 1996, **16**, 501.
94. Leamon, C. P.; Low, P. S. *Proc Natl Acad Sci U S A* 1991, **88**, 5572.
95. Coney, L. R.; Tomassetti, A.; Carayannopoulos, L.; Frasca, V.; Kamen, B. A.; Colnaghi, M. I.; Zurawski, V. R. J. *Cancer Res* 1991, **51**, 6125.
96. Weitman, S. D.; Lark, R. H.; Coney, L. R.; Fort, D. W.; Frasca, V.; Zurawski, V. R.; Kamen, B. A. *Cancer Res* 1992, **52**, 3396.
97. Ross, J. F.; Chaudhuri, P. K.; Ratnam, M. *Cancer* 1994, **73**, 2432.
98. Weitman, S. D.; Frazier, K. M.; Kamen, B. A. *J Neurol Oncol* 1994, **21**, 107.
99. Boerman, O. C.; van Niekerk, C. C.; Makkink, K.; Hanselaar, T. G.; Kenemans, P.; Poels, L. G. *Int J Gynecol Pathol* 1991, **10**, 15.
100. Garin-Chesa, P.; Campbell, I.; Saigo, P. E.; Lewis, J. L.; Old, L. J.; Rettig, W. J. *Am J Pathol* 1993, **142**, 557.
101. Mattes, M. J.; Major, P. P.; Goldenberg, D. M.; Dion, A. S.; Hutter, R. V. P.; Klein, K. M. *Cancer Res Suppl* 1990, **50**, 880S.
102. Weitman, S. D.; Weiberg, A. G.; Coney, L. R.; Zurawski, V. R.; Jennings, D. S.; Kamen, B. A. *Cancer Res* 1992, **52**, 6708.
103. Toffoli, G.; Cernigoi, C.; Russo, A.; Gallo, A.; Bagnoli, M.; Boiocchi, M. *Int J Cancer* 1997, **74**, 193.
104. Holm, J.; Hansen, S. I.; Hoier-Madsen, M.; Bostad, L. *Biochem J* 1991, **280**, 267.
105. Berenson, J. R.; Yang, J.; Mickel, R. A. *Oncogene* 1989, **4**, 1111.
106. Orr, R. B.; Kriesler, A. R.; Kamen, B. A. *J Natl Cancer Inst* 1995, **87**, 299.
107. Parker, N.; Turk, M. J.; Westrick, E.; Lewis, J. D.; Low, P. S.; Leamon, C. P. *Anal Biochem* 2005, **338**, 284.
108. Grapp, M.; Wrede, A.; Schweizer, M.; Huwel, S.; Galla, H. J.; Snaidero, N.; Simons, M.; Buckers, J.; Low, P. S.; Urlaub, H.; Gartner, J.; Steinfeld, R. *Nat Commun* 2013, **4**, 2123.
109. Patrick, T. A.; Kranz, D. M.; van Dyke, T. A.; Roy, E. J. *J Neurooncol* 1997, **32**, 111.
110. Kennedy, M. D.; Jallad, K. N.; Low, P. S.; Ben-Amotz, D. *Pharm Res* 2003, **20**, 5, 714–719.
111. Han, W.; Zaynagetdinov, R.; Yull, F. E.; Polosukhin, V. V.; Gleaves, L. A.; Tanjore, H.; Young, L. R.; Peterson, T. E.; Manning, H. C.; Prince, L. S.; Blackwell, T. S. *Am J Respir Cell Mol Biol* 2014, **6**, 6.
112. Assaraf, Y. G.; Leamon, C. P.; Reddy, J. A. *Drug Resist Updat* 2014, **17**, 89.
113. Zimmerman, J. *Gastroenterology* 1990, **99**, 964.
114. Morshed, K. M.; Ross, D. M.; McMartin, K. E. *J Nutr* 1997, **127**, 1137.
115. Birn, H.; Selhub, J.; Christensen, E. I. *Am J Physiol* 1993, **264**, C302.
116. Birn, H.; Nielsen, S.; Christensen, E. I. *Am J Physiol* 1997, **272**, F70.
117. Turk, M. J.; Breur, G. J.; Widmer, W. R.; Paulos, C. M.; Xu, L. C.; Grote, L. A.; Low, P. S. *Arthritis Rheum* 2002, **46**, 1947.
118. Nakashima-Matsushita, N.; Homma, T.; Yu, S.; Matsuda, T.; Sunahara, N.; Nakamura, T.; Tsukano, M.; Ratnam, M.; Matsuyama, T. *Arthritis Rheum* 1999, **42**, 1609.
119. Shen, J.; Putt, K. S.; Visscher, D. W.; Murphy, L.; Cohen, C.; Singhal, S.; Sandusky, G.; Feng, Y.; Dimitrov, D. S.; Low, P. S. *Oncotarget* 2015, **30**, 30.
120. Sun, J. Y.; Shen, J.; Thibodeaux, J.; Huang, G.; Wang, Y.; Gao, J.; Low, P. S.; Dimitrov, D. S.; Sumer, B. D. *Laryngoscope* 2014, **124**, 4.

121. Shen, J.; Hilgenbrink, A. R.; Xia, W.; Feng, Y.; Dimitrov, D. S.; Lockwood, M. B.; Amato, R. J.; Low, P. S. *J Leukoc Biol* 2014, **96**, 563.
122. Jager, N. A.; Westra, J.; Golestani, R.; van Dam, G. M.; Low, P. S.; Tio, R. A.; Slart, R. H.; Boersma, H. H.; Bijl, M.; Zeebregts, C. J. *J Nucl Med* 2014, **55**, 1945.
123. Shen, J.; Chelvam, V.; Cresswell, G.; Low, P. S. *Mol Pharm* 2013, **10**, 1918.
124. Kularatne, S. A.; Belanger, M. J.; Meng, X.; Connolly, B. M.; Vanko, A.; Suresch, D. L.; Guenther, I.; Wang, S.; Low, P. S.; McQuade, P.; Trotter, D. G. *Mol Pharm* 2013, **10**, 3103.
125. Gent, Y. Y.; Weijers, K.; Molthoff, C. F.; Windhorst, A. D.; Huisman, M. C.; Smith, D. E.; Kularatne, S. A.; Jansen, G.; Low, P. S.; Lammertsma, A. A.; van der Laken, C. J. *Arthritis Res Ther* 2013, **15**, 1, R37.
126. Henne, W. A.; Rothenbuhler, R.; Ayala-Lopez, W.; Xia, W.; Varghese, B.; Low, P. S. *Mol Pharm* 2012, **9**, 1435.
127. Henne, W. A.; Kularatne, S. A.; Ayala-Lopez, W.; Doorneweerd, D. D.; Stinnette, T. W.; Lu, Y.; Low, P. S. *Bioorg Med Chem Lett* 2012, **22**, 709.
128. Ayala-Lopez, W.; Xia, W.; Varghese, B.; Low, P. S. *J Nucl Med* 2010, **51**, 768.
129. Xia, W.; Hilgenbrink, A. R.; Matteson, E. L.; Lockwood, M. B.; Cheng, J. X.; Low, P. S. *Blood* 2009, **113**, 438.
130. Matteson, E. L.; Lowe, V. J.; Prendergast, F. G.; Crowson, C. S.; Moder, K. G.; Morgenstern, D. E.; Messmann, R. A.; Low, P. S. *Clin Exp Rheumatol* 2009, **27**, 253.
131. Low, P. S.; Henne, W. A.; Doorneweerd, D. D. *Acc Chem Res* 2008, **41**, 120.
132. Varghese, B.; Haase, N.; Low, P. S. *Mol Pharm* 2007, **4**, 679.
133. Poh, S.; Chelvam, V.; Low, P. S. *Nanomedicine* 2015, **10**, 1439.
134. Lu, Y.; Wollak, K. N.; Cross, V. A.; Westrick, E.; Wheeler, L. W.; Stinnette, T. W.; Vaughn, J. F.; Hahn, S. J.; Xu, L. C.; Vlahov, I. R.; Leamon, C. P. *Clin Immunol* 2014, **150**, 64.
135. Piscaer, T. M.; Muller, C.; Mindt, T. L.; Lubberts, E.; Verhaar, J. A.; Krenning, E. P.; Schibli, R.; De Jong, M.; Weinans, H. *Arthritis Rheum* 2011, **63**, 1898.
136. Mathias, C. J.; Wang, S.; Lee, R. J.; Waters, D. J.; Low, P. S.; Green, M. A. *J Nucl Med* 1996, **37**, 1003.
137. Mathias, C. J.; Wang, S.; Waters, D. J.; Turek, J. J.; Low, P. S.; Green, M. A. *J Nucl Med* 1998, **39**, 1579.
138. Leamon, C. P.; Low, P. S. *Drug Discov Today* 2001, **6**, 44.
139. Campbell, I. G.; Jones, T. A.; Foulkes, W. D.; Trowsdale, J. *Cancer Res* 1991, **51**, 5329.
140. Wang, S.; Lee, R. J.; Mathias, C. J.; Green, M. A.; Low, P. S. *Bioconjug Chem* 1996, **7**, 56.
141. Ilgan, S.; Yang, D. J.; Higuchi, T.; Zareneyrizi, F.; Bayham, H.; Yu, D.; Kim, E. E.; Podoloff, D. A. *Cancer Biother Radiopharm* 1998, **13**, 427.
142. Guo, W.; Hinkle, G. H.; Lee, R. J. *J Nucl Med* 1999, **40**, 1563.
143. Linder, K. E.; Wedeking, P.; Ramalingam, K.; Nunn, A. D.; Tweedle, M. F. *Soc Nucl Med Proc 47th Annual Meeting* 2000, **41**, 119P.
144. Leamon, C. P.; Parker, M. A.; Vlahov, I. R.; Xu, L. C.; Reddy, J. A.; Vetzel, M.; Douglas, N. *Bioconjug Chem* 2002, **13**, 6, 1200–1210.
145. Wang, S.; Luo, J.; Lantrip, D. A.; Waters, D. J.; Mathias, C. J.; Green, M. A.; Fuchs, P. L.; Low, P. S. *Bioconjug Chem* 1997, **8**, 673.

146. Fisher, R. E.; Siegel, B. A.; Edell, S. L.; Oyesiku, N. M.; Morgenstern, D. E.; Messmann, R. A.; Amato, R. J. *J Nucl Med* 2008, **49**, 899.
147. Konda, S. D.; Aref, M.; Brechbiel, M.; Wiener, E. C. *Invest Radiol* 2000, **35**, 50.
148. Lee, J. W.; Lu, J. Y.; Low, P. S.; Fuchs, P. L. *Bioorg Med Chem* 2002, **10**, 2397.
149. Citro, G.; Szczylik, C.; Ginobbi, P.; Zupi, G.; Calabretta, B. *Br J Cancer* 1994, **69**, 463.
150. Li, S.; Huang, L. *J Liposome Res* 1997, **7**, 63.
151. Li, S.; Huang, L. *J Liposome Res* 1998, **8**, 239.
152. Li, S.; Deshmukh, H. M.; Huang, L. *Pharm Res* 1998, **15**, 1540.
153. Leopold, L. H.; Shore, S. K.; Newkirk, T. A.; Reddy, R. M.; Reddy, E. P. *Blood* 1995, **85**, 2162.
154. Ward, C. M.; Acheson, N.; Seymour, L. M. *J Drug Target* 2000, **8**, 119.
155. Lu, J. Y.; Lowe, D. A.; Kennedy, M. D.; Low, P. S. *J Drug Target* 1999, **7**, 43.
156. Leamon, C. P.; Low, P. S. *J Biol Chem* 1992, **267**, 24966.
157. Leamon, C. P.; Low, P. S. *Biochem J* 1993, **291**, 855.
158. Leamon, C. P.; Low, P. S. *J Drug Target* 1994, **2**, 101.
159. Kranz, D. M.; Patrick, T. A.; Brigle, K. E.; Spinella, M. J.; Roy, E. J. *Proc Natl Acad Sci U S A* 1995, **92**, 9057.
160. Cho, B. K.; Roy, E. J.; Patrick, T. A.; Kranz, D. M. *Bioconjug Chem* 1997, **8**, 338.
161. Kranz, D. M.; Manning, T. C.; Rund, L. A.; Cho, B. K.; Gruber, M. M.; Roy, E. J. *J Control Release* 1998, **53**, 77.
162. Rund, L. A.; Cho, B. K.; Manning, T. C.; Holler, P. D.; Roy, E. J.; Kranz, D. M. *Int J Cancer* 1999, **83**, 141.
163. Amato, R. J.; Shetty, A.; Lu, Y.; Ellis, R.; Low, P. S. *J Immunother* 2013, **36**, 268.
164. Lee, R. J.; Low, P. S. *J Biol Chem* 1994, **269**, 3198.
165. Lee, R. J.; Low, P. S. *Biochim Biophys Acta* 1995, **1233**, 134.
166. Vogel, K.; Wang, S.; Lee, R. J.; Chmielewski, J.; Low, P. S. *J Am Chem Soc* 1996, **118**, 1581.
167. Rui, Y.; Wang, S.; Low, P. S.; Thompson, D. H. *J Am Chem Soc* 1998, **120**, 11213.
168. Gabizon, A.; Horowitz, A. T.; Goren, D.; Tzemach, D.; Mandelbaum-Shavit, F.; Qazen, M. M.; Zalipsky, S. *Bioconjug Chem* 1999, **10**, 289.
169. Wang, B.; Galliford, C. V.; Low, P. S. *Nanomedicine (Lond)* 2014, **9**, 2, 313–330.
170. Zhang, Y.; Kohler, N.; Zhang, M. *Biomaterials* 2002, **23**, 1553.
171. Oyewumi, M. O.; Mumper, R. J. *Bioconjug Chem* 2002, **13**, 1328.
172. Oyewumi, M. O.; Mumper, R. J. *Int J Pharm* 2003, **251**, 85.
173. Gottschalk, S.; Cristiano, R. J.; Smith, L. C.; Woo, S. L. C. *Gene Ther* 1994, **1**, 185.
174. Mislick, K. A.; Baldeschwieler, J. D.; Kayyem, J. F.; Meade, T. J. *Bioconjug Chem* 1995, **6**, 512.
175. Douglas, J. T.; Rogers, B. E.; Rosenfeld, M. E.; Michael, S. I.; Feng, M.; Curiel, D. T. *Nat Biotechnol* 1996, **14**, 1574.
176. Leamon, C. P.; Weigl, D.; Hendren, R. W. *Bioconjug Chem* 1999, **10**, 947.
177. Guo, W.; Lee, R. J. *AAPS PharmSci* 1999, **1**, 19.
178. Reddy, J. A.; Dean, D.; Kennedy, M. D.; Low, P. S. *J Pharm Sci* 1999, **88**, 1112.
179. Reddy, J. A.; Low, P. S. *J Control Release* 2000, **64**, 27.

180. Reddy, J. A.; Abburi, C.; Hofland, H.; Howard, S. J.; Vlahov, I.; Wils, P.; Leamon, C. P. *Gene Ther* 2002, **9**, 1542.

181. Hofland, H. E.; Masson, C.; Iginla, S.; Osetinsky, I.; Reddy, J. A.; Leamon, C. P.; Scherman, D.; Bessodes, M.; Wils, P. *Mol Ther* 2002, **5**, 739.

182. Liu, S.; Edwards, D. S.; Barrett, J. A. *Bioconjug Chem* 1997, **8**, 621.

183. Reddy, J. A.; Xu, L. C.; Parker, N.; Vetzel, M.; Leamon, C. P. *J Nucl Med* 2004, **45**, 857.

184. Palmer, E.; Scott, J.; Symanowski, J. *J Nucl Med Meet Abstr* 2013, **54**, 400.

185. Maurer, A. H.; Elsinga, P.; Fanti, S.; Nguyen, B.; Oyen, W. J.; Weber, W. A. *J Nucl Med* 2014, **55**, 701.

186. Reddy, J. A.; Westrick, E.; Vlahov, I.; Howard, S. J.; Santhapuram, H. K.; Leamon, C. P. *Cancer Chemother Pharmacol* 2006, **58**, 229.

187. Leamon, C. P.; Reddy, J. A.; Vlahov, I. R.; Vetzel, M.; Parker, N.; Nicoson, J. S.; Xu, L. C.; Westrick, E. *Bioconjug Chem* 2005, **16**, 803.

188. Naumann, R. W.; Coleman, R. L.; Burger, R. A.; Sausville, E. A.; Kutarska, E.; Ghamande, S. A.; Gabrail, N. Y.; DePasquale, S. E.; Nowara, E.; Gilbert, L.; Gersh, R. H.; Teneriello, M. G.; Harb, W. A.; Konstantinopoulos, P. A.; Penson, R. T.; Symanowski, J. T.; Lovejoy, C. D.; Leamon, C. P.; Morgenstern, D. E.; Messmann, R. A. *J Clin Oncol* 2013, **31**, 35, 4400–4406.

189. Kamen, B. A.; Wang, M. T.; Streckfuss, A. J.; Peryea, X.; Anderson, R. G. W. *J Biol Chem* 1988, **263**, 13602–13609.

190. Lu, Y.; Low, P. S. *Cancer Immunol Immunother* 2002, **51**, 153.

191. Amato, R. J.; Shetty, A.; Lu, Y.; Ellis, P. R.; Mohlere, V.; Carnahan, N.; Low, P. S. *J Immunother* 2014, **37**, 237.

20

PROTEIN AND PEPTIDE CONJUGATES FOR TARGETING THERAPEUTICS AND DIAGNOSTICS TO SPECIFIC CELLS

Barlas Büyüktimkin, John Stewart, Jr., Kayann Tabanor, Paul Kiptoo, and Teruna J. Siahaan

Department of Pharmaceutical Chemistry, The University of Kansas, Lawrence, KS, USA

20.1 INTRODUCTION

In the past decade, there has been a trend to develop drugs to target specific types of cells in certain diseases (i.e., cancers and autoimmune diseases) and a group of patients to improve the outcome of the treatments. Thus, there is a need to elucidate the intricate details of the development and progression of a disease as well as to ascertain the different patient populations that will respond well to certain treatments. Progress in modern chemistry and biology has allowed scientists to better pinpoint these intricate differences between normally functioning cells and "diseased" cells. Often, the differences between the diseased and normal cells are due to changes in cell surface receptors, shifts in cytokine or hormone production, differences in cell migration properties, changes in pH in cell compartments, and changes in intracellular endosome trafficking properties. A targeted therapy is designed to exploit the differences between diseased and normal cells to provide selectivity and better drug efficacy for patients. In this case, the drug is guided to specific cells at the site of action for increasing drug efficacy and lowering side effects [1, 2]. To accomplish

Drug Delivery: Principles and Applications, Second Edition. Edited by Binghe Wang, Longqin Hu, and Teruna J. Siahaan.

these goals, peptides and proteins (e.g., antibodies) have been conjugated with drugs to direct the drugs to the diseased cells with upregulated cell surface receptors (Fig. 20.1a).

(a)

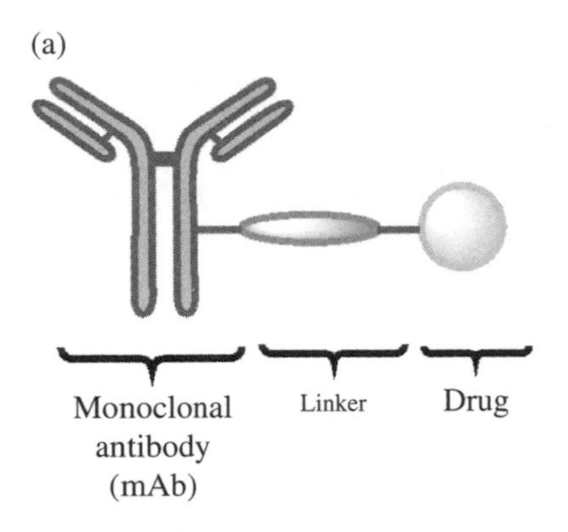

(b)

Enzyme cleavable linker

Spacer

Monomethyl auristatin E, MMAE

Adcetrus® (brentuximab vedotin)

(c)

Kadcyla® (trastuzumab emtansine, T-DM1)

FIGURE 20.1 (a) The general structure of an antibody–drug conjugate (ADC) with the monoclonal antibody conjugated to the drug via a linker. (b) Adcetris (brentuximab vedotin) is a conjugate between anti-CD30 mAb and monomethyl auristatin E (MMAE) conjugated via a linker that can be cleaved by an enzyme. (c) Kadcyla (trastuzumab emtansine, T-DM1) is a conjugate between anti-CD33 mAb and merstansine (DM1) via a stable thioether linker. (d) Mylotarg (gemtuzumab ozogamicin) is a conjugate between anti-CD33 mAb and calicheamicin, which is connected via a hydrazone–disulfide linker. (e) BR96-doxorubicin (BR-96 DOX) is a conjugate between anti-Lewis Y mAb and DOX via a hydrazone–thioether linker. (f) The huC242-SPDB-DM4 compound is a conjugate between huC242 mAb and DM4.

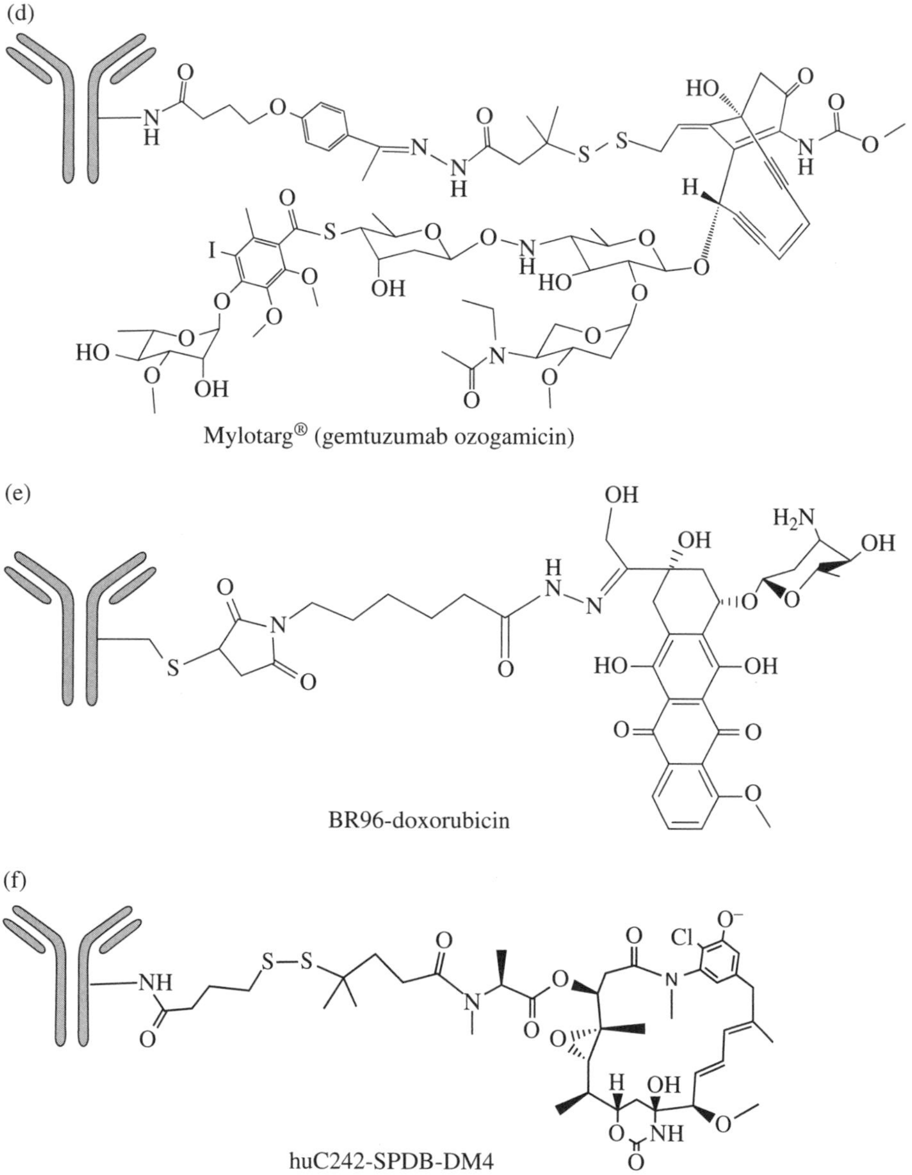

(d)

Mylotarg® (gemtuzumab ozogamicin)

(e)

BR96-doxorubicin

(f)

huC242-SPDB-DM4

FIGURE 20.1 (*Continued*)

Recently, some attention has been paid to developing peptide/protein conjugates with drugs in which the peptides or proteins serve as guiding molecules that bind to a specific receptor on the cell surface of diseased cells [3]. There have been successes in developing antibody–drug conjugates (ADCs) such as Adcetris®

(brentuximab vedotin) and Kadcyla® (trastuzumab emtansine, T-DM1) to reach patients (Fig. 20.1b–c). Adcetris was approved in August 2011 for the treatment of systemic anaplastic large-cell lymphoma (ALCL) and Hodgkin's lymphoma (HL) [4, 5]. Adcetris utilizes a monoclonal antibody (mAb) called "brentuximab" that binds to CD30 proteins on lymphoma cells to deliver the cytotoxic drug monomethyl auristatin E (MMAE), a potent cytotoxic agent that inhibits tubulin polymerization that causes intratumoral vascular damage [6]. Similarly, trastuzumab is used in Kadcyla to deliver mertansine (DM1) to metastatic breast cancer cells by targeting cell surface receptors called "human epidermal growth factor receptor 2" (Her2 or CD33) [7]. A cytotoxic maytansoid, mertansine (DM1), was conjugated to a trastuzumab using a noncleavable thioether linker (Fig. 20.1c). Trastuzumab is an mAb that preferentially targets HER2, an antigen highly overexpressed on some metastatic breast cancers. However, there are still many challenges in developing ADC because they are new types of drugs with limited track records. Mylotarg® (gemtuzumab ozogamicin, Fig. 20.1d) was approved by the FDA in 2000 to treat acute myeloid leukemia (AML) [8]; it was withdrawn from the market with the possibility of reentry because no clear benefits to patients were seen. There was also an increase in patient deaths during the clinical trials. The development of many conjugates has been relatively slow because of several factors, including (i) the lack of prior knowledge in developing these types of therapeutics, (ii) the challenging nature of working with proteins and peptides, (iii) the difficulty of producing a consistent active pharmaceutical ingredient (API) as a mixture, and (iv) the difficulty in formulating and analyzing the drug candidates.

Understanding the mechanism of disease development and progress is one of many important steps in developing therapeutic agents from drug conjugates. An ideal conjugate takes advantage of the carrier molecule specificity such as the specificity of an mAb to target a unique receptor or an antigen that is highly expressed on the surface of target disease cells (e.g., malignant cancer cells) (Fig. 20.2). After binding to the surface receptor, the conjugate is typically internalized and trafficked to intracellular organelles such as early endosomes and late endosomes (Fig. 20.2). While the receptor may be recycled to the cell surface from the early sorting endosomes, the conjugates that are trapped in the late endosomes can later be degraded in the lysosomes to release the active drug. The active drug can escape from the lysosomes to modulate the target protein or DNA, leading to the desired therapeutic response (e.g., cell death in cancer therapy). To develop safe and effective macromolecule conjugates, consideration should be taken to ensure target specificity of the carrier molecule, potency of the drug, use of an appropriate linker that allows the release of the drug at the intended site, and a good pharmacokinetic profile and plasma stability of the conjugate in systemic circulation. This chapter provides an overview of the development of drug conjugate technologies to improve the delivery of a drug conjugate to a specific tissue or cell type, ultimately providing greater efficacy and safety than the parent drug.

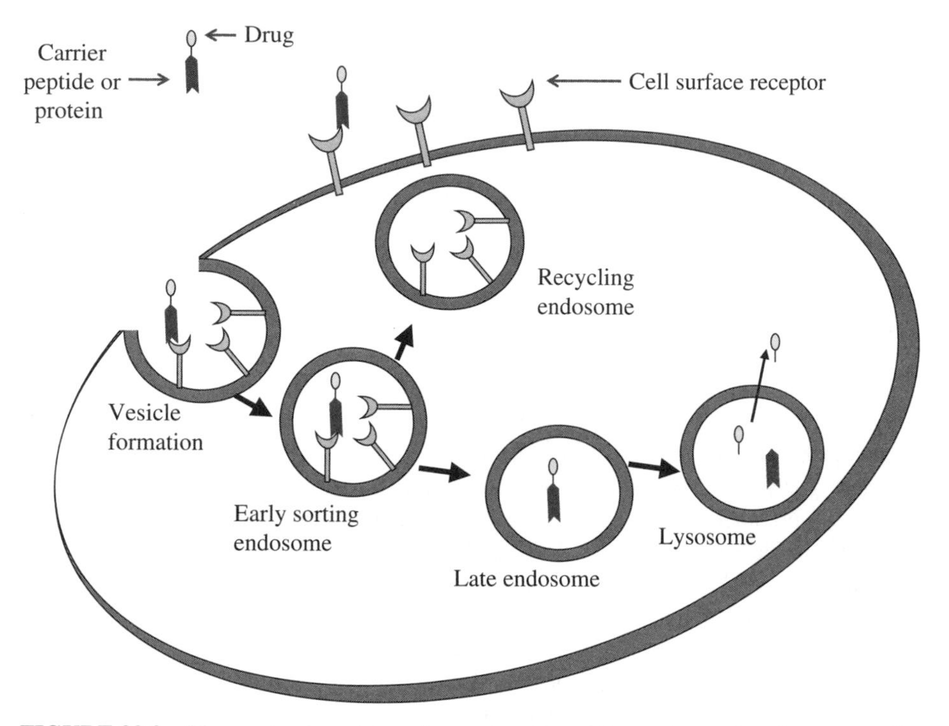

FIGURE 20.2 The mechanism of targeting and uptake of the conjugate into the target cells. The carrier molecule binds to the target receptor on the cell surface followed by endocytosis into the cytoplasm inside the early endosomes. From the early endosomes, the receptors can move into recycling endosomes to carried back to the cell surface. The conjugates are released from the receptor and are trapped inside the late endosomes. Finally, the conjugates are degraded in the lysosomes to release the free drugs.

20.2 RADIOLABELED ANTIBODIES FOR CANCER TREATMENT

Although the idea of using mAbs to target various diseases dates back more than a century, the development of mAb technology by Kohler and Milstein in 1975 spurred several pharmaceutical and biotechnology companies to actively develop ADCs that were selective and efficacious [9]. The approval in the United States of two radioimmunoconjugates for cancer treatment stimulated the development of ADCs [10]. Radioimmunotherapy takes advantage of the exquisite selectivity of an antibody to its target and the intrinsic properties of a radionuclide to deliver a therapeutic dose of radiation to malignant cells. Radionuclides such as ^{131}I, ^{90}Y, and ^{177}Lu, which emit high-energy β electrons, are widely used in radioimmunotherapy [11–13]. The radionuclide selected for most cancer therapy depends on the size of the tumor being treated. High-energy beta emitters (e.g., ^{90}Y) are used to treat larger tumors, while ^{131}T and ^{177}Lu are medium-energy emitters suitable for treating smaller tumors [12]. Because of the inherent ability of radiolabeled antibodies to accumulate at neoplastic sites, the buildup of mAbs in normal cells is limited, leading to less nonselective killing of normal cells (crossfire effect) [13]. ^{131}I-tositumomab (Bexar®) and ^{90}Y-ibritumomab tiuxetan (Zevalin®) are two radioimmunoconjugates currently approved by the FDA; they both target CD20 on the cancer cells of non-Hodgkin's lymphoma (NHL) [11, 14]. ^{90}Y-epratuzumab teraxetan, which is specific for CD22-expressing cancer cells, is being investigated for the treatment of aggressive non-Hodgkin's lymphoma [11, 15].

20.3 ANTIBODY–DRUG CONJUGATE

The high specificity of mAbs has successfully allowed direct toxic anticancer drugs to kill cancer cells by utilizing the expression of upregulated antigens or receptors on the surface of cancer cells over the normal cells. For a proof of concept, early ADCs were developed by conjugating mAbs to conventional and approved anticancer drugs. Unfortunately, ADCs containing drugs such as KS1/4-desacetylvinblastine hydrazide [16] and BR96-doxorubicin (BR96-DOX) [17] were unsuccessful in a clinical setting. It has been shown that in cancer therapy the potency of the drug is important; therefore, the limited success of the early ADCs was due to the low cytotoxic potency of the conjugated drugs (e.g., DOX and vinca alkaloids) [18]. Thus, highly potent cytotoxic agents have been used in recent ADCs, including DNA-damaging agents (i.e., calicheamicin and duocarmycin) and microtubule inhibitors (i.e., auristatins, taxanes, and maytansinoids) [19]. Another important factor is the limited number of conjugation sites on the mAbs, which leads to inefficiency and poor accumulation of cytotoxic drug inside the target cells [18].

Overall, the failures in developing ADCs are primarily due to low potency of the conjugated drugs, no clear difference in expressions of target molecules (i.e., antigens) on normal vs. cancer cells, premature release of the drugs due to instability of the linker, and immunogenicity of the conjugate [5]. To overcome these shortcomings, efforts have been focused on several improvements, including (i) the development of efficient conjugation methods, (ii) the development of linker technologies to improve stability, (iii) the development of formulations to decrease ADC aggregation for lowering potential immunogenicity, and (iv) the use of highly potent cytotoxic drugs to lower the dose needed and minimize side effects [5]. These improvements have led to the development of ADCs such as Adcetris and Kadcycla® that have had some success in treating patients.

The successful ADCs that reach the market as cancer therapeutics consist of a very specific antibody conjugated to a highly cytotoxic drug via a relatively stable linker. In Kadcyla, trastuzumab is very specific for upregulated Her-2 receptors and found to be more highly regulated on the surface of cancer cells than normal cells. Gemtuzumab, an anti-CD33 mAb (P67.6), is utilized to deliver Mylotarg, a very toxic calicheamicin, to treat AML [8]. Prior to its use as a targeting agent, the radiolabeled anti-CD33 mAb was used during marrow irradiation prior to marrow transplantation for AML because 80% of the patients have high expression of CD33 in their myeloid lineage cells [8]. Furthermore, CD33-positive cells internalize the ^{125}I-labeled P67.6 anti-CD33 mAb, which is another reason to use it to target drugs to myeloid cancer cells [8].

20.3.1 Sites of Conjugation on mAbs, Linkers, and Drugs

The most common conjugation sites used to connect the mAb and the active drug via the linker are side chains of the cysteine and lysine residues and the carbohydrate groups. Typically, there are around 80 lysine residues on an mAb; however, not all of the lysine residues are available for conjugation. The amino group of lysine is most commonly conjugated to the linker via an amide bond (see Kadcyla and Mylotarg, Fig. 20.1). The free thiol group of cysteine can be conjugated to the linker via a thioether bond (Adcetris). The carbohydrates at the CH2 domain on mAbs (e.g., anti-CD33) have also been used as conjugation sites [20].

Most of the conjugates incorporate a conditionally stable linker between the targeting agent and the drug. Because of this, linker technologies have become an important factor that should be taken into account in designing effective conjugates. Important considerations for developing ADCs are the density of target antigen, the turnover rate of the antigen, the potency of the drug, and the drug loading known as drug-to-antibody ratio (DAR) [20]. Linkers have been shown to affect the selectivity, pharmacokinetics, and therapeutic index, as well as the overall success of the conjugate. An ideal linker must be stable during transport through the systemic circulation but labile enough to allow efficient release of the drug at the target site such as inside the lysosomes of cancer cells [5, 18, 19, 21–24]. Therefore, various types of linkers have been designed to provide a balance of both stability and labile properties.

The structure of the linker can regulate the kinetics of its degradation to release the drug. Figure 20.3 shows that the degradation in compound 2 was slower than in compound 1; this is presumably due to the effect of steric hindrance of the gem-dimethyl group of the alpha-carbon to the disulfide bond in compound 2 compared to the mono-methyl group in compound 1. The release of the drug inside the cells can be produced by reduction or disulfide exchange by glutathione. The release of the drug in compound 3 (Fig. 20.3) is slower than in compound 2 because the thioether bond in 3 is more stable than the disulfide bond in 2.

Acid-sensitive cleavable linkers have been developed to provide drug release upon a change to acidic pH in lysosomes (Fig. 20.2). The most commonly used acid-sensitive linker is the hydrazone linker shown in compound 4 (Fig. 20.3). The hydrazone linker is relatively stable in blood at pH ~ 7.4, but cleaves in acidic endosomes (pH 5.0–6.5) and lysosomes (pH 4.5–5.0) to release the toxic drug in the intracellular space [24]. An early example of the use of a hydrazone linker was a conjugate of DOX to BR96 mAb to make BR96-DOX (see Fig. 20.1e). DOX kills cancer cells upon entering the nucleus, and intercalates with DNA for inhibition of cell cycles [2]. The ketone of DOX was modified to a hydrazone group, then linked to an amide bond with maleimide-hexanoic acid, and the BR96 mAb was coupled to the linker with DOX via the cysteine residue to make BR96-DOX (Fig. 20.1).

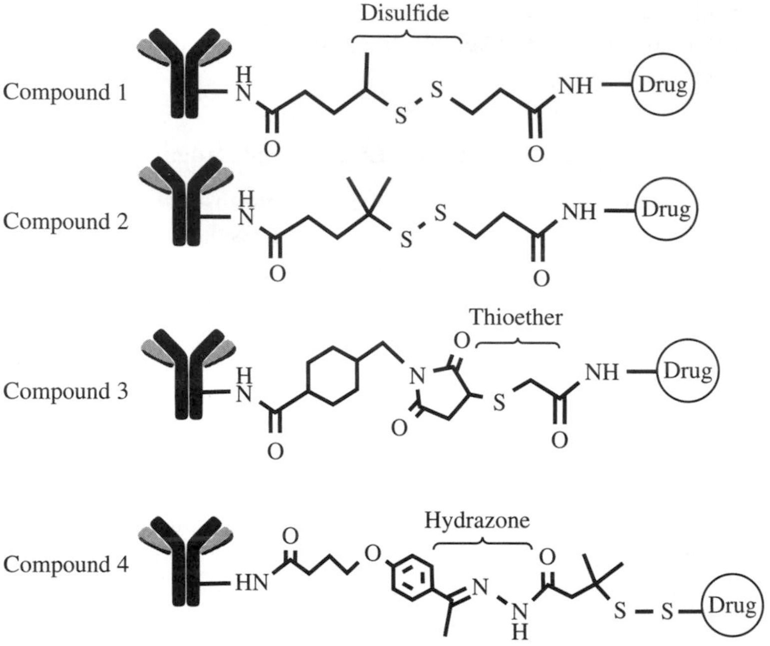

FIGURE 20.3 Examples of drug conjugates with various linkers to control the drug release. Compounds 1 and 2 contain a disulfide linker with monomethyl and gem-dimethyl group at the alpha-carbon of the disulfide bond, respectively. The drug release is faster in compound 1 than in compound 2. Compound 3 has a stable thioether linkage and the drug release in compound 3 is slower than in compound 2. Compound 4 contains a hydrazone–disulfide linker to control the drug release. The drug release from the hydrazone linker is acidic pH-sensitive as in the lysosomes.

BR96 mAb binds to the Lewis-Y receptor, which is overexpressed on many carcinomas [2, 17]. Preclinical findings showed that BR96-DOX had better antitumor efficacy than free DOX; however, BR96-DOX was unsuccessful in phase II clinical trials for patients with metastatic breast cancer. This failure was mainly attributed to a side effect of gastrointestinal toxicity because the Lewis-Y antigen is also expressed on some normal tissues such as gastric mucosa, small intestine, and pancreas. It was suggested that the gastrointestinal cells expressing Lewis-Y antigens may act as an "antigen sink" and prevent much of the drug from reaching the desired target cancer cells [17].

A disulfide bond is considered a reducing linker (Fig. 20.3) because it relies on the reducing environment with the presence of millimolar concentrations of glutathione inside the target cells for the drug release compared to micromolar concentrations in the bloodstream [24, 25]. As in Mylotarg, a disulfide bond was used to conjugate a maytansinoid derivative DM4 to humanized C242 (huC242) mAb via the lysine residues to make huC242-SPDB-DM4 (Fig. 20.1f) [26]; the huC242 mAb targets CanAg antigen on cancer cells (i.e., colorectal, pancreas, and lung cancers). The drug

DM4, as in Kadcyla, functions as a cell antimitotic and binds to tubulin for inhibiting microtubule assembly [27]. It is proposed that the conjugates are endocytosed into cancer cells by CaAg and degraded in lysosomes to release DM4. As previously mentioned, the drug release via disulfide bond reduction can be controlled by methylation of the carbon alpha to the disulfide bond using mono or di-methylation [26]. This modification allows the drug to pass from targeted tumor cells into neighboring cells, creating a "bystander effect" [26]. Modification of the alpha-methyl carbon of a disulfide bond linker may affect the plasma half-life and the drug release from the conjugates [28].

During research to develop Mylotarg, calicheamicin was conjugated to the mAb using two methods: (i) the lysine residues using an amide bond and (ii) carbohydrates on the CH2 domain. Conjugation via the carbohydrate group utilizes a hydrazone group in which the alcohol group of the sugar moiety was oxidized to produce an aldehyde that can be reacted with hydrazine to make a hydrazone bond [20]. The amide bond via lysine conjugation is very stable; thus, the release of the drug is through reduction of the disulfide bond [29]. The carbohydrate conjugate was more selective toward CD33-expressing HL60 cells (IC_{50} < 0.006 ng/ml) compared to Raji cells (0.79 ng/ml). In addition, the carbohydrate conjugate was 7000 times more potent than the amide conjugate toward HL-60 cells [20]. Nine of ten mice with tumors were free of tumors when treated with the carbohydrate conjugate, while none of the mice treated with lysine–amide conjugate were tumor free [20]. In AML pediatric patients, the carbohydrate conjugate inhibits tumor growth better than the lysine–amide conjugate [20]. This supports the idea that the linker design is an important factor for the success of the conjugate and that the therapeutic index of the conjugate can be increased by optimizing the linker [20]. Although there is a possibility that conjugation to the lysine residue of mAb could affect the binding affinity to CD33 antigen, the lysine conjugate of calicheamicin still binds effectively to target CD33 protein [20]. Thus, the lower activity of the lysine conjugate is presumably due to the inefficiency of the disulfide bond in releasing the drug compared to the hydrazone group. Therefore, *in vitro*, *ex vivo*, and *in vivo* efficacy results were consistent in showing that the carbohydrate conjugate was better than the amide conjugate [20].

Mylotarg utilizes both hydrazone and disulfide functional groups (bifunctional linker) for its linker to conjugate calicheamicin to anti-CD33 IgG4 mAb (compound 4, Fig. 20.3) [30, 31]. Calicheamicin is a cytotoxic natural product shown to bind to the minor groove of DNA, causing sequence-specific cleavage of double-stranded DNA [20, 29]. Two different functional groups were used to control the stability of Mylotarg with the hope of hydrolysis of hydrazine in acidic lysosomes and reductive cleavage of the disulfide bond to release the active drug. The electronic properties and steric hindrance present in the hydrazone–disulfide linker allow it to control the rate of hydrolysis, which can lead to optimal clinical results [20]. In phase II trials, 30% of patients receiving Mylotarg experienced complete remission [32].

For hP67.h conjugate derivatives, various hydrazone spacers with aromatic derivatives (Fig. 20.4) were designed to alter the physicochemical properties of hydrazone to control the rate of drug release in the plasma and cell lysosomes [29]. The different hydrazone aromatic spacers have been shown to have different rates of

	% Hydrolysis		In vitro toxicity		% Tumor/control
	pH 7.4	pH 4.5	IC_{50} (ng cal/ml)	Selectivity	
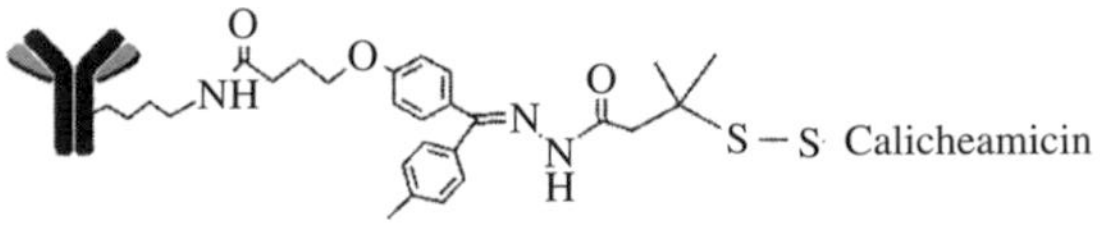 Compound 5 (S–S Calicheamicin)	1	25	99	3	ND
Compound 6 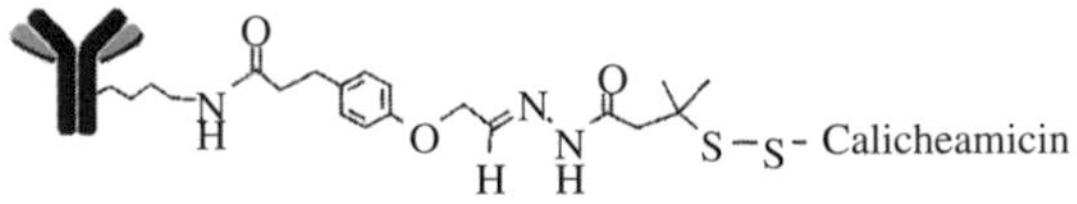 (S–S- Calicheamicin)	90	100	2.2	125	ND
Compound 7 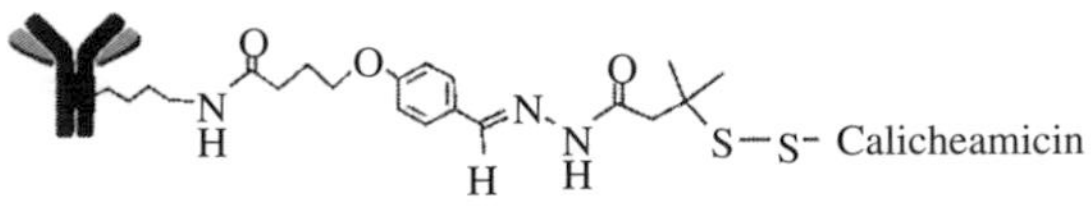 (S–S- Calicheamicin)	5	60	0.52	2900	12
Compound 8 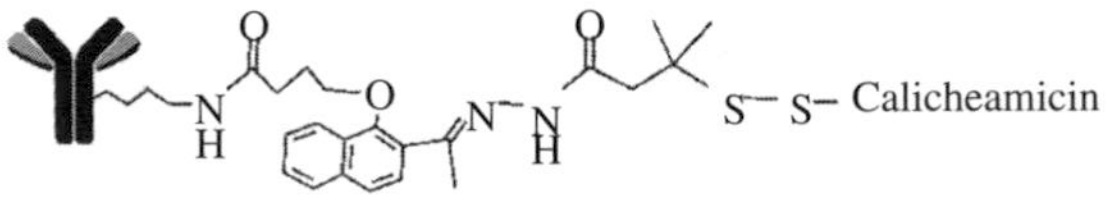 (S–S– Calicheamicin)	0	16	0.061	5000	36
Compound 9 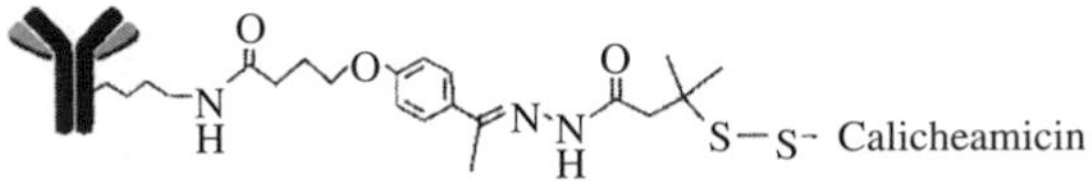 (S–S- Calicheamicin)	6	97	0.040	6400	0

FIGURE 20.4 Compounds 5–9 are conjugates of mAb and calicheamicin with different hydrazone linkers. The effect of different hydrazone derivatives on drug release at pH 7.4 and 4.5 to mimic the blood and lysosomes, respectively. The modified linkers affect the IC_{50} and selectivity of the ADC in cell toxicity.

release at pH 4.5, which mimics the lysosome pH, and at pH 7.4, as in the systemic circulation or extracellular space. For example, when compounds 5 and 9 were incubated at 37°C for 24 hours in buffer, compound 5 showed 1% hydrolysis at pH 7.4 and 25% at pH 4.5. In contrast, compound 9 had 6% hydrolysis at pH 7.4 and 97% at pH 4.5 [29]. In most cases, with some exceptions, the rate of hydrolysis at pH 4.5 can be correlated with the *in vitro* cytotoxicity and selectivity of the hP67.h conjugate when compared to a control hCTM01 conjugate in HL-60 target cells [29]. These linker modification studies resulted in an optimized conjugate called Mylotarg, which can suppress HL-60 tumors when given three times at 50 μg cal/kg while the unconjugated mixture was not active [29].

The reaction to make calicheamicin–anti-CD33 mAb conjugates can cause protein aggregations that lower the recovery of the desired product. This aggregation is presumably due to the change in the physicochemical properties of the ADCs compared to those of the parent mAb [10]. Different additives can be used to improve the conjugation reaction to prevent aggregations and improve product recovery [10]. For example, conjugation using the lysine residues lowers the total positive charges on the mAb and increases the surface hydrophobicity due to addition of hydrophobic spacers and drugs (e.g., calicheamicin). The original reaction to produce the ADC of calicheamicin utilized 25% DMF as one of the solvents; this reaction produced a DAR of 2.9 with 50% aggregates and 26% recovery [10]. To overcome the problem with aggregation and product recovery, additives were included in the reaction mixture. For example, 25% *t*-butanol produced a DAR of 3.6 with 10% aggregate and 54% recovery [10]. One of the best additives was a mixture of 25% propylene glycol and 80 mM octanoic acid, and this additive produced a DAR of 3.6 and 80% recovery with only 3% aggregates [10]. Some linkers incorporate polar moieties such as a polyethylene glycols or sulfate groups to improve conjugate reaction recovery and water solubility, decrease systemic clearance, and increase drug load on the ADCs while limiting aggregation or loss of receptor affinity [33, 34].

Some linkers can be rapidly hydrolyzed or reduced in plasma to release the drug prior to reaching the target site. One way to improve stability is by introducing a peptide linker that is a substrate for lysosomal proteases such as cathepsin or plasmin. In such a case, the enzymes cleave the peptide to release the free active drug inside the targeted cell [35]. Conjugates with peptide linkers are expected to have better serum stability since the proteases that recognize the peptide sequence are not active in the bloodstream due to the high pH and inhibition by serum protease inhibitors [35]. Adcetris (Fig. 20.1) utilizes valine–citrulline because it has good plasma stability and can be efficiently hydrolyzed by cathepsin B inside the cell [6, 36]. The use of a valine–citrulline linker increased the half-life of Adcetris in human plasma by nearly a hundred fold compared to the MMAE linked to the antibody by a hydrazone linker [6]. In phase II clinical trials, HL and ALCL patients treated with Adcetris showed high response rates exceeding 70 and 80%, respectively, as well as complete remission rates of 34 and 57% for HL and ALCL, respectively. It was also found that Adcetris can be administered with a maximum tolerated dose (MTD) of 1.8 mg/kg every 3 weeks [37].

A stable thioether linker offers another alternative approach to controlling the drug release; it is used in Adcetris [6, 36] and Kadcyla [38, 39] (Fig. 20.1). The good plasma stability of Adcetris could be due in part to the more stable thioether bond that connects the antibody and the linker [6, 36]. Because the thioether is a stable linker, it was speculated that drug release in the lysosome relies on mAb degradation. It was found that Kadcyla (T-DM1) has a half-life of 134 hours, and the released DM1 drug was conjugated to a lysine residue and this modified drug molecule was still active [38]. After a single dose in Sprague–Dawley rats, T-DM1 was found to be the major circulating compound, suggesting its systemic stability. The *in vivo* catabolites found were DM1, (*N*-maleimidomethyl) cyclohexane-1-carboxylate-DM1 (MCC-DM1), and lysine–MCC–DM1 [39]. In the plasma of patients treated with a single dose of T-DM1, the intact T-DM1 was found to be a major component with a small amount of released drugs [39]. The tissue and kidney levels of the conjugate were lower than in the blood, and the lowest concentration was in the brain as expected [39]. Most of the conjugate and metabolites were excreted in the feces (80%) with small fraction in the urine [39]. Because of the complexity of the molecules, the traditional method of identifying small drug molecules has to be modified for the conjugates [39]. In a randomized phase II clinical study, patients with previously untreated metastatic breast cancer had a higher response to Kadcyla (T-DM1) than to the standard care with a chemotherapeutic agent, docetaxel [40]. T-DM1 was found to lower the "by-stander effect" because of its charges that prevent it from diffusing into neighboring cells [40].

20.4 NON-ANTIBODY-BASED PROTEIN–DRUG CONJUGATES

Besides antibodies, other biologically active proteins such as transferrin (Tf), epidermal growth factor (EGF), and low-density lipoprotein (LDL) have been investigated as drug conjugates to target specific cells as potential therapeutic agents to treat cancer and autoimmune diseases. However, their development and successful applications have been impeded by several challenges, including low receptor affinity, insufficient stability, immunogenicity, and poor bioavailability. In contrast to most proteins, albumin is a versatile and robust protein carrier for drug targeting. Albumin has an effective diameter of 7.2 nm, and has been known to undergo extravasation specifically through blood vessels of tumor tissues but not normal tissues [41]. After extravasation, albumin accumulates in tumor tissues and reduces its clearance [42].

A conjugate of methotrexate (MTX) and human serum albumin (MTX–HSA, Fig. 20.5a) has been used to improve the therapeutic potential of methotrexate in the treatment of rheumatoid arthritis (RA) [41]. MTX has been used to treat patients suffering from RA; however, this drug has severe side effects. In comparative studies between MTX–HSA and MTX, the MTX–HSA conjugate was more effective in preventing the progression of collagen-induced arthritis in mice [41]. Other conjugates using albumin to target drugs include albumin–DOX [43] and albumin–paclitaxel nanoparticles (abraxane) [44].

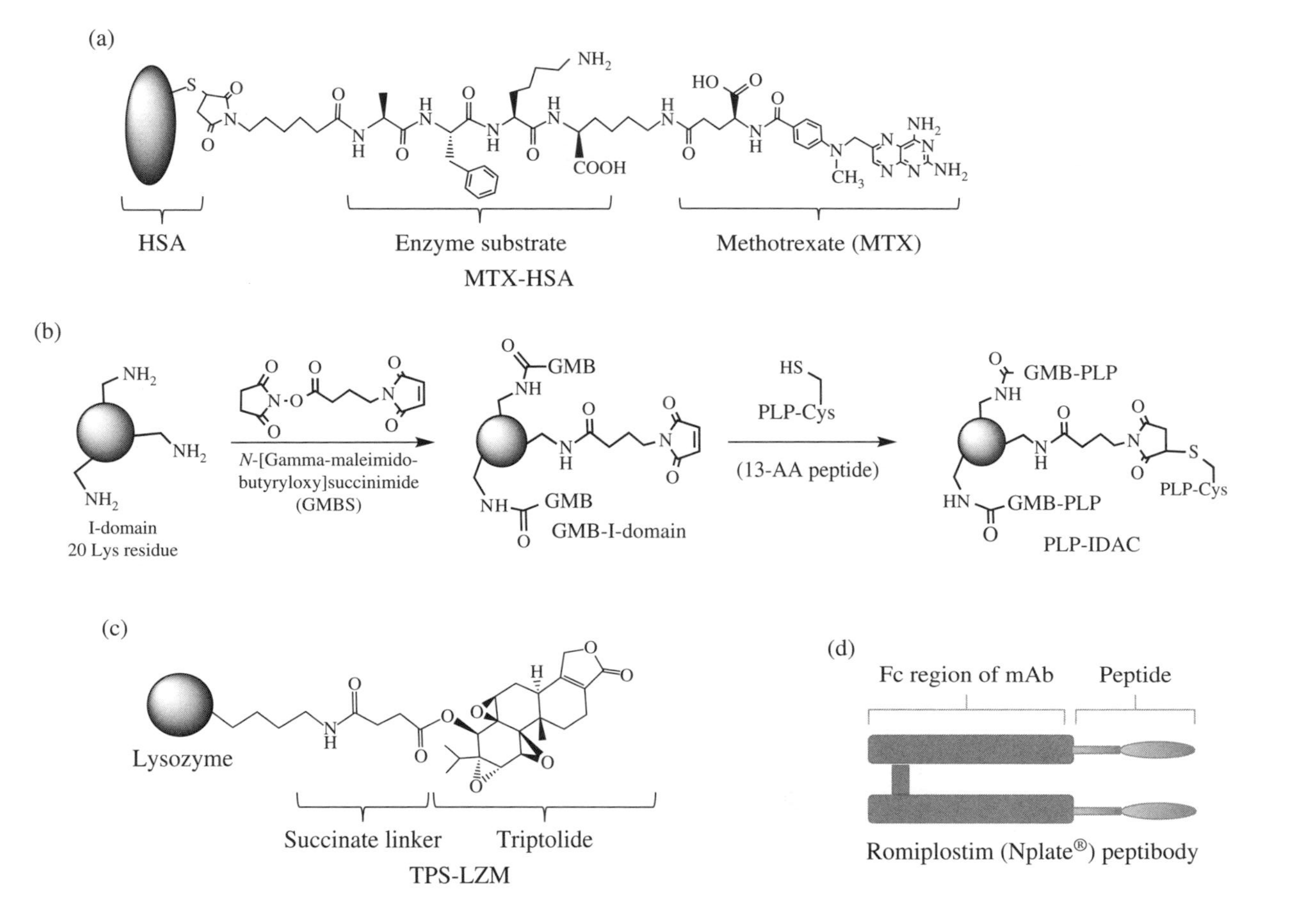

FIGURE 20.5 The structures of non-antibody protein conjugates including (a) MTX-HSA, (b) PLP-IDAC with the reactions to make the conjugate, (c) TPS-LZM, and (d) peptibody.

The I-domain (ID) was conjugated to antigenic peptides to give I-domain–antigen conjugates (IDACs, Fig. 20.5b). The IDAC molecules were used to deliver antigenic peptides to antigen-presenting cells (APCs) with the goal of altering the commitment of naïve T cells in autoimmune diseases from inflammatory to regulatory and suppressor T cells. These IDAC molecules suppressed the autoimmune diseases in the animal models of autoimmune diseases. A peptide from proteolipid protein ($PLP_{139-151}$) was conjugated to the lysine residues on the ID to make PLP–IDAC molecules. Because the ID has 20 lysine residues, these lysine residues were derivatized by reacting them with *N*-(γ-maleimidobutyryloxy)succinimide ester (GMBS) to make a maleimide derivative of the ID called GMB-ID. The maleimide groups were subsequently reacted with the thiol group of the cysteine residue located on the C-terminus of $PLP_{139-151}$ to give PLP–IDAC as a mixture of conjugates [45, 46]. As prophylactic and vaccine-like treatments, PLP–IDAC molecules effectively suppressed experimental autoimmune encephalomyelitis (EAE) in the mouse model, a model for multiple sclerosis (MS) [46, 47]. Cytokine studies indicated that PLP–IDAC molecules suppress EAE by enhancing the production of suppressor and regulatory T cells (T-reg) while suppressing Th17 cells [46, 47].

Lysozyme is an endogenous low-molecular-weight protein that has been used as a carrier molecule for drugs such as triptolide, naproxen, captopril, DOX, and sulfamethoxazole. Lysozyme–drug conjugates, such as triptolide–lysozyme (TPS–LZM, Fig. 20.5c), are normally used for drug delivery to the kidneys because of lysozyme's specific renal uptake and biodegradability [48]. Lysozyme belongs to a class of biologically active proteins in the circulatory system called low-molecular-weight glomerular proteins (LMWPS). They are small-molecular-weight ($MW < 30{,}000\,Da$) proteins that can be filtered at the glomerulus and reabsorbed in the renal tubules *via* receptor-mediated endocytosis [48]. In the evaluation of TPS–LZM conjugate, an ester linkage was used to make a 1 : 1 TPS–LZM conjugate [49, 50]. The renal targeting efficiency of the TPS–LZM is higher than that of the drug alone with prolonged residence times [49]. Compared to a physical mixture of triptolide and lysozyme, the conjugate was more effective in reversing the progression of *in vivo* renal ischemia–reperfusion injury at a very low drug concentration [49]. Furthermore, the conjugate is safer since it induces less hepatotoxicity than triptolide alone.

An increased expression of ICAM-1 on immune and endothelial cells has been reported in several diseases, including autoimmune diseases, malignancies, and neurological disorders. As such, ligands that bind to ICAM-1 are attractive drug-targeting agents. The ID protein derived from the α-subunit of leukocyte function-associated antigen-1 (LFA-1) has been used to target molecules to APC to control in autoimmune diseases [51]. The fluorescein-5′-isothiocyanate (FITC)-ID has been shown to selectively bind to intercellular adhesion molecule-1 (ICAM-1) receptors on lymphocytes and is also internalized into cells (i.e., Raji cells) [45].

20.5 PEPTIBODY

Peptibody is a chimeric between a protein and a peptide; an example is the fusing of the Fc domain of an antibody with a bioactive peptide (Fig. 20.5d) [52–54]. The formation of a peptibody is an attractive way to improve peptide delivery, efficacy, and cellular uptake. Peptides suffer from rapid renal clearance due to poor metabolic stability, but these problems can be overcome by forming a peptibody. A peptibody protects the peptide from plasma degradation and increases systemic circulation time by preventing glomerular filtration of the molecule due to its large size or hydrodynamic radius [53]. Furthermore, grafting these bioactive peptides to the Fc region of IgG improves their pharmacokinetic properties by utilizing the FcRn recycling process. The prolonged half-life of a peptibody occurs because the unbound IgG is degraded at the lysosomes at low pH while the peptibody is returned to the cell surface and released again [54].

The advantage of peptibody formation is that it eliminates the need to separately synthesize and chemically conjugate the peptide to the Fc region and the normal recombinant protein expression methods can be used to produce peptibodies. Romiplostim (Nplate®) is the only FDA-approved peptibody on the market for the treatment of chronic immune thrombocytopenic purpura (ITP). This drug is the Fc fragment of IgG1 fused to a thrombopoietin (TPO) agonist peptide (14 amino acid). The peptide binds to a receptor called AF12505, which stimulates platelet production [55]. Here, two molecules of the receptor AF12505 are fused to bind two peptide fragments on Nplate peptibody. This idea was developed because it was found that the dimeric form of the TPO agonist peptide was more active than the monomeric form [56]. This is the first design of a peptibody to simultaneously increase receptor-binding affinity and extend the circulation half-life of the peptide.

The peptide sequence of diphtheria toxin has been fused with interleukin-2 (IL-2) to make denileukin diftitox (Ontak®), which is used to treat cutaneous T-cell lymphoma (CTCL) [1]. Other toxins such as ricin, RNAse, and *Pseudomonas* exotoxin have been conjugated to mAbs and are being investigated for cancer treatments [57–59].

20.6 PROTEIN CONJUGATES FOR DIAGNOSTICS

The use of antibodies in noninvasive imaging is one of the most active research fields for diagnostic purposes. Optical imaging that utilizes fluorescence and bioluminescence imaging probes has emerged as a preferred method of diagnostics over radioisotope-based imaging [60]. This is due mainly to the low cost and portability of imaging instruments as well as the absence of ionizing radiation. Once excited at a specific wavelength, fluorescent molecules (e.g., green fluorescent proteins or GFPs) have been used to observe gene production and location of specific proteins in cells [60]. In one study, fluorescent protein rhodamine green (RhodG) was conjugated to herceptin (antibody to HER2) to give the herceptin–rhodG conjugate for detecting tumor metastasis [60, 61]. Under normal conditions, HER2 is expressed at basal levels, but it is overexpressed in a number of epithelial malignancies, including breast, ovarian, gastric, prostate, and other cancers. Thus, herceptin–rhodG conjugate was able to be detected in HER2-positive tumors with high specificity and sensitivity [60, 61].

Initially, the widespread use of optical imaging encountered difficulties in achieving a high target-to-background ratio arising from their permanently "On" property [62].

However, recent advances have led to development of fluorophores that can be activated on-site through a number of techniques such as electron transfer, target-specific enzymatic activation, or removal of a quencher. The development of fluorescence resonance energy transfer (FRET) has further suppressed the nonspecific activation characteristic [63]. FRET-based probes include pairing between a fluorophore and a quencher conjugated to a specific targeting agent. A rhodamine fluorophore TAMRA and a quencher QSY7 have been linked to avidin to make Av–TM-Q7 conjugate for targeting the D-galactose receptor. Similarly, TAMRA and QSY7 were conjugated to trastuzumab mAb to make Traz–TM-Q7 conjugate, and this conjugate has been shown to detect HER2 receptors on tumor cells (Fig. 20.6) [63]. TAMRA does not emit fluorescence when it is in close proximity to QSY7 in

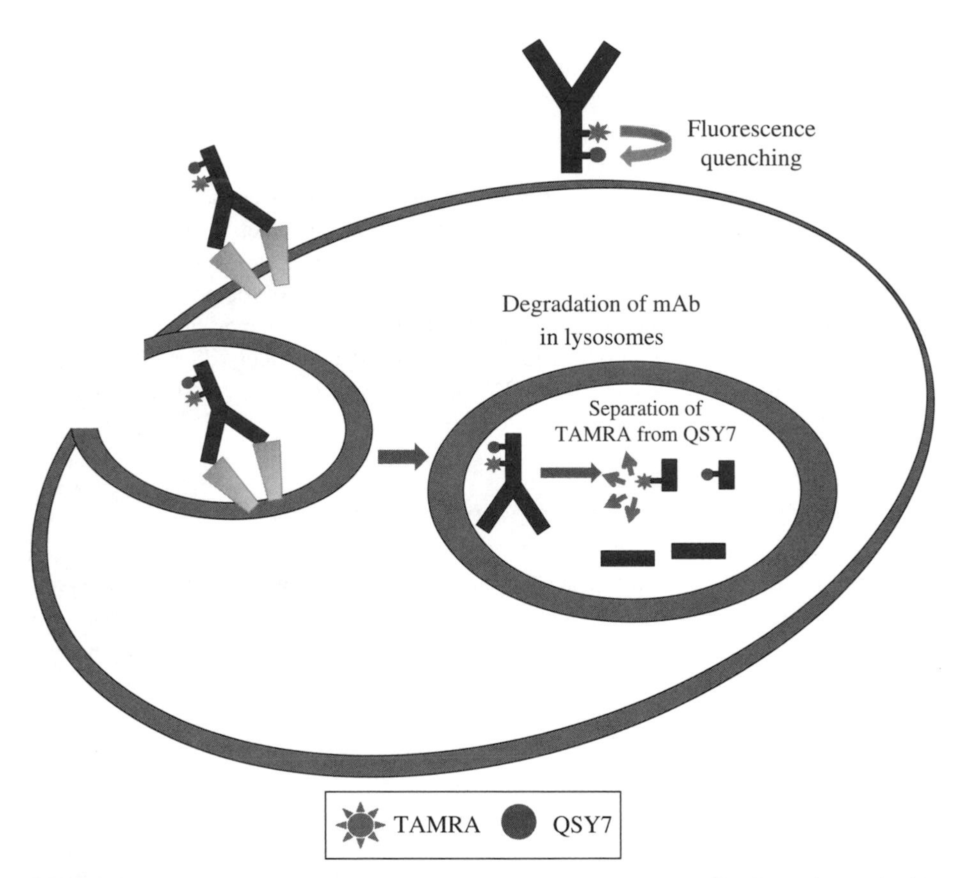

FIGURE 20.6 The mechanism of cancer cell detection using an antibody conjugated with TAMRA and QSY7 that binds to antigens on the surface of cancer cell. The antibody will be taken up by the cancer cells via receptor-mediated endocytosis and processed through early and late endosomes. In the early and late endosomes, the intact antibody with TAMRA and QSY7 has very low fluorescence intensity due to quenching of TAMRA fluorescence by QSY7. After degradation of the antibody in the lysosomes, the TAMRA is separated from the quencher QSY7 to produce fluorescence in the cancer cells.

the conjugates, but once internalized, the conjugates are degraded in the lysosomes to separate TAMRA and QSY7 molecules (Fig. 20.6). When TAMRA is chemically and physically separated from QSY7 molecules as a quencher, it appears as bright fluorescent intracellular dots in target cells. These probes have been used to detect small tumors in the mouse model [63].

Infrared imaging is particularly attractive in live animal imaging due to its deep penetrating ability, high sensitivity, and low tissue autofluorescence in the range of infrared imaging [64]. These imaging techniques are becoming more attractive in clinical situations where "see and treat" strategies are needed in disease management. A method that would aid a surgeon in real time would be welcomed for effective surgical resection in patients with locally invasive prostate cancer. Long-term cancer control in these patients can be achieved if adequate margins are obtained at surgery. Noninvasive and highly targeted imaging probes that have few false positives and are activated only at targeted sites would be more instrumental in achieving better patient outcomes. In this case, a near-infrared dye, indocyanine green (ICG), was conjugated to mAbs such as daclizumab (Dac), panitumumab (Pan), and trastuzumab (Tra) to produce Dac–ICG, Pan–ICG, and Tra–ICG, respectively. Dac–ICG, Pan–ICG, and Tra–ICG conjugates have been shown to detect CD25, Her2, and Her1 antigens on tumor cells in *in vitro* and *in vivo* systems [64]. When the ICG is conjugated to the mAb, it has a weak fluorescence and will emit fluorescence when it is cleaved from the protein. In this case, the mAb conjugates were bound and internalized to the cell surface antigens (CD25, Her2, or Her1) on the tumor cells. Then, ICG is released inside the tumor cells after degradation in the lysosomes [64].

20.7 PEPTIDE–DRUG CONJUGATES

Peptides are generally defined as having a chain of less than 50 amino acids; they may exhibit a high degree of secondary structure, but they may not have tertiary structure. Compared to proteins such as mAbs, peptides can be synthesized relatively easily with straightforward structure manipulations. However, peptides tend to have lower affinities to the target proteins on the cell surface compared to antibodies. Peptides also have rapid rate of renal clearance compared to antibodies because the kidneys filter molecules with cut-off size of 50 kDa [65]. In addition, peptides normally have low half-lives, ranging from minutes to a few hours. To combat against renal clearance and exposure to exopeptidases, polyethylene glycol (PEG) has been attached to the N- and C-termini of peptides [66]. A variety of peptides have been investigated for delivery of drugs to specific cells, including cell-penetrating peptides (CPPs), cell adhesion peptides, and receptor-specific peptides.

Highly cationic CPPs contain a predominant population of either the arginine or lysine residues and thus can cross the cell lipid bilayer without the need for a specific receptor [67–69]. Although the mechanism of cellular uptake of CPPs has not been fully elucidated, they have been use to deliver drugs into different cells [70, 71]. Trans-activator of transcription (TAT) peptides, penetratin, and VP22 are examples of CPPs that have been used to deliver biologically active proteins [67, 72]. A TAT_{47-57} peptide (YGRKKRRQRRR) is derived from the human immunodeficiency virus type 1 (HIV-1), which is involved in cellular transduction [73, 74]. Both endosomal and non-endosomal (energy-independent) mechanisms have been reported to be the mechanisms of uptake of CPPs [75, 76]. CPP conjugates that enter the cell through non-endosomal pathways are able to deliver the drugs directly into the cytoplasm [72, 77]. On the other hand, conjugates that enter via the endosomal route risk being trapped in intracellular vesicular compartments; in this particular case, additional formulation to facilitate release into the cytosol is needed [72, 77].

CPPs (e.g., TAT, penetratin, and the D-form of octaarginine (r8)) have been conjugated to DOX for targeting to tumor xenografts *in vivo* [70]. Fluorescence imaging studies showed that all these CPPs had significantly higher accumulation in liver, kidney, lung, and spleen; however, r8 had preferential accumulation in tumor xenografts [70]. Furthermore, DOX-r8 (Fig. 20.7a) conjugate was effective in tumor suppression with no significant weight loss compared to free DOX [70]. Another application of a CPP–drug conjugate has been illustrated using a δPKC inhibitor, KAI-9803, which is used in stroke patients. Here, the compound was conjugated to TAT_{47-57} via a disulfide bond where the conjugate showed a reduction in apoptotic cell death after ischemia [78]. Although CPPs have provided success in delivering drugs into cells, they have relatively low target-organ specificity and their accumulation depends mainly on high vasculature organs or macrophage-like cells [70].

Cell adhesion molecules such as integrin (e.g., $\alpha_V\beta_3$, $\alpha_V\beta_5$, and LFA-1) and immunoglobulin (e.g., ICAM-1, VCAM-1) receptors are known to undergo endocytosis and offer alternative proteins on the cell surface as targets for drug delivery [79]. The upregulation of cell adhesion molecules on the surface of leukocytes and endothelial cells has been correlated with inflammation and some cancers (i.e., lung and pancreatic cancers) [80, 81]. RGD peptides that are selective for $\alpha_v\beta_3$ and $\alpha_v\beta_5$ receptors have been used as targets for delivering drugs to tumor cells [79, 82]. DOX was conjugated to a bicyclic RGDC4 peptide to make DOX-RGDC4 conjugate (Fig. 20.7b); it has excellent efficacy compared to free DOX in suppressing the growth of human breast carcinoma in the mouse tumor model [82]. DOX-RGDC4 improved the survival of mice compared to DOX alone. The low systemic toxicity of DOX-RGDC4 conjugate compared to DOX suggests that the conjugate targets the tumor cells using upregulated $\alpha_v\beta_3$ and $\alpha_v\beta_5$ receptors [82].

A conjugate of an ICAM-1 peptide cIBR with MTX (MTX–cIBR, Fig. 20.7c) was used to deliver MTX and lower its side effects; MTX–cIBR was aimed at delivering MTX to LFA-1-bearing activated immune cells to suppress RA [83]. cIBR peptide has been shown to bind to the I-domain of LFA-1 ($\alpha_L\beta_2$ integrin) and is internalized into LFA-1-expressing cells (i.e., T cells and HL-60) via receptor-mediated endocytosis [84–86]. MTX–cIBR conjugate was effective in suppressing RA in the collagen-induced arthritis (CIA) mouse model [87] and rat adjuvant arthritis model [83]. *In vitro*, MTX–cIBR has lower toxicity than MTX alone, but MTX–cIBR was more selective than MTX for LFA-1-expressing cells than cells without LFA-1 expression. In *in vivo* studies, MTX–cIBR in has low toxicity in the rat adjuvant model of RA and has different mechanisms of action than MTX in suppressing RA [83].

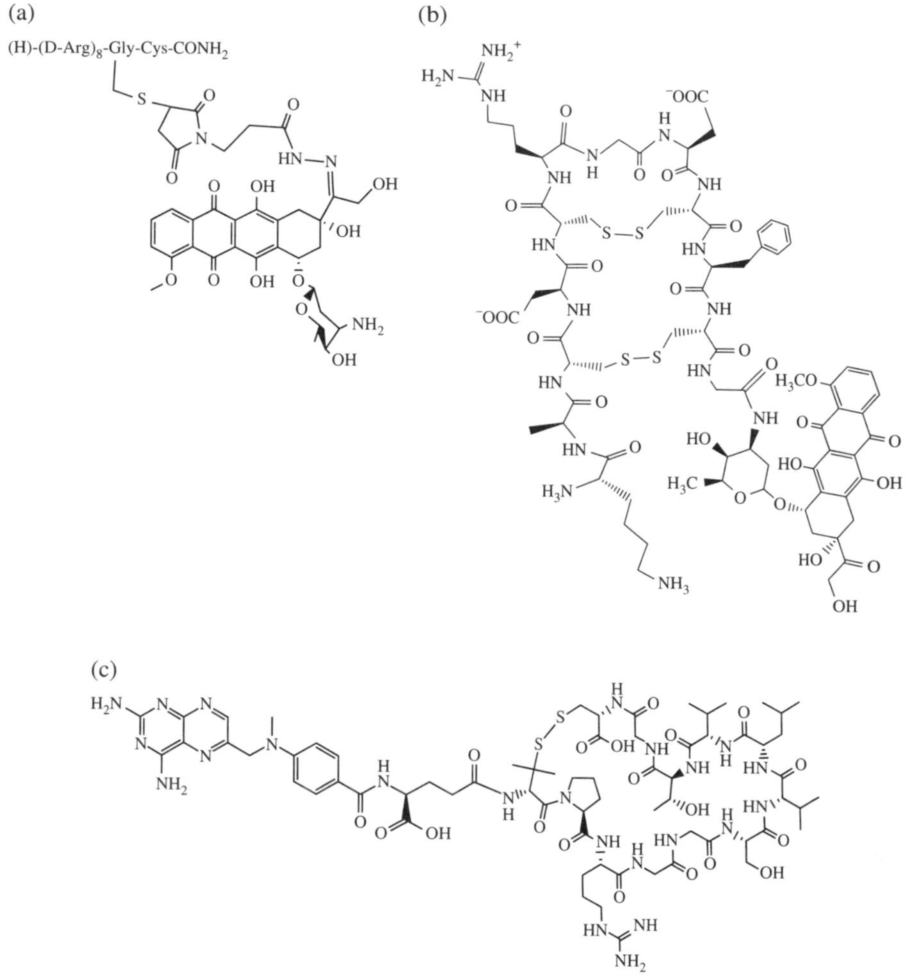

(a)

(H)-(D-Arg)$_8$-Gly-Cys-CONH$_2$

(b)

(c)

FIGURE 20.7 The structures of drug–peptide conjugates including (a) DOX-r8, (b) DOX-RGDC4, and (c) MTX-cIBR.

The LABL peptide derived from the ID of LFA-1 has been shown to target antigenic peptides from PLP [88, 89], glutamic acid decarboxylase (GAD) [90], and collagen-II (CII) [91] to APC to control autoimmune diseases such as EAE, type-1 diabetes, and RA [51, 92]. In this case, LABL peptide was conjugated via a linker to PLP, GAD, and CII peptides to make bifunctional peptide inhibitor (BPI) molecules called PLP–BPI, GAD–BPI and CII–BPI (Fig. 20.8). Each respective BPI molecule has been shown to suppress a different target autoimmune disease, depending on the antigenic peptide used. For example, PLP–BPI has been shown to selectively suppress EAE disease in the mouse model (a model for MS) [88, 89]. Similarly, the GAD–BPI and CII–BPI molecules have been shown to effectively suppress T1D and RA, respectively [87, 90]. The cytokine studies indicate that BPI molecules

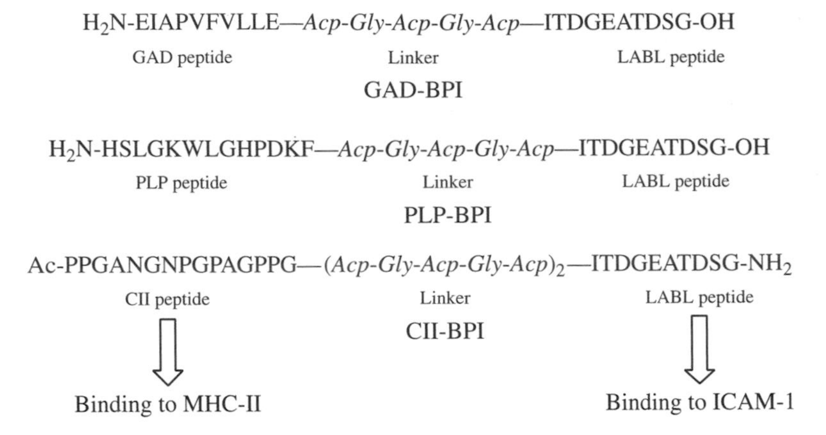

FIGURE 20.8 The sequence of BPI molecule, which is a conjugate between an antigenic peptide (GAD, PLP, or CII peptide) and LABL peptide that are connected via a linker. GAD-BPI has been shown to suppress Type 1 diabetes in the non-obese diabetes mouse model. PLP-BPI suppresses the experimental autoimmune encephalomyelitis (EAE) in the mouse model as treatment, prophylactic, and vaccine. CII-BPI suppresses rheumatoid arthritis in collagen-induced rheumatoid arthritis. *Acp*, amino caproic acid; *Gly*, glycine.

suppress inflammatory responses such as suppression of Th17 and possibly stimulate regulatory (e.g., T-reg) and suppressor (Th2) responses [51]. The hypothesis is that BPI molecules simultaneously bind to major histocompatibility-II (MHC-II) and ICAM-1 receptors on the surface of APC to alter the balance of inflammatory-to-regulatory T cells by blocking the formation of the immunological synapse [51, 92]. It is interesting to find that PLP–BPI can suppress EAE when administered as a vaccine [93, 94], a prophylactic [89], or a treatment [88]. In addition to suppressing the disease development, PLP–BPI has been shown to prevent blood–brain barrier leakiness in EAE mice [93].

20.8 CHALLENGES IN ANALYZING CONJUGATES

The analysis of peptide conjugates is more straightforward than analyzing protein conjugates because peptide conjugates consist of a single entity. The chemistry of conjugation can be incorporated during peptide synthesis and, thus, the drug can be conjugated in one functional group within the peptide. In contrast, most protein conjugates are heterogeneous mixtures of products because the drugs are randomly conjugated to a specific functional groups such as an amine group of the lysine residues in protein [95, 96]. The heterogeneity of the conjugates makes them challenging to analyze. Conjugation of maytansines and calicheamicin to the side chain amino group of the lysine residues of the mAb generates a wide range of species; for example, trastuzumab emtansine is a mixture with DAR of 0–8 or an average of 3.5 [95, 97, 98]. Similarly, auristatins conjugated to the mAb via reduced interchain disulfides gave an average DAR of 4, which is the clinically preferred value [95, 99]. Some efforts have been carried out to limit the heterogeneity of the product of conjugation; for example, a lysine residue can be mutated to a cysteine residue or unnatural amino acid such a phenyl-methyl ketone derivative of phenyl alanine residue [100, 101] (Fig. 20.9) for selective conjugation to the linker. As previously described, the cysteine residue can be conjugated to the linker via maleimide group and the methyl-phenyl ketone can be connected to the linker using a hydrazone group. Recently, homogeneous ADCs with a defined DAR have been produced through engineered cysteine residues at specific sites [102, 103]. Despite efforts to develop nearly homogeneous conjugates, the dynamic nature of the mAb introduces heterogeneity. For instance, the conjugated drug can be partially released *in vivo*, resulting in changes in DAR values [102].

The composition of the conjugates can affect the clinical safety and efficacy. Therefore, sensitive and reliable analytical methods for characterization of the conjugates should be developed for evaluating the heterogeneity of the conjugates. One of the most common analytical techniques used for analysis of conjugates is UV/VIS spectroscopy. Typically, the A_{max} of proteins or antibodies is around 280 nm, which is a different A_{max} than that of the conjugated drug. Using the absorbance ratio between the drug and mAb, the DAR can be calculated [96]. Reversed-phase, ion-exchange, size-exclusion, and hydrophobic interaction chromatography (HIC) have been used to characterize DAR of ADC. Separation of macromolecule conjugates by HIC relies on hydrophobic interactions between the protein and stationary phase. Most of the conjugated drugs (e.g., paclitaxel, DM1, calicheamicin, and DOX) are

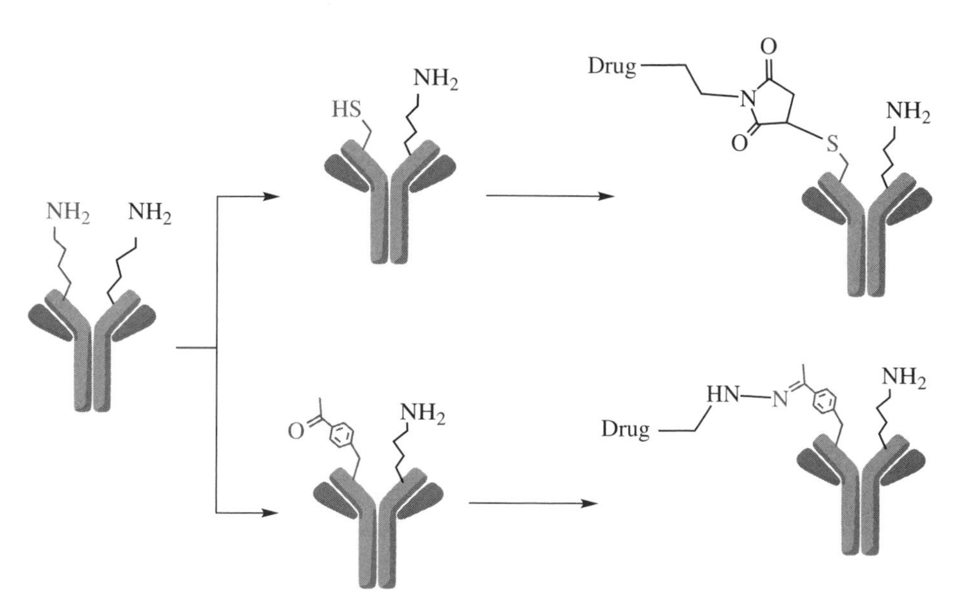

FIGURE 20.9 The mutation residue in the monoclonal antibody-to-cysteine residue or unnatural amino acid for site-specific and homogeneous drug conjugation.

very hydrophobic; as a result, the conjugates have increased hydrophobicity compared to the parent protein or mAb [96, 97]. HIC allows the analysis of various components of the conjugates based on their different hydrophocities. Because HIC utilizes non-denaturing conditions at neutral pH with decreasing salt gradient, elution of drug-loaded conjugates can be achieved with the use of a low concentration of an organic modifier [96]. Charge-based separations (e.g., ion-exchange chromatography (IEC), or iso-electric focusing gel electrophoresis (IEF)) can also be used to determine drug distribution on conjugates. For example, the attachment of drugs to the lysine residues on Mylotarg reduces the net positive charge by one for each conjugation, and this change is reflected in the retention time of each conjugate population on IEC [96].

One of the challenges encountered during manufacturing and storage of these conjugates is the likelihood of aggregate formation because conjugation of drug to antibody can affect the physicochemical properties of the ADC compared to the parent mAb. Clinically, the presence of these aggregates can elicit anti-therapeutic response (ATA), which is known to cause infusion-associated reactions, change in product pharmacokinetics, and reduced drug exposure [96]. To analyze for the extent of aggregation and fragmentation, size-exclusion chromatography (SEC) has been used to monitor the stability of the conjugate. As with HIC, SEC is done under non-denaturing conditions, and can also be used to analyze various aggregates of the intact conjugates. In T-MD1 (Kadcyla), the DM1 was conjugated via lysine residues with DAR of 3.5, causing the loss of positive charges on the mAb and increasing the hydrophobicity. As predicted, the results from Tmab and T-DM1 differential scanning calorimetry (DSC) produced two major transitions where the first transition is for the melting of the CH2 domain and the second is for the melting of the CH3 domain. The first transition (CH2 domain) of T-DM1 (63.8°C) was lower than that of Tmab (68.2°C), while the second transition (CH3 domain) did not show any difference [104]. The change in the first transition temperature is due to the fact that the majority of drug conjugations were at the CH2 domain; this is because the CH2 domain is flexible and the lysine residues in this domain are accessible. Incubation of T-DM1 and Tmab for 70 days at 40°C showed 11% aggregation of T-DM1 while no aggregation of Tmab was found.

Liquid chromatography–mass spectrometry (LCMS) methods are standard means employed in the analysis of conjugates. In the most contemporary LCMS methods, separation is achieved through the use of organic solvents by electrospray ionization electrospray ionization coupled with time-of-flight (TOF) or triple quadrupole mass detectors [105, 106]. These methods can be used to characterize drug-load distribution of T-DM1. This technique enhances mass accuracy as well as improves resolution and facilitates determination of molecular masses of conjugates with varying DAR. Mass detection in combination with peptide mapping analysis has been used to study DAR and map sites of conjugation [105]. A combination of enzyme digestion and LCMS can identify the sites of drug conjugation on the proteins. For example, tryptic digest and LCMS have been used to identify the sites of conjugation of FITC-labeled and PLP peptides on the lysine residue on the ID protein [46, 107].

Other bioanalytical assays have also been developed to detect the conjugates and their metabolites during pharmacokinetics, efficacy, and safety studies. The total antibody, antibody–drug conjugate, and free drug released from antibody conjugates in biological fluids were normally determined using enzyme-linked immunosorbent assay (ELISA) or its modified forms [97]. A combination of affinity capture using an immobilized biotinylated target antigen on paramagnetic streptavidin beads and LCMS has also been used to detect the conjugates and their components in biological systems [102].

20.9 CONCLUSIONS

There have been some successes in developing ADCs that reach patients, including brentuximab vedotin (Adcetris) and trastuzumab emtansine (Kadcyla). Although gentuzumab ozogamicin (Mylotarg) was withdrawn with possible reentry, the knowledge gained in developing this ADC is valuable for future development of new ADCs and other protein conjugates. Although most current successes are antibody conjugates, it is predicted that other types of targeting molecules such as other proteins and peptides may reach patients in the future. The current area of therapy for conjugates is mostly in cancer treatment; however, there are other areas of therapeutics such as autoimmune diseases that are being investigated for many different conjugates. A future strategy to consider is development of conjugates with dual activities to potentially modulate two or more receptors because, for many immune disorders, many effector cells require modulation of two or more activated receptors or signals. BPI and IDAC molecules that target two cell surface receptors have been shown to suppress autoimmune diseases. Conjugates of multiple peptides onto proteins, liposomes, and polymers may be investigated for therapeutics and diagnostics in the future. The hope is that conjugates will provide a new avenue for the development of better drugs.

REFERENCES

1. Olsen E., Duvic M., Frankel A., Kim Y., Martin A., Vonderheid E., Jegasothy B., Wood G., Gordon M., Heald P., Oseroff A., Pinter-Brown L., Bowen G., Kuzel T., Fivenson D., Foss F., Glode M., Molina A., Knobler E., Stewart S., Cooper K., Stevens S., Craig F., Reuben J., Bacha P., Nichols J., *J Clin Oncol* 2001, **19**, 376–388.
2. Trail P. A., Willner D., Lasch S. J., Henderson A. J., Hofstead S., Casazza A. M., Firestone R. A., Hellstrom I., Hellstrom K. E., *Science* 1993, **261**, 212–215.
3. Walsh G., *Nat Biotechnol* 2010, **28**, 917–924.
4. Younes A., Bartlett N. L., Leonard J. P., Kennedy D. A., Lynch C. M., Sievers E. L., Forero-Torres A., *N Engl J Med* 2010, **363**, 1812–1821.
5. Senter P. D., Sievers E. L., *Nat Biotechnol* 2012, **30**, 631–637.
6. Doronina S. O., Toki B. E., Torgov M. Y., Mendelsohn B. A., Cerveny C. G., Chace D. F., DeBlanc R. L., Gearing R. P., Bovee T. D., Siegall C. B., Francisco J. A., Wahl A. F., Meyer D. L., Senter P. D., *Nat Biotechnol* 2003, **21**, 778–784.

7. Niculescu-Duvaz I., *Curr Opin Mol Ther* 2010, **12**, 350–360.
8. Appelbaum F. R., Matthews D. C., Eary J. F., Badger C. C., Kellogg M., Press O. W., Martin P. J., Fisher D. R., Nelp W. B., Thomas E. D., Bernstein I. D., *Transplantation* 1992, **54**, 829–833.
9. Kohler G., Milstein C., *Nature* 1975, **256**, 495–497.
10. Hollander I., Kunz A., Hamann P. R., *Bioconjug Chem* 2008, **19**, 358–361.
11. Jackson M. R., Falzone N., Vallis K. A., *Clin Oncol (R Coll Radiol)* 2013, **25**, 604–609.
12. Steiner M., Neri D., *Clin Cancer Res* 2011, **17**, 6406–6416.
13. David K. A., Milowsky M. I., Kostakoglu L., Vallabhajosula S., Goldsmith S. J., Nanus D. M., Bander N. H., *Clin Genitourin Cancer* 2006, **4**, 249–256.
14. Goldsmith S. J., *Semin Nucl Med* 2010, **40**, 122–135.
15. Linden O., Hindorf C., Cavallin-Stahl E., Wegener W. A., Goldenberg D. M., Horne H., Ohlsson T., Stenberg L., Strand S. E., Tennvall J., *Clin Cancer Res* 2005, **11**, 5215–5222.
16. Petersen B. H., DeHerdt S. V., Schneck D. W., Bumol T. F., *Cancer Res* 1991, **51**, 2286–2290.
17. Tolcher A. W., Sugarman S., Gelmon K. A., Cohen R., Saleh M., Isaacs C., Young L., Healey D., Onetto N., Slichenmyer W., *J Clin Oncol* 1999, **17**, 478–484.
18. Flygare J. A., Pillow T. H., Aristoff P., *Chem Biol Drug Des* 2013, **81**, 113–121.
19. Sievers E. L., Senter P. D., *Annu Rev Med* 2013, **64**, 15–29.
20. Hamann P. R., Hinman L. M., Beyer C. F., Lindh D., Upeslacis J., Flowers D. A., Bernstein I., *Bioconjug Chem* 2002, **13**, 40–46.
21. Leal M., Sapra P., Hurvitz S. A., Senter P., Wahl A., Schutten M., Shah D. K., Haddish-Berhane N., Kabbarah O., *Ann N Y Acad Sci* 2014, **1321**, 41–54.
22. Trail P. A., Bianchi A. B., *Curr Opin Immunol* 1999, **11**, 584–588.
23. Xie H., Blattler W. A., *Expert Opin Biol Ther* 2006, **6**, 281–291.
24. Ducry L., Stump B., *Bioconjug Chem* 2010, **21**, 5–13.
25. Ranson M., Sliwkowski M. X., *Oncology* 2002, **63** Suppl 1,17–24.
26. Erickson H. K., Park P. U., Widdison W. C., Kovtun Y. V., Garrett L. M., Hoffman K., Lutz R. J., Goldmacher V. S., Blattler W. A., *Cancer Res* 2006, **66**, 4426–4433.
27. Krop I. E., Beeram M., Modi S., Jones S. F., Holden S. N., Yu W., Girish S., Tibbitts J., Yi J. H., Sliwkowski M. X., Jacobson F., Lutzker S. G., Burris H. A., *J Clin Oncol* 2010, **28**, 2698–2704.
28. Xie H., Audette C., Hoffee M., Lambert J. M., Blattler W. A., *J Pharmacol Exp Ther* 2004, **308**, 1073–1082.
29. Hamann P. R., Hinman L. M., Hollander I., Beyer C. F., Lindh D., Holcomb R., Hallett W., Tsou H. R., Upeslacis J., Shochat D., Mountain A., Flowers D. A., Bernstein I., *Bioconjug Chem* 2002, **13**, 47–58.
30. Larson R. A., Boogaerts M., Estey E., Karanes C., Stadtmauer E. A., Sievers E. L., Mineur P., Bennett J. M., Berger M. S., Eten C. B., Munteanu M., Loken M. R., Van Dongen J. J., Bernstein I. D., Appelbaum F. R., Mylotarg Study G., *Leukemia* 2002, **16**, 1627–1636.
31. Sievers E. L., Larson R. A., Stadtmauer E. A., Estey E., Lowenberg B., Dombret H., Karanes C., Theobald M., Bennett J. M., Sherman M. L., Berger M. S., Eten C. B., Loken

M. R., van Dongen J. J., Bernstein I. D., Appelbaum F. R., Mylotarg Study G., *J Clin Oncol* 2001, **19**, 3244–3254.

32. Larson R. A., Sievers E. L., Stadtmauer E. A., Lowenberg B., Estey E. H., Dombret H., Theobald M., Voliotis D., Bennett J. M., Richie M., Leopold L. H., Berger M. S., Sherman M. L., Loken M. R., van Dongen J. J., Bernstein I. D., Appelbaum F. R., *Cancer* 2005, **104**, 1442–1452.
33. Burke P. J., Senter P. D., Meyer D. W., Miyamoto J. B., Anderson M., Toki B. E., Manikumar G., Wani M. C., Kroll D. J., Jeffrey S. C., *Bioconjug Chem* 2009, **20**, 1242–1250.
34. King H. D., Dubowchik G. M., Mastalerz H., Willner D., Hofstead S. J., Firestone R. A., Lasch S. J., Trail P. A., *J Med Chem* 2002, **45**, 4336–4343.
35. Koblinski J. E., Ahram M., Sloane B. F., *Clin Chim Acta* 2000, **291**, 113–135.
36. Francisco J. A., Cerveny C. G., Meyer D. L., Mixan B. J., Klussman K., Chace D. F., Rejniak S. X., Gordon K. A., DeBlanc R., Toki B. E., Law C. L., Doronina S. O., Siegall C. B., Senter P. D., Wahl A. F., *Blood* 2003, **102**, 1458–1465.
37. Fanale M. A., Forero-Torres A., Rosenblatt J. D., Advani R. H., Franklin A. R., Kennedy D. A., Han T. H., Sievers E. L., Bartlett N. L., *Clin Cancer Res* 2012, **18**, 248–255.
38. Polson A. G., Calemine-Fenaux J., Chan P., Chang W., Christensen E., Clark S., de Sauvage F. J., Eaton D., Elkins K., Elliott J. M., Frantz G., Fuji R. N., Gray A., Harden K., Ingle G. S., Kljavin N. M., Koeppen H., Nelson C., Prabhu S., Raab H., Ross S., Slaga D. S., Stephan J. P., Scales S. J., Spencer S. D., Vandlen R., Wranik B., Yu S. F., Zheng B., Ebens A., *Cancer Res* 2009, **69**, 2358–2364.
39. Shen B. Q., Bumbaca D., Saad O., Yue Q., Pastuskovas C. V., Khojasteh S. C., Tibbitts J., Kaur S., Wang B., Chu Y. W., LoRusso P. M., Girish S., *Curr Drug Metab* 2012, **13**, 901–910.
40. Hurvitz S. A., Dirix L., Kocsis J., Bianchi G. V., Lu J., Vinholes J., Guardino E., Song C., Tong B., Ng V., Chu Y. W., Perez E. A., *J Clin Oncol* 2013, **31**, 1157–1163.
41. Kratz F., *J Control Release* 2008, **132**, 171–183.
42. Noguchi Y., Wu J., Duncan R., Strohalm J., Ulbrich K., Akaike T., Maeda H., *Jpn J Cancer Res* 1998, **89**, 307–314.
43. Di Stefano G., Kratz F., Lanza M., Fiume L., *Dig Liver Dis* 2003, **35**, 428–433.
44. Desai N., Trieu V., Yao Z., Louie L., Ci S., Yang A., Tao C., De T., Beals B., Dykes D., Noker P., Yao R., Labao E., Hawkins M., Soon-Shiong P., *Clin Cancer Res* 2006, **12**, 1317–1324.
45. Manikwar P., Tejo B. A., Shinogle H., Moore D. S., Zimmerman T., Blanco F., Siahaan T. J., *Theranostics* 2011, **1**, 277–289.
46. Manikwar P., Buyuktimkin B., Kiptoo P., Badawi A. H., Galeva N. A., Williams T. D., Siahaan T. J., *Bioconjug Chem* 2012, **23**, 509–517.
47. Buyuktimkin B., Manikwar P., Kiptoo P. K., Badawi A. H., Stewart J. M., Jr., Siahaan T. J., *Mol Pharm* 2013, **10**, 297–306.
48. Zhou P., Sun X., Zhang Z., *Acta Pharma Sin B* 2014, **4**, 37–42.
49. Zheng Q., Gong T., Sun X., Zhang Z. R., *Arch Pharm Res* 2006, **29**, 1164–1170.
50. Zheng Q., Ye L., Gao R., Han J., Xiong M., Zhao D., Gong T., Zhang Z., *Biomed Chromatogr* 2007, **21**, 724–729.
51. Manikwar P., Kiptoo P., Badawi A. H., Buyuktimkin B., Siahaan T. J., *Med Res Rev* 2012, **32**, 727–764.

52. Mezo A. R., McDonnell K. A., Low S. C., Song J., Reidy T. J., Lu Q., Amari J. V., Hoehn T., Peters R. T., Dumont J., Bitonti A. J., *Bioconjug Chem* 2012, **23**, 518–526.
53. Shimamoto G., Gegg C., Boone T., Queva C., *MAbs* 2012, **4**, 586–591.
54. Wu B., Sun Y. N., *J Pharm Sci* 2014, **103**, 53–64.
55. Molineux G., Newland A., *Br J Haematol* 2010, **150**, 9–20.
56. Cwirla S. E., Balasubramanian P., Duffin D. J., Wagstrom C. R., Gates C. M., Singer S. C., Davis A. M., Tansik R. L., Mattheakis L. C., Boytos C. M., Schatz P. J., Baccanari D. P., Wrighton N. C., Barrett R. W., Dower W. J., *Science* 1997, **276**, 1696–1699.
57. Hursey M., Newton D. L., Hansen H. J., Ruby D., Goldenberg D. M., Rybak S. M., *Leuk Lymphoma* 2002, **43**, 953–959.
58. Kreitman R. J., *Curr Pharm Des* 2009, **15**, 2652–2664.
59. Pastan I., Hassan R., FitzGerald D. J., Kreitman R. J., *Annu Rev Med* 2007, **58**, 221–237.
60. Kaur S., Venktaraman G., Jain M., Senapati S., Garg P. K., Batra S. K., *Cancer Lett* 2012, **315**, 97–111.
61. Koyama Y., Hama Y., Urano Y., Nguyen D. M., Choyke P. L., Kobayashi H., *Clin Cancer Res* 2007, **13**, 2936–2945.
62. Bremer C., Bredow S., Mahmood U., Weissleder R., Tung C. H., *Radiology* 2001, **221**, 523–529.
63. Ogawa M., Kosaka N., Longmire M. R., Urano Y., Choyke P. L., Kobayashi H., *Mol Pharm* 2009, **6**, 386–395.
64. Ogawa M., Kosaka N., Choyke P. L., Kobayashi H., *Cancer Res* 2009, **69**, 1268–1272.
65. Sato A. K., Viswanathan M., Kent R. B., Wood C. R., *Curr Opin Biotechnol* 2006, **17**, 638–642.
66. Werle M., Bernkop-Schnurch A., *Amino Acids* 2006, **30**, 351–367.
67. Lindgren M., Hallbrink M., Prochiantz A., Langel U., *Trends Pharmacol Sci* 2000, **21**, 99–103.
68. Lindgren M., Langel U., *Methods Mol Biol* 2011, **683**, 3–19.
69. Lindgren M. E., Hallbrink M. M., Elmquist A. M., Langel U., *Biochem J* 2004, **377**, 69–76.
70. Nakase I., Konishi Y., Ueda M., Saji H., Futaki S., *J Control Release* 2012, **159**, 181–188.
71. Holm T., Johansson H., Lundberg P., Pooga M., Lindgren M., Langel U., *Nat Protoc* 2006, **1**, 1001–1005.
72. Wang F., Wang Y., Zhang X., Zhang W., Guo S., Jin F., *J Control Release* 2014, **174**, 126–136.
73. Vives E., Brodin P., Lebleu B., *J Biol Chem* 1997, **272**, 16010–16017.
74. Bechara C., Sagan S., *FEBS Lett* 2013, **587**, 1693–1702.
75. Magzoub M., Graslund A., *Q Rev Biophys* 2004, **37**, 147–195.
76. Wagstaff K. M., Jans D. A., *Curr Med Chem* 2006, **13**, 1371–1387.
77. Nasrollahi S. A., Taghibiglou C., Azizi E., Farboud E. S., *Chem Biol Drug Des* 2012, **80**, 639–646.
78. Miyaji Y., Walter S., Chen L., Kurihara A., Ishizuka T., Saito M., Kawai K., Okazaki O., *Drug Metab Dispos* 2011, **39**, 1946–1953.

79. Dunehoo A. L., Anderson M., Majumdar S., Kobayashi N., Berkland C., Siahaan T. J., *J Pharm Sci* 2006, **95**, 1856–1872.
80. Lee S. J., Benveniste E. N., *J Neuroimmunol* 1999, **98**, 77–88.
81. Seidel M. F., Keck R., Vetter H., *J Histochem Cytochem* 1997, **45**, 1247–1253.
82. Arap W., Pasqualini R., Ruoslahti E., *Science* 1998, **279**, 377–380.
83. Majumdar S., Anderson M. E., Xu C. R., Yakovleva T. V., Gu L. C., Malefyt T. R., Siahaan T. J., *J Pharm Sci* 2012, **101**, 3275–3291.
84. Anderson M. E., Siahaan T. J., *Pharm Res* 2003, **20**, 1523–1532.
85. Anderson M. E., Tejo B. A., Yakovleva T., Siahaan T. J., *Chem Biol Drug Des* 2006, **68**, 20–28.
86. Zimmerman T., Oyarzabal J., Sebastian E. S., Majumdar S., Tejo B. A., Siahaan T. J., Blanco F. J., *Chem Biol Drug Des* 2007, **70**, 347–353.
87. Büyüktimkin B., Kiptoo P., Siahaan T. J., *Clin Cell Immunol* 2014, **5**, 273–281.
88. Kobayashi N., Kiptoo P., Kobayashi H., Ridwan R., Brocke S., Siahaan T. J., *Clin Immunol* 2008, **129**, 69–79.
89. Kobayashi N., Kobayashi H., Gu L., Malefyt T., Siahaan T. J., *J Pharmacol Exp Ther* 2007, **322**, 879–886.
90. Murray J. S., Oney S., Page J. E., Kratochvil-Stava A., Hu Y., Makagiansar I. T., Brown J. C., Kobayashi N., Siahaan T. J., *Chem Biol Drug Des* 2007, **70**, 227–236.
91. Büyüktimkin B., Kiptoo P., Siahaan T. J., *Clin Cell Immunol* **5**, 273–282.
92. Badawi A. H., Siahaan T. J., *Clin Immunol* 2012, **144**, 127–138.
93. Badawi A. H., Kiptoo P., Wang W. T., Choi I. Y., Lee P., Vines C. M., Siahaan T. J., *Neuropharmacology* 2012, **62**, 1874–1881.
94. Buyuktimkin B., Wang Q., Kiptoo P., Stewart J. M., Berkland C., Siahaan T. J., *Mol Pharm* 2012, **9**, 979–985.
95. Gorovits B., Alley S. C., Bilic S., Booth B., Kaur S., Oldfield P., Purushothama S., Rao C., Shord S., Siguenza P., *Bioanalysis* 2013, **5**, 997–1006.
96. Wakankar A., Chen Y., Gokarn Y., Jacobson F. S., *MAbs* 2011, **3**, 161–172.
97. Alley S. C., Anderson K. E., *Curr Opin Chem Biol* 2013, **17**, 406–411.
98. Lewis Phillips G. D., Li G., Dugger D. L., Crocker L. M., Parsons K. L., Mai E., Blattler W. A., Lambert J. M., Chari R. V., Lutz R. J., Wong W. L., Jacobson F. S., Koeppen H., Schwall R. H., Kenkare-Mitra S. R., Spencer S. D., Sliwkowski M. X., *Cancer Res* 2008, **68**, 9280–9290.
99. DiJoseph J. F., Armellino D. C., Boghaert E. R., Khandke K., Dougher M. M., Sridharan L., Kunz A., Hamann P. R., Gorovits B., Udata C., Moran J. K., Popplewell A. G., Stephens S., Frost P., Damle N. K., *Blood* 2004, **103**, 1807–1814.
100. Xiao H., Chatterjee A., Choi S. H., Bajjuri K. M., Sinha S. C., Schultz P. G., *Angew Chem Int Ed Engl* 2013, **52**, 14080–14083.
101. Wals K., Ovaa H., *Front Chem* 2014, **2**, 15.
102. Xu K., Liu L., Saad O. M., Baudys J., Williams L., Leipold D., Shen B., Raab H., Junutula J. R., Kim A., Kaur S., *Anal Biochem* 2011, **412**, 56–66.
103. Junutula J. R., Bhakta S., Raab H., Ervin K. E., Eigenbrot C., Vandlen R., Scheller R. H., Lowman H. B., *J Immunol Methods* 2008, **332**, 41–52.
104. Wakankar A. A., Feeney M. B., Rivera J., Chen Y., Kim M., Sharma V. K., Wang Y. J., *Bioconjug Chem* 2010, **21**, 1588–1595.

105. Wagner-Rousset E., Janin-Bussat M. C., Colas O., Excoffier M., Ayoub D., Haeuw J. F., Rilatt I., Perez M., Corvaia N., Beck A., *MAbs* 2014, **6**, 173–184.
106. Wang L., Amphlett G., Blattler W. A., Lambert J. M., Zhang W., *Protein Sci* 2005, **14**, 2436–2446.
107. Manikwar P., Zimmerman T., Blanco F. J., Williams T. D., Siahaan T. J., *Bioconjug Chem* 2011, **22**, 1330–1336.

21

DRUG DELIVERY TO THE LYMPHATIC SYSTEM

QIUHONG YANG AND LAIRD FORREST
Department of Pharmaceutical Chemistry, The University of Kansas, Lawrence, KS, USA

21.1 INTRODUCTION

The lymphatic system is a subsystem of the circulatory system central to maintaining tissue homeostasis and immunofunction [1, 2], including removing excess extravascular fluid, presenting to the immune system foreign bodies and pathogens, maturation of lymphocytes, and absorption of lipid-soluble nutrients from the digestive tract. The lymphatic system consists of an extensive network of lymphatic vessels, lymph nodes, lymph, lymphatic organs (e.g., spleen and thymus), and lymphoid tissues (e.g., tonsils and Peyer's patches).

The lymphatic system has distinctive physiological functions in lipid absorption and microparticulate uptake, which could be harnessed as an alternative route for the delivery of pharmaceuticals to enhance their bioavailability and efficacy. Drug agents that would benefit from lymphatic concentration, such as vaccines and anticancer chemotherapeutics, could be engineered for preferential uptake into the lymphatic system to improve efficacy with reduced systemic distribution. In addition, with the increase in resolution and specificity of imaging technologies, there is the potential to monitor tumor cells that migrate through the lymphatic system and accumulate in the lymph nodes, which could detect very early micrometastatic disease. Two approaches are generally proposed to modify drug molecules for targeted accumulation of the lymphatic system. One approach is to chemically tailor the structures of the drug molecules and synthesize "prodrugs" with enhanced

Drug Delivery: Principles and Applications, Second Edition. Edited by Binghe Wang, Longqin Hu, and Teruna J. Siahaan.

lipophilicity, primarily used to improve gastric lymphatic absorption [3, 4], and the other is to develop particulate drug carriers to encapsulate drug molecules. With recent advances in materials science and technology, a variety of models of drug delivery to the lymphatic system have been reported in the literature, including drug–polymer conjugates [5–10], drug-loaded emulsions [11–19], liposomes [20–44], solid lipid nanoparticles (SLNs) [45–51], nanostructured lipid carriers [52–58], polymeric micelles [59–62], polymeric nanoparticles [63–69] and microparticles [70–75], carbon nanotubes [76, 77], and nanocapsules [78]. In this chapter, we first give a brief description of the lymphatic system, followed by a characterization of physicochemical properties of typical drug carriers and their influence on lymphatic uptake and transport, and then review the recent development of drug carriers for the lymphatic system delivery with a special focus on nano-scaled particles and their clinical applications in chemotherapy (Table 21.1). In the end of this chapter, we will compare alternative administration routes for lymphatic targeting.

21.2 ANATOMY AND PHYSIOLOGY OF THE LYMPHATIC SYSTEM

21.2.1 Lymph

The blood vessels and capillaries are not fluid-tight; a portion of blood volume will be pushed into the surrounding tissues, primarily through the capillary walls, due to the hydrostatic pressure differential. The plasma osmotic pressure is greater than the interstitial space, but the hydrostatic pressure is sufficient to cause a net outward flow of about 10% of the total volume passing through the capillary network. This excess fluid forms the interstitial fluid. These interstitial fluids bath the living tissue cells and then diffuse into the small, fenestrated, and blind-ended lymphatic capillaries. The osmotic pressure of fluids in the lymphatic venules is higher than the surrounding interstitial tissues, but the hydrostatic pressure difference drives fluids into the venules [91]. Once the intestinal fluids transverse the lymphatic capillaries, they are called "lymph," a clear interstitial fluid that has comparable composition to blood plasma but with a larger number of white blood cells, mostly lymphocytes, and approximately 50% reduced protein content [92, 93].

21.2.2 Lymphatic Vessels

Lymphatic vessels are a family of four distinct vessels, including lymphatic capillaries, lymphatic collecting vessels, lymphatic trunks, and lymphatic ducts. The lymphatic capillaries are the smallest lymphatic vessels, and they are widely distributed throughout the body except for the avascular tissues and the central nervous system. They consist of a single layer of nonfenestrated endothelial cells that are connected to the extracellular matrix by elastic anchoring filaments to form the porous capillary walls with numerous clefts [94, 95]. Being in close proximity to the blood capillaries, the lymphatic capillaries primarily function by continuously removing interstitial fluids. As the "cleft-like" intercellular junctions (Fig. 21.1) along the surface of the

TABLE 21.1 Examples of Anticancer Drug-Loaded Nanoparticles That Have Demonstrated Enhanced Chemotherapeutic Efficacies in Tumor Models

Formulations	Encapsulated Drugs	Tumor Models	References
Drug-polymer conjugates	Interferon α2	Breast	[5]
	Doxorubicin	Breast	[6]
	Cisplatin	Breast, head, and neck	[79, 80]
Emulsions	Paclitaxel	Breast	[17]
	Pirarubicin	Gastric	[18]
Liposomes	Paclitaxel	Ovarian, melanoma	[20, 81]
	Doxorubicin	Breast, lung, melanoma, gastric, ovarian, KB oral carcinoma, and Kaposi's sarcoma	[21–23, 26, 29, 82–86]
	Muramyl tripeptide phosphoethanolamine	Ovarian	[30]
	DNA	Ovarian	[31]
	Melphalan	Mammary adenocarcinoma	[87]
	Cisplatin	Colon and lung	[32, 88]
	Combination of vitamin E analogue and 9-nitro camptothecin	Lung metastasis	[89]
	Combination of vitamin E analogue and paclitaxel	Lung metastasis	[90]
Solid lipid nanoparticles	Paclitaxel	Non-small-cell lung	[51]
Nanostructured lipid carriers	Combination of Celecoxib with doxetaxel	Non-small-cell lung	[58]
	Isoliquiritigenin	Liver and sarcoma	[54]
	Camptothecin	Melanoma	[56]
Polymeric micelles	Cisplatin	Tongue	[59]
	Vinorelbine	Breast	[60]
	(1,2 Diaminocyclohexane) platinum (II)	Gastric	[61]
Poly (lactic-*co*-glycolic) nanoparticles	Paclitaxel	Ovarian	[67]
Poly (lactic-*co*-glycolic)/poly (L-lactic acid) microspheres	Paclitaxel	Intraperitoneal carcinomatosis, subcutaneous carcinoma, and lung	[70, 72, 73]
Chitosan nanoparticles	Camptothecin	Liver and soft tissues	[64]
Carbon nanotubes	Gemcitabine	Pancreatic	[76]

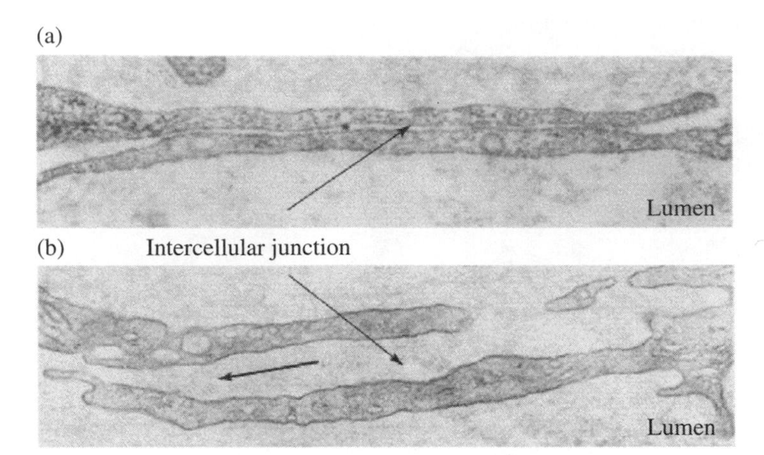

FIGURE 21.1 Microscopic view of the lymphatic capillary illustrating a closed intercellular junction (a) and an open intercellular junction (b) allowing luminal access to molecules along the pathway indicated by the arrow. McLennan et al. [96], pp. 89–96. Reproduced with permission of Elsevier.

lymphatic capillaries are very thin, with an estimated diameter between 10 and 60 μm [97], they facilitate the absorption of lymph fluid that contains protein molecules, cellular debris, large foreign particles, and pathogens from the interstitial space. After the absorption, the lymph fluid drains from the initial lymphatic capillaries into the collecting vessels, followed by filtration through lymph nodes and then is transported to the lymphatic trunks, driven by the combined effect of skeletal muscle action, changes in thoracic pressure and pulsation of nearby arteries, as shown in Figure 21.2. During such one-way flow, the semilunar valves in lymphatic vessels serve to spare lymph from flowing backward under the circumstances of skeleton muscle relaxation. The lymphatic trunks eventually converge together to form two major lymphatic vessels, the right and left (thoracic) lymphatic ducts, from which lymph fluid reenters the circulatory system. The right lymphatic duct drains the upper right quadrant of the body, while the left lymphatic duct drains the lymph in the remaining three quarters of the body. These two ducts connect to the right and left subclavian veins, respectively, which return blood from the superior vena cava to the right atrium [98].

Lacteals are a type of specialized lymphatic capillaries located in the central section of small intestine villi that absorb fats or fatty acids, cholesterol, and soluble vitamins from the intestinal tract. Lacteals join lymphatic capillaries from the large intestine and that of mucosa and submucosa to form collecting lymphatic vessels for the transportation of nutrients through the lymphatic duct into the subclavian vein.

21.2.3 Lymph Nodes

Lymph nodes, as shown at the upper right side of Figure 21.2, are small, oval-shaped lymphatic organs that contain macrophages and lymphocytes to filter foreign particulates, pathogens, viruses, and bacteria carried in the lymph fluid, either

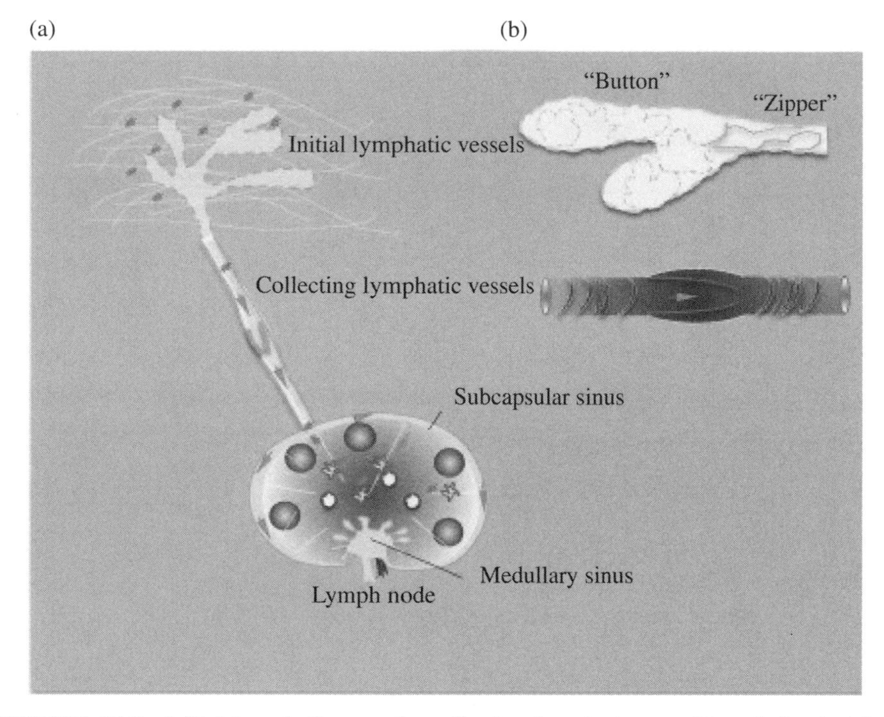

FIGURE 21.2 Initial lymphatic vessels, collecting lymphatic vessels, and the draining lymph node. (a) Schematic figure of lymphatic system, including initial lymphatic vessels, collecting lymphatic vessels, and the draining lymph node. (b) Initial lymphatic vessels and collecting lymphatic vessels. The initial lymphatic vessels have discontinuous cell junctions (Button pattern). Collecting lymphatics have continuous junction molecules and basement membrane and organized smooth muscle cell coverage. Organized smooth muscle coverage in collecting lymphatics allows phasic lymphatic contractions to propel lymph through lymphatic vessels. Liao and Weid [2], pp. 83–89. Reproduced with permission of Elsevier.

by phagocytosis or via the immune response [92]. Except for the central nervous system, the lymph nodes are encapsulated in most connective tissues and are widely distributed in the body along the lymph vessels routes, especially in the neck cervical, axillary, groin inguinal, vertebral column, and intestinal mesenteric.

The clinical significance of lymph nodes in cancer metastasis has been demonstrated in multiple types of human cancers, including breast [99–101], colon [102–104], prostate [105–107], bladder [108], head, and neck cancers [109–112]. These studies together demonstrate that regional lymph node metastasis is strongly associated with the diminished survival in cancer patients. In the process of tissue invasion by tumors (Fig. 21.3), malignant cells first secrete cytokines to induce lymphangiogenesis—the formation of new lymphatic vessels—around the tumor periphery. After tumor cells invade the adjacent lymphatic capillaries through the clefts formed by lymphatic endothelial cells, they travel with the lymph fluid into the first draining lymph node, prior to the systemic dissemination into distant tissues and

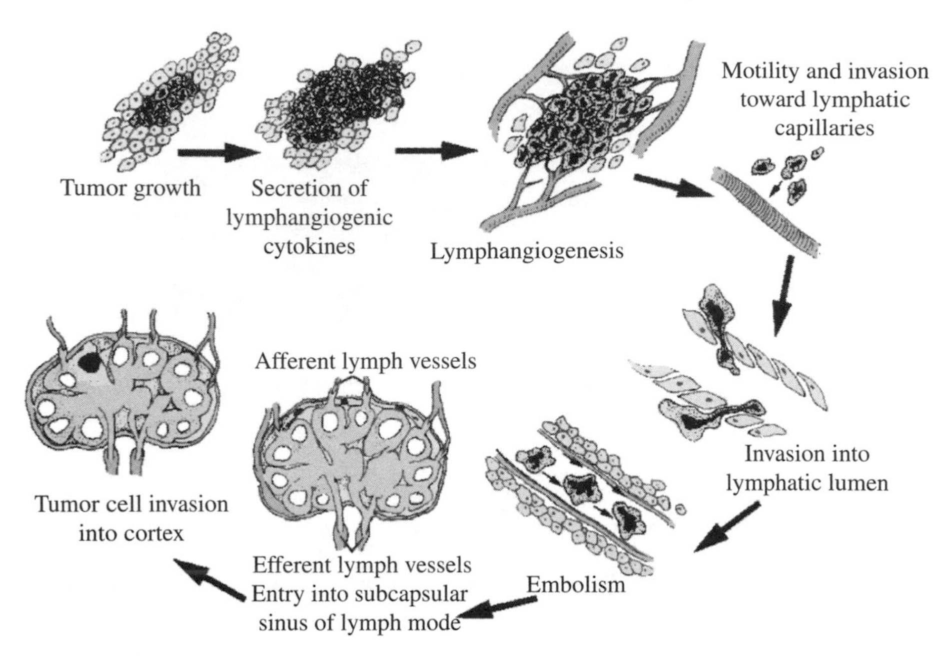

FIGURE 21.3 Tumor-associated lymphatic vessels serve as a route for lymph node metastasis. Nathanson [113], pp. 413–423. Reproduced with permission of John Wiley & Sons. (*See insert for color representation of the figure.*)

organs. A strategy to address lymphatic metastasis is loco-regional chemotherapy, which delivers drugs into the lymph nodes and lymphatic vessels draining the tumor. Doing so could increase intranodal drug concentration [114, 115] and result in superior therapeutic effects with less morbidity.

21.2.4 Lymph Organs

The spleen is the largest lymphoid organ in adults and is located in the upper left quadrant of the abdomen. The spleen consists of white pulp tissues (lymphocytes and macrophages) and red pulp tissues (large venous sinuses). Its primary function is to eliminate both foreign substances and damaged or old erythrocytes from the blood. Also in the case of hemorrhage, the spleen acts as an oxygen-rich blood reservoir to compensate for a loss of red blood cells. The thymus is another lymphatic organ, which is located in the lower neck and extends in the mediastinum over the heart. The thymus is large and highly active in infants and atrophies throughout childhood until it is mostly replaced by adipose tissue in young adults. The thymus is responsible for the maturation and selection of self-tolerate immature lymphocytes known as T-lymphocytes during the neonate period and early childhood, although the thymus is still active to a limited extent throughout life. T-lymphocytes later leave the thymus, travel through the lymphatic and the blood systems, and migrate to the other lymphatic organs.

21.3 INFLUENCE OF PHYSICOCHEMICAL CHARACTERISTICS OF DRUG CARRIERS ON LYMPHATIC UPTAKE AND TRANSPORT

Multiple factors, including the physiological state of the interstitial space, immunological state, the lymph flow rate, the administration routes, the distance between the injection site and the lymph nodes, and the physicochemical parameters of the drug carrier/particles, may affect the effectiveness of lymphatic drainage and transport of a drug carrier/particle. However, numerous studies have established that the physicochemical characteristics of a drug carrier play predominant roles in determining its efficiency, and these characteristics lie in three key aspects: particle size, surface charge, and hydrophobicity [14, 27, 65, 116–118].

21.3.1 Size

Particle size is one of the strongest determinants of tissue, lymphatic, and circulatory distribution, particularly after subcutaneous (SC) administration. Lymphatic uptake and node accumulation are greatest for particle sizes between 10 and 80 nm (Fig. 21.4a and b) [14, 65, 68, 116, 118]. Larger particles over 100 nm cannot easily traverse through the interstitium, and most of them remain in the injection site until phagosomal clearance (Fig. 21.4c). At the other end of spectra, particles smaller than

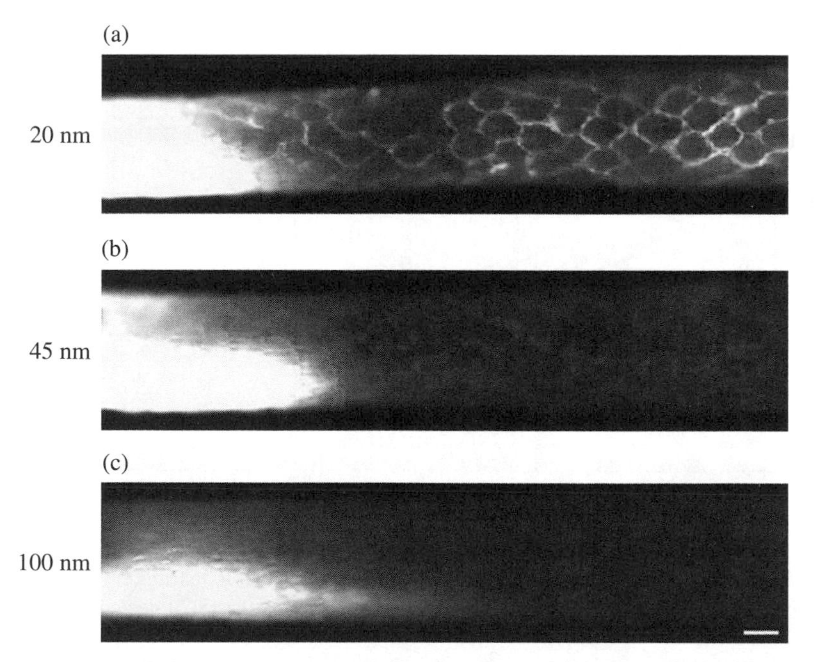

FIGURE 21.4 Comparison of nanoparticle uptake into the initial lymphatics. Fluorescence microlymphangiography of the lymphatic capillary network in mouse tail skin after 90 minutes infusion with fluorescent nanoparticles of: (a) 20, (b) 45 and (c) 100 nm diameter. Reddy et al. [119], pp. 26–34. Reproduced with permission of Elsevier.

10 nm are more likely to diffuse into the blood capillaries and subsequently enter the systemic circulation. In rats administered with liposomes subcutaneously, there is an inverse correlation between liposome size and lymphatic uptake from the injection site over a range of 40–400 nm [120] (Fig. 21.5a). Mass diffusion plays a role in the lymphatic uptake; smaller nanoparticles more readily traverse the interstitium after SC injection and are absorbed by the initial lymphatic capillaries. These smaller particles are less likely to be phagocytized by the macrophages within nodal sinuses, and they can consequently pass through the nodes instead of being filtered out. Larger liposomes are filtered more efficiently into the nodal sinuses, leading to greater accumulation in the lymph nodes (Fig. 21.5b). This observed "trade-off" between lymphatic drainage versus lymph node accumulation is also observed after intraperitoneal (IP) injection [33].

Antigen-presenting cells (APCs), primarily dendritic cells (DCs) that present in the lymph nodes, also play active roles in particle phagocytosis [119]. After intradermal (ID) injection in mice of DC-targeting polymeric nanoparticles, fluorescence microlymphangiography (Fig. 21.6) of the lymph nodes 120 hours post-injection shows that 20-nm particles not only exhibit the most rapid uptake from the injection site (Fig. 21.4a) but also achieve the highest nodal accumulation over the entire time of microlymphangiography due to a greater APC internalization of the smaller particles. The optimal size range to achieve both efficient lymphatic uptake and lymph node retention is thought to lie between 20 and 45 nm.

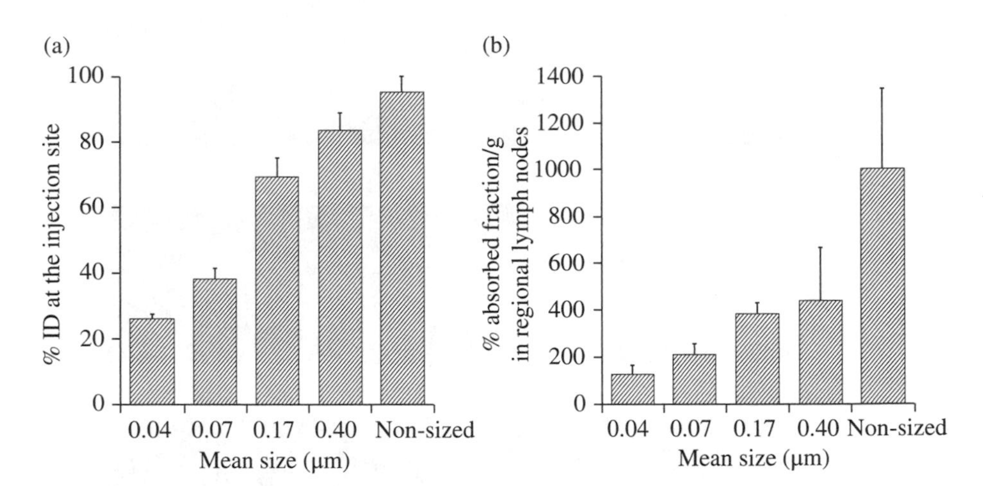

FIGURE 21.5 Influence of size on the pharmacokinetics and biodistribution of SC-administered liposomes. A single dose of liposomes of varying sizes was SC injected into the dorsal side of the foot of rats. Levels of radioactivity were determined 52 hours post injection. (a) Percentage of injected dose recovered from the SC injection site. (b) Percentage of the lymphatically absorbed fraction per gram recovered from the regional lymph nodes. Oussoren et al. [120], pp. 261–272. Reproduced with permission of Elsevier.

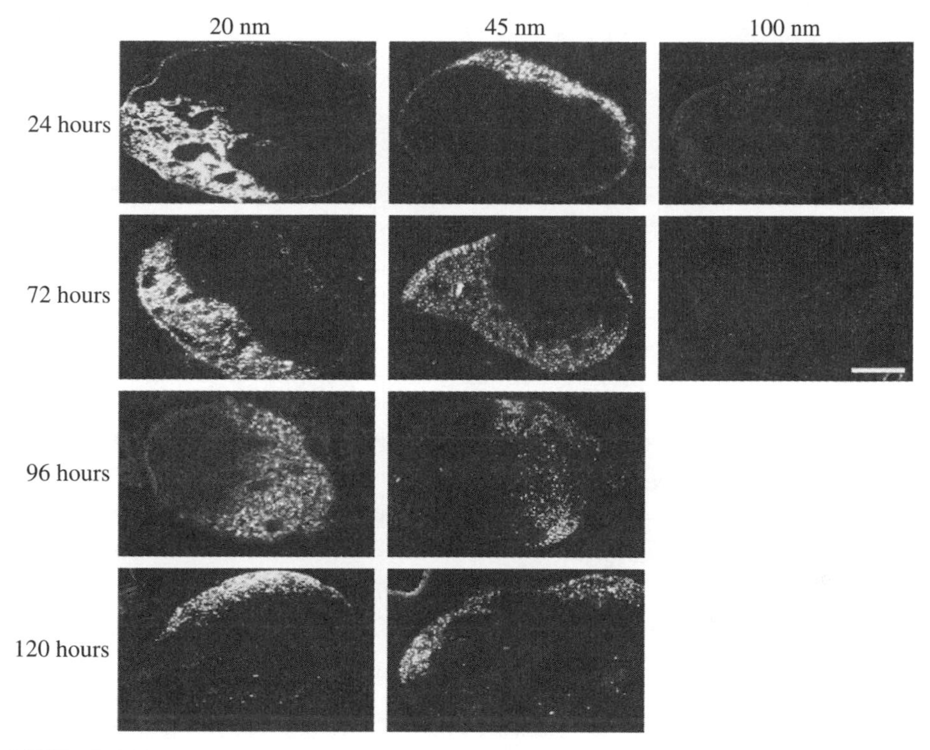

FIGURE 21.6 Lymph node retention of nanoparticles. Shown are sections from draining lymph nodes following the interstitial injections into the mouse tail with 20, 45, and 100 nm PPS nanoparticles. Nanoparticles were present at all time points for 20 and 45 nm nanoparticles, but 100 nm particles were not seen in the lymph nodes. Reddy et al. [119], pp. 26–34. Reproduced with permission of Elsevier.

21.3.2 Surface Charge

Surface charge and chemical function groups also affect the particle uptake, but these effects are less uniform across particle types than size. For small uniformal liposomes of phosphatidylcholine–cholesterol given SC, the liposome localization in the lymph nodes followed the order of negative > positive > neutral (Fig. 21.7) [121]. In drug carrier systems such as dendrimers [122] and poly-(lactic-*co*-glycolic acid) (PLGA)-based nanoparticles [65], negative charges on the carrier surfaces have been related to faster lymphatic migration and greater nodal retention than the positive or neutral drug carriers. Such a trend could be attributed to a slightly negatively charged environment in the interstitial matrix caused by glycosaminoglycans (largely hyaluronan) [123]. Anionic drug carriers administered by SC injection first enter the interstitial region underlying the skin dermis, and then they are propelled from the interstitium into the lymphatic capillaries because they do not interact strongly with the glycosaminoglycans, due to the charge–charge

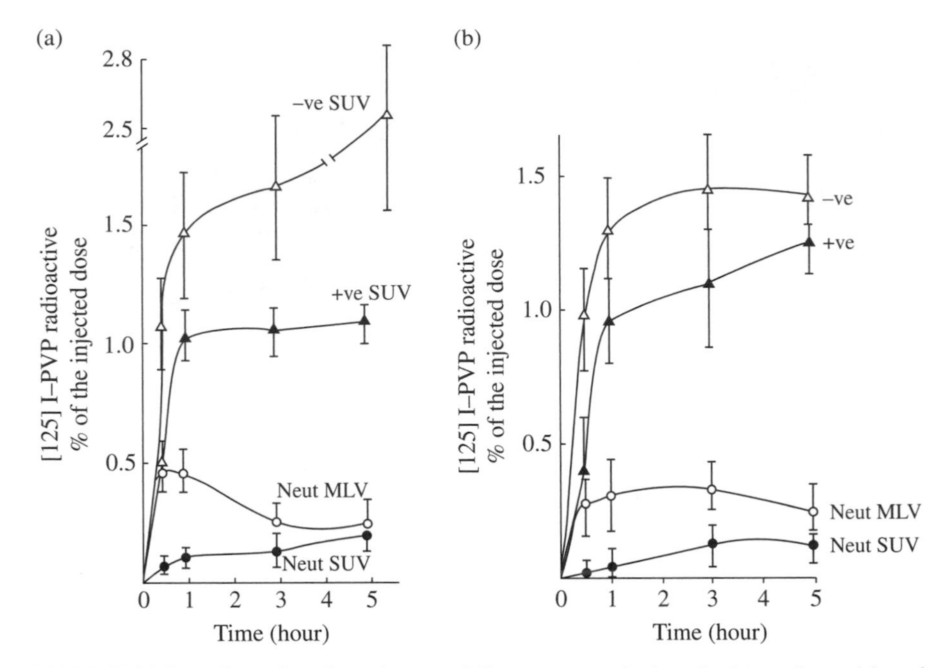

FIGURE 21.7 Effect of surface charges of liposomes on the lymphatic uptake and lymph nodal retention. Radiolabeled liposomes with negative, positive, or neutral charges were SC injected into rats footpad, and the percentage of the injected dose in both (a) primary lymph nodes and (b) secondary regional lymph nodes was calculated. Patel et al. [121], pp. 76–86. Reproduced with permission of Elsevier.

repulsion. Conversely, the electrostatic attraction between the cationic particles and the negatively charged matrix hinders the particles from traversing interstitium. The negative surface charges on particles have also been speculated to trigger the phagocytosis in the lymph nodes, which results in the enhanced regional lymph node retention.

An opposite relation between surface charge and lymphatic uptake has been observed in a pharmacokinetics study of differently charged liposomes in rats after intramuscular (IM) injection [42]. When neutral methotrexate (MTX)-encapsulated liposomes are modified via incorporating stearylamine or dicetyl phosphate to render positive or negative charges, respectively, on the particle surfaces, the positively charged MTX-liposomes are found to yield the highest nodal accumulation at 24 hours post injection, followed by the negatively and neutrally charged MTX-liposomes. However, zidovudine (ZDV)-loaded liposomes showed a different distribution after SC administration [35]. Negatively charged liposomes showed greater lymphatic uptake compared with the positively charged liposomes, in line with other liposomal systems [65, 121, 122]. The difference in the nodal drug accumulation of the MTX-liposomes and ZDV-liposomes has not been clearly elucidated yet, but it could be in part due to the different injection routes.

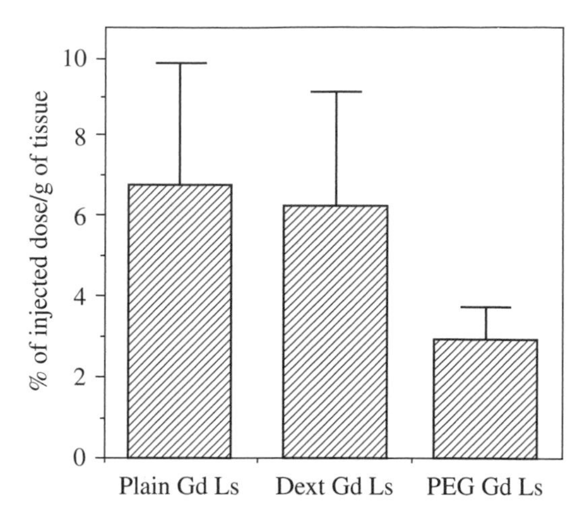

FIGURE 21.8 The axillary lymph nodal accumulation of plain liposomes (plain Gd Ls), dextran-modified liposomes (Dext Gd Ls), and PEG-modified liposomes (PEG Gd Ls). Trubetskoy et al. [36], pp. 31–37. Reproduced with permission of Elsevier.

21.3.3 Hydrophobicity

Opsonin molecules are able to nonspecifically bind foreign bodies and signal phagocytosis, and there is a known preference for hydrophobic rather than hydrophilic surfaces [124]. Therefore, the modification of particles by adding hydrophilic coating polymer, such as dextran, polyethylene glycol (PEG), poloxamers, and poloxamines block copolymers, could be used to prevent macrophages recognition and avoid nonspecific sequestration by the reticuloendothelial system (RES) after intravenous (IV) injection, and thus lead to prolonged systemic circulation. However, the increased stealth properties conferred by surface modification reduce internalization of the particles by phagocytic cells in lymph nodes, which in turn results in less nodal accumulation and faster lymph drainage. ^{111}In radiolabeled Gd liposomes [36] with a 5-kDa PEG coating accumulate 50% less in the axillary lymph node than the plain or 6-kDa dextran-modified liposomes (Fig. 21.8) 2 hours after the SC injection. Various approaches [27, 28, 65, 125–127] have been reported to tailor the extent of surface hydrophobicity/hydrophilicity via optimizing the amount [27, 28, 65, 127] or chain length [125, 126] of the PEG polymer. Dioleoyltrimethylammoniumpropanes (DOTAP) cationic liposomes modified with 1 or 5 mol% DSPE-PEG2000 drain much faster in the lymph nodes than unmodified liposomes [27] (Fig. 21.9).

21.4 CARRIERS FOR LYMPHATIC DRUG DELIVERY

From the earlier discussion it is clear that to achieve targeted drug lymphatic delivery a desired carrier should possess following features:

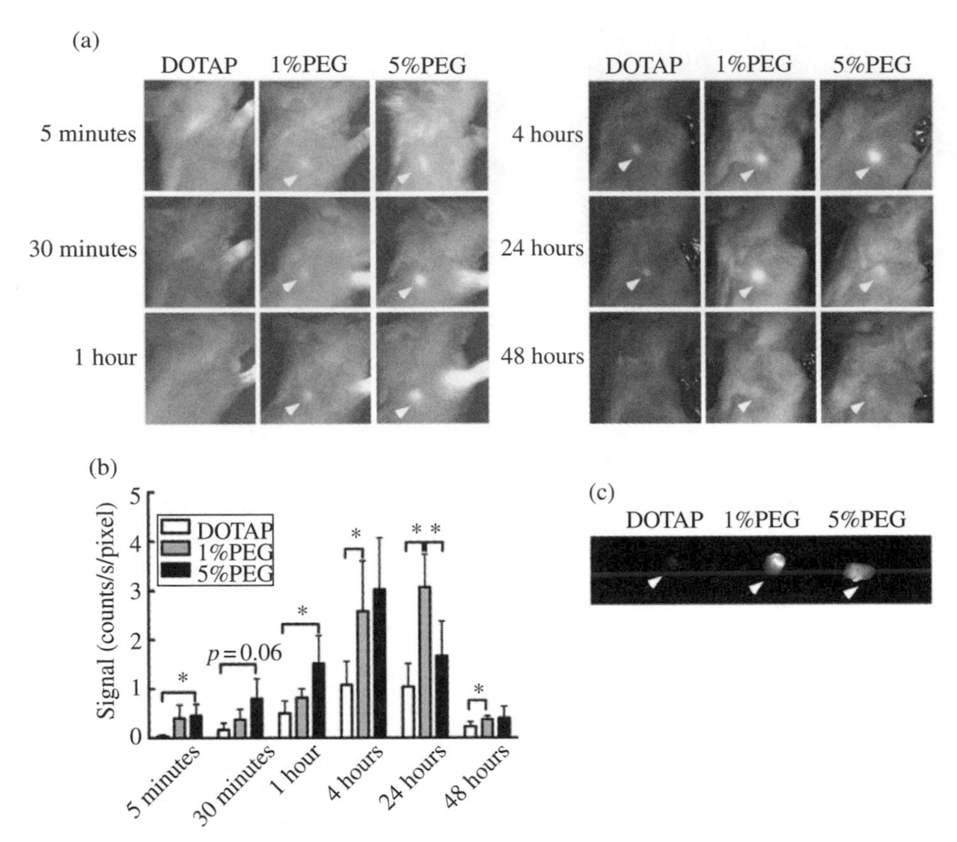

FIGURE 21.9 The lymphatic trafficking of fluorescent liposomes after SC injection into forepaws of Balb/c mice. (a) The accumulation of different liposomes in draining lymph nodes (LNs) from 5 minutes to 48 hours post injection. (b) The fluorescent intensity of LN images measured with the maestro software. Data are shown as mean ± SE ($n=5$), $^*p<0.05$; $^{**}p<0.01$. (c) The fluorescent image of draining LNs at 24 hours. White arrows indicate draining LNs. Zhuang et al. [27], pp. 135–142. Reproduced with permission of Elsevier. (*See insert for color representation of the figure.*)

- Low or nontoxicity to healthy tissues and organs;
- High colloidal, chemical, and biological stability;
- Rapid drainage into the lymphatic capillaries from the injection site;
- High lymph nodal retention;
- Efficient delivery and release of therapeutic agents to the tissues or organs of interest.

In the process of creating optimal drug carriers, nanoscaled particles, especially those of being lipid- and polymer-based, have distinct advantages over alternative models. These advantages, such as tunable size control and feasible surface modifications with targeting molecules, have attracted intense interests.

21.4.1 Liposomes

A liposome is a spherical self-enclosed vesicle with a lipid bilayer comprising amphiphilic monomers (e.g., phosphalipids). The average diameter of small unilamellar liposomes ranges from 25 to 100 nm [128]. As shown in Figure 21.10, liposomes possess a structure with the long hydrocarbon chains of phospholipids lining up against one another to form a membrane and the charged hydrophilic groups on both sides orienting toward aqueous medium. Such a unique structure makes the liposomes especially suitable to encapsulate water-soluble molecules in the hydrophilic core and to trap lipophilic molecules within the bilayer. The structures of liposomes impart high colloidal stability, rendering the nanoparticles less sensitive to external environment change than micellular structures or emulsions. Additionally, polymer coatings on the surface can further sterically stabilize liposomes for prolonged blood retention.

Drug-encapsulated liposomes are typically prepared by a four-step film-rehydration method shown in Figure 21.11, which consists of (i) dissolving the drug and phospholipid (e.g., phosphatidylcholine or phosphatidylethanolamine) or cholesterol in an organic solvent such as chloroform or methanol, (ii) removal of the solvent by rotary evaporation under reduced pressure to form a uniform and thin film, (iii) hydration of the film in an aqueous medium, and (iv) size reduction via sonication or extrusion.

Liposomes are generally considered as a promising drug carrier for lymphatic delivery, attributed to their biocompatibility, tunable size, and dual-loading capacity for both hydrophilic and hydrophobic drugs. Listed in Table 21.2 [128] are a variety of liposomal formulations of anticancer drugs that either have been approved by the drug administration agencies or are currently in the advanced stages of clinical trials. For example, Doxil®, the first FDA-approved nanodrug, is a doxorubicin-encapsulated liposomal formulation coated with PEG. Doxil has been used as a first-line anticancer drug administered via IV infusion to treat a series of cancers including Kaposi's sarcoma, breast cancer, gastric cancer, ovarian cancer, and head and neck squamous cell carcinoma (HNSCC) [26, 82–85, 130].

Unmodified and polymer-coated liposomes drain into the lymphatic capillaries and then are internalized by macrophages in the regional lymph nodes and tumor cells. As most tumor cells, tumor lymphatic and tumor-associated macrophages express specific membrane-associated proteins such as receptors, membrane transporters, and adhesion molecules, conjugation of targeting motifs to the surface of liposomes could specifically target the encapsulated drugs to the tumor lymphatics via receptor-mediated interaction and consequently enhance the therapeutics efficacy. Table 21.3 lists some frequently used ligands for targeting the liposomes.

LyP-1, for example, is a cyclic nonapeptide (CGNKRTRGC) that possesses specific binding ability with the p32/gC1q receptor overexpressed in tumor cells, tumor lymphatics, and tumor-associated macrophages [131–133]. Yan et al. [21, 22] synthesized LyP-1-conjugated PEG-DSPE and prepared a lymphatic-targeting liposomal formulation of doxorubicin. The targeted doxorubicin liposomes has demonstrated superior chemotherapeutic efficacy in animal models implanted with lymphatic metastatic breast and

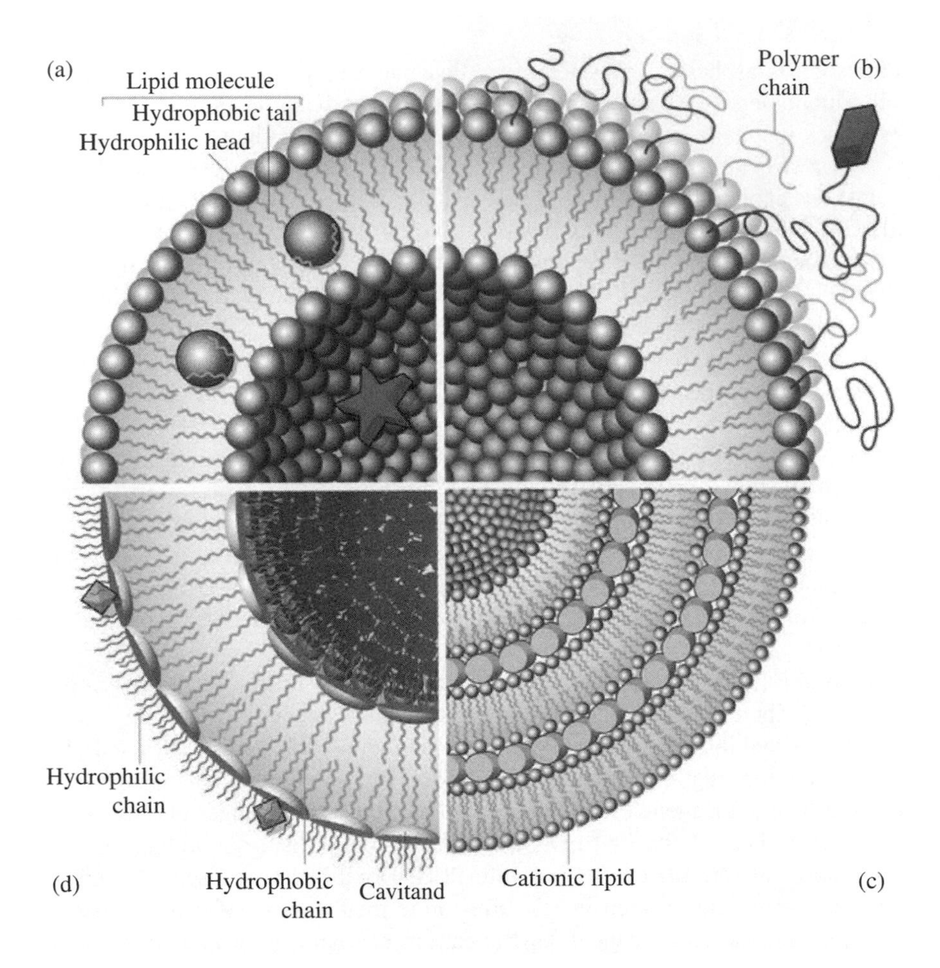

FIGURE 21.10 Typical structure of liposomes. (a) Simple liposomes are vesicles that have a shell consisting of a lipid bilayer. A liposome can trap hydrophobic guest molecules a few nanometres in diameter (spheres) within the hydrophobic bilayer and hydrophilic guests up to several hundred nanometres (star) in its larger interior. (b) In "stealth" liposomes developed for drug-delivery applications, the lipid bilayer contains a small percentage of polymer lipids. Peptides (rectangle) that target specific biological targets may also be attached to the polymers. (c) Most cationic liposome–DNA complexes have an onion-like structure, with DNA (rods) sandwiched between cationic membranes. (d) Kubitschke et al. report liposomes in which the bilayer assembles from cavitands—vase-shaped molecules—to which the authors attached hydrophobic and hydrophilic chains. The cavitands can trap Angström-sized guest compounds (diamonds) in their hydrophobic cavities. These vesicles can therefore encapsulate guest molecules of different sizes in the cavitands, the bilayer and the liposome's interior. Safinya and Kai [129], pp. 372–374. Reproduced with permission of Macmillan Publishers Ltd.

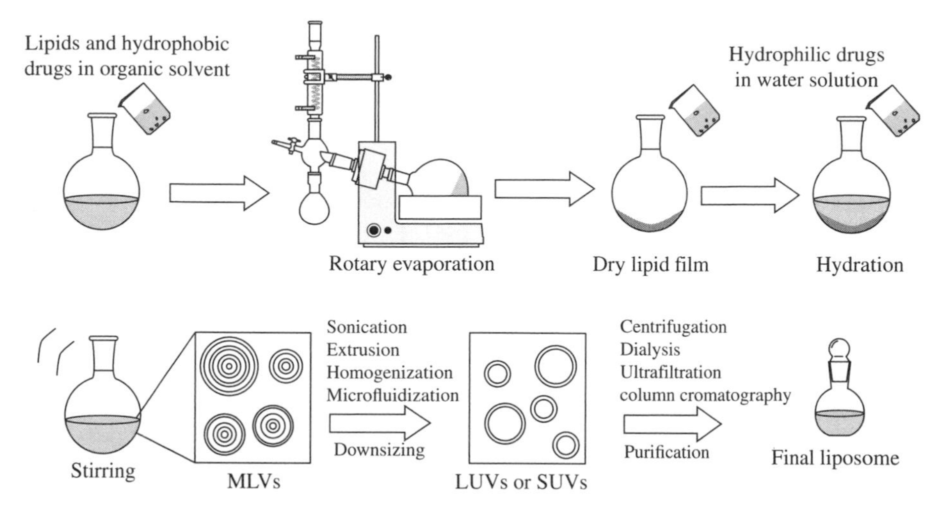

FIGURE 21.11 Representation of liposomes production by lipid hydration followed by size-reduction process. Lopes et al. [128].

lung tumors, in part by suppressing lymph node metastases and disrupting tumor lymphatics (Fig. 21.12). To evaluate the specific targeting ability of LyP-1-PEG liposomes to breast tumor lymphatics, fluorescein-encapsulated LyP-1-PEG liposomes were first subcutaneously injected into nude mice with lymph node metastasis. Two metastatic-draining lymph nodes were then excised, sectioned, and separately stained with two lymphatic vessel markers—LYVE-1 and podoplanin. Visualized by an immunofluorescence analysis shown in Figure 21.13, the targeted liposomes were colocalized with LYVE-1 and podoplanin because of the specific binding. Pharmacokinetic profiles of untargeted and LyP-1-PEG liposomes shown in Figure 21.14 indicated a significant enhancement in the uptake of the liposomes by the metastatic lymph nodes after the LyP-1 modification.

Surface modification with antigen presenting cell–specific antibodies is another approach to enhance the lymphatic delivery of liposomes, and coupling a nonspecific IgG to PEG-bearing liposomes has been found to increase its retention in both the primary and secondary lymph nodes [24]. A closer examination of the immuno-PEG-liposomes structure revealed that IgG chains orient randomly on the liposomes surface and expose its Fc region to facilitate the recognition by Fc receptors on the surface of macrophages. Immunoliposome-mediated lymphatic targeting to specific tumor cells, on the other hand, could be enhanced by attaching Fab fragments of monoclonal antibodies that are directed against specific antigen present on the cancer cells [28, 29]. Bestman-Smith et al. [28] reported an elevated accumulation of the anti-HLA-DR-coupled PEG-liposomes in lymph nodes over the plain liposomes or PEG-liposomes.

In addition to ligand-attached liposomes, cationic liposomes prepared from DOTAP and DOPC have been shown to preferentially target tumor endothelial cells and vessels [27, 31, 81, 134]. However, the mechanism underlying the selective targeting has not yet been fully elucidated.

TABLE 21.2 Approved and Emerging Liposome-Encapsulated Anticancer Drugs[a]

Product Name	Entrapped Drug	Lipid Composition	Therapeutic Indication	Status[b]
Doxil/Caelyx®	Doxorubicin	HSPC/CHOL/DSPE-PEG2000	Kaposi's sarcoma, recurrent ovarian, multiple myeloma, and metastatic breast cancer	A
Myocet®	Doxorubicin	EPC/CHOL	Metastatic breast cancer	A[c]
DaunoXome®	Daunorubicin	DSPC/CHOL	Kaposi's sarcoma	A
DepoCyt®	Cytarabine	DOPC/DPPG/CHOL/TRIOLEIN	Lymphomatous meningitis	A
SPI-077®	Cisplatin	HSPC/CHOL/DSPE-PEG2000	Ovarian cancer	P II
Lipoplatin®	Cisplatin	DPPG/SPC/CHOL/DSPE-PEG2000	Lung cancer	P III
Aroplatin®	Bis-Neodecanoate diaminocyclohexane platinum	DMPC/DMPG	Colorectal, lung, and pancreatic cancer	P II
LEP-ETU®	Paclitaxel	DOPC/CHOL/CARDIOLIPIN	Breast, lung, and ovarian cancer	P II
EndoTAG-1®	Paclitaxel	DOPC/DOTAP	Breast, pancreatic, and hepatic cancer	P II
ThermoDox®	Doxorubicin	DPPC/MSPC/DSPE-PEG2000	Bone metastasis, breast, and hepatocellular cancer	P II
Marqibo®	Vincristine	DSPC/CHOL	Non-Hodgkin's lymphoma, acute lymphoblastic leukemia, and Hodgkin's lymphoma	P III
OSI-211® (NX211)	Lurtotecan	HSPC/CHOL	Ovarian cancer and small-cell lung cancer	P III
LE-SN38®	Irinotecan metabolite SN38	DSPC/CHOL	Colorectal and lung cancer	P II
INX-0076®	Topotecan	Sphingomyelin/CHOL	Ovarian and small-cell lung cancer	P II
Alocrest®	Vinorelbine	Sphingomyelin/CHOL	Non-small-cell lung cancer and breast cancer	P I
Oncolipin®	Interleukin 2	DMPC	Kidney cancer	P II
OSI-7904L®	Thymidylate synthase inhibitor	HSPC/CHOL	Colorectal cancer	P II
CPX-351	Cytarabine and daunorubicin	DSPC/DSPG/CHOL	Acute myeloid leukemia	P II
CPX-1	Irinotecan and floxuridine	DSPC/DSPG/CHOL	Advanced colorectal cancer	P II

[a] © 2013 Lopes, Giuberti, Rocha, Ferreira, Leite, Oliveira. Originally published in [128] under CC BY 3.0 license. Available from: http://dx.doi.org/ 10.5772/55290.

[b] A, approved; P I, phase I study; P II, phase II study; P III, phase III study.

[c] Approved by EMA.

CHOL, cholesterol; DMPC, dimyristoyl phophatidylcholine; DOPC, dioleylphosphatidylcholine; DOTAP, dioleytrimethylammoniumpropane; DSPC, distearoylphosphatidylcholine; DSPE-PEG2000, distearoylphosphatidylethanolaminepolyethyleneglycol2000; DSPG, distearoylphosphatidylglycerol; EPC, egg, phosphatidylcholine; HSPC, hydrogenated soy phosphatidylcholine; MSPC, myristoylstearoylphosphatidylcholine; SPC, soy phosphatidylcholine.

TABLE 21.3 Targeting Ligands Frequently Used for Targeted Liposomes

Targeting Ligands	Function	References
LyP-1	Recognizes mitochondrial protein 32 overexpressed in tumor cells and tumor lymphatics	[21, 22]
Monoclonal antibody	Binds specific virus surface	[24, 28, 29, 37]
Mannose	Interacts with mannose receptors in macrophage tissues	[35]
VEGF	Binds vascular endothelial growth factor receptors (VEGFR) secreted on tumor cells	[23]
Sialyl LewisX	Targets E-selectin expressed on tumor vascular endothelial cells	[32]

21.4.2 Lipid-Based Emulsions and Nanoparticles

Considering the effective intestinal lymphatic transport of lipophilic drug molecules and fatty acids, novel lipid-based nanoparticles (Fig. 21.15) including self-micro-emulsifying drug delivery systems, solid lipid nanoparticles, and nanostructured lipid carriers have the potential for lymphatic delivery of poorly water-soluble drugs.

21.4.2.1 Self-Emulsifying Drug Delivery System Self-emulsifying drug delivery system (SMEDDS) (Fig. 21.15, left) is as an isotropic mixture of lipophilic drug molecules, natural or synthetic oils, and solid or liquid surfactants (Table 21.4). When introduced into an aqueous phase, such as water or physiological media, the SMEDDS spontaneously forms a drug-containing oil in water (o/w) emulsion with a particle size between 20 and 100 nm under gentle agitation. One distinct advantage of SMEDDS over the other lipid-based carriers is that, with the optimal choice of components, concentration, and ratio, SMEDDS can be easily prepared with nearly 100% drug-loading efficiency after gentle introduction into aqueous media. These components can be formulated as a soft gelatin capsule that will release and self-emulsify in the stomach or small intestine [135, 136]. Marketed soft gelatin capsule of SMEDDS formulations include Norvir® (Ritonavir) from Abbott, Neoral® (Cyclosporine A), Fortovase® (Saquinavir) from Roche, and Convulex® from Pharmacia.

SMEDDS formulations have greatly enhanced the bioavailability of poorly water-soluble drugs, including carvedilol [12], raloxifene [16], phenytoin [19] and sirolimus [15]. After oral administration, SMEDDS emulsify into subnanoscale o/w emulsions, driven by the peristaltic motility in the gastrointestinal tract, and then are absorbed by lymphatic pathways. Sun et al. [15] proposed a positive correlation ($r = 0.9998$) between the oil content and the lymphatic transport of sirolimus-loaded SMEDDS formulations.

21.4.2.2 Solid Lipid Nanoparticles and Nanostructured Lipid Nanoparticles Solid lipid nanoparticles (SLNs) are spherical particulates that are typically 50–150 nm in

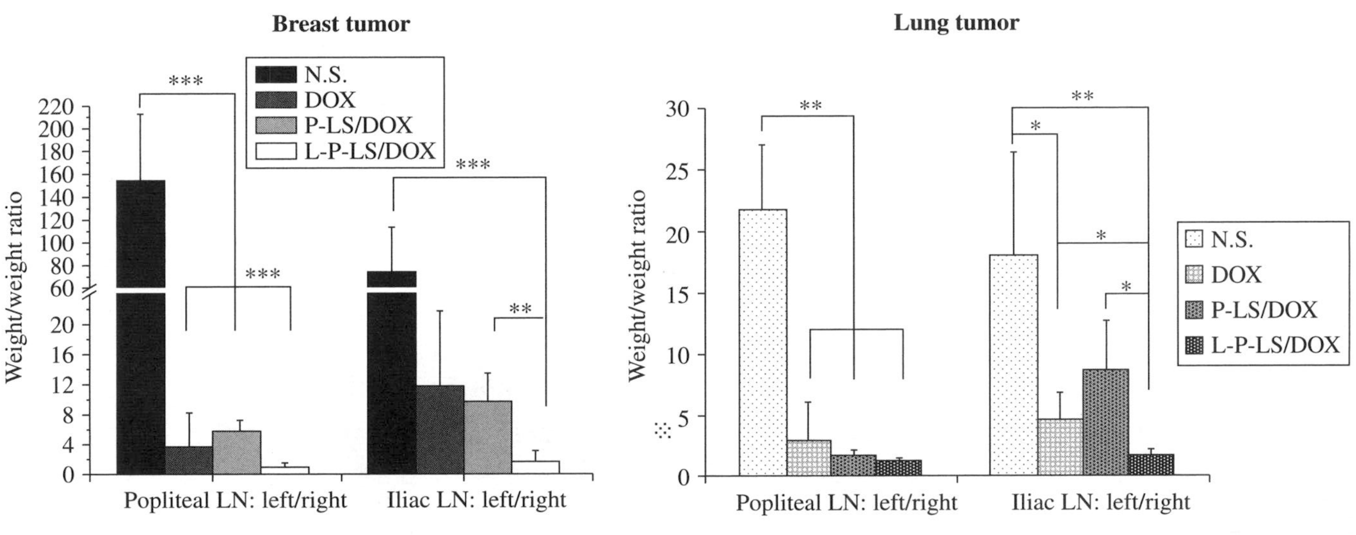

FIGURE 21.12 Weight ratios of popliteal and iliac lymph nodes (LNs) on the left side (tumor inoculation and metastasis side) to the corresponding LNs at the right side of breast or lung tumor models after SC injection of normal saline (N.S.), doxorubicin (DOX) solution, untargeted PEG liposomes (P-LS/DOX), and LyP-1-PEG liposomes (L-P-LS/DOX). Compared with untargeted liposomes, LyP-1-PEG liposomes exhibited significantly enhanced inhibition of popliteal and iliac LNs metastases in breast tumor model but only iliac LN metastasis in lung tumor model (*$p<0.05$, **$p<0.01$, ***$p<0.001$). Yan et al. [21], pp. 118–125. *J Control Release*; Yan et al. [22], p. 415103. Reproduced with permission of IOP Publishing.

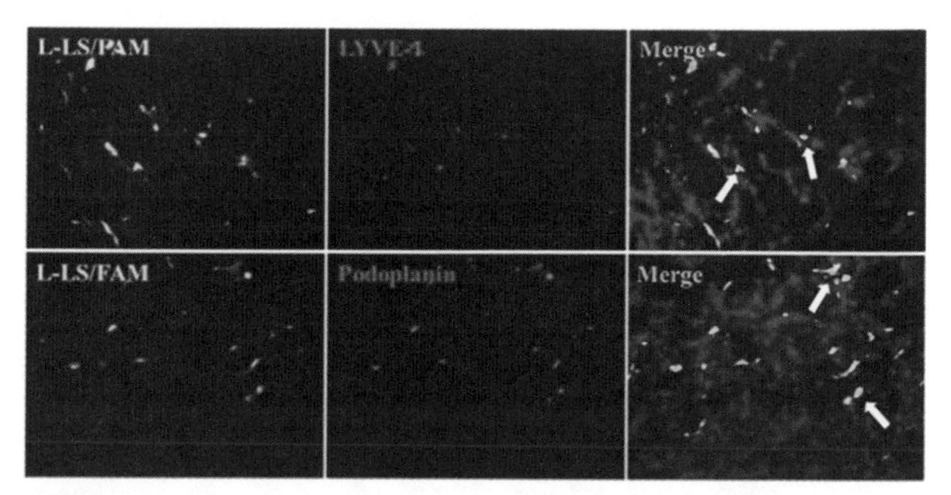

FIGURE 21.13 Specific targeting of LyP-1-PEG liposomes to breast tumor lymphatics. Fluorescent LyP-1-PEG liposomes were colocalized with lymphatic vessel markers (shown for LYVE-1 and podoplanin arrows indicate the colocalization in metastatic LNs, indicating the specific binding ability of LyP-1-PEG liposomes to tumor lymphatics. Yan et al. [22], p. 415103. Reproduced with permission of IOP Publishing. (*See insert for color representation of the figure.*)

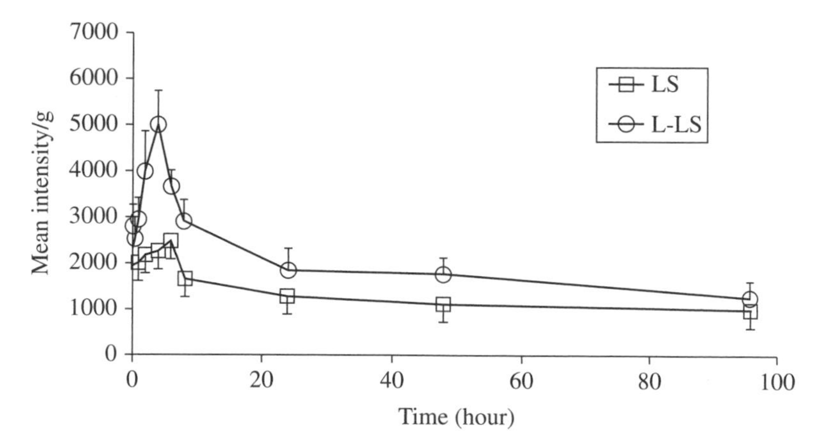

FIGURE 21.14 Pharmacokinetic profiles of fluorescein-loaded untargeted and LyP-1-PEG liposomes in metastatic popliteal lymph nodes (LNs) of MDA-MB-435 tumor models after the SC injection. The AUC of targeted liposomes (L-LS) in metastatic popliteal LNs was 53.9% higher than that of untargeted liposomes (LS) ($n = 3$, Mean ± SD), verifying the active targeting ability of LyP-1-PEG liposomes metastatic LNs. Yan et al. [22], p. 415103. Reproduced with permission of IOP Publishing.

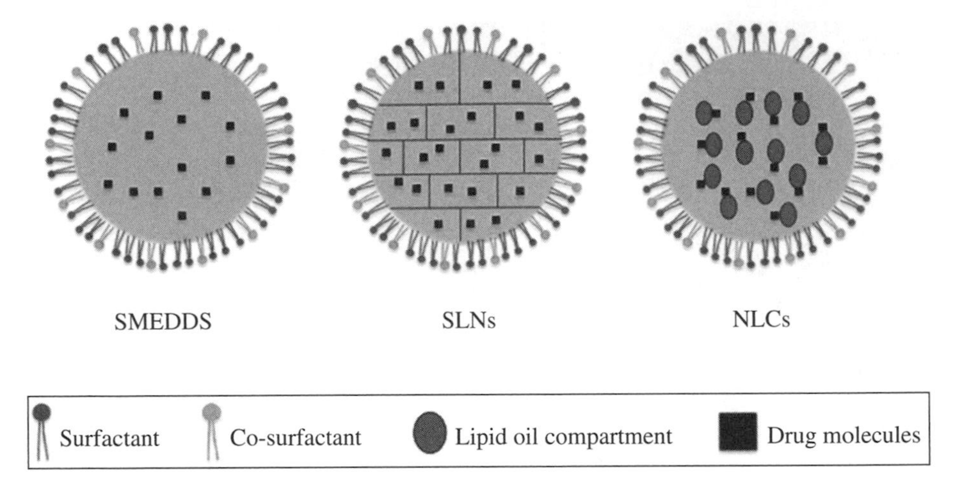

FIGURE 21.15 Schematics of three types of lipid-based drug carriers.

diameter comprising a solid core with a melting point above body temperature (Fig. 21.15 middle). SLNs are generally prepared from physiological lipids (e.g., waxes, fatty acids, steroids, and mono-, di-, or triglyceride mixtures) and biocompatible surfactants (e.g., Tween 80, Poloxamer 188, Poloxamine 908, and sodium dodecyl sulfate) as emulsifiers [137, 138]. SLNs can be prepared from low-cost ingredients on a large scale, and the physiological lipids in SLNs can remain in the solid state at both room and body temperature, thus improving drug stability and releasing drug molecules in a controlled manner to achieve prolonged *in vivo* therapeutic effects. Procedures generally used for the preparation of SLNs include high shear homogenization and ultrasound, high-pressure homogenization, solvent emulsification/evaporation, and dilution of microemulsions. The size and surface charge of SLNs are greatly influenced by the ingredient composition and manufacture procedures [50, 138–140]. For example, the average sizes of tripalmitin SLNs are 28 and 124 nm when they are prepared by the solvent evaporation method and melt homogenization method, respectively [138].

SLN formulations have been demonstrated to predominantly drain into the lymphatic system rather than systemic circulation system after oral administration. Using a radiolabeled SLNs, Bargoni et al. reported [141] an approximately 100-fold higher radioactivity in the lymph compared to blood. SLNs prepared with a Compritol® 888 lipid mixture as a core material were found to be superior to other lipid-based SLNs for oral lymphatic delivery of MTX [50]. The percentage lymphatic uptake of MTX at different time points was compared in drug solution (PBS pH 7.4), and four SLN formulations were prepared with Compritol 888, stearic acid, monostearin, and tristearin, respectively. As shown in Figure 21.16, while all the SLNs exhibited notably enhanced lymphatic uptake compared to PBS solution, Compritol 888–based SLNs showed the highest MTX concentration in the lymphatic region.

TABLE 21.4 List of Lipids and Surfactants Typically Used for SMEDDS Formulations[a]

Excipients	Function	Chemical Structure
Labrafac® CC	Oil (HLB[b] = 1)	Medium-chain triglycerides of caprylic and capric acids (C8–C10)
Capmul® MCM	Oil	Glyceryl monocaprylate (C8)
Soybean oil	Oil	Alpha-linolenic acid, linoleic acid, oleic acid (C16–C18)
Lauroglycol® FCC	Oil and solubilizer (HLB = 4)	Propylene glycol laurate (C12)
Labrasol®	Surface-active agent (HLB = 14)	Mono-, di-, and triglycerides and mono- and di-fatty esters of polyethylene glycol. The predominant fatty acids are caprylic/capric (C8–C10)
Tween 20	Surface-active agent (HLB = 16.7)	Polyoxyethylene (20) sorbitan monolaurate
Tween 60	Surface active agent (HLB = 14.9)	Polyoxyethylene (20) sorbitan monostearate
Tween 60	Surface-active agent (HLB = 15.0)	Polyoxyethylene (20) sorbitan monooleate
Acconon® CC-6	Surface-active agent (HLB = 12.5)	Polyoxyethylene (6) caprylic/capric glycerides
Plurol® oleique	Coemulsifier/solubilizer (HLB = 10)	Polyglyceryl-6 dioleate (C18)
Transcutol®	Solubilizer	Purified diethylene glycol monoethyl ether

[a] Reproduced from Refs. [16, 19] with permission.
[b] HLB, hydrophilic–liphophilic balance.

SLNs formulation offers a versatile platform for lymphatic delivery of poorly water-soluble drugs. However, due to a nearly perfect solid lattice structure in the lipid core of SLNs, drug molecules could be expelled out during the storage. This problem can be overcome by adjusting the lipid composition with the addition of liquid lipids such as medium-chain triglyceride (MCT) to form a new generation of lipid-based nanoparticles—nanostructured lipid carriers (NLCs) [52–58, 142, 143]. The resultant imperfect or amorphous lipid matrix of the NLCs (Fig. 21.15 right) provides space for accommodating the drug molecules to enhance drug-encapsulation efficiency and storage stability, with the minimum incidence of drug expulsion. For example, incorporation of 13.9% of MCT oil into the lipid core has been found to increase the entrapment efficiency of Amitone B to 84.7% for NLSs when compared with 52.2% for SLNs [53]. Moreover, significantly enhanced chemotherapeutic activity of isoliquiritigenin (ISL)-loaded NLCs has been recently reported in animals implanted with sarcoma and liver tumors [54]. The ISL-NLCs were administered IP at 10, 20, and 40 mg/kg dose levels, and the suppression rates for tumor growth increased from 75.70% to 83.90% in sarcoma-bearing mice and from 71.49% to 85.62% in liver tumor–bearing mice.

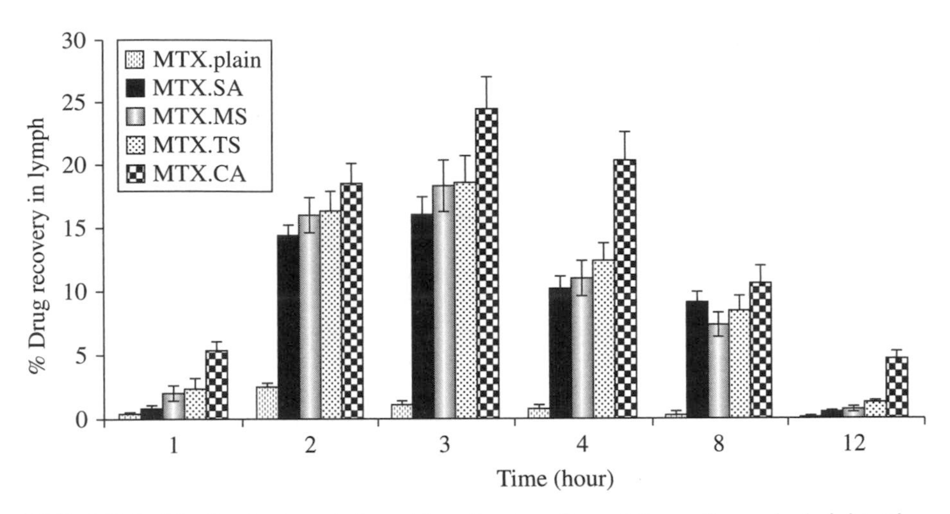

FIGURE 21.16 *In vivo* lymphatic uptake of various formulations after oral administration of MTX PBS solution (MTX-Plain), MTX-loaded SLNs prepared from stearic acid (MTX-SA), monostearin (MTX-MS), tristearin (MTX-TS), and Compritol 888 (MTX-CA). Paliwal et al. [50], pp. 184–191. Reproduced with permission of Elsevier.

21.4.3 Polymer-Based Carriers

Numerous therapeutic agents have been conjugated or encapsulated into polymer-based nanoparticles for targeted and sustained lymphatic delivery to inhibit tumor growth and metastases. The widely utilized polymers are categorized into natural polymers such as dextran [7, 8], hyaluronic acid (HA) [6, 9, 10], and synthetic polymers including poly (ethylene glycol)-block-poly (3-caprolactone) (PEG-PCL) [62], poly (L-lactic acid) [70], poly (lactide-*co*-glycolide) (PLGA) [65, 66, 144], poly (hexylcyanoacrylate) nanoparticles (PHCA), and poly (methylmethacrylate) (PMMA) [145].

21.4.3.1 Natural Polymers Polysaccharide-based nanoparticles, such as dextran and HA (Fig. 21.17), have gained popularity in developing controlled-release drug delivery system, attributed to their biocompatibility, biodegradability, broad range of physicochemical properties, and versatility of modification by simple chemical conjugations with drug molecules.

The feasibility of mitomycin C–dextran conjugate (MMC-D) as a lymphotropic drug delivery system has been earlier demonstrated by Takakura et al. [7]. Anticancer antibiotic, mitomycin C, was first conjugated to dextran with molecular weights of 10–500 kDa via ε-aminocaproic acid as a spacer, and then administered by IM injection to rats inoculated with leukemia cells. In contrast to the free drug and MMC-D prepared with the lowest molecular weight (10 kDa), larger MMC-Ds afforded significant higher drug concentration in the thoracic lymph as well as substantially

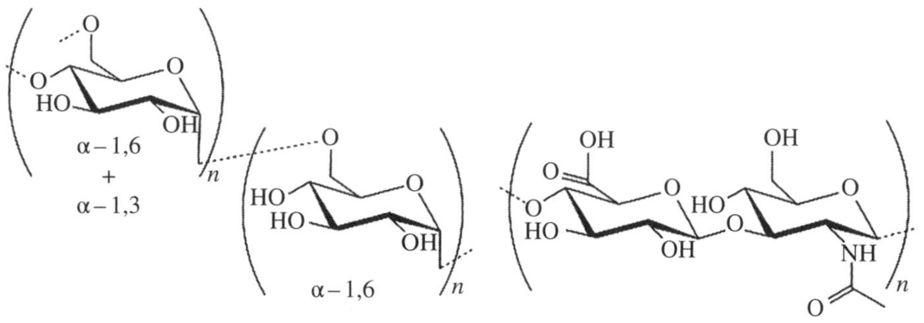

Dextran Hyaluronan

FIGURE 21.17 Structures of Dextran and Hyaluronan.

enhanced accumulation in the regional lymph nodes, resulting in more efficient suppression of tumor growth and metastasis.

Hyaluronic acid (HA) is another type of natural polysaccharide that consists of alternating D-glucuronic acid and *N*-acetyl D-glucosamine. The HA is present throughout the body with particularly high concentrations in the connective tissues. In the process of clearance by the lymphatic system, the HA is catabolized in the lymph nodes by receptor-mediated endocytosis and lysosomal degradation. Some invasive tumors such as breast, head, and neck tumors, preferentially uptake the HA over normal tissues due to their surface overexpression of the HA receptor CD44. Therefore, conjugation of chemotherapeutic agents to HA could provide an efficacious approach to treating lymphatic metastases. Cai et al. synthesized two HA-cytotoxic drug nanoconjugates, HA-cisplatin [10, 79, 80, 146] and HA-doxorubicin [6], which have both exhibited *in vivo* sustained-release profiles and enhanced drug retention by lymph nodes in locally aggressive metastatic tumor models of head and neck squamous cell carcinoma and breast cancer. Consequently, intralymphatic delivery of HA-cisplatin and HA-doxorubicin significantly inhibited tumor progression and led to the increased survival rates when compared with the conventional chemotherapy.

21.4.3.2 Synthetic Polymeric Nanoparticles In aqueous solution, biodegradable amphiphilic block copolymers such as PEG-PCL or PEG-poly (amino acid) self-assemble into the core-shell structured polymeric micelles with a size of 20–100 nm, as shown in Figure 21.18. The description, characterization of polymeric micelles, and the recent developments in functional biodegradable micelles for safe and efficient cancer chemotherapy have been extensively reviewed by Croy et al. [148] and Deng et al. [149]. In the past decade, polymeric micelles have generated great interest for sustained delivery of poorly water-soluble anticancer drugs attributed to their inherent physiochemical features. For example, a core formed from poly (3-caprolactone) or poly (amino acid) provides a hydrophobic environment for accommodating lipophilic drug molecules, while the presence of PEG in the hydrophilic shell spares micelles from aggregation and also reduces the fast clearance of the micelles by RES.

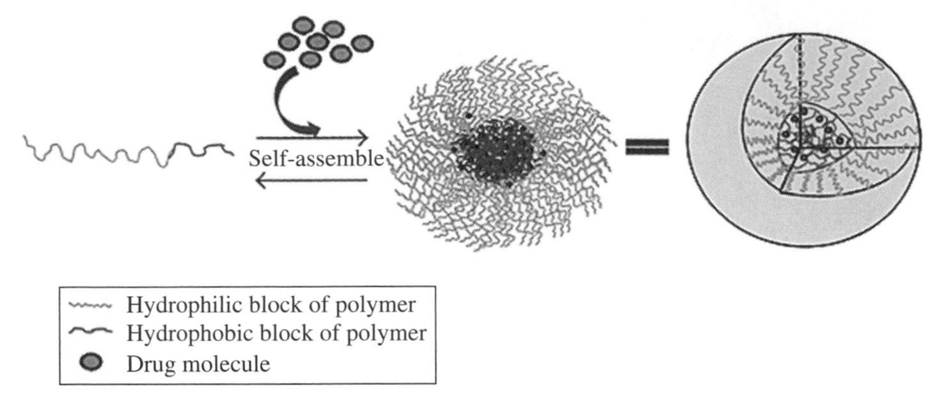

FIGURE 21.18 Self-assembly of polymeric micelles encapsulated with poorly water-soluble anticancer drugs. Xu et al. [147], p. 15.

Drug-loaded polymeric micelles generally enter the primary tumor via passive diffusion through the blood capillaries after IV administration. Surface modification of the polymeric micelles with targeting ligands can enhance their binding with tumor lymphatic vessels through the receptor-mediated interaction to achieve active lymphatic targeting. LyP-1 peptide, as mentioned in Section 21.4.1, has been identified as a ligand for p32/gC1qR that are overexpressed in highly metastatic tumor cells and tumor lymphatics [131]. Wang et al. [62] developed a LyP-1 peptide-conjugated PEG-PCL micelles (LyP-1-PM) with a size of around 30 nm. The specific binding of the LyP-1-PM with highly metastatic tumor cells (MDA-MB-435S) and lymphatic endothelial cells was first observed by flow cytometry and laser confocal microscopy, and then its *in vivo* lymphatic-targeting capability was investigated in a breast tumor model. After IV injection of the fluorescein-loaded LyP-1-PM into the breast tumor–bearing mice, *in vivo* fluorescent imaging (Fig. 21.19) only showed good colocalization of the LyP-1-PM with the lymph vessel marker (LYVE-1) but not with the blood vessel marker (CD31). In addition, targeted delivery of artemisinin-encapsulated LyP-1-PM remarkably enhanced the antitumor efficacy and reduced the systemic toxicity, indicating the potential of LyP-1-targeting polymeric micelles as specific drug carriers for lymphatic delivery.

PLGA-based drug carriers have also been used for lymphatic targeting. This biocompatible and biodegradable polymer has been approved by the FDA for use in humans as therapeutic device, and it has been extensively studied as a sustained release system to deliver chemotherapeutic agents [65, 66, 71, 144]. Drug-incorporated PLGA microspheres or sub-100 nm nanoparticles could be readily prepared by emulsification solvent evaporation methods [150–154], and it has been reported that surface coating with poloxamer or poloxamine block copolymers could remarkably enhance the maximal lymphatic uptake from 6 to 17% after the SC injection [69]. Paclitaxel-loaded PLGA (PTX-PLGA) microspheres have been demonstrated to effectively inhibit tumor growth in animal models bearing IP carcinomatosis [70] and SC carcinoma [72]. Liu et al. [73] further incorporated PTX-PLGA into a biodegradable gelatin sponge that enabled a continuous release of

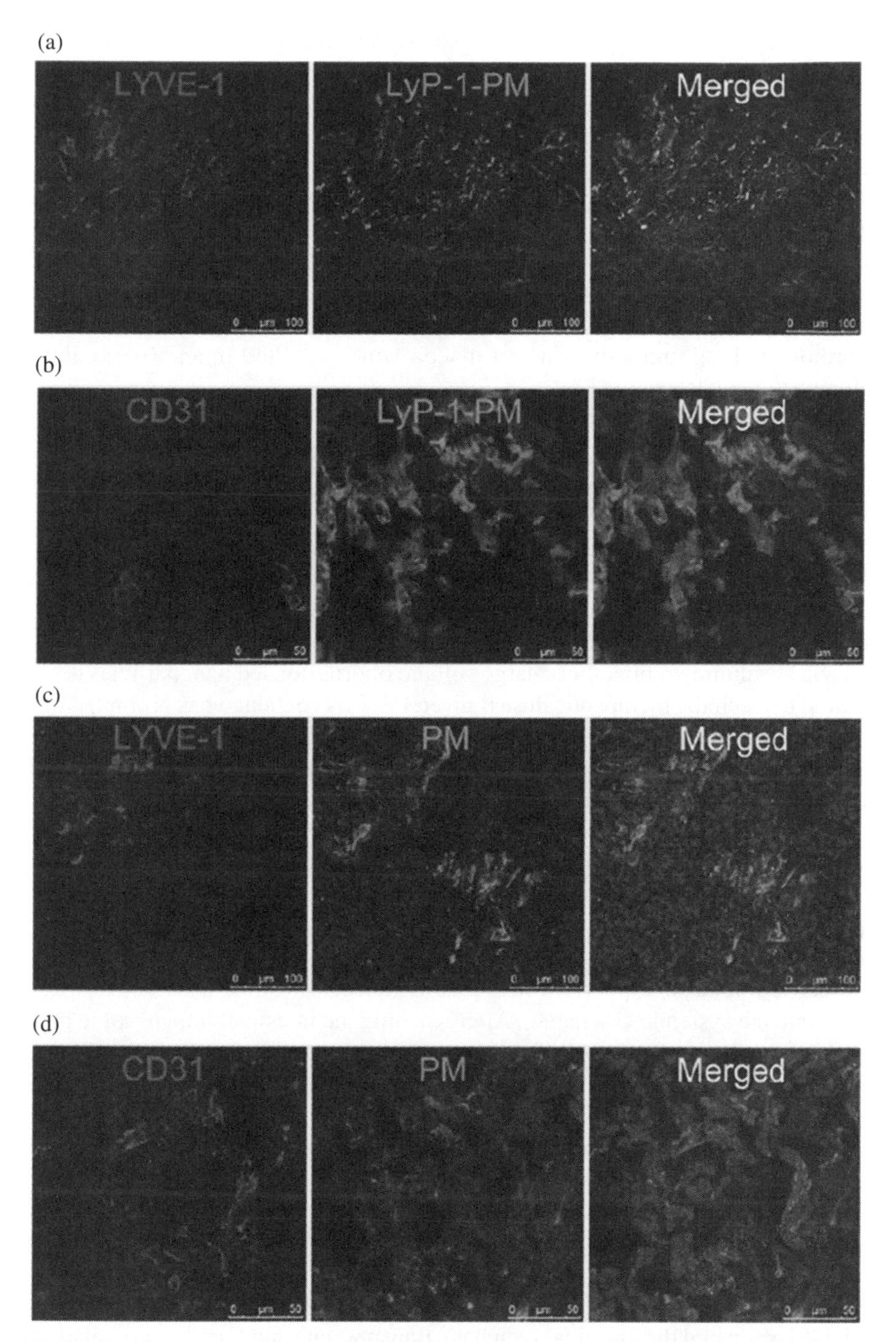

FIGURE 21.19 The targeted delivery of LyP-1-PM to tumor lymphatics in MDA-MB-435S tumor-bearing nude mice by immunofluorescence technique. LyP-1-PM colocalized with the (a) lymphatic endothelial marker (LYVE-1), but not with (b) blood vessel markers (CD31). On the contrary, PM had good colocalization with (d) CD31, but not with (c) LYVE-1. Nuclei were counterstained with Hoechst 33258. Wang et al. [62], pp. 2646–2657. Reproduced with permission of American Chemical Society. (*See insert for color representation of the figure.*)

PTX into the lymphatics with subsequent accumulation in the regional lymph nodes. The PTX-PLGA sponge was implanted into the pleural space of rats, and its retention in mediastinal lymph nodes was compared with that of the IV PTX solution. During the given 28-day experimental time, an over 400-fold increase in PTX exposure at the ipsilateral mediastinal lymph nodes and approximately a 100-fold increase at the contralateral side were observed in the group treated with IP PTX-PLGA sponge. Attributed to the improved drug distribution into the lymphatic system, the PTX-PLGA sponge also exhibited the enhanced antitumor efficacy in an orthotopic lung tumor–bearing rat model, which resulted in an 80% inhibition of lymphatic metastasis.

21.5 ADMINISTRATION ROUTES FOR LYMPHATIC DELIVERY

Conventional cytotoxic drugs or drug-loaded carriers are generally administered by IV infusion. The limitations of IV administration routes include rapid clearance by the RES system, premature drug leakage, and loss in nontarget tissues that leads to the occurrence of side effects. Moreover, the lymphatic system is not highly accessible via IV administration, thus a large volume of drug-loaded nanoparticles is often required to reach the therapeutic dose if given IV. This challenge has prompted many explorations of alternative routes such as intestinal and pulmonary delivery, and local parenteral injections, including SC, IP, and IM administration, to achieve efficient lymphatic targeting. Listed in Table 21.5 are some advantages that these alternative routes offer for lymphatic delivery.

21.5.1 Intestinal

The intestinal lymphatic system is a critical route through which fat-soluble vitamins, food-derived lipids, and poorly water-soluble molecules can be absorbed and transported into the systemic circulation. After reaching the intestinal lumen, some highly lipophilic drugs (log $P_{oil/water} > 5$), fatty acids (e.g., long-chain triglyceride), and molecules with high molecular weights can transit across the intestinal absorptive cells (enterocytes) only with the assistance of lipoproteins (e.g., chylomicrons and VLDL) due to their inability to diffuse across the blood capillaries [3]. The lipoprotein-bound drug molecules thus preferentially enter the lymphatic system via the highly permeable lymphatic capillaries that underline the enterocytes in close proximity (Fig. 21.20). Therefore, intestinal administration of lipid-based drug carriers (i.e., SMEDDS, SLNs, and NLCs) can be exploited as an efficient drug delivery strategy to improve the oral bioavailability and can prevent the first-pass metabolism in the liver. Alex et al. [48] compared the intestinal lymphatic transport rate and lymph accumulation of lopinavir in Compritol 888–based SLNs formulation versus in conventional suspension with 0.5% (w/v) of methyl cellulose, and they found that drug-loaded SLNs exhibited a 4.9-fold higher lymph accumulation 6 hours post administration (Fig. 21.21) compared with the drug suspension. Another pharmacokinetics study of idarubicin-loaded SLNs [161] observed the predominant transmucosal transport of the drug SLNs into

TABLE 21.5 Summary of Primary Improvements of Drug-Carriers Administered via Alternative Routes Over Conventional Therapy

Routes of Administration	Improvement Over Conventional Therapy	References
Intestinal	• Increased intestinal permeability • Prolonged gastric residence time • Increased drug localization in intestinal lymph • Augmented and prolonged therapeutic efficacy	[38, 39, 46, 48–50, 155]
Pulmonary	• Prolonged drug deposition and residency in lung • Improved penetration to lymphatic vessels • Enhanced suppression of primary mammary tumor and reduced lung and lymph node micrometastasis	[40, 51, 58, 89, 90, 156, 157]
Subcutaneous	• Increased accessibility to lymphatic system • Accelerated lymphatic drainage and prolonged lymph node retention • Enhanced therapeutic efficacy of chemotherapy and augmented vaccine-induced immune responses	[21, 22, 24, 27, 28, 34–36, 65, 72, 79, 120, 158, 159]
Intraperitoneal	• Selective absorption from the peritoneal cavity via lymphatics and enhanced lymphatic channel uptake • Increased retention by the diaphragm, liver, spleen, pelvic lymph node, renal lymph node, and thoracic lymph node • Pronounced antitumor efficacy of chemotherapy and gene therapy against ovarian tumor to achieve prolonged survival against ovarian tumor	[20, 29–31, 33, 41, 44, 54, 67, 70, 73, 145],
Intramuscular	• Enhanced long-lasting serum titer of interferon liposomes • Enhanced drug retention in regional lymph node	[7, 42, 43]

the lymph after intraduodenal administration, which resulted in a 1.5-fold increased bioavailability.

21.5.2 Pulmonary

Pulmonary delivery allows the administered drug molecules to directly reach trachea and lungs, which are abundant in lymphatic vessels and a variety of lymph nodes (Fig. 21.22). Drug molecules and foreign particles (<500 nm) that are deeply deposited in the alveolar region are mostly removed by alveolar macrophages and drain into the lymphatic system [163, 164]. Therefore, pulmonary administrations such as aerosol inhalation and intratracheal instillation can effectively target drug-loaded nanoparticles to the pulmonary lymphatics.

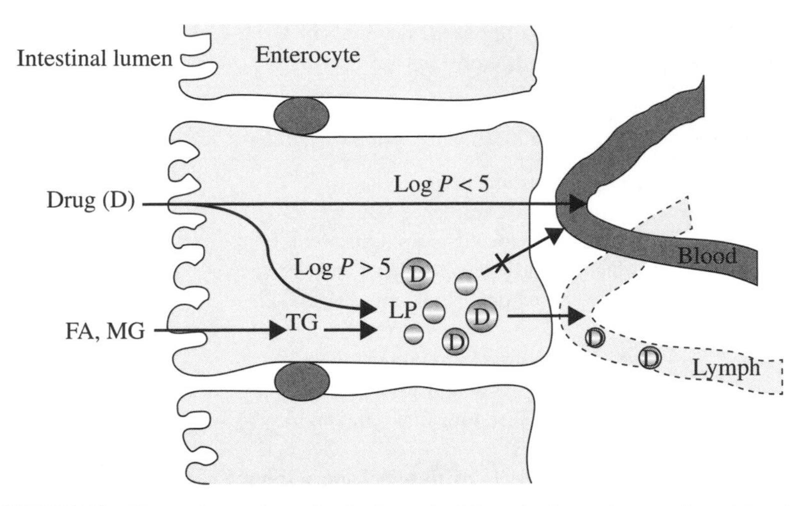

FIGURE 21.20 Drug absorption via the intestinal lymphatic system and portal vein. FA, fatty acid; LPs, lipoproteins; MG, monoglyceride; TG, triglyceride. Trevaskis et al. [160], pp. 702–716. Reproduced with permission of Elsevier.

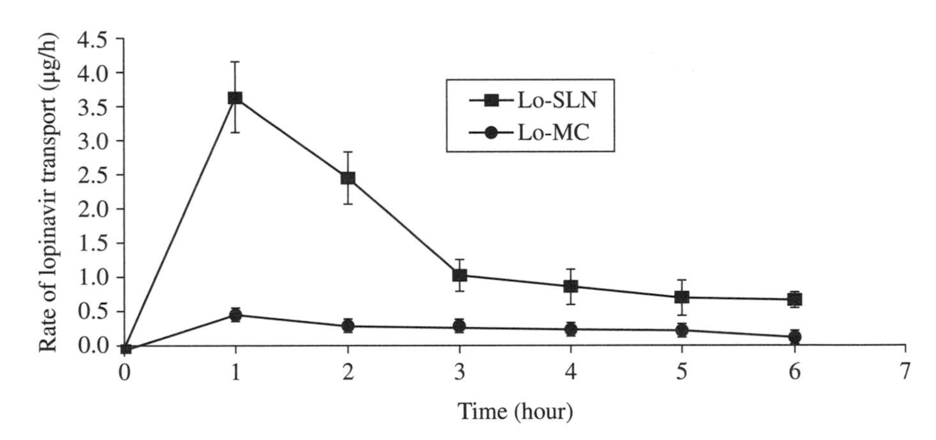

FIGURE 21.21 Rate of intestinal lymphatic uptake of lopinavir. Lo-MC, conventional lopinavir suspension; Lo-SLN, Lopinavir SLNs formulation ($p < 0.001$). Alex et al. [48], pp. 11–18. Reproduced with permission of Elsevier.

Prolonged residency and preferential uptake of SLNs into the lymphatic system have been visualized by whole-body gamma camera imaging of rats administered with radiolabeled SLNs via pulmonary administration [156, 165]. For example, when compared with the IV injection, the pulmonary route achieved substantially enhanced SLNs accumulation and longer retention in the lung (Fig. 21.23). Considering the predominant biodistribution of nanoparticles in the lymphatic system, direct delivery of chemotherapeutic drugs via the pulmonary route provides an alternative over IV infusion for chemotherapy of lymphatically metastatic lung cancer.

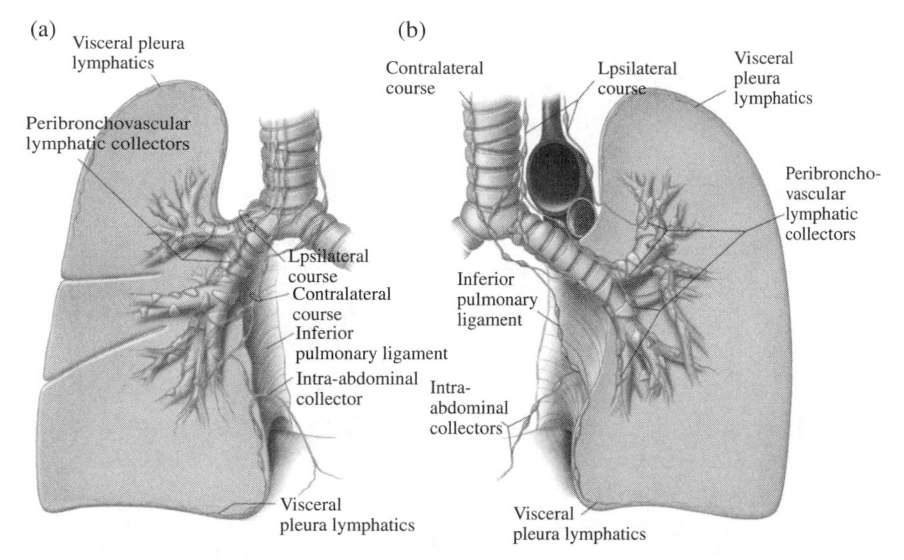

FIGURE 21.22 (a) Visceral pleura lymphatics and peribronchial vascular lymphatic collectors of the right lung. (b) Visceral pleura lymphatics and peribronchial vascular lymphatic collectors of the left lung. Riquet [162], pp. 619–638. Reproduced with permission of Elsevier.

To demonstrate the efficacy of drug delivery via the pulmonary route, Videira et al. [51] prepared a PTX-loaded SLNs formulation, which was nebulized into an aerosol for pulmonary delivery into mice with lung metastases. The PTX-SLNs aerosol exhibited substantially superior metastasis inhibition over the conventional IV PTX solution. Moreover, vitamin E analogue (α-TEA) and camptothecin derivative (9-NC) were formulated [89] into dilauroylphosphatidylcholine liposomes and administered as aerosols into lung metastasis, bearing Balb/c mice, either alone as an individual treatment or together as a combination treatment. Compared with untreated controls and individual treatments, combination of the α—TEA and 9—NC was found to be significantly more efficient at inhibiting tumor growth and reducing the metastases to lung and lymph nodes (Fig. 21.24). A similar enhanced therapeutic efficacy of combination treatment against lung tumor metastasis has been observed in α—TEA/PTX liposomal aerosol formulations [90]. In addition, aerosolized celecoxib-encapsulated NLCs in combination with IV doxetaxel solution [58] showed significant reduction in non-small-cell lung tumor growth. Therefore, sequential inhalation of liposomal-formulated chemotherapeutic agents could represent a promising lymphatic-targeting strategy for selective regional delivery to the deep lung region against lung cancer.

21.5.3 Subcutaneous

Considering the ready access that the SC route provides to the lymphatic system [166, 167] (Fig. 21.25), SC injection of drug-loaded carriers under the dermis is another effective approach to targeting therapeutics to the lymphatic system. Upon

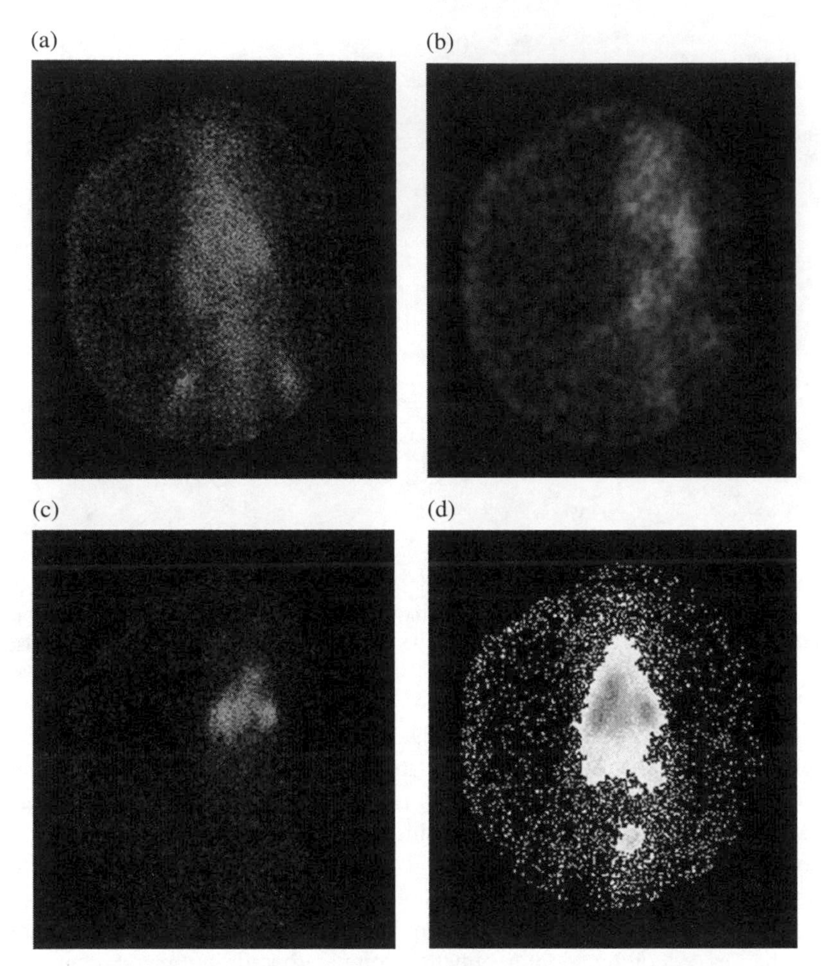

FIGURE 21.23 Gamma scintigraphy photographs of rats receiving ^{99m}Tc-amikacin-loaded SLNs (a) IV after 0.5 hour, (b) IV after 6 hours, (c) pulmonary after 0.5 hour, and (d) pulmonary after 6 hours. Varshosaz et al. [165], p. 8. (*See insert for color representation of the figure.*)

SC injection, drug-loaded carriers that possess proper physicochemical characteristics (size [120], surface charge [65], [121–124], hydrophobicity [27, 36]) are preferentially absorbed from the injection site to the interstitial area beneath the dermis followed by the lymphatic capillaries and then drain with peripheral lymph into the connecting lymph nodes.

Three administration routes—IV, IP, and SC—were compared using radiolabeled etoposide SLNs as a carrier model in Dalton's lymphoma-bearing mice [168]. Comparatively, the accumulation of SLNs in tumor was of the order SC > IP > IV. At 24 hours post injection, the SC route rendered a 59- and 8-fold higher etoposide uptake than the IV and IP routes, respectively (Fig. 21.26). These data are consistent in principle with what was reported in a pharmacokinetics and lymphatic uptake

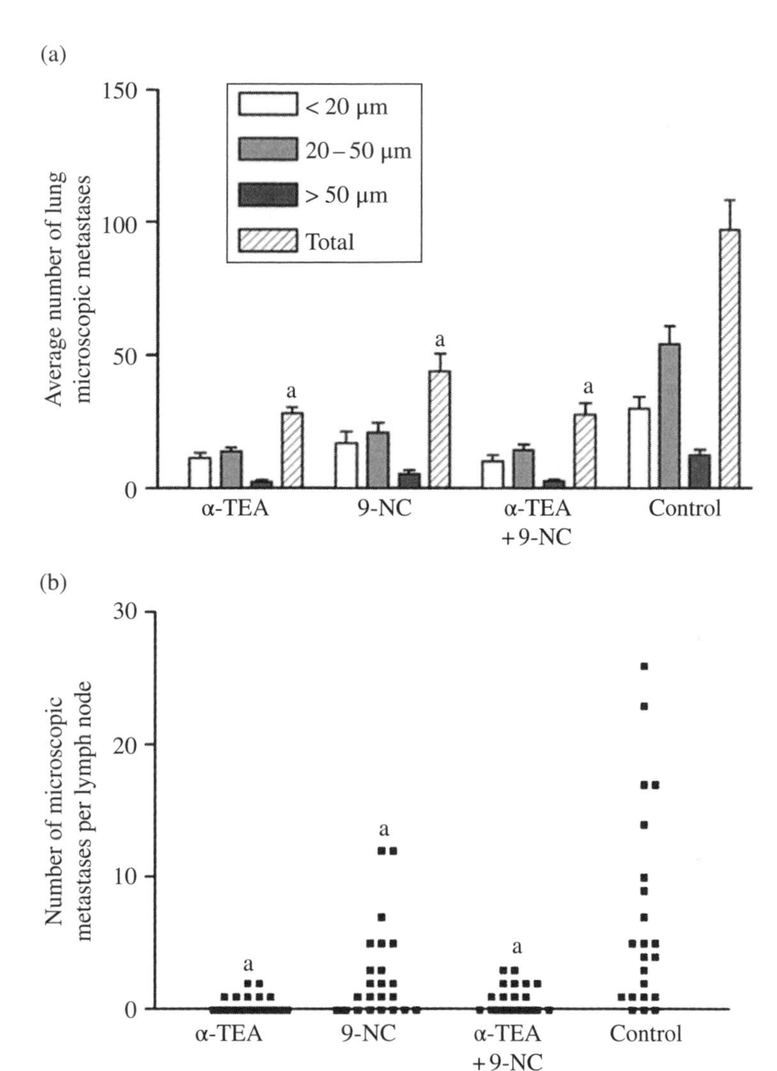

FIGURE 21.24 Aerosolized α-TEA, 9-NC, and combination treatments inhibited lung and lymph node metastasis. The number of fluorescent microscopic metastases (a) on the surface of the left lung lobe or (b) on the surface of individual lymph nodes were determined ([a]significantly different from control; $p<0.05$). Lawson et al. [89], pp. 421–431. Reproduced with permission of Springer.

study in rats administered with PEGylated polylysine dendrimers [159]. Retention of the 13.4-nm PEGylated dendrimers in popliteal and iliac nodes after IV administration was only approximately 10% of the levels reached after SC administration.

Using a lymphatically metastasized HNSCC model, Cai et al. [79] reported that the SC administration of HA-cisplatin (HA-Pt) nanoconjugate could significantly improve antitumor efficacy and survival rate of HNSCC xenografts in Nu/Nu mice

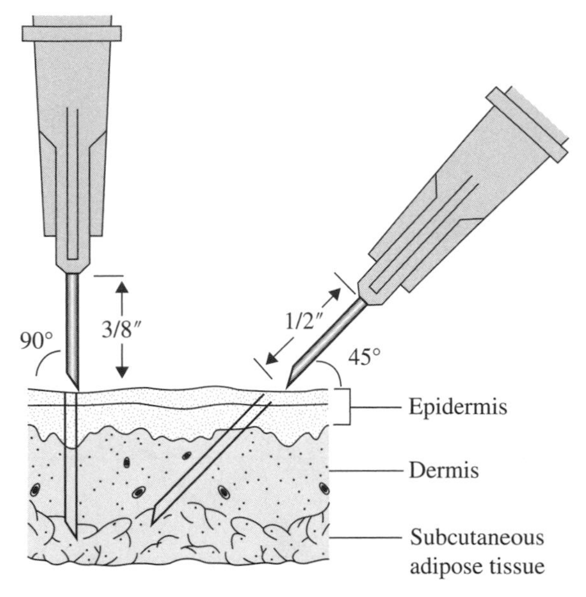

FIGURE 21.25 Diagrammatic representation of the subcutaneous injection site. Lammon [166]. Reproduced with permission of Elsevier.

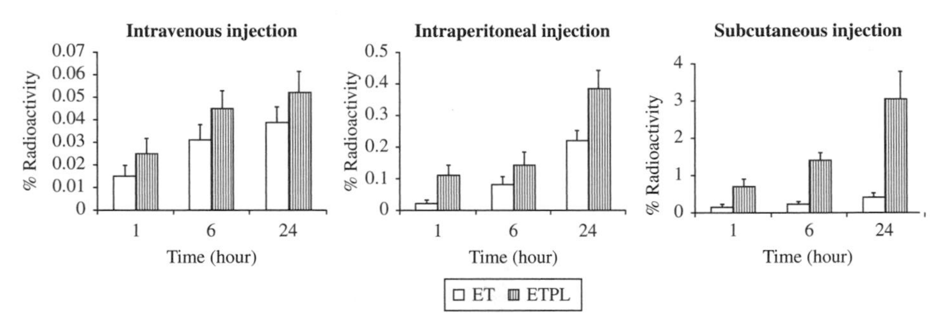

FIGURE 21.26 Tumor concentration of free ^{99m}Tc-etoposide and ^{99m}Tc-ETPL SLNs after injecting in Dalton's lymphoma-bearing mice via different routes. Each value is the mean ± S.D. of three experiments. ET, etoposide; ETPL, etoposide SLNs. Adapted from Reddy et al. [168], pp. 185–198. Reproduced with permission of Elsevier.

when compared with IV cisplatin (CDDP) and IV HA-Pt. Tumor growth of the mice treated with SC HA-Pt was substantially inhibited due to the increased drug accumulation in the primary tumor and the cervical lymph node. By contrast, the average tumor burden of animals in either IV CDDP group or IV HA-Pt group reached 1000 mm^3 by week 8 after the tumor cell implantation (Fig. 21.27a). In addition, approximately 60% of the mice in SC HA-Pt group were cured within 6 weeks without disease recurrence, and the survivors lived throughout the study (Fig. 21.27b and c), whereas 100% of the mice in the other treatment or control groups were euthanized by week 10 (Fig. 21.27b and d).

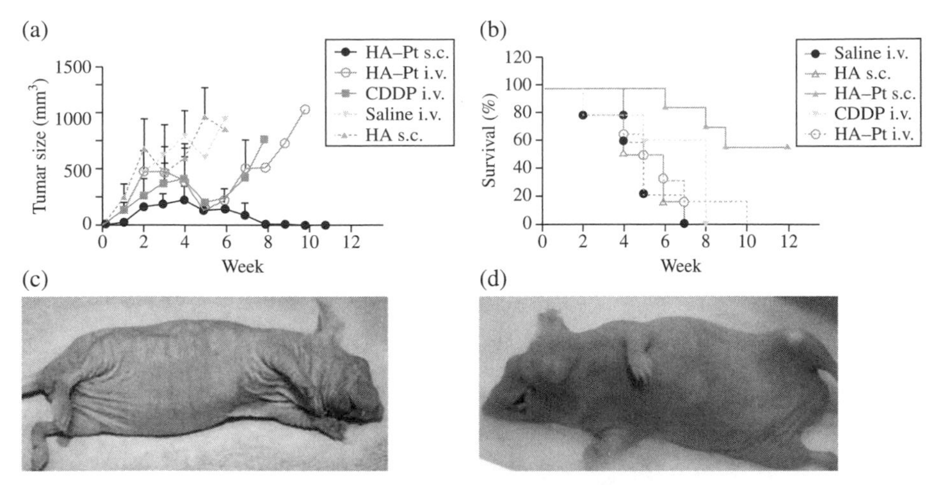

FIGURE 21.27 Measurement of tumor size (a) and survival curves (b) of female Nu/Nu mice administered with IV saline, SC HA, equivalent doses of IV CDDP, SC HA–Pt, and IV HA–Pt (3.5 mg/kg on Pt basis, $5 \leq N \leq 7$). Representative photographs of animals in (c) SC HA–Pt group at 12 weeks or (d) IV CDDP group at 8 weeks. Cai et al. [79], pp. 237–245. Reproduced with permission of Future Science Ltd.

21.5.4 Intraperitoneal

A major reason for failure of chemotherapy against colorectal cancer is due to cancer metastases within the abdominal and pelvic regions as a local recurrence of the primary tumor, which is usually considered as a terminal condition. Local delivery of chemotherapeutic agents results in a higher drug concentration with a prolonged residence of the drug in the peritoneal cavity, thus improving the antitumor efficiency while reducing the systemic toxicities. Upon IP injection, small molecules with molecular weights of less than 20 kDa are mostly absorbed through the peritoneal capillaries into the systemic circulation [33, 169], causing an insufficient local retention. Larger molecules or particulates, on the contrary, are preferentially cleared by the lymphatic drainage through the stomata on the subdiaphragmic surface that connects to the lymphatic vessels [170]. In order to extend the residence time of the chemotherapeutic agents, a number of drug delivery approaches such as liposomes, NLCs, and PLGA/PLA microspheres have been investigated in tumor models of ovarian, liver, and lung tumors.

The lymphatic uptake of IP-injected Lipusu®, a commercialized liposomal formulation of PTX in rat ovarian cancer xenografts [20] resulted in higher drug exposure in the tumor and pelvic lymph nodes compared to the free PTX solution (Fig. 21.28). However, there was no statistically significant difference in tumor weight reduction between the two formulations. Comparatively, a PLGA-formulated PTX markedly enhanced both of the lymphatic targeting and tumor inhibition [67]. After IP administration to ovarian cancer xenografts, the drug concentration in pelvic lymph nodes of PTX-PLGA-treated rats at 48 hours post injection was 20-fold higher

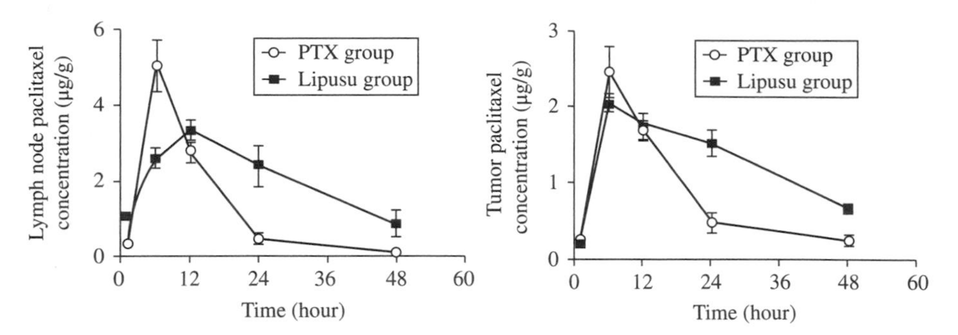

FIGURE 21.28 Paclitaxel lymph node (left) and tumor (right) distribution versus time after IP injection of 5 mg/kg paclitaxel (PTX) or Lipusu. Data represent the mean ± SD of three animals. Ye et al. [20], pp. 200–206. Reproduced with permission of Elsevier.

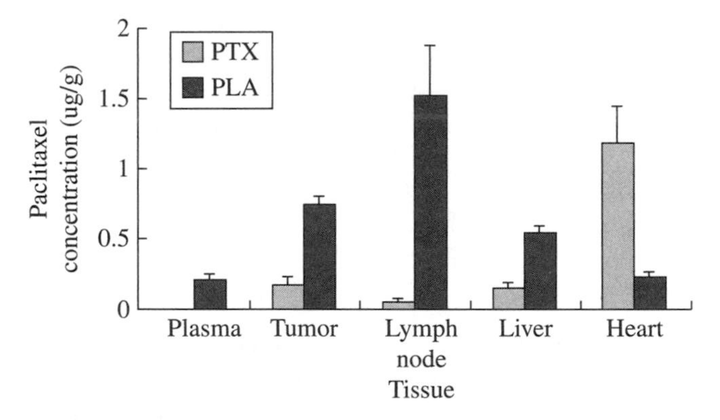

FIGURE 21.29 Comparison of paclitaxel concentration in all tissues collected at 48 hours after paclitaxel loading nanoparticle (PLA) and PTX IP administration at 5 mg/kg. Lu et al. [67], pp. 175–181. Reproduced with permission of Springer.

than the PTX solution–treated animals, and the drug exposure in tumor also increased by three fold (Fig. 21.29). Moreover, PTX-PLGA significantly reduced the tumor weight and ascites volume (Table 21.6), indicating the enhanced chemotherapeutic efficacy toward the ovarian cancer.

21.6 LYMPHATIC-TARGETING VACCINATION

The development of therapeutic subunit vaccines with purified antigens has gained much attention recently as they can elicit a focused immune response while avoiding the risks associated with the live attenuated pathogens [171]. A successful delivery of the subunit vaccine typically requires efficient drainage of the antigen/adjuvant to reach the secondary lymphoid organs, where the circulation of naïve T cells is restricted and T cells are activated by foreign antigens presented by APCs [172, 173].

TABLE 21.6 Therapeutic Effect for Ovarian Cancer in F344 Rats Xenografts (Mean ± SD)[a]

Experiment Group	Tumor Weight (g)	Ascite Volume (ml)	*P*
Saline	15.65±0.80	**85.30±3.50**	
Empty PLGA	15.85±0.75	**86.55±2.50**	
PTX-PLGA	4.55±0.11	**3.55±0.50**	<0.001
PTX solution	10.13±0.52	**30.45±1.55**	<0.001

[a] Reproduced from Ref. [67] with permission.
The dose is 5 mg/kg, and the drug was given weekly for 5 weeks.

Among lipid-based NPs, a unique class, called interbilayer crosslinked multilamellar vesicles (ICMVs), has been found to accumulate in the subcapsular sinus of the lymph nodes and preferentially localize with macrophages at 2 weeks post SC injection [174]. This property of ICMVs, therefore, lends themselves to an efficient vaccine carrier to deliver to a recombinant *Plasmodium vivax* circumsporozoite antigen, VMP001, and to generate strong humoral and cellular immune responses with enhanced titers and prolonged antibody response. A novel strategy, called "albumin-hitchhiking," was recently exploited to target the subunit vaccine to lymph nodes [175], where a fatty acid–based lipophilic albumin-binding domain was linked to a peptide antigen via a solubilizing PEG spacer to construct the molecular vaccines (Fig. 21.30a) with CpG DNAs as the adjuvants. Visualized by *in vivo* fluorescent imaging, the albumin-binding CpG accumulated in the lymph nodes up to 3 days after SC injection, leading to a 12-fold increase in lymph node accumulation compared with the unmodified CpG. The targeted molecular vaccines have been demonstrated to be able to elicit a 30-fold increase in T-cell priming and at the same time significantly regress the large-TC tumors (Fig. 21.30b) and slow the tumor growth of melanomas (Fig. 21.30c).

The synthetic particle vaccines mimic microbes such as viruses or bacteria. After SC or IM injection, the particles drain into the lymph nodes, where they interact with DCs in lymph nodes, inducing T-cell activation and differentiation to elicit the immune response. However, because the SC or muscular tissue is not as abundant with immune-competent cells as the dermis, lymph nodes or afferent lymphatic conducts, immune response in these tissues may be relatively modest in magnitude compared with the lymph nodes. To deliver vaccine to deep lymph nodes within a shortened period of time, injection routes that target immune-privileged anatomical sites may be used [176] to benefit from the proximity of the APCs and T cells. For example, a biodistribution and immunogenicity study of alum-adsorbed protein vaccine has shown that intralymphatic (IN) injection greatly enhanced the humoral and cellular immune response, resulting in an approximately 100-fold increase in lymph node accumulation of protein vaccine compared with SC injection [177]. In a phase I clinical trial in patients with metastatic melanoma [178], four patients were directly injected with a therapeutic vaccine that comprises a plasmid and two peptides corresponding to Melan A and tyrosinase. In the patients' lymph nodes the plasmid priming-peptide booted regimen lead to an overall immune response rate of

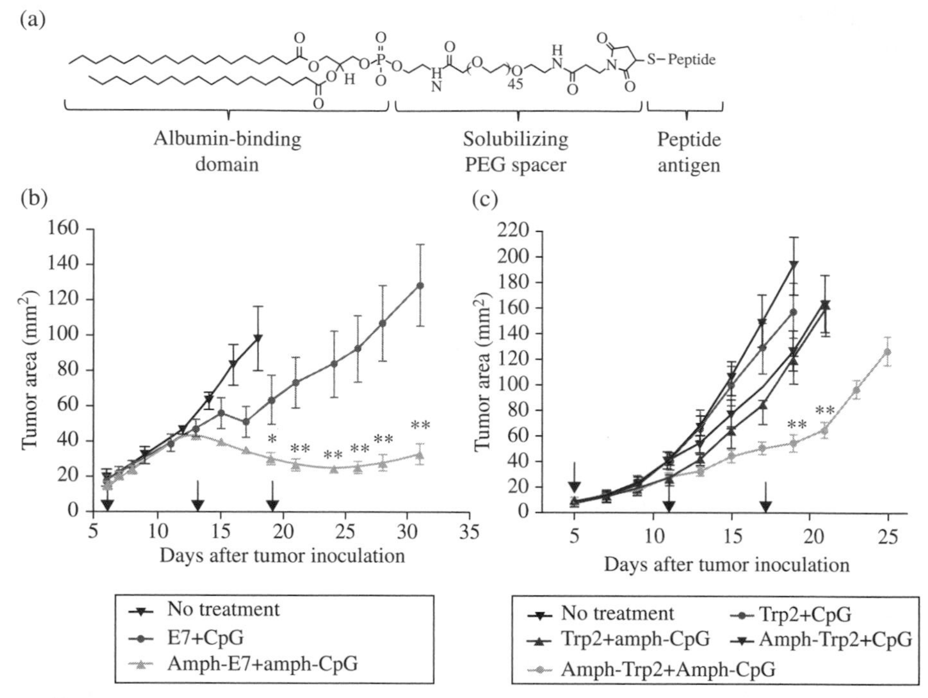

FIGURE 21.30 (a) Structure of albumin-binding molecular vaccine. (b and c) Tumor growth in C57BL/6 mice (n=8 per group) inoculated with 3×10^5 TC-1 (b) or B16F10 (c) tumor cells and vaccinated with CpG plus E7 peptide or Trp2 peptide (10 μg prime, 20 μg boost), respectively, on days indicated by arrows. Statistically significant differences between soluble and amph-vaccines are indicated by asterisks: **$P<0.01$, *$P<0.05$ by one-way ANOVA with Bonferroni post-test. Data show mean ± s.e.m. of 2–4 independent experiments. NS, not significant. Liu et al. [175], pp. 519–522. Reproduced with permission of Macmillan Publishers Ltd.

50% along with evident tumor regression. These encouraging examples demonstrate a potent improvement in vaccination by lymphatic targeting.

21.7 CONCLUSIONS

Running parallel to blood capillaries in the body tissues, the lymphatic system is critical in balancing the tissue fluids, maintaining metabolism and function of the immune system. In particular, the lymphatic system plays active roles in cancer progression, hence it provides novel targets for anticancer treatments. Localized delivery of chemotherapeutic agents to the lympatics will spare the systemic toxicities and suppress the tumor metastases. Favorable recognitions of foreign particulates by the macrophages in the lymph nodes with their subsequent removal from the lymphatic vessels have led to a new approach to targeting the lymphatic system by developing nanoscale drug carriers. A variety of nanoparticles for lymphatic delivery have been exploited, including liposomes, SLNs/NLCs, natural polymers, and polymeric nanoparticles. The uptake and distribution of the drug carriers in the lymphatic system have been observed to be highly dependent on their size, surface charge, hydrophobicity, and the administration routes. Moreover, modification of the nanoparticles surface with ligands that bind with specific markers for lymphatic endothelium has led to a significant enhancement in lymphatic uptake of drug carriers, resulting in substantially augmented treatment efficacy.

REFERENCES

1. Maby-El Hajjami, H. and T.V. Petrova, Developmental and pathological lymphangiogenesis: from models to human disease. *Histochem Cell Biol*, 2008. **130**(6): p. 1063–78.
2. Liao, S. and P.Y. von der Weid, Lymphatic system: an active pathway for immune protection. *Semin Cell Dev Biol*, 2015. **38**: p. 83–9.
3. Yanez, J.A., et al., Intestinal lymphatic transport for drug delivery. *Adv Drug Deliv Rev*, 2011. **63**(10–11): p. 923–42.
4. White, K.L., et al., Lymphatic transport of Methylnortestosterone undecanoate (MU) and the bioavailability of methylnortestosterone are highly sensitive to the mass of coadministered lipid after oral administration of MU. *J Pharmacol Exp Ther*, 2009. **331**(2): p. 700–9.
5. Kaminskas, L.M., et al., PEGylation of interferon α2 improves lymphatic exposure after subcutaneous and intravenous administration and improves antitumour efficacy against lymphatic breast cancer metastases. *J Control Release*, 2013. **168**(2): p. 200–8.
6. Cai, S., et al., Localized doxorubicin chemotherapy with a biopolymeric nanocarrier improves survival and reduces toxicity in xenografts of human breast cancer. *J Control Release*, 2010. **146**(2): p. 212–18.
7. Takakura, Y., et al., Enhanced lymphatic delivery of mitomycin C conjugated with dextran. *Cancer Res*, 1984. **44**(6): p. 2505–10.
8. Kim, J., et al., Lymphatic delivery of 99mTc-labeled dextran acetate particles including cyclosporine A. *J Microbiol Biotechnol*, 2008. **18**(9): p. 1599–605.
9. Jeong, Y.I., et al., Cisplatin-incorporated hyaluronic acid nanoparticles based on ion-complex formation. *J Pharm Sci*, 2008. **97**(3): p. 1268–76.
10. Cai, S., et al., Intralymphatic chemotherapy using a hyaluronan-cisplatin conjugate. *J Surg Res*, 2008. **147**(2): p. 247–52.
11. Wu, H., et al., Examination of lymphatic transport of puerarin in unconscious lymph duct-cannulated rats after administration in microemulsion drug delivery systems. *Eur J Pharm Sci*, 2011. **42**(4): p. 348–53.
12. Singh, B., et al., Optimized nanoemulsifying systems with enhanced bioavailability of carvedilol. *Colloids Surf B Biointerfaces*, 2013. **101**: p. 465–74.
13. Sha, X., et al., Self-microemulsifying drug-delivery system for improved oral bioavailability of probucol: preparation and evaluation. *Int J Nanomedicine*, 2012. **7**: p. 705–12.
14. Khullar, O.V., et al., Nanoparticle migration and delivery of Paclitaxel to regional lymph nodes in a large animal model. *J Am Coll Surg*, 2012. **214**(3): p. 328–37.

15. Sun, M., et al., Intestinal absorption and intestinal lymphatic transport of sirolimus from self-microemulsifying drug delivery systems assessed using the single-pass intestinal perfusion (SPIP) technique and a chylomicron flow blocking approach: linear correlation with oral bioavailabilities in rats. *Eur J Pharm Sci*, 2011. **43**(3): p. 132–40.
16. Thakkar, H., et al., Formulation and characterization of lipid-based drug delivery system of raloxifene-microemulsion and self-microemulsifying drug delivery system. *J Pharm Bioallied Sci*, 2011. **3**(3): p. 442–8.
17. Liu, R., et al., Prevention of nodal metastases in breast cancer following the lymphatic migration of paclitaxel-loaded expansile nanoparticles. *Biomaterials*, 2013. **34**(7): p. 1810–19.
18. Yoshimura, K., et al., Evaluation of endoscopic pirarubicin-Lipiodol emulsion injection therapy for gastric cancer. *Gan To Kagaku Ryoho*, 1996. **23**(11): p. 1519–22.
19. Atef, E. and A.A. Belmonte, Formulation and in vitro and in vivo characterization of a phenytoin self-emulsifying drug delivery system (SEDDS). *Eur J Pharm Sci*, 2008. **35**(4): p. 257–63.
20. Ye, L., et al., Antitumor effect and toxicity of Lipusu in rat ovarian cancer xenografts. *Food Chem Toxicol*, 2013. **52**: p. 200–6.
21. Yan, Z., et al., LyP-1-conjugated PEGylated liposomes: a carrier system for targeted therapy of lymphatic metastatic tumor. *J Control Release*, 2012. **157**(1): p. 118–25.
22. Yan, Z., et al., LyP-1-conjugated doxorubicin-loaded liposomes suppress lymphatic metastasis by inhibiting lymph node metastases and destroying tumor lymphatics. *Nanotechnology*, 2011. **22**(41): p. 415103.
23. Herringson, T.P. and J.G. Altin, Effective tumor targeting and enhanced anti-tumor effect of liposomes engrafted with peptides specific for tumor lymphatics and vasculature. *Int J Pharm*, 2011. **411**(1–2): p. 206–14.
24. Moghimi, M. and S.M. Moghimi, Lymphatic targeting of immuno-PEG-liposomes: evaluation of antibody-coupling procedures on lymph node macrophage uptake. *J Drug Target*, 2008. **16**(7): p. 586–90.
25. Jain, S., A.K. Tiwary, and N.K. Jain, PEGylated elastic liposomal formulation for lymphatic targeting of zidovudine. *Curr Drug Deliv*, 2008. **5**(4): p. 275–81.
26. Akamo, Y., et al., Chemotherapy targeting regional lymph nodes by gastric submucosal injection of liposomal adriamycin in patients with gastric carcinoma. *Jpn J Cancer Res*, 1994. **85**(6): p. 652–8.
27. Zhuang, Y., et al., PEGylated cationic liposomes robustly augment vaccine-induced immune responses: role of lymphatic trafficking and biodistribution. *J Control Release*, 2012. **159**(1): p. 135–42.
28. Bestman-Smith, J., et al., Sterically stabilized liposomes bearing anti-HLA-DR antibodies for targeting the primary cellular reservoirs of HIV-1. *Biochim Biophys Acta*, 2000. **1468**(1–2): p. 161–74.
29. Vingerhoeds, M.H., et al., Immunoliposome-mediated targeting of doxorubicin to human ovarian carcinoma in vitro and in vivo. *Br J Cancer*, 1996. **74**(7): p. 1023–9.
30. Malik, S.T., et al., Therapy of human ovarian cancer xenografts with intraperitoneal liposome encapsulated muramyl-tripeptide phosphoethanolamine (MTP-PE) and recombinant GM-CSF. *Br J Cancer*, 1991. **63**(3): p. 399–403.
31. Lee, M.J., et al., Intraperitoneal gene delivery mediated by a novel cationic liposome in a peritoneal disseminated ovarian cancer model. *Gene Ther*, 2002. **9**(13): p. 859–66.

32. Hirai, M., et al., Novel and simple loading procedure of cisplatin into liposomes and targeting tumor endothelial cells. *Int J Pharm*, 2010. **391**(1–2): p. 274–83.
33. Hirano, K. and C.A. Hunt, Lymphatic transport of liposome-encapsulated agents: effects of liposome size following intraperitoneal administration. *J Pharm Sci*, 1985. **74**(9): p. 915–21.
34. Oussoren, C. and G. Storm, Lymphatic uptake and biodistribution of liposomes after subcutaneous injection: III. Influence of surface modification with poly(ethyleneglycol). *Pharm Res*, 1997. **14**(10): p. 1479–84.
35. Kaur, C.D., M. Nahar, and N.K. Jain, Lymphatic targeting of zidovudine using surface-engineered liposomes. *J Drug Target*, 2008. **16**(10): p. 798–805.
36. Trubetskoy, V.S., et al., Controlled delivery of Gd-containing liposomes to lymph nodes: surface modification may enhance MRI contrast properties. *Magn Reson Imaging*, 1995. **13**(1): p. 31–7.
37. Torchilin, V.P., et al., Targeted delivery of diagnostic agents by surface-modified liposomes. *J Control Release*, 1994. **28**(1–3): p. 45–58.
38. Perrie, Y., et al., Liposome (Lipodine)-mediated DNA vaccination by the oral route. *J Liposome Res*, 2002. **12**(1–2): p. 185–97.
39. Ling, S.S., et al., Enhanced oral bioavailability and intestinal lymphatic transport of a hydrophilic drug using liposomes. *Drug Dev Ind Pharm*, 2006. **32**(3): p. 335–45.
40. Koshkina, N.V., et al., Distribution of camptothecin after delivery as a liposome aerosol or following intramuscular injection in mice. *Cancer Chemother Pharmacol*, 1999. **44**(3): p. 187–92.
41. Parker, R.J., K.D. Hartman, and S.M. Sieber, Lymphatic absorption and tissue disposition of liposome-entrapped [14C]Adriamycin following intraperitoneal administration to rats. *Cancer Res*, 1981. **41**(4): p. 1311–17.
42. Kim, C.K. and J.H. Han, Lymphatic delivery and pharmacokinetics of methotrexate after intramuscular injection of differently charged liposome-entrapped methotrexate to rats. *J Microencapsul*, 1995. **12**(4): p. 437–46.
43. Rutenfranz, I., A. Bauer, and H. Kirchner, Pharmacokinetic study of liposome-encapsulated human interferon-gamma after intravenous and intramuscular injection in mice. *J Interferon Res*, 1990. **10**(3): p. 337–41.
44. Zavaleta, C.L., et al., Use of avidin/biotin-liposome system for enhanced peritoneal drug delivery in an ovarian cancer model. *Int J Pharm*, 2007. **337**(1–2): p. 316–28.
45. Kuo, Y.C. and H.F. Ko, Targeting delivery of saquinavir to the brain using 83-14 monoclonal antibody-grafted solid lipid nanoparticles. *Biomaterials*, 2013. **34**(20): p. 4818–30.
46. Chalikwar, S.S., et al., Formulation and evaluation of Nimodipine-loaded solid lipid nanoparticles delivered via lymphatic transport system. *Colloids Surf B Biointerfaces*, 2012. **97**: p. 109–16.
47. Alex, A., et al., Enhanced delivery of lopinavir to the CNS using Compritol-based solid lipid nanoparticles. *Ther Deliv*, 2011. **2**(1): p. 25–35.
48. Aji Alex, M.R., et al., Lopinavir loaded solid lipid nanoparticles (SLN) for intestinal lymphatic targeting. *Eur J Pharm Sci*, 2011. **42**(1–2): p. 11–18.
49. Baek, J.S., et al., Solid lipid nanoparticles of paclitaxel strengthened by hydroxypropyl-beta-cyclodextrin as an oral delivery system. *Int J Mol Med*, 2012. **30**(4): p. 953–9.

50. Paliwal, R., et al., Effect of lipid core material on characteristics of solid lipid nanoparticles designed for oral lymphatic delivery. *Nanomedicine*, 2009. **5**(2): p. 184–91.
51. Videira, M., A.J. Almeida, and A. Fabra, Preclinical evaluation of a pulmonary delivered paclitaxel-loaded lipid nanocarrier antitumor effect. *Nanomedicine*, 2012. **8**(7): p. 1208–15.
52. Sun, M., et al., Quercetin-nanostructured lipid carriers: characteristics and anti-breast cancer activities in vitro. *Colloids Surf B Biointerfaces*, 2014. **113**: p. 15–24.
53. Luan, J., et al., Design and characterization of Amoitone B-loaded nanostructured lipid carriers for controlled drug release. *Drug Deliv*, 2013. **20**(8): p. 324–30.
54. Zhang, X.Y., et al., Preparation of isoliquiritigenin-loaded nanostructured lipid carrier and the in vivo evaluation in tumor-bearing mice. *Eur J Pharm Sci*, 2013. **49**(3): p. 411–22.
55. Shete, H., et al., Long chain lipid based tamoxifen NLC. Part II: pharmacokinetic, biodistribution and in vitro anticancer efficacy studies. *Int J Pharm*, 2013. **454**(1): p. 584–92.
56. Hsu, S.H., et al., Formulation design and evaluation of quantum dot-loaded nanostructured lipid carriers for integrating bioimaging and anticancer therapy. *Nanomedicine (Lond)*, 2013. **8**(8): p. 1253–69.
57. Bondi, M.L., et al., Nanostructured lipid carriers-containing anticancer compounds: preparation, characterization, and cytotoxicity studies. *Drug Deliv*, 2007. **14**(2): p. 61–7.
58. Patel, A.R., et al., Efficacy of aerosolized celecoxib encapsulated nanostructured lipid carrier in non-small cell lung cancer in combination with docetaxel. *Pharm Res*, 2013. **30**(5): p. 1435–46.
59. Endo, K., et al., Tumor-targeted chemotherapy with the nanopolymer-based drug NC-6004 for oral squamous cell carcinoma. *Cancer Sci*, 2013. **104**(3): p. 369–74.
60. Qin, L., et al., Polymeric micelles for enhanced lymphatic drug delivery to treat metastatic tumors. *J Control Release*, 2013. **171**(2): p. 133–42.
61. Rafi, M., et al., Polymeric micelles incorporating (1,2-diaminocyclohexane)platinum (II) suppress the growth of orthotopic scirrhous gastric tumors and their lymph node metastasis. *J Control Release*, 2012. **159**(2): p. 189–96.
62. Wang, Z., et al., LyP-1 modification to enhance delivery of artemisinin or fluorescent probe loaded polymeric micelles to highly metastatic tumor and its lymphatics. *Mol Pharm*, 2012. **9**(9): p. 2646–57.
63. Luo, G., et al., LyP-1-conjugated nanoparticles for targeting drug delivery to lymphatic metastatic tumors. *Int J Pharm*, 2010. **385**(1–2): p. 150–6.
64. Zhou, L., et al., In vivo antitumor and antimetastatic activities of camptothecin encapsulated with N-trimethyl chitosan in a preclinical mouse model of liver cancer. *Cancer Lett*, 2010. **297**(1): p. 56–64.
65. Rao, D.A., et al., Biodegradable PLGA based nanoparticles for sustained regional lymphatic drug delivery. *J Pharm Sci*, 2010. **99**(4): p. 2018–31.
66. Hawley, A.E., L. Illum, and S.S. Davis, Preparation of biodegradable, surface engineered PLGA nanospheres with enhanced lymphatic drainage and lymph node uptake. *Pharm Res*, 1997. **14**(5): p. 657–61.
67. Lu, H., et al., Paclitaxel nanoparticle inhibits growth of ovarian cancer xenografts and enhances lymphatic targeting. *Cancer Chemother Pharmacol*, 2007. **59**(2): p. 175–81.

68. Manolova, V., et al., Nanoparticles target distinct dendritic cell populations according to their size. *Eur J Immunol*, 2008. **38**(5): p. 1404–13.
69. Hawley, A.E., L. Illum, and S.S. Davis, Lymph node localisation of biodegradable nanospheres surface modified with poloxamer and poloxamine block co-polymers. *FEBS Lett*, 1997. **400**(3): p. 319–23.
70. Liggins, R.T., et al., Paclitaxel loaded poly(L-lactic acid) microspheres for the prevention of intraperitoneal carcinomatosis after a surgical repair and tumor cell spill. *Biomaterials*, 2000. **21**(19): p. 1959–69.
71. Choi, H.S., et al., Preparation and characterization of fentanyl-loaded PLGA microspheres: in vitro release profiles. *Int J Pharm*, 2002. **234**(1–2): p. 195–203.
72. Azouz, S.M., et al., Prevention of local tumor growth with paclitaxel-loaded microspheres. *J Thorac Cardiovasc Surg*, 2008. **135**(5): p. 1014–21.
73. Liu, J., et al., Translymphatic chemotherapy by intrapleural placement of gelatin sponge containing biodegradable Paclitaxel colloids controls lymphatic metastasis in lung cancer. *Cancer Res*, 2009. **69**(3): p. 1174–81.
74. Tawde, S.A., et al., Formulation and evaluation of oral microparticulate ovarian cancer vaccines. *Vaccine*, 2012. **30**(38): p. 5675–81.
75. Coppi, G. and V. Iannuccelli, Alginate/chitosan microparticles for tamoxifen delivery to the lymphatic system. *Int J Pharm*, 2009. **367**(1–2): p. 127–32.
76. Yang, F., et al., Magnetic functionalised carbon nanotubes as drug vehicles for cancer lymph node metastasis treatment. *Eur J Cancer*, 2011. **47**(12): p. 1873–82.
77. Yang, D., et al., Hydrophilic multi-walled carbon nanotubes decorated with magnetite nanoparticles as lymphatic targeted drug delivery vehicles. *Chem Commun (Camb)*, 2009(29): p. 4447–9.
78. Nassar, T., et al., High plasma levels and effective lymphatic uptake of docetaxel in an orally available nanotransporter formulation. *Cancer Res*, 2011. **71**(8): p. 3018–28.
79. Cai, S., et al., Carrier-based intralymphatic cisplatin chemotherapy for the treatment of metastatic squamous cell carcinoma of the head & neck. *Ther Deliv*, 2010. **1**(2): p. 237–45.
80. Cohen, M.S., et al., A novel intralymphatic nanocarrier delivery system for cisplatin therapy in breast cancer with improved tumor efficacy and lower systemic toxicity in vivo. *Am J Surg*, 2009. **198**(6): p. 781–6.
81. Strieth, S., et al., Paclitaxel encapsulated in cationic liposomes increases tumor microvessel leakiness and improves therapeutic efficacy in combination with Cisplatin. *Clin Cancer Res*, 2008. **14**(14): p. 4603–11.
82. Tejada-Berges, T., et al., Caelyx/Doxil for the treatment of metastatic ovarian and breast cancer. *Expert Rev Anticancer Ther*, 2002. **2**(2): p. 143–50.
83. Frenkel, V., et al., Delivery of liposomal doxorubicin (Doxil) in a breast cancer tumor model: investigation of potential enhancement by pulsed-high intensity focused ultrasound exposure. *Acad Radiol*, 2006. **13**(4): p. 469–79.
84. Prescott, L.M., Doxil offers hope to KS sufferers. *J Int Assoc Physicians AIDS Care*, 1995. **1**(11): p. 43–4.
85. O'Brien, M.E., et al., Reduced cardiotoxicity and comparable efficacy in a phase III trial of pegylated liposomal doxorubicin HCl (CAELYX/Doxil) versus conventional doxorubicin for first-line treatment of metastatic breast cancer. *Ann Oncol*, 2004. **15**(3): p. 440–9.

86. Pan, X.Q., H. Wang, and R.J. Lee, Antitumor activity of folate receptor-targeted liposomal doxorubicin in a KB oral carcinoma murine xenograft model. *Pharm Res*, 2003. **20**(3): p. 417–22.
87. Khato, J., E.R. Priester, and S.M. Sieber, Enhanced lymph node uptake of melphalan following liposomal entrapment and effects on lymph node metastasis in rats. *Cancer Treat Rep*, 1982. **66**(3): p. 517–27.
88. Vaage, J., et al., Therapy of a xenografted human colonic carcinoma using cisplatin or doxorubicin encapsulated in long-circulating pegylated stealth liposomes. *Int J Cancer*, 1999. **80**(1): p. 134–7.
89. Lawson, K.A., et al., Novel vitamin E analogue and 9-nitro-camptothecin administered as liposome aerosols decrease syngeneic mouse mammary tumor burden and inhibit metastasis. *Cancer Chemother Pharmacol*, 2004. **54**(5): p. 421–31.
90. Latimer, P., et al., Aerosol delivery of liposomal formulated paclitaxel and vitamin E analog reduces murine mammary tumor burden and metastases. *Exp Biol Med (Maywood)*, 2009. **234**(10): p. 1244–52.
91. Leak, L.V., The structure of lymphatic capillaries in lymph formation. *Fed Proc*, 1976. **35**(8): p. 1863–71.
92. Watson, R., *Anatomy and physiology for nurses*. 13th ed. 2011, Edinburgh/New York: Baillière Tindall/Elsevier. vi, 367 p.
93. Yoffey, J.M. and F.C. Courtice, *Lymphatics, lymph and the lymphomyeloid complex*. 1970, London/New York: Academic Press. xviii, 942 p.
94. Casley-Smith, J.R. and H.W. Florey, The structure of normal small lymphatics. *Q J Exp Physiol Cogn Med Sci*, 1961. **46**: p. 101–6.
95. Leak, L.V., Electron microscopic observations on lymphatic capillaries and the structural components of the connective tissue-lymph interface. *Microvasc Res*, 1970. **2**(4): p. 361–91.
96. McLennan, D.N., C.J.H. Porter, and S.A. Charman, Subcutaneous drug delivery and the role of the lymphatics. *Drug Discov Today Technol*, 2005. **2**(1): p. 89–96.
97. Charman, W.N. and V.J. Stella, *Lymphatic transport of drugs*. 1992, Boca Raton: CRC Press. 331 p.
98. Mohrman, D.E. and L.J. Heller, *Cardiovascular physiology*. 4th ed. 1997, New York: McGraw-Hill, Health Professions Division. xi, 254 p.
99. Sakorafas, G.H., J. Geraghty, and G. Pavlakis, The clinical significance of axillary lymph node micrometastases in breast cancer. *Eur J Surg Oncol*, 2004. **30**(8): p. 807–16.
100. Donovan, C.A., et al., Correlation of breast cancer axillary lymph node metastases with stem cell mutations. *JAMA Surg*, 2013. **148**(9): p. 873–8.
101. Cox, C.E., et al., Significance of sentinel lymph node micrometastases in human breast cancer. *J Am Coll Surg*, 2008. **206**(2): p. 261–8.
102. Ito, Y., et al., Characterization of a novel lymph node metastasis model from human colonic cancer and its preclinical use for comparison of anti-metastatic efficacy between oral S-1 and UFT/LV. *Cancer Sci*, 2010. **101**(8): p. 1853–60.
103. Kojima, T., et al., In vivo biological purging for lymph node metastasis of human colorectal cancer by telomerase-specific oncolytic virotherapy. *Ann Surg*, 2010. **251**(6): p. 1079–86.
104. Markl, B., et al., The clinical significance of lymph node size in colon cancer. *Mod Pathol*, 2012. **25**(10): p. 1413–22.

105. Cheng, L., et al., Cell proliferation in prostate cancer patients with lymph node metastasis: a marker for progression. *Clin Cancer Res*, 1999. **5**(10): p. 2820–3.
106. Bastide, C., et al., A Nod Scid mouse model to study human prostate cancer. *Prostate Cancer Prostatic Dis*, 2002. **5**(4): p. 311–15.
107. Datta, K., et al., Mechanism of lymph node metastasis in prostate cancer. *Future Oncol*, 2010. **6**(5): p. 823–36.
108. Guzzo, T.J., et al., Impact of adjuvant chemotherapy on patients with lymph node metastasis at the time of radical cystectomy. *Can J Urol*, 2010. **17**(6): p. 5465–71.
109. Goldenberg, D., et al., Cystic lymph node metastasis in patients with head and neck cancer: an HPV-associated phenomenon. *Head Neck*, 2008. **30**(7): p. 898–903.
110. Beasley, N.J., et al., Intratumoral lymphangiogenesis and lymph node metastasis in head and neck cancer. *Cancer Res*, 2002. **62**(5): p. 1315–20.
111. Moore, B.A., et al., Lymph node metastases from cutaneous squamous cell carcinoma of the head and neck. *Laryngoscope*, 2005. **115**(9): p. 1561–7.
112. Veness, M.J., et al., Cutaneous head and neck squamous cell carcinoma metastatic to cervical lymph nodes (nonparotid): a better outcome with surgery and adjuvant radiotherapy. *Laryngoscope*, 2003. **113**(10): p. 1827–33.
113. Nathanson, S.D., Insights into the mechanisms of lymph node metastasis. *Cancer*, 2003. **98**(2): p. 413–23.
114. Yokoyama, J., et al., Impact of lymphatic chemotherapy targeting metastatic lymph nodes in patients with tongue cancer (cT3N2bM0) using intra-arterial chemotherapy. *Head Neck Oncol*, 2012. **4**: p. 64.
115. Yokoyama, J., et al., A feasibility study of lymphatic chemotherapy targeting sentinel lymph nodes of patients with tongue cancer (cT3N0M0) using intra-arterial chemotherapy. *Head Neck Oncol*, 2012. **4**: p. 60.
116. Oussoren, C. and G. Storm, Liposomes to target the lymphatics by subcutaneous administration. *Adv Drug Deliv Rev*, 2001. **50**(1–2): p. 143–56.
117. Nishioka, Y. and H. Yoshino, Lymphatic targeting with nanoparticulate system. *Adv Drug Deliv Rev*, 2001. **47**(1): p. 55–64.
118. Hawley, A.E., S.S. Davis, and L. Illum, Targeting of colloids to lymph nodes: influence of lymphatic physiology and colloidal characteristics. *Adv Drug Deliv Rev*, 1995. **17**(1): p. 129–48.
119. Reddy, S.T., et al., In vivo targeting of dendritic cells in lymph nodes with poly(propylene sulfide) nanoparticles. *J Control Release*, 2006. **112**(1): p. 26–34.
120. Oussoren, C., et al., Lymphatic uptake and biodistribution of liposomes after subcutaneous injection. II. Influence of liposomal size, lipid composition and lipid dose. *Biochim Biophys Acta*, 1997. **1328**(2): p. 261–72.
121. Patel, H.M., K.M. Boodle, and R. Vaughan-Jones, Assessment of the potential uses of liposomes for lymphoscintigraphy and lymphatic drug delivery. Failure of 99m-technetium marker to represent intact liposomes in lymph nodes. *Biochim Biophys Acta*, 1984. **801**(1): p. 76–86.
122. Kaminskas, L.M. and C.J. Porter, Targeting the lymphatics using dendritic polymers (dendrimers). *Adv Drug Deliv Rev*, 2011. **63**(10–11): p. 890–900.
123. Schmid-Schonbein, G.W., Microlymphatics and lymph flow. *Physiol Rev*, 1990. **70**(4): p. 987–1028.

124. Patel, H.M., Serum opsonins and liposomes: their interaction and opsonophagocytosis. *Crit Rev Ther Drug Carrier Syst*, 1992. **9**(1): p. 39–90.
125. Moghimi, S.M., The effect of methoxy-PEG chain length and molecular architecture on lymph node targeting of immuno-PEG liposomes. *Biomaterials*, 2006. **27**(1): p. 136–44.
126. Christy, N., *Targeting of colloids to the lymphatic system*, 1992, University of Nottingham, Nottingham.
127. Sanjula, B., et al., Effect of poloxamer 188 on lymphatic uptake of carvedilol-loaded solid lipid nanoparticles for bioavailability enhancement. *J Drug Target*, 2009. **17**(3): p. 249–56.
128. Lopes, S.C.d.A., et al., Liposomes as carriers of anticancer drugs, in *Cancer treatment—conventional and innovative approaches*, L. Rangel, Editor. 2013, Rijeka: InTech.
129. Safinya, C.R.a.E. and K. Kai, Materials chemistry: liposomes derived from molecular vases. *Nature*, 2012. **489**(7416): p. 372–4.
130. Soundararajan, A., et al., [(186)Re]Liposomal doxorubicin (Doxil): in vitro stability, pharmacokinetics, imaging and biodistribution in a head and neck squamous cell carcinoma xenograft model. *Nucl Med Biol*, 2009. **36**(5): p. 515–24.
131. Laakkonen, P., et al., A tumor-homing peptide with a targeting specificity related to lymphatic vessels. *Nat Med*, 2002. **8**(7): p. 751–5.
132. Laakkonen, P., et al., Antitumor activity of a homing peptide that targets tumor lymphatics and tumor cells. *Proc Natl Acad Sci U S A*, 2004. **101**(25): p. 9381–6.
133. Fogal, V., et al., Mitochondrial/cell-surface protein p32/gC1qR as a molecular target in tumor cells and tumor stroma. *Cancer Res*, 2008. **68**(17): p. 7210–18.
134. Thurston, G., et al., Cationic liposomes target angiogenic endothelial cells in tumors and chronic inflammation in mice. *J Clin Invest*, 1998. **101**(7): p. 1401–13.
135. Zhang, P., et al., Preparation and evaluation of self-microemulsifying drug delivery system of oridonin. *Int J Pharm*, 2008. **355**(1–2): p. 269–76.
136. Pouton, C.W., Lipid formulations for oral administration of drugs: non-emulsifying, self-emulsifying and "self-microemulsifying" drug delivery systems. *Eur J Pharm Sci*, 2000. **11**(Suppl 2): p. S93–8.
137. Muller, R.H., K. Mader, and S. Gohla, Solid lipid nanoparticles (SLN) for controlled drug delivery—a review of the state of the art. *Eur J Pharm Biopharm*, 2000. **50**(1): p. 161–77.
138. Mehnert, W. and K. Mader, Solid lipid nanoparticles: production, characterization and applications. *Adv Drug Deliv Rev*, 2001. **47**(2–3): p. 165–96.
139. Helgason, T., et al., Effect of surfactant surface coverage on formation of solid lipid nanoparticles (SLN). *J Colloid Interface Sci*, 2009. **334**(1): p. 75–81.
140. Jores, K., et al., Investigations on the structure of solid lipid nanoparticles (SLN) and oil-loaded solid lipid nanoparticles by photon correlation spectroscopy, field-flow fractionation and transmission electron microscopy. *J Control Release*, 2004. **95**(2): p. 217–27.
141. Bargoni, A., et al., Solid lipid nanoparticles in lymph and plasma after duodenal administration to rats. *Pharm Res*, 1998. **15**(5): p. 745–50.
142. Shidhaye, S.S., et al., Solid lipid nanoparticles and nanostructured lipid carriers—innovative generations of solid lipid carriers. *Curr Drug Deliv*, 2008. **5**(4): p. 324–31.
143. Selvamuthukumar, S. and R. Velmurugan, Nanostructured lipid carriers: a potential drug carrier for cancer chemotherapy. *Lipids Health Dis*, 2012. **11**: p. 159.

144. Niu, C., et al., Doxorubicin loaded superparamagnetic PLGA-iron oxide multifunctional microbubbles for dual-mode US/MR imaging and therapy of metastasis in lymph nodes. *Biomaterials*, 2013. **34**(9): p. 2307–17.
145. Maincent, P., et al., Lymphatic targeting of polymeric nanoparticles after intraperitoneal administration in rats. *Pharm Res*, 1992. **9**(12): p. 1534–9.
146. Cai, S., et al., Pharmacokinetics and disposition of a localized lymphatic polymeric hyaluronan conjugate of cisplatin in rodents. *J Pharm Sci*, 2010. **99**(6): p. 2664–71.
147. Xu, W., P. Ling, and T. Zhang, Polymeric micelles, a promising drug delivery system to enhance bioavailability of poorly water-soluble drugs. *J Drug Deliv*, 2013. **2013**: p. 15.
148. Croy, S.R. and G.S. Kwon, Polymeric micelles for drug delivery. *Curr Pharm Des*, 2006. **12**(36): p. 4669–84.
149. Deng, C., et al., Biodegradable polymeric micelles for targeted and controlled anticancer drug delivery: promises, progress and prospects. *Nano Today*, 2012. **7**(5): p. 467–80.
150. Murakami, H., et al., Preparation of poly(DL-lactide-co-glycolide) nanoparticles by modified spontaneous emulsification solvent diffusion method. *Int J Pharm*, 1999. **187**(2): p. 143–52.
151. Govender, T., et al., PLGA nanoparticles prepared by nanoprecipitation: drug loading and release studies of a water soluble drug. *J Control Release*, 1999. **57**(2): p. 171–85.
152. Mainardes, R.M. and R.C. Evangelista, PLGA nanoparticles containing praziquantel: effect of formulation variables on size distribution. *Int J Pharm*, 2005. **290**(1–2): p. 137–44.
153. Wang, Y.M., et al., Preparation and characterization of poly(lactic-co-glycolic acid) microspheres for targeted delivery of a novel anticancer agent, taxol. *Chem Pharm Bull (Tokyo)*, 1996. **44**(10): p. 1935–40.
154. Jalil, R. and J.R. Nixon, Microencapsulation using poly(L-lactic acid). I: microcapsule properties affected by the preparative technique. *J Microencapsul*, 1989. **6**(4): p. 473–84.
155. Sarmento, B., et al., Oral insulin delivery by means of solid lipid nanoparticles. *Int J Nanomedicine*, 2007. **2**(4): p. 743–9.
156. Videira, M.A., et al., Lymphatic uptake of pulmonary delivered radiolabelled solid lipid nanoparticles. *J Drug Target*, 2002. **10**(8): p. 607–13.
157. Videira, M.A., et al., Lymphatic uptake of lipid nanoparticles following endotracheal administration. *J Microencapsul*, 2006. **23**(8): p. 855–62.
158. Mishra, H., et al., Evaluation of solid lipid nanoparticles as carriers for delivery of hepatitis B surface antigen for vaccination using subcutaneous route. *J Pharm Pharm Sci*, 2010. **13**(4): p. 495–509.
159. Kaminskas, L.M., et al., PEGylation of polylysine dendrimers improves absorption and lymphatic targeting following SC administration in rats. *J Control Release*, 2009. **140**(2): p. 108–16.
160. Trevaskis, N.L., W.N. Charman, and C.J. Porter, Lipid-based delivery systems and intestinal lymphatic drug transport: a mechanistic update. *Adv Drug Deliv Rev*, 2008. **60**(6): p. 702–16.
161. Zara, G.P., et al., Pharmacokinetics and tissue distribution of idarubicin-loaded solid lipid nanoparticles after duodenal administration to rats. *J Pharm Sci*, 2002. **91**(5): p. 1324–33.
162. Riquet, M., Bronchial arteries and lymphatics of the lung. *Thorac Surg Clin*, 2007. **17**(4): p. 619–38.

163. McIntire, G.L., et al., Pulmonary delivery of nanoparticles of insoluble, iodinated CT X-ray contrast agents to lung draining lymph nodes in dogs. *J Pharm Sci*, 1998. **87**(11): p. 1466–70.

164. Leak, L.V. and V.J. Ferrans Lee, Lymphatics and lymphoid tissue, in *The lung: scientific foundations*, R.G. Crystal, West, J.B., Cherniack, N.S. and Weibel, E.R., Editors. 1991, New York: Raven Press. p. 779–86.

165. Varshosaz, J., et al., Biodistribution of amikacin solid lipid nanoparticles after pulmonary delivery. *BioMed Res Int*, 2013. **2013**: p. 8.

166. Lammon, C.B., *Clinical nursing skills*. 1995, Philadelphia: W.B. Saunders.

167. Moffett, V.S., *Human physiology: foundations and frontiers*. 2nd ed. 1993, St. Louis: Mosby, Incorporated. 880.

168. Harivardhan Reddy, L., et al., Influence of administration route on tumor uptake and biodistribution of etoposide loaded solid lipid nanoparticles in Dalton's lymphoma tumor bearing mice. *J Control Release*, 2005. **105**(3): p. 185–98.

169. Flessner, M.F., et al., Peritoneal absorption of macromolecules studied by quantitative autoradiography. *Am J Physiol*, 1985. **248**(1 Pt 2): p. H26–32.

170. Negrini, D., et al., Distribution of diaphragmatic lymphatic stomata. *J Appl Physiol (1985)*, 1991. **70**(4): p. 1544–9.

171. Guy, B., The perfect mix: recent progress in adjuvant research. *Nat Rev Microbiol*, 2007. **5**(7): p. 505–17.

172. Johansen, P., et al., Lympho-geographical concepts in vaccine delivery. *J Control Release*, 2010. **148**(1): p. 56–62.

173. Moon, J.J., B. Huang, and D.J. Irvine, Engineering nano- and microparticles to tune immunity. *Adv Mater*, 2012. **24**(28): p. 3724–46.

174. Moon, J.J., et al., Enhancing humoral responses to a malaria antigen with nanoparticle vaccines that expand Tfh cells and promote germinal center induction. *Proc Natl Acad Sci U S A*, 2012. **109**(4): p. 1080–5.

175. Liu, H., et al., Structure-based programming of lymph-node targeting in molecular vaccines. *Nature*, 2014. **507**: p. 519–22.

176. Senti, G., P. Johansen, and T.M. Kundig, Intralymphatic immunotherapy. *Curr Opin Allergy Clin Immunol*, 2009. **9**(6): p. 537–43.

177. Martinez-Gomez, J.M., et al., Intralymphatic injections as a new administration route for allergen-specific immunotherapy. *Int Arch Allergy Immunol*, 2009. **150**(1): p. 59–65.

178. Ribas, A., et al., Intralymph node prime-boost vaccination against Melan A and tyrosinase for the treatment of metastatic melanoma: results of a phase 1 clinical trial. *Clin Cancer Res*, 2011. **17**(9): p. 2987–96.

22

THE DEVELOPMENT OF CANCER THERANOSTICS: A NEW EMERGING TOOL TOWARD PERSONALIZED MEDICINE

HONGYING SU[1,2], YUN ZENG[2,3], GANG LIU[2], AND XIAOYUAN CHEN[4]

[1] *Department of Chemical Engineering, Kunming University of Science and Technology, Kunming, China*

[2] *State Key Laboratory of Molecular Vaccinology and Molecular Diagnostics & Center for Molecular Imaging and Translational Medicine, School of Public Health, Xiamen University, Xiamen, China*

[3] *Sichuan Key Laboratory of Medical Imaging, North Sichuan Medical College, Nanchong, China*

[4] *Laboratory of Molecular Imaging and Nanomedicine (LOMIN), National Institute of Biomedical Imaging and Bioengineering (NIBIB), National Institutes of Health (NIH), Bethesda, MD, USA*

22.1 INTRODUCTION

Cancer is the second leading cause of death in the United States and accounts for approximately one in every four deaths. Statistics from the National Cancer Institute (NCI) showed that more than 1590 Americans die of cancer everyday, and a total of 1,660,290 new cases would suffer from cancer in the United States in 2013 [1]. Current treatment of cancer involves several techniques including surgery, radiotherapy, chemotherapy, hyperthermia, immunotherapy, stem cell therapy, and combinations thereof [2]. Approximately 50% of human cancer treatments rely on the use of

Drug Delivery: Principles and Applications, Second Edition. Edited by Binghe Wang, Longqin Hu, and Teruna J. Siahaan.

chemotherapy, especially for metastatic cancers. However, the nonspecific distribution of conventional chemotherapeutic drugs in the human body may cause cytotoxicity to both normal and cancer cells, leading to dose-related side effects and inadequate drug concentrations at the tumor site. The effectiveness of chemotherapy is also limited by the selection of drug-resistant cells expressing the multidrug-resistance (MDR) phenotype [3–5]. Successful cancer management depends on accurate and early diagnostics, along with specific treatment protocols. The idea of monitoring the medical treatment process timely in different individuals leads to the concept of theranostics [6, 7].

Theranostics, a portmanteau of therapeutics and diagnostics, is related to but different from traditional imaging and therapeutics. It examines a diverse set of cancer biomarkers, identifies suitable molecular targets to treat, determines the pharmacokinetics and pharmacodynamics of a drug, and tailors the personalized therapeutic strategies. The term "theranostics" was first used by the PharmaNetics' president and CEO John Funkhouser [8] in describing his company's business model in developing diagnostic tests directly linked to the application of a specific therapeutics. Currently, it is defined as a treatment strategy that combines diagnostics with therapeutics on a single platform, aiming to monitor the response to treatment and increase drug efficacy and safety [6]. Theranostics would, therefore, be a key part of personalized medicine and require considerable advances in predictive medicine. In the past decade, theranostics is growing rapidly along with the advances and collaborations in pharmacy, molecular imaging, medicine, biology, nanotechnology, and chemistry (Fig. 22.1) [9].

Recent progress in nanotechnology offers an opportunity to draw diagnosis and therapy closer for cancer theranostics. A new term "nanotheranostics" has been

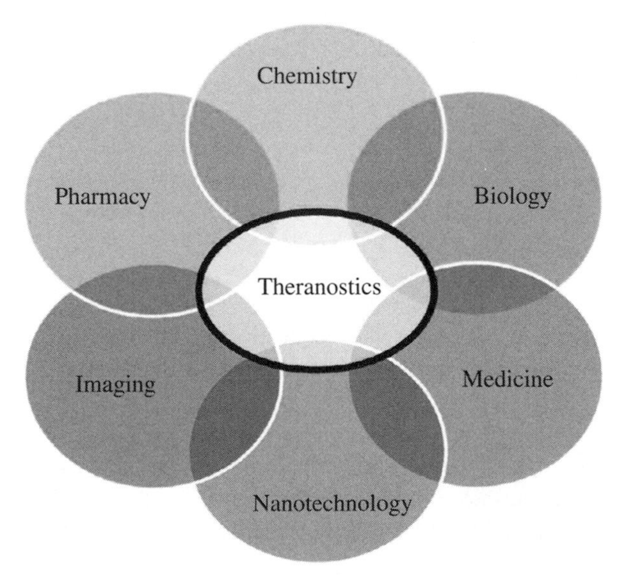

FIGURE 22.1 Schematic representation of the highly interdisciplinary field of theranostics; the future of theranostics depends on multidisciplinary cooperation between medical scientists, molecular biologists, chemists, materials scientists, physicists, and imaging specialists. Lammers et al. [9], pp. 1029–1038. Reproduced with permission of American Chemical Society.

proposed to describe these nanoplatforms using combined imaging probes and therapeutic agents together for simultaneous tumor detection and imaging-guided drug delivery and release [10–12]. The most promising advantages of nanoparticle-based therapeutics, diagnostics, and theranostics are their potential to accumulate (or be targeted) at the tumor site and reduce the possible numerous untoward side effects by using both passive- and active-targeting strategies. The nanometric size of these biomaterials precludes them from being readily cleared through the kidneys, thereby extending the circulation time in the bloodstream depending on their surface functionalization characteristics. Nanoscopic platforms that combine therapeutic agents, diagnostic imaging capabilities, and molecular targeting are emerging as the next generation of multifunctional nanomedicine to improve the treatment outcome for cancer patients [9, 13–15]. Work in "nanotheranostics" can be significant for being cutting edge and clinically translatable, and the community has grown rapidly.

22.2 IMAGING-GUIDED DRUG DELIVERY AND THERAPY

Molecular imaging allows the *in vivo* characterization and measurement of biological processes at the cellular and molecular level. Along with the development of molecular and cell biology techniques, molecular imaging has emerged as an indispensable tool for diagnosis and therapy of many diseases by understanding their pathological mechanism at cellular and molecular levels [16–20]. Several imaging modalities are involved in molecular imaging (as shown in Fig. 22.2) [21, 22], including magnetic resonance imaging (MRI), X-ray computed tomography (CT), positron emission tomography (PET), single-photon emission computed tomography (SPECT), fluorescence imaging (FLI), and ultrasound (US) imaging. By exploiting specific molecular probes or contrast media, these powerful diagnostic imaging techniques can be used to target and detect early-stage cancer and provide a rapid approach for treatment evaluation. Medical imaging has already become an important tool in the processes of drug discovery and development that improve the efficiency of drug screening and investigate the pharmacology and safety of novel drugs [23].

However, each imaging modality mentioned earlier has its own advantages and disadvantages in terms of spatial resolution, sensitivity, and penetration depth [24]. For example, US is safe and of relatively low cost, but its spatial resolution is poor compared to CT and MRI. The CT is a classical anatomical imaging modality and deals with the visualization of organs or tissues, such as skeletal structure and the lung. The MRI offers sensitive detection of soft tissue pathologies and conveys valuable information related to physiological process. Nuclear imaging (i.e., PET and SPECT) is a sensitive tool for diagnosing diseases at an early stage, analyzing drug biodistribution and assessing their efficacy in patients, but the spatial resolution of PET is lower than US, CT, and MRI. Optical imaging technologies, including fluorescence and bioluminescence imaging, are highly sensitive and easily accessible at limited depths of a few millimeters. Combination of two or more imaging modalities may thus improve the overall outcome of diagnosis and therapeutic efficacy monitoring.

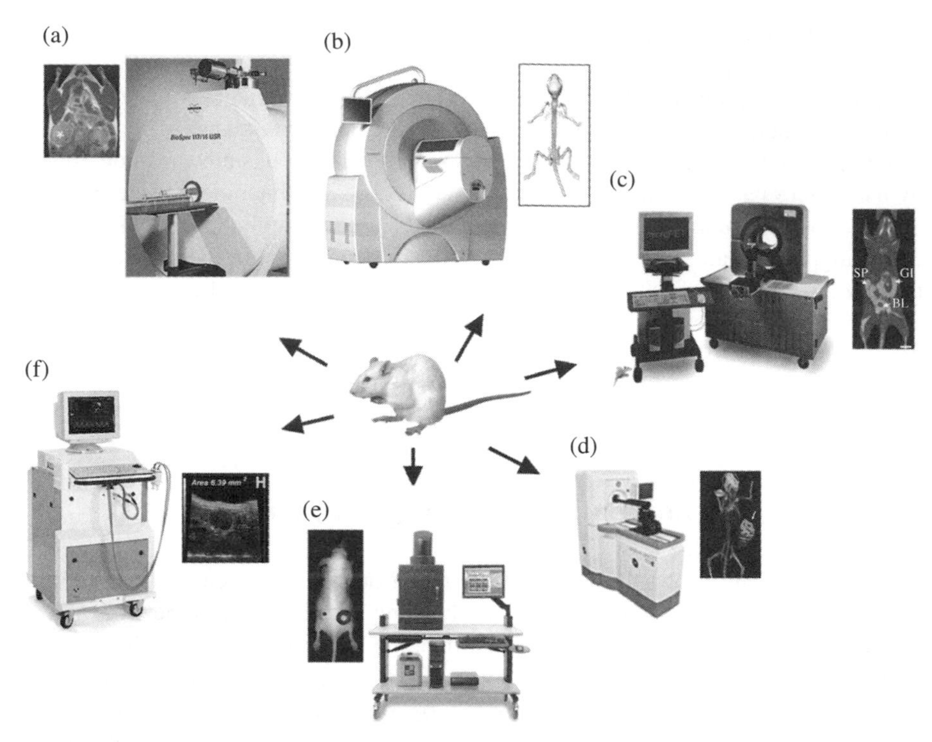

FIGURE 22.2 Typical molecular imaging instruments and images representative of each modality: (a) MRI, (b) computed tomography, (c) positron emission tomography (PET), (d) single-photon emission computed tomography (SPECT), (e) optical imaging, and (f) ultrasound. Janib et al. [21], pp. 1052–1063. Reproduced with permission of Elsevier.

With the recent advances in nanotechnology, nanoparticle-based contrast agents are quickly becoming valuable and potentially transformative tools for enhancing medical diagnostics for a wide range of *in vivo* imaging modalities. Compared with conventional molecular-scale contrast agents, nanoparticles (NPs) promise improved abilities for *in vivo* detection and potentially enhanced targeting efficiencies through longer engineered circulation times, designed clearance pathways, and multimeric-binding capacities [25–27]. The emergence of nanotechnology has offered an opportunity to draw diagnosis and therapy closer. Nanoparticle-based imaging and therapy have been investigated separately, and understanding of them has now evolved to a point enabling the birth of NP-based theranostics, which can be defined as nanoplatforms that can co-deliver therapeutic and imaging functions. Among all these imaging modalities, MRI, PET, SPECT, and optical imaging (bioluminescence and fluorescence) are currently available for nanoplatform-based theranostics. Combinations of imaging probes with therapeutic agents lead to a multifunctional platform that makes early detection of tumors and timely monitoring of the medical treatment process possible. In this chapter, we will focus on the theranostic platforms based on nanoparticles and their applications.

22.3 OPTICAL IMAGING-BASED THERANOSTICS

Among these molecular imaging modalities mentioned earlier, optical imaging (bioluminescence and fluorescence) is less expensive and suitable primarily for small animal studies: it has already become the most commonly used approach for *in vitro* and *ex vivo* applications in molecular and cellular biology. A fundamental issue in optical imaging is its poor tissue penetration (1–2 cm) and significant background due to tissue autofluorescence and light absorption in the visible light region (395–600 nm) [16, 28, 29]. Progress in optical imaging strategies was made recently with the development of near-infrared (NIR) fluorochromes and targeted bioluminescence probes. Currently, FLI and bioluminescence imaging (BLI) have become two of the most extensively studied optical imaging techniques. One major advantage of FLI is that multiple fluorescence probes with different emission spectra can be used for multiplexed molecular imaging. In FLI with probes, excitation light illuminates the subject and the emission light is collected at different wavelengths. The optical signals can provide molecular and cellular information of biological tissues related to cancer metabolism and biochemistry [30, 31]. Bioluminescence can be imaged as deep as several centimeters within tissue and allows at least organ-level resolution. This form of optical imaging modality has been proven to be a powerful methodology in research on monitoring transgene expression, progression of infection, tumor growth and metastasis, gene therapy, and so on. In the following section we will focus on discussing the developments and applications of NIR FLI and BLI used for optical image-guided cancer diagnosis and treatment.

22.3.1 NIR Fluorescence Imaging

Traditional fluorescence modality suffers from poor tissue penetration and significant background due to tissue autofluorescence and light absorption by proteins, heme groups, and even water [22, 28, 32]. NIR FLI possesses enhanced light penetration depth through living tissues because the absorbance spectra for all bimolecular materials reach minima in the NIR region (700–1000 nm) [29, 33, 34]. Therefore, NIR FLI has a high sensitivity and offers a unique advantage for the imaging of pathophysiological changes at the cellular and molecular level, which is most suitable for rapid and cost-effective preclinical evaluation in small animal models for *in vivo* imaging. Low levels of autofluorescence from organisms and tissues in the NIR spectral range can minimize background interference. NIR imaging has a high sensitivity and offers a unique advantage for the imaging of small pathophysiological changes, which is most suitable for *in vivo* imaging.

In recent years, NIR FLI coupled with novel fluorescent probes have received particular attention. Highly sensitive and efficient NIR fluorescent probes can be functionalized with targeting moieties and therapeutic agents, providing new potentials for clinical diagnostics and therapies and will undoubtedly play a critical role in cancer therapy [35–38]. NIR probes can be generally divided into two categories: inorganic and organic molecules [36, 39]. Inorganic NIR imaging probes are mainly associated with nanoparticles (NPs) such as quantum dots (QDs) and upconversion NPs.

Organic NIR fluorochromes are mainly small molecules with low molecular weight. Two kinds of organic fluorochromes are commercially available, including cyanine and Alexa Fluor dyes.

Traditional organic NIR dyes have low toxicity, weak photodegradation, and functional groups allowing convenient chemical modification. Recently, newly developed NIR dyes with improved photophysical properties, lower cost, and suitability for large-scale chemical synthesis make them attractive candidates for FLI. One of the best known cyanine dyes is Cy5.5, which can be detected *in vivo* at sub-nanomolar concentrations and has been proven to be a promising contrast agent for *in vivo* demarcation of tumors by several groups [40–42]. This NIR dye has an absorbance maximum at 675 nm and emission maximum at 694 nm. Several reactive Cy5.5 dyes modified with active groups are now commercially available and can be easily conjugated to some macromolecules (proteins, nuclear acids, dendrimers, etc.) or nanocarriers. Talanov and coworkers initially developed a type of dendrimer-based bifunctional NPs by adding covalently attached Gd-(III)-DTPA chelates and units of the Cy5.5 dye to polyamidoamine (PAMAM) dendrimers [42]. The potential of the resulting NPs as an MRI/FLI dual-modality imaging agent was demonstrated *in vivo* by efficient visualization of sentinel lymph nodes in mice.

Compared with organic dyes and fluorescent proteins, NIR probes based on NPs offer several advantages: (i) higher photo stability than organic dye molecules, allowing continuous imaging of cell or tissue over an extended period of time, (ii) separate luminescent excitation and emission peaks resulting in minimization of the high autofluorescence background of tissue, (iii) bright photoluminescence, improving the sensitivity of *in vivo* imaging, and (iv) multicolor QDs nanocrystals of different sizes can be excited by a single wavelength, which makes it possible to track a panel of molecular markers simultaneously [43, 44].

An emerging new class of NP-based NIR probes for *in vivo* FLI is quantum dots (QDs). QDs, also known as semiconductor nanocrystals, are a special class of NPs, ranging from 2 to 10 nm in diameter, which are single crystals composed of elements from periodic groups of II–VI (CdSe) or III–V (InP) [45–48]. QDs have unique optical and electronic properties: size-tunable emission, extreme brightness, resistance against photobleaching, and simultaneous excitation of multiple fluorescence colors. These properties ensure that they are widely used in biological applications as fluorescent probes [49–53]. NIR emitting QDs, also called type II QDs, including PbSe [54], CdTe/CdSe (core/shell) [55, 56], CuInSe [57], and $InAs_xP_{1-x}$/InP/ZnSe QDs [58], have been developed. Chen et al. reported an integrin-targeted NIR QDs for *in vivo* targeting and imaging of integrin $\alpha_v\beta_3$ (Fig. 22.3) [59]. In this study, CdTe/ZnS QD705 (emission maximum at 705 nm) was labeled by arginine-glycine-aspartic acid (RGD) peptide. NIR imaging after injection of QD705-RGD showed that the accumulation of QDs resulted in increasing fluorescence intensity in integrin expressing U87MG tumor. At 6 hours post injection, the tumor signal intensity reached its maximum. This work demonstrated that RGD peptide–labeled QDs can specifically target integrin $\alpha_v\beta_3$ in living mice, which opens up new perspectives for integrin-targeted optical imaging and shows great potential as QD-based theranostics.

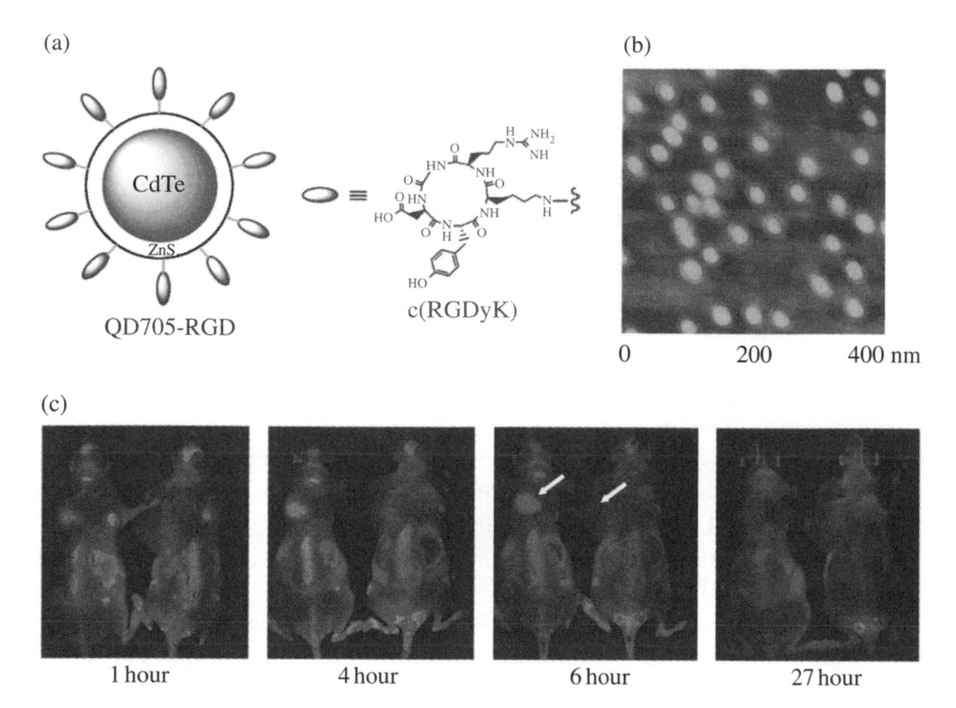

FIGURE 22.3 RGD peptide-labeled QD705 for NIR fluorescence imaging of tumor vasculature: (a) Schematic illustration of the peptide-labeled QDs (QD705-RGD), (b) AFM of QD705-RGD deposited on a silicon wafer and (c) *in vivo* NIR fluorescence imaging of U87MG tumor-bearing mice injected with QD705-RGD (left) and QD705 (right), the tumor signal intensity reached its maximum at 6-hour postinjection with QD705-RGD (pointed by white arrows). Cai et al. [59], pp. 669–676. Reproduced with permission of American Chemical Society. (*See insert for color representation of the figure.*)

Upconversion (UC) NPs are another new generation of fluorophores, which can convert long-wavelength radiation to a short-wavelength one *via* a nonlinear optical process. Mechanisms of the UC process are based on the sequential absorption of two or more photons by metastable, long-lived energy states, leading to highly excited states, which trigger the UC emission. NIR-to-NIR UC luminescence shows higher penetration depth because of the "optical transmission window" for biological tissues. Excitation and emission in the NIR range not only allows a deep light penetration and reduced photodamage but also produces low autofluorescence and light scattering. Tm^{3+}-doped UC NPs, which were reported to have NIR-to-NIR UC luminescence, are expected to be used as NIR probes for tumor imaging [60–62]. Recently, Chen et al. also reported a water-soluble core/shell (a-$NaYbY_4$: Tm^{3+})/CaF_2 NIR-to-NIR UC NPs with the coating of hyaluronic acid (HA) for deep tissue imaging [61]. Whole-animal imaging of BALB/c mice injected via a tail vein with these HA-coated NPs showed the high-contrast photoluminescence imaging of deep tissues by NIR-to-NIR UC NPs.

Despite the excellent optical properties of NPs-based fluorescent probes, *in vivo* biological applications of this class of contrast agents are limited due to several fundamental problems and technical barriers, such as the potential cytotoxicity of their heavy metal component (i.e., Cd, Se), surface-coated ligands (CTAB, for example) or nanoparticle aggregation [63–65], small-scale and expensive preparation, difficulties in reproducibility and comparability as well as quantification [66].

22.3.2 Bioluminescence Imaging

BLI is another widely used optical imaging modality developed over the past decade, which involves the detection of light signatures emitted from living organisms in the visible or NIR regions. Bioluminescence is a chemiluminescent reaction based on enzymatic light production of luciferases in the presence of substrate luciferins under physiological conditions [67, 68]. Luciferases are a family of photoproteins isolated from marine organisms, prokaryotes, or insects. In the luciferase reaction, light in the region of 400–620 nm is emitted when luciferase acts on the appropriate luciferin substrate. Some traditional luciferase reactions depend on the presence of oxygen and ATP, and BLI has been used to detect ATP. Currently, novel luciferase–luciferin systems that do not require exogenous substrate supplies for light emissions have been established.

BLI offers advantages including high sensitivity, low noise, and nontoxicity, which make it a powerful noninvasive methodology for real-time analysis of ongoing biological processes at the molecular level. With the development of novel luciferase–luciferin systems coupled with targeting agents, BLI has been used widely in studies of tumor responses to chemotherapeutics, transgene expression, tumor growth and metastasis, and gene and cell therapy [69–72]. For instance, as most of the antitumor drugs cause tumor cell death primarily by the induction of apoptosis [73, 74], real-time imaging of the apoptosis process is important to assess the therapeutic efficacy and toxicity of drugs. BLI has been proven to be a promising imaging modality to evaluate tumor growth and regression in response to therapy due to its relatively high sensitivity at the single-cell level *in vivo* [75]. Chen's group did some pioneer researches aiming to visualize the dynamics of apoptotic process with temporal BLI using an apoptosis-specific bioluminescence reporter gene [76]. In this study, both human head and neck squamous carcinoma UM-SCC-22B cells and murine breast cancer 4T1 cells were genetically modified with a caspase-3–specific cyclic firefly luciferase reporter gene (pcFluc-DEVD). After treatment with doxorubicin, bioluminescence signal changes in cells and tumor models were noninvasively acquired at different time points and correlated with caspase-3 activity.

Compared with the excessive effort in luciferase engineering, only a few studies have focused on the modification of luciferins. Recently, conjugation of luciferin onto nanocarriers was investigated in order to make controlled BLI at specific locations of interest possible. One remarkable technique that enables such control is the use of light to manipulate compounds that are photoactive (or photocaged) in various biological systems [77–79]. Xing et al. recently developed a system for the controlled uncaging of D-luciferin and BLI that is based on photocaged NIR-to-UV upconversion NPs [80]. In this study, D-luciferin caged with a 1-(2-nitrophenyl)ethyl group was conjugated onto the surface of Tm/Yb co-doped $NaYF_4$ core-shell UCNPs (silica-UCNPs), dissociation of D-luciferin molecules from the surface of the nanoparticle can be triggered upon NIR light irradiation. This D-luciferin/UCNPs conjugate have advantages of low cellular damage due to less exposure to high-intensity UV and the deep tissue penetration attributable to NIR light irradiation, which may offer new candidates for BLI probes for noninvasive *in vivo* monitoring of drug delivering and tumor detection.

22.3.3 Gold Nanoparticle as a Theranostics Platform

Gold nanoparticles (AuNPs) are one of the most promising candidates for drug delivery and molecular diagnostics because of their size and shape-dependent optical properties as well as their excellent biocompatibility [81]. These types of NPs can absorb light energy and then scatter specific types of diagnostic/therapeutic signals, for example, US, heat, Raman, or fluorescence signals, a phenomenon commonly known as localized surface plasmon resonance (LSPR) [82]. These AuNPs can also function as theranostic NPs on their own including imaging-based detection, photothermal therapy (PTT), chemical therapy, and drug delivery.

AuNPs, also known as colloidal gold, are usually a suspension of nanosized gold particles in water, which were discovered by Faraday in 1857 when he observed a range of colors originating from some unidentified particles in a gold chloride solution after reducing and stabilizing agents were added [83]. With the development of inorganic nanoparticle synthesis and surface modification, well-dispersed AuNPs with different sizes and shapes can be fabricated. AuNPs in the forms of spheres [84, 85], rods [86, 87], and cages [88, 89], have all been used as contrast media in preclinical investigations. The strong semicovalent interaction between thiol and Au is favored to load functional entities onto the surface of Au NPs [90, 91], and a number of therapeutic drugs have been loaded in such a manner. Prethiolated DNA oligos, for instance, have long been used to stabilize AuNPs, and the resulting conjugates have been investigated as gene therapy agents [85].

Due to their unique LSPR feature, AuNPs are also promising candidate materials as energy transducers for PTT. AuNPs accumulated specifically at the tumor site can convert the energy of laser irradiation into heat and kill adjacent tumor cells. Such a treatment paradigm is active only within the limited illumination area, with minimized normal tissue damage. AuNPs in the form of nanorod, nanocage, or nanoshell with absorption in the NIR region were reported to be suitable for PTT. Xia et al. demonstrated recently that PEG-coated Au nanocages can accumulate in a U87MG xenograft model, and when exposed to NIR light they can increase the tumor surface temperature to 54°C within 2 minutes [92].

Two obvious disadvantages limit the applications of AuNPs as a theranostic nanoplatform. These are as follows: (i) The high cost of production may cast a shadow over applications that otherwise have encouraging clinical perspectives. (ii) The stability of conjugates based on the thio-Au chemistry under a reducing environment can be an issue, such as glutathione (GSH) that is of high concentration in a living subject. Considering that more and more investigations are to be performed *in vivo*, new chemistries that result in more stable conjugates can be highly favorable.

22.4 MRI-BASED THERANOSTICS

MRI is among the best noninvasive diagnostic modalities today in clinical imaging for assessing anatomy and function of tissues [16, 17]. MRI offers several advantages including excellent temporal and spatial resolution, lack of exposure to radiation, rapid *in vivo* acquisition of images, and a long effective imaging window. However, it is much less sensitive compared to nuclear medicine or FLI, and more than 40% of all MRI examinations rely on the administration of a contrast agent [93]. High MRI sensitivity is required, especially for the accurate diagnosis of early-stage cancers, and searching for ultrasensitive contrast agents has drawn a lot of attention during the past decade [94]. The majority of the clinically used MRI contrast agents contain paramagnetic or superparamagnetic complexes that induce an increase in the longitudinal (T_1) or transverse (T_2) relaxation. Currently, there are two kinds of MRI contrast media available in both basic and clinical applications depending on T_1 or T_2 relaxation enhancement [95, 96]. Complexes of paramagnetic metals (Gd^{3+} or Mn^{2+}) are widely used as longitudinal (T_1) contrast agents. However, the administration of Gd-based MRI contrast agents can generate serious side effects such as nephrogenic systemic fibrosis (NSF) in some patients [97], leading to increased interest in iron oxide–based T_2 imaging probes or manganese-enhanced MRI (MEMRI) [98, 99]. As a typical T_2 contrast agent, superparamagnetic iron oxide (SPIO) NPs are strong enhancers of proton relaxation with superior T_2 shortening effect, and can be used at low concentration [100–102]. Commercial SPIO-based MRI contrast agents such as Feridex® and Resovist® are available and used widely in clinical MRI imaging. With the help of SPIO contrast agents, MRI has made great progress in studying gene delivery, cell trafficking, drug delivery, tumor diagnosis, and many other fields [103–106].

Iron oxide NPs are iron oxide particles with diameters between about 1 and 100 nm. As the diameter of iron oxide NPs is decreased to less than 20 nm, they display superparamagnetic behaviors at room temperature. These types of NPs are called SPIO NPs, which are widely used as a platform in theranostics, mainly because (i) SPIO NPs have the most prominent transverse relaxation time T_2/T_2^* among those available MRI contrast agents [107]. They can shorten T_2 relaxation time and bring negative contrast, resulting in hypointense images to monitor small pathological changes and drug delivery *in vivo*; (ii) SPIO NPs have a large surface area that is favored for carrying various biomolecules and drugs. With the development of nanotechnology, SPIO NPs possessing controllable size, morphology, compositions, magnetizations, relaxivities, and surface chemistry are now available [108, 109]; (iii) due to their potential in hyperthermia, SPIO NPs can play an attractive imaging/therapy dual function in cancer treatment [110]; and (iv) SPIO NPs exhibit excellent biosafety because they can be degraded and metabolized into the serum Fe pool to form hemoglobin or to enter other metabolic processes [24].

Relaxivity is a measurement of the ability of an MRI contrast agent to influence either longitudinal relaxation time T_1 ($r_1 = 1/T_1$) or transverse relaxation time T_2 ($r_2 = 1/T_2$). The relaxivity of SPIO was reported to be influenced by its particle size, compositions, and strength of the external magnetic field. Aggregates of multiple

SPIO nanocrystals were found to have much stronger T_2 effect than single ones, and the T_2 relaxation rate would enhance dramatically with an increased particle size [104, 105, 111].

Clusters of SPIO represent a new class of ultrasensitive MRI contrast agents. With further surface functionalization by targeting agents, optical dyes, or permeation enhancers, their applications in imaging, *in vivo* cell tracking, early detections of cancers, targeted drug delivery, and gene therapy have been widely investigated. Taking targeted drug delivery as an example, SPIO nanoparticle–based nanoplatforms have been intensively studied as vehicles of antitumor drugs. Generally, multiple SPIO NPs are loaded, along with therapeutics, into polymer-based matrices labeled with targeting moieties [112–114]. Various formulations of SPIO NPs have been developed for theranostic applications. A good example of SPIO aggregates as theranostics is reported by Nasongkla et al. in 2006 [104]. They loaded doxorubicin (DOX) and a cluster of SPIO NPs simultaneously into the hydrophobic cores of poly(ethylene glycol)-poly(D,L-lactide) (PEGPLA) micelles (Fig. 22.4). In addition, a cyclic arginine-glycine-aspartic acid (cRGD) ligand as the targeting moiety was conjugated onto the micelle surface to target the integrin $\alpha_v\beta_3$ overexpressed on tumors or angiogenic endothelial cells. The integrated capability of the probe to be used as MRI-imaging agents and cancer-targeting drug delivery vehicles makes it a promising candidate for future cancer diagnosis and therapy. Another advantage provided by nanoparticle-based drug carriers is that they offer an improved efficacy of chemotherapy by circumventing multidrug resistance (MDR), because the encapsulated drugs are resistant to drug efflux [115–117].

22.5 NUCLEAR IMAGING-BASED THERANOSTICS

Nuclear imaging is a noninvasive imaging technique producing images by detecting radiation from different parts of the body after a very low dose of radioactive tracer material is administered. In contrast with imaging techniques (i.e., MRI and CT) that mainly show anatomy information, nuclear imaging can provide quantitative functional information about normal tissues or disease conditions at the molecular and cellular levels. Depending on the properties of the radiotracer applied, various aspects of biochemical processes (i.e., tissue blood flow and metabolism, expression of cell receptors in normal and abnormal cells, and cell trafficking and homing) can be targeted and visualized under nuclear imaging scans. The advantage of assessing the function of an organ is that it helps physicians make a diagnosis and plan present or future treatments for the part of the body being evaluated, which make nuclear imaging a sensitive tool for diagnosing diseases at an early stage, analyzing drug biodistribution and assessing the efficacy in living subjects.

Current nuclear medicine imaging techniques include PET and SPECT, which are based on, respectively, positron- and gamma-emitting radionuclides for the generation of signal. Both external and internal radiotherapy can be directed by diagnostic PET and SPECT, and diagnostic imaging and radiotherapy are merging into theranostics, resulting in more personalized medicine. Both PET and SPECT have distinct

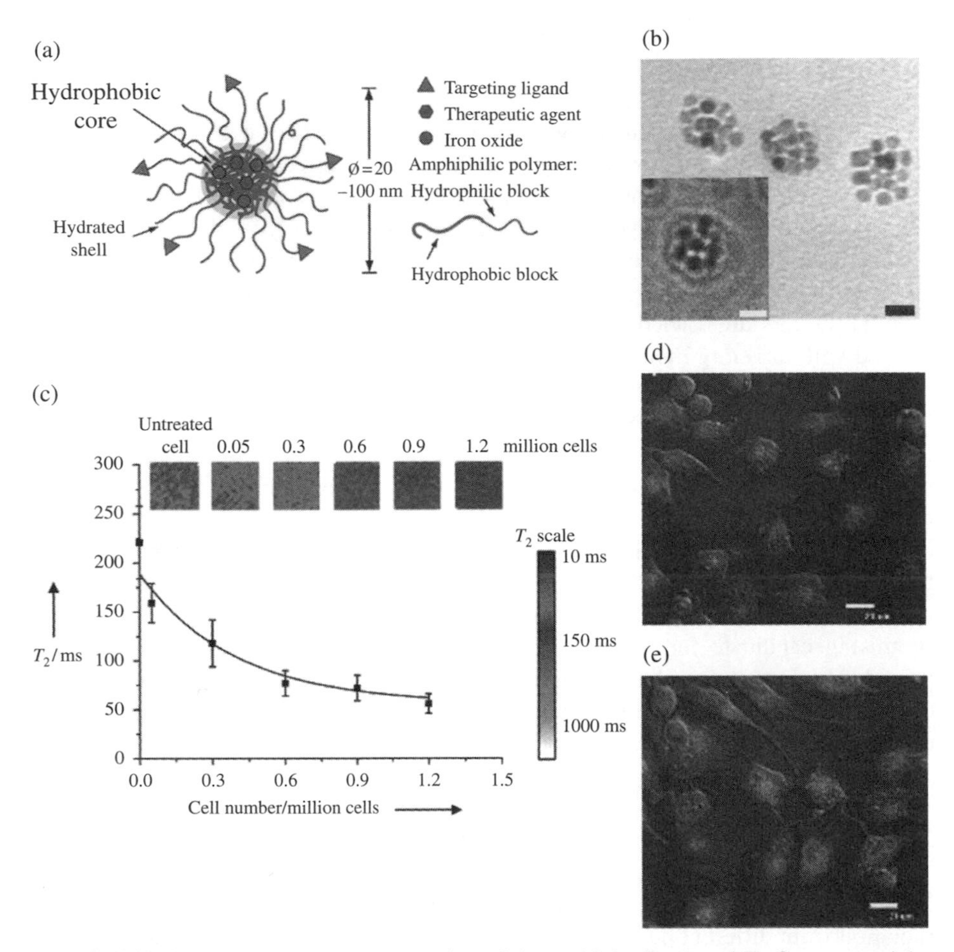

FIGURE 22.4 (a) Schematic representation of doxorubicin (Dox) and Fe_3O_4 nanoparticle-loaded PEG-PLA micelles, with cRGD peptide conjugated on the micelle surface. (b) TEM image of cRGD-DOXO-SPIO-loaded polymeric micelles (scale bars: 20 nm). (c) T_2 values of SLK cells treated with 16% cRGD-DOXO-SPIO micelles as a function of cell number. (d, e) Confocal laser scanning microscopy of SLK cells treated with 10 and 16% cRGD-DOXO-SPIO micelles. Nasongkla et al. [104], pp. 2427–2430. Reproduced with permission of American Chemical Society. (*See insert for color representation of the figure.*)

advantages and disadvantages that make them useful for detection in different conditions. For example, SPECT can use more than one radiotracer at a time, and the longer half-life of these tracers makes SPECT more readily available compared with PET. However, PET can provide more physiological information and is more sensitive than SPECT by a factor of 2–3 [118].

A radiopharmaceutical used in nuclear imaging is either a radionuclide alone or radionuclides attached to a macromolecule or particle, and the radionuclides used include fluorine-18, indium-111, iodine-123, and so on. After being introduced into the body by injection or swallowing, the radiotracer will accumulate in the organ or tissue of interest. These tracers are not dyes or medicines, and they have no side effects. The amount of radiation a patient receives in a typical nuclear medicine scan tends to be very low. To date, PET scans used in diagnosis of local recurrence and metastatic sites of various cancers and evaluation of treatment response are mainly based on fluorine-18-fluorodeoxyglucose ([^{18}F]FDG). This imaging technique is also known as FDG-PET. Radiopharmaceutical [^{18}F]FDG is a glucose analog labeled with a radionuclide [fluorine-18] at the 2′ position, and it can be used to assess the differences between cancer and normal cells in glucose metabolism. Cancer cells, particularly those from aggressive tumors, proliferate more rapidly than normal cells and consume considerably larger amounts of glucose. However, [^{18}F]FDG is not a target-specific PET tracer and cannot provide enough information about tumor biology and/or its vulnerability to potential treatments. Searching for the selective biologic radiotracers that will yield specific biochemical information and allow for noninvasive molecular imaging has drawn a lot of attention. Recently, several non-[^{18}F]FDG PET tracers for specific tumor biology processes, and their preclinical and clinical applications were reported. Small molecules or biomolecules such as peptides, affibodies, antibodies, aptamers, and oligonucleotides can be labeled with PET isotopes and have been evaluated recently for their potential as diagnostic imaging agents [119].

Another newly developed PET tracer is based on radiolabeled NPs, which has enormous potential for clinical applications and has several advantages [120, 121]: (i) maximize the binding affinity via multimeric receptor-specific biomolecules based on the multivalency principle, (ii) biodistributions of NPs labeled with radiotracers can be monitored precisely by SPECT or PET, (iii) dual-modality imaging combined functional nuclear imaging and anatomical CT, MRI, or other imaging modality is allowed, and (iv) therapeutic properties can be achieved as NPs can be used as a carrier of therapeutic agents. For instance, targeting moieties (i.e., antibodies, peptides, or any molecule with biological activity) can be linked to the surface of radiolabeled NPs (QDs, SIPO, AuNPs, silica NPs, or single-walled carbon nanotube (SWCNT), leading to dual-modality imaging agents with highly efficient tumor-targeting capabilities. Chen et al. developed a type of surface functionalized mesoporous silica (mSiO_2) NPs for actively targeted PET imaging and drug delivery in 4T1 murine breast tumor-bearing mice (Fig. 22.5) [122]. In this study, mSiO_2 nanoparticle was conjugated to human/murine chimeric IgG1 monoclonal antibody (TRC105, binding to both human and murine CD105) and ^{64}Cu labeled (*S*)-2-(4-isothiocyanatobenzyl)-1,4,7-triazacyclononane-1,4,7-triacetic acid (*p*-SCN-Bn-NOTA) to form the ^{64}Cu-NOTA-mSiO_2-PEG-TRC105 nanoconjugate. Since the tumor vasculature in 4T1 tumor tissue expresses high level of CD105, ^{64}Cu-NOTA-mSiO_2-PEG-TRC105 nanoconjugate exhibited excellent target specificity (approximately twofold enhancement) compared to that of passive targeting based on the enhanced permeability and retention (EPR) effect. The capability of enhanced tumor-targeted delivery of DOX by DOX-loaded NOTA-mSiO_2-PEG-TRC105 nanoconjugate was also demonstrated in 4T1 tumor-bearing mice upon intravenous injection, using TRC105 as the targeting ligand and mSiO_2 as the drug carrier, which holds great potential for future image-guided drug delivery and targeted cancer therapy.

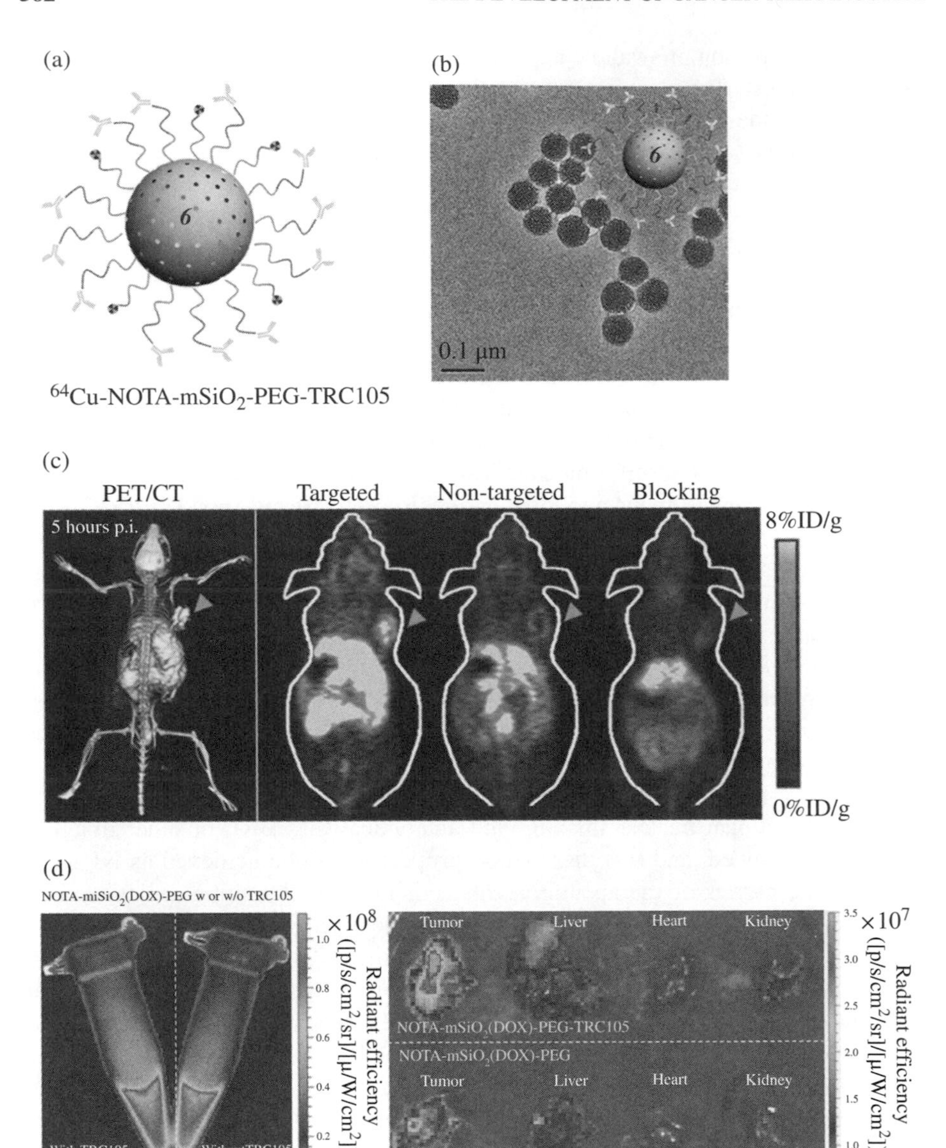

FIGURE 22.5 (a) Schematic illustration of ^{64}Cu-NOTA-mSiO$_2$-PEG-TRC105 nanoconjugate. (b) TEM image of NOTA-mSiO$_2$-PEG-TRC105 in PBS solution. (c) Representative PET/CT and PET images of mice at 5 hours postinjection. Tumors are indicated by yellow arrowheads. (d) Fluorescence images of DOX-loaded nanocomposite in PBS solution and *ex vivo* optical image of major organs at 0.5 hour after intravenous injection of DOX-loaded nanocomposite. Chen et al. [122], pp. 9027–9039. Reproduced with permission of American Chemical Society. (*See insert for color representation of the figure.*)

However, the spatial resolution of PET is much lower than those anatomy-based imaging modalities, such as US, CT, and MRI, the increasing availability of PET and SPECT fused/coregistered with CT or MRI for precise anatomic localization, coupled with the discovery of a multitude of new biochemical targets that characterize a specific disease, has led to tremendous interest in molecular imaging in oncology [123]. Clinical decisions based on nuclear imaging studies are changing cancer patient management by the combination of the functional information from PET/SPECT to the conventional anatomical information obtained from CT or MRI [124].

22.6 ULTRASOUND-BASED THERANOSTIC PLATFORM

Ultrasound (US) is an oscillating sound pressure wave with a frequency greater than 20 kilohertz (20 kHz), which can be used in object detections, cleaning, medical imaging (ultrasonography), and many other different fields. As a medical tool, US is used for clinical diagnostic imaging and therapeutic purposes. Taking advantages of its low cost, portability, and excellent safety, US imaging has become one of the most widely used diagnostic tools in modern medicine.

Ultrasound imaging, also known as ultrasonography, is a medical diagnostic imaging modality that provides anatomic images using the reflection and scattering of acoustic waves generated and received by an acoustic transducer, widely used for routine screening examinations of breast, abdomen, tendons, and other soft tissues with real-time tomographic images. US imaging is usually performed without contrast agents; sufficient information can be obtained for right diagnosis in routine and emergency examinations. However, contrast-enhanced US imaging is necessary for situations such as the characterization of liver lesions, the assessment of therapy response and biological processes at the molecular level [125–127]. There are two main classes of US contrast agents: microbubble and nonmicrobubble-based contrast agents. Microbubbles are gas–liquid emulsions, usually having a diameter of about 1–4 μm, which have evolved from free gas bubbles in solution over bubbles stabilized by surfactants to bubbles encapsulated with a shell of phospholipids, proteins, or polymers. Due to the very high echogenic response caused by the gaseous core, microbubbles are the most commonly used contrast agents for contrast-enhanced US imaging [125, 128].

In theory, microbubbles are potential candidates as theranostic platforms, given their propensity to be visualized *in vivo* with extremely high sensitivity, their ability to load therapeutic moieties into or onto their shell, and the possibility of improving drug delivery by passively or actively targeted modifications [129]. Progresses have been recently made on targeted microbubbles as US molecular imaging contrast agent [130]. Leong-Poi et al. did some pioneering study in this regard by using cationic microbubbles complex plasmid DNA encoding for vascular endothelial growth factor-165 (VEGF(165)) to induce therapeutic arteriogenesis in chronically ischemic skeletal muscle [131]. The result showed that the vascular endothelial growth factor (VEGF) expression in ischemic hind limbs in rats was enhanced, leading to the increase of microvessel density and improvement of microvascular blood flow. However, only very few studies have evaluated the possibilities of microbubbles for theranostic applications.

Besides these diagnostic applications, US was widely used for therapeutic purposes, which can be dated back to before US imaging. As the US waves can be focused onto very small volumes and greatly increasing their intensity, US applied for physical therapy was named high-intensity focused ultrasound (HIFU). Therapeutic US is based on the thermal effects of HIFU, and have been proven to be a promising surgical tool for noninvasive tumor treatment under the guidance and therapy control by magnetic resonance imaging (MRI), diagnostic US, or other imaging modality [132–134]. For instance, MR-guided HIFU ablation is clinically applied for the treatment of deep-seated tumors. In this procedure, HIFU exposures can target more accurately to special organ or tumor site with the help of high-resolution anatomical images provided by an MRI, and destroy the unwanted tissues by heating it to about 60°C [135, 136].

Photoacoustic (PA) imaging is a hybrid modality combining the high spatial resolution of US imaging with the high-contrast and spectroscopic-based specificity of optical imaging, which shows great potential as a real-time molecular imaging technique for biomedical applications [137, 138]. The PA imaging relies on the PA effect and images using laser-induced US. In theory, the PA imaging is based on the excitation of endogenous tissue chromophores or exogenous contrast agents induced by a short-pulsed laser beam, leading to the thermoelastic expansion of the tissue and thus wideband ultrasonic emission. The emitted ultrasonic waves are then captured by an ultrasonic transducer and utilized for image reconstruction [139–141].

Biomedical applications of contrast-enhanced PA molecular imaging in cancer researches have attracted many researchers' attentions, as the background signal can be suppressed via wavelength selection and specific targeting of molecules expressed on the tumor tissue [138]. For instance, RGD-conjugated single-walled carbon nanotubes labeled with indocyanine green dyes were used to make integrin expression on the neovasculature of glioblastoma visualized by PA molecular imaging [142]. Similarly, PA imaging contrast media based on gold nanorods or NPs have been used to image and identify epidermal growth factor receptor–expressed cancer cells [143]. However, clinical applications of PA molecular imaging are now still underexplored because of the lack of available clinical system and some other issues to be addressed.

22.7 MULTIMODALITY IMAGING-BASED THERANOSTIC PLATFORM

Molecular imaging is now playing a key role in drug discovery at the preclinical level, clinical diagnosis, and treatment of many diseases. However, among all these molecular imaging techniques, no single modality is perfect and sufficient to provide physicians with all the necessary information. It is thus natural to consider multimodality imaging [107, 144–146]. By combining new imaging media with multimodality-imaging instruments that merge structural information and functional data, physicians can perform multiple functional-imaging assays simultaneously with anatomic analyses [17]. For example, MRI is among the best noninvasive methodologies today in clinical medicine for assessing anatomy and the function of tissues. However, it is much less sensitive than nuclear medicine or optical imaging in monitoring small tissue lesions, molecular or cellular activities. Combination of MRI with other imaging techniques can overcome this drawback and offer synergistic advantages beyond single modality; coregistration with MRI images provides the anatomic landscape for localizing the functional or molecular data generated by optical, PA imaging, or/and PET. Multimodality imaging such as MRI/optical, MRI/PET has already demonstrated promising prospects in tumor detection or therapy. In this section, we will introduce briefly the application of several multimodality-imaging techniques in cancer theranostics.

22.7.1 PET/CT

As nuclear medicine imaging can offer quantitative functional information of normal and diseased tissues, and CT scans provide high-resolution anatomical information, hybrid systems combining these two imaging modalities can provide both functional and anatomical data in the same scanning session [147–149]. Combining the imaging techniques of PET/CT or SPECT/CT has further expanded the utility and accuracy of nuclear imaging particularly in the field of oncology [123, 150]. With the help of the combined-modality PET/CT or SPECT/CT devices, functional processes can be localized precisely within the body to an anatomically identified or, in some instances, as yet unidentifiable structural alteration. These devices provide additional information that improves diagnostic accuracy and impacts on patient management.

Townsend et al. developed the prototype of PET/CT dual-modality imaging in 2000, by integrating FDG-PET and CT to obtain an image from the fusion of hardware of CT and PET [124, 151]. Since then, PET/CT fusion imaging has revolutionized many fields of medical diagnosis, by adding precise anatomic localization to functional imaging, which was lacking from single PET imaging. Diagnoses of tumors, surgical planning, cancer staging, and the assessment of therapy response have all been changing rapidly under the influence of PET/CT availability. Consequently many diagnostic imaging procedures and centers have gradually abandoned conventional PET devices and substituted them with PET/CT systems, although the combined/hybrid device is considerably more expensive. Radiolabeled NPs for PET/CT fusion imaging were developed aiming to maximize the binding affinity and evaluate noninvasively the biodistribution or tumor-targeting efficacy of the labeled NPs. For example, Weissleder et al. have reported ^{18}F labeled NPs for *in vivo* PET/CT imaging [152]. In this study, ^{18}F doped cross-linked iron oxide (CLIO) NPs were synthesized to image the liver and blood pool of a mouse by PET/CT imaging.

Currently, different types of hybrid systems with PET/CT capability are now commercially available [153], the only obstacle to a wider dissemination of PET/CT fusion imaging is the difficulty and cost of producing and transporting the radiopharmaceuticals, which are usually extremely short-lived. For instance, the half-life of [^{18}F]FDG is only 2 hours, and its production requires a very expensive cyclotron as well as a production line. The research on development of new and improved tracers is constantly growing.

22.7.2 MRI/Optical

NIR FLI possesses enhanced light penetration depth through living tissues because the absorbance spectra for all biomolecules reach minima in the NIR region (700–900 nm) [29, 33, 34]. Therefore, NIR FLI has a high sensitivity and offers a unique advantage for the imaging of pathophysiological changes at the cellular and molecular level, which is most suitable for rapid and cost-effective preclinical evaluation in small-animal *in vivo* imaging. The combination of MRI and NIR imaging modalities can provide useful information with high spatial resolution and good sensitivity, which requires the availability of dual-modality probes. For instance, Wang et al. have conjugated NIR fluorescent dye Cy5.5 with amphiphilic polyethylenimine (PEI), which was used for encapsulation of multiple SPIO NPs [154]. These MRI/NIR dual-modality probes can label MCF-7/Adr cells with high efficiency; accumulation of kyl-PEI25-Cy5.5/SPIO nanocomposites in the cytoplasm resulted in the MR signal darkening and strong NIR fluorescence of these labeled cells *in vitro*.

Recently, Xu et al. developed a multifunctional MRI/NIR nanostructure with integrated upconversion luminescence (UCL) emission, paramagnetic property, and strong NIR absorption, for dual-model imaging and imaging combined with drug delivery [113]. They encapsulated hydrophobic UCNPs, iron oxide NPs (IONPs), and DOX together by using amphiphilic poly(styrene-block-allyl alcohol) (PS_{16}-b-PAA_{10}) to form a UC-IO@Polymer-DOX complex (Fig. 22.6). *In vivo* UCL/MRI images of nanocomposite-injected nude mice show strong UCL signals and dramatic darkening effect predominately in the liver tissue, owning to the high uptake of UC-IO@Polymer nanocomposite in the reticuloendothelial system (RES). Magnetism-targeted drug delivery was demonstrated by incubating Hela cells with UC-IO@Polymer-DOX complex in the presence of a magnetic field. The UCL/FL/MR visibility will enable us to monitor the delivery of drug and targeted therapy in a non-invasive and dynamic mode with the combination of anatomical, functional, and molecular information.

22.7.3 MRI/PET

Another widely studied dual-modality imaging technique based on the combination of PET and MRI. PET is among these systems to open new horizons for multimodality molecular imaging by offering simultaneous morphologic, functional, and molecular information of a living system [155, 156]. The combined PET/MRI systems were first reported by Shao et al. in the late 1990s [157–159]. Shortly after that, PET/MRI dual-modality imaging was reported to be used in oncology of small-animal tumor studies. Nowadays, integrated PET/MRI scanners are commercially available for clinical use, and searching for multimodality agents has drawn considerable attention. Yang et al. reported a multifunctional SPIO nanocarrier for targeted drug delivery and PET/MRI dual-modality imaging of tumors with integrin $\alpha_v\beta_3$ expression [160]. In this study, cRGD peptides and ^{64}Cu chelators (NOTA) were conjugated onto the distal ends of PEG arms wrapped on the surface of SPIO NPs, and an anticancer drug, DOX, was conjugated onto the PEGylated SPIO nanocarriers *via*

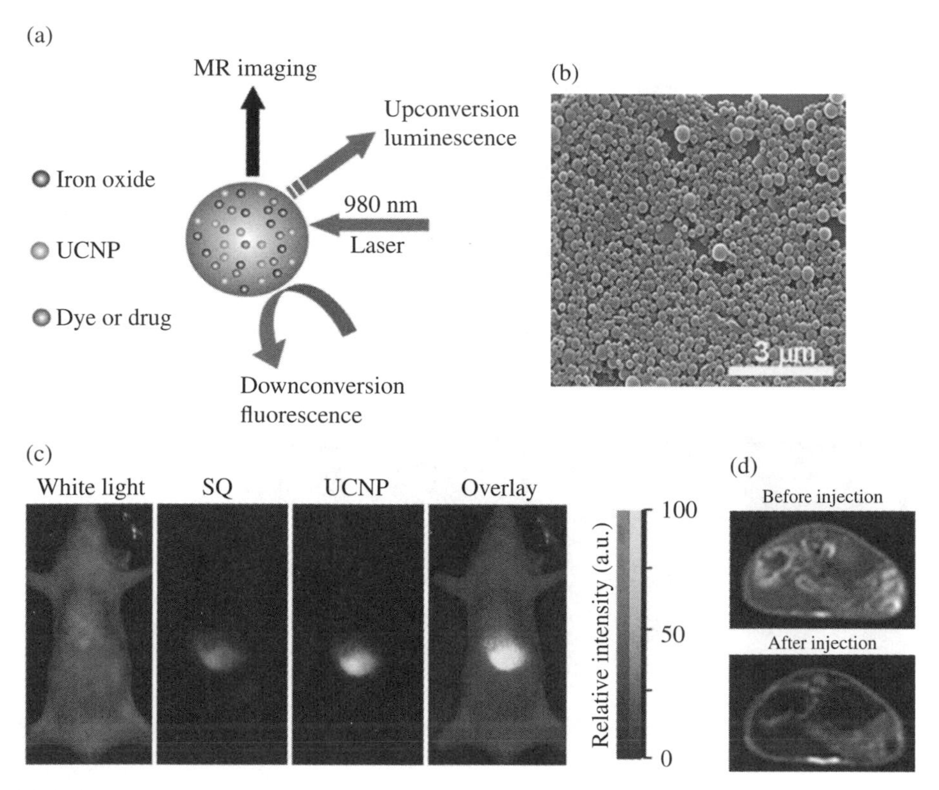

FIGURE 22.6 (a) Schematic illustration for UC-IO@Polymer nanocomposite. (b) SEM image of UC-IO@Polymer. (c) UCL (green) and FL (red) images of Squaraine (SQ) dye–loaded nanocomposites (UC-IO@Polymer-SQ) obtained by the Maestro *in vivo* imaging system. (d) T_2-weighted MR images of mice before and post injection of UC-IO@Polyme-SQ. Xu et al. [113], pp. 9364–9373. Reproduced with permission of Elsevier. (*See insert for color representation of the figure.*)

pH-sensitive bonds to achieve pH-responsive drug release. These SPIO nanocarriers had a hydrodynamic diameter of 68 ± 2 nm, with similar *in vitro* MRI r_2 relaxivity to Feridex, and pH-sensitive drug release behavior. *In vivo* PET images and biodistribution analyses showed that these cRGD-conjugated SPIO nanocarriers had a much higher level of tumor accumulation compared to cRGD-free SPIO (Fig. 22.7). These multifunctional SPIO nanocarriers have the potential for combined tumor-targeting drug delivery and PET/MR imaging thereby making cancer theranostics possible.

22.8 CONCLUSION AND FUTURE PERSPECTIVES

Theranostics combine therapeutic and diagnostic agents into a single platform for simultaneous diagnosis and treatment of many diseases. Along with the advantages of noninvasive medical imaging techniques based on PET, MRI, and optical imaging and

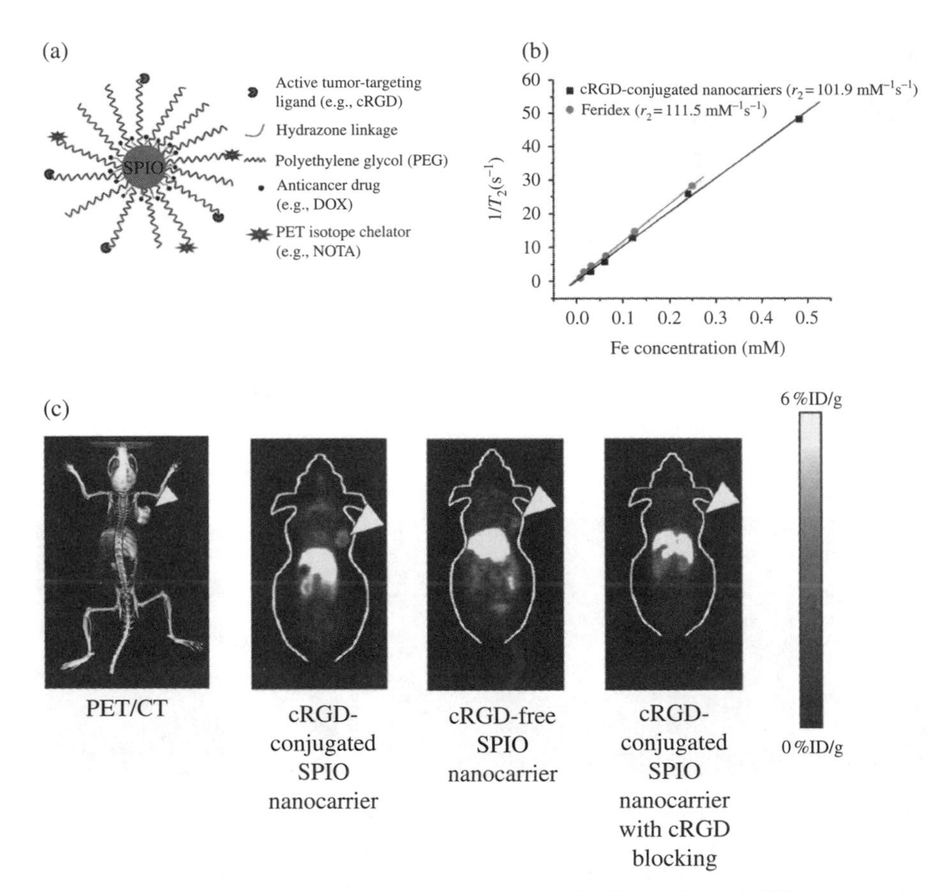

FIGURE 22.7 (a) Illustration of the multifunctional cRGD-conjugated SPIO nanocarriers for combined tumor-targeting drug delivery and PET/MR imaging. (b) T_2 relaxation rates ($1/T_2$, s^{-1}) as a function of iron concentration (mM) for both cRGD-conjugated SPIO nanocarriers and commercial Feridex. (c) PET images of U87MG tumor-bearing mice at 0.5 hour post injection of ^{64}Cu-labeled SPIO nanocarriers (cRGD-conjugated, cRGD-free, and cRGD-conjugated with a blocking dose of cRGD). Yang et al. [160], pp. 4151–4160. Reproduced with permission of Elsevier. (*See insert for color representation of the figure.*)

their ability to reveal cellular/molecular processes and disease mechanisms present in physiologically authentic environments, imaging-guided drug delivery assisted by molecular imaging techniques has become a key part of personalized medicine.

As the rapid development of nanotechnology proceeds, a variety of nanotheranostic systems combining imaging agents and therapeutics have been explored. For example, nanoparticle-based imaging probes such as QDs, AuNPs, SPIO, and radiolabeled NPs with superior physicochemical property and passive- and/or active-targeting abilities have been reported. On the other hand, multimodality imaging enables the combination of anatomical, functional, and molecular information by combining images from different modalities taken at the same point, which has emerged as a strategy that combines the strengths of different modalities and yields a hybrid imaging platform with characteristics superior to those of any of its constituents considered alone. We have all the reasons to believe that nanoscopic platforms that combine therapeutic agents and molecular-targeting ligands with nanoparticle-based imaging moieties are emerging as the next generation of multifunctional nanomedicine to improve the therapeutic outcome of drug therapy.

In this chapter, we have highlighted several theranostic nanoplatforms based on different imaging modalities with great potential as cancer theranostics. However, although considerable efforts have been made by the development of theranostic nanoplatforms and approaches, theranostic NPs have yet to be employed in a clinical setting. To date, most research on nanoparticle-based theranostic platforms has mainly focused on the nanostructures design, but their clinical applications have not been sufficiently considered. Many factors need to be optimized to design advanced NPs for cancer theranostics, including biocompatibility, pharmacokinetics, *in vivo* targeting efficacy, and cost effectiveness. For instance, NP-based imaging probes consisting of heavy metal components (i.e., Cd, Se) and surface-coated ligands suffer from the potential cytotoxicity related to their surface charge, the dosage, and duration of exposure. In addition, polymeric NPs enable targeted tumor-imaging/therapy and also facilitate monitoring the therapeutic effect; however, numerous concerns on safety and efficacy still remain. Consequently, to date, nanoparticle platforms have not shown significant potential for cancer theranostics in clinical settings. Moreover, a number of issues on the safety and efficacy of the theranostic nanostructures should be addressed. With that goal in mind, extensive discussion and multidisciplinary cooperation among clinicians, biologists, engineers, and materials scientists are required.

ACKNOWLEDGMENTS

Address correspondence to gangliu.cmitm@xmu.edu.cn and Shawn.Chen@nih.gov. This work was supported partially by the National Basic Research Program of China (973 Program) (2013CB733802 and 2014CB744503); the National Natural Science Foundation of China (NSFC) (81101101, 51273165, 81371596, 81422023, and 51503090); the Key Project of Chinese Ministry of Education (212149); the Fundamental Research Funds for the Central Universities (2013121039); the Program for New Century Excellent Talents in University (NCET-13-0502); and the Scientific Research Start-up Fund of Kunming University of Science and Technology (KKSY201305089).

REFERENCES

1. Siegel, R.; Naishadham, D.; Jemal, A. *CA Cancer J Clin* 2013, **63**, 11–30.
2. Fernandez-Fernandez, A.; Manchanda, R.; McGoron, A. *Appl Biochem Biotechnol* 2011, **165**, 1628–1651.
3. Wang, X.; Wang, Y.; Chen, Z. G.; Shin, D. M. *Cancer Res Treat* 2009, **41**, 1–11.
4. Gottesman, M. M.; Fojo, T.; Bates, S. E. *Nat Rev Cancer* 2002, **2**, 48–58.

5. Szakacs, G.; Paterson, J. K.; Ludwig, J. A.; Booth-Genthe, C.; Gottesman, M. M. *Nat Rev Drug Discov* 2006, **5**, 219–234.
6. Pene, F.; Courtine, E.; Cariou, A.; Mira, J.-P. *Crit Care Med* 2009, **37**, Supplement 1, 50–58.
7. Kelkar, S. S.; Reineke, T. M. *Bioconjug Chem* 2011, **22**, 1879–1903.
8. Funkhouser, J. *Curr Drug Discov* 2002, **2**, 17–19.
9. Lammers, T.; Aime, S.; Hennink, W. E.; Storm, G.; Kiessling, F. *Acc Chem Res* 2011, **44**, 1029–1038.
10. Lammers, T.; Kiessling, F.; Hennink, W. E.; Storm, G. *Mol Pharm* 2010, **7**, 1899–1912.
11. Mura, S.; Couvreur, P. *Adv Drug Deliv Rev* 2012, **64**, 1394–1416.
12. Sun, D. *Mol Pharm* 2010, **7**, 1879.
13. Sumer, B.; Gao, J. *Nanomedicine* 2008, **3**, 137–140.
14. Choi, K. Y.; Liu, G.; Lee, S.; Chen, X. *Nanoscale* 2012, **4**, 330–342.
15. Riehemann, K.; Schneider, S. W.; Luger, T. A.; Godin, B.; Ferrari, M.; Fuchs, H. *Angew Chem Int Ed* 2009, **48**, 872–897.
16. Weissleder, R.; Mahmood, U. *Radiology* 2001, **219**, 316–333.
17. Herschman, H. R. *Science* 2003, **302**, 605–608.
18. Hoffman, J. M.; Gambhir, S. S. *Radiology* 2007, **244**, 39–47.
19. Kircher, M. F.; Hricak, H.; Larson, S. M. *Mol Oncol* 2012, **6**, 182–195.
20. Rudin, M.; Weissleder, R. *Nat Rev Drug Discov* 2003, **2**, 123–131.
21. Janib, S. M.; Moses, A. S.; MacKay, J. A. *Adv Drug Deliv Rev* 2010, **62**, 1052–1063.
22. Massoud, T. F.; Gambhir, S. S. *Genes Dev* 2003, **17**, 545–580.
23. Willmann, J. K.; van Bruggen, N.; Dinkelborg, L. M.; Gambhir, S. S. *Nat Rev Drug Discov* 2008, **7**, 591–607.
24. Hua, A. *Adv Drug Deliv Rev* 2011, **63**, 772–788.
25. Hahn, M.; Singh, A.; Sharma, P.; Brown, S.; Moudgil, B. *Anal Bioanal Chem* 2011, **399**, 3–27.
26. Wickline, S. A.; Lanza, G. M. *Circulation* 2003, **107**, 1092–1095.
27. Na, H. B.; Song, I. C.; Hyeon, T. *Adv Mater* 2009, **21**, 2133–2148.
28. Ntziachristos, V. *Annu Rev Biomed Eng* 2006, **8**, 1–33.
29. Frangioni, J. V. *Curr Opin Chem Biol* 2003, **7**, 626–634.
30. Libutti, S. K.; Choyke, P.; Choy, G. *Mol Imaging* 2003, **2**, 303–312.
31. Becker, A.; Hessenius, C.; Licha, K.; Ebert, B.; Sukowski, U.; Semmler, W.; Wiedenmann, B. *Nat Biotechnol* 2001, **19**, 327–331.
32. Quek, C.-H.; Leong, K. W. *Nanomaterials* 2012, **2**, 92–112.
33. Chen, X.; Conti, P. S.; Moats, R. A. *Cancer Res* 2004, **64**, 8009–8014.
34. He, X.; Wang, K.; Cheng, Z. *Wiley Interdiscip Rev Nanomed Nanobiotechnol* 2010, **2**, 349–366.
35. Paganin-Gioanni, A.; Bellard, E.; Paquereau, L.; Ecochard, V.; Golzio, M.; Teissié, J. *Radiol Oncol* 2010, **44**, 142–148.
36. Luo, S.; Zhang, E.; Su, Y.; Cheng, T.; Shi, C. *Biomaterials* 2011, **32**, 7127–7138.
37. Huang, H.-C.; Barua, S.; Sharma, G.; Dey, S. K.; Rege, K. *J Control Release* 2011, **155**, 344–357.

38. Kobayashi, H.; Ogawa, M.; Alford, R.; Choyke, P. L.; Urano, Y. *Chem Rev* 2009, **110**, 2620–2640.
39. Rao, J.; Dragulescu-Andrasi, A.; Yao, H. *Curr Opin Biotechnol* 2007, **18**, 17–25.
40. Weissleder, R.; Tung, C.-H.; Mahmood, U.; Bogdanov, A. *Nat Biotechnol* 1999, **17**, 375–378.
41. Veiseh, O.; Sun, C.; Gunn, J.; Kohler, N.; Gabikian, P.; Lee, D.; Bhattarai, N.; Ellenbogen, R.; Sze, R.; Hallahan, A.; Olson, J.; Zhang, M. *Nano Lett* 2005, **5**, 1003–1008.
42. Talanov, V. S.; Regino, C. A. S.; Kobayashi, H.; Bernardo, M.; Choyke, P. L.; Brechbiel, M. W. *Nano Lett* 2006, **6**, 1459–1463.
43. Gao, X.; Chan, W. C. W.; Nie, S. *J Biomed Opt* 2002, **7**, 532–537.
44. Chan, W. C. W.; Maxwell, D. J.; Gao, X.; Bailey, R. E.; Han, M.; Nie, S. *Curr Opin Biotechnol* 2002, **13**, 40–46.
45. Vossmeyer, T.; Katsikas, L.; Giersig, M.; Popovic, I. G.; Diesner, K.; Chemseddine, A.; Eychmueller, A.; Weller, H. *J Phys Chem* 1994, **98**, 7665–7673.
46. Schmidt, M. E.; Blanton, S. A.; Hines, M. A.; Guyot-Sionnest, P. *Phys Rev B* 1996, **53**, 12629–12632.
47. Murray, C. B.; Norris, D. J.; Bawendi, M. G. *J Am Chem Soc* 1993, **115**, 8706–8715.
48. Peng, X.; Wickham, J.; Alivisatos, A. P. *J Am Chem Soc* 1998, **120**, 5343–5344.
49. Bruchez, M.; Moronne, M.; Gin, P.; Weiss, S.; Alivisatos, A. P. *Science* 1998, **281**, 2013–2016.
50. Jamieson, T.; Bakhshi, R.; Petrova, D.; Pocock, R.; Imani, M.; Seifalian, A. M. *Biomaterials* 2007, **28**, 4717–4732.
51. Zhang, H.; Yee, D.; Wang, C. *Nanomedicine* 2008, **3**, 83–91.
52. Byers, R. J.; Hitchman, E. R. *Prog Histochem Cytochem* 2011, **45**, 201–237.
53. Cassette, E.; Helle, M.; Bezdetnaya, L.; Marchal, F.; Dubertret, B.; Pons, T. *Adv Drug Deliv Rev* 2013, **65**, 719–731.
54. Pietryga, J. M.; Werder, D. J.; Williams, D. J.; Casson, J. L.; Schaller, R. D.; Klimov, V. I.; Hollingsworth, J. A. *J Am Chem Soc* 2008, **130**, 4879–4885.
55. Kim, S.; Fisher, B.; Eisler, H.-J.; Bawendi, M. *J Am Chem Soc* 2003, **125**, 11466–11467.
56. Kim, S.; Lim, Y. T.; Soltesz, E. G.; De Grand, A. M.; Lee, J.; Nakayama, A.; Parker, J. A.; Mihaljevic, T.; Laurence, R. G.; Dor, D. M.; Cohn, L. H.; Bawendi, M. G.; Frangioni, J. V. *Nat Biotechnol* 2004, **22**, 93–97.
57. Allen, P. M.; Bawendi, M. G. *J Am Chem Soc* 2008, **130**, 9240–9241.
58. Kim, S.-W.; Zimmer, J. P.; Ohnishi, S.; Tracy, J. B.; Frangioni, J. V.; Bawendi, M. G. *J Am Chem Soc* 2005, **127**, 10526–10532.
59. Cai, W.; Shin, D.-W.; Chen, K.; Gheysens, O.; Cao, Q.; Wang, S. X.; Gambhir, S. S.; Chen, X. *Nano Lett* 2006, **6**, 669–676.
60. Nyk, M.; Kumar, R.; Ohulchanskyy, T. Y.; Bergey, E. J.; Prasad, P. N. *Nano Lett* 2008, **8**, 3834–3838.
61. Chen, G.; Shen, J.; Ohulchanskyy, T. Y.; Patel, N. J.; Kutikov, A.; Li, Z.; Song, J.; Pandey, R. K.; Ågren, H.; Prasad, P. N.; Han, G. *ACS Nano* 2012, **6**, 8280–8287.
62. Zhou, J.; Sun, Y.; Du, X.; Xiong, L.; Hu, H.; Li, F. *Biomaterials* 2010, **31**, 3287–3295.
63. Wang, Y.; Hu, R.; Lin, G.; Roy, I.; Yong, K.-T. *ACS Appl Mater Interfaces* 2013, **5**, 2786–2799.

64. Lewinski, N.; Colvin, V.; Drezek, R. *Small* 2008, **4**, 26–49.
65. Derfus, A. M.; Chan, W. C. W.; Bhatia, S. N. *Nano Lett* 2003, **4**, 11–18.
66. Resch-Genger, U.; Grabolle, M.; Cavaliere-Jaricot, S.; Nitschke, R.; Nann, T. *Nat Methods* 2008, **5**, 763–775.
67. Greer, L. F.; Szalay, A. A. *Luminescence* 2002, **17**, 43–74.
68. Ozawa, T.; Yoshimura, H.; Kim, S. B. *Anal Chem* 2012, **85**, 590–609.
69. Weissleder, R.; Pittet, M. J. *Nature* 2008, **452**, 580–589.
70. Huang, N. F.; Okogbaa, J.; Babakhanyan, A.; Cooke, J. P. *Theranostics* 2012, **2**, 346–354.
71. Zhang, F.; Zhu, L.; Liu, G.; Hida, N.; Lu, G. M.; Eden, H. S.; Niu, G.; Chen, X. Y. *Theranostics* 2011, **1**, 302–309.
72. Maguire, C. A.; Bovenberg, M. S.; Crommentuijn, M. H. W.; Niers, J. M.; Kerami, M.; Teng, J.; Sena-Esteves, M.; Badr, C. E.; Tannous, B. A. *Mol Ther Nucleic Acids* 2013, **2**, e99.
73. Johnstone, R. W.; Ruefli, A. A.; Lowe, S. W. *Cell* 2002, **108**, 153–164.
74. Fesik, S. W. *Nat Rev Cancer* 2005, **5**, 876–885.
75. Laxman, B.; Hall, D. E.; Bhojani, M. S.; Hamstra, D. A.; Chenevert, T. L.; Ross, B. D.; Rehemtulla, A. *Proc Natl Acad Sci U S A* 2002, **99**, 16551–16555.
76. Niu, G.; Zhu, L.; Ho, D. N.; Zhang, F.; Gao, H. K.; Quan, Q. M.; Hida, N.; Ozawa, T.; Liu, G.; Chen, X. Y. *Theranostics* 2013, **3**, 190–200.
77. Umeda, N.; Ueno, T.; Pohlmeyer, C.; Nagano, T.; Inoue, T. *J Am Chem Soc* 2010, **133**, 12–14.
78. Orange, C.; Specht, A.; Puliti, D.; Sakr, E.; Furuta, T.; Winsor, B.; Goeldner, M. *Chem Commun* 2008, 1217–1219.
79. Nguyen, A.; Rothman, D. M.; Stehn, J.; Imperiali, B.; Yaffe, M. B. *Nat Biotechnol* 2004, **22**, 993–1000.
80. Yang, Y.; Shao, Q.; Deng, R.; Wang, C.; Teng, X.; Cheng, K.; Cheng, Z.; Huang, L.; Liu, Z.; Liu, X.; Xing, B. *Angew Chem Int Ed* 2012, **51**, 3125–3129.
81. Jeong, E.; Jung, G.; Hong, C.; Lee, H. *Arch Pharm Res* 2014, **37**, 53–59.
82. Hu, M.; Chen, J.; Li, Z.-Y.; Au, L.; Hartland, G. V.; Li, X.; Marquez, M.; Xia, Y. *Chem Soc Rev* 2006, **35**, 1084–1094.
83. Faraday, M. *Philos Trans R Soc Lond* 1857, **147**, 145–181.
84. Daniel, M.-C.; Astruc, D. *Chem Rev* 2003, **104**, 293–346.
85. Rosi, N. L.; Giljohann, D. A.; Thaxton, C. S.; Lytton-Jean, A. K. R.; Han, M. S.; Mirkin, C. A. *Science* 2006, **312**, 1027–1030.
86. Oyelere, A. K.; Chen, P. C.; Huang, X.; El-Sayed, I. H.; El-Sayed, M. A. *Bioconjug Chem* 2007, **18**, 1490–1497.
87. Huang, X.; El-Sayed, I. H.; Qian, W.; El-Sayed, M. A. *Nano Lett* 2007, **7**, 1591–1597.
88. Skrabalak, S. E.; Chen, J.; Sun, Y.; Lu, X.; Au, L.; Cobley, C. M.; Xia, Y. *Acc Chem Res* 2008, **41**, 1587–1595.
89. Xia, Y.; Li, W.; Cobley, C. M.; Chen, J.; Xia, X.; Zhang, Q.; Yang, M.; Cho, E. C.; Brown, P. K. *Acc Chem Res* 2011, **44**, 914–924.
90. Nuzzo, R. G.; Zegarski, B. R.; Dubois, L. H. *J Am Chem Soc* 1987, **109**, 733–740.
91. Love, J. C.; Estroff, L. A.; Kriebel, J. K.; Nuzzo, R. G.; Whitesides, G. M. *Chem Rev* 2005, **105**, 1103–1170.

92. Chen, J.; Glaus, C.; Laforest, R.; Zhang, Q.; Yang, M.; Gidding, M.; Welch, M. J.; Xia, Y. *Small* 2010, **6**, 811–817.
93. Caravan, P.; Ellison, J. J.; McMurry, T. J.; Lauffer, R. B. *Chem Rev* 1999, **99**, 2293–2352.
94. Yan, G.; Zhuo, R. *Chin Sci Bull* 2001, **46**, 1233–1237.
95. Hermann, P.; Kotek, J.; Kubicek, V.; Lukes, I. *Dalton Trans* 2008, 3027–3047.
96. Yan, G.; Robinson, L.; Hogg, P. *Radiography* 2007, **13**, Supplement 1, 5–19.
97. Sieber, M. A.; Steger-Hartmann, T.; Lengsfeld, P.; Pietsch, H. *J Magn Reson Imaging* 2009, **30**, 1268–1276.
98. Pan, D.; Caruthers, S. D.; Senpan, A.; Schmieder, A. H.; Wickline, S. A.; Lanza, G. M. *Wiley Interdiscip Rev Nanomed Nanobiotechnol* 2011, **3**, 162–173.
99. Bellin, M.-F. *Eur J Radiol* 2006, **60**, 314–323.
100. Wang, Y.-X.; Hussain, S.; Krestin, G. *Eur Radiol* 2001, **11**, 2319–2331.
101. Bulte, J. W. M.; Kraitchman, D. L. *NMR Biomed* 2004, **17**, 484–499.
102. Thorek, D. J.; Chen, A.; Czupryna, J.; Tsourkas, A. *Ann Biomed Eng* 2006, **34**, 23–38.
103. Liu, G.; Wang, Z.; Lu, J.; Xia, C.; Gao, F.; Gong, Q.; Song, B.; Zhao, X.; Shuai, X.; Chen, X.; Ai, H.; Gu, Z. *Biomaterials* 2011, **32**, 528–537.
104. Nasongkla, N.; Bey, E.; Ren, J. M.; Ai, H.; Khemtong, C.; Guthi, J. S.; Chin, S. F.; Sherry, A. D.; Boothman, D. A.; Gao, J. M. *Nano Lett* 2006, **6**, 2427–2430.
105. Lu, J.; Ma, S.; Sun, J.; Xia, C.; Liu, C.; Wang, Z.; Zhao, X.; Gao, F.; Gong, Q.; Song, B.; Shuai, X.; Ai, H.; Gu, Z. *Biomaterials* 2009, **30**, 2919–2928.
106. de Vries, I. J. M.; Lesterhuis, W. J.; Barentsz, J. O.; Verdijk, P.; van Krieken, J. H.; Boerman, O. C.; Oyen, W. J. G.; Bonenkamp, J. J.; Boezeman, J. B.; Adema, G. J.; Bulte, J. W. M.; Scheenen, T. W. J.; Punt, C. J. A.; Heerschap, A.; Figdor, C. G. *Nat Biotechnol* 2005, **23**, 1407–1413.
107. Louie, A. *Chem Rev* 2010, **110**, 3146–3195.
108. Gupta, A. K.; Gupta, M. *Biomaterials* 2005, **26**, 3995–4021.
109. Xie, J.; Liu, G.; Eden, H. S.; Ai, H.; Chen, X. *Acc Chem Res* 2011, **44**, 883–892.
110. Yallapu, M. M.; Othman, S. F.; Curtis, E. T.; Gupta, B. K.; Jaggi, M.; Chauhan, S. C. *Biomaterials* 2011, **32**, 1890–1905.
111. Ai, H.; Flask, C.; Weinberg, B.; Shuai, X.; Pagel, M. D.; Farrell, D.; Duerk, J.; Gao, J. M. *Adv Mater* 2005, **17**, 1949–1952.
112. Kievit, F. M.; Wang, F. Y.; Fang, C.; Mok, H.; Wang, K.; Silber, J. R.; Ellenbogen, R. G.; Zhang, M. *J Control Release* 2011, **152**, 76–83.
113. Xu, H.; Cheng, L.; Wang, C.; Ma, X.; Li, Y.; Liu, Z. *Biomaterials* 2011, **32**, 9364–9373.
114. Liu, G.; Gao, J.; Ai, H.; Chen, X. *Small* 2013, **9**, 1533–1545.
115. Jabr-Milane, L. S.; van Vlerken, L. E.; Yadav, S.; Amiji, M. M. *Cancer Treat Rev* 2008, **34**, 592–602.
116. Malam, Y.; Loizidou, M.; Seifalian, A. M. *Trends Pharmacol Sci* 2009, **30**, 592–599.
117. Dong, X.; Mumper, R. J. *Nanomedicine* 2010, **5**, 597–615.
118. Rahmim, A.; Zaidi, H. *Nucl Med Commun* 2008, **29**, 193–207.
119. Jacobson, O.; Chen, X. *Pharmacol Rev* 2013, **65**, 1214–1256.
120. Hong, H.; Zhang, Y.; Sun, J.; Cai, W. *Nano Today* 2009, **4**, 399–413.

121. Welch, M. J.; Hawker, C. J.; Wooley, K. L. *J Nucl Med* 2009, **50**, 1743–1746.
122. Chen, F.; Hong, H.; Zhang, Y.; Valdovinos, H. F.; Shi, S.; Kwon, G. S.; Theuer, C. P.; Barnhart, T. E.; Cai, W. *ACS Nano* 2013, **7**, 9027–9039.
123. Schillaci, O.; Simonetti, G. *Cancer Biother Radiopharm* 2004, **19**, 1–10.
124. Beyer, T.; Townsend, D. W.; Brun, T.; Kinahan, P. E.; Charron, M.; Roddy, R.; Jerin, J.; Young, J.; Byars, L.; Nutt, R. *J Nucl Med* 2000, **41**, 1369–1379.
125. Lanza, G. M.; Wickline, S. A. *Prog Cardiovasc Dis* 2001, **44**, 13–31.
126. Deshpande, N.; Needles, A.; Willmann, J. K. *Clin Radiol* 2010, **65**, 567–581.
127. Kiessling, F.; Fokong, S.; Bzyl, J.; Lederle, W.; Palmowski, M.; Lammers, T. *Adv Drug Deliv Rev* 2014, **72**, 15–27.
128. Kiessling, F.; Fokong, S.; Koczera, P.; Lederle, W.; Lammers, T. *J Nucl Med* 2012, **53**, 345–348.
129. Ferrara, K.; Pollard, R.; Borden, M. *Annu Rev Biomed Eng* 2007, **9**, 415–447.
130. Klibanov, A. *J Nucl Cardiol* 2007, **14**, 876–884.
131. Leong-Poi, H.; Kuliszewski, M. A.; Lekas, M.; Sibbald, M.; Teichert-Kuliszewska, K.; Klibanov, A. L.; Stewart, D. J.; Lindner, J. R. *Circ Res* 2007, **101**, 295–303.
132. Al-Bataineh, O.; Jenne, J.; Huber, P. *Cancer Treat Rev* 2012, **38**, 346–353.
133. Kennedy, J. E. *Nat Rev Cancer* 2005, **5**, 321–327.
134. Kennedy, J. E.; ter Haar, G. R.; Cranston, D. *Br J Radiol* 2003, **76**, 590–599.
135. Voogt, M. J.; Trillaud, H.; Kim, Y. S.; Mali, W. P. T. M.; Barkhausen, J.; Bartels, L. W.; Deckers, R.; Frulio, N.; Rhim, H.; Lim, H. K.; Eckey, T.; Nieminen, H. J.; Mougenot, C.; Keserci, B.; Soini, J.; Vaara, T.; Köhler, M. O.; Sokka, S.; Bosch, M. A. J. *Eur Radiol* 2012, **22**, 411–417.
136. Merckel, L.; Bartels, L.; Köhler, M.; den Bongard, H. J. G. D.; Deckers, R.; Mali, W. T. M.; Binkert, C.; Moonen, C.; Gilhuijs, K. A.; den Bosch, M. A. J. *Cardiovasc Intervent Radiol* 2013, **36**, 292–301.
137. Wang, L. V.; Hu, S. *Science* 2012, **335**, 1458–1462.
138. Wilson, K. E.; Wang, T. Y.; Willmann, J. K. *J Nucl Med* 2013, **54**, 1851–1854.
139. Wang, L. V. *Med Phys* 2008, **35**, 5758–5767.
140. Beard, P. *Interface Focus* 2011, **1**, 602–631.
141. Zackrisson, S.; van de Ven, S. M. W. Y.; Gambhir, S. S. *Cancer Res* 2014, **74**, 979–1004.
142. Zerda, A. d. l.; Liu, Z.; Bodapati, S.; Teed, R.; Vaithilingam, S.; Khuri-Yakub, B. T.; Chen, X.; Dai, H.; Gambhir, S. S. *Nano Lett* 2010, **10**, 2168–2172.
143. Luke, G.; Yeager, D.; Emelianov, S. *Ann Biomed Eng* 2012, **40**, 422–437.
144. Su, H.; Liu, Y.; Wang, D.; Wu, C.; Xia, C.; Gong, Q.; Song, B.; Ai, H. *Biomaterials* 2013, **34**, 1193–1203.
145. Lee, S.; Chen, X. *Mol Imaging* 2009, **8**, 87–100.
146. Cherry, S. R. *Annu Rev Biomed Eng* 2006, **8**, 35–62.
147. Costa, D. C.; Visvikis, D.; Crosdale, I.; Pigden, I.; Townsend, C.; Bomanji, J.; Prvulovich, E.; Lonn, A.; Ell, P. J. *Nucl Med Commun* 2003, **24**, 351–358.
148. Israel, O.; Keidar, Z.; Iosilevsky, G.; Bettman, L.; Sachs, J.; Frenkel, A. *Semin Nucl Med* 2001, **31**, 191–205.
149. Townsend, D.; Cherry, S. *Eur Radiol* 2001, **11**, 1968–1974.

150. Townsend, D. W.; Carney, J. P. J.; Yap, J. T.; Hall, N. C. *J Nucl Med* 2004, **45**, Supplement 1, 4–14.

151. Townsend, D. W. *J Nucl Med* 2001, **42**, 533–534.

152. Devaraj, N. K.; Keliher, E. J.; Thurber, G. M.; Nahrendorf, M.; Weissleder, R. *Bioconjug Chem* 2009, **20**, 397–401.

153. Townsend, D. W.; Beyer, T.; Blodgett, T. M. *Semin Nucl Med* 2003, **33**, 193–204.

154. Wang, D.; Su, H.; Liu, Y.; Wu, C.; Xia, C.; Sun, J.; Gao, F.; Gong, Q.; Song, B.; Ai, H. *Chin Sci Bull* 2012, **57**, 4012–4018.

155. Judenhofer, M. S.; Wehrl, H. F.; Newport, D. F.; Catana, C.; Siegel, S. B.; Becker, M.; Thielscher, A.; Kneilling, M.; Lichy, M. P.; Eichner, M.; Klingel, K.; Reischl, G.; Widmaier, S.; Rocken, M.; Nutt, R. E.; Machulla, H.-J.; Uludag, K.; Cherry, S. R.; Claussen, C. D.; Pichler, B. J. *Nat Med* 2008, **14**, 459–465.

156. Yankeelov, T. E.; Peterson, T. E.; Abramson, R. G.; Garcia-Izquierdo, D.; Arlinghaus, L. R.; Li, X.; Atuegwu, N. C.; Catana, C.; Manning, H. C.; Fayad, Z. A.; Gore, J. C. *Magn Reson Imaging* 2012, **30**, 1342–1356.

157. Shao, Y.; Cherry, S. R.; Farahani, K.; Meadors, K.; Siegel, S.; Silverman, R. W.; Marsden, P. K. *Phys Med Biol* 1997, **42**, 1965–1970.

158. Shao, Y.; Cherry, S. R.; Farahani, K.; Slates, R.; Silverman, R. W.; Meadors, K.; Bowery, A.; Siegel, S.; Marsden, P. K.; Garlick, P. B. *IEEE Trans Nucl Sci* 1997, **44**, 1167–1171.

159. Slates, R. B.; Farahani, K.; Shao, Y.; Marsden, P. K.; Taylor, J.; Summers, P. E.; Williams, S.; Beech, J.; Cherry, S. R. *Phys Med Biol* 1999, **44**, 2015.

160. Yang, X.; Hong, H.; Grailer, J. J.; Rowland, I. J.; Javadi, A.; Hurley, S. A.; Xiao, Y.; Yang, Y.; Zhang, Y.; Nickles, R. J.; Cai, W.; Steeber, D. A.; Gong, S. *Biomaterials* 2011, **32**, 4151–4160.

23

INTRACELLULAR DELIVERY OF PROTEINS AND PEPTIDES

Can Sarisozen[1] and Vladimir P. Torchilin[1,2]

[1] *Center for Pharmaceutical Biotechnology and Nanomedicine, Northeastern University, Boston, MA, USA*

[2] *Department of Biochemistry, Faculty of Science, King Abdulaziz University, Jeddah, Saudi Arabia*

23.1 INTRODUCTION

Since the first introduction of recombinant therapeutic protein human insulin [1], Eli Lilly & Co.'s Humulin in 1982 (The FDA's New Drug Application Approval Database), more than 200 protein products have been marketed. After more than 30 years of recombinant insulin's first introduction, the number and the market of protein therapeutics expand exponentially. The sales of peptide and protein-based therapeutics were \$94 billion in 2007, which corresponded to approximately 15% of the global pharmaceutical industry [2]. Global market size of the recombinant therapeutic proteins including the monoclonal antibody (mAb)-based products had increased to \$99 billion in 2009. Since then, the global sales of peptide and protein therapeutics recorded \$108, \$113, and \$125 billion for 2010, 2011, and 2012, respectively (source: La Merie Publishing, Germany). And finally, Humira® (adalimumab, Abbott and Eisai), a recombinant fully humanized mAb, became one of the biggest selling products of all time and recorded annual sales of almost \$10 billion in 2012 (Abbott annual sales report 2012). The increase in the peptide and protein therapeutics can be attributed to faster clinical development and FDA approval time compared with small-molecule drugs. Indeed, between 1980 and 2002, FDA approval times were more than 1 year faster for the 33 protein therapeutics than 294 small-molecule

Drug Delivery: Principles and Applications, Second Edition. Edited by Binghe Wang, Longqin Hu, and Teruna J. Siahaan.

drugs [3]. Moreover, the time and coverage of patent protection for protein drugs can be much greater than the small molecules. Between 2006 and 2010, 21 out of 99 approved new molecular entities were biopharmaceuticals [2]. But these financial advantages are not enough by themselves to fully explain the current attraction and increase in the peptide and protein therapeutics.

Peptide and protein therapeutics have several advantages over small-molecule drugs. Most of the peptides and proteins possess highly specific and effective functions due to their complex molecular structure. At the same time, they are highly potent therapeutics that possess low possibility of interference with biological processes. Different classifications could be done for peptide- and protein-based therapeutics. They can be grouped into their molecular types as protein receptors, mAbs, mAb fractions and bispecific antibodies, blood factors, protein immunotoxins, protein–drug conjugates, bone morphogenetic proteins, growth factors, hormones, interferons, interleukins, and enzymes [4]. Alternatively, a classification based upon their function and activity [5] has been proposed in recent years.

Insulin is generally the first protein therapeutic that comes to mind for the treatment of diseases caused by a deficient protein. However, enzymes regardless of the different classifications represent a wide and important group of the therapeutic proteins. Therapeutic enzymes in general include antitumor enzymes acting by catabolism of amino acids necessary for tumor growth, enzymes for the replacement of digestive enzymes, enzymes for thrombolytic therapy, antibacterial and antiviral enzymes, and anti-inflammatory enzymes [6]. But more importantly certain diseases can only be treated by the administration of exogenous enzymes. Lysosomal storage diseases (LSD) are a group of disorders that involve loss of function in more than 40 enzymes in the lysosomes including proteases, lipases, sulfatases, or proteins important in their synthesis or trafficking [7]. When these enzymes are missing or not functioning properly, their substrates accumulate in the lysosomes, and that *storage* eventually damages the organs. One important example of the many LSD is Gaucher's disease—a chronic disease of lipid metabolism [8]. It is caused by β-glucocerebrosidase (or acid β-glucosidase) enzyme deficiency—an enzyme that degrades glucocerebroside (also called glucosylceramide). The storage and accumulation of this lipid in macrophages causes enlarged liver and spleen, bone lesions, and neurological conditions [9]. Other types of common LSD include Fabry's disease caused by deficiency in the activity of α-galactosidase A that hydrolyzes globotriaosylceramide [10] and Pompe disease caused by the deficiency of acid α-glucosidase that results in the intracellular accumulation of glycogen [11].

Protein- and peptide-based drugs (partially summarized in Table 23.1) are currently among the most effective treatment options for various diseases and conditions including cancer, diabetes, hemophilia, immunodeficiencies, rheumatoid arthritis, neutropenia, embolisms, and allograft rejections. But the advantages such as high specificity, low interference with normal processes in the body, improved efficacy, and greater safety that make the protein and peptide drugs unique, effective, and favorable over conventional small-molecule therapeutics are also associated with their complex macromolecular structure. Unlike small molecules, peptides and especially proteins possess secondary, tertiary, and quaternary structures as well as primary structure. All these structures, which are required for peptide and protein

TABLE 23.1 Examples of Peptide and Protein Therapeutics on the Market

Protein	Clinical Use	References
Protein therapeutics for protein deficiencies and abnormalities		
Insulin and derivatives	Diabetes mellitus	[1, 12]
Growth hormone, somatotropin	Growth failure that is caused by growth hormone deficiency, Turner syndrome	[13]
Factor VIII	Hemophilia A	[14]
Factor IX	Hemophilia B	[15]
β-Glucocerebrosidase	Gaucher's disease	[8, 16]
Alglucosidase-α	Pompe disease	[17]
Agalsidase-β	Fabry disease	[18]
α-1-Proteinase inhibitor	α-1-Antitrypsin deficiency	[19]
Protein therapeutics augmenting an existing pathway		
Erythropoietin, darbepoetin-α	Treatment of anemia	[20]
Filgrastim, pegfilgrastim	Neutropenia	[21]
Salmon calcitonin	Postmenopausal osteoporosis	[22]
Tissue plasminogen activator	Myocardial infarction, acute ischemic stroke, pulmonary embolism	[23]
Protein therapeutics with novel activity		
Botulinum toxin type A and B	Dystonia, cosmetic uses	[24]
Human deoxyribonuclease I	Cystic fibrosis	[25]
Humanized or chimeric mAb		
Bevacizumab (binds VEGFA)	Colorectal and non-small-cell lung cancer	[26]
Cetuximab (binds EGFR)	Colorectal and head and neck cancer	[27]
Trastuzumab (binds HER2/Neu)	Breast cancer	[28]
Rituximab (binds CD20)	Various non-Hodgkin's lymphomas	[29]
Adalimumab (binds TNFα)	Rheumatoid arthritis, Crohn's disease	[30]
Basiliximab and daclizumab (binds the α chain of CD25)	Prophylaxis against allograft rejection	[31]
Fusion proteins		
Abatacept	Rheumatoid arthritis	[32]
Alefacept	Chronic plaque psoriasis	[33]

activity, also are subjected to many different degradation and instability pathways both *in vitro* and *in vivo*. Hydrolysis, oxidation, isomerization, disulfide exchange, and racemization of peptides and proteins cause chemical instability and rapid degradation [34]. Physical instability includes denaturation, nonspecific and undesirable adsorption to surfaces, and aggregation depending on the temperature, pH, salt type and concentration, preservatives, and surfactant content of the solution [35]. All these mechanisms lead to rapid inactivation of the peptides and proteins, eventually elimination from the circulation due to enzymatic degradation, nonspecific organ and tissue accumulation, and uptake by the mononuclear phagocyte system (MPS). Moreover, the immune response to the administered peptide or protein is also a critical process to keep in mind. Problems and mechanisms given earlier could be solved with different approaches detailed in this chapter, but if the activity of the peptide or protein therapeutic depends on its antigenic part, the solution is not as easy.

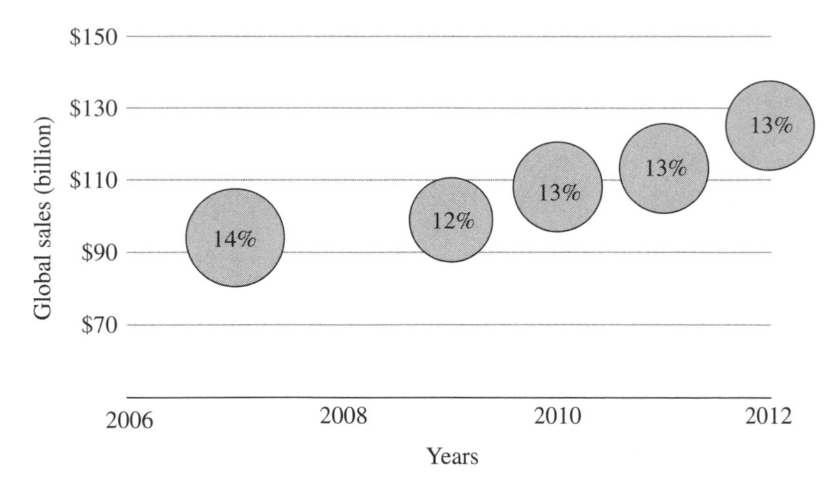

FIGURE 23.1 Global sales of the peptide and protein therapeutics between 2007 and 2012. The center points of the circles represent the approximate global sales values, while the diameter of the circles (the percentage values) corresponds to the ratio in the wholesale therapeutics market. Even though the sales numbers of peptide and protein drugs are going higher every year, their percentage in the global pharmaceutics market remains similar, which indicates a stable share despite of the current developments. Data combined from [2, 36] and various La Merie (http://www.lamerie.com) financial and public reports.

There are also challenges related to the action mechanism of the peptides and proteins. Some peptide and protein drugs including antibodies interact with the cell extracellularly and exert their action by binding to cell surface receptors. However, a big portion of the peptide and protein drugs have to enter cells because their targets are in the cellular compartments. In this case, the impermeability of the cell membrane to peptides and especially proteins limits and in some cases prevents their action. To successfully deliver peptides and proteins to their site of action, which is mostly inside of the target cell, two main obstacles have to be overcome. First, the peptide and protein therapeutics need to be protected until they reach their site of action; second they need to be delivered intracellularly into the cells. The latter part is the main focus of this chapter (Fig. 23.1).

23.2 INTRACELLULAR DELIVERY STRATEGIES OF PEPTIDES AND PROTEINS

The need to deliver peptide and protein therapeutics inside the cell to exert their activity is mainly prevented by cellular membrane. To enhance the intracellular delivery of large, membrane-impermeable molecules, two main invasive methods have been used in the previous decades [37]. In the electroporation technique, transient pores are formed in the cell membrane by brief electric pulse of high field strength [38]. This method is mainly used for DNA or RNA delivery into cells for transformation and expression of gene products [39] or intracellular antibody delivery [37a, 38]. In the microinjection technique, by its definition, peptides or proteins are injected into single cells, which limits the use of this technique when high number of cells are needed [40]. Both of these techniques are invasive in nature and usually damage the cell membrane.

Successful noninvasive peptide and protein delivery depends on nontoxic carriers or vectors, which can efficiently deliver the macromolecular drug intracellularly to exert their therapeutic action inside the cytoplasm or onto nucleus or other specific organelles, such as lysosomes, mitochondria, or endoplasmic reticulum. To achieve this purpose, viral vectors have been used effectively. Since the viruses have highly evolved machinery for cell entry, they are mainly used for gene delivery applications [41]. Viral-based vectors use the infection pathway to enter the cell without the expression of the viral genes. On the other hand, disadvantages like strong immune response, possibility of chromosomal insertion and proto-oncogene activation, toxicity, and likelihood of contamination with the live virus limit their effective use [41a, 42].

Nonviral vectors for gene delivery make use of naturally occurring or synthetic materials to deliver the gene of interest to the target cells. The compounds used to manufacture nonviral vectors usually do not elicit an immune reaction and are less toxic. Additional functionalities on nonviral vectors improve their specificity toward the target sites. They are relatively easy to produce and can be used for repetitive administration. Among the most popular and well-investigated peptide and protein drug delivery systems are nano-sized carriers like liposomes, nanoparticles, micelles, polyplexes, and lipoplexes.

23.3 CONCEPTS IN INTRACELLULAR PEPTIDE AND PROTEIN DELIVERY

23.3.1 Longevity in the Blood

To better understand the intracellular peptide and protein delivery process, it is important to have a background on basic concepts related to nano-sized delivery systems, their fate upon administration, their interaction with the cells, and finally their internalization steps. All of these are important and have a critical effect on successful peptide and protein delivery. As mentioned earlier, before internalization and even interaction with the target cells, peptide and protein therapeutics have to be formulated in a way so that stability and delivery to the target site can be ensured. In 1979 Maeda et al. demonstrated that styrene maleic acid (SMA)–conjugated anticancer protein neocarzinostatin (SMANCS) was able to accumulate higher in the tumor tissue than their unconjugated (NCS) derivatives [43]. They showed that proteins that are bigger than 40 kDa could not only selectively accumulate in the tumor areas but also stay there for prolonged time periods. After 7 years of comprehensive research they build up the enhanced permeation and retention (EPR) effect theory, which since then became one of the golden standards in drug delivery [44]. The EPR effect is now a well-established phenomenon that the endothelial lining of the blood vessel wall in tumors, infarcts, and inflammation sites becomes more permeable as compared to the normal state [45]. The increase in the spacing between the endothelial cells causes a leaky structure that allows large molecules and particles to leave the vascular bed and accumulate in the interstitial space in such areas. The cut-off size in these permeabilized vasculature sites varies from case to case (100–800 nm) but usually falls between the range of 100 and 500 nm. Large molecules with appropriate sizes can slowly accumulate in pathological sites with leaky vasculature, hence the term "enhanced permeation." This accumulation works especially well with tumors because of the lack of lymphatic drainage, meaning that the molecules and particles that are accumulated (*leaked*) will continue to stay in the tumor site [44b, 46].

To benefit from the EPR effect, it is clear that the large molecules or particles need to stay in the circulation for prolonged time period to provide sufficient level of accumulation in the target site. But peptides and proteins rapidly get eliminated from the circulation mainly due to the renal filtration, enzymatic degradation, nontarget organ/tissue accumulation, and most importantly uptake by the MPS. In addition, plain nanocarriers for peptide and protein delivery are usually recognized as foreign particles by the immune system of the body. These carriers rapidly interact and are coated with plasma proteins mainly due to their hydrophobic surfaces. This process is called opsonization and helps macrophages of the MPS to uptake these particles, thus causing clearance from the circulation. To help better accumulation of peptide and protein drugs or drug-loaded pharmaceutical carriers in the target site and increase their interaction with the target cells due to the larger number of passages through the target with blood, prolonged circulation is required.

Modification of peptide or protein therapeutics as well as nano-sized peptide/protein carriers with certain synthetic water-soluble polymers to mask their surface is one of the usual and widely accepted approaches to increase their circulation time in the blood [47]. For this masking purpose, polyethylene glycol (PEG) is the most popular polymer and modification with PEG is called PEGylation [48]. Other types of biocompatible, soluble, and hydrophilic polymers for steric protection are poly(2-methyl-2-oxazoline) [49], phosphatidyl polyglycerols [50], poly(acryloyl morpholine) [51], poly(acryl amide), poly(vinyl pyrrolidone) [52], and poly(vinyl alcohol) [53], but PEG still remains the first choice. It has been showed that modification of polypeptides with water-soluble polymers slows down their renal filtration, and PEGylated L-asparaginase has a circulation time of 5.7 days, significantly longer comparing to 1.2 days for the nonmodified enzyme [54]. There are numerous examples of long-circulating PEGylated polypeptides and proteins for therapeutic applications (also reviewed by Eliason in Refs. [55]) [56]. Benefits of PEGylation, and thus slowing down the opsonization, clearance by MPS, and hindering the contact between the nanocarriers and phagocytic cells, were also shown for nano-sized peptide and protein delivery systems. Even though it is best demonstrated with liposomes [57] and is currently used in clinical conditions (Doxil® and Caelyx®) [58], PEGylation is used to prepare a variety of long-circulating carriers like nanoparticles [59], micelles [60], dendrimers [61], and solid lipid nanoparticles [62]. In conclusion, to achieve successful intracellular peptide and protein delivery *in vivo*, EPR effect-mediated stabilization and PEGylation is the first step before interactions at cellular level. Without proper stabilization of the peptide and protein therapeutics, intracellular delivery of them will be a weak attempt.

23.3.2 Cellular Uptake Pathways

Intracellular transport of the peptide and protein therapeutics has been one of the main problems in drug delivery. Several biological barriers need to be overcome for successful and effective intracellular delivery. Lipophilic nature of the biological membranes, especially the cell membrane, prevents peptides and proteins from directly reaching the cytosol. There are a number of different ways that peptides and proteins are internalized into cells, and success of intracellular delivery usually relates to these internalization pathways. The main intracellular entry mechanism of peptides and proteins as well as nano-sized peptide/protein deliver systems is endocytosis. Mechanistically there are two broad types of endocytosis: phagocytosis or pinocytosis. In addition, endocytosis can be divided into three categories (fluid-phase, adsorptive, and receptor-mediated) in terms of kinetics [63]. Phagocytosis is restricted to specialized mammalian cells including macrophages, monocytes, and neutrophils, basically cell types involved in large particle internalization processes. Pinocytosis, on the other hand, occurs in all types of cells and often considered to be synonymous with endocytosis in general. Morphologically there are four distinct pinocytosis pathways described (Fig. 23.2): clathrin-mediated endocytosis (CME), caveolae-mediated endocytosis, clathrin- and caveolae-independent endocytosis, and macropinocytosis [63, 64].

Clathrin-mediated endocytosis (CME) is a highly regulated energy-dependent process. It has been previously referred to as receptor-mediated endocytosis [65], but after realization of involvement of receptor–ligand interactions in most pinocytic uptake pathways, the term "clathrin-mediated endocytosis" has been adopted. The first requirement of CME is the ligand binding to the specific cell membrane receptor. Then ligand–receptor complexes cluster into coated pits on the plasma membrane

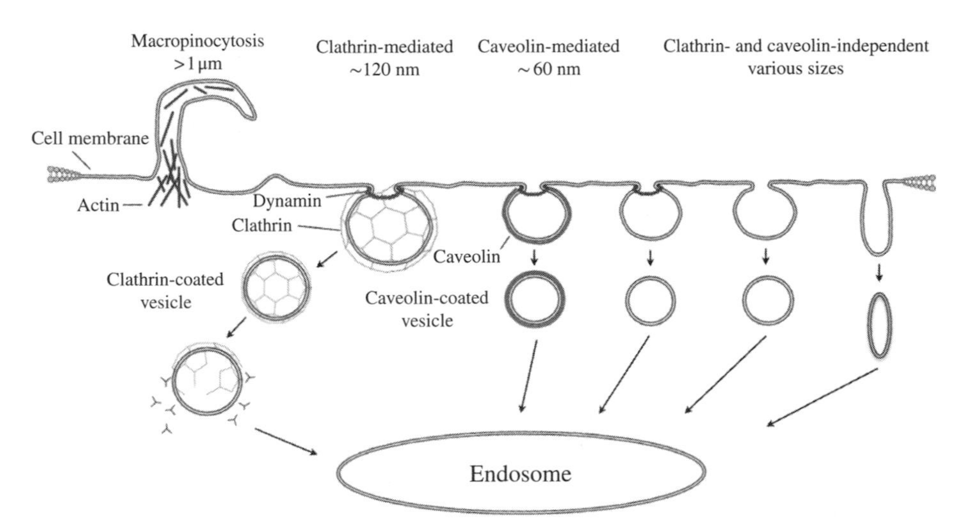

FIGURE 23.2 Different pinocytosis pathways. Compared to the other endocytosis pathways, the vesicles of the macropinocytosis are significantly larger. Some clathrin- and caveolae-independent pathways also require dynamin.

[66]. These pits were formed mainly by the assembly of clathrin, a cytosolic coat protein. Coated pits carrying receptor–ligand complexes invaginate and pinch off to form endocytic vesicles covered with clathrin. Three-legged structure of clathrin is consisted of clathrin heavy chains and light chains [67]. Along with the assembly proteins, clathrin molecules form a closed cage structure coating the endocytic vesicles, which carry the concentrated receptor–ligand complexes into the cell (Fig. 23.2) [68]. Dynamin, a GTPase, also plays an important regulatory role in the CME and other trafficking events at the cell surface [69]. It has been observed that assembled dynamin helices become constricted, and this causes the pinching off and severs invaginated pits upon GTP hydrolysis [70]. Also it is believed to be responsible of sending the formed vesicles into the cytosol [71].

The clathrin-coated vesicles are in size of approximately 120 nm (100–150 nm range). Following the vesicle formation and pinching off, the clathrin coating depolymerizes and forms early endosomes. These early endosomes fuse with each other and/or preformed endosomes to form late endosomes, which eventually form lysosomes. The pH in the lumen of early endosomes is around 5.9–6.0 [68b], which means that peptides and proteins or their carriers that enter the cells via this pathway will experience a high drop in the pH. The pH further drops to 5 during the formation of late endosomes and lysosomes.

Caveolae-mediated endocytosis involves the hydrophobic membrane domains that are rich in cholesterol and glycosphingolipids called caveolae. Even though they are described as flask-shaped invaginations, they can also be in the forms of flat, tubular, or even detached vesicles. Caveolae are present on the membrane of many cells and are involved in several cellular processes. Their structure and formation is associated with a cholesterol-binding protein called caveolin. Caveolin inserts cholesterol as a loop into the inner plasma membrane and self-associates to form a coat on the surface of membrane invaginations. It has been showed that caveolae are static structures of the plasma membrane but following a signaling cascade that results in tyrosine-phosphorylation their internalization can be triggered. This signaling cascade can be activated by phosphatase inhibitors, activation of albumin receptor gp60, or simian virus (SV40) fragments.

One of the main advantages of the caveolae-mediated endocytosis is the possibility of bypassing the lysosomal entrapment. It is believed to be a nonacidic and nondigestive internalization route, because caveolae usually does not subject to pH drop [72]. Also this internalization pathway is believed to be advantageous for the DNA delivery. But the size of caveolae are rather small (~50–60 nm in diameter), thus they can carry only small amounts of fluids into the cells. Moreover, their internalization times are slow, being $t_{1/2} > 20$ minutes. Due to these characteristics of the caveolae-mediated internalization pathway, it has been suggested that large or significant amounts of peptide or protein uptake through this pathway is unlikely.

Clathrin- and caveolae-independent endocytosis pathways remain poorly understood, and they are still described only in negative terms. Clathrin- and caveolae-independent pathways are generally the internalization pathways that cannot be collected into large groups. Clathrin-independent endocytosis has been showed in neuroendocrine cells, involving the neuron-specific isoform of dynamin [73].

It has been also shown that two proteins with a similar topology to caveolin-1, the main protein responsible for caveolae-dependent endocytosis, are able to form caveolae-like structures [74]. These proteins are called flotillin 1 and flotillin 2, and the coassembly of these proteins causes the formation of microdomains in the plasma membrane that appear to bud into cell. These observations suggest a flotillin 1– and flotillin 2–specific endocytic pathway [75]. But the rate at which flotillin microdomains bud into cell is much more slower than clathrin-coated vesicles [74c, 76], indicating flotillin-coated vesicles are not abundant in the cytoplasm and the endocytosis rate and amount through this pathway is limited, like caveolae-dependent endocytosis.

RhoA (a small GTPase)-dependent endocytosis is another clathrin- and caveolae-independent endocytosis mechanism [77], even though this pathway requires dynamin. It was first shown for internalization of β-chain of the interleukin-2 receptor. Clathrin- and caveolae-independent endocytosis mechanisms that do not require dynamin can be grouped into CDC42- and Arf6-regulated internalization [78]. CDC42 dynamin–independent pathway seems to be the main route for the nonclathrin, noncaveolar uptake of cholera toxin B (CtxB), ricin, and the *Helicobacter pylori* vacuolating toxin (VacA) [79]. Arf family GTPase Arf6 plays not-explicitly-established but potent role in clathrin- and caveolae-independent endocytosis of some proteins like class I major histocompatibility complex molecules (MHC I), β1 integrin, carboxy-peptidase E (CPE), and E-cadherin by involving actin remodeling [80].

Another intriguing example of clathrin- and caveolae-independent endocytosis is related to a virus that caused an epidemic in Asia during 2002–2003, SARS coronavirus. Severe acute respiratory syndrome coronavirus (SARS-CoV) was first believed to enter cells through direct fusion with the plasma membrane. But in 2008, Wang et al. reported that SARS-CoV enters the cells via translocation of its functional receptor angiotensin-converting enzyme 2 (ACE2) from the cell membrane to endosomes [81]. Authors showed that entry mechanism was clathrin- and caveolae-independent, but involved cholesterol and sphingolipid-rich lipid raft microdomains in the membrane.

Macropinocytosis refers to the special clathrin-, caveolae-, and dynamin-independent endocytosis pathway. These vesicles generated by actin-driven envagination of the plasma membrane, different from the invagination that occurs during clathrin- and caveolae-dependent endocytosis. Cell surface ruffling formed by a linear band of outward-directed actin polymerization [82] near the membrane surface is the first step in macropinocytosis. These membrane ruffles become longer, then close onto the cell membrane and form large vacuoles called macropinosomes (Fig. 23.2) [83]. Macropinocytosis is involved in many functions, especially as an efficient route for internalization of large volume of solutes because of comparatively massive size of macropinosomes, as large as 5 μm in diameter. Macropinosomes do not have a coating that can be found in clathrin-dependent endocytosis, and they do not concentrate receptor as in clathrin- and caveolae-dependent endocytosis.

In macrophages macropinosomes move into the center of the cell and then completely merge into the lysosomal compartment, but in human cells it has been showed that macropinosomes do not interact with the endocytic compartments and do not fuse with the lysosomes. Moreover, macropinosomes are thought to be inherently leaky vesicles compared with endosomes. These characteristics of the macropinocytosis pathway provide several advantages especially for intracellular peptide and protein delivery due to the avoidance of lysosomal degradation and ease of escape from the leaky vesicles.

23.3.3 Endosomal Escape

Almost all proteins that are identified and encoded by the human genome function in the cells. Thus, peptides and proteins that have therapeutic applications need to be delivered intracellularly. When the cellular uptake methods described earlier are evaluated, it can be seen that almost all these pathways lead their substrates to endosomes. Early endosomal encapsulation is a common starting point of the internalized solutes in the cell. Most of the times the journey leads them in a one-way street to late endosomes, and matured late endosomes eventually fuse with lysosomes. The pH change that internalized peptides and proteins experience during their endosomal fate starts with a sudden drop to pH 5.9–6.0 in early endosomes and ends with pH 5.0 in the lysosomes. Along with the pH change, peptides and proteins get subjected to enzymatic degradation in the lysosomes. All these steps that peptide and protein therapeutics are subjected to cause instability and degradation, and eventually prevent their function in the cell. In conclusion, other than specific purposes, peptide and protein drugs need to be able to escape from the endosomes whether they are to be delivered intracellularly as conjugates or in the delivery systems. So far several approaches have been applied to allow the peptides and proteins to escape from endosomal encapsulation pathway and to protect them from degradation. In this chapter commonly used and established mechanisms that have been proposed for endosomal escape are discussed.

Some peptides that are soluble in water but also have an affinity to bind the lipid bilayers could be used to form holes in the endosome membrane. The main mechanism of pore formation is called barrel-stave model [84]. In this mechanism, transmembrane pores are formed in the interior regions of the membrane by bundles of amphipathic α-helices. But instead of a single monomer binding to the membrane, more peptide monomers are needed to facilitate the formation of a pore. Although this mechanism is mainly used by viruses such as picornaviridae, parvoviridae, and reoviridae to enter the cell after endosomal escape [85], synthetic amphipathic peptide GALA with the repeated amino acid sequence (glutamic-alanine-leucine-alanine) is suggested to use this mechanism [86].

pH buffering effect or proton sponge effect is another endosomal escape mechanism successfully used for peptide and protein delivery into the cells. The proton sponge effect is mediated by high buffering capacity agents that have the flexibility to swell when protonated. One of the polymeric carriers that can facilitate this mechanism is polyethyleneimine (PEI). Upon PEI-mediated entry into the cell, the polymer acts in the endosomes as a sponge that adsorbs protons due to its high buffering capacity and its primary, secondary, and tertiary amine groups. This protonation causes an influx of H^+ and Cl^- ions and water into the endosome and eventually leads to swelling and bursting of the endosomes because of the osmotic pressure [87]. PEI has been used successfully for delivery of peptide-proteins and genes into the cells. PEI-containing polyplexes have been successful for *in vivo* gene delivery to a variety of tissues [88], and the success behind the high gene transfer activity is believed to be due to efficient endosomal escape by proton sponge effect. PEI is not the only polymer that has the buffering capacity and ability to escape from endosomes. Polymers containing crowded histidines (imidazoles), morpholinos, and polyamidoamine (PAMAM) polymers also use this mechanism to facilitate the endosomal escape [89].

PEI can be synthesized in different lengths, be branched or linear, and possess a capability of protonation of amino group at every third position. This latter feature gives PEI a high positive charge density at physiological pH and permits the condensation of negatively charged macromolecules (DNA, siRNA, ODN) into dense particles by electrostatic interactions. Combined with its proton sponge effect and endosomal escape abilities, PEI has a high efficacy for intracellular delivery of peptides and proteins. Positive charges of the PEI complexes interact with negatively charged components of cell membranes and thus trigger cellular uptake of the complexes, but they also cause interaction with blood components and opsonization, leading to rapid clearance from the blood circulation. As a result, the PEI/DNA complexes are cleared from circulation in a few minutes and accumulate mainly in RES organs such as the liver and spleen [90]. Moreover the positive charge of the PEI is usually linked to high toxicity, especially for the high molecular weight PEI [91]. These drawbacks limited the therapeutic use of PEI for macromolecule delivery *in vivo*.

Recently our laboratory developed a novel nonviral gene delivery vector, a micelle-like nanoparticle (MNP) suitable for systemic application [92]. MNPs were engineered by condensing plasmid DNA with a chemical conjugate of phospholipid with PEI (PLPEI) and then coating the complexes with an envelope of lipid monolayer additionally containing PEG–phosphoethanolamine (PEG–PE), resulting in spherical "hard-core" nanoparticles loaded with DNA. We have shown that these MNP formulations were able to protect DNA from enzymatic degradation, give resistance to salt-induced aggregation, and reduce the toxicity. Moreover we have shown that after 1 hour of injection to mice, the amount of MNP formulations still in the circulation was 4 times higher than the only PEI/DNA polyplexes, due to the PEGylation. Intravenous injection of MNP loaded with plasmid DNA encoding for the green fluorescent protein (GFP) resulted in an effective transfection of a distal tumor. When comparable dose of PEI/DNA complex was alone injected, we observed the death of animals only after 30 minutes from respiratory failure.

We have also investigated the potential of MNP systems for intracellular siRNA delivery. These MNP formulations combined the favorable properties of the low molecular weight PEI 1.8 kDa (positive charge, nucleic acid condensation, endosomal escape) with PEGylated nanocarriers (long circulation in the blood) [93]. At a nitrogen/phosphate ratio of 10, siRNA binding was complete. MNPs protected the loaded siRNA from RNase activity and serum degradation and were found to be nontoxic for different range of cell lines. The GFP was successfully downregulated in the cells by these formulations carrying GFP-targeted siRNA, indicating that intracellular peptide delivery was achieved efficiently. Further evaluation of these systems was carried out with a therapeutic siRNA downregulating P-glycoprotein (P-gp), an efflux membrane protein that is overexpressed in multidrug-resistant (MDR) cancer cells [94]. Dioleoylphosphatidylethanolamine-PEI (DOPE-PEI) conjugate was used to prepare the MNPs, and P-gp overexpressing cells were successfully transfected with these conjugates. After P-gp silencing, the cells were treated with doxorubicin, and significantly higher therapeutic effect was achieved. Most recently we have also evaluated the lipid structure effect on the transfection and intracellular delivery of the macromolecules. When PEI was conjugated to dipalmitoylphosphatidylethanolamine (DPPE), DOPE or phosphocholine (PC), the physicochemical properties and siRNA-binding capacities of the formulations were not affected. But we have shown that the transfection efficacy, which indicated the intracellular delivery of the peptide, was significantly higher with DOPE conjugates [95].

For successful endosomal escape, DOPE is widely used as a helper lipid for its fusogenic property. The role of DOPE as a helper lipid is attributed to its endosomolytic activity [96]. The ability of DOPE to destabilize endosomal membranes is based on its small head group area and a large hydrocarbon area, which causes a tendency to adopt an inverted hexagonal lipid phase as the H_{II}. This structure of DOPE favors a nonbilayer structure with a cone shape that facilitates the destabilization of endosomal membrane, leading the peptide or protein therapeutic or its carrier to escape the endosome [97]. Both dioleoylphosphatidylcholine (DOPC) and DPPE share very similar structures with DOPE. But the cone-shaped head group of DOPE displays a high tendency to form inverted hexagonal phase, especially at acidic pH in the endosomes while the head groups of DOPC or DPPE do not [97a, 98].

The cationic lipids have been shown to destabilize endosome membranes and thus lead to endosomal escape of the peptide and protein drugs. Electrical interaction between the cationic lipids and the negatively charged endosomal membranes results in the formation of ion-pair that promotes the formation of inverted hexagonal phase (H_{II}) and disruption of the endosomal membrane [97c, 99]. H_{II} phase is an intermediate structure when two bilayers start to fuse each other. This fusion causes destabilization in both bilayers, meaning that both the endosomal membrane and the peptide–protein carrier with the cationic lipid are destabilized in the process. This destabilization provides endosomal escape as well as peptide or protein release from the carrier or complex [97c, 100]. This mechanism is successfully used for intracellular oligonucleotide delivery of siRNA molecules [101].

Photochemical disruption of the endosomal membrane is called photochemical internalization (PCI) and involves a light-mediated endosomal escape and intracellular delivery of peptides and proteins. In this mechanism a number of photosensitizers that can bind to and localize in the membrane of the endosomes and lysosomes are used [102]. Following the interaction of the peptide and protein complexes or carriers with the cells, these carriers were internalized by endosomes. After the exposure to the light, photosensitizers induce the formation of free radicals with a short lifetime. Due to the effect of these free radicals on unsaturated fatty acids within the endosomal membrane, endosome membrane is disrupted. The photosensitizers used for intracellular peptide and protein delivery include meso-tetraphenylporphine

($TPPS_{2a}$) [103], $TPPS_4$, disulfonated aluminum phthalocyanine ($AlPcS_{2a}$), and zinc phthalocyanine dendrimer.

In an attempt to facilitate the endosomal escape, mimicking the viral mechanisms was also used. Most of the viruses have peptides that can undergo conformational changes in the low pH environment such as endosome lumen. Hemagglutinin (HA2), influenza virus coat peptide, is a well-established example for this kind of fusogenic peptides [87b]. At neutral pH HA2 subunit is existed as the nonhelical hydrophilic coil conformation due to the charge repulsion of ionized glutamic and aspartic acid residues. But when the HA2-containing delivery system of protein conjugate/complex internalized through endosomal uptake, HA2 subunit adopts a helical conformation because of the protonation of acid residues, and the new structure leads to fusion with the endosomal membrane [104].

There are other fusogenic peptides that have been developed and synthesized for efficient endosomal escape of peptides and proteins. GALA and its cationic counterpart KALA peptide have been shown to have fusogenic properties along with their pore formation effect. They have been used for intracellular delivery of genes [105] and siRNA [106]. Other types of fusogenic peptides will be detailed in the flowing sections of this chapter.

Protein kinases are another regulator group that regulate endocytosis [107]. Recently, ur Rehman et al. reported data related to intracellular delivery of oligonucleotides by branched polyethylenimine polymers (BPEIs) [108] and suggested that protein kinase A (PKA) inhibition could be used to modulate the intracellular delivery of the peptide delivery systems. They showed that inhibition of PKA activity modulates the intracellular routing of the BPEI polyplexes and prevents trafficking into late endosomes/lysosomes and thus prevents the peptide degradation. Researchers proposed a new compartment to which the cargo reroutes after the inhibition of PKA, and they reported two- to threefold transfection efficiency. The endosomal escape strategies mentioned earlier involve first the endosomal encapsulation via endocytic uptake pathways and then use of vectors that help the cargo to escape from the endosomes. PKA inhibition and rerouting the clathrin-mediated pathway to bypass the lysosomal fusion could be used as another promising strategy for intracellular peptide and protein delivery.

Despite the strategies given earlier for an efficient endosomal escape of the peptides and proteins or their carrier systems, there are still problems and dilemmas to be overcome. One of those dilemmas is related to the PEGylation of the peptide and protein carriers. Even though the clear advantages of PEGylation have been reviewed in this chapter, paradoxically PEGylation also causes a steric hindrance for the vectors that are used for endosomal escape and prevents them to freely interact with the endosomal/lysosomal membranes. Thus reduced interaction causes insufficient endosomal escape and eventually can decrease the intracellular delivery of the loaded peptide and protein. This effect of PEG is called the "PEG dilemma." Most ideal peptide and protein carrier system should have the PEG coating layer for long circulation times and decreased opsonization *in vivo*, but this PEG layer should also dissociate from the carrier surface at the right place and time. After a peptide or protein carrier reaches its target site and is internalized by the cells, the covering PEG chain becomes unnecessary and prevents the endosomal escape by fusogenic peptides and other types of vectors. The reversible PEGylation strategy is able to combine the advantages of both PEGylation and endosomal escape. Stimuli-triggered removal of PEG coating has allowed the design of smart multifunctional systems. There are different stimuli-responsive cleavage mechanisms for PEG chains. pH-sensitive bonds and polymers as well as enzymatic cleavage for PEG chain removal at the low pH levels and high matrix metalloproteinase (MMP)–expressed cells (i.e., cancer cells) have been used successfully for intracellular delivery [109].

23.4 PEPTIDE AND PROTEIN DELIVERY TO LYSOSOMES

Endocytosis, meaning internalization into early endosomes, maturation of the late endosomes, and eventually fusing with the lysosomes, was most of the time seen as the archenemy of effective and successful intracellular peptide and protein delivery because of the degradation due to highly acidic environment and enzymatic activity. But in case of a group of diseases called LSD, this pathway is the only option for successful treatment. With more than 40 inherited disorders, LSD caused by mutations in the proteins are critical for proper lysosomal function [8, 16a, 110]. Lysosomes take their substrates with various mechanisms and they degrade the substrates with complex machinery. When one or more lysosomal proteins that are involved in the regulation of this degradation process are absent or mutated, progressive accumulation of its substrate occurs. The accumulation of the undigested molecules eventually starts to alter cellular mechanisms. Even though LSD are not getting the attention as they deserve, the combined incidence of the LSD is increasing from 1 in 8000 [111] in 1999 to 1 in 5000 [112] in 2006 in live births. After the discovery of intracellular enzymes and proteins targeted into the lysosomes via mannose-6-phosphate receptor-mediated pathway [113], enzyme replacement therapy became available to treat the LSD. So far, this treatment is the only approved treatment option for not only type I Gaucher's disease but also for the LSD in general [7, 114]. But still there are some drawbacks of this therapy that require more effective and enhanced intracellular and lysosome-targeted delivery. The treatment is life long, and its effectiveness is mainly dose dependent. It has been shown that to achieve adequate substrate degradation and cleaning from the affected tissues and organs, usually higher doses are required [16a, 115]. Moreover protein delivery is invasive and time consuming, in addition to the inability of the proteins to cross the blood–brain barrier, which is important for treating the LSD as most of the diseases affect the central nervous system.

Gaucher's disease is the most common LSD and characterized by a deficiency of the lysosomal enzyme glucocerebrosidase (GlcCerase) [116]. Patients diagnosed with Gaucher's disease have 5–25% of the normal GlcCerase activity, and this deficiency causes accumulation of the GlcCerase substrate glucocerebroside in the lysosomes, primarily in the macrophages [117]. Enzyme replacement therapy, despite its disadvantages, is the only approved therapy option for Gaucher disease. In our lab, we were also interested in an effective intracellular enzyme delivery in recent years. We have demonstrated that, liposome-based enzyme therapy can be achieved by using liposomes specifically targeted to lysosomes. To achieve lysosomal delivery, we have modified the liposome surface with lysosomotropic octadecyl-rhodamine B (Rh) and showed that liposome-loaded model marker FITC-dextran was significantly increased in HeLa cells [118]. In the light of the results, most recently we investigated the possibility of successful and effective active enzyme Velaglucerase alfa (VPRIV™) delivery to lysosomes with Rh-modified liposomes [119]. Confocal laser scanning microscopy studies with plain and Rh-modified liposomes loaded with FITC-dextran proved the significant targeting and localization of Rh-modified liposomes in the lysosomes. Non-Rh-modified plain liposomes showed significantly lower lysosomal localization even after 20 hours of incubation with the incubation of monocyte-derived macrophages (MDM) with inhibited GlcCerase activity. To determine the GlcCerase activity, we have used the Gaucher's fibroblasts that have lower level of or no GlcCerase compared with the normal fibroblasts. We have shown that Rh-modified lysosomotropic liposomes strongly improve the lysosomal accumulation of liposomal glucocerebrosidase both in nonphagocytic Gaucher's fibroblasts and phagocytic MDMs. The intralysosomal accumulation of liposomal VPRIV in the cells treated with Rh-modified liposomes was 68% higher relative to the cells treated with nonmodified liposomes or free VPRIV, thus it can be concluded that such an improvement in the intracellular enzyme delivery can shift the balance from pathological to normal cell types.

A large variety of small molecules have been identified, which specifically target and accumulate in lysosomes. Among them, neutral red (NR) and rhodamine B (RhB) are routinely used for the visualization of lysosomes and other acidic organelles in live cells. Meerovich et al. [120] had synthesized different ligands based on NR and RhB with short- and long-PEG spacers suitable for introduction into the lipid bilayer and compared their ability to enhance the lysosomal delivery of these ligand-modified liposomes loaded with the model compound FITC-dextran. They reported that derivatives of NR showed low or no colocalization of liposomal dextran with antibody-marked lysosomes, while samples of cells incubated with (RhB) derivatives showed elevated colocalization of liposomal load in comparison to plain dextran-loaded liposomes under the same condition. Liposomes modified with two of the synthesized ligands—RhB DSPE-PEG2k-amide and 6-(3-(DSPE-PEG2k)-thioureido) RhB—enhance lysosomal delivery of model drug load *in vitro*, in some cases, higher than commercially available RhB octadecyl ester and can be used for further investigation and potentially for intracellular protein delivery for the treatment of the LSD.

23.5 RECEPTOR-MEDIATED INTRACELLULAR DELIVERY OF PEPTIDES AND PROTEINS

23.5.1 Transferrin Receptor–Mediated Delivery

Transferrins belong to a family of iron-binding glycoproteins that have been classified into three main groups. Serum transferrin (Tf) is found in blood and other mammalian fluids such as lymph and cerebrospinal fluid, bile, colostrum, and amniotic fluid [121]. Other members are ovotransferrin, which can be found in avian and reptilian egg white, and lactoferrin in mammalian milk, tears, and saliva [122]. The Tf as well as its other family members is a single-chain glycoprotein containing 679 amino acids with a molecular weight of approximately 80 kDa. The main function of the Tf is related to iron transport from the intake sites into the cells and tissues. Iron is needed for many redox reactions in the organism as a cofactor. But Fe^{3+} ions cannot enter the cells readily because of insoluble hydroxide complex formation, and even they have important roles in cell metabolism events such as heme synthesis, they are not available for the cells. The Tf binds two iron atoms per Tf molecule, and the iron-loaded Tf is called holo-Tf, while the iron-free Tf is called apo-Tf. Cells can uptake iron with various mechanisms but the principal pathway for mammalian cells is by the receptor-mediated uptake of holo-tranferrin [123]. Uptake process is triggered by the binding of Tf to its specific cell surface receptor (TfR). After the binding, CME pathway starts and Tf–TfR complex (with the loaded iron) is encapsulated in the endosomal compartment. At the low pH values in endosomes, binding of iron to its carrier weakens and iron is released from the protein. Immediately Fe^{3+} is reduced by the divalent metal transporter (DMT1) on the endosomal membrane and released into the cytoplasm as Fe^{2+} [124]. After the iron is separated from the Tf–TfR complex, the protein and receptor complex are routed to the exocytic vesicles, which carries the complex to the cell surface. At physiologic pH, apo-Tf disassociates from TfR receptor and returns to circulation (Fig. 23.3). ATP is necessary for sustained

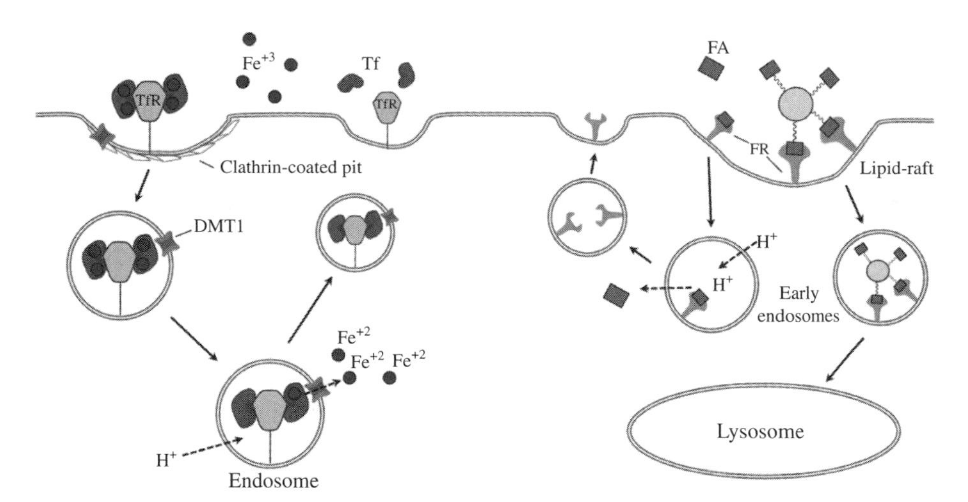

FIGURE 23.3 The cellular uptake pathways of Tf and FA via their receptors. Holo-Tf binds to the TfR on the cell surface and complexes localize in clathrin-coated pits, which invaginate to initiate endocytosis. Acidification of the endosome results in a decrease in pH that stimulates a conformational change in Tf and its subsequent release of iron. The iron is then transported out of the endosome into the cytosol by DMT1. Apo-Tf remains bound to the TfR1 while in the endosome and is only released once the complex reaches the cell surface and neutral pH. Clathrin-independent endocytosis of FA differs according to the FA–FR interaction. Univalent complexes go through a different cycle and lower pH drop compared with the multivalent complexes, which reach lysosomes rapidly.

turnover of TfR-mediated endocytosis [123c, 125]. Molecular mechanisms of TfR expression are highly regulated, and iron-regulatory proteins control the rate of translation [126]. Although iron is essential for all living cells and TfR is expressed in many cell and tissue types, normal cells in their resting state only need small amounts of iron for normal cell functions, and TfR is not overexpressed in normal cells. TfR is overexpressed in the cells that need large amounts of iron due to their increased metabolism and growth, that is, cancer cells [127]. Thus, it is not surprising that most of the Tf- and TfR-related intracellular delivery was focused on cancer.

There are a number of studies related to direct Tf or anti-TfR antibody conjugates with small-molecule anticancer drugs such as doxorubicin (DOX) [128], cisplatin [129], chlorambucil [130], or mitomycin C [131]. With the same approach, Tf-toxic protein conjugates were also used for intracellular protein delivery. One of these proteins is ricin, a highly toxic type II ribosome-inactivating protein extracted form *Ricinus communis*. A chain of the protein (RTA) is responsible for blocking ribosomal activity and the B chain (RTB) for binding to the cell surface [132]. RTA inactivates the ribosomes and thus leads to the inhibition of protein synthesis and death of the cell. Even though the toxic protein part is the RTA, it lacks the cell binding without the RTB and is less toxic because of the limited intracellular entry. To enhance the intracellular delivery of only the toxic protein part, human Tf was conjugated to RTA and found to be 10,000-fold more toxic than the RTA alone [133]. The efficacy of the Tf–RTA conjugate was also investigated on the MCF7 3D multicellular cancer cell spheroid model and was found to inhibit the growth completely at 0.0175 and 0.035 nM immunotoxin concentration [134]. Tf is not the only ligand conjugated to the RTA for TfR-mediated intracellular delivery. Murine monoclonal TfR IgG1 antibody 7D3 and 454A12 were also conjugated to RTA for intracellular protein delivery and was found to be effective both *in vitro* and *in vivo* [135].

Another protein directly conjugated to Tf for TfR-mediated intracellular delivery was the diphtheria toxin (DT), which blocks protein synthesis. It was found that Tf conjugation to DT [136] or its mutant form CRM107 increased the intracellular delivery of DT or its mutant through TfR-mediated endocytosis [137].

Moreover, ribonucleases are another group of proteins that were intracellularly delivered using Tf. In order to increase intracellular delivery to cancer cells bovine pancreatic ribonuclease A [138] and human pancreatic ribonuclease [139] were conjugated to Tf, and their cytotoxicity was found to increase due to increased intracellular delivery.

Direct conjugation of the proteins to Tf or anti-TfR antibodies was reported to successfully deliver the proteins intracellularly in preclinical models but most of these studies were completed almost two decades ago and no significant clinical successes have been demonstrated. Direct conjugation was definitely not the only option for TfR-mediated intracellular peptide and protein delivery. Tf and/or anti-TfR antibodies can be used to modify peptide and protein carriers, that is, liposomes, micelles, or nanoparticles for enhanced cytosolic delivery. Liposomes are among the most popular and well-investigated carriers for peptide and protein delivery. Tf-modified liposomes were used to deliver macromolecules, for example, the tumor suppressor gene p53 into the DU145 prostate cancer cells [140]. Moreover, different liposome formulations with different TfR-targeted ligand such as human Tf or scFv antibody fragment against TfR were successfully used to deliver genes into the cells [141]. These examples indicate that TfR-targeted peptide and protein carriers formulated with effective endosomal escape vectors summarized in the related section of this chapter could be used for effective intracellular peptide and protein delivery both for *in vivo* and *in vitro* applications.

23.5.2 Folate Receptor–Mediated Delivery

Folic acid (FA) is a vitamin that is needed for one-carbon reactions, especially for nucleotide bases. Because it is involved in the nucleic acid synthesis, FA is consumed by the proliferating cells in large amounts. Cellular entry of FA (oxidized form of folate) is regulated by three transporters: the reduced folate carrier [142], proton-coupled folate transporter [143], and the folate receptor (FR) [144]. The first two of these transporters are found in all cells and act as the primary folate uptake pathway into the cells. But these transporters do not have an affinity to the folate conjugates, that is, folate-conjugated proteins or folate-conjugated carriers. The FR is found in the polarized epithelial cells and activated macrophages [145] but the receptor is generally absent in normal human tissues [146]. The low concentrations of the reduced folate carrier or proton-coupled folate transporter on the membranes are usually enough to internalize FA for normal metabolic reactions. But like TfR, FR is overexpressed in cancer cells too, probably due to the increased nucleotide synthesis caused by higher proliferation rates [147]. This overexpression makes the FR an ideal target for cancer cells. Moreover, the FR can be further used for tumor-specific target *in vivo*, since the accessible FR on normal cells is limited to macrophages and the proximal kidney tubules [148]. When the difference in the FR levels of the normal and malignant tissues of same origin was investigated, the difference in the expression levels of FR was found significantly higher in the malignant tissues such as ovary, uterus, and brain [147b]. This selectivity makes the FR receptors' natural ligand FA a popular targeting moiety for targeting therapeutics to the cancer cells. The affinity of FR to its ligand FA is high with K_d values between 0.1 and 1 nM [149]. In addition to this high affinity, FA is a small molecule with low immunogenicity and high stability in different solvents [150], which allows it to be handled and used easily for different applications.

FA internalization via FR is a receptor-mediated endocytosis uptake. Early reports suggest that, after binding of FA to FR that are clustered in invaginated caveolae, the caveolae stay attached to the plasma membrane, and FA is released into the cytosol from these still-membrane-attached vacuoles [151], a mechanisms named potocytosis. But more recent studies indicate that FRs cluster on the membrane within lipid rafts, and these rafts invaginate into the cytosol to form early endosomes [152]. The FA is separated from its receptors in these early endosomes, and the FR is moved to another separate endosome that carries it to the cell surface [153].

When the FA and FR are used for the intracellular peptide and protein delivery applications, there are two main strategies: (i) peptides or proteins can be directly conjugated to FA or an anti-FR Ab and (ii) peptide and protein carriers such as liposomes, micelles, nanoparticles, dendrimers, etc. can be modified with a ligand for FR targeting. The internalization pathway given earlier is used for monovalent FA or its conjugates. When the FA is conjugated to peptide or protein carriers, the intracellular fate of the carrier changes. Since the FA or related ligands in the surface of these carriers are more than one, the interaction with the clustered FR on the membrane results in a different intracellular pathway, which leads the endosomal captured carriers into an old enemy, lysosomes. When the monovalent FA conjugates are internalized by the cells, the pH values that they are subjected are found to be around 6.8–6.9. On the other hand, multivalent FA conjugates experience much lower pH values (around 5.0) due to the lysosomal encapsulation [154]. This important difference should be kept in mind for the FR-mediated intracellular peptide and protein delivery methods.

An important group of FA conjugated proteins are toxins. Different conjugation strategies were discussed in depth in [155]. The advantage of the FA-protein conjugates is the bypassing of the possible lysosomal encapsulation; thus, enzymes do not rapidly destroy the conjugates and the proteins. Toxic proteins such as plant-derived momordin and bacteria-derived pseudomonas exotoxin A (PE38) were used for effective intracellular delivery to cancer cells [156]. Both of those proteins do not contain cell-binding domains, thus their entry to the cells is impeded. The IC_{50} values of both toxins were found to be higher than 10^{-5} M when they were used as free proteins. But when momordin was conjugated to FA, its cytotoxicity and intracellular delivery to the FR overexpressing cells were increased; almost 1000-fold decrease in the IC_{50} value was achieved in HeLA and KB cells. The protein conjugation chemistry to FA also plays an important role in the final activity of the intracellularly delivered proteins. For example, disulfide- and amide-linked folate conjugates of gelonin, a protein toxin, bind the FR with the same affinity but their activity of protein synthesis inhibition differs by more than 225-fold, with the disulfide-conjugated protein being more effective [157].

It's clear that FR targeting with an FA-conjugated protein increases the intracellular delivery of the protein. But even under the optimum conditions, the *in vivo* use of these conjugated proteins or peptides is limited because of their short circulation time and low ability to reach the tumor site. Peptide and protein carriers such as liposomes, micelles, or polymer complex systems can be used to eliminate these drawbacks. But it should be noted that the endocytic pathway of the multivalent carriers involves the endosomal encapsulation and lysosomal fusion, thus proper endosomal escape methodology has to be used for effective intracellular delivery. For example, PEGylated long-circulating protein complexes modified with folate were used for intracellular delivery, and increased intracellular delivery of the protein was achieved due to the folate-mediated endocytosis [158]. But when these folate-modified polymer complexes were prepared with PEI instead of PLL, proton sponge effect and thus endosomal escape caused by PEI significantly increased the intracellular delivery of caspase-3 and enhanced the apoptosis ratios in the KB cells [159]. Folate-modified liposomal systems can also be used for intracellular peptide and protein delivery. FR-targeted liposomes prepared for gene delivery were also found successful due to the enhanced intracellular uptake through FR-mediated endocytosis, and when an endosomal escape vector such as DOPE was included in the formulations, gene delivery efficacy increased significantly [160].

23.6 TRANSMEMBRANE DELIVERY OF PEPTIDES AND PROTEINS

Plasma membrane of the cell is one of the most important and also effective barriers for internalization of the peptides and proteins. In general, a molecule should be nonpolar and smaller than 500 Da in size to permeate through the membrane. To overcome this limitation, many strategies including the ones that are summarized in this chapter like TfR- or FR-mediated intracellular delivery methods have been used to enhance the therapeutic peptide and protein uptake into the cells. However, there are limitations of these strategies such as insufficient uptake or endosomal encapsulation as discussed earlier. Since Ryser and Hancock's discovery of increased cellular uptake of albumin by histones and cationic polyamines [161], many other peptide and protein structures that have the ability to efficiently pass through the plasma membrane have been discovered. The proof of protein transduction into the cells was described by Green [162] and Frankel [163] in 1988, independently from each other, after discovering the ability of HIV-1 transactivator of transcription (TAT) protein to cross the cell membranes in a receptor-independent but concentration-dependent manner. After this discovery, in 1994 Fawell et al. demonstrated the increased intracellular delivery of functional proteins with the peptide fragments derived from HIV-1 TAT protein [164]. These studies have provided a new opportunity for overcoming the cellular barrier for intracellular peptide and protein delivery based on the use of certain proteins and peptides that contain the so-called protein transduction domains (PTDs) or cell-penetrating peptides (CPPs), usually less than 20 amino acids and highly rich in basic residues. Subsequently, this property of translocation was found in many other peptides and proteins, and this new class of intracellular peptide and protein delivery vectors has been one of the most efficient choices for protein and peptide delivery into the cells. In 2003, "membrane translocating sequence (MTS)" was proposed [165] as another term to describe these peptide sequences, but in this chapter CPP will be used since it is the most popular and self-explanatory option.

23.6.1 Well Studied Classes of CPPs for Peptide and Protein Delivery

Of all the different CPPs reported over the past 25 years (more than 100 peptide sequences varying from 5 to 40 amino acids in length) [166], only a handful of them are well studied and investigated. CPP can be divided into two classes: (i) peptides with a high degree of amphipathicity such as model amphipathic peptide (MAP) [167], transportan [168], and Pep-1 [169] where the charge contribution originates from lysine (Lys) residues and (ii) low amphipathic peptides with arginine (Arg)-rich structure such as penetratin [170], TAT peptide [171], and repeating units of arginine residues, most notably R8 and R9 [172]. Most of the known CPPs are not cell type or tissue specific, and the intracellular entrance depends on the sequences of positively charged amino acids, especially Arg and Lys. Here we will give short description of the most often used CPPs for intracellular peptide and protein delivery.

Penetratin is the minimal PTD of Antp, the homeodomain of Antennapedia, and contains the 16-mer peptide present in the third helix of the homeodomain. Penetratin was found to be more efficient for the transfer of the small proteins than large proteins of 100 residues or more [173]. Transportan is a chimeric 27-amino-acid-long CPP that contains peptide sequence from N-terminus of neuropeptide galanin linked through a lysine residue to mastoparan [168], and it was found to be more effective for crossing the epithelial layers [174]. Transportan has been used for intracellular delivery of siRNA [175], peptide nucleic acids [176], and for protein with a wide range of sizes between 30 and 150 kDa [177]. VP22 is another CPP and the major structural component of herpes simplex virus type 1 (HSV-1) [178]. This protein is synthesized in the infected cells, then penetrates into the surrounding cells, and concentrates in the nuclei and binds chromatin [178, 179]. This peptide has been used for intracellular delivery of proteins due to its membrane translocation [178, 180]. MAP is a designed 18-mer amphipathic peptide [167] with the given sequence in Table 23.2. Internalization efficiencies of different peptides designed on the basis of MAP suggest that amphipathic character of the peptide is the main responsible factor for effective intracellular delivery since MAP promotes the internalization of the cargo but its nonamphipathic derivatives do not [183]. MAP has the fastest uptake and the highest cargo delivery efficiency when compared to transportan, TAT, and penetratin [184].

TAT peptide (TATp), the most frequently used CPP, is derived from the transcriptional activator protein encoded by human immunodeficiency virus type 1 (HIV-1) [185]. Depending on the different strains of the HIV virus, the length of the TAT protein can be different, ranging from 86 to 102 amino acids. It had been shown in 1994 that proteins containing TAT 37–72 peptide could effectively deliver different large proteins such as β-galactosidase, horseradish peroxidase, and RNAse A intracellularly [164]. But it was in 1997 when Vives et al. [171] characterized the arginine-rich segment (Table 23.2) of the protein (positions 48–60) that is mainly responsible for the effective translocation, even though other reported sequences such as residues 47–57 are also responsible for the protein transduction [181].

23.6.2 Cellular Uptake Mechanisms of CPPs

Different cellular uptake mechanisms are proposed for different CPPs and their cargos. Despite their common cationic property, the intracellular delivery properties of CPPs differ significantly, and the mechanism of CPP accumulation in the cytoplasm is not fully understood. It seems clear that two types of intracellular uptake coexist, but differ dramatically in terms of the efficiency of accumulation and, therefore, in possible applications. In addition to the CPP electrostatic interactions and hydrogen bonding that are responsible for the direct transduction and penetration of small molecules through the lipid bilayer, energy-dependent macropinocytosis is a primary endocytic pathway responsible for CPP-mediated intracellular delivery of peptides and proteins or carriers (Fig. 23.4). But for any of these uptake mechanisms to be effective, the first step is the necessity of direct contact of the positively charged CPP with the negatively charged components of the cellular membrane.

Direct penetration of CPPs, in general, is possible when high concentrations of amphipathic CPPs such as transportan or MAP analogues were used [186]. In the

TABLE 23.2 Some of the Known CPPs and Their Sequences

CPP	Origin	Sequence	Number of Residues	Reference
TATp	HIV-TAT protein	YGRKKRRQRRR	11	[181]
Penetratin (Antp)	Antennapedia *Drosophila melanogaster*	RQIKIWFQNRRMKWKK	16	[170]
MAP	Model amphipathic peptides, chimeric	KLALKLALKALKAALKLA	18	[167]
Transportan	Galanin-mastoparan	GWTLNSAGYLLGKINLKALAALAKKIL	27	[168]
VP22	HSV-1	DAATATRGRSAASRPTERPRAPAR-SASRPRRPVD	34	[182]

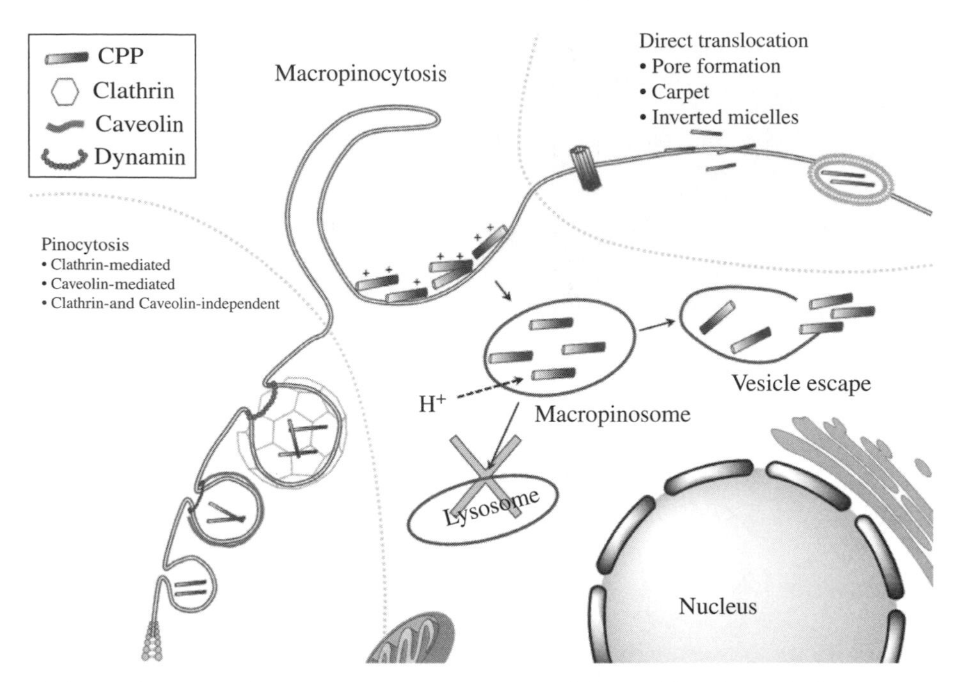

FIGURE 23.4 Possible cellular uptake pathways of CPPs. After the CPP (i.e., TATp)-enclosed macropinosomes enter the cytoplasm; the pH decrease stimulates the leakage of CPP-conjugated macromolecules from the vesicle. Structures are not proportional.

inverted micelle model there is an interaction in addition to ionic one between the hydrophobic residues of the CPP such as tryptophan and the hydrophobic residues of the cell membrane. It is suggested that this interaction of CPP with the cell membrane causes a disturbance in the membrane, thus leads to formation of inverted hexagonal structures or "inverted micelles," in which the CPPs would be trapped until the destabilization of these micelles and their cytoplasmic release. But the most important limitation in this model for intracellular peptide and protein delivery is the size of the large CPP conjugates, due to the small compartment size of these inverted micelle structures. Pore formation is another possible direct penetration mechanism for the CPP uptake, which can occur by barrel-stave or toroidal model [187], which is explained in the general cellular uptake pathways. In both mechanisms, CPPs need to be present at higher than their threshold concentrations. Carpet-like model involves the interactions between the cationic CPP and the negatively charged phospholipids that cause strong association of the CPP on the membrane and eventually phospholipid reorganization [188].

Different studies have suggested diversity of mechanisms other than the direct penetration for the translocation of CPPs. Clathrin-mediated intracellular uptake is one of these pathways where the TATp shows colocalization with Tf [189]. Moreover, caveolin-mediated endocytosis is also suggested for CPP internalization, but the rate of internalization is very slow: it takes several hours to reach the plateau [190]. The cellular uptake mechanism of CPPs remains difficult to determine, probably because there is no "the" mechanism for cellular uptake by CPPs. While working with CPPs, several important issues require careful consideration starting from the CPP-induced cytotoxicity. Even though there are not so many studies on the cytotoxic effects of CPPs, it is known that the nonadherent cell types are more subjected to CPP-mediated cytotoxicity than the adherent ones, probably due to the larger surface area exposed to the CPPs [191]. Another important thing to keep in mind is the possible effect of the covalently conjugated fluorophore groups to the CPPs. Many of the popular fluorescent dyes have very different structures and characteristics, which can affect the internalization of the CPPs, either negatively or positively. Different partition and diffusion coefficients should be taken into account during the selection of the fluorescent dye to investigate the CPP internalization studies. For example, the fluorescence of FITC could be reduced by 70% at the lysosomal pH of 5 [192]. Different fluorophores attached to the same CPP can cause different localization in cell compartments or different endosomal escape of the conjugate [193].

The internalization of TAT peptide was first thought to be direct penetration, because the internalization seemed to occur both at 37 and 4°C. But in 2003, Lundberg et al. proved that this observation was caused by the fixation procedure used during the evaluation [189b], which led to a reexamination of the mechanisms of TAT internalization. Wadia et al. [194] demonstrated that TAT-fusion protein entered the cell via endocytic pathway, circumvented lysosomal degradation. As of now, more than 95% of the TAT-conjugated peptides and proteins are reported to enter the cells via macropinocytosis [195]. It looks like that the lipid raft–mediated uptake and macropinocytosis are the key mechanisms of the uptake of TAT-conjugated large cargoes [195, 196]. It can be concluded that more than one mechanism works for CPP-mediated intracellular delivery of small and large molecules. Individual CPPs or CPP-conjugated to small molecules are internalized into cells via electrostatic interactions and hydrogen bonding, while CPP-conjugated to large molecules occur via the energy-dependent macropinocytosis. However, in both cases, the direct contact between the CPPs and the negative residues on cell surface is a prerequisite for the successful transduction.

Very recently, Hirose et al. proposed that the direct penetration of CPPs into the cells is a valid mechanism but depends on several requirements. They used 12-arginine peptide (R12) conjugated to Alexa 488 fluorescent dye and concluded that this conjugate enters the cells by direct penetration at specific sites where small particle-like cell surface structures are formed. Moreover, when compared to shorter R4-Alexa conjugates, they suggest that if long arginine-containing CPP couples to a small hydrophobic cargo, direct internalization occurs at specific sites of the membrane that are competent to induce multivesicular structures [197].

23.6.3 CPP-Mediated Delivery of Peptides and Proteins

Since traversal through cellular membranes represents a major barrier for efficient delivery of macromolecules into cells, cell-penetrating peptides may serve to ferry various macromolecules into mammalian cells *in vitro* and *in vivo*. The use of peptides and protein domains with amphipathic sequences for drug and gene delivery across cellular membranes is gaining increasing attention. Hitching of therapeutic peptides and proteins onto CPPs may circumvent conventional limitations by allowing the transport of these compounds into a wide variety of cells *in vitro* and *in vivo*. Especially in recent years, a growing number of reports related to intracellular peptide or protein delivery into cancer cells have been published.

The first report related to *in vivo* use of protein delivery into the tissues was published by Dowdy's group in 1999 [198]. Since then many applications of CPP-mediated protein delivery into cancer cells have been reported. Among them, restoring p53 gene function in the cancer cells is one of the most studied. Half of the human tumors lack p53 gene activity, a gene that suppresses the tumor growth, and restoring the function of this gene can inhibit the growth and proliferation of the cancer cells. Harbour et al. found that when N-terminal of the p53 peptide was fused with the TAT peptide, the increased accumulation of the peptide in the cells and the decreased binding of its negative regulator HDM2 occur, and this leads to increased p53 activity and preferential killing of the tumor cells [199]. In a similar approach, TAT peptide–mediated intracellular delivery of a peptide from the C-terminal of p53 gene was shown to be a useful approach due to increased internalization of the peptide when applied *in vivo* [200]. The full-length p53 protein was also conjugated to TAT peptide to restore the p53 functionality in cancer cells [201]. Although the increased p53 activity in the malignant cells was achieved, normal cells too were affected by the treatment, which raises the concern about selectivity. When p27, a tumor suppressor protein, was synthesized as a conjugate with TAT peptide, it caused increased apoptosis in tumor cells [202]. P16 tumor suppressor peptide conjugated to penetratin [203] or TAT peptide [204] caused similar effects and cell cycle arrest of the cancer cells. Another successful TAT-mediated delivery of tumor suppressor is the TAT-VHL peptide, which efficiently reduced the cell proliferation by 80% and inhibited the invasiveness in the renal cell carcinomas when injected intraperitoneally to nude mice [205].

CPP-mediated delivery of proteins and peptides to regulate apoptosis was also reported. To induce apoptosis and facilitate the antitumor effect, several CPPs linked to different apoptosis regulatory proteins including proapoptotic smac peptide [206], Bcl-2 family proteins [207], S100 family proteins [208], and activating transcription factor 2 [209].

TAT PTD was also used for intracellular delivery of a biologically active neuroprotectant Bcl-xL in cerebral ischemia. TAT PTD and Bcl-xL fusion protein (called as PTDHA-Bcl-xL) resulted in robust protein transduction in cultures and also delivered the protein across the blood–brain barrier. Similarly, TAT PTD was fused to an artificial cytoprotective protein FNK, obtained from Bcl-xL by site-directed mutagenesis; the TAT PTD-FNK fusion protein protected cultured neuronal cells against glutamate-induced excitotoxicity and staurosporine-induced apoptosis. When administered i.p. in gerbils, it reduced ischemic injury of hippocampal CA1 neurons [210].

The erythroid-related disorders (ERDs) represent a large group of hematological diseases, which in most cases are attributed either to the deficiency or malfunction of biosynthetic enzymes or oxygen transport proteins. Most recently, Papadopoulou and Tsiftsoglou developed a new approach for treating ERDs with the help of CPP-mediated intracellular protein delivery [211]. They first produced the genetically engineered human CPP–mediated protein of interest, which is missing or mutated in the patient. Then they isolate the target cells from the patient blood and ex vivo transduce the cells with the CPP-mediated protein followed by readministration of the transduced cells back into the same patient.

Fusion of dominant negative forms of Ras or PI3K to TAT peptide for inflammatory response inhibition is another example of use of CPPs in the treatment of diseases such as asthma [212]. TAT-mediated protein delivery also showed potential as a therapeutic and prophylactic vaccine. Exogenous proteins cannot enter the cytosol and access the MHC class I processing pathway. Therefore, it is difficult to design a protein-based vaccine that induces class I-restricted cytotoxic T-lymphocyte (CTL) response. However, after conjugating the antigenic protein to TATp (49–57), such as TAT–ovalbumin conjugate, the conjugate was processed by antigen presenting cells, resulting in effective killing of the target cells by antigen-specific CTLs [213].

CPPs were also used for successful intracellular delivery of siRNA. CPP-mediated siRNA delivery could be achieved by different strategies. It was found that intracellular delivery and efficacy of siRNA could be enhanced when different CPPs were covalently attached, especially with the disulfide bond due to the easy intracellular cleavage and separation from the CPP [175a, 214]. Moreover noncovalent CPP-siRNA complexes can be prepared using the interaction between the negatively charged siRNA and highly cationic CPPs [215].

23.6.4 CPP-Modified Carriers for Intracellular Delivery of Peptides and Proteins

CPPs can also enhance the intracellular delivery of nanocarriers, which are loaded with peptides or proteins. Liposomes are the most investigated carriers for this purpose, and CPP modification of liposomes has been shown to be a promising option for intracellular delivery. TAT peptide (47–57)–modified liposomes could be delivered intracellularly in different cells, such as murine Lewis lung carcinoma (LLC) cells, human breast tumor BT20 cells, and rat cardiac myocyte H9C2 cells [57c]. The liposomes were tagged with TAT peptide via the spacer, *p-nitrophenylcarbonyl*–PEG–PE (*pNP*-PEG-PE), at the density of a few hundreds of TAT peptide per single liposome vesicle. It was shown that the cells treated with liposomes, in which TAT peptide–cell interaction was hindered either by direct attachment of TAT peptide to the liposomes surface or by the long PEG grafts on the liposome surface shielding the TAT moiety, did not show TAT–liposome internalization; however, the preparations of TAT–liposomes, which allowed for the direct contact of TAT peptide residues with cells, displayed an enhanced uptake by the cells. This suggested that the translocation of TAT peptide (TATp)–liposomes into cells requires direct free interaction of TAT peptide with the cell surface. Further studies on the intracellular trafficking of rhodamine-labeled TATp–liposomes loaded with FITC-dextran revealed that TATp–liposomes remained intact inside the cell cytoplasm within 1 hour of translocation, after 2 hour they migrated into the perinuclear zone, and at 9 hour the liposomes disintegrated there [216].

Although of considerable clinical potential, CPPs also have a few important drawbacks and limitations. First, they have the undesirable characteristic of nonspecificity and can enter any cell they come in contact with. This lack of selectivity affects the risk of drug-induced toxic effect on normal tissues. Second, the *in vivo* stability of these peptides is at risk until they reach their target. These peptides can be enzymatically cleaved by plasma enzymes and thus need to be sterically protected [217]. Recently, another approach has been suggested for this problem, which proposed that CPPs be incorporated into "smart" nanocarrier delivery platforms [218]. Thus, during the first phase of nanocarrier delivery, the nonspecific CPP function is sterically protected ("shielded") by a polymer or targeting antibody. Upon accumulation in the target, the protective moiety attached to the surface of the carrier via a stimulus-sensitive bond will detach under local environmental conditions to reveal the CPP and affect targeted delivery. There are a large number of different stimuli, especially in the cancer tissue environment that can be used to trigger the action or de-shielding (For a recent review please refer to Ref. [219].) For example, Kale et al. formulated a PEGylated liposomal delivery system [109b, 220] for the plasmid pGFP with TATp conjugated to the surface of the particles along with long, pH-sensitive PEG blocks to act as a peptide shield. The liposomes that reached tumor sites (aided by the EPR effect) lost their PEG coating in the low pH tumor environment, exposing the underlying TAT peptides, which then mediated the transport into the tumor cells. A triple functional liposomal carrier has also been designed, with its membrane decorated with the anticancer 2C5 mAb, TAT peptide, and a "shielding" pH-sensitive PEG block [221].

23.7 CONCLUSION

Thus, our current knowledge provides some promising approaches on how to deliver peptide and/or protein-based drugs not only to the site of disease but also inside the target cell for enhanced therapy. For the successful and effective intracellular delivery of peptide and protein drugs, the necessity of endosomal escape or bypassing the endosomal encapsulation is one of the most important strategies. Most of the examples and methods that are summarized in this chapter mainly focuses on *in vitro* studies and are on preclinical stage. But the fact that new delivery platform technologies that have matured into combination of potentially useful strategies provides optimism for a wide range of therapeutic applications, which should eventually pave the way for their combined usefulness in the clinic for intracellular protein and peptide delivery.

REFERENCES

1. Goeddel, D. V.; Kleid, D. G.; Bolivar, F.; Heyneker, H. L.; Yansura, D. G.; Crea, R.; Hirose, T.; Kraszewski, A.; Itakura, K.; Riggs, A. D., Expression in *Escherichia coli* of chemically synthesized genes for human insulin. *Proc Natl Acad Sci U S A* 1979, **76** (1), 106–10.
2. Walsh, G., Biopharmaceutical benchmarks 2010. *Nat Biotechnol* 2010, **28** (9), 917–24.

3. Reichert, J. M., Trends in development and approval times for new therapeutics in the United States. *Nat Rev Drug Discov* 2003, **2** (9), 695–702.
4. (a) Carter, P. J., Introduction to current and future protein therapeutics: a protein engineering perspective. *Exp Cell Res* 2011, **317** (9), 1261–9; (b) Nicolaides, N. C.; Sass, P. M.; Grasso, L., Advances in targeted therapeutic agents. *Expert Opin Drug Discov* 2010, **5** (11), 1123–40.
5. Leader, B.; Baca, Q. J.; Golan, D. E., Protein therapeutics: a summary and pharmacological classification. *Nat Rev Drug Discov* 2008, **7** (1), 21–39.
6. (a) Torchilin, V., Intracellular delivery of protein and peptide therapeutics. *Drug Discov Today Technol* 2008, **5** (2–3), e95–103; (b) Vellard, M., The enzyme as drug: application of enzymes as pharmaceuticals. *Curr Opin Biotechnol* 2003, **14** (4), 444–50.
7. Schultz, M. L.; Tecedor, L.; Chang, M.; Davidson, B. L., Clarifying lysosomal storage diseases. *Trends Neurosci* 2011, **34** (8), 401–10.
8. Brady, R. O.; Pentchev, P. G.; Gal, A. E.; Hibbert, S. R.; Dekaban, A. S., Replacement therapy for inherited enzyme deficiency. Use of purified glucocerebrosidase in Gaucher's disease. *N Engl J Med* 1974, **291** (19), 989–93.
9. Zhao, H.; Keddache, M.; Bailey, L.; Arnold, G.; Grabowski, G., Gaucher's disease: identification of novel mutant alleles and genotype-phenotype relationships. *Clin Genet* 2003, **64** (1), 57–64.
10. (a) Garman, S. C.; Garboczi, D. N., The molecular defect leading to Fabry disease: structure of human alpha-galactosidase. *J Mol Biol* 2004, **337** (2), 319–35; (b) Connock, M.; Juarez-Garcia, A.; Frew, E.; Mans, A.; Dretzke, J., A systematic review of the clinical effectiveness and cost-effectiveness of enzyme replacement therapies for Fabry's disease and mucopolysaccharidosis type 1. *Health Technol Assess* 2006, **10** (20), 130.
11. Geel, T. M.; McLaughlin, P. M.; de Leij, L. F.; Ruiters, M. H.; Niezen-Koning, K. E., Pompe disease: current state of treatment modalities and animal models. *Mol Genet Metab* 2007, **92** (4), 299–307.
12. (a) Keen, H.; Glynne, A.; Pickup, J. C.; Viberti, G. C.; Bilous, R. W.; Jarrett, R. J.; Marsden, R., Human insulin produced by recombinant DNA technology: safety and hypoglycaemic potency in healthy men. *Lancet* 1980, **2** (8191), 398–401; (b) Skyler, J. S.; Cefalu, W. T.; Kourides, I. A.; Landschulz, W. H.; Balagtas, C. C.; Cheng, S. L.; Gelfand, R. A., Efficacy of inhaled human insulin in type 1 diabetes mellitus: a randomised proof-of-concept study. *Lancet* 2001, **357** (9253), 331–5; (c) Hirsch, I. B., Insulin analogues. *N Engl J Med* 2005, **352** (2), 174–83.
13. (a) Hardin, D. S., Treatment of short stature and growth hormone deficiency in children with somatotropin (rDNA origin). *Biologics* 2008, **2** (4), 655–61; (b) Lee, P. A.; Savendahl, L.; Oliver, I.; Tauber, M.; Blankenstein, O.; Ross, J.; Snajderova, M.; Rakov, V.; Pedersen, B. T.; Christesen, H. T., Comparison of response to 2-years' growth hormone treatment in children with isolated growth hormone deficiency, born small for gestational age, idiopathic short stature, or multiple pituitary hormone deficiency: combined results from two large observational studies. *Int J Pediatr Endocrinol* 2012, **2012** (1), 22.
14. (a) Lusher, J. M.; Arkin, S.; Abildgaard, C. F.; Schwartz, R. S., Recombinant factor VIII for the treatment of previously untreated patients with hemophilia A. Safety, efficacy, and development of inhibitors. Kogenate Previously Untreated Patient Study Group. *N Engl J Med* 1993, **328** (7), 453–9; (b) Mannucci, P. M.; Mancuso, M. E.; Santagostino, E., How we choose factor VIII to treat hemophilia. *Blood* 2012, **119** (18), 4108–14.

15. Roth, D. A.; Kessler, C. M.; Pasi, K. J.; Rup, B.; Courter, S. G.; Tubridy, K. L.; Recombinant Factor, I. X. S. G., Human recombinant factor IX: safety and efficacy studies in hemophilia B patients previously treated with plasma-derived factor IX concentrates. *Blood* 2001, **98** (13), 3600–6.
16. (a) Barton, N. W.; Brady, R. O.; Dambrosia, J. M.; Di Bisceglie, A. M.; Doppelt, S. H.; Hill, S. C.; Mankin, H. J.; Murray, G. J.; Parker, R. I.; Argoff, C. E.; Grewal, R. P.; Yu, K.-T., Replacement therapy for inherited enzyme deficiency—macrophage-targeted glucocerebrosidase for Gaucher's disease. *N Engl J Med* 1991, **324** (21), 1464–70; (b) Grabowski, G. A.; Hopkin, R. J., Enzyme therapy for lysosomal storage disease: principles, practice, and prospects. *Annu Rev Genomics Hum Genet* 2003, **4**, 403–36; (c) Torchilin, V. P., Immobilised enzymes as drugs. *Adv Drug Deliv Rev* 1988, **1** (3), 270.
17. (a) van der Ploeg, A. T.; Clemens, P. R.; Corzo, D.; Escolar, D. M.; Florence, J.; Groeneveld, G. J.; Herson, S.; Kishnani, P. S.; Laforet, P.; Lake, S. L.; Lange, D. J.; Leshner, R. T.; Mayhew, J. E.; Morgan, C.; Nozaki, K.; Park, D. J.; Pestronk, A.; Rosenbloom, B.; Skrinar, A.; van Capelle, C. I.; van der Beek, N. A.; Wasserstein, M.; Zivkovic, S. A., A randomized study of alglucosidase alfa in late-onset Pompe's disease. *N Engl J Med* 2010, **362** (15), 1396–406; (b) Nicolino, M.; Byrne, B.; Wraith, J. E.; Leslie, N.; Mandel, H.; Freyer, D. R.; Arnold, G. L.; Pivnick, E. K.; Ottinger, C. J.; Robinson, P. H.; Loo, J. C.; Smitka, M.; Jardine, P.; Tato, L.; Chabrol, B.; McCandless, S.; Kimura, S.; Mehta, L.; Bali, D.; Skrinar, A.; Morgan, C.; Rangachari, L.; Corzo, D.; Kishnani, P. S., Clinical outcomes after long-term treatment with alglucosidase alfa in infants and children with advanced Pompe disease. *Genet Med* 2009, **11** (3), 210–19.
18. Schiffmann, R.; Kopp, J. B.; Austin, H. A., 3rd; Sabnis, S.; Moore, D. F.; Weibel, T.; Balow, J. E.; Brady, R. O., Enzyme replacement therapy in Fabry disease: a randomized controlled trial. *JAMA* 2001, **285** (21), 2743–9.
19. Dirksen, A.; Dijkman, J. H.; Madsen, F.; Stoel, B.; Hutchison, D. C.; Ulrik, C. S.; Skovgaard, L. T.; Kok-Jensen, A.; Rudolphus, A.; Seersholm, N.; Vrooman, H. A.; Reiber, J. H.; Hansen, N. C.; Heckscher, T.; Viskum, K.; Stolk, J., A randomized clinical trial of alpha(1)-antitrypsin augmentation therapy. *Am J Respir Crit Care Med* 1999, **160** (5 Pt 1), 1468–72.
20. (a) Silver, M.; Corwin, M. J.; Bazan, A.; Gettinger, A.; Enny, C.; Corwin, H. L., Efficacy of recombinant human erythropoietin in critically ill patients admitted to a long-term acute care facility: a randomized, double-blind, placebo-controlled trial. *Crit Care Med* 2006, **34** (9), 2310–16; (b) Nissenson, A. R.; Swan, S. K.; Lindberg, J. S.; Soroka, S. D.; Beatey, R.; Wang, C.; Picarello, N.; McDermott-Vitak, A.; Maroni, B. J., Randomized, controlled trial of darbepoetin alfa for the treatment of anemia in hemodialysis patients. *Am J Kidney Dis* 2002, **40** (1), 110–18; (c) Ghali, J. K.; Anand, I. S.; Abraham, W. T.; Fonarow, G. C.; Greenberg, B.; Krum, H.; Massie, B. M.; Wasserman, S. M.; Trotman, M. L.; Sun, Y.; Knusel, B.; Armstrong, P.; Study of Anemia in Heart Failure Trial, G., Randomized double-blind trial of darbepoetin alfa in patients with symptomatic heart failure and anemia. *Circulation* 2008, **117** (4), 526–35.
21. (a) Kaczmarski, R. S.; Mufti, G. J., Low-dose filgrastim therapy for chronic neutropenia. *N Engl J Med* 1993, **329** (17), 1280–1; (b) Bedell, C., Pegfilgrastim for chemotherapy-induced neutropenia. *Clin J Oncol Nurs* 2003, **7** (1), 55–6, 63–4.
22. Chesnut, C. H., 3rd; Silverman, S.; Andriano, K.; Genant, H.; Gimona, A.; Harris, S.; Kiel, D.; LeBoff, M.; Maricic, M.; Miller, P.; Moniz, C.; Peacock, M.; Richardson, P.; Watts, N.; Baylink, D., A randomized trial of nasal spray salmon calcitonin in postmenopausal women with established osteoporosis: the prevent recurrence of osteoporotic fractures study. PROOF Study Group. *Am J Med* 2000, **109** (4), 267–76.

23. (a) Katzan, I. L.; Furlan, A. J.; Lloyd, L. E.; Frank, J. I.; Harper, D. L.; Hinchey, J. A.; Hammel, J. P.; Qu, A.; Sila, C. A., Use of tissue-type plasminogen activator for acute ischemic stroke: the Cleveland area experience. *JAMA* 2000, **283** (9), 1151–8; (b) Colman, E.; Hedin, R.; Swann, J.; Orloff, D., A brief history of calcitonin. *Lancet* 2002, **359** (9309), 885–6.

24. Jankovic, J.; Brin, M. F., Therapeutic uses of botulinum toxin. *N Engl J Med* 1991, **324** (17), 1186–94.

25. Fuchs, H. J.; Borowitz, D. S.; Christiansen, D. H.; Morris, E. M.; Nash, M. L.; Ramsey, B. W.; Rosenstein, B. J.; Smith, A. L.; Wohl, M. E., Effect of aerosolized recombinant human DNase on exacerbations of respiratory symptoms and on pulmonary function in patients with cystic fibrosis. The Pulmozyme Study Group. *N Engl J Med* 1994, **331** (10), 637–42.

26. (a) Yang, J. C.; Haworth, L.; Sherry, R. M.; Hwu, P.; Schwartzentruber, D. J.; Topalian, S. L.; Steinberg, S. M.; Chen, H. X.; Rosenberg, S. A., A randomized trial of bevacizumab, an anti-vascular endothelial growth factor antibody, for metastatic renal cancer. *N Engl J Med* 2003, **349** (5), 427–34; (b) Ferrara, N.; Hillan, K. J.; Gerber, H. P.; Novotny, W., Discovery and development of bevacizumab, an anti-VEGF antibody for treating cancer. *Nat Rev Drug Discov* 2004, **3** (5), 391–400.

27. Blick, S. K.; Scott, L. J., Cetuximab: a review of its use in squamous cell carcinoma of the head and neck and metastatic colorectal cancer. *Drugs* 2007, **67** (17), 2585–607.

28. Vogel, C. L.; Cobleigh, M. A.; Tripathy, D.; Gutheil, J. C.; Harris, L. N.; Fehrenbacher, L.; Slamon, D. J.; Murphy, M.; Novotny, W. F.; Burchmore, M.; Shak, S.; Stewart, S. J.; Press, M., Efficacy and safety of trastuzumab as a single agent in first-line treatment of HER2-overexpressing metastatic breast cancer. *J Clin Oncol* 2002, **20** (3), 719–26.

29. Keating, M. J.; O'Brien, S.; Albitar, M.; Lerner, S.; Plunkett, W.; Giles, F.; Andreeff, M.; Cortes, J.; Faderl, S.; Thomas, D.; Koller, C.; Wierda, W.; Detry, M. A.; Lynn, A.; Kantarjian, H., Early results of a chemoimmunotherapy regimen of fludarabine, cyclophosphamide, and rituximab as initial therapy for chronic lymphocytic leukemia. *J Clin Oncol* 2005, **23** (18), 4079–88.

30. Olsen, N. J.; Stein, C. M., Drug therapy—new drugs for rheumatoid arthritis. *N Engl J Med* 2004, **350** (21), 2167–79.

31. (a) Vincenti, F.; Kirkman, R.; Light, S.; Bumgardner, G.; Pescovitz, M.; Halloran, P.; Neylan, J.; Wilkinson, A.; Ekberg, H.; Gaston, R.; Backman, L.; Burdick, J., Interleukin-2-receptor blockade with daclizumab to prevent acute rejection in renal transplantation. Daclizumab Triple Therapy Study Group. *N Engl J Med* 1998, **338** (3), 161–5; (b) Nashan, B.; Moore, R.; Amlot, P.; Schmidt, A. G.; Abeywickrama, K.; Soulillou, J. P., Randomised trial of basiliximab versus placebo for control of acute cellular rejection in renal allograft recipients. CHIB 201 International Study Group. *Lancet* 1997, **350** (9086), 1193–8.

32. Genovese, M. C.; Becker, J. C.; Schiff, M.; Luggen, M.; Sherrer, Y.; Kremer, J.; Birbara, C.; Box, J.; Natarajan, K.; Nuamah, I.; Li, T.; Aranda, R.; Hagerty, D. T.; Dougados, M., Abatacept for rheumatoid arthritis refractory to tumor necrosis factor alpha inhibition. *N Engl J Med* 2005, **353** (11), 1114–23.

33. Ellis, C. N.; Krueger, G. G.; Alefacept Clinical Study, G., Treatment of chronic plaque psoriasis by selective targeting of memory effector T lymphocytes. *N Engl J Med* 2001, **345** (4), 248–55.

34. Goolcharran, C.; Khossravi, M.; Borchardt, R. T.; Frokjaer, S.; Hovgaards, L., *Chemical pathways of peptide and protein degradation*. Taylor and Francis: London, 2000.

35. Chi, E. Y.; Krishnan, S.; Randolph, T. W.; Carpenter, J. F., Physical stability of proteins in aqueous solution: mechanism and driving forces in nonnative protein aggregation. *Pharm Res* 2003, **20** (9), 1325–36.
36. Walsh, G., Biopharmaceutical benchmarks 2014. *Nat Biotechnol* 2014, **32** (10), 992–1000.
37. (a) Chakrabarti, R.; Wylie, D. E.; Schuster, S. M., Transfer of monoclonal antibodies into mammalian cells by electroporation. *J Biol Chem* 1989, **264** (26), 15494–500; (b) Arnheiter, H.; Haller, O., Antiviral state against influenza virus neutralized by microinjection of antibodies to interferon-induced Mx proteins. *EMBO J* 1988, **7** (5), 1315–20.
38. Lukas, J.; Fau - Bartek, J.; Strauss, M., Efficient transfer of antibodies into mammalian cells by electroporation. *J Immunol Methods* 1994, **170**, 255–9.
39. Miller, J. F.; Dower, W. J.; Tompkins, L. S., High-voltage electroporation of bacteria: genetic transformation of Campylobacter jejuni with plasmid DNA. *Proc Natl Acad Sci U S A* 1988, **85** (3), 856–60.
40. (a) Bar-Sagi, D.; Feramisco, J. R., Microinjection of the ras oncogene protein into PC12 cells induces morphological differentiation. *Cell* 1985, **42** (3), 841–8; (b) Komarova, Y.; Peloquin, J.; Borisy, G., Microinjection of protein samples. *CSH Protoc* 2007, **2007**, pdb. prot4657.
41. (a) Thomas, C. E.; Ehrhardt, A.; Kay, M. A., Progress and problems with the use of viral vectors for gene therapy. *Nat Rev Genet* 2003, **4** (5), 346–58; (b) Tomanin, R.; Scarpa, M., Why do we need new gene therapy viral vectors? Characteristics, limitations and future perspectives of viral vector transduction. *Curr Gene Ther* 2004, **4** (4), 357–72.
42. Wang, T.; Upponi, J. R.; Torchilin, V. P., Design of multifunctional non-viral gene vectors to overcome physiological barriers: dilemmas and strategies. *Int J Pharm* 2012, **427** (1), 3–20.
43. Maeda, H.; Takeshita, J.; Kanamaru, R., A lipophilic derivative of neocarzinostatin. A polymer conjugation of an antitumor protein antibiotic. *Int J Pept Protein Res* 1979, **14** (2), 81–7.
44. (a) Matsumura, Y.; Maeda, H., A new concept for macromolecular therapeutics in cancer chemotherapy: mechanism of tumoritropic accumulation of proteins and the antitumor agent smancs. *Cancer Res* 1986, **46** (12 Pt 1), 6387–92; (b) Maeda, H.; Wu, J.; Sawa, T.; Matsumura, Y.; Hori, K., Tumor vascular permeability and the EPR effect in macromolecular therapeutics: a review. *J Control Release* 2000, **65** (1–2), 271–84; (c) Maeda, H., Vascular permeability in cancer and infection as related to macromolecular drug delivery, with emphasis on the EPR effect for tumor-selective drug targeting. *Proc Jpn Acad Ser B Phys Biol Sci* 2012, **88** (3), 53–71.
45. (a) Torchilin, V. P.; Klibanov, A. L.; Huang, L.; O'Donnell, S.; Nossiff, N. D.; Khaw, B. A., Targeted accumulation of polyethylene glycol-coated immunoliposomes in infarcted rabbit myocardium. *FASEB J* 1992, **6** (9), 2716–19; (b) Torchilin, V. P., Polymer-coated long-circulating microparticulate pharmaceuticals. *J Microencapsul* 1998, **15** (1), 1–19; (c) Palmer, T. N.; Caride, V. J.; Caldecourt, M. A.; Twickler, J.; Abdullah, V., The mechanism of liposome accumulation in infarction. *Biochim Biophys Acta* 1984, **797** (3), 363–8; (d) Jain, R. K., Transport of molecules, particles, and cells in solid tumors. *Annu Rev Biomed Eng* 1999, **1**, 241–63.
46. (a) Iyer, A. K.; Khaled, G.; Fang, J.; Maeda, H., Exploiting the enhanced permeability and retention effect for tumor targeting. *Drug Discov Today* 2006, **11** (17–18), 812–18; (b) Maeda, H., The enhanced permeability and retention (EPR) effect in tumor vasculature:

the key role of tumor-selective macromolecular drug targeting. *Adv Enzyme Regul* 2001, **41**, 189–207; (c) Maeda, H.; Sawa, T.; Konno, T., Mechanism of tumor-targeted delivery of macromolecular drugs, including the EPR effect in solid tumor and clinical overview of the prototype polymeric drug SMANCS. *J Control Release* 2001, **74** (1–3), 47–61.

47. Torchilin, V. P., How do polymers prolong circulation time of liposomes? *J Liposome Res* 1996, **6** (1), 99–116.
48. (a) Roberts, M. J.; Bentley, M. D.; Harris, J. M., Chemistry for peptide and protein PEGylation. *Adv Drug Deliv Rev* 2002, **54** (4), 459–76; (b) Veronese, F. M.; Pasut, G., PEGylation, successful approach to drug delivery. *Drug Discov Today* 2005, **10** (21), 1451–8.
49. Woodle, M. C.; Engbers, C. M.; Zalipsky, S., New amphipatic polymer lipid conjugates forming long-circulating reticuloendothelial system-evading liposomes. *Bioconjug Chem* 1994, **5** (6), 493–6.
50. Maruyama, K.; Okuizumi, S.; Ishida, O.; Yamauchi, H.; Kikuchi, H.; Iwatsuru, M., Phosphatidyl polyglycerols prolong liposome circulation in-vivo. *Int J Pharm* 1994, **111** (1), 103–7.
51. Monfardini, C.; Schiavon, O.; Caliceti, P.; Morpurgo, M.; Harris, J. M.; Veronese, F. M., A branched monomethoxypoly(ethylene glycol) for protein modification. *Bioconjug Chem* 1995, **6** (1), 62–9.
52. Torchilin, V. P.; Levchenko, T. S.; Whiteman, K. R.; Yaroslavov, A. A.; Tsatsakis, A. M.; Rizos, A. K.; Michailova, E. V.; Shtilman, M. I., Amphiphilic poly-N-vinylpyrrolidones: synthesis, properties and liposome surface modification. *Biomaterials* 2001, **22** (22), 3035–44.
53. Takeuchi, H.; Kojima, H.; Toyoda, T.; Yamamoto, H.; Hino, T.; Kawashima, Y., Prolonged circulation time of doxorubicin-loaded liposomes coated with a modified polyvinyl alcohol after intravenous injection in rats. *Eur J Pharm Biopharm* 1999, **48** (2), 123–9.
54. (a) Asselin, B. L., The three asparaginases—comparative pharmacology and optimal use in childhood leukemia. *Adv Exp Med Biol* 1999, **457**, 621–9; (b) Abuchowski, A.; Kazo, G. M.; Verhoest, C. R.; Vanes, T.; Kafkewitz, D.; Nucci, M. L.; Viau, A. T.; Davis, F. F., Cancer-therapy with chemically modified enzymes.1. Antitumor properties of polyethylene glycol-asparaginase conjugates. *Cancer Biochem Biophys* 1984, **7** (2), 175–86.
55. Eliason, J. F., PEGylated proteins in immunotherapy of cancer. In *Delivery of protein and peptide drugs in cancer*, Torchilin, V. P., Ed. Imperical College Press: London, 2006, 111–26.
56. Greenwald, R. B.; Choe, Y. H.; McGuire, J.; Conover, C. D., Effective drug delivery by PEGylated drug conjugates. *Adv Drug Deliv Rev* 2003, **55** (2), 217–50.
57. (a) Klibanov, A. L.; Maruyama, K.; Torchilin, V. P.; Huang, L., Amphipathic polyethyleneglycols effectively prolong the circulation time of liposomes. *FEBS Lett* 1990, **268** (1), 235–7; (b) Maruyama, K.; Yuda, T.; Okamoto, A.; Ishikura, C.; Kojima, S.; Iwatsuru, M., Effect of molecular weight in amphipathic polyethyleneglycol on prolonging the circulation time of large unilamellar liposomes. *Chem Pharm Bull (Tokyo)* 1991, **39** (6), 1620–2; (c) Allen, T. M.; Hansen, C.; Martin, F.; Redemann, C.; Yauyoung, A., Liposomes containing synthetic lipid derivatives of poly(ethylene glycol) show prolonged circulation half-lives invivo. *Biochim Biophys Acta* 1991, **1066** (1), 29–36; (d) Papahadjopoulos, D.; Allen, T. M.; Gabizon, A.; Mayhew, E.; Matthay, K.; Huang, S. K.; Lee, K. D.; Woodle, M. C.; Lasic, D. D.; Redemann, C., Sterically stabilized liposomes: improvements in pharmacokinetics and antitumor therapeutic efficacy. *Proc Natl Acad Sci U S A* 1991, **88** (24), 11460–4.

58. (a) Gabizon, A. A., Pegylated liposomal doxorubicin: metamorphosis of an old drug into a new form of chemotherapy. *Cancer Invest* 2001, **19** (4), 424–36; (b) Cianfrocca, M. E.; Kaklamani, V. G.; Rosen, S. T.; Von Roenn, J. H.; Rademaker, A.; Smith, D. A.; Rubin, S. D.; Meservey, C.; Uthe, R.; Gradishar, W. J., Phase I trial of pegylated liposomal doxorubicin and lapatinib in the treatment of metastatic breast cancer (MBC): final results. *J Clin Oncol* 2012, **30** (15-Suppl), 610; (c) O'Shaughnessy, J. A., Pegylated liposomal doxorubicin in the treatment of breast cancer. *Clin Breast Cancer* 2003, **4** (5), 318–28; (d) Symon, Z.; Peyser, A.; Tzemach, D.; Lyass, O.; Sucher, E.; Shezen, E.; Gabizon, A., Selective delivery of doxorubicin to patients with breast carcinoma metastases by stealth liposomes. *Cancer* 1999, **86** (1), 72–8.

59. (a) Gref, R.; Minamitake, Y.; Peracchia, M. T.; Trubetskoy, V.; Torchilin, V.; Langer, R., Biodegradable long-circulating polymeric nanospheres. *Science* 1994, **263** (5153), 1600–3; (b) Gref, R.; Domb, A.; Quellec, P.; Blunk, T.; Muller, R. H.; Verbavatz, J. M.; Langer, R., The controlled intravenous delivery of drugs using PEG-coated sterically stabilized nanospheres. *Adv Drug Deliv Rev* 2012, **64**, 316–26.

60. (a) Weissig, V.; Whiteman, K. R.; Torchilin, V. P., Accumulation of protein-loaded long-circulating micelles and liposomes in subcutaneous Lewis lung carcinoma in mice. *Pharm Res* 1998, **15** (10), 1552–6; (b) Torchilin, V. P., Micellar nanocarriers: pharmaceutical perspectives. *Pharm Res* 2007, **24** (1), 1–16.

61. (a) Gajbhiye, V.; Kumar, P. V.; Tekade, R. K.; Jain, N. K., PEGylated PPI dendritic architectures for sustained delivery of H(2) receptor antagonist. *Eur J Med Chem* 2009, **44** (3), 1155–66; (b) Kojima, C.; Kono, K.; Maruyama, K.; Takagishi, T., Synthesis of polyamidoamine dendrimers having poly(ethylene glycol) grafts and their ability to encapsulate anticancer drugs. *Bioconjug Chem* 2000, **11** (6), 910–17; (c) Bhadra, D.; Bhadra, S.; Jain, S.; Jain, N. K., A PEGylated dendritic nanoparticulate carrier of fluorouracil. *Int J Pharm* 2003, **257** (1–2), 111–24; (d) Kaminskas, L. M.; Boyd, B. J.; Karellas, P.; Krippner, G. Y.; Lessene, R.; Kelly, B.; Porter, C. J. H., The impact of molecular weight and PEG chain length on the systemic pharmacokinetics of PEGylated poly L-lysine dendrimers. *Mol Pharm* 2008, **5** (3), 449–63.

62. (a) Fundaro, A.; Cavalli, R.; Bargoni, A.; Vighetto, D.; Zara, G. P.; Gasco, M. R., Non-stealth and stealth solid lipid nanoparticles (SLN) carrying doxorubicin: pharmacokinetics and tissue distribution after i.v. administration to rats. *Pharmacol Res* 2000, **42** (4), 337–43; (b) Almeida, A. J.; Souto, E., Solid lipid nanoparticles as a drug delivery system for peptides and proteins. *Adv Drug Deliv Rev* 2007, **59** (6), 478–90; (c) Chen, D. B.; Yang, T. Z.; Lu, W. L.; Zhang, Q., In vitro and in vivo study of two types of long-circulating solid lipid nanoparticles containing paclitaxel. *Chem Pharm Bull* 2001, **49** (11), 1444–7.

63. Khalil, I. A.; Kogure, K.; Akita, H.; Harashima, H., Uptake pathways and subsequent intracellular trafficking in nonviral gene delivery. *Pharmacol Rev* 2006, **58** (1), 32–45.

64. (a) Lamaze, C.; Schmid, S. L., The emergence of clathrin-independent pinocytic pathways. *Curr Opin Cell Biol* 1995, **7** (4), 573–80; (b) Conner, S. D.; Schmid, S. L., Regulated portals of entry into the cell. *Nature* 2003, **422** (6927), 37–44.

65. (a) Subtil, A.; Hemar, A.; Dautryvarsat, A., Rapid endocytosis of interleukin-2 receptors when clathrin-coated pit endocytosis is inhibited. *J Cell Sci* 1994, **107**, 3461–8; (b) Parton, R. G.; Joggerst, B.; Simons, K., Regulated internalization of caveolae. *J Cell Biol* 1994, **127** (5), 1199–215.

66. Huang, F.; Khvorova, A.; Marshall, W.; Sorkin, A., Analysis of clathrin-mediated endocytosis of epidermal growth factor receptor by RNA interference. *J Biol Chem* 2004, **279** (16), 16657–61.

67. (a) Brodsky, F. M.; Chen, C. Y.; Knuehl, C.; Towler, M. C.; Wakeham, D. E., Biological basket weaving: formation and function of clathrin-coated vesicles. *Annu Rev Cell Dev Biol* 2001, **17**, 517–68; (b) Kirchhausen, T., Adaptors for clathrin-mediated traffic. *Annu Rev Cell Dev Biol* 1999, **15**, 705–32.

68. (a) Takei, K.; Haucke, V., Clathrin-mediated endocytosis: membrane factors pull the trigger. *Trends Cell Biol* 2001, **11** (9), 385–91; (b) Maxfield, F. R.; McGraw, T. E., Endocytic recycling. *Nat Rev Mol Cell Biol* 2004, **5** (2), 121–32.

69. (a) Sever, S.; Damke, H.; Schmid, S. L., Garrotes, springs, ratchets, and whips: putting dynamin models to the test. *Traffic* 2000, **1** (5), 385–92; (b) Yan, L.; Ma, Y.; Sun, Y.; Gao, J.; Chen, X.; Liu, J.; Wang, C.; Rao, Z.; Lou, Z., Structural basis for mechanochemical role of Arabidopsis thaliana dynamin-related protein in membrane fission. *J Mol Cell Biol* 2011, **3** (6), 378–81; (c) Hinshaw, J. E., Dynamin and its role in membrane fission. *Annu Rev Cell Dev Biol* 2000, **16**, 483–519.

70. (a) Hinshaw, J. E.; Schmid, S. L., Dynamin self-assembles into rings suggesting a mechanism for coated vesicle budding. *Nature* 1995, **374** (6518), 190–2; (b) Sweitzer, S. M.; Hinshaw, J. E., Dynamin undergoes a GTP-dependent conformational change causing vesiculation. *Cell* 1998, **93** (6), 1021–9.

71. Stowell, M. H.; Marks, B.; Wigge, P.; McMahon, H. T., Nucleotide-dependent conformational changes in dynamin: evidence for a mechanochemical molecular spring. *Nat Cell Biol* 1999, **1** (1), 27–32.

72. Ritter, T. E.; Fajardo, O.; Matsue, H.; Anderson, R. G.; Lacey, S. W., Folate receptors targeted to clathrin-coated pits cannot regulate vitamin uptake. *Proc Natl Acad Sci U S A* 1995, **92** (9), 3824–8.

73. Artalejo, C. R.; Elhamdani, A.; Palfrey, H. C., Sustained stimulation shifts the mechanism of endocytosis from dynamin-1-dependent rapid endocytosis to clathrin- and dynamin-2-mediated slow endocytosis in chromaffin cells. *Proc Natl Acad Sci U S A* 2002, **99** (9), 6358–63.

74. (a) Hansen, C. G.; Nichols, B. J., Molecular mechanisms of clathrin-independent endocytosis. *J Cell Sci* 2009, **122** (Pt 11), 1713–21; (b) Frick, M.; Bright, N. A.; Riento, K.; Bray, A.; Merrified, C.; Nichols, B. J., Coassembly of flotillins induces formation of membrane microdomains, membrane curvature, and vesicle budding. *Curr Biol* 2007, **17** (13), 1151–6; (c) Glebov, O. O.; Bright, N. A.; Nichols, B. J., Flotillin-1 defines a clathrin-independent endocytic pathway in mammalian cells. *Nat Cell Biol* 2006, **8** (1), 46–54.

75. Langhorst, M. F.; Reuter, A.; Jaeger, F. A.; Wippich, F. M.; Luxenhofer, G.; Plattner, H.; Stuermer, C. A., Trafficking of the microdomain scaffolding protein reggie-1/flotillin-2. *Eur J Cell Biol* 2008, **87** (4), 211–26.

76. Bauer, M.; Pelkmans, L., A new paradigm for membrane-organizing and -shaping scaffolds. *FEBS Lett* 2006, **580** (23), 5559–64.

77. Lamaze, C.; Dujeancourt, A.; Baba, T.; Lo, C. G.; Benmerah, A.; Dautry-Varsat, A., Interleukin 2 receptors and detergent-resistant membrane domains define a clathrin-independent endocytic pathway. *Mol Cell* 2001, **7** (3), 661–71.

78. Mayor, S.; Pagano, R. E., Pathways of clathrin-independent endocytosis. *Nat Rev Mol Cell Biol* 2007, **8** (8), 603–12.

79. (a) Sabharanjak, S.; Sharma, P.; Parton, R. G.; Mayor, S., GPI-anchored proteins are delivered to recycling endosomes via a distinct cdc42-regulated, clathrin-independent pinocytic pathway. *Dev Cell* 2002, **2** (4), 411–23; (b) Llorente, A.; Rapak, A.; Schmid, S. L.;

van Deurs, B.; Sandvig, K., Expression of mutant dynamin inhibits toxicity and transport of endocytosed ricin to the Golgi apparatus. *J Cell Biol* 1998, **140** (3), 553–63; (c) Gauthier, N. C.; Monzo, P.; Kaddai, V.; Doye, A.; Ricci, V.; Boquet, P., Helicobacter pylori VacA cytotoxin: a probe for a clathrin-independent and Cdc42-dependent pinocytic pathway routed to late endosomes. *Mol Biol Cell* 2005, **16** (10), 4852–66.

80. (a) Kalia, M.; Kumari, S.; Chadda, R.; Hill, M. M.; Parton, R. G.; Mayor, S., Arf6-independent GPI-anchored protein-enriched early endosomal compartments fuse with sorting endosomes via a Rab5/phosphatidylinositol-3'-kinase-dependent machinery. *Mol Biol Cell* 2006, **17** (8), 3689–704; (b) D'Souza-Schorey, C.; Chavrier, P., ARF proteins: roles in membrane traffic and beyond. *Nat Rev Mol Cell Biol* 2006, **7** (5), 347–58.
81. Wang, H.; Yang, P.; Liu, K.; Guo, F.; Zhang, Y.; Zhang, G.; Jiang, C., SARS coronavirus entry into host cells through a novel clathrin- and caveolae-independent endocytic pathway. *Cell Res* 2008, **18** (2), 290–301.
82. Hewlett, L. J.; Prescott, A. R.; Watts, C., The coated pit and macropinocytic pathways serve distinct endosome populations. *J Cell Biol* 1994, **124** (5), 689–703.
83. Swanson, J. A., Shaping cups into phagosomes and macropinosomes. *Nat Rev Mol Cell Biol* 2008, **9** (8), 639–49.
84. Ehrenstein, G.; Lecar, H., Electrically gated ionic channels in lipid bilayers. *Q Rev Biophys* 1977, **10** (01), 1–34.
85. (a) Brabec, M.; Schober, D.; Wagner, E.; Bayer, N.; Murphy, R. F.; Blaas, D.; Fuchs, R., Opening of size-selective pores in endosomes during human rhinovirus serotype 2 in vivo uncoating monitored by single-organelle flow analysis. *J Virol* 2005, **79** (2), 1008–16; (b) Schober, D.; Kronenberger, P.; Prchla, E.; Blaas, D.; Fuchs, R., Major and minor receptor group human rhinoviruses penetrate from endosomes by different mechanisms. *J Virol* 1998, **72** (2), 1354–64.
86. Parente, R. A.; Nir, S.; Szoka, F. C., Jr., Mechanism of leakage of phospholipid vesicle contents induced by the peptide GALA. *Biochemistry* 1990, **29** (37), 8720–8.
87. (a) Akinc, A.; Thomas, M.; Klibanov, A. M.; Langer, R., Exploring polyethylenimine-mediated DNA transfection and the proton sponge hypothesis. *J Gene Med* 2005, **7** (5), 657–63; (b) Cho, Y. W.; Kim, J. D.; Park, K., Polycation gene delivery systems: escape from endosomes to cytosol. *J Pharm Pharmacol* 2003, **55** (6), 721–34; (c) Behr, J.-P., The proton sponge: a trick to enter cells the viruses did not exploit. *Chimia* 1997, **51** (1–2), 34–6.
88. (a) Goula, D.; Remy, J. S.; Erbacher, P.; Wasowicz, M.; Levi, G.; Abdallah, B.; Demeneix, B. A., Size, diffusibility and transfection performance of linear PEI/DNA complexes in the mouse central nervous system. *Gene Ther* 1998, **5** (5), 712–17; (b) Boletta, A.; Benigni, A.; Lutz, J.; Remuzzi, G.; Soria, M. R.; Monaco, L., Nonviral gene delivery to the rat kidney with polyethylenimine. *Hum Gene Ther* 1997, **8** (10), 1243–51; (c) Coll, J. L.; Chollet, P.; Brambilla, E.; Desplanques, D.; Behr, J. P.; Favrot, M., In vivo delivery to tumors of DNA complexed with linear polyethylenimine. *Hum Gene Ther* 1999, **10** (10), 1659–66.
89. (a) Asayama, S.; Hamaya, A.; Sekine, T.; Kawakami, H.; Nagaoka, S., Aminated poly(L-histidine) as new pH-sensitive DNA carrier. *Nucleic Acids Symp Ser (Oxf)* 2004, (**48**), 229–30; (b) Midoux, P.; Pichon, C.; Yaouanc, J. J.; Jaffres, P. A., Chemical vectors for gene delivery: a current review on polymers, peptides and lipids containing histidine or imidazole as nucleic acids carriers. *Br J Pharmacol* 2009, **157** (2), 166–78; (c) Tian, W. D.; Ma, Y. Q., Insights into the endosomal escape mechanism via investigation of dendrimer-membrane interactions. *Soft Matter* 2012, **8** (23), 6378–84.

90. Neu, M.; Fischer, D.; Kissel, T., Recent advances in rational gene transfer vector design based on poly(ethylene imine) and its derivatives. *J Gene Med* 2005, **7** (8), 992–1009.

91. (a) Thomas, M.; Klibanov, A. M., Enhancing polyethylenimine's delivery of plasmid DNA into mammalian cells. *Proc Natl Acad Sci U S A* 2002, **99** (23), 14640–5; (b) Thomas, M.; Ge, Q.; Lu, J. J.; Chen, J. Z.; Klibanov, A. M., Cross-linked small polyethylenimines: while still nontoxic, deliver DNA efficiently to mammalian cells in vitro and in vivo. *Pharm Res* 2005, **22** (3), 373–80.

92. Ko, Y. T.; Kale, A.; Hartner, W. C.; Papahadjopoulos-Sternberg, B.; Torchilin, V. P., Self-assembling micelle-like nanoparticles based on phospholipid-polyethyleneimine conjugates for systemic gene delivery. *J Control Release* 2009, **133** (2), 132–8.

93. Navarro, G.; Sawant, R. R.; Essex, S.; Tros de Ilarduya, C.; Torchilin, V. P., Phospholipid-polyethylenimine conjugate-based micelle-like nanoparticles for siRNA delivery. *Drug Deliv Transl Res* 2011, **1** (1), 25–33.

94. Navarro, G.; Sawant, R. R.; Biswas, S.; Essex, S.; Tros de Ilarduya, C.; Torchilin, V. P., P-glycoprotein silencing with siRNA delivered by DOPE-modified PEI overcomes doxorubicin resistance in breast cancer cells. *Nanomedicine (Lond)* 2012, **7** (1), 65–78.

95. Navarro, G.; Essex, S.; Sawant, R. R.; Biswas, S.; Nagesha, D.; Sridhar, S.; de Ilarduya, C. T.; Torchilin, V. P., Phospholipid-modified polyethylenimine-based nanopreparations for siRNA–mediated gene silencing: implications for transfection and the role of lipid components. *Nanomed: Nanotechnol, Biol Med* 2014, **10**, 411–19.

96. Zhou, X.; Huang, L., DNA transfection mediated by cationic liposomes containing lipopolylysine: characterization and mechanism of action. *Biochim Biophys Acta* 1994, **1189** (2), 195–203.

97. (a) Farhood, H.; Serbina, N.; Huang, L., The role of dioleoyl phosphatidylethanolamine in cationic liposome mediated gene transfer. *Biochim Biophys Acta* 1995, **1235** (2), 289–95; (b) Fasbender, A.; Marshall, J.; Moninger, T. O.; Grunst, T.; Cheng, S.; Welsh, M. J., Effect of co-lipids in enhancing cationic lipid-mediated gene transfer in vitro and in vivo. *Gene Ther* 1997, **4** (7), 716–25; (c) Hafez, I. M.; Cullis, P. R., Roles of lipid polymorphism in intracellular delivery. *Adv Drug Deliv Rev* 2001, **47** (2–3), 139–48.

98. Zuhorn, I. S.; Bakowsky, U.; Polushkin, E.; Visser, W. H.; Stuart, M. C.; Engberts, J. B.; Hoekstra, D., Nonbilayer phase of lipoplex-membrane mixture determines endosomal escape of genetic cargo and transfection efficiency. *Mol Ther* 2005, **11** (5), 801–10.

99. Xu, Y.; Szoka, F. C., Jr., Mechanism of DNA release from cationic liposome/DNA complexes used in cell transfection. *Biochemistry* 1996, **35** (18), 5616–23.

100. Ewert, K. K.; Ahmad, A.; Evans, H. M.; Safinya, C. R., Cationic lipid-DNA complexes for non-viral gene therapy: relating supramolecular structures to cellular pathways. *Expert Opin Biol Ther* 2005, **5** (1), 33–53.

101. (a) Semple, S. C.; Akinc, A.; Chen, J.; Sandhu, A. P.; Mui, B. L.; Cho, C. K.; Sah, D. W.; Stebbing, D.; Crosley, E. J.; Yaworski, E.; Hafez, I. M.; Dorkin, J. R.; Qin, J.; Lam, K.; Rajeev, K. G.; Wong, K. F.; Jeffs, L. B.; Nechev, L.; Eisenhardt, M. L.; Jayaraman, M.; Kazem, M.; Maier, M. A.; Srinivasulu, M.; Weinstein, M. J.; Chen, Q.; Alvarez, R.; Barros, S. A.; De, S.; Klimuk, S. K.; Borland, T.; Kosovrasti, V.; Cantley, W. L.; Tam, Y. K.; Manoharan, M.; Ciufolini, M. A.; Tracy, M. A.; de Fougerolles, A.; MacLachlan, I.; Cullis, P. R.; Madden, T. D.; Hope, M. J., Rational design of cationic lipids for siRNA delivery. *Nat Biotechnol* 2010, **28** (2), 172–6; (b) Semple, S. C.; Klimuk, S. K.; Harasym, T. O.; Dos Santos, N.; Ansell, S. M.; Wong, K. F.; Maurer, N.; Stark, H.; Cullis, P. R.; Hope, M. J.; Scherrer, P., Efficient encapsulation of antisense oligonucleotides in lipid vesicles using

ionizable aminolipids: formation of novel small multilamellar vesicle structures. *Biochim Biophys Acta* 2001, **1510** (1–2), 152–66.

102. Prasmickaite, L.; Hogset, A.; Berg, K., Evaluation of different photosensitizers for use in photochemical gene transfection. *Photochem Photobiol* 2001, **73** (4), 388–95.

103. Ndoye, A.; Dolivet, G.; Hogset, A.; Leroux, A.; Fifre, A.; Erbacher, P.; Berg, K.; Behr, J. P.; Guillemin, F.; Merlin, J. L., Eradication of p53-mutated head and neck squamous cell carcinoma xenografts using nonviral p53 gene therapy and photochemical internalization. *Mol Ther* 2006, **13** (6), 1156–62.

104. (a) Mahato, R. I., Non-viral peptide-based approaches to gene delivery. *J Drug Target* 1999, **7** (4), 249–68; (b) Mahat, R. I.; Monera, O. D.; Smith, L. C.; Rolland, A., Peptide-based gene delivery. *Curr Opin Mol Ther* 1999, **1** (2), 226–43; (c) Martin, M. E.; Rice, K. G., Peptide-guided gene delivery. *AAPS J* 2007, **9** (1), E18–29.

105. (a) Kogure, K.; Moriguchi, R.; Sasaki, K.; Ueno, M.; Futaki, S.; Harashima, H., Development of a non-viral multifunctional envelope-type nano device by a novel lipid film hydration method. *J Control Release* 2004, **98** (2), 317–23; (b) Kogure, K.; Akita, H.; Yamada, Y.; Harashima, H., Multifunctional envelope-type nano device (MEND) as a non-viral gene delivery system. *Adv Drug Deliv Rev* 2008, **60** (4–5), 559–71; (c) Kogure, K.; Akita, H.; Harashima, H., Multifunctional envelope-type nano device for non-viral gene delivery: concept and application of Programmed Packaging. *J Control Release* 2007, **122** (3), 246–51; (d) Kakudo, T.; Chaki, S.; Futaki, S.; Nakase, I.; Akaji, K.; Kawakami, T.; Maruyama, K.; Kamiya, H.; Harashima, H., Transferrin-modified liposomes equipped with a pH-sensitive fusogenic peptide: an artificial viral-like delivery system. *Biochemistry* 2004, **43** (19), 5618–28; (e) Li, W.; Nicol, F.; Szoka, F. C., Jr., GALA: a designed synthetic pH-responsive amphipathic peptide with applications in drug and gene delivery. *Adv Drug Deliv Rev* 2004, **56** (7), 967–85; (f) Min, S. H.; Lee, D. C.; Lim, M. J.; Park, H. S.; Kim, D. M.; Cho, C. W.; Yoon do, Y.; Yeom, Y. I., A composite gene delivery system consisting of polyethylenimine and an amphipathic peptide KALA. *J Gene Med* 2006, **8** (12), 1425–34.

106. Hatakeyama, H.; Ito, E.; Akita, H.; Oishi, M.; Nagasaki, Y.; Futaki, S.; Harashima, H., A pH-sensitive fusogenic peptide facilitates endosomal escape and greatly enhances the gene silencing of siRNA-containing nanoparticles in vitro and in vivo. *J Control Release* 2009, **139** (2), 127–32.

107. (a) Liberali, P.; Ramo, P.; Pelkmans, L., Protein kinases: starting a molecular systems view of endocytosis. *Annu Rev Cell Dev Biol* 2008, **24**, 501–23; (b) Korolchuk, V.; Banting, G., Kinases in clathrin-mediated endocytosis. *Biochem Soc Trans* 2003, **31** (Pt 4), 857–60; (c) Pelkmans, L.; Fava, E.; Grabner, H.; Hannus, M.; Habermann, B.; Krausz, E.; Zerial, M., Genome-wide analysis of human kinases in clathrin- and caveolae/raft-mediated endocytosis. *Nature* 2005, **436** (7047), 78–86.

108. ur Rehman, Z.; Hoekstra, D.; Zuhorn, I. S., Protein kinase A inhibition modulates the intracellular routing of gene delivery vehicles in HeLa cells, leading to productive transfection. *J Control Release* 2011, **156** (1), 76–84.

109. (a) Auguste, D. T.; Furman, K.; Wong, A.; Fuller, J.; Armes, S. P.; Deming, T. J.; Langer, R., Triggered release of siRNA from poly(ethylene glycol)-protected, pH-dependent liposomes. *J Control Release* 2008, **130** (3), 266–74; (b) Kale, A. A.; Torchilin, V. P., Enhanced transfection of tumor cells in vivo using "Smart" pH-sensitive TAT-modified pegylated liposomes. *J Drug Target* 2007, **15** (7–8), 538–45; (c) Li, W.; Huang, Z.; MacKay, J. A.; Grube, S.; Szoka, F. C., Jr., Low-pH-sensitive poly(ethylene glycol) (PEG)-stabilized plasmid nanolipoparticles: effects of PEG chain length, lipid composition

and assembly conditions on gene delivery. *J Gene Med* 2005, **7** (1), 67–79; (d) Simoes, S.; Moreira, J. N.; Fonseca, C.; Duzgunes, N.; de Lima, M. C., On the formulation of pH-sensitive liposomes with long circulation times. *Adv Drug Deliv Rev* 2004, **56** (7), 947–65; (e) Zhu, L.; Wang, T.; Perche, F.; Taigind, A.; Torchilin, V. P., Enhanced anti-cancer activity of nanopreparation containing an MMP2-sensitive PEG-drug conjugate and cell-penetrating moiety. *Proc Natl Acad Sci U S A* 2013, **110**, 17047–52.

110. Brady, R. O.; Tallman, J. F.; Johnson, W. G.; Gal, A. E.; Leahy, W. R.; Quirk, J. M.; Dekaban, A. S., Replacement therapy for inherited enzyme deficiency. Use of purified ceramidetrihexosidase in Fabry's disease. *N Engl J Med* 1973, **289** (1), 9–14.
111. Meikle, P. J.; Hopwood, J. J.; Clague, A. E.; Carey, W. F., Prevalence of lysosomal storage disorders. *JAMA* 1999, **281** (3), 249–54.
112. Fuller, M.; Meikle, P. J.; Hopwood, J. J., Epidemiology of lysosomal storage diseases: an overview. In *Fabry disease: perspectives from 5 years of FOS*, Mehta, A.; Beck, M.; Sunder-Plassmann, G., Eds. PharmaGenesis: Oxford, 2006, Chapter 2.
113. Desnick, R. J.; Schuchman, E. H., Enzyme replacement and enhancement therapies: lessons from lysosomal disorders. *Nat Rev Genet* 2002, **3** (12), 954–66.
114. (a) Platt, F. M.; Boland, B.; van der Spoel, A. C., The cell biology of disease: lysosomal storage disorders: the cellular impact of lysosomal dysfunction. *J Cell Biol* 2012, **199** (5), 723–34; (b) Brady, R. O., Enzyme replacement for lysosomal diseases. *Annu Rev Med* 2006, **57**, 283–96.
115. Grabowski, G. A.; Leslie, N.; Wenstrup, R., Enzyme therapy for Gaucher disease: the first 5 years. *Blood Rev* 1998, **12** (2), 115–33.
116. Grabowski, G. A., Phenotype, diagnosis, and treatment of Gaucher's disease. *Lancet* 2008, **372** (9645), 1263–71.
117. Jmoudiak, M.; Futerman, A. H., Gaucher disease: pathological mechanisms and modern management. *Br J Haematol* 2005, **129** (2), 178–88.
118. Koshkaryev, A.; Thekkedath, R.; Pagano, C.; Meerovich, I.; Torchilin, V. P., Targeting of lysosomes by liposomes modified with octadecyl-rhodamine B. *J Drug Target* 2011, **19** (8), 606–14.
119. Thekkedath, R.; Koshkaryev, A.; Torchilin, V. P., Lysosome-targeted octadecyl-rhodamine B-liposomes enhance lysosomal accumulation of glucocerebrosidase in Gaucher's cells in vitro. *Nanomedicine (Lond)* 2013, **8** (7), 1055–65.
120. Meerovich, I.; Koshkaryev, A.; Thekkedath, R.; Torchilin, V. P., Screening and optimization of ligand conjugates for lysosomal targeting. *Bioconjug Chem* 2011, **22** (11), 2271–82.
121. Gomme, P. T.; McCann, K. B.; Bertolini, J., Transferrin: structure, function and potential therapeutic actions. *Drug Discov Today* 2005, **10** (4), 267–73.
122. Qian, Z. M.; Li, H.; Sun, H.; Ho, K., Targeted drug delivery via the transferrin receptor-mediated endocytosis pathway. *Pharmacol Rev* 2002, **54** (4), 561–87.
123. (a) Dautry-Varsat, A.; Ciechanover, A.; Lodish, H. F., pH and the recycling of transferrin during receptor-mediated endocytosis. *Proc Natl Acad Sci U S A* 1983, **80** (8), 2258–62; (b) Andrews, N. C., Iron homeostasis: insights from genetics and animal models. *Nat Rev Genet* 2000, **1** (3), 208–17; (c) Morgan, E. H., Cellular iron processing. *J Gastroenterol Hepatol* 1996, **11** (11), 1027–30; (d) Lieu, P. T.; Heiskala, M.; Peterson, P. A.; Yang, Y., The roles of iron in health and disease. *Mol Aspects Med* 2001, **22** (1–2), 1–87; (e) Goswami, T.; Rolfs, A.; Hediger, M. A., Iron transport: emerging roles in health and disease. *Biochem Cell Biol* 2002, **80** (5), 679–89.

124. (a) Tabuchi, M.; Yoshimori, T.; Yamaguchi, K.; Yoshida, T.; Kishi, F., Human NRAMP2/DMT1, which mediates iron transport across endosomal membranes, is localized to late endosomes and lysosomes in HEp-2 cells. *J Biol Chem* 2000, **275** (29), 22220–8; (b) Watkins, J. A.; Altazan, J. D.; Elder, P.; Li, C. Y.; Nunez, M. T.; Cui, X. X.; Glass, J., Kinetic characterization of reductant dependent processes of iron mobilization from endocytic vesicles. *Biochemistry* 1992, **31** (25), 5820–30; (c) Nunez, M. T.; Gaete, V.; Watkins, J. A.; Glass, J., Mobilization of iron from endocytic vesicles. The effects of acidification and reduction. *J Biol Chem* 1990, **265** (12), 6688–92.

125. Morgan, E. H., Mechanisms of iron transport into rat erythroid cells. *J Cell Physiol* 2001, **186** (2), 193–200.

126. (a) Casey, J. L.; Hentze, M. W.; Koeller, D. M.; Caughman, S. W.; Rouault, T. A.; Klausner, R. D.; Harford, J. B., Iron-responsive elements: regulatory RNA sequences that control mRNA levels and translation. *Science* 1988, **240** (4854), 924–8; (b) Hentze, M. W.; Caughman, S. W.; Rouault, T. A.; Barriocanal, J. G.; Dancis, A.; Harford, J. B.; Klausner, R. D., Identification of the iron-responsive element for the translational regulation of human ferritin mRNA. *Science* 1987, **238** (4833), 1570–3; (c) Leibold, E. A.; Munro, H. N., Cytoplasmic protein binds in vitro to a highly conserved sequence in the 5' untranslated region of ferritin heavy- and light-subunit mRNAs. *Proc Natl Acad Sci U S A* 1988, **85** (7), 2171–5.

127. (a) Daniels, T. R.; Delgado, T.; Helguera, G.; Penichet, M. L., The transferrin receptor part II: targeted delivery of therapeutic agents into cancer cells. *Clin Immunol* 2006, **121** (2), 159–76; (b) Richardson, D. R.; Ponka, P., The molecular mechanisms of the metabolism and transport of iron in normal and neoplastic cells. *Biochim Biophys Acta* 1997, **1331** (1), 1–40.

128. (a) Seymour, G. J.; Walsh, M. D.; Lavin, M. F.; Strutton, G.; Gardiner, R. A., Transferrin receptor expression by human bladder transitional cell carcinomas. *Urol Res* 1987, **15** (6), 341–4; (b) Callens, C.; Moura, I. C.; Lepelletier, Y.; Coulon, S.; Renand, A.; Dussiot, M.; Ghez, D.; Benhamou, M.; Monteiro, R. C.; Bazarbachi, A.; Hermine, O., Recent advances in adult T-cell leukemia therapy: focus on a new anti-transferrin receptor monoclonal antibody. *Leukemia* 2008, **22** (1), 42–8; (c) Ng, P. P.; Dela Cruz, J. S.; Sorour, D. N.; Stinebaugh, J. M.; Shin, S. U.; Shin, D. S.; Morrison, S. L.; Penichet, M. L., An anti-transferrin receptor-avidin fusion protein exhibits both strong proapoptotic activity and the ability to deliver various molecules into cancer cells. *Proc Natl Acad Sci U S A* 2002, **99** (16), 10706–11; (d) Ng, P. P.; Helguera, G.; Daniels, T. R.; Lomas, S. Z.; Rodriguez, J. A.; Schiller, G.; Bonavida, B.; Morrison, S. L.; Penichet, M. L., Molecular events contributing to cell death in malignant human hematopoietic cells elicited by an IgG3-avidin fusion protein targeting the transferrin receptor. *Blood* 2006, **108** (8), 2745–54; (e) Wang, X.; Yang, L.; Chen, Z. G.; Shin, D. M., Application of nanotechnology in cancer therapy and imaging. *CA Cancer J Clin* 2008, **58** (2), 97–110; (f) Lubgan, D.; Jozwiak, Z.; Grabenbauer, G. G.; Distel, L. V., Doxorubicin-transferrin conjugate selectively overcomes multidrug resistance in leukaemia cells. *Cell Mol Biol Lett* 2009, **14** (1), 113–27.

129. (a) Wu, L.; Wu, J.; Zhou, Y.; Tang, X.; Du, Y.; Hu, Y., Enhanced antitumor efficacy of cisplatin by tirapazamine-transferrin conjugate. *Int J Pharm* 2012, **431** (1–2), 190–6; (b) Elliott, R. L.; Stjernholm, R.; Elliott, M. C., Preliminary evaluation of platinum transferrin (Mptc-63) as a potential nontoxic treatment for breast-cancer. *Cancer Detect Prev* 1988, **12** (1–6), 469–80.

130. Beyer, U.; Roth, T.; Schumacher, P.; Maier, G.; Unold, A.; Frahm, A. W.; Fiebig, H. H.; Unger, C.; Kratz, F., Synthesis and in vitro efficacy of transferrin conjugates of the anticancer drug chlorambucil. *J Med Chem* 1998, **41** (15), 2701–8.

131. Tanaka, T.; Shiramoto, S.; Miyashita, M.; Fujishima, Y.; Kaneo, Y., Tumor targeting based on the effect of enhanced permeability and retention (EPR) and the mechanism of receptor-mediated endocytosis (RME). *Int J Pharm* 2004, **277** (1–2), 39–61.

132. (a) Frankel, A. E.; Bugge, T. H.; Liu, S. H.; Vallera, D. A.; Leppla, S. H., Peptide toxins directed at the matrix dissolution systems of cancer cells. *Protein Pept Lett* 2002, **9** (1), 1–14; (b) Sandvig, K.; van Deurs, B., Endocytosis, intracellular transport, and cytotoxic action of Shiga toxin and ricin. *Physiol Rev* 1996, **76** (4), 949–66.

133. Raso, V.; Basala, M., A highly cytotoxic human transferrin-ricin A chain conjugate used to select receptor-modified cells. *J Biol Chem* 1984, **259** (2), 1143–9.

134. Chignola, R.; Foroni, R.; Franceschi, A.; Pasti, M.; Candiani, C.; Anselmi, C.; Fracasso, G.; Tridente, G.; Colombatti, M., Heterogeneous response of individual multicellular tumour spheroids to immunotoxins and ricin toxin. *Br J Cancer* 1995, **72** (3), 607–14.

135. (a) Recht, L. D.; Griffin, T. W.; Raso, V.; Salimi, A. R., Potent cytotoxicity of an antihuman transferrin receptor-ricin A-chain immunotoxin on human glioma cells in vitro. *Cancer Res* 1990, **50** (20), 6696–700; (b) Engebraaten, O.; Hjortland, G. O.; Juell, S.; Hirschberg, H.; Fodstad, O., Intratumoral immunotoxin treatment of human malignant brain tumors in immunodeficient animals. *Int J Cancer* 2002, **97** (6), 846–52.

136. (a) Trowbridge, I. S.; Domingo, D. L., Anti-transferrin receptor monoclonal antibody and toxin-antibody conjugates affect growth of human tumour cells. *Nature* 1981, **294** (5837), 171–3; (b) O'Keefe, D. O.; Draper, R. K., Characterization of a transferrin-diphtheria toxin conjugate. *J Biol Chem* 1985, **260** (2), 932–7.

137. (a) Greenfield, L.; Johnson, V. G.; Youle, R. J., Mutations in diphtheria toxin separate binding from entry and amplify immunotoxin selectivity. *Science* 1987, **238** (4826), 536–9; (b) Johnson, V. G.; Wilson, D.; Greenfield, L.; Youle, R. J., The role of the diphtheria toxin receptor in cytosol translocation. *J Biol Chem* 1988, **263** (3), 1295–300.

138. Rybak, S. M.; Saxena, S. K.; Ackerman, E. J.; Youle, R. J., Cytotoxic potential of ribonuclease and ribonuclease hybrid proteins. *J Biol Chem* 1991, **266** (31), 21202–7.

139. Suzuki, M.; Saxena, S. K.; Boix, E.; Prill, R. J.; Vasandani, V. M.; Ladner, J. E.; Sung, C.; Youle, R. J., Engineering receptor-mediated cytotoxicity into human ribonucleases by steric blockade of inhibitor interaction. *Nat Biotechnol* 1999, **17** (3), 265–70.

140. Xu, L.; Frederik, P.; Pirollo, K. F.; Tang, W. H.; Rait, A.; Xiang, L. M.; Huang, W.; Cruz, I.; Yin, Y.; Chang, E. H., Self-assembly of a virus-mimicking nanostructure system for efficient tumor-targeted gene delivery. *Hum Gene Ther* 2002, **13** (3), 469–81.

141. (a) Xu, L.; Pirollo, K. F.; Chang, E. H., Transferrin-liposome-mediated p53 sensitization of squamous cell carcinoma of the head and neck to radiation in vitro. *Hum Gene Ther* 1997, **8** (4), 467–75; (b) Zhang, X.; Koh, C. G.; Yu, B.; Liu, S.; Piao, L.; Marcucci, G.; Lee, R. J.; Lee, L. J., Transferrin receptor targeted lipopolyplexes for delivery of antisense oligonucleotide g3139 in a murine k562 xenograft model. *Pharm Res* 2009, **26** (6), 1516–24; (c) Xu, L.; Tang, W. H.; Huang, C. C.; Alexander, W.; Xiang, L. M.; Pirollo, K. F.; Rait, A.; Chang, E. H., Systemic p53 gene therapy of cancer with immunolipoplexes targeted by anti-transferrin receptor scFv. *Mol Med* 2001, **7** (10), 723–34; (d) Xu, L.; Huang, C. C.; Huang, W. Q.; Tang, W. H.; Rait, A.; Yin, Y. Z.; Cruz, I.;

Xiang, L. M.; Pirollo, K. F.; Chang, E. H., Systemic tumor-targeted gene delivery by anti-transferrin receptor scFv-immunoliposomes. *Mol Cancer Ther* 2002, **1** (5), 337–46.

142. Matherly, L. H.; Hou, Z.; Deng, Y., Human reduced folate carrier: translation of basic biology to cancer etiology and therapy. *Cancer Metastasis Rev* 2007, **26** (1), 111–28.
143. (a) Zhao, R.; Min, S. H.; Wang, Y.; Campanella, E.; Low, P. S.; Goldman, I. D., A role for the proton-coupled folate transporter (PCFT-SLC46A1) in folate receptor-mediated endocytosis. *J Biol Chem* 2009, **284** (7), 4267–74; (b) Shin, D. S.; Zhao, R.; Fiser, A.; Goldman, I. D., Role of the fourth transmembrane domain in proton-coupled folate transporter function as assessed by the substituted cysteine accessibility method. *Am J Physiol Cell Physiol* 2013, **304** (12), C1159–67.
144. Antony, A. C., The biological chemistry of folate receptors. *Blood* 1992, **79** (11), 2807–20.
145. (a) Nakashima-Matsushita, N.; Homma, T.; Yu, S.; Matsuda, T.; Sunahara, N.; Nakamura, T.; Tsukano, M.; Ratnam, M.; Matsuyama, T., Selective expression of folate receptor beta and its possible role in methotrexate transport in synovial macrophages from patients with rheumatoid arthritis. *Arthritis Rheum* 1999, **42** (8), 1609–16; (b) Turk, M. J.; Breur, G. J.; Widmer, W. R.; Paulos, C. M.; Xu, L. C.; Grote, L. A.; Low, P. S., Folate-targeted imaging of activated macrophages in rats with adjuvant-induced arthritis. *Arthritis Rheum* 2002, **46** (7), 1947–55.
146. Weitman, S. D.; Lark, R. H.; Coney, L. R.; Fort, D. W.; Frasca, V.; Zurawski, V. R., Jr.; Kamen, B. A., Distribution of the folate receptor GP38 in normal and malignant cell lines and tissues. *Cancer Res* 1992, **52** (12), 3396–401.
147. (a) Parker, N.; Turk, M. J.; Westrick, E.; Lewis, J. D.; Low, P. S.; Leamon, C. P., Folate receptor expression in carcinomas and normal tissues determined by a quantitative radioligand binding assay. *Anal Biochem* 2005, **338** (2), 284–93; (b) Ross, J. F.; Chaudhuri, P. K.; Ratnam, M., Differential regulation of folate receptor isoforms in normal and malignant-tissues in-vivo and in established cell-lines—physiological and clinical implications. *Cancer* 1994, **73** (9), 2432–43.
148. (a) Low, P. S.; Henne, W. A.; Doorneweerd, D. D., Discovery and development of folic-acid-based receptor targeting for imaging and therapy of cancer and inflammatory diseases. *Acc Chem Res* 2008, **41** (1), 120–9; (b) Sudimack, J.; Lee, R. J., Targeted drug delivery via the folate receptor. *Adv Drug Deliv Rev* 2000, **41** (2), 147–62; (c) Pan, X.; Lee, R. J., Tumour-selective drug delivery via folate receptor-targeted liposomes. *Expert Opin Drug Deliv* 2004, **1** (1), 7–17.
149. Hilgenbrink, A. R.; Low, P. S., Folate receptor-mediated drug targeting: from therapeutics to diagnostics. *J Pharm Sci* 2005, **94** (10), 2135–46.
150. Reddy, J. A.; Low, P. S., Folate-mediated targeting of therapeutic and imaging agents to cancers. *Crit Rev Ther Drug Carrier Syst* 1998, **15** (6), 587–627.
151. (a) Anderson, R. G.; Kamen, B. A.; Rothberg, K. G.; Lacey, S. W., Potocytosis: sequestration and transport of small molecules by caveolae. *Science* 1992, **255** (5043), 410–11; (b) Matveev, S.; Li, X.; Everson, W.; Smart, E. J., The role of caveolae and caveolin in vesicle-dependent and vesicle-independent trafficking. *Adv Drug Deliv Rev* 2001, **49** (3), 237–50.
152. Yang, J.; Chen, H.; Vlahov, I. R.; Cheng, J. X.; Low, P. S., Evaluation of disulfide reduction during receptor-mediated endocytosis by using FRET imaging. *Proc Natl Acad Sci U S A* 2006, **103** (37), 13872–7.

153. (a) Chatterjee, S.; Smith, E. R.; Hanada, K.; Stevens, V. L.; Mayor, S., GPI anchoring leads to sphingolipid-dependent retention of endocytosed proteins in the recycling endosomal compartment. *EMBO J* 2001, **20** (7), 1583–92; (b) Paulos, C. M.; Reddy, J. A.; Leamon, C. P.; Turk, M. J.; Low, P. S., Ligand binding and kinetics of folate receptor recycling in vivo: impact on receptor-mediated drug delivery. *Mol Pharmacol* 2004, **66** (6), 1406–14.

154. Yang, J.; Chen, H.; Vlahov, I. R.; Cheng, J. X.; Low, P. S., Characterization of the pH of folate receptor-containing endosomes and the rate of hydrolysis of internalized acid-labile folate-drug conjugates. *J Pharmacol Exp Ther* 2007, **321** (2), 462–8.

155. Reddy, J. A.; Leamon, C. P.; Low, P. S., Folate-mediated delivery of protein and peptide drugs into tumors. In *Delivery of protein and peptide drugs in cancer*, Torchilin, V. P., Ed. Imperial College Press: London, 2006, 183–204.

156. (a) Leamon, C. P.; Low, P. S., Cytotoxicity of momordin-folate conjugates in cultured human cells. *J Biol Chem* 1992, **267** (35), 24966–71; (b) Leamon, C. P.; Pastan, I.; Low, P. S., Cytotoxicity of folate-Pseudomonas exotoxin conjugates toward tumor cells. Contribution of translocation domain. *J Biol Chem* 1993, **268** (33), 24847–54; (c) Leamon, C. P.; Low, P. S., Selective targeting of malignant cells with cytotoxin-folate conjugates. *J Drug Target* 1994, **2** (2), 101–12.

157. Atkinson, S. F.; Bettinger, T.; Seymour, L. W.; Behr, J. P.; Ward, C. M., Conjugation of folate via gelonin carbohydrate residues retains ribosomal-inactivating properties of the toxin and permits targeting to folate receptor positive cells. *J Biol Chem* 2001, **276** (30), 27930–5.

158. Kim, S. H.; Jeong, J. H.; Joe, C. O.; Park, T. G., Folate receptor mediated intracellular protein delivery using PLL-PEG-FOL conjugate. *J Control Release* 2005, **103** (3), 625–34.

159. Cho, K. C.; Jeong, J. H.; Chung, H. J.; Joe, C. O.; Kim, S. W.; Park, T. G., Folate receptor-mediated intracellular delivery of recombinant caspase-3 for inducing apoptosis. *J Control Release* 2005, **108** (1), 121–31.

160. (a) Reddy, J. A.; Dean, D.; Kennedy, M. D.; Low, P. S., Optimization of folate-conjugated liposomal vectors for folate receptor-mediated gene therapy. *J Pharm Sci* 1999, **88** (11), 1112–18; (b) Turk, M. J.; Reddy, J. A.; Chmielewski, J. A.; Low, P. S., Characterization of a novel pH-sensitive peptide that enhances drug release from folate-targeted liposomes at endosomal pHs. *Biochim Biophys Acta* 2002, **1559** (1), 56–68.

161. Ryser, H. J.; Hancock, R., Histones and basic polyamino acids stimulate the uptake of albumin by tumor cells in culture. *Science* 1965, **150** (3695), 501–3.

162. Green, M.; Loewenstein, P. M., Autonomous functional domains of chemically synthesized human immunodeficiency virus tat trans-activator protein. *Cell* 1988, **55** (6), 1179–88.

163. Frankel, A. D.; Pabo, C. O., Cellular uptake of the tat protein from human immunodeficiency virus. *Cell* 1988, **55** (6), 1189–93.

164. Fawell, S.; Seery, J.; Daikh, Y.; Moore, C.; Chen, L. L.; Pepinsky, B.; Barsoum, J., Tat-mediated delivery of heterologous proteins into cells. *Proc Natl Acad Sci U S A* 1994, **91** (2), 664–8.

165. Lundberg, P.; Langel, U., A brief introduction to cell-penetrating peptides. *J Mol Recognit* 2003, **16** (5), 227–33.

166. Lindgren, M.; Langel, U., Classes and prediction of cell-penetrating peptides. *Methods Mol Biol* 2011, **683**, 3–19.

167. Oehlke, J.; Scheller, A.; Wiesner, B.; Krause, E.; Beyermann, M.; Klauschenz, E.; Melzig, M.; Bienert, M., Cellular uptake of an alpha-helical amphipathic model peptide with the potential to deliver polar compounds into the cell interior non-endocytically. *Biochim Biophys Acta* 1998, **1414** (1–2), 127–39.

168. Pooga, M.; Hallbrink, M.; Zorko, M.; Langel, U., Cell penetration by transportan. *FASEB J* 1998, **12** (1), 67–77.

169. Henriques, S. T.; Castanho, M. A., Translocation or membrane disintegration? Implication of peptide-membrane interactions in pep-1 activity. *J Pept Sci* 2008, **14** (4), 482–7.

170. Derossi, D.; Joliot, A. H.; Chassaing, G.; Prochiantz, A., The 3rd helix of the Antennapedia homeodomain translocates through biological-membranes. *J Biol Chem* 1994, **269** (14), 10444–50.

171. Vives, E.; Brodin, P.; Lebleu, B., A truncated HIV-1 Tat protein basic domain rapidly translocates through the plasma membrane and accumulates in the cell nucleus. *J Biol Chem* 1997, **272** (25), 16010–17.

172. Futaki, S.; Suzuki, T.; Ohashi, W.; Yagami, T.; Tanaka, S.; Ueda, K.; Sugiura, Y., Arginine-rich peptides—an abundant source of membrane-permeable peptides having potential as carriers for intracellular protein delivery. *J Biol Chem* 2001, **276** (8), 5836–40.

173. Derossi, D.; Chassaing, G.; Prochiantz, A., Trojan peptides: the penetratin system for intracellular delivery. *Trends Cell Biol* 1998, **8** (2), 84–7.

174. (a) Ross, M. F.; Filipovska, A.; Smith, R. A.; Gait, M. J.; Murphy, M. P., Cell-penetrating peptides do not cross mitochondrial membranes even when conjugated to a lipophilic cation: evidence against direct passage through phospholipid bilayers. *Biochem J* 2004, **383** (Pt. 3), 457–68; (b) Lindgren, M. E.; Hallbrink, M. M.; Elmquist, A. M.; Langel, U., Passage of cell-penetrating peptides across a human epithelial cell layer in vitro. *Biochem J* 2004, **377** (Pt 1), 69–76.

175. (a) Muratovska, A.; Eccles, M. R., Conjugate for efficient delivery of short interfering RNA (siRNA) into mammalian cells. *FEBS Lett* 2004, **558** (1–3), 63–8; (b) Turner, J. J.; Jones, S.; Fabani, M. M.; Ivanova, G.; Arzumanov, A. A.; Gait, M. J., RNA targeting with peptide conjugates of oligonucleotides, siRNA and PNA. *Blood Cells Mol Dis* 2007, **38** (1), 1–7.

176. (a) Shiraishi, T.; Nielsen, P. E., Peptide nucleic acid (PNA) cell penetrating peptide (CPP) conjugates as carriers for cellular delivery of antisense oligomers. *Artif DNA PNA XNA* 2011, **2** (3), 90–9; (b) Pooga, M.; Soomets, U.; Hallbrink, M.; Valkna, A.; Saar, K.; Rezaei, K.; Kahl, U.; Hao, J. X.; Xu, X. J.; Wiesenfeld-Hallin, Z.; Hokfelt, T.; Bartfai, T.; Langel, U., Cell penetrating PNA constructs regulate galanin receptor levels and modify pain transmission in vivo. *Nat Biotechnol* 1998, **16** (9), 857–61; (c) Pooga, M.; Soomets, U.; Bartfai, T.; Langel, U., Synthesis of cell-penetrating peptide-PNA constructs. *Methods Mol Biol* 2002, **208**, 225–36.

177. Pooga, M.; Kut, C.; Kihlmark, M.; Hallbrink, M.; Fernaeus, S.; Raid, R.; Land, T.; Hallberg, E.; Bartfai, T.; Langel, U., Cellular translocation of proteins by transportan. *FASEB J* 2001, **15** (8), 1451–3.

178. Phelan, A.; Elliott, G.; O'Hare, P., Intercellular delivery of functional p53 by the herpesvirus protein VP22. *Nat Biotechnol* 1998, **16** (5), 440–3.

179. Lobanov, V. A.; Zheng, C.; Babiuk, L. A.; van Drunen Littel-van den Hurk, S., Intracellular trafficking of VP22 in bovine herpesvirus-1 infected cells. *Virology* 2010, **396** (2), 189–202.

180. Dilber, M. S.; Phelan, A.; Aints, A.; Mohamed, A. J.; Elliott, G.; Smith, C. I.; O'Hare, P., Intercellular delivery of thymidine kinase prodrug activating enzyme by the herpes simplex virus protein, VP22. *Gene Ther* 1999, **6** (1), 12–21.

181. Schwarze, S. R.; Hruska, K. A.; Dowdy, S. F., Protein transduction: unrestricted delivery into all cells? *Trends Cell Biol* 2000, **10** (7), 290–5.

182. Elliott, G.; O'Hare, P., Intercellular trafficking and protein delivery by a herpesvirus structural protein. *Cell* 1997, **88** (2), 223–33.

183. (a) Wolf, Y.; Pritz, S.; Abes, S.; Bienert, M.; Lebleu, B.; Oehlke, J., Structural requirements for cellular uptake and antisense activity of peptide nucleic acids conjugated with various peptides. *Biochemistry* 2006, **45** (50), 14944–54; (b) Scheller, A.; Oehlke, J.; Wiesner, B.; Dathe, M.; Krause, E.; Beyermann, M.; Melzig, M.; Bienert, M., Structural requirements for cellular uptake of alpha-helical amphipathic peptides. *J Pept Sci* 1999, **5** (4), 185–94.

184. (a) Zorko, M.; Langel, U., Cell-penetrating peptides: mechanism and kinetics of cargo delivery. *Adv Drug Deliv Rev* 2005, **57** (4), 529–45; (b) Hallbrink, M.; Floren, A.; Elmquist, A.; Pooga, M.; Bartfai, T.; Langel, U., Cargo delivery kinetics of cell-penetrating peptides. *Biochim Biophys Acta* 2001, **1515** (2), 101–9; (c) Floren, A.; Mager, I.; Langel, U., Uptake kinetics of cell-penetrating peptides. *Methods Mol Biol* 2011, **683**, 117–28.

185. Giacca, M., The HIV-1 Tat protein: a multifaceted target for novel therapeutic opportunities. *Curr Drug Targets Immune Endocr Metabol Disord* 2004, **4** (4), 277–85.

186. (a) Duchardt, F.; Fotin-Mleczek, M.; Schwarz, H.; Fischer, R.; Brock, R., A comprehensive model for the cellular uptake of cationic cell-penetrating peptides. *Traffic* 2007, **8** (7), 848–66; (b) Kosuge, M.; Takeuchi, T.; Nakase, I.; Jones, A. T.; Futaki, S., Cellular internalization and distribution of arginine-rich peptides as a function of extracellular peptide concentration, serum, and plasma membrane associated proteoglycans. *Bioconjug Chem* 2008, **19** (3), 656–64.

187. Yang, L.; Harroun, T. A.; Weiss, T. M.; Ding, L.; Huang, H. W., Barrel-stave model or toroidal model? A case study on melittin pores. *Biophys J* 2001, **81** (3), 1475–85.

188. Shai, Y., Mechanism of the binding, insertion and destabilization of phospholipid bilayer membranes by alpha-helical antimicrobial and cell non-selective membrane-lytic peptides. *Biochim Biophys Acta* 1999, **1462** (1–2), 55–70.

189. (a) Console, S.; Marty, C.; Garcia-Echeverria, C.; Schwendener, R.; Ballmer-Hofer, K., Antennapedia and HIV transactivator of transcription (TAT) "protein transduction domains" promote endocytosis of high molecular weight cargo upon binding to cell surface glycosaminoglycans. *J Biol Chem* 2003, **278** (37), 35109–14; (b) Lundberg, M.; Wikstrom, S.; Johansson, M., Cell surface adherence and endocytosis of protein transduction domains. *Mol Ther* 2003, **8** (1), 143–50.

190. Rothbard, J. B.; Jessop, T. C.; Lewis, R. S.; Murray, B. A.; Wender, P. A., Role of membrane potential and hydrogen bonding in the mechanism of translocation of guanidinium-rich peptides into cells. *J Am Chem Soc* 2004, **126** (31), 9506–7.

191. (a) Sugita, T.; Yoshikawa, T.; Mukai, Y.; Yamanada, N.; Imai, S.; Nagano, K.; Yoshida, Y.; Shibata, H.; Yoshioka, Y.; Nakagawa, S.; Kamada, H.; Tsunoda, S. I.; Tsutsumi, Y., Comparative study on transduction and toxicity of protein transduction domains. *Br J Pharmacol* 2008, **153** (6), 1143–52; (b) Saar, K.; Lindgren, M.; Hansen, M.; Eiriksdottir, E.; Jiang, Y.; Rosenthal-Aizman, K.; Sassian, M.; Langel, U., Cell-penetrating peptides: a comparative membrane toxicity study. *Anal Biochem* 2005, **345** (1), 55–65.

192. Ohkuma, S.; Poole, B., Fluorescence probe measurement of the intralysosomal pH in living cells and the perturbation of pH by various agents. *Proc Natl Acad Sci U S A* 1978, **75** (7), 3327–31.

193. (a) Szeto, H. H.; Schiller, P. W.; Zhao, K.; Luo, G., Fluorescent dyes alter intracellular targeting and function of cell-penetrating tetrapeptides. *FASEB J* 2005, **19** (1), 118–20; (b) Srinivasan, D.; Muthukrishnan, N.; Johnson, G. A.; Erazo-Oliveras, A.; Lim, J.; Simanek, E. E.; Pellois, J. P., Conjugation to the cell-penetrating peptide TAT potentiates the photodynamic effect of carboxytetramethylrhodamine. *PLoS One* 2011, **6** (3), e17732; (c) Mishra, A.; Lai, G. H.; Schmidt, N. W.; Sun, V. Z.; Rodriguez, A. R.; Tong, R.; Tang, L.; Cheng, J.; Deming, T. J.; Kamei, D. T.; Wong, G. C., Translocation of HIV TAT peptide and analogues induced by multiplexed membrane and cytoskeletal interactions. *Proc Natl Acad Sci U S A* 2011, **108** (41), 16883–8.

194. Wadia, J. S.; Stan, R. V.; Dowdy, S. F., Transducible TAT-HA fusogenic peptide enhances escape of TAT-fusion proteins after lipid raft macropinocytosis. *Nat Med* 2004, **10** (3), 310–15.

195. van den Berg, A.; Dowdy, S. F., Protein transduction domain delivery of therapeutic macromolecules. *Curr Opin Biotechnol* 2011, **22** (6), 888–93.

196. (a) Fittipaldi, A.; Ferrari, A.; Zoppe, M.; Arcangeli, C.; Pellegrini, V.; Beltram, F.; Giacca, M., Cell membrane lipid rafts mediate caveolar endocytosis of HIV-1 tat fusion proteins. *J Biol Chem* 2003, **278** (36), 34141–9; (b) Ferrari, A.; Pellegrini, V.; Arcangeli, C.; Fittipaldi, A.; Giacca, M.; Beltram, F., Caveolae-mediated internalization of extracellular HIV-1 tat fusion proteins visualized in real time. *Mol Ther* 2003, **8** (2), 284–94; (c) Nakase, I.; Niwa, M.; Takeuchi, T.; Sonomura, K.; Kawabata, N.; Koike, Y.; Takehashi, M.; Tanaka, S.; Ueda, K.; Simpson, J. C.; Jones, A. T.; Sugiura, Y.; Futaki, S., Cellular uptake of arginine-rich peptides: roles for macropinocytosis and actin rearrangement. *Mol Ther* 2004, **10** (6), 1011–22; (d) Kaplan, I. M.; Wadia, J. S.; Dowdy, S. F., Cationic TAT peptide transduction domain enters cells by macropinocytosis. *J Control Release* 2005, **102** (1), 247–53.

197. Hirose, H.; Takeuchi, T.; Osakada, H.; Pujals, S.; Katayama, S.; Nakase, I.; Kobayashi, S.; Haraguchi, T.; Futaki, S., Transient focal membrane deformation induced by arginine-rich peptides leads to their direct penetration into cells. *Mol Ther* 2012, **20** (5), 984–93.

198. Schwarze, S. R.; Ho, A.; Vocero-Akbani, A.; Dowdy, S. F., In vivo protein transduction: delivery of a biologically active protein into the mouse. *Science* 1999, **285** (5433), 1569–72.

199. Harbour, J. W.; Worley, L.; Ma, D.; Cohen, M., Transducible peptide therapy for uveal melanoma and retinoblastoma. *Arch Ophthalmol* 2002, **120** (10), 1341–6.

200. (a) Tang, X.; Molina, M.; Amar, S., p53 short peptide (p53pep164) regulates lipopolysaccharide-induced tumor necrosis factor-alpha factor/cytokine expression. *Cancer Res* 2007, **67** (3), 1308–16; (b) Snyder, E. L.; Meade, B. R.; Saenz, C. C.; Dowdy, S. F., Treatment of terminal peritoneal carcinomatosis by a transducible p53-activating peptide. *PLoS Biol* 2004, **2** (2), E36.

201. Ryu, J.; Lee, H. J.; Kim, K. A.; Lee, J. Y.; Lee, K. S.; Park, J.; Choi, S. Y., Intracellular delivery of p53 fused to the basic domain of HIV-1 Tat. *Mol Cells* 2004, **17** (2), 353–9.

202. (a) Parada, Y.; Banerji, L.; Glassford, J.; Lea, N. C.; Collado, M.; Rivas, C.; Lewis, J. L.; Gordon, M. Y.; Thomas, N. S. B.; Lam, E. W.-F., BCR-ABL and interleukin 3 promote haematopoietic cell proliferation and survival through modulation of cyclin D2 and p27Kip1 expression. *J Biol Chem* 2001, **276** (26), 23572–80; (b) Nagahara, H.;

Vocero-Akbani, A. M.; Snyder, E. L.; Ho, A.; Latham, D. G.; Lissy, N. A.; Becker-Hapak, M.; Ezhevsky, S. A.; Dowdy, S. F., Transduction of full-length TAT fusion proteins into mammalian cells: TAT-p27Kip1 induces cell migration. *Nat Med* 1998, **4** (12), 1449–52.

203. Fujimoto, K.; Hosotani, R.; Miyamoto, Y.; Doi, R.; Koshiba, T.; Otaka, A.; Fujii, N.; Beauchamp, R. D.; Imamura, M., Inhibition of pRb phosphorylation and cell cycle progression by an antennapedia-p16(INK4A) fusion peptide in pancreatic cancer cells. *Cancer Lett* 2000, **159** (2), 151–8.

204. (a) Gius, D. R.; Ezhevsky, S. A.; Becker-Hapak, M.; Nagahara, H.; Wei, M. C.; Dowdy, S. F., Transduced p16INK4a peptides inhibit hypophosphorylation of the retinoblastoma protein and cell cycle progression prior to activation of Cdk2 complexes in late G1. *Cancer Res* 1999, **59** (11), 2577–80; (b) Ezhevsky, S. A.; Nagahara, H.; Vocero-Akbani, A. M.; Gius, D. R.; Wei, M. C.; Dowdy, S. F., Hypo-phosphorylation of the retinoblastoma protein (pRb) by cyclin D:Cdk4/6 complexes results in active pRb. *Proc Natl Acad Sci U S A* 1997, **94** (20), 10699–704.

205. Datta, K.; Sundberg, C.; Karumanchi, S. A.; Mukhopadhyay, D., The 104-123 amino acid sequence of the beta-domain of von Hippel-Lindau gene product is sufficient to inhibit renal tumor growth and invasion. *Cancer Res* 2001, **61** (5), 1768–75.

206. (a) Fulda, S.; Wick, W.; Weller, M.; Debatin, K. M., Smac agonists sensitize for Apo2L/TRAIL- or anticancer drug-induced apoptosis and induce regression of malignant glioma in vivo. *Nat Med* 2002, **8** (8), 808–15; (b) Yang, L.; Mashima, T.; Sato, S.; Mochizuki, M.; Sakamoto, H.; Yamori, T.; Oh-Hara, T.; Tsuruo, T., Predominant suppression of apoptosome by inhibitor of apoptosis protein in non-small cell lung cancer H460 cells: therapeutic effect of a novel polyarginine-conjugated Smac peptide. *Cancer Res* 2003, **63** (4), 831–7.

207. (a) Holinger, E. P.; Chittenden, T.; Lutz, R. J., Bak BH3 peptides antagonize Bcl-xL function and induce apoptosis through cytochrome c-independent activation of caspases. *J Biol Chem* 1999, **274** (19), 13298–304; (b) Letai, A.; Bassik, M. C.; Walensky, L. D.; Sorcinelli, M. D.; Weiler, S.; Korsmeyer, S. J., Distinct BH3 domains either sensitize or activate mitochondrial apoptosis, serving as prototype cancer therapeutics. *Cancer Cell* 2002, **2** (3), 183–92.

208. Makino, E.; Sakaguchi, M.; Iwatsuki, K.; Huh, N. H., Introduction of an N-terminal peptide of S100C/A11 into human cells induces apoptotic cell death. *J Mol Med (Berl)* 2004, **82** (9), 612–20.

209. Bhoumik, A.; Gangi, L.; Ronai, Z., Inhibition of melanoma growth and metastasis by ATF2-derived peptides. *Cancer Res* 2004, **64** (22), 8222–30.

210. (a) Cao, G.; Pei, W.; Ge, H.; Liang, Q.; Luo, Y.; Sharp, F. R.; Lu, A.; Ran, R.; Graham, S. H.; Chen, J., In vivo delivery of a Bcl-xL fusion protein containing the TAT protein transduction domain protects against ischemic brain injury and neuronal apoptosis. *J Neurosci* 2002, **22** (13), 5423–31; (b) Asoh, S.; Ohsawa, I.; Mori, T.; Katsura, K.; Hiraide, T.; Katayama, Y.; Kimura, M.; Ozaki, D.; Yamagata, K.; Ohta, S., Protection against ischemic brain injury by protein therapeutics. *Proc Natl Acad Sci U S A* 2002, **99** (26), 17107–12.

211. Papadopoulou, L. C.; Tsiftsoglou, A. S., The potential role of cell penetrating peptides in the intracellular delivery of proteins for therapy of erythroid related disorders. *Pharmaceuticals (Basel)* 2013, **6** (1), 32–53.

212. Myou, S.; Leff, A. R.; Myo, S.; Boetticher, E.; Meliton, A. Y.; Lambertino, A. T.; Liu, J.; Xu, C.; Munoz, N. M.; Zhu, X., Activation of group IV cytosolic phospholipase A2 in

human eosinophils by phosphoinositide 3-kinase through a mitogen-activated protein kinase-independent pathway. *J Immunol* 2003, **171** (8), 4399–405.

213. (a) Moy, P.; Daikh, Y.; Pepinsky, B.; Thomas, D.; Fawell, S.; Barsoum, J., Tat-mediated protein delivery can facilitate MHC class I presentation of antigens. *Mol Biotechnol* 1996, **6** (2), 105–13; (b) Kim, D. T.; Mitchell, D. J.; Brockstedt, D. G.; Fong, L.; Nolan, G. P.; Fathman, C. G.; Engleman, E. G.; Rothbard, J. B., Introduction of soluble proteins into the MHC class I pathway by conjugation to an HIV tat peptide. *J Immunol* 1997, **159** (4), 1666–8.

214. (a) Davidson, T. J.; Harel, S.; Arboleda, V. A.; Prunell, G. F.; Shelanski, M. L.; Greene, L. A.; Troy, C. M., Highly efficient small interfering RNA delivery to primary mammalian neurons induces MicroRNA-like effects before mRNA degradation. *J Neurosci* 2004, **24** (45), 10040–6; (b) Ganguly, S.; Chaubey, B.; Tripathi, S.; Upadhyay, A.; Neti, P. V.; Howell, R. W.; Pandey, V. N., Pharmacokinetic analysis of polyamide nucleic-acid-cell penetrating peptide conjugates targeted against HIV-1 transactivation response element. *Oligonucleotides* 2008, **18** (3), 277–86; (c) Shiraishi, T.; Nielsen, P. E., Enhanced delivery of cell-penetrating peptide-peptide nucleic acid conjugates by endosomal disruption. *Nat Protoc* 2006, **1** (2), 633–6; (d) Moschos, S. A.; Jones, S. W.; Perry, M. M.; Williams, A. E.; Erjefalt, J. S.; Turner, J. J.; Barnes, P. J.; Sproat, B. S.; Gait, M. J.; Lindsay, M. A., Lung delivery studies using siRNA conjugated to TAT(48–60) and penetratin reveal peptide induced reduction in gene expression and induction of innate immunity. *Bioconjug Chem* 2007, **18** (5), 1450–9.

215. (a) Crombez, L.; Morris, M. C.; Dufort, S.; Aldrian-Herrada, G.; Nguyen, Q.; Mc Master, G.; Coll, J.-L.; Heitz, F.; Divita, G., Targeting cyclin B1 through peptide-based delivery of siRNA prevents tumour growth. *Nucleic Acids Res* 2009, **37** (14), 4559–69; (b) Simeoni, F.; Morris, M. C.; Heitz, F.; Divita, G., Insight into the mechanism of the peptide-based gene delivery system MPG: implications for delivery of siRNA into mammalian cells. *Nucleic Acids Res* 2003, **31** (11), 2717–24; (c) Mo, R. H.; Zaro, J. L.; Shen, W. C., Comparison of cationic and amphipathic cell penetrating peptides for siRNA delivery and efficacy. *Mol Pharm* 2012, **9** (2), 299–309.

216. Torchilin, V. P.; Levchenko, T. S.; Rammohan, R.; Volodina, N.; Papahadjopoulos-Sternberg, B.; D'Souza, G. G., Cell transfection in vitro and in vivo with nontoxic TAT peptide-liposome-DNA complexes. *Proc Natl Acad Sci U S A* 2003, **100** (4), 1972–7.

217. Grunwald, J.; Rejtar, T.; Sawant, R.; Wang, Z.; Torchilin, V. P., TAT peptide and its conjugates: proteolytic stability. *Bioconjug Chem* 2009, **20** (8), 1531–7.

218. (a) Torchilin, V., Multifunctional and stimuli-sensitive pharmaceutical nanocarriers. *Eur J Pharm Biopharm* 2009, **71** (3), 431–44; (b) Sawant, R. M.; Hurley, J. P.; Salmaso, S.; Kale, A.; Tolcheva, E.; Levchenko, T. S.; Torchilin, V. P., "SMART" drug delivery systems: double-targeted pH-responsive pharmaceutical nanocarriers. *Bioconjug Chem* 2006, **17** (4), 943–9.

219. Zhu, L.; Torchilin, V. P., Stimulus-responsive nanopreparations for tumor targeting. *Integr Biol (Camb)* 2013, **5** (1), 96–107.

220. Kale, A. A.; Torchilin, V. P., "Smart" drug carriers: PEGylated TATp-modified pH-sensitive liposomes. *J Liposome Res* 2007, **17** (3–4), 197–203.

221. Koren, E.; Apte, A.; Jani, A.; Torchilin, V. P., Multifunctional PEGylated 2C5-immunoliposomes containing pH-sensitive bonds and TAT peptide for enhanced tumor cell internalization and cytotoxicity. *J Control Release* 2012, **160** (2), 264–73.

24

VACCINE DELIVERY: CURRENT ROUTES OF ADMINISTRATION AND NOVEL APPROACHES

NEHA SAHNI*, YUAN CHENG*, C. RUSSELL MIDDAUGH, AND DAVID B. VOLKIN

Department of Pharmaceutical Chemistry, Macromolecule Vaccine Stabilization Center, The University of Kansas, Lawrence, KS, USA

24.1 INTRODUCTION

A variety of different vaccine dosage forms are commercially available for human use including (i) liquid solutions stored in glass vials and/or prefilled syringes (PFS) administered by injection (or in limited cases nasally), (ii) lyophilized powders requiring reconstitution with diluent prior to injection or oral delivery, and (iii) liquid formulations filled into plastic tubes, or lyophilized powders placed into capsules or tablets, for oral administration. How and why were these different vaccine presentations designed and developed? Some of the relevant considerations include patient safety and efficacy (route of administration, ensuring best immune responses, etc.), pharmaceutical properties (vaccine potency and stability), manufacturing capacity (ability to reproducibly make vaccine at large scale in a GMP facility), and commercial feasibility (cost, distribution including the need for a vaccine cold chain, patient/medical convenience, etc.). These interrelated variables are typically summarized as a "target product profile (TPP)" during the clinical development of a new vaccine candidate. The TPP is a strategic work plan

* Contributed equally as first authors.

Drug Delivery: Principles and Applications, Second Edition. Edited by Binghe Wang, Longqin Hu, and Teruna J. Siahaan.

that not only defines the components and rationale behind a vaccine dosage form (safety, efficacy, stability, patient delivery, etc.) but also coordinates all of these activities within a vaccine company during research, clinical trials, manufacturing, and regulatory approvals [1].

The stabilization, formulation, and delivery of a vaccine antigen (or multiple antigens), resulting in a defined pharmaceutical dosage form that can be distributed through a vaccine cold chain and administered to patients as a prophylactic medicine, are key components of the TPP [2–4]. Since vaccines typically contain macromolecular antigens or microorganisms with complex and often delicate three-dimensional structures, vaccines are labile and usually require careful maintenance of the cold chain, that is, refrigerators and even freezers, for their distribution and storage [2]. To achieve the goal of developing a stable vaccine formulation, a series of experimental steps are typically performed including (i) evaluating the physicochemical properties of the vaccine antigen; (ii) developing stability-indicating analytical assays including potency tests; (iii) selecting pharmaceutical excipients [5] (to enhance stability, solubility, and ensure tonicity), and in some cases as discussed in more detail later; and (iv) evaluating and selecting adjuvants to boost *in vivo* immune responses [6]. These activities are coordinated to design, develop, and produce a vaccine dosage form (containing a combination of antigen, adjuvant, and excipients) that is stable during manufacturing, long-term storage, and distribution in the vaccine cold chain and can be easily and appropriately administered to the target patient population.

The focus of this chapter is not only to review how currently available commercial vaccines are formulated and administered to patients but also how new vaccines may be administered in the future using novel delivery devices and formulation technologies. Most vaccines are administered by injection. The parenteral route of administration includes a wide variety of approaches such as intravenous (IV) injection, which deliver drugs directly into the circulation system for the most rapid onset of action (not typically used with vaccines), and intramuscular (IM) and subcutaneous (SC) injections, which can provide slower and more sustained release of drugs [7, 8]. Most vaccines are delivered via the IM or SC route, but newer approaches include intradermal (ID) delivery to target the natural environment of many immune cells such as macrophages and dendritic cells as well as novel experimental approaches to better direct and target antigen to specific immune cells (e.g., intranodal lymph delivery). Parenteral administration of vaccines by needle and syringe is the standard, well-accepted approach, but needle-free injection technologies are also of increasing interest. As shown in Figure 24.1, ID, SC, and IM injections result in vaccines being delivered to different biological tissues, as illustrated with a liquid dye solution delivered by a traditional needle–syringe and needle-free injections (the latter being developed to eliminate needlestick injuries and minimize medical waste [10, 11]). By delivery of a vaccine to either the skin, SC tissue, or muscle, as discussed later in more detail, both the quality and quantity of the immune responses as well as the tolerability of the vaccine injection can be modulated.

Despite the remarkable success in protecting the general public with vaccines administered by injection, certain drawbacks do exist including the need to maintain

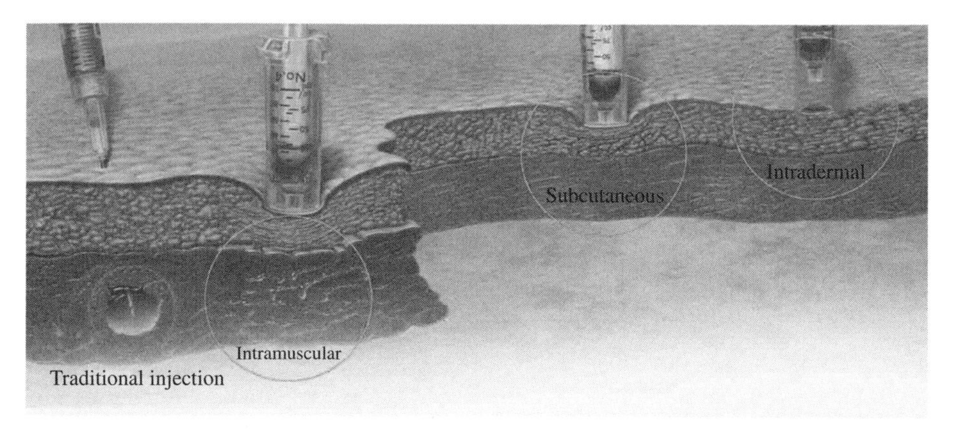

FIGURE 24.1 Intramuscular (IM), subcutaneous (SC), and intradermal (ID) delivery by traditional needle injection and newer needle-free delivery systems. Figure is used with permission from Bioject Inc. from Ref. [9]. Courtesy of Bioject Inc.

sterile conditions during administration, the requirement of special skills to administer injections, possible pain upon injection, lack of patient compliance due to fear of needles, and generation of medical waste requiring needle disposal. By utilizing alternative routes of administration, not only can these concerns be addressed, but potentially better immune responses and protection against disease are possible by better mimicking natural infections. For example, many viruses spread by oral–fecal transmission or by airborne droplets, thus triggering immune response in the gut and nasal cavities, respectively. Therefore, oral and nasal administration of a vaccine could be very desirable, and as discussed later, several vaccines have been successfully developed employing these routes of administration. Finally, many new vaccine delivery technologies are being evaluated both in the laboratory and the clinic to provide more targeted delivery of vaccines to specific tissues, using either injection (ID, IM, SC) or oral, nasal, or even buccal or aerosol delivery.

24.2 PARENTERAL ADMINISTRATION OF VACCINES

24.2.1 Currently Available Vaccines and Devices for Intramuscular and Subcutaneous Delivery

The IM route of injection is preferred for many commercial vaccines, especially for those containing inactivated or purified antigens and formulated with aluminum adjuvants. IM injection permits more sustained presentation of antigens *in vivo* [12] and avoids or minimizes local irritation and inflammation [13]. As shown in Table 24.1, examples of intramuscularly administered vaccines include tetanus, diphtheria, and pertussis, hepatitis A virus (HAV), hepatitis B virus (HBV), and Hib vaccines, as well as their various combination products. In addition, aluminum-adjuvant containing pneumococcal conjugate vaccines (PCV) [15] are administered intramuscularly to minimize local irritation as recommended by the Centers for Disease

Control and Prevention (CDC). Historically, IM vaccine injections were given in the buttocks, but their fat layers do not contain specific immune cells to produce potent immune response [16]. Most IM injections are now given in the deltoid or the anterolateral aspect of the thigh since it has a richer blood supply [17].

Inactivated influenza and polio vaccines, which contain inactivated viruses without adjuvant, can be administered by either intramuscularly or subcutaneously since they induce minimal local inflammatory reactions [18] (Table 24.1). Live attenuated viral vaccines including measles, mumps, and rubella (MMR) and varicella zoster virus (VZV) and their quadrivalent combination (MMRV) along with several polysaccharide vaccines (i.e., pneumococcal and meningococcal vaccines) are administered

TABLE 24.1 List of US-Approved Vaccines and Their Dose and Route of Administration[a]

Vaccine	Adjuvant	Dose	Route
Diphtheria, tetanus, pertussis (DTaP, DT, Tdap, Td)	Aluminum	0.5 ml	IM
Haemophilus influenzae type b (Hib)	Aluminum	0.5 ml	IM
Hepatitis A (Hep A)	Aluminum	≤18 year: 0.5 ml ≥19 year: 1.0 ml	IM
Hepatitis B (Hep B)	Aluminum	<19 year: 0.5 ml ≥20 year: 1.0 ml	IM
Human papillomavirus (HPV)	Aluminum	0.5 ml	IM
Influenza, inactivated	None	6–35 month: 0.25 ml ≥3 year: 0.5 ml	IM
Measles, mumps, and rubella (MMR)	None	0.5 ml	SC
Meningococcal conjugate (MCV)	None	0.5 ml	IM
Meningococcal polysaccharide (MPSV)	None	0.5 ml	SC
Pneumococcal conjugate (PCV)	Aluminum	0.5 ml	IM
Pneumococcal polysaccharide (PPSV)	None	0.5 ml	IM or SC
Polio, inactivated (IPV)	None	0.5 ml	IM or SC
Varicella (Var)	None	0.5 ml	SC
Zoster (Zos)	None	0.65 ml	SC
DTap/HepB/IPV (Pediarix)	Aluminum	0.5 ml	IM
DTap/IPV/Hib (Pentacel)	Aluminum	0.5 ml	IM
DTap/IPV (Kinrix)	Aluminum	0.5 ml	IM
Hib/HepB (Comvax)	Aluminum	0.5 ml	IM
MMRV (ProQuad)	None	≤12 year: 0.5 ml	SC
HepA/HepB (Twinrix)	Aluminum	≥18 year: 1.0 ml	IM
Influenza, inactivated: Fluzone ID	None	18–64 year: 0.1 ml	ID
Rotavirus vaccine: Rotarix	None	1.0 ml	Oral
Rotavirus vaccine, pentavalent: RotaTeq	None	2.0 ml	Oral
Typhoid vaccine	None	Capsule	Oral
Polio vaccine, live attenuated	None	0.5 ml (2 drops/dose)	Oral
Adenovirus type 4 and type 7 vaccine	None	Tablet	Oral
Influenza: FluMist	None	0.2 ml	Intranasal

[a] Table was adapted from Immunization Action Coalition [14].
ID, intradermal; IM, intramuscular; SC, subcutaneous.

subcutaneously. There are no general regulatory guidelines for the use of specific routes of administration (IM or SC) for vaccine administration. This is established by clinical trials for each vaccine and is then recommended per manufacturer's instructions. In general, injections through the SC route can be associated with increased rates of local reactions. This is probably due to the poorer drainage system in SC tissue that causes longer retention of vaccine antigen [15]. The SC route has been associated with formation of abscesses and granulomas in certain cases [19]. In addition, SC administration has resulted in suboptimal efficacy with several vaccines such as hepatitis B and rabies vaccines [20, 21].

The selection of optimal needle length and gauge is of utmost importance during IM and SC injection of vaccines to avoid pain and undesirable local side effects and for efficient delivery [21] as summarized in Table 24.2. In addition, the angle of injection impacts the depth of needle penetration. SC injections are administered at a 45° angle into the thigh of infants or in the upper outer triceps for adults with 5/8″ (23–25 gauge) long needles. IM injections are administered at an angle of 90° into

TABLE 24.2 Injection Site and Needle Size for Intramuscular and Subcutaneous Administration of Vaccines[a]

Injection Site and Needle Size (Choose the injection site that is appropriate to the person's age and body mass)

Subcutaneous (SC) Injection (Use a 23–25 gauge needle)

Age	**Needle Length**	**Injection Site**
Infants (1–12 months)	5/8″	Fatty tissue over anterolateral thigh muscle
Children 12 months or older, adolescents, and adults	5/8″	Fatty tissue over anterolateral thigh muscle or fatty tissue over triceps

Intramuscular (IM) Injection (Use a 22–25 gauge needle)

Age	**Needle Length**	**Injection Site**
Newborns (first 28 days)	5/8″[b]	Anterolateral thigh muscle
Infants (1–12 months)	1″	Anterolateral thigh muscle
Toddlers (1–2 years)	1–1¼″ 5/8–1″[b]	Anterolateral thigh muscle or deltoid muscle of arm
Children and teens (3–18 years)	5/8–1″[b] 1–1¼″	Deltoid muscle of arm or anterolateral thigh muscle
Adults 19 years or older		
Male or female less than 130 lbs	5/8–1″[b]	Deltoid muscle of arm
Female 130–200 lbs Male 130–260 lbs	1–1½″	Deltoid muscle of arm
Female 200+ lbs Male 260+ lbs	1½″	Deltoid muscle of arm

[a] Table was obtained from the Immunization Action Coalition [14].

[b] 5/8″ needle may be used for patients weighing less than 130 lbs (<60 kg) for IM injection in the deltoid muscle only if the skin is stretched tight, the subcutaneous tissue is not bunched, and the injection is made at a 90° angle [14].

the thigh or deltoid muscle [10], and the needle length varies from 5/8 to 1½″ (22–25 gauge) as it should be long enough to reach the muscle mass (Table 24.2).

For vaccines filled and packaged into glass vials, administration includes the use of a disposable plastic syringe with a luer lock connected to disposable injection needles (Fig. 24.2a). These components are affordable, presterilized, and ready for use. Typically, for a liquid vaccine formulation, two steps are required: removal of the liquid from the vial (through the rubber stopper) followed by administration to the patient. According to the CDC, there is no need to change the needle between drawing the vaccine from vial and administration unless the needle is contaminated or damaged. For a lyophilized formulation, an additional initial step is required to add diluent to the vaccine vial for reconstitution. Diluents are either provided by the vaccine manufacturer or acquired separately (e.g., water for injection).

The use of vaccines packaged into glass PFS has several advantages over vaccines filled into vials (Fig. 24.2b). For example, PFS packaging reduces the risk associated with microbial contamination (i.e., eliminating the need for removing a vaccine dose from the vial with a disposable syringe) and also is more convenient to use with improved accuracy of the delivered dose. The PFS configuration comes with either a luer lock to connect a disposable needle or is available with a staked needle already attached to the syringe. PFS dosage forms have the potential to encourage self-administration by patients at home, hence reducing the cost of injection and promoting patient compliance [25]. The commercial quadrivalent human papillomavirus vaccine (Gardasil, Merck & Co.) is one example of vaccine packaged in a PFS

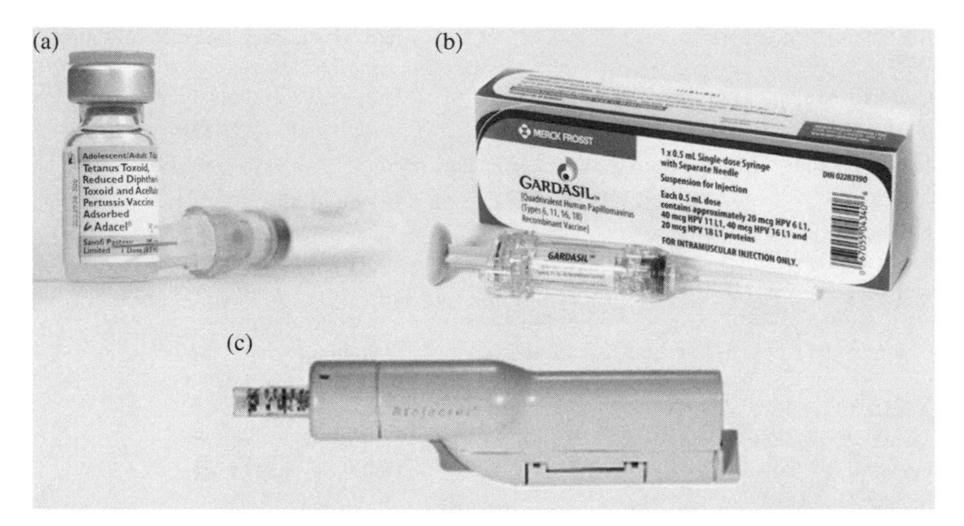

FIGURE 24.2 Devices for parenteral administration of vaccines. (a) Disposable needle and syringe for IM/SC delivery of a vaccine. The image was acquired from Free Stock Photos courtesy CDC/Amanda Mills [22]. (b) Prefilled syringe vaccine against human papilloma virus (Gardasil) for IM delivery. Reproduced with permission from The Liberty Beacon [23]. (c) Needle-free device Biojector® 2000 (Bioject Inc.) for IM/SC delivery. Plotkin [24]. Reproduced with permission from Elsevier.

(Fig. 24.2b) for IM delivery. PFS can come with a device enabling an autodisable function as part of single-dose delivery [26]. These devices cover or protect the needle after use resulting in additional safety and reduction of medical waste.

Unfortunately, improper injection practices can lead to blood-borne infections and transmission of infectious agents such as human immunodeficiency virus (HIV), HBV, and HCV. According to OSHA, approximately 0–3.62 needlestick injuries per 100,000 vaccination occur every year in the United States [27]. To minimize and prevent the risks related to IM/SC delivery, not only are autodisable devices becoming more widely available with needle and syringes, but needle-free delivery devices are also available including multiuse nozzle jet injectors and single-use disposable cartridge jet injectors. Jet injectors are noninvasive devices that are specifically designed for the delivery of liquid formulations. They work by accelerating liquid formulations to a speed that can penetrate into the skin. Needleless delivery devices have the potential to provide faster and cost-effective vaccination programs with minimal pain. Multiuse nozzle jet injectors can be used for multiple patients with accurate delivery doses, while cartridge jet injectors are single-use devices with disposable cartridges [28]. Multiuse nozzle jet injectors still have safety and health risks associated with them that can be overcome using disposable cartridges to prevent cross-contamination between patients. According to the CDC, these injectors should be able to provide 600 injections per hour in large immunization campaigns [13, 29].

Handheld jet injectors are devices used to deliver a vaccine through a nozzle with high pressure [30] (Fig. 24.2c). These devices deliver the vaccine with the use of a narrow, high-velocity fluid jet that penetrates the skin in a fraction of a second. They can be gas propelled or spring powered. For example, the Biojector® 2000 (Bioject Inc., Tustin, CA) is a gas-powered injector approved for ID, IM, and SC use (Fig. 24.2c), while the Bioject ZetaJet® is spring-powered device employed for SC and IM delivery. Other available devices include the Injex® (Equidyne Systems, Inc., Tustin, CA) [13] and PharmaJet® Stratis (PharmaJet, Inc., Golden, CO) [31]. These devices also can have autodisable function to prevent refilling ensuring single use only. Additional disposable jet injectors are under development including Imojet® and Imule® for typhoid and tetanus toxoid, respectively [13], and Vitajet™ for influenza [32]. Studies have shown that these jet injectors were able to produce similar or equivalent immune responses compared to traditional needle and syringe delivery. Some studies have shown needle-free delivery that elicits enhanced immune responses in comparison to standard IM immunization [33, 34]. The widespread use of jet injector devices has been hindered by certain weakness including high operational costs, difficulty in maintenance, and, for some patients, increased discomfort associated with administration [35].

24.2.2 Currently Available Intradermal Vaccines and Associated Delivery Devices

Due to the complexity of traditional ID delivery techniques, ID vaccination has had limited application to date with modern vaccines. ID vaccines were traditionally administered by the Mantoux technique, which uses a hypodermic needle and syringe.

This technique is difficult to perform and inconsistent in delivering well-defined vaccine doses. The smallpox vaccine and worldwide disease eradication were accomplished primarily by ID administration of the vaccine using a bifurcated needle. The mass vaccination program against smallpox was started in the 1950s by the World Health Organization (WHO). The disease was abolished as a part of an extensive worldwide vaccination program in 1980 [36, 37]. The vaccine was introduced into the prongs of the bifurcated needle and then administered to the skin. This method was very simple to use and the needles were easily sterilizable and required minimum maintenance [38]. Newer cell culture versions of the traditional smallpox vaccine have been developed and manufactured for biodefense purposes and are administered with similar bifurcated needles [39].

With recent advances in our understanding of the skin immune system along with advances in ID delivery technologies, ID vaccination is emerging as a promising alternative to traditional IM and SC immunizations for other vaccines. The dermal layer contains a high concentration of immune cells that can be important for initiating and enhancing immune responses for many vaccines. ID vaccination has numerous attractive advantages over IM/SC immunization: (i) it is less invasive, reducing administration-associated injuries and infections; (ii) it causes less pain and stress, favoring patient compliance, particularly in children; and (iii) it usually requires a smaller dose to elicit potent immune responses, providing a dose-sparing strategy.

New delivery devices have allowed for more reproducible ID vaccinations. Fluzone® Intradermal (Sanofi Pasteur, Inc.), marketed in 2011, is the first FDA-approved ID influenza vaccine [7]. Fluzone® Intradermal is administered through a prefilled microinjection system (Soluvia®, Becton Dickinson) that directly delivers the vaccine to the dermis layer of the skin [7] (see Fig. 24.3a). The Soluvia device contains a 1.5 mm long ultrathin needle and a syringe that can hold 100–200 µl of liquid [42]. An appropriate insertion depth is ensured by a depth limiter integrated in the microinjection system. Due to the use of these microneedles, Soluvia™ is associated with minimal discomfort [43–46]. Clinical studies have shown that dose-sparing vaccination by Fluzone Intradermal can elicit immune responses comparable to those induced by standard IM injection [47]. Intradermal rabies vaccine (IDRV) is another example of an ID delivered vaccine. It was first approved by the WHO for postexposure prophylaxis in developing countries in 1992 [48]. IDRV is administered to the lymphatic drainage sites in the upper arm through a beveled needle and syringe. This ID regimen results in reduced vaccine usage and fewer hospital visits compared to the IM regimen [48].

24.2.3 Novel Devices for Parenteral Injection

There have been significant technical developments in the area of needle-free devices to deliver vaccine antigens either intramuscularly or subcutaneously. A number of these devices are being tested in preclinical and clinical phases. For example, CELLECTRA®, MedPulser®, and Elgen® electroporation devices by Inovio are being developed for IM delivery and have been tested to deliver plasmid DNA vaccine

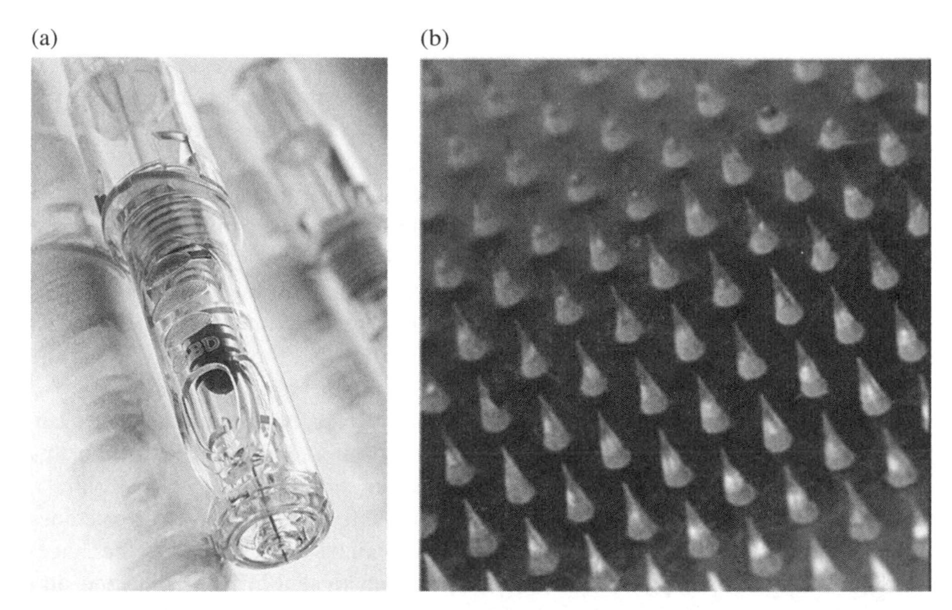

FIGURE 24.3 Devices for intradermal delivery of vaccines. (a) BD SoluviaTM prefillable microinjection system currently marketed for use with influenza vaccine (Fluzone® Intradermal). Reproduced with permission from BD [40]. (b) TheraJect VaxMat® dissolvable microneedle arrays currently under development. Kommareddi [41], pp. 1021–1027. Reproduced with permission from John Wiley & Sons, Inc.

candidates into the muscle to elicit strong immune response. LectraJet® HS (D'Antonio Consultants International, Inc., East Syracuse, NY) is another device under development with autodisable function allowing single use only [13, 49]. PowderJect (PowderJect Pharmaceuticals PLC, now a part of Novartis) is an example of a jet injector under development to deliver dry powder vaccines. This technology uses pressured gas to accelerate vaccine particles to a high velocity that can penetrate into the skin. To improve the efficiency of skin penetration, vaccine antigens are often absorbed onto the surface of high-density carriers. Gold is the most commonly used carrier for powder injectors because of its high density and good biocompatibility. Biodegradable materials, such as high-density sugar particles, have also been evaluated as carriers for powder injectors [50, 51]. This technology has been tested in animals for an influenza vaccine [52]. The PowderJect XR device [53] has also been tested with hepatitis B surface antigen (HBsAg)-coated gold particles in mice [54].

There have been numerous efforts recently to develop novel ID devices for influenza vaccines [55]. A study from van Damme et al., for example, evaluated the safety and efficacy of a novel microneedle device, MicronJet® (NanoPass), for ID influenza vaccination in healthy adults [56]. This device can deliver vaccines directly into the skin when used with a standard syringe. A marketed influenza vaccine (α-RIX®, GlaxoSmithKline (GSK) Biologicals) was used as the test antigen in this study. The results suggest that dose-sparing ID vaccination by MicronJet® elicits similar immunogenic responses to the full-dose IM vaccination [56]. In another example, a candidate plasmid DNA-based herpes simplex virus type 2 (HSV-2) vaccine was examined [57]. Studies in mice showed that an HSV-2 vaccine based on gene fragments, when delivered intradermally by a noninvasive technology designated particle-medicated epidermal delivery (PMED), can elicit specific immune responses that protect the animals from lethal infectious challenge.

Another emerging ID delivery technology is the potential use of microneedle arrays. These devices consist of an adhesive patch with a large number of attached microneedles (see Fig. 24.3b). Microneedle arrays can be classified into three general groups: solid microneedles, hollow microneedles, and dissolvable microneedles [35]. Solid microneedle arrays can be used to perforate the skin before applying vaccines to the skin surface. This pretreatment can improve the permeability of the skin and thereby the immunogenicity of vaccine antigens [58]. Alternatively, solid microneedle arrays with antigen coated on the surface of the microneedles can directly deliver antigens to a defined depth of the skin. Studies have shown this approach can elicit protective immune responses in certain cases [59]. In contrast, hollow microneedle arrays are potentially more effective in delivering vaccines across skin barriers. Injection flow rate and dosing can be well controlled by connection to a syringe or pump. A recent study with an inactivated influenza vaccine suggested that vaccines delivered by hollow microneedle arrays can elicit immune responses comparable to those induced by the standard IM method [56].

A general concern with microneedle arrays is the breakage of microneedles in the skin. To address this problem, microneedles made of dissolvable materials, such as sugars and polylactic acids (PLA), are being developed [60–62]. An example of a dissolvable, biodegradable microneedle array is the TheraJect VaxMAT® (Fig. 24.3b). The tips of these microneedles are made of a mixture of antigens, trehalose, and sodium carboxymethyl cellulose. The microneedles begin to dissolve on the order of minutes after piercing the skin [41]. *In vivo* studies in mice have shown protective immunity against monovalent influenza strains when the vaccine is delivered using microneedle patches [41]. Another example is vaccination against measles using microneedle patchs [63]. These microneedle patches have also been recently tested against influenza virus [64] and compared directly to IM and SC injections. The study showed that dissolvable microneedles produced much higher immune responses [64, 65]. The influenza vaccine in these studies was coated onto microneedle surfaces or encapsulated into the needle made of a biodegradable polymer.

24.2.4 Novel Formulations and Delivery Approaches for Parenteral Injection

The stratum corneum is the skin's main barrier to the outside world and poses the greatest challenge for ID delivery. A few abrasive methods that can remove stratum corneum have been shown to enhance the immune responses induced by ID immunization [66, 67]. Other approaches that can improve the permeability of the skin, such as thermal ablation, ultrasound, electroporation, and chemical enhancers, have also been shown to facilitate the ID delivery of vaccines [68]. Alternatively, the use of adjuvants and delivery systems is being assessed to improve immune responses to vaccines upon ID injection. Novel molecular adjuvant molecules that can more efficiently cross skin barriers, such as bacterial enterotoxins and Toll-like receptor (TLR) ligands, are primarily being examined for this purpose [69]. Cholera toxin (CT) and *Escherichia coli* heat-labile toxin (LT), and their related mutated forms, are two of the most extensively studied bacterial enterotoxin-based adjuvant systems [70]. The ADP-ribosyltransferase activity of these toxins may play an important role in eliciting strong immune response [71]. For example, the ID administration of IpaB and IpaD antigens with *E. coli* heat-labile double mutant enterotoxin (dmLT) using microneedles prevented *Shigella* infection [72]. TLR ligands are another class of adjuvant molecules that mediate the innate and adaptive immune responses against pathogens. The TLR9 ligand polynucleotide-based CpG adjuvant given via the ID route has been evaluated for immunization in preclinical studies [73, 74].

Conventional particulate adjuvants such as aluminum salts and oil-in-water emulsions are not typically used for ID vaccine delivery due to challenges in transport across skin barriers. Nonetheless, these particulate carriers have the potential to improve immune responses by prolonging the residence of vaccine antigens at the injection site by a depot effect [75], facilitating antigen uptake by antigen-presenting cells (APCs) [76], and retaining antigens in close proximity to these immune cells [77]. Particulate carriers prepared with lipids or surfactants have been explored for the delivery of ID vaccines. Transfersome® (prepared with a mixture of phosphatidylcholine, Span 80, and ethanol) has been evaluated with HBsAg and has shown comparable IgG titers, and significantly increased secretory IgA titers, compared to the same dose of aluminum salt-adjuvanted HBsAg delivered via IM injection [78]. Other delivery systems, such as PLA nanoparticles, have also been investigated for ID delivery, although limited success has so far been achieved [79, 80]. The failure has largely been attributed to their inability to transport across skin barriers. This is supported by the observations that reduction in the size of particulate carriers (or perforation of the skin) improves immune responses elicited by nanoparticle-delivered antigens [80, 81].

Alternative sites of vaccination are another way of improving immune responses produced by parenteral administration. For example, vaccination by the intranodal route has also shown great potential in delivering antigens with adjuvants directly to lymph nodes, since an effective immune response is thought to be best produced when an antigen travels to, and is presented to, the lymph node to activate both cellular and humoral responses. Recent studies have shown improved potency of DNA, RNA, and cell-based vaccine candidates when delivered via the intranodal route [82, 83]. Intranodal delivery has also been tested in phase I clinical trials for the treatment of metastatic melanoma [84]. The direct administration of plasmid DNA vaccines into the lymph nodes has shown greater immunogenicity in comparison to parenteral routes [85]. In terms of related immunotherapy treatments, allergic rhinitis is a major health problem typically treated by immunotherapy given subcutaneously [86]. This treatment involves multiple injections to control allergic symptoms although only small amounts of antigen reach the lymph nodes. These studies showed improved results with lower dose of antigen when delivered via intralymph nodes. Beust et al. showed with bee venom allergen that a lower antigen dose combined with CpG oligonucleotides as adjuvant could be directly administered to the lymph nodes resulting in greater immune responses [87]. In general, improving delivery of vaccine antigens through the intranodal route lies in the efficient presentation of antigens and avoiding rapid clearing of these antigens by lymphatic drainage. The encapsulation of the vaccine antigen in biopolymers can help avoid degradation of antigen and produce efficient immune responses when delivered directly into the lymph nodes [88]. A nontoxic tracer is injected peripherally to locate the site of injection in the lymph node. The drainage of tracer dye followed by SC injection enables the visualization of the lymph nodes as discussed in the study performed by Jewell et al. [83]. Vaccine antigens can also be designed to be delivered into lymph nodes using albumin hitchhiking approach [89]. This approach consists of an antigen modified with lipophilic albumin binding domain that helps in delivering an antigen directly to the lymph nodes.

24.3 ORAL DELIVERY OF VACCINES

24.3.1 Currently Available Orally Administered Vaccines

The mucosa lining the gastrointestinal (GI) tract is the major replication and infection sites of many pathogens. Thus, unlike parenteral immunization, oral vaccination can better mimic the natural route of infection and induce both systemic and mucosal immunity. From a vaccine administration point of view, orally delivered vaccines offer ease of administration and lower costs compared to injections. It has been demonstrated that administration of the vast majority of inactivated or recombinant vaccines by the oral route does not elicit protective immune responses, often due to loss and instability of various vaccine antigens [90]. The only examples of successful oral vaccination are with live attenuated vaccines that naturally infect through the gut. One key limitation and technical challenge is to develop a formulation of a live vaccine with sufficient storage stability, or in the case of cholera vaccine, an inactivated bacterial vaccine formulated with or without mucosal adjuvant [24], which at the same time can survive exposure to gastric acid in the stomach.

Examples of marketed orally administered vaccines are listed in Table 24.1. The oral polio vaccine (OPV) has been the most widely used orally delivered vaccine. In the 1950s, live attenuated poliovirus vaccine (IPV) was administered against poliomyelitis in humans, but it was not initially licensed due to its neurotropic effects [24]. This led to the development of different, more attenuated strains of poliovirus resulting in the currently licensed trivalent OPV developed by Sabin. OPV contains live, attenuated poliovirus types 1, 2, and 3 with stabilizers such as magnesium chloride [91]. The recommended dose for OPV is 0.5 ml, which is filled into a disposable plastic tube and given as single dose into the oral cavity. More recently, RotaTeq® (Merck & Co., Inc.) was developed and approved as an orally administered pentavalent live attenuated viral vaccine for the prevention of rotavirus gastroenteritis in infants (see Fig. 24.4a). RotaTeq is formulated in a ready-to-use liquid formulation in a vaccination series consisting of three doses administered at 4- to 10-week intervals starting at 6–12 weeks of ages [24]. A combination of different excipients was

(a)

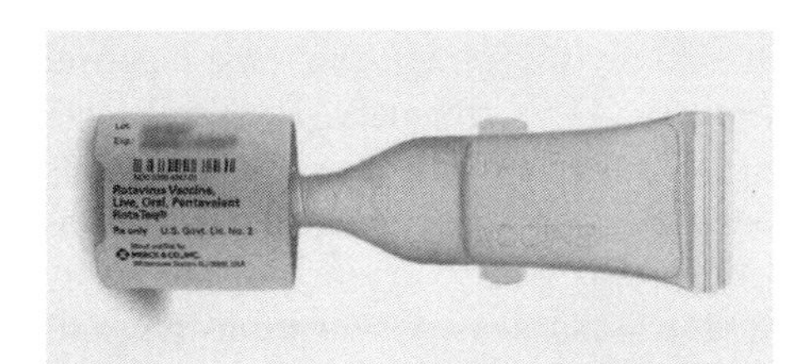

(b)

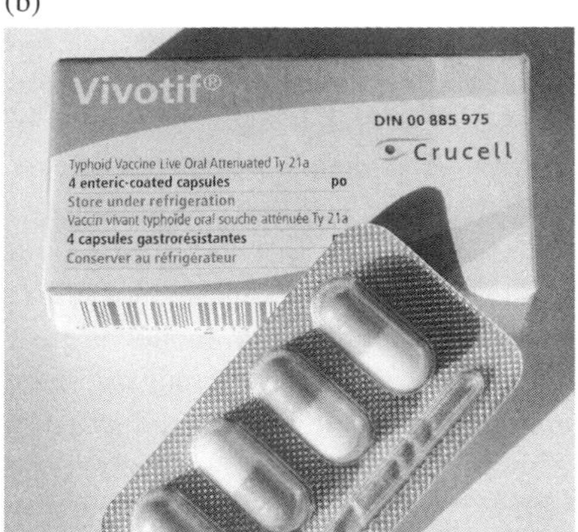

FIGURE 24.4 Different dosage forms for oral administration of vaccines. (a) Liquid formulation in a plastic tube for oral delivery for RotaTeq®, a live, attenuated pentavalent rotavirus vaccine (Merck & Co., Inc.). This figure is used with permission from CDC from Ref. [92]. (b) Enteric coated capsules containing lyophilized powder of live, attenuated bacteria for Vivotif®, a typhoid vaccine marketed by Crucell. This image was acquired from Wikipedia, The Free Encyclopedia [93].

identified to both stabilize the vaccine during long-term storage as a liquid formulation and provide acid-neutralizing capacity to protect the virus from gastric acid during oral administration [94, 95]. Rotarix® (GSK) is another vaccine that is administered orally for the prevention of rotavirus gastroenteritis caused by G1 and non-G1 types [96].

Tablet and capsule formulations have also been developed for oral administration of certain vaccines. For example, an adenovirus vaccine is administered orally and licensed for use in the military. Initially, Wyeth Laboratories supplied the adenovirus vaccine but ceased manufacturing in the 1990s. A manufacturing license was issued to Teva by the FDA in 2011. There are two strains of adenovirus in the vaccine (Ad4 and Ad7) that are lyophilized, formulated as enteric-coated tablets resistant to stomach acid, and administered for dissolution in the upper gastrointestinal tract [24]. In another example, Vivotif® (Berna Biotech, Ltd.) is an orally administered typhoid vaccine containing the attenuated strain Salmonella typhi Ty21a [77, 24]. Vivotif is delivered as either a liquid formulation (after reconstitution of a lyophilized vaccine) or as enteric-coated capsules containing the lyophilized formulation (see Fig. 24.4b). The lyophilized bacteria formulated in capsules have higher storage stability than corresponding liquid formulations [97].

24.3.2 Novel Formulations and Delivery Approaches for Oral Administration

One active area of research to facilitate oral administration of new vaccines is the development of novel mucosal adjuvants. Currently known mucosal adjuvants can be categorized into three general groups: bacterial enterotoxins, TLR ligands, and non-TLR immunostimulants. CT and LT are two bacterial enterotoxin adjuvants that can facilitate potent immune responses but are often associated with serious enterotoxicity. Many studies have been performed to identify nontoxic mutants of CT and LT. One successful example is the double mutant of LT, designated as LT(R192G/L211A). This double mutant (dmLT) retains adjuvanticity but exhibits no enterotoxicity in mice [98]. Recently, orally delivered dmLT was tested in phase I human clinical trials and was shown to be safe and enhanced immunogenicity [99]. TLR agonists represent another major class of potentially useful adjuvants for mucosal vaccination and include monophosphoryl lipid A (TLR4 agonist) [100] and CpG (TLR9 agonist) [101]. Recent studies suggest that the mechanism of TLR agonists may involve the direct activation of B cells or other APCs [102]. A variety of non-TLR immunostimulants have also been identified as potential mucosal adjuvants [102]. These molecules function by activating the innate sensors on various types of immune cells. Examples include the NKT cell activator α-galactosylceramide, recombinant IL-1 family cytokines, dectin-1 agonists, mast cell activators, and M-cell targeting ligands [102].

Several oral vaccine candidates are currently under development using novel adjuvants. One example is an oral vaccine against enterotoxigenic *E. coli* (ETEC) [103], a frequent cause of traveler's diarrhea. Early studies suggested that immunity in the small intestine is of particular importance for the prevention of ETEC-induced diarrhea [104]. Current studies have been focused on identifying protective antigens that can effectively elicit immunity in the small intestine against ETEC. Based on the pathogenic mechanisms of ETEC, virulence factors including heat-labile enterotoxin and the small polypeptide heat-stable enterotoxin and colonization factors have been selected as potential antigens for ETEC vaccination. One clinical study showed that a combination of recombinant CT (a homologue of ETEC heat-labile enterotoxin) and formalin-inactivated *E. coli* expressing major colonization factors can provide effective protection against traveler's diarrhea [103]. Another example of oral vaccine candidates under development are malaria antigens including AMA1, MSP1, and the erythrocyte surface antigen [105, 106]. These antigens have poor immunogenicity in the absence of adjuvants. One study showed that the fusion of AMA1 and MSP1 antigens with CT subunit B can significantly enhance their immunogenicity when delivered orally. These fusion antigens have been shown to induce both systemic and mucosal immune responses [107].

A wide variety of pharmaceutical delivery systems are also being evaluated to facilitate oral vaccination. Synthetic micro- and nanoparticles are the most widely studied since these carriers have the potential to protect encapsulated vaccine antigens from the harsh environment in the gastrointestinal tract and to facilitate antigen uptake in the intestines [108]. Biocompatible materials, such as polylactic-*co*-glycolic acid (PLGA), PLA, and chitosan, have been used to prepare synthetic particulate carriers [109–112]. The compatibility of synthetic particulate carriers with various types of antigens has been examined, and protein antigens have been successfully encapsulated into particulate carriers without extensively losing structural and immunologic integrity [113–115]. It has also been shown that synthetic particulate carriers loaded with DNA plasmids and inactivated bacteria can also induce both mucosal and systemic immune responses [116–118]. To improve the bioavailability of oral vaccines, strategies have been developed to facilitate the transport of particulate carriers across intestinal mucosa such as modifying the surface properties of particulate carriers. For example, PEG or chitosan modification can improve paracellular transport by enhancing the electrostatic interaction of particulate carriers with cell membranes or tight junctions [119, 120]. Another strategy is to conjugate particulate carriers with ligands that can specifically target cells involved in antigen uptake such as intestinal epithelial or M cells [121, 122]. Commonly used ligands include various types of lectins and the binding subunit of *E. coli* heat-labile toxin.

Other pharmaceutical carrier systems that have been evaluated for oral delivery of vaccines include liposomes, virosomes, emulsions, and immune-stimulating complexes (ISCOMs). Liposomes are artificial vesicles composed of a phospholipid shell and an aqueous core. Antigens of various sizes and properties can be incorporated into liposomes, while the surface properties of liposomes can be modified to better control the release of the antigen or to target specific types of cells. Liposomes have been used to deliver protein and DNA vaccine antigens via the oral route [123, 124]. Virosomes are a special type of liposomes composed of not only phospholipids but also viral membrane proteins that can enhance the immune responses induced by their antigen payloads [125]. For example, immunopotentiating reconstituted influenza virosomes (IRIVs) contain membrane proteins of influenza virus and lipids derived from viral, egg, or synthetic lipids [126]. Emulsions are another type of carrier consisting of mainly two types, water-in-oil and oil-in-water, which are used for the delivery of hydrophilic and hydrophobic substances, respectively [127]. Nanoemulsion technologies can prepare emulsions with good colloidal stability ranging in size from 20 to 200 nm [90]. Nanoemulsions are similar in size to natural pathogens, potentially facilitating their uptake by M cells and subsequent presentation by APCs, and have been studied for the delivery of mucosal vaccines [90]. Finally, ISCOMs are spherical structures formed by a mixture of the immunostimulatory fractions from Quillaja saponaria (Quil A), cholesterol, phospholipids, and cell membrane proteins. It has been shown that ISCOMs can facilitate the uptake of antigens from the GI tract and the induction of both systemic and mucosal immunity [128].

The use of biological delivery systems is another area of active research to improve oral vaccinations including bacteria, viruses, viruslike particles (VLPs), and plant cells. Live attenuated strains of orally transmissible bacteria, such as *Salmonella*, *E. coli*, *Shigella*, *Listeria*, and *Lactobacillus*, have been used as biological carriers to orally deliver vaccine antigens [129]. Genes encoding heterologous antigens can be directly integrated into the carrier chromosomes or carried in the form of plasmids. Such bacterial carriers can enter APCs through phagocytosis and release the antigen-encoding DNA into the cytosol. The antigens are then expressed and presented on the surface of APCs. One pitfall of bacterial carriers is the immunogenicity of the carrier itself. It has been shown that immunity against the bacterial carriers will predominate over time, which can compromise the immunity against the vaccine antigens or use of the carrier for other vaccines [130]. Orally transmissible viruses, including adenoviruses, poxviruses, influenza viruses, herpesviruses, and polioviruses, have all been used to orally deliver vaccine antigens [129]. In this approach, antigen-encoding genes are directly inserted into viral genomes and delivered into host cells via viral infection. For safety reasons, these viral carriers are either attenuated or engineered to eliminate the ability to cause diseases and transmission among humans while retaining the ability to infect human cells. One recent study from Steel et al. examined the use of adenovirus vectors for oral vaccination against breast cancers [131]. Another study investigated the safety and immunogenicity of an adenovirus vector vaccine for H5N1 influenza [132]. VLPs have been explored as vectors for oral vaccination. Heterologous antigens can be coexpressed with viral capsids to form a chimeric VLP that presents antigens on the surface [133]. VLPs have also been used to deliver encapsulated DNA vaccine antigens for oral delivery [134]. Finally, vaccine antigens expressed and bioencapsulated in plant cells are protected from the acidic stomach environment. Encapsulated vaccine antigens are released upon the digestion of plant cells by microbes in the intestines. Studies showed that plant cell-delivered oral vaccines can induce both systemic and mucosal immune response and protect against bacterial, viral, and protozoan pathogens or toxin challenge [135]. Plant cell delivery systems avoid the denaturing encapsulation process required for the preparation of polymeric particulate carriers, which often causes significant structural damage to vaccine antigens. Harvested leafs containing vaccine antigens are lyophilized, powdered, and packaged into capsules. A number of plant cell-based oral vaccines are currently under development such as plague, cholera, and malaria vaccines [136].

Finally, there are alternative sites of vaccination available to improve immune responses produced by oral administration. For example, both sublingual and buccal delivery have been gaining interest. For sublingual administration, a drug or vaccine is placed under the tongue, while buccal administration involves placement between the gums and the cheek. These routes have an advantage over GI routes since they avoid degradation by enzymes present in the GI tract. Sublingual and buccal delivery of vaccines may enhance mucosal immunity, a current limitation of the parenteral route [137]. There are now many studies exploring the potential of administering live attenuated virus vaccines delivered by the sublingual route. For example, a live attenuated influenza virus vaccine has been administered to mice through the sublingual route [138], along with several other recombinant virus [139] and bacterial candidate vaccines [140]. A candidate vaccine containing antigens from *Helicobacter pylori* and adjuvanted with dmLT has been evaluated by sublingual immunization in animals [141]. The sublingual route has also been tested for respiratory pathogens such as respiratory syncytial virus (RSV) [142], severe acute respiratory syndrome (SARS) virus [137, 143], and adenovirus (Ad5)-based vaccine candidate for Ebola, the latter in mice and guinea pigs [144]. Sublingual allergy treatment is the only currently approved commercial immunotherapy that is administered through the sublingual route [145]. The buccal route of administration has not been studied as extensively for vaccination. A drug to treat candidiasis is given through the buccal route [146], and there are several buccal patches available for the slow release of drugs in the oral cavity [147].

24.4 NASAL AND AEROSOL DELIVERY OF VACCINES

24.4.1 Currently Available Nasally Administered Vaccines

One of the most common ways that pathogens enter the human body is through mucous membranes of the nose and lungs [148]. Thus, vaccination through the nasal route is an attractive mechanism for delivery of antigen since it mimics some natural infections. The intranasal route is potentially preferred over the oral one for inducing mucosal immune responses due to the need for lower doses of antigen and better antigen stability [149]. Immunization through the nasal route can elicit both mucosal and systemic immune responses [150]. Recent studies in rodents have focused on the role of nasal-associated lymphoid tissue (NALT) for the induction of immune responses upon nasal administration [151]. NALT is the inductive site in the nasal cavity for mucosal immunity and consists of B- and T-cell follicles with specialized M cells. The immune response at this site consists of secretory IgA and innate immunity [152] and has been widely targeted while developing vaccine candidates for delivery by the nasal route.

The only commercially available licensed intranasal vaccine is a live virus influenza vaccine. Flu is a seasonal disease caused by influenza virus, which belongs to Orthomyxoviridae family [153], and results in high levels of morbidity and mortality worldwide each year. Currently, there are three types of licensed vaccines available for protection against influenza infections: live, attenuated virus, inactivated virus, and recombinant HA protein-based vaccines [154, 155]. The live attenuated vaccine marketed under the trade name FluMist™ is given by the intranasal route [156, 157]. Recently, the FDA approved FluMist Quadrivalent (MedImmune, Inc.), consisting of two influenza A strains and two influenza B strains [156]. For nasal delivery of FluMist [158], a glass PFS holds vaccine with an attached nasal delivery device, Accuspray™ (BD Technologies) (see Fig. 24.5a). This device permits an accurate delivery dose with ease of administration. The dose can be divided for delivery into each nostril. This vaccine has recently been shown to be especially effective with children [158].

Although not used for commercial vaccines, liquid formulations of drugs, either aqueous solutions, suspensions, or emulsions [159], are commonly delivered by nasal drops. Many investigational studies looking at nasal vaccination in animals utilize nasal drops. Nasal drops provide a simple way to deliver drugs to the nose but have the disadvantage of lacking dose precision. The drops can also be delivered using a pipette or a glass dropper that could be inserted into the nostril for delivery. One potential problem is partial reexpulsion of the dose through a sneeze or runny nose. Several other devices can also be used such as squeeze bottles and squirt tubes. The squeeze bottles are employed to deliver multiple doses of over-the-counter drugs such as decongestants, but extra care and use of preservatives are required to avoid contamination by microorganisms.

24.4.2 Novel Devices and Formulations for Nasal Administration

For small-molecule drugs, nasal droppers have been widely replaced by metered-dose spray pumps or single-dose pipettes. These devices have not been widely applied to vaccination to date. Metered-dose devices are commonly used to deliver

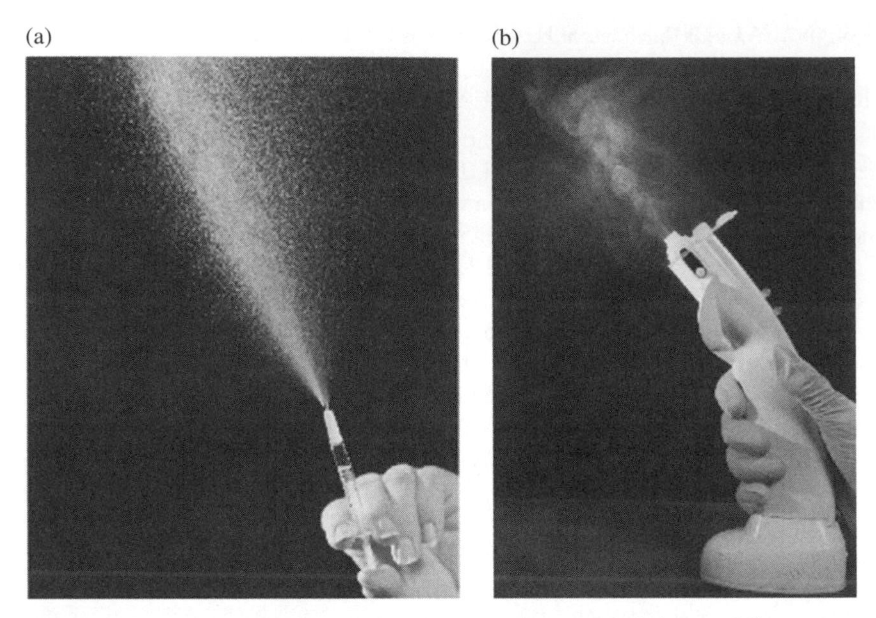

FIGURE 24.5 Devices for intranasal delivery. (a) Intranasal delivery of live attenuated influenza vaccine FluMist™. The vaccine is filled into a glass prefilled syringe with an attached nasal delivery device, Accuspray™ (Becton, Dickinson and Company). (b) Nebulizer (AerovectRx, Inc.) utilizing battery powered piezoelectric energy to drive an aerosol from a disposable drug cartridge. Plotkin [24]. Reproduced with permission from Elsevier.

decongestants/saline and a precise amount of dose from 25 to 200 μl per spray [160]. Several multidose spray pumps are available in market for topical steroids (GSK), osteoporosis (Novartis), and induction of labor (Sigma-Tau Pharmaceuticals, Inc.) [160, 161]. They provide higher reproducibility for the desired dose but require priming and overfill to ensure dose conformity for multiple uses. Single-dose spray devices can be used to overcome these limitations. Other single-dose devices currently being marketed are Imitrex (GSK) and Zomig (Pfeiffer/Aptar) for migraine delivered through the nasal route [162]. In addition, many new devices are currently being developed for nasal delivery. One example is the OptiNose breath-powered devices, involving bidirectional nasal delivery using posterior connection between the nasal passages [151, 163]. We can expect expansion in the use of such devices to deliver vaccines potentially in the near future.

There are also several formulation approaches under development to facilitate vaccination by the nasal route. First, new vaccine antigens are under development, which can be administered intranasally. Studies have shown that the nasal administration of recombinant adenovirus encoding tetanus toxin C (Tc) fragment produced high titers of anti-Tc antibodies after a single dose [164]. DNA vaccines have also been developed for intranasal delivery [157]. Several animal studies have been performed testing DNA-based vaccine antigens with chitosan through the nasal route, for example, against the nucleocapsid of severe acute respiratory syndrome coronavirus (SARS-CoV) [165, 166]. Mucosal immunization of chitosan nanoparticles with plasmid DNA encoding the major surface antigen of HBV in Balb/c mice produced high antibody titers against the viral antigen [167]. Other studies have evaluated adjuvants to enhance immune responses upon nasal immunization including CT B, which produces larger mucosal immune responses in respiratory and genitovaginal tracts [151, 168]. Another example is an oil-in-water emulsion formulation (size smaller than 400 nm) for a mucosal smallpox vaccine [37]. In mice, immunization with nanoemulsions through the nasal route produced high levels of both mucosal and systemic antibodies. Similar formulations are also being tested for a candidate gp 120 vaccine against HIV [169].

In addition, several different types of polymer carriers are being evaluated to enhance mucosal immunity via nasal administration. The two such polymers are poly(lactide-*co*-glycolides) and PLA [170]. Nanoprecipitation and simple emulsification can also be used to encapsulate antigens [171]. Tetanus toxoid, influenza, and HBsAg are among the several antigens that have been encapsulated in PLG and PLA polymers. This technique has been associated with certain limitations due to antigen degradation as a consequence of organic solvent exposure during the encapsulation process. Another approach is the adsorption of vaccine antigens to the surface of polymeric microparticles, often referred to as polymeric lamellar substrate particles (PLSP) [172]. This bypasses exposure of antigens to organic solvents. Several other polymers have also been evaluated including polycaprolactone, polystyrene benzyltrimethylammonium chloride, and poly(methyl methacrylate) (PMMA). PMMA nanoparticles have recently been used for the intranasal delivery of an HIV-1 Tat protein vaccine [173].

Several natural biopolymers have also been employed such as *N*-trimethyl chitosan and sodium alginate. Chitosan-based nanoparticles can be prepared by self-assembly resulting in 160 nm sized particles [174]. Chitosan formulations with vaccine antigens have included studies with recombinant toxins, DNA-based plasmids, and filamentous hemagglutinin [175, 176]. A powder formulation with chitosan nanoparticles was also tested in mice for an influenza subunit antigen through intranasal route [177] and produced higher immunogenic response in comparison to the liquid formulation [178]. Similar studies including human clinical trials with a Norwalk virus VLP have been performed [179]. Chitosan-based formulations may increase the adsorption in the nasal cavity by opening tight junctions protecting the paracellular pathway [180]. Sodium alginate is another natural biopolymer used for intranasal administration of drugs and vaccines [157, 181, 182]. Encapsulation in alginate microspheres using divalent or trivalent ions provides a nontoxic and potentially less expensive delivery system [183]. A study employing delivery of alginate-encapsulated tetanus toxoid in rabbits showed higher levels of antitetanus toxoid IgG in serum upon intranasal administration [184].

Several different types of liposomes have also been evaluated as carriers to enhance mucosal immunity via nasal administration. Liposomes consist of a phospholipid bilayer outer shell with an aqueous core [185]. A variety of antigens can be adsorbed onto the surface of liposomes due to their simultaneous hydrophobic and hydrophilic character. Liposomes can be prepared using reversed-phase evaporation process by dissolving phospholipids in an organic solvent [186]. Studies with surface-modified forms of liposomes including oligomannose-coated liposomes have shown cellular and humoral immune responses through nasal delivery [157, 187]. The intranasal administration of liposomes with membrane proteins of *Bordetella pertussis* was shown to produce IgA and IgG antibodies in the lungs of mice [188]. A liposomal formulation of *Yersinia pestis* has also shown promising results through intranasal delivery [151]. Several lipopeptide-based liposomes have been employed for intranasal delivery of vaccines [189], including virosomes consisting of viral glycoprotein envelopes with higher affinity toward the mucosal surfaces of the respiratory tract [190]. The glycoprotein envelope increases the uptake of vehicles containing antigens by M cells and APCs. Virosomes have been tested for the intranasal delivery of several antigens such as influenza, HIV, and DNA-based vaccines. Influenza virosomes with empty influenza virus shells were tested for the trivalent inactivated influenza vaccine in humans and animal models [191, 192]. Finally, ISCOMs, which are composed of approximately 40 nm colloidal saponin particles, can be used as vaccine delivery systems by stimulating higher immune responses and higher uptake by dendritic cells [189]. The two most important components of ISCOMs are cholesterol and saponin [193]. ISCOMs have been used in combination with several vaccine antigens (e.g., hepatitis B, influenza, and herpes simplex virus type I) to induce local and systemic immune response through intranasal delivery [194].

24.4.3 Devices and Delivery Systems for Aerosol Administration of Vaccines

For aerosol delivery, there are many devices currently available for administration of drugs, but their applicability to vaccines remains investigational. For devices to deliver drug powders, many different powder/aerosol technologies are available and are classified into three general categories: pressurized metered-dose inhalers (pMDIs), dry powder inhalers (DPIs), and nebulizers [195]. pMDIs consist of drug formulations that are packed under pressure with propellant in a canister with valve to allow precise dose delivery. Some pMDI products have been banned due to their chlorofluorocarbon (CFC) propellants [196]. New types of DPIs have been developed based on hydrofluoroalkane (HFA). The first such pMDI to deliver the topical steroid beclomethasone dipropionate (BDP) was approved for allergic rhinitis [197]. There is a nitrogen gas-driven pMDI manufactured by Impel NeuroPharma currently being evaluated in clinical trials [198]. For DPIs, currently available drugs include Aerolizer (Novartis, Basel, Switzerland), Handihaler (Boehringer, Ingelheim, Germany), Diskhaler (GSK, Middlesex, United Kingdom), and Turbuhaler (AstraZeneca, London, United Kingdom). A topical steroid (dexamethasone cipecilate) is being marketed in Japan by Nippon Shinyaku Co., Ltd. for allergic rhinitis [195]. Finally, nebulizers are used to generate small particle sizes and deliver larger doses with minimal patient skill. There are different kinds of nebulizers: VibrENT pulsation membrane nebulizer, Aeroneb Solo vibrating mesh nebulizer, ViaNase atomizer, and the Impel nitrogen-driven atomizer. The marketed nebulizers include pulsation membrane (by PARI) and vibrating mesh nebulizer (by Aerogen) for nasal polyps [199, 200]. Handheld mechanical (from Kurve Tech) nebulizer devices are under clinical trials for the treatment of Alzheimer's disease [201]. AeroEclipse (Trudell Medical International, London, ON, Canada) and HaloLite (Medic-Aid

Limited, West Sussex, United Kingdom) have been developed as breath-actuated nebulizers [159, 202]. The mesh nebulizers provide a potential advantage over other jet nebulizers by delivering more labile biomolecules without loss in activity [203]. Nebulizers can also be coupled with software to control the dose-dependent drug deposition based on patient's inhalation pattern. Two examples of commercially available softwares are Activaero AKITA® and Philips I-neb® nebulizer systems [204]. One example of a nebulizer device is shown in Figure 24.5b.

Many vaccine studies have been performed involving pulmonary delivery of nebulized live attenuated vaccines including measles [205], BCG [206], and rubella [207, 208]. In addition, there have been major efforts to develop devices for aerosol delivery of vaccines, especially measles vaccine. The aerosol-based measles vaccination has been shown to elicit better humoral and cellular immune response compared to the traditional SC route [205]. These devices have also gained interest for mass vaccination campaigns against measles virus [209]. Inhalable dry powder formulations are also being developed for measles vaccines. This formulation can eliminate the reconstitution step of the lyophilized vaccine with water [210]. Dry powder measles formulations are primarily being developed using spray drying [211]. These formulations can be reconstituted for injections or used as inhalable aerosol powders. The particle size of droplets can also be controlled by spray drying to optimize their potential use for mucosal delivery of measles vaccines [212].

24.5 CONCLUSIONS

This chapter reviewed the different routes of administration for commercial vaccines and explored the potential utility of new vaccine delivery devices and formulation technologies for future vaccines. Most vaccines are administered by parenteral injection, in order of frequency, by IM, SC, and ID delivery. Vaccine dosage forms for parenteral administration by needle and syringe include liquid solutions, lyophilized powders followed by reconstitution, or liquid solutions packaged in PFS. New devices are actively being developed to improve parenteral injections including special needles for ID delivery and needle-free technologies. Administering vaccines by the oral or nasal routes is of great interest not only to avoid the use of needles but also to better mimic natural infections and provide better protective immune responses, and several commercial vaccine products are available. Not only must these commercial vaccine dosage forms be optimized for the targeted route of administration, but the vaccine antigens must be stabilized and formulated to ensure long-term storage stability including distribution through the vaccine cold chain. New delivery devices and formulation technologies are currently being assessed in preclinical and clinical studies to better target vaccine antigens to mucosal surfaces by oral, nasal or even aerosol delivery. There are still many technical challenges to the design, development, and production of novel delivery devices and formulation technologies for the complex, labile biological molecules and microorganisms used as vaccine antigens. The opportunities to improve human health by protecting against infectious diseases by vaccination, however, are well worth the effort and investment.

REFERENCES

1. Lambert, W.J., Considerations in developing a target product profile for parenteral pharmaceutical products. *AAPS PharmSciTech*, 2010. **11**(3): p. 1476–81.
2. Kumru, O.S., et al., Vaccine instability in the cold chain: mechanisms, analysis and formulation strategies. *Biologicals*, 2014. **42**: p. 237–59.
3. Volkin, D.B. and C.R. Middaugh, Vaccines as physically and chemically well-defined pharmaceutical dosage forms. *Expert Rev Vaccines*, 2010. **9**(7): p. 689–91.
4. Hem, S.L., et al., Preformulation studies—the next advance in aluminum adjuvant-containing vaccines. *Vaccine*, 2010. **28**(31): p. 4868–70.
5. Kamerzell, T.J., et al., Protein-excipient interactions: mechanisms and biophysical characterization applied to protein formulation development. *Adv Drug Deliv Rev*, 2011. **63**(13): p. 1118–59.
6. Reed, S.G., M.T. Orr, and C.B. Fox, Key roles of adjuvants in modern vaccines. *Nat Med*, 2013. **19**(12): p. 1597–608.
7. Kagan, L., et al., The role of the lymphatic system in subcutaneous absorption of macromolecules in the rat model. *Eur J Pharm Biopharm*, 2007. **67**(3): p. 759–65.
8. Poland, G.A., et al., Determination of deltoid fat pad thickness. Implications for needle length in adult immunization. *JAMA*, 1997. **277**(21): p. 1709–11.
9. Bioject Inc., *The Bioject advantage*. Available from website: http://www.bioject.com/pdf/BiojectAdvantage.pdf (accessed September 26, 2015).
10. Petousis-Harris, H., Vaccine injection technique and reactogenicity—evidence for practice. *Vaccine*, 2008. **26**(50): p. 6299–304.
11. Logomasini, M.A., R.R. Stout, and R. Marcinkoski, Jet injection devices for the needle-free administration of compounds, vaccines, and other agents. *Int J Pharm Compd*, 2013. **17**(4): p. 270–80.
12. Gupta, R.K., Aluminum compounds as vaccine adjuvants. *Adv Drug Deliv Rev*, 1998. **32**(3): p. 155–72.
13. Giudice, E.L. and J.D. Campbell, Needle-free vaccine delivery. *Adv Drug Deliv Rev*, 2006. **58**(1): p. 68–89.
14. Immunization Action Coalition, *Administering vaccines: dose, route, site and needle size*. Available from website: http://www.immunize.org/catg.d/p3085.pdf (accessed August 20, 2014).
15. Cook, I.F., D. Pond, and G. Hartel, Comparative reactogenicity and immunogenicity of 23 valent pneumococcal vaccine administered by intramuscular or subcutaneous injection in elderly adults. *Vaccine*, 2007. **25**(25): p. 4767–74.
16. Michaels, L. and R.W. Poole, Injection granuloma of the buttock. *Can Med Assoc J*, 1970. **102**(6): p. 626–8.
17. Zuckerman, J.N., The importance of injecting vaccines into muscle. Different patients need different needle sizes. *BMJ*, 2000. **321**(7271): p. 1237–8.
18. Kashiwagi, Y., et al., Inflammatory responses following intramuscular and subcutaneous immunization with aluminum-adjuvanted or non-adjuvanted vaccines. *Vaccine*, 2014. **32**(27): p. 3393–401.
19. Turner, P.V., et al., Administration of substances to laboratory animals: routes of administration and factors to consider. *J Am Assoc Lab Anim Sci*, 2011. **50**(5): p. 600–13.

20. Shaw, F.E., Jr., et al., Effect of anatomic injection site, age and smoking on the immune response to hepatitis B vaccination. *Vaccine*, 1989. **7**(5): p. 425–30.
21. Groswasser, J., et al., Needle length and injection technique for efficient intramuscular vaccine delivery in infants and children evaluated through an ultrasonographic determination of subcutaneous and muscle layer thickness. *Pediatrics*, 1997. **100**(3 Pt 1): p. 400–3.
22. Free Stock Photos, *A vaccine and syringe*. CDC/Amanda Mills acquired from Public Health Image Library Website. Available from website: http://www.freestockphotos.biz/stockphoto/16366 (accessed September 26, 2015).
23. Staff, T., *The truth about HPV and gardasil*. 2013. Available from website: www.thelibertybeacon.com/2013/02/04/the-truth-about-hpv-and-gardasil/ (accessed September 26, 2015).
24. Plotkin, S.A., W.A. Orenstein, and P.A. Offit, *Vaccines*. 6th ed. St. Louis, MO: Elsevier Saunders, 2013.
25. Badkar, A., et al., Development of biotechnology products in pre-filled syringes: technical considerations and approaches. *AAPS PharmSciTech*, 2011. **12**(2): p. 564–72.
26. Drain, P.K., C.M. Nelson, and J.S. Lloyd, Single-dose versus multi-dose vaccine vials for immunization programmes in developing countries. *Bull World Health Organ*, 2003. **81**(10): p. 726–31.
27. de Perio, M.A., *Needlestick injuries among employees at a retail pharmacy chain—nationwide*. Cincinnati, OH: CDC, 2012. Available from website: http://www.cdc.gov/niosh/hhe/reports/pdfs/2011-0063-3154.pdf (accessed September 26, 2015).
28. Simon, J.K., et al., Safety, tolerability, and immunogenicity of inactivated trivalent seasonal influenza vaccine administered with a needle-free disposable-syringe jet injector. *Vaccine*, 2011. **29**(51): p. 9544–50.
29. Weniger, B.G., *New High Speed Jet Injectors for Mass Vaccination: Pros and Cons of DCJIs Versus MUNJIs*. WHO Initiative for Vaccine Research: Global Vaccine Research Forum; Montreux, Switzerland 2004.
30. Jackson, L.A., et al., Safety and immunogenicity of varying dosages of trivalent inactivated influenza vaccine administered by needle-free jet injectors. *Vaccine*, 2001. **19**(32): p. 4703–9.
31. Weniger, B.G., *Needle-free injection technology*. Golden, CO: PharmaJet, 2005. Available from website: http://www.sustpro.com/upload/22867/documents/124/70412017962013_15.pdf (accessed September 26, 2015).
32. Parent du Chatelet, I., et al., Clinical immunogenicity and tolerance studies of liquid vaccines delivered by jet-injector and a new single-use cartridge (Imule): comparison with standard syringe injection. Imule Investigators Group. *Vaccine*, 1997. **15**(4): p. 449–58.
33. Mitragotri, S., Current status and future prospects of needle-free liquid jet injectors. *Nat Rev Drug Discov*, 2006. **5**(7): p. 543–8.
34. Mitragotri, S., Immunization without needles. *Nat Rev Immunol*, 2005. **5**(12): p. 905–16.
35. Kis, E.E., G. Winter, and J. Myschik, Devices for intradermal vaccination. *Vaccine*, 2012. **30**(3): p. 523–38.
36. Selgelid, M.J., Bioterrorism and smallpox planning: information and voluntary vaccination. *J Med Ethics*, 2004. **30**(6): p. 558–60.

37. Bielinska, A.U., et al., A novel, killed-virus nasal vaccinia virus vaccine. *Clin Vaccine Immunol*, 2008. **15**(2): p. 348–58.
38. Behbehani, A.M., The smallpox story: life and death of an old disease. *Microbiol Rev*, 1983. **47**(4): p. 455–509.
39. Kennedy, R.B., I. Ovsyannikova, and G.A. Poland, Smallpox vaccines for biodefense. *Vaccine*, 2009. **27**(Suppl 4): p. D73–9.
40. BD, *BD Soluvia™ prefillable microinjection system*. Available from website: http://www.bd.com/pharmaceuticals/products/microinjection.asp (accessed November 21, 2015).
41. Kommareddy, S., et al., Dissolvable microneedle patches for the delivery of cell-culture-derived influenza vaccine antigens. *J Pharm Sci*, 2012. **101**(3): p. 1021–7.
42. Laurent, A., et al., Echographic measurement of skin thickness in adults by high frequency ultrasound to assess the appropriate microneedle length for intradermal delivery of vaccines. *Vaccine*, 2007. **25**(34): p. 6423–30.
43. Alarcon, J.B., et al., Preclinical evaluation of microneedle technology for intradermal delivery of influenza vaccines. *Clin Vaccine Immunol*, 2007. **14**(4): p. 375–81.
44. Holland, D., et al., Intradermal influenza vaccine administered using a new microinjection system produces superior immunogenicity in elderly adults: a randomized controlled trial. *J Infect Dis*, 2008. **198**(5): p. 650–8.
45. Leroux-Roels, I., et al., Seasonal influenza vaccine delivered by intradermal microinjection: a randomised controlled safety and immunogenicity trial in adults. *Vaccine*, 2008. **26**(51): p. 6614–19.
46. Van Damme, P., et al., Evaluation of non-inferiority of intradermal versus adjuvanted seasonal influenza vaccine using two serological techniques: a randomised comparative study. *BMC Infect Dis*, 2010. **10**: p. 134.
47. Gorse, G.J., et al., Intradermally-administered influenza virus vaccine is safe and immunogenic in healthy adults 18–64 years of age. *Vaccine*, 2013. **31**(19): p. 2358–65.
48. Verma, R., et al., Intra-dermal administration of rabies vaccines in developing countries: at an affordable cost. *Hum Vaccin*, 2011. **7**(7): p. 792–4.
49. LectraJet. *LectraJet injection systems*. Available from website: http://www.dantonioconsultants.com/prod_ji_human.htm#A (accessed September 26, 2015).
50. Maa, Y.F., et al., Hepatitis-B surface antigen (HBsAg) powder formulation: process and stability assessment. *Curr Drug Deliv*, 2007. **4**(1): p. 57–67.
51. Schiffter, H., J. Condliffe, and S. Vonhoff, Spray-freeze-drying of nanosuspensions: the manufacture of insulin particles for needle-free ballistic powder delivery. *J R Soc Interface*, 2010. **7**(Suppl 4): p. S483–500.
52. Chen, D., et al., Epidermal immunization by a needle-free powder delivery technology: immunogenicity of influenza vaccine and protection in mice. *Nat Med*, 2000. **6**(10): p. 1187–90.
53. Fuller, D.H., P. Loudon, and C. Schmaljohn, Preclinical and clinical progress of particle-mediated DNA vaccines for infectious diseases. *Methods*, 2006. **40**(1): p. 86–97.
54. Chen, D., et al., Epidermal powder immunization induces both cytotoxic T-lymphocyte and antibody responses to protein antigens of influenza and hepatitis B viruses. *J Virol*, 2001. **75**(23): p. 11630–40.
55. Young, F. and F. Marra, A systematic review of intradermal influenza vaccines. *Vaccine*, 2011. **29**(48): p. 8788–801.

56. Van Damme, P., et al., Safety and efficacy of a novel microneedle device for dose sparing intradermal influenza vaccination in healthy adults. *Vaccine*, 2009. **27**(3): p. 454–9.
57. Braun, R.P., et al., Multi-antigenic DNA immunization using herpes simplex virus type 2 genomic fragments. *Hum Vaccin*, 2008. **4**(1): p. 36–43.
58. Henry, S., et al., Microfabricated microneedles: a novel approach to transdermal drug delivery. *J Pharm Sci*, 1998. **87**(8): p. 922–5.
59. Prausnitz, M.R. and R. Langer, Transdermal drug delivery. *Nat Biotechnol*, 2008. **26**(11): p. 1261–8.
60. Ito, Y., et al., Feasibility of microneedles for percutaneous absorption of insulin. *Eur J Pharm Sci*, 2006. **29**(1): p. 82–8.
61. Ito, Y., et al., Self-dissolving microneedles for the percutaneous absorption of EPO in mice. *J Drug Target*, 2006. **14**(5): p. 255–61.
62. Miyano, T., et al., Sugar micro needles as transdermic drug delivery system. *Biomed Microdevices*, 2005. **7**(3): p. 185–8.
63. Edens, C., et al., Measles vaccination using a microneedle patch. *Vaccine*, 2013. **31**(34): p. 3403–9.
64. Norman, J.J., et al., Microneedle patches: usability and acceptability for self-vaccination against influenza. *Vaccine*, 2014. **32**(16): p. 1856–62.
65. Weldon, W.C., et al., Microneedle vaccination with stabilized recombinant influenza virus hemagglutinin induces improved protective immunity. *Clin Vaccine Immunol*, 2011. **18**(4): p. 647–54.
66. Gill, H.S., et al., Selective removal of stratum corneum by microdermabrasion to increase skin permeability. *Eur J Pharm Sci*, 2009. **38**(2): p. 95–103.
67. Guebre-Xabier, M., et al., Immunostimulant patch containing heat-labile enterotoxin from *Escherichia coli* enhances immune responses to injected influenza virus vaccine through activation of skin dendritic cells. *J Virol*, 2003. **77**(9): p. 5218–25.
68. Kim, Y.C. and M.R. Prausnitz, Enabling skin vaccination using new delivery technologies. *Drug Deliv Transl Res*, 2011. **1**(1): p. 7–12.
69. Bal, S.M., et al., Advances in transcutaneous vaccine delivery: do all ways lead to Rome? *J Control Release*, 2010. **148**(3): p. 266–82.
70. Williams, J., et al., Hepatitis A vaccine administration: comparison between jet-injector and needle injection. *Vaccine*, 2000. **18**(18): p. 1939–43.
71. Snider, D.P., The mucosal adjuvant activities of ADP-ribosylating bacterial enterotoxins. *Crit Rev Immunol*, 1995. **15**(3–4): p. 317–48.
72. Heine, S.J., et al., Intradermal delivery of Shigella IpaB and IpaD type III secretion proteins: kinetics of cell recruitment and antigen uptake, mucosal and systemic immunity, and protection across serotypes. *J Immunol*, 2014. **192**(4): p. 1630–40.
73. Scharton-Kersten, T., et al., Transcutaneous immunization with bacterial ADP-ribosylating exotoxins, subunits, and unrelated adjuvants. *Infect Immun*, 2000. **68**(9): p. 5306–13.
74. Beignon, A.S., et al., Immunization onto bare skin with synthetic peptides: immunomodulation with a CpG-containing oligodeoxynucleotide and effective priming of influenza virus-specific CD4+ T cells. *Immunology*, 2002. **105**(2): p. 204–12.
75. Wilson-Welder, J.H., et al., Vaccine adjuvants: current challenges and future approaches. *J Pharm Sci*, 2009. **98**(4): p. 1278–316.

76. Perrie, Y., et al., Vaccine adjuvant systems: enhancing the efficacy of sub-unit protein antigens. *Int J Pharm*, 2008. **364**(2): p. 272–80.
77. Schlosser, E., et al., TLR ligands and antigen need to be coencapsulated into the same biodegradable microsphere for the generation of potent cytotoxic T lymphocyte responses. *Vaccine*, 2008. **26**(13): p. 1626–37.
78. Mishra, D., et al., Elastic liposomes mediated transcutaneous immunization against hepatitis B. *Vaccine*, 2006. **24**(22): p. 4847–55.
79. Mattheolabakis, G., et al., Transcutaneous delivery of a nanoencapsulated antigen: induction of immune responses. *Int J Pharm*, 2010. **385**(1–2): p. 187–93.
80. Young, S.L., et al., Transcutaneous vaccination with virus-like particles. *Vaccine*, 2006. **24**(26): p. 5406–12.
81. Bal, S.M., et al., Microneedle-based transcutaneous immunisation in mice with *N*-trimethyl chitosan adjuvanted diphtheria toxoid formulations. *Pharm Res*, 2010. **27**(9): p. 1837–47.
82. Senti, G., P. Johansen, and T.M. Kundig, Intralymphatic immunotherapy. *Curr Opin Allergy Clin Immunol*, 2009. **9**(6): p. 537–43.
83. Jewell, C.M., S.C. Lopez, and D.J. Irvine, In situ engineering of the lymph node microenvironment via intranodal injection of adjuvant-releasing polymer particles. *Proc Natl Acad Sci U S A*, 2011. **108**(38): p. 15745–50.
84. Ribas, A., et al., Intra-lymph node prime-boost vaccination against Melan A and tyrosinase for the treatment of metastatic melanoma: results of a phase 1 clinical trial. *Clin Cancer Res*, 2011. **17**(9): p. 2987–96.
85. Maloy, K.J., et al., Intralymphatic immunization enhances DNA vaccination. *Proc Natl Acad Sci U S A*, 2001. **98**(6): p. 3299–303.
86. Hylander, T., et al., Intralymphatic allergen-specific immunotherapy: an effective and safe alternative treatment route for pollen-induced allergic rhinitis. *J Allergy Clin Immunol*, 2013. **131**(2): p. 412–20.
87. von Beust, B.R., et al., Improving the therapeutic index of CpG oligodeoxynucleotides by intralymphatic administration. *Eur J Immunol*, 2005. **35**(6): p. 1869–76.
88. Andorko, J.I., et al., Intra-lymph node injection of biodegradable polymer particles. *J Vis Exp*, 2014. **83**: p. e50984.
89. Liu, H., et al., Structure-based programming of lymph-node targeting in molecular vaccines. *Nature*, 2014. **507**(7493): p. 519–22.
90. Woodrow, K.A., K.M. Bennett, and D.D. Lo, Mucosal vaccine design and delivery. *Annu Rev Biomed Eng*, 2012. **14**: p. 17–46.
91. Furesz, J., Developments in the production and quality control of poliovirus vaccines—historical perspectives. *Biologicals*, 2006. **34**(2): p. 87–90.
92. Hibbs, B.F., E.R. Miller, and T. Shimabukuro, Notes from the field: rotavirus vaccine administration errors—United States, 2006–2013. *MMWR Morb Mortal Wkly Rep*, 2014. **63**(04): p. 81.
93. Wikipedia, The Free Encyclopedia. July 8, 2014. Available from website: http://en.wikipedia.org/w/index.php?title=Ty21a&oldid=616746979 (accessed November 21, 2015).
94. Clark, H.F., et al., Safety, immunogenicity and efficacy in healthy infants of G1 and G2 human reassortant rotavirus vaccine in a new stabilizer/buffer liquid formulation. *Pediatr Infect Dis J*, 2003. **22**(10): p. 914–20.

95. Buckland, B.C., The process development challenge for a new vaccine. *Nat Med*, 2005. **11**(4 Suppl): p. S16–19.
96. Dennehy, P.H., Rotavirus vaccines: an overview. *Clin Microbiol Rev*, 2008. **21**(1): p. 198–208.
97. Ohtake, S., et al., Room temperature stabilization of oral, live attenuated *Salmonella enterica* serovar Typhi-vectored vaccines. *Vaccine*, 2011. **29**(15): p. 2761–71.
98. Norton, E.B., et al., Characterization of a mutant *Escherichia coli* heat-labile toxin, LT(R192G/L211A), as a safe and effective oral adjuvant. *Clin Vaccine Immunol*, 2011. **18**(4): p. 546–51.
99. El-Kamary, S.S., et al., Safety and immunogenicity of a single oral dose of recombinant double mutant heat-labile toxin derived from enterotoxigenic *Escherichia coli*. *Clin Vaccine Immunol*, 2013. **20**(11): p. 1764–70.
100. Fox, C.B., et al., Synthetic and natural TLR4 agonists as safe and effective vaccine adjuvants. *Subcell Biochem*, 2010. **53**: p. 303–21.
101. Huang, C.F., et al., Effect of neonatal sublingual vaccination with native or denatured ovalbumin and adjuvant CpG or cholera toxin on systemic and mucosal immunity in mice. *Scand J Immunol*, 2008. **68**(5): p. 502–10.
102. Lawson, L.B., E.B. Norton, and J.D. Clements, Defending the mucosa: adjuvant and carrier formulations for mucosal immunity. *Curr Opin Immunol*, 2011. **23**(3): p. 414–20.
103. Svennerholm, A.M., From cholera to enterotoxigenic *Escherichia coli* (ETEC) vaccine development. *Indian J Med Res*, 2011. **133**: p. 188–96.
104. Sanchez, J. and J. Holmgren, Virulence factors, pathogenesis and vaccine protection in cholera and ETEC diarrhea. *Curr Opin Immunol*, 2005. **17**(4): p. 388–98.
105. Greenwood, B.M., et al., Malaria. *Lancet*, 2005. **365**(9469): p. 1487–98.
106. Maher, B., Malaria: the end of the beginning. *Nature*, 2008. **451**(7182): p. 1042–6.
107. Davoodi-Semiromi, A., et al., Chloroplast-derived vaccine antigens confer dual immunity against cholera and malaria by oral or injectable delivery. *Plant Biotechnol J*, 2010. **8**(2): p. 223–42.
108. Lowe, P.J. and C.S. Temple, Calcitonin and insulin in isobutylcyanoacrylate nanocapsules: protection against proteases and effect on intestinal absorption in rats. *J Pharm Pharmacol*, 1994. **46**(7): p. 547–52.
109. Martin-Banderas, L., et al., Functional PLGA NPs for oral drug delivery: recent strategies and developments. *Mini Rev Med Chem*, 2013. **13**(1): p. 58–69.
110. Jain, A.K., et al., PEG-PLA-PEG block copolymeric nanoparticles for oral immunization against hepatitis B. *Int J Pharm*, 2010. **387**(1–2): p. 253–62.
111. Nayak, B., et al., Formulation, characterization and evaluation of rotavirus encapsulated PLA and PLGA particles for oral vaccination. *J Microencapsul*, 2009. **26**(2): p. 154–65.
112. Li, L., et al., Potential use of chitosan nanoparticles for oral delivery of DNA vaccine in black seabream *Acanthopagrus schlegelii* Bleeker to protect from *Vibrio parahaemolyticus*. *J Fish Dis*, 2013. **36**(12): p. 987–95.
113. Fattal, E., et al., Biodegradable microparticles for the mucosal delivery of antibacterial and dietary antigens. *Int J Pharm*, 2002. **242**(1–2): p. 15–24.
114. Allaoui-Attarki, K., et al., Protective immunity against *Salmonella typhimurium* elicited in mice by oral vaccination with phosphorylcholine encapsulated in poly(DL-lactide-*co*-glycolide) microspheres. *Infect Immun*, 1997. **65**(3): p. 853–7.

115. Jones, D.H., et al., Orally administered microencapsulated *Bordetella pertussis* fimbriae protect mice from *B. pertussis* respiratory infection. *Infect Immun*, 1996. **64**(2): p. 489–94.
116. Esparza, I. and T. Kissel, Parameters affecting the immunogenicity of microencapsulated tetanus toxoid. *Vaccine*, 1992. **10**(10): p. 714–20.
117. Challacombe, S.J., D. Rahman, and D.T. O'Hagan, Salivary, gut, vaginal and nasal antibody responses after oral immunization with biodegradable microparticles. *Vaccine*, 1997. **15**(2): p. 169–75.
118. Kim, S.Y., et al., Oral immunization with Helicobacter pylori-loaded poly(D, L-lactide-*co*-glycolide) nanoparticles. *Helicobacter*, 1999. **4**(1): p. 33–9.
119. Patel, V.F., F. Liu, and M.B. Brown, Advances in oral transmucosal drug delivery. *J Control Release*, 2011. **153**(2): p. 106–16.
120. Bagan, J., et al., Mucoadhesive polymers for oral transmucosal drug delivery: a review. *Curr Pharm Des*, 2012. **18**(34): p. 5497–514.
121. Gref, R., et al., Surface-engineered nanoparticles for multiple ligand coupling. *Biomaterials*, 2003. **24**(24): p. 4529–37.
122. Foster, N. and B.H. Hirst, Exploiting receptor biology for oral vaccination with biodegradable particulates. *Adv Drug Deliv Rev*, 2005. **57**(3): p. 431–50.
123. Liu, J., et al., Oral vaccination with a liposome-encapsulated influenza DNA vaccine protects mice against respiratory challenge infection. *J Med Virol*, 2014. **86**(5): p. 886–94.
124. Wang, D., et al., Liposomal oral DNA vaccine (mycobacterium DNA) elicits immune response. *Vaccine*, 2010. **28**(18): p. 3134–42.
125. Felnerova, D., et al., Liposomes and virosomes as delivery systems for antigens, nucleic acids and drugs. *Curr Opin Biotechnol*, 2004. **15**(6): p. 518–29.
126. Zurbriggen, R., et al., IRIV-adjuvanted hepatitis A vaccine: in vivo absorption and biophysical characterization. *Prog Lipid Res*, 2000. **39**(1): p. 3–18.
127. Bagwe, R.P., et al., Improved drug delivery using microemulsions: rationale, recent progress, and new horizons. *Crit Rev Ther Drug Carrier Syst*, 2001. **18**(1): p. 77–140.
128. Mowat, A.M., et al., Oral vaccination with immune stimulating complexes. *Immunol Lett*, 1999. **65**(1–2): p. 133–40.
129. Azizi, A., et al., Enhancing oral vaccine potency by targeting intestinal M cells. *PLoS Pathog*, 2010. **6**(11): p. e1001147.
130. Wells, J.M. and A. Mercenier, Mucosal delivery of therapeutic and prophylactic molecules using lactic acid bacteria. *Nat Rev Microbiol*, 2008. **6**(5): p. 349–62.
131. Steel, J.C., et al., Oral vaccination with adeno-associated virus vectors expressing the Neu oncogene inhibits the growth of murine breast cancer. *Mol Ther*, 2013. **21**(3): p. 680–7.
132. Gurwith, M., et al., Safety and immunogenicity of an oral, replicating adenovirus serotype 4 vector vaccine for H5N1 influenza: a randomised, double-blind, placebo-controlled, phase 1 study. *Lancet Infect Dis*, 2013. **13**(3): p. 238–50.
133. Ma, Y. and J. Li, *Vesicular stomatitis* virus as a vector to deliver virus-like particles of human norovirus: a new vaccine candidate against an important noncultivable virus. *J Virol*, 2011. **85**(6): p. 2942–52.

134. Takamura, S., et al., DNA vaccine-encapsulated virus-like particles derived from an orally transmissible virus stimulate mucosal and systemic immune responses by oral administration. *Gene Ther*, 2004. **11**(7): p. 628–35.
135. Daniell, H., et al., Plant-made vaccine antigens and biopharmaceuticals. *Trends Plant Sci*, 2009. **14**(12): p. 669–79.
136. Kwon, K.C., et al., Oral delivery of human biopharmaceuticals, autoantigens and vaccine antigens bioencapsulated in plant cells. *Adv Drug Deliv Rev*, 2013. **65**(6): p. 782–99.
137. Shim, B.S., et al., Sublingual delivery of vaccines for the induction of mucosal immunity. *Immune Netw*, 2013. **13**(3): p. 81–5.
138. Song, J.H., et al., Sublingual vaccination with influenza virus protects mice against lethal viral infection. *Proc Natl Acad Sci U S A*, 2008. **105**(5): p. 1644–9.
139. Kim, S.H., et al., Mucosal vaccination with recombinant adenovirus encoding nucleoprotein provides potent protection against influenza virus infection. *PLoS One*, 2013. **8**(9): p. e75460.
140. Amuguni, J.H., et al., Sublingually administered *Bacillus subtilis* cells expressing tetanus toxin C fragment induce protective systemic and mucosal antibodies against tetanus toxin in mice. *Vaccine*, 2011. **29**(29–30): p. 4778–84.
141. Sjokvist Ottsjo, L., et al., A double mutant heat-labile toxin from *Escherichia coli*, LT(R192G/L211A), is an effective mucosal adjuvant for vaccination against Helicobacter pylori infection. *Infect Immun*, 2013. **81**(5): p. 1532–40.
142. Kim, S., et al., Dual role of respiratory syncytial virus glycoprotein fragment as a mucosal immunogen and chemotactic adjuvant. *PLoS One*, 2012. **7**(2): p. e32226.
143. Shim, B.S., et al., Sublingual immunization with recombinant adenovirus encoding SARS-CoV spike protein induces systemic and mucosal immunity without redirection of the virus to the brain. *Virol J*, 2012. **9**: p. 215.
144. Choi, J.H., et al., A single sublingual dose of an adenovirus-based vaccine protects against lethal Ebola challenge in mice and guinea pigs. *Mol Pharm*, 2012. **9**(1): p. 156–67.
145. Kraan, H., et al., Buccal and sublingual vaccine delivery. *J Control Release*, 2014. **190**: p. 580–92.
146. Gilhotra, R.M., et al., A clinical perspective on mucoadhesive buccal drug delivery systems. *J Biomed Res*, 2014. **28**(2): p. 81–97.
147. Wong, C.F., K.H. Yuen, and K.K. Peh, Formulation and evaluation of controlled release Eudragit buccal patches. *Int J Pharm*, 1999. **178**(1): p. 11–22.
148. Brandtzaeg, P., Humoral immune response patterns of human mucosae: induction and relation to bacterial respiratory tract infections. *J Infect Dis*, 1992. **165**(Suppl 1): p. S167–76.
149. Vajdy, M. and D.T. O'Hagan, Microparticles for intranasal immunization. *Adv Drug Deliv Rev*, 2001. **51**(1–3): p. 127–41.
150. Mestecky, J., et al., Routes of immunization and antigen delivery systems for optimal mucosal immune responses in humans. *Behring Inst Mitt*, 1997. **98**: p. 33–43.
151. Sharma, S., et al., Pharmaceutical aspects of intranasal delivery of vaccines using particulate systems. *J Pharm Sci*, 2009. **98**(3): p. 812–43.
152. Whaley, K.J. and L. Zeitlin, Preventing transmission: plant-derived microbicides and mucosal vaccines for reproductive health. *Vaccine*, 2005. **23**(15): p. 1819–22.

153. Nayak, D.P., et al., Influenza virus morphogenesis and budding. *Virus Res*, 2009. **143**(2): p. 147–61.

154. He, W., et al., Molecular basis of live-attenuated influenza virus. *PLoS One*, 2013. **8**(3): p. e60413.

155. Leroux-Roels, I. and G. Leroux-Roels, Current status and progress of prepandemic and pandemic influenza vaccine development. *Expert Rev Vaccines*, 2009. **8**(4): p. 401–23.

156. Jabbal-Gill, I., Nasal vaccine innovation. *J Drug Target*, 2010. **18**(10): p. 771–86.

157. Zaman, M., S. Chandrudu, and I. Toth, Strategies for intranasal delivery of vaccines. *Drug Deliv Transl Res*, 2013. **3**(1): p. 100–9.

158. Belshe, R.B., et al., Live attenuated versus inactivated influenza vaccine in infants and young children. *N Engl J Med*, 2007. **356**(7): p. 685–96.

159. Djupesland, P.G., Nasal drug delivery devices: characteristics and performance in a clinical perspective-a review. *Drug Deliv Transl Res*, 2013. **3**(1): p. 42–62.

160. Vidgren, M.T. and H. Kublik, Nasal delivery systems and their effect on deposition and absorption. *Adv Drug Deliv Rev*, 1998. **29**(1–2): p. 157–77.

161. Berger, W.E., J.W. Godfrey, and A.L. Slater, Intranasal corticosteroids: the development of a drug delivery device for fluticasone furoate as a potential step toward improved compliance. *Expert Opin Drug Deliv*, 2007. **4**(6): p. 689–701.

162. Rapoport, A. and P. Winner, Nasal delivery of antimigraine drugs: clinical rationale and evidence base. *Headache*, 2006. **46**(Suppl 4): p. S192–201.

163. Djupesland, P.G., et al., Breath actuated device improves delivery to target sites beyond the nasal valve. *Laryngoscope*, 2006. **116**(3): p. 466–72.

164. Shi, Z., et al., Protection against tetanus by needle-free inoculation of adenovirus-vectored nasal and epicutaneous vaccines. *J Virol*, 2001. **75**(23): p. 11474–82.

165. Raghuwanshi, D., et al., Dendritic cell targeted chitosan nanoparticles for nasal DNA immunization against SARS CoV nucleocapsid protein. *Mol Pharm*, 2012. **9**(4): p. 946–56.

166. Shim, B.S., et al., Intranasal immunization with plasmid DNA encoding spike protein of SARS-coronavirus/polyethylenimine nanoparticles elicits antigen-specific humoral and cellular immune responses. *BMC Immunol*, 2010. **11**: p. 65.

167. Khatri, K., et al., Plasmid DNA loaded chitosan nanoparticles for nasal mucosal immunization against hepatitis B. *Int J Pharm*, 2008. **354**(1–2): p. 235–41.

168. Holmgren, J. and C. Czerkinsky, Mucosal immunity and vaccines. *Nat Med*, 2005. **11**(4 Suppl): p. S45–53.

169. Bielinska, A.U., et al., Nasal immunization with a recombinant HIV gp120 and nano-emulsion adjuvant produces Th1 polarized responses and neutralizing antibodies to primary HIV type 1 isolates. *AIDS Res Hum Retroviruses*, 2008. **24**(2): p. 271–81.

170. Pillai, O. and R. Panchagnula, Polymers in drug delivery. *Curr Opin Chem Biol*, 2001. **5**(4): p. 447–51.

171. Makadia, H.K. and S.J. Siegel, Poly lactic-*co*-glycolic acid (PLGA) as biodegradable controlled drug delivery carrier. *Polymers (Basel)*, 2011. **3**(3): p. 1377–97.

172. Jabbal-Gill, I., et al., Polymeric lamellar substrate particles for intranasal vaccination. *Adv Drug Deliv Rev*, 2001. **51**(1–3): p. 97–111.

173. Bettencourt, A. and A.J. Almeida, Poly(methyl methacrylate) particulate carriers in drug delivery. *J Microencapsul*, 2012. **29**(4): p. 353–67.

174. Lee, K.Y., et al., Preparation of chitosan self-aggregates as a gene delivery system. *J Control Release*, 1998. **51**(2–3): p. 213–20.
175. Illum, L., et al., Chitosan as a novel nasal delivery system for vaccines. *Adv Drug Deliv Rev*, 2001. **51**(1–3): p. 81–96.
176. van der Lubben, I.M., et al., Chitosan for mucosal vaccination. *Adv Drug Deliv Rev*, 2001. **52**(2): p. 139–44.
177. Amidi, M., et al., *N*-trimethyl chitosan (TMC) nanoparticles loaded with influenza subunit antigen for intranasal vaccination: biological properties and immunogenicity in a mouse model. *Vaccine*, 2007. **25**(1): p. 144–53.
178. Bacon, A., et al., Carbohydrate biopolymers enhance antibody responses to mucosally delivered vaccine antigens. *Infect Immun*, 2000. **68**(10): p. 5764–70.
179. El-Kamary, S.S., et al., Adjuvanted intranasal Norwalk virus-like particle vaccine elicits antibodies and antibody-secreting cells that express homing receptors for mucosal and peripheral lymphoid tissues. *J Infect Dis*, 2010. **202**(11): p. 1649–58.
180. Cui, F., F. Qian, and C. Yin, Preparation and characterization of mucoadhesive polymer-coated nanoparticles. *Int J Pharm*, 2006. **316**(1–2): p. 154–61.
181. Augst, A.D., H.J. Kong, and D.J. Mooney, Alginate hydrogels as biomaterials. *Macromol Biosci*, 2006. **6**(8): p. 623–33.
182. Singh, M. and D. O'Hagan, The preparation and characterization of polymeric antigen delivery systems for oral administration. *Adv Drug Deliv Rev*, 1998. **34**(2–3): p. 285–304.
183. Coelho, J.F., et al., Drug delivery systems: advanced technologies potentially applicable in personalized treatments. *EPMA J*, 2010. **1**(1): p. 164–209.
184. Tafaghodi, M., S.A. Sajadi Tabassi, and M.R. Jaafari, Induction of systemic and mucosal immune responses by intranasal administration of alginate microspheres encapsulated with tetanus toxoid and CpG-ODN. *Int J Pharm*, 2006. **319**(1–2): p. 37–43.
185. Heurtault, B., B. Frisch, and F. Pons, Liposomes as delivery systems for nasal vaccination: strategies and outcomes. *Expert Opin Drug Deliv*, 2010. **7**(7): p. 829–44.
186. Kersten, G.F. and D.J. Crommelin, Liposomes and ISCOMS as vaccine formulations. *Biochim Biophys Acta*, 1995. **1241**(2): p. 117–38.
187. Ishii, M. and N. Kojima, Mucosal adjuvant activity of oligomannose-coated liposomes for nasal immunization. *Glycoconj J*, 2010. **27**(1): p. 115–23.
188. Brownlie, R.M., et al., Stimulation of secretory antibodies against *Bordetella pertussis* antigens in the lungs of mice after oral or intranasal administration of liposome-incorporated cell-surface antigens. *Microb Pathog*, 1993. **14**(2): p. 149–60.
189. Gregory, A.E., R. Titball, and D. Williamson, Vaccine delivery using nanoparticles. *Front Cell Infect Microbiol*, 2013. **3**: p. 13.
190. Vacher, G., et al., Recent advances in mucosal immunization using virus-like particles. *Mol Pharm*, 2013. **10**(5): p. 1596–609.
191. De Magistris, M.T., Mucosal delivery of vaccine antigens and its advantages in pediatrics. *Adv Drug Deliv Rev*, 2006. **58**(1): p. 52–67.
192. Kammer, A.R., et al., A new and versatile virosomal antigen delivery system to induce cellular and humoral immune responses. *Vaccine*, 2007. **25**(41): p. 7065–74.
193. Cox, J.C., A. Sjolander, and I.G. Barr, ISCOMs and other saponin based adjuvants. *Adv Drug Deliv Rev*, 1998. **32**(3): p. 247–71.

194. Morein, B., et al., ISCOM: a delivery system for neonates and for mucosal administration. *Adv Vet Med*, 1999. **41**: p. 405–13.
195. Lu, D., Recent development in aerosol devices for pulmonary vaccine delivery. *Beijing Da Xue Xue Bao*, 2012. **44**(5): p. 683–7.
196. Hankin, C.S., et al., Medical costs and adherence in patients receiving aqueous versus pressurized aerosol formulations of intranasal corticosteroids. *Allergy Asthma Proc*, 2012. **33**(3): p. 258–64.
197. Meltzer, E.O., et al., Safety and efficacy of once-daily treatment with beclomethasone dipropionate nasal aerosol in subjects with perennial allergic rhinitis. *Allergy Asthma Proc*, 2012. **33**(3): p. 249–57.
198. Hoekman, J.D. and R.J. Ho, Enhanced analgesic responses after preferential delivery of morphine and fentanyl to the olfactory epithelium in rats. *Anesth Analg*, 2011. **113**(3): p. 641–51.
199. Moller, W., et al., Nasally inhaled pulsating aerosols: lung, sinus and nose deposition. *Rhinology*, 2011. **49**(3): p. 286–91.
200. Vecellio, L., et al., Deposition of aerosols delivered by nasal route with jet and mesh nebulizers. *Int J Pharm*, 2011. **407**(1–2): p. 87–94.
201. Craft, S., et al., Intranasal insulin therapy for Alzheimer disease and amnestic mild cognitive impairment: a pilot clinical trial. *Arch Neurol*, 2012. **69**(1): p. 29–38.
202. Dolovich, M., New propellant-free technologies under investigation. *J Aerosol Med*, 1999. **12**(Suppl 1): p. S9–17.
203. Hatley, R. and J. Pritchard, Comment on: devices and formulations for pulmonary vaccination. *Expert Opin Drug Deliv*, 2013. **10**(11): p. 1593–5.
204. Chan, J.G., et al., Advances in device and formulation technologies for pulmonary drug delivery. *AAPS PharmSciTech*, 2014. **15**(4): p. 882–97.
205. Wong-Chew, R.M., et al., Immunogenicity of aerosol measles vaccine given as the primary measles immunization to nine-month-old Mexican children. *Vaccine*, 2006. **24**(5): p. 683–90.
206. Cohn, M.L., C.L. Davis, and G. Middlebrook, Airborne immunization against tuberculosis. *Science*, 1958. **128**(3334): p. 1282–3.
207. Lu, D. and A.J. Hickey, Pulmonary vaccine delivery. *Expert Rev Vaccines*, 2007. **6**(2): p. 213–26.
208. Ganguly, R., et al., Rubella immunization of volunteers via the respiratory tract. *Infect Immun*, 1973. **8**(4): p. 497–502.
209. Coates, A.L., et al., How many infective viral particles are necessary for successful mass measles immunization by aerosol? *Vaccine*, 2006. **24**(10): p. 1578–85.
210. Burger, J.L., et al., Stabilizing formulations for inhalable powders of live-attenuated measles virus vaccine. *J Aerosol Med Pulm Drug Deliv*, 2008. **21**(1): p. 25–34.
211. Ohtake, S., et al., Heat-stable measles vaccine produced by spray drying. *Vaccine*, 2010. **28**(5): p. 1275–84.
212. Wang, S.H., et al., Dry powder vaccines for mucosal administration: critical factors in manufacture and delivery. *Curr Top Microbiol Immunol*, 2012. **354**: p. 121–56.

25

DELIVERY OF GENES AND OLIGONUCLEOTIDES

Charles M. Roth

Department of Chemical & Biochemical Engineering, Department of Biomedical Engineering, Rutgers, The State University of New Jersey, Piscataway, NJ, USA

25.1 INTRODUCTION

The concept of gene therapy, originally conceived in the early 1970s as the introduction of genetic material to treat single-gene disorders, was expected to revolutionize the practice of medicine [1]. Over 40 years later, implementation has been slower than expected, but a broader spectrum of genetic therapeutic possibilities has emerged and is being pursued, with tremendous potential for having a significant impact in a number of recalcitrant disorders. In addition to the ability to restore gene function via gene transfer, the repertoire of nucleic acid therapeutics has expanded markedly to include exon skipping and other forms of gene correction, the use of antisense oligonucleotides and short interfering RNA (siRNA) for gene silencing, and both microRNA (miRNA) mimics and antagonists ("antagomirs") to alter the regulatory activities related to miRNA [2, 3]. These methods are ubiquitous in the research laboratory, as they provide the tools to answer a variety of questions about gene function and biological mechanisms. In similar fashion, RNA interference screens are becoming more widespread tools in the identification and validation of drug targets for pharmaceutical development [4]. Clinical translation has been slower, with essentially no US FDA approvals for gene transfer to date. One aptamer was approved in 2004 [5], and two antisense drugs were approved, one each in 1998 [6] and 2013 [7]. Another antisense therapeutic with orphan drug designation is seeing

Drug Delivery: Principles and Applications, Second Edition. Edited by Binghe Wang, Longqin Hu, and Teruna J. Siahaan.

clinical use under the named patient program. These limited clinical successes highlight both the feasibility of nucleic acid therapeutics and the barriers hindering more rapid clinical implementation.

Clinical translation of gene modification approaches into nucleic acid therapeutics presents a number of challenges. Some of these relate to the nature of the therapeutic and its pharmacodynamics and safety profile. Yet, the biggest challenge is delivery. Nucleic acids are not significantly present in the circulation where they are rapidly degraded by nucleases, and even chemically modified nucleic acids exhibit less than optimal pharmacokinetic profiles. As a result, the formulation of nucleic acids with delivery vehicles (*vectors*) that can provide protection, transport to the site of action, and release is needed to advance nucleic therapeutics into a wider range of applications.

Viruses have evolved over tens of thousands of years to infect mammalian organisms including humans. However, safety issues have plagued the development of viral gene therapies [8]. Nonetheless, there is much to be learned from virus structure and mechanisms of infection toward the development of synthetic, nonviral delivery vectors. However, the delivery barriers and challenges are formidable. As a result, despite considerable progress in understanding the barriers and developing multifunctional materials for delivery, the need remains for continued research into nucleic acid delivery mechanisms and development of vectors that are efficient, safe, and appropriate for current and emerging therapeutic opportunities. This chapter highlights the major mechanisms of delivery as well as best practices in delivery vector design and development.

25.2 SYSTEMIC DELIVERY BARRIERS

Nucleic acids encounter a number of formidable obstacles between the site of administration and the site of action. In this light, it is perhaps not surprising that the solitary-approved aptamer therapeutic and the first antisense therapeutic were for ocular applications, where the privileged environment of the eye simplifies the delivery problem. Nonetheless, the majority of genes and oligonucleotides currently under development are formulated for systemic administration. Without a carrier, even chemically modified oligonucleotides exhibit extensive plasma protein binding and are not distributed extensively to organs other than the kidneys and liver [9]. Partially as a result, there is increasing emphasis on the use of carriers for nucleic acid delivery. A carrier must avoid filtration or immune cell recognition, protecting the nucleic acid cargo until it reaches the tissues [10–12]. Furthermore, it may facilitate entry into the cell, intracellular trafficking, and eventually unpackage, or release, of the nucleic acid so that it may find its target mRNA, in the case of oligonucleotides, or be transcribed in the case of plasmid DNA [13]. Many of the issues encountered in nucleic acid delivery are common to other forms of drug delivery, but two that are relatively unique to nucleic acids are the high susceptibility to degradation (depending on chemistry) and the polyelectrolyte nature of nucleic acids.

A wide variety of nucleases exist in plasma as well as in the cytosol of cells that serve to cleave nucleic acids. Single-stranded natural phosphodiester DNA is so extremely susceptible to exonuclease degradation that it is not used even for *in vitro* antisense experiments, being replaced initially by phosphorothioates and subsequently by 2′-*O*-methyl, 2′-*O*-methoxyethyl, and related chemical modifications [14]. Similarly, siRNAs are recognized and rapidly degraded by serum nucleases [15] and plasmid DNA by DNases present in serum as well as the cytosol [16, 17], with increased stability observed for supercoiled as opposed to linear DNA. Only for siRNAs have unmodified chemistries been utilized in human clinical trials, and in this case also the trend is toward chemically modified oligonucleotides [18].

The anionic charge on nucleic acids presents a particular challenge to formulating them for delivery. Charged species do not permeate cell membranes by passive means, and in fact polyanionic species are expected to experience electrostatic repulsion for the negatively charged cell membranes and extracellular matrix (ECM) components. Furthermore, they are apt to bind to proteins in plasma as well in ECM. These reasons contribute to the motivation for nucleic acid delivery vector development. Most carriers for nucleic acid delivery are built upon cationic polymers or lipids, as these provide a means to bind to the nucleic acid as well as to neutralize its charge. Unfortunately, net positive polyelectrolyte complexes, as often result from the association of nucleic acid and carrier, may also associate with biological macromolecules such as glycosaminoglycans [19], leading to sequestration and reducing efficiency of transport.

25.2.1 Viruses: Learning from Nature

Evolution has endowed viruses with the ability to deliver a nucleic acid cargo, their own genome, to host cells. Various classes of viruses exist, whose sizes, shapes, and structural components represent a significant diversity. Nonetheless, they share several common features. They all possess the ability to bind and package nucleic acids into an efficiently compacted structure, which is then protected from the exterior environment by a lipid bilayer. Viruses all have the means to enter cells via mechanisms that restrict entry to a defined tropism. They often carry bioactive elements that facilitate unpackaging of the nucleic acids within the infected cell, entry of DNA into the nucleus, and integration of viral DNA into the host genome [8, 20]. In addition, they are all nanoparticles, ranging in characteristic dimensions from 20 to 300 nm.

Given that viruses have had many thousands of years to evolve their delivery capabilities, why not simply use viruses for delivery? Indeed, viruses are still the preferred route for clinical gene therapy protocols. To date, synthetic materials have not been developed that match the efficiency of their viral counterparts with respect to delivering genes into cells in a disease setting. However, there are as yet no FDA-approved gene therapies, and a number of serious safety issues have arisen over the years [8]. Furthermore, recombinant viruses used for gene therapy are severely limited in the size of the cargo they can deliver. As such, nonviral delivery approaches have also been explored extensively in laboratory and preclinical models, and these may see increasing opportunities for clinical evaluation. In the case of oligonucleotide-based therapeutics, there is a clearer advantage toward the use of engineered delivery systems. Rather than trying to transduce a cell with a single copy of a gene, the goal is to deliver a sufficient number of oligonucleotides to create a thermodynamic driving force for either direct binding to the target mRNA (antisense oligonucleotides, ribozymes) or incorporation into the RISC complex for presentation (siRNA, miRNA) [21]. This is more easily achieved when the oligonucleotides are self-assembled into a delivery construct rather than genetically encoded into a virus.

Thus, in the remainder of this chapter, the focus is primarily on nonviral (engineered) nucleic acid delivery systems. Yet, we are mindful in the construction of nucleic acid delivery vectors of the mechanisms that viruses have evolved for cellular delivery, and we try to incorporate these as design elements as much as possible. Viruses self-assemble into homogeneous particles that incorporate nucleic acids and protect them until their eventual release in the cell. They are able to co-opt cellular receptors to mediate entry, typically through viral envelope proteins. Furthermore, they have elements that direct intracellular trafficking, mediating escape from endosomes where they are part of the entry route, and, for genes, unlocking entry into the cell nucleus.

25.2.2 Materials for Nucleic Acid Delivery

The two main classes of materials used for delivery of genes or oligonucleotides are polymers and lipids. These may be either naturally derived, purely synthetic, or hybrid (i.e., formed by chemical modification of natural materials). Biological molecules also have been utilized either as carriers or to provide recognition elements to aid in the delivery to a particular type of cell or tissue.

Some of the earliest materials employed to aid in nucleic acid delivery were cationic polymers, such as poly-L-lysine (PLL) and poly(ethyleneimine) (PEI) [22–25]. These polymers have high charge densities, and PEI exhibits other favorable attributes including branched as well as linear architectures and the presence of multiple types of amines to provide pH buffering [26]. Dendrimers provide extremely high charge densities and present functional groups at the exterior that can be modified to tailor their properties [22, 27–30]. Key variables relating to cationic polymers are their molecular weight and architecture (branching pattern) [31]. Higher molecular weight cationic polymers have greater avidity for binding to anionic nucleic acids due to cooperative binding interactions [32, 33]. This can be an advantage in reducing the amount of polymer needed to facilitate nucleic acid delivery. However, complexes formed from higher molecular weight polyelectrolytes tend to be more difficult to *unpackage* or dissociate, once the carrier has delivered its cargo inside of the cell [24, 34]. Furthermore, it has been observed extensively that higher molecular weight polycations exhibit greater cytotoxicity [35, 36]. Thus, the lowest molecular weight polycation with favorable delivery properties should be utilized, or in many instances degradable polymers have been employed as a solution to this issue [37].

Lipids were soon found to provide a complementary class of materials for nucleic acid delivery. Due to their amphiphilic nature, lipids self-assemble into higher-order structures and have moieties suitable for interactions both with nucleic acids and cellular membranes. Mixtures of cationic lipids with neutral or zwitterionic lipids are used to tune the charge density, association with nucleic acids, and interactions with biological membranes, often by creating phase instabilities [38]. Cholesterol has also proven a useful excipient in stabilizing liposomes [39].

For both polymer-based and lipid-based carriers, interaction with serum proteins is a major issue. The binding of anionic serum proteins such as albumin to carriers promotes aggregation and opsonization, leading to rapid clearance and poor distribution. This problem can be ameliorated by shielding the nanoparticles in a hydrophilic polymer coating that is not favorable for protein binding. This type of shielding layer is most frequently formed using poly(ethylene oxide) (PEO or PEG), which is used extensively in drug delivery and biomaterial applications for similar purposes. The main design variables are the length (molecular weight) of PEG and its density of grafting onto the polymers or lipids within the nanoparticle [40, 41]. In a systematic study of these two variables in the use of PEI-PEG for siRNA delivery, it was determined that the molecular weight of PEG must be at least 2000 Da to ensure stable complex formation, and a PEG molecular weight of 5000 Da proved optimal for siRNA-mediated gene silencing in cells [42].

25.2.3 Characterization of Nanoparticles

Generally, polymers or lipids assemble spontaneously with nucleic acids to form nanostructures [43–45]. At this size scale, association of entities cannot be observed or assumed and thus must be measured in some manner. The size and charge of the resulting structures are critical attributes characterizing their suitability for biological or *in vivo* use, and related assays providing information on stability to environmental conditions or biochemical challenge are also useful.

The first test that is typically performed to assess whether there is an interaction between a polymer or lipid and a nucleic acid is the electrophoretic gel shift assay. Varying proportions of the carrier and nucleic acid are mixed to form complexes, and a small aliquot of each is pipetted into the wells of an electrophoresis gel and run together. Compared to a control of nucleic acid only, which will migrate readily in an agarose gel (whose density will depend on the size of the nucleic acid being investigated), complexes of greater size will migrate more slowly in the gel, and indeed nanostructures will be too large to migrate at all in the gels employed for this assay [46]. The nucleic acids are detected using any of a variety of dyes that are appropriate for the nucleic acid chemistry (e.g., DNA vs. RNA; double-stranded vs. single-stranded). Sometimes, the dyes can be used without electrophoresis in a dye exclusion assay. If the nucleic acid is shielded from the solution, including dye, by incorporation into a physical complex, then the dye will be unable to bind nucleic acid and produce fluorescence, whereas nucleic acid in solution (not in a complex) will be bound by the dye, producing fluorescence (Fig. 25.1a) [23, 47].

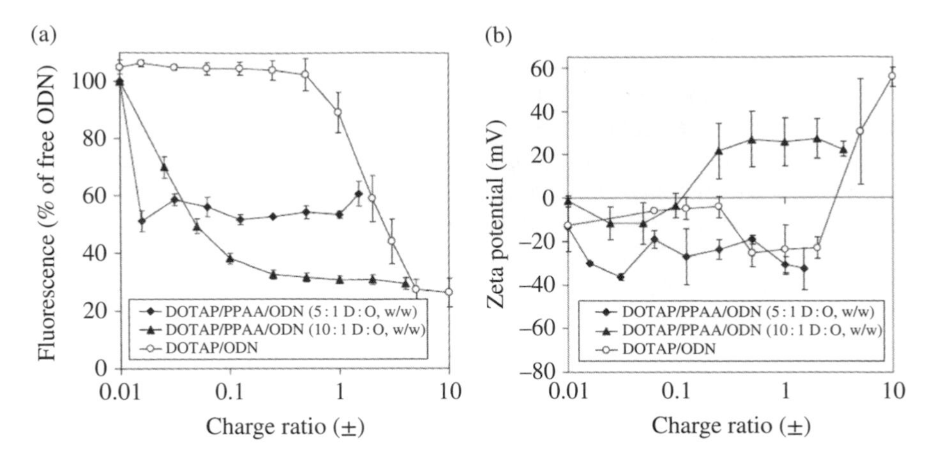

FIGURE 25.1 Characterization of ternary complexes for oligonucleotide delivery. (a) DNA accessibility in complexes of the cationic liposomes DOTAP with poly(propylacrylic acid) (PPAA) and oligodeoxynucleotide (ODN). A diluted solution of the single-stranded DNA-binding reagent OliGreen was added to the complexes, and the decrease of OliGreen fluorescence was measured at excitation and emission wavelengths of 485 and 530 nm, respectively. Means of three separate experiments ± SEM are shown. (b) Zeta potentials of DOTAP/PPAA/ODN complexes. Aliquots (120 µl) of complexes were diluted with 1.5 ml HBS before their zeta potentials were determined by phase analysis light scattering. Means of two separate experiments ± SEM are shown.

Once it is established that a complex has been formed, properties of the complexes such as size and charge are ascertained. There are a large number of techniques available for measuring particle size. Perhaps the simplest is dynamic light scattering (DLS), which can be used on particles in solution and provides not just an average size but a distribution. However, most DLS instruments require approximately a milliliter of solution, and a particle shape (e.g., spherical) must be assumed; furthermore, interference may be encountered with serum-containing media. Techniques that provide a more direct view of particle morphology include transmission electron microscopy (TEM) and atomic force microscopy (AFM). The latter techniques are very useful but may not be available in all labs and require immobilization of the particles and thus may be challenging or impossible to perform in the natural environment of the particles.

Assessment of the charge on particles can be made by zeta potential measurement, which is also performed in solution and can frequently be performed in tandem with DLS size measurements. The net zeta potential on the particles will reflect the overall charge from the relative composition of positive and negative charges within the particles, which may be different from the relative composition of positive and negative charges in solution. This can be particularly useful in assessing changes to particles as environmental conditions changes, such as pH, or as additional moieties are added to a particle surface. For example, the adsorption of anionic polymers onto lipoplexes has been evaluated in this way [48–50] (Fig. 25.1b).

Several common biophysical assays of polymer or lipid–nucleic acid complexes utilize these assays in conjunction with chemical or biochemical challenge. For instance, a number of groups are interested in developing pH-sensitive carriers that alter their size, membrane interactions, or release characteristics under low pH environments encountered in tumors or the endosomal compartments of cells. The pH responsiveness of the carriers can thus be evaluated biophysically in a controlled environment before testing in cells or animals. Similarly, it has been observed that delivery vectors must be sufficiently stable to not fall apart in serum or upon encountering ECM molecules, yet sufficiently labile to allow unpackaging or dissociation of the nucleic acid from its carrier. The effect of various polymer compositions on the stability and unpackaging can be assessed, for example, in a heparin dissociation assay (Fig. 25.2). In this assay, complexes are exposed to heparin or similar ECM molecules, with the stability of the complexes measured by gel shift or dye exclusion techniques [23, 51].

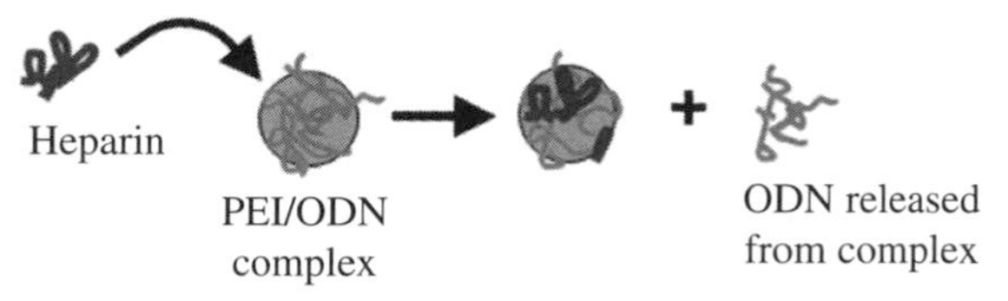

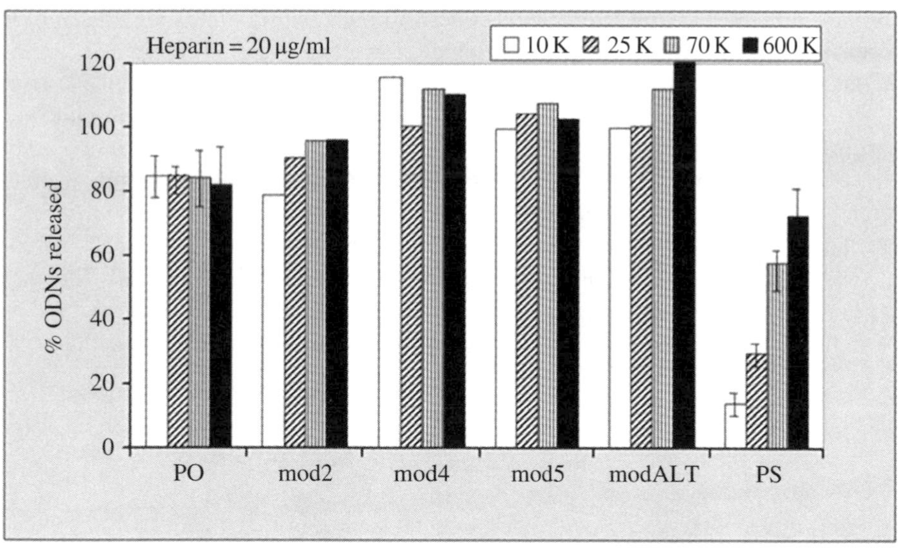

FIGURE 25.2 Heparin dissociation assay. The anionic polyelectrolyte heparin is added to polymer–nucleic acid complexes. The release of nucleic acid is a measure of the unpackaging capability of the complexes. In the example shown here, oligodeoxynucleotides (ODNs) with varying backbones—phosphodiester (PO), phosphorothioate (PS), or partially PS (with mod2, for example, indicating that two linkages on each end are phosphorothioate, and modALT, indicating alternating PO and PS linkages throughout)—were incubated with PEI of varying molecular weight. Results show that PS ODNs are resistant to heparin-induced dissociation compared to PO or mixed-backbone ODNs.

25.2.4 Targeted Delivery of Nucleic Acids

Drugs that are administered systemically travel through the body and distribute to varying extents across all tissues, but they are generally intended to act on only one tissue. A major role of a drug delivery vehicle can thus be to direct the drug preferentially to the desired tissue, not only to increase the concentration of drug at the site of action but also to reduce the concentration on other tissues, which often induces undesired side effects. This principle of targeted drug delivery is applicable to nucleic acid-based drugs as well.

There are two main ideas related to drug targeting. The most common is *active targeting*, the first step of which is to first identify a biomarker (receptor) that is overexpressed on the target cell type relative to other tissues. A ligand is obtained that combines affinity and specificity for the receptor. The ligand is attached, often covalently but sometimes via simple physical adsorption, to the carrier (Fig. 25.3). An interesting variation for oligonucleotide therapeutics is the creation of a multifunctional nucleic acid consisting of an aptamer selected as the ligand along with an siRNA or other bioactive oligonucleotide component [52]. The carrier should be one that has limited nonspecific interactions with nontarget cells so that the ligand–receptor interaction becomes the dominant one governing distribution and cellular uptake of the carrier with its nucleic acid cargo. Alternatively, the ligand may be conjugated directly to the cargo [53]; this is more feasible for oligonucleotides than for plasmid DNA. Despite tremendous interest in nanoparticle-mediated delivery in recent years, advances in molecular conjugates for active targeting continue including advancement of GalNAc into clinical trials [54, 55].

A common misconception about targeted delivery is that the ligand-endowed carrier has the ability to "find" its target among all the tissues and cells in the body. However, this is not the case. Rather, the ligand–receptor interaction merely enables retention when the ligand-bearing carrier and receptor-expressing cell do have an encounter. As such, it is essential that part of the targeting strategy must include mechanisms that allow the carrier to come into contact with the target cell. This may explain why RGD peptides, which target many subfamilies of integrins and in particular $\alpha_v\beta_3$ integrins that tend to be overexpressed on tumor neovasculature as

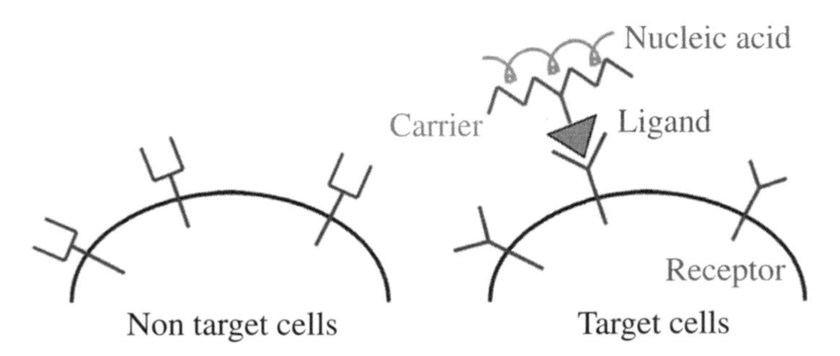

FIGURE 25.3 Active targeted delivery. Through a binding agent (shown) or via direct conjugation, a ligand is able to shuttle its nucleic acid cargo selectively to cells bearing the corresponding biological receptor, with comparatively less uptake in cells absent the receptor.

well as on many tumor cells, have been successful ligands for targeting tumors despite the fact that integrins are expressed widely in the vasculature and across other tissues.

The targeted delivery concept has been applied most frequently to cancer, where the cytotoxic mechanism of action places a premium on selective delivery to cancer cells. Often, tumor targeting relies on *passive targeting* instead of or in addition to active targeting. Passive targeting involves exploiting a nonspecific feature of the carrier for selective delivery to the target tissue. For tumors, selective uptake of particles in the nanoscale range of roughly 20–400 nm can be achieved via the *enhanced permeability and retention* (EPR) effect, which results from a combination of the leakiness of the immature tumor vasculature and incomplete lymphatic drainage in a developing tumor [56, 57]. An example of such an approach is the cyclodextrin-based nanoparticles that incorporate transferrin ligands presented at the end of PEG chains, developed by the Davis group [58, 59]. This delivery system was utilized in the first siRNA nanomedicine introduced in human clinical trials and resulted in demonstrable gene silencing in humans [60, 61].

Some noncancerous organs may also be amenable to a combination of passive and active targeting. The liver is a major site of nanoparticle accumulation within the space of Disse created among the liver sinusoidal fenestrations. This acts to retain nanoparticles for sufficient time that they may be phagocytosed by Kupffer cells (liver macrophages) or enter hepatocytes, which can be facilitated by active targeting of the asialoglycoprotein receptor (ASGPR), which is expressed exclusively on hepatocytes [62]. This approach forms the basis of the "Dynamic PolyConjugate" approach, in which *N*-acetylgalactosamine is used as the ligand for ASGPR targeting, PEG provides shielding, and the backbone polymer degrades in the low pH environment in such a way as to promote endosomal escape [63]. This technology is currently in Phase II clinical trials for hepatitis B [64]. More generally, a preponderance of clinical siRNA drug development efforts are being directed toward liver diseases, due to the preferential accumulation of nanoparticle carriers in this organ [18].

25.3 CELLULAR DELIVERY BARRIERS

Even after overcoming systemic barriers and entering the target cells, a vector must still make its cargo available for expression (gene delivery) or to silence its target mRNA (antisense or siRNA). Two key barriers in achieving this task are escape from endosomes and unpackaging of the carrier from the nucleic acid (Fig. 25.4).

25.3.1 Endosomal Escape

Nucleic acids with their carriers typically enter cells via some form of endocytosis [65]. Endosomal trafficking leads to fusion with lysosomes in which the contents are degraded. Exocytosis may also act to remove vectors and their nucleic acids from cells [66]. In order for therapeutic nucleic acids to exert their activity, they must escape the endosome. Even for highly optimized siRNA formulations currently in clinical trials,

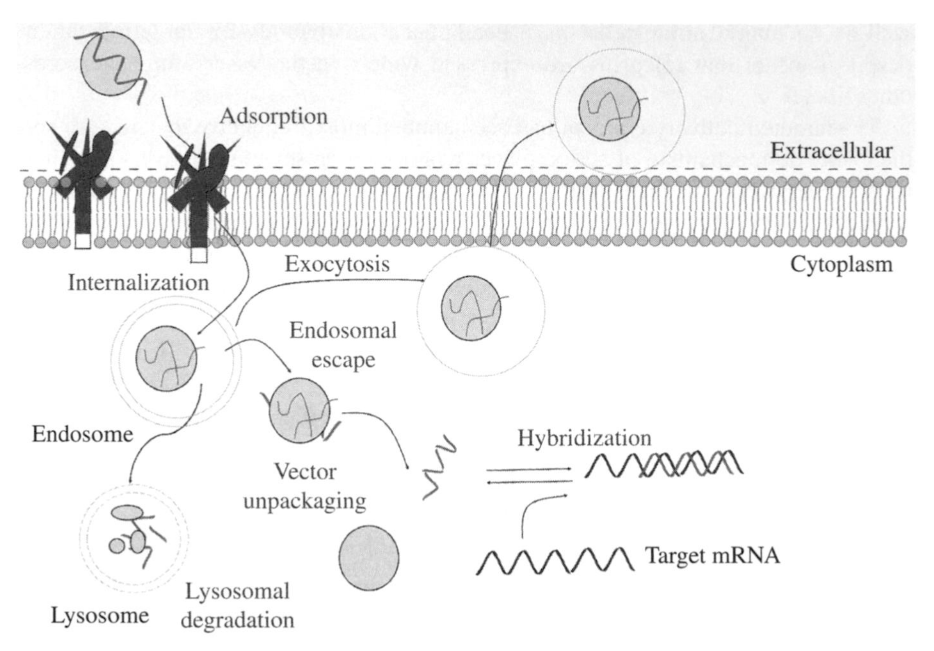

FIGURE 25.4 Intracellular delivery barriers. Nucleic acid vectors typically enter cells via endocytosis or via micropinocytosis leading eventually to fusion with the endosomal pathway. Escape from the endosomes before exocytosis or degradation in lysosomes is necessary for the nucleic acid to exert its biological activity, as is dissociation (unpackaging) from the carrier.

for example, the efficiency of escape may be only 1–2% [62]. Thus, most delivery systems are now engineered with a mechanism to promote endosomal escape. There exist a number of strategies to achieve endosomal escape, including membrane fusion, osmotic expansion, and the use of amphipathic peptides or synthetic polymers that respond to the low pH environment within endosomes.

Some liposomal systems are designed to disrupt endosomal membranes. For example, cationic lipids have been developed with architectures such that ion pairing with anionic membrane phospholipids would compromise the phase stability of the endosomal membrane and thus promote membrane leakiness and endosomal escape of the contents [67]. A common strategy employed with polymeric delivery systems is to exploit the *proton sponge effect* for osmotic destabilization [68]. For polymers such as branched PEI that possess a mixture of primary, secondary, and tertiary amines, acidification of the endosome drives titration of some of the unprotonated amines. This absorption of protons leads to a net flux of protons into the endosomes, and this is accompanied by a flux of chloride or other anions to maintain electroneutrality. The ion flux causes an increase in osmotic pressure within the endosomes leading to membrane destabilization. Recent mechanistic studies suggest that this may not lead to a complete burst of the endosomes but rather creation of transient pores through which the nucleic acid cargo can escape [69].

The other broad strategy for endosomal escape involves designing a material that responds to the low pH of the endosome in such a way as to promote membrane destabilization. This approach is inspired by the action of many viruses, such as influenza, in which surface proteins undergo a conformational change at endosomal pH to one that triggers insertion into the endosomal membrane [70]. The most direct translation of this mechanism is the use of peptides derived from viruses or that mimic the action of viruses. One class of peptides is designed to be alpha-helical, with one face of the helix possessing cationic amino acid residues that complex with nucleic acids and the other face lipophilic for membrane interactions [71, 72]. An example of this is the KALA peptide, which is actually a 30-amino-acid peptide with repeating motifs of the KALA sequence. Arginine-rich peptides mimicking a motif found in the HIV TAT protein have also been utilized [73].

An alternate route to exploiting the low pH environment of the endosome is to mimic the action of pH-sensitive amphipathic peptides in a synthetic polymer. For example, it has been observed that the poly(alkylacrylic acid) polymers exhibit pH-dependent membrane lytic activity, as measured using a red blood cell hemolysis assay [74–76]. Among these, poly(propylacrylic acid) (PPAA) is lytic to membranes selectively at endosomal pH of approximately 5 but not at physiological pH. For this reason, PPAA has been combined with lipoplexes and shown to improve the endosomal escape and net biological activity of delivered genes [74, 77] and antisense oligonucleotides [48]. By grafting onto the PPAA either PEG or more sophisticated poly(alkylene oxides), composite carriers are formed with the ability to avoid serum attack, enter cells and promote intracellular release, and achieve gene silencing with both antisense oligonucleotides and siRNA [49, 50, 78]. This approach has also been utilized to silence oncogenes in tumors [79].

25.3.2 Vector Unpackaging

A major role of the carrier is to protect the nucleic acid cargo during the delivery process, and to do this a stable complex must be formed. However, to perform its biological function, the nucleic acid must be released inside of the cell, a process termed *unpackaging*. Early generations of vectors carried high charge densities that are difficult to dissociate, and it has been found that tuning the charge density by acetylating the amines on a cationic polymer can be an effective means to promote unpackaging [80]. Unfortunately, high degrees of acetylation can also abrogate the pH buffering capabilities of polymers, which may be responsible for their ability to escape endosomes [47, 80]. A simple biophysical assay often used to evaluate unpackaging involves challenging a vector with the anionic macromolecule heparin, and after a period of time, measuring the release of nucleic acid [42]. The results of this assay have been shown to correlate with trends in gene silencing data for antisense oligonucleotides [24].

An alternative approach to the unpackaging problem is to have a polymer that degrades at the target site. For instance, polymers with disulfide groups generally exhibit reduction in the low pH environment of the endosomes, releasing their cargo [81]. This approach can help to overcome both the unpackaging and cytotoxicity associated with cationic polymers such as PEI and PAMAM [82–84]. The use of intracellularly reducible chemistries can produce vectors that promote disassembly and endosomal release, as in the case of the Dynamic PolyConjugates [63] and similar materials [85].

25.4 CURRENT AND FUTURE APPROACHES TO NUCLEIC ACID DELIVERY

A wide variety of polymer and lipid chemistries has been explored as potential carriers for therapeutic nucleic acids without a consensus emerging as to the chemical structures best suited for the task. Formulations have advanced to the clinic from both main camps: (i) highly charged, primarily hydrophilic polymers, and (ii) amphiphilic lipids or lipid-like structures.

25.4.1 Vectors in the Clinic

The first nucleic acids to be introduced into humans were plasmid DNA molecules, which are often administered intramuscularly, and without a carrier, for DNA vaccines and in targeting myogenic disorders such as muscular dystrophy. Cationic lipid/DNA complexes have been used as vaccine adjuvants in clinical trials [86] and were investigated in some immunotherapy trials for delivering IL-2 [87]. Furthermore, a cationic lipid formulation of a plasmid encoding two genes reached Phase III clinical trials before being discontinued [88].

Initial efforts to introduce oligonucleotides into humans were performed in the absence of a carrier, and indeed two antisense therapies have been approved based on naked oligonucleotides, albeit one of them administered directly into the privileged environment of the eye. The development of antisense therapeutics has emphasized extensive chemical modification of the oligonucleotides to improve the stability and other properties of the drug, rather than implementation of delivery systems being developed in research laboratories. A few isolated trials have utilized delivery systems for antisense oligonucleotides, including a Phase I study of an oral formulation to assess bioavailability [89], convection-enhanced delivery in the brain for glioblastoma [90], and a liposomal formulation investigated for cancer [91].

With the more recent discovery and development of siRNAs as therapeutics, delivery systems are being utilized in the clinical development of many molecules going into humans. The first siRNA to enter clinical trials with a nanoparticle was CALA-001, targeting RRM2 in solid tumors. CALA-001 is formulated using a hydrophilic polymer, cyclodextrin, coformulated for self-assembly with siRNA, PEG-adamantane and transferrin-PEG-adamantane. The pharmacokinetics and toxicity in humans of this construct were found to be predictable from animal (rodent and primate) studies [92]. Since then, several other polymer-based siRNA formulations have entered the clinic [93]. These range from one of the earliest carriers, PEI, to the more sophisticated, rationally designed Dynamic PolyConjugate system, which combines the elements of ligand targeting, PEG stabilization, and carrier degradation to promote endosomal escape and siRNA release [63, 94].

In parallel, stable liposome formulations termed stable nucleic acid lipid particles (SNALPs) have achieved demonstrable delivery of siRNA and gene silencing in humans [95]. Several clinical trials are underway that feature variations of the SNALP delivery technology. Other cationic liposome formulations featuring modified lipid structures have also been utilized in siRNA clinical trials [96].

25.4.2 Combinatorial Chemistry Approaches

Given the essentially infinite chemical design space for nucleic acid carriers at both molecular (e.g., PEI-PEG-ligand) and aggregated (e.g., self-assembled, 150 nm diameter nanoparticle at 2 : 1 charge ratio) forms, convergence upon the most effective carriers, or design principles in terms of chemistry or form, for various applications has been slow. One approach to accelerate the carrier discovery process is to employ combinatorial chemistry to produce a library of candidates that can be screened with a series of biological assays. One notable implementation of this strategy has featured the generation of lipidoids by parallel combination of a set of epoxide-terminated alkyl chains (for amphiphilicity) with a set of amines in varying architectures (for charge) [97]. A number of the generated molecules were able to form stable complexes with siRNA and mediate effective gene silencing in a luciferase-based screen *in vitro*. The ability of the candidate vectors generated in this fashion to mediate delivery and silencing *in vivo* varied widely, highlighting the challenges of developing carriers systematically for therapeutic use. This concept has been extended to libraries of additional diversity, for example, lipopeptides [98]. Similar approaches have been used to identify materials suitable for nonviral gene delivery and to form composite theranostic devices for imaging and delivery [99, 100]. In conjunction with mechanistic modeling and/or multivariate statistical analysis, the use of combinatorial libraries may help elucidate the physicochemical determinants of effective nucleic acid delivery materials [101], which can aid in the design process as current combinatorial polymer libraries still only sample a very minute portion of design space.

25.4.3 Polymer–Lipid Nanocomposites

As there seems to be no clear winner in the race between polymer- and lipid-based nucleic acid delivery vectors, efforts to combine the best features from both types of molecules are emerging. Indeed the lipidoids represent an example of polymers generated with partially lipid-like properties. Similarly, charged segments based on PEI and similar cations have been conjugated to lipids and other amphiphilic molecules, producing novel micellar or liposomal particles with favorable properties for nucleic acid complexation and cell membrane interactions [46, 102]. Alternatively, combinations of lipids and polymers can be used to generate composite structures. For instance, polymers modified with groups for serum stabilization, endosomal release, and active targeting may be coated onto the surface of lipoplexes to produce effective hybrid vectors [49, 50, 74, 78, 79, 81, 103–105]. In some cases, lipids and polymers may be simply mixed with nucleic acids and allowed to self-assemble into a hybrid nanoparticle with favorable properties [106, 107], whereas in other cases the lipid and polymer are conjugated together [108].

25.5 SUMMARY AND FUTURE DIRECTIONS

Considerable progress has been achieved toward the creation of vectors that are suitable for nucleic acid delivery *in vivo*. Early efforts demonstrated that cationic polymers and lipids can form nanostructures with nucleic acids via self-assembly and that these nanoparticles can enter cells and deliver nucleic acids in biologically active form, albeit with low efficiency. Subsequent research has identified a number of key barriers to nucleic acid delivery, including serum stability, clearance and biodistribution, endosomal sequestration, and vector unpackaging. Understanding of these barriers has provided the design rules for the development of increasingly sophisticated carriers, commonly consisting of highly functionalized polymers and/or combinations of molecules with distinct roles in overcoming the existing delivery barriers. Especially for siRNA delivery, these carriers are now undergoing extensive evaluation in clinical trials, and it is reasonable to expect that the next several years will see the first FDA approval of a nucleic acid-based nanomedicine.

With the experience gained from past and current efforts to deliver nucleic acids clinically, we are likely to see increasing emphasis on more sophisticated nucleic acid therapeutics, employing combinations of agents or exploiting novel mechanisms. Dozens of clinical trials have been initiated using DNA vaccines [109], which consist of plasmid DNA that can be injected subcutaneously or intramuscularly but with low resulting immunogenicity. A variety of novel delivery approaches are currently being employed including bombardment of gold nanoparticles onto which are loaded the plasmid DNA, electroporation, and dermal patches possibly incorporating the plasmid in nanoparticles. The growing understanding of the regulatory roles of microRNAs in biology and in human disease opens a new category of therapeutic possibilities, including the use of miRNA mimics with siRNA-like chemistries to restore hypoactive miRNA functions as well as oligonucleotide “antagomiRs” that hybridize with and block hyperactive or oncogenic miRNAs [110]. Future therapeutic formulations may combine multiple antisense oligonucleotides [111], multiple forms of nucleic acid such as siRNA and plasmid DNA [112], or combinations of nucleic acids and small-molecule drugs [113]. With these myriad possibilities, the prospects for nucleic acid therapeutics remain bright, and nucleic acid delivery technologies represent a key cornerstone of their development and clinical application.

REFERENCES

1. Friedmann, T.; Roblin, R. *Science* 1972, **175**, 949.
2. Krutzfeldt, J.; Rajewsky, N.; Braich, R.; Rajeev, K. G.; Tuschl, T.; Manoharan, M.; Stoffel, M. *Nature* 2005, **438**, 685.

3. Farooqi, A. A.; Rehman, Z. U.; Muntane, J. *Onco Targets Ther* 2014, **7**, 2035.
4. Schenone, M.; Dancik, V.; Wagner, B. K.; Clemons, P. A. *Nat Chem Biol* 2013, **9**, 232.
5. Que-Gewirth, N. S.; Sullenger, B. A. *Gene Ther* 2007, **14**, 283.
6. Crooke, S. T. *Antisense Nucleic Acid Drug Dev* 1998, **8**, vii.
7. Lee, R. G.; Crosby, J.; Baker, B. F.; Graham, M. J.; Crooke, R. M. *J Cardiovasc Transl Res* 2013, **6**, 969.
8. Sheridan, C. *Nat Biotechnol* 2011, **29**, 121.
9. Geary, R. S.; Watanabe, T. A.; Truong, L.; Freier, S.; Lesnik, E. A.; Sioufi, N. B.; Sasmor, H.; Manoharan, M.; Levin, A. A. *J Pharmacol Exp Ther* 2001, **296**, 890.
10. Juliano, R.; Alam, M. R.; Dixit, V.; Kang, H. *Nucleic Acids Res* 2008, **36**, 4158.
11. Watts, J. K.; Corey, D. R. *J Pathol* 2012, **226**, 365.
12. Castanotto, D.; Rossi, J. J. *Nature* 2009, **457**, 426.
13. Roth, C. M. *Biotechnol Prog* 2008, **24**, 23.
14. Sanghvi, Y. S. *Curr Protoc Nucleic Acid Chem* 2011, **Chapter 4**, Unit 4.1.1.
15. Bartlett, D. W.; Davis, M. E. *Biotechnol Bioeng* 2007, **97**, 909.
16. Lechardeur, D.; Sohn, K. J.; Haardt, M.; Joshi, P. B.; Monck, M.; Graham, R. W.; Beatty, B.; Squire, J.; O'Brodovich, H.; Lukacs, G. L. *Gene Ther* 1999, **6**, 482.
17. Crook, K.; McLachlan, G.; Stevenson, B. J.; Porteous, D. J. *Gene Ther* 1996, **3**, 834.
18. Kanasty, R.; Dorkin, J. R.; Vegas, A.; Anderson, D. *Nat Mater* 2013, **12**, 967.
19. Hanzlikova, M.; Ruponen, M.; Galli, E.; Raasmaja, A.; Aseyev, V.; Tenhu, H.; Urtti, A.; Yliperttula, M. *J Gene Med* 2011, **13**, 402.
20. Roth, C. M.; Yarmush, M. L. *Annu Rev Biomed Eng* 1999, **1**, 265.
21. Roth, C. M. *Biophys J* 2005, **89**, 2286.
22. Sun, X.; Zhang, N. *Mini Rev Med Chem* 2010, **10**, 108.
23. Sundaram, S.; Viriyayuthakorn, S.; Roth, C. M. *Biomacromolecules* 2005, **6**, 2961.
24. Sundaram, S.; Lee, L. K.; Roth, C. M. *Nucleic Acids Res* 2007, **35**, 4396.
25. Putnam, D.; Gentry, C. A.; Pack, D. W.; Langer, R. *Proc Natl Acad Sci U S A* 2001, **98**, 1200.
26. Lemkine, G. F.; Demeneix, B. A. *Curr Opin Mol Ther* 2001, **3**, 178.
27. Patil, M. L.; Zhang, M.; Betigeri, S.; Taratula, O.; He, H.; Minko, T. *Bioconjug Chem* 2008, **19**, 1396.
28. Patil, M. L.; Zhang, M.; Minko, T. *ACS Nano* 2011, **5**, 1877.
29. Wang, D.; Kopeckova, J. P.; Minko, T.; Nanayakkara, V.; Kopecek, J. *Biomacromolecules* 2000, **1**, 313.
30. Yoo, H.; Juliano, R. L. *Nucleic Acids Res* 2000, **28**, 4225.
31. Pack, D. W.; Hoffman, A. S.; Pun, S.; Stayton, P. S. *Nat Rev Drug Discov* 2005, **4**, 581.
32. Liu, G.; Molas, M.; Grossmann, G. A.; Pasumarthy, M.; Perales, J. C.; Cooper, M. J.; Hanson, R. W. *J Biol Chem* 2001, **276**, 34379.
33. Nayvelt, I.; Thomas, T.; Thomas, T. J. *Biomacromolecules* 2007, **8**, 477.
34. Forrest, M. L.; Meister, G. E.; Koerber, J. T.; Pack, D. W. *Pharm Res* 2004, **21**, 365.
35. Fischer, D.; Bieber, T.; Li, Y.; Elsasser, H. P.; Kissel, T. *Pharm Res* 1999, **16**, 1273.
36. Kunath, K.; von Harpe, A.; Fischer, D.; Petersen, H.; Bickel, U.; Voigt, K.; Kissel, T. *J Control Release* 2003, **89**, 113.

37. Breunig, M.; Lungwitz, U.; Liebl, R.; Goepferich, A. *Proc Natl Acad Sci U S A* 2007, **104**, 14454.
38. Hafez, I. M.; Maurer, N.; Cullis, P. R. *Gene Ther* 2001, **8**, 1188.
39. Zelphati, O.; Uyechi, L. S.; Barron, L. G.; Szoka, F. C., Jr. *Biochim Biophys Acta* 1998, **1390**, 119.
40. Alakhov, V. Y.; Ochietti, B.; Guerin, N.; Vinogradov, S. V.; St-Pierre, Y.; Lemieux, P.; Kabanov, A. V. *J Drug Target* 2002, **10**, 113.
41. Bromberg, L.; Deshmukh, S.; Temchenko, M.; Iourtchenko, L.; Alakhov, V.; Alvarez-Lorenzo, C.; Barreiro-Iglesias, R.; Concheiro, A.; Hatton, T. A. *Bioconjug Chem* 2005, **16**, 626.
42. Mao, S.; Neu, M.; Germershaus, O.; Merkel, O.; Sitterberg, J.; Bakowsky, U.; Kissel, T. *Bioconjug Chem* 2006, **17**, 1209.
43. Creusat, G.; Zuber, G. *Chembiochem* 2008, **9**, 2787.
44. Davis, M. E.; Chen, Z. G.; Shin, D. M. *Nat Rev Drug Discov* 2008, **7**, 771.
45. Fenske, D. B.; Cullis, P. R. *Expert Opin Drug Deliv* 2008, **5**, 25.
46. Sparks, S. M.; Waite, C. L.; Harmon, A. M.; Nusblat, L. M.; Roth, C. M.; Uhrich, K. E. *Macromol Biosci* 2011, **11**, 1192.
47. Waite, C. L.; Sparks, S. M.; Uhrich, K. E.; Roth, C. M. *BMC Biotechnol* 2009, **9**, 38.
48. Lee, L. K.; Williams, C. L.; Devore, D.; Roth, C. M. *Biomacromolecules* 2006, **7**, 1502.
49. Peddada, L. Y.; Harris, N. K.; Devore, D. I.; Roth, C. M. *J Control Release* 2009, **140**, 134.
50. Mishra, S.; Vaughn, A. D.; Devore, D. I.; Roth, C. M. *Integr Biol (Camb)* 2012, **4**, 1498.
51. Zelphati, O.; Szoka, F. C. *Proc Natl Acad Sci U S A* 1996, **93**, 11493.
52. Mai, J.; Huang, Y.; Mu, C.; Zhang, G.; Xu, R.; Guo, X.; Xia, X.; Volk, D. E.; Lokesh, G. L.; Thiviyanathan, V.; Gorenstein, D. G.; Liu, X.; Ferrari, M.; Shen, H. *J Control Release* 2014, **187**, 22.
53. Rajur, S. B.; Roth, C. M.; Morgan, J. R.; Yarmush, M. L. *Bioconjug Chem* 1997, **8**, 935.
54. Meade, B. R.; Gogoi, K.; Hamil, A. S.; Palm-Apergi, C.; Berg, A.; Hagopian, J. C.; Springer, A. D.; Eguchi, A.; Kacsinta, A. D.; Dowdy, C. F.; Presente, A.; Lonn, P.; Kaulich, M.; Yoshioka, N.; Gros, E.; Cui, X. S.; Dowdy, S. F. *Nat Biotechnol* 2014, **32**, 1256.
55. Nair, J. K.; Willoughby, J. L.; Chan, A.; Charisse, K.; Alam, M. R.; Wang, Q.; Hoekstra, M.; Kandasamy, P.; Kel'in, A. V.; Milstein, S.; Taneja, N.; O'Shea, J.; Shaikh, S.; Zhang, L.; van der Sluis, R. J.; Jung, M. E.; Akinc, A.; Hutabarat, R.; Kuchimanchi, S.; Fitzgerald, K.; Zimmermann, T.; van Berkel, T. J.; Maier, M. A.; Rajeev, K. G.; Manoharan, M. *J Am Chem Soc* 2014, **136**, 16958.
56. Ranganathan, R.; Madanmohan, S.; Kesavan, A.; Baskar, G.; Krishnamoorthy, Y. R.; Santosham, R.; Ponraju, D.; Rayala, S. K.; Venkatraman, G. *Int J Nanomedicine* 2012, **7**, 1043.
57. Bae, Y. H.; Park, K. *J Control Release* 2011, **153**, 198.
58. Heidel, J. D.; Yu, Z.; Liu, J. Y.; Rele, S. M.; Liang, Y.; Zeidan, R. K.; Kornbrust, D. J.; Davis, M. E. *Proc Natl Acad Sci U S A* 2007, **104**, 5715.
59. Bellocq, N. C.; Pun, S. H.; Jensen, G. S.; Davis, M. E. *Bioconjug Chem* 2003, **14**, 1122.
60. Davis, M. E. *Mol Pharm* 2009, **6**, 659.
61. Davis, M. E.; Zuckerman, J. E.; Choi, C. H.; Seligson, D.; Tolcher, A.; Alabi, C. A.; Yen, Y.; Heidel, J. D.; Ribas, A. *Nature* 2010, **464**, 1067.

62. Gilleron, J.; Querbes, W.; Zeigerer, A.; Borodovsky, A.; Marsico, G.; Schubert, U.; Manygoats, K.; Seifert, S.; Andree, C.; Stoter, M.; Epstein-Barash, H.; Zhang, L.; Koteliansky, V.; Fitzgerald, K.; Fava, E.; Bickle, M.; Kalaidzidis, Y.; Akinc, A.; Maier, M.; Zerial, M. *Nat Biotechnol* 2013, **31**, 638.
63. Rozema, D. B.; Lewis, D. L.; Wakefield, D. H.; Wong, S. C.; Klein, J. J.; Roesch, P. L.; Bertin, S. L.; Reppen, T. W.; Chu, Q.; Blokhin, A. V.; Hagstrom, J. E.; Wolff, J. A. *Proc Natl Acad Sci U S A* 2007, **104**, 12982.
64. Sebestyen, M. G.; Wong, S. C.; Trubetskoy, V.; Lewis, D. L.; Wooddell, C. I. *Methods Mol Biol* 2015, **1218**, 163.
65. Juliano, R. L.; Ming, X.; Carver, K.; Laing, B. *Nucleic Acid Ther* 2014, **24**, 101.
66. Sahay, G.; Querbes, W.; Alabi, C.; Eltoukhy, A.; Sarkar, S.; Zurenko, C.; Karagiannis, E.; Love, K.; Chen, D.; Zoncu, R.; Buganim, Y.; Schroeder, A.; Langer, R.; Anderson, D. G. *Nat Biotechnol* 2013, **31**, 653.
67. Semple, S. C.; Akinc, A.; Chen, J.; Sandhu, A. P.; Mui, B. L.; Cho, C. K.; Sah, D. W.; Stebbing, D.; Crosley, E. J.; Yaworski, E.; Hafez, I. M.; Dorkin, J. R.; Qin, J.; Lam, K.; Rajeev, K. G.; Wong, K. F.; Jeffs, L. B.; Nechev, L.; Eisenhardt, M. L.; Jayaraman, M.; Kazem, M.; Maier, M. A.; Srinivasulu, M.; Weinstein, M. J.; Chen, Q.; Alvarez, R.; Barros, S. A.; De, S.; Klimuk, S. K.; Borland, T.; Kosovrasti, V.; Cantley, W. L.; Tam, Y. K.; Manoharan, M.; Ciufolini, M. A.; Tracy, M. A.; de Fougerolles, A.; MacLachlan, I.; Cullis, P. R.; Madden, T. D.; Hope, M. J. *Nat Biotechnol* 2010, **28**, 172.
68. Boussif, O.; Lezoualc'h, F.; Zanta, M. A.; Mergny, M. D.; Scherman, D.; Demeneix, B.; Behr, J. P. *Proc Natl Acad Sci U S A* 1995, **92**, 7297.
69. ur Rehman, Z.; Hoekstra, D.; Zuhorn, I. S. *ACS Nano* 2013, **7**, 3767.
70. Kalani, M. R.; Moradi, A.; Moradi, M.; Tajkhorshid, E. *Biophys J* 2013, **105**, 993.
71. Wyman, T. B.; Nicol, F.; Zelphati, O.; Scaria, P. V.; Plank, C.; Szoka, F. C., Jr. *Biochemistry* 1997, **36**, 3008.
72. Niidome, T.; Wakamatsu, M.; Wada, A.; Hirayama, T.; Aoyagi, H. *J Pept Sci* 2000, **6**, 271.
73. El-Sayed, A.; Futaki, S.; Harashima, H. *AAPS J* 2009, **11**, 13.
74. Jones, R. A.; Cheung, C. Y.; Black, F. E.; Zia, J. K.; Stayton, P. S.; Hoffman, A. S.; Wilson, M. R. *Biochem J* 2003, **372**, 65.
75. Yessine, M. A.; Lafleur, M.; Meier, C.; Petereit, H. U.; Leroux, J. C. *Biochim Biophys Acta* 2003, **1613**, 28.
76. Murthy, N.; Robichaud, J. R.; Tirrell, D. A.; Stayton, P. S.; Hoffman, A. S. *J Control Release* 1999, **61**, 137.
77. Kyriakides, T. R.; Cheung, C. Y.; Murthy, N.; Bornstein, P.; Stayton, P. S.; Hoffman, A. S. *J Control Release* 2002, **78**, 295.
78. Mishra, S.; Peddada, L. Y.; Devore, D. I.; Roth, C. M. *Acc Chem Res* 2012, **45**, 1057.
79. Peddada, L. Y.; Garbuzenko, O. B.; Devore, D. I.; Minko, T.; Roth, C. M. *J Control Release* 2014, **194**, 103.
80. Gabrielson, N. P.; Pack, D. W. *Biomacromolecules* 2006, **7**, 2427.
81. Murthy, N.; Campbell, J.; Fausto, N.; Hoffman, A. S.; Stayton, P. S. *J Control Release* 2003, **89**, 365.
82. Forrest, M. L.; Koerber, J. T.; Pack, D. W. *Bioconjug Chem* 2003, **14**, 934.
83. Lei, Y.; Wang, J.; Xie, C.; Wagner, E.; Lu, W.; Li, Y.; Wei, X.; Dong, J.; Liu, M. *J Gene Med* 2013, **15**, 291.

84. Islam, M. A.; Park, T. E.; Singh, B.; Maharjan, S.; Firdous, J.; Cho, M. H.; Kang, S. K.; Yun, C. H.; Choi, Y. J.; Cho, C. S. *J Control Release* 2014, **193**C, 74.
85. Guidry, E. N.; Farand, J.; Soheili, A.; Parish, C. A.; Kevin, N. J.; Pipik, B.; Calati, K. B.; Ikemoto, N.; Waldman, J. H.; Latham, A. H.; Howell, B. J.; Leone, A.; Garbaccio, R. M.; Barrett, S. E.; Parmar, R. G.; Truong, Q. T.; Mao, B.; Davies, I. W.; Colletti, S. L.; Sepp-Lorenzino, L. *Bioconjug Chem* 2014, **25**, 296.
86. Lay, M.; Callejo, B.; Chang, S.; Hong, D. K.; Lewis, D. B.; Carroll, T. D.; Matzinger, S.; Fritts, L.; Miller, C. J.; Warner, J. F.; Liang, L.; Fairman, J. *Vaccine* 2009, **27**, 3811.
87. O'Malley, B. W., Jr.; Li, D.; McQuone, S. J.; Ralston, R. *Laryngoscope* 2005, **115**, 391.
88. Bedikian, A. Y.; Richards, J.; Kharkevitch, D.; Atkins, M. B.; Whitman, E.; Gonzalez, R. *Melanoma Res* 2010, **20**, 218.
89. Tillman, L. G.; Geary, R. S.; Hardee, G. E. *J Pharm Sci* 2008, **97**, 225.
90. Jaschinski, F.; Rothhammer, T.; Jachimczak, P.; Seitz, C.; Schneider, A.; Schlingensiepen, K. H. *Curr Pharm Biotechnol* 2011, **12**, 2203.
91. Zhang, C.; Pei, J.; Kumar, D.; Sakabe, I.; Boudreau, H. E.; Gokhale, P. C.; Kasid, U. N. *Methods Mol Biol* 2007, **361**, 163.
92. Zuckerman, J. E.; Gritli, I.; Tolcher, A.; Heidel, J. D.; Lim, D.; Morgan, R.; Chmielowski, B.; Ribas, A.; Davis, M. E.; Yen, Y. *Proc Natl Acad Sci U S A* 2014, **111**, 11449.
93. Bouchie, A. *Nat Biotechnol* 2012, **30**, 1154.
94. Wong, S. C.; Klein, J. J.; Hamilton, H. L.; Chu, Q.; Frey, C. L.; Trubetskoy, V. S.; Hegge, J.; Wakefield, D.; Rozema, D. B.; Lewis, D. L. *Nucleic Acid Ther* 2012, **22**, 380.
95. Coelho, T.; Adams, D.; Silva, A.; Lozeron, P.; Hawkins, P. N.; Mant, T.; Perez, J.; Chiesa, J.; Warrington, S.; Tranter, E.; Munisamy, M.; Falzone, R.; Harrop, J.; Cehelsky, J.; Bettencourt, B. R.; Geissler, M.; Butler, J. S.; Sehgal, A.; Meyers, R. E.; Chen, Q.; Borland, T.; Hutabarat, R. M.; Clausen, V. A.; Alvarez, R.; Fitzgerald, K.; Gamba-Vitalo, C.; Nochur, S. V.; Vaishnaw, A. K.; Sah, D. W.; Gollob, J. A.; Suhr, O. B. *N Engl J Med* 2013, **369**, 819.
96. Schultheis, B.; Strumberg, D.; Santel, A.; Vank, C.; Gebhardt, F.; Keil, O.; Lange, C.; Giese, K.; Kaufmann, J.; Khan, M.; Drevs, J. *J Clin Oncol* 2014, **32**, 4141.
97. Love, K. T.; Mahon, K. P.; Levins, C. G.; Whitehead, K. A.; Querbes, W.; Dorkin, J. R.; Qin, J.; Cantley, W.; Qin, L. L.; Racie, T.; Frank-Kamenetsky, M.; Yip, K. N.; Alvarez, R.; Sah, D. W.; de Fougerolles, A.; Fitzgerald, K.; Koteliansky, V.; Akinc, A.; Langer, R.; Anderson, D. G. *Proc Natl Acad Sci U S A* 2010, **107**, 1864.
98. Dong, Y.; Love, K. T.; Dorkin, J. R.; Sirirungruang, S.; Zhang, Y.; Chen, D.; Bogorad, R. L.; Yin, H.; Chen, Y.; Vegas, A. J.; Alabi, C. A.; Sahay, G.; Olejnik, K. T.; Wang, W.; Schroeder, A.; Lytton-Jean, A. K.; Siegwart, D. J.; Akinc, A.; Barnes, C.; Barros, S. A.; Carioto, M.; Fitzgerald, K.; Hettinger, J.; Kumar, V.; Novobrantseva, T. I.; Qin, J.; Querbes, W.; Koteliansky, V.; Langer, R.; Anderson, D. G. *Proc Natl Acad Sci U S A* 2014, **111**, 3955.
99. Vu, L.; Ramos, J.; Potta, T.; Rege, K. *Theranostics* 2012, **2**, 1160.
100. Barua, S.; Joshi, A.; Banerjee, A.; Matthews, D.; Sharfstein, S. T.; Cramer, S. M.; Kane, R. S.; Rege, K. *Mol Pharm* 2009, **6**, 86.
101. Alabi, C. A.; Love, K. T.; Sahay, G.; Yin, H.; Luly, K. M.; Langer, R.; Anderson, D. G. *Proc Natl Acad Sci U S A* 2013, **110**, 12881.

102. Navarro, G.; Sawant, R. R.; Essex, S.; Tros de Ilarduya, C.; Torchilin, V. P. *Drug Deliv Transl Res* 2011, **1**, 25.

103. Cheung, C. Y.; Murthy, N.; Stayton, P. S.; Hoffman, A. S. *Bioconjug Chem* 2001, **12**, 906.

104. Cheung, C. Y.; Stayton, P. S.; Hoffman, A. S. *J Biomater Sci Polym Ed* 2005, **16**, 163.

105. Kiang, T.; Bright, C.; Cheung, C. Y.; Stayton, P. S.; Hoffman, A. S.; Leong, K. W. *J Biomater Sci Polym Ed* 2004, **15**, 1405.

106. Xue, H. Y.; Narvikar, M.; Zhao, J. B.; Wong, H. L. *Pharm Res* 2013, **30**, 572.

107. Yessine, M. A.; Meier, C.; Petereit, H. U.; Leroux, J. C. *Eur J Pharm Biopharm* 2006, **63**, 1.

108. Metselaar, J. M.; Bruin, P.; de Boer, L. W.; de Vringer, T.; Snel, C.; Oussoren, C.; Wauben, M. H.; Crommelin, D. J.; Storm, G.; Hennink, W. E. *Bioconjug Chem* 2003, **14**, 1156.

109. Ferraro, B.; Morrow, M. P.; Hutnick, N. A.; Shin, T. H.; Lucke, C. E.; Weiner, D. B. *Clin Infect Dis* 2011, **53**, 296.

110. Zhang, Q.; Ran, R.; Zhang, L.; Liu, Y.; Mei, L.; Zhang, Z.; Gao, H.; He, Q. *J Control Release* 2015, **197**, 208.

111. Ferrari, N.; Seguin, R.; Renzi, P. *Future Med Chem* 2011, **3**, 1647.

112. Francis, S. M.; Taylor, C. A.; Tang, T.; Liu, Z.; Zheng, Q.; Dondero, R.; Thompson, J. E. *Mol Ther* 2014, **22**, 1643.

113. Li, J.; Wang, Y.; Zhu, Y.; Oupicky, D. *J Control Release* 2013, **172**, 589.

INDEX

Drug Delivery: Principles and Applications, Second Edition. Edited by Binghe Wang, Longqin Hu, and Teruna J. Siahaan.